Streptococcus A

E. coli OI57:H7

Influenza A

AIDS

Hantavirus

Vancomycin−resistant
Enterococcus

Dengue

Pandemic
cholera

Lyme
disease

Drug-resistant
pneumococcus

Cryptosporidiosis

Venezuelan
hemorrhagic fever

Microorganisms
in
our World

Microorganisms in our World

Ronald M. Atlas, PhD
University of Louisville
Louisville, Kentucky

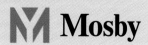

St. Louis Baltimore Berlin Boston Carlsbad Chicago London Madrid
Naples New York Philadelphia Sydney Tokyo Toronto

Mosby
Dedicated to Publishing Excellence

Publisher: James M. Smith
Editor: Robert J. Callanan
Developmental Editor: Diana Lyn Laulainen
Project Manager: Carol Sullivan Weis
Production Editor: Florence Achenbach
Design Managers: Betty Schulz, Sheilah Barrett
Design: Jeanne Wolfgeher
Cover Designer: E. Rohne Rudder
Art: Pagecrafters, Inc.
Cover Illustrations: Alterimage, Image Bank. *Front,* Yoav Levy, Phototake; AM Siegelman, Visuals Unlimited. *Back,* Brownie Harris, Tony Stone Worldwide; David A. Wagner, Phototake.

Case studies and anatomy appendix written by Randall W. Oelerich, MD.

Printed in the United States of America

Composition by Graphic World, Inc.
Printing/Binding by Von Hoffmann Press, Inc.

Mosby-Year Book, Inc.
11830 Westline Industrial Drive
St. Louis, Missouri 63146

Library of Congress Cataloging in Publication Data

Atlas, Ronald M.
   Microorganisms in our world   /   Ronald M. Atlas.
      p.   cm.
   Includes bibliographical references and index.
   ISBN 08-8016-7804-8
   1. Microbiology. 2. Medical microbiology. I. Title.
QR 41.2.A842   1994
576—dc20   94-31334
                                                CIP

95   96   97   98   99   /   9   8   7   6   5   4   3   2

PREFACE

The field of microbiology is fascinating. Most people associate microorganisms only with disease. However, cheese, wine, and sourdough bread, for example, are products of microbial metabolism. The quality of our environment also depends on the activities of microorganisms. *Microorganisms in our World* will introduce the student to the wide world of microbiology. A text written by a microbiologist, *Microorganisms in Our World* gives the perspective of microbiology as a broad relevant field and of microorganisms as important to human health and the functioning of planet Earth.

THE BOOK

The text is written in a manner that is easy to read. Difficult topics are thoroughly explained so that they can be easily understood. Technical terms are explained as they are introduced and a conceptual framework of microbiology is presented. Students will learn how to communicate microbiological information. They will be able to understand the numerous news reports related to microbiology, especially those about relationships between microorganisms and human health. The depth of presentation is limited to the essentials but provides enough information to develop an understanding of a topic. The information presented represents a distillation of endless volumes of scientific studies. Facts are presented that reflect the latest state of knowledge. Accuracy is not sacrificed for brevity or simplicity. The breadth of the field of microbiology is covered at an appropriate level for students.

Topics are organized in a logical yet flexible manner. Coverage of medical microbiology emphasizes microorganisms and how they cause disease. The sense of discovery is used to entice students into wanting to learn more. Microbiology is an exciting field, and students should develop enthusiasm toward scientific discovery. They should never be bored by the information they are learning. The text contains numerous boxes with items that will interest students. The relevance of microbiology in our daily lives and selected careers is presented so that today's pragmatic students know why they are studying a particular topic and want to learn about microorganisms.

The diversity of the microbial world is shown with relevant examples of how microorganisms affect our daily lives. The interactions of microorganisms with humans are highlighted, especially as they relate to human disease. The properties of microorganisms that are responsible for their ability to cause disease and permit prevention, diagnosis, and treatment of disease are emphasized.

ORGANIZATION

Microorganisms in our World is organized into six units. The first three units focus on the principles of microbiology, including cell structure, cellular metabolism, microbial genetics, and growth. Unit Four and Unit Five focus on the medical applications of microbiology—microorganisms as disease-causing entities and the body's defenses against infection. The final unit of the text discusses microbiological applications to industry and the environment.

Microorganisms and Microbiology (Unit One) introduces the student to the field of microbiology and the techniques that are used by microbiologists in research and clinical settings. Chapter 1 (Unity of the Microbial World) introduces the field of microbiology, giving insight into the relevance of microorganisms of human health and how microbiologists work—the scientific method. Chapter 2 (Diversity of the Microbial World) continues by introducing the student to the vast breadth of the microbial world. Chapter 3 (Science of Microbiology: Methods for Studying Microorganisms) discusses the techniques that microbiologists use to do their work. The various techniques of light and electron microscopy are explained in addition to coverage of culturing, aseptic technique, and plating. Unit One concludes with Chapter 4 (Chemistry for the Microbiologist)—an introduction for students who have not had a basic chemistry course and a convenient review for those who have.

The text continues with coverage of **Cellular and Molecular Microbiology** (Unit Two). Chapter 5 (Cell Structure) introduces the student to the general structure of prokaryotic cells and compares and contrasts them to the structure of eukaryotic cells. Chapter 6 (Cellular Metabolism) explains the energetics of microorganisms—cellular metabolism. Chapter 7 (Microbial Genetics: Replication and Expression of Genetic Information) discusses microbial genetics and is immediately followed by an introduction to biotechnology in Chapter 8 (Genetic Recombination and Recombinant DNA Technology).

Unit Three focuses on **Microbial Growth and Its Control.** The unit begins with a study of growth in viruses in Chapter 9 (Viral Replication), and Chapter 10 (Bacterial Reproduction and Growth of Microorganisms) focuses on the growth of bacteria. Chapter 11 (Control of Microbial Growth and Death) discusses how bacterial growth can be minimized or stopped using techniques such as pasteurization and sterilization and antimicrobial agents, including preservatives, disinfectants, and antibiotics.

In **Microorganisms and Human Diseases** (Unit Four) the content of the book moves from the principles-based coverage of microbiology in the first three units to more applied topics showing how microorganisms affect our lives. Chapter 12 (Microorganisms and Human Diseases) introduces Koch's postulates, the virulence factors of pathogenic microorganisms, and transmission routes of infectious agents. The body's immune response to infection is explained first in Chapter 13 (Nonspecific Host Defenses against Microbial Infections: The Immune Response), followed by Chapter 14 (Specific Host Defenses against Microbial Infections: The Immune Response). Chapter 14 (Immune Response and Human Disease) discusses topics that include immunization, immunodeficiencies, autoimmunity, hypersensitivity reactions, and transplantation. Chapter 15 (Diagnosis of Human Disease) introduces the student to methodology used in the clinical microbiology laboratory. These methods include skin testing, culturing, immunofluorescence, gene probes, and the polymerase chain reaction. This unit concludes with a discussion of the uses of antimicrobial agents in Chapter 17 (Treatment of Infectious Diseases).

The **Infectious Disease** unit (Unit Five) is comprised of six chapters. Diseases are organized by the causative microorganisms and then divided into the system's of the body that they affect. This allows the student to understand commonalities of viral infection and pathogenesis of viral disease, separate from the treatment of bacterial diseases. This unit begins with two chapters covering viral diseases—Chapter 18 (Viral Diseases of the Respiratory, Gastrointestinal, and Genital Tracts) and Chapter 19 (Viral Diseases of the Central Nervous, Cardiovascular, and Lymphatic Systems). Three chapters follow,

focusing on the bacterial diseases—Chapter 20 (Bacterial Diseases of the Respiratory and Gastrointestinal Tracts), Chapter 21 (Bacterial Diseases of the Central Nervous, Cardiovascular, and Lymphatic Systems), and Chapter 22 (Bacterial Diseases of the Urinary Tract, Genital Tract, Skin, Eyes, Ears, and Oral Cavity). The unit concludes with Chapter 23 (Diseases Caused by Eukaryotic Organisms).

The final unit of the text, **Applied and Environmental Microbiology** (Unit Six), explains the importance of microbiology in regard to industrial and environmental applications. Chapter 24 (Industrial Microbiology) covers various topics, including fermentation, production of antibiotics, and the recovery of mineral resources. Chapter 25 (Environmental Microbiology) details the role of microorganisms in solid waste disposal, water treatment, biodegradation of pollutants, and other selected applications.

Because prerequisites for this course vary from school to school and an understanding of the anatomy and physiology of the human body is important to understanding microbiology, an Anatomy Appendix has been added to the text. This illustrated appendix offers a brief overview of the structure and function of each of the eleven body systems, including a section that details the importance of each system from a microbiological perspective.

FEATURES

Many features have been added to *Microorganisms in our World* to make the content accessible and relevant to the student. The strong emphasis on disease-causing microorganisms will be especially interesting to students entering health careers and general education students. The application of microbiological principles to human health is important to the reader.

BOXED MATERIAL

Boxed asides have been added to the text to make the microbiological principles and applications more relevant to the students. Four types of boxes appear in the text; each will focus on a specific area as detailed below:

Newsbreak: Features important advances and current events in the field of microbiology.

Highlight: Expands coverage on certain topics that are important in the study of microbiology.

Historical Perspective: Shows the relevance of the past of microbiology.

Methodology: Shows students how microbiologists apply scientific methods and approaches in the study of microorganisms.

CASE STUDIES

Since many students will be entering the health professions and disease applications are interesting to them, clinical case studies have been added to the end of each of the chapters in Unit Five. Written by a medical doctor, these case studies provide the story of a patient's illness from the initial assessment at the health care facility through laboratory findings and diagnosis, culminating with a discussion of treatment and course of the disease. Because most students have not taken a pathophysiology course, italicized remarks in the case studies interpret the health care provider's thought process. The student is able to understand why certain tests have been ordered, what clues the tests offer the physician, and how a final diagnosis was confirmed.

ILLUSTRATIONS

The full-color illustrations are accessible, highlighting key information that is covered in the text. Effort has been made to ensure that the illustrations support the narrative and are easily understood.

A color-coding system is maintained throughout the text whereby specific structures always appear in a specific color. This consistency will enable today's visual learners to readily identify the identical structures in different microorganisms. The key to the color coding is presented below.

In addition, selected colorization of electron micrographs is used to highlight the full extent of structures rather than relying on labels and lead lines that may obstruct important parts of an image. The same color code applies to the micrographs.

In the metabolism chapter, dual panel illustrations present the metabolic pathways in two ways. The actual chemical structures are shown on the left for the instructor who wants detail, while on the right, a conceptual pathway using only words is shown. This allows the instructor the flexibility to use or not use the detail provided based on the objectives of the course.

Finally, the use of real-life photographs throughout the book, in conjunction with an illustration or separately, allows the student to see the relevance of microbiology in everyday life.

PEDAGOGY

Several learning aids have been added to the text to help students learn the content. They include:

Chapter Outlines Show students the big picture of the topic they are studying and helps them plan study time.

Chapter Objectives Detail essential information that a student should understand from the chapter.

Bold-faced Terms Emphasize key terms within the chapter.

Concept Checks Brief synopses of content at the end of many major sections.

Tables Information is summarized in many tables in each chapter to help the student organize and learn the material.

Chapter Summaries Organize and condense chapter contents for easy access.

Review Questions Allow students to test themselves on key concepts of the chapter.

Critical Thinking Questions Foster critical thinking skills by asking students to answer questions as if they were microbiologists attempting to solve a problem.

KEY TO COLOR CODE OF CHEMICALS AND STRUCTURES			
Color	Chemical	Structure	Microorganism
	Protein, lipoprotein	Viral capsid, bacterial pili, flagella	Virus
	Peptidoglycan	Bacterial cell wall	Bacteria
	Carbohydrate glycoprotein, lipopolysaccharide	Bacterial outer membrane, glycocalyx, capsule	
	DNA	Bacterial chromosome, plasmas, chloroplasts	
	RNA, ATP	Ribosomes, nucleus	
	Lipid, phospholipid	Membranes, mitochondria	Eukaryotes

Further Readings A selection of relevant articles and books suggesting sources of further information. Each reference offers a brief synopsis of the article.

Microbiology CD-ROM Tutorial This innovative microbiology tutorial supplements lecture and text material.

ANCILLARIES

In addition to the text, an ancillary package has been assembled to aid in teaching. Selected items will be provided to qualified adopters using *Microorganisms in our World.* The ancillaries include:

Instructor's Manual and Testbank Written by James Parsons of Bloomsburg University, this manual will prove invaluable in planning your course. It features chapter outlines, key terms, lecture outlines, teaching tips, and audiovisual references. The test bank contains over 750 multiple choice questions to aid the instructor in evaluating student progress.

Computerized Instructor's Manual and Testbank The book above in an electronic format that instructors can access through their word processing software. This ancillary is available in IBM and Macintosh versions.

Computerized Testbank The testbank is also available on Diploma IV, a test-generating software program. This ancillary is available in IBM and Macintosh versions.

Transparency Acetates A set of 200 full-color transparency acetates chosen from the illustrations in the text and relabeled with large, bold lettering to enhance projection in the classroom.

Slide Set The same images in the transparency set are also provided in 2 × 2 slides.

Infectious Disease Slide Set A set of fifty 2 × 2 slides of photomicrographs and clinical conditions.

Study Guide Written by William Wellnitz of Augusta College, the Study Guide includes various learning activities to increase students understanding and retention.

Laboratory Manual This comprehensive laboratory manual emphasizes general principles and health science applications of microbiology. An Instructor's Manual accompanies the lab manual.

ACKNOWLEDGMENTS

Microorganisms in our World is the result of the effort of many individuals. Some individuals informally shared ideas. Others formally reviewed drafts of the manuscript and illustrations, ensuring clarity and accuracy. Thanks to the reviewers listed below:

Robert K. Alico, *Indiana University of Pennsylvania*
Jerry Allen, *Northwestern State University of Louisiana*
Delia Anderson, *University of Southern Mississippi*
Carolyn H. Bohach, *University of Idaho*
Dan Brannan, *Abilene Christian University*
Laura Bukovsan, *State University of New York, Oneonata*
Michael Burke, *North Dakota State School of Science*
John Clausz, *Carroll College*
Don C. Dailey, *Austin Peay State University*
Richard H. Davis, *City College of San Francisco*
Jim De Kloe, *Solano County Community College*
Warren R. Erhardt, *Daytona Beach Community College*
Katharine B. Gregg, *West Virginia Wesleyan College*
Eileen Gregory, *Rollins College*
Eddie P. Hill, *Macalaster College*
Diana Kaftan, *Diablo Valley College*
Steve Karr, *Carson-Newman College*
Anne C. Lund, *Hampden-Sydney College*
John C. Makemson, *Florida International University*
RGE Murray, *University of Western Ontario*
Brian Nummer, *Tennessee Tech University*
Lawrence Parks, *University of Louisville*
Joseph Pearson, *Scottsdale Community College*
Helen K. Pigage, *US Air Force Academy*
Maxine Plummer, *Bucks County Community College*
Laurie Richardson, *Florida International University*
Barbara B. Rundell, *Rosary College*
Paul Small, *Eureka College*
Mary L. Taylor, *Portland State University*
Gerald Tritz, *Kirksville College of Osteopathic Medicine*
William R. Wellnitz, *Augusta College*
Anne Wolffe, *Laramie Community College*
Brian Yamamoto, *Kauai Community College*

Case Studies and Anatomy Appendix
Randall W. Oelerich, MD

CONTENTS IN BRIEF

CONTENTS

HISTORICAL PERSPECTIVE

METHODOLOGY

UNIT ONE

Microorganisms

and Microbiology

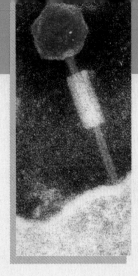

CHAPTER 1

Unity of the Microbial World

PREVIEW TO CHAPTER 1

In this chapter we will:
- Study unifying characteristics of microorganisms.
- Define the scope of the science of microbiology.
- Describe the attributes of microorganisms.
- Distinguish microorganisms from plants and animals.
- Examine the structural organization of microorganisms.
- Identify major characteristics of viruses, prokaryotic eubacteria and archaebacteria, and eukaryotic fungi, algae, and protozoa.
- Gain insight into the relevance of microorganisms to human health and well-being.
- Learn the following key terms and names:

algae	microscope
archaebacteria	multicellular
bacteria	nucleus
cell	organelles
cellular metabolism	pathogens
DNA (deoxyribonucleic acid)	plasma membrane
eubacteria	prokaryotic cells
eukaryotic cells	protozoa
fungi	tissues
microbiology	unicellular
microorganism	viruses

MICROORGANISMS

Many of us equate the terms *microorganisms* and *germs*. This is not surprising since we continuously battle microorganisms to maintain our health. Unseen microorganisms exert a powerful influence on our lives. By learning about microorganisms you will be able to understand why we sometimes suffer infections and disease. You also will be able to see how modern medical practices attempt to control and to eliminate disease-causing microorganisms and to treat infectious diseases when they do occur. At the same time you may be surprised that, while the general layman's view is that all microorganisms are harmful to human health, most microorganisms do not cause disease. Our skin, hair, tongue, intestines, and other body parts are literally swarming with bacteria that do us no harm. In fact, they are advantageous, or even necessary to us. Life on earth depends on microorganisms that have an enormous capacity to degrade and to recycle materials.

WHAT IS A MICROORGANISM?

As you begin your study of **microbiology**—the science that deals with microorganisms—you may ask, exactly what is a microorganism and how do microorganisms differ from other organisms? As implied by the word **microorganism** (from the Greek word *micro*, meaning small), microorganisms are very small life forms—so small that individual microorganisms usually cannot be seen without magnification (FIG. 1-1). These microorganisms include the viruses, bacteria, fungi, algae, and protozoa. They are the organisms that microbiologists study (FIG. 1-2).

To describe size, scientists use the metric system measure of length that is based on a meter (a meter is approximately a yard). Fractions of a meter are described by using prefixes such as milli (one thousandth), micro (one millionth), and nano (one billionth). A millimeter (mm) is a thousandth of a meter

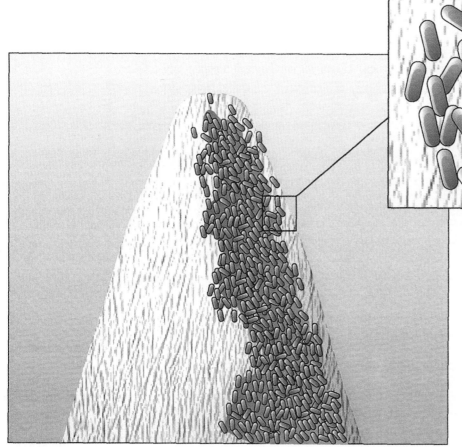

FIG. 1-1 Microorganisms are extremely small, so they are invisible to the naked eye. Most bacteria are 0.5 to 2 μm in length. A million bacteria can fit on the tip of a pin. Individual bacteria can only be seen by using a microscope to magnify the image.

FIG. 1-2 Students in an introductory microbiology laboratory class get their first view of the microbial world while peering through their microscopes.

NEWSBREAK

FRENCH REVOLUTION LEADS TO THE METRIC SYSTEM

The metric system was introduced during the French Revolution, in part because of concerns with differences in the measuring systems used in various parts of France but equally to demonstrate that a new governmental regime was in charge. The fundamental unit of measuring length is the metric system, the meter (m), was defined as one ten-millionth the distance from the earth's north pole to the equator. The official measurement was made in 1799 geometrically between Dunkirk and Barcelona. A platinum meter was deposited in the Archives of the Republic as the official standard measure of length. Although French proponents of the reform in weights and measures sought a system based on some natural universal unit to be used throughout the world, England rejected the metric system as impractical. Jefferson, although enthusiastic about such a change for the United States, rejected the French system and tried to develop his own system of measures. Thus, while the rest of the world uses the metric system, the United States and Britain do not generally use this system of measure, except for scientists—who, the world over, use the metric system.

or approximately 0.004 inches; a micrometer (μm) is a millionth of a meter or approximately 0.000004 inches, and a nanometer (nm) is a billionth of a meter or approximately 0.000000004 inches.

The smallest object that can be seen with the naked eye is about a tenth of a millimeter (0.1 mm), which is equal to 100 micrometers (100 μm). Most microorganisms are smaller than this. Largest bacterial cells, for example, are only 5 μm in length (0.005 mm), and small bacterial cells are about 0.1 μm in length (much less than a millionth of a meter). Therefore we generally must magnify the images of most bacteria about 1,000 times just to be able to see them. The viruses are smaller yet. The largest viruses are almost 0.1 μm and the smallest viruses are 0.01 μm (less than a millionth of a meter). Thus the images of viruses must be magnified 10,000 to 100,000 times to be seen.

Microorganisms are defined by their small size, since they generally are invisible to the naked eye and can only be seen with a microscope.

ORGANIZATIONAL STRUCTURE OF MICROORGANISMS

Cells of Living Organisms

The **cell** is the fundamental unit of all living systems, including all living microorganisms. Cells have a boundary layer, called the **plasma membrane**, that separates the living cell from the external surroundings. The plasma membrane controls the flow of materials into and out of the cell. It permits the cell to maintain the highly organized state that is a major characteristic of living systems. If the plasma membrane is damaged and fails to regulate the flow of materials, the cell dies and life ceases. To protect this essential structure, many cells have a rigid cell wall surrounding the plasma membrane.

Cells are the fundamental functional units of living organisms.

All cells have a plasma membrane that separates the living system from the nonliving surroundings.

Cells carry out the essential functions of life, including processing of energy and materials for growth and replication of hereditary information for reproduction. Cells contain the molecule **DNA (deoxyribonucleic acid)**, which is the universal substance of living cells that passes hereditary information to offspring cells. The DNA of a cell specifies the potential characteristics of that cell and is often called the "master molecule of life" because it directs the activities of the cell. In all living cells there is a flow of information from the DNA through molecules of RNA (ribonucleic acid) to direct the synthesis of proteins. The synthesis of proteins occurs at numerous ribosomes within the cell. Proteins are the action molecules of life, catalyzing chemical reactions within living cells. By specifying the proteins that a cell makes, the information contained in the DNA directs **cellular metabolism,** the process in which living cells utilize energy and transform materials into the structures necessary for growth and reproduction. All living cells must carry out metabolism to sustain life, one of the essential functions of which is to generate ATP (adenosine triphosphate) as a central currency of cellular energy. Thus cells uniformly have structures that regulate the flow of materials, the storage and expression of genetic information, and the ability to carry out metabolism (FIG. 1-3).

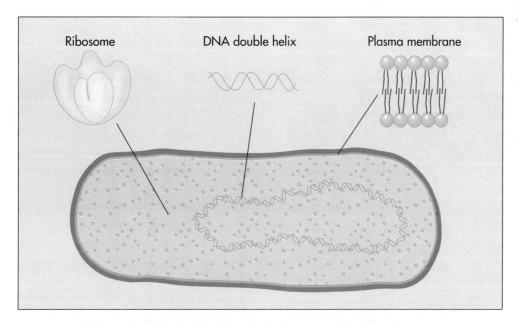

FIG. 1-3   The cell is the fundamental unit of all living organisms. It is separated from its surroundings by a plasma membrane that regulates the flow of materials into and out of the cell. The organism's hereditary information is contained within the cell in molecules of DNA. Proteins that catalyze the metabolism of the living organism are synthesized at ribosomes within the cell.

DISCOVERY OF MICROORGANISMS

Considering their small size, it is no wonder that the existence of microorganisms was not recognized until only a few centuries ago. The advent of the microscope, which is an instrument used to enlarge objects and images, permitted us to see them. Nor is it surprising that we still have a lot to learn about microorganisms. The invention of the microscope occurred at the end of the sixteenth century, essentially at the same time as the introduction of the telescope. As some scientists looked upward and outward toward the stars with telescopes, others began to search inward with microscopes.

The first microscopes were simple ground glass lenses that magnified images of previously unseen objects. By the late seventeenth century, microscopes permitted magnifications of several hundred times, making it possible to discover the microbial world. Among the first to observe the previously invisible microbial world was Robert Hooke, an English scientist. His detailed drawings of fungi, made in 1667, reflect hours of tedious observations (FIG. A).

Antonie van Leeuwenhoek, an amateur scientist and official winetaster of Delft, Holland, made the first recorded observations of bacteria in 1670 (FIG. B). Van Leeuwenhoek's interest in microscopes was probably related to the use of magnifying glasses by drapers to examine fabrics. His hobby was microscopy and he made over 100 microscopes, each consisting of a simple glass lens (FIG. C). These microscopes were little more than magnifying glasses, each capable of magnifying an image about 300 times, so that bacteria could barely be seen as fuzzy images. Leeuwenhoek must have had great patience and persistence to squint through the lens of his handheld microscope at dimly lighted specimens

A

B

Robert Hooke made microscopic observations and provided the earliest descriptions of many fungi. Various species of fungi can clearly be identified in his drawings made from 1635 to 1703 and recorded in his book, *Micrographia*.

Antonie van Leeuwenhoek (1632-1723), here seen holding one of his microscopes, opened the door to the hidden world of microorganisms when he described bacteria. Although he was only an amateur scientist, Leeuwenhoek's keen interest in optics and his diligence allowed him to make this important discovery.

and record drawings of microorganisms as he did. He thought the bacteria he observed were little animals because they moved, therefore he called them "animalcules." Van Leeuwenhoek's observation of bacteria (animalcules) set the stage for the development of the field of microbiology (FIG. D).

Although not a professional scientist, van Leeuwenhoek asked questions and performed experiments. For example, after observing animalcules in rainwater from his garden, van Leeuwenhoek decided to test whether microorganisms came from heaven or whether they came from earthly sources. He washed a porcelain bowl in fresh rainwater, set it out in his garden during a storm, and observed no microorganisms in the freshly collected rainwater sample. After allowing the water to sit for a few days, he observed numerous microorganisms in the sample. He concluded that a few microorganisms in the sample had multiplied and that "life begets life—even for animalcules."

Although his observations stimulated much controversy, no one at that time made a serious attempt to repeat or to extend them. Van Leeuwenhoek's animalcules remained mere oddities of nature to the scientists of the day. It was not until two centuries later that the significance of the observations made by van Leeuwenhoek became evident, when the science of microbiology began to flourish and to develop into the vibrant field of scientific study it is today (FIG. E).

C

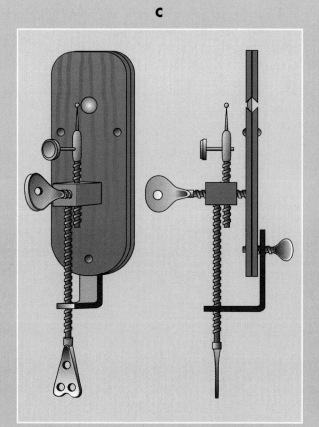

Leeuwenhoek's microscopes were little more than magnifying glasses.

D

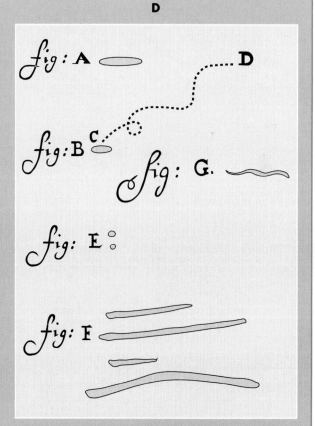

Colorized reproduction of Leeuwenhoek's sketches of bacteria from the human mouth illustrate several common types of bacteria, including rods and cocci. *A*, A motile *Bacillus*; *B*, *Selenomonas sputigena*; *C* and *D*, the path of *S. sputigena's* motion; *E*, micrococci; *F*, *Leptothrix buccalis*; *G*, a spirochete.

Continued.

THE DISCOVERY OF MICROORGANISMS—CONT'D

E

1641–1700	1701–1740	1741–1780	1781–1820	1821–1840	1841–1860	1861–1880

Boyle's law of gases

Hooke's Micrographia

van Leeuwenhoek observes bacteria and protozoa

Malpighi describes plant and animal anatomy

Newton's theory of gravitation

Fahrenheit constructs mercury thermometer

Montagu introduces smallpox vaccination

Micheli publishes work on fungi

Linnaeus describes classification system for living organisms

Osmosis described

Benjamin Franklin shows lightning to be electricity

Hydrogen, oxygen, and nitrogen discovered

Spallanzani disputes theory of spontaneous generation

Watt perfects the steam engine

Lavoisier produces a table of chemical elements

Jenner introduces vaccination

Discovery of UV rays

Dalton's atomic theory

Wave theory of light

Avogadro proposes molecular theory of matter

Faraday's electric motor

Ohm's law of current

Schwann, Kützing, and Latour report yeasts cause fermentation

Development of centrifuge

Liebig studies on biochemistry

Joule defines first law of thermodynamics

Clausius outlines second law of thermodynamics

Bunsen invents gas burner

Darwin's theory of natural selection

Pasteur studies fermentation

Lister begins practice of antiseptic surgery

Periodic table of elements

Koch describes bacteria as cause of disease; develops methods for pure culture of bacteria

Cohn discovers endospores of bacteria

Development of autoclave

Time line of some scientific discoveries important to the development of microbiology. The study of microorganisms, which began in the late 1600s with the observations of Leeuwenhoek, developed in the mid-nineteenth century into a true science through the studies of Robert Koch and Louis Pasteur. Major developments have been made in the twentieth century, including the discovery of antibiotics and the molecular basis of heredity. Today microbiology is a flourishing field of science.

1881–1900	1901–1920	1921–1940	1941–1960	1961–1980	1981–1990	1991–

Zeiss constructs modern microscope

Virology begins with discovery of plant and animal viruses

Bergey's Manual of Determinative Bacteriology

Beadle and Tatum isolate mutants

Breaking of the genetic code

AIDS epidemic begins

New emerging diseases: Hantavirus, Cryptosporidium, reemerging tuberculosis

Gram stain developed

Ehrlich studies chemotherapy

Fleming discovers penicillin

Use of antibiotics in medicine begins

DNA recombination studies lead to genetic engineering

Biotechnology develops as industry based on genetic engineering

Koch's postulates

Virus-malignancy link discovered

Kluyver and Van Niel work on comparative microbial metabolism

Avery shows bacterial transformation

Smallpox eliminated by worldwide immunization

Miniaturized serological kits introduced for rapid diagnosis of disease

Multidrug-resistant diseases

Pasteur inoculates against rabies

Discovery of bacteriophage

Invention of electron microscope

Discovery that actinomycetes produce antibiotics

Recognition of archaebacteria

Petri's culture dish

Influenza pandemic kills millions

Stanley crystallizes a virus

Antipolio vaccine

First outbreak of Legionnaire's disease

Gene probes developed for identification of microorganisms

Beijerinck studies nitrogen fixation

Birdseye deep-freezes food

Introduction of DPT vaccine

Watson and Crick's DNA double helix

Measles, mumps, rubella vaccine

Winogradsky studies autotrophy

Mosquito-malaria link discovered

Cells have genetic information in the form of DNA that directs the form and function of the cell and passes hereditary information from one generation to the next.

Cells carry out metabolism that transforms energy and materials for growth and reproduction.

Unicellular and Multicellular Organisms

Living organisms are composed of one or more cells. The human body, for example, is composed of trillions of cells. Many microorganisms are **unicellular,** meaning that the entire organism is composed of a single cell. Most bacteria, for example, are unicellular. Each bacterial cell comprises the entire organism and is capable of independently carrying out metabolism for growth and reproduction to form progeny (offspring). Some microorganisms are **multicellular,** meaning that they are composed of many cells. Most multicellular microorganisms—such as the fungi that can often be seen as fuzzy filaments growing on bread—can be viewed as aggregates of cells with each individual cell having the properties of the entire organism. Each individual cell retains the capacity of reproducing to form the entire organism.

Microorganisms do not exhibit the advanced structural organization of plants and animals that form specialized groups of cells called **tissues.** Microorganisms lack tissues. The cells of a plant or animal tissue function together as integrated units. In animals, for example, we find epithelial tissue covering the external body surface, connective tissue binding together and supporting other tissues, muscle tissue moving parts of the animal's body, and nerve tissue transmitting messages from one part of the animal to another. The cells of these differentiated tissues usually lose the ability to generate the entire organism.

Microorganisms are more simply organized than plants and animals.

Microorganisms lack differentiated tissues.

Prokaryotic and Eukaryotic Cells

Among all living organisms there are two fundamentally different types of cells: **prokaryotic cells** (from the Greek word meaning before a nucleus) and **eukaryotic cells** (from the Greek word meaning having a nucleus) (FIG. 1-4). Both prokaryotic cells and eukaryotic cells carry out all the essential life functions: exchange of materials with the environment, energy processing, and reproduction. However, the structures of these types of cells, discussed in detail in Chapter 5, are quite different (Table 1-1).

A prokaryotic cell has a much simpler internal structure than a eukaryotic cell. Eukaryotic cells have numerous membrane-bound compartments, called **organelles,** that perform specialized functions. Prokaryotic cells do not contain membrane-bound organelles. Of prime importance is the fact that the DNA (the substance that encodes the hereditary information of the cell) is not separated within a specialized organelle from the rest of the cell contents in a prokaryotic cell. The DNA of a eukaryotic cell, in contrast, is contained within an organelle called the **nucleus.**

TABLE 1-1		
Comparison of Prokaryotic and Eukaryotic Cells		
	PROKARYOTIC CELL	**EUKARYOTIC CELL**
Plasma membrane	Present	Present
Internal organelles (membrane bound)	Absent	Present
Genetic (hereditary) molecule	DNA as a single circular bacterial chromosome not enclosed in a nucleus	DNA as multiple linear chromosomes enclosed within a nucleus
Site of energy (ATP) generation	Cytoplasm and, in some cases, plasma membrane or photosynthetic membranes	Cytoplasm and mitochondria or chloroplasts

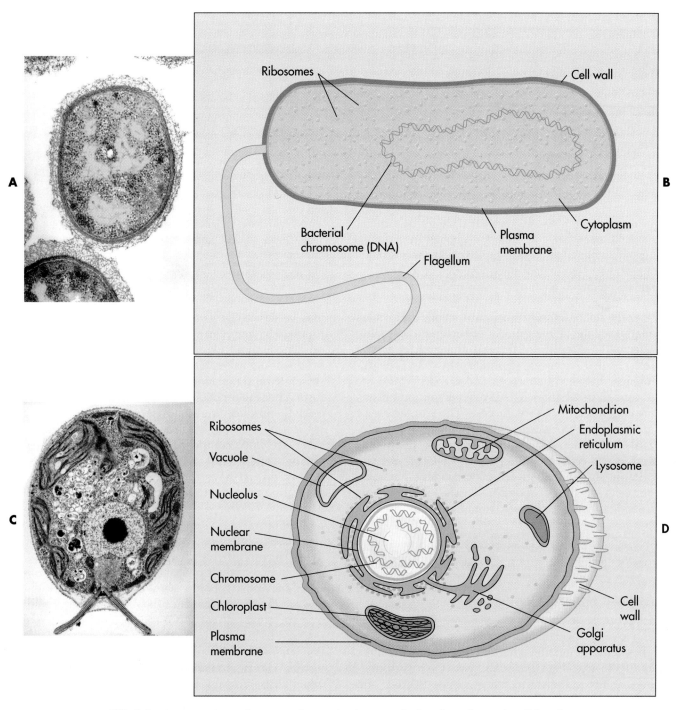

FIG. 1-4 A comparison of structural organization reveals that the eukaryotic cell has far more internal organization than the prokaryotic cell; the membrane-bound organelles found in eukaryotic cells do not occur in prokaryotic cells. **A,** Colorized micrograph of a prokaryotic cell of the bacterium *Pseudomonas aeruginosa* (32,400×). **B,** Drawing of a prokaryotic cell. **C,** Colorized micrograph of a eukaryotic cell of the green alga *Chlamydomonas reinhardtii* (6,750×). **D,** Drawing of an algal eukaryotic cell.

TABLE 1-2

Cellular Structures of Different Microorganisms

ORGANISM	INTERNAL ORGANIZATION	SITE OF ATP GENERATION	CELL WALL
PROKARYOTIC CELLS			
Eubacteria	No internal organelles*	Plasma membrane or in cytoplasm	Plasma membrane protected by a surrounding cell wall with unique chemical structure
Archaebacteria	No internal organelles	Plasma membrane or in cytoplasm	Plasma membrane protected by a surrounding cell wall with chemical structure different than eubacterial cell walls
EUKARYOTIC CELLS			
Fungi	Numerous internal organelles	Mitochondria and cytoplasm	Plasma membrane generally protected by surrounding cell wall
Algae	Numerous internal organelles	Chloroplasts and in mitochondria and cytoplasm	Plasma membrane generally protected by surrounding cell wall
Protozoa	Numerous internal organelles	Mitochondria and cytoplasm	Plasma membrane generally not protected by surrounding cell wall

*Although prokaryotic cells lack membrane-bound organelles, a few specialized eubacteria (for example photosynthetic bacteria) have internal membranes; these membranes are derived from and often continuous with the plasma membrane.

Some microorganisms are prokaryotic and others are eukaryotic (Table 1-2). All **eubacteria** (commonly referred to simply as bacteria) and **archaebacteria** (often called Archaea) are prokaryotic cells. (The differences between eubacteria and archaebacteria are discussed in Chapter 3.) All other living organisms are composed of eukaryotic cells. Three of the major groups of microorganisms—**fungi, algae,** and **protozoa**—are composed of eukaryotic cells. Plant cells and animal cells, including human cells, are eukaryotic. The difference between prokaryotic and eukaryotic cells sets the eubacteria and archaebacteria apart from all other organisms in a fundamental way.

All eubacteria and archaebacteria are prokaryotic cells that lack membrane-bound organelles; all other living organisms are composed of eukaryotic cells that contain membrane-bound organelles.

Acellular Nonliving Viruses

So far we have considered organisms that have cells and clearly meet all the criteria of living systems. These criteria include a high degree of organization, ability to exchange materials with their surroundings, ability to transform energy, ability to grow, and ability to reproduce independent progeny. Some microorganisms, however, do not have cells and do not meet all these criteria. Viruses are acellular (noncellular) microorganisms, meaning that they have neither prokaryotic nor eukaryotic cells. By some criteria, these organisms give the appearance of being alive, but by many other criteria they clearly are nonliving entities.

Viruses lack the fundamental structure of living systems. Viruses have a genetic molecule—which may be either DNA or RNA (ribonucleic acid)—surrounded by a protein coat (FIG. 1-5). Although the viral genetic molecule is capable of directing viral reproduction, viruses do not have the cellular support structures and metabolic machinery necessary to perform other life functions. They depend on living host cells to produce the materials necessary for their replication. Although some viruses are surrounded by an envelope composed of a membrane that they obtain from a host cell in which they replicate, viruses do not have functional plasma membranes separating them from their surroundings. They have no means of carrying out independent metabolism or replication.

Viruses rely entirely on the metabolic activities of living cells to provide energy and materials for their replication. On their own, viruses are inanimate objects. They passively interact with their environment and are unable to reproduce themselves. They do not transform energy, carry out metabolism, or actively respond to their environment. All of these are essential characteristics of living systems. Therefore viruses can be considered as nonliving. However, when viruses are able to enter (infect) living cells, the viral nucleic acid molecule has the capability of directing the replication of the complete virus. Within the confines of a living cell, the genetic information of the viral nucleic acid takes over control of the metabolic activities of that cell. Many microbiologists, therefore, view viruses as genetic extensions of the host cells in which they replicate.

Viruses can carry out the functions characteristic of living organisms only within living cells.

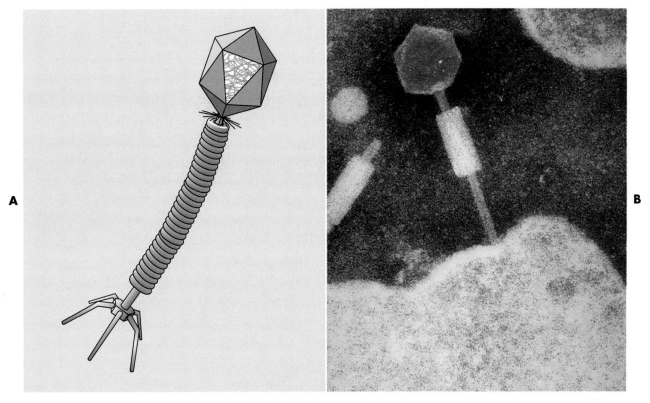

FIG. 1-5 A, A virus has a protein coat (capsid) surrounding a nucleic acid hereditary molecule (RNA or DNA *[green]*). **B,** Colorized micrograph of a bacteriophage *(blue)* that infects and replicates within cells of the bacterium *Chromatium violaceum (tan).*

HIGHLIGHT

PRACTICAL SIGNIFICANCE OF ORGANIZATIONAL DIFFERENCES AMONG MICROORGANISMS

Organizational differences between the acellular viruses, prokaryotic eubacteria and archaebacteria, and eukaryotic fungi, algae, and protozoa are of more than just scientific interest. They also have great practical importance. The use of many antibiotics (drugs that adversely affect microorganisms) to treat human diseases caused by bacteria, for example, depends on the ability of the antibiotic to act selectively against the prokaryotic cells of the invading bacteria without killing the eukaryotic human cells at the same time. Antibiotics, such as penicillin, are widely used to treat bacterial infections, such as syphilis, because they specifically target prokaryotic cells. If you were to develop pneumonia caused by a particular species of bacterium, a physician could prescribe any number of antibiotics that could kill the infecting bacteria and help cure you of the disease. It is more difficult to use antibiotics against disease-causing fungi and protozoa, such as malaria, because fungi and protozoa, like humans, are composed of eukaryotic cells, so that what is toxic to them is also often toxic to the human cells.

Likewise, it is hard to find antibiotics that can be used to treat viral diseases because viruses replicate within the confines of the living cells of the organism they infect. Chemicals that inhibit viral replication generally do so by killing the host cell within which the virus replicates. This prevents the therapeutic use of these chemicals. No one wants to use a drug that kills the patient at the same time it kills the disease-causing microorganism. So, when you have the common cold—a disease caused by a virus—a physician generally can only tell you to rest, stay warm, and drink plenty of fluids. This prescription enables your body's own defenses to fight the invading viruses.

Without question we have learned a great deal about microorganisms in the last century. This knowledge now enables us to control microorganisms for human good—at least to some extent. We have developed methods for preventing microbial attack on foods and other materials. Refrigeration, freezing, canning, and other methods prevent microbial growth and, thereby, preserve foods. We paint wood surfaces and keep fabrics dry to prevent deterioration due to microbial growth. Microorganisms are used for the treatment of sewage and municipal wastes, ridding the world of waste materials.

HIGHLIGHT

A PRACTICAL VIEW OF MICROBIOLOGY

Microbiological researchers at university, government, and industrial laboratories carry out studies that reveal the fundamental nature of microorganisms and the practical applications of that knowledge. Understanding the genetics and metabolism of microorganisms facilitates the beneficial uses of microorganisms.

Health care workers treat individuals with diseases caused by microorganisms and try to prevent the spread of infectious diseases. Vaccination is used to prevent many once deadly diseases. Antiseptics are used to preclude microorganisms from infecting wounds. Antibiotics are administered to cure patients of infecting microorganisms.

The clinical microbiology laboratory helps physicians diagnose diseases and determine the most effective treatments (FIG. A). The clinical microbiologist and staff receive samples from physicians and run various tests to identify the underlying causes of disease in patients. The clinical microbiology laboratory also tests the effectiveness of various drugs against disease-causing microorganisms.

The food industry maintains strict quality control to preserve foods and to protect against the foodborne spread of disease. Technicians in the food industry frequently test batches of foods to ensure safe levels of microorganisms. Canning, refrigeration, freezing, and other methods are routinely used to prevent food spoilage.

Many foods and beverages are produced using microorganisms, including bread (yeasts are used for leavening to cause the dough to rise), cheese (bacteria are used to convert milk into various cheeses), wine (yeasts are used to convert grapes into this alcoholic beverage), beer (yeasts are used to convert grains into this alcoholic beverage), and numerous other products. We all have enjoyed eating or drinking the foods produced by microorganisms (FIG. B).

Farmers use agricultural practices that control the spread of plant pathogens (microorganisms that cause diseases of plants). Agricultural extension service workers help farmers plan appropriate control measures to ensure maximal crop production. Fungicides are often used to prevent fungal diseases that destroy crops.

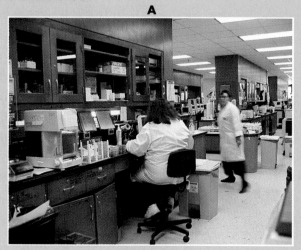

The clinical microbiology laboratory helps diagnose the causes of infectious diseases and aids the physician in determining appropriate therapies.

Microorganisms are cultivated to produce alcoholic beverages such as beer. Copper kettles often are used for preparing the substrate for yeast fermentation (mash production). The fermentation often is carried out in open fermentors. Bubbles of carbon dioxide in the fermentor accompany ethanol production.

MICROORGANISMS AND DISEASE

Our knowledge of how disease-causing microorganisms (pathogens) spread has permitted us to reduce the incidence of many diseases. Improved sanitation practices, based on our understanding of how disease causing microorganisms are transmitted to humans, have greatly reduced the incidence of diseases such as plague and gastrointestinal infections. Disinfection of water and proper cooking and handling of foods has greatly reduced the incidence of disease such as typhoid that were once widespread. Sanitary practices in medicine introduced in the latter half of the nineteenth century are now routinely used to protect patients and medical staff against infections (FIG. 1-6).

Some microorganisms can also be used as biological control agents to replace chemical pesticides.

Microorganisms are used to degrade wastes and pollutants (FIG. C). Sewage treatment plants, compost piles, and septic tanks use microorganisms to decompose wastes into products that can be accommodated by the environment. Sanitary engineers are important in seeing that these microbial processes run effectively. Microorganisms are also used to bioremediate polluted sites, such as shorelines contaminated by oil spills.

Microorganisms are used in industry to produce antibiotics, vaccines, solvents, and numerous other products of economic value (FIG. D). Numerous technicians oversee the daily production operation and ensure the quality of the final product. The biotechnology industry, which uses microorganisms for the production of these products, has been rapidly growing. The ability to genetically modify microorganisms now permits the production of many new products and undoubtedly will lead to the production of many more.

Microorganisms are used to degrade wastes and pollutants to maintain and restore environmental quality. An activated sludge treatment facility has tanks in which microorganisms degrade wastes. Extensive aeration and agitation maintain aerobic conditions that favor complete degradation of organic compounds by microorganisms.

C

D

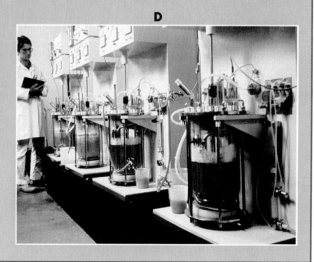

Microorganisms are grown in fermentors to produce pharmaceuticals. Small-scale fermentors are used in the research and development phase of pharmaceutical development.

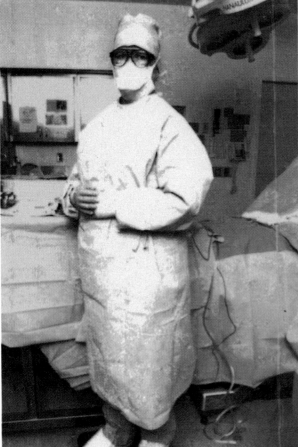

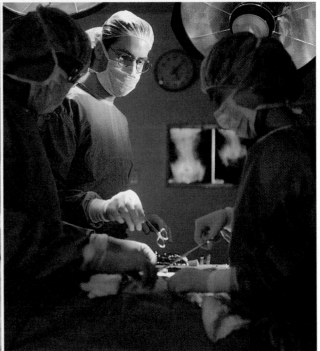

FIG. 1-6 **A,** Florence Nightingale revolutionized nursing practice during the Crimean War when she introduced sanitary practices that help control the spread of infectious diseases. **B,** Today nurses wear masks, gloves, and gowns during surgery and other medical procedures to prevent infections. **C,** Surgery is performed in clean operating theaters that are relatively free of microorganisms.

A mere century ago, 5 of every 10 deaths in the United States were due to microbial infections (FIG. 1.7). Tuberculosis, pneumonia, and gastrointestinal infections were the three leading killers, with diphtheria and bronchitis also listed among the top 10. Today, the only remaining infectious disease caused by microorganisms in this list is pneumonia, a disease that occurs when the defense systems of the body fail—for example, as a consequence of another disease. Although tuberculosis, gastrointestinal infections, bronchitis, and diphtheria still occur, and the number of cases of tuberculosis has recently risen, the percentage of deaths due to these diseases has dropped from over 20% in 1900 to less than 1% of total deaths today.

The reduced mortality from infectious diseases is largely due to the discovery and widespread medical uses of antibiotics and vaccines that have enabled contemporary medicine to reduce the impact of infectious diseases. Antibiotics such as penicillin were not available in medical practice until after World War II. They are now routinely used to control microbial in-

NEWSBREAK

ORIGINS OF VACCINATION

Lady Mary Wortley Montagu, wife of the English ambassador to Turkey (1716-1718), described a crude form of inoculation against smallpox practiced there. In September, old women insert some of "the matter of the best sort of smallpox" into a vein. After 8 days, those so inoculated develop a fever, remain ill for 2 or 3 days, recover within 8 days, and get no pox marks. Lady Montagu tried to persuade the English to adopt this practice, but without success. Edward Jenner got the idea for the technique of vaccination against smallpox from a young girl, a milkmaid it is said, who told him that she would not catch smallpox because she had had cowpox. Jenner is credited with the introduction of vaccination in Europe.

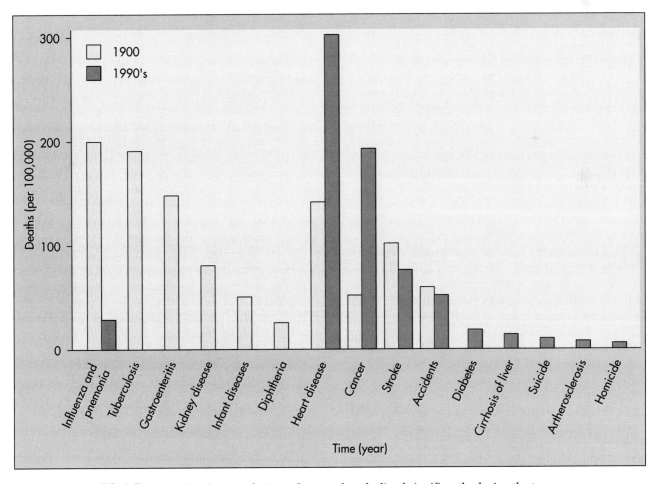

FIG. 1-7 Mortality due to infectious deseases has declined significantly during the twentieth century because of an increased understanding of the roles of microorganisms in human disease and the introduction of practices that reduce the occurrence and effects of microbial infections. Deaths due to many once deadly diseases such as diphtheria, gastroenteritis, and even tuberculosis are very low (near 0 per 100,000) in the 1990s.

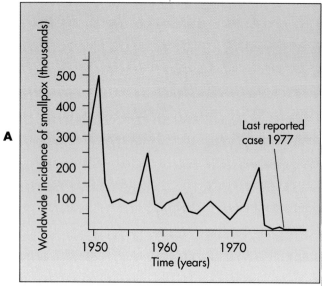

A

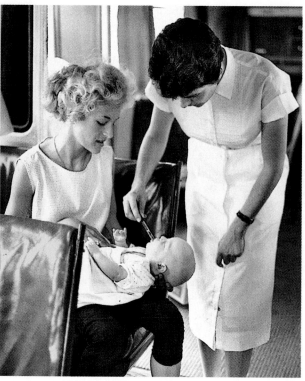

B

FIG. 1-8   A, Smallpox has been eliminated through vaccination. The vaccination program that eliminated smallpox began in the 1700s with the work of Lady Mary Montagu and Edward Jenner and became a major effort of the United Nations World Health Organization after World War II. **B,** Vaccination is used today to prevent many diseases.

fections. Clinical microbiology laboratories assist physicians by rapidly identifying pathogens in infected individuals and determining which antibiotics can be used to control that disease-causing microorganism.

Vaccines are extremely important in preventing disease. An amazing accomplishment of vaccination is the elimination of smallpox. Once 80% of all people on earth could expect to contract smallpox. Now the world is free of this disease. Three centuries of vaccination against smallpox and a massive 20-year public health effort by the World Health Organization finally were victorious over the smallpox virus (FIG. 1-8). The last natural case of smallpox occurred in Somalia, Africa, in 1977. For the first and only time, an infectious disease was eliminated from earth. Effective vaccination programs have also reduced the occurrences of other diseases, such as polio, measles, diphtheria, and whooping cough.

Microbiologists continue to search for new vaccines (substances that stimulate the body's defenses, rendering an individual immune to disease) and antibiotics and to use the techniques of molecular biology to engineer genetically new microorganisms that can produce useful products such as antibiotics and vaccines. Such genetically engineered microorganisms will help meet our future technological needs and improve human health and well-being.

The reduced impact on humankind of diseases caused by microorganisms is due in part to advances in our scientific knowledge of microorganisms and their interactions with humans.

Against the tide of enormous advancement in our understanding of the microorganisms has come the emergence of new infectious diseases, including acquired immunodeficiency syndrome (AIDS). AIDS has shown that diseases caused by microorganisms can still radically change the lives of infected individuals and also of entire societies. Daily, we read in the newspapers, hear on the radio, and see on television reports about the AIDS virus and other diseases caused by microorganisms. Illness and death spark public concern, and the medical aspects of microbiology usually are of the most interest to many of us.

So far, scientists have been unable to develop an effective means of preventing AIDS or drugs that will rid patients of the human immunodeficiency virus (HIV) that causes it once an infection occurs (FIG. 1-9). The Centers for Disease Control estimate the incidence of infection with HIV in the United States in 1995 to be 3 in 1,000, with the infection more common in males than females. The World Health Organization estimates that in Africa the incidence in 1995 will be much higher, with 1 in 10 men and women infected with the AIDS causing HIV virus. The cost of treating AIDS patients in New York City alone is running well over $1 billion per year. Only changes in behavioral patterns that reduce the risk of infection with HIV hold any immediate promise for reducing the incidence of this disease. These include the avoidance of unprotected sex with individuals who may carry HIV and avoidance of exposure to contaminated blood through transfusions or shared

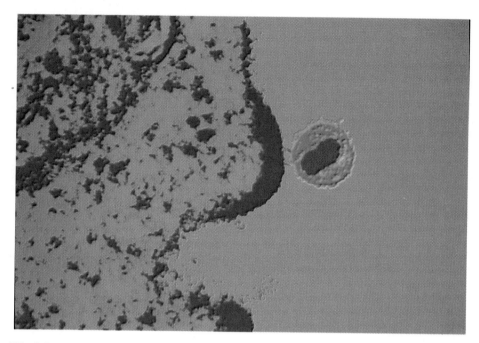

FIG. 1-9 Colorized micrograph of the human immunodeficiency virus (HIV) *(blue)*, that causes AIDS, on the surface of a cell *(red)*.

intravenous needles. If behavioral changes do not occur, and if no cure or preventive is found, the impact of this disease on society will be staggering.

AIDS is just the latest major outbreak of disease to afflict humankind. All societies, both contemporary and ancient, have suffered from the ravages of dis-eases caused by microorganisms (FIG. 1-10). Plague, tuberculosis, polio, smallpox, and various other dis-eases caused by microorganisms followed explorers, traders, missionaries, immigrants, or soldiers. Mi-croorganisms have had a significant impact on the course of human civilization, including warfare, reli-

FIG. 1-10 During the Middle Ages there were major epidemics of plague. The disease followed the trade routes, as did the rats carrying the flea vectors infected with *Yersinia pestis*.

gion, migration of populations, art, agriculture, science, and technology. Major disease outbreaks, often plague, altered the course of history many times and contributed to the decline of empires. Widespread disease in ancient Athens led to that city–state's loss to its rival Sparta, during the Peloponnesian War. This prevented Athens from establishing a unified

Greek empire. When the central government of the Roman Empire was weakened by the debilitating effects of plague, it could not consolidate its power over its vast empire and the influence of Rome declined.

Population patterns have often been determined by the actions of microorganisms. A fungal infection of potatoes in Ireland in the mid-nineteenth century caused widespread famine. This led to the migration of large numbers of Irish people to the United States. European settlers in the New World thought that the diseases with which they infected the natives were God's way of showing approval for their colonization. Entire tribes of native North Americans were eradicated due to outbreaks of smallpox. These outbreaks were sometimes initiated when Europeans gave them smallpox infected blankets.

Microorganisms have had major impacts on human health and society.

BENEFICIAL USES OF MICROORGANISMS

We have long depended on microorganisms for the production of many of the foods and beverages we enjoy. Even early societies produced beer, wine, and bread using microorganisms. We have built sewage treatment facilities where microorganisms are grown to degrade our wastes. We have also been able to greatly reduce the incidence of many diseases through the use of antibiotics and vaccines produced by microorganisms (FIG. 1-11). Indeed, the vaccines,

FIG. 1-11 **A,** Some viral vaccines are produced by culture in eggs or tissure culture. **B,** Vaccines are often produced by growing microorganisms and harvesting them.

A

 B

antibiotics, and other pharmaceuticals that are used to prevent or to treat many human diseases are microbial products.

Many microorganisms benefit humans, producing foods we eat, removing wastes we produce, and making antibiotics and vaccines for use in medicine.

Many agricultural practices encourage the beneficial metabolic activities of microorganisms and discourage their harmful effects. Spacing of crops, the development of disease resistant plant varieties, and the use of chemical pesticides aim to limit infections due to plant pathogens. They also limit crop damage due to insects and other nonmicrobial pests. Crop rotation uses plants such as alfalfa and soybeans that are associated with nitrogen-fixing bacteria. These bacteria can convert atmospheric nitrogen (N_2) into forms of nitrogen that can be used by plants. Plant growth depends on the forms of nitrogen provided by these bacteria or on the addition of nitrogen fertilizers by farmers. Without the microbial activities that transform decaying dead material to substances that can be reused by the living, life on earth would soon cease. We could not survive without microorganisms.

Microorganisms are essential for maintaining the balance of nature and sustaining life on earth.

Clearly the microorganisms are complex and difficult to understand—powerful but small, necessary yet sometimes deadly. Through the science of micro-

NEWSBREAK

BIOTECHNOLOGY TO CURE GLOBAL WARMING

Microorganisms may have a role in reversing global warming. They may be used to remove carbon dioxide from the atmosphere. Japanese researchers are investigating the possibility of using genetically engineered algae to remove carbon dioxide from the atmosphere and maintain levels of atmospheric carbon dioxide that will not cause global warming. Already, Japanese researchers have found an alga that has ten times the carbon dioxide fixing capacity of trees. This would be a novel employment of biotechnology for the maintenance of global environmental quality. The idea is to genetically engineer algae with high photosynthetic capacities that produce polymers that are not easily biodegraded. In this way, carbon dioxide would be drawn out of the atmosphere and incorporated into compounds that are not easily degraded back to carbon dioxide.

biology we have increased standards of living, improved agricultural crop yields, helped maintain environmental quality, and decreased the incidence of disease. By studying the microorganisms, we hope to develop the necessary understanding to control microorganisms for the benefit of the future of humankind.

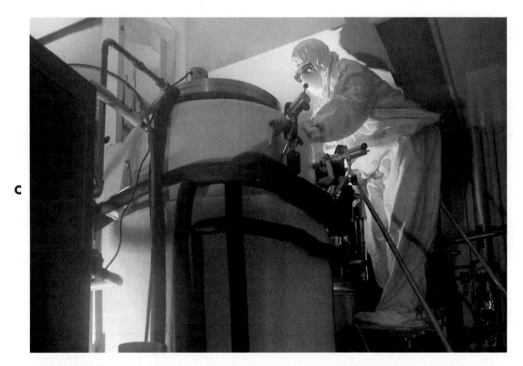

FIG. 1-11 cont'd C, Some vaccines are produced by growing recombinant vaccine strains in fermentors; the fermentor in this photograph is used to produce hepatitis B vaccine employing a recombinant strain of *Escherichia coli* containing the viral genes.

SUMMARY

Microorganisms (pp. 3-13)

- Microorganisms are very small life forms.
- The observation of microorganisms depends on the use of microscopes.
- Antonie van Leeuwenhoek made the first recorded microscopic observations of bacteria.

The Organizational Structure of Microorganisms (pp. 5-13)

- Even microorganisms, the smallest forms of life, are highly organized living systems.
- The cell is the fundamental unit of all living systems.
- The plasma membrane acts as a barrier that controls the flow of material into and out of the cell.
- Cells contain hereditary information in DNA and carry out metabolism that transforms energy and materials.
- Many microorganisms are unicellular (single celled organisms).
- Some microorganisms are multicellular (composed of many cells).
- Bacteria are cellular organisms, with simple cells that lack organelles.
- Bacterial cells are prokaryotic.
- Fungi, algae, protozoa, plants, and animals (including humans) all have eukaryotic cells.

- Eukaryotic cells have organelles, including a nucleus that houses the genetic information.
- Organelles are membrane-bound compartments within a cell.
- Viruses are not made of cells.
- Viruses only replicate within living cells.

Importance of Microorganisms to Humankind (pp. 14-21)

- Some microorganisms cause disease.
- Diseases caused by microorganisms have a major impact on individuals and human society.
- Today's reduced rates of sickness and death are a direct result of our ability to control the growth of microorganisms and our basic understanding of the involvement of microorganisms in infectious diseases.
- Most microorganisms are nonpathogens that serve useful industrial and essential environmental functions.
- Microorganisms are used to make various foods, beverages, and other useful products.
- Microorganisms are essential for life to continue on earth.

CHAPTER REVIEW

REVIEW QUESTIONS

1. What kinds of organisms are studied by microbiologists?
2. What is a microorganism?
3. How can you distinguish living from nonliving microorganisms?
4. How would you define a living organism?
5. What are the arguments for viruses being living organisms? Against?
6. Why is microbiology considered a relatively new field of science?
7. Describe the fundamental organizational structure of living organisms.

8. What are the differentiating characteristics of prokaryotic and eukaryotic cells?
9. What is the difference between unicellular and multicellular organisms? Which microorganisms are unicellular and which are multicellular?
10. How are microorganisms distinguished from higher plants and animals?
11. What are the characteristics that make eubacteria similar to archaebacteria? What are the characteristics that differentiate them?
12. What are the characteristics that make fungi, algae, and protozoa similar? What are the characteristics that differentiate them?

CRITICAL THINKING QUESTIONS

1. How have microorganisms affected your life? Consider the adverse effects microorganisms may have had, such as infectious diseases you, your friends, and your relatives may have had, and the benefits you may have experienced, such as foods and beverages you may have enjoyed.
2. What would life be like without microorganisms? What would be the consequences of eliminating all pathogens from earth? What changes would occur in society if there were no infectious diseases?

3. If you were sending a mission to Mars to explore for life, what would you look for? How would you know whether life existed or not? How would you recognize the difference between an intelligent computer and a living organism?
4. Within the last few years, a new bacterium has been isolated from the intestinal tracts of surgeonfish in the Red Sea. These bacteria are so large that they are visible to the unaided eye. What does this mean with respect to the definition of microorganisms?

READINGS

Local and National Newspapers and News Magazines.

Frequent articles about microorganisms, diseases caused by microorganisms, and biotechnology.

De Kruif P: 1926. *Microbe Hunters.* NY, Harcourt, Brace and Co.

Inspiring stories of the lives and works of some important historical figures in microbiology.

Dixon B: 1994. *Power unseen: how microbes rule the world.* New York, W.H. Freeman.

A well-written book on the microbial perspective highlighting the importance of microbiology.

Dobell C (ed.): 1932. *Antony van Leeuwenhoek and His Little Animals,* London, Constable and Co.

Classic volume describing the life and times of Antony van Leeuwenhoek; includes Leeuwenhoek's drawings of his observations.

Gest H: 1988. *The World of Microbes,* Madison, Wisconsin, Science Tech Publications.

This explanation of microbiology is written for the average nonscientist to read and is spiced with many historical anecdotes, drawings, diagrams, and useful appendices.

Henderson DA: 1976. The eradication of smallpox, *Scientific American* 235(4):25-33.

Describes the campaign that ultimately led to the worldwide elimination of this deadly disease.

Hopkins DR: 1983. *Princes and Peasants: Smallpox in History,* Chicago, University of Chicago Press.

The history of smallpox and smallpox's influence on history with chapters on the disease in Europe, China, Japan and the Pacific, India, Africa, Latin America, and North America. Interesting insights into smallpox, its impact on societies, and primitive treatments.

Latour B: 1988. *The Pasteurization of France,* Cambridge, Harvard University Press.

Study of science and society in nineteenth century France.

Lederberg J (ed.): 1992. *Encyclopedia of Microbiology,* 4 volumes, San Diego, Academic Press.

A comprehensive four-volume reference covering all major areas of microbiology.

McEvedy C: 1988. The bubonic plague, *Scientific American* 258(2):118 123.

Discusses the factors involved in the rise and fall of incidence of bubonic plague.

Moberg CL and ZA Cohn: 1991. Rene Jules Dubos, *Scientific American* 264(5):66-67, 70-74.

A description of the life and accomplishments of a noted microbiologist and philosopher.

Phaff HJ: 1986. My life with yeasts, *Annual Review of Microbiology* 40:1-28.

This professional autobiography of a foremost scientist in this field emphasizes his interest in the ecology and taxonomy of yeasts.

Postgate J: 1992. *Microbes and Man,* Cambridge, England; Cambridge University Press.

A well-written book on the practical aspects of microbiology showing how microorganisms affect our daily lives.

Reid R: 1975. *Microbes and Men,* New York, Saturday Review Press.

Script of a BBC television series that portrayed several important historical events in microbiology, including the discoveries of Pasteur, Koch, and Fleming.

Scientific American: 1988. What science knows about AIDS. 259(4): a single topic issue.

Ten articles detailing the history, molecular biology, epidemiology, and the clinical, cellular, therapeutic, and social status of AIDS.

Thomas L: 1979 & 1980. *The Lives of a Cell* and *The Medusa and the Snail,* New York, Bantam Books.

These writings of a physician/scientist/philosopher make science an exciting, exhilarating human activity.

CHAPTER 2

Diversity of the Microbial World

PREVIEW TO CHAPTER 2

In this chapter we will:
- Learn how microorganisms are classified.
- See how microorganisms evolved.
- Gain an overview of the microbial world by examining the fundamental characteristics of the various types of microorganisms.
- Examine characteristics of the acellular, nonliving microorganisms—viruses, viroids, and prions.
- Examine characteristics of the prokaryotic microorganisms—archaebacteria and eubacteria.
- Examine characteristics of the eukaryotic microorganisms—fungi, algae, and protozoa.
- See that the structural differences between acellular, prokaryotic, and eukaryotic microorganisms have practical importance in medical practice.
- Learn the following key terms and names:

algae	fungi
Archaea	Gram-negative bacteria
bacteria	Gram-positive bacteria
bacteriophage	Monera
Bergey's Manual	morphology
binomial name	mycelia
DNA hybridization	prions
endospores	Protista
endosymbiotic theory of	protozoa
evolution	species
eubacteria	taxonomy
eukaryotes	viroids
evolution	virus
evolutionary relatedness	yeasts
filamentous fungi	
five-Kingdom	
classification system	

The microbial world is extremely diverse. Some organisms have common features that permit grouping into common categories. Scientists use a hierarchical (ranking system) organizational structure in which organisms are classified (grouped) according to their degrees of similarity (FIG. 2-1). Classification systems are used to systematically establish the characteristics of different groups of organisms so that specific types of organisms can be described and identified. The broadest groups in a classification system are called Kingdoms and the smallest are called species. Related species are grouped into genera. For example, *Mycobacterium tuberculosis* and *Mycobacterium leprae* are both in the genus *Mycobacterium.*

The science of identifying and classifying organisms is called **taxonomy,** and scientists who study the classification of organisms are called **taxonomists.** Arguments among taxonomists are often quite heated. Some taxonomists tend to classify many similar organisms into large groups. Others favor small groups that emphasize even minor differences between organisms. Interpreting the evidence rests on personal judgement on the relative importance given certain criteria. Is shape, color, the ability to metabolize a given compound in a particular way, or something else the most important characteristic that should be used to distinguish one microbial species from another? Many modern taxonomists use evolu-

tionary relationships—determined by molecular biological analyses—as the basis for establishing order in classifying organisms.

> **Classification systems arrange living organisms into groups based on common factors and show how they are related.**

It has been difficult to classify microorganisms based on how they are related to each other in terms of their evolutionary relationships. We use fossils that were preserved in the Earth's crust to piece together evolutionary relationships of plants and animals. Microorganisms lack structures such as bones and woody tissues that are preserved as fossils. There is only a limited fossil record that provides a basis for comparing ancient forms with contemporary forms of microorganisms. Direct examination of the ancestors of today's microorganisms is for the most part impossible. As a result, the classification of microorganisms generally has been based on relationships inferred from observed similarities and differences in *morphology* (form and structure) and *physiology* (functions) among actual living microorganisms.

The **species** is the fundamental grouping used in classifying all organisms. In higher organisms the species is defined by the ability to interbreed and produce fertile offspring. Organisms that can interbreed are considered members of the same species.

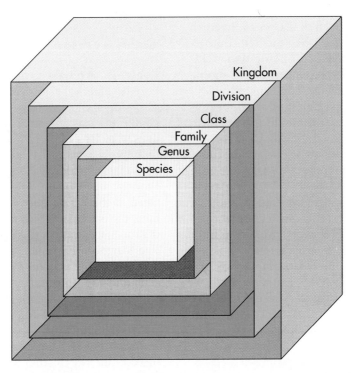

FIG. 2-1 The system used for the classification of living organisms employs hierarchical levels of organization: the highest level of organization (most encompassing) is the Kingdom; the species is the fundamental level of organization.

Those that cannot sexually reproduce fertile off-spring are defined as members of different species. Many microorganisms, however, reproduce asexually, with a single cell dividing to form progeny. Therefore, defining microbial species based on the ability to reproduce sexually is not meaningful.

Microbiologists have been hard pressed to find a working definition of a microbial species. Often, they consider a species as a group of microorganisms that have similar characteristics and that are significantly different from those of other groups. Thus, a bacterial species is a group of very closely related bacteria. Many bacterial species that are only genetically distantly related, however, have evolved similar morphological and physiological characteristics, making this definition ambiguous. DNA molecules (hereditary molecules) of all members of the same bacterial species have a very high degree of homology (similarity) to each other, and therefore modern classification systems increasingly rely on molecular analyses of DNA and RNA.

EARLY CLASSIFICATION SYSTEMS

Early classification systems date from the time of Aristotle, before the existence of microorganisms was known. These early systems recognized only two major groups—the plant and the animal Kingdoms—which could be easily distinguished based on motility. All organisms that moved were classified as animals and all those that remained stationary were called plants. In the eighteenth century, just after the microbial world was discovered, the noted Swedish botanist, Carl Linnaeus, established the first comprehensive classification system of all living things. He recognized that some microorganisms were motile and others remained stationary. Linnaeus placed all microorganisms into a single genus, which he appropriately named *Chaos* because he could not establish objective criteria for distinguishing among them.

Gradually, attempts were made to change *Chaos* into an orderly classification system. As is the case for higher organisms, identifying and naming a microorganism requires a system of classification in which the critical characteristics are systematically established that distinguish one microorganism from another. But because many microorganisms look exactly alike, we are usually unable to simply look at a microorganism and identify it. In most cases, to identify a microorganism, it is necessary to consider not only its form (morphology) but also its metabolism (physiology) and its evolutionary history as revealed in its genetics (its molecular sequences).

In 1866, Ernst Heinrich Haeckel proposed a classification system based on inferred evolutionary relationships among species. This was the first attempt to include evolution theory within a classification system. **Evolution** is the process of change that results

from the interaction between the genetic information and the environment of organisms of a species. Evolution results in the formation of new organisms that are better adapted to the environment. Modern classification systems attempt to reflect the gradual evolutionary changes that have resulted in the appearance of new organisms. Such **phylogenetic classification systems** (from the Greek *phylo,* meaning kind or type, and *geny,* meaning origin) categorize organisms based on their evolutionary relatedness.

A phylogenetic classification system takes into account evolutionary relatedness.

Haeckel's system contained three Kingdoms: Protista, Animalia, and Plantae. Haeckel proposed that the microorganisms—bacteria, fungi, algae, and protozoa—all belonged to one primary Kingdom, which he called the **Protista.** The Protista were characterized by their simple structural organization, that is, by their lack of specialized tissues. Haeckel believed that the Protista was the first Kingdom to evolve and that both the Plantae and Animalia evolved from the Protista. They both had tissues. The Plantae had specialized tissues and were photosynthetic, using light energy for growth. The Animalia also had specialized tissues but were heterotrophic, meaning that they used organic compounds for energy and growth. Although this system was somewhat naive, it did show a great deal of insight. Its recognition of microorganisms as a single Kingdom distinct from plants and animals is appealing to microbiologists even today.

MODERN CLASSIFICATION SYSTEMS

Robert H. Whittaker, in 1969, proposed a modification of Haeckel's system that became widely accepted by biologists (FIG. 2-2). Whittaker tried to organize his classification system along evolutionary lines, which he inferred from observed characteristics of living organisms. Whittaker proposed a five-Kingdom classification system: Animalia, Plantae, Fungi, Protista, and Monera. The microorganisms constitute three of the Kingdoms. The Kingdom **Monera** contains the bacteria, which are separated from all other organisms based on their prokaryotic cell structure. The Kingdom **Protista** contains the unicellular algae and protozoa based on the fact they have eukaryotic cells and are unicellular. The fungi, some of which are unicellular and some of which are multicellular, are in the Kingdom **Fungi,** based on their lack of specialized tissues.

According to Whittaker, the Monera (bacteria—prokaryotic organisms) were the first organisms on Earth and the Protista (protists—eukaryotic unicellular organisms) evolved directly from the Monera. Whittaker proposed that fungi, plants, and animals evolved from the protists via three separate direc-

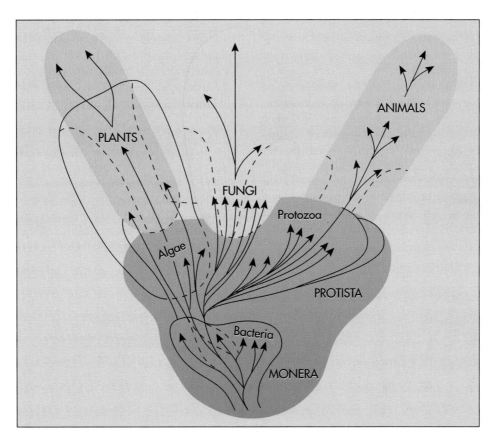

FIG. 2-2   Robert H. Whittaker in 1969 proposed a five-Kingdom classification system that emphasized evolutionary divergence from a presumed common ancestral prokaryotic bacterium. The lines of evolution followed nutritional strategies. Microorganisms occupy three of the five Kingdoms in this system.

tions of evolution. These evolutionary lines were based on differences in how the organisms met their nutritional needs. According to Whittaker's hypotheses, the fungi evolved as the most complex multicellular organisms that obtained their nutrients by absorption, that is, by taking up the chemicals that they needed for growth and reproduction. Animals evolved based on the ability to ingest other organisms to meet their nutritional needs. Plants evolved based on their photosynthetic capacity for self-feeding, that is, their ability to synthesize organic compounds from inorganic compounds using light as an energy source. Among the plants, he included the multicellular algae, such as the structurally complex brown and red algae. Whittaker's system, like the earlier ones, however, lacked direct evidence to support its validity.

> **The five-Kingdom classification system of Whittaker is based on evolution of species from the prokaryotes (bacteria) along strategies for obtaining nutrition.**

In the early 1970s, developments in the field of molecular biology provided the first tools for directly examining the evolutionary relationships among microorganisms. Methods were developed that permitted determination of the degree of relatedness of the DNA from two different organisms. DNA contains the genetic information that is passed to progeny cells generation after generation. It is the primary substance of heredity. DNA provides a molecular record of relatedness that is far more accurate than the fossil record for tracing the path of evolutionary change for any organism. Additionally, RNA can be used to infer evolutionary relatedness among organisms. RNA is made by directly using the information in the DNA. Therefore it can be used to infer hereditary and, hence, evolutionary relatedness.

> **A precise determination of microbial relatedness can be made based on DNA or RNA nucleotide sequences.**

Carl Woese, in the 1970s, analyzed RNA to explore the evolution of microorganisms. Woese reasoned that the RNA of ribosomes was changed only relatively slowly during the evolution of new organisms. Ribosomes are the sites where proteins are made within all cells. So, ribosomal RNA (rRNA) should not have changed much as a result of small evolutionary steps. Rather, rRNA should change significantly only as a result of major steps in evolution. Before Woese's studies, it was generally accepted that all bacteria were closely related because they all were prokaryotic cells. Woese's analyses of rRNA molecules revealed that the bacteria actually fell into two distinct and only distantly related groups. One of these groups is as distantly related to organisms with eukaryotic cells as it is to the other group of bacteria.

Woese proposed a new and radically different classification system that defines these three groups as three primary Kingdoms for all living things: *Archaebacteria (Archaea), Eubacteria,* and *Eukaryotes* (FIG. 2-3). According to Woese's theory, three separate paths of evolution from a common progenitor cell produced three different types of cells: the archaebacterial cell, the eubacterial cell, and the eukaryotic cell. Further studies on the physiological characteris-

tics of the organisms designated by Woese as archaebacteria appear to substantiate his proposal. The chemical composition of the plasma membranes of archaebacteria is totally different from all other cells. Additionally, the archaebacteria have metabolic capabilities not found in any other organisms.

There are three distinct lines of cellular evolution that have led to the archaebacteria, eubacteria, and eukaryotes.

With regard to the evolution of the modern eukaryotic cell, with its multiple organelles, Woese accepted an idea put forth a few years earlier by Lynn Margulis—the **endosymbiotic theory of evolution.** Margulis proposed that some bacterial cells had be-

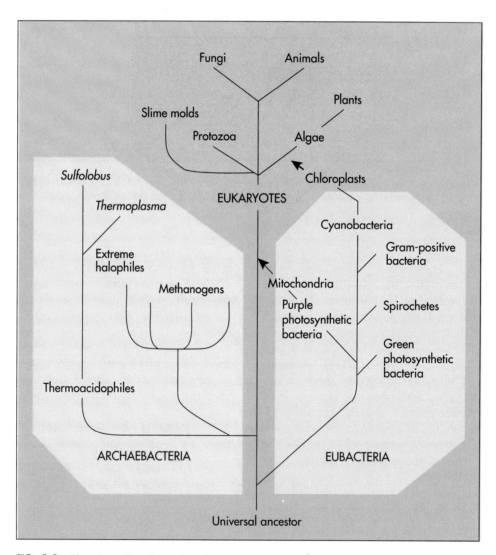

FIG. 2-3 The three-Kingdom classification system proposed by Carl Woese was developed using the modern techniques of molecular biology and, in particular, the examination of the RNA macromolecules of ribosomes. Unlike previous classification systems, the analysis of conserved gene products permits a direct assessment of genetic and, thus, evolutionary relatedness. Based on rRNA analyses, Woese found that there were three primary lines of evolution leading to the archaebacteria (Archaea), eubacteria, and eukaryotes. Modern eukaryotes have cells that incorporated mitochondria (derived from purple photosynthetic bacteria) and chloroplasts (derived from cyanobacteria); mitochondria and chloroplasts became organelles of eukaryotic cells by endosymbiosis.

gun to live within the predecessors of modern eukaryotic cells in a mutually beneficial (symbiotic) relationship that probably helped each meet its metabolic energy and nutritional requirements (FIG. 2-4). Living together, the prokaryotic and eukaryotic organisms could help each other live under conditions where either might not be able to live independently. Eventually the endosymbiotic bacteria lost their capacity for independent existence and developed into two of the cellular organelles (mitochondria and chloroplasts) involved in energy transformations that are found in eukaryotic cells today.

Endosymbiosis accounts for the acquisition of organelles by eukaryotes.

Woese was not concerned with the further evolution of the eukaryotic cell. He believed that the origins of the different types of cells, as revealed by the molecular record of evolution retained within DNA and rRNA molecules, should be used as the primary criteria in defining Kingdoms. Most microbiologists who concentrate their studies on organisms composed of one or relatively few cells readily accept this concept. However, many biologists who study higher organisms have difficulty in accepting Woese's system because it places all organisms with eukaryotic cells—including fungi, protozoa, algae, plants, and animals—into a single Kingdom, the eukaryotes. It is

not surprising that lumping humans into the same Kingdom as fungi and algae meets some resistance and it is likely that further modifications to Woese's system will be made to subdivide the eukaryotes.

Classification systems continue to develop based on our abilities to examine the characteristics of organisms. Classification of living organisms has changed from systems based on first observational glimpses at the microbial world to systems based on detailed molecular analyses. In the progression of classification systems, we see a quest to change chaos into order and to establish a system of classification that reflects evolutionary relatedness.

Increased technological ability to examine the characteristic properties of organisms has permitted the establishment of more detailed and accurate classification systems for living things.

Interestingly, none of these classification systems considers the viruses. The viruses, perhaps as they should be, are treated as nonliving entities. An examination of the genetic molecules (RNA or DNA) of viruses indicates that they probably evolved from their respective host cells. They probably did not evolve in a hereditary lineage from one virus to the next. Hence, it is appropriate to classify viruses in relation to their host cells, for example, as a tobacco mosaic virus or a human immunodeficiency virus.

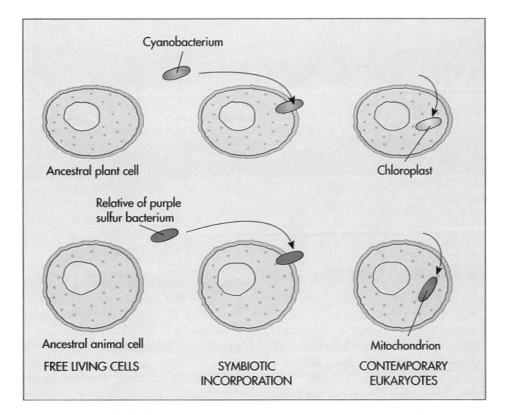

Cyanobacterium

Ancestral plant cell

Chloroplast

Relative of purple sulfur bacterium

Ancestral animal cell

Mitochondrion

FREE LIVING CELLS   SYMBIOTIC INCORPORATION   CONTEMPORARY EUKARYOTES

FIG. 2-4 According to endosymbiotic theory the chloroplasts and mitochondria of contemporary eukaryotic cells evolved from prokaryotic cells living within ancestral eukaryotic cells that lacked these organelles, providing photosynthetic and respiratory capabilities, respectively.

EVOLUTION OF MICROORGANISMS

We do not know exactly how the first living organism developed. However, it is possible to show in laboratory experiments that some chemicals that could have accumulated in the primitive atmosphere of Earth spontaneously collect into spheres when wet with water (see Figure). These spheres, called *micelles*, resemble the cells that are the fundamental organizational units of all living systems. A micelle is separated from the surrounding environment by a chemical boundary layer. Its structure allows restricted exchange of materials with the surroundings while permitting the maintenance of a high degree of internal organization. It is likely that about 3.6 billion years ago the chemicals that accumulated within a micelle, including nucleic acid molecules

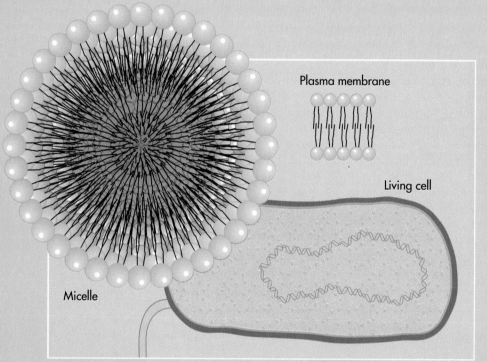

A micelle that forms spontaneously when certain chemicals interact with water resembles the structure of a living cell.

SURVEY OF MICROORGANISMS

ACELLULAR NONLIVING MICROORGANISMS

Viruses, viroids, and prions are acellular (noncellular), nonliving microorganisms that can replicate within the confines of a living cell of a compatible host organism (Table 2-1). It is this capacity for multiplying (replicating one's own structure) within a host cell that gives them their "lifelike character" and distinguishes them from other nonliving chemical combinations of molecules. A **host cell** is a compatible cell of a living organism within which a specific virus, viroid, or prion is capable of being replicated.

Viruses, viroids, and prions are obligate intracellular parasites that do not have an independent capacity to carry out life functions. When these acellular microorganisms are replicated within the cells of host organisms, they use the living cells' metabolic functions. Often in doing so they disrupt normal cellular functions, producing diseases in those organisms. Each virus, for example, produces characteristic symptoms when it replicates within a host organism, as typified by the characteristic mosaic pattern on the leaves of a tobacco plant when tobacco mosaic virus

that can transmit hereditary information, permitted essential life functions to occur. These functions were the ability to process materials and energy and to reproduce. Thus the first living microorganism could have evolved on Earth. Paleobiologists believe they have found fossilized imprints of microbial communities that existed 3.5 to 3.8 billion years ago in western Australia and Greenland.

Of the chemicals that may have accumulated in a micelle, RNA (ribonucleic acid) appears to have had a key role. RNA, like the hereditary molecule of all contemporary living organisms, DNA (deoxyribonucleic acid), can encode genetic information. Also some RNA molecules can act as catalysts for chemical reactions, a role most often played by protein catalysts (enzymes) in living cells. In 1993, scientists at the Scripps Research Institute, made an artificial RNA molecule. When they mixed it with proteins it began to reproduce. As long as they continued to supply proteins, the RNA molecule churned out copies of itself.

In a similar manner the first microorganism probably used RNA as catalyst and guiding template and energy from the organic compounds that accumulated spontaneously in the primordial atmosphere or on the Earth's surface. Generally speaking, organic compounds are substances that contain carbon and hydrogen. With these compounds a living cell could have carried out chemical reactions. (Collectively these reactions are called metabolism.) From these reactions the living cell could obtain energy and transform substances into the materials needed to survive and reproduce. There would have been no molecular oxygen (O_2) in the primitive atmosphere. Therefore this first microorganism would have been an anaerobe, that is, an organism that grows and reproduces without using molecular oxygen.

Reproduction of this first microbial cell produced other living cells. Each of the new cells also reproduced, forming new cells with hereditary information that could be passed on to their progeny. In this continuous chain of descendents, errors or changes in the replication of DNA occasionally occurred so that some of the new cells received somewhat differing hereditary characteristics. Thus new microorganisms evolved. Some of the microorganisms that evolved could synthesize complex organic compounds from the carbon dioxide in the atmosphere. Some of these microorganisms were photosynthetic and were able to use light energy to make organic compounds. The metabolism of these microorganisms gradually changed conditions in the environment so that other organisms could evolve.

The first photosynthetic microorganisms were able to grow only in the absence of oxygen. Later—probably 2 billion years ago, based on geologic evidence—microorganisms evolved that could produce molecular oxygen from water. This oxygen-producing photosynthesis made possible other ways for cells to obtain energy, including aerobic respiration (a life-supporting process that uses oxygen). Over the next 0.6 billion years, the pace of evolution apparently quickened. While many microorganisms became extinct, an astonishing number of new organisms developed, including many types of plants, animals, and microorganisms.

TABLE 2-1	
Some Characteristics of Acellular, Nonliving Microorganisms	
MICROORGANISM	**DESCRIPTION**
Virus	Highly structured; contains DNA or RNA as the genetic informational molecule surrounded by a protein coat; may have an additional outer lipid-containing structure called an envelope; replicates within specific hosts (bacteriophage replicate in bacterial cells, plant viruses replicate in plant cells, and animal viruses replicate in animal cells), using the synthetic capabilities of the host cells.
Viroid	Infectious RNA molecule that can be replicated within specific plant cells
Prion	Infectious protein molecule that can be replicated within specific animal cells

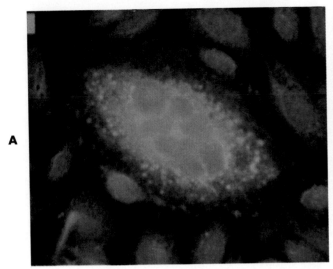 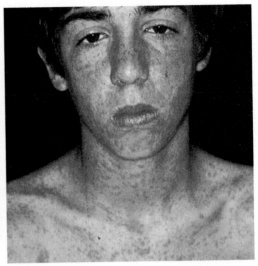

A   **B**

FIG. 2-5   **A,** Micrograph of cells infected with measles viruses. **B,** A child with a typical measles rash.

(TMV) infects the cells of that plant. We easily recognize the red rash of measles when the measles virus replicates within certain human cells (FIG. 2-5).

> Viruses, viroids, and prions are nonliving acellular microorganisms that can only replicate within living cells.

Viruses

A **virus** is made up of two essential parts: a central genetic nucleic acid molecule and a protein coat called a **capsid** (see FIG. 1-5). The capsid surrounds and protects the viral nucleic acid. The capsid gives the virus a characteristic shape. It also helps establish the specificity of the virus for a particular host cell. Some viruses also have an external **viral envelope** made up of plasma membrane that they acquire from host cells within which they replicate. Unlike prokaryotic and eukaryotic cells, viruses contain only one type of nucleic acid—RNA or DNA—as their hereditary molecule. Some viruses are RNA viruses and others are DNA viruses.

> Viruses are composed of a central genetic core— DNA or RNA—surrounded by a protein coat called a capsid.

Viruses are not capable of carrying out metabolism independent of a host cell. They replicate only in specific host cells. For example, viruses that replicate within a bacterial cell do not replicate within the cells of other organisms. This specificity, as well as other characteristics of viruses, is the basis for believing that they evolved from host cells rather than beginning as primitive or independent entities.

Interactions between the viral capsid and the outer layer of the host cell determine whether the viral nucleic acid will be able to enter the host cell. Viral replication within host cells causes numerous plant and animal diseases, including, in humans, the common cold, influenza, AIDS, herpes, measles, mumps, chickenpox, yellow fever, hepatitis, and many more.

There are three groups of viruses. **Animal viruses** infect and replicate within animal cells, **plant viruses** within plant cells, and **bacteriophage** or **phage** within bacterial cells. Viruses that infect bacterial cells cannot infect and replicate within cells of other organisms. Often the specificity is even greater than this. A virus that infects a tomato plant generally cannot infect other plants. This results in greater specificity of viral diseases. A phage that infects the bacterium *Escherichia coli* often cannot infect other bacteria or even other subtypes of *E. coli*. It is possible to utilize this specificity to identify bacteria, a procedure called phage typing.

Viroids

Unlike viruses, **viroids** are composed exclusively of RNA that contains their genetic information. They have no structures surrounding their genetic molecules. Inside a suitable host cell the RNA of a viroid is capable of initiating its own replication. This sometimes manifests as disease symptoms in the host organism. It was discovered in the early 1970s that viroids cause potato spindle tuber disease, chrysanthemum stunt, citrus exocortis and a few other plant diseases. Thus far they appear to affect only plants.

> Viroids are infectious RNA molecules found in plants.

Prions

The most recently discovered and least understood microorganisms are the **prions** (for proteinaceous infectious particles). What is so unusual about prions is that they seem to be composed only of protein. Like viroids, these "organisms" have no structures. They are only individual protein molecules that contain the information that codes for their replication when they infect a suitable host cell. They are properly called infectious proteins.

Unlike cellular and even other acellular organisms, prions do not store their genetic (hereditary) material in nucleic acid molecules. This presents a problem in understanding how prions replicate. Prions do not fit into our current understanding of how genetic information in nucleic acid molecules is replicated and how it can determine the specific structural and functional characteristics of each organism. Scientists do not know how a protein can direct its replication, and thus they do not understand how prions replicate.

Prions are infectious proteins.

Prions can cause diseases of the nervous system. They were discovered during the search for the cause of scrapie, an infectious and usually fatal disease of sheep. This disease was known to be caused by an agent that could pass through a filter that could trap bacteria and larger organisms. Therefore it was believed to be caused by a virus. No virus could be found, and eventually scrapie was shown to be caused by an infectious protein, that is, a prion. Prions have been found to cause some exotic human diseases such as kuru, a disease restricted to certain tribes of New Guinea that practice cannibalism. Some scientists have also hypothesized that prions may cause various degenerative nervous disorders, including Alzheimer's disease, that up until now have had no known cause.

PROKARYOTES

Microorganisms with prokaryotic cells include the archaebacteria and the eubacteria (Table 2-2). The prokaryotic cell is structurally not as complex as the eukaryotic cell. However, don't be misled by this statement. All living systems are extremely complex. Eubacteria and archaebacteria, like all other living organisms, must meet their energy and material needs through their metabolic activities, replicate their hereditary (genetic) information, and reproduce. Failure to accomplish any or all of these tasks results in death.

The distinction between archaebacteria and eubacteria is relatively new. Before 1980 all organisms with prokaryotic cells were called bacteria. However, as we discussed earlier, beginning in the late 1970s, scientists discovered that there were two distinct lines of evolutionary descent among the prokaryotes. The distinctions between these two groups will be considered later. One group was given the name **archaebacteria,** implying that these were early evolutionary forms. The archaebacteria have subsequently been called by some the *Archaea* or *archaeobacteria*. The second group was given the name **eubacteria.** Most often these prokaryotes are simply called bacteria.

Naming Bacteria

Like all other living organisms, each bacterial species is given a formal name that distinguishes it from all other organisms (Table 2-3). The formal names of bacteria and all other organisms are given in Latin. Latin is used because it was the classical language of science at the time when formal names were first given to organisms on a systematic basis. When typed or handwritten, species names are underlined. In print, species names are italicized, for example, the name *Streptococcus pneumoniae*. (This bacterial species sometimes causes pneumonia in humans.)

The formal name of a species is a **binomial name,** indicating that it has two parts. The use of binomial

TABLE 2-2			
Some Characteristics of Prokaryotic Microorganisms			
MICROORGANISM	**PLASMA MEMBRANE**	**CELL WALL**	**GROWTH**
Eubacteria (usually referred to simply as bacteria)	Composed largely of phospholipid molecules that have an ester bond linking the fatty acids to the glycerol	Most surrounded by a rigid cell wall composed of a unique molecule called peptidoglycan that has a carbohydrate (glycan) portion and a proteinlike (peptide) portion that contains amino acids not typically found in other organisms	Diverse, including growth on organic compounds, inorganic compounds, and light energy for different species
Archaebacteria (also called archaeobacteria or Archaea)	Composed of lipid molecules that have an ether bond that links the molecules together	Some surrounded by a rigid cell wall composed of unique molecules that resemble but are not the same as peptidoglycan; others have a protein coat or a complex carbohydrate	Many species grow at extreme conditions, such as in high temperature, high acidity, and saline environments

names was introduced by Carl Linnaeus. The first part of the binomial name, for example, *Streptococcus*, is the genus name. A **genus** is a group of closely related species. The first letter of the genus name is capitalized. The second part of the binomial name,

for example, *pneumoniae* is the **species epithet.** The species epithet is written in all lower case letters. The **species name** must always contain the genus name and the species epithet. It is permissible, though, to abbreviate the genus name. In this case the genus is

	TABLE 2-3			
	Some Names of Bacteria and Their Characteristics.			
NAME	**MORPHOLOGY**		**PHYSIOLOGY**	
	SHAPE	**STAINING**	**GROWTH**	**PATHOGENICITY**
Bacillus anthracis	Large rod-shaped cells	Gram positive	Aerobic growth; forms heat-resistant endospores	Causes anthrax
B. cereus	Large rod-shaped cells	Gram positive	Aerobic growth; forms heat-resistant endospores	Nonpathogen or food poisoning
B. subtilis	Large rod-shaped cells	Gram positive	Aerobic growth; forms heat-resistant endospores	Nonpathogen
Bordetella pertussis	Short rod-shaped cells	Gram negative	Fastidious (demanding growth requirements)	Causes pertussis (whooping cough)
Chlamydia trachomatis	Small spherical (coccal) cells	Gram negative	Grows only within mammalian cells	Causes a sexually transmitted disease and trachoma
Clostridium botulinum	Large rod-shaped cells	Gram positive	Anaerobic growth; forms heat-resistant endospores	Causes botulism
C. perfringens	Large rod-shaped cells	Gram positive	Anaerobic growth; forms heat-resistant endospores	Causes gas gangrene and food poisoning
C. tetani	Large rod-shaped cells	Gram positive	Anaerobic growth; forms heat-resistant endospores	Causes tetanus
Corynebacterium diphtheriae	Rod-shaped cells; cells remain attached at angles	Gram positive	Fastidious (demanding growth requirements)	Causes diphtheria
Enterobacter aerogenes	Short rod-shaped cells	Gram negative	Aerobic and anaerobic growth	Nonpathogen or opportunistic pathogen
Escherichia coli	Short rod-shaped cells	Gram negative	Aerobic and anaerobic growth	Nonpathogen that lives in human gut; commonly used in microbial genetics; some strains pathogenic
Haemophilus influenzae	Short rod-shaped cells	Gram negative	Fastidious; requires blood factors	Causes upper respiratory infections and meningitis
Klebsiella pneumoniae	Short rod-shaped cells	Gram negative	Aerobic and anaerobic growth	Causes pneumonia
Lactobacillus acidophilus	Short rod-shaped cells, forms chains of cells	Gram positive	Produces lactic acid	Nonpathogen
Legionella pneumophila	Short rod-shaped cells	Gram negative	Fastidious; requires iron and cysteine	Causes Legionnaire's disease and Pontiac fever
Micrococcus luteus	Spherical (coccal) cells occurring in packets	Gram positive	Produces a yellow pigment	Nonpathogen
Mycobacterium leprae	Rod-shaped cells	Acid fast	Slow grower	Causes leprosy
M. tuberculosis	Rod-shaped cells	Acid fast	Slow grower	Causes tuberculosis

indicated by a single capital letter followed by a period—for example, *S. pneumoniae.*

We sometimes designate a subspecies or type to specify some significant characteristics of a particular microorganism. For example, a particular strain of *S. pneumoniae* of the American Type Culture Collection (ATCC) that is resistant to treatment with penicillin is designated *S. pneumoniae* ATCC 35088. Also the pathovar (PV) or serovar (SV) may be noted by letters or numbers, for example, *Escherichia coli* O157:H7, a strain of this common species that is associated with serious dysentery.

NAME	MORPHOLOGY		PHYSIOLOGY	
	SHAPE	STAINING	GROWTH	PATHOGENICITY
Neisseria gonorrhoeae	Large coccal-shaped cells that occur as pairs	Gram negative	Fastidious	Causes gonorrhea
N. meningitidis	Large coccal-shaped cells that occur as pairs	Gram negative	Fastidious	Causes meningitis
Proteus vulgaris	Short rod-shaped cells	Gram negative	Aerobic and anaerobic growth	Nonpathogen or opportunistic pathogen
Pseudomonas aeruginosa	Rod-shaped cells that occur singly	Gram negative	Aerobic growth	Opportunistic pathogen; important in skin infections
Rickettsia rickettsii	Rod-shaped cells	Gram negative	Obligate intracellular parasite	Causes Rocky Mountain spotted fever, transmitted by ticks
R. typhi	Rod-shaped cells	Gram negative	Obligate intracellular parasite	Causes typhus, transmitted by body lice
Salmonella typhi	Rod-shaped cells	Gram negative	Aerobic and anaerobic growth	Causes typhoid fever
Serratia marcescens	Rod-shaped cells	Gram negative	Produces bright red pigment	Opportunistic pathogen
Shigella dysenteriae	Rod-shaped cells	Gram negative	Aerobic and anaerobic growth	Causes dysentery
Spirillum volutans	Helical-shaped cells	Gram negative	At low oxygen concentration	Nonpathogen
Staphylococcus aureus	Coccal-shaped cells that occur in grape-like clusters	Gram positive	Aerobic growth	Causes various diseases, including food poisoning, boils, and upper respiratory infections
Streptococcus pneumoniae	Coccal-shaped cells that occur in chains	Gram positive	Produces lactic acid	Causes pneumonia
Streptococcus pyogenes	Coccal-shaped cells that occur in chains	Gram positive	Produces lactic acid	Causes upper respiratory infections, rheumatic fever, and scarlet fever
Treponema pallidum	Helical-shaped cells that are wound around flagella that are attached to the ends of the cells	Gram negative	Fastidious	Causes syphilis
Vibrio cholerae	Curved, comma-shaped rods	Gram negative	Aerobic and anaerobic growth	Causes cholera
Yersinia pestis	Short rod-shaped cells	Gram negative	Fastidious	Causes plague, transmitted by fleas

Characteristics Used in Classifying and Identifying Bacteria

Many different characteristics are used in classifying and identifying bacteria (Table 2-4). These include general tests that are applied for virtually all bacteria and very specialized tests that are used to identify specific bacterial strains. Often completing the tests to identify a bacterial species, for example, for clinical laboratory identification, will take 1 or more days. Modern identification techniques are reducing this time to less than a day and even to just a few minutes for the identification of some disease-causing bacteria.

The differing morphologies (shapes and structures) of bacteria are used for classification and identification. Microscopic observations are used to view the shapes of bacterial cells. Bacteria are described as *cocci* if the cells are spherical and as *rods* if the cells are shaped like cylinders (FIG. 2-6.) Other shapes include curved rods, spirals, and others. The arrangement pattern of the cells, derived from the number and planes of division, is also an important characteristic that distinguishes one species from another. The cells of some bacteria, such as *Escherichia coli,* occur singly. Other species have cells that occur as pairs, tetrads, chains, or irregular clusters.

Specific staining reactions are also used to describe and classify bacteria. The **Gram stain** is a specific staining procedure, described in Chapter 3; it is especially important in describing and classifying bacteria. The Gram stain reaction is determined by the specific chemical composition of the cell wall, a structure that in most bacterial cells surrounds and protects the cell. Some bacterial species stain blue-purple by the Gram stain procedure and are called **Gram-positive bacteria;** the other bacterial species stain pink-red and are called **Gram-negative bacteria** (see FIG. 3-13). The bacterium *E. coli,* which occurs in the human intestine and is the most widely studied bacterial species, is a Gram-negative rod-shaped bacterium that occurs as single cells. *Streptococcus pneumoniae,* on the other hand, is a Gram-positive cocci that occurs in pairs and sometimes causes pneumonia.

The arrangement of **flagella,** which are filamentous structures involved in the movement of cells, also is important in classification. Some bacterial cells, such as those of *E. coli,* have flagella, called **peritrichous flagella,** that surround the cell. Other bacteria, such as *Pseudomonas aeruginosa,* have one or more flagella, called **polar flagella,** originating from one end of the bacterial cell.

Other special structures are also used to distinguish specific bacteria. For example, the presence of **endospores,** which are dormant heat-resistant structures, is important in classifying some bacteria. *Clostridium* and *Bacillus* species have endospores that make them particularly difficult to kill by heating. This is a particular problem in food canning where heat is used to kill microorganisms that can spoil food and cause disease.

Although morphological and staining characteristics are useful in describing and identifying bacteria, alone they are insufficient to discriminate among bacteria. Too many bacteria look alike. **Growth characteristics,** such as the optimal temperature for growth, and **metabolic characteristics,** such as the ability to produce acid from glucose, are important for classifying and identifying bacteria. Even when examining only a limited number of bacteria, such as those that cause human infections and that are seen in the clinical microbiology laboratory, at least 20 such metabolic characteristics generally must be determined to identify a particular bacterial species (FIG. 2-7).

TABLE 2-4	
Some Characteristics Used for Classifying and Identifying Bacteria	
CRITERION	**EXAMPLE**
Cellular Characteristics	
Morphology	Cell shape, cell size, arrangement of cells, arrangement of flagella, capsule, endospores
Staining reactions	Gram stain, acid-fast stain
Growth and nutritional characteristics	Appearance in liquid culture, colonial morphology, pigmentation, energy sources, carbon sources, nitrogen sources, fermentation products, modes of metabolism (autotrophic, heterotrophic, fermentative, respiratory)
Biochemical characteristics	Cell wall constituents, pigment biochemicals, storage inclusions, antigens, RNA molecules
Physiological and ecological characteristics	Temperature range and optimum, oxygen relationships, pH tolerance range, osmotic tolerance, salt requirement and tolerance, antibiotic sensitivity
Genetic characteristics	DNA mole% G + C, DNA hybridization

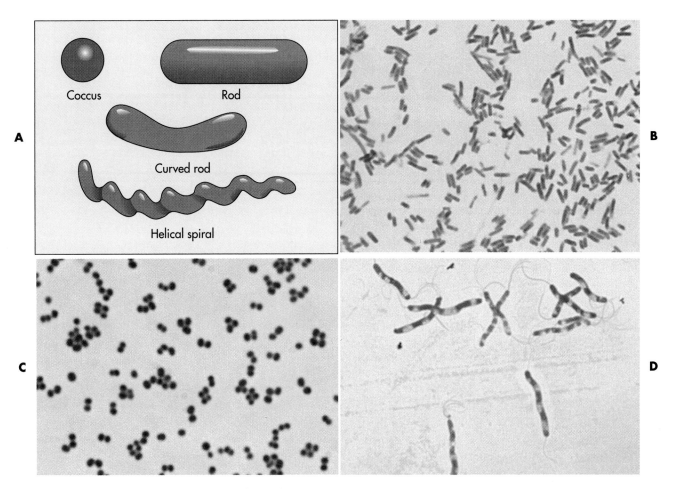

FIG. 2-6 **A,** The cells of each bacterial species have characteristic shapes. The most common shapes of bacterial cells are rods (cylindrical structures), cocci (spheres), and helical spiral (curved cells). **B,** Micrograph showing the rod-shaped cells of *Escherichia coli,* which lives in the human intestines (1,800×). **C,** Micrograph showing the coccal-shaped cells of *Staphylococcus epidermidis,* which lives on the human skin. (1,800×). **D,** Micrograph showing the spiral-shaped cells of the aquatic bacterium *Spirillum volutans.*

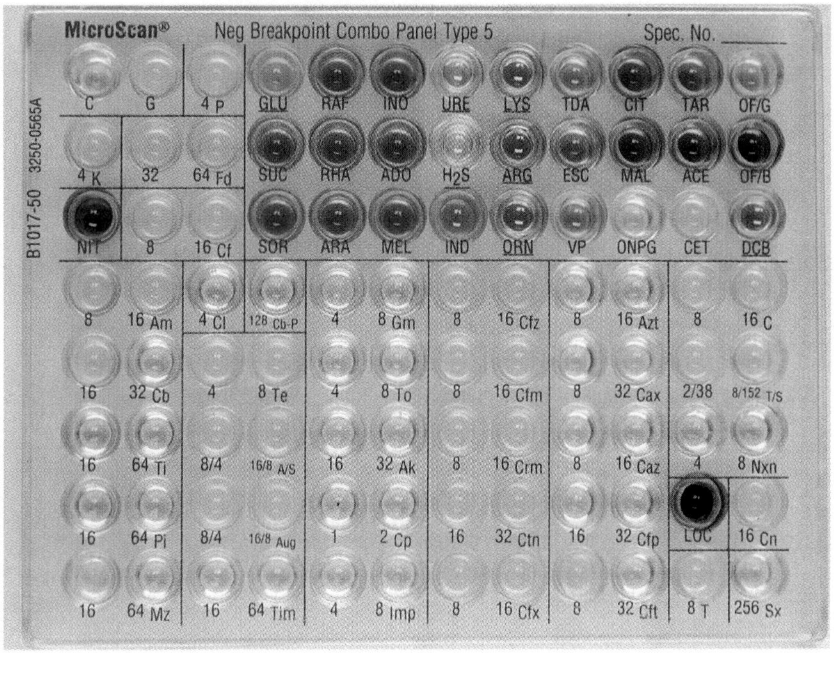

FIG. 2-7 Microtiter plate used in clinical microbiology laboratory for determination of metabolic characteristics of isolated bacteria. The color reactions indicate utilization of specific substances.

METHODOLOGY

DNA Hybridization: A Method for Analyzing Genetic Relatedness

One method used to assess genetic relatedness among organisms is called DNA hybridization (see Figure). To understand how this DNA hybridization method permits scientists to determine the relatedness of two organisms, we must first recognize that (1) DNA is composed of a series of individual molecules called nucleotides that are linked together in a particular order to establish the informational content of each organism, much in the way that the letters of our alphabet are linked together to form words and sentences, (2) DNA is composed of two complementary strands, and (3) DNA is the universal hereditary substance of all living organisms. The question of relatedness can be explored by determining what sequences of nucleotides the organisms have in common, both in terms of the specific nucleotides and their particular order (sequence) within the DNA molecule.

In the process of DNA hybridization, a radioisotope or a fluorescent dye is used to label or tag the DNA from one of the organisms. Then the DNA from both organisms is converted from double strands to single strands. These labelled and nonlabelled single strands are incubated together, allowing the DNA to reform double strands. Where the labelled DNA matches the nonlabelled DNA, the amount of radioactivity or dye measured will determine the extent to which the DNA from the two organisms combined.

Using this technique, scientists can determine directly the relatedness of organisms from bacteria to mammals. The formation of hybrid double-stranded DNA, that is, a double-stranded DNA molecule in which each strand came from a different organism, is a measure of the similarity, called homology, between the DNA of the two organisms. DNA molecules from organisms that are closely related, such as members of the same species, have a high degree of homology. Because evolution is based on changes in hereditary information and because DNA contains the hereditary information, DNA hybridization provides a tool for directly analyzing evolutionary relatedness for all organisms. No longer is it necessary to infer relationships based solely on appearances or metabolic functions. Using DNA hybridization, microbiologists can determine the relatedness of bacterial species.

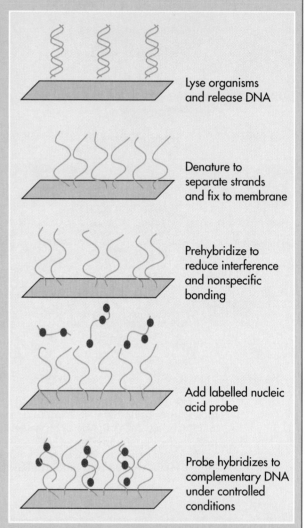

Lyse organisms and release DNA

Denature to separate strands and fix to membrane

Prehybridize to reduce interference and nonspecific bonding

Add labelled nucleic acid probe

Probe hybridizes to complementary DNA under controlled conditions

In DNA hybridization procedures for gene probe detection, cells are lysed to release double-stranded DNA. The DNA is denatured to convert it to single-stranded target DNA. The single-stranded DNA is affixed to a membrane. A prehybridization solution is used to prevent nonspecific binding to the membrane. A labelled nucleic acid probe (gene probe) is added. (The label may be a dye or a radioactive element.) The labelled probe hybridizes to complementary regions (if any) of the target DNA.

Additionally, **genetic characteristics** are employed in modern classification systems. DNA from different bacteria is compared to reveal the degree of similarity. In some clinical identifications, species are identified using DNA hybridization. Often, short segments of DNA, called *gene probes*, are used to determine the presence or absence of specific diagnostic genes. This is a powerful technique for identifying specific microorganisms

Many morphological, metabolic, and genetic characteristics are used for describing, classifying, and identifying bacteria.

TABLE 2-5
Some of the Major Groups of Bacteria Described in *Bergey's Manual*

GROUP	DESCRIPTION
Spirochetes	Very slender rods that are helically coiled around a central axial filament; includes the bacteria that cause syphilis and Lyme disease
Gram-negative aerobic rods and cocci	Bacteria that have a cell wall structure that results in their staining pink-red by the Gram stain procedure, are cylindrical or spherical in shape, and obtain their energy by respiration
Gram-negative facultatively anaerobic rods	Bacteria that have a cell wall structure that results in their staining pink-red by the Gram stain procedure and obtain their energy either by respiration or anaerobic fermentation
Gram-negative anaerobic bacteria	Bacteria that have a cell wall structure that results in their staining pink-red by the Gram stain procedure and obtain their energy by anaerobic fermentation
Rickettsias and chlamydias	Bacteria that are obligate intracellular parasites, that is, capable of reproducing only within host cells; includes the bacteria that cause Rocky Mountain spotted fever, trachoma, and urinary tract infections
Phototrophic bacteria	Photosynthetic bacteria that derive their energy from light; contain pigmented molecules that are used in photosynthesis; includes the purple and green sulfur bacteria, which do not produce oxygen during photosynthesis, and cyanobacteria, which do produce oxygen during photosynthesis
Gram-positive cocci	Bacteria that have a cell wall structure that results in their staining blue-purple by the Gram stain procedure and that are spherical; include the streptococci and staphylococci
Endospore-forming rods and cocci	Bacteria that form heat-resistant bodies called endospores within their cells; include the bacteria that cause gas gangrene, botulism, tetanus, and anthrax
Gram-positive, nonspore-forming rod-shaped bacteria	Bacteria that have a cell wall structure that results in their staining blue-purple by the Gram stain procedure, are cylindrical in shape, and do not form endospores
Mycobacteria	Bacteria that have a cell wall structure that results in their staining pink-red by the acid-fast staining procedure; include the bacteria that cause tuberculosis and leprosy
Actinomycetes and related organisms	Bacteria that form branching filaments and those that are closely related based on cell wall structure
Mycoplasmas	Bacteria that lack a cell wall; include bacteria that can cause atypical pneumonia that is not treatable with penicillin
Archaebacteria	Physiologically unique bacteria distantly related to other prokaryotes that include methane producers, bacteria that grow at low pH and high temperature, and bacteria that require high salt concentrations

Major Groups of Bacteria

Information on the characteristics of bacterial genera and species is updated periodically and published in a book entitled ***Bergey's Manual of Determinative Bacteriology*** (Table 2-5). *Bergey's Manual* is the standard reference for descriptions of microorganisms—the "bible" of bacterial taxonomy. Examining *Bergey's Manual* quickly reveals that relatively few bacterial species cause disease, even though these may be the ones about which we are most commonly concerned. Most bacteria perform metabolic activities that maintain the ecological balance of the Earth. Only a small portion of the numerous bacterial species described

in *Bergey's Manual* are ever seen in the clinical laboratory or ever discussed in introductory microbiology courses. In this book we will be able to examine only a small portion of the diverse bacteria that exist in nature.

> **Beyond their common prokaryotic cell structure, bacteria exhibit extreme diversity of form and function and in their metabolic characteristics.**

Several of the major groups of bacteria described in *Bergey's Manual* are distinguished primarily on morphological characteristics, namely: cell shapes (rods, cocci, curved, or filament forming); spore production (endospores, or other spores); staining reactions (color after staining according to the Gram stain procedure—Gram negative or Gram positive); and motility (nonmotile, motile with peritrichous flagella surrounding the cell, motile with polar flagella projecting from the end of the cell, motile by gliding). Other major bacterial groups are defined based on their metabolism, in particular how they generate energy in the form of ATP (photosynthetic [using light

HIGHLIGHT

MICROVIEW OF BACTERIAL DIVERSITY

Gram-positive cocci include the genus *Staphylococcus*, which typically form grape-like clusters and the genus *Streptococcus*, which occur in pairs or chains. Species of *Staphylococcus* commonly occur on skin surfaces where they live without causing disease. *Staphylococcus aureus*, however, is a potential human pathogen, infecting wounds, membranes, and linings and also causing food poisoning. Some members of *Streptococcus* are also human pathogens. For example, rheumatic fever is caused by *Streptococcus pyogenes*. Several *Streptococcus* species are also responsible for the formation of dental caries, which cause tooth decay.

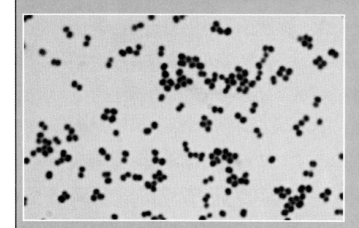

Micrograph of *Staphylococcus aureus*. This bacterium forms grape-like clusters of cells. It causes several human infections, including boils of the skin.

The two most important genera of **endospore-forming bacteria**, *Bacillus* and *Clostridium*, are Gram-positive rods. *Bacillus* species can grow in the presence of air, whereas *Clostridium* species are obligately anaerobic. Food spoilage by *Bacillus* and *Clostridium* species is of great economic importance. Several *Clostridium* species are important human pathogens. For example, *Clostridium botulinum* is the causative agent of botulism, *Clostridium tetani* causes tetanus, and *Clostridium perfringens* causes gas gangrene.

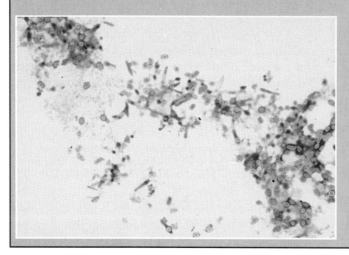

Micrograph of a *Clostridium* species after endospore staining. The spores appear green and the bacterial cells are stained red. The spores of this bacterium are heat resistant, and can survive in boiling water for hours.

energy], chemolithotrophic [using inorganic compounds], heterotrophic [using organic compounds]), and whether the metabolism is anaerobic (not using molecular oxygen), aerobic (using molecular oxygen), or facultatively anaerobic (capable of aerobic and anaerobic metabolism). Yet others are defined based on combined morphological and physiological characteristics. These characteristics are determined by grouping cultures of bacteria and making observations on those cultures and the cells that grow.

Text continued on p. 51.

The **Gram-positive nonspore-forming, rod-shaped bacteria** include bacteria that produce lactic acid. *Lactobacillus* are Gram-positive nonspore-forming rods that occur in chains. The lactobacilli are extremely important in the dairy industry. Cheese, yogurt, and many other fermented products are made by the metabolic activities of *Lactobacillus* species. They also inhabit regions of the human body, including teeth.

Micrograph of *Lactobacillus* species in a vaginal smear. Lactobacilli colonize many body surfaces. The lactic acid they produce helps protect against infections with disease-causing microorganisms.

The **Gram-negative facultatively anaerobic rods** include intestinal enteric bacteria. They are motile by means of peritrichous flagella and often live in the human intestinal tract. Much of what we know about bacterial metabolism and bacterial genetics has been elucidated in studies using *Escherichia coli. E. coli* is employed as an indicator of fecal contamination in environmental microbiology. The genera *Salmonella* and *Shigella* contain many species, many of which are important human pathogens. In particular, typhoid fever and various gastrointestinal upsets are caused by *Salmonella* species and bacterial dysentery is caused by *Shigella*.

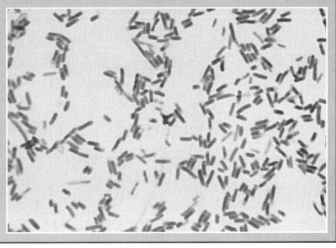

Micrograph showing the rod-shaped cells of *Escherichia coli*. This bacterium lives in the human gut and is the most commonly studied.

Continued.

The **Gram-negative aerobic rods** encompass a metabolically diverse group of bacteria; all of which can carry out aerobic metabolism. *Pseudomonas* is an aerobic motile bacterial species that is rod-shaped with polar flagella. Many *Pseudomonas* species are nutritionally versatile and capable of degrading many natural and synthetic organic compounds. Some *Pseudomonas* species are plant and animal pathogens. *P. aeruginosa,* can be a human pathogen and is commonly isolated from wound, burn, and urinary tract infections.

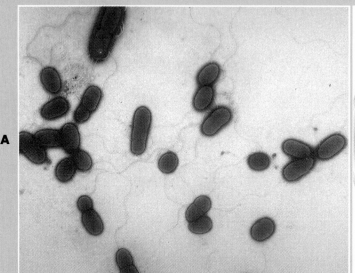
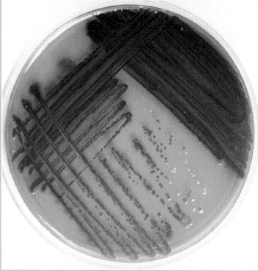

Various pseudomonads degrade environmental pollutants. Others cause human and plant diseases. **A,** Colorized micrograph of *Pseudomonas* sp. **B,** *Pseudomonas aeruginosa* growing on an agar plate.

Azotobacter, Rhizobium, and *Bradyrhizobium* are Gram-negative aerobic rods that are capable of fixing atmospheric nitrogen. This is a metabolic function unique to a select group of bacteria. *Rhizobium* species infect leguminous plant roots, where they cause the formation of tumorous growths called nodules within which they live in a mutually beneficial relationship. Within nodules, *Rhizobium* cells are irregularly shaped.

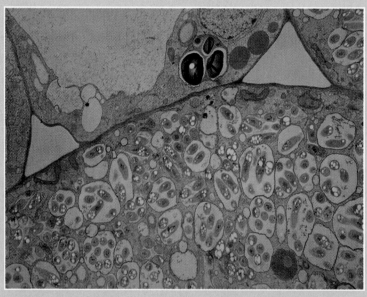

Colorized micrograph of sections showing *Bradyrhizobium japonicum (purple)* within a nodule of a soybean plant cell. This bacterium fixes atmospheric nitrogen and provides nitrogen-containing nutrients to the plant.

The Gram-negative aerobic rod-shaped *Agrobacterium* produces tumorous growths on infected plants. These growths are known as galls. *Agrobacterium tume-* *faciens* causes galls of many different plants and is an extremely important plant pathogen, causing large economic losses in agriculture.

A tree with crown gall caused by tumorous growth at base of the tree.

The genus *Neisseria* is a representative example of the **Gram-negative cocci.** *Neisseria gonorrhoeae* causes gonorrhea and *Neisseria meningitidis* causes bacterial meningitis. These bacteria tend to form relatively large cocci that typically look like kidney beans. The cells occur in pairs (diplococci).

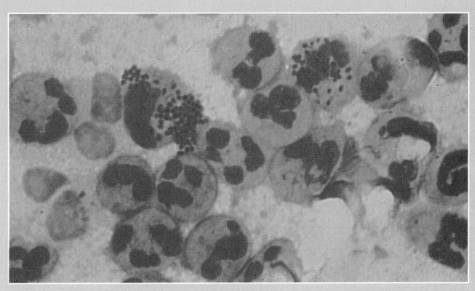

Micrograph of intracellular, Gram-negative diplococci. The presence of these bacteria in a urethral discharge is diagnostic for gonorrhea and in a vaginal discharge is presumptive for gonorrhea.

Continued.

The **helical** and **curved bacteria** group are helically curved rods that may have less than one complete turn (comma-shaped) to many turns (helical). *Campylobacter fetus* is a curved bacterium that frequently is the cause of gastrointestinal infections in infants. *Bdellovibrio*, which also is curved, has the unique characteristic of being able to penetrate and reproduce within the cells of other bacteria.

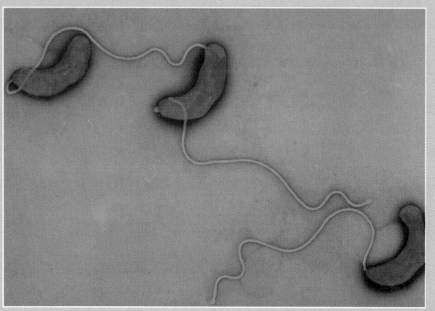

Colorized micrograph of the comma-shaped bacterium *Bdellovibrio bacteriovorus*. This bacterium reproduces within cells of the bacterium *Escherichia coli*.

The **spirochetes** are helically coiled rods, with one or more central axial fibril(s) wound around each cell. Many spirochetes are human pathogens. Several members of the genus *Treponema*, for example, are human pathogens. *Treponema pallidum* causes syphilis, which is a sexually transmitted disease. *Treponema pertenue* causes yaws. *Borrelia burgdorferi*, another spirochete, causes Lyme disease.

Micrograph of the spirochete *Treponema pallidum* (*green helical-shaped cells*) after fluorescent antibody staining. This bacterium causes syphilis.

The **budding** and/or **appendaged bacteria** are grouped together because they produce cell appendages. Several of these bacteria reproduce by budding. This involves separating a portion of the cell to form a new progeny cell. Many of the appendaged bacteria grow well at low nutrient concentrations. *Caulobacter*, for example, can grow in very dilute concentrations of organic matter in lakes and even is able to grow in distilled water. Some adhere to surfaces of other cells via their appendages.

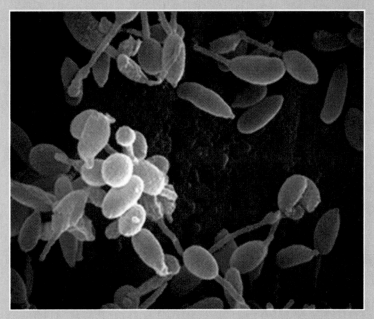

Colorized micrograph of the stalk-forming bacterium *Hyphomicrobium*. The stalks *(pink)* are part of the cells *(purple)*.

The **sheathed bacteria** comprise bacteria whose cells occur within a filamentous structure known as a sheath. The formation of a sheath enables these bacteria to attach themselves to solid surfaces. It also affords protection against predators and parasites. *Sphaerotilus natans* is a sheathed bacterium that is often referred to as the sewage fungus. This organism normally occurs in polluted flowing waters, such as sewage effluents, where it may be present in high concentrations just below sewage outfalls.

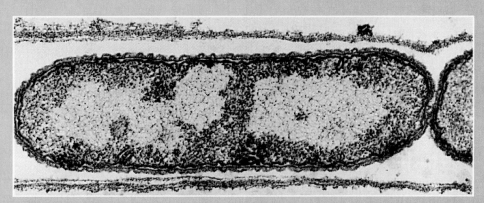

Colorized micrograph of the filamentous bacterium *Sphaerotilus natans*. Cells of this bacterium are enclosed within a sheath *(orange)*. It occurs abundantly in rivers below sewage outfalls and is called the "sewage fungus" because of its filamentous (fungal-like) appearance.

Continued.

Some bacteria are grouped based on their *gliding motility* on solid surfaces. These bacteria lack the specialized structures—flagella—that other bacteria use to propel themselves. The myxobacteria are gliding bacteria that have a unique feature. Under appropriate conditions they aggregate to form *fruiting bodies*. These fruiting bodies represent a stage in the complex reproductive process carried out by these bacteria. The fruiting bodies of myxobacteria occur on decaying plant material, on the bark of living trees, or on animal dung, appearing as highly colored slimy growths that may extend above the surface of the substrate.

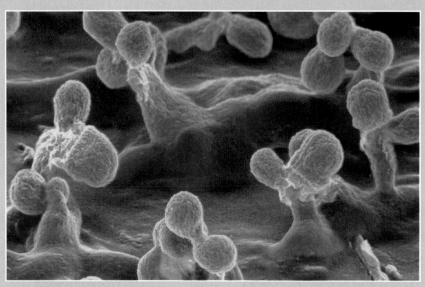

Colorized micrograph of the fruiting myxobacterium *Stigmatella auranticaca*, formed by the aggregation of many thousands of this bacterium.

Actinomycetes are bacteria that resemble fungi in appearance because they form filamentous growths. The production of antibiotics by actinomycetes, such as *Streptomyces griseus*, is extremely important in the pharmaceutical industry. Many previously fatal diseases are now easily controlled by using antibiotics produced by actinomycetes. Other actinomycetes are human pathogens.

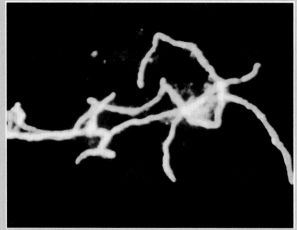
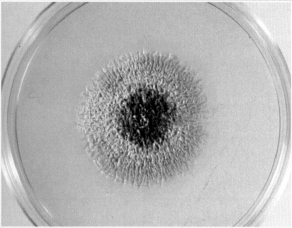

A, Micrograph of the actinomycete *Actinomyces israelii* showing hyphae formation. This bacterium is a human pathogen. **B,** *Streptomyces griseus* growing on an agar plate.

Mycobacteria are acid-fast, meaning that stained cells resist decolorization with acid alcohol, thus remaining red. This genus contains several important human pathogens, including *Mycobacterium tuberculosis* (causative agent of tuberculosis) and *Mycobacterium leprae* (causative agent of Hansen disease [leprosy]).

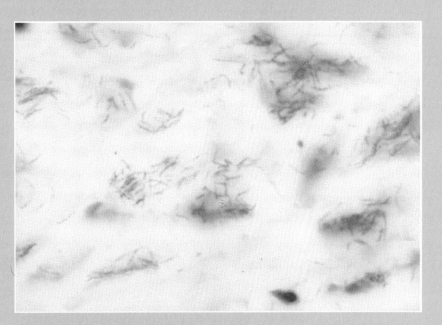

Micrograph of *Mycobacterium tuberculosis* in a sputum sample of an individual with tuberculosis. The appearance of red rods after acid-fast staining indicates the presence of mycobacteria and is diagnostic of tuberculosis.

The **coryneform bacteria** are defined by a characteristic irregular appearance of the cells. They tend to show incomplete separation following cell reproduction. When coryneforms reproduce, the progeny cells do not completely separate from one another (called snapping division) and form groups resembling "Chinese characters" when viewed under the microscope. Many species of *Corynebacterium* are plant or animal pathogens. For example, *Corynebacterium diphtheriae* is the causative agent of diphtheria.

Micrograph of the Gram-positive pleomorphic rod-shaped bacterium *Corynebacterium diphtheriae*. This bacterium causes diphtheria.

Continued.

The **rickettsias** and **chlamydias** are obligate intracellular parasites, that is, they can only reproduce within living host cells. Rickettsias are unable to produce sufficient amounts of metabolic energy to support their reproduction. They obtain energy from the host cells in which they grow. Most rickettsias that cause diseases in humans are transmitted by fleas, ticks, and lice. For example, *Rickettsia rickettsii* is transmitted by ticks and causes Rocky Mountain spotted fever.

Chlamydia reproduction is characterized by a change from a small, rigid-walled infectious form (elementary body) into a larger, thin-walled noninfectious form (initial body). *Chlamydia* cause human respiratory and urogenital tract diseases. They also cause conjunctivitis and trachoma. In birds they cause respiratory diseases and generalized infections. For example, the disease psittacosis, parrot fever, is caused by *Chlamydia psittaci.*

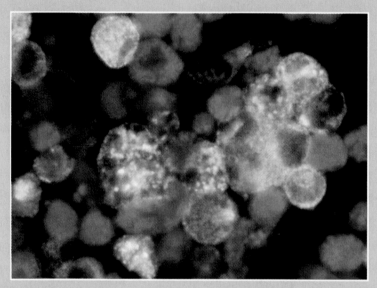

Micrograph showing fluorescent-antibody-stained inclusions formed by clumps of *Chlamydia trachomatis* (*light yellow*) within infected cells (*green*).

The **mycoplasmas** differ from other bacteria in that they lack a cell wall. They are the smallest organisms capable of self-reproduction. Several members of this genus cause diseases in humans. For example, atypical pneumonia is caused by *Mycoplasma pneumoniae. Ureaplasma* causes a sexually transmissible disease.

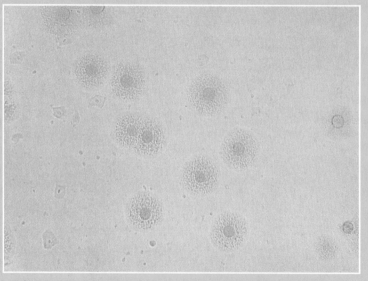

Colonies of *Mycoplasma hominis* with characteristic "fried-egg" appearance.

The **photosynthetic bacteria** are distinguished from other bacterial groups by their ability to use light energy to obtain cellular energy. Some photosynthetic bacteria carry out photosynthesis without the production of oxygen. Such anaerobic (nonoxygen-requiring) photosynthetic bacteria include the purple nonsulfur bacteria, purple sulfur bacteria, green sulfur bacteria, and green flexibacteria.

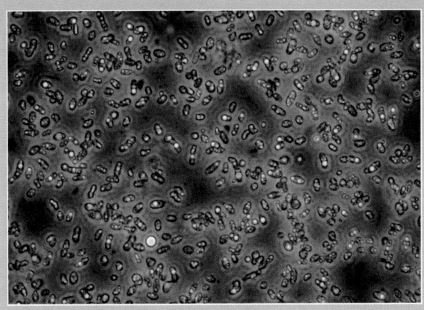

Micrograph of the purple sulfur bacterium *Chromatium* species. These bacteria deposit sulfur granules within their cells that are iridescent and appear multicolored.

The **cyanobacteria,** or blue-green bacteria (formerly called blue-green algae), carry out a type of photosynthesis that resembles that of higher plants. It results in the production of molecular oxygen.

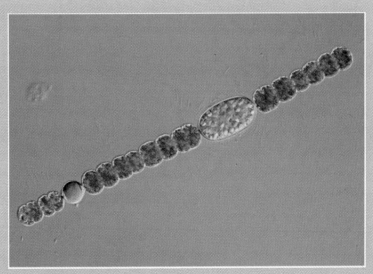

Micrograph of the cyanobacterium *Anabaena cylindrica* showing vegetative cells and a heterocyst (enlarged cell) in which nitrogen fixation occurs.

Continued.

Chemolithotrophic bacteria use inorganic compounds to generate ATP. The metabolic transformations of inorganic compounds mediated by these organisms cause global-scale cycling of various elements between the air, water, and soil. *Thiobacillus thiooxidans* is a chemolithotroph that uses sulfur to generate its energy. It is often found in association with waste coal heaps. The metabolic activities of this organism form sulfuric acid from the sulfur in coal, producing acid mine drainage, a serious ecological problem associated with some coal mining operations. Some chemolithotrophic bacteria, called *nitrifying bacteria,* convert ammonia (NH_4^+) to nitrate (NO_3^-). This conversion is important for the global cycling of nitrogen. This chemical change, however, causes leaching of nitrate into groundwater, which alters soil fertility. Nitrate also can cause a life-threatening disease of human infants. In human infants, nitrate can block the ability of hemoglobin to transport oxygen. Infants that drink water with too much nitrate die of "blue baby syndrome."

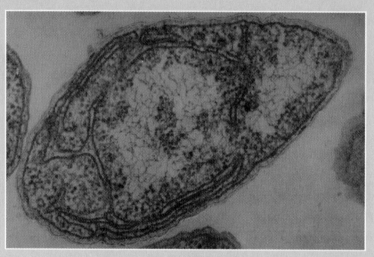

Colorized micrograph of membrane systems (*tan*) of the nitrifying bacterium *Nitrosomonas europea.*

The **archaebacteria,** or **Archaea,** represent a distinct evolutionary lineage of prokaryotes. Although they are prokaryotes, they have unique characteristics that distinguish them from eubacteria. Archaea have unusual physiological properties that permit many of them to live in extreme environments such as boiling hot springs, concentrated sulfuric acid, and salt lakes. The thermophilic archaebacteria grow only at temperatures above 85° C. The acidophilic archaebacteria grow at pH values of less than 2. The halophilic archaebacteria grow at high salt concentrations such as in brines that have sodium chloride concentrations of 15%. The methanogenic archaebacteria are very strict anaerobes that grow only in the absence of air and produce methane.

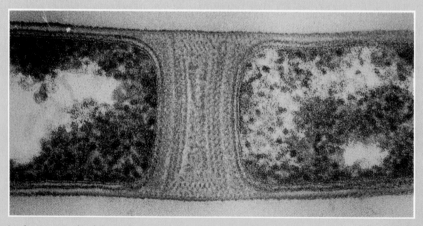

Colorized micrograph of the archaebacterium, *Methanospirillum hungatei* showing cells within a protein sheath (*orange*).

TABLE 2-6		
Some Characteristics of Eukaryotic Microorganisms		
MICROBIAL GROUP	**DESCRIPTION**	
Fungi	Nonphotosynthetic (heterotrophic) metabolism; cells usually have cell walls, which typically contain chitin; some form spores for reproduction; may reproduce sexually or asexually; generally nonmotile; unicellular (yeasts) or multicellular filamentous forms (molds); no differentiated tissues	
Algae	Photosynthetic metabolism; contain chlorophylls and accessory photosynthetic pigments such as carotenoids; no differentiated tissues	
Protozoa	Nonphotosynthetic (heterotrophic) metabolism; cells typically lack cell walls; usually unicellular; generally motile; no differentiated tissues	

EUKARYOTES

The eukaryotic microorganisms include the fungi, algae, and protozoa (Table 2-6). These microorganisms, like the higher plants and animals, have eukaryotic cells. They evolved along different lines of descent, apparently based on how they obtain nutrition. The algae carry out photosynthesis, obtaining energy from light and carbon from inorganic carbon dioxide for cell growth. The fungi absorb organic nutrients that they use to generate cellular energy and cell constituents. The protozoa tend to engulf nutrients, sometimes growing on other cells.

Fungi

Like the bacteria, the **fungi** are extremely diverse. Unlike bacteria, however, fungi are composed of eukaryotic cells. Most fungi have cell walls, which most often contain chitin, the substance that makes up insect skeletons and crab shells. These cell walls help protect the cells against physical damage and chemical attack. Some fungi—**yeasts**—are primarily unicellular (FIG. 2-8). Others, called **filamentous fungi** or **molds,** form tube-like filaments called **hyphae** (FIG.

2-9). Some hyphae are coenocytic, meaning they lack cross-walls to separate cells; coencytic hyphae are mutinucleate.

Hyphae, which are composed of many cells, can form integrated masses called **mycelia.** Mycelia are the visible structures seen when molds grow on bread and other substrates. In some cases, elongation of hyphae occurs without forming separate cells. Long, multinucleate, fungal hyphae develop. More commonly, separate cells are formed by branches and crosswalls as the hyphae grow. The crosswalls are called **septa.** Even when crosswalls form, cellular materials flow through pores in the septa.

Fungi include the molds, which are filamentous organisms, and single-celled organisms called yeasts.

Fungi obtain their energy from the metabolism of organic compounds. They generally absorb nutrients from their surroundings, often from plant materials. In nature, fungi are very important decomposers. They cause, for example, the decay of dead logs. Un-

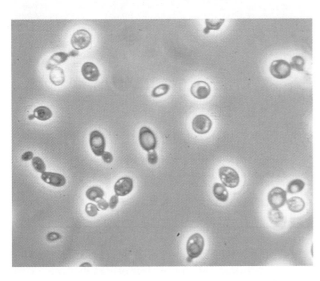

FIG. 2-8 Micrograph of *Saccharomyces cerevisiae* showing budding to produce progeny. (1,075×).

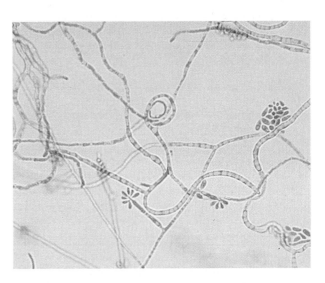

FIG. 2-9 Micrograph of the fungus *Exophiala jeanselmei* that, like other molds, forms long filamentous filaments of intertwined mycelia.

FIG. 2-10 Brown rot of an apricot.

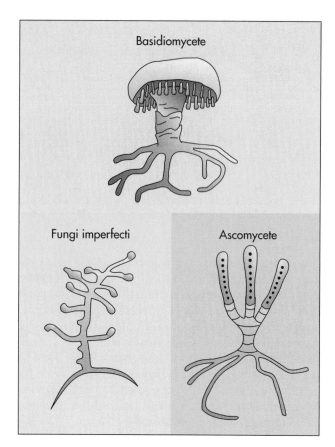

FIG. 2-11 The classification of fungi is based largely on spore formation. Various types of spores and fruiting bodies are formed by different fungi. Basidiomycetes form spores on basidia; ascomycetes form spores within asci; fungi imperfecti do not form sexual reproductive spores.

fortunately many are also plant pathogens, causing great losses of agricultural crops. We sometimes use chemical fungicides to protect many cultivated plants (FIG. 2-10). A few also cause human diseases such as athlete's foot, histoplasmosis, coccidioidomycosis, and yeast *(Candida)* infections.

The classification of fungi is based primarily on their sexual reproductive spores (FIG. 2-11). Some sexual spores of fungi are formed within a specialized structure known as the **ascus.** Such spores are called **ascospores** and the fungi that produce them are called **ascomycetes.** Another major group of fungi, the **basidiomycetes,** produce sexual spores on a specialized structure known as the *basidium.* A mushroom is such a basidium. The spores produced by basidomycetes are called **basidiospores.** Other fungi, known as the **deuteromycetes** or **fungi imperfecti,** have no known sexual reproductive phase. As far as we know, they are restricted to asexual means of reproduction. The fungi imperfecti include *Penicillium* and *Aspergillus,* two of the more common fungi that we may observe growing on foods such as bread.

> **Fungi are composed of eukaryotic cells, may be unicellular or multicellular, often form filaments, usually have cell walls, and typically form reproductive spores.**
>
> **Fungi are classified primarily based on their sexual means of reproduction.**

To distinguish among species of yeasts, we employ a few morphological (structural) observations and numerous metabolic characteristics. The procedure for identifying the unicellular yeasts is very similar to that employed for identifying bacteria. For the filamentous fungi, on the other hand, we rely almost entirely on morphological observations. We use the same basic approach used in identifying plants. Mushroom-producing fungi, for example, are identified based on visual appearance (FIG. 2-12). Great

FIG. 2-12 The ink cap mushroom *Coprinus atramentarius.*

care should be taken in identifying mushrooms. Some mushrooms are edible but others that may look quite similar are deadly poisonous. Even experts have sometimes been fooled and have died after eating mushrooms that they improperly identified.

Algae

Algae are eukaryotic photosynthetic microorganisms that contain chlorophyll and utilize light energy to generate their chemical energy. They are the only eukaryotic photosynthetic microorganisms. As such, they are the microorganisms most closely related to the plants. They also are able to produce oxygen (O_2) from water. Many of the characteristics used in the classification of algae have been adapted from botanists. These characteristics tend to be those that can be observed. They are generally not metabolic characteristics that must be experimentally determined. The major groups of algae are identified by their characteristic colored pigments and cell morphologies. The major groups of algae include the green algae, euglenoids, golden and yellow-green algae, cryptomonads, and dinoflagellates.

The **diatoms** are unicellular algae that have cell walls containing silicon. They are quite beautiful when seen through the microscope (FIG. 2-13). The euglenoid algae, dinoflagellates, and many green algae are also unicellular. Some green algae are multicellular but do not form structures nearly as complex as the brown and red algae. As examples, the green alga *Volvox* forms a multicellular colonial aggregation of cells and the green alga *Spirogyra* produces a multicellular filament (FIG. 2-14).

Algae are photosynthetic eukaryotic microorganisms.

Most algae are nonpathogens.

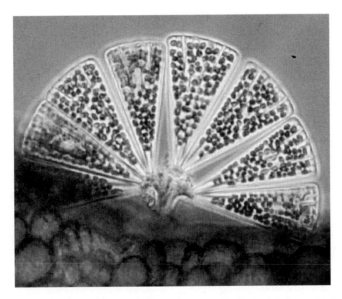

FIG. 2-13   Micrograph of the marine diatom *Licmophora*. (320×).

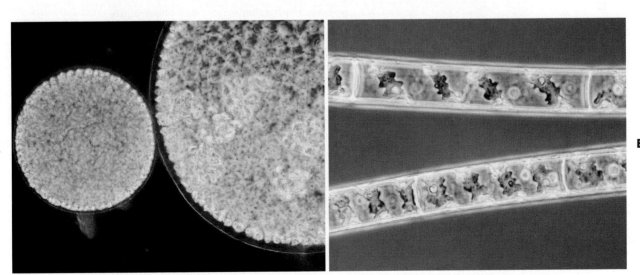

FIG. 2-14   **A,** Micrograph of the green alga *Volvox aureus.* (175×). **B,** Micrograph showing detail of spiral chloroplasts of the green alga *Spirogyra* species. (525×).

Some organisms that have traditionally been classified as algae, namely the red and brown algae, produce complex macroscopic multicellular structures. The kelps, for example, are brown algae that often reach lengths of 50 meters. Kelps have holdfasts, which are rootlike structures that make them plant-like. They are clearly not microorganisms and have been appropriately reclassified as plants instead of algae.

Protozoa

The **protozoa** for the most part are unicellular, non-photosynthetic eukaryotic organisms. They generally lack a cell wall. Many of the protozoa are motile, meaning they are able to move. Since at one time all organisms that moved were considered animals, protozoa have been studied largely by zoologists. The protozoa have heterotrophic metabolism, meaning that they obtain cellular energy from organic substances such as proteins. Protozoa tend to engulf their food sources, a characteristic that makes them similar to higher animals. Many of the characteristics used in classifying protozoa are analogous to the morphological characteristics used in describing animals.

> **Protozoa are eukaryotic heterotrophic microorganisms that are classified on the basis of morphological characteristics, especially their means of locomotion.**

Motility is a major characteristic traditionally used in classifying protozoa. It is still very important in the latest classification system proposed in 1980. Some protozoa, the **Sarcodina,** form extensions of their cells known as **pseudopodia,** or **false feet** (FIG. 2-15). By forming pseudopodia, these protozoa move from place to place increasing their chances of encountering food. This group includes the genus *Amoeba*. *Amoeba* is an amorphous organism that moves and engulfs food particles by extending its cytoplasm. Unlike the Sarcodina, the cells of the **Ciliophora** are covered by numerous hairlike projections, called **cilia.** Cilia beat continuously, either propelling the protozoan cell or moving food to the protozoan so that it can be engulfed. *Paramecium* is a common example of a ciliate protozoan (FIG. 2-16). The cilia of *Paramecium* propel it from one place to another and drive food toward its "mouthlike" region.

Yet other protozoa, the **Mastigophora,** have flagella that emanate from one end of the protozoan cell. They propel the cell. *Giardia* is a flagellate protozoan that infects the human gastrointestinal tract and causes severe diarrhea. It is becoming an increasing health problem in the United States (FIG. 2-17).

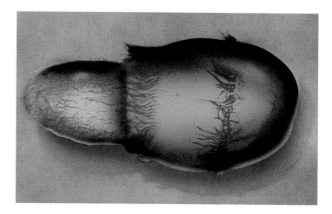

FIG. 2-16 Colorized micrograph of the protozoan *Didinium (green)* consuming the protozoan *Paramecium (yellow).*

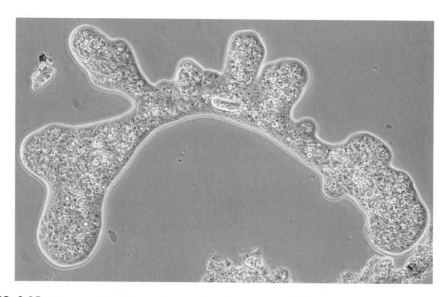

FIG. 2-15 Micrograph of the protozoan *Amoeba proteus*, which moves by extending its cytoplasm to form false feet. (225×).

FIG. 2-17 Colorized micrograph of *Giardia lamblia*, a protozoan that is one of the most common causes of gastrointestinal infections.

Another protozoan group that traditionally has been recognized is the **Sporozoa.** Members of the Sporozoa generally are nonmotile, produce spores during their life cycles, and grow only on or in a host organism from which they derive nutrition.

In the protozoan classification system proposed in 1980, the Sporozoa were reclassified into several smaller groups based on morphological characteristics. Many of these characteristics can only be observed with an electron microscope. Various other protozoa were also placed into these smaller groups regardless of their means of locomotion. For example, *Plasmodium*—the protozoan genus that includes the species that cause malaria—was placed into the group Apicomplexa because it produces a small umbrella-like structure at the end of the cell.

The classification of protozoa now includes electron microscopic observations of previously unseen morphological characteristics.

SUMMARY

Evolution of Microorganisms (pp. 25-30)
- Microorganisms evolved from abiotically formed micelles.
- As microorganisms evolved they altered their surroundings, permitting new forms of life to develop.

Classifying Microorganisms (pp. 25-30)
Early Classification Systems (p. 26)
- Early classification systems were based on observable characteristics of microorganisms, such as shape and motility.
- The species is the fundamental unit used in classifying all organisms.
Modern Classification Systems (pp. 26-29)
- Modern classification systems attempt to reflect evolutionary relationships.
- The contemporary classification system of Woese, which is based on genetic analyses at the molecular level, recognizes three primary Kingdoms: archaebacteria (primitive specialized prokaryotes), eubacteria (common prokaryotes or "true bacteria"), and eukaryotes (all other cellular organisms, including plants, animals, protozoa, algae, and fungi).

Survey of Microorganisms (pp. 30-55)
Acellular Nonliving Microorganisms—Viruses, Viroids, and Prions (pp. 30-33)
- The acellular microorganisms include the viruses, viroids, and prions.
- Viruses have two parts: a protein coat (the capsid) and a hereditary molecule, which may be DNA or RNA.
- Viroids are composed of only RNA.

- Prions are infectious proteins.
Prokaryotes—Eubacteria and Archaebacteria (p. 33-50)
- The eubacteria and archaebacteria are prokaryotic cells.
- Like other organisms, bacteria are named using a binomial system of genus and species.
- The major groups of bacteria are distinguished based on morphological, metabolic, and genetic characteristics.
- The bacteria are described in *Bergey's Manual,* which is revised periodically.
Eukaryotes (pp. 51-55)
Fungi (pp. 51-52)
- The fungi are heterotrophic eukaryotic microorganisms.
- Fungal cells commonly have cell walls.
- Some fungi form multicellular filamentous growths called mycelia and others are unicellular (yeasts).
- The fungi are classified into major groups based on their modes of reproduction and, in particular, the reproductive spores they produce.
Algae (pp. 53-54)
- The algae are photosynthetic eukaryotic microorganisms.
- The major groups of algae are defined based on their characteristic pigments and cell morphologies.
Protozoa (pp. 54-55)
- The protozoa are heterotrophic microorganisms.
- The cells of protozoa generally lack cell walls.
- The major groups of protozoa are classified based on their means of motility or on specialized morphological characteristics.

CHAPTER REVIEW

REVIEW QUESTIONS

1. Why is it hard to classify microorganisms by their evolutionary relationships?
2. Compare and contrast the classification systems of Linnaeus, Haeckel, Whittaker, and Woese.
3. How do viroids differ from viruses?
4. What are prions? Why are prions different from other living organisms?
5. What are the main characteristics of organisms in each of the Kingdoms in the five-Kingdom system of classification?
6. How are bacteria named?
7. What is *Bergey's Manual*?
8. How do humans differ from bacteria?
9. What are some of the morphological and physiological characteristics that have been used to classify bacteria?
10. What are the main features of the endosymbiont theory of evolution?
11. What are the modern methods used to classify organisms?
12. What were the characteristics of the first microorganisms?
13. How could life have evolved on Earth? Could life have evolved on other planets? If so, how?
14. What are the differences between classification and identification. How can clinical laboratories identify pathogens based on only 20 tests?
15. Why has the proposal by Carl Woese for three primary Kingdoms (eubacteria, archaebacteria, and eukaryotes) upset many traditional plant and animal biologists?

CRITICAL THINKING QUESTIONS

1. You have recently discovered a new microorganism. What criteria would you use to help classify it?
2. You would like to develop a new system of classification that groups all the eukaryotic microorganisms into one category. What characteristics are fundamentally the same for all eukaryotic microorganisms? What characteristics are fundamentally different for the major groups of eukaryotic microorganisms?
3. Compare methods of categorizing organisms. Are any of them better than the others or are they just different ways of looking at the same thing?
4. Protozoologists classified many organisms as protozoa that have traditionally been considered as fungi or as algae. These organisms include the slime molds and the euglenoid algae. How would you resolve the disputes between mycologists and protozoologists over the slime molds and between phycologists and protozoologists over the unicellular euglenoid algae?
5. Should the photosynthetic "blue-greens" be considered cyanobacteria or should they be considered blue-green algae?

READINGS

Balows A, HG Truper, M Dworkin, W Harder, K-H Schleifer: 1992. *The Prokaryotes*, ed. 2, New York, Springer-Verlag.

A comprehensive set of volumes on the prokaryotes that is a valuable reference for information on the properties of specific bacteria.

Cairns-Smith AG: 1985. The first organisms, *Scientific American* 252(6):90-100.

Presents an alternative theory for the origin of life, arguing that clay, not primordial soup, provided the fundamental materials from which life arose.

Chang ST and PG Miles: 1984. A new look at cultivated mushrooms, *BioScience* 34:358-362.

A fascinating account of the rapidly growing industry that produces these fungi.

Courtenay B and HH Burdsall Jr: 1982. *A Field Guide to Mushrooms and Their Relatives*, New York, Van Nostrand Reinhold.

Excellent identification manual, illustrating and describing more than 350 species of ascomycetes and basidiomycetes.

Dickerson R: 1978. Chemical evolution and the origin of life, *Scientific American* 239(3):70-86.

Explains the chemical changes thought to have occurred during the evolution of life.

Diener TO: 1982. Viroids: minimal biological systems, *BioScience* 32:38-44.

Describes the studies that established the existence of viroids.

Dyer BD and R Obar (eds.): 1985. *The Origin of Eukaryotic Cells*, New York, Van Nostrand Reinhold.

Considers the question of what initiated the development of mitochondria, chloroplasts, and organelles of motility and espouses a belief in the endosymbiont theory to explain these processes.

Fox GE et al: 1980. The phylogeny of prokaryotes, *Science* 209:457-463.

Describes the relationships of the prokaryotes, particularly the Archaebacteria.

Holt J (ed.): 1994. *Bergey's Manual of Determinative Bacteriology*, ed. 9, Baltimore, Williams & Wilkins.

The most recent volume of Bergey's Manual describes the characteristics of bacteria and provides valuable diagnostic tables to aid in the identification of bacteria.

Horgan J: 1991. In the beginning... *Scientific American* 264(2):116-125.

Reviews recent theories inspired by Stanley Miller's 1953 experiments on the origin of life; discusses when, where, and how life began.

Joyce GF: 1992. Directed molecular evolution, *Scientific American* 267(6):90-97.

An interesting examination of the underlying molecular basis of evolution.

Krieg NR and J Holt (eds.): 1984. Bergey's Manual *of Systematic Bacteriology*, 4 Volumes. Baltimore, Williams & Wilkins.

The comprehensive volumes of Bergey's Manual *is an invaluable resource, giving the descriptions of bacteria.*

Margulis L and KV Schwartz: 1988. *Five Kingdoms: An Illustrated Guide to the Phyla of Life on Earth,* New York, W.H. Freeman.

A catalog of the diversity of the world's living organisms, describing with profuse illustrations all major groups of organisms—100 phyla belonging to the five Kingdoms proposed by Robert Whittaker.

Margulis L: 1984. *Early Life,* Boston, Jones & Bartlett.

Presents the author's views on the origins of life.

Margulis L: 1982. *Early Life,* Boston, Science Books International.

Short treatise on the evolution of life on earth, presenting an overview of the genesis of living organisms and their life processes.

Margulis L: 1981. *Symbiosis in Cell Evolution,* San Francisco, W.H. Freeman.

Extensive arguments for the endosymbiotic theory of eukaryotic origins, that mitochondria, chloroplasts, and other organelles were acquired by eukaryotes from symbiotes.

Margulis L: 1970. *Origin of Eukaryotic Cells; Evidence and Research Implications for a Theory of the Origin and Evolution of Microbial, Plant, and Animal Cells on the Precambrian Earth,* New Haven, Connecticut, Yale University Press.

Presents the author's views on the origin and evolution of cells based on the endosymbiotic origin of specific eukaryotic organelles.

Palleroni NJ: 1983. The taxonomy of bacteria, *BioScience* 33:370-377.

Reveals the methods by which bacteria are classified.

Prusiner SB: 1984. Prions, *Scientific American* 251(3):50-59.

Introduction to prions, disease-causing agents that contain no DNA or RNA and yet are able to reproduce within cells.

Schopf JW (ed.): 1983. *Earth's Earliest Biosphere: Its Origin and Evolution,* Princeton University Press.

Articles by experts on the evolution of early forms of life and the geochemical evidence of the kinds of metabolism they conducted.

Schopf JW: 1978. The evolution of the earliest cells, *Scientific American* 239(3):110-138.

Argues that the earliest cells gave rise to the oxygen in the atmosphere on which modern life depends.

Shapiro JA: 1988. Bacteria as multicellular organisms, *Scientific American* 258(6):82-89.

Discusses how bacteria grown in the laboratory can form complex multicellular communities, differentiating into various cell types and forms.

Skerman VBD: 1967. Order out of confusion: a case for a systematic approach to microbiology, *Impact of Science on Society* 17:225-244.

Reviews the history of microbiology in order to show the general basis on which the description of microorganisms is founded and how it is constantly changing.

Skerman VBD, V McGowan, PHA Sneath: 1989. *Approved Lists of Bacterial Names,* Washington, D.C.; American Society for Microbiology.

Includes all valid names of bacteria that have been adequately described and, if cultivatable, for which there is a type, neotype, or reference strain available.

Whittaker RH and L Margulis: 1978. Protist classification and the Kingdoms of organisms, *Biosystems* 10(1-2):3-18.

Some insights on the five-Kingdom system.

Whittaker RH: 1969. New concepts of kingdoms of organisms, *Science* 163:150-160.

Whittacker's proposed modifications in 1969 of Haeckel's system became widely accepted for the five-Kingdom classification system.

Woese CR: 1981. Archaebacteria, *Scientific American* 244(6):98-122.

Insights on early evolution and the archaebacteria and the author's revolutionary new three-Kingdom classification system based on genetic analyses.

CHAPTER 3

Science of Microbiology: Methods for Studying Microorganisms

PREVIEW TO CHAPTER 3

In this chapter we will:
- Examine the scientific method that scientists use in their investigations.
- See how the scientific method was used to disprove the theory of spontaneous generation.
- Learn how the studies of Louis Pasteur began the science of microbiology.
- Examine how the microscope is used to view microorganisms.
- Learn about the principles of microscopy that determine how large an object can be magnified and how small an object can be seen.
- Learn about the different types of microscopes—their applications and limitations—including light and electron microscopes.
- Learn how microbiologists grow pure cultures of microorganisms so that they can study their physiological properties.
- Learn the following key terms and names:

acid-fast staining	negative staining
agar	objective lens
aseptic technique	ocular lens
autoclave	oil immersion objective
bright-field microscope	Louis Pasteur
colony	Petri plate
compound light	phase contrast microscope
microscope	pure culture
controlled experiments	resolving power
culture	scanning electron
dark-field microscope	microscope (SEM)
differential medium	scientific method
differential staining	selective media
procedures	simple staining
electron microscope	procedures
endospore staining	spread plate method
procedure	staining
Gram stain procedure	sterile
Gram-negative bacteria	sterilization procedures
Gram-positive bacteria	streak plate method
hypothesis	theory of spontaneous
immunofluorescence	generation
microscopy	transmission electron
magnification	microscope (TEM)
microscope	

Suppose you saw a piece of rotting meat covered with wormlike fly larvae. You ask whether these larvae, called maggots, are always found on rotting meat. You search out other pieces of rotting meat and see whether all of them are covered with maggots. If you note that maggots are present every time you observe a piece of rotting meat, you could generalize that when meat rots, maggots appear. Such generalizing from specific observations is called **inductive reasoning.** You could also begin with a generalization. By reasoning from it, you could arrive at a specific conclusion. This process is called **deductive reasoning.** In reaching a conclusion by deductive reasoning, one essentially says that if this happens, then that will happen; if I observe this, then I will observe that. If I see a piece of meat rotting, then I will find maggots on the meat. This type of "if-then" reasoning is essential in science.

However, not all conclusions reached by such deductive reasoning are correct. For example, you could erroneously conclude, based on the above ob-servations, that maggots arise spontaneously from rotting meat. In fact, the repeated observation of maggots on rotting meat led in part to the *theory of spontaneous generation.* This theory held that living organisms could arise spontaneously from nonliving matter. The belief that maggots spring forth by spontaneous generation from decaying meat persisted in various forms from before the time of Aristotle in the fourth century B.C. to the late seventeenth century. Almost all scientists and philosophers until the seventeenth century believed that living animals could be generated from nonliving matter. In this case, the theory of spontaneous generation was logical but wrong. Demonstrating the fallacy of the theory of spontaneous generation required the use of an approach known as the **scientific method** (FIG. 3-1).

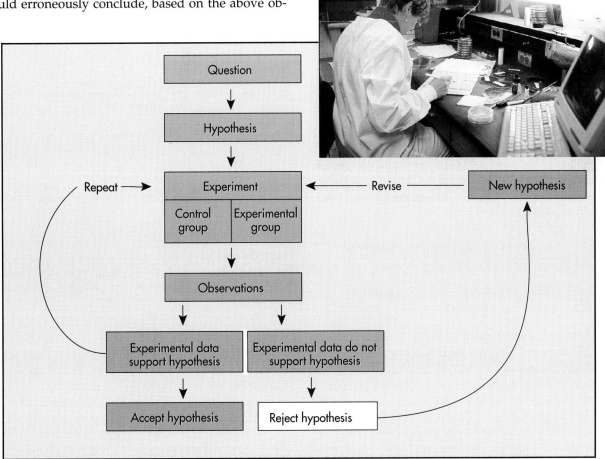

FIG. 3-1 In the scientific method, a hypothesis is proposed that can be tested. Experiments typically are run to determine whether predictions based on the hypothesis are accurate. The hypothesis is rejected if the predictions are not validated by experimental observations (data). An alternate new hypothesis may then be proposed and new experiments performed to assess its validity. When predictions based on a hypothesis are experimentally confirmed, the hypothesis is accepted.

The philosophy for this method of inquiry was developed by the English philosopher Francis Bacon in the seventeenth century. The scientific method relies on observations and deductive reasoning. It also demands that conclusions be subjected to thorough testing to develop objective evidence that can be evaluated before their credibility is accepted.

Scientific inquiry depends on systematic observation and tests.

In the scientific method a scientist first poses a question, for example, where do the maggots observed on rotting meat come from? The scientist then proposes a tentative answer to that question, for example, the maggots arise spontaneously. This tentative answer is called a **hypothesis.** Then the scientist tests the validity of the hypothesis by making *systematic observations.* Scientists often run **controlled experiments** to determine whether the tentative answers (hypotheses) are correct. In a controlled experiment, scientists attempt to hold all factors constant except the ones under study. By properly designing their experiments, scientists manipulate conditions so that they can control the factors that may cause a particular phenomenon. This way they can explore the cause-and-effect relationships among phenomena in the universe.

The design of a controlled experiment includes a **control group** and an **experimental group.** The control group serves as the reference. It has a set of conditions that the scientist does not vary. In contrast, in the experimental group, the scientist varies some factor or factors. By comparing the experimental group to the control group, the scientist is able to determine the effect(s) of the factor(s) that is varied on some other parameter(s).

Testing hypotheses through controlled experiments is the basis of scientific inquiry.

SHOWING THAT MAGGOTS DO NOT ARISE SPONTANEOUSLY FROM DECAYING MEAT

At about the same time that the scientific method was developed, Francesco Redi devised a controlled experiment to test the validity of the hypothesis that maggots arise spontaneously from rotting meat (FIG. 3-2). Redi was an Italian poet and physician. He placed a piece of loosely woven cloth over one piece of meat. This prevented insects from reaching the meat. He left a second piece of meat uncovered as a control. The presence or absence of the cloth was the only difference between the two pieces of meat. Redi hypothesized that if the maggots really arose spontaneously from the meat they would be trapped inside the cloth. He reasoned that if the maggots reached the meat from some other source, the cloth would

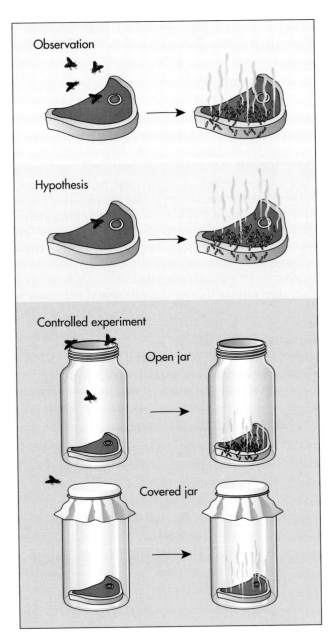

FIG. 3-2 Francisco Redi demonstrated that maggots do not arise spontaneously from rotting meat. When he covered the meat with cheesecloth, no maggots developed in the meat.

prevent the maggots from getting to the meat. Then maggots would be observed only on the outside of the cloth.

Redi observed that as the meat rotted, maggots appeared on uncovered meat. Maggots only appeared on the outside of the cloth of the covered meat. Thus Redi showed that maggots do not arise from within the decaying meat. It was important that the experiment conducted by Redi included a control that showed that the meat would rot and that maggots would appear on the meat in the absence of the cloth covering. The design of the experiment permitted him to determine by direct observation that the maggots came from outside the meat. The outcome of this experiment disproved the theory of spontaneous generation for large multicellular organisms such as insects. Other experiments by other scientists later demonstrated that maggots appear when eggs deposited by adult flies hatch.

It was significant that the experiments performed by Redi to test his hypothesis were repeatable by other scientists. You could easily repeat his controlled experiment today. Your experiment would also show that maggots do not arise spontaneously from rotting meat. Objectivity and the ability to repeat an experiment are essential parts of the scientific method. Reporting of the results of all experiments must be free of subjectivity, or bias, on the part of the scientist. Scientists always question the design and accuracy of experiments performed by their colleagues and the conclusions that are drawn from their experiments. No idea can be put forward with the expectation that its validity will not be challenged. As new information is obtained and experiments are repeated independently, scientists may gain confidence that accepted explanations are correct. Alternatively, they may determine that a once accepted hypothesis is incorrect and that a new hypothesis to explain the available data is needed.

Scientific experiments must be repeatable.

SHOWING THAT MICROORGANISMS DO NOT ARISE SPONTANEOUSLY DURING DECAY AND FERMENTATION

Scientists can only accept or reject hypotheses based on the evidence at hand. When Redi, for example, demonstrated in 1668 that maggots do not arise spontaneously, scientists did not know that microorganisms existed. It was not until 1670, 2 years after Redi performed his experiment, that Antonie van Leeuwenhoek peered through a microscope and recorded the first observations of bacteria "swimming" in rainwater and other liquids. The observation of microorganisms meant that new questions could be asked, new hypotheses put forth, and new controlled experiments performed to test these hypotheses.

In the eighteenth century, scientists using microscopes began to ask where the large numbers of microorganisms came from that they were now able to observe "swimming" in vats of spoiling wine and pots of rotting beef broth. What had been learned about maggots was not immediately applied to microorganisms. Many scientists, including numerous noted chemists, held that the rotting of meat (decay) and the conversion of sugar to alcohol (fermentation) as occurs when the sugar in grapes is made into the alcohol in wine were strictly chemical processes. They believed that the microorganisms seen in fer-

menting grape juice or rotting beef broth arose by spontaneous generation. Again, scientists had to demonstrate by using the scientific method that living organisms, in this case, living microorganisms, do not arise by spontaneous generation during decay and fermentation.

Scientists had a difficult time performing controlled experiments with microorganisms they could not see with the naked eye. Woven cloth could not prevent very small microorganisms from entering fermenting grape juice or decaying beef broth. Microorganisms could not be eliminated from these liquids without taking steps that also potentially altered the chemistry of the liquid and the air over it. Hence, the validity of experiments aimed at showing that, like maggots, microorganisms do not arise spontaneously was difficult to establish.

Lazzaro Spallanzani, an eighteenth century Italian priest, was among the first to use the scientific method to investigate whether or not microorganisms arise by spontaneous generation. Spallanzani hypothesized that microorganisms do not arise spontaneously. He hypothesized that, like maggots, microorganisms arise from the reproduction of living organisms. He performed controlled experiments to test his hypothesis.

In his controlled experiments Spallanzani placed various kinds of seeds in a flask. He then sealed the flask by covering the mouth with a stopper. By sealing the flask he prevented any microorganisms that were not already in the flask from reaching the liquid. Spallanzani then boiled the liquid in one flask for 45 minutes to kill any living microorganisms it contained. As a control Spallanzani did not boil the liquid in another otherwise similarly prepared flask. A few days later Spallanzani collected samples from each flask and observed them under a microscope. He saw many microorganisms in the samples from the flask that had not been boiled. He saw no micro-

HIGHLIGHT

Spontaneous Generation of Frogs and Mice

The ancient Greek philosopher Aristotle said that anything that becomes humid and any humid thing that dries produces animals. He taught that insects and worms came from dewdrops and slime, that mice were generated by dank soil, and that eels and fish sprang forth from sand, mud, and putrefying algae. In the seventeenth century, others expanded this belief into the theory of spontaneous generation and created recipes for producing frogs from the mud of ponds, eels from river waters, and mice from straw.

Control   Schwann's experiment   Latour and Kützing experiment

FIG. 3-3 Several scientists, including Latour, Kützing, and Schwann, attempted to disprove the theory of spontaneous generation. They used heat and sealed vessels, which chemists argued destroyed or eliminated oxygen, the "vital force of life." Thus these experiments did not definitively disprove the theory of spontaneous generation.

ganisms in the samples from the flask that he had boiled. The liquid that had been boiled in the sealed flask remained clear, free of microorganisms, and unspoiled indefinitely. Spallanzani concluded that boiling killed the microorganisms and that new microorganisms could not arise spontaneously from the nonliving liquid in a sealed flask.

However, nineteenth century advocates of the theory of spontaneous generation of microorganisms assailed Spallanzani's experiments. They claimed that by sealing the flasks Spallanzani had eliminated molecular oxygen. They argued that the oxygen in air was an essential *"life force"* necessary for the spontaneous formation of microorganisms. Thus, Spallanzani's experimental design was criticized. His opponents claimed that his conclusions were invalid. Chemists held fast to the view that changes in organic chemicals occurred by strictly chemical processes and were not brought about by living organisms.

In the 1830s, several scientists proposed hypotheses and conducted experiments to demonstrate that microorganisms do not arise spontaneously during the course of the decay of organic matter. Like Spallanzani, these scientists used boiling to kill the microorganisms in liquids contained in flasks. However, they did not seal the flasks to prevent microorganisms from entering the liquid broth. Theodor Schwann in Germany placed a flame at the mouth of an open flask to kill microorganisms that might be in the air, thereby preventing microorganisms from entering the flask. Charles Cagniard de Latour in France and Friedrich Kützing in Germany used cotton plugs to prevent microorganisms from entering the flasks and reaching boiled broths (FIG. 3-3).

These experiments showed that when microorganisms were prevented from entering the liquid in the flask, decay and/or fermentation did not occur. Neither were microorganisms observed in the broth when their entry was prevented, whereas microorganisms always were seen in the broths when microorganisms could enter the flasks. However, these experiments were also subject to criticism. Chemists claimed each experiment destroyed or eliminated some essential component called the *"life force"* in the air. They said air was needed for the decay of the organic matter and the spontaneous generation of microorganisms. Remember that such criticism is part of the scientific method. Scientists always ask how do you know that your results are correct, that your interpretation of the experimental results is the best possible one. Only experimental results and interpretations that withstand intense criticism and review become accepted in science.

The results of an experiment are carefully and critically reviewed before their validity is accepted by the scientific community.

PASTEUR AND THE FINAL REFUTATION OF THE THEORY OF SPONTANEOUS GENERATION

In 1862, **Louis Pasteur** (FIG. 3-4) designed a controlled experiment that overcame all the criticisms of the chemists, thereby disproving spontaneous generation. Pasteur was a Frenchman who had been trained as a chemist. This provided him with the experience needed to use the scientific method in studies on microorganisms. It also gave him some degree of credibility with other professional scientists. Pasteur believed that the origin of microorganisms had

FIG. 3-4 Louis Pasteur (1822-1895) began as a chemist but soon became a pioneer microbiologist. Pasteur's work emcompassed pure research and many areas of applied science that produced several important practical discoveries. Among his many accomplishments, Pasteur discredited the theory of spontaneous generation, introduced vaccination to treat rabies, and solved industrial problems related to the production and spoilage of foods.

to be understood before microbiology could be established as an experimental science. He also held that "one cannot expect the doctrine of spontaneous generation to be abandoned as long as serious argument can be presented in its favor."

When he turned his attention to microorganisms, Pasteur initially asked two questions: Are microorganisms needed for decay and/or fermentation to occur?; and do microorganisms arise spontaneously during decay and/or fermentation? Pasteur hypothesized that microorganisms cause decay and fermentation and that microorganisms do not arise by spontaneous generation. By using controlled experiments, Pasteur left flasks containing boiled beef broth or boiled grape juice open to the air and showed that decay of the beef broth and fermentation of the grape juice did not occur unless microorganisms could reach the liquid from the outside.

For these experiments Pasteur designed a flask that could be left completely open to the air and still prevent microorganisms from reaching the broth contained within the flask (FIG. 3-5). He made glass flasks with several curves in the neck of each flask. Microorganisms would settle with dust particles in the depressions of the neck and never reach the liquid within the flask. The neck of the flask was curved to look like a "swan's neck" and the flask was therefore called a **swan-neck flask.** As a control Pasteur used a flask with a straight neck of the same length. The only difference between the flasks he used in his experiments was the shape of the neck of the flask.

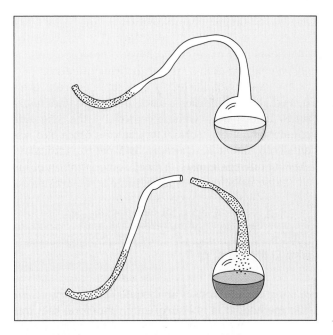

FIG. 3-5 To discredit the theory of spontaneous generation, Pasteur used various shapes in the design of his swan-necked flasks. Pasteur boiled the liquid containing nutrients to kill any microorganisms that were already there. He then left the flasks open to the air. The curved necks of the flasks trapped dust particles, preventing them from carrying microorganisms to the liquid broth growth medium so that the broth remained clear and free of microorganisms (*top*). These experiments demonstrate that spontaneous generation does not occur. When he broke the necks of some of the flasks (*bottom*), dust carried a microorganism into the broth growth medium. The microorganism grew in the broth, making it turbid (cloudy).

Chemists could not claim that this difference affected the *"life force."*

In his experiments, Pasteur found that liquids subjected to boiling remained free of living microorganisms in the flasks with the swan necks. Liquids boiled in the same manner soon swarmed with microorganisms in the flasks with the straight necks. He concluded that, although air could freely enter the flask, the shape of the swan neck of the flask prevented airborne microorganisms from entering the liquid.

As an additional control to demonstrate that the microorganisms that cause decay and fermentation do not arise spontaneously, Pasteur broke off the neck of one of the swan-neck flasks. This allowed microorganisms to enter the medium in the flask that had, until that point, remained free of living microorganisms. Shortly thereafter, Pasteur observed numerous microorganisms in the broth contained in that flask. This demonstrated that the medium in the flask could in fact support microbial life if microorganisms were introduced. The flask with the intact swan neck remained free of microorganisms. He concluded that the microorganisms in the broth came from living microorganisms that were floating unseen in the air. Pasteur had shown that microorganisms are like other forms of life—that living cells come only from the reproduction of pre-existing living cells. Pasteur's swan-neck flasks, now sealed to prevent the evaporation of water, are still free of microorganisms and on exhibit at the Pasteur Institute in Paris.

The results of Pasteur's experiments were challenged. His experiments had to be repeated many times before they were accepted by the scientific community. This is characteristic of science. No scientist, no matter how famous, escapes the critical review of other scientists. A final demonstration of his experiments before a tribunal of other scientists was convened to critically evaluate his disproof of the spontaneous generation of microorganisms. At this time Pasteur declared: "There is no condition known today in which you can affirm that microscopic beings came into the world without germs, without parents like themselves. Those who allege it have been the sport of illusions, of ill-made experiments filled with errors that they have not been able to perceive." Referring to the liquids in his swan-neck flasks he concluded: "I have kept from them the one thing that is above the power of humans to make; I have kept from them the germs which float in the air. I have kept them from life."

The results of Pasteur's experiments discredited the theory of spontaneous generation. They established once and for all that living microorganisms are responsible for the chemical changes that occur during fermentation and decay and that microorganisms do not arise spontaneously. With the success of his experiments, Pasteur proclaimed: "No more shall spontaneous generation rear its ugly head!"

Pasteur's refutation of the theory of spontaneous generation was a critical milestone in the development of microbiology as a scientific discipline. It demonstrated that scientists could successfully use the scientific method for examining microorganisms. It also showed that microorganisms are like other living organisms, that only living organisms give rise to other living organisms. It further revealed the importance of microorganisms in bringing about chemical changes. It was the activities of the microorganisms that transformed grape juice into wine and caused meat to decay. Pasteur went on to study many other aspects of microorganisms and to help pioneer the birth of microbiology as a scientific discipline.

With the studies of Pasteur, microbiology emerged as a scientific discipline.

MICROSCOPY

Making observations is an essential part of the scientific method. It is not surprising, therefore, that microbiology did not develop as a field of science until the necessary instruments and methods were developed for observing microorganisms. Unlike organisms that could be described and studied by the unassisted eye, microorganisms could not be studied until instruments were available with sufficient magnifying power so that they could be seen. The **microscope** is an instrument for producing enlarged images of objects that are too small to be seen unaided. Microscopic observations generally provide information only about the morphology of microorganisms and not their physiological characteristics. The light microscope can be used for viewing objects as small as bacteria. The electron microscope extends the

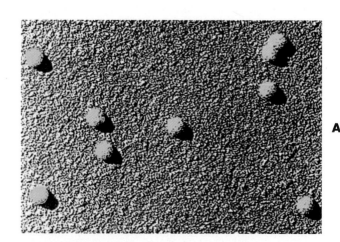

A

FIG. 3-6   A, Colorized transmission electron micrograph of polioviruses observed through an electron microscope.

B

FIG. 3-6, cont'd **B,** Micrograph of a bacterial biofilm viewed by confocal fluorescence scanning microscopy. Color indicates depth within specimen (purple nearest surface and red deepest in biofilm).

range to even smaller objects, such as viruses and large molecules (FIG. 3-6).

By using microscopes to magnify their images, microorganisms can be observed and studied.

PRINCIPLES OF LIGHT MICROSCOPY

Magnification

Because the observation of microorganisms through the microscope is fundamental to the study of microbiology, we should consider some of the principles of microscopy. Let us first discuss how magnification is achieved. The apparent size of an object viewed directly by the eye depends on its distance from the eye. As an object is brought nearer to the eye, the apparent size of the object increases. However, if the object is too close to the eye, the eye no longer can form a clear image. Placing a curved glass lens between the object and the eye can restore the sharpness of the image. This is because the lens changes the direction (angle) of the light rays reaching the eye. The bending of the light rays by a glass lens is called **refraction.** It forms an image of the object that is larger than

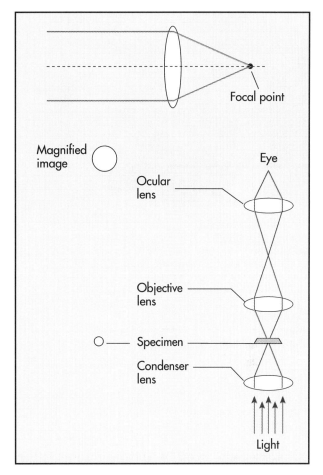

FIG. 3-7 A convex-convex lens theoretically focuses light at a single focal point. However, red light (long wavelength light) focuses more distantly from the lens than blue light (short wavelength light), causing chromatic aberrations. Also, light rays passing through the periphery of the lens (axial rays) focus more distantly from the lens than light rays that pass through the center of the lens (marginal rays), causing spherical aberrations. Various lenses greatly inprove the performance of the microscope lens by correcting for these aberrations. Modern microscopes use several such lenses to produce a high-quality magnified image.

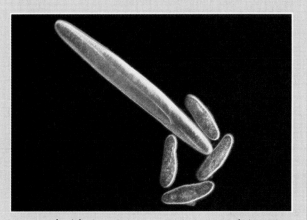

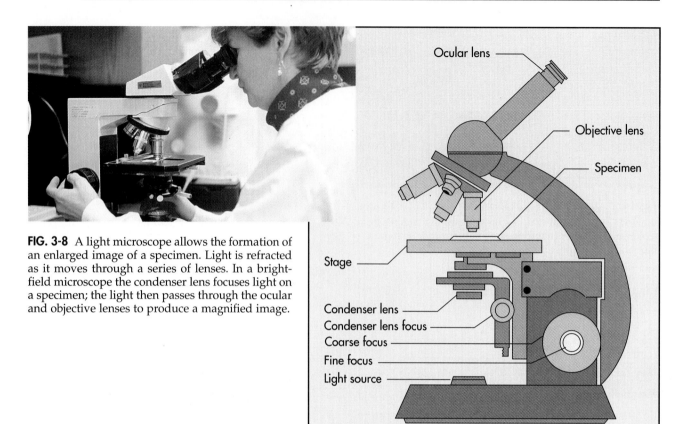

FIG. 3-8 A light microscope allows the formation of an enlarged image of a specimen. Light is refracted as it moves through a series of lenses. In a bright-field microscope the condenser lens focuses light on a specimen; the light then passes through the ocular and objective lenses to produce a magnified image.

the object itself (FIG. 3-7). This is because the light reaching the eye from the top to the bottom of the object is at a greater angle than it would be if the lens had not caused the light to bend.

> **The microscope produces an enlarged (magnified) image of a specimen through the bending (refraction) of light rays by curved glass lenses.**

The **compound light microscope** has multiple lenses that refract light and achieve magnification of the image of a specimen (FIG. 3-8). The lenses can be moved to achieve focus. Light, typically from an incandescent bulb source, is focused onto a specimen by a **condenser lens.** The specimen is normally on a clear glass microscope slide that can be moved by a mechanical stage. The light passes through the specimen and enters the **objective lens,** which refracts the light and begins the magnification process. The light then passes through the **ocular lens,** which further refracts the light and completes the magnification process. The light enters the eye and the magnified image is seen.

The degree of magnification (enlargement of the image) that can be achieved with a compound microscope is the product of the individual magnifying powers of the ocular and objective lenses. The total magnifying power of a microscope is calculated by multiplying the power of the objective by the power of the ocular. The magnifying power obtained with a

$100\times$ objective and a $10\times$ ocular, for example, would be $1,000\times$. The most commonly used light microscopes have objective lenses with powers of $10\times$, $40\times$, and $100\times$ and an ocular lens of $10\times$ so that one can obtain magnifying powers of 100, 400, and 1,000 times with them.

> **The magnifying power of a compound microscope is calculated by multiplying the magnifying powers of the ocular and objective lenses.**

Resolution

One could well ask, why don't we simply continue to increase the magnifying capacity of the lenses to make more and more powerful microscopes? Why not combine a $100\times$ objective lens with a $100\times$ ocular lens so that an image could be magnified $10,000\times$? The answer is that such an image, while larger, would not show additional details of the specimen. **Resolution** is the ability to distinguish detail in the object that is viewed. In biological specimens, resolution is equated with the ability to see structures of the organism.

Resolving power is a distance measure, sometimes called the resolving distance, that is defined as the closest spacing between two points at which they can still be seen clearly as separate entities. We measure the resolving power of a microscope in units called nanometers (nm), which are billionths of a meter (10^{-9} m). Objects that are closer than the resolving power of the microscope cannot be seen as separate and distinct, and objects that are smaller than the resolving power cannot be seen at all.

> **The ability to see detail is determined by the resolving power and not the magnifying power of the microscope.**
>
> **Resolution describes the amount of detail that can be seen in an image.**

The resolving power of the light microscope is about 200 nm. This is much better than the 0.1 mm resolving power of the unaided eye. Because all cellular organisms, including bacteria, are larger than 100 nm, the light microscope permits the observation of the microbial world. Most viruses and other acellular microorganisms, which are all smaller than 100 nm, however, cannot be seen with the light microscope.

The resolving power (R) of a microscope is approximately one half the wavelength of the light (λ) that is used to illuminate the specimen divided by the numerical apertures (NAs) of the objective lens and the condenser lens (FIG. 3-9).

$$R = \lambda/(NA_{objective\ lens} + NA_{condenser\ lens})$$

The numerical aperture represents the amount of light from an object that actually enters a lens. The greater the amount of light that can enter the lens, the higher the numerical aperture. The higher the numerical aperture, the smaller R and hence the better the resolving power. Since the numerical apertures of the objective lens and the condenser can be about the same, the formula for resolving power normally is written as:

$$R = \lambda/(2 \times NA)$$

For visible light, wavelength is seen as color. Blue light has a short wavelength and red light has a long wavelength. The best resolution of a light microscope can be achieved by using a blue light source (short wavelength light) to illuminate the specimen and a lens with a high numerical aperture.

> **Resolution depends on the wavelength of light (λ) and the numerical aperture (NA) of the lens.**
>
> **The best resolving power of a light microscope occurs when short wavelength light and a high numerical aperture objective lens are used.**

To achieve a high numerical aperture, we insert a drop of clear oil between the specimen and the objective lens. This requires a special type of objective lens called an **oil immersion objective.** Using oil improves the resolving power of the microscope because the oil allows light to pass in a straight line from the specimen to the objective lens. The typical oil immersion objective has a numerical aperture of 1.25 to 1.30, whereas the numerical aperture of a non-oil immersion lens is less than 1. As a rule, the useful magnification of a microscope above which greater detail cannot be seen because of the limitation to resolving power is 1,000 times the numerical aperture of the objective lens being used. It is possible, using an oil immersion lens, to achieve magnifications of just over 1,000×, which is essential for viewing bacteria.

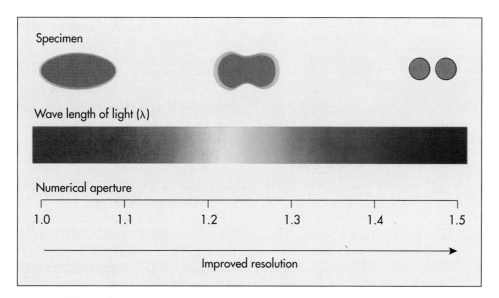

Specimen

Wave length of light (λ)

Numerical aperture

1.0   1.1   1.2   1.3   1.4   1.5

Improved resolution

FIG. 3-9 Microscopy depends on the ability to see detail, that is, to resolve distinct points. At low resolution, structures blur; at greater resolution, more detail can be observed. Resolution depends, in part, on the wavelength of light. Blue light, which has a short wavelength (380 nm), gives superior resolution to red light, which has a long wavelength (750 nm).

Contrast and Staining

Staining (dyeing) is used to increase the contrast between the specimen and the background. This usually is necessary to create adequate contrast between the specimen and the background so the specimen can be seen. Most microorganisms are colorless and cannot be seen readily without being stained. Many types of staining procedures, which employ dyes of differing colors, are used for different purposes in microbiology. Most staining procedures for light microscopic observation of microorganisms begin with the transfer of a suspension of microorganisms to a glass microscope slide (FIG. 3-10). The microorganisms are spread as a thin film across the slide that is allowed to dry in the air. Then the slide is quickly passed through a flame to *fix* the cells to the slide. Fixing preserves the shape of the cells and prevents them from being washed off during staining. A stain is added and allowed to penetrate and to react with the cells on the slide for a period of time. Excess stain is then rinsed from the slide and the specimen is ready for viewing under the microscope.

Staining creates contrast between a specimen and its background so it can be seen.

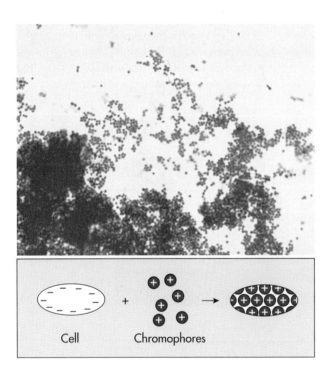

FIG. 3-10 In a simple staining procedure, microorganisms are affixed to a glass slide and stained with an appropriate dye (colored chromophore). This increases the contrast between the cells and the background so they can easily be seen using a light microscope. Because the outer layer of a cell is negatively charged, a positively charged stain chromophore is attracted to the cell; this is the basis of positive staining procedures.

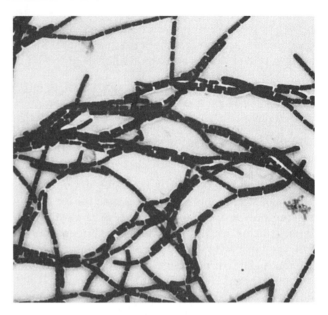

FIG. 3-11 Light micrograph of the bacterium *Bacillus cereus* after simple positive staining with carbolfuchsin. (1,400×). The cells of the rod-shaped bacterium appear red in contrast to the clear background.

Simple staining procedures use a single stain. A stain is a salt comprised of a positively charged ion and a negatively charged ion. If the colored portion of the stain is positively charged, it will be attracted to microbial cells, which have negative charges. Staining with a positively charged dye is called **positive staining**. Stained cells that look dark or colored against a light or clear background will be seen. Methylene blue, for example, will stain microbial cells blue by positive staining. This stain is useful for seeing the shapes of bacterial cells (FIG. 3-11). If the colored portion of the stain is repelled by the negatively charged microorganisms, clear unstained microorganisms will be seen against a dark background. Such staining with a negatively charged dye is called **negative staining**. India ink and nigrosin, which have negatively charged black particles, are examples of dyes that can be used for negative staining (FIG. 3-12). Like positive staining, negative staining is used to reveal the shapes of microorganisms.

Differential staining procedures use multiple stains to distinguish different cell structures and/or cell types. In differential staining procedures, specific types of microorganisms and/or particular structures of a microorganism exhibit different affinities for certain stains. By using multiple stains and washing steps in a differential staining procedure, structural differences between microbial species can be revealed. These differences aid in the classification and identification of microbial species.

The **Gram stain procedure** is the most widely used differential staining procedure in bacteriology

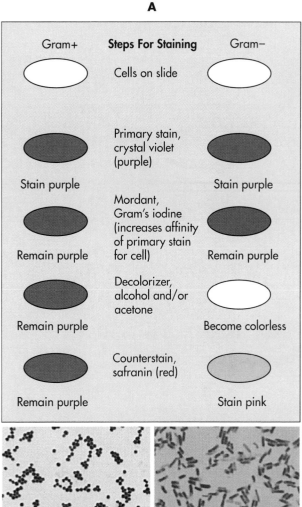

FIG. 3-12 Light micrograph of the bacterium *Klebsiella pneumoniae* after simple negative staining with India ink. The cells appear clear against a dark background because the ink does not penetrate the cells.

today (FIG. 3-13). It was developed in 1884 by the Danish physician Hans Christian Gram. Gram was trying to develop a method for seeing bacteria within mammalian tissues. He failed at that task but instead discovered a method for the diagnostic differentiation of bacterial species based on the way they are stained by his procedure. This Gram staining procedure begins with primary staining with crystal violet, which stains all bacterial cells blue-purple. Then Gram's iodine is applied as a mordant. A mordant is a substance that increases the affinity of the primary stain for the bacterial cells. The cells are rinsed with acetone-alcohol or another decolorizing agent to try to wash out the primary stain. A red counterstain (safranin) is then applied, which stains the bacteria that were decolorized in the previous step so that they can be easily seen.

The decolorization step of the Gram stain procedure is the critical step. It differentiates bacterial species based on their cell wall structure. The cell wall structure of certain bacteria does not permit decolorization. These bacteria remain blue-purple and are called **Gram-positive bacteria.** Gram-positive bacteria appear blue-purple at the end of the Gram stain procedure. *Staphylococcus* and *Streptococcus* are examples of Gram-positive coccal-shaped (round) bacteria. The remaining bacterial species, called **Gram-negative bacteria,** are decolorized and appear red-pink following counterstaining at the completion of the Gram stain procedure. *Escherichia coli,* which occurs in huge numbers in the human intestine, is a Gram-negative bacterial species.

Gram-positive bacteria remain blue-purple after the Gram stain procedure; Gram-negative bacteria appear red-pink.

A

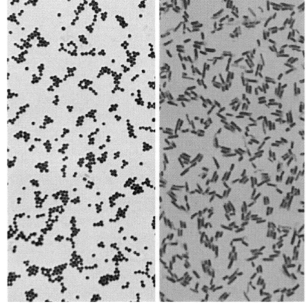

B

C

FIG. 3-13 **A,** The Gram stain procedure is widely used to differentiate major groups of bacteria. Gram-positive bacteria stain blue and Gram-negative bacteria stain pink-red by this staining procedure. Gram-positive and Gram-negative bacteria stain purple with the primary stain. The primary stain is removed from Gram-negative cells by the decolorizer and they are then stained pink by the counterstain. Gram-positive cells retain the primary stain and remain purple. **B,** Cells of the Gram-positive bacterium *Staphylococcus aureus* appear as blue-purple cocci. **C,** Cells of the Gram-negative bacterium *Escherichia coli* appear as pink-red rods.

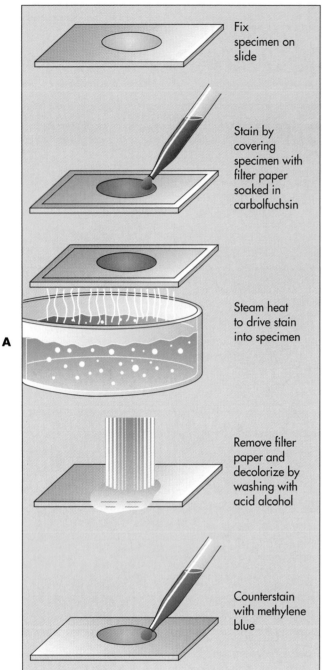

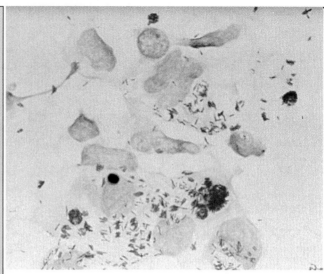

FIG. 3-14   A, The acid-fast staining procedure uses the red dye carbolfuchsin to stain cells. Acid-fast bacteria, such as mycobacteria, retain this stain when treated with acid alcohol. Nonacid-fast bacteria and tissue cells are decolorized by acid alcohol; they are counterstained with methylene blue. **B,** Acid-fast mycobacteria, including *Mycobacterium leprae* (the cause of leprosy), appear red after this staining procedure, as shown in this light micrograph.

The acid-fast bacteria probably retain the carbolfuchsin because of waxy chemicals that occur in their cell walls. These chemicals are not found in nonacid-fast bacterial cells. To complete the acid-fast procedure, cells are counterstained with methylene blue and observed under the microscope. Acid-fast bacteria appear red and nonacid-fast bacteria appear blue. The red mycobacteria can thus be easily recognized even in a sputum sample containing numerous other blue bacterial cells.

> **Acid-fast staining is used to identify *Mycobacterium*, which are acid-fast and appear red after staining.**
>
> **Nonacid-fast bacteria appear blue after acid-fast staining.**

The **endospore staining procedure** is used to reveal specifically the presence or absence of endospores (FIG. 3-15). Endospores are heat-resistant resting bodies produced by only a few bacteria, such as species of the genera *Clostridium* and *Bacillus*. In the endospore staining procedure, the primary stain malachite green is driven into the endospores by steaming. Water is then used as a decolorizing agent. Water will wash the malachite green from the cells but not from the endospores. Next, safranin, a counterstain, is used to stain the cells. When viewed under the microscope, endospores—if they are present—appear as green spheres. They are either within the red-pink cells that produced them or free if those cells have died.

Acid-fast staining is another differential staining procedure frequently used in bacteriology (FIG. 3-14). The acid-fast stain procedure is especially useful in identifying members of the bacterial genus *Mycobacterium* and is important in identifying the causative organisms of tuberculosis *(M. tuberculosis)* and leprosy *(M. leprae)*. In the acid-fast staining procedure, the red stain carbolfuchsin is used as a primary stain. Next, acid-alcohol is used as a decolorizer. The acid-alcohol will remove the red stain from bacteria, such as *Escherichia coli,* that are not acid-fast. The acid-fast mycobacteria will remain red.

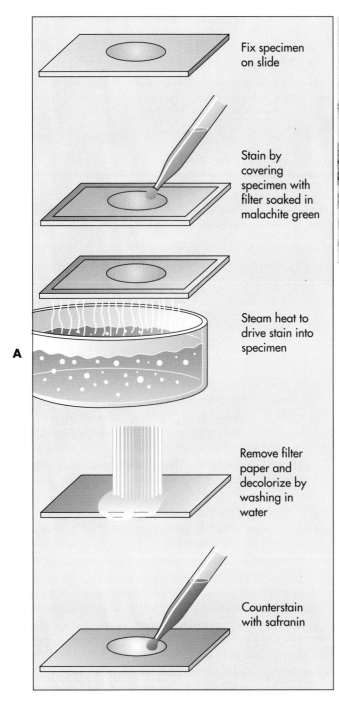

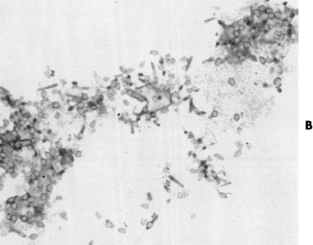

FIG. 3-15 **A,** The endospore staining procedure involves primary staining with malachite green. This stain is retained by endospores but washed out of cells with water. Cells are then counterstained with the red dye safranin. The spores appear green and the cells pink-red after staining by this procedure. **B,** Micrograph of *Clostridium tetani* after endospore staining. The spores appear green and the bacterial cells are stained red.

Fix specimen on slide

Stain by covering specimen with filter soaked in malachite green

Steam heat to drive stain into specimen

Remove filter paper and decolorize by washing in water

Counterstain with safranin

TYPES OF LIGHT MICROSCOPES

Just as there are different staining procedures that are used for different purposes, there are different types of microscopes (Table 3-1, p. 72). These microscopes differ in magnification and resolution capacities. Some have specialized applications, and others have widespread applications in microbiology.

Bright-field Microscope

Virtually every microbiology laboratory has a **bright-field microscope,** so named because the field of view is brightly illuminated (FIG. 3-16). The typical bright-

FIG. 3-16 The bright-field microscope is used routinely by students of microbiology and practicing microbiologists, such as the technician in the clinical microbiology laboratory seen here.

TABLE 3-1			
Comparison of Various Types of Microscopes			
TYPE OF MICROSCOPE	**MAXIMUM USEFUL MAGNIFICATION**	**RESOLUTION**	**DESCRIPTION**
Bright-field	1,500×	100-200 nm	Extensively used for the visualization of microorganisms; usually necessary to stain specimens for viewing
Dark-field	1,500×	100-200 nm	Used for viewing live microorganisms, particularly those with characteristic morphology; staining not required; specimen appears bright on a dark background
Fluorescence	1,500×	100-200 nm	Uses fluorescent staining; useful in many diagnostic procedures for identifying microorganisms
Phase contrast	1,500×	100-200 nm	Used to examine structures of living microorganisms; does not require staining
TEM (transmission electron microscope)	500,000-1,000,000×	0.1 nm	Used to view ultrastructure of microorganisms, including viruses; much greater resolving power and useful magnification than can be achieved with light microscopy
SEM (scanning electron microscope)	10,000-100,000×	1-10 nm	Used for showing detailed surface structures of microorganisms; produces a three-dimensional image

field microscope is a compound microscope that has two lenses that contribute to the magnification of the image. The lenses are actually sets of several individual lenses that are designed to minimize distortions of light, called *aberrations,* that would interfere with the clear observation of a specimen. Besides its compound lenses, the modern bright-field microscope also has a built-in light source, a condenser lens that focuses the light on the specimen, and a mechanical stage that holds and permits controlled movement of the specimen.

Fluorescence Microscope

The **fluorescence microscope** is specifically designed for use with fluorescent stains (FIG. 3-17). These stains emit light of a different wavelength (color)

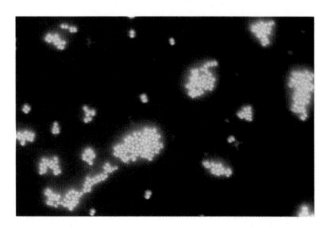

FIG. 3-17 Micrograph of clusters of staphylococci in blood after staining with acridine orange; the cells of this bacterium fluoresce orange.

than the wavelength of light used to illuminate the specimen. For example, when fluorescein isothiocyanate (FITC) is illuminated with blue or ultraviolet light, it emits an apple-green light. The fluorescence microscope has one filter that sets the wavelength of light used to illuminate the specimen and another filter that determines the wavelength of light that is viewed. As with the bright-field microscope, the specimens usually are stained. Some bacteria, such as *Pseudomonas,* contain fluorescent pigments and can be viewed without staining.

Microscopy that uses fluorescent dyes is known as fluorescence microscopy.

Fluorescence microscopy is important because fluorescent dyes can be chemically linked to antibodies to produce **fluorescent-conjugated antibodies.** Antibodies are molecules that take part in animal defense systems against invading microorganisms. They have very specific chemical targets with which they react. This means that scientists can design and create fluorescent-conjugated antibodies that will react only with specific microorganisms to the exclusion of all others. By attaching a fluorescent dye to an antibody, the presence of a particular target can be visualized (FIG. 3-18). If the target is a bacterial cell, this forms the basis for **immunofluorescence microscopy** (differential staining using fluorescent-conjugated antibodies). Many pathogens, including the bacterium *Treponema pallidum,* which causes syphilis, can be quickly and positively identified by immunofluorescence microscopy without having to grow cultures in the laboratory. Diagnosis of disease is accomplished rapidly so that appropriate therapy can be initiated as quickly as possible.

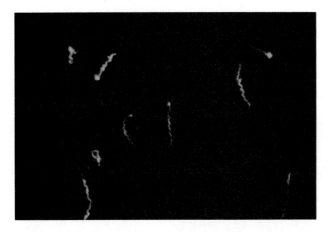

FIG. 3-18 Light micrograph of the spirochete *Treponema pallidum (green)* after fluorescent antibody staining. This bacteria causes syphilis.

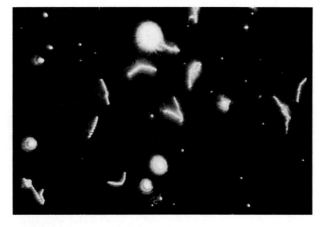

FIG. 3-20 Light micrograph of the bacterium *Treponema pallidum*, which causes syphilis, viewed by dark-field microscopy.

Dark-field Microscope

The **dark-field microscope** is designed to enhance the contrast between the specimen and the background without the use of staining. Most staining procedures kill microorganisms. Using the dark-field microscope we can view live as well as dead specimens. The dark-field microscope has a special condenser that does not permit light to be transmitted directly through the specimen and into the objective lens (FIG. 3-19). This special dark-field condenser focuses light on the specimen at an oblique angle, such that the only light entering the objective is reflected off the object under observation. Thus only light that reflects off the specimen will be seen. In the absence of a specimen the entire field will appear black. Bacteria viewed with a dark-field microscope appear very bright on a black background (FIG. 3-20). The contrast between the specimen and the background is sufficient to permit easy visualization of bacteria and other cellular microorganisms.

Phase Contrast Microscope

The **phase contrast microscope** also provides a means for achieving adequate contrast for visualizing microorganisms without the necessity of staining. The design of this microscope relies on the fact that light passing through a cell is slowed by the difference in density between it and the surrounding medium. Differences in brightness are created and this contrast is exactly what is needed to visualize specimens (FIG. 3-21). Consequently, with the phase contrast microscope, living organisms can be clearly

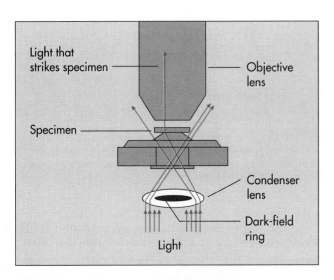

FIG. 3-19 Diagram of a dark-field microscope showing the path of light. The dark-field ring in the condenser blocks the direct passage of light through the specimen and into the objective lens. Only light that is reflected off a specimen will enter the objective lens and be seen.

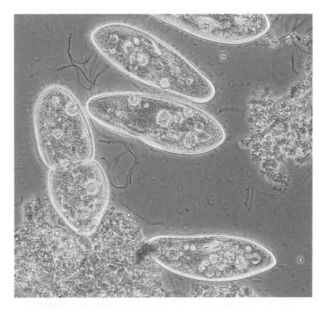

FIG. 3-21 Light micrograph of the protozoan *Paramecium caudatum* viewed by phase contrast microscopy. (264×).

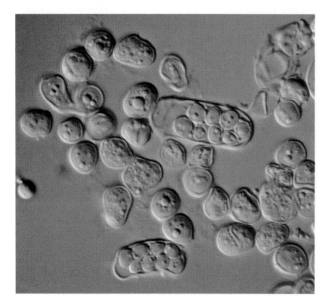

FIG. 3-22 Micrograph of the yeast *Schizosaccharomyces* viewed by Nomarski interference microscopy. (12,000.×).

observed in great detail without staining them. This permits the study of their movements and their appearances in their natural surroundings.

Interference Microscope

There are several types of **interference microscopes,** such as the Nomarski interference microscope, that are now used in many microbiology laboratories. Like the phase contrast microscope these interference microscopes work on the principle that light from two light waves can be combined to increase or to decrease brightness. These microscopes are designed so that a beam of light can be split. One part of the light passes through the specimen and another part goes around the specimen. The beams are then recombined in a specialized prism and viewed. The result is a colored image that has a three-dimensional appearance (FIG. 3-22).

ELECTRON MICROSCOPY

An **electron microscope** uses an electron beam instead of visible light. Because an electron beam has a much shorter wavelength than the wavelengths of visible light, the electron microscope can achieve superior resolution. The electron microscope, therefore, permits higher useful magnifications than can be achieved with light microscopy. Consequently, viruses and even the smallest structures of a bacterial cell can be seen by using an electron microscope.

> **Electron microscopes use electron beams instead of visible light. Electron microscopes achieve better resolution and higher useful magnification than light microscopes.**

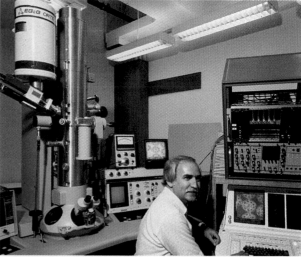

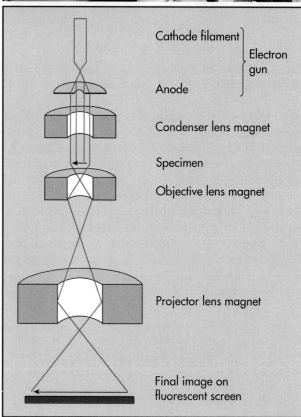

FIG. 3-23 The transmission electron microscope (TEM) allows the visualization of the fine detail of the microbial cell. The TEM uses an electron beam and electromagnets instead of the light source and glass lenses used in light microscopy.

In the **transmission electron microscope** (TEM), the electron beam is transmitted through the specimen (FIG. 3-23). The other type of electron microscope is called the **scanning electron microscope** (SEM). In the SEM an electron beam is scanned across the surface of the specimen (FIG. 3-24). In either type of electron microscope, the electron beam must be transmitted through a vacuum. Otherwise the elec-

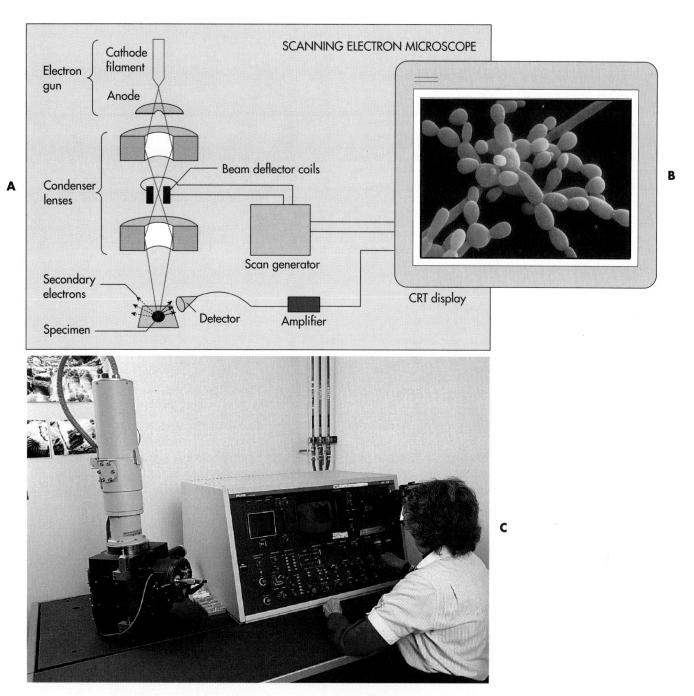

FIG. 3-24   A, The scanning electron microscope (SEM) is used for viewing surface structures and their three-dimensional spatial relationships. An electron beam is scanned across a specimen. The electrons emitted from the surface of the specimen determine the intensity of the image. The relative lengths of the scans across the specimen and the cathode ray tube (CRT) display determine the magnification. **B,** Colorized micrograph of the fungus *Candida albicans* viewed using a SEM. (6,300×). **C,** Photograph of a scanning electron microscope.

trons would collide with the molecules in air and scatter. This produces a blurred image, if any image is produced at all. Much of the apparent complexity and bulk of an electron microscope is the result of the need to produce a high vacuum system for the transmission of the electron beam. Also, because we can

not view the electron beam directly, we view the magnified image of an electron microscope on a phosphorescent screen and sometimes make a permanent image on photographic film. The images produced by electron microscopes are called electron micrographs or just micrographs.

Transmission Electron Microscope

The principles of operation of the **transmission electron microscope** (TEM) are very similar to those of the compound light microscope. Instead of a light source, the TEM uses an electron beam. In place of glass lenses, the TEM uses a series of electromagnets to focus the electron beam and to magnify the image of the specimen. The electron microscope has a theoretical resolution of approximately 0.1 nm. This is about a thousand times better resolving power than can be achieved by using light microscopy. Consequently, the useful magnification for an electron microscope is higher than 500,000×. Magnifications this high permit the visualization of all microorganisms, including viruses. The internal structures of microorganisms can also be seen in fine detail (FIG. 3-25).

> The transmission electron microscope (TEM) permits very high magnification (>500,000×) and superior resolution (0.1 nm).

> The TEM permits the visualization of the detailed structures of all microorganisms, including viruses.

There are some special problems in viewing biological specimens with the electron microscope. The specimens must be killed before being placed in a high vacuum chamber and exposed to a high voltage electron beam. Because of the high magnifications achieved with the electron microscope, there is a great potential for creating artifacts that could be mistakenly viewed as real structures in electron mi-

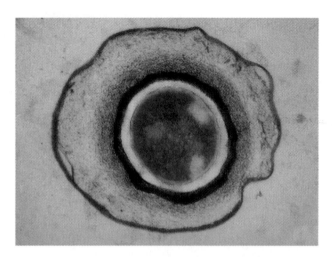

FIG. 3-25 Colorized transmission electron micrograph of a thin section of an endospore of *Bacillus sphaericus*.

crographs. An artifact is the appearance of something in an image or micrograph caused by the magnification system or procedures used for preparing specimens.

> An artifact is not a true representation of the features of the specimen on view.

To minimize artifacts, special preparation procedures must be used in electron microscopy (FIG. 3-26). Biological specimens containing water cannot simply be placed under high vacuum because the water would boil. The integrity of the organisms

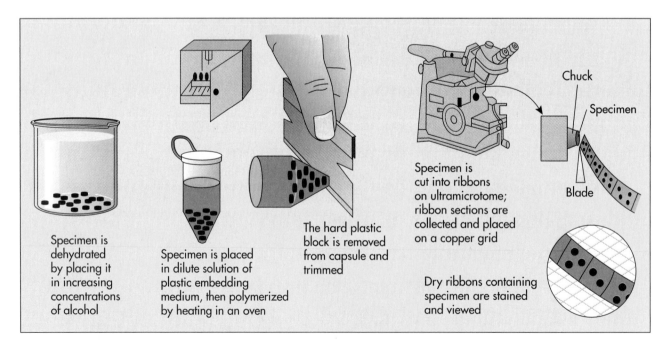

Specimen is dehydrated by placing it in increasing concentrations of alcohol

Specimen is placed in dilute solution of plastic embedding medium, then polymerized by heating in an oven

The hard plastic block is removed from capsule and trimmed

Specimen is cut into ribbons on ultramicrotome; ribbon sections are collected and placed on a copper grid

Dry ribbons containing specimen are stained and viewed

Chuck
Specimen
Blade

FIG. 3-26 Extensive preparation of a specimen is generally needed for viewing by transmission electron microscopy. Water must be removed; this dehydration of the specimen usually is achieved using alcohol. Many specimens must be cut into thin sections before they can be viewed in the electron microscope. Sectioning is accomplished by placing the specimen in a plastic and then cutting the sections with an ultramicrotome.

HIGHLIGHT

Mesosomes: Real Structures or Artifacts?

Sometimes, when thin sections of bacteria are viewed by transmission electron microscopy, extensive folds of membranes are seen within the cell (see Figure). These membranes are called mesosomes. Careful observations of these mesosomes indicate that they are continuous with the plasma membrane. Having observed mesosomes in many routine electron microscopic preparations, several independent investigators tried to identify their functions. Although many functions were hypothesized for mesosomes, establishing their real function proved elusive. For each proposed function, cells were found without mesosomes that carried out the same functions as cells with mesosomes. For example, because mesosomes were often observed in the region where a bacterial cell divides when it reproduces, one proposed function was that they were involved in cell division. However, mesosomes are not always seen when bacterial cells divide, nor, when they are present, do they always occur near the site of cell division.

After considering the observations made in many studies, it became obvious to the scientists who had first described mesosomes that the evidence for the existence of mesosomes had to be reevaluated. These scientists asked why mesosomes were only sometimes observed and why their absence didn't result in the loss of some function. They hypothesized that mesosomes might be an artifact. Based on numerous observations

they found that mesosomes were almost always found to be attached or closely associated with DNA within the cell. They reasoned that the DNA might be shrinking due to dehydration during preparation for electron microscopy. Also, since bacterial DNA is attached to the plasma membrane, such DNA shrinkage might pull part of the plasma membrane inward. If this occurred, they hypothesized, it would create an artifact that appeared as a mesosome.

To see if this were the case, cells were frozen in liquid nitrogen and then exposed to X radiation (X-rays) to break up the DNA before the cells were dehydrated during preparation for electron microscopic viewing. When this procedure was followed, no mesosomes were observed. Mesosomes, however, were observed in control specimens in which the DNA had not been broken up by X radiation prior to dehydration. This suggested that these observed structures were artifacts of preparation for electron microscopic observation, rather than real structures of the bacterial cell.

After years of investigation, the evidence, therefore, supports the hypothesis that mesosomes are artifacts and not real structures. This is a good example of why scientists must always repeat their observations, test their hypotheses, and continue to evaluate their interpretations until they exhaust all possible alternative explanations for their observations.

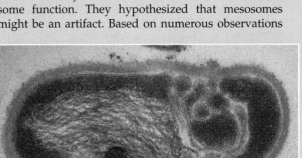

Colorized transmission electron micrograph of a thin-sectioned bacterium, *Corynebacterium* species, shows the invaginated membrane *(tan)* that has been called a mesosome.

would be destroyed. Therefore, before viewing a microorganism with a transmission electron microscope, it is necessary to preserve (fix) and to dehydrate (remove water from) the specimen. Water is usually removed by transferring the specimen through a series of alcohol baths of increasing alcohol concentrations. Inadequate fixation or insufficient dehydration will result in the formation of artifacts.

Additionally, as surprising as it may seem, whole microorganisms often are too thick to view with a transmission electron microscope. To see the fine detail that is possible with the high resolving power of the TEM, it is normally necessary to slice microor-

ganisms into thin sections or to fracture them into pieces before trying to observe their detailed structures. Microorganisms can be sliced by first embedding them in plastic. A specialized instrument called a *microtome*, equipped with a glass or diamond knife, is used to cut a thin film containing the sliced microorganisms. To break them into pieces, microorganisms can be frozen in liquid nitrogen and then bludgeoned with a dull instrument, such as a hammer. Here again there is a possibility that artifacts will be created in processing the specimen. Therefore great caution must be observed in interpreting electron micrographs.

Caution must be used in interpreting electron micrographs because the techniques used in their preparation/creation have a high potential for producing artifacts.

Scanning Electron Microscope

The operation and design of the **scanning electron microscope** (SEM) are quite different from those of the TEM. In SEM, a narrowly focused beam of electrons is rapidly scanned across the surface of the specimen. The primary scanning electron beam knocks electrons out of the specimen surface. These secondary electrons are collected by a detector, amplified, and used to generate an image on a cathode ray tube (CRT) screen. The image on the CRT screen has a three-dimensional appearance due to the shadowing effects of the sweep beam (FIG. 3-27). The primary beam that scans across the specimen is synchronized with a beam that scans across the CRT display so that the image on the CRT screen is an accurate magnified representation of the specimen. A typical scanning electron microscope is capable of

FIG. 3-27 Colorized scanning electron micrograph of the diatom *Achnanthes exigna*. (875×). The frustule that forms the outer shell of the diatom is composed of silicon dioxide and has two overlapping halves.

achieving a useful magnification of 10,000× to 100,000× with a resolution of 10 to 20 nm. The SEM is used primarily for viewing surface details.

The scanning electron microscope (SEM) is used primarily for viewing surface structures.

PURE CULTURE METHODS

Many microorganisms look exactly alike. Examining their physiological characteristics, therefore, is essential for studying and identifying microbial species. In macroorganisms the metabolism of a single organism can be measured. For example, the rate of respiratory production of carbon dioxide by a human athlete can be measured. However, the metabolism of a single microbial cell doesn't transform sufficient material for the scientist to measure. A population of identical microorganisms, on the other hand, will transform measurable amounts of materials. Microbiologists culture (grow) colonies of microorganisms that contain millions of individuals. This way sufficient material to study and determine their metabolic characteristics is provided (FIG. 3-28).

A **pure culture** of bacteria is a population of identical bacteria all derived by asexual reproduction from a single bacterial cell. Bacteria most often reproduce asexually by a process called *binary fission,* in which a single cell divides into two equal-sized identical progeny cells. Scientists, thus, can grow pure bacterial cultures starting with a single bacterial cell (FIG. 3-29). Cultures of bacteria contain millions of identical bacterial cells. These cultures can be used by scientists to study the metabolism and growth characteristics of the individual unseen bacterium. Cultures grown in the laboratory can also be used to study the rates of microbial growth under different environmental conditions.

Microbial growth is essentially synonymous with cell reproduction. The growth of a bacterium, for example, is measured as an increase in cell number due

FIG. 3-28 Microbial cultures are grown in pure culture so they can be studied and used for various purposes.

to reproduction. Using cultures, scientists can observe under what conditions a microorganism will reproduce. For example, they can determine the temperature range over which the microorganism can grow and the temperature at which the microorganism grows fastest. The ability to produce new cells is often used to determine whether or not a bacterium is alive. Living or viable bacteria will reproduce, whereas nonviable bacteria will not.

A pure culture of a microorganism, a population of identical bacteria, can be grown to observe the characteristic properties of that microorganism.

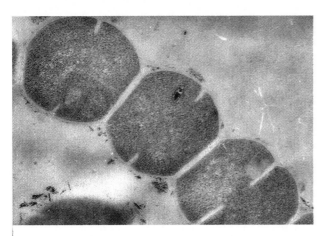

FIG. 3-29 Colorized transmission electron micrograph showing the reproduction of *Streptococcus pyogenes* by binary fission. The inward growth of the septum *(purple)* divides the parent cell to produce two equal-sized progeny cells. Other components of the cell have already been synthesized and segregated into the new cells.

EARLY DEVELOPMENT OF PURE CULTURE METHODS

Many of the methods for culturing bacteria were developed in the late nineteenth century in Germany by Robert Koch and his assistants. They developed simple methods for the isolation and maintenance of pure cultures of microorganisms. They grew cultures on the surfaces of solid media containing the nutrients needed to support the growth of the microorganisms. Before then, microorganisms were grown only in liquid broths, making it very difficult to separate one microorganism from another. By growing cultures on solid media, individual cells could be separated from other cells. This facilitated the isolation of the progeny of a single cell for the establishment of pure cultures. It also permitted observation of populations of single species of microorganisms.

At first, Koch grew bacteria on solid fruits and vegetables, such as slices of boiled potato, but many bacteria cannot grow on such substances. Koch developed a way of solidifying liquid broths that could support the growth of a greater variety of microorganisms. Initially he used gelatin and later an algal extract called **agar** as the solidifying agent. The suggestion for using agar originated with the wife of one of the investigators at Koch's Institute, Frau Fanny Hesse, who had seen her mother using agar to make jellies. Agar is particularly useful because it melts at 100° C but does not resolidify until it cools to 42° C; agar also resists proteases. Tubes or flasks containing agar can be placed into a boiling water bath to melt the agar. Agar can be poured into other containers while it is still liquid. Bacteria can even be added to the agar while it is liquid at 42° C and survive. Once agar has cooled and solidified it will remain solid at the temperatures used for the culturing of bacteria. Most bacteria that are human pathogens are grown at 37° C. Other bacteria from nature often are grown at lower or higher temperatures.

> **Agar is a useful medium for the culture of bacteria because it melts at 100° C and resolidifies at 42° C, can be poured, can have bacteria added when liquid, and remains solid at typical bacterial growth temperatures.**

One of Koch's students, Richard J. Petri, suggested placing the solidified media into a new type of container. This container was circular in shape and had an overlapping lid that prevented microorganisms in the air from contaminating the cultures placed into the container. The basic design has become known as the **Petri plate.** It is used in virtually all microbiological laboratories in essentially the same design described by Petri (FIG. 3-30). If all microorganisms in the medium in a Petri plate are killed by heating, the plate will remain free of microorganisms, that is sterile, until the lid of the plate is lifted. If the lid is removed for only a very short time in a clean room, specific microorganisms can be placed on the surface

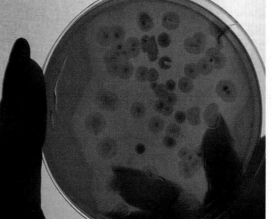

FIG. 3-30 Petri plates are routinely used for culturing microorganisms. **A,** The Petri plate consists of two overlapping halves that permit air to enter but prevent contamination by extraneous microorganisms. **B,** Microorganism grown on a solid medium within the Petri plate, forming visible colonies.

of the solid medium without much risk of contamination by other microorganisms in the air.

It is in this manner that microorganisms can be transferred. They can be transferred from some source, such as a blood or urine specimen collected from a patient, into a Petri plate with the assurance that all microorganisms in the plate came from that source. Once introduced, microorganisms reproduce by using the nutrients in the medium in the Petri plate to support their metabolism. If a single microbial species is introduced into the Petri plate, its reproduction will produce a pure culture. There will be no other microbial species. The Petri plate, with solidified agar, thus permits the establishment of pure cultures of microorganisms. Most microbiological studies, including those used to identify the microorganisms causing a particular disease, require pure cultures for study to establish that a particular effect—such as a disease—is caused by a specific microorganism.

> **Pure culture methods using agar media and Petri plates are fundamental techniques used in virtually all microbiology laboratories in the world today.**

Sterilization

To establish pure cultures of microorganisms, the media, containers, and all the implements used for manipulating microorganisms must initially be free of living microorganisms. An environment that is totally free of all living organisms is **sterile.** Sterilization procedures are used to kill or to remove all living microorganisms from a specified area. Since living organisms, including microorganisms, cannot arise by spontaneous generation, a sterile area will remain sterile unless a living microorganism enters that area naturally from somewhere else or until a scientist intentionally introduces a microorganism.

There are several ways of sterilizing the liquids, containers, and instruments used in pure culture procedures (FIG. 3-31). These methods include: (1) filtration to remove microorganisms; (2) exposure to elevated temperatures (usually over 100° C), (3) exposure to toxic chemicals (such as ethylene oxide), and (4) exposure to ionizing radiation (such as gamma radiation) to kill microorganisms. These methods are discussed in detail in Chapter 11. Sterilization procedures are absolutely essential if pure cultures are later to be obtained for clinical diagnosis, scientific investigation, or industrial use. It also is essential to sterilize instruments and materials for other purposes, such as in surgical procedures where it is critical to avoid the introduction of microorganisms into the body, thereby preventing infection and disease.

Removal of microorganisms from liquids and gases by filtration generally is accomplished by pas-

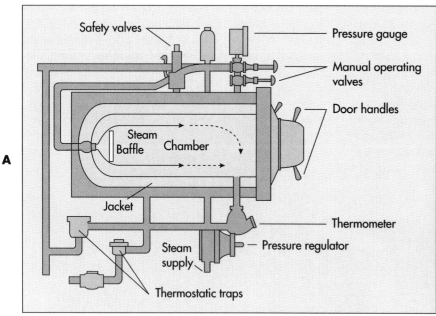

A

B

FIG. 3-31 **A,** Diagram of an autoclave. This instrument is routinely used for sterilization of media and other items in the microbiology laboratory. In an autoclave, steam is introduced under pressure into a chamber containing the material to be sterilized. The pressure generally is adjusted to 15 lb/in² so that a temperature of 121° C (250° F) is reached. Valving of the autoclave permits rapid entry of steam from a preheated jacket into the chamber and the subsequent slow exhausting of steam from the chamber; this process permits rapid heating of the material and prevents liquids from boiling out of their containers, as would happen if the pressure was suddenly reduced. **B,** A technician is loading an autoclave with media for sterilization—a routine operation in most microbiology laboratories.

sage of the substance through a filter with 0.2 to 0.45 μm diameter pores. Most bacteria are trapped on the filter, but viruses and some very small bacteria may pass through it. When viruses were first discovered they were referred to as filterable agents because they passed through bacteriological filters (filters with pore sizes less than 0.2 μm). Clean rooms, such as operating theaters, often employ high efficiency particulate air filters (HEPA filters) to remove microorganisms from the air. HEPA filters remove particulate material larger than about 0.3 μm from the air.

Heat sterilization at a temperature that kills all microorganisms, including their heat-resistant endospores, is often used to eliminate unwanted microorganisms. Incineration can be used to destroy medical or other microbiological wastes. Incineration (burning) is also used to sterilize inoculation loops that are frequently used in the microbiology laboratory to transfer cultures of microorganisms. Dry heat sterilization requires high temperatures and long exposure periods to kill all of the microorganisms in a sample. Exposure in an oven for 2 hours at 170° C (328° F) is generally used for the dry heat sterilization of glassware and other laboratory items.

Much time is spent in the preparation and sterilization of media for the bacteriology laboratory. Culture media preparation usually employs an autoclave that uses steam under pressure for sterilization. An **autoclave** is an instrument that exposes substances to steam at elevated temperatures. Steam has a high penetrating power and a much higher heat capacity than dry heat. Thus it is very effective at killing microorganisms. Generally, exposure for 15 minutes at 121° C, achieved by using a pressure of 15 lb/in² (SI equivalent = 103.4 kPa), is used to sterilize microbiological culture media.

Exposure to ionizing radiation is useful in sterilizing materials that are destroyed by heat, such as plastics. Many of the Petri plates used in microbiology laboratories are radiation sterilized by using gamma radiation. Radiation exposure can be used to increase the shelf life of various foods and to kill foodborne pathogens such as *Clostridium botulinum*, the bacterial species that causes botulism. Exposure to radiation does not leave any residual radioactivity in the food, but the method is still controversial. Labels are placed on foods that have been radiation treated to inform consumers.

> **Microorganisms can be killed or removed by several methods to achieve sterility. A sterile environment contains no living microorganisms.**

Aseptic Technique

Once substances are sterilized, they stay sterile as long as they remain within containers that do not permit living microorganisms to enter. Scientists, however, often must transfer materials from one con-

tainer to another. To maintain sterility, they use methods that do not permit the accidental entry of living microorganisms during the transfer process. These methods are collectively called **aseptic technique.** For example, when pouring a sterile liquid from one vessel to another sterile vessel, the openings of the vessels often are passed through the flame of a bunsen burner. This is done immediately after they are opened to kill any microorganisms that may be there and to create convection currents away from the mouth of the tube so that microorganisms in the air don't enter the tube. Then the liquid is quickly poured from one vessel to the other. Any contact with surfaces that contain microorganisms that could contaminate the liquid is avoided. After the liquid has been poured, the openings of the vessels are again flamed to reensure that no microorganisms are there that could enter the sterile liquid. Finally the vessels are resealed. Minimizing the time that the vessels are open and working in a clean area are important parts of aseptic technique and limit the possibility that microorganisms will enter the vessel accidentally.

Aseptic technique also must be used to establish a pure culture. The scientist must ensure that only the desired microorganism enters the culture vessel. To transfer microorganisms from one place to another, microbiologists often use a device called a transfer loop (FIG. 3-32). A transfer loop is a piece of wire with a handle on one end and a small circle of wire on the other end. The metal wire usually is made of

FIG. 3-32 Aseptic transfer procedures are essential for preventing contamination of bacterial cultures and for ensuring that the microorganisms being cultured do not escape into the laboratory.

nichrome or platinum. It can be heated red hot by placing it into an electric heater or the flame of a bunsen burner to kill all the microorganisms on it. The circular wire loop can then be cooled and touched to an already established pure microbial culture. It can also be dipped into a liquid containing microorganisms to pick up living microbial cells.

By avoiding contact with any surfaces containing other microorganisms and by working quickly, the living microorganisms on the loop can be transferred to a new location. For example, a bacterial culture can be moved onto a Petri plate containing sterile medium, without contamination with other living microorganisms. When the transfer loop is touched to the surface of the solid sterile medium some of the living microorganisms fall off. The transfer loop is then heated red hot again to kill any remaining microorganisms. Aseptic transfer technique not only ensures that a pure sample of the microorganism is transferred, it also protects the scientist from contact with the culture. This is especially critical when the culture contains microorganisms that can cause disease. Thus aseptic technique protects microorganism and microbiologist alike from contamination.

> **Aseptic technique is used to prevent the contamination of a pure culture of a microorganism with extraneous microorganisms and to prevent human contact with potentially dangerous microorganisms.**

ISOLATION OF PURE CULTURES

Pure cultures of microorganisms can be isolated and maintained using aseptic transfer techniques and sterile media. The isolation of a pure bacterial culture involves separating bacteria into individual cells. These cells are then allowed to reproduce to form a colony on the surface of a plate of a solid medium (FIG. 3-33). A **colony,** which contains millions of cells

and can be seen visually, develops from the asexual reproduction of a single cell. Each colony contains a population of a single bacterial species. Separating this colony from all other bacteria results in the formation of a pure culture.

Several methods are used to isolate pure cultures of microorganisms. The **streak plate method** is the most commonly used method. In this method a transfer loop is dipped into a substance containing bacterial cells. The source can be a liquid, such as blood, urine, or water, or a solid, such as soil (FIG. 3-34). The bacterial cells that adhere to the loop are picked up. The loop is gently dragged across the solid agar surface of the nutrient medium. The cells are thus deposited. The surface of the medium has been solidified by the addition of agar. In the initial portions of this continuous streak, many cells will be deposited close together. When the plates are incubated, these cells will reproduce to form colonies. Because the cells are deposited close together, the colonies will overlap. The cells in the overlapping colonies do not originate from a single cell. The colony may therefore contain more than one bacterial species. As the transfer loop is streaked further across the plate, there are progressively fewer cells still on the loop that can be deposited on the plate. Often, the microorganisms are dragged across part of the plate, the loop is resterilized, and a new streak is made at an angle to the original streak. Bacterial cells in the latter portions of the streak will be deposited further apart. The reproduction of these cells will give rise to separate, well-isolated colonies. Because each well-isolated colony arises from the reproduction of a single bacterium, it is a pure culture. The isolated colonies can then be transferred, using a sterile inoculating loop and aseptic transfer technique, onto a fresh medium in a separate vessel and grown as pure cultures.

Pure cultures can also be isolated by dripping a small volume of a dilute liquid suspension of microorganisms onto the center of an agar plate. The liquid suspension is then spread over the surface of the agar with a sterile glass rod. This method for isolating pure cultures is called the **spread plate method** (FIG. 3-35). By spreading the suspension over the plate, individual cells are separated from the other cells in the suspension. Well-isolated colonies develop in regions where cells are deposited far apart on the surface of the medium.

In the **pour plate method,** dilute suspensions of microorganisms are added to melted agar that has been cooled to approximately 42° to 45° C (FIG. 3-36). Since agar is a liquid until it cools below 42° C and since most bacteria are not killed by exposure to 45° C for several minutes, living bacteria can be mixed with the liquid agar medium and then poured into sterile Petri plates using aseptic technique. When the agar cools below 42° C and solidifies, individual bacterial

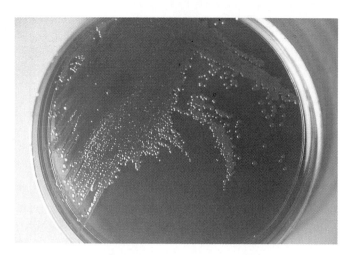

FIG. 3-33 Colonies of *Vibrio parahaemolyticus* growing in a Petri plate on thiosulfate citrate-bile salts-sucrose (TCBS) agar.

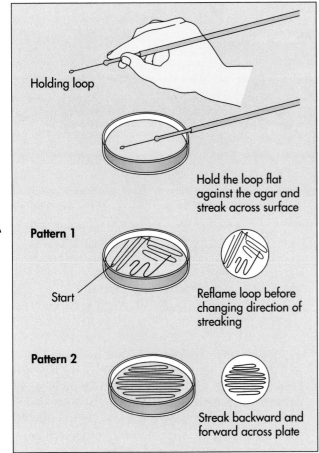

Holding loop

Hold the loop flat against the agar and streak across surface

Pattern 1

Start

Reflame loop before changing direction of streaking

Pattern 2

Streak backward and forward across plate

A

B

FIG. 3-34 **A,** Streaking for the isolation of pure cultures, showing two different streaking patterns. In this procedure, a culture is diluted by drawing a loopful of the organism across a medium. For a dilute culture, only a single streak may be used (Pattern 2). For more concentrated cultures, a second streak is drawn across the first streaks so that some cells are picked up and further diluted; several additional streaks are made to ensure sufficient dilution so that only single cells are deposited at a given location (Pattern 1). The growth of each isolated individual cell results in the formation of a discrete colony. **B,** Streak plate of *Vibrio cholerae* on thiosulfate citrate-bile salts-sucrose (TCBS) agar.

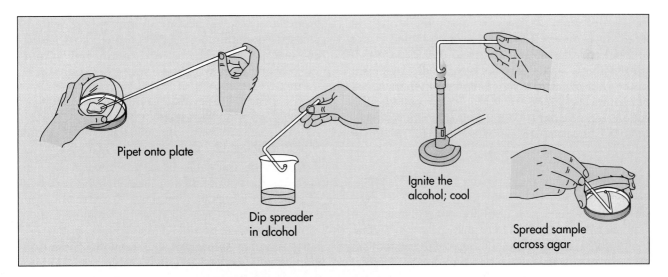

Pipet onto plate

Dip spreader in alcohol

Ignite the alcohol; cool

Spread sample across agar

FIG. 3-35 The spread plate technique for isolating and enumerating microorganisms. *1,* A sample is aseptically pipetted onto an agar medium; *2,* a spreading rod is sterilized by dipping in alcohol and flaming; *3,* the sterile rod is used to spread the suspension over the surface of the medium.

cells become trapped at locations throughout the medium. Although the medium holds bacteria in place, it is soft enough to permit reproduction of bacteria and the formation of discrete isolated colonies. As with the other isolation methods, individual col-

onies are then picked up and transferred to separate vessels where they can be grown as pure cultures.

The streak plate, spread plate, and pour plate methods are used to isolate pure cultures of microorganisms.

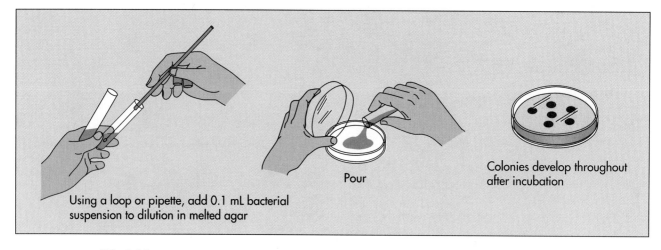

Using a loop or pipette, add 0.1 mL bacterial suspension to dilution in melted agar

Pour

Colonies develop throughout after incubation

FIG. 3-36 The pour plate technique for isolating and enumerating microorganisms. A sample of a known dilution is mixed with a liquefied agar medium that has been cooled to 45° to 50° C and poured into a Petri plate. After incubation the numbers of colonies that develop are counted and the concentration of microorganisms in the original suspension is calculated.

CONDITIONS AND MEDIA FOR LABORATORY GROWTH OF MICROORGANISMS

To culture microorganisms, microbiologists must establish the conditions necessary for microbial growth in the laboratory. They must provide suitable environmental conditions and the necessary nutrients so that the microorganism can carry out its metabolism and reproduce. Under optimal growth conditions, most microorganisms reproduce very rapidly. Visible colonies of many bacteria, for example, develop in less than 24 hours if the culture is incubated under optimal conditions. The ability to obtain pure cultures within a matter of hours is especially important in the clinical microbiology laboratory where speed of identification is essential. Many microbial identifications can be completed in less than a day. This permits the physician to begin appropriate treatment quickly.

Incubators

Temperature is one of the most important environmental factors to control for optimal microbial growth. Microorganisms only survive within certain temperature limits. Rates of microbial metabolism are greatly affected by temperature. Each microbial species has an optimal growth temperature, which is the temperature at which that microbial species reproduces at its fastest rate. Incubators are used to grow bacteria in the laboratory. Within an **incubator** the temperature can be controlled. The temperature is usually set near the optimal growth temperature of the microorganism of interest.

Incubators are temperature control chambers in which microbial cultures are grown.

As with temperature, different microbial species will grow best at different concentrations of molecular oxygen (O_2). **Obligate aerobes** require molecular oxygen for growth and will not grow in its absence. To grow in the laboratory, microbiologists must use procedures that assure the absence of molecular oxygen. Like temperature and oxygen, other environmental factors can also be controlled for culturing microorganisms. For example, some pathogenic microorganisms grow best in an atmosphere with increased concentrations of carbon dioxide. Scientists can use special incubators that control carbon dioxide concentration to cultivate such microorganisms.

Scientists control environmental conditions to grow pure cultures of microorganisms in the laboratory.

Culture Media

To culture microorganisms, microbiologists must ensure that the culture medium contains the variety of organic and inorganic nutrients that are required for microbial metabolism. All organisms require carbon, nitrogen, oxygen, hydrogen, phosphorus, sulfur, and various other substances for growth. These substances must be available in a usable chemical form to meet the nutritional requirements of a particular microorganism and to permit that organism to grow. Not all microorganisms have the same nutritional requirements. In fact, the specific nutritional requirements for different microorganisms vary greatly.

Commonly used culture media contain protein that have been digested with enzymes or acids and/or carbohydrates as growth substances that a microorganism can utilize. These media also generally contain sources of nitrogen—such as ammonium nitrate, phosphate, and sulfate. Additionally, magnesium, sodium, potassium, and chloride ions are needed to meet the inorganic metabolic needs of the microorganism. Because of the difficulty in defining the specific nutritional requirements of individual microbial species, microbiologists often use **complex media,** which are media that contain various substances whose precise chemical compositions are unknown. Such complex culture media will support the growth of many different types of microorganisms that require organic compounds as their source of energy. Many complex media contain beef extract, peptones, and yeast extract. In some cases, scientists must add specific compounds to get a microorganism to grow in the laboratory. Clinical microbiologists often incorporate blood into media that are designed for the culture of disease-causing microorganisms to provide necessary but unspecified nutrients for the growth of these microorganisms. Some disease-causing microorganisms, called **fastidious microorganisms,** are nutritionally demanding. Some require factors in blood or other specific substances to support their growth. They often are difficult to culture in the laboratory.

A growth medium contains the necessary nutrients to support the nutritional requirements of cultured microorganisms.

Different types of media are needed for the cultivation of different microorganisms.

Fastidious microorganisms have specific and demanding growth requirements.

Scientists sometimes design media to prevent the growth of unwanted microorganisms or to permit the differentiation of specific types of microorganisms. Media are called **selective media** if they favor the growth of specific microorganisms and **differential media** if they permit the recognition of specific types of microorganisms; in some cases a medium is both differential and selective (Table 3-2). Selective media usually are designed to allow some microorganisms to grow while preventing growth of others.

TABLE 3-2
Some Differential and Selective Media

MEDIUM	DESCRIPTION
MacConkey Agar MacConkey agar is a differential plating medium for the selection and recovery of Enterobacteriaceae and related enteric Gram-negative rods. Bile salts and crystal violet are included to inhibit the growth of Gram-positive bacteria and some fastidious Gram-negative bacteria. Lactose is the sole carbohydrate.	Lactose-fermenting bacteria produce colonies that are varying shades of red because of the conversion of the neutral red indicator dye (red below pH 6.8) from the production of mixed acids. Colonies of non-lactose-fermenting bacteria appear colorless or transparent.

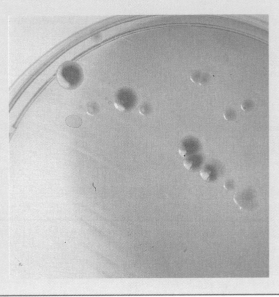

Colonies of *Escherichia coli* growing on MacConkey agar showing characteristic colony morphology and coloration. *E. coli* frequently causes urinary tract infections and also is a common cause of bacterial meningitis in infants.

Continued.

TABLE 3-2—cont'd
Some Differential and Selective Media

MEDIUM	DESCRIPTION

Eosin Methylene Blue (EMB) Agar

EMB agar is a differential plating medium that can be used in place of MacConkey agar in the isolation and detection of the Enterobacteriaceae and related coliform rods from specimens with mixed bacteria. The aniline dyes (eosin and methylene blue) in this medium inhibit Gram-positive and fastidious Gram-negative bacteria. They also combine to form a precipitate at acid pH, thus also serving as indicators of acid production.

Escherichia coli growing on eosin-methylene blue (EMB) agar showing characteristic colonies with green metallic sheen.

Desoxycholate-citrate (DCA) Agar

DCA agar is a differential plating medium used for the isolation of members of the Enterobacteriaceae from mixed cultures. The medium contains about three times the concentration of bile salts (sodium desoxycholate) of MacConkey agar, making it most useful in selecting species of *Salmonella* from specimens overgrown or heavily contaminated with coliform bacteria or Gram-positive organisms. Sodium and ferric citrate salts in the medium retard the growth of *Escherichia coli*. Lactose is the sole carbohydrate, and neutral red is the pH indicator and detector of acid production.

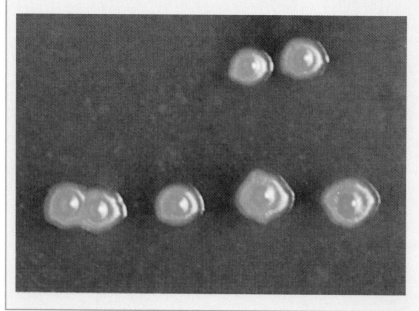

Colonies of *Salmonella arizonae* growing on DCA. This bacterium produces characteristic pink colonies on this medium.

MEDIUM	DESCRIPTION

HEKTOEN ENTERIC (HE) AGAR

HE agar is devised as a direct plating medium for fecal specimens to increase the yield of species of *Salmonella* and *Shigella* from the heavy numbers of normal microbiota. The high bile salt concentration of this medium inhibits the growth of all Gram-positive bacteria and retards the growth of many stains of coliforms.

Acids may be produced from three carbohydrates, and acid fuchsin reacting with thymol blue produces a yellow color when the pH is lowered. Sodium thiosulfate is a sulfur source, and H_2S gas is detected by ferric ammonium citrate, producing a black precipitate.

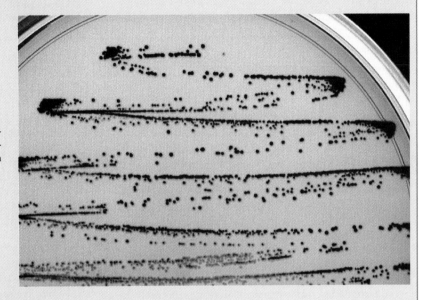

Salmonella enteritidis growing on Hektoen enteric (HE) agar showing characteristic black colonies due to production of hydrogen sulfide.

XYLOSE LYSINE DESOXYCHOLATE (XLD) AGAR

XLD agar is less inhibitory to the growth of coliform bacteria. It was designed to detect *Shigella* species in feces after enrichment in Gram-negative broth. Bile salts in relatively low concentration make this medium less selective than the other media included in this table. Three carbohydrates are available for acid production, and phenol red is the pH indicator. Lysine-positive organisms, such as most *Salmonella enteriditis* strains, produce initial yellow colonies from xylose utilization and delayed red colonies from lysine decarboxylation. The H_2S detection system is similar to that of HE agar.

Salmonella typhimurium on xylose lysine desoxycholate agar (XLD) showing the black growth due to production of hydrogen sulfide.

For example, the stain methylene blue is more toxic to Gram-positive bacteria than to Gram-negative bacteria. By incorporating 0.5% methylene blue into a culture medium, the growth of Gram-positive bacteria can be inhibited. This will not interfere with the growth of Gram-negative bacteria. Eosin methylene blue (EMB) agar contains methylene blue and, therefore, is a selective medium. It is frequently used for the selective culture of Gram-negative bacteria, such as *Escherichia coli,* from samples, such as stool specimens, that contain both Gram-negative and Gram-positive bacteria. EMB agar is also a differential medium because colonies of different bacterial

species growing on it will appear different based on the type of metabolism that they carry out. The bacterium *E. coli,* which metabolizes lactose, produces colonies with a green metallic sheen when growing on EMB agar. These lactose-utilizing bacteria can be easily differentiated from any other bacteria growing on EMB agar.

Selective media contain substances that favor the growth of one type of microorganism and discourage the growth of other types.

Differential media contain substances that permit recognition of different types of microorganisms.

SUMMARY

Scientific Method and Development of the Science of Microbiology (pp. 59-64)

- All fields of science, including microbiology, rely on the scientific method in which a scientist asks a question, proposes a hypothesis, makes systematic observations, and conducts controlled experiments to test the validity of that hypothesis.
- Scientists use deductive reasoning ("if-then" thinking) and test their predictions by conducting experiments.
- Controlled experiments are designed to yield unambiguous answers regarding whether a hypothesis is correct or incorrect.
- The results of experiments must be repeatable, and interpretations of experimental results become accepted only after they have been critically evaluated.
- The struggle to disprove the theory of spontaneous generation represents a good example of the scientific method at work.
- Louis Pasteur's experiments with swan-neck flasks provided the final refutation of the theory of spontaneous generation.

Microscopy (pp. 64-78)

Principles of Light Microscopy (pp. 65-70)

- The usefulness of a particular type of microscopy depends on the ability to produce a magnified image of a microorganism that (a) is large enough to be seen; (b) has not been distorted by the method of preparation or viewing; (c) has sufficient resolution to distinguish structures of interest; and (d) has sufficient contrast so that the organisms and the structures of interest can be distinguished from the surrounding background.
- The light microscope focuses light onto and through a specimen and produces a magnified image; images can be magnified 1,000 times and structures larger than 100 nm can be resolved.
- Magnification is useful only if detail can still be resolved. The resolving power of the microscope (the amount of detail that can be seen) determines the useful magnification.
- The resolving power of the light microscope depends on the wavelength of the illuminating light source

and the numerical aperture of the objective lens. The shorter the wavelength of light and the higher the numerical aperture of the objective lens, the better the resolving power.
- Staining increases the contrast between a specimen and its background so it can be seen under the microscope.
- The Gram stain procedure is a widely used differential staining procedure that distinguishes between species that stain red-pink (Gram-negative bacteria) and species that stain blue-purple (Gram-positive bacteria).
- Other differential staining procedures include the acid-fast staining procedure and the endospore staining procedure.

Types of Light Microscopes (pp. 71-74)

- The bright-field microscope is most widely used in microbiology.
- Fluorescent microscopes are used to observe microorganisms that have been stained with a fluorescent dye.
- Certain immunofluorescent procedures, in which fluorescent dyes are coupled with specific antibodies, are widely used for the specific diagnosis of infectious diseases.
- The dark-field microscope relies on the reflection of light striking the specimen at an oblique angle.
- The phase contrast and interference microscopes enhance contrast without the need to stain the specimen.

Electron Microscopy (pp. 74-78)

- The short wavelength of an electron beam permits better resolution and hence higher useful magnifications using electron microscopes than can be achieved with the light microscope.
- There are two types of electron microscopes: the transmission electron microscope (TEM) and the scanning electron microscope (SEM).
- Both TEM and SEM require extensive procedures to prepare the specimen for viewing, while at the same time preventing the production of artifacts.
- The TEM is used to see finely detailed structures. The SEM is most useful for the observation of surface structures.

Pure Culture Methods (pp. 78-88)

- The scientific study of microorganisms requires the growth of large numbers of organisms in pure culture, where the metabolism and other characteristics of living microorganisms can be observed.
- Obtaining and maintaining pure cultures necessitates the elimination of all microorganisms from the growth medium (sterilization), the separation of the microorganism that is being cultured from a mixture of microorganisms (isolation), the movement of the microorganism from one place to another without contamination (aseptic transfer technique), and the growth of the microorganisms in the laboratory.
- Streak plate, spread plate, and pour plate methods are isolation techniques designed so that individual microbial cells are sufficiently separated on the surface of solidified agar so that when they reproduce they develop into well-isolated colonies.
- Colonies are macroscopic clumps of microbial cells.
- A culture medium contains the nutrients that are required by the microorganism for growth. Media used to grow microorganisms in the laboratory generally contain a source of carbon, nitrogen, phosphorus, iron, magnesium, sulfur, sodium, potassium, and chloride. Microorganisms require all of these for metabolism and reproduction.
- In some cases selective and/or differential media are employed to isolate specific microbial species. Such media are designed to suppress the growth of some microorganisms and to permit the recognition of others.

CHAPTER REVIEW

REVIEW QUESTIONS

1. Compare and contrast inductive and deductive reasoning.
2. Describe the rise and fall of the theory of spontaneous generation.
3. What is the scientific method? How are hypotheses tested by scientists?
4. Describe the advantages of the bright-field microscope and what kinds of observations are made using it.
5. Define resolution. What is its importance to microscopy? How does wavelength affect resolution?
6. Why is staining usually done?
7. What is the difference between a simple staining procedure and a differential one?
8. Describe the steps in the Gram stain.
9. What advantages does the dark-field microscope provide?
10. What advantages does the phase-contrast microscope provide?
11. What advantages does the fluorescence microscope provide?
12. What is a pure culture? Why was the ability to isolate a pure culture important to the development of microbiology as a science?
13. How do clinical microbiologists culture bacteria? Why does it often take over a day to positively diagnose a disease?
14. What is a selective medium? How is such a medium used to help identify microorganisms in the case of a gastrointestinal tract infection?
15. What is a differential medium? How is such a medium used to help identify microorganisms in the case of a respiratory tract infection?

CRITICAL THINKING QUESTIONS

1. You are working in the clinical microbiology laboratory when a blood sample from a critically ill patient arrives. The physician wants to know as soon as possible whether there are any disease-causing microorganisms in the blood and, if so, the identity of the infecting agent. What would you do?
2. There is an outbreak of a mysterious illness in the southwestern United States. Several individuals who are native American Indians have already died. Someone proposes that the deaths are due to a viral infection. How would you go about proving or disproving that hypothesis? Assuming it was correct, how would you go about finding the source of the disease?
3. There is an outbreak of an illness in Philadelphia. Several individuals who attended a convention suddenly died and several others are ill and have been hospitalized. Someone proposes that the deaths are due to a terrorist attack that introduced a poison into the food that was served at a banquet. Another hypothesis is that a pathogenic bacterium has spread through the air-conditioning cooling system. How would you go about proving or disproving these hypotheses?
4. You are admitted to nursing school. You are able to attend all the scheduled classes but because you are also raising a family you are unable to attend all the laboratory sessions—particularly when they have extended open hours. You therefore decide to purchase a microscope so that you can examine specimens at home and not fall behind in your studies. How would you go about purchasing a microscope that would permit you to examine human tissue specimens and also prepared stained slides of microorganisms? What features would you consider most important in comparing various microscopes?

READINGS

Aldrich HC and WJ Todd (eds.): 1986. *Ultrastructural Techniques for Microorganisms,* New York, Plenum Press.
> *Important, useful, and current techniques for microscopy written by experts in their fields.*

Atlas RM: 1993. *Handbook of Microbiological Media,* Boca Raton, CRC Press.
> *Gives the recipes and uses of 2,700 media employed for the culture of microorganisms.*

Balows A: 1991. *Manual of Clinical Microbiology,* ed. 5, American Society for Microbiology, Washington, D.C.
> *A definitive reference for the methods used by clinical microbiologists.*

Bradbury S: 1984. *An Introduction to the Optical Microscope,* New York, Oxford University Press.
> *Well-illustrated handbook that explains basic concepts, parts, and use of the microscope.*

Dubos RJ: 1976. *Louis Pasteur: Freelance of Science,* N Y, Scribner's.
> *This well-known microbiologist and philosopher provides an insightful view into the work of Louis Pasteur.*

Farley J: 1978. The social, political, and religious background to the work of Louis Pasteur, *Annual Review of Microbiology* 32:143-154.
> *Questions how extrascientific factors influenced the debate over spontaneous generation during the period dominated by Louis Pasteur and how Pasteur dealt with these factors.*

Gerhardt P, RGE Murray, WA Wood, NR Krieg (eds.): 1993. *Methods for General Bacteriology,* Washington, ASM Press.
> *A guide to the methods employed by microbiologists for the study of microorganisms.*

Hayat MA: 1986. *Basic Techniques for Transmission Electron Microscopy,* New York, Academic Press.
> *Discusses chemical fixation, rinsing, dehydrating, embedding, sectioning, staining, support films, and specific preparation methods for TEM studies.*

Howells MR, J Kirz, W Sayre: 1991. X-ray microscopes, *Scientific American* 264(2):88-95.
> *X-ray microscopes render three-dimensional images of cells in their natural state at 10 times the resolution of optical microscopes.*

Laudan L: 1981. *Science and Hypothesis: Historical Essays on Scientific Methodology,* Boston, D. Reidel.
> *Examines how the method of hypothesis came to be important in the philosophy of science and an integral part of the scientific method.*

Morris R: 1984. *Dismantling the Universe: the Nature of Scientific Discovery,* New York, Simon and Schuster.
> *Discusses the nature of scientific discovery and its relation to other kinds of human creativity.*

Norris JR and DW Ribbons (eds.): 1969—. *Methods in Microbiology,* New York, Academic Press.
> *An ongoing series of volumes containing comprehensive reviews of specific methods employed in microbiology.*

Oldroyd DR: 1986. *The Arch of Knowledge: An Introductory Study of the History of the Philosophy of Science,* London, Methuen.
> *Traces chronologically the historical development of ideas relating to the nature of scientific knowledge and the ways in which it is acquired.*

Olson AJ and DS Goodsell: 1992. Visualizing biological molecules, *Scientific American* 267(5):76-81.
> *An article describing the modern methods that can be used to reveal the specific molecules that make up living systems.*

Palm LC and HAM Snelders (eds.): 1982. *Antoni van Leeuwenhoek, 1632-1723: Studies on the Life and Work of the Delft Scientist Commemorating the 350th Anniversary of His Birth,* Amsterdam, Rodopi.
> *Discusses van Leeuwenhoek's microscopes, his views of the world, his ideas on spontaneous generation and his contribution to biology.*

Shih G and R Kessel: 1982. *Living Images: Biological Microstructures Revealed by Scanning Electron Microscopy,* Boston, Science Books International.
> *This book presents numerous SEM photomicrographs of microorganisms with clear explanations of what is pictured.*

Sieburth JM: 1975. *Microbial Seascapes,* Baltimore, University Park Press.
> *A magnificent collection of micrographs of microorganisms showing the beauty of the microbial world. Includes a discussion of the principles of microscopy.*

Tweney RD et al. (eds.): 1981. *On Scientific Thinking,* New York, Columbia University Press.
> *Selections from the original writings of Bacon, Descartes, Newton, Locke, and modern thinkers on scientific thinking and creativity.*

Wickramasinghe HK: 1989. Scanned-probe microscopes, *Scientific American* 261(4):98-105.
> *A new generation of microscopes uses sharp probes to "feel" the contours of atoms.*

Williams RC: 1978. Spectroscopes, telescopes, microscopes, *Annual Review of Microbiology* 32:1-18.
> *An eminent microscopist tells the story of his lifelong involvement with optical methods, following his work with hydrogen atoms, the stars, metal films, viruses, and enzymes.*

Yoshii Z, J Tokunaga, J Tawara: 1976. *Atlas of Scanning Electron Microscopy,* Baltimore, Williams & Wilkins.
> *A beautiful collection of scanning electron micrographs of microorganisms, with a discussion of scanning electron microscopy.*

van Zuylen J: 1981. On the microscopes of Antoni van Leeuwenhoek, *J Microsc.* 121(3):309-328.
> *A historical perspective of the microscopes used by Leeuwenhoek to first observe microorganisms.*

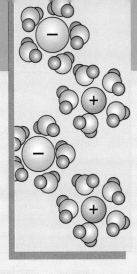

CHAPTER 4

Chemistry for the Microbiologist

PREVIEW TO CHAPTER 4

In this chapter we will:

- Review some fundamental chemical principles essential for understanding microbial structure and function.
- Discuss the structure of the atom and its relationship to the chemical properties of elements.
- Examine how atoms form chemical compounds by exchanging or sharing electrons to establish chemical bonds that hold molecules together.
- See why water is essential for life and how carbon forms the backbone of the organic chemicals that comprise living systems.
- Examine the types of chemical reactions that occur between molecules that allow cells to obtain energy and materials for growth and reproduction.
- Discuss the essential role of enzymes in the chemical reactions that constitute the metabolism of cells.
- Examine the major types of chemicals that make up living systems and learn about their properties, which we later will relate to microbial structure and function.
- Learn the following key terms and names:

acid	isomers
activation energy	lipids
amino acids	macromolecules
anion	molecule
atom	nucleic acids
ATP (adenosine triphosphate)	nucleotides
	oxidation
base	oxidation-reduction reactions
buffer	
carbohydrates	peptide bond
cation	pH
coenzyme	phosphodiester bond
covalent bonds	phospholipids
deoxyribonucleic acid (DNA)	polymers
	polypeptide chain
electrons	polysaccharides
enzymes	reduction
functional groups	ribonucleic acid (RNA)
hydrogen bond	substrate
ionic bond	

Early biological studies centered on the observation of living organisms—what they looked like, where they lived, what they ate. Naturalists cataloged the species of plants and animals in a region, recorded their distributions, and observed their appearances and behaviors. Early microbiologists continued in this tradition, looking at microorganisms and describing their morphologies and movements. Antonie van Leeuwenhoek, for example, recorded the shapes and movements of the "animalcules" he observed. Such microscopic observations revealed the existence of the living microbial world, but gave only limited insight into how microorganisms interact with their environment, or how they obtain the matter and energy needed to sustain life.

In the first half of the nineteenth century chemists developed a fundamental understanding of matter—the physical material of the universe. With the recognition that all matter in the universe has certain unifying chemical and physical properties and that living organisms are manifestations of their underlying chemical composition and the chemical reactions that they carry out, the fields of chemistry, physics, and biology began to be drawn together. Biologists soon recognized that to understand living organisms they had to investigate the chemistry of life. Microbiologists realized that the scientific understanding of the microorganisms, what they are and what they can do, necessitates the understanding of their underlying chemistry. So they incorporated chemistry as an integral part of the field of microbiology. To understand the chemistry of living systems, it is necessary to learn the "language" that chemists use for communicating information about chemicals. It is necessary to become conversant with the chemical terms and principles that are applied to the descriptions of microorganisms and their activities.

CHEMICAL ELEMENTS

An **element** is the fundamental unit of a chemical that cannot be broken down further without destroying the properties of that pure chemical substance. There are 92 different naturally occurring elements—such as carbon, hydrogen, nitrogen, and oxygen. Chemists have assigned each element a **chemical symbol** that is a one- or two-letter abbreviation of its English or Latin name. The chemical symbol for the element hydrogen is H, oxygen is O, carbon is C, and so forth. The same chemical symbol is used regardless of the element's name in the language of the country in which it is being used. Thus, even though nitrogen is called *azoto* in Italy and *stickstoff* in Germany, its chemical symbol is always N. These symbols for the chemical elements form the "alphabet" of the language of chemistry. Biologists, generally, are only concerned with the 26 elements that form the major components of living systems. The most abundant elements in living systems are carbon (C), hydrogen (H), oxygen (O), nitrogen (N), phosphorus (P), sulfur (S), sodium (Na), potassium (K), magnesium (Mg), calcium (Ca), iron (Fe), and chlorine (Cl). Of these, carbon is the element that forms the backbone of all molecules that comprise living organisms.

STRUCTURE OF ATOMS

The smallest unit of an element that still retains the chemical properties of that element is called an **atom** (Greek, meaning that which cannot be cut). Atoms were thought to be the smallest particles into which matter could be divided. It was not until the twentieth century that physicists showed that atoms are composed of yet smaller subatomic particles. Chemists subsequently discovered that the number of an atom's constituent subatomic particles deter-

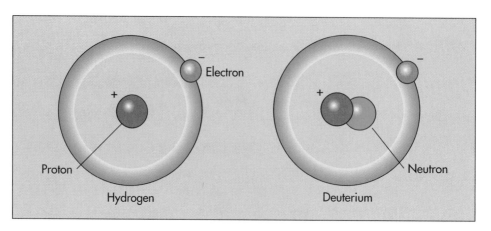

FIG. 4-1 Atoms are the fundamental units of chemical elements. They are composed of subatomic particles (negatively charged electrons, positively charged protons, and, with the exception of hydrogen, neutrally charged neutrons).

mines the characteristic properties of the atoms of different elements, such as their capacities to combine with other atoms.

These subatomic particles of atoms carry a positive or a negative charge, or no charge (FIG. 4-1). Positively charged particles are called **protons,** uncharged neutral particles are called **neutrons,** and negatively charged particles are called **electrons.** The atom is organized with the protons and neutrons in a central region called the **nucleus.** The electrons move in regions of space around the nucleus.

The nucleus of an atom has a net positive charge because it contains positively charged protons. However, because the number of protons in each nucleus is equal to the number of electrons, the total positive charge of the nucleus's protons equals the total negative charge of the electrons. Therefore each atom has a net charge of zero. An atom is said to be neutral. As discussed later, the chemical properties of atoms, which allow them to participate in chemical reactions, depend on the number and arrangements of their electrons.

> **Atoms, which are the smallest units of elements, contain positively charged protons, uncharged neutrons, and negatively charged electrons; an atom has a net electronic charge of zero because it contains equal numbers of protons and electrons.**

Ions

The number of electrons moving around the nucleus of an atom can increase or decrease. Some atoms have a tendency to lose one or more electrons. Other atoms tend to gain electrons. An atom that has lost or gained an electron is called an **ion.** It is no longer neutral. When a sodium (Na) atom loses an electron, it becomes a positively charged sodium ion (Na^+). Such a positively charged ion is called a **cation.** Other examples of cations are the potassium ion (K^+), magnesium ion (Mg^{2+}), and calcium ion (Ca^{2+}). Atoms of hydrogen can lose an electron and become a stable positive ion (H^+). The formation of hydrogen ions is very important because this is what causes acidity in the watery solutions that are an integral part of biological systems.

When a chlorine (Cl) atom gains an electron, it becomes a negatively charged chloride ion (Cl^-). Such a negatively charged ion is called an **anion.** Other examples of anions are the iodide ion (I^-) and sulfide ion (S^{2-}). Notice that the symbol for an ion is the chemical abbreviation followed by a superscript designating the ion's number of positive (+) or negative (−) charges.

> **An ion is an atom that has gained or lost one or more electrons.**
>
> **Atoms that gain electrons form negatively charged ions (anions) and atoms that lose electrons form positively charged ions (cations).**

Atomic Number and Weight

Chemists have assigned each element a unique atomic number. The **atomic number** of an element is determined by the number of protons in its nucleus. No two elements have the same number of protons. Therefore each chemical element has a different atomic number.

> **The atomic number of an atom of an element is the number of protons in the nucleus.**

The **atomic weight** of an element is the total number of protons and neutrons in each atom of that element. Each proton and each neutron has one unit of atomic weight. Electrons contribute only negligibly to the weight of an atom. Therefore the atomic weight is calculated by adding only the numbers of protons and neutrons.

> **The atomic weight of an atom of an element is the sum of the numbers of protons and neutrons in its nucleus.**

Isotopes

Isotopes of an element have varying numbers of neutrons. All isotopes of a given element have the same number of protons in their nuclei. Hence, they all have the same atomic number. Their atomic weights differ because they have different numbers of neutrons. The most abundant isotope of carbon (^{12}C), for example, has six protons and six neutrons. Another isotope of carbon (^{14}C) has six protons and eight neutrons.

Many isotopes are stable. They do not change spontaneously into other atomic forms. Some isotopes, though, have unstable combinations of pro-

NEWSBREAK

Beginning of Chemistry and the End of the Chemist Lavoisier

It was the eighteenth century when the science of chemistry emerged from the tradition of alchemy. Antoine Lavoisier introduced quantitative methods in chemistry by demonstrating that substances increase in weight during combustion. He began the use of chemical equations and distinguished between chemical elements and compounds. Lavoisier also explained the process of fermentation and respiration by showing the role of oxygen and carbon dioxide. Despite these brilliant accomplishments, Lavoisier was beheaded by the Revolutionaries during the French Revolution in 1794 because he was a part of the tax-collecting bureau of the monarchy.

tons and neutrons in their nuclei. Such isotopes are called **radioactive isotopes.** A radioactive isotope breaks down or decays by giving off subatomic particles and energy (radiation). For example, carbon-14 (^{14}C) is a radioisotope of carbon because it has too many neutrons in its nucleus to be stable. The instability within the nucleus of ^{14}C results in the breaking apart of a neutron. Energy and a beta particle (an electron formed by the decomposition of a neutron) are emitted from the nucleus. Radioactive isotopes, such as ^{14}C, are useful for labelling biological substances because beta particles can be easily detected. Caution must be used, however, whenever handling radioisotopes because the energy they give off can damage biological systems.

> **Atoms of the same element that have different numbers of neutrons are called isotopes.**

Electron Arrangements and Chemical Reactivity

The protons and neutrons in the nucleus determine the atomic weight of an atom, but electrons of the atom determine its chemical properties. It is the electrons that actually participate in chemical reactions. Each element's atoms differ from those of all other elements in the number and arrangement of their electrons. Electrons move in regions of space around the nucleus.

The regions of space where electrons are likely to be found are called **shells.** Each electron shell represents an energy level. Shells closest to the nucleus have the lowest energy. Shells furthest from the nucleus have the greatest energy. Each shell has a maximal number of electrons that it can hold. The further away from the nucleus a shell is located, the greater the number of electrons that can occupy that shell (FIG. 4-2). The shell closest to the nucleus can hold only two electrons. Electrons first occupy the shells closest to the nucleus. Only after these shells are

filled do electrons occupy the shells with higher energy levels.

The outermost shell is called the **valence shell.** The number of electrons that can occupy the valence electron shell establishes in large part the capacity of that atom to combine with other atoms. The basic principle of atomic reactivity is that when its outermost electron shell is completely full an atom is stable. It will not react with other atoms. By interacting, atoms gain, lose, or share electrons to fill their outer shells. The outer electron shells of the atoms of the major elements in biological systems—hydrogen, carbon, nitrogen, oxygen, phosphorus, and sulfur— are all incomplete. The atoms of these elements, therefore, readily react with other atoms to achieve stable outer electron shells.

> **The fullness of the outer electron shell (valence shell) of an atom determines the capacity of that atom to combine with other atoms.**
>
> **Atoms react with each other by losing, gaining, or sharing electrons to fill their outer electron shells.**

MOLECULES AND CHEMICAL BONDS

When elements combine with each other they form compounds. A **compound** is a specific combination of elements in which the elements are present in a fixed and unvarying proportion. For example, water (H_2O) is a compound that has a fixed 2:1 proportion of two elements: hydrogen and oxygen. Because the proportion of elements in compounds never varies, they are distinct from **mixtures.** In a mixture two or more elements can be present in different and varying proportions. A mixture can be separated by physical means, such as filtering or sorting. A compound cannot be split into its component parts by such means.

A **molecule** is the simplest form of a compound that still retains the properties of that compound. A molecule is formed when atoms combine with each other. Chemists write **molecular formulas** to describe how many and which specific atoms form a molecule. For example, the molecular formula for water (H_2O) communicates the fact that this molecule is formed when two atoms of hydrogen and one atom of oxygen combine. Likewise the molecular formula for glucose ($C_6H_{12}O_6$) tells us that this sugar is formed by combining six carbon atoms, twelve hydrogen atoms, and six oxygen atoms. If the atoms of elements are the "letters" of the chemical alphabet, then the molecules of compounds are the "words" in the language of the chemist. Like atoms, molecules have specific physical properties, such as density. Molecules also have chemical properties, such as the ability to react with other molecules.

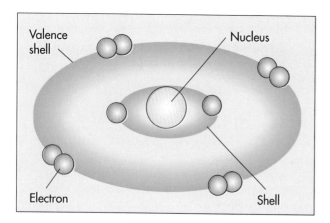

FIG. 4-2 The electrons of an atom are arranged in shells. Each shell represents a different energy level.

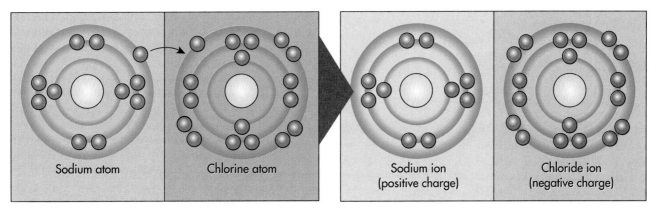

FIG. 4-3 The formation of ionic bonds involves the loss and gain of electrons. In the formation of NaCl, the chlorine atom gains an electron to fill its outer electron shell and the sodium atom loses an electron, so that all the remaining electron shells are filled. After the formation of the ionic bond, the sodium ion has a positive charge and the chloride ion a negative charge.

Chemical bonds are formed between atoms when atoms combine by transferring or sharing electrons. Stable bonds occur when atoms establish complete outer valence shells. The chemical bonds of a molecule hold together the constituent atoms that make up that molecule. The number of bonds that an atom can form depends on the number of electrons required by that atom to fill its outer electron shell. A carbon atom, for example, has four electrons in its outer electron shell that can hold a maximum of eight electrons. A carbon atom, therefore, can establish up to four bonds with other elements.

> **Stable chemical bonds occur when atoms fill their outer shells with electrons.**
>
> **Molecules—the fundamental units of compounds—are specific combinations of atoms formed when atoms form chemical bonds. They form these bonds by sharing or transferring electrons.**
>
> **A molecular formula specifies the numbers and kinds of atoms of elements that are bonded together to form a molecule of a compound.**

Three types of chemical bonds can form between atoms. **Ionic bonds** are based on attractions of ions with opposite electronic charges. **Covalent bonds** are based on sharing of electrons. **Hydrogen bonds** are based on interactions of hydrogen atoms with weak opposing electronic charges. Each type of bond is important in establishing and determining the properties of the molecules that make up living systems.

Ionic Bonds

Two ions with different charges can be held together by the mutual attraction of these charges. These are called electrostatic forces. A chemical bond based on such electrostatic forces is called an **ionic bond.** This is the type of bonding that holds sodium and chloride ions together in table salt (NaCl) (FIG. 4-3). Similarly, positively charged hydrogen ions can form ionic bonds with negatively charged chloride ions to form hydrochloric acid. The atoms of certain other elements similarly can lose or gain electrons and thereby establish ionic bonds.

> **In an ionic bond, two ions with different charges are held together by the mutual attraction of the opposite charges of the two ions.**

Ionic bonds readily dissociate (break apart) in water. This is because of the interactions with water molecules. Hydrochloric acid, for example, readily dissociates in water into H^+ and Cl^-. The concentration or relative amount of H^+ formed by such dissociation of acids is what determines the acidity of a solution.

> **Ionic bonds typically dissociate in water.**

Covalent Bonds

Covalent bonds are formed when atoms are held together by sharing electrons. Many of the molecules in living systems are based on the ability of carbon atoms to form covalent bonds. The covalent bonds between carbon atoms is what holds together the molecules that make up the structures of all living systems. The outer shell of carbon contains four electrons and is completed by the addition of four more electrons. Carbon atoms can form four covalent bonds. A carbon atom, for example, can combine with four hydrogen atoms to form methane (CH_4) (FIG. 4-4). Methane is a simple **organic compound,**

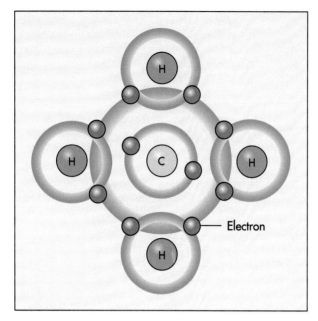

FIG. 4-4   The formation of the covalent bond involves the sharing of electrons. In methane, each hydrogen atom shares an electron with a carbon atom, completing the orbitals of the carbon and hydrogen atoms.

that is, a molecule that contains carbon and hydrogen. In the methane molecule, the carbon atom shares four electrons to complete its outer shell. It shares one electron with each of the four hydrogen atoms. A hydrogen atom has one electron and can hold two in its only shell. Each hydrogen atom completes its shell by sharing one electron from the carbon atom. A carbon atom can also form a covalent bond with another carbon atom, as well as with hydrogen atoms to form a chain of carbon atoms linked to each other. A chain of carbon atoms with hydrogen

atoms attached to the carbon atoms is called a hydrocarbon. Similarly, carbon forms covalent bonds with other atoms to establish the large and complex molecules of living systems.

> Stable covalent bonds are formed when atoms completely fill their outer electron shells as a result of sharing electrons with other atoms.

> The number of covalent bonds that a particular atom can form depends on the number of electrons in its outer electron shell and the number of electrons needed to complete that shell.

Water is an essential molecule for life. When water (H_2O) forms from the elements hydrogen and oxygen, the outer electron shells in both elements reach a stable configuration (FIG. 4-5). The oxygen atom initially has six electrons in the outer electron shell that can hold eight electrons. It completes its outer shell by sharing two electrons—one with each of the two hydrogen atoms. The hydrogen atoms each share an electron with oxygen so that they completely fill their outer electron shells.

In most cases only one pair of electrons is shared to form a **single bond.** A covalent single bond is represented as a line ($—$). In some cases, atoms share two pairs of electrons. This gives rise to a **double bond,** which is expressed as two lines ($=$). Double bonds occur most frequently when carbon is double bonded to carbon ($C=C$) or when carbon is double bonded to oxygen ($C=O$). They are found in many biologically important molecules. Carbon dioxide (CO_2), for example, is the molecule from which plants, algae, and most photosynthetic bacteria obtain the carbon to build cellular structures. It contains two double bonds between carbon and oxygen ($O=C=O$).

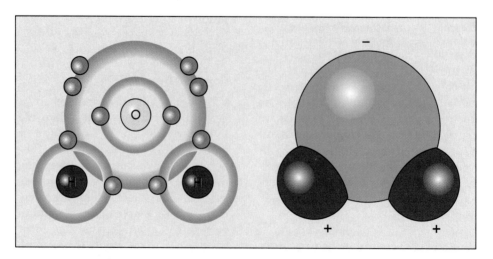

FIG. 4-5   The spatial arrangement of atoms in a water molecule results in a dipole moment because of unequal distribution of electrons between hydrogen and water. As a result, water is a good polar solvent because water can surround both positively and negatively charged ions.

Three pairs of electrons can form a **triple bond.** To form a triple bond, three electron pairs are shared between two atoms. A triple covalent bond is expressed as three single lines ($\equiv$). Molecular nitrogen (N_2) is an example of a biologically important molecule with a triple bond ($N\equiv N$) This triple bond structure is very stable and difficult to break. Although it constitutes 78% of the atmosphere, molecular nitrogen cannot be used by most organisms in their metabolism. A few bacterial species, however, are able to use molecular nitrogen. Such species are called nitrogen-fixing bacteria. They incorporate the nitrogen atoms from N_2 into proteins and other chemicals that make up their cellular structures. These nitrogen-fixing bacteria are extremely important because they form nitrogen-containing nutrients that can be used by other bacteria and higher organisms.

> **By sharing one, two, or three pairs of electrons, atoms can form single, double, and triple covalent bonds.**

Covalent bonds are strong. A relatively large amount of energy is required to break them. Atoms held together by covalent bonds generally do not dissociate in water as do ionic bonds. The covalent bonds between carbon atoms are strong enough to form the backbones of the major molecules of living systems. The fact that carbon atoms can form four single covalent bonds is important. This allows carbon to form backbone chains of covalently bonded carbon molecules. It also allows carbon to bond with other atoms, such as hydrogen, oxygen, or nitrogen. Covalent carbon-carbon bonds provide much of the stability needed to establish the very large molecules essential to the operation and reproduction of microorganisms and other living organisms. Such large molecules are called **macromolecules** and include DNA and proteins.

> **Biochemists have concluded that of all the naturally occurring elements only carbon atoms can form the bonds that will hold together the large molecules of living systems.**

Hydrogen Bonds

When hydrogen forms a covalent bond with atoms of oxygen or nitrogen, the relatively large positive nucleus of these larger atoms attracts the hydrogen electron more strongly than the single hydrogen proton. This establishes **polarity** within the molecule. This means that one end of the molecule has a positive charge and the other a negative charge. The positively charged end of the molecule is the end with the hydrogen atoms. The positively charged end can be attracted to the negatively charged end of another molecule. In this way a **hydrogen bond** is formed. When molecules of water (H_2O), for example, come close to each other, a hydrogen atom of one of the water molecules is attracted to the negatively charged

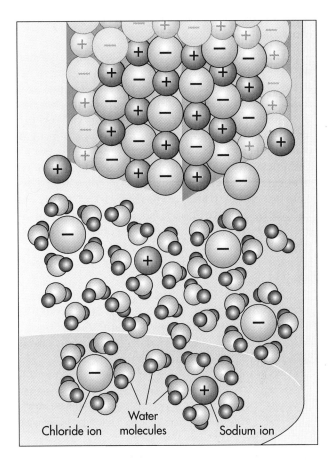

FIG. 4-6 Water forms hydrogen bonds with polar compounds. This allows water to act as a solvent. Most polar substances readily dissolve in water.

oxygen atoms of another (FIG. 4-6). The result is a lattice of water molecules that are held together by these hydrogen bonds established by charge interactions. Such hydrogen bonds do not link atoms together as strongly as do covalent bonds.

> **A hydrogen bond is formed when a hydrogen atom that is covalently bonded to an oxygen or a nitrogen atom is attracted to a polar atom in another molecule.**

A hydrogen bond has only about 5% of the strength of a covalent bond. Although such hydrogen bonds are weak, they are important because they hold different molecules together. They help establish the three-dimensional structures of large molecules by forming weak bonds between atoms with long chains of covalently bonded atoms. They also establish important chemical properties of various molecules. For example, the capacity of water to dissolve many substances is due to water's polarity and its capacity to form hydrogen bonds with ions and polar molecules. Hydrogen bonds are also important in the formation of helical molecules.

> **Hydrogen bonds are relatively weak bonds that help establish important properties of molecules.**

Isomers and the Three-dimensional Structures of Molecules

Molecules that contain identical types and numbers of atoms, but that have different arrangements of those atoms, are called **isomers**. The isomers of a molecule can have very different properties because the ability of each isomer to interact with other chemicals depends in part on the precise position of its constituent atoms in three-dimensional space. For example, glucose and fructose—both of which have the molecular formula $C_6H_{12}O_6$—are isomers that have different chemical properties (FIG. 4-7). Some bacteria can use glucose but not fructose as a source of energy for their metabolism. Other isomers can be distinguished based on how they rotate light: those rotating light to the left are called L-isomers and those rotating light to the right are called D-isomers (L stands for levorotary [left turning] and D for dextrorotary [right turning]). The amino acids that make up proteins are all L-amino acids; D-amino acids occur only in rare and special molecules of living organisms, such as the cell walls of bacteria.

Molecules with different arrangements of the same atoms (isomers) often have different chemical and physical properties.

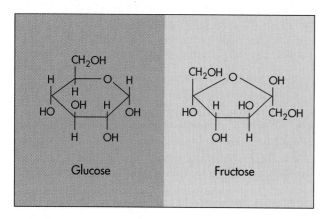

CH₂OH

Glucose

Fructose

FIG. 4-7 The sugars glucose and fructose are isomers. They have the same elemental composition but their atoms are arranged differently.

Functional Groups

When certain atoms are bound together and behave chemically as a unit that is part of a larger molecule, they are called a functional group. **Functional groups** act in the same way regardless of the molecules in which they occur (FIG. 4-8). Functional groups determine some of the characteristics, such as solubility, and chemical reactivity of the molecules to which they are attached.

Bonding of specific atoms that behave as a unit form a functional group within a molecule.

Functional groups have specific chemical properties, even when they are attached to very different molecules.

The bonding of an oxygen atom to a hydrogen atom forms a polar **hydroxyl (-OH) functional group**. The hydroxyl group has polarity because shared electrons of a covalent bond are drawn closer to the oxygen atom than to the hydrogen atom. This gives the hydroxyl group a slightly negative oxygen atom and a slightly positive hydrogen atom. Alcohols such as ethanol and glycerol are organic molecules that have hydroxyl functional groups. They tend to be relatively soluble in water. Alcohols dissolve in water because of the hydrogen bonding between water molecules and the hydroxyl group of the alcohol.

The bonding of a nitrogen atom with two hydrogen atoms forms a polar functional group, called an **amino (-NH₂) functional group**. The bonding of carbon with oxygen and a hydroxyl group forms a polar **carboxyl (-COOH) functional group**. Amino acids, which are the building blocks of proteins, have both amino and carboxyl functional groups.

Functional groups are important because they permit molecules to react with each other in predictable ways. In some cases, reactions between functional groups form bonds that link molecules together. This is how small molecules can be joined to form large molecules. Proteins, for example, are large molecules that are assembled by linking smaller amino acid molecules. Other functional groups have characteristic properties and engage in chemical reactions that are essential for sustaining microorganisms and other living systems.

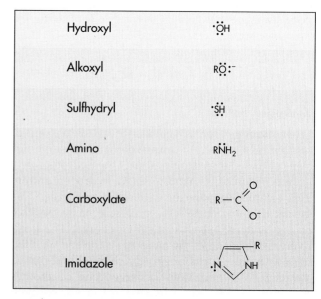

Hydroxyl	·ÖH
Alkoxyl	RÖ:⁻
Sulfhydryl	·ṢH
Amino	RṄH₂
Carboxylate	R—C(=O)O⁻
Imidazole	(imidazole ring structure)

FIG. 4-8 The chemical characteristics of a substance are determined in large part by its functional groups.

So far we have considered molecules as if they were fixed structures that remain stationary and do not change. In reality, molecules are in constant motion. They possess **kinetic energy,** the energy of motion. The faster molecules move, the more kinetic energy they have. When moving molecules collide, there may be sufficient kinetic energy to break bonds apart. When this occurs, the atoms can form new bonds. This permits new arrangements of molecules to form. During each chemical reaction, existing chemical bonds are broken and new chemical bonds form to yield different molecules. When molecules react and form new molecules, the combinations of atoms get rearranged. The total kinds and number of atoms always remain unchanged. Atoms can be neither formed nor destroyed in any chemical reaction. This **conservation of matter** is a fundamental law governing the universe.

> Kinetic energy, the energy of motion, is used to break chemical bonds.

> The law of conservation of matter states that atoms cannot be created or destroyed in chemical reactions.

CHEMICAL EQUATIONS

The conservation of matter always applies to all chemical reactions, including the chemical reactions occurring in living systems. There must be a balance between what goes into a chemical reaction, the **reactants,** and what is produced by that reaction, the **products.** The **chemical equation** describes the relationship between the reactants and products. The reactants are shown on the left side and the products are shown on the right side of the equation. If elements are the "letters" of chemistry and molecules are the "words," then chemical equations are the "sentences" in the language of chemistry. A chemical equation identifies the reactants and products by name or chemical formula. It permits chemists to describe the changes that occur during chemical reactions.

In a **balanced chemical equation,** the number of atoms of each element in the reactant molecules must equal the number of product molecules of that element. For example, the equation for the reaction of sodium chloride (NaCl) in water to form sodium (Na$^+$) and chloride (Cl$^-$) ions is written:

$$NaCl \longrightarrow Na^+ + Cl^-$$

The numbers of sodium and chlorine atoms on both sides of the equation are the same. The equation is properly balanced. Water is required for this reaction to occur. It is not shown in the equation because it is not changed or transformed in the process of the reaction. When a substance acts as a solvent and does not participate directly in the reaction, it is not shown within the equation. Sometimes, however, the solvent is indicated above the arrow to show that its presence is necessary for the reaction to occur.

$$NaCl \xrightarrow{\text{water}} Na^+ + Cl^-$$

> Chemical equations show the changes that occur during a chemical reaction.

> The chemical equation shows the balance between reactants and products.

EQUILIBRIUM

Virtually all chemical reactions are reversible. In a reversible reaction, the reactants can become the products and the products the reactants. The direction of the reaction depends in part on the concentrations of reactants and products. The likelihood of molecules colliding with sufficient kinetic energy to break bonds depends on the relative abundances of molecules with sufficient kinetic energies to react. Concentrations of reactants and products are expressed in units called moles. A **mole** is a measure of the number of molecules (6×10^{23} molecules). A mole is equal to the weight in grams of the molecular weight of a molecule. Thus, 1 mole of water weighs 18 grams because the molecular weight of water is 18, the sum of the molecular weights of two ^1H atoms and one ^{16}O atom.

The greater the concentration of reactants, the greater the opportunity for collisions to occur and, hence, the faster the forward reaction. The greater the concentration of products, the faster the reverse reaction. As more and more product molecules form, fewer and fewer reactant molecules remain. This lowers the rate of the forward reaction. As the concentration of product molecules increases, they will collide more frequently with each other than before. In some cases the reaction is reversed so that the original reactants are reformed. Eventually, a balance—called **equilibrium**—is achieved between the reactants and products of the forward and reverse reactions. At equilibrium the rates of the forward and reverse reactions are equal. This does not mean that the amounts of the products and reactants are equal. In this state there is no net change in the concentrations of reactants and products even though the molecular reactions are still continuing.

> Chemical reactions are reversible and chemical reactions tend toward a state of equilibrium.

> At equilibrium, the rates of the forward and the reverse reactions are equal and there is no further net change in the concentrations of reactants and products.

ENERGY AND CHEMICAL REACTIONS

There is only a finite amount of energy in the universe. This energy cannot be created or destroyed. Energy, however, can be converted from one form to another. The various forms of energy include the chemical energy stored in molecular bonds (*potential* or *stored energy*), kinetic energy (energy of motion), electrical energy (energy produced by movement of electrons), and radiant energy (heat or light energy) from the sun. During chemical reactions chemical bonds are broken and new bonds are formed. Energy is transformed during these reactions. In a chemical reaction there always is a net balance between the energy required to break chemical bonds, the energy released by the new bonds that are formed, and the energy—such as heat energy—that is exchanged with the surroundings.

Energy is neither created nor destroyed in chemical reactions; however, energy can be converted from one form to another.

The products of chemical reactions can end up with either less or more energy than the reactants had. Some chemical reactions release energy. Others require the input of energy. Energy-requiring reactions can occur only when extra energy enters into the reaction. The extra energy must come from some other system.

Chemical reactions involve energy changes.

Most often the energy needed to drive energy-requiring chemical reactions in living systems is supplied by **ATP (adenosine triphosphate).** ATP is called an energy-rich or high-energy compound. It contains chemical bonds that can release a relatively large amount of energy (FIG. 4-9). The release of this energy from the ATP molecule can be coupled to energy-requiring reactions. In this fashion, energy-releasing reactions drive the energy-requiring reactions of cell growth, movement, and transport. ATP serves almost universally in biological systems as the energy source for energy-requiring reactions. Cellular processes requiring energy most likely depend on the use of ATP. As such, ATP can be termed the *universal currency of energy* in biological systems.

ATP has a central role in the flow of energy through living systems.

TYPES OF CHEMICAL REACTIONS

Enzymatic Reactions

For a chemical reaction to occur, the reactant molecules must collide with sufficient kinetic energy to bring about the reaction. The amount of energy needed is called the **activation energy.** The activation energy is the amount of energy needed to start the reaction. It is not the amount consumed or released by the breaking and forming of chemical bonds. A chemical reaction can occur only when the energy of activation is provided to start the reaction. Chemists often heat chemicals with a Bunsen burner to provide the energy needed for chemical reactions to occur. But cells operate at nearly constant temperatures. Cells employ other methods to overcome the energy barrier to starting a chemical reaction presented by the activation energy.

Biological systems depend on enzymes to lower the activation energy of a chemical reaction (FIG. 4-10). **Enzymes** are proteins that act as biological catalysts; some RNA molecules, called ribozymes, can similarly act as biological catalysts. A catalyst speeds

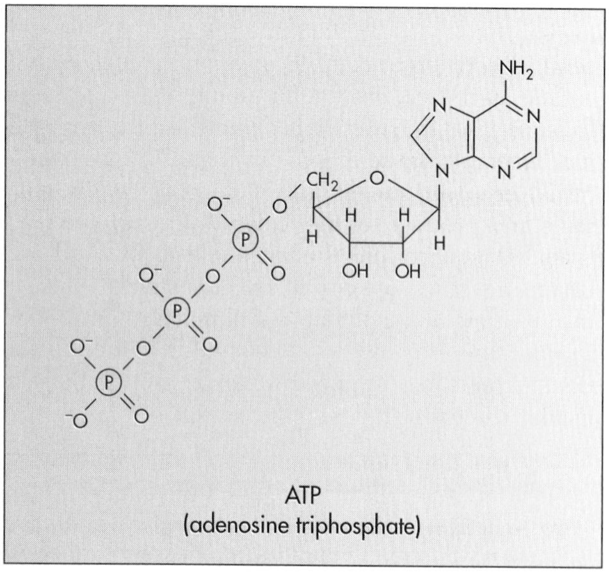

FIG. 4-9 ATP is a compound with high-energy phosphate bonds.

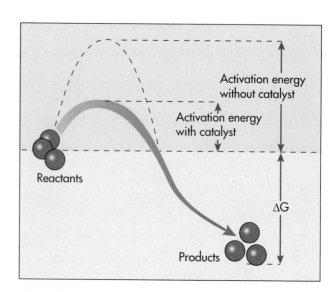

FIG. 4-10 An input of energy called the activation energy is needed to start a chemical reaction. A catalyst lowers the activation energy. In biological systems, enzymes serve as the catalysts to lower the activation energy.

NEWSBREAK

Catalysis by Cell-Free Enzymes

Jons Berzelius discovered the principle of catalysis, based on the recognition that certain substances have a catalytic force that enables them to decompose other compounds and rearrange the decomposed products without being chemically altered in the process. The German chemist Eduard Buchner received a Nobel Prize in 1907 for his demonstration of an alcoholic fermentation without yeast cells. He showed that sugar could be changed into alcohol by soluble proteins from yeasts. He called the active proteins "zymase," from which we derive the term enzyme. Today many cell free enzymes are used, for example, in detergents to help remove proteins and in the production of high-fructose corn syrup sweetener for soft drinks.

up a reaction without being consumed in that reaction. At a given temperature, a catalyzed reaction proceeds more rapidly than an uncatalyzed reaction. Because catalysts are not consumed in the reaction, enzymes theoretically may continue to function indefinitely. Since they can be reused, only small amounts of enzymes are often required.

Without enzymes, chemical reactions would not occur fast enough within a cell to support life functions. The rapid rates at which chemical reactions occur in living systems are possible because of the role played by enzymes in lowering the activation energy. Each microbial cell must possess many enzymes, thousands in fact, to carry out the essential metabolic activities involved in its growth and reproduction.

An enzyme works by binding with a molecule called the **substrate.** An enzyme can bond only to a specific substrate. The degree of substrate specificity exhibited by enzymes reflects the fact that the enzyme and the substrate must fit together in a specific way (FIG. 4-11). The precision of fit between enzyme

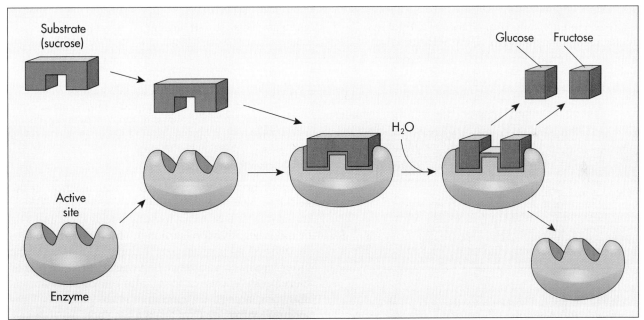

FIG. 4-11 **A,** The fit between the enzyme and the substrate to form an enzyme-substrate complex has been likened to that of a lock and key. Actually, this interaction modifies the three-dimensional structure of the enzyme so that the substrate induces its fit to the enzyme. The precision of fit is responsible for the high degree of specificity of enzymes for particular substrates. **B,** Model showing the fit between the polysaccharide component of a bacterial cell wall *(yellow substrate)* and the active site of the enzyme lysozyme that catalyzes the breakdown of the bacterial cell wall.

and substrate molecules permits the establishment of exactly the right spatial orientation so that the numerous chemical reactions of an organism can occur with greater speed.

> Enzymes are proteins that act as biological catalysts.

> Enzymes are highly specific, both in terms of their substrates and the reactions they catalyze.

Oxidation-reduction Reactions

Oxidation-reduction reactions are based on the transfers of electrons between molecules (FIG. 4-12). **Oxidation** is the process of removing one or more electrons from an atom or molecule. **Reduction** is the process of adding one or more electrons to an atom or molecule. Oxidation and reduction are coupled because they involve the simultaneous removal of an electron from one substance and the addition of that electron to another. Often in biological systems a proton or hydrogen ion (H^+) is transferred with the electron during oxidation-reduction reactions. Oxidation reactions that involve the removal of an electron and hydrogen ion are called *dehydrogenation reactions*. Reduction reactions that involve the addition of an electron and hydrogen ion are called *hydrogenation reactions*.

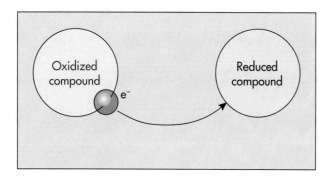

FIG. 4-12 Oxidation-reduction reactions involve an exchange of electrons. The substance that accepts an electron becomes reduced and the substance that donates an electron becomes oxidized.

An example of biological oxidation-reduction reactions is shown in the following equation:

$$X_{reduced} + NAD^+ \longrightarrow X_{oxidized} + NADH + H^+$$

In this example, molecule X serves as the source of two electrons and two protons that are transferred to a molecule of nicotinamide adenine dinucleotide (NAD^+). Molecule X becomes oxidized (two electrons are removed) and NAD^+ becomes reduced (two electrons are added). At the same time, two protons are removed from molecule X. One of these is added

to NAD^+ (so that the reduced form of NAD is written as NADH) and the other proton (H^+) is released into the medium.

> Oxidation is the removal of electrons and reduction is the addition of electrons.

> The oxidation of one substance must always be coupled with the reduction of another.

Oxidation-reduction reactions are important in the metabolism of cells for energy changes and the formation of new cellular material. Oxidation-reduction reactions can release energy that is used to form ATP. For example, during cellular respiration the oxidation of glucose to form carbon dioxide is coupled with the reduction of oxygen to form water. This provides the energy to drive ATP production. Living organisms use the energy released from such oxidation-reduction reactions for growth, reproduction, and other life processes—such as movement.

Oxidation-reduction reactions also are used by living systems to store energy. Thus, when carbon dioxide is reduced to glucose during photosynthesis, energy is stored within the organic molecules of the organism. This stored energy can later be released when organisms oxidize sugars during cellular respiration.

> Oxidation-reduction reactions can release energy and are used to fuel cellular reactions, to store energy, and for biosynthesis of the macromolecules of the cell.

Acid-base Reactions

Another important type of chemical reaction is the acid-base reaction. An **acid** is a substance that dissociates into one or more hydrogen ions (H^+) and one or more negative ions (anions). Thus an acid can also be defined as a proton (H^+) donor. A **base,** on the other hand, is a substance that dissociates into one or more positive ions (cations), plus one or more anions that can accept or combine with protons. Thus sodium hydroxide (NaOH) is a base because in water it dissociates to release hydroxyl ions (OH^-), which have a strong attraction for protons. Bases that produce hydroxyl ions are among the most important proton acceptors.

> Acids increase the concentration of hydrogen ions in solution.

> Bases decrease the concentration of hydrogen ions in solution.

The amount of H^+ in a solution is expressed by a logarithmic pH scale that ranges from 0 to 14 (FIG. 4-13). The **pH** of a solution is the negative logarithm to the base 10 of the hydrogen ion concentration.

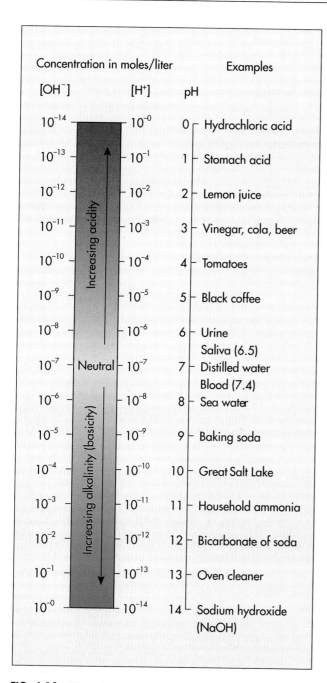

FIG. 4-13 The pH scale showing pH values of some common substances.

Measuring pH

It often is critical to determine the pH of a solution. Microorganisms grow only within certain pH ranges. Adjusting the pH of a growth medium is essential to provide a favorable condition for microbial growth. Adjusting or maintaining the pH of many foods so as to ensure unfavorable conditions for microbial growth is used as a food preservation method. *Clostridium botulinum*, the bacterium that causes a fatal form of food poisoning called botulism, for example, will not grow at low pH (acidic conditions). Acidic foods such as red tomatoes are unsuitable for the growth of this deadly bacterium. Other foods such as low-acid yellow tomatoes provide favorable conditions for growth of *C. botulinum* when they are preserved by canning. Therefore higher temperatures are used when canning yellow tomatoes to protect against botulism.

Several methods are used to determine pH. The pH can be determined electronically using a pH meter. The probe of a pH meter is placed into a solution and the instrument responds to the hydrogen ion concentration by registering the pH. pH meters are simple to use and rapidly provide information on the pH of a solution. Various chemical pH indicators are also used to determine pH (see Table). These indicators change color at different pH values. A drop of a solution can be added to the pH indicator and the color reveals the approximate pH.

Color Reactions of Some pH Indicators

INDICATOR	pH RANGE	COLOR	
		Acid	**Base**
Cresol Red	1.0-2.2	Red	Yellow
Thymol Blue	1.2-2.8	Red	Yellow
Bromphenol Blue	3.0-4.8	Yellow	Blue
Methyl Orange	3.1-4.4	Red	Yellow
Bromcresol Green	4.0-5.6	Yellow	Blue
Methyl Red	4.4-6.2	Red	Yellow
Bromcresol Purple	5.2-6.8	Yellow	Purple
Bromthymol Blue	6.0-7.6	Yellow	Blue
Phenol Red	6.4-8.0	Yellow	Red
Phenolphthalein	8.0-10.0	Colorless	Red
Thymolphthalein	9.4-10.6	Colorless	Blue
Alizarin Yellow	10.0-12.1	Yellow	Red

$$pH = -\log[H^+]$$

The greater the hydrogen ion concentration the lower the pH. Because the pH scale is logarithmic, a change of one whole pH unit represents a tenfold change from the previous concentration of hydrogen ions. Thus a solution with a pH 1 has 10 times more H^+ ions than one with a pH 2 and 100 times more H^+ ions than a solution with a pH 3. Acidic solutions contain more H^+ ions than OH^- ions and have a pH lower than 7. Basic or alkaline solutions have more OH^- ions than H^+ ions and a pH higher than 7. In pure water the concentrations of H^+ and OH^- are equal and the pH is 7. This pH level is called **neutral.**

> **pH describes the concentration of hydrogen ions.**
>
> **Acidic solutions have pH values less than 7; basic solutions have pH values greater than 7; water is neutral (pH = 7).**

When an acid reacts with a base, there is a reaction between the hydrogen ions produced by the acid and the hydroxyl ions produced by the base. Acid-base reactions result in the formation of water and a salt. For example, when hydrochloric acid reacts with sodium hydroxide, the products are sodium chloride and water.

$$HCl + NaOH \longrightarrow NaCl + H_2O$$

$$H^+ + Cl^- + Na^+ + OH^- \longrightarrow Na^+ + Cl^- + H_2O$$

If the amounts of acid and base are balanced, all the free hydrogen ions react with all the free hydroxyl ions. This is known as a **neutralization reaction** because it results in a neutral solution of the salt. The hydrogen ion and hydroxyl ion concentrations are balanced and thus achieve a neutral pH of 7.

As living organisms take up nutrients, carry out chemical reactions, and excrete wastes, they may change the balance of acids and bases. This change may occur both within their cells and in the surrounding solution. When bacteria are grown in laboratory medium, for example, some of the chemicals produced by their metabolism are acids that can alter the pH of the medium. Unchecked, the pH of the medium would become too acidic for the bacteria to live. To prevent this, microbiologists add pH buffers to the culture medium. A **buffer** limits the change of pH by reacting with acids or bases to form neutral

salts. Phosphate buffer containing K_2HPO_4 and KH_2PO_4 is often used to maintain a pH near 7.0 in culture media.

Condensation and Hydrolysis Reactions

Condensation reactions involve the bonding of two molecules into one. Condensation reactions are very important in forming the large molecules of living systems. In a **condensation reaction,** a hydrogen ion (H^+) removed from one functional group of a molecule and a hydroxyl ion (OH^-) from another combine to form a molecule of water (H_2O). The component molecules are joined by a covalent bond (FIG. 4-14). For example, glucose molecules combine into larger molecules containing multiple glucose subunits. Large molecules formed from the bonding of many subunit molecules are called **polymers.** The polymers produced by condensation reactions may contain millions of individual subunit molecules, called **monomers.** These monomers may or may not be identical. As a rule polymers are less soluble and more stable (long-lived) than monomers. Polymers are important components of many biological structures (Table 4-1).

The reverse reaction is **hydrolysis.** A hydrolytic reaction breaks down polymers into their component monomers (FIG. 4-15). Covalent bonds between parts of molecules are broken and H^+ and OH^- ions from water become attached to the component subunit molecules. Hydrolysis reactions, such as the hydrolysis of ATP, are important for the extraction of energy from molecules. They yield the energy needed to support energy-requiring reactions in cells. Hydrolysis reactions also produce the small molecules that are used by cells for the synthesis of the large molecules that make up the structures of organisms.

> **Condensation and hydrolysis reactions permit the linking and breaking apart of molecules, including some of the very large molecules of living systems.**

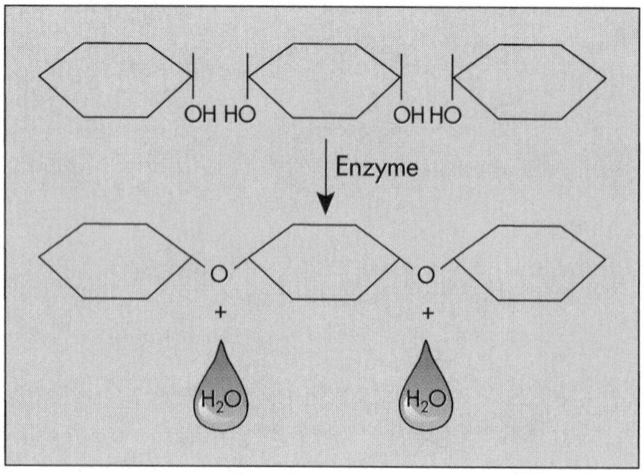

FIG. 4-14 Polymers are formed when smaller chemical units join together in a condensation reaction.

TABLE 4-1

POLYMER	MONOMER SUBUNITS	BIOLOGICAL FUNCTION
		Chemical Compositions of Some Biologically Important Polymers
Polysaccharides	Monosaccharides	Structural components that support and protect cells; nutrient storage within cells
Cellulose	Glucose	Structural component of cell walls of algal and plant cells; protection of cell
Starch	Glucose	Storage of carbon as nutrient source within some algal and plant cells
Glycogen	Glucose	Storage of carbon as nutrient source within some protozoan and animal cells
Chitin	N-acetylglucosamine	Structural component of cell walls of many fungal cells; protection of cell; structural component of skeletons or shells of some animals
Peptidoglycan	N-acetylglucosamine and N-acetylmuramic acid	Structural component of cell walls of most bacterial cells; protection of cell
Nucleic Acids	Nucleotides	Storage, transmission, and utilization of cellular hereditary information
DNA	Deoxyribonucleotides	Storage and transmission of hereditary information from one generation to the next in living cells and some viruses
RNA	Ribonucleotides	Structural component of ribosomes; transfer of genetic information from DNA for use in directing protein synthesis in living cells; hereditary informational molecule in some viruses
Proteins	L-Amino acids	Structural component of almost all cell structures; enzymes that catalyze cellular chemical reactions; transport of most chemicals into and out of cells; protective coat of viruses

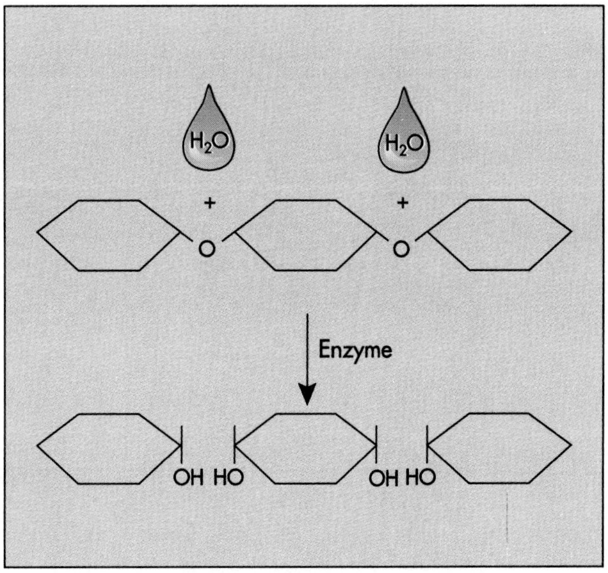

FIG. 4-15 Large molecules are broken down into smaller molecules in hydrolysis reactions.

A common feature of all living systems is that they are based on carbon atoms. Organic molecules that contain carbon form the essential components of all living organisms. Carbon, hydrogen, oxygen, and nitrogen atoms comprise 99% of the mass of living organisms. Carbon atoms are able to establish strong bonds with these atoms, as well as with other carbon atoms. Therefore carbon is well suited for uniting the atoms of living systems into stable macromolecules. Microorganisms are composed of various organic macromolecules representing four major classes of chemicals: carbohydrates, lipids, proteins, and nucleic acids. In addition, microorganisms are also composed largely of water. All living systems depend on the availability of water and various other inorganic molecules, such as carbon dioxide and phosphate.

WATER

Of the inorganic compounds of living systems, water is without doubt the most abundant and important. Life cannot exist in the absence of water. Water usually comprises over 75% of the weight of a living cell. Water serves as a solvent that permits the dissociation of chemicals, allowing chemical reactions of many molecules to occur within living cells that produce numerous new combinations of molecules.

Water's structural and chemical properties make it particularly suitable for living cells (FIG. 4-16). The hydrogen (H^+) and hydroxyl (OH^-) portions of water (HOH) can split apart and later rejoin. This enables water to participate as a reactant or a product in many chemical reactions. Water molecules, for example, are involved in many chemical reactions as an important source of the hydrogen and oxygen atoms that are incorporated into the numerous organic compounds that make up living cells.

Because the oxygen region of the water molecule has a slightly negative charge and the hydrogen region has a slightly positive charge, water has a **polar** nature. The polarity of water means that many charged or polar substances dissolve in water by dissociating into individual molecules. Molecules dissolved in water are called **solutes.** The negatively charged part of the water molecule is attracted to the positively charged part of the solute molecule. At the same time the positively charged part of the water molecule is attracted to the negatively charged part of the solute molecule. Solid NaCl, for example, dissolves in water by dissociating into the positively charged sodium ions (Na^+) and chloride ions (Cl^-). The positive sodium ions are attracted to the negatively charged oxygen atom of water. The negative chloride ions are attracted to the positively charged

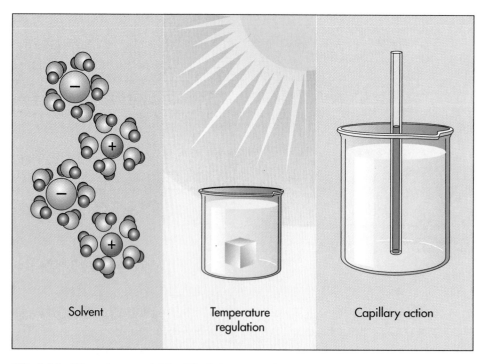

Solvent · Temperature regulation · Capillary action

FIG. 4-16 Water is essential for life. It has critical roles in life's functions, including its ability to regulate temperature and to act as a solvent, that are based largely on its ability to form hydrogen bonds.

hydrogen atoms of water. Thus the Na^+ and Cl^- ions of solid NaCl are separated by the water molecule and table salt dissolves in water.

This polarity of the water molecule also means that hydrogen bonds are formed between nearby water molecules. The hydrogen bonds between water molecules make water an excellent temperature regulator. Cells are mostly water and live surrounded by water. Water readily maintains a constant temperature and tends to protect cells from sudden environmental temperature changes. Also, a great deal of heat energy is required to separate water molecules—held together by hydrogen bonds—from each other to form water vapor, that is, to convert liquid water into gaseous steam. Water exists in the liquid state at temperatures of 0° to 100° C. Liquid water is available on most of the Earth's surface and readily available for use as a solvent.

CARBOHYDRATES

Carbohydrates are a large and diverse group of organic compounds. This group includes sugars and compounds such as starch that are derived from sugars. Each sugar molecule has a fixed ratio of carbon:hydrogen:oxygen of 1:2:1. Therefore, all **carbohydrates** have the same basic chemical formula—$C_n(H_2O)_n$, where n is a whole number equal to or greater than 3.

Carbohydrates include the **monosaccharides** (saccharide is the Greek word for sugar). Monosaccharides are simple sugars with three to seven carbon atoms (FIG. 4-17). Monosaccharides may be linked to form larger carbohydrate molecules. A **disaccharide** contains two monosaccharide units, an **oligosaccharide** contains three to ten monosaccharide units, and a **polysaccharide** contains more than ten monosaccharide units. Monosaccharides with five or more carbon atoms tend to form ring structures when dissolved in water. Thus, within cells, **pentoses,** which have five carbon atoms, and **hexoses,** which have six carbon atoms, form molecules with ring structures. Pentoses and hexoses are biologically significant. They both serve as energy sources and as the structural backbones of large informational molecules. Deoxyribose is a pentose found in deoxyribonucleic acid (DNA), the genetic material of the cell. Ribose, another pentose, is found in ribonucleic acid (RNA). RNA is the molecule used to transfer genetic information within cells for the expression of genetic information. Glucose is a common hexose and the main energy-supplying molecule of living cells.

Disaccharides are formed when two monosaccharides join in a condensation reaction (FIG. 4-18). For example, molecules of two monosaccharides, glucose and fructose, combine to form a molecule of the disaccharide sucrose (table sugar). Sucrose is the form in which carbohydrates are transported through plants. The disaccharide lactose is formed by the bonding of one glucose and one galactose subunit. Lactose occurs in the milk of mammals. The bond linking the monosaccharides in these disaccharides is a type of covalent bond known as a glycosidic bond. In a **glycosidic bond** an oxygen atom forms a bridge between two carbon atoms.

CH$_2$OH

Glucose

FIG. 4-17   A monosaccharide is the fundamental chemical unit of carbohydrates.

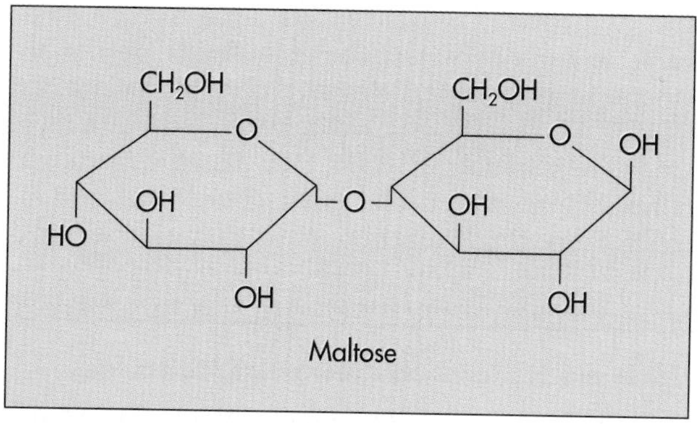

CH$_2$OH   CH$_2$OH

OH   OH   OH

HO

OH   OH

Maltose

FIG. 4-18   A disaccharide is composed of two monosaccharides.

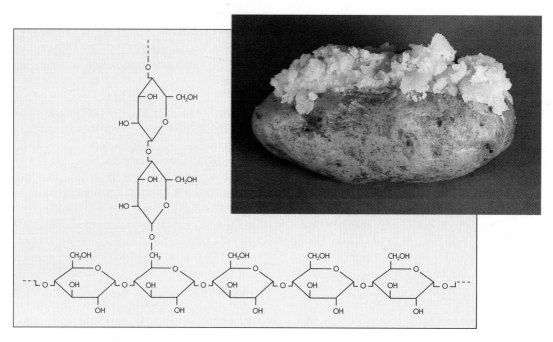

FIG. 4-19 A polysaccharide is composed of numerous monosaccharides.

Many monosaccharide units can likewise be linked to form **polysaccharides** (FIG. 4-19). Polysaccharides, like starch and glycogen, are composed of many units of glucose that are linked together. They function as important carbon and energy reserves in bacteria, plants, and animals. Other polysaccharides, such as cellulose, function as structural supports as in the cell walls of algal and plant cells.

Carbohydrates serve as sources of energy for cells and make up key structures, such as the walls that surround some cells.

LIPIDS

Like carbohydrates, **lipids** are organic compounds composed of atoms of carbon, hydrogen, and oxygen. Lipids, however, are mostly made up of carbon and hydrogen. They have very little oxygen compared to carbohydrates. Therefore lipids are nonpolar and **hydrophobic.** Being hydrophobic means that they do not readily dissolve in water. Although most lipids are insoluble in water, they dissolve readily in nonpolar solvents such as ether, chloroform, and alcohol. Some lipids function in the storage and transport of energy. Others are key components of membranes, protective coats, and other structures of cells.

Lipids are Hydrophobic Nonpolar Molecules

Many lipids have fatty acid components (FIG. 4-20). A **fatty acid molecule** consists of a carboxyl (–COOH) functional group attached to the end of a

long hydrocarbon chain composed only of carbon and hydrogen atoms. Thus fatty acids contain a highly hydrophobic hydrocarbon chain, usually 16 to 18 carbon atoms long, and a carboxyl functional group that is highly hydrophilic. Being hydrophilic means that it is attracted to water molecules. This gives fatty acids interesting chemical properties, such as the ability of part of the fatty acid molecule to associate with water molecules while the other part is pushed away. The carboxyl functional group can donate hydrogen ions in a chemical reaction with the alcohol group of another molecule. In this way fatty acids can combine with alcohols such as glycerol to form fats.

Fats consist of fatty acids bonded to the 3-carbon alcohol glycerol (FIG. 4-21). In the fat molecule the fatty acid is usually stretched out like a flexible tail. A fat molecule is formed when a molecule of glycerol combines with one, two, or three fatty acid molecules to form a **monoglyceride, diglyceride,** or **triglyceride,** respectively. The chemical bond formed between a fatty acid and an alcohol group of glycerol is called an **ester linkage.** Plants and animals store lipids as triglycerides. Glycerides are the most abundant lipids and the richest source of energy in the human body. They are insoluble in water and tend to clump into fat globules.

Complex lipids have additional components such as phosphate, nitrogen, or sulfur, or small hydrophilic carbon compounds such as sugars. For example, the cell wall of *Mycobacterium tuberculosis,* the bacterium that causes tuberculosis, is distinguished

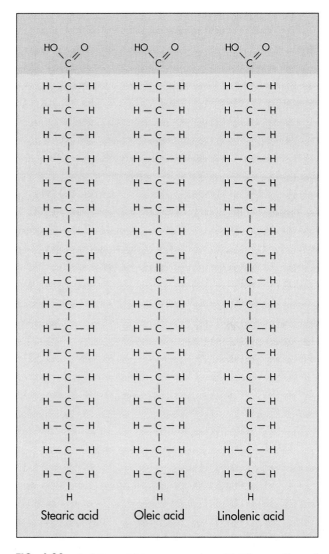

FIG. 4-20 A fatty acid is an organic acid. The portion of the fatty acid with the functional carboxylic acid group is polar and hydrophilic (attracted to water), whereas the remaining hydrocarbon portion is hydrophobic (repelled by water).

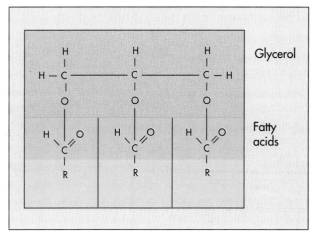

FIG. 4-21 A triglyceride is composed of glycerol and three fatty acids.

by the presence of abundant glycolipids (carbohydrates that are joined to lipids). These glycolipids give the bacterium a wax-like covering that contributes to its distinctive acid-fast staining characteristic.

Phospholipids are complex lipids made up of glycerol, two fatty acids, and a phosphate functional group (FIG. 4-22). Phospholipids are the major chemical component of biological membranes, including the plasma membrane. Their molecules contain both hydrophobic and hydrophilic portions. This enables phospholipids to aggregate into bilayers in which the hydrophobic components of each layer interact with each other and the hydrophilic components are exposed to the aqueous interior or exterior of the cell. The chemical properties of phospholipids make them effective structural components of a cell's plasma membrane. Water soluble (polar) substances are un-

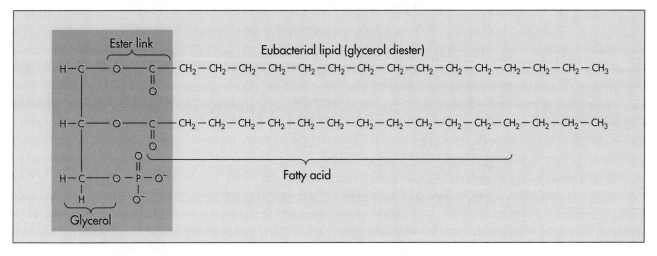

FIG. 4-22 Phospholipids are composed of glycerol linked to two fatty acids and a phosphate group.

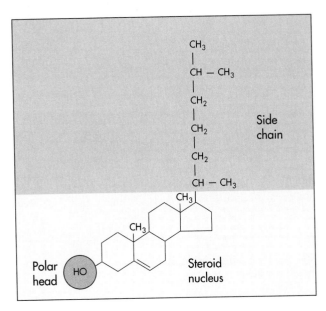

FIG. 4-23 A steroid is a nonpolar lipid with four rings. Cholesterol is an example of a steroid.

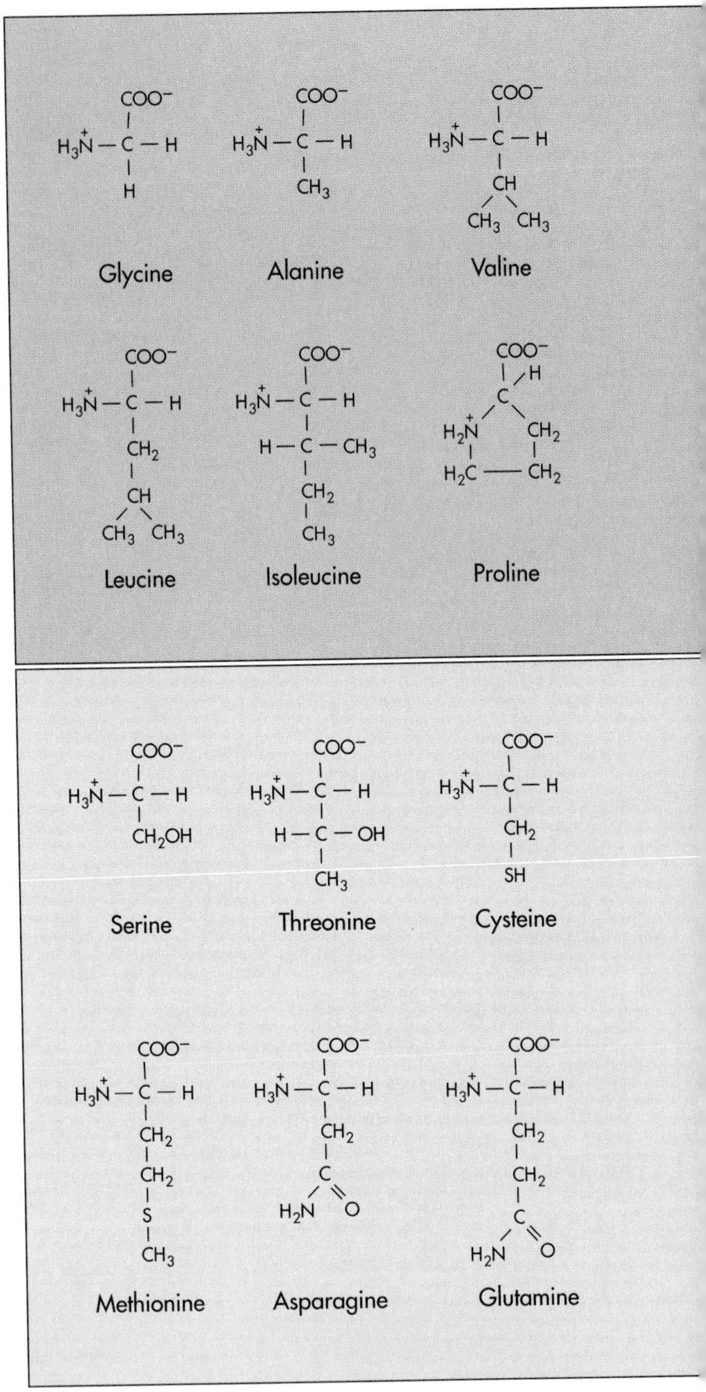

able to flow through the hydrophobic fatty acid portion of the bilayer. Phospholipids, thus, enable the plasma membrane to restrict the flow of materials into and out of the cell.

> Phospholipids, which have hydrophilic and hydrophobic portions, form an integral part of the plasma membrane.

Steroids are also lipids but they are structurally very different from the lipids described previously. Cholesterol is a steroid compound that contains a —OH group, making it a **sterol** (FIG. 4-23). Cholesterol is an important component of the plasma membrane of eukaryotic animal cells. Other eukaryotic cells such as fungal cells contain different sterols in their plasma membranes. Cholesterol and other sterols wedge between phospholipids in the plasma membrane, maintaining membrane fluidity. Cholesterol and other sterols generally are absent from the plasma membrane of a prokaryotic cell.

PROTEINS

Proteins are large molecules made up of hundreds or thousands of amino acid subunits. **Amino acids** are the building blocks of proteins. An amino acid contains at least one carboxyl (—COOH) functional group and one amino (—NH₂) functional group attached to the same carbon atom. This carbon atom is

FIG. 4-24 The structural formulas of 20 common amino acids. Each is an L-α-amino acid. The structures differ in the other constituents.

called the alpha- carbon (α-carbon). There are only 20 amino acids naturally found in proteins (FIG. 4-24). Amino acids exist in mirror images called **stereoisomers.** They are designated as either L or D forms (FIG. 4-25). The amino acids found in proteins are always **L-amino acids.** Also attached to the alpha-carbon is a

Phenylalanine   Tyrosine   Tryptophan

Lysine   Arginine   Histidine

Aspartate   Glutamate

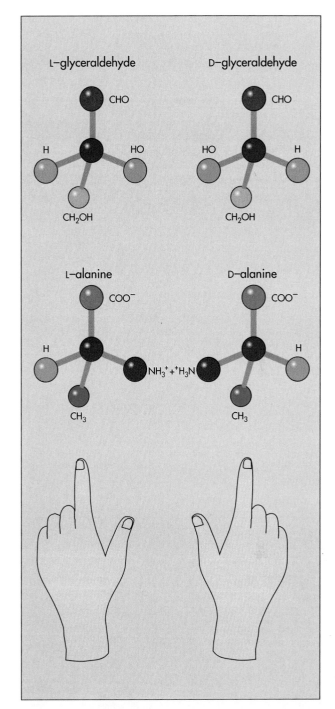

FIG. 4-25   An asymmetric carbon allows molecules to exist in two different forms (isomers) that are mirror images: The D and L forms of gylceraldehyde molecules and their relation to L-alanine. In determining the absolute configuration of a carbohydrate, the D designation is used to indicate that the groups H, CHO, and OH, in that order, are situated in a clockwise fashion about the asymmetric carbon atom, when the CH_2OH group is directed away from the viewer. The designation L is used if the order is counterclockwise. In determining the configurations of amino acids, we still use glyceraldehyde as the reference, with the NH_2 group substituting for OH and the COOH group substituting for CHO.

side or **R group.** These R groups are the amino acid's distinguishing factors. The R group can be a hydrogen atom, an unbranched or branched carbon chain, or cyclic ring structure. It can also contain functional groups—such as the sulfhydryl (—SH), hydroxyl (—OH), or additional carboxyl or amino groups.

The L-amino acids of a protein molecule are linked by covalent bonds. These are called peptide bonds. A **peptide bond** forms between the amino group of one amino acid and the carboxyl group of another. The bonding of two amino acids by a peptide bond forms a dipeptide. Three or more amino acids linked by peptide bonds form a **polypeptide chain** (FIG. 4-26).

A protein is composed of a chain of L-amino acids held together by peptide linkages.

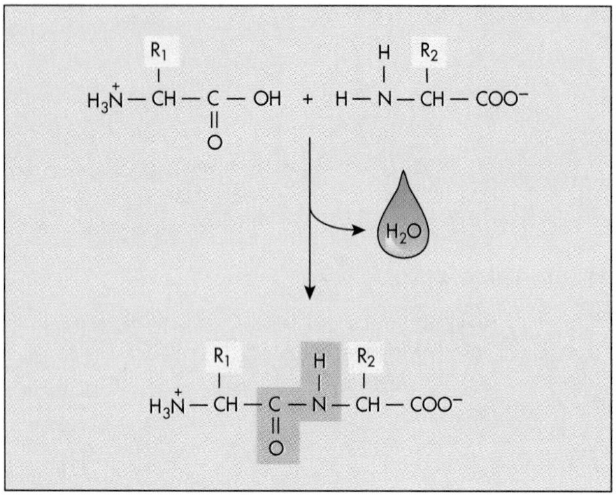

FIG. 4-26   A polypeptide has a free amino end and a free carboxyl end.

Protein Structure

Proteins have very highly organized three-dimensional structures (FIG. 4-27). Both the number and order of the specific amino acids within the polypeptide chain are important. They establish the structure and functional properties of protein molecules. Proteins have different lengths, different quantities of the various amino acid subunits, and different specific sequences in which the amino acids are bonded. Hence, the number of proteins is practically endless. Every living cell produces many different proteins.

There are only 20 different L-amino acids found in proteins, and virtually every protein contains the same amino acids. Yet each different protein has a unique sequence of amino acids. This sequence of amino acids forms the **primary structure** of a protein. The primary structure influences the three-dimensional shape of a protein, its function, and how it will interact with other substances. Alterations in amino acid sequences can have profound metabolic effects. For example, a single incorrect amino acid in a blood protein can produce the deformed hemoglobin molecule characteristic of sickle cell anemia.

The primary structure of a protein is the sequence of amino acids in its polypeptide chain.

The primary structure of a polypeptide determines how the molecule can fold and twist. The po-

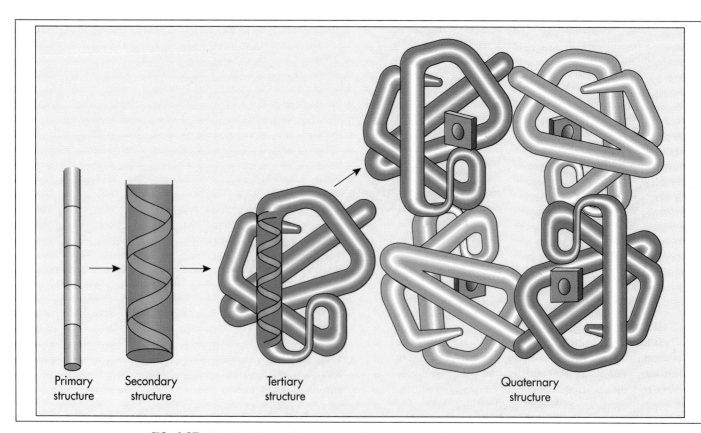

Primary structure   Secondary structure   Tertiary structure   Quaternary structure

FIG. 4-27   Proteins have primary, secondary, tertiary, and quaternary structures.

sitioning of the R groups of the amino acids is dictated by the primary structure of the peptide chain. The R group position forces the polypeptide to twist and fold in a specific way. The term *secondary structure* refers to the helical or extended protein structures that result when different amino acids are positioned close enough to allow hydrogen bonding to occur. Most often, hydrogen bonds form between every fourth amino acid. They hold the chain in a specific structure, called the α-helix, in which a helical coil is wound about its own axis. In other cases the chain is almost fully extended and hydrogen bonds form between different chains. These bonds hold many chains side by side in a sheetlike structure. In the β-sheet or pleated sheet, the chain of amino acids in the polypeptide folds back and forth on itself. R groups are thus exposed that can undergo extensive hydrogen bonding. The R groups of some amino acids tend to favor helical patterns; others tend to favor sheetlike patterns.

The secondary structure of a protein is stabilized by hydrogen bonding between the amino acids of the polypeptides.

Most helically coiled chains become further folded into some characteristic shape. The folding of polypeptide chains is called **tertiary structure.** Folding of a helical polypeptide accomplishes two things. The polypeptide becomes a unique shape that is compatible with a specific biological function, and the folding process converts the molecule into its most chemically stable form. The tertiary structure is based on interactions between various R groups of specific amino acids. Hydrophobic R groups associate with each other at the interior of folded chains. Hydrophilic R groups assume exterior positions where they can form weak bonds with other polar R groups or with water. The highly nonpolar regions of the polypeptide are brought close together by tertiary folding. They contribute stability to the folded structure by preventing the penetration of water into these regions. In addition, sulfhydryl groups (—SH) on two amino acid subunits can form a covalent, disulfide bond (—S—S—). This bond further stabilizes the folding of the protein molecule, contributing to the tertiary structure of the protein.

The actual shape of a protein is the result of the interactions of the polar covalent bonds of the peptide linkage and the combined interactions of the polar and nonpolar side chains of the individual amino acids.

Some proteins consist of more than one polypeptide chain. Their structures are even more complex. In some cases, the polypeptide chains are linked by disulfide bridges. For example, the antibodies that help protect the human body against disease are composed of four peptide chains that are linked by disulfide bonds. Such proteins have a quaternary structure. The **quaternary structure** describes the arrangement in space of multiple peptide chains when they make up the structure of a protein.

The three-dimensional shape formed by the secondary, tertiary, and quaternary structure of a protein is essential for the function of all proteins, including those that act as enzymes. The sequence of the amino acids and the three-dimensional shape they assume determines where a substrate can bind and the catalytic properties of the active site. Enzymes with different three-dimensional shapes at their active sites catalyze different metabolic reactions.

Primary, secondary, tertiary, and quaternary structures contribute to the three-dimensional shape of a protein that is essential for its proper functioning.

The ability of a protein to function as an enzyme and catalyze chemical reactions depends on its three-dimensional shape.

Denaturation of Proteins

If the three-dimensional structure of a protein is disrupted, the protein is **denatured** (FIG. 4-28). Denatu-

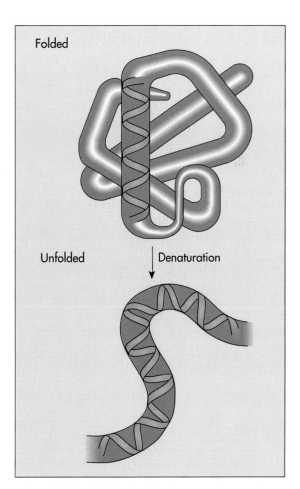

FIG. 4-28 Denaturation occurs when the three-dimensional shape of a protein is altered so that it no longer functions properly. Heat, salt, and various other factors can disrupt the tertiary structure of a protein, denaturing it.

ration occurs when there is a change in the three-dimensional structure of the protein that results in the loss of proper function of the protein. If the protein is an enzyme, denaturation results in the loss of catalytic capacity. The critical three-dimensional configuration of a protein can be disrupted without breaking the covalent peptide bonds of the polypeptide chain. Exposure to high temperatures, typically above 60° C, or certain chemical agents can disrupt the hydrogen bonds, sulfhydryl bonds, and hydrophobic interactions on which secondary, tertiary, and quaternary structures are based. This is one reason that high temperatures can be used to kill microorganisms. High salt or high H^+ concentrations can also denature proteins. They alter the weak bond interactions that maintain the structure of the protein molecule.

> Proteins lose their functional capabilities when they are denatured because denaturation disrupts critical three-dimensional structures.

NUCLEIC ACIDS

Nucleic acids are polymers composed of monomeric **nucleotides.** Each nucleotide has three different parts: a nitrogen-containing base, a pentose, and a phosphate group (FIG. 4-29). These three parts of the nucleotide are joined by covalent bonds. The nitrogen-containing base is either adenine, guanine, cytosine, thymine, or uracil. Adenine and guanine are double-ring structures. Collectively they are referred to as **purines.** Thymine, uracil, and cytosine are smaller, single-ring structures. They are called **pyrimidines.**

A single strand of a nucleic acid consists of nucleotide units strung into long chains. A phosphate bridge connects their sugars. Bases stick out to the side. The type of bond that links the nucleotides in a nucleic acid is called a **phosphodiester bond** (FIG. 4-30). The backbone of the nucleic acid molecule is always the same. It consists of alternating sugar and phosphate units. The nitrogenous bases attached to the sugar portion of the nucleotides vary in the chain. Hence, when chemists refer to a specific sequence of nucleotides in a nucleic acid, they really are describing the sequence of nitrogenous bases. The sequence of bases in a DNA or RNA molecule carries the genetic information necessary to produce the proteins required by the organism.

> Nucleic acids are polymers composed of nucleotide monomers held together by phosphodiester linkages.

DNA

All cells contain hereditary material called genes. Each gene is a segment of a **deoxyribonucleic acid (DNA)** molecule. Genes determine all hereditary traits. They control all the potential activities that take place within living cells. When a cell divides, its hereditary information is passed on to the next generation. This transfer of information is possible because of DNA's unique structure. DNA contains four

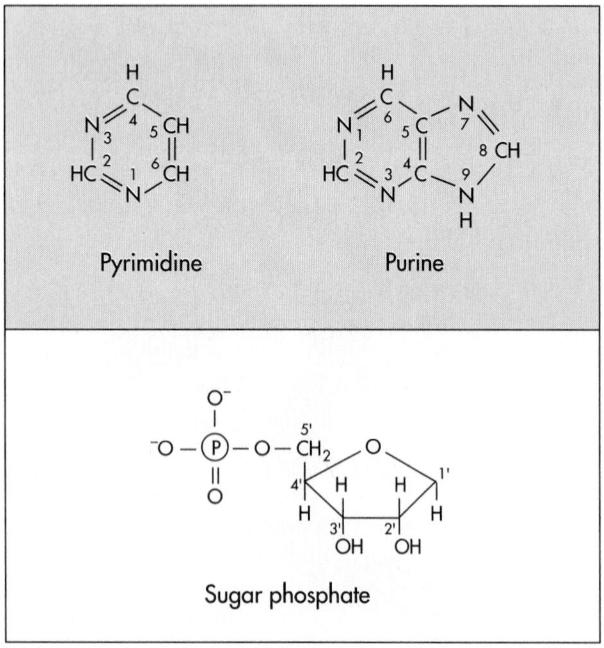

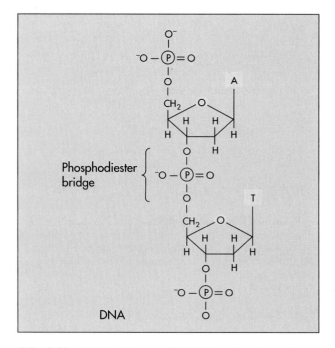

FIG. 4-29 The structural components of a nucleotide—the basic unit of the informational macromolecules of living systems. The carbon atoms of the sugar are assigned sequential number primes.

FIG. 4-30 Nucleotides are linked by phosphate diester bonds to form dimers and polymeric units (A, adenine; T, thymine). The linked nucleotides form long chains in DNA and RNA macromolecules.

nucleotides that each contain one of four nucleic acid bases: adenine (A), guanine (G), cytosine (C), and thymine (T). The ordering of the nucleotides determines the hereditary information contained in DNA. Nucleotides are held together by phosphodiester bonds between deoxyribose, the sugar found in the backbone of DNA. There are two strands of DNA held together by hydrogen bonds to form a coiled molecule called a **double helix.** Hydrogen bonds can form between adenine (A) and thymine (T) or between guanine (G) and cytosine (C).

The four nucleic acid bases in DNA are adenine, guanine, cytosine, and thymine.

RNA

Ribonucleic acid (RNA) differs from DNA in several respects. The five-carbon sugar in the RNA nucleotide is ribose. Ribose has one more oxygen atom than does the deoxyribose in DNA. One of RNA's bases is uracil (U) instead of thymine. Whereas DNA is normally double stranded, RNA is usually single stranded.

The four nucleic acid bases in RNA are adenine, guanine, cytosine and uracil.

Other Nucleotides

The nucleotide **adenosine triphosphate (ATP)** is the principal energy-carrying molecule of all cells. It stores the chemical energy released by some chemical reactions. It also provides the energy for energy-requiring reactions when needed. ATP consists of adenine, ribose, and three phosphate groups. ATP is called a high-energy molecule because it releases a large amount of usable energy upon hydrolysis of a phosphate group. The product is **adenosine diphosphate (ADP).** The production and utilization of ATP are essential to the bioenergetics of the cell. All of the metabolic pathways of microorganisms are involved in producing or consuming ATP.

Nucleotides also serve as coenzymes. A **coenzyme** is a temporary carrier of substances such as electrons. During metabolism, coenzymes transport hydrogen atoms and electrons. Nicotinamide adenine dinucleotide (NAD^+) and flavin adenine dinucleotide (FAD) are two of these coenzymes. These coenzymes gain electrons and hydrogen during some chemical reactions and donate them during other chemical reactions. Much of the metabolism of microorganisms involves chemical reactions requiring ATP and/or coenzymes. This accounts for the importance of these nucleotide molecules in the chemical reactions of living systems.

SUMMARY

Organization of Matter (pp. 92-98)
Chemical Elements (p. 92)
- An element is the fundamental unit of a chemical.
- Chemists use letter abbreviations (symbols) for the chemical elements (H, C, O, N, and so forth).

Structure of Atoms (pp. 92-94)
- All matter is made of atoms that are composed of a central nucleus, containing protons and neutrons, and electrons moving around the nucleus.
- The atoms of each element have a unique number of protons.
- The number of protons is the atomic number of an element.
- The sum of protons and neutrons is the atomic weight of an element.
- Electrons are arranged in electron shells of differing energy levels.
- Each shell can contain a fixed maximum number of electrons.
- An atom is most stable, and therefore least reactive, when its outermost shell is either completely full or completely empty.

Molecules and Chemical Bonds (pp. 94-98)
- Atoms combine to form molecules, which are fixed combinations of elements in which atoms are held together by chemical bonds formed when atoms transfer or share electrons.

- There are three principal types of chemical bonds: ionic, covalent, and hydrogen bonds.
- When atoms form ions by gaining or losing electrons, they can interact to form ionic bonds based on charge interactions.
- Covalent bonds are strong bonds formed when two atoms share electrons.
- If atoms share electrons equally they form a nonpolar covalent bond.
- If atoms share electrons unequally they form a polar covalent bond in which one of the atoms has a relative negative charge and the other atom has a relative positive charge.
- A hydrogen bond is the attraction between charged atoms of different molecules or distant parts of a large molecule that result from polar covalent bonds.
- Isomers are molecules that contain different arrangements of the same types and numbers of atoms, which gives the molecules different properties.
- Functional groups are specific combinations of atoms that act the same no matter to which molecules they are attached. They determine the characteristics, solubilities, and reactivity of the molecules.

Chemical Reactions (pp.99-105)
- In chemical reactions, molecules combine, break up, or transfer atoms or electrons.

- As a result of chemical reactions, new molecules are formed and energy can be released from the bonds of molecules to support the energy-requiring activities of living systems.

Chemical Equations (p. 99)
- Chemical equations are symbolic representations of what changes take place in a chemical reaction.

Equilibrium (p. 99)
- Chemical reactions are reversible and tend to move toward a state of equilibrium.

Energy and Chemical Reactions (p. 100)
- Chemical reactions involve energy changes.

Types of Chemical Reactions (pp. 100-105)
- Enzymes are proteins that act as catalysts.
- An enzyme lowers the activation energy, that is, the energy needed to start a reaction. By lowering the activation energies of chemical reactions, enzymes permit chemical reactions to occur rapidly at temperatures where living systems can maintain their structural integrity and organization.
- The activity of an enzyme depends on its three-dimensional shape, which allows it to bind with a substrate and catalyze a chemical reaction.
- Oxidation is the loss of electrons; reduction is the gain of electrons.
- Oxidation-reduction reactions are always coupled, so that as one atom or molecule is oxidized another atom or molecule is reduced.
- Some of the chemical reactions in cells are acid-base reactions.
- Acids donate hydrogen ions; bases accept hydrogen ions.
- The concentration of hydrogen ions in a solution is measured as pH.
- The pH of a solution = $-\log[H^+]$.
- Condensation reactions in which small molecules combine to form larger molecules are very important for the biosynthesis of the molecules that make up the structures of cells.
- Most large biological molecules are synthesized by linking many smaller subunit molecules; chains of subunits are connected by covalent bonds through condensation reactions.
- Hydrolysis reactions that break down molecules into smaller molecules are important for releasing energy and for forming the subunit molecules that are used for biosynthesis.

Molecules of Living Systems (pp. 106-115)
Water (pp. 106-107)
- The properties of water allow it to act as a solvent for polar molecules and to modulate temperature; these properties are critical for supporting life processes.

Carbohydrates (pp. 107-108)
- Carbohydrates include sugars, and large polysaccharides, such as starch and cellulose.
- Sugars (monosaccharides and disaccharides) are used for storage of energy and for the construction of other molecules.
- Starch and glycogen are polysaccharides that serve for long-term energy storage in eukaryotic cells.
- Cellulose and related polysaccharides form the cell walls of bacteria, fungi, and other organisms.

Lipids (pp. 108-110)
- Lipids are water-insoluble molecules of diverse chemical structure, and include oils, fats, phospholipids, and steroids.
- Lipids are used for energy storage and as the principal component of cell membranes (phospholipids).

Proteins (pp. 110-114)
- Proteins are chains of amino acids linked together by peptide bonds.
- The function of each protein is determined by the sequence of amino acids in the chain.
- The three-dimensional shapes of proteins, determined by their primary structure (covalent bonds) and by higher order interactions (weak hydrogen bonds and hydrophobic interactions), are especially important in establishing the specificity of roles played by enzymes.
- When the three-dimensional structure of a protein is disrupted, it cannot function.

Nucleic Acids (pp. 114-115)
- Nucleic acids are chains of nucleotides.
- Each nucleotide is composed of a phosphate group, a sugar group, and a nitrogen-containing base.
- The two types of nucleic acids are deoxyribonucleic acid (DNA) and ribonucleic acid (RNA).
- Other nucleotides include energy carrier molecules (ATP) and coenzymes.

CHAPTER REVIEW

REVIEW QUESTIONS

1. Describe the four levels of protein structure.
2. Describe the chemical composition of carbohydrates, lipids, proteins, and nucleic acids.
3. Why are most enzymes active only over a particular range of temperatures? Why are most enzymes not active at temperatures above 100° C?
4. What are the most common elements in living systems? What are their chemical symbols?
5. Compare and contrast covalent, ionic, and hydrogen bonding.
6. What are the differences between atoms, ions, and isotopes.

7. Describe how the chemical nature of water affects its properties relative to living systems. Why is water essential for life? What properties of water contribute to functions in living systems?
8. What is the difference between a monosaccharide and a polysaccharide? A peptide and a polypeptide? A nucleotide and nucleic acid?
9. What are the main properties of enzymes?
10. What are the differences between enzymes and coenzymes? Why are both needed for living systems?
11. What role does ATP have in living systems?
12. What role does DNA have in living systems?

CRITICAL THINKING QUESTIONS

1. Describe the different roles that are played by covalent and noncovalent bonds in biological systems. Why do living cells require molecules with both covalent and noncovalent bonds?
2. All living cells are based on organic molecules that contain carbon atoms. Why is this so? Why couldn't hydrogen, which can establish only a single covalent bond with another atom, form the basis for organic molecules? Could life have evolved based on silicon, which like carbon can form four covalent bonds with other atoms?
3. In oxidation-reduction reactions some substance becomes oxidized and another becomes reduced. If one substance causes another one to become reduced, the first substance is called a reducing agent. Will a reducing agent be oxidized or reduced in the process? What do you think an oxidizing agent does? Will an oxidizing agent be oxidized or reduced?

READINGS

Alberts B et al: 1995. *Molecular Biology of the Cell, ed.3*, New York, Garland Publishing.
  Clear descriptions of the biologically important molecules with good illustrations.
Breed A et al: 1982. *Through the Molecular Maze*, Los Altos, California; Kaufmann.
  Clear, concise guide to basic chemical concepts.
Bretscher MS: 1985. The molecules of the cell membrane, *Scientific American* 253(4):100-108
  Well-illustrated discussion of the structure and function of the cell membrane.
Doolittle RF: 1985. Proteins, *Scientific American* 243(4):88-99.
  Describes the roles proteins play in cell functions.
Fruton JS: 1972. *Molecules and Life: Historical Essays on the Interplay of Chemistry and Biology*, New York, Wiley-Interscience.
  Covers the period 1800 to 1950, from the work of Lavoisier to the discovery of the DNA double helix, with emphasis on ferments and enzymes, the nature of proteins, nucleic acids, intracellular respiration, and pathways of biochemical change, demonstrating the interplay between concepts and experiments.
Karplus M and JA McCammon: 1986. The dynamics of proteins, *Scientific American* 254(4):42-51.
  Computer simulations describe the incessant motion of proteins and the role of that motion in protein function.

Lehninger AL, DL Nelson, MM Cox: 1993. *Principles of Biochemistry*, ed. 2, New York, Worth Publishers.
  An advanced biochemistry text that covers all aspects of the relationship of chemistry to living systems.
Richards FM: 1991. The protein folding problem, *Scientific American* 264(1):54-57, 60-63.
  The chemistry of proteins and the tertiary structure is discussed in this technical article.
Sackheim GI: 1985. *Chemistry for the Health Sciences*, New York, Macmillan.
  Easy-to-read, basic information.
Scientific American: 1985. The molecules of life, 253(4):a single topic issue.
  A comprehensive view of what is known about biologically important molecules, with individual articles on DNA, RNA, proteins, etc.
Sharon N: 1980. Carbohydrates, Scientific American 243(5):90-116.
  Describes the structures of carbohydrates and the diverse roles they assume in organisms.
Weinberg R: 1985. The molecules of life, *Scientific American* 253(4):48-57.
  Survey of the new techniques and discoveries of molecular biology.
Zubay G: 1993. *Biochemistry*, ed. 3, Dubuque, Iowa; William C. Brown.
  An advanced biochemistry text that covers all aspects of the relationship of chemistry to living systems.

UNIT TWO

Cellular and

Molecular

Microbiology

CHAPTER 5

Cell Structure

PREVIEW TO CHAPTER 5

In this chapter we will:
• Examine the structures of the cell and see how the chemical composition of each structure relates to the life processes it carries out.
• Discuss the relationships between form and function and see that even minor, seemingly trivial, differences in chemical structure can have a profound influence on the ability of a cell to survive.
• Compare prokaryotic and eukaryotic cells and examine how each meets the essential requirements for sustaining life.
• Explore the practical implications of the structural differences between eukaryotic and prokaryotic cells, such as why certain antibiotics can be used to kill the bacteria causing infections without also killing human cells.
• Learn the following key terms and names:

bacterial chromosome
capsule
cell wall
chemiosmosis
chemosensors
chloroplasts
chromosomes
cyst
cytosis
diffusion
endoplasmic reticulum
endospores
endotoxin
exocytosis
facilitated diffusion
flagella
glycocalyx
Golgi apparatus
Gram-negative bacterial cell wall
Gram-positive bacterial cell wall
group translocation
lipopolysaccharide (LPS)
lysosomes

magnetotaxis
nucleus
osmosis
outer membrane
passive diffusion
peptidoglycan
periplasmic space
peritrichous flagella
permeases
phagocytosis
phagolysosome
phagosome
plasma membrane
plasmids
polar flagella
porins
protonmotive force
protoplast
ribosomes
slime layer
spheroplast
spore
thylakoids
trophozoite
vegetative cells

All living organisms are composed of cells, which are the fundamental units of all living systems. A cell is a self-contained system capable of independently carrying out metabolism. Cells also are the units of reproduction for living organisms. They house the hereditary information that is passed from one generation to the next. Cells come only from pre-existing cells.

All cells have certain common functional and structural properties:

1. Each cell has a plasma membrane that surrounds it. The plasma membrane forms a boundary layer between the living cell and its surroundings. The plasma membrane regulates the passage of materials into and out of the cell.

2. Each cell contains a fluid substance, called the cytoplasm. Chemical reactions take place in the cytoplasm. These reactions transform the energy and material needed for cell growth and reproduction. The cytoplasm consists of a solution, called the cytosol, and various particulate structures.

3. Each cell contains a copy of the hereditary information stored in molecules of DNA. The genetic molecules of DNA direct the activities of the cell and pass hereditary information to new cells formed as a result of cellular reproduction.

4. Each cell has thousands of small particles, called ribosomes, where proteins are made. The actual transfer of genetic information from DNA involves the formation of another informational molecule, RNA. Several different types of RNA are involved in the transfer process: messenger RNA (mRNA), transfer RNA (tRNA), and ribosomal RNA (rRNA). Messenger RNA carries the genetic information to the ribosomes where that information is used to direct the synthesis of proteins. There are structural proteins, regulatory proteins, and enzymatic proteins. Enzymes are proteins that are the action molecules catalyzing the metabolic functions of the cell.

5. Each cell utilizes energy from ATP, the "universal energy currency" of living cells. Cells carry out metabolism through which they generate ATP and cell constituents for growth and reproduction.

Every living cell is surrounded by a plasma membrane, contains a fluid called cytoplasm that has a solution portion called cytosol, contains hereditary information in molecules of DNA, processes genetic information, employs RNA intermediates to form proteins at ribosomes, and uses ATP as the cellular form of energy.

PLASMA MEMBRANE: MOVEMENT OF MATERIALS INTO AND OUT OF CELLS

The **plasma membrane** that surrounds the cell acts as a semipermeable barrier, regulating the flow of materials into and out of the cell. Its ability selectively to control which materials enter and which leave the cell is essential, since exchanges with the external surroundings must be selective and restricted to maintain life functions. Cells grow and reproduce by acquiring energy and materials from their surroundings and they also must discharge wastes into those same surroundings. The plasma membrane regulates the flow of materials into and out of the cell. It also allows the maintenance of the highly organized dynamic state that is characteristic of living systems.

The plasma membrane regulates the flow of materials into and out of the cell.

The plasma membrane, which is almost always a lipid bilayer with various proteins distributed within it, has a limited capacity to handle the essential exchanges of materials with the cell's surroundings. This is because of the plasma membrane's limited surface area and, hence, limited passageways through which materials may pass. Most cells are very small—less than 100 μm in diameter. The size of a cell is limited by the relationship between the cell's surface area and its volume. If a cell grows too large in volume, the surface area of the plasma membrane becomes small compared to the volume. This is because volume of a sphere increases much more rapidly than surface area as the diameter increases. If the diameter increases by a factor of 10, the volume increases by a factor of 1,000, but the surface area increases only by a factor of 100.

This does not provide sufficient surface area for substances to move across. Then rates of exchange between a cell and its surroundings are not fast enough to meet the needs of the cell. On the other hand, cells must be large enough to house their genetic information and proteins that are required for metabolism and reproduction. The smallest cells are about 0.1 μm in diameter. Recently, a new bacterium has been discovered that is 0.5 mm (50 μm) long. However, most bacterial cells have diameters of 0.2 to 2 μm.

Enclosed within the plasma membrane is the cytoplasm. Cytoplasm is a semifluid substance containing various embedded cell structures and solutions of chemicals. The fluid portion of the cytoplasm is called the **cytosol.** It contains various dissolved substances such as amino acids and sugars. The concentrations of chemicals within the cytosol are very different from those of the outside environment. The cytosol receives raw materials from the external environment that pass through the plasma membrane. Enzymatic reactions within the cytosol then degrade them to yield usable energy and new substances that are used for the synthesis of new cellular materials. If the plasma membrane breaks, the cytosol leaks out and the cell dies.

> **Life depends on the integrity of the plasma membrane and its ability to act as a semipermeable barrier, regulating the flow of materials into and out of the cell.**

STRUCTURE OF THE PLASMA MEMBRANE

The plasma membranes of most prokaryotic and eukaryotic cells are composed of phospholipid and protein. The chemical nature of a phospholipid explains how it contributes to the ability of the plasma membrane to regulate the flow of materials into and out of the cell. A phospholipid molecule has two parts. It has a phosphate portion, which is hydrophilic and attracted to water. It also has a fatty acid (lipid) portion, which is hydrophobic and repelled by water. When phospholipids are surrounded by water, hydrophobic interactions cause their fatty acid tails to move away from water and to cluster. When a thin cross section of a cell is viewed with a transmission electron microscope, the plasma membrane appears like a railroad track completely encircling the cell (FIG.

5-1). The dark rail-like portions of the membrane correspond to the electron-dense hydrophilic phosphate portions. The clear space between the "rails" corresponds to the lipid portions of the phospholipid molecules.

The basic structure of the plasma membranes of eubacterial and eukaryotic cells is a lipid bilayer. In this structure, the hydrophobic fatty acids of the phospholipid are sandwiched in the middle of the bilayer. The hydrophilic phosphate portions of one layer of phospholipid molecules interact with the water outside the cell and those of the other layer interact with the cytosol within the cell. The individual phospholipid molecules within the bilipid layer move about. They slide sideways, spin in place, and flex their fatty acid tails. These movements prevent the lipids from packing tightly together. They also impart fluidity to the membrane. This is important because it allows some molecules to pass through the plasma membrane without destroying its integrity.

> **The plasma membranes of most cells contain phospholipids that contribute to their ability to act as a semipermeable barrier.**

Whether or not molecules pass through the plasma membrane is determined by their size and polarity. Nonpolar molecules, which dissolve easily in lipids, pass through the phospholipid portion of the plasma membrane more readily than do polar substances, which do not dissolve in lipids. Relatively small uncharged molecules, such as oxygen (O_2), nitrogen (N_2), and hydrogen (H_2), usually pass through the plasma membrane easily. Ions and other charged molecules pass through the membrane only

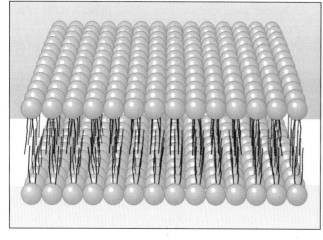
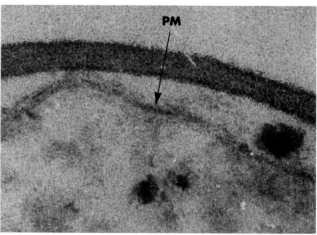

A | | B

FIG. 5-1 **A,** The typical plasma membrane structure of eubacterial and eukaryotic cells is a lipid bilayer, as illustrated here, showing the orientations of the hydrophilic (*tan spheres*) and hydrophobic (*black*) ends of phospholipids that make up this structure. The hydrophilic portions (phosphate groups) occur near the water inside and outside the cell. The hydrophobic portions (formed from fatty acids) are sequestered in the interior of the membrane. **B,** Colorized electron micrograph of the plasma membrane (*PM*) of the bacterium *Bacillus subtilis* reveals the characteristic railroad track appearance of this lipid bilayer.

very slowly. Their electrostatic charges prevent them from interacting with the lipids in the middle of the bilayer. Large molecules with high molecular weights, such as proteins, cannot pass through the phospholipid portion of the plasma membrane.

However, proteins that compose part of the plasma membrane provide passageways through which polar and relatively large molecules can pass. Proteins are distributed in a patchlike or mosaic pattern in the plasma membrane. Some of the proteins span the bilayer and others are partially embedded in it. The partially embedded proteins are exposed only to the internal cytosol or the external surrounding environment. Since some of these proteins move in the plane of the membrane, scientists have proposed the fluid mosaic model for the plasma membrane. This model explains most of the functional and structural aspects of the plasma membrane (FIG. 5-2). The proteins that span the membrane create channels that connect the outside of the cell with the inside. The passage of some, but not all, materials across the plasma membrane is thus allowed.

Proteins form channels and serve as carriers to move materials across the plasma membrane.

Several antimicrobial agents work because they disrupt the structure of the plasma membrane, causing the death of the cell. These include the polymyxin antibiotics that destroy the phospholipid bilayer of the plasma membrane. Leakage of intracellular contents then occurs, followed by cell death. Various alcohols also destroy plasma membranes and are commonly used to kill microorganisms, such as when alcohol is applied before puncturing the skin with a hypodermic syringe.

COMPARISON OF ARCHAEBACTERIAL AND EUBACTERIAL PLASMA MEMBRANES

The plasma membranes of archaebacterial cells have a lipid composition that is fundamentally different from all other organisms (FIG. 5-3). The chemical composition of the archaebacterial plasma membrane indicates that they evolved as a separate Kingdom. It also shows that archaebacterial cells are only distantly related to eubacterial and eukaryotic cells.

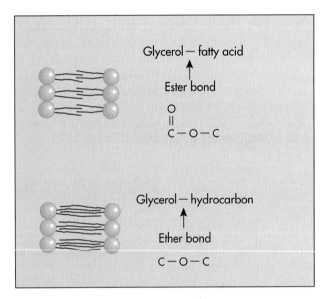

FIG. 5-3 Unlike the lipids that make up the plasma membranes of eubacterial cells, which are glycerol diesters (they have ester bonds formed by the reaction of glycerol and a fatty acid), major lipids of archaebacteria are glycerol diethers, diglycerol tetraethers, and tetrapentacyclic diglycerol tetraethers (they have ether bonds formed by the reaction of a hydrocarbon with glycerol). The glycerol diethers, like the phospholipids (glycerol diesters) of eubacterial and eukaryotic cells, form a lipid bilayer. Diglycerol tetraethers form a single layer with the two glycerols on the outside.

In particular, the presence of ether linkages in the lipids of the archaebacterial plasma membranes is unique in the biological world. Despite their chemical uniqueness, archaebacterial plasma membranes are functionally similar to those of eubacterial and eukaryotic cells. They also serve as selectively permeable barriers. The unique chemical structure permits this membrane to function in extreme environments, such as the very hot, very salty, and very acidic environments where some archaebacteria live. Under these conditions, phospholipids in the plasma membranes of other organisms would not be able to maintain the integrity of the permeability barrier.

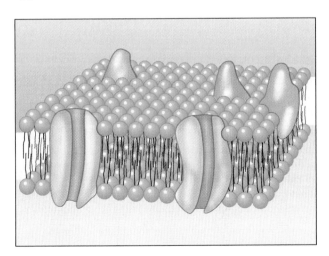

FIG. 5-2 The fluid mosaic model of membrane structure accounts for the facts that proteins (*blue*), as well as lipids (*beige* and *black*), comprise an integral part of membranes and that the structure is dynamic as opposed to static. Some proteins extend through the membrane (integral proteins) and others are associated with one side or the other (peripheral proteins) and do not completely transverse the membrane.

COMPARISON OF EUBACTERIAL AND EUKARYOTIC PLASMA MEMBRANES

Eukaryotic cells have sterols in their membranes. Prokaryotic cells generally do not. Sterols make the eukaryotic membrane stronger. They permit eukaryotes to survive without cell walls in environments where bacteria cannot survive unless they have other protective structures, such as a bacterial cell wall. Only a few bacteria, such as *Mycoplasma,* have sterols in their plasma membranes. These bacteria are restricted to certain environments. Sterols are an integral part of the eukaryotic plasma membranes of disease-causing eukaryotic microorganisms such as pathogenic fungi. They are often the targets of antimicrobial drugs. Amphotericin B, for example, is used to treat some fungal infections. It reacts with the specific sterols in fungal plasma membranes, such as ergosterol, disrupting membrane function and causing death of the infecting fungi. Amphotericin B has a lesser affinity for the sterols in human cells, such as cholesterol, which is why it can be used therapeutically in individuals with fungal infections.

The plasma membranes of eukaryotic cells contain sterols, whereas those of most prokaryotic cells do not.

TRANSPORT ACROSS THE PLASMA MEMBRANE

The structure of the plasma membrane permits the selective movement of substances into and out of living cells by several different mechanisms (Table 5-1). Different cells have different mechanisms for transporting specific substances into and out of the cell. The specific transport capabilities of a cell determine what substances it can use for growth and reproduction.

Diffusion

The concentrations of many substances differ inside and outside a cell so that there is a *concentration gradient,* that is, a region of high relative concentration on one side of the membrane and a region of low relative concentration on the other side. Unless prevented from doing so, substances will move from regions where they are in relatively high concentration to regions where they are in relatively low concentration. This process of moving from high to low concentration is called **diffusion.** To understand how diffusion works, let us consider a simple example. If a drop of red dye is added to a pool of water, it initially can be seen as a red spot where the dye is in high concentration. Soon, however, the dye begins to disperse, spreading out by diffusion from the initial region of high concentration, forming a concentration gradient from the region of highest to the region of lowest concentration of the dye. The diffusion of the dye continues until the dye is equally distributed throughout the water. If unrestricted, all chemicals will naturally move toward a state of equal concentration.

Diffusion is the movement of molecules down a concentration gradient, moving from a region of relatively high concentration to a region of relatively low concentration.

TABLE 5-1	
Comparison of Transport Mechanisms Across the Plasma Membrane	
TRANSPORT PROCESS	**DESCRIPTION**
Passive diffusion	Movement of substances across the plasma membrane due to a concentration gradient; movement is from the region of high to low relative concentration at a rate proportional to the concentration difference
Facilitated diffusion	Movement of substances across the plasma membrane due to a concentration gradient; movement is from the region of high to low relative concentration at a rate that is higher than expected based on the concentration difference; movement of substance is aided by permeases in the membrane, which are protein carriers that help move substances across the membrane; the action of permeases accounts for the relatively high rate of movement
Active transport	An energy-requiring transport mechanism mediated by permeases that allows substances to move against a concentration gradient, that is, from a region of relatively low to a region of relatively high concentration; ATP often supplies the energy for active transport
Group translocation	Occurs only in prokaryotic cells; substances are chemically modified as they move across the membrane; energy-requiring process with energy supplied by phosphoenolpyruvate
Cytosis	Occurs only in eukaryotic cells; the plasma membrane wraps around a substance, enclosing it and releasing it when the plasma membrane subsequently unwraps it

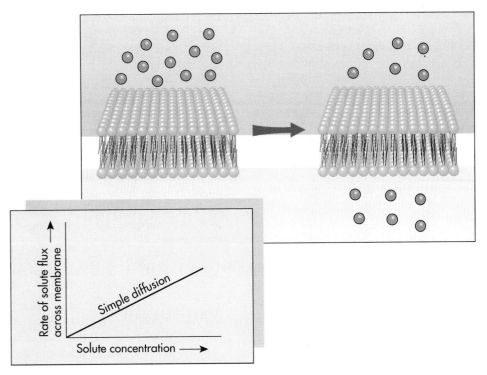

FIG. 5-4 Diffusion across a membrane occurs when substances pass through the membrane and when there is a favorable concentration gradient; this type of transport represents the downhill flow of a substance along a concentration gradient from high to low.

Some substances move across the plasma membrane and enter or exit the cell by a process called **passive diffusion,** so named because it does not require an input of energy or the use of special carriers in the membrane (FIG. 5-4). Substances will tend to move by passive diffusion from a region of higher concentration to one of lower concentration at a rate that is determined by the concentration gradient. The greater the concentration difference across the plasma membrane, the more rapid the rate of diffusion. Many small molecules (including water) can enter and leave cells by passive diffusion. These substances are able to diffuse into and out of cells because there is a concentration gradient across the membrane. The plasma membrane does not prevent their movement. The structure of the plasma membrane, however, restricts movement of large and charged molecules, including ions such as H^+ and Na^+.

The rate of passive diffusion into and out of a cell depends on the relative concentrations of a substance on opposing sides of the membrane.

In some cases, substances move more rapidly across the plasma membrane than can be explained by the concentration gradient alone because of a process called **facilitated diffusion** that increases the rate of passage across the membrane. Like passive diffusion, facilitated diffusion involves the movement of chemicals down a concentration gradient. Faster movement occurs because of proteins that span the plasma membrane and selectively increase the permeability of the membrane for specific substances. Such proteins are called **permeases.** Permeases act as carriers, making it easier for a substance to move through the membrane. Permeases are highly specific and will carry only selected molecules across the plasma membrane. A substance is not chemically modified by a permease as it moves across the membrane. The permease simply picks up the substance at one side of the membrane and pulls it through the membrane. The substance is released unchanged at the other side.

Facilitated diffusion involves permeases that carry substances across the plasma membrane and thereby increase the rate of diffusion when concentration gradients are small.

Osmosis

While water can move across the plasma membrane through channels without restriction, the movement of many substances dissolved in water is restricted. Substances dissolved in water are called *solutes.* When a solute cannot move across the plasma membrane in response to a concentration gradient, water will move across the membrane. This occurs because the presence of solute changes the concentration of water. If there is a high concentration of solute on one side of the membrane the water concentration is correspondingly lower. In this case, water will move across the membrane from the region of high water concentration to the region of low water concentration. Such movement of water across a semipermeable membrane is called **osmosis.**

Osmosis is the movement of water across a semipermeable membrane due to diffusion.

In a medium where the solute concentration is higher outside the cell than inside the cell, water will flow out of the cell. The cell will shrink. This process is called **plasmolysis** (FIG. 5-5). The reverse will occur if the cell is in a medium where the solute concentration is lower outside the membrane than inside the cell. In this latter case, water will flow into the cell. The cell will expand. This is often characteristic

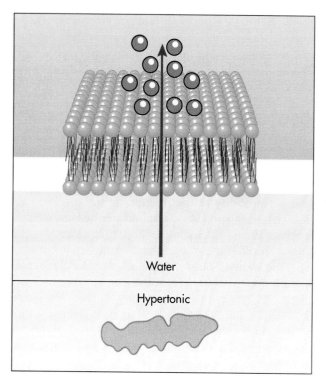

Water

Hypertonic

FIG. 5-5 Cells respond to osmotic pressure because water can move across the plasma membrane by osmosis and cause the cell to expand or shrink. Under isotonic conditions (equal concentrations of solute on both sides of the plasma membrane), cell shape is maintained. Under hypertonic conditions (higher solute concentration outside cell), the cell loses water and shrivels. Under hypotonic conditions (lower solute concentration outside cell), water moves into the cell, pressing the cell to expand; because the cell has very limited ability to increase volume without new synthesis of cell wall and plasma membrane components, osmotic pressure increases and the cell lyses due to osmotic shock.

of microorganisms because the concentrations of chemicals within the cell normally are in considerably higher concentrations than in the surrounding solution.

If the flow of water into the cell is unrestricted, the plasma membrane will burst. Pressure is exerted on the plasma membrane by the water entering the cell. This pressure is called **osmotic pressure.** Rupture of the cell is called *cell lysis.* When this rupture is the result of excessive osmotic pressure, the cell is said to have died due to **osmotic shock.** To survive, microorganisms have developed various strategies, such as having a rigid cell wall surrounding the plasma membrane, for preventing cell lysis due to excessive osmotic pressure.

Cell lysis will occur if the flow of water into a cell due to osmosis is not controlled.

Active Transport

Diffusion only allows the movement of chemicals down a concentration gradient. Substances can be moved across the plasma membrane against a concentration gradient by a process called **active transport.** To reverse the direction of movement of diffusion, active transport requires an energy input. In active transport, specific membrane proteins (permeases) act as carriers. Energy from ATP or another source is used to move substances across the plasma membrane against a concentration gradient. Many substances that do not freely diffuse through the plasma membrane, such as most sugars, amino acids, ions, and the like, can pass through the membrane by active transport. They can be concentrated to over 1,000 times that of the solution surrounding the cell.

The active transport process can be likened to a pump that uses energy to move water uphill, that is, against an energy gradient. As with any pump, active transport requires that energy be expended and work be performed. In a eukaryotic cell, ATP generated by the metabolism of the cell is used to drive the uptake of substances by active transport. In bacteria, the energy for driving the pump usually comes from the potential energy of a concentration gradient. This gradient is formed by the expulsion of hydrogen ions (protons) from the cell across the membrane. This energy is called the *protonmotive force.* The potential energy of the concentration gradient of protons is used to drive the uptake of nutrients by active transport into bacterial cells.

In active transport, cells use metabolic energy and protein carriers in the plasma membrane to move substances across the plasma membrane against a concentration gradient.

Group Translocation

Group translocation is a form of energy-requiring transport that uses groups of enzymes and alters the chemical as it moves across the plasma membrane. This process occurs only in bacteria. The fact that the transported substance is chemically changed distinguishes group translocation from active transport. Additionally, ATP is not the source of energy for group transport. Phosphoenolpyruvate (PEP) is used by *Escherichia coli* to move glucose and other substances into the cell. The glucose is changed to glucose phosphate in this group transport process.

> In group translocation, bacteria use multiple enzymes and phosphoenolpyruvate as the source of energy to pump substances across the plasma membrane and simultaneously alter them.

Cytosis

Cytosis allows substances to move into or out of cells without passing through the hydrophobic internal portion of the plasma membrane. Only eukaryotic cells are capable of cytosis. In cytosis the plasma membrane wraps around a substance. The plasma membrane engulfs that substance to form a membrane-bound sphere called a *vesicle*. A vesicle now can open on the other side of the membrane and release that substance. Alternatively, it can separate from the plasma membrane and remain as an intact vesicle.

The prefixes *endo-* (into) and *exo-* (out of) indicate whether the substance is entering or leaving a cell (FIG. 5-6). **Endocytosis** refers to the movement of materials into the cell. **Exocytosis** denotes movement out of the cell. Both endo- and exo-cytosis are important in moving some substances in bulk into and out of cells. These substances include large and polar molecules that cannot move across the plasma membrane by other processes. In some instances one cell can even engulf and ingest another cell by cytosis. The process of engulfing and ingesting cells or other solid materials is called **phagocytosis** (FIG. 5-7). Many protozoa obtain their nutrients by ingesting bacteria by phagocytosis. The ability of certain white blood cells to engulf microbial cells by phagocytosis, and then later to digest and kill them, is a very important part of our body's defense against infections.

> Eukaryotic cells can move substances into and out of the cell by cytosis.

> Phagocytosis is the engulfment by one cell of another cell.

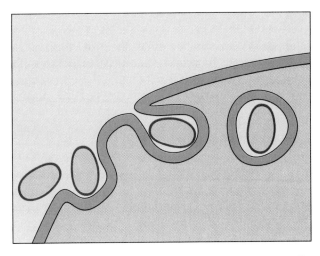

FIG. 5-6   In cytosis, which only occurs in eukaryotic cells, a substance is transported into or out of the cell without actually passing through the membrane. This is important for the movement of large substances and large quantities of substances into and out of the cell.

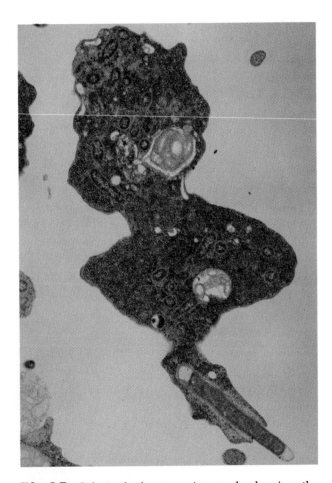

FIG. 5-7   Colorized electron micrograph showing the phagocytic capture of a rod-shaped bacterium *(purple)* by the soil protozoan *Vahlkampfia*. Many protozoa graze on bacteria as their food sources.

The plasma membrane defines the outer limit of the cell. Many cells have external cellular structures that are critical to their survival. Some of these structures, such as the bacterial cell wall, protect the plasma membrane from physical disruption. Others, such as flagella, move the cell in search of food or to escape hostile environments.

CELL WALL

Because the plasma membrane is a selective barrier that maintains cell viability, it is necessary to protect this structure. Prokaryotic plasma membranes are relatively fluid. Bacterial cells tend to expand under the force of osmotic pressure until they burst. The plasma membrane must be protected by an external structure against excessive expansion. The cell wall provides such protection (FIG. 5-8). Only bacterial cells growing in regions of high external solute concentration, such as in the fluids that occur in the human lungs or the high sugar concentrations of some laboratory solutions, can survive with a defective or missing cell wall.

Protection of the bacterial cell against osmotic shock is the main function of the bacterial cell wall.

The cell wall also establishes the shape of a bacterial cell. Bacteria occur as spheres called *cocci,* cylinders called *rods* or *bacilli,* and helical shapes called *spirilla,* as well as other diverse forms (see FIG. 1-4). The *morphology* (shape) of a bacterial cell is an important characteristic used in its identification. Each bacterial genus has a characteristic shape. For example, *Staphylococcus, Streptococcus,* and *Neisseria* are all cocci, whereas *Escherichia, Enterobacter, Bacillus,* and

Clostridium are all rods, and *Rhodospirillum* and *Spirillum* are all helical cells (spirilla).

The shape of a bacterial cell affects its ability to survive and grow in different environments. Cocci are round and can usually survive more severe drying than can rod- or spiral-shaped bacterial cells. Hence, many of the bacteria living on the surface of human skin are cocci. For example, staphylococci can withstand the dry conditions that often exist on the skin. Rod-shaped bacteria, on the other hand, have more surface exposed per unit volume than cocci. Therefore they can take up nutrients from dilute solutions more readily than cocci. Most bacteria that live in lakes, rivers, and oceans are rod shaped.

The bacterial cell wall protects the cell against osmotic shock and gives the cell its shape.

The cell walls of most eubacteria consist in large part of peptidoglycan. Peptidoglycan is a polymer that is part peptide and part carbohydrate. Its chemical structure gives it the physical strength to withstand osmotic pressure and prevent lysis. Peptidoglycan is a biochemically unique substance that is found only in eubacterial cell walls. Many archaebacterial and eukaryotic cells also have cell walls. However, the chemical composition of their walls is different. They do not contain peptidoglycan. Archaebacterial cells have a substance called pseudopeptidoglycan, plant cells have cell walls made of cellulose, and many fungal cell walls contain chitin.

Chemical Composition of the Bacterial Cell Wall

The peptidoglycan polymer of the eubacterial cell wall has a peptide portion and a glycan, or sugar,

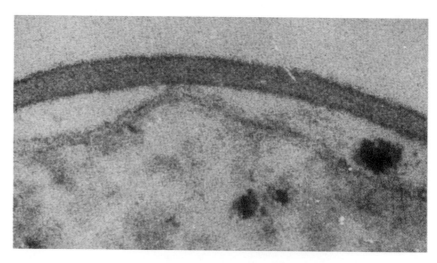

FIG. 5-8 Colorized electron micrograph of the cell wall of the bacterium *Bacillus subtilis.* This cell wall *(purple)* completely surrounds and protects the plasma membrane *(tan).*

portion (FIG. 5-9). The peptide portion is composed of amino acids connected by peptide linkages. Some of these amino acids are found only in the peptidoglycan of the bacterial cell wall. For example, D- and L-amino acids occur in peptidoglycan whereas only L isomers of amino acids occur in protein. The glycan portion that forms the backbone of the molecule is a complex polysaccharide composed of two amino sugar molecule subunits, N-acetylglucosamine (NAG) and N-acetylmuramic acid (NAM). The sugar subunits are linked so that they form an al-

ternating and repeating unit—NAG-NAM-NAG-NAM and so forth.

Short peptide chains are covalently linked to some of the N-acetylmuramic acid groups. These peptides hang like tails from the glycan backbone of this structure. They are linked together by additional short peptides. This reduces the flexibility of the entire peptidoglycan molecule and provides the desired rigidity. They also form bridges between adjacent peptidoglycan polymers. The cell wall is a multilayered sheet, linked together into one functional unit.

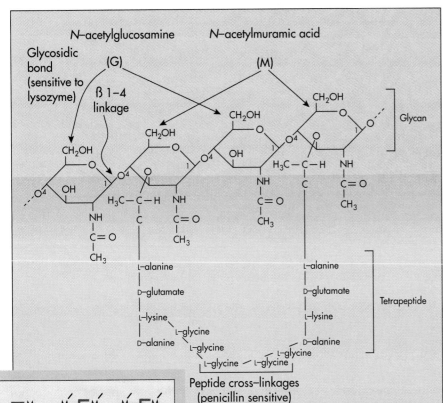

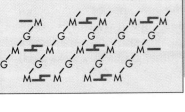

FIG. 5-9 Peptidoglycan is the backbone chemical of the bacterial cell wall; it is composed of repeating alternating units of N-acetylglucosamine (NAG) and N-acetylmuramic acid (NAM) and has cross-linked, short peptide chains, some of which have unusual amino acids such as D-alanine. The cross-linkages provide the needed structural support of the wall and are characteristic for specific bacterial genera.

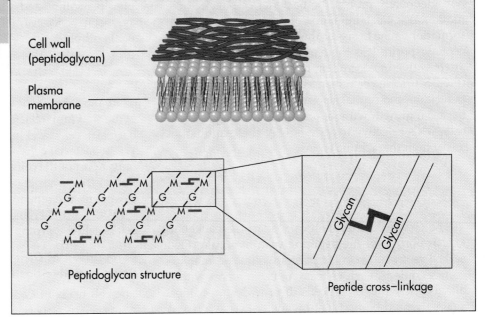

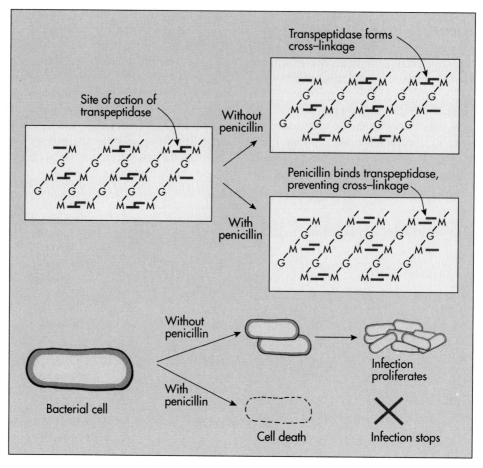

FIG. 5-10 The mode of action of penicillin involves inhibition of the formation of the normal cross-linkages in the peptidoglycan layer of the bacterial cell wall. Penicillin forms an inactive complex with transpeptidase, a key enzyme in cell-wall sythesis, so that the peptide cross-linkages do not form.

Enzymes are involved in cross-linkage of peptides during cell wall formation. Bacterial cell walls are weakened by substances that inhibit these enzymes. When this happens, the cell wall is defective. It cannot adequately protect the bacterial cell against osmotic shock. Such inhibition of peptide cross-linkage formation is the basis for the action of penicillin and cephalosporin. It is the reason that these antibiotics are effective in controlling many bacterial infections (FIG. 5-10). Since human cells lack peptidoglycan, penicillins and cephalosporins will selectively inhibit bacterial growth without adversely affecting human cells. This selective toxicity is why these antibiotics are of high therapeutic value.

Penicillins and cephalosporins work by blocking formation of peptide cross-linkages in the bacterial cell wall.

The enzyme lysozyme breaks apart the backbone glycan portion of the peptidoglycan molecule (FIG. 5-11). Lysozyme occurs as part of various normal human secretions, such as tears and saliva and within phagocytic white blood cells. It provides protection against would-be bacterial invaders. Lysozyme and cell wall–inhibiting antibiotics prevent the growth and/or kill many species of bacteria. However, these agents will not attack the few species that do not have peptidoglycan-containing cell walls. For example, penicillin and cephalosporin are ineffective against *Mycoplasma*, a bacterial genus that lacks a cell wall entirely.

The enzyme lysozyme will destroy the peptidoglycan of the bacterial cell wall and kill bacteria.

In the laboratory, scientists sometimes use lysozyme to remove bacterial cell walls, so they can lyse bacterial cells and recover their internal contents. When bacterial cells are treated with lysozyme, all or only a part of the cell wall may be destroyed. If a portion of the bacterial cell wall remains after lysozyme treatment, the remaining cell is called a **spheroplast**. If the cell wall is removed completely, the remaining cell is called a **protoplast** (see FIG. 5-11).

Comparison of Gram-positive and Gram-negative Bacterial Cell Walls

The cell walls of Gram-negative and Gram-positive bacteria are structurally different (FIG. 5-12). This

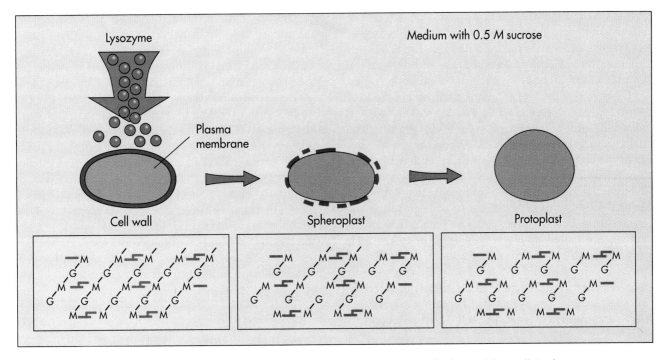

FIG. 5-11 Lysozyme cleaves the glycan portion of the peptidoglycan. The wall is degraded but as long as there is an osmotic support (such as a 0.5 M sucrose solution), the cells do not lyse (rupture). Spheroplasts have some intact wall, whereas protoplasts have none. If the osmotic support is removed, both spheroplasts and protoplasts lyse.

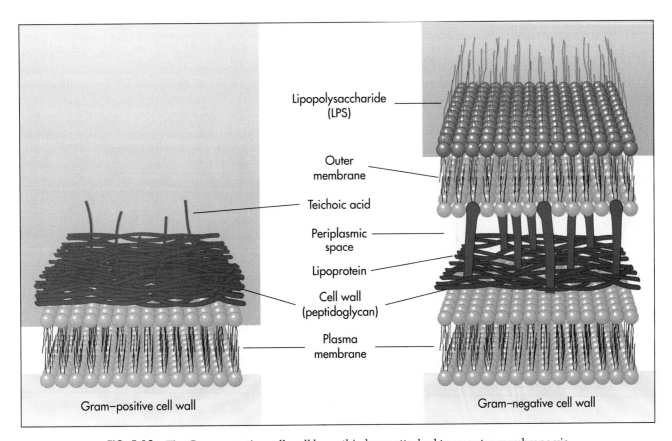

FIG. 5-12 The Gram-negative cell wall has a thin layer attached to an outer membrane via lipoproteins. The outer membrane contains phospholipid on its inner surface and lipopolysaccharide (LPS) on its outer surface. The space between the outer membrane and the plasma membrane is called the periplasmic space.

difference in cell wall structure is critical. It affects the survival of bacteria in nature and the effectiveness of antibiotics against bacterial pathogens. It is also of further practical importance because of the use of the Gram stain in identifying bacterial species.

Over 90% of the Gram-positive cell wall is made of peptidoglycan (see FIG. 5-8). This provides a strong, relatively thick protective layer that protects the plasma membrane from lysis by osmotic shock. Besides peptidoglycan, Gram-positive cell walls contain teichoic acids, which are acidic polysaccharides. Teichoic acids make the Gram-positive cell wall acidic. This is important because cells produce enzymes that break down peptidoglycan. These enzymes are called *autolysins*. At relatively low pH, however, autolysins do not degrade the cell wall. Autolysins are normally involved in cell wall growth and restructuring. Teichoic acids regulate autolytic activity so that autolysins can perform their necessary function without causing self-destruction (autolysis) of the cell.

Gram-negative cell walls are far more complex. Two distinct structures make up the Gram-negative wall: a peptidoglycan layer and an outer membrane (FIG. 5-13). The peptidoglycan layer of the Gram-negative cell is very thin, often comprising only 10% or less of the cell wall. It does not contain teichoic acids as in Gram-positive bacterial cell walls. The peptidoglycan layer and outer membrane are held together by lipoproteins (lipid linked to protein molecules).

The region between the plasma membrane and the outer membrane is called the **periplasmic space,** or **periplasmic gel.** Some substances get trapped within the periplasmic gel. Various chemical reactions occur there. The periplasmic gel occurs only in Gram-negative bacterial cells. Additionally, there are regions where the plasma membrane and the outer membrane are attached to one another.

The **outer membrane** is located at the outer extremity of the Gram-negative cell wall. The outer membrane, like the plasma membrane, is a semipermeable structure. The outer membrane restricts the passage of materials based largely on pore size. Many harmful chemicals are excluded from the cell by this structure. The region is unique to Gram-negative bacteria. The outer membrane is a phospholipid bilayer that contains a unique lipopolysaccharide (LPS). LPS is called an *endotoxin* because it is part of the bacterial cell and can cause a toxic reaction in mammals. The symptoms of some infections caused by Gram-negative bacteria are the direct result of the action of endotoxin. Endotoxin causes fever, lysis of red blood cells, and coagulation of blood in capillaries when it is released from dead Gram-negative bacterial cells and enters the circulatory system.

> The cell walls of Gram-positive and Gram-negative bacterial cells are chemically different.

> The Gram-negative bacterial cell wall is a complex structure that has an outer membrane containing the endotoxin lipopolysaccharide (LPS).

The outer membrane can render a Gram-negative bacterial cell resistant to antibiotics. It has channels made of proteins called **porins.** Small polar molecules freely pass through porins. The outer membrane blocks the passage of very large molecules, such as lysozyme, or molecules that have both polar and nonpolar ends, such as penicillin. Some antibiotics cannot cross the outer membrane. They cannot reach their targets at the cell wall's peptidoglycan layer or gain entry into the cell itself. Consequently, for treating infections caused by Gram-negative bacteria, physicians often must select different antibiotics that can cross the outer membrane of the Gram-negative bacterial cell wall and can work against other targets within the cell.

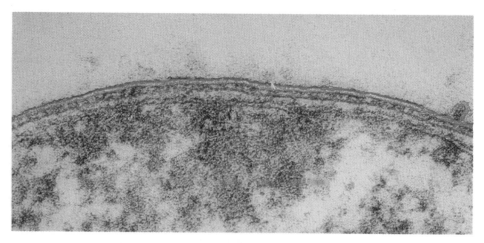

FIG. 5-13 Colorized electron micrograph of the cell wall of the Gram-negative bacterium *Escherichia coli*. (154,000×). The outer membrane *(red)* encloses the peptidoglycan *(purple)*. The entire cell wall surrounds the plasma membrane *(tan)*.

Bacterial Capsules and Glycocalyces

In addition to a cell wall and an outer membrane, some bacteria form another protective structure called a **capsule** (FIG. 5-14). A capsule is a surface layer that surrounds the cell outside of the cell wall. It is almost always composed of polysaccharides. Negative staining with India ink and phase-contrast microscopy readily reveal the presence of a capsule in bacteria having this structure. Capsules occur only in some Gram-positive and Gram-negative bacterial species, such as the pneumonia-causing pathogens *Streptococcus pneumoniae, Klebsiella pneumoniae,* and *Haemophilus influenzae.*

The capsule is especially important in protecting bacterial cells against phagocytosis by some eukaryotic cells. When most bacterial cells lacking a capsule infect human tissues, they are engulfed by phagocytic white blood cells and digested. Phagocytic white blood cells are one of the major defenses against infections. Bacterial cells with capsules are much more difficult for phagocytic white blood cells to engulf. Bacteria lacking a capsule are easily destroyed by such cells in the lungs. The presence of a capsule, therefore, can be a major factor in determining the virulence of a bacterium, that is, the ability of that bacterium to cause disease. Capsule-producing bacteria such as *Streptococcus pneumoniae* and *Klebsiella pneumoniae* are relatively resistant to the phagocytic white blood cells that protect the lungs against infections, which is why many of the bacteria that cause pneumonia have capsules. Capsules also protect the bacteria they surround from desiccation (drying) and virus attack.

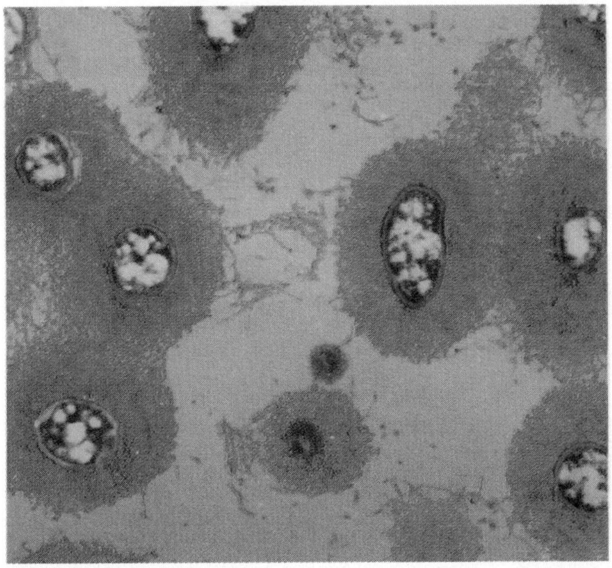

> A capsule is a polysaccharide layer that surrounds some bacterial cells external to their cell walls and protects the bacterial cell against phagocytosis.

In some bacteria, the polysaccharide surface layer is less firmly attached to the bacterial cell and is called a **slime layer.** *Pseudomonas aeruginosa,* a bacterium that produces a significant slime layer, is a particular problem for patients with cystic fibrosis (FIG. 5-15). These bacteria become entrapped in the mucus of the respiratory tract of individuals with cystic fibrosis and their slime layer contributes to difficulty in breathing.

The **glycocalyx** is a less organized structure than the capsule (FIG. 5-16). The glycocalyx, like the cap-

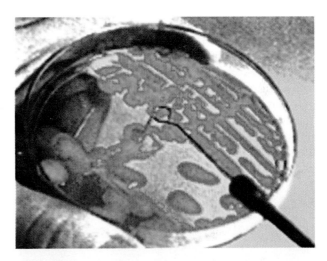

FIG. 5-15 Abundant mucoid material produced by a strain of *Pseudomonas aeruginosa* isolated from a patient with cystic fibrosis. Note adherence of slime with bacterial cells to the inoculating loop.

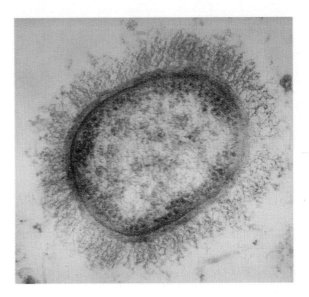

FIG. 5-14 Colorized electron micrograph showing capsules *(red)* of the bacterium *Alcaligenes faecalis.* The capsule surrounds and protects the cell.

FIG. 5-16 Colorized electron micrograph of the glycocalyx *(red)* of a Gram-negative bacterium. (58,500×).

sule, is often made of polysaccharide and surrounds some bacterial cells. The glycocalyx was not observed until very recently. This structure has a very high water content. This makes it difficult to view by light microscopy. It also makes it easy to destroy during the dehydration steps used in preparing specimens for viewing with the electron microscope.

The glycocalyx allows the bacterial cell to attach to solid surfaces. By attaching to surfaces, bacterial cells maintain themselves in a location that has favorable conditions for growth and survival. Some bacteria in aquatic habitats seem to attach to rocks through their glycocalyces. These bacteria obtain nutrients from water as it flows by. They do not have to expend energy in search of food. Other pathogenic bacteria adhere to the animal tissues they invade via a glycocalyx. This enables them to obtain nutrients from the animal.

Some bacteria adhere to the surfaces of teeth via their glycocalyces. These bacteria produce a glycocalyx that consists of sticky polysaccharides made from sucrose. This extensive polysaccharide material attaches the bacteria to the tooth surface. It also entraps other oral bacterial cells. This layer of polysaccharide and entrapped bacterial cells is called **dental plaque** (FIG. 5-17). When dental plaque secures bacteria to the surfaces of teeth, it also traps lactic acid produced by the adhering bacteria. The trapped lactic acid eats away at the enamel surface of teeth and may result in dental caries, that is, tooth decay. Treatment of teeth with fluorides helps make the enamel more resistant to the action of lactic acid, reducing the frequency of tooth decay.

The glycocalyx allows bacteria to attach to surfaces.

PILI

Some bacterial cells produce short (less than 1 μm) hairlike projections made of protein (FIG. 5-18). Such projections are called **pili** or **fimbriae.** Pili are involved in specific attachment (adhesion) processes. For example, pili permit some bacteria to adhere to rocks in flowing rivers. Pili also permit some bacteria to attach to cells and to grow on the surfaces of the cells. Sometimes there is sufficient growth to cause disease in the host organism. For example, *Vibrio cholerae* attaches to the surfaces of cells lining the human gastrointestinal tract via its pili, where it reproduces, causing the disease cholera (FIG. 5-19).

A special type of pilus called the *F pilus* or *sex pilus* is involved in mating between bacteria. During mating, DNA is passed from a donor to a recipient bacterial cell. In the absence of an F pilus, mating between bacterial cells cannot occur. F pili are found exclusively on the donor cells that donate DNA during this process. Recipient cells lack F pili.

Pili are hairlike surface projections that attach bacterial cells to other cells and surfaces.

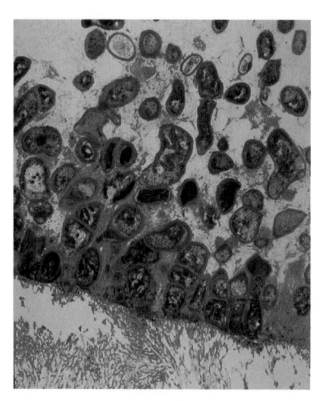

FIG. 5-17 Glucan *(red)* production allows bacteria to adhere to teeth, forming dental plaque. This colorized electron micrograph shows bacterial cells and the dextran matrix of plaque. (10,200×).

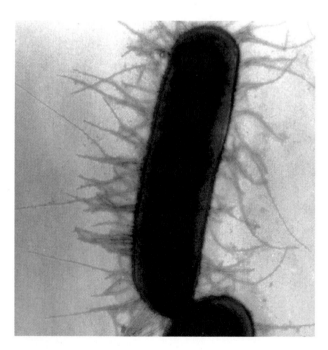

FIG. 5-18 Colorized electron micrograph of pili *(blue)* emanating from the surface of a cell of *Escherichia coli.* (16,400×).

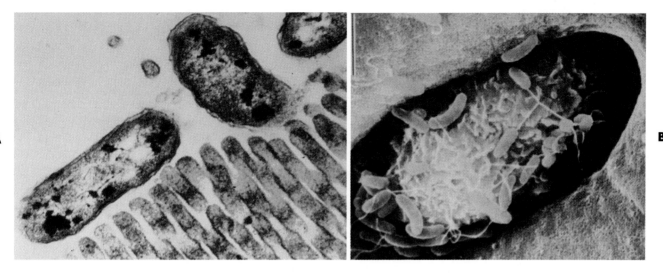

FIG. 5-19   A, Colorized electron micrograph of *Vibrio cholerae* adhering to the cell lining *(tan)* of the gastrointestinal tract via its pili. **B,** Colorized scanning electron micrograph of *Vibrio cholerae* attachment to cells in human ileal mucosa via pili *(blue).*

BACTERIAL FLAGELLA

Flagella are structures that project from the cell surface and propel the cell. Bacteria with flagella are able to move. Some bacteria have only one attached flagellum, whereas others have numerous flagella. The arrangement of the flagella on the bacterial surface is characteristic of a particular genus and is an important diagnostic characteristic used in classifying bacteria (FIG. 5-20). In some bacteria such as *Pseudomonas,* one or more flagella emanate from one end, or pole, of the cell. In other bacteria such as *Spirillum,* flagella may project from both ends of the cell. Whether they emerge from one or both ends of the cell, such flagella are called **polar flagella.** In other bacteria, such as those of the bacterial genus *Proteus,* flagella surround the entire bacterial cell. Flagella that occur around a bacterial cell are called **peritrichous flagella.**

Flagella are structures that propel bacterial cells.

Polar flagella emanate from the ends of cells.

Peritrichous flagella emanate from points all around the cell.

Bacterial flagella extend from the plasma membrane as relatively long, rigid helixes, usually several

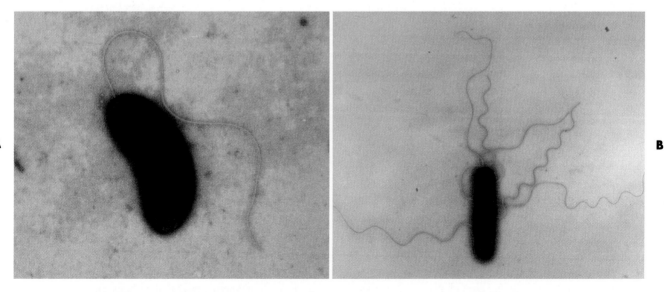

FIG. 5-20   A, Colorized electron micrograph of a *Vibrio* sp. shows that a single polar flagellum *(blue)* emanates from the end of the cell. (25,500×). **B,** Colorized electron micrograph of a *Salmonella* sp. shows that peritrichous flagella *(blue)* arise anywhere on the cell. (20,000×).

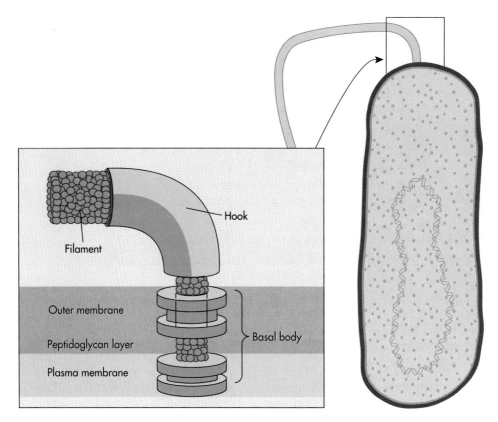

FIG. 5-21   The flagellum is anchored to the cell via a hook and basal body structure. There are rings that join the flagellum to the outer and plasma membranes of a Gram-negative cell. This structure permits the flagellum to rotate. The energy for rotation comes from the protonmotive force.

times longer than the length of the cell. Each bacterial flagellum consists of a single tubelike filament composed of a protein called flagellin. An individual flagellum does not grow from the base, like an animal hair; rather, it grows from its tip. Flagellin molecules formed within the cell pass up through the hollow core of the flagellum and add on at the tip.

Each flagellum is attached to the bacterial cell by a hook and basal body (FIG. 5-21). The basal body is a complex structure consisting of a small central rod that passes through a system of rings. In Gram-negative bacteria, there are two pairs of rings: an outer pair of rings in the cell wall and an inner pair of rings in the plasma membrane. Gram-positive bacteria, which lack the outer lipopolysaccharide layer, only have an inner pair of rings attached to the plasma membrane. The hook is a tubular structure that passes between and extends just beyond the rings of the basal body. The hook has a wider diameter than the filament of the flagellum so that the filament fits into the hook and is thereby attached to the bacterial cell.

The attached flagellum can spin within the rings. It spins much like the shaft of an electric motor spinning within surrounding magnetic coils. The structure of the bacterial flagellum allows it to spin. The

propeller-like rotation of the bacterial flagellum propels the bacterial cell from place to place. This adaptive feature increases the bacterium's ability to obtain nutrition or to escape from hostile microenvironments. Bacterial flagella propel the cell by rotating.

Bacterial flagella rotate to propel the cell.

The bacterial flagellum provides some bacteria with a mechanism for swimming toward or away from particular chemicals. This behavior is known as **chemotaxis** (FIG. 5-22). Some bacteria move toward certain chemicals (*attractants*) and away from others (*repellents*). Sensory structures associated with the cell wall or plasma membrane of the cell, called *chemosensors*, detect certain chemicals and signal the flagella to respond. The chemosensors permit the detection of concentration gradients as the bacterium moves so that the bacterium can detect whether it is moving toward a region of higher or lower concentration of a specific substance.

Some bacteria exhibit chemotaxis in which they move in response to chemical concentration gradients; chemotaxis indicates that some bacteria can sense and respond to their chemical environment.

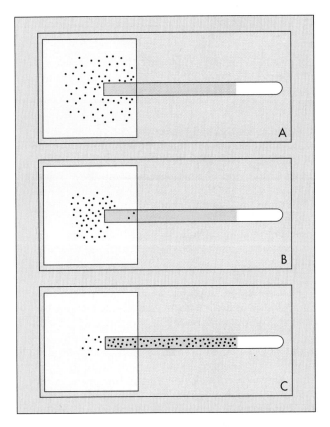

FIG. 5-22 Chemotactic behavior is readily demonstrated and measured by placing the tip of a thin capillary tube containing an attractant solution in a suspension of motile *Escherichia coli* bacteria. The suspension is placed on a slide, under a cover slip. **A,** At first, the bacteria are distributed at random throughout the suspension. **B,** After 20 minutes, they have congregated at the mouth of the capillary tube. **C,** After about an hour, many cells have moved up into the capillary tube. If the capillary tube had contained a repellant, few bacteria, if any, would have entered. Using this technique, it is possible to show which chemicals attract bacteria and which do not.

When bacteria move, they periodically change direction. They do not reach their destination by swimming in a single straight line. Their movement consists of a series of *runs* (straight line movements) and *tumbles* (turning motions). Counterclockwise rotation of the flagella produces runs. Tumbling occurs when the flagella rotate in a clockwise direction. The direction of flagella rotation, and hence the length of a run, is determined by the interactions of the chemosensors with attractants or repellents in the cell's plasma membrane, or in the periplasmic gel. An increasing concentration of attractant, for example, interacts with the chemosensors to decrease the frequency of tumbling. A decreasing concentration of attractant causes increasing tumbling, and hence shorter runs. The reverse is true for the interactions with repellents, where increasing concentration of the repellent causes increased tumbling. The net effect of this process is an overall movement toward an attractant or away from a repellent.

EUKARYOTIC FLAGELLA AND CILIA

In contrast to the rather simple structure of the bacterial flagellum, the flagella and cilia of eukaryotic microorganisms are larger, far more complex, and operate by entirely different mechanisms. **Eukaryotic flagella** also propel cells. They originate from the polar region of the eukaryotic cell (FIG. 5-23). **Cilia** generally are also involved in cell movement. They surround the cell and are more numerous and shorter than flagella (FIG. 5-24). Unlike the rigid bacterial flagella that rotate, cilia and flagella of eukaryotic cells undulate in a flexible wave-like motion to propel the cell.

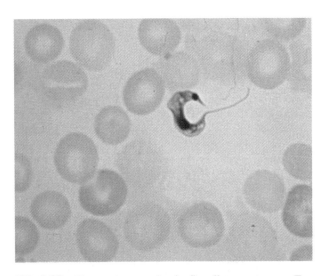

FIG. 5-23 Photomicrograph of a flagellate protozoan *Trypanosoma cruzi*, which is a human pathogen.

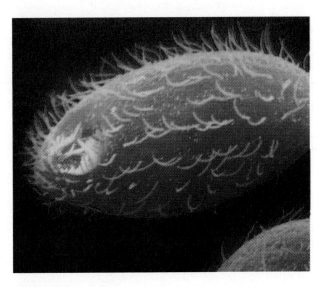

FIG. 5-24 Colorized scanning electron micrograph of the ciliate protozoan *Tetrahymena*.

Magnetotaxis: Specialized Movement of Bacteria in Response to Magnetic Fields

A few bacteria can sense and respond to magnetic fields. The movement of bacteria in response to a magnetic field is known as *magnetotaxis*. These bacteria orient their movements in relation to the Earth's magnetic field. The magnetotactic bacteria were discovered by Richard Blakemore while viewing the movement of some Gram-negative bacteria obtained from sulfide-rich lake mud. He observed that the bacteria were all moving in the same direction across the microscope's field of view. Even when he turned the microscope around or moved it to another location in the room these bacteria continued to swim in the same geographic direction.

The bacteria did not appear to be swimming in response to any particular chemical stimulus. They continued to swim in one direction even in a darkened room. Blakemore hypothesized that the bacteria might be orienting their movements to the Earth's magnetic field. To test his hypothesis, Blakemore placed a magnet adjacent to the microscope slide. He observed that the bacteria instantly began to swim toward the end of the magnet that attracts the north-seeking end of a compass needle. When he turned the magnet around, the bacteria also turned around. They began to swim back toward the same end of the magnet that they had been previously moving toward. These observations clearly showed that these bacteria could orient the direction of their movements in response to a magnetic field.

Blakemore then hypothesized that if he collected bacteria from the Southern Hemisphere they would show the opposite responses to magnetic fields. When he examined magnetotactic bacteria from New Zealand, in the Southern Hemisphere, they swam toward the south pole and away from the north pole of a magnet. Magnetotactic bacteria from the Northern and Southern Hemispheres swim in opposite directions in response to a magnetic field.

Blakemore continued his studies by observing the cells of these bacteria with an electron microscope. He observed dense particles within the bacterial cells of these bacteria. He did not observe such particles in bacteria that did not exhibit magnetotactic behavior (see Figure). Analysis of these particles showed that they were composed of iron oxide (Fe_3O_4), a magnetic metal compound. The iron oxide is stored in structures, called *magnetosomes*, that act like magnets and allow bacterial cells containing them to respond to magnetic fields and, thus in their natural environments, to orient themselves along the Earth's magnetic field. This enables magnetotactic aquatic bacteria to point their cells downward toward sediments where nutrients are more abundant and where oxygen concentrations are favorable for their growth.

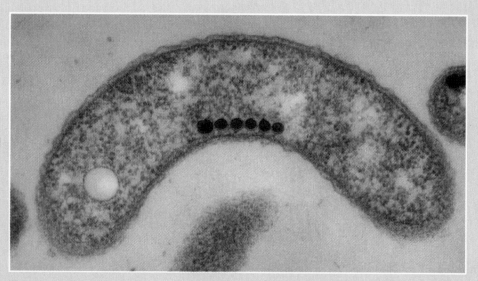

Colorized electron micrograph of a section of *Aquaspirillum magnetotacticum* shows characteristic magnetosomes (*red granules*) that allow this bacterium to respond to a magnetic field. (42,900×).

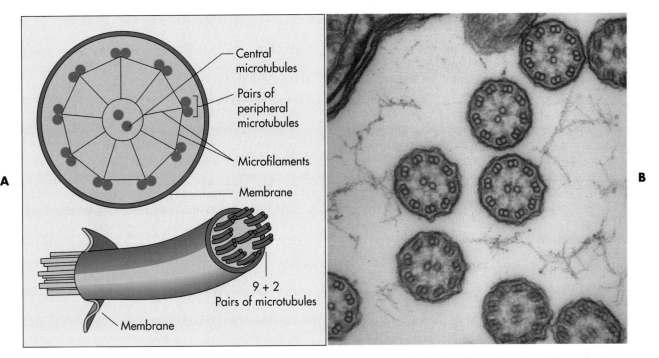

FIG. 5-25 **A,** The structure of the eukaryotic flagellum and cilia has nine pairs of peripheral microtubules surrounding a central pair of microtubules. The microtubules are connected by microfilaments. The peripheral microtubules slide past the central microtubules, causing the flagellum or cilia to bend. **B,** Electron micrograph of a cross section of the cilia of the ciliate protozoan *Mesodinium.* (98,800×).

Both eukaryotic flagella and cilia consist of a series of microtubules (FIG. 5-25). These microtubules are hollow cylinders composed of proteins surrounded by a membrane. The protein that makes up the microtubule is called tubulin. The arrangement of microtubules in eukaryotic flagella and cilia is known as the "9 + 2" system. It consists of nine pairs of peripheral microtubules that form a circle around two single central microtubules. Attached to each of the peripheral pairs of microtubules are molecules of a protein called dynein. Dynein is involved in the conversion of chemical energy from ATP into the mechanical energy of flagellar movement. The movement of flagella or cilia is based on a sliding microtubule mechanism. The peripheral pairs of microtubules slide past each other, resulting in bending of the flagella or cilia. This causes the flagella or cilia to bend and propel cells with a whiplike motion.

Eukaryotic flagella flex to propel cells based on the sliding movement of microtubules.

Cilia may also be involved in moving materials such as food particles past the cell surface while the eukaryotic cell remains stationary. Paramecia use such movement of cilia to obtain food. Some human tissues have cells with cilia. For example, the cells lining the human respiratory tract have cilia that beat

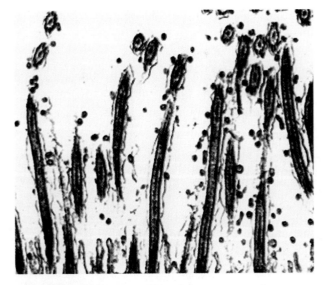

FIG. 5-26 Colorized micrograph of the cilia (*tan*) lining the human respiratory tract. These cilia help prevent viruses, such as those shown in blue, from entering the body via the respiratory tract.

with an upward wave-like motion. They sweep mucus outward and prevent particles such as bacteria and viruses trapped in the mucus from reaching the lungs. Cilia thus help to defend the human respiratory tract against microbial infections (FIG. 5-26).

DNA is sometimes called the "master molecule" of all living cells. It contains the hereditary information that is passed from generation to generation. DNA determines the potential metabolic and morphological characteristics of the cell. It also regulates the expression of genetic information. The total genetic material of an organism is called the **genome.** The genome is divided into functional segments, or genes. Each **gene** specifies the amino acid sequences of specific proteins or regulates when particular proteins or RNA molecules are to be made. The DNA in a bacterial cell typically contains about 6,000 genes. Passage of replica copies of the DNA to progeny cells during reproduction transmits the hereditary information from one generation to the next.

The genetic information **(genotype),** which is stored in the DNA, is expressed through the production of functional proteins. It controls the **phenotype** (appearance) of the cell, that is, the cell's morphology and metabolism. In this way, the genetic information determines the properties and structural characteristics of the cell. The actual steps in the expression and transmission of genetic information, which are two essential life functions, will be considered in detail when we examine microbial genetics in Chapter 7. Here, we will only discuss the way in which genetic information is stored in prokaryotic and eukaryotic cells and the structures involved in the expression of this information.

BACTERIAL CHROMOSOME

In the bacterial cell, almost all the genetic information is contained within a single DNA macromolecule called the **bacterial chromosome.** The cytoplasmic region occupied by the bacterial chromosome is referred to as the *nucleoid region.* This region is not actually surrounded by a covering or membrane. It can move within the cell (FIG. 5-27).

The bacterial chromosome is composed virtually exclusively of DNA. The DNA forms a double helix—two strands wound around each other, giving the appearance of a spiral staircase with two handrails. The bacterial chromosome is a closed loop of DNA. If this circular bacterial chromosome were to be cut open, the linear DNA molecule formed would be approximately 1 mm (1,000 μm) long. Since most bacterial cells are less than 5 μm in length, bacterial DNA must be extensively coiled into a twisted molecule. This twisted molecule is called supercoiled DNA. Supercoiled DNA is very compact and fits within the bacterial cell. This molecule houses all the essential information that determines the structural and functional characteristics of the bacterial cell.

> The bacterial chromosome is circular and houses the essential genetic information of the cell.

PLASMIDS

Some bacterial cells also contain one or more small circular macromolecules of DNA that store additional specialized information. These are known as plasmids. **Plasmids** contain a limited amount of specific genetic information. This information supplements the essential genetic information contained in the bacterial chromosome. Plasmids contain only 1% to 5% as much DNA as is in the bacterial chromosome, that is, about 20 genes. Nevertheless, the sup-

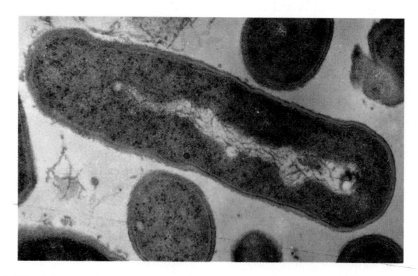

FIG. 5-27 Colorized electron micrograph of a section of the bacterium *Mycobacterium phlei* showing the nucleoid region *(green)* within the cytoplasm where the bacterial chromosome occurs.

plemental genetic information contained in plasmids can be quite important. They can establish such characteristics as resistance to antibiotics and tolerance to heavy metals. The gene products of plasmids may permit the survival of the bacterium under conditions that are normally unfavorable for growth and survival. Having a plasmid that codes for resistance to an antibiotic, for example, can mean 100% survival for a bacterium. Plasmids can be transferred from one bacterial cell to another, sometimes even from one bacterial species to another. Thus plasmids conferring antibiotic resistance can become prevalent among bacteria surviving in areas where antibiotics are widely used, such as within hospitals. For this reason physicians limit the use of antibiotics to situations where they will be especially useful in treating a specific disease.

> **Plasmids are small circular DNA molecules that contain supplemental genetic information that can enhance survival potential, including resistance to antibiotics.**

NUCLEUS OF THE EUKARYOTIC CELL

The DNA of eukaryotic cells is contained in a specialized organelle called the **nucleus.** The nucleus is segregated from the rest of the cell because it is surrounded by two membranes, the **nuclear envelope** (FIG. 5-28). The nuclear envelope is rather porous so that exchange of molecules such as RNA and pro-

teins can easily occur between the cytoplasm and the fluid within the nucleus. This fluid is called the **nucleoplasm.** DNA and associated proteins are suspended in the nucleoplasm. The nucleus also contains a region called the **nucleolus** where ribosomes are assembled.

> **DNA in eukaryotic cells is contained within the nucleus, an organelle surrounded by a nuclear envelope that separates the nucleoplasm from the cytoplasm.**

The critical functions that occur within the nucleus are the processing and transmission of hereditary information. When a eukaryotic cell is about to divide, the DNA and its associated proteins fold and twist into condensed structures called **chromosomes.** Chromosomes can be observed through a light microscope (see FIG. 5-28). Chromosomes pass the hereditary information from one generation to the next. Eukaryotic chromosomes are linear and generally occur in pairs. Each individual chromosome in the pair has virtually the same genes. Thus there generally is a duplication of genetic information in eukaryotic cells that is not found in bacterial cells. Most eukaryotic cells are *diploid* (having two copies of each gene), whereas prokaryotic cells are *haploid* (having one copy of each gene).

When a nondividing eukaryotic cell is stained for observation through a light microscope, all that can

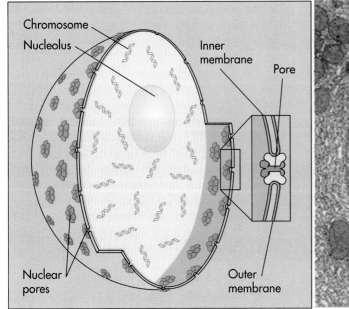
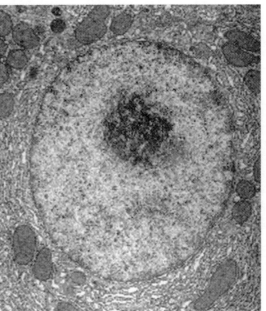

A

B

FIG. 5-28 **A,** The nucleus that contains the hereditary information in a eukaryotic cell is surrounded by two membranes: an inner and an outer membrane. The nucleolus within the nucleus is the site where ribosomal subunits are made. There are pores in the membranes through which materials can move, including messenger RNA that carries information from the DNA within the nucleus to the ribosomes in the cytoplasm. **B,** Colorized micrograph of the nucleus of a eukaryotic cell shows the double membrane structure and the pores of this organelle that contains the chromosomes of eukaryotic cells. (8,400×).

FIG. 5-29 **A,** A nucleosome showing how DNA is wrapped around histones (basic proteins), establishing coiling of DNA within the nucleus of eukaryotic cells. **B,** Colorized micrograph of a region of a chromosome showing the beadlike appearance of nucleosomes.

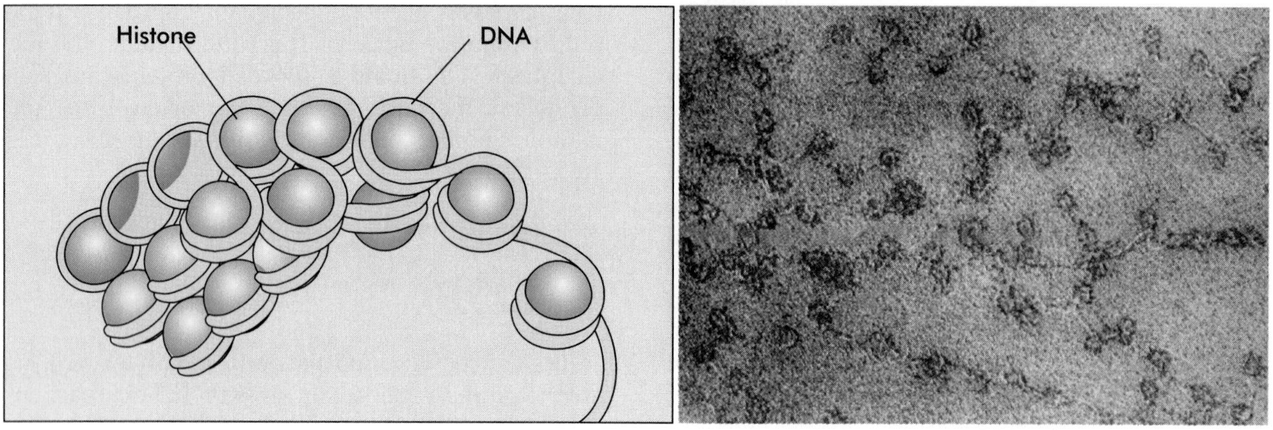

be seen within the nucleus is a grainy material, called *chromatin.* Chromatin is composed of strung-out DNA and its associated proteins. The DNA of the chromatin directs the metabolic activities of the eukaryotic cell. Some proteins of the chromatin, called **histones,** maintain the highly coiled shape of the DNA macromolecule. The DNA wound around the histones forms subunits of chromatin known as *nucleosomes* (FIG. 5-29). Each nucleosome is composed of about 200 nucleotides coiled around several different histones. When viewed by electron microscopy, nucleosomes appear as spherical particles, like beads on a string.

> **Chromatin consists of DNA and associated proteins, including histones that help maintain the shape of the DNA molecule.**

> **DNA in chromatin is coiled around the histones to form nucleosomes.**

RIBOSOMES

The expression of genetic information requires the formation of proteins—particularly enzymes that are the action molecules that determine what a cell is and what it does. All cells use the genetic information stored in their DNA to direct the production of specific proteins. First, the information in the DNA is transferred to RNA in a process called *transcription.* Following transcription, some RNA molecules, called *messenger RNA (mRNA),* carry the genetic information to the ribosomes in the cytoplasm of the cell.

Ribosomes are the structures within a cell where proteins are made. It is at the ribosomes that protein synthesis, *translation,* occurs. During translation the information in RNA is used to direct the synthesis of a specific protein. Thousands of different proteins are continuously needed to support the metabolism of a cell. A typical prokaryotic cell may have 10,000 or more ribosomes. Eukaryotic cells contain considerably more.

> **Ribosomes are the structures within cells where protein synthesis occurs.**

Functional ribosomes are made up of two subunits, a larger and a smaller one. Each subunit is made up of specific RNA molecules and a group of different proteins. Protein synthesis can occur only when the subunits are joined. In eukaryotic cells the ribosomal subunits are produced within the nucleus. They then are exported through the pores of the nuclear envelope to the cytoplasm where they are assembled into functional ribosomes.

Ribosomes of bacteria are different from those of eukaryotic cells. Bacterial cells have **70S ribosomes.** The ribosomes in the cytoplasm of eukaryotic cells are **80S ribosomes** (FIG. 5-30). (Eukaryotic cells also have 70S ribosomes within specialized organelles involved in ATP generation: mitochondria and chloroplasts.) "S" stands for Svedburg units. Svedburg units represent a combined measure of molecular weight and shape derived from the rate of sedimentation in an ultracentrifuge. An ultracentrifuge is a centrifuge that spins at very high speeds, in excess of 25,000 revolutions per minute.

> **Bacterial cells have 70S ribosomes.**

> **Eukaryotic cells have 80S ribosomes in the cytoplasm.**

The difference between eukaryotic and prokaryotic ribosomes is significant. Various antibiotics discriminate between 70S and 80S ribosomes. Antibiotics such as erythromycin, tetracyclines, and many others are effective for treating human bacterial in-

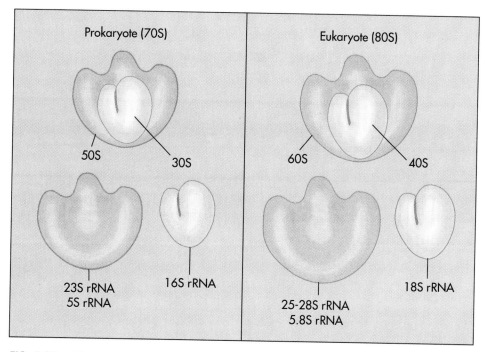

FIG. 5-30 A basic difference between prokaryotic and eukaryotic cells is the nature of the ribosomes in the cytoplasm. The prokaryotic cell has 70S ribosomes composed of 30S and 50S subunits. The 30S subunit contains about 21 proteins and a 16S rRNA molecule, having approximately 1,540 nucleotides; the 50S subunit is composed of approximately 34 proteins, a 23S rRNA, having approximately 2,900 nucleotides, and a small 5S rRNA species having only about 120 nucleotides. A eukaryotic cell has 80S ribosomes in its cytoplasm composed of 60S and 40S subunits. The 40S subunit contains proteins and an 18S rRNA, and the larger 60S subunit has proteins, 25S to 28S rRNA, and 5.8S rRNA. These differences form the basis for the specificity of action of some antibiotics that inhibit protein synthesis.

fections because they are able to block protein synthesis by specifically binding with the 70S ribosomes of bacterial cells. These antibiotics do not bind to 80S eukaryotic ribosomes in the cytoplasm. They do not disrupt protein synthesis at the 80S ribosomes of human cells. Therefore they can be used to selectively inhibit protein synthesis of infecting bacteria and can be used therapeutically in the treatment of human in-

fections caused by pathogenic bacteria. (A few adversely affect a patient because human cells have 70S ribosomes within mitochondria and these have more limited therapeutic uses.) Once again, a fundamental difference between prokaryotic and eukaryotic cell structure provides the opportunity for selective drug action against infecting bacteria while leaving eukaryotic human cells untouched.

CELLULAR STRUCTURE AND ENERGY TRANSFORMATIONS

Besides having mechanisms for storing and expressing genetic information, cells must have a means of obtaining energy to carry out life functions. Cells do not create energy. Cells transform the energy they obtain from light or chemicals to produce usable cellular energy stored as molecules of ATP. Many of the chemical reactions that occur during metabolism are involved in generating ATP. Cells must generate sufficient ATP for their maintenance, growth, and reproduction, or they will die. The metabolic reactions that generate ATP in prokaryotic or eukaryotic cells

can occur in either of two locations. They can occur within the cytosol or in association with specialized membranes, or at both sites.

Living systems depend on their ability to generate ATP, the universal currency of energy transfer within the cell.

The mechanism by which ATP is generated at membranes was proposed in 1965 by Peter Mitchell. Initially the scientific community rejected Mitchell's hypothesis for ATP generation but later accepted his

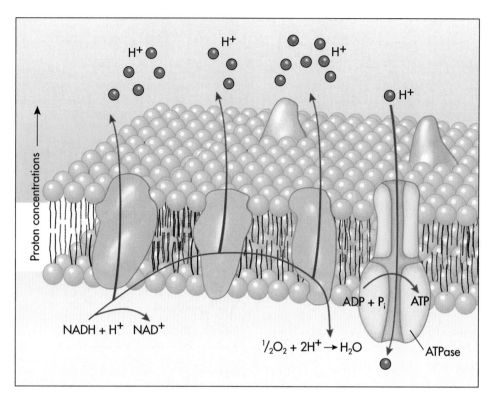

FIG. 5-31 The protonmotive force drives the formation of ATP by chemiosmosis. Protons are extruded across the membrane to establish a proton gradient. The membrane is impermeable to protons, which can only move back across the membrane through protein channels associated with ATPase. As the protons move by diffusion through these channels, energy is transferred to form ATP.

idea and awarded him a Nobel Prize in 1978. Mitchell reasoned that transport of substances across the membrane and the semipermeable properties of biological membranes have a special role in the generation of ATP by many organisms. He proposed that protons (hydrogen ions) are extruded across membranes as a result of oxidation-reduction reactions that occur when electrons are transferred between compounds embedded within the membrane. The metabolism of the cell is thus connected to the establishment of a proton gradient across the membrane (FIG. 5-31). The concentration difference between hydrogen ions on the opposing sides of the membrane represents a potential energy called the **protonmotive force.** The protonmotive force can be used to drive the formation of ATP. This process for the generation of ATP is called **chemiosmosis.** In chemiosmosis, energy from metabolic oxidations is used to move protons (H⁺) across a membrane, establishing a concentration gradient. Diffusion of H⁺ back across the membrane is coupled to the synthesis of ATP.

Chemiosmosis is the process in which the protonmotive force (potential energy across a membrane based on a concentration gradient of protons) is used to generate ATP.

SITES OF CHEMIOSMOTIC GENERATION OF ATP IN PROKARYOTIC CELLS

In bacterial respiration the plasma membrane is an important site of chemiosmotic generation of ATP. Protons are moved across the plasma membrane out from the cell. The concentration gradient of protons establishes the protonmotive force that is used to generate ATP. When protons diffuse back across the plasma membrane, ATP is generated by chemiosmosis. In some specialized groups of bacteria, internal membranes are similarly involved with chemiosmotic generation of ATP (FIG. 5-32). Such specialized bacteria include the nitrifying bacteria and the photosynthetic bacteria. Nitrifying bacteria are bacteria that oxidize inorganic nitrogen-containing compounds to generate ATP. Photosynthetic bacteria use light energy to generate ATP. In photosynthetic bacteria the internal membranes are the sites where light energy is converted to chemical energy in the form of ATP during photosynthesis. These specialized photosynthetic membranes can be simple extensions of the plasma membrane (as in the purple sulfur bacteria), cylindrically shaped vesicles (as in the green photosynthetic bacteria), or an extensive multilayered system of membranes (as in the cyanobacteria) (FIG. 5-33).

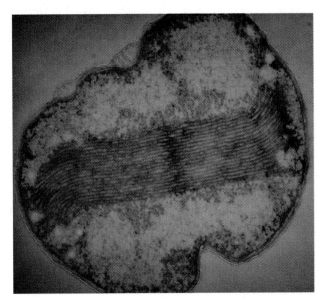

FIG. 5-32 Colorized electron micrograph of a section of the nitrifying bacterium *Nitrosococcus oceanus* showing internal membrane *(tan)*.

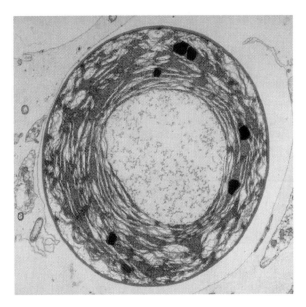

FIG. 5-33 Colorized electron micrograph of the photosynthetic bacterium *Prochloron didemni* reveals that it has extensive internal membranes that have photosynthetic pigments *(green)*. (6,400×). These membranes are the sites of chemiosmotic generation of ATP by this bacterium, which derives the energy for ATP formation from light energy.

SITES OF CHEMIOSMOTIC GENERATION OF ATP IN EUKARYOTIC CELLS

Mitochondria

Mitochondria are important sites of chemiosmotic ATP generation in respiring eukaryotic cells (FIG. 5-34). These organelles appear to have arisen as a result of endosymbiotic evolution and, like prokaryotic cells, contain 70S ribosomes and a circular DNA macromolecule. Some yeast cells have as few as two mitochondria per cell. Other eukaryotic cells have many more. Many animal cells, for example, contain 1,000 mitochondria. A mitochondrion is a double membrane–bounded compartment. It has an interior membrane that forms a separate compartment and an outer membrane that acts as the boundary between the mitochondrion and the cell cytosol. The establishment of a proton gradient across the inner membrane is used for the chemiosmotic generation of ATP.

The inner membrane of the mitochondrion has many folds or convolutions. These folds are called *cristae*. They increase the surface area. Particles of ATPase, the enzyme that catalyzes the chemiosmotic generation of ATP, are bound to the cristae. Protein

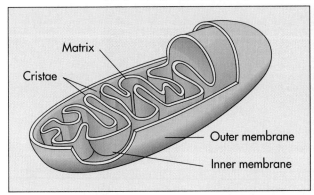

Matrix

Cristae

Outer membrane

Inner membrane

A

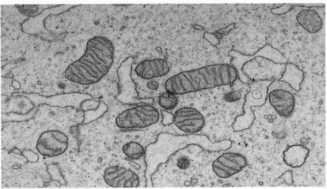

B

FIG. 5-34 **A,** A mitochondrion is the site of ATP generation by chemiosmosis in eukaryotic cells. There are two distinct membranes and extensive folding of the internal membrane. Protons are pumped across the inner membrane into the space between the inner and outer membranes. This establishes the protonmotive force that drives the formation of ATP. **B,** Colorized electron micrograph of mitochondria *(tan)* of a human cell. (11,900×).

channels in the outer membrane allow passage of any molecule of molecular weight less than approximately 10,000. This permeability permits the ATP produced within the mitochondria to move into the cytosol, where it is used in energy-requiring reactions.

Mitochondria are the sites of chemiosmotic generation of ATP in eukaryotic cells.

Chloroplasts

Chloroplasts are the structures where photosynthetic generation of ATP occurs in algal and plant cells (FIG. 5-35). They are quite similar in many ways to mitochondria in structure and function. They appear to have arisen as a result of endosymbiotic evolution. Chloroplasts are composed of extensively invaginated membranes, contain 70S ribosomes, and have a circular DNA macromolecule—as do the mitochondria. Like the mitochondrion, the chloroplast contains an outer membrane that separates the organelle from the cytosol and an inner membrane.

Chloroplasts also have an additional complex internal membranous system, known as the **thylakoids.** Thylakoids are sac-like membranous vesicles that contain various photosynthetic pigments. These pigments include the chlorophylls that are the primary photosynthetic pigments involved in the conversion of light energy to cellular chemical energy. Groups of thylakoids may be organized to form densely stacked piles called *grana.*

The fluid within the chloroplast, called the *stroma*, is where the fixation of carbon dioxide occurs during photosynthesis. The thylakoid membranes of the chloroplasts are highly impermeable to protons, so that protons moved across the membrane cannot freely diffuse back across the membrane. The proton gradient and associated protonmotive force established across thylakoid membranes is used to drive the synthesis of ATP by chemiosmosis when protons move through ATPase channels in the membranes. This chemiosmotic generation of ATP at the thylakoid membranes of chloroplasts is analogous to the synthesis of ATP in the mitochondria.

Chloroplasts are the structures where photosynthetic chemiosmotic generation of ATP and carbon fixation occur in eukaryotic algal and plant cells.

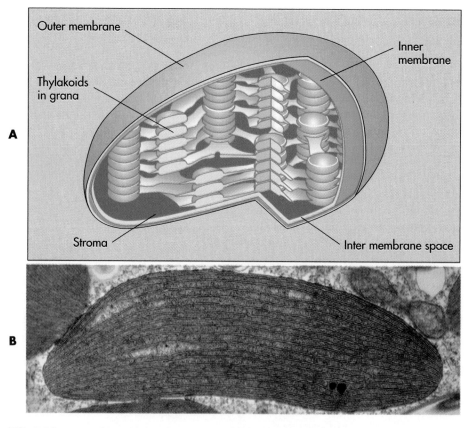

FIG. 5-35 **A,** A chloroplast is the site of ATP generation by chemiosmosis in eukaryotic photosynthetic cells. Light energy is trapped by the chlorophyll in the chloroplast. This initiates a process in which water is converted into oxygen (O_2) and protons (H^+) There are two distinct membranes. Protons are pumped inward across the inner membrane. This establishes the protonmotive force that drives the formation of ATP. The return flow of protons into the space between the inner and outer membranes passes through a protein channel associated with ATPase. **B,** Colorized electron micrograph of a chloroplast *(green)* of the alga *Euglena proxima.* (28,900×).

In addition to generating ATP as a usable form of cellular energy, cells must move substances from one place to another within the cell. In prokaryotes this is relatively simple. The lack of internal membrane-bound organelles means that substances can freely mix within the cell's cytoplasm. To compensate for the lack of internal compartments, some functions occur extracellularly or within the cytoplasmic membranes of the prokaryotic cells that occur within the organelles of eukaryotic cells. Some bacterial cells form relatively insoluble reserve materials, such as the fatlike molecule poly-β-hydroxybutyric acid (PHB), the phosphate-rich polyphosphate (also called volutin or metachromatic granules), and ele-

mental sulfur granules. Because they are nonpolar molecules and hence do not readily dissolve in cytosol, these reserve materials naturally segregate from the other components within the cell and clump, forming granules (FIG. 5-36).

CYTOMEMBRANE SYSTEM OF EUKARYOTIC CELLS

In marked contrast to the prokaryotic cell, the eukaryotic cell is filled with membranous organelles involved with the processing and storage of materials within the cell (Table 5-2, p. 148). These membrane-bound organelles physically separate different chemicals so that they do not mix haphazardly within the cytoplasm. The extensive internal membrane system of eukaryotic cells also permits the segregation of chemical reactions that might interfere with one another. Organelles furthermore separate different chemical reactions in space and time. Molecules being used and produced are passed from one organelle to another in specific reaction sequences. However, this segregation increases the need to coordinate and manage the functions of the cell's subunit organelles.

Many of the organelles of the eukaryotic cell are linked together. They function in a coordinated manner through what is collectively called the *cytomembrane system.* The cytomembrane system includes the nuclear membrane, the endoplasmic reticulum, the Golgi apparatus, lysosomes, vacuoles, and the cytoskeleton. The endoplasmic reticulum isolates, packages, and transports proteins and other substances. The Golgi apparatus is used for the modification and distribution of materials and to produce various membrane-bound vesicles and vacuoles. These membrane-bound vesicles and vacuoles are involved in the storage, processing, and transport of materials. The movement of materials through this cytomembrane system is highly organized. Functioning together, the outer nuclear membrane, the endoplasmic reticulum, the Golgi bodies, and the secretory vesicles perform a sequence of coordinated activities that move and process materials (often proteins) through the cytoplasm to the exterior of the cell.

> Eukaryotic cells contain a network of membranous organelles that are involved in the storage and processing of materials.

Endoplasmic Reticulum

As part of their cytomembrane system, eukaryotic cells contain an extensive membranous network known as the **endoplasmic reticulum** (ER) (FIG. 5-37). The endoplasmic reticulum is composed of a se-

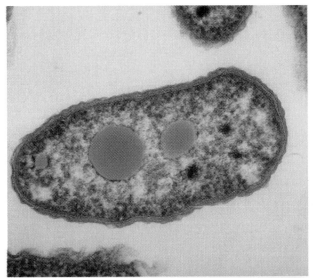

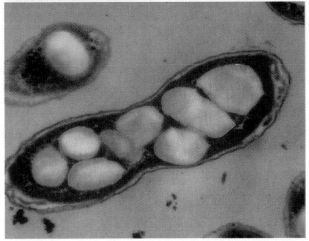

FIG. 5-36 **A,** Polyphosphate *(gold)* accumulates in some bacterial cells such as *Pseudomonas aeruginosa* (colorized electron micrograph; 44,100×). **B,** The polyhydroxybutyrate inclusions *(pink)* of a *Vibrio* species nearly fill the cells of this bacterium (colorized electron micrograph; 32,800×).

TABLE 5-2

Comparison of Structures in Prokaryotic and Eukaryotic Cells

STRUCTURE	FUNCTION	PROKARYOTIC CELLS		EUKARYOTIC CELLS				
		ARCHAEBACTERIA	EUBACTERIA	FUNGI	ALGAE	PROTOZOA	PLANTS	ANIMALS
Plasma membrane	Semipermeable barrier; regulation of substances moving into and out of cell	+	+	+	+	+	+	+
Cell wall (with peptidoglycan)	Protects cell against osmotic shock or physical damage	−	+	−	−	−	−	−
Cell wall (without peptidoglycan)	Protects cell against osmotic shock or physical damage	+	−	+	+	−	+	−
Flagella* that lack microtubules and rotate	Cell movement by rotation	+	+	−	−	−	−	−
Flagella† with microtubules that undulate	Cell movement by flexion	−	−	+	+	+	+/−	+
Cilia†	Cell movement; movement of materials	−	−	−	−	+	+/−	+
Nucleoid	Region of DNA concentration; heredity control	+	+	−	−	−	−	−
Nucleus	Membrane-bound organelle containing DNA; region of heredity control	−	−	+	+	+	+	+
Nucleolus	Formation of ribosomal subunits	−	−	+	+	+	+	+
Bacterial chromosome	Circular molecule that contains genome (hereditary information)	+	+	−	−	−	−	−
Chromosomes	Linear molecules that contain genomes; DNA stores the hereditary information; protein establishes structure of the chromosome essential for gene expression	−	−	+	+	+	+	+
Ribosome	Translation of genetic information carried by mRNA into proteins; protein synthesis	+	+	+	+	+	+	+
Endoplasmic reticulum	Processing and transport of proteins and other substances through cell; communication of chemicals and coordination of functions within cell	−	−	+	+	+	+	+
Golgi body	Processing of substances and packaging of materials into vesicles for export from cell	−	−	+	+	+	+	+
Lysosome	Containment of digestive enzymes; controlled degradation of substances	−	−	+	+	+	+	+
Cytoskeleton	Organization and support of organelles within cell	−	−	+	+	+	+	+
Mitochondrion	Respiratory chemiosmotic generation of ATP	−	−	+	+	+	+	+
Chloroplast	Photosynthetic chemiosmotic generation of ATP	−	−	−	+	−	+	−
Endospore	Survival; heat resistance	−	+	−	−	−	−	−

*Simple protein structure. †Complex structure.

ries of interconnected membranous tubes and sacs. These tubes and sacs form compartments within the cytoplasm. This membranous network appears to provide a communication system that helps coordinate the metabolic activities of the cell. The ER shows two distinct morphologies when examined by electron microscopy. One type of ER (rough ER) looks "rough" because ribosomes are attached to it. The other type, smooth ER, appears smooth because it is not associated with ribosomes. The attachment of ribosomes to the endoplasmic reticulum permits the coordinated synthesis and sequestering of certain proteins. Proteins made at these ribosomes can be sent through the channels of the endoplasmic reticulum to other organelles within the cell for storage, use, modification, or export. Prokaryotic cells have no analogous membrane structure to which ribosomes can attach. They have no system comparable to the endoplasmic reticulum to coordinate movement of materials within the cell.

> The endoplasmic reticulum is an extensive membrane network that coordinates the synthesis, segregation, and movement of proteins in the eukaryotic cytoplasm.

Golgi Apparatus

The **Golgi apparatus** is a series of flattened membranous sacs. It forms a continuous network with the rough endoplasmic reticulum (FIG. 5-38). Substances are packaged at the Golgi and transported to the plasma membrane or organelles within the cell. Vesicles, derived from the endoplasmic reticulum, carry protein and lipids to the Golgi apparatus where these materials can be chemically incorporated into membrane-bound vesicles.

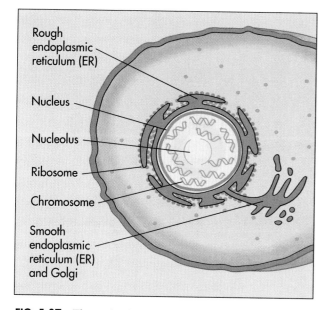

FIG. 5-37   The endoplasmic reticulum (ER) is an extensive membrane network that runs throughout the eukaryotic cell. Regions of the ER that have attached ribosomes are called rough ER; those lacking ribosomes are called smooth ER. These names are derived from the appearances of the ER when viewed by electron microscopy.

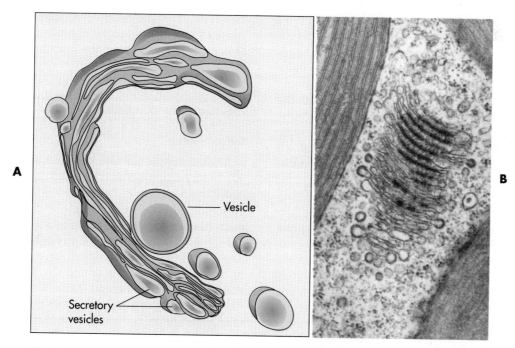

FIG. 5-38   **A,** The Golgi apparatus is involved in the packaging of substances for export from the cell. Secretory vesicles are formed at the Golgi apparatus that carry substances out of the cell. **B,** Colorized electron micrograph of the Golgi apparatus (*tan*) of the alga *Euglena*. (30,400×).

Repackaging of materials into special **secretory vesicles** occurs in the Golgi apparatus. The secretory vesicles move to the plasma membrane where they release their contents through exocytosis. Such a process is important for the construction of structures external to the plasma membrane, such as the cell walls of eukaryotic cells. Digestive enzymes that are used extracellularly (outside the cell) are also packaged at the Golgi apparatus. They are subsequently secreted from the eukaryotic cell. Thus materials are packaged and routed by the Golgi apparatus for distribution to specific sites.

The Golgi apparatus is involved in packaging and distribution of materials.

Lysosomes

Lysosomes are specialized membrane-bound organelles produced at the Golgi apparatus. The lysosomes contain various digestive enzymes that are used within the cell. It is essential that these enzymes do not mix freely with the contents of the cytoplasm. They could digest the substances found there. The lysosome membrane is impermeable to the outward movement of digestive enzymes. It also is resistant to their action. This segregation of digestive enzymes within the lysosome prevents these enzymes from digesting many of the cell's structural components. A lysosome can fuse with a membranous sac, bringing together the contents of each, thus digesting selected materials.

Lysosomes segregate and store digestive enzymes.

When a white blood cell captures a bacterium by phagocytosis, it encloses the bacterial cell in a membranous sac, called a **phagocytic vesicle** or **phagosome** (FIG. 5-39). This membranous vesicle, containing the bacterial cell, fuses with a lysosome. A single structure called a *phagolysosome* is formed. Within the phagolysosome, the lysosomal enzymes can digest the bacterial cell without simultaneously digesting the eukaryotic cell. The remains of the bacterial cell after such digestion are small enough to be transported across the membrane surrounding the phagolysosome into the cytoplasm. Once there, they can be used in the metabolism of the cell. Some bacteria, such as *Salmonella typhi, Legionella pneumophila,* and *Mycobacterium tuberculosis,* are resistant to lysosomal digestive enzymes and survive within the phagolysosome. The ability to resist digestion within phagocytic cells is important to the abilities of these pathogens to resist body defenses and to cause human disease.

Vacuoles

Various types of membrane-bound sacs serve different purposes within the cells of eukaryotic micro-

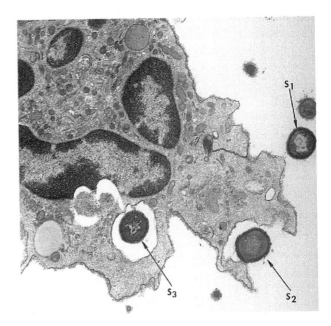

FIG. 5-39 Colorized electron micrograph showing the phagocytosis of the bacterium *Streptococcus pyogenes* by a human polymorphonuclear leukocyte, which is a white blood cell that helps defend the body against infection. (15,000×). One bacterial cell is free (S_1); one is in the process of being phagocytized (S_2); and one is within the phagosome (S_3).

organisms. These membrane-bound sacs are called **vacuoles.** One type of vacuole segregates reserve materials from other cytoplasmic constituents of eukaryotic cells. This is the **storage vacuole.** Other vacuoles fuse with the plasma membrane to move materials out of the cell by exocytosis. In other cases, endocytosis of a food particle results in the formation of a vacuole that fuses with lysosomes. This establishes a digestive vacuole within which the ingested food is digested to small molecules. These small molecules can diffuse across the membrane of the vacuole into the cytoplasm.

Cytoskeleton

The eukaryotic cell also has a cytoskeleton (FIG. 5-40). The **cytoskeleton** is composed of flexible strands called microfilaments (which are about 5 nm thick) and microtubules (which are about 25 nm thick). The microfilaments and microtubules are organized into a three-dimensional network through the cytoplasm. The cytoskeleton is involved in the support and movement of membrane-bound structures, including the plasma membrane and the various organelles of the eukaryotic cell. It helps the cell maintain or change its shape.

Microfilaments contain a protein that allows them to contract. They are involved in the movement of materials within cells. For example, microfilaments move chloroplasts in response to changes in the posi-

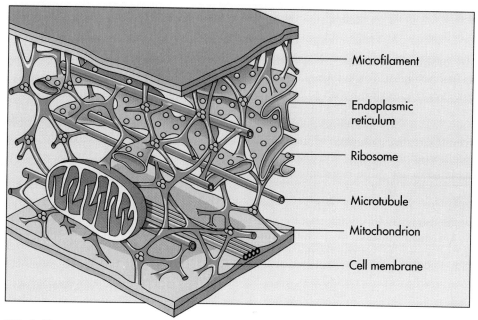

FIG. 5-40   The cytoskeleton is a complex network that links the organelles of the eukaryotic cell. Organelles are attached to microfilaments of the cytoskeleton.

tion of the sun. During division of a eukaryotic cell, microtubules pull chromosomes apart. Microtubules thus aid in the distribution of chromosomes to progeny cells. Microtubules are also involved in cellular movement. For example, microtubules located just beneath the surface of the plasma membrane apparently provide the basis for vesicle movement during endo- and exo-cytosis. The movement of microtubules permits the extension of the cytoplasmic membrane to form the "false feet" (pseudopodia) used by some protozoa, like *Amoeba*. These protozoa move by extending their cytoplasm in a particular direction as they continuously change shape (FIG. 5-41).

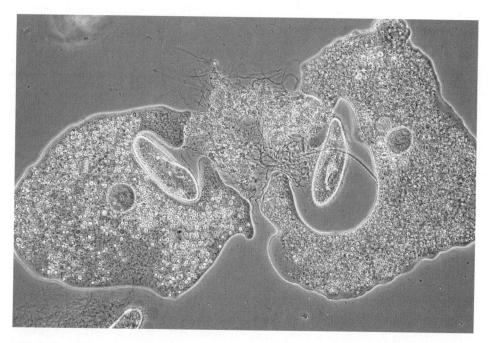

FIG. 5-41   Colorized micrograph of the protozoan *Amoeba* consuming the ciliate protozoan *Paramecium*.

Some organisms produce specialized cells that enhance their survival potential. These specialized cells are called **spores**. Spores may be involved in reproduction, dispersal, or the ability of the organism to withstand adverse environmental conditions. They enhance the overall survival potential of the organism. For example, spores involved in the dispersal of microorganisms usually are quite resistant to desiccation (drying), enabling some microorganisms to survive extended passage through the air. Many spores are *dormant* (nongrowing) with minimal metabolism. They are distinguished from the normal *vegetative cells* of an organism, which are the cells involved in growth.

BACTERIAL ENDOSPORES

The **bacterial endospore** is a heat-resistant spore formed within the cells of a few bacterial genera (FIG. 5-42). The bacterial endospore is highly resistant to elevated temperatures, desiccation (drying), and radiation. It can survive at temperatures as high as 100° C for extended periods, whereas bacterial cells are killed by brief exposures to such high temperatures. Placing an endospore in boiling water, even for several hours, often doesn't kill it.

Several structural and chemical factors contribute to the relative resistance of endospores to conditions that kill vegetative cells. The endospore is a complex multilayered structure that contains a spore coat and a cortex. The cortex contains calcium dipicolinate, but little water, which contributes to its ability to survive under adverse conditions.

> **Bacterial endospores are highly heat-resistant structures that permit survival under adverse conditions.**

Endospores are formed when conditions are unfavorable for the continued growth of bacteria, such as when nutrients are not available. Dormant endospores are formed under starvation conditions. In the first observable stage of spore formation (*sporogenesis*), a newly replicated bacterial chromosome and a small portion of cytoplasm are isolated by an inward growth of the plasma membrane. This forms a structure called the *spore septum* (FIG. 5-43). The spore septum becomes a double-layered membrane that surrounds the chromosome and cytoplasm. A structure, called a *forespore,* that is entirely enclosed as a separate compartment within the original cell is thus produced. Thick layers of peptidoglycan and calcium dipicolinate are laid down between the two membrane layers. The presence of calcium dipicolinate contributes to the heat resistance of the spore.

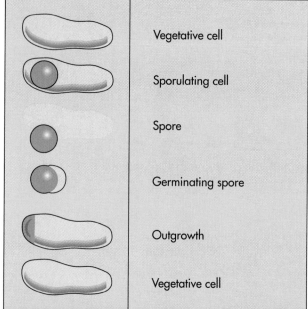

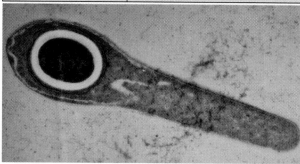

A

B

FIG. 5-42 **A,** The cycle of endospore formation, release, and germination. The endospore is a multilayered structure that is heat resistant. It is formed within a vegetative cell and released when the cell lyses. Subsequently, an endospore can germinate under favorable conditions to form new vegetative cells. **B,** Colorized transmission electron micrograph of an endospore of *Clostridium tetani* shows the complex multilayers of this heat-resistant body within a cell.

Then a thick *spore coat* of protein forms around the outside membrane for additional protection. When the endospore matures, the vegetative cell wall lyses and the endospore is released.

Only a few bacterial genera form endospores. The most important endospore producers are members of two genera, *Bacillus* and *Clostridium*. *Bacillus* and *Clostridium* are defined as Gram-positive rods that form endospores. *Bacillus* is aerobic, growing in the presence of oxygen. *Clostridium* is obligately anaerobic, growing only in the absence of oxygen. Some members of these genera are human pathogens, such as *B. anthracis* (cause of anthrax), *C. perfringens* (cause of gas gangrene), and *C. tetani* (cause of tetanus). The

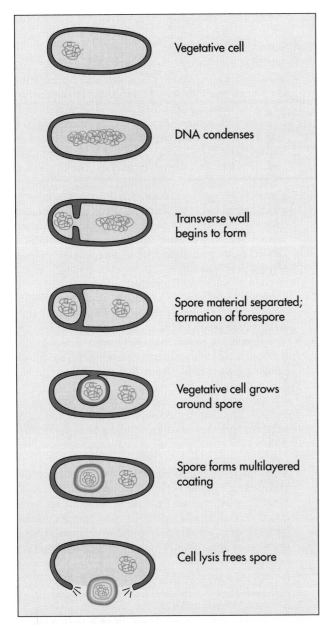

Vegetative cell

DNA condenses

Transverse wall begins to form

Spore material separated; formation of forespore

Vegetative cell grows around spore

Spore forms multilayered coating

Cell lysis frees spore

FIG. 5-43 The formation of an endospore is a complex process that occurs in stages. The spore contains a copy of the bacterial chromosome, which is separated from the rest of the cell by several layers that contain peptidoglycan. There is also a layer containing calcium dipicolinate that contributes to the heat resistance of the spore.

endospores of these bacteria can survive in soil for extended periods, even years. Contamination of a wound with endospores (for example, those of *C. tetani*) can lead to tetanus—often fatal in unvaccinated individuals.

The ability of *Clostridium* and *Bacillus* species to form endospores presents special problems for the food industry. The food industry must employ special processes that rely on heat to prevent spoilage of products. Some endospores can withstand boiling for more than 1 hour. To ensure that such endospores are

killed, it is necessary to heat liquids to a temperature greater than 120° C and hold that temperature for at least 15 minutes. It is especially important to use sufficient heating to kill endospores when canning foods. The endospore-producer *Clostridium botulinum*, the bacterium that causes botulism, sometimes contaminates foods. If the endospores are not killed during canning, this bacterium can grow under the anaerobic conditions in a canned food and produce sufficient toxin to kill anyone who eats that food (antitoxin is available but must be quickly administered).

FUNGAL SPORES

Some fungi produce spores that enhance their survival under conditions of low moisture, high temperature, or lack of nutrients. These spores are not as heat resistant as bacterial endospores. Various fungal spores perform different functions, including reproduction. (Bacterial endospores are not a means of reproduction.) Fungal spores involved in the dispersal of microorganisms usually are quite resistant to desiccation, and the production of such spores is an important adaptive feature that permits the survival of fungi for long periods of time during transport in the air. The spread of many fungi depends on the successful transport of fungal spores from one place to another. Unfortunately, many of these fungi are plant pathogens and cause great agricultural damage as a result of their ability to move effectively from field to field.

PROTOZOAN CYSTS

Some protozoa can form survival structures. The active feeding stage of a protozoan is called a **trophozoite.** Trophozoites are the vegetative stage of several parasitic and free-living protozoa. Trophozoites often cannot withstand harsh environmental conditions. To cope with such conditions many protozoa secrete a thick, resistant covering. They develop into a resting stage called a **cyst.**

A protozoan can produce many cysts, which are readily dispersed in the environment. Since these cysts represent the infectious form of pathogenic protozoa, they represent a high potential for disease transmission. This was clearly evident in Milwaukee, Wisconsin, in the summer of 1993 when there was a major outbreak of the waterborne disease caused by the ingestion of cysts of the protozoa *Cryptosporidium*. Similarly, there have been major outbreaks of infections in cities like St. Petersberg, Russia, due to ingestion of cysts of the protozoan *Giardia*.

The process of encystment represents a type of cell differentiation. Cysts often aid in the spread of disease pathogens.

SUMMARY

Cells p. 121

- Every living organism is made up of one or more cells. The cell is the fundamental unit of living systems.
- Each new cell arises only from cells that already exist.

Plasma Membrane: Movement of Materials into and out of Cells (pp. 121-127)

- All cells are surrounded by plasma membranes, contain cytoplasm, have DNA and ribosomes, and utilize energy from ATP. The plasma membrane is a semipermeable barrier that separates the organism (living system) from the surrounding environment, serving as the boundary layer between the fluid around the cell and the cytoplasm contained within the cell. The plasma membrane controls the flow of materials into and out of the cell, allowing some substances to pass and preventing others from entering or exiting the cell.

Structure of the Plasma Membrane (pp. 122-123)

- The plasma membrane of eubacterial and eukaryotic cells is composed of a phospholipid bilayer and proteins.
- The fluid mosaic model explains the structure of the plasma membrane.

Comparison of Archaebacterial and Eubacterial Plasma Membranes (p. 123)

- Archaebacterial plasma membranes have ether linkages.
- Eubacterial plasma membranes have phospholipids with ester linkages.
- Eubacterial and archaebacterial plasma membranes serve identical functions.

Comparison of Eubacterial and Eukaryotic Plasma Membranes p. 124)

- Plasma membranes of eukaryotic cells contain sterols.
- Plasma membranes of eubacterial cells generally lack sterols.

Transport Across the Plasma Membrane (pp. 124-127)

- Some substances enter and leave the cell by diffusion, moving from regions of high concentration on one side of the membrane to regions of low concentration on the other side of the membrane. When water moves by diffusion, the process is called osmosis. The rate of movement by simple diffusion is dictated by the concentration gradient across the membrane.
- Some substances move across the plasma membrane at accelerated rates—called facilitated diffusion—due to the action of permeases. Active transport uses energy to transport substances through the plasma membrane against a concentration gradient. In eukaryotic cells, substances can enter or leave the cell by cytosis.
- Group translocation modifies substrates as they move across the plasma membrane.
- Phagocytosis occurs when a eukaryotic cell engulfs another cell.

Cell Surface (pp. 128-139)

Cell Wall (pp. 128-132)

- Most bacterial cells have a rigid cell wall surrounding the plasma membrane that protects the cell against lysis due to osmotic shock. Cell walls give bacteria their characteristic shapes. Peptidoglycan makes the cell wall rigid, but if the peptidoglycan is not properly formed or if it is digested, the cell wall is defective and cannot protect the cell from lysis. The Gram-positive cell wall is composed almost entirely of peptidoglycan, a unique biochemical that does not occur in other organisms. The Gram-negative cell wall has a relatively thin layer of peptidoglycan and also additional material, including lipopolysaccharide (LPS).
- LPS is known as endotoxin and can cause adverse reactions in humans.

Bacterial Capsules and Glycocalyces (pp. 133-134)

- Some bacteria have a specialized structure, called a capsule, that surrounds the cell and makes such cells relatively resistant to phagocytosis.
- The glycocalyx, which surrounds some bacterial cells, is responsible for the ability of some bacteria to attach to living tissues and inanimate surfaces.

Pili (p. 134)

- Pili are short hairlike projections that are involved in various attachment processes, such as the mating of bacteria and the attachment of bacteria to human cells.

Bacterial Flagella (pp. 135-137)

- Many bacterial cells are motile by means of flagella. Peritrichous flagella occur all around a bacterial cell. Polar flagella emanate from the ends of the bacterial cell. Bacterial flagella rotate to propel the cell. Motile bacterial cells are able to respond to chemicals, moving toward some and away from others, by a process known as chemotaxis.
- Some very specialized bacteria navigate along magnetic fields by magnetotaxis.

Eukaryotic Flagella and Cilia (pp. 137-139)

- The flagella and cilia of eukaryotic cells undulate in a wave-like motion.

Storage and Expression of Genetic Information (pp. 140-143)

Bacterial Chromosome (p. 140)

- In bacterial cells the DNA is not segregated from the rest of the cell. All of the essential genetic information of a bacterial cell is contained within a single bacterial chromosome.
- The bacterial chromosome is a circular loop of double helical DNA.

Plasmids (pp. 140-141)

- Some bacterial cells have plasmids, which are genetic elements that contain ancillary information.

Nucleus of the Eukaryote Cell (pp. 141-142)

- In eukaryotic cells the DNA is contained within a nucleus.

- Chromosomes of eukaryotic cells are linear.
- Eukaryotic cells often contain multiple pairs of chromosomes.
- The DNA of a eukaryotic chromosome is wound around specialized proteins called histones, which establish the tightly coiled structure of the chromosome.

Ribosomes (pp. 142-143)
- Ribosomes are the sites of protein synthesis.
- The ribosomes of prokaryotic cells are 70S. The ribosomes in the cytoplasm of eukaryotic cells are 80S. Various antibiotics, such as tetracyclines and erythromycin, bind specifically to 70S ribosomes, thereby selectively inhibiting bacterial protein synthesis.

Cellular Structure and Energy Transformations (pp. 143-146)

- Pumping of hydrogen ions across a membrane to establish a hydrogen ion gradient forms the basis of chemiosmotic ATP generation. Mitochondria and chloroplasts are organelles of eukaryotic cells where ATP is generated.

Segregation and Movement of Materials within Cells (pp. 147-151)

- Eukaryotic cells contain various membrane-bound organelles that establish a compartmentalization of function and permit many different and sophisticated specialized activities to occur within an integrated framework. Eukaryotic cells have an extensive internal cytomembrane system that includes the endoplasmic reticulum, the Golgi apparatus, lysosomes, vacuoles and the cytoskeleton.

Survival Through the Production of Spores and cysts (pp. 152-153)

- Members of a few bacterial genera, most notably species of *Bacillus* and *Clostridium*, produce a specialized structure, the endospore, that has great survival value.
- Some endospores can withstand boiling for more than 1 hour.
- Most fungi produce spores that facilitate their dissemination.
- Some protozoa produce resistant resting stages called cysts.

CHAPTER REVIEW

REVIEW QUESTIONS

1. Eubacteria and archaebacteria are prokaryotes. Describe their similarities and differences.
2. Compare and contrast simple diffusion, facilitated diffusion, and active transport.
3. Draw and label the parts of a typical eubacterial cell. Describe the function of each part.
4. Draw and label the parts of a typical fungal cell. Describe the function of each part.
5. Draw and label the parts of a typical algal cell. Describe the function of each part.
6. Describe the cell walls of Gram-positive and Gram-negative bacteria.
7. How do bacteria move? How do they respond to their chemical environment?
8. Describe the structure and function of cilia, flagella, and pseudopodia.
9. Why are endospores known as resting cells?
10. What advantages does having endospores bring to bacteria?
11. What are functions of the bacterial plasma membrane?

CRITICAL THINKING QUESTIONS

1. Why have organelles been hypothesized to be necessary for eukaryotic cells and not for prokaryotic cells? How has the finding of a bacterium with cells of 0.5 mm altered this view?
2. How do bacteria adapt to their environment? How do they find places where they can grow?
3. Eukaryotic and prokaryotic cells have several different structures. How do these differences relate to medical practice and, in particular, the use of drugs to treat diseases caused by microorganisms? Why is it generally easier to cure a patient of a bacterial disease than a protozoan disease?
4. Since eukaryotic human cells have 70S ribosomes in their mitochondria, how can drugs that target the 70S ribosomes—such as erythromycin—be used to treat bacterial infections?
5. How long can endospore-forming bacteria survive? Could there be a million-year-old bacterium that still could be grown in the laboratory?
6. Why are archaebacteria considered to be descendants of prokaryotes that evolved early in the history of life on earth? How do the physiological properties of archaebacteria allow them to survive in extreme environments?

READINGS

Adler J: 1976. The sensing of chemicals by bacteria, *Scientific American* 234(4):40-47.

Interesting description of how bacteria respond to chemical stimuli, moving toward some chemicals and away from others.

Allen RD: 1987. The microtubule as an intracellular engine, *Scientific American* 256(2):42-49.

Video-enhanced microscopy is used to show how microtubules move organelles around the cell.

Becker WM and DW Deamer: 1991. *World of the Cell*, Menlo Park, California; Benjamin-Cummings.

An easy-to-read book with outstanding illustrations of cell structures.

Berg HC: 1975. How bacteria swim, *Scientific American* 233(2):36-44.

Describes how cells move, the way in which flagella propel bacteria, and how all this was discovered.

Blakemore RP and RB Frankel: 1981. Magnetic navigation in bacteria, *Scientific American* 245(6):58-67.

Discusses the surprising discovery that some bacteria respond to magnetic fields and can navigate by magnetotaxis.

Costerton JW, GG Geesey, K-J Cheng: 1978. How bacteria stick, *Scientific American* 238(1):86-95.

Considers revelations about how bacteria attach to substances, focusing on the role of the glycocalyx; includes spectacular micrographs.

DeDuve C: 1984 and 1985. A Guided Tour of the Living Cell, New York, W.H. Freeman.

These two short volumes produce an excellent introduction to the cell. Illustrated with freeze-fracture electron micrographs.

Drlica K and M Riley (eds.): 1990. *The Bacterial Chromosome*, Washington, D.C.; American Society for Microbiology.

Reviews background and current research into the structure and function of the bacterial chromosome, includes genetic and physical maps, DNA configuration, and global responses to stress, among other topics.

Ferris FG and TJ Beveridge: 1985. Functions of bacterial cell surface structures, *Bioscience* 35:172-176.

Reviews the significance of bacterial cell wall features.

Hill WE et al. (eds.): 1990. *The Ribosome: Structure, Function, and Evolution*, Washington, D.C.; American Society for Microbiology.

Explores the state of our knowledge and recent research on the ribosome in eukaryotic and prokaryotic organisms. Includes how ribosomes and antibiotics interact and the evolution of the ribosome.

Koch AL: 1990. Growth and form of the bacterial cell wall, *American Scientist* 78:327-341.

An informative review of the formation of bacterial cell walls.

Kornberg RD and A Klug: 1981. The nucleosome, *Scientific American* 244(2):55-72.

Detailed description of the structure of the nucleosome.

Marchalonis JJ: 1988. *The Lymphocytes: Structure and Function*, New York, Marcel Dekker.

Explores the biochemical nature and molecular biological and functional properties of lymphocytes in immune recognition.

Neidhardt FC, JL Ingraham, M Schaechter: 1990. *Physiology of the Bacterial Cell: A Molecular Approach*, Sunderland, Massachusetts; Sinauer Associates.

An extensive review of prokaryotic cell structures and functions.

Rothman J: 1985. The compartmental organization of the Golgi apparatus, *Scientific American* 253(3):74-89.

A close-up view of Golgi structure and function, especially of the sacs of the Golgi, which form three biochemically distinct processing units. (Offprint No. 1563).

Shapiro JA: 1988. Bacteria as multicellular organisms, *Scientific American* 258(6):82-89.

Bacteria in the laboratory form multicellular communities that take specific shapes and follow programmed movements.

Weber K and M Osborn. 1985. The molecules of the cell matrix, *Scientific American* 253(4):110-120.

New techniques in biochemistry and microscopy are used to probe the structure and function of the cytoskeleton.

CHAPTER 6

Cellular Metabolism

PREVIEW TO CHAPTER 6

In this chapter we will:
- Study the chemical reactions mediated by enzymes that constitute the metabolism of a cell.
- Learn about the diverse strategies of metabolism that enable cells to sustain life functions.
- Examine how a cell meets its energy needs by generating ATP.
- Compare various forms of autotrophic and heterotrophic metabolism.
- See the chemical reactions that occur in various metabolic pathways, including fermentation, respiration, photosynthesis, and chemolithotrophy.
- Learn the following key terms and names:

activation energy
active site
adenosine triphosphatase
  (ATPase)
aerobic respiration
alcoholic fermentation
anabolic pathway
anaerobic respiration
autotrophic metabolism
butanediol fermentation
butanol fermentation
butyric acid fermentation
Calvin cycle
capnophiles
catabolic pathway
cellular metabolism
chemiosmosis
chemoautotrophic
  metabolism
chemolithotrophic
  metabolism
coenzyme
electron transport chain

Embden-Meyerhof
  pathway
enzymes
heterolactic fermentation
heterotrophic metabolism
homolactic fermentation
Krebs cycle
lactic acid fermentation
Methyl Red (MR) test
microaerophiles
mixed-acid fermentation
nitrogen fixation
nitrogenase
photoautotrophic
  metabolism
propionic acid
  fermentation
protonmotive force
respiration
substrate-level
  phosphorylation
substrates
Voges-Proskauer test

The evolution of living cells began with the use of the energy released from a chemical bond. Early organisms evolved that were able to convert this energy into ATP. Proteins evolved diverse catalytic functions, making possible the retention of a larger portion of the chemical bond energy available in abiotic (nonliving) organic molecules. Breaking a series of chemical bonds in successive steps enabled this to occur and allowed cells to carry out metabolism.

Metabolic processes are believed to have evolved in the essentially oxygen-free atmosphere that characterized the atmosphere of the Earth during the time life began on Earth. Primitive life forms are thought to have obtained chemical energy by breaking down organic molecules formed by non-metabolic reactions.

The totality of all of the chemical reactions that a cell carries out is called **cellular metabolism.** Through the process of cellular metabolism, cells bring about chemical changes through which they obtain energy and materials for growth and reproduction. Energy is required for living things to sustain life processes. Cells can store energy or use it for the synthesis of new molecules by controlling the status of chemical bonds. Adenosine triphosphate (ATP) is the central chemical in the energy transformations of cellular metabolism. ATP cannot be stored for long periods of time and therefore must be continually made. Within a living cell, the flow of energy involves the formation and consumption of ATP. Cellular metabolism transforms the energy stored in light or chemicals into ATP. Cellular metabolism also transforms starting materials into the numerous carbon-containing chemicals that make up the structural and functional components of the cell.

> **Energy and materials are transformed within living cells through a complex integrated network of chemical reactions that collectively constitute the metabolism of the cell, or cellular metabolism.**

ROLE OF ENZYMES

Cellular metabolism is based on chemical reactions catalyzed by enzymes. **Enzymes** are biological catalysts, which are substances produced by cells that accelerate the rates of chemical reactions. Almost all biological catalysts are proteins but a very few are RNA. Virtually every step in cell metabolism involves an enzyme. Enzymes increase the rates of a cell's chemical reactions by more than a million times.

Energy is required for a chemical reaction to occur. Enzymes bind to molecules and bonds in such a way that the energy required to initiate a chemical reaction, called the **activation energy**, is lowered (FIG. 6-1). This lowering of the activation energy is critical because it permits reactions to occur at life-support-

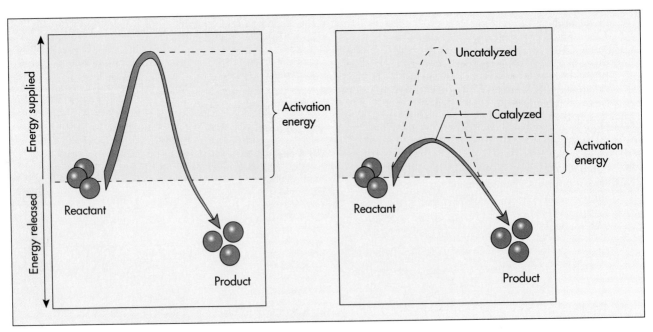

FIG. 6-1 An input of energy, called the activation energy, is needed to start a chemical reaction. A catalyst lowers the activation energy. In biological systems, enzymes serve as the catalysts to lower the activation energy.

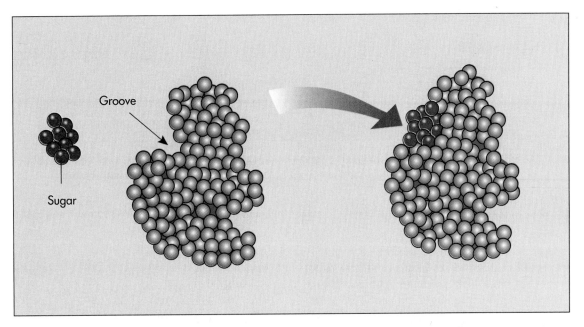

FIG. 6-2   The fit between the enzyme and the substrate to form an enzyme-substrate complex has been likened to that of a lock and key. Actually, this interaction modifies the three-dimensional structure of the enzyme so that the substrate induces its fit to the enzyme. The precision of fit is responsible for the high degree of specificity of enzymes for particular substrates.

ing temperatures. The result is that the reaction occurs more rapidly than it otherwise would at those temperatures.

An input of energy—the activation energy—is needed to initiate a chemical reaction.

Enzymes are biological catalysts that lower the activation energy so that chemical reactions can occur within living cells.

Some molecules can bind to a particular enzyme so that the enzyme can catalyze a chemical reaction. Such molecules are called the **substrates** of that enzyme (FIG. 6-2). Enzymes can bind to substrate molecules because the three-dimensional shape of the enzyme fits the substrate molecule, sort of like a lock fits a key. Enzymes exhibit great specificity in the reactions that they catalyze. When an enzyme binds to a substrate it forms an enzyme-substrate complex. The enzyme-substrate complex then breaks down, releasing the enzyme and the product(s) of the reaction. The enzyme is not consumed in the overall reaction and can continue to act as a catalyst.

Substrate + Enzyme $\rightleftarrows$ Enzyme-substrate complex $\rightleftarrows$ Enzyme + Product

Different enzymes are needed to bring about reactions that transform even very similar chemical compounds. Thousands of enzymes involved in the metabolic reactions of each cell are necessary for cellular growth and reproduction. Enzymes catalyzing key reactions in metabolic pathways govern whether molecules are degraded as an ATP-generating energy source or converted for use in biosynthesis.

The actual site of the enzyme that is responsible for its catalytic action is called the **active site.** Because protein shapes are not rigid, when some enzymes bind to their substrate, there may be a slight alteration in the shape of the enzyme molecule so that there is a good fit between substrate and enzyme. The fit is essential for the enzymatic reaction to occur. If the shape of the enzyme is altered so that it can no longer function as a catalyst, the enzyme is said to be denatured. Heating and certain chemicals can denature enzymes, destroying their catalytic activities.

Enzymes exhibit a high degree of substrate specificity.

The enzymes a particular cell synthesizes will determine which chemical reactions occur in cellular metabolism of that cell.

COENZYMES AND OXIDATION-REDUCTION REACTIONS

Many of the chemical reactions in cellular metabolism are oxidation-reduction reactions in which electrons and protons are exchanged between molecules. Many of these reactions are used to extract energy from organic compounds. In these reactions, electrons and protons often are transferred to a molecule

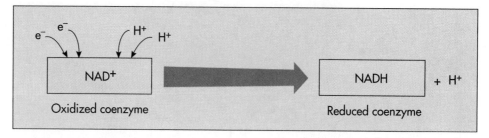

FIG. 6-3 The reduction of the oxidized coenzyme NAD^+ to the reduced coenzyme NADH + H^+ is a critical reaction that often is coupled with the oxidation of substrates within a cell. This reaction can be written as $NAD^+ \rightarrow$ NADH.

called a coenzyme (FIG. 6-3). A **coenzyme** is an organic molecule that serves as a carrier of electrons and/or protons during metabolism. The coenzyme NAD^+ (nicotinamide adenine dinucleotide) is the common temporary holder of electrons and protons in many metabolic pathways. An example of such a reaction is:

$$malate + NAD^+ \longrightarrow oxaloacetate + NADH + H^+$$

The reduced coenzyme NADH formed in this reaction can then donate an electron and a proton to another molecule so that the coenzyme is reoxidized. Other important coenzymes used in cellular metabolism are NADP (nicotinamide adenine dinucleotide phosphate) and FAD (flavin adenine dinucleotide).

ATP AND CELLULAR ENERGY

A central concern of cellular metabolism is the flow of energy. All of the activities of living organisms use energy. Living systems can neither create nor destroy energy. Rather, living systems transform energy, capturing energy from one source and using that energy to drive the essential chemical reactions that enable cells to carry out the life functions of growth and reproduction. Some cells capture energy directly from sunlight; others obtain energy from the oxidation ("burning") of organic or inorganic chemicals. Regardless of the source of energy, all cells employ the same strategy of carrying out metabolic

reactions that transfer energy to molecules of ATP. Then ATP serves as the molecule for transferring energy within the cell. A growing cell of the bacterium *Escherichia coli* must synthesize approximately 2.5 million molecules of ATP per second to support its energy needs.

> **During metabolism, energy is transferred to and stored within molecules of ATP.**

> **ATP is the universal energy carrier of all living cells.**

In particular, the energy from ATP is used to drive the energy-requiring reactions of biosynthesis that are needed for cellular growth and reproduction. ATP contains a phosphate functional group joined to the rest of the molecule by a high-energy bond (FIG. 6-4). Breaking this bond yields an inorganic phosphate group (P_i) and a molecule of adenosine diphosphate (ADP) and releases a large amount of energy—approximately 7300 calories for every 6.023×10^{23} molecules (mole) of ATP converted to ADP + P_i. This energy can be used to drive the cell's energy-requiring chemical reactions.

> **In cellular metabolism, energy-requiring reactions occur because they are coupled with reactions that cleave ATP.**

The formation of ATP from ADP + P_i is an energy-requiring reaction. This reaction cannot occur spontaneously. When coupled to a spontaneously occur-

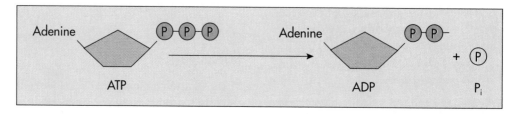

FIG. 6-4 Adenosine triphosphate (ATP) is a compound with high-energy phosphate bonds. When ATP is converted to adenosine diphosphate (ADP) a high-energy phosphate bond is cleaved, releasing about 7.5 kcal/mole that can be used to drive other chemical reactions.

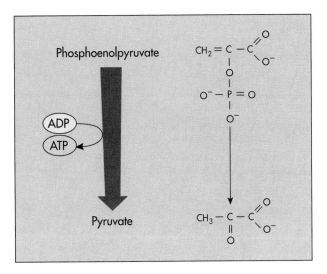

FIG. 6-5 In substrate-level phosphorylation an energy yield reaction directly provides the energy for the generation of ATP; this occurs via a coupled reaction.

ring, energy-releasing reaction, however, the synthesis of ATP from ADP + P_i does take place, because energy is now available to drive the synthesis of ATP. The generation of ATP by coupling energetically favorable reactions to the synthesis of ATP from ADP plus either an organic or inorganic phosphate source is called **substrate-level phosphorylation** (FIG. 6-5).

In other cellular reactions the chemical formation of ATP is driven by a diffusion force in a process known as **chemiosmosis** (FIG. 6-6). Protons are

pumped out of prokaryotic cells across the plasma membrane or the membrane-bound organelles (mitochondria and chloroplasts) of eukaryotic cells as a result of oxidation-reduction reactions that transport electrons and protons through membrane-embedded carriers. As the proton concentration across the membrane becomes higher on one side than the other, the protons on that side of the membrane are driven back across the membrane. The force exerted by these protons to drive them back across the membrane is called the **protonmotive force.** The passage of these protons through specific channels in the membrane provides the energy for the formation of ATP from ADP + P_i.

METABOLIC PATHWAYS AND CARBON FLOW

The chemical reactions of metabolism occur in sequences. In each sequence the product of one chemical reaction becomes the substrate for the enzyme that catalyzes the next reaction. The overall ordered sequences of enzyme-catalyzed chemical reactions are called **biochemical pathways** or **metabolic pathways.** The sequential steps between the starting substrate molecule(s) and the end product(s) constitute the **intermediary metabolism** of the cell.

> The enzymatically mediated metabolic reactions of a cell proceed via a series of small discrete steps that establish a metabolic pathway.

Metabolic pathways that involve the breakdown (degradation) of organic molecules are said to be

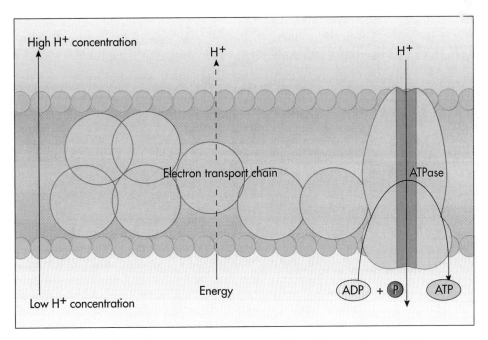

FIG. 6-6 In chemiosmosis the metabolic reactions of the cell are used to establish a proton gradient across a membrane. This proton gradient exerts a force called the protonmotive force that is used to generate ATP.

catabolic (meaning to break down). A catabolic pathway is one in which larger molecules are split into smaller ones. Such pathways can be used to obtain energy from organic chemicals, for example. The processes that expend energy to synthesize organic molecules are said to be **anabolic** (meaning to build up). The catabolic and anabolic pathways of a cell are interconnected so that the substrates used to feed the cell can be changed into the molecules that make up a living cell.

> **Catabolism means degradative process and anabolism means biosynthetic process.**

Cells must be able to process matter from available starting material into their own structural and functional components. As with energy, the chemical reactions of living systems can neither create nor destroy matter. Cells obtain matter from their surroundings and the chemical reactions the cells perform can change the combinations of atoms within that matter to form new molecules. Thus the various available starting substrate molecules are transformed by cellular metabolism into the many different macromolecules of the cell. These macromolecules include, among others, proteins for enzymes, lipids for membranes, carbohydrates for various structures such as cell walls, and nucleic acids for the storage and expression of genetic information.

Carbon is the backbone atom of all organic chemicals. In terms of carbon flow, the basic strategy of the cell is to form relatively small molecules that can act as the basis for the carbon skeletons of larger macromolecules (FIG. 6-7). When a microorganism uses an organic substrate, like glucose, to generate ATP, it follows a catabolic pathway to break that molecule down into smaller compounds; these smaller compounds then act as building blocks—called **precursors**—for the biosynthesis of macromolecules. Then the microorganism uses an anabolic pathway to transform small molecules into larger molecules.

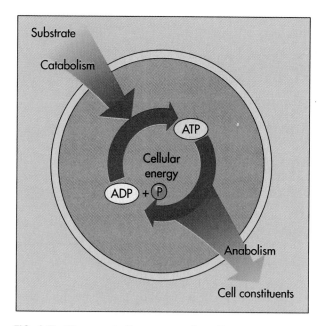

FIG. 6-7   The metabolic strategy of a cell is to break down substances into smaller compounds via catabolic pathways. The small compounds that are formed are used as the substrates for biosynthetic reactions in anabolic pathways.

AUTOTROPHIC AND HETEROTROPHIC METABOLISM

Microorganisms exhibit differing strategies of metabolism for meeting their common needs. These needs include synthesizing ATP and transforming carbon-containing molecules into the macromolecules that constitute the cells of the microorganism. Two distinct modes of microbial metabolism have evolved for accomplishing these tasks: **autotrophy** (meaning self-feeding) and **heterotrophy** (meaning other-feeding).

Autotrophic Metabolism

Microorganisms with **autotrophic metabolism** are called **autotrophs** (Table 6-1). The cellular metabo-

TABLE 6-1	
Types of Autotrophic Microbial Metabolism Used to Generate ATP	
TYPE OF METABOLISM	**DESCRIPTION**
Oxygenic photosynthesis	Uses two connected photosystems and results in evolution of oxygen, as well as generation of ATP; carried out by algae and cyanobacteria, which gain reducing power (H^+) from photolysis of water
Anoxygenic photosynthesis	Uses one photosystem and does not result in evolution of oxygen; carried out by anaerobic photosynthetic bacteria, e.g., green and purple sulfur bacteria, and under some conditions by cyanobacteria, which gain reducing power (H^+) from H_2S or organic compounds
Chemoautotrophic (chemolithotrophic)	Uses oxidation of inorganic compounds such as sulfur, nitrite, nitrate, and hydrogen to establish a protonmotive force across a membrane that results in generation of ATP by chemiosmosis

lism of autotrophic organisms uses inorganic carbon dioxide as a source of carbon for the biosynthesis of the molecules of the cell. Also, autotrophic metabolism almost always generates ATP from the oxidation of inorganic compounds or through the conversion of light energy to chemical energy—not from organic compounds. Photoautotrophic (photosynthetic) microorganisms use light energy and chemoautotrophic (chemolithotrophic) microorganisms use the energy derived from oxidation of inorganic compounds to supply energy needed for synthesis of ATP.

> **Carbon for the macromolecules of autotrophic microorganisms originates from inorganic carbon dioxide.**

> **Autotrophic metabolism does not use organic compounds for the generation of ATP but rather captures light energy or energy from the oxidation of inorganic chemicals; the cellular carbon of autotrophs comes from carbon dioxide.**

Heterotrophic Metabolism

In **heterotrophic metabolism** the generation of ATP is based on the use of an organic substrate molecule (Table 6-2). The conversion of the organic substrate to end products occurs via a metabolic pathway that releases sufficient energy to be coupled with the synthesis of ATP. The catabolic pathway involves reactions that break down an organic molecule into smaller molecules. Besides using organic compounds to provide energy for ATP generation, heterotrophs obtain their cellular carbon from organic substrates. In heterotrophic metabolism, organic compounds are broken down into smaller molecules, called **intermediary metabolites**, that subsequently are used for biosynthesis.

> **Heterotrophic metabolism uses organic chemicals to supply the energy for ATP generation; the cellular carbon of heterotrophs also comes from organic compounds.**

Heterotrophic metabolism occurs by either of two processes: respiration or fermentation. Respiration links the metabolism of an organic substrate with the utilization of an inorganic compound. **Respiration** is defined as a type of metabolism involving oxidation-reduction reactions where the final electron acceptor that completes the metabolism is an inorganic molecule. Often the inorganic compound is molecular oxygen (O_2) so that the process is dependent on air (aerobic respiration) but, in some cases, respiration occurs in the absence of air (anaerobic respiration). ATP is formed during respiration both by substrate-level phosphorylation and by chemiosmosis.

Fermentation is an anaerobic catabolic process that releases energy from sugars or other organic compounds. During fermentation the final electron acceptor is an organic molecule. Only substrate-level phosphorylation is used to generate ATP in a fermentation pathway. During fermentation, hydrogen ions and electrons are transferred from NADH to pyruvic acid, which is turned into various end products. Various microorganisms are able to ferment different substrates; the end products depend on the particular microorganism, the substrate, and the activity of the enzymes that are present (Table 6-3, p. 164).

> **Respiration requires an inorganic substance, often molecular oxygen, to complete the metabolism of an organic substrate; ATP is generated by both substrate level phosphorylation and chemiosmosis during respiration.**

TABLE 6-2	
Types of Heterotrophic Microbial Metabolism Used to Generate ATP	
TYPE OF METABOLISM	**DESCRIPTION**
Respiration	Uses complete oxidation of organic compounds, requiring an external electron acceptor to balance oxidation-reduction reactions used to generate ATP; much of the ATP is formed as a result of chemiosmosis based on the establishment of a proton gradient across a membrane
Aerobic respiration	Uses oxygen as the terminal electron acceptor in the membrane-bound pathway that establishes the proton gradient for chemiosmotic ATP generation
Anaerobic respiration	Uses compounds other than oxygen, e.g., nitrate or sulfate, as the terminal electron acceptor in the membrane-bound pathway that establishes the proton gradient for chemiosmotic ATP generation
Fermentation	Does not require an external electron acceptor, achieving a balance of oxidation-reduction reactions using the organic substrate molecule; various fermentation pathways produce different end products

TABLE 6-3	
Types of Fermentative Metabolism	
FERMENTATION PATHWAY	**END PRODUCTS**
Homolactic acid	Lactic acid
Heterolactic acid	Lactic acid + ethanol + CO_2
Ethanolic	Ethanol + CO_2
Propionic acid	Propionic acid + CO_2
Mixed acid	Ethanol + acetic acid + lactic acid + succinic acid + formic acid + H_2 + CO_2
Butanediol	Butanediol + CO_2
Butyric acid	Butyric acid + butanol + acetone + CO_2

Fermentation does not require an inorganic substance to utilize an organic substrate for generating ATP as a form of usable cellular energy and materials for growth; fermentation does not require air; ATP is generated exclusively by substrate level phosphorylation.

Microorganisms can be grouped into categories based on their requirement for or intolerance of oxygen. **Aerobes** grow in the presence of air that contains molecular oxygen. They generally grow only by respiratory metabolism. Other microorganisms, called **microaerophiles,** grow only at reduced concentrations of molecular oxygen. Such organisms require oxygen for growth but only at concentrations (about 5%) lower than atmospheric levels (20%). Generally, microaerophilic organisms will not grow in air. Some microaerophiles prefer to grow at elevated carbon dioxide concentrations (5% to 10%) and are called **capnophiles.** *Campylobacter jejuni,* a bacterium that causes gastroenteritis and diarrhea, is capnophilic (FIG. 6-8).

Facultative anaerobes can grow in the presence or absence of air. Many facultative anaerobes, such as *Escherichia coli,* switch between respiration and fermentation, depending on the availability of molecular oxygen; they usually carry out fermentative metabolism in the absence of oxygen and respiration in the presence of oxygen. Other facultative anaerobes, such as streptococci, carry out only a fermentative metabolism whether or not oxygen is present.

Other bacteria are **anaerobes** and grow only in the absence of air. **Obligate anaerobes,** such as *Clostridium* species, can only carry out fermentative metabolism. Only a few groups of bacteria and protozoa are obligate anaerobes. Some of these, such as *Clostridium botulinum,* the bacterium that causes botulism, are very sensitive to oxygen and even a brief exposure to oxygen will kill them.

FIG. 6-8 Colonies of *Campylobacter jejuni* after 48 hours of incubation on Campy blood agar in an atmosphere with reduced oxygen concentration.

METABOLIC PATHWAYS

The metabolism of a cell occurs via a specific series of chemical reactions, called **metabolic pathways,** in which energy is transformed to generate ATP. A metabolic pathway has discrete steps between a starting substance (substrate molecule) and the products of the chemical reactions (end products). All of the ATP-generating strategies involve metabolic pathways consisting of multiple discrete enzyme-catalyzed steps. These steps operate in a specific sequence to convert an initial substrate or substrates into the end product or products of the pathway, accompanied by the formation of ATP. Several central metabolic pathways have key roles in the metabolism of microbial cells. The reactions that lead to ATP generation involve various intermediary metabolites linked together in a series of small steps to form unified metabolic pathways. The intermediary metabolites and the cellular energy are used to build the molecules of new cells for growth and reproduction.

RESPIRATION

A respiration pathway has three distinct phases (FIG. 6-9). These three phases for the respiratory metabolism of glucose are: phase 1, glycolysis; phase 2, Krebs cycle; and phase 3, electron transport chain. The overall respiration pathway results in the formation of carbon dioxide from the organic substrate

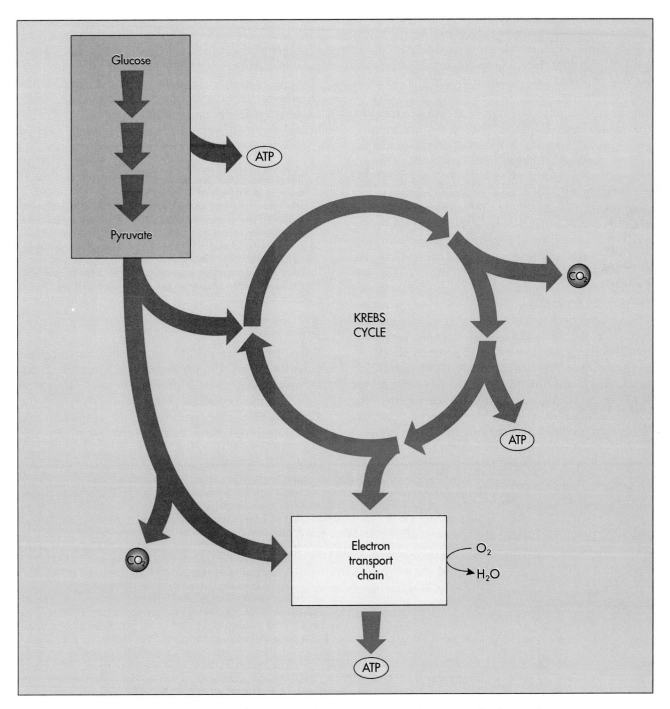

FIG. 6-9 Respiration occurs in three distinct metabolic pathways. In the first pathway, called glycolysis, carbohydrates are broken down to pyruvate. Then in the Krebs cycle the pyruvate is further metabolized to carbon dioxide. Reduced coenzymes formed during glycolysis and the Krebs cycle are used to generate a protonmotive force for chemiosmotic generation of ATP in the third and final phase of respiration.

molecule. In aerobic respiration, water is also produced as a result of the reduction of oxygen. The classic equation that summarizes aerobic respiration using glucose as substrate is:

$$C_6H_{12}O_6 + 6O_2 \longrightarrow 6CO_2 + 6H_2O$$
$$\underset{\text{glucose}}{} \quad \underset{\text{oxygen}}{} \quad \underset{\substack{\text{carbon} \\ \text{dioxide}}}{} \quad \underset{\text{water}}{}$$

In the first phase of a respiration pathway, the organic substrate is converted into small organic molecules. For example, glucose can be converted to pyruvate in a pathway called **glycolysis.** The word glycolysis comes from the Greek, *glyco,* meaning sweet (sugar), plus *lysis,* meaning to break. For each molecule of glucose that passes through this catabolic pathway, the cell acquires a few ATP molecules by substrate-level phosphorylation. It also produces reduced coenzyme (NADH). This means that glycolysis captures about 2% of the available chemical energy of glucose within the ATP molecules that are formed.

The metabolites produced in this first stage of respiration, for example pyruvate, next enter the **Krebs cycle** (named for its discoverer, Hans Krebs). In the Krebs cycle these small organic compounds are further oxidized, producing inorganic carbon dioxide and generating reduced coenzyme (NADH and FADH$_2$). Whereas glucose and other carbohydrates are initially broken down to pyruvate via glycolysis, other classes of chemicals are converted via different metabolic pathways into small organic molecules that can also enter the second stage of respiration—the Krebs cycle. Proteins are broken down into amino acids, which are further converted to form various small organic acids that can enter the Krebs cycle. Lipids are converted to fatty acids, which are broken down into products that can enter the Krebs cycle.

In the third and final stage of respiration, reduced coenzyme molecules formed during glycolysis and the Krebs cycle are reoxidized and ATP is generated by chemiosmosis. This reoxidation of reduced coenzymes involves the transport of electrons through a series of membrane-bound carriers to the terminal electron acceptor. This transfer of electrons is called **electron transport** and the carriers are referred to as the **electron transport chain.** The process of electron transport through the membrane-embedded carriers also results in the expulsion of hydrogen ions (protons) across the membrane. Because substances move by diffusion from regions of high to low concentration, the difference of proton concentration across the membrane exerts a force. This protonmotive force, caused by the concentration gradient of protons across the membrane, is used to generate ATP by chemiosmosis.

> **Respiration involves three distinct phases: the catabolic breakdown of a substrate by glycolysis with the generation of some ATP and reduced coenzyme (NADH), the complete oxidation of the intermediary metabolite to produce CO$_2$ in the Krebs cycle with the generation of additional ATP and reduced coenzymes (NADH + FADH), and the reoxidation of the reduced coenzymes involving an electron transport chain with the resultant generation of ATP by chemiosmosis.**

Glycolysis

Glycolysis is the initial stage of all the main pathways of carbohydrate metabolism. It occurs within the cytoplasm in both prokaryotic and eukaryotic cells. The most common pathway of glycolysis is the **Embden-Meyerhof pathway** (FIG. 6-10). In this pathway the 6-carbon molecule glucose is first con-

FIG. 6-10 The Embden-Meyerhof pathway of glycolysis is a central metabolic pathway in various eukaryotic and prokaryotic cells for the conversion of carbohydrates to pyruvate and the formation of ATP. Each red ball represents a carbon atom. In the Embden-Meyerhof pathway a molecule of glucose is converted to two molecules of pyruvate, with the net production of two molecules of ATP and two molecules of reduced coenzyme NADH. In the initial step in the Embden-Meyerhof pathway, glucose is converted to glucose-6-phosphate. ATP is used here rather than produced. The reactive glucose-6-phosphate is then modified and converted to fructose-1,6-bisphosphate. Again, ATP is consumed rather than produced so that now the net ATP balance is minus two. Fructose-1,6-bisphosphate is enzymatically split into two phosphate-containing molecules: glyceraldehyde-3-phosphate (PGA) and dihydroxyacetone phosphate (DHAP). DHAP is converted into PGA so that now there are two molecules of PGA, and beyond this point in the pathway each step occurs twice for each molecule of glucose metabolized. Next, the PGA molecule loses 2 electrons—becoming more oxidized—and the electrons are transferred to NAD$^+$ to form NADH. Also in this reaction an additional phosphate group is added to form 1,3-bisphosphoglycerate. This is a substrate level phosphorylation reaction in which the generation of ATP does not depend on chemiosmosis. At this point in the glycolytic pathway there is an exact balance between the amount of ATP consumed and the amount of ATP generated.

Now net ATP production begins. The 1,3-bisphosphoglycerate is used to convert ADP to ATP, thus balancing the original energy investment of ATP. The resulting molecule is rearranged to produce phosphoenolpyruvate (PEP). PEP is transformed into pyruvate, the end product of the glycolytic pathway. It is in this last reaction that a phosphate group is added to ADP to form additional ATP, yielding the net gain of two ATP molecules for each glucose molecule degraded via this pathway.

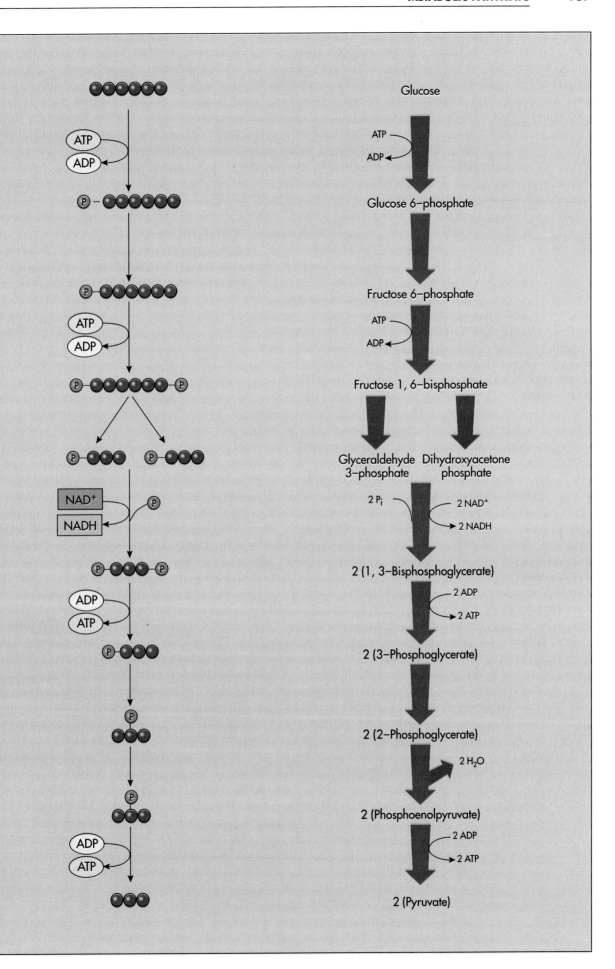

verted to the phosphate-containing compound fructose 1,6 bisphosphate. This conversion requires an input of energy so that two ATP molecules are consumed in the reactions that bring about this conversion. The energy that has been transferred to fructose 1,6 bisphosphate subsequently is used to form ATP. Fructose 1,6 bisphosphate, which like glucose has 6-carbon atoms, is broken down into molecules that have three carbon atoms. This results in the production of two molecules of pyruvate (see FIG. 6-10). It also results in the net production of two molecules of NADH and the net synthesis of two ATP molecules so that the overall equation for the Embden-Meyerhof pathway of glycolysis can be written as:

$$\text{glucose} + 2\,\text{ADP} + 2\,\text{P}_i + 2\,\text{NAD}^+ \longrightarrow$$
$$2\,\text{pyruvate} + 2\,\text{NADH} + 2\,\text{ATP}$$

where P_i stands for inorganic phosphate.

In the Embden-Meyerhof pathway of glycolysis, glucose is partially broken down into pyruvate, two NADH molecules are formed, and the energy released leads to a net yield of two ATP.

Krebs Cycle

The second phase of respiratory metabolism occurs when the pyruvate generated by glycolysis feeds into the **Krebs cycle,** which is also known as the **tricarboxylic acid** or **citric acid cycle** (FIG. 6-11). In prokaryotic cells, the Krebs cycle occurs within the cytoplasm. In eukaryotic cells, this metabolic process occurs within mitochondria. In the series of chemical reactions that make up the Krebs cycle, the potential chemical energy stored in intermediate compounds derived from pyruvate is released in a series of oxidation-reduction reactions. As a result of the reactions of the Krebs cycle the pyruvate molecules formed during glycolysis are oxidized to form carbon dioxide. Thus at the end of the Krebs cycle, six carbon dioxide molecules are produced for each 6-carbon glucose molecule metabolized.

To begin the Krebs cycle, pyruvate produced by glycolysis is split and a fragment of it is attached to coenzyme A (CoA). The combined molecule is called *acetyl-CoA.* Acetyl-CoA then enters the Krebs cycle, initiating a series of reactions that release electrons and protons (H$^+$). NADH is generated during several reactions of the Krebs cycle. Another coenzyme, flavin adenine dinucleotide (FAD), is also reduced to FADH$_2$. One of the reactions of the Krebs cycle is also directly coupled with the substrate-level generation of a high-energy phosphate-containing compound called guanidine triphosphate (GTP). GTP can be converted to ATP, and, for accounting purposes, the GTP generated in this reaction is counted as ATP in determining the net synthesis of ATP during respiration.

The net reaction of metabolism through the Krebs cycle, starting with the pyruvate generated from glucose, can be written as:

$$2\,\text{pyruvate} + 2\,\text{ADP} + 2\,\text{FAD} + 8\,\text{NAD}^+ \longrightarrow$$
$$6\,\text{CO}_2 + 2\,\text{ATP} + 2\,\text{FADH}_2 + 8\,\text{NADH}$$

At the end of the Krebs cycle, the cell has converted all of the substrate carbon of the glucose molecule to carbon dioxide. There also has been a net synthesis of four ATP molecules—the production of ten reduced coenzyme molecules as NADH and the generation of two reduced coenzyme molecules as FADH$_2$.

In the Krebs cycle, intermediary metabolites, such as pyruvate, are completely oxidized to CO$_2$ with the production of reduced coenzymes and some ATP.

The Krebs cycle and glycolytic pathways have important roles within the overall respiratory generation of ATP. They also occupy a central place in the flow of carbon through the cell. As a result of its function of supplying small biochemical molecules for biosynthetic pathways, the Krebs cycle is rarely completed. Some of the intermediates are siphoned out of the cycle, especially into amino acid biosynthesis, and so some of the intermediary metabolites of this pathway must be continuously resynthesized to continue the Krebs cycle. In many microorganisms, only part of the substrate is completely oxidized for driving the synthesis of ATP, and the remainder is used for biosynthesis. Similarly, the reduced coenzymes generated in this pathway can be used for generating ATP or for the synthesis of the reduced coenzyme NADPH (reduced nicotinamide adenine dinucleotide phosphate) for use in biosynthesis.

Electron Transport Chain and Chemiosmotic Generation of ATP

The reduced coenzyme molecules that are generated during glycolysis and the Krebs cycle can be reoxidized and the energy stored in them used to generate additional ATP (FIG. 6-12). The energy-requiring synthesis of ATP from ADP in respiration is driven largely by chemiosmosis. As electrons pass down an electron transport chain, some of their carrier molecules extrude protons from one side of the membrane to the other. Some of the compounds in the electron transport chain can accept and transfer whole hydrogen atoms (that is, a proton and an electron), whereas others can accept only electrons. When only electrons are accepted, the protons (hydrogen ions) must go somewhere. In bacteria they are extruded to the outside of the plasma membrane. Since the phospholipid portions of plasma membranes are normally impermeable to protons, this pumping establishes a

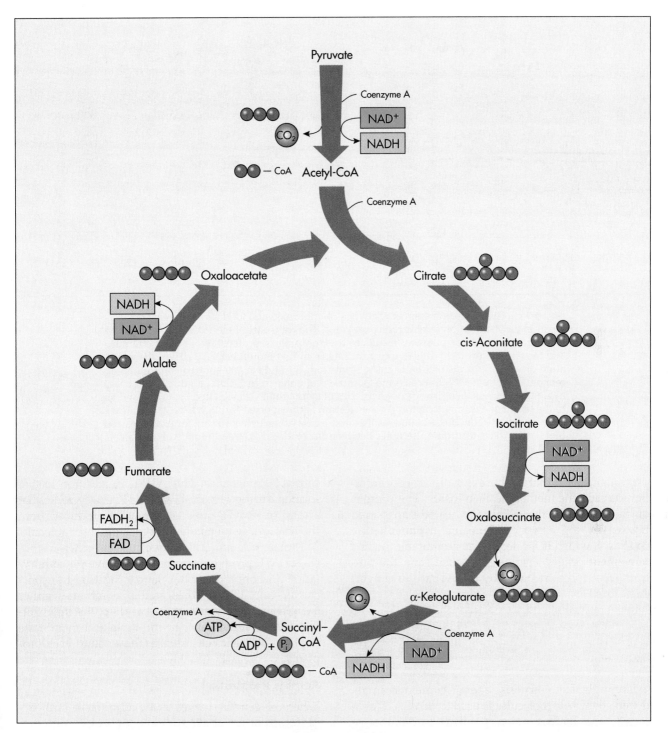

FIG. 6-11 The Krebs cycle is a metabolic pathway central to respiratory metabolism and provides a critical link between the metabolism of the different classes of macromolecules. The metabolism of pyruvate through the tricarboxylic acid Krebs cycle results in the generation of ATP and reduced coenzymes and the formation of CO_2. The oxidation of pyruvate takes place in two stages: the decarboxylation of pyruvate, that is, CO_2 removal, to form acetyl-CoA, and the subsequent oxidation of the acetyl-CoA to form CO_2. When the pathway is completed, the intermediate carboxylic acids are regenerated and continue to cycle throughout the reactions of the Krebs cycle. Each carbon atom in the molecules of the pathway is represented as a red ball.

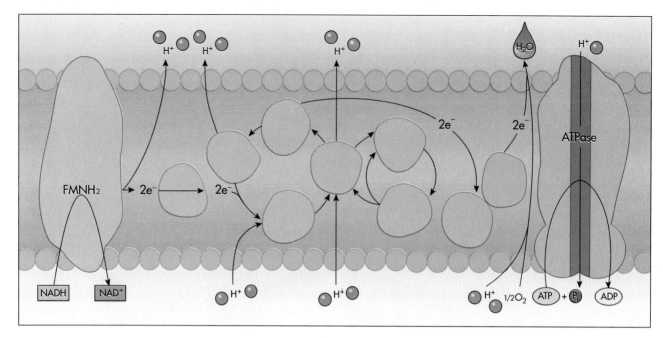

FIG. 6-12 The electron transport chain is a membrane-embedded series of reactions that result in the reoxidation of reduced coenzymes. The transport of electrons through the cytochrome chain of this pathway results in the pumping of protons across the membrane, and the return flow of hydrogen ions resulting from this proton gradient drives the generation of ATP. Electrons that enter the system from NADH are transported through flavin mononucleotide (FMN) to coenzyme Q; those that enter from $FADH_2$ go directly to coenzyme Q. Electrons then flow through a series of cytochromes, designated cyt *b, c, a,* and *a₃,* to the terminal electron acceptor. As the electron is transported through each carrier, there is an oxidation-reduction reaction, so that in the case of the cytochromes, for example, iron (Fe) within the cytochrome alternates between the oxidized Fe^{3+} and reduced Fe^{2+} states.

proton gradient, that is, there is a higher concentration outside the membrane than inside. The protons on the outside of the membrane can be transported inward by the enzyme **adenosine triphosphatase (ATPase),** which is located in the membrane. As this movement occurs, energy is released from the protonmotive force and used by the ATPase to convert ADP to ATP.

> **Chemiosmosis is based on pumping protons across a membrane and using the energy released by their diffusive return (protonmotive force) to generate ATP.**

Through chemiosmosis, energy contained in reduced coenzyme molecules is used to drive ATP synthesis. The hydrogen ion gradient (protonmotive force) across a membrane and chemiosmosis drives the formation of ATP. The reduced coenzyme NADH contains more stored chemical energy than the reduced coenzyme $FADH_2$. For each NADH molecule, three ATP molecules can be synthesized during oxidative phosphorylation, compared to only two ATP molecules for each $FADH_2$. The ten NADH molecules generated during glycolysis and the Krebs cycle, therefore, can be converted to 30 ATP molecules during oxidative phosphorylation. The two $FADH_2$ molecules generated during the Krebs cycle can generate

four ATP molecules. This ATP is in addition to that formed during glycolysis and the Krebs cycle, so that a total of 38 ATP molecules may be produced from the respiratory metabolism of each glucose molecule.

Prokaryotic and eukaryotic cells use chemiosmosis for ATP production in both oxidative phosphorylation and photophosphorylation. Electron transport carriers and ATPase are located in membranes, either the plasma membrane of prokaryotes, the inner mitochondrial membrane, or the thylakoid membrane of photosynthetic cells, such as those found in chloroplasts.

Aerobic Respiration

When oxygen (O_2) serves as the terminal electron acceptor, as in the above example, the respiratory metabolism is called **aerobic respiration.** The oxygen is reduced to form water in this process. The overall reaction for the aerobic respiratory metabolism of glucose can be written as:

$$glucose + 6\,O_2 + 38\,ADP + 38\,P_i \longrightarrow$$
$$6\,CO_2 + 6\,H_2O + 38\,ATP$$

Aerobic respiration (respiration in which molecular oxygen serves as the terminal electron acceptor) is remarkably efficient. The initial breakdown of glucose by glycolysis yields only two ATP molecules per molecule of glucose. In aerobic respiration, an additional 36 ATP molecules can be formed. Hence, approximately 40% of the chemical bond energy of the glucose that is broken down in aerobic respiration is recovered as high-energy-containing ATP.

Anaerobic Respiration

Some cells can use substances other than oxygen as the terminal electron acceptor in a respiratory pathway (FIG. 6-13). Bacteria such as *Pseudomonas* and *Bacillus* can use a nitrate ion (NO_3^-) as a terminal electron acceptor. This NO_3^- is then reduced to nitrite ion (NO_2^-), nitrous oxide (N_2O), or nitrogen gas (N_2).

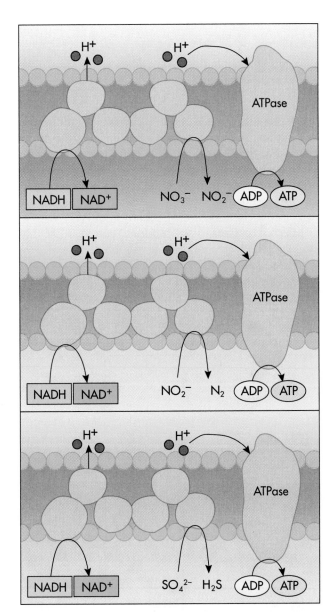

FIG. 6-13 Anaerobic respiration in which nitrate, nitrite, or sulfate serves as the terminal electron acceptor has specific electron carriers. The electron transfer results in the formation of nitrite from nitrate, molecular nitrogen from nitrite, or hydrogen sulfide from sulfate.

HIGHLIGHT

A MICROBIAL EXPLANATION FOR GHOSTS

Under anaerobic conditions, where there is no oxygen, nitrate, or sulfate to serve as terminal electron acceptors for respiration, some microorganisms may still be able to carry out respiration using phosphate as the terminal electron acceptor. The use of phosphate (PO_4^{3-}) as the terminal electron acceptor results in the production of phosphine (PH_3), a reactive green-glowing gas. Conditions where this might occur include areas of extensive organic matter decomposition, such as might be encountered around cemeteries where the graves are not sealed. Mass burial sites were often used in villages where cemeteries were located on hillsides and numerous bodies were placed daily in times of famine and epidemic outbreaks of disease. Imagine the myths that might arise when villagers heard a rumbling noise and saw a green glow rising from the graves on the hillside.

Other bacteria, such as *Desulfovibrio*, use sulfate (SO_4^{2-}) as the terminal electron acceptor to form hydrogen sulfide (H_2S). When a molecule other than oxygen, such as nitrate or sulfate, serves as the terminal electron acceptor, the metabolic pathway is called **anaerobic respiration.** As the name implies, anaerobic respiration does not require the presence of air (oxygen). Anaerobic respiration is often less efficient than aerobic respiration, producing about one third of the ATP made by aerobes. This is because a complete Krebs cycle often does not function in the absence of molecular oxygen and because two rather than three ATP molecules are generated from each NADH by chemiosmosis.

> In the absence of molecular oxygen, nitrate can serve as the terminal electron acceptor with the production of molecular nitrogen and water.

> In the absence of both oxygen and nitrate, sulfate can serve as the terminal electron acceptor, becoming reduced to hydrogen sulfide and water in the process.

LIPID AND PROTEIN CATABOLISM

Besides glucose and other carbohydrates, microorganisms also utilize lipids and proteins as substrates for metabolism. Because of their high molecular weights these substances can not enter the cell unless they are first broken down outside the cell (extracellularly). Some microorganisms produce extracellular enzymes for the breakdown of proteins and lipids, as well as for attacking polysaccharides such as cellulose, chitin, and starch.

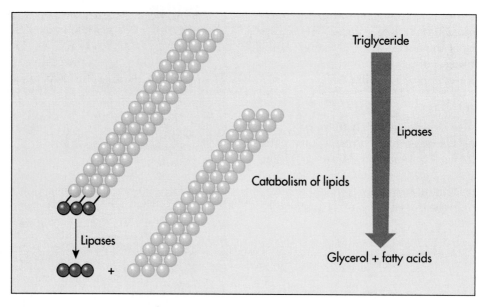

FIG. 6-14 Lipids are broken down by lipases into glycerol and fatty acids. The fatty acids are further metabolized by β-oxidation to smaller fatty acids and acetate.

Various microorganisms produce extracellular enzymes, called *lipases*, that break down lipids (fats) into their fatty acid and glycerol components. These components enter the cell, where they are further metabolized (FIG. 6-14). Many microorganisms convert glycerol into dihydroxyacetone phosphate, one of the intermediates formed during glycolysis. The dihydroxyacetone phosphate enters the glycolytic pathway and is further metabolized to CO_2. Fatty acids are catabolized by *beta-oxidation*. In this process, carbon fragments of long chains of fatty acids are removed two at a time and acetyl-CoA is formed. A fatty acid with sixteen carbons yields eight molecules of acetyl-CoA in seven cleavage steps. As the molecules of acetyl-CoA form, they enter the Krebs cycle, as do the acetyl-CoA molecules formed by the oxidation of pyruvate. In this process, reduced coenzymes are produced and their subsequent reoxidation results in the chemiosmotic generation of ATP.

Microorganisms also produce extracellular enzymes, called *proteases*. Proteases break down proteins into short polypeptides and amino acids (FIG. 6-15). These small subunits enter the cell and are further attacked. Amino acids can be enzymatically

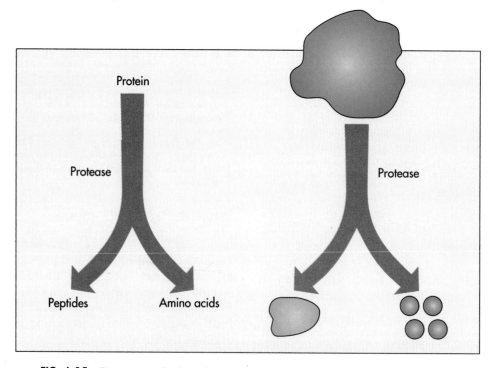

FIG. 6-15 Proteins are broken down by proteases into peptides and amino acids.

deaminated, that is, the amino group can be removed and converted to inorganic ammonia. Removal of an amino group from an amino acid produces a carboxylic acid. The carboxylic acids produced in this manner can enter the glycolytic pathway or the Krebs cycle. Further catabolism results in decarboxylation with the production of CO_2. ATP is generated as a result of the protonmotive force formed by the establishment of a proton gradient across the plasma membrane of a prokaryotic cell or the mitochondrial membranes of eukaryotic cells.

FERMENTATION

In fermentation pathways the organic substrate molecule is converted to another organic molecule that serves as the terminal electron acceptor. There is no net change in the oxidation state of the products relative to the starting substrate molecule. The oxidized products are exactly counterbalanced by reduced products, and thus the required oxidation-reduction balance is always achieved. Such a metabolic pathway can occur in the absence of air because there is no requirement for oxygen or any other inorganic compound to serve as the terminal electron acceptor to balance a change in the oxidation state of the organic molecule. Some microorganisms can carry out fermentation in the presence of air, simply ignoring the presence of oxygen, whereas others can carry out fermentation only in the absence of oxygen.

Fermentation is an anaerobic process that does not require an external electron acceptor to balance the oxidation-reduction reactions in the pathway.

In a fermentation pathway, substrate-level phosphorylation converts ADP to ATP. Unlike respiration, fermentation does not involve an electron transport chain or chemiosmosis for generating ATP. Rather, the synthesis of ATP in fermentation is largely restricted to the amount formed during glycolysis. Fermentation yields far less ATP per substrate molecule than respiration. This is because during a fermentation pathway the organic substrate molecule is not completely oxidized to carbon dioxide. As a result, not as much energy can be released from the substrate molecule to drive the synthesis of ATP.

Fermentation generates ATP by substrate-level phosphorylation.

Fermentation yields far less ATP per substrate molecule utilized than respiration.

The energy obtained from the complete oxidation of glucose to carbon dioxide and water by respiration is more than 10 times greater than that obtained when glucose is metabolized by fermentation. Because more ATP can be generated per molecule of substrate, fewer substrate molecules must be metabolized during respiration than during fermentation to achieve equivalent growth. From both the viewpoints of the energy needs of living organisms and conservation of available organic nutrient resources, respiration is more favorable than fermentation. Organisms that have the metabolic capability to carry out both types of metabolism will generally use the energetically more favorable respiration pathway, when conditions permit, and will rely on fermentation only when there is no available external electron acceptor.

The initial metabolic steps in the fermentation of a carbohydrate are identical to those in respiration. It begins with glycolysis. If the microorganism uses the Embden-Meyerhof glycolytic pathway, it generates two pyruvate molecules, two reduced coenzyme NADH molecules, and two ATP molecules for each molecule of glucose as a result of the glycolytic breakdown of glucose.

In respiration the reoxidation of the coenzymes occurs in the electron transport chain and requires an external electron acceptor. In fermentation the reoxidation of NADH to NAD$^+$ depends on the reduction of the pyruvate molecules formed during glycolysis to balance the oxidation-reduction reactions. Different microorganisms have developed different pathways for utilizing the pyruvate to reoxidize the reduced coenzyme. These pathways have different ter-

NEWSBREAK

INTESTINAL YEAST INFECTIONS CAUSE INTOXICATION

Some people in Japan have been reported to suffer from a strange form of alcoholism apparently related to a yeast infection in their intestines. These people had not consumed any alcoholic beverages but nevertheless frequently appeared to be inebriated. The evidence indicates that yeast growing in their intestines is carrying out an alcoholic fermentation. This fermentation produced sufficient alcohol to bring about this state of drunkenness. Reduction of the yeast population in their intestines by treatment with antibiotics eliminated this state of intoxication. Thus there is a cure for such infectious intoxication.

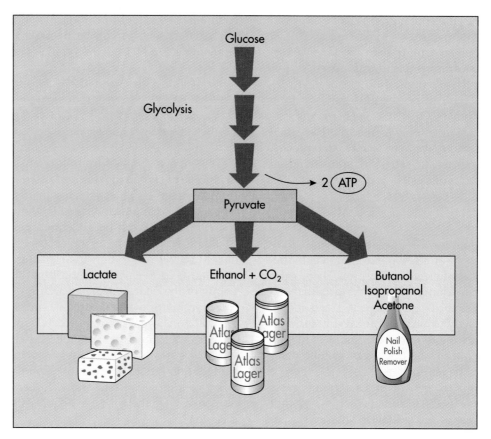

FIG. 6-16 There are various fermentation pathways that produce different end products.

minal sequences and hence the various fermentation pathways result in the formation of varying end products (FIG. 6-16).

A complete fermentation pathway (1) begins with a substrate, (2) includes glycolysis, and (3) terminates with the formation of end products. There is no net change in the oxidative state of the coenzymes during the overall fermentation pathway, and the coenzyme does not appear in the overall fermentation equation. There are several different fermentation pathways that are generally named for the characteristic end products that are formed (see Table 6-3). Many of the products of fermentation pathways have commercial value. As will become apparent from the discussion of the specific fermentation pathways, numerous foods, beverages, and other products we frequently use are the products of microbial fermentations.

Ethanolic Fermentation

In the **ethanolic fermentation,** or **alcoholic fermentation,** pyruvate is converted to ethanol and carbon dioxide. Ethanolic fermentation is carried out by many yeasts, such as *Saccharomyces cerevisiae,* but by relatively few bacteria. The ethanolic fermentation pathway can be written as:

glucose + 2 ADP + 2 P_i $\longrightarrow$ 2 ethanol + 2 CO_2 + 2 ATP

This fermentation pathway is very important in food and industrial microbiology. It is used to produce beer, wine, and distilled spirits (FIG. 6-17). For example, the carbohydrates in grains are converted to ethanol by yeasts in the production of beer and spirits. The sugars in grapes are the substrates for ethanol production by the wine-producing yeasts. *Saccharomyces cerevisiae,* also known as baker's yeast, is used in bread making; the carbon dioxide released during ethanolic fermentation causes bread dough to rise in a process called *leavening* (FIG. 6-18). The ethanol is driven from the bread during baking, which, according to the United States Environmental Protection Agency, is an important source of air pollution that must be controlled. The ethanol produced by *S. cerevisiae* is also used as a fuel to augment gasoline in a product known popularly as *gasohol.*

Lactic Acid Fermentation

Lactic acid fermentation is carried out by bacteria that—by virtue of these metabolic reactions—are classified as the lactic acid bacteria. This fermentation pathway produces lactic acid as an end product. When the Embden-Meyerhof scheme of glycolysis is used in the lactic acid fermentation pathway, the overall pathway is a **homolactic fermentation** because the only end product formed is lactic acid. Homolactic fermentation is carried out by *Streptococcus*

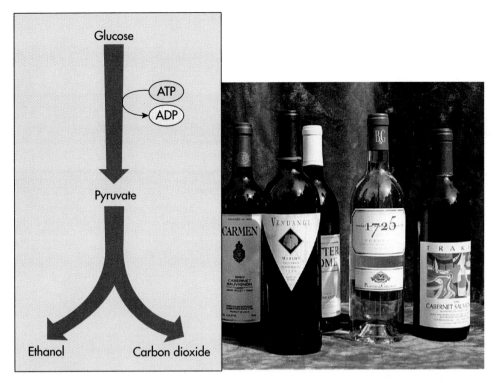

FIG. 6-17 The ethanolic fermentation pathway results in the formation of ethanol and CO_2. The fermentation of carbohydrates to these end products forms the basis of the beer, wine, and spirits industries.

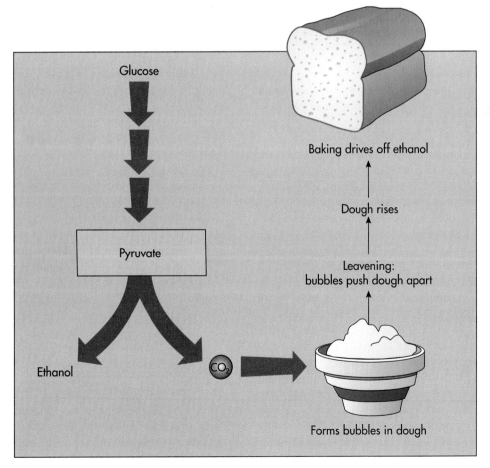

FIG. 6-18 Bread is made by an alcoholic fermentation. The CO_2 from the ethanolic fermentation causes the bread to rise.

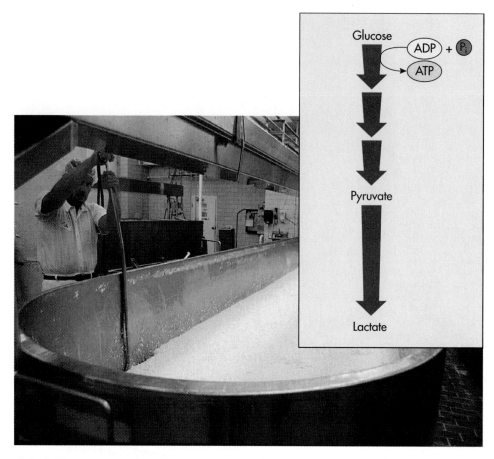

FIG. 6-19 The homolactic acid fermentation pathway results in the production of lactate (lactic acid). This fermentation pathway is the basis for cheese production.

and *Lactobacillus* species (FIG. 6-19). The homolactic acid fermentation pathway can be written as:

$$\text{glucose} + 2\text{ ADP} + 2\text{ P}_i \longrightarrow 2\text{ lactic acid} + 2\text{ ATP}$$

Streptococci, even though they are metabolically obligate anaerobes, live on human tooth surfaces and produce lactic acid that is held against the tooth by dental plaque. The acid can gradually eat through the enamel of the tooth and create cavities. The lactic acid produced by *Lactobacillus* species in the vaginal tract helps protect against sexually transmitted pathogens. *Lactobacillus* species are the initial colonizers of the human intestinal tract and occur in the human digestive tract, aiding in the digestion of milk and dairy products. Many adults cannot digest the carbohydrates in milk and so *L. acidophilus* is added to various commercial milk products, such as acidophilus milk, to aid those who cannot tolerate milk. The enzymes produced by *L. acidophilus* convert milk sugars to products that do not accumulate and cause gastrointestinal upset. The homolactic acid fermentation pathway is important in the dairy industry because it is responsible for the souring of milk. It is used in the production of many types of cheese, yogurt, and other dairy products. Buttermilk and sour cream are made by using different strains of lactic

acid bacteria as starter cultures and different parts of whole milk as the starting substrate.

In some bacteria, however, a different pathway of glycolysis—the Entner-Doudoroff pathway—is used. Such bacteria are termed *heterolactic* because both ethanol and carbon dioxide are produced in addition to lactic acid. The ethanol and carbon dioxide come from the glycolytic portion of the pathway. This fermentative pathway is carried out by *Leuconostoc* and various *Lactobacillus* species and is responsible for the production of sauerkraut (FIG. 6-20). Several types of sausage are produced by allowing meat to undergo heterolactic acid fermentation during curing. The overall reaction for the **heterolactic fermentation** can be written as:

$$\text{glucose} + \text{ADP} + \text{P}_i \longrightarrow$$
$$\text{lactic acid} + \text{ethanol} + \text{CO}_2 + \text{ATP}$$

Propionic Acid Fermentation

Another fermentation pathway of interest is the **propionic acid fermentation pathway.** This metabolic sequence carried out by the propionic acid bacteria produces propionic acid and carbon dioxide. The bacterial genus *Propionibacterium*, which contributes to acne, is defined as Gram-positive rods that produce propionic acid from the metabolism of carbohy-

FIG. 6-20 The heterolactic acid fermentation pathway results in the production of lactate (lactic acid), ethanol, and CO_2. This fermentation pathway is the basis for sauerkraut production.

HIGHLIGHT

PRODUCTION OF CHEESE

Cheeses are produced by microbial fermentations. Cheese consists of milk curds that have been separated from the liquid portion of milk (whey). The curdling of milk is accomplished by using the enzyme rennin and lactic acid bacterial starter cultures. Rennin is obtained from calf stomachs or by microbial production.

Cheeses are classified as: (1) soft, if they have a high water content (50% to 80%); (2) semihard, if the water content is about 45%; and (3) hard, if they have a low water content (less than 40%). Cheeses are also classified as unripened, if they are produced by a single-step fermentation, or as ripened, if additional microbial growth is required during maturation of the cheese to achieve the desired taste, texture, and aroma. Processed cheeses are made by blending various cheeses to achieve a desired product. If the water content is elevated during processing, thereby diluting the nutritive content of the product, the product is called a "processed cheese food" rather than a cheese.

The natural production of cheeses involves a lactic acid fermentation with various mixtures of *Streptococcus* and *Lactobacillus* species used as starter cultures to initiate fermentation. The flavors of different cheeses result from the use of different microbial starter cultures, varying incubation times and conditions, and the inclu-

sion or omission of secondary microbial species late in the fermentation process.

Ripening of cheeses involves additional enzymatic transformations after formation of the cheese curd, using enzymes produced by lactic acid bacteria or enzymes from other sources. Unripened cheeses do not require additional enzymatic transformations. Cottage cheese and cream cheese are produced by using a starter culture similar to the one used for the production of cultured buttermilk and are soft cheeses that do not require ripening. Sometimes a cheese is soaked in brine to encourage development of selected bacterial and fungal populations during ripening. Limburger is a soft cheese produced in this manner. During ripening, the curds are softened by proteolytic and lipolytic enzymes and the cheese acquires its characteristic aroma. The production of Parmesan cheese also involves brine curing.

Various fungi are also used in the ripening of different cheeses. The unripened cheese is normally inoculated with fungal spores and incubated in a warm moist room to favor the growth of filamentous fungi. For example, blue cheeses are produced by using *Penicillium* species, Roquefort cheese is produced by using *P. roqueforti*, and camembert and brie are produced by using *P. camemberti* and *P. candidum*.

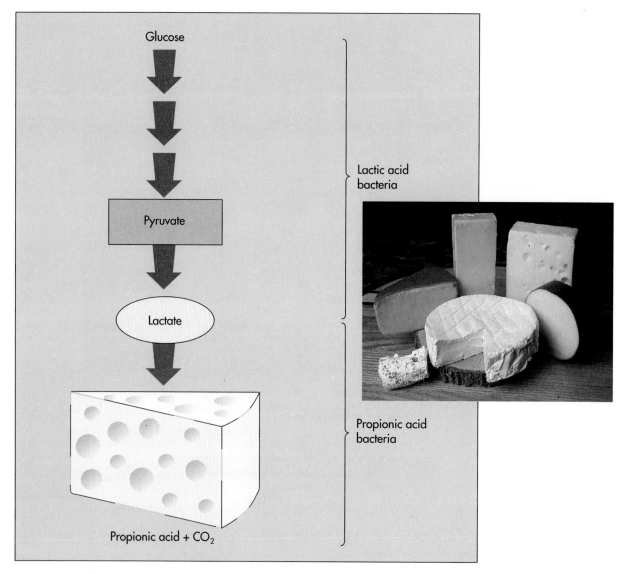

FIG. 6-21   The propionic acid fermentation produces propionic acid and carbon dioxide. The carbon dioxide produced in this fermentation forms the holes in Swiss cheese.

drates (FIG. 6-21). These bacteria have the ability to carry out this fermentation pathway beginning with lactic acid as the substrate. The ability to utilize the end product of another fermentation pathway is quite unusual, but it allows species of *Propionibacterium* to carry out late fermentation during the production of cheese.

Lactic acid bacteria convert the initial substrates in milk to lactic acid. Propionic acid bacteria then convert the lactic acid to propionic acid and carbon dioxide. Propionic acid bacteria begin their fermentation only after the cheese curd has formed through the action of lactic acid bacteria. The release of carbon dioxide during this late fermentation forms gas bubbles in the semisolid cheese curd, forming the holes in Swiss cheese. The propionic acid formed during this

fermentation also gives Swiss cheese its characteristic flavor.

Mixed-acid Fermentation

The **mixed-acid fermentation** pathway is carried out by *Escherichia coli*, as well as hundreds of other bacterial species. In the mixed-acid fermentation pathway the pyruvate formed during glycolysis is converted to various products—ethanol, acetate (acetic acid), formate (formic acid), lactate (lactic acid), molecular hydrogen (H_2), and carbon dioxide (CO_2). It is called the mixed-acid fermentation because of these many products. The proportions of the products vary, depending on the bacterial species.

Mixed-acid fermentation can be detected by the **Methyl Red (MR) test,** which is based on the color

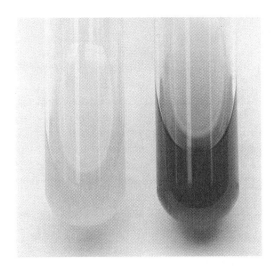

FIG. 6-22 The Methyl Red test, which detects the production of acid, is useful for differentiating various bacterial species, including *Enterobacter aerogenes* and *Escherichia coli.* Negative test result for *Enterobacter aerogenes* is on left. Positive test result for *Escherichia coli* is on right.

FIG. 6-23 The Voges-Proskauer test, which detects the production of acetoin, is useful for differentiating various bacterial species, including *Enterobacter aerogenes* and *Escherichia coli.* Negative test result for *Escherichia coli* is on left. Positive test result for *Enterobacter aerogenes* is on right.

reaction of the pH indicator Methyl Red (FIG. 6-22). This test is one of several typically used in clinical identification systems, including miniaturized commercial identification systems used for the identification of bacteria, such as *E. coli*, which cause urinary tract and other infections.

Butanediol Fermentation

Some bacteria, such as members of the bacterial genus *Enterobacter* and *Klebsiella,* carry out a **butanediol fermentation** pathway. An intermediary product in the butanediol fermentation pathway is acetoin (acetyl methyl carbinol). One of the classic diagnostic tests used for separating *E. coli* from *Enterobacter aerogenes* is the **Voges-Proskauer test,** which detects the presence of acetoin (FIG. 6-23). *E. coli* does not carry out a butanediol pathway, whereas *Enterobacter aerogenes* does. Thus *Enterobacter aerogenes* is Voges-Proskauer positive and *E. coli* is Voges-Proskauer negative.

It is important to distinguish between these organisms because *E. coli* is used as an indicator of human fecal contamination in assessing water quality and safety. The Methyl Red test is also used for separating *Enterobacter aerogenes* from *E. coli*. It detects very low pH resulting from high amounts of acid production. Because *Enterobacter aerogenes* channels part of its substrate into the neutral fermentation end product, butanediol, it does not produce as much acid and thus does not lower the pH as much as *E. coli*, which channels all of its substrate into the mixed-acid fermentation pathway. Thus *E. coli* shows a positive Methyl Red test, whereas *Enterobacter aerogenes* yields the opposite reaction.

Butanol Fermentation

In yet another pathway, members of the genus *Clostridium* carry out a **butanol fermentation.** Different species of *Clostridium* form various end products via this fermentation pathway, with pyruvate being converted either to acetone and carbon dioxide, propanol and carbon dioxide, butyrate, or butanol. Several of these products are good solvents. Chaim Weizmann, a Polish-born microbiologist working in Britain, discovered the butanol fermentation pathway in time to allow the British to produce acetone for the manufacture of munitions for World War I. Their ability to do so was instrumental in the success of the Allied forces. Today acetone for nail polish remover often is produced by microbial fermentation via this pathway.

PHOTOSYNTHETIC METABOLISM

Some early organisms evolved a different way of generating ATP by **photoautotrophic metabolism,** or **photosynthesis.** Instead of obtaining energy for ATP synthesis by breaking chemical bonds, these organisms, called **photoautotrophs,** developed the ability to use light as the source of energy that moves protons across membranes of their cells to create a protonmotive force. Photoautotrophs also obtain their carbon from inorganic CO_2.

> **In photosynthesis, light energy is captured and used to generate ATP.**

In photosynthesis, sunlight strikes pigments embedded within the membranes of the cell. In this first stage of photosynthesis, electrons within special re-

FIG. 6-24 Micrograph of the cyanobacterium *Aphanizomenon flos-aquae.*

action centers of certain pigments are excited, meaning that they have absorbed energy from the light. These excited electrons pass from the reaction-center pigments through a series of proteins embedded within the membrane, eventually reaching and activating channels that enable the transport of protons across the membrane. These channels retain the electrons and export the protons across the membrane. The protonmotive force generated in this process is used to produce ATP by chemiosmosis. As a result of electron transport, the reduced coenzyme NADPH also is produced; NADPH is later used to produce sugars from carbon dioxide.

Photoautotrophs

Organisms that use light as a source of energy and carbon dioxide as their chief source of carbon are called **photoautotrophs.** The photoautotrophs include photosynthetic bacteria (green sulfur and purple sulfur bacteria, and cyanobacteria), algae, and green plants (FIG. 6-24). In the photosynthetic reactions of cyanobacteria, algae, and plants, the electrons of water are used to reduce carbon dioxide, and oxygen gas (O_2) is given off (FIG. 6-25). Because this photosynthetic process produces O_2, it is sometimes

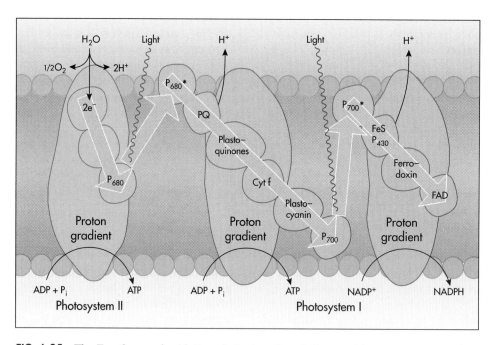

FIG. 6-25 The Z pathway of oxidative photophosphorylation combines two separate photoactivation steps (photosystems) into a unified pathway. These are the excitation of P_{680} to P_{680^*} and the excitation of P_{700} to P_{700^*}. The P_{680} has a sufficiently positive reduction potential to use H_2O as an electron donor. The resulting P_{680^*} is at a considerably more negative reduction potential, such that the resulting electrons can "fall down" a potential gradient in which protons are extruded across the membrane and form a proton gradient (protonmotive force) to generate ATP. The electrons are passed to the P_{700} complex, which when excited is at a potential more negative than that of the NADP$^+$/NADPH redox pair and is thereby capable of reducing NADP$^+$ to NADPH. The pathway is called the Z *pathway* because when these reactions are plotted as a function of reduction potential the resulting figure resembles a Z. The protons originate from the photolysis of water to establish the protonmotive force.

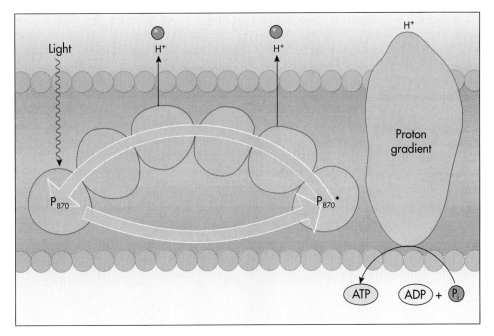

FIG. 6-26 Cyclic photophosphorylation of anaerobic photosynthetic bacteria uses P_{870}. This pathway generates the proton gradient needed to drive the formation of ATP and does not produce reduced coenzyme.

called **oxygenic photosynthesis.** The light-trapping pigment required by green plants, algae, and cyanobacteria is the green compound **chlorophyll** *a*. It is located in the thylakoids of chloroplasts in algae and green plants and in the thylakoids that form a part of an elaborate internal membrane structure in cyanobacteria.

> **Algae and cyanobacteria have chlorophyll** *a*, **which absorbs light energy.**
> **Oxygen is formed from water during oxygenic photosynthesis.**

Besides the cyanobacteria, there are several other families of photosynthetic prokaryotes that are classified according to the way they reduce carbon dioxide. These bacteria cannot use electrons from water to reduce carbon dioxide and cannot carry on photosynthesis when oxygen is present, that is, they must have an anaerobic environment (FIG. 6-26). Their photosynthetic process does not produce oxygen gas, and therefore this type of photosynthesis is called **anoxygenic photosynthesis.**

The green sulfur and purple sulfur bacteria are photoautotrophic microorganisms that carry out anoxygenic photosynthesis. The chlorophyll used by these photosynthetic bacteria is *bacteriochlorophyll*, which absorbs light of longer wavelengths than that absorbed by chlorophyll *a*. The green sulfur bacteria use sulfur, sulfur compounds, or hydrogen gas to reduce carbon dioxide and form organic compounds. Applying the energy from light to bacteriochloro-

phyll and the appropriate enzymes, these bacteria oxidize sulfur to sulfuric acid, hydrogen sulfide to sulfur, or hydrogen gas to water. The purple sulfur bacteria also use sulfur, sulfur compounds, or hydrogen gas to reduce carbon dioxide. They are distinguished from the green sulfur bacteria on the basis of their biochemistry and morphology.

> **Oxygen is not produced during anoxygenic photosynthesis.**
> **Anaerobic photosynthetic sulfur bacteria oxidize hydrogen sulfide (H_2S) so that they can reduce carbon dioxide to organic matter.**

Photoheterotrophs

Photoheterotrophs use light as a source of energy and can convert CO_2 to sugars, but they also require organic compounds as sources of carbon for growth. Their organic compound sources include alcohols, fatty and other organic acids, and carbohydrates. Among the photoheterotrophs are the green and purple nonsulfur bacteria.

Photosystems and ATP Generation

Chlorophyll and various other colored pigmented compounds are light-trapping pigments that are organized into clusters of 200 to 300 molecules. These compounds harvest or collect light energy in one of their chemical bonds. When this happens, an electron becomes excited and reaches a higher energy level. That extra energy is released and transferred to a neighboring pigment when the electron returns to a lower energy level. The high energy electron is transferred to an electron acceptor. The capture of light energy and the transfer of electrons and energy occurs via a system called a **photosystem.** These systems consist of pigment molecules that absorb light energy and a series of molecules that alternately accept and donate electrons and protons to form a chain of oxidation-reduction reactions through which electrons and protons are passed. The transfers of protons establishes a protonmotive force that is used for the chemiosmotic generation of ATP.

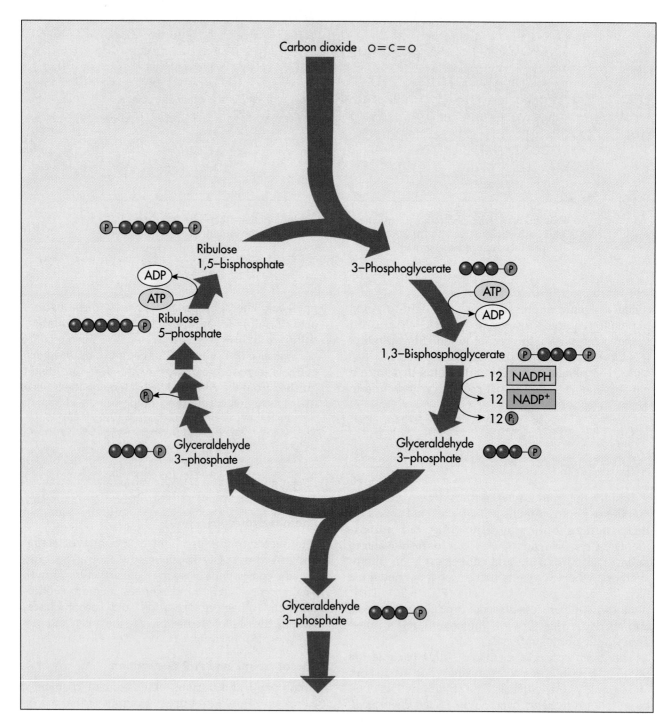

FIG. 6-27 The Calvin, or carbon reduction, cycle is the main metabolic pathway used by autotrophs for the conversion of carbon dioxide to organic carbohydrates. The pathway, which is active in photoautotrophs and chemolithotrophs, requires the input of carbon dioxide, ATP (energy), and NADPH (reducing power).

Calvin Cycle and CO₂ Fixation

Autotrophic microorganisms carry out a metabolic sequence of reactions known as the **Calvin cycle.** In the Calvin cycle, carbon dioxide is used to form organic matter. The conversion of CO_2 to organic matter requires a significant input of ATP and reduced coenzyme (NADPH).

The Calvin cycle is a complex series of reactions that actually represents three slightly different but fully integrated metabolic sequences (FIG. 6-27). It effectively takes three turns of the Calvin cycle to synthesize one molecule of the organic product of this metabolic pathway, which is glyceraldehyde 3-phosphate. Because glyceraldehyde 3-phosphate contains three carbon atoms, the Calvin cycle is sometimes referred to as a C_3 *pathway.* The glyceraldehyde 3-phosphate molecules that are formed during the Calvin cycle can further react to form glucose and polysaccharides of glucose, such as starch and cellulose. It takes six turns of the Calvin cycle to form a 6-carbon carbohydrate, such as glucose. The net input of energy—as ATP—and reducing power—as NADPH—required for the conversion of carbon dioxide to glucose is 18 ATP and 12 NADPH molecules.

> **In the Calvin cycle, carbon dioxide is reduced to form organic compounds for glucose synthesis.**

In photoautotrophs the ATP and NADPH (energy and reducing power) to drive the Calvin cycle come from the light reactions of photosynthesis. In chemolithotrophs (discussed below) the needed ATP and reduced coenzymes come from inorganic compounds. The Calvin cycle itself is known as a "light-independent" or "dark reaction" because, although it requires ATP and NADPH, it does not require any light reactions.

CHEMOAUTOTROPHIC METABOLISM

Some bacteria evolved the metabolic capacity to use inorganic substances as substrates to generate ATP for cellular energy (Table 6-4). For example, some bacteria use reduced sulfur compounds, such as iron sulfide, to generate reducing power and cellular energy. Bacteria that obtain energy in this way are called **chemoautotrophs** or **chemolithotrophs,** from the Greek, meaning obtaining nourishment from stones. These organisms do not require an organic compound or light as a source of energy. They obtain all their energy by oxidizing an inorganic compound. These bacteria have electron transport chains and establish a protonmotive force across membranes, which is used to generate ATP by chemiosmosis. Only a few genera of bacteria obtain their energy from chemoautotrophic metabolism, and all other living organisms depend on them to provide the continuous cycling of materials that are needed for growth.

> **Chemoautotrophic metabolism, also called chemolithotrophic metabolism, uses inorganic compounds to generate ATP.**

Sulfur Oxidation

Various sulfur compounds can be oxidized by chemolithotrophs to meet their energy needs. The chemolithotrophic activities of sulfur-oxidizing microorganisms received considerable attention when it was found that a highly productive submarine area off the Galapagos Islands is supported by the productivity of chemolithotrophs growing on reduced sulfur released from thermal vents in the ocean floor (FIG. 6-28). It is unusual to find an ecological system driven by chemolithotrophic metabolism. Some sulfur-oxidizing chemolithotrophic bacteria, such as *Thiobacillus thiooxidans,* can oxidize large amounts of reduced sulfur compounds with the formation of sulfate. The sulfur-oxidizing activities of this bacterium are detrimental in nature because they result in the formation of acid mine drainage; however, they are beneficial for mineral recovery processes and are used for the recovery of copper and uranium, as well.

TABLE 6-4	
Some Examples of Chemolithotrophic Metabolism	
REACTION	**BACTERIA**
$H_2 + \frac{1}{2}O_2 \rightarrow H_2O$	*Alcaligenes eutrophus*
$NO_2^- + \frac{1}{2}O_2 \rightarrow NO_3^-$	*Nitrobacter winogradsky*
$NH_4^+ + 1\frac{1}{2}O_2 \rightarrow NO_2^- + H_2O + 2H^+$	*Nitrosomonas europaea*
$S^\circ + 1\frac{1}{2}O_2 + H_2O \rightarrow H_2SO_4$	*Thiobacillus denitrificans*
$S_2O_3^{2-} + 2O_2 + H_2O \rightarrow 2SO_4^{2-} + 2H^+$	*Sulfolobus acidocaldarius*
$2Fe^{2+} + 2H^+ + \frac{1}{2}O_2 \rightarrow 2Fe^{3+} + H_2O$	*Thiobacillus ferrooxidans*
$CO + O_2 + 2H^+ \rightarrow CO_2 + H_2O$	*Hydrogenomonas carboxydovorans*

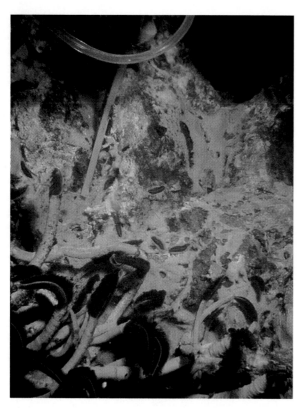

FIG. 6-28 The tube worms *(Riftia pachyptila)* that grow extensively near deep sea thermal vents have no gut. They have extensive internal populations of sulfur-oxidizing chemolithotrophic bacteria that produce the nutrients used by these animals for sustenance. The red-brown color of the worms is due to a form of hemoglobin that supplies oxygen and hydrogen sulfide to the chemolithotrophic bacteria within the tissues of the tube worms. Microbial mats of *Beggiatoa* grow between strands of the tube worms at the Guaymas Basin vent site (Gulf of California) at a depth of 2,010 meters.

Nitrification

Nitrifying bacteria oxidize either ammonium or nitrite ions. Bacteria, such as *Nitrosomonas*, oxidize ammonia to nitrite (FIG. 6-29). Other bacteria, such as *Nitrobacter*, oxidize nitrite to nitrate. Because the chemolithotrophic oxidation of reduced nitrogen compounds yields relatively little energy, chemolithotrophic bacteria carry out extensive transformations of nitrogen in soil and aquatic habitats to synthesize their required ATP. The activities of these bacteria are important in soil because the alteration of the oxidation state radically changes the mobility of these nitrogen compounds in the soil column. Nitrifying bacteria lead to decreased soil fertility because positively charged ammonium ions bind to negatively charged soil clay particles, whereas the negatively charged nitrite and nitrate ions do not bind and are therefore leached from soils by rainwater.

NITROGEN FIXATION

The evolution of a mechanism for converting atmospheric nitrogen into reduced nitrogen compounds such as ammonia was a major event in the progress and development of cellular metabolism. It is this process, called **nitrogen fixation,** that makes nitrogen available for incorporation into proteins. This is critical because, while carbohydrates and lipids can be synthesized from photosynthetic products based on CO_2 fixation, proteins and nucleic acids cannot be synthesized, because they contain nitrogen. Therefore life could not have persisted and expanded in the early oceans unless a means of replacing organic nitrogen compounds evolved. The process of nitro-

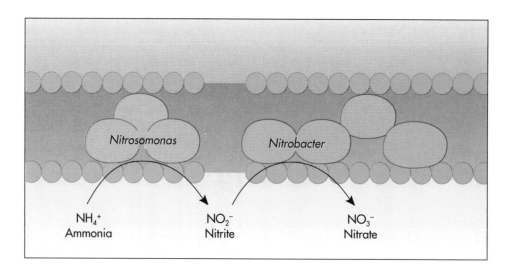

FIG. 6-29 Nitrifying bacteria are chemolithotrophs that oxidize inorganic nitrogen compounds to generate ATP. Some, such as *Nitrosomonas,* oxidize ammonium ions (NH_4^+) to nitrite ions (NO_2^-) *(left);* others, such as *Nitrobacter,* oxidize nitrite ions to nitrate ions (NO_3^-) *(right).* These reactions take place within specialized membranes that intrude within the cytoplasm of nitrifying bacteria.

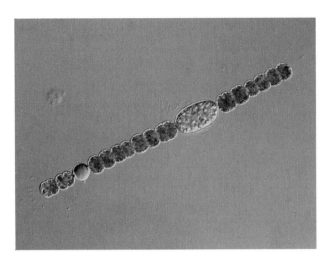

FIG. 6-30 Micrograph of the cyanobacterium *Anabaena cylindrica* showing vegetative cells and a heterocyst (enlarged cell) in which nitrogen fixation occurs. (400 ×).

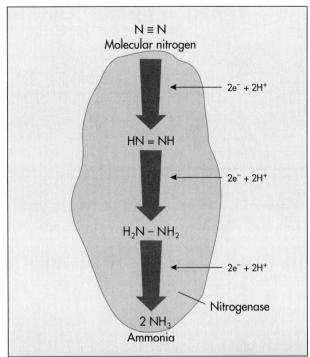

gen fixation, the incorporation of nitrogen atoms from N_2 gas into protein, requires the breaking of an $N \equiv N$ triple bond. This is a very strong bond that is extremely difficult to break.

In the biological fixation of nitrogen, the triple bond of molecular nitrogen is enzymatically broken by **nitrogenase**. This is a complex enzyme system. An iron-containing compound such as *ferredoxin* first obtains electrons from the breakdown of organic molecules or from photosynthetic light reactions and carries them to a protein, *nitrogen reductase*, which channels them to another protein, *dinitrogenase*. With the transfer of six electrons and the use of twelve ATP and four water molecules, nitrogenase converts nitrogen gas into two molecules of ammonia.

Nitrogen-fixing bacteria, called *Rhizobium* and *Bradyrhizobium*, live mutualistically in the nodules on the roots of legume plants (FIG. 6-30). Within the nodule, leghemoglobin, a protein produced by the plant, provides controlled amounts of oxygen so that aerobic energy-yielding metabolism can be carried out without inactivating nitrogenase, which is sensitive to oxygen exposure. When growing alone, *Rhizobium* and *Bradyrhizobium* require oxygen for their metabolism and are unable to fix nitrogen. When they live within root nodules, *Rhizobium* and *Bradyrhizobium* survive in this oxygen-free environment by utilizing the metabolites of the plant.

Other nitrogen-fixing bacteria, such as *Azotobacter* and *Beijerinckia*, are free living. Nitrogen-fixing cyanobacteria have specialized cells, called **heterocysts**, that contain the nitrogenase (FIG. 6-31). The heterocyst provides protection for nitrogenase against molecular oxygen, which is produced photosynthetically by cyanobacteria and which denatures nitrogenase.

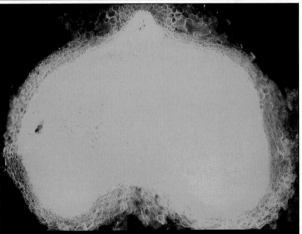

FIG. 6-31 Bacterial cells of *Bradyrhizobium japonicum* within a nodule of a soybean produce nitrogenase, which results in the conversion of molecular nitrogen to ammonia.

METHANOGENESIS

Some archaebacteria are able to use hydrogen and carbon dioxide to generate the ATP and molecules that compose their cellular structures. The metabolism of these archaebacteria produces methane and they are therefore called **methanogens** (FIG. 6-32). Other methanogens use fatty acids instead of carbon dioxide for the production of methane. Hydrogen gas, carbon dioxide, and fatty acids were available at the time life evolved on Earth. The methanogenic ar-

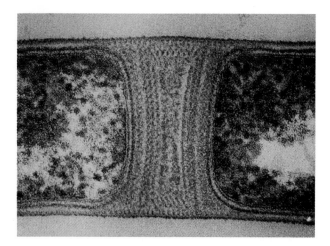

FIG. 6-32 Colonized micrograph of *Methanospirillum hungatei* cells within a sheath *(orange)*. (84,000×). The cells are separated by a cell spacer.

chaebacteria may have been among the first organisms to carry out cellular metabolism. The methanogenic archaebacteria are strict anaerobes. They not only do not use oxygen in their metabolism, they are killed by exposure to oxygen. Methanogens could have grown on the compounds available in the primitive atmosphere of the Earth. Descendants of these archaebacteria still carry out anaerobic methane production today.

The metabolism of methane-producing archaebacteria involves a series of oxidation-reduction reactions. In these reactions, electrons and protons are transferred from one coenzyme to another (FIG. 6-33). The oxidation-reduction reactions of the coenzymes establish an electron chain through which the electrons move. This movement of electrons is coupled with the pumping of protons (hydrogen ions) across a membrane. The return flow of protons by diffusion across the membranes of these archaebacterial cells provides the energy needed to drive the synthesis of ATP by chemiosmosis. This process of chemiosmosis is able to drive the formation of ATP because when protons are pumped out of the cell, a low concentration of protons exists within the cell. This causes diffusion to force the flow of protons from outside the cell back in. The membrane is generally impermeable to protons except that protons can pass through special proton-transporting channels that cross the plasma membrane. The passage of protons through these channels releases energy that results in the synthesis of ATP from ADP and inorganic phosphate (P_i).

Anaerobic methane-producing archaebacteria (methanogens) use chemiosmosis, involving an electron transport chain and proton movement across a membrane, to generate ATP.

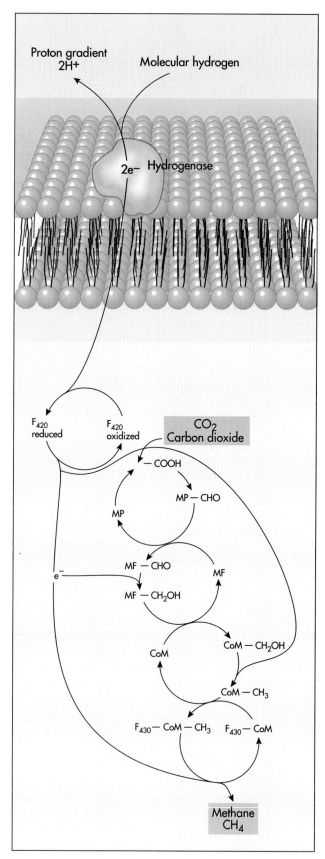

FIG. 6-33 The production of methane by methanogens involves several unique coenzymes and oxidation-reduction reactions to establish a protonmotive force for chemiosmotic ATP generation.

NEWSBREAK

EXPLOSIONS THAT DESTROY HOUSES TRACED TO METHANE FROM LANDFILL

Methane is produced at sites where organic matter decomposes and creates anaerobic conditions. Sanitary landfills represent such sites, and methane production at landfills is so extensive that it may be trapped and used as a fuel. In some cases, however, such methane production represents a serious problem, particularly for nearby houses. Several houses in Kentucky and other states that were located near landfills have been blown off their foundations because methane from a nearby landfill seeped through the ground into their basements. The methane was ignited by the pilot lights of the water heater or furnace systems. The explosions moved the houses several feet away from their original locations. These incidents serve as important reminders to keep houses a safe distance from landfills.

The production of methane by methanogens has several practical consequences. Methane is found in natural gas. It is a flammable gas. Methane seeping from landfills, where it is produced by methanogens degrading the waste deposits, sometimes enters nearby houses and causes explosions. Some sewage treatment facilities collect the methane that is formed during anaerobic decomposition of wastes and use it as a source of fuel for generating heat and electricity. Some communities are supplied with a portion of their natural gas (methane) from this metabolic process. In the future, methane produced by microorganisms may be used as a fuel for automobiles.

SUMMARY

Process of Metabolism Within a Cell (pp. 158-164)

- Cells exhibit various strategies for converting chemical and light energy into the energy stored within ATP, the central currency of energy of the cell. These processes of cellular metabolism also transform starting materials into the organic chemicals that make up the cell's structural and functional components.

Role of Enzymes (pp. 158-159)

- Enzymes are proteins that act as biological catalysts to accelerate the rates of chemical reactions by lowering the activation energy necessary for the reaction to occur. Different enzymes are needed to catalyze different reactions, each cell having thousands of enzymes, each enzyme binding only specific substrates to its active site.

Coenzymes and Oxidation-reduction
Reactions (pp. 159-160)

- An oxidation-reduction or redox reaction involves the transfer of an electron from a donor, which is oxidized, to an acceptor, which is reduced. All redox reactions must be balanced. Cellular metabolism generates reducing power to convert substrate materials into the more reduced molecules of the cell by coupling oxidation reactions with the reduction of coenzymes.

ATP and Cellular Energy (pp. 160-161)

- All cells carry out metabolic reactions that transfer energy to ATP. ATP requires energy-releasing metabolic reactions for its formation. ATP's stored energy is used to drive energy-requiring biosynthetic reactions. Breaking an ATP high-energy bond yields 7300

calories per mole of energy. Substrate level phosphorylation is the generation of ATP from ADP + P_i coupled to energetically favorable reactions. The formation of ATP can also be driven by a protonmotive force in chemiosmosis.

Metabolic Pathways and Carbon Flow (pp. 161-162)

- The metabolic pathways utilized in ATP generation involve various intermediary metabolites linked together in a series of small steps to form unified biochemical pathways. In a catabolic pathway, larger molecules are split into smaller ones. Cells form relatively small molecules that can act as the basis for the carbon skeletons of larger macromolecules that are synthesized in anabolic (biosynthetic) pathways.

Autotrophic and Heterotrophic
Metabolism (pp. 162-164)

- The synthesis of ATP can be achieved autotrophically—through the oxidation of inorganic substrates or through the conversion of light energy to chemical energy—or may be generated heterotrophically through the utilization of organic substrates.

Metabolic Pathways (pp. 164-187)

Respiration (pp. 165-171)

- The Embden-Meyerhof pathway of glycolysis converts the 6-carbon molecule glucose into two molecules of the 3-carbon molecule pyruvate, plus two molecules of reduced coenzyme and two molecules of ATP.
- Glycolysis is the first step in all pathways of carbohydrate metabolism. Its product, pyruvate, feeds into

the Krebs cycle and is converted to carbon dioxide with a net production of four ATP molecules. The Krebs cycle is not always completed; its intermediary products are siphoned out of the cycle and so must be continuously resynthesized.

- During oxidative phosphorylation, electrons from NADH and $FADH_2$ are transferred through an electron transport chain, which includes a series of oxidation-reduction reactions of membrane-bound carrier molecules and the reduction of a terminal electron acceptor. Chemiosmosis provides the energy for ATP production as a result of this process.

- An external electron acceptor is required to complete respiratory metabolic pathways. In aerobic respiration, oxygen is the terminal electron acceptor. Aerobic respiration is an efficient generator of ATP that comes from chemiosmosis.

- Protonmotive force is the potential energy gradient across a membrane established when protons are pumped across the membrane. Energy released when protons move back across the membrane by diffusion is coupled with the energy-requiring conversion of ADP to ATP. Generation of ATP using protonmotive force is called chemiosmosis.

Lipid and Protein Catabolism (pp. 171-173)

- Lipases break down fats into their fatty acid and glycerol components, which are further metabolized in the cell. Fatty acids are catabolized by beta-oxidation in which carbon fragments of long chains of fatty acids are removed two at a time and acetyl-CoA is formed. ATP is generated chemiosmotically by the reoxidation of reduced coenzymes.

- Proteases break down proteins into short polypeptides and amino acids. Amino acids are enzymatically deaminated, producing carboxylic acid, which can enter either the glycolytic pathway or the Krebs cycle. ATP is generated as a result of the protonmotive force across the plasma membrane of prokaryotes or the mitochondrial membranes of eukaryotes.

Fermentation (pp. 173-179)

- In fermentation, organic substrate molecules are used to generate ATP by substrate level phosphorylation. Organic molecules formed as products of fermentative metabolism serve as terminal electron acceptors. The amount of ATP is limited to that formed during glycolysis, yielding far less ATP per substrate molecule than respiration.

- All fermentation pathways are anaerobic. A complete fermentation pathway begins with a substrate, includes glycolysis and reoxidation of the coenzyme, and terminates in the formation of end products.

- Ethanolic or alcoholic fermentation converts pyruvate to ethanol and carbon dioxide by yeasts, such as *Saccharomyces cerevisiae.* It is used to produce beer, wine, and distilled liquor.

- Lactic acid fermentation produces lactic acid as an end product. Homolactic fermentation uses the Embden-Meyerhof pathway of glycolysis and produces only lactic acid. It is used in the production of dairy products such as cheese and yogurt. It is carried out by streptococci and lactobacilli. Hetero-

lactic fermentation produces ethanol and carbon dioxide and is carried out by *Leuconostoc* and *Lactobacillus* species.

- Propionic acid fermentation is carried out by propionic acid bacteria and produces propionic acid and carbon dioxide. This pathway is used in the production of Swiss cheese, giving it the characteristic holes and flavor.

- Mixed-acid fermentation yields ethanol, acetic acid, formic acid, hydrogen, and carbon dioxide. This pathway is carried out by members of the Enterobacteriaceae, including *E. coli.* It can be detected by the Methyl Red test.

- In the butanediol fermentation pathway, *Klebsiella* species produce butanediol. An intermediary metabolite in this pathway, acetoin, can be detected by the Voges-Proskauer test, which distinguishes *E. coli* from *Enterobacter aerogenes* for water quality testing.

- The butanol fermentation pathway is carried out by members of the genus *Clostridium;* the end products of this pathway can be acetone and carbon dioxide, propanol and carbon dioxide, butyrate or butanol.

Photosynthetic Metabolism (pp. 179-183)

- Photoautotrophs, which include the photosynthetic bacteria, algae, and green plants, use light as their energy source and carbon dioxide as their carbon source. In oxygenic photosynthesis the electrons of water reduce carbon dioxide, and oxygen gas is given off. Chlorophyll and bacteriochlorophyll are the light-trapping pigments.

- Photoheterotrophs use light as the energy source and CO_2 and organic compounds as the carbon source. Green and purple nonsulfur bacteria are photoheterotrophs.

- In photosynthetic microorganisms the flow of electrons—initiated when a chlorophyll molecule is energetically excited by absorbing light energy—establishes a protonmotive force across a membrane during the process of photophosphorylation.

- Photosystems are light-trapping pigments organized into clusters of 200 to 300 molecules. These pigments harvest light energy in their chemical bonds, causing electrons to become excited and reach a higher energy level. That energy is released and transferred to a neighboring pigment when the electron returns to a lower energy level.

- Electron transport systems consist of a series of molecules that alternately accept and donate electrons from and to their neighboring molecules. The electrons are those expelled from a photosystem. These transfers accomplish oxidation-reduction reactions with the release of energy.

- The photosynthetic metabolism of cyanobacteria, algae, and plants releases oxygen atoms split from water into the atmosphere.

- The Calvin cycle is carried out by most autotrophic microorganisms in which carbon dioxide is reduced to form organic matter. This requires reducing power in the form of reduced coenzyme NADPH and ATP. The product of this pathway is glyceraldehyde 3-phosphate.

Chemoautotrophic Metabolism (pp. 183-184)

- Chemoautotrophs (chemolithotrophs) are bacteria that can combine inorganic substances such as sulfur or nitrogen with oxygen to generate ATP for cellular energy via aerobic respiration. Such bacteria play important roles in mineral cycling, for example, the conversion of ammonia to nitrite and nitrate, and hydrogen sufite to sufate.
- Chemoautotrophic microorganisms couple the oxidation of an inorganic compound with the reduction of a suitable coenzyme. They use chemiosmosis to generate ATP. Important mineral cycling reactions are the result of chemoautotrophic metabolism.
- Regardless of the mode of metabolism the strategies are the same: synthesize ATP, reduce coenzyme (NADPH) and small precursor molecules to serve as the building blocks of macromolecules, and then use the energy, reducing power, and precursor molecules to synthesize the macromolecular constituents of the organism.

Nitrogen Fixation (pp. 184-185)

- Nitrogen fixation is the conversion of atmospheric nitrogen into reduced nitrogen-containing compounds such as ammonia. Nitrogen fixation is carried out only by members of a few bacterial genera.
- Nitrogenase is the enzyme that converts molecular nitrogen into reduced nitrogen compounds.
- Some nitrogen fixing bacteria live in symbiotic association with plants in specialized structures called nodules. Others have structures called heterocysts where nitrogen fixation occurs.

Methanogenesis (pp. 185-187)

- Methanogens are strictly anaerobic archaebacteria that produce methane as a result of anaerobic respiration. The oxidation-reduction reactions of the coenzymes in the metabolism of methanogens establishes an electron chain through which electrons move. This movement of electrons is coupled with the moving of protons across a membrane, and the protonmotive force is used for ATP synthesis by chemiosmosis.

CHAPTER REVIEW

REVIEW QUESTIONS

1. Define metabolism.
2. What are the differences between catabolism and anabolism?
3. What is an enzyme?
4. What factors influence the activity of enzymes?
5. Describe the chemiosmotic theory of ATP generation.
6. Explain what an oxidation-reduction reaction is.
7. Compare and contrast autotrophy and heterotrophy.
8. Describe what occurs during each of the phases of respiratory metabolism.
9. Describe the similarities and differences between aerobic and anaerobic respiration.
10. What are the starting substrates and end products of glycolysis?
11. What are the light and dark reactions of photosynthesis?
12. What are some commercially useful products of fermentative processes?
13. Compare photoautotrophic and chemolithotrophic metabolism.
14. What is nitrogen fixation? Why is it an important process?
15. What are the differences between fermentation and respiration?

CRITICAL THINKING QUESTIONS

1. What is the significance of the origin of oxygen-producing photosynthesis? How could the origin of this type of metabolism have altered the course of evolution?
2. Lactobacilli and streptococci are microorganisms that normally inhabit the gastrointestinal and vaginal tracts. These organisms typically carry out a homolactic acid fermentation. What is the significance of this observation? How could it relate to resistance to disease? What would happen if these lactic acid fermenters were eliminated from the body?
3. All fermentation reactions have the same common feature with respect to regeneration of oxidation-reduction capability. What is this common feature? How do different fermentation pathways achieve this goal?
4. It is often said that life on Earth depends on the input of solar radiation and that animals and other heterotrophic organisms always depend on green plants. Is this statement accurate? How might you modify this statement to account for the roles of photosynthetic bacteria and algae? How has the discovery of thermal vent regions with biologically productive communities altered the validity of the dependence of life on solar input and photosynthesis?

READINGS

Dawes EA: 1986. *Microbial Energetics*, Glasgow, Blackie.
  Discusses microbial metabolism and respiration and how the principles of energy relate to microbial growth.
Erickson LE and DY-C Fung: 1988. *Handbook on Anaerobic Fermentations*, New York, Dekker.
  Thorough review of fermentative metabolism.
Gottschalk G: 1986. *Bacterial Metabolism*, New York, Springer-Verlag.
  Review of bacterial physiology and metabolic pathways.
Harold FM: 1986. *The Vital Force: A Study of Bioenergetics*, New York, W. H. Freeman.
  Explains the principles of bioenergetics with emphasis on membrane-mediated events.
Hellingwerf KJ and WN Konigs: 1985. The energy flow in bacteria: the main free energy intermediates and their regulatory role, *Advances in Microbial Physiology* 26:125-154.
  Detailed discussion of metabolic pathways and their control in bacteria.
Hinkle P and R McCarty: 1978. How cells make ATP, *Scientific American* 238(23):104-123.
  Reviews how the chemiosmotic theory explains ATP formation in chloroplasts and mitochondria.
Latruffe N: 1988. *Dynamics of Membrane Proteins and Cellular Energies*, Berlin, Springer-Verlag.
  Reviews research methodology and analysis of cell membrane proteins, metabolism, and transport.

Lehninger AL, DL Nelson and MM Cox: 1993. *Principles of Biochemistry*, ed. 2, New York, Worth Publishers.
  An advanced biochemistry text that covers all aspects of the relationship of chemistry to living systems.
Moody MM: 1986. Microorganisms and iron limitation, *Bioscience* 36: 618-623.
  Discusses siderophores in bacteria and their roles in microbial growth and transport.
Rosen BP and S Silver (eds.): 1987. *Ion Transport in Prokaryotes*, San Diego, Academic Press.
  Discusses the role of ion channels and active biological transport in microbial metabolism.
Stryer L: 1988. *Biochemistry*, New York, W. H. Freeman.
  Excellent text covering all aspects of biochemistry and metabolism.
Watson JD, N Hopkins, J Roberts, J Steitz, A Weiner: 1987. *Molecular Biology of the Gene*, Menlo Park, California; Benjamin/Cummings.
  A comprehensive treatise covering molecular biology and metabolism.
Youvan D and B Marrs: 1987. Molecular mechanisms of photosynthesis, *Scientific American* 256(6): 42-50.
  A discussion of the molecular basis of ATP generation by photosynthesis and how spectroscopy, X-ray crystallography, and molecular genetics have helped us understand the process.
Zubay G: 1993. *Biochemistry*, ed. 3, Dubuque, Iowa; W. C. Brown.
  An advanced biochemistry text that covers all aspects of the relationship of chemistry to living systems.

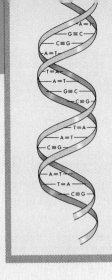

CHAPTER 7

Microbial Genetics: Replication and Expression of Genetic Information

In 1928 a British microbiologist, Frederick Griffith, was trying to develop a vaccine against pneumonia. He was working with two different strains of the causative bacterium *Streptococcus pneumoniae* (FIG. 7-1). One strain was pathogenic, killing the mice injected with it. The other strain was nonpathogenic. The two strains differed in appearance when viewed under the microscope. The nonpathogenic strain appeared rough and was not surrounded by a capsule. The pathogenic strain appeared smooth, surrounded by a polysaccharide capsule. When Griffith injected heat-killed cells of this smooth, pathogenic strain of *S. pneumoniae* into a mouse, the mouse survived because the dead bacteria were unable to establish an infection in the mouse. However, when he injected a mouse with living cells of the rough nonpathogenic strain, together with dead smooth bacteria, knowing that neither of them could cause disease alone, the mouse died. Unlike the live, rough bacteria he injected, the bacteria he isolated from the dead mouse appeared smooth and surrounded by a capsule.

This was a most puzzling observation. Griffith reasoned that genetic material from the heat-killed bacteria had somehow entered the living nonpathogens and transformed them into pathogenic bacteria. He postulated that heat could kill the pathogenic cells without destroying the substance containing their hereditary information, which included instructions

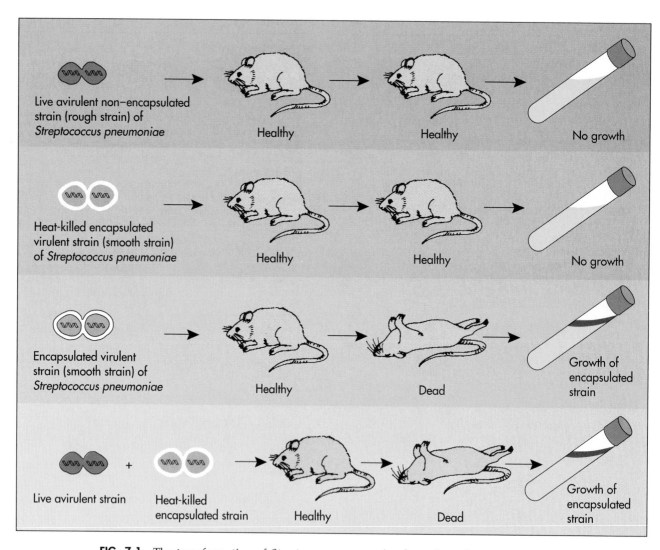

FIG. 7-1 The transformation of *Streptococcus pneumoniae* shows how the properties of a bacterial strain can be altered by a hereditary substance (later identified as DNA). When cells of *S. pneumoniae* are heat killed they leak DNA, which can be picked up by other cells and incorporated into the genetic information of those cells. In this manner, avirulent (nonpathogenic) strains of *S. pneumoniae* that lack the gene for capsule production (virulence factor that contributes to their ability to cause fatal disease) can acquire the gene (DNA) that encodes for capsule production. When this occurs, an avirulent noncapsule-producing strain of *S. pneumoniae* is transformed into a virulent strain that produces a capsule.

on how to cause infection and disease. Griffith had, in fact, observed the movement of hereditary material from one cell to another. The chemical that transmitted the hereditary information for causing disease leaked from the dead pathogens and was picked up by the living bacteria, transforming them into pathogens when it became part of their hereditary material.

Other scientists then began to investigate the specific chemical substance that caused the transformation of a nonpathogen to a pathogen. They were looking for the molecular basis for heredity. Chemical analyses narrowed the possible hereditary molecules to either proteins or nucleic acids. Most scientists hypothesized that proteins were the basis of heredity because their essential roles in metabolism were known. The specific chemical nature of the transforming material observed by Griffith, however, remained a puzzle until 1944 when Oswald Avery and his co-workers were able to demonstrate the chemical nature of the substance that transformed nonpathogenic *S. pneumoniae* to pathogenic *S. pneumoniae*.

Avery hypothesized that a nucleic acid, deoxyribonucleic acid (DNA), rather than protein was the hereditary molecule. He designed experiments to prove this. In Avery's experiments the transforming principle of *S. pneumoniae,* which had been shown to be predominantly DNA with a trace of protein, was treated sequentially with an enzyme that destroys protein and an enzyme that destroys DNA (FIG. 7-2). Avery observed that the protein-destroying enzyme did not affect the ability of the material to transform nonpathogenic *S. pneumoniae* into pathogenic *S. pneumoniae,* whereas treatment with the DNA-destroying enzyme eliminated such transformation. Based on these observations, Avery concluded that the transforming principle must be DNA.

Despite this quite convincing demonstration, the scientific community was not ready to accept that DNA was the universal hereditary molecule. Most scientists remained convinced that proteins would eventually be shown to be the basis of heredity for organisms other than bacteria. Another set of experiments conducted with bacteriophage (viruses that replicate within bacterial cells), however, added convincing evidence that nucleic acids, not proteins, are the source of hereditary information. These experiments, conducted in 1952 by Alfred Hershey and Martha Chase, examined the replication of bacteriophage T2. Although bacteriophage are not living cells, they were known to contain DNA and protein, making them good simple models to examine whether it is protein or DNA that carries hereditary information.

Hershey and Chase used two different radioactive labels to track the movement of protein and DNA separately (FIG. 7-3). Most proteins contain sulfur but none contain phosphorus. Thus the radioactive isotope ^{35}S can be used to label the bacteriophage protein. DNA contains phosphorus but no sulfur, so they used the radioactive isotope ^{32}P to label the bacteriophage DNA. Thus Hershey and Chase cleverly devised a method for following both the DNA and protein components of bacteriophage T2. When they added bacteriophage that had been labelled with ^{35}S to a culture of growing cells of the bacterium *Escherichia coli,* they observed that the ^{35}S label remained outside of the bacterial cells. Thus protein did not enter the bacterial cells. In contrast, when they similarly added ^{32}P-labelled bacteriophages, the ^{32}P label entered the interior of the bacterial cells. This indicated that DNA was the material that entered the cells and therefore must be the substance that carried the hereditary information. The progeny bacteriophages produced from the replication of the original

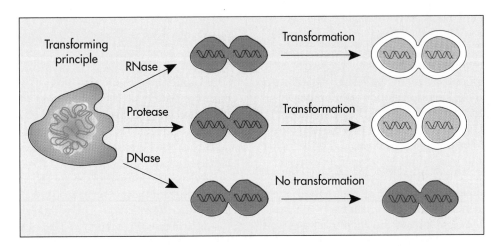

FIG. 7-2 To prove that the hereditary substance was DNA, enzymes that degrade proteins were added to cell extracts. These enzymes did not eliminate transformation, showing that the substance was not a protein. In contrast, the addition of a DNA-destroying enzyme eliminated transformation.

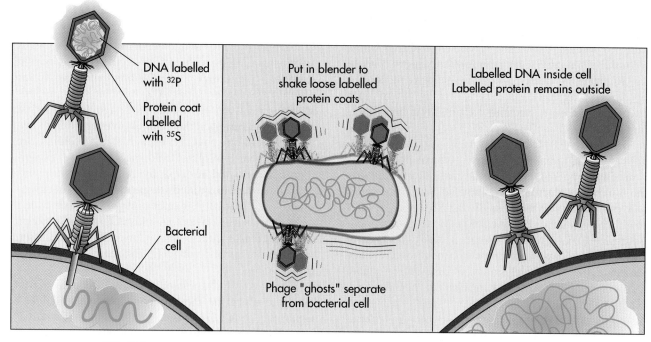

FIG. 7-3 Hershey and Chase demonstrated that nucleic acids are the hereditary substances of viruses. In their experiments ^{32}P was used to label nucleic acids and ^{35}S was used to label proteins. The ^{35}S remained outside of the host cell, whereas the ^{32}P entered the cell. This indicated that the ^{32}P-labelled nucleic acid carried the hereditary information.

bacteriophage contained ^{32}P and not ^{35}S, indicating further that the hereditary material passed from one generation to the next, was, in fact, DNA. Although subsequent experiments have shown that another nucleic acid (ribonucleic acid [RNA]) sometimes is the hereditary substance for viruses, it was now firmly established that DNA is the hereditary molecule for many viruses and all living cells.

DNA is the substance that transmits the hereditary information of many viruses and all cellular organisms.

STRUCTURE OF DNA

NUCLEOTIDES—BUILDING BLOCKS OF THE GENETIC CODE

To understand how DNA stores and transmits hereditary information, it is necessary to examine the chemical structure of this molecule. DNA is a large, high-molecular-weight molecule, called a *macromolecule*. It is composed of many subunits called **nucleotides** (FIG. 7-4). Each nucleotide subunit of DNA has three parts: deoxyribose (a 5-carbon sugar), phosphate, and one of four **nitrogenous bases** (sometimes referred to as **nucleic acid bases**). The four different nitrogenous bases that occur in DNA are *adenine* (A), *thymine* (T), *guanine* (G), and *cytosine* (C).

These four nucleotides are like an "alphabet" that makes up the genetic code. They establish the first important property of DNA as the chemical basis for heredity—the ability to encode the genetic information. This is achieved by linking the nucleotides in a specific order—much as the letters of the alphabet are joined to form words.

The hereditary information is coded by the order in which the four different nucleotides occur within the DNA macromolecule.

CHAINS OF NUCLEOTIDES—DIRECTIONALITY OF DNA

Individual nucleotides are linked to form a long chain consisting of several million nucleotides. The bonds holding the nucleotides together are covalent and hence strong. This is important for the long-term stability of the hereditary macromolecule. Within this chain, nucleotides are locked together in order, thereby establishing the sequences that encode the genetic information. Once encoded in the chain of DNA, the information remains intact unless acted on by a destructive force, such as certain chemicals or radiation.

The chemical bonds holding the chains of nucleotides together are called **3'-5' phosphodiester bonds** (FIG. 7-5). They are so-named because phos-

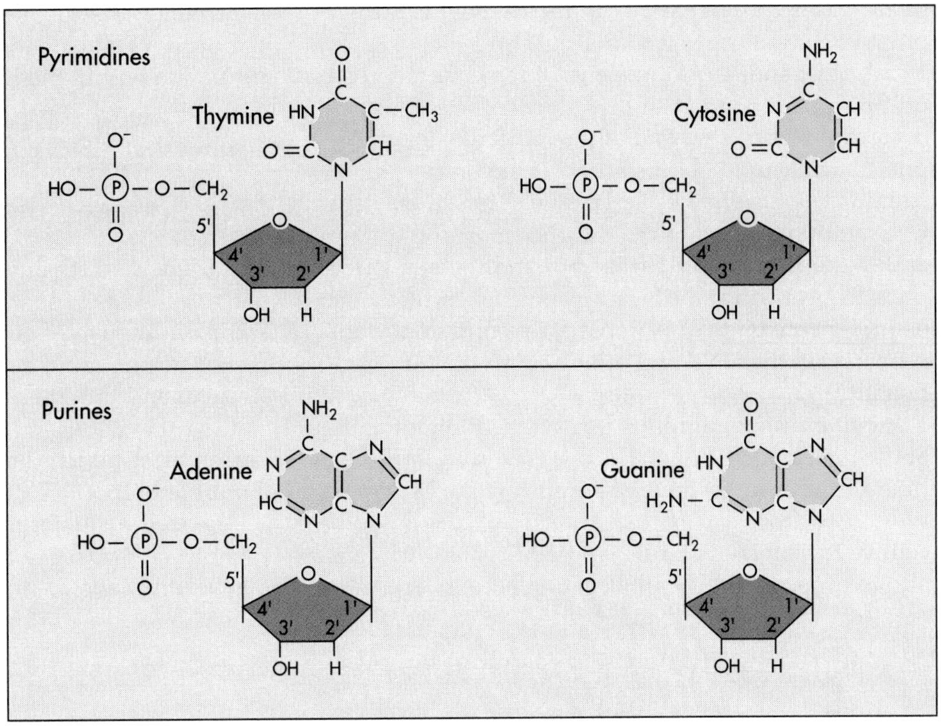

FIG. 7-4 Four different deoxyribonucleotides comprise the subunit molecules of DNA. These have differing nucleic acid bases: thymine, cytosine, adenine, and guanine.

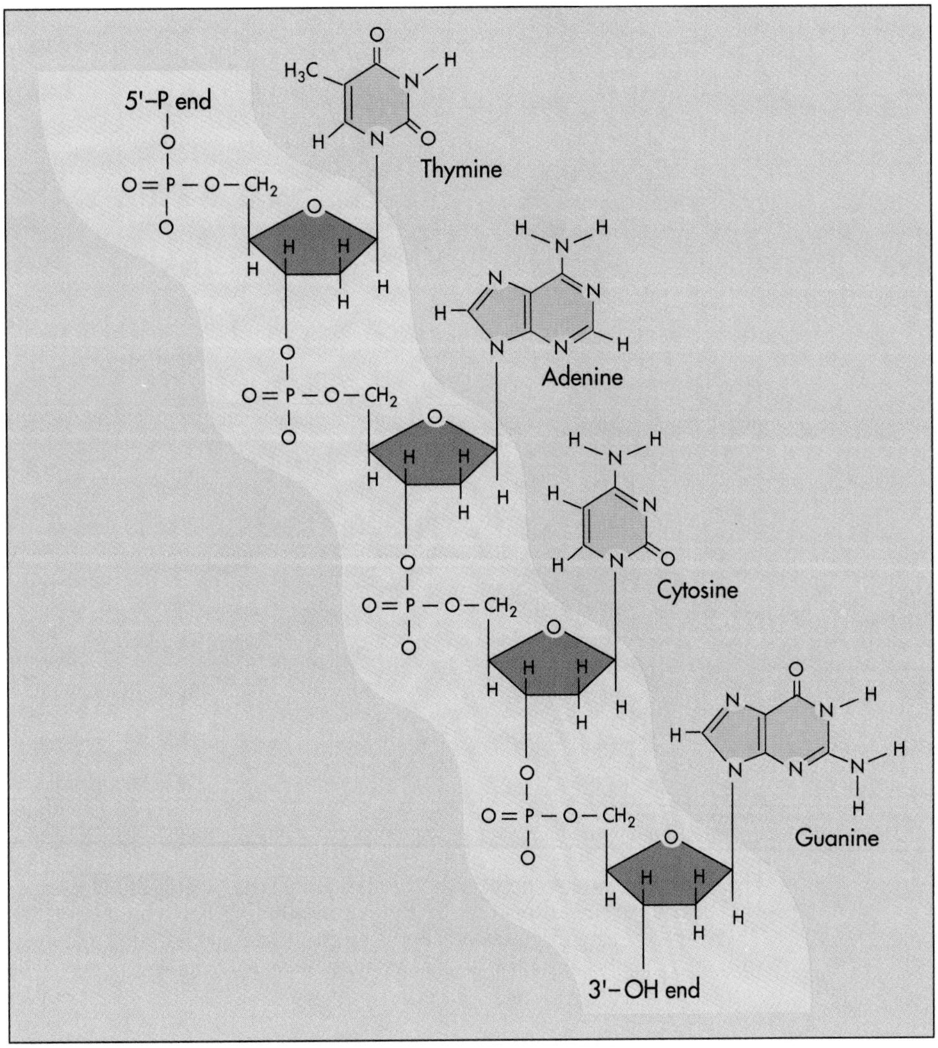

FIG. 7-5 Nucleotides are joined together by phosphodiester bonds between the 3'-OH and 5'-P positions. There is a free 5'-P at one end of the polynucleotide chain and a free 3'-OH at the other end.

phate forms a bridge between the number 3-carbon of one deoxyribose sugar molecule and the number 5-carbon of another. At one end of the chain there is no phosphate bonded to the 3-carbon of a deoxyribose, and at the other end of the molecule, the phosphate attached to the 5-carbon is not involved in forming a phosphodiester linkage. Thus there is a free hydroxyl group at the 3-carbon position at one end of the chain (**3′-OH free end**) and there is a free phosphate group at the 5-carbon position at the other end (**5′-P free end**). This imparts a second important property on the DNA macromolecule—that of directionality.

Having different groups free at the different ends of the molecule distinguishes one end from the other (like left from right) thereby permitting the molecule to be read from a particular direction. This is especially critical for a molecule whose purpose is to store and to transmit genetic information because it estab-

lishes the basis for the correct direction for reading the order of nucleotides that encode the genetic information within the DNA macromolecule.

The chains of DNA macromolecules are different at either end, which allows directional recognition.

DNA DOUBLE HELIX—COMPLEMENTARITY

There are two chains of nucleotides in the DNA macromolecule. The two polynucleotide chains that comprise the DNA double helix run in opposite directions—one chain runs from the 3′-OH to the 5′-P free end and the complementary chain runs from the 5′-P to the 3′-OH free end. These complementary chains twist together to form an arrangement called a **double helix** (FIG. 7-6).

The two chains of the DNA double helix are held together by hydrogen bonding between complementary chains. Two of the nitrogenous bases (C and T)

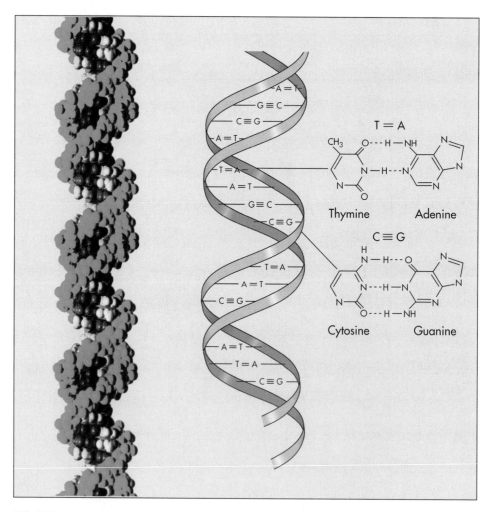

FIG. 7-6 The double helix is the fundamental structure of the DNA macromolecule. The two strands are held together by hydrogen bonding between complementary base pairs. There are three hydrogen bonds between the base pairs guanine and cytosine and two hydrogen bonds between the base pairs adenine and thymine.

DISCOVERING THE STRUCTURE OF DNA

In the early 1950s, James Watson, who had only recently received a doctoral degree from Indiana University, teamed with Francis Crick, a Cambridge University researcher, and pieced together the available data to determine the chemical structure of the DNA molecule, showing how DNA could store and transmit hereditary information (see Figure).

Watson and Crick reasoned that the double-ringed adenine and guanine molecules were probably paired with the single-ringed thymine and cytosine molecules along the entire length of DNA. This would fit the observation made by Erwin Chargaff in the United States in the late 1940s that in any given DNA macromolecule the amount of adenine present is always equal to the amount of thymine and the amount of cytosine is always equal to the amount of guanine. It also was consistent with the observations of Maurice Wilkins and Rosalind Franklin at Cambridge University in England, who, using X-ray diffraction methods, showed that DNA was long and thin, with a uniform diameter. If this was not the case, then DNA would bulge where the two double rings were paired and narrow where two single rings were paired, something that was never seen in the X-ray diffraction images of Wilkins and Franklin.

The problem Watson and Crick had was that the structural formulas for the nucleic acid bases guanine and thymine that were in the organic chemistry books at that time were wrong. Because they were trying to build a model with the wrong structural representations for these molecules, they were unable to get the molecules to fit in a way that would be stabilized by hydrogen bonding between their atoms. Only after getting the structural forms of guanine and thymine corrected during a chance meeting with a visiting scientist were Watson and Crick able to build the correct model of DNA. Once they had the accurate structures, they began shifting the bases in and out of various pairing possibilities.

They next used a plumb line and a measuring stick to determine the relative positions of all of the atoms in a single nucleotide. By assuming a helical shape like a spiral staircase, it became clear that the locations of the atoms in one nucleotide would automatically generate the position of the other. They had constructed a model of the DNA double helix. The model Watson and Crick constructed also showed how the chains could be separated and how each could code for a new complementary chain, thereby unraveling the mystery of how a chemical could pass hereditary information to the next generation with such fidelity.

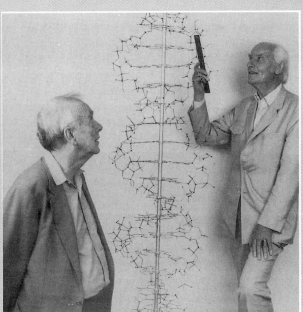

A, James Watson at age 23 and Francis Crick at age 34 developed a model for the structure of DNA while working at the Cavendish Laboratory at Cambridge University, England. The model explained how DNA can transmit hereditary information. They announced their discovery of the molecular structure of DNA in 1953 and shared the Nobel prize for medicine in 1962 along with Maurice Wilkins. **B,** In 1993, on the fortieth anniversary of their discovery, they again posed with their model of DNA, which has proven to be correct.

in DNA are single-ring structures called **pyrimidines** and the other two (A and G) are double-ring structures called **purines.** The charge interactions between purines and pyrimidines allow them to form weak hydrogen bonds (FIG. 7-7). Chemically, the most stable hydrogen bonding occurs when guanine forms three hydrogen bonds with cytosine and when adenine forms two hydrogen bonds with thymine. The proper alignment to form these hydrogen bonds occurs only when the sugar-phosphate backbones of the two DNA chains run in opposing directions and are twisted together to form the double helix.

DNA, which stores and transmits cellular hereditary information, is a double helical molecule.

The hydrogen bonding of A to T and C to G is called **base pairing.** It is this complementarity that establishes the basis for the double helical arrangement of DNA and for the accurate replication of the DNA macromolecule. This is essential for passage of hereditary information from one generation to the next. It also means that in the double helical DNA molecule, the amount of adenine is always the same as the amount of thymine, and the amount of guanine is always the same as the amount of cytosine (A = T and G = C).

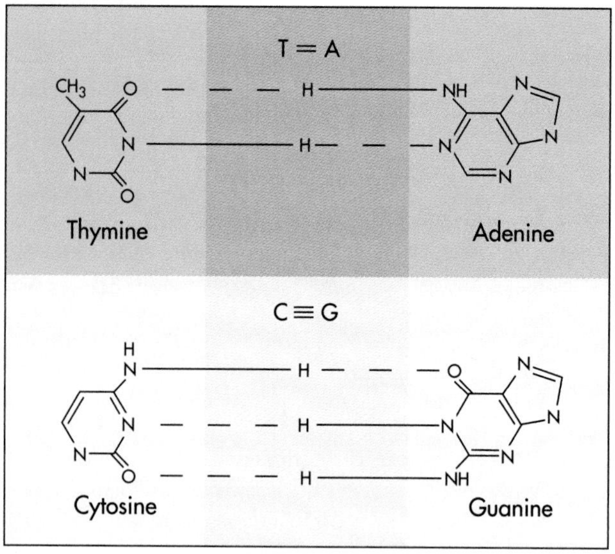

FIG. 7-7 Hydrogen bonding occurs between nucleotide base pairs. Adenine forms two hydrogen bonds with thymine. Guanine forms three hydrogen bonds with cytosine.

Base pairing occurs between complementary nucleotides—adenine pairs with thymine and guanine pairs with cytosine.

REPLICATION OF DNA

When a cell divides, its hereditary information is passed to the next generation. Replication of the hereditary information involves synthesizing new DNA molecules that have the same nucleotide sequences as those of the parental organism. The transfer of hereditary information is possible because DNA has a unique chemical structure in which the two chains of the DNA double helix are **complementary** in nucleotide sequence. Wherever a G is found in one chain, a C is found in the other, and wherever a T is present in one chain, its complementary chain will have an A. A nucleotide sequence of ATCG in one chain has a corresponding sequence of TAGC in the other chain. The nucleotide sequence in one chain specifies the sequence in the other. The information in DNA is, thus, accurately replicated so that an exact copy is passed from one generation to the next.

The order of nucleotides in each chain of a double helical DNA molecule specifies the order of nucleotides in the new complementary chains.

SEMICONSERVATIVE DNA REPLICATION

The process by which a double helical DNA molecule is copied to form a duplicate DNA macromolecule is called **semiconservative replication.** It is so named because during replication each of the chains of nucleotides in the DNA being replicated remains intact. The two chains of nucleotides in the double-stranded DNA molecule are conserved—and a new, complementary chain is assembled for each one. Each of the conserved parental DNA chains serves as the template that specifies the sequence of nucleotides in the newly synthesized strands.

Semiconservative replication was demonstrated experimentally by Matthew Meselson and Franklin Stahl at the California Institute of Technology in 1958 (FIG. 7-8). They grew a culture of *Escherichia coli* in a medium in which the sole source of nitrogen was the heavy isotope ^{15}N. The heavy nitrogen was incorporated into the nucleotides of DNA during bacterial reproduction, so that the DNA of these bacteria became heavier than usual. They then transferred these bacteria to a medium containing the normal lighter isotope ^{14}N. At various time intervals they collected cells and analyzed the DNA to determine if it was "heavy" (^{15}N label), light (^{14}N label), or intermediate (mixture of ^{15}N and ^{14}N label). For these analyses they used an ultracentrifuge—an instrument that spins its contents at high speed—which caused materials to separate out according to their different densities.

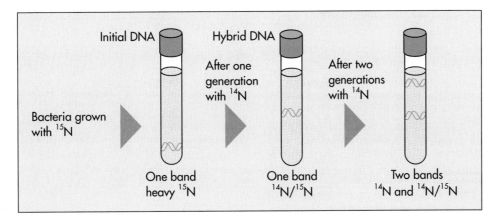

FIG. 7-8 The semiconservative nature of DNA replication was demonstrated by labelling DNA in one generation by the incorporation of heavy nitrogen (^{15}N) and following the fate of this tagged DNA from one generation to the next, using density gradient ultracentrifugation. The location of the bands obtained by ultracentrifugation, that is, the distance that the DNA moves, which is a function of the molecular weight of the DNA, permitted the tracking of the fate of the heavy DNA when the cells were grown in the presence of normal light nitrogen (^{14}N). The banding pattern obtained in these experiments, which is illustrated in the figure, proved that DNA replication occurs by a semiconservative method.

Denser molecules move farther than lighter molecules in cesium chloride density gradient centrifugation, so DNA containing ^{15}N moves a greater distance than DNA containing only ^{14}N. The movement is such that bands of DNA can be distinguished corresponding to light, heavy, and intermediate DNA.

Initially Meselson and Stahl detected only one band. This band corresponded to heavy DNA in which both chains of the DNA contained the ^{15}N label. After sufficient time for one complete round of DNA replication, again only one band of DNA was detected, but now the band was at an intermediate level between all-light isotope and all-heavy isotope DNA. This intermediate band was exactly what was predicted by the hypothesis that DNA replication is semiconservative. Each DNA double helix had one chain from the parental DNA that contained the heavy ^{15}N isotope and one newly synthesized chain that contained only the light ^{14}N isotope. Also as predicted, after sufficient time for a second round of DNA replication, Meselson and Stahl observed two bands of DNA, one intermediate and the other light. This occurred because when the intermediate DNA containing one light and one heavy chain replicated, it contributed one heavy chain to form another intermediate DNA macromolecule and one light chain to form a new all-light DNA macromolecule. This experiment confirmed that DNA replication is semiconservative as suggested by the Watson-Crick model of the DNA double helix.

DNA replication is semiconservative, producing two "half-old, half-new" DNA macromolecules every time the DNA is duplicated.

STEPS IN DNA REPLICATION

Unwinding the DNA Double Helix—Replication Forks

The first step in semiconservative DNA replication is to pull apart a portion of the DNA helix. This enables each of the chains to act as a template (pattern) to direct the synthesis of a new complementary chain of nucleotides. This can occur because hydrogen bonds are relatively weak. Thus the two chains can separate without breaking apart the covalently linked nucleotides of the chains, which would destroy the information encoded within them. This establishes the basis for one chain serving as a template for the synthesis of a new chain of DNA with a sequence of nucleotides that is exactly complementary.

The chains do not entirely separate before DNA replication. Rather, a localized region of the DNA unwinds because the two parental DNA chains are pulled apart by specific enzymes. This creates a region of two single strands and provides space for individual nucleotides to align opposite their complementary bases for the synthesis of new chains. This region of localized DNA synthesis is called a **replication fork** (FIG. 7-9, p. 201). At the replication fork, enzymes link nucleotides to form a new DNA strand that is complementary to the original template DNA.

The DNA double helix unwinds to form a replication fork where DNA synthesis occurs.

In eukaryotic cells, multiple replication forks form at different locations. Simultaneous synthesis of different portions of the DNA is thus made possible. In a bacterial cell, DNA replication is initiated at only

METHODOLOGY

POLYMERASE CHAIN REACTION (PCR)

Understanding the mechanism of DNA replication has enabled scientists to develop a method for replicating segments of DNA in the laboratory from slight traces that otherwise might be too small to analyze. The method, called polymerase chain reaction (PCR), has enormous practical importance because it allows rapid amplification of trace DNA by making many additional copies through replication of specific DNA sequences that can then be detected with great sensitivity (see Figure). This permits the detection of even rare genes. Already, PCR has permitted extremely sensitive detection of the AIDS-causing virus in blood, which is essential for the protection of the blood supply. The impact on microbiology and molecular biology is enormous and PCR has become one of the most widely used methods in science. In recognition of the importance of PCR, the 1993 Nobel Prize in chemistry was awarded to Kerry Mullis, who discovered this method.

PCR is based on the following facts about DNA replication: DNA serves as a template for its own replication; the DNA double helix separates into two chains for replication; a pool of free nucleotides provides the nucleotides for the synthesis of new chains; DNA polymerase catalyzes the formation of the new chains; DNA polymerase adds only to the 3'-free OH end of a nucleotide chain; and DNA polymerase requires a short chain of nucleotides (oligonucleotide) to serve as a primer to initiate DNA replication.

To accomplish the replication of DNA outside of living cells by using PCR, a source of template DNA is added along with a pool of free nucleotides and a DNA polymerase. Also added are short oligonucleotide primers that are complementary to the nucleotide sequences flanking the region of the DNA that is to be replicated. These primers define the region of DNA that is replicated by providing the 3'-OH free ends onto which the DNA polymerase can add nucleotides.

PCR procedure uses heat to provide energy for breaking the hydrogen bonds to separate the chains of the DNA double helix. Heating to 95° C will break the hydrogen bonds without breaking the covalent bonds that link the nucleotides in the chains. Once the chains are separated the reaction is cooled, for example, to 40° C, which allows hydrogen bonds to form between the oligonucleotide primers and their complementary regions of the template DNA. The temperature is then raised to approximately 72° C to allow the DNA polymerase to quickly add nucleotides.

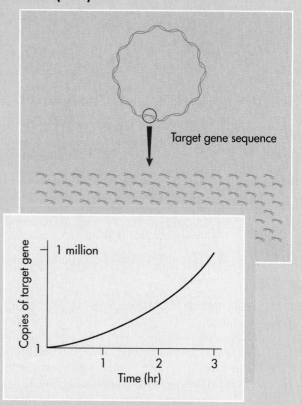

The polymerase chain reaction (PCR) is an *in vitro* method for replicating DNA. A target nucleotide sequence is copied repeatedly so that a million copies can be made in less than an hour.

The DNA polymerase used in PCR, called *Taq* polymerase, comes from a bacterium—*Thermus aquaticus*—that lives in hot springs; it is not denatured at high temperatures. Thus this DNA polymerase can withstand repeated exposure to 95° C. This is critical because in PCR the temperature is repeatedly cycled to separate the chains of the DNA double helix, to bind the primers to the template DNA, and to allow the DNA polymerase to synthesize new strands. Each cycle lasts only a few minutes. The effect of repeated cycling is to exponentially increase the number of copies of a defined segment of the DNA. Within an hour a single copy of a gene can be amplified to a million copies. PCR technology has applications for research and diagnosis and is fast becoming a standard procedure in biotechnology and medical diagnostic laboratories.

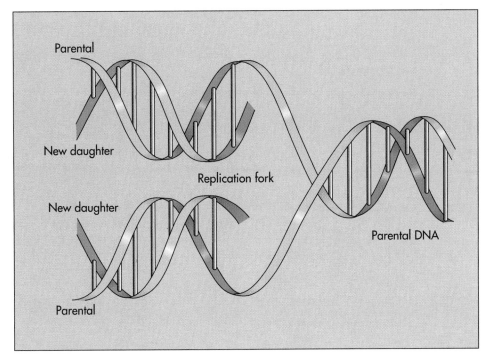

FIG. 7-9   During DNA replication, enzymes separate the two strands of DNA in a localized region called the replication fork. At this site, new nucleotides align opposite base pairs and new strands of DNA are synthesized.

one site, with two replication forks moving from the initiation site in opposite directions around the circular bacterial chromosome. As the replication forks move around the bacterial chromosome, an enzyme—**DNA gyrase**—twists the DNA. This enzyme is unique to bacteria and hence a potential site for the action of an antimicrobial agent. In fact, a new class of antibacterial agents, the **quinolones,** have been discovered that interfere with DNA gyrase. By preventing the formation of replication forks in bacterial cells, quinolones block bacterial reproduction and, hence, can be used to treat bacterial infections. The quinolone ciprofloxacin, for example, is useful in treating *Pseudomonas* infections.

> **DNA gyrase untwists the DNA of the bacterial chromosome.**
>
> **Quinolones are antibacterial agents that inhibit DNA gyrase.**

Formation of a New Chain of Nucleotides—DNA Polymerase

Free nucleotides within the cell in association with DNA polymerase are positioned opposite their complementary nucleotides in the template. This process of aligning complementary nucleotides (A opposite T and C opposite G) is called base pairing. The order of

the nucleotides is specified by the template DNA. After the nucleotides are aligned by base pairing, an enzyme called **DNA polymerase** links the nucleotides by forming phosphodiester bonds. The action of DNA polymerase can be likened to a zipper where the teeth of the zipper are initially aligned and progressively linked together in a continuous motion.

DNA polymerase adds nucleotides to the free 3'-OH end of an existing nucleotide chain of nucleotides (FIG. 7-10). Because DNA polymerase adds nucleotides only to the 3'-OH free end, the direction of DNA synthesis is 5'-P → 3'-OH. Since the two chains of the double helical DNA molecule are antiparallel (one running from the 5'-P → 3'-OH free end and the other running from the 3'-OH → 5'-P free end) this indicates that the synthesis of the two complementary DNA chains must proceed in opposite directions.

One DNA chain can be continuously synthesized. It is the chain that runs in the appropriate direction for the continuous addition of new free nucleotides to the free 3'-OH end. This is the **continuous** or **leading strand of DNA.** Its synthesis occurs simultaneously with the unwinding of the double helical molecule and progresses toward the replication fork.

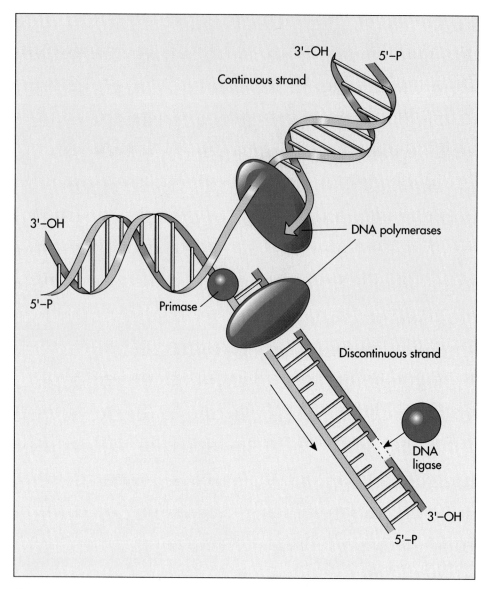

FIG. 7-10   DNA polymerases add nucleotides only to the 3′-OH ends of the newly synthesized DNA polynucleotide chains. One chain is elongated continuously along the direction of formation of the replication fork. The other strand is synthesized as discontinuous segments (Okazaki fragments) that are then joined together by DNA ligase.

The other strand of DNA, however, cannot be synthesized continuously. This is because it runs 3′-OH to 5′-P but DNA polymerase only adds nucleotides in the 5′-P to 3′-OH direction. The initiation of its synthesis can begin only after the double helix has undergone some unwinding. Synthesis of this strand involves formation of short DNA fragments (called Okazaki fragments after the husband and wife team that discovered them) in the direction opposite the direction in which the parent DNA unwinds. Because it is synthesized discontinuously and only after synthesis of the continuous strand has begun, it is called the **discontinuous** or **lagging strand of DNA.** The short DNA fragments of the discontinuous strand are joined together by enzymes called **ligases.** The combined action of DNA polymerase and DNA ligase, thus, accomplishes the synthesis of both complementary strands of DNA during replication.

To make a complementary copy of DNA, the double helix is pulled apart to form a replication fork, complementary nucleotides are aligned by base pairing, and phosphodiester linkages are formed by DNA polymerase.

Replication of DNA should always produce exact copies of the hereditary information. Errors, however, sometimes occur. Such errors are called mutations. A **mutation** is any change in the sequence of nucleotides within DNA. Mutations can involve the addition, deletion, or substitution of nucleotides. Even a simple change, such as the deletion or addition of a single nucleotide, can greatly alter the characteristics of an organism. Once they occur, these changes in the DNA are heritable and are passed from one generation to the next. Mutations introduce genetic variability that makes evolutionary change possible. They also sometimes increase the virulence of pathogens and make some microorganisms resistant to antibiotics.

> **Mutations are stable heritable changes in the nucleotide sequences of DNA.**

Types of Mutations

There are several types of mutations (FIG. 7-11). One type of mutation, **base substitution,** occurs when one pair of nucleotide bases in the DNA is replaced by another pair of nucleotides. A **deletion mutation** involves removal of one or more nucleotide base pairs from the DNA. An **insertion mutation** involves the addition of one or more base pairs. Even though they may represent minor changes in the sequence of nucleotides, mutations can have major effects, sometimes proving lethal to the progeny (offspring or descendants) of the organism.

Sometimes a mutation results in the death of the microorganism or its inability to reproduce. This is called a **lethal mutation.** In other cases, the mutation alters the nutritional requirements for the progeny of a microorganism. Such a mutation is called a **nutritional mutation.** Often, nutritional mutants will be unable to synthesize essential biochemicals, such as amino acids. **Auxotrophs** are nutritional mutants that require growth factors that are not needed by the parent **(prototroph)** strain.

Replica plating is a method frequently used to detect auxotrophs (FIG. 7-12). In this method, bacterial cells are grown on a master plate and then transferred to sterile plates by repeatedly stamping a pad over the master plate and pressing the pad into plates with media of differing composition. The distribution of microbial colonies should be replicated exactly on each new plate. If a colony is unable to grow on the minimal media, which lacks a specific growth factor, but will grow on the complete medium, this indicates that nutritional mutants, or auxotrophs, are occurring. This method allows an investigator to screen a large number of bacteria for mutations.

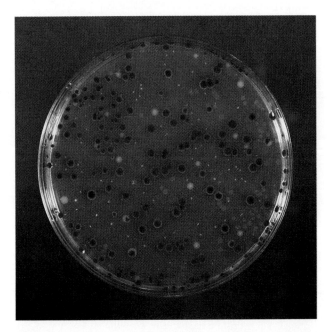

FIG. 7-11 Plate showing growth of *Serratia marcescens.* The wild type colonies are red and the mutant colonies are gray.

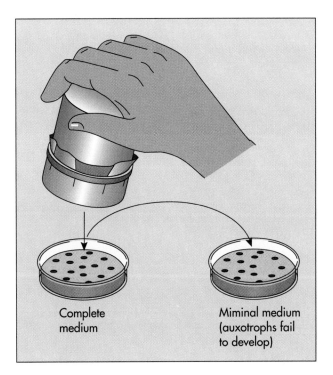

Complete medium

Miminal medium (auxotrophs fail to develop)

FIG. 7-12 Replica plating is used to identify mutants by transferring identical colonies to different types of media and comparing the colonies that develop on the respective plates. This method is critical in identifying auxotrophic mutants. All colonies develop on a complete medium that satisfies the nutritional needs of both the parental and mutant strains. Colonies of the auxotrophic mutant fail to develop on a minimal medium lacking the specific nutritional growth factors required by the mutant.

203

METHODOLOGY

AMES TEST

The fact that microorganisms are susceptible to chemical mutagens can be used to determine the ability of various chemicals to increase the rate of mutation, or mutagenicity. In the Ames test procedure, a strain of the bacterium *Salmonella typhimurium* is the test organism for determining chemical mutagenicity (see Figure). This bacterial strain is an auxotroph (nutritional mutant) that requires the amino acid histidine. These organisms are exposed to increasing amounts of the chemical being tested on a solid growth medium that lacks histidine. Normally, the bacteria cannot grow and in the absence of a chemical mutagen no colonies can develop. If the chemical is a mutagen, lethal mutations will occur in the areas of high chemical concentration and no growth will occur in these areas. At lower chemical concentrations along the concentration gradient, however, fewer mutations will occur and some of the cells may revert through mutation to nutritional types that do not require histidine for growth. Such mutants are able to grow and produce visible bacterial colonies on the medium. The appearance of bacterial colonies, therefore, demonstrates that the chemical is a mutagen and the absence of colonies indicates that it is not.

The Ames test procedure also is used to screen chemicals to determine if they are potential cancer-causing agents, or carcinogens. The theoretical basis for this use of the Ames test is that nearly all carcinogens that act directly by attacking DNA are also mutagens. Rather than screening chemicals directly for carcinogenicity, Bruce

Ames and his co-workers thought it would be better to test them first for mutagenicity. They recognized, though, that some chemicals are chemically modified in the body and, in particular, some chemicals are inadvertently transformed into carcinogens in the liver in an apparent effort by the body to detoxify these chemicals. Therefore, in testing for potential carcinogenicity, once a chemical has been found to be mutagenic, it is incubated with a preparation of rat liver enzymes to simulate what normally occurs in the liver. Various concentrations of this preparation are then incubated with the *Salmonella* auxotroph to determine whether any products that would cause mutations are formed. The chemicals that do not produce mutations are assumed to be noncarcinogenic or are carcinogens that are not detected by this procedure. Those shown to be mutagenic are subjected to further testing.

Although the Ames test does not positively establish whether a chemical causes cancer, determining whether a chemical has mutagenic activity is useful in screening large numbers of chemicals for potential mutagens, because it is highly probable that a chemical that is a mutagen is also a carcinogen. Since this test can be completed in 24 hours, rapid identification of a mutagen is possible. Today, the use of this bacterial assay greatly simplifies the task of screening many potentially dangerous chemicals, permitting us to recognize potentially carcinogenic compounds. Using bacteria in the Ames assay also allows scientists to avoid animal testing in many cases.

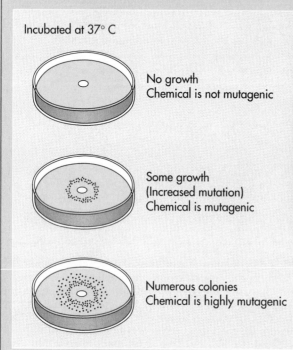

Incubated at 37° C

No growth
Chemical is not mutagenic

Some growth
(Increased mutation)
Chemical is mutagenic

Numerous colonies
Chemical is highly mutagenic

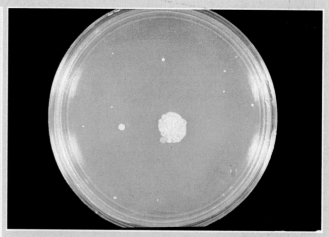

The Ames test procedure is used to screen for mutagens and potential carcinogens. The auxotrophic strain used in this procedure, generally a histidine-requiring mutant of *Salmonella typhimurium*, will not grow on a minimal medium. Mutants that revert to the prototrophic wild type will grow on this medium. The number of colonies that develop after exposure to a chemical indicates the effect of that chemical on mutation rate and therefore its degree of mutagenicity. The development of many colonies indicates that the chemical is highly mutagenic.

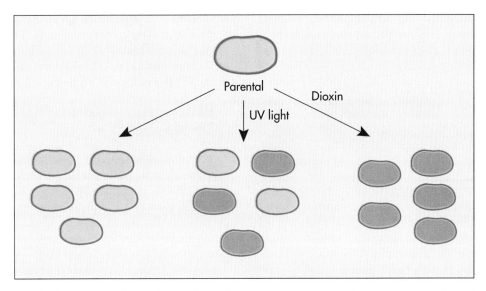

FIG. 7-13 Various chemical and physical agents increase rates of mutation. UV light and dioxin are mutagens that cause formation of mutants *(pink cells).*

FACTORS AFFECTING RATES OF MUTATION

Naturally occurring rates of mutation are relatively low, about one in a million times the rate of DNA replication. Various physical and chemical agents, however, can modify the nucleotides within DNA, increasing the rate of mutation (FIG. 7-13). Agents that increase the rates of mutation are called **mutagens.**

Exposure to high-energy radiation such as X-rays can cause mutations. Such high-energy ionizing radiation produces breaks in the DNA molecule. The time and intensity of exposure determines the number of lethal mutations that occur. Exposure to gamma radiation, such as that emitted by an isotope of cobalt, ^{60}Co, can be used for sterilizing objects, including plastic Petri plates, because sufficient exposure results in lethal mutations and the death of all exposed microorganisms. Gamma radiation from ^{60}Co has also been used to kill microorganisms on the surfaces of some foods, thereby delaying the spoilage of the food.

Exposure to ultraviolet light also can result in **thymine dimer** formation, that is, the covalent linkage of one thymine to another thymine. Covalent linkages are formed between two thymines on the same strand of DNA. A thymine dimer cannot act as a template for DNA polymerase, and the occurrence of such dimers, therefore, prevents the proper functioning of DNA polymerase. Exposure to ultraviolet light can cause lethal mutations and is sometimes used to kill microorganisms in sterilization procedures. Sometimes ultraviolet light is placed above a work surface when it is not in use to maintain sterility of the surface. Also, air is sometimes passed over ultraviolet lights in hospitals treating patients with respiratory diseases to kill airborne pathogens.

Radiation causes lethal mutations and can be used for sterilization of some materials.

EXPRESSION OF GENETIC INFORMATION

The expression of genetic information involves using information encoded within DNA to direct the synthesis of proteins. In a cell, DNA stores and transmits the complete hereditary information called the **genotype.** Proteins mediate functional activity—called the **phenotype**—that is, the actual appearance and activities of the organism. For example, proteins (enzymes) catalyze all the metabolic activities of cells and produce the observable characteristics that distinguish one microorganism from another.

Proteins mediate functional activity (phenotype); DNA mediates the informational capacity (genotype) of the cell.

The sequence of nucleotides within the DNA molecule ultimately codes for the sequence of amino acids in proteins. Because proteins are the "action" molecules within cells and nucleic acids dictate the formation of proteins, it is the ordering of nucleotides within DNA that provides the information for establishing, controlling, and reproducing cell structure and function. By specifying and regulating protein synthesis, the genetic informational macromolecules define and control the metabolic capabilities of microorganisms.

> **The sequence of nucleotides within a cell's DNA determines the sequence of amino acids in its protein molecules.**

GENES

DNA is divided into functional sequences known as **genes.** Each gene codes for a specific function. Some genes code for the synthesis of RNA and proteins, respectively determining the sequences of nucleotides and amino acids in these macromolecules. Such genes are known as **structural genes.** Other genes, called **regulatory genes,** act to determine when structural genes are actually expressed. Regulatory genes exert control over the activities of the cell by turning the expression of structural genes on or off. Together,

structural and regulatory genes constitute the genotype, that is, the complete genetic informational capacity of an organism. Genes determine all hereditary traits, and they control all the potential activities that can take place within living cells.

> **Genes are sequences of DNA that have specific functions.**

Bacterial cells usually have a single set of genes and are said to be **haploid.** In contrast, eukaryotic microorganisms generally have pairs of matching chromosomes and are **diploid,** having two copies of each gene during at least part of their life cycles. Some eukaryotic microorganisms are haploid during part of their life cycle. When both copies of the gene are identical, the cell is **homozygous.** When the corresponding copies of the gene differ, the cell is **heterozygous.** For example, eukaryotic microorganisms with homozygous genes for color may appear red or white; those with heterozygous genes for red and

HISTORICAL PERSPECTIVE

ONE GENE—ONE POLYPEPTIDE

At the turn of the twentieth century, a physician named Archibald Garrod was studying human metabolic disorders that seemed to run in families. He hypothesized that the specific units of inheritance must function through the synthesis of specific enzymes. This hypothesis received strong support from the studies reported by George Beadle and Edward Tatum in 1941 on the relationship between genes—the units of inheritance—and the metabolism of the bread mold *Neurospora crassa* (see Figure). Prototrophic strains of *N. crassa* can grow on a minimal medium containing only sucrose, mineral salts, and biotin. From these substances, *N. crassa* can synthesize all of its other nutritional requirements, including the amino acids needed for making proteins. This fungus made an excellent choice as the experimental organism to study gene function because much was already known about some of its metabolic pathways. In particular each of the enzymatic steps in the biosynthetic pathway for making the amino acid arginine was known.

Beadle and Tatum designed experiments to detect mutations that produced auxotrophs that could not synthesize arginine. They then analyzed these mutants to see which metabolic step in the synthesis of arginine was affected. To increase the frequency of mutations, Beadle and Tatum X-rayed *Neurospora* spores. The prog-

eny were allowed to grow on a complete medium, which contained all necessary metabolites, including arginine. Next, they tested the abilities of the progeny to grow on a minimal medium lacking amino acids to see if mutations had occurred that resulted in the inability to synthesize arginine. An auxotrophic mutant that could not synthesize arginine would grow on the complete medium but not on the minimal medium.

Beadle and Tatum hypothesized that if genes (hereditary units) controlled the production of specific enzymes, they could detect mutants that could not make different specific enzymes and so could be blocked at different steps in the metabolic pathway for arginine biosynthesis. In fact, Beadle and Tatum identified and isolated many such mutants.

Analysis of cell extracts of mutants revealed a different defective enzyme in each mutant strain. For each enzyme in the arginine biosynthetic pathway, Beadle and Tatum were able to isolate a mutant strain with a defective form of the enzyme needed for that step. They thus provided evidence favoring the "one gene, one enzyme" hypothesis. Other scientists continued to study the relationship between genes and proteins, modifying the original hypothesis slightly, reaching the conclusion that genes code for polypeptides, the chains of amino acids that make up proteins.

white may appear red or white if one gene is preferentially expressed (dominant) over the other (recessive) or pink if both genes are simultaneously expressed (codominant).

RNA Synthesis

The transfer of information from DNA to protein is accomplished in two stages (FIG. 7-14). The information in the DNA molecule is initially used to direct the synthesis of ribonucleic acid (RNA) molecules in a process called transcription. Transcription is so named because the information in the DNA is effectively copied or transcribed into RNA. In the second stage, called translation, the RNA directs the synthesis of proteins. Translation is so named because the order of nucleotides in RNA is translated into the order of amino acids of proteins. Transcription and translation occur in tandem in prokaryotic cells.

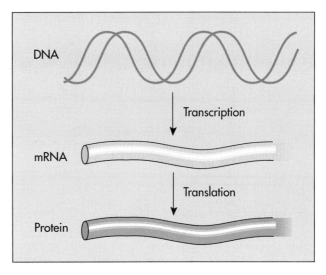

FIG. 7-14 The information in DNA is transferred to RNA during transcription. The RNA then directs the synthesis of proteins during translation.

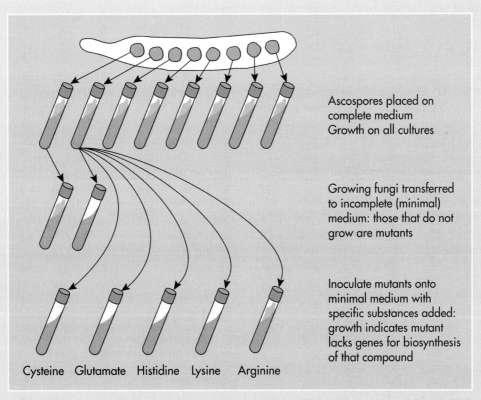

Ascospores placed on complete medium
Growth on all cultures

Growing fungi transferred to incomplete (minimal) medium: those that do not grow are mutants

Inoculate mutants onto minimal medium with specific substances added: growth indicates mutant lacks genes for biosynthesis of that compound

Cysteine   Glutamate   Histidine   Lysine   Arginine

The one gene–one hypothesis was experimentally demonstrated by Beadle and Tatum. They studied the fungus *Neurospora*, which forms spores that have single sets of genes. They were able to collect individual spores and to culture them so that they could observe any changes in the genes of the fungus that altered its nutritional requirements. Mutants, which were unable to carry out complete biosynthetic pathways, could grow on complete media but not on minimal media. The minimal media lacked the nutrients required for growth that the mutants could no longer synthesize. Growth on minimal media with specific compounds added, such as vitamins and amino acids, enabled them to determine which compounds the mutant fungi could not synthesize. In this manner they were able to identify the genetic changes that occurred and to associate specific genes with specific metabolic activities.

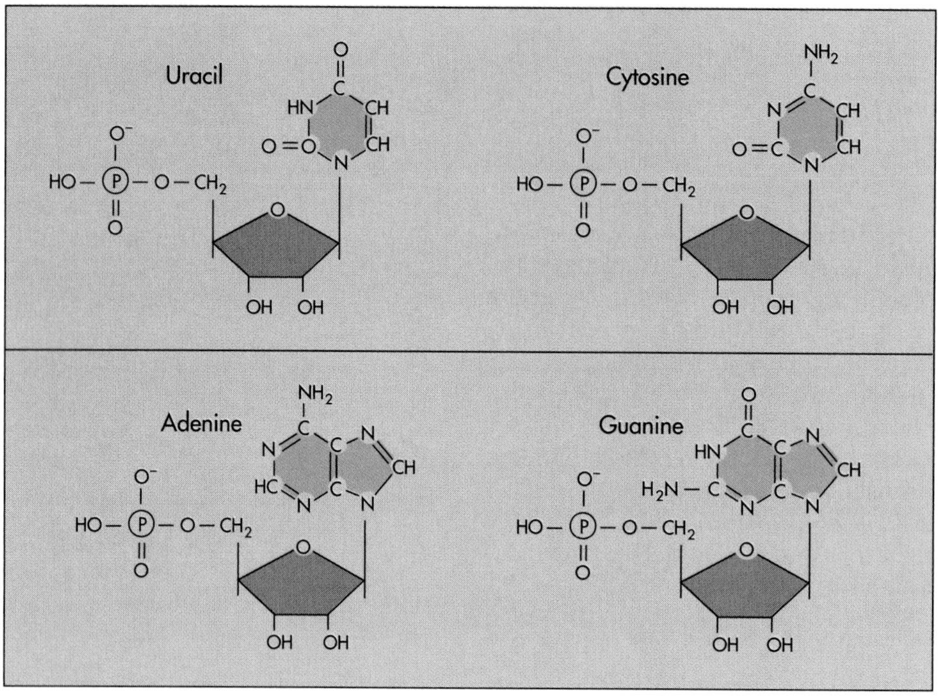

FIG. 7-15   Ribonucleic acid (RNA) is composed of ribonucleotides that have the sugar ribose, a phosphate group, and one of four nucleic acid bases: uracil, cytosine, adenine, or guanine.

However, in eukaryotic cells, transcription occurs in the nucleus and translation subsequently occurs on the ribosomes in the cytoplasm. Therefore, in eukaryotic cells, transcription and translation are separated in time and space.

> **RNA acts as an informational mediator between the DNA where genetic information is stored and the proteins that functionally express that information.**

Ribonucleic Acid (RNA)—Functions and Types

RNA differs from DNA in several respects (FIG. 7-15). The 5-carbon sugar in the RNA nucleotide is ribose, which has one more oxygen atom than the deoxyribose in DNA. One of RNA's nitrogenous bases is uracil instead of thymine. Also, RNA is usually single-stranded.

> **RNA contains ribose, adenine, uracil, cytosine, and guanine.**

Three different kinds of RNA occur in all living cells. They are ribosomal RNA, messenger RNA, and transfer RNA. Each of these RNA molecules has different functions and different physical-chemical properties (Table 7-1). **Ribosomal RNA** (rRNA) is an important structural component of ribosomes, the sites where proteins are synthesized within cells. **Messenger RNA (mRNA)** carries the information from the DNA molecule to the ribosome. At the ribosomes, protein synthesis actually occurs when the information encoded in the mRNA molecule is used to specify the sequence of amino acids that comprise the

protein. The mRNA contains the genetic information of DNA in a single-stranded molecule complementary in base sequence to a portion of the base sequence of DNA. The mRNA sequence directs incorporation of amino acids into a growing polypeptide in the process called translation, which takes place on the surface of the ribosome. **Transfer RNA** (tRNA) molecules help align amino acids during protein synthesis in the order specified by an mRNA molecule. Transfer RNAs form a bridge between mRNA and an amino acid, thereby transferring the genetic information carried by the mRNA to the amino acid sequence of the polypeptide. Working together, the three types of molecules convert genetic information from the language of nucleotides to the language of amino acids, the building blocks of proteins.

> **Three types of RNA molecules (rRNA, mRNA, and tRNA) are involved in the transfer of information from DNA to proteins.**

Transcription

Transcription is the process by which information stored in the DNA molecule is used to code for the synthesis of RNA (FIG. 7-16). During transcription, RNA nucleotides in association with RNA polymerase pair with complementary DNA bases. The base pairs between DNA and RNA are: thymine (DNA) and adenine (RNA), adenine (DNA) and uracil (RNA), guanine (DNA) and cytosine (RNA), and cytosine (DNA) and guanine (RNA). Once the RNA bases are properly aligned in the order speci-

TABLE 7-1

Characteristics of Various Types of RNA

TYPE OF RNA	ABBREVIATION	SEDIMENTATION COEFFICIENT	FUNCTION
Messenger RNA	mRNA	6—50S	Carries genetic information from DNA to the ribosomes where the information is used to direct the synthesis of polypeptides
Transfer RNA	tRNA	4S	Carries amino acids to the ribosomes and assists in the translation of the information carried by mRNA
Prokaryotic Ribosomal RNA	rRNA	5S 16S 23S	The major structural components of ribosomes that interact with mRNA and tRNA to ensure proper synthesis of polypeptides; there are three major types of rRNAs with differing sedimentation coefficients
Eukaryotic Ribosomal RNA	rRNA	5.8S 18S 28S	The major structural components of ribosomes that interact with mRNA and tRNA to ensure proper synthesis of polypeptides; there are three major types of rRNAs with differing sedimentation coefficients

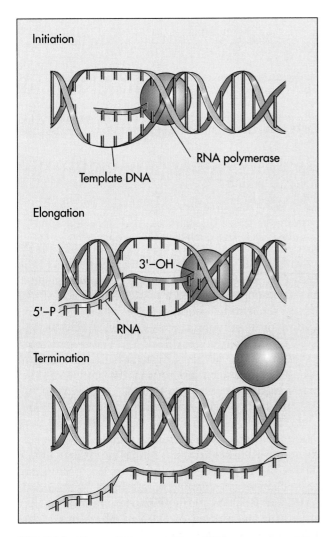

FIG. 7-16 Transcription produces RNA using one of the strands of DNA as a template. The formation of the RNA is catalyzed by RNA polymerase. After initiation the RNA is elongated by this enzyme until termination.

fied by the DNA molecule, the enzyme **RNA polymerase** links the bases together.

The synthesized molecule of RNA is antiparallel, that is it runs 5'-P to 3'-OH if the chain of DNA that serves as the template runs 3'-OH to 5'-P. Only one chain of the DNA serves as a template for the synthesis of a particular RNA molecule. This DNA chain coding for the synthesis of RNA is known as the **template strand.** Both chains of the DNA can serve as template strands in different regions, and the term *template strand* is applied only to the specific region of the DNA that is being transcribed.

Initiation and Termination of Transcription

The transfer of information from DNA to RNA requires that transcription begin and stop at precise locations. Both prokaryotes and eukaryotes have multiple initiation sites along the DNA molecule for transcription (FIG. 7-17). The specific site where RNA polymerase initially binds to the DNA and, thus,

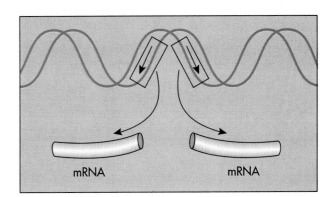

FIG. 7-17 Transcription can occur at multiple sites. This permits the formation of several proteins simultaneously.

where transcription begins is called the **promoter region.** There are different promoters for the initiation of transcription of different genes. The promoter determines the site of transcription initiation and which of the two DNA strands is to serve as the sense strand for transcription in that region.

Transcription begins in the promoter region of DNA where RNA polymerase binds.

The RNA polymerase moves along the DNA template, synthesizing a new strand of RNA until a specific termination sequence is reached. Then the RNA polymerase falls off the sense strand of the DNA and transcription stops.

Synthesis of mRNA in Prokaryotic and Eukaryotic Cells—Split Genes

There is a fundamental difference in how mRNA is formed in prokaryotic and eukaryotic cells. The sequence of nucleotides in the messenger RNA molecule of prokaryotic cells corresponds exactly with the sequence of nucleotides in the DNA. In contrast, the RNA molecules of eukaryotic microorganisms are generally extensively modified after transcription from the DNA to form mRNA (FIG. 7-18).

Unlike the genes of bacterial cells, the DNA sequences coding for RNA molecules in eukaryotic cells is not continuous. There are intervening DNA sequences, called **introns,** between the nucleotide sequences that actually constitute the gene. The sequences that constitute the gene are called **exons.** Hence, the genes of eukaryotic cells are said to be **split genes.** When DNA is transcribed the RNA initially contains both introns and exons. This precursor of messenger RNA in eukaryotes, known as **hnRNA (heterogeneous nuclear RNA),** is then subjected to substantial post-transcriptional modification within the nucleus. This post-transcriptional modification removes the introns and forms a messenger RNA that only has exons.

PROTEIN SYNTHESIS—TRANSLATION OF THE GENETIC CODE

mRNA and the Genetic Code

The information in mRNA specifies the sequence of amino acids in the protein made during **translation.** The translation of the information in the mRNA molecule, that is, reading mRNA, is a directional process. Messenger RNA is read in a 5'-P → 3'-OH direction. The polypeptide that is made is synthesized from the amino terminal to the carboxyl terminal end. Here, we see the importance of having a mechanism for recognizing direction in informational macromolecules. Just as we have the convention for reading the English language from left to right, the correct interpretation of the information stored in the mRNA

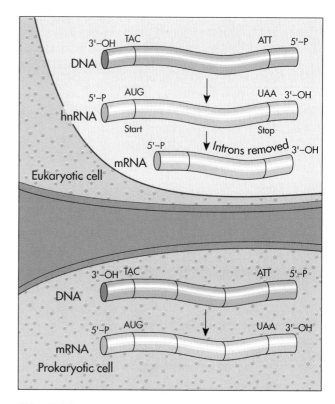

FIG. 7-18 In eukaryotic cells the primary transcript (hnRNA) is extensively modified within the nucleus to produce mRNA. The conversion of hnRNA to mRNA involves the removal of introns. Each mRNA usually encodes only a single gene. In prokaryotic cells the primary transcript serves as the mRNA. It often encodes several genes.

molecule requires that it be read from the 5'-P → 3'-OH free end.

The information in mRNA specifies the sequence of amino acids in a protein during translation.

Within mRNA, three sequential nucleotides are used to code for a given amino acid. The genetic code, therefore, is termed a **triplet code.** Each of the triplet nucleotide sequences is called a **codon.** Adding or deleting nucleotides causes mutations because the codon changes (FIG. 7-19). The genetic code is said to be degenerate because more than one codon can specify the same amino acid. The mRNA molecule is read one codon at a time. In other words, the three nucleotides that specify a single amino acid are read together. There are no spaces between the codons that are read. Therefore establishing a reading frame is critical for extracting the proper information. Adding or deleting a single base pair in the DNA changes the reading frame of the transcribed messenger RNA. Such **frame-shift mutations** can result in the misreading of large numbers of codons (FIG. 7-20), thus producing proteins that are inactive because they have the wrong amino acid sequence.

Three nucleotides constitute a codon that specifies an amino acid.

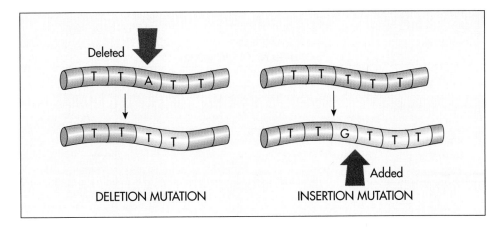

FIG. 7-19 Deletion mutations occur when one or more nucleotides are omitted during DNA replication. Addition mutations occur when one or more nucleotides are added during DNA replication.

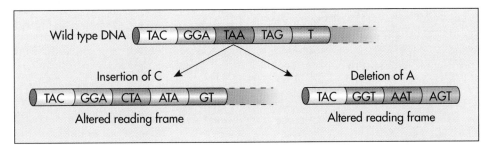

FIG. 7-20 A frame shift mutation results from nucleotide deletions or additions.

Each codon consists of three nucleotides. There are four different nucleotides, making possible 64 codons. The four different nucleotides in different three-base combinations lead to $4 \times 4 \times 4 = 64$ possibilities (Table 7-2, p. 212). The genetic language, which is almost universal, can therefore be said to have four letters in the alphabet and 64 words in the dictionary, each word containing three letters.

Proteins in biological systems normally contain only 20 L-amino acids. Thus there are many more codons than are strictly needed for the translation of genetic information into functional proteins. More than one codon can code for the insertion of the same amino acid into the polypeptide chain. Consequently, a silent mutation can occur in which there is an alteration in the nucleotide sequence without a change in the amino acid sequence. In such cases, changes in genotype would not be reflected in changes in phenotype.

Additionally, there are three codons that do not code for any amino acid. These codons have been referred to as **nonsense codons.** Actually, nonsense codons serve a very important function, acting like the period at the end of a sentence and signalling termination of synthesis of a polypeptide chain.

More than one codon can code for the same amino acid. The genetic code is degenerate.

tRNA and Polypeptide Formation

Translation of the genetic code into protein molecules occurs at the ribosomes. Ribosomes provide the spatial framework and structural support for aligning the translational process of protein synthesis. Distortion of the proper configuration of the ribosome can prevent proper information exchange and expression of the genetic information. This forms the basis for the action of many antibiotics, such as erythromycin.

Transfer RNA (tRNA) attaches to amino acids and brings them to the ribosomes. There is a specific tRNA for each amino acid. In addition to bringing the amino acids to the ribosomes, the tRNA also properly aligns them during translation (FIG. 7-21). Each tRNA molecule contains a specific **anticodon,** a three-base nucleotide sequence that is complementary to the three-base nucleotide sequence of the codon. The pairing of the codons of the mRNA molecules with the anticodons of the tRNA molecules determines the order of amino acid sequence in the polypeptide chain. The third base of the anticodon does not always properly recognize the third base of

TABLE 7-2						
Codons of the Genetic Code (mRNA shown in 5'-P → 3'-OH direction)						
	SECOND NUCLEIC ACID					
First Nucleic Acid (5'-P End)	**U**	**C**	**A**	**G**		**Third Nucleic Acid (3'-OH End)**
U	UUU UUC Phenylalanine / UUA UUG Leucine	UCU UCC UCA UCG Serine	UAU UAC Tyrosine / UAA UAG STOP	UGU UGC Cysteine / UGA STOP / UGG Tryptophan		U C A G
C	CUU CUC CUA CUG Leucine	CCU CCC CCA CCG Proline	CAU CAC Histidine / CAA CAG Glutamine	CGU CGC CGA CGG Arginine		U C A G
A	AUU AUC AUA Isoleucine / AUG Methionine	ACU ACC ACA ACG Threonine	AAU AAC Asparagine / AAA AAG Lysine	AGU AGC Serine / AGA AGG Arginine		U C A G
G	GUU GUC GUA GUG Valine	GCU GCC GCA GCG Alanine	GAU GAC Aspartate / GAA GAG Glutamate	GGU GGC GGA GGG Glycine		U C A G

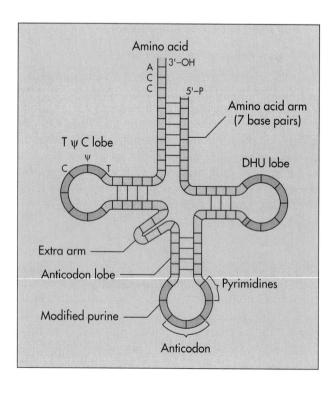

FIG. 7-21 All tRNA molecules have a characteristic four-lobe structure that results from internal base pairing of some of the nucleotides. Each lobe of the tRNA molecule has a distinct function. Several of the lobes are characterized by the inclusion of unusual nucleotides. These nucleotides are formed by enzymatic modification of the nucleotides directly coded for by the DNA; that is, the DNA does not have additional nucleotides that directly call for the insertion of nucleic acid bases other than adenine, uracil, cytosine, and guanine into the RNA. One of the lobes, designated the *DHU* or *D lobe*, contains dihydrouracil (DHU). This lobe binds to the enzyme involved in forming the peptide during translation. The TψC lobe contains the sequence ribothymine (T), pseudouracil (ψ), and cytosine (C). This lobe binds to the ribosome. A third lobe, which also contains modified purines, is designated the *anticodon lobe* because it is complementary to the region of the mRNA, the codon, that specifies the amino acid to be incorporated during protein synthesis. The 3'-OH end always has the terminal sequence ACC, which is where the amino acid binds. This terminal sequence is usually referred to as the *CCA end*, reading from the 5'-P end of the tRNA molecule.

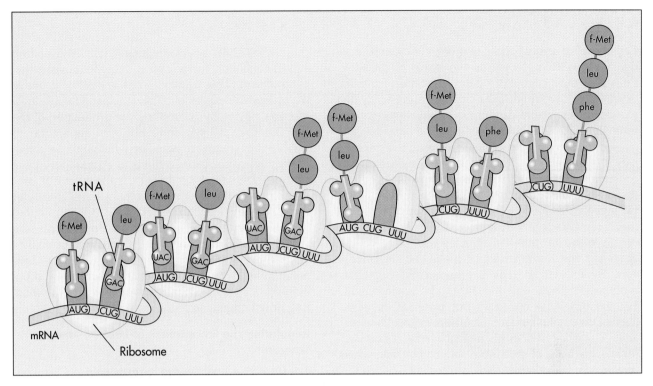

FIG. 7-22 During protein synthesis the codons of the mRNA are translated into an amino acid sequence at the ribosome. Each codon of the mRNA matches an anticodon of a tRNA so that the proper amino acid sequence is formed. The start codon AUG specifies the insertion of formyl methionine (f-Met) at the peptidyl site. A second amino acid is aligned at the aminoacyl site by the pairing of a tRNA with the codon. Formyl methionine is transferred to the amino acid at the aminoacyl site with the formation of a peptide bond. The mRNA then moves along the ribosome so that the tRNA with its two attached amino acids moves to the peptidyl site. A new amino acid is aligned at the aminoacyl site, again by pairing of the appropriate tRNA with the codon. The two amino acids are transferred to the amino acid with the formation of a new peptide bond so that the peptide chain now has three amino acids. The process is repeated over and over to form the long polypeptide chain of amino acids joined by peptide bonds in the sequence specified by the mRNA.

the messenger RNA codon. The first and second bases of the codon sequence are therefore more important in matching the codon with the anticodon. As a result, a codon may pair with more than one anticodon that differs only in the third base position, a phenomenon called **wobble.**

Transfer RNA brings amino acids to the ribosomes and properly aligns them during translation.

Forming the Polypeptide

During translation, transfer RNA molecules bring individual amino acids to be sequentially inserted into the polypeptide chain (FIG. 7-22). The codon of the messenger RNA that specifies where the synthesis of a polypeptide is initiated is called the **start codon.** The tRNAs arrive in the order specified by the codons in the mRNA as the mRNA moves across the surface of a ribosome. When tRNA molecules arrive at the ribosome, the proper anticodon pairs with its

matching codon of mRNA. The amino acid is thus aligned so that it can be covalently bound to a growing peptide chain. After a peptide bond is established between amino acids already in the polypeptide chain and the newly aligned amino acid, the messenger RNA then moves along the ribosome by three nucleotides. The movement of messenger RNA, transfer RNA, and the growing polypeptide chain along the ribosome is known as **translocation.** The process is repeated over and over, resulting in the elongation of the polypeptide chain. Eventually, one of the nonsense codons appears on the mRNA as it moves across the ribosome. Since no tRNA molecule pairs with the nonsense codon, the translational process is terminated and the polypeptide is physically released from the ribosome.

Translocation is the movement of messenger RNA, transfer RNA, and the polypeptide chain along the ribosome.

Cells have structural genes that encode the information for specific polypeptide sequences of proteins. Cells also have regulatory genes that code for gene expression. It would be inappropriate and energy depleting for the entire genome to be expressed at one time. By controlling which genes of the organism are to be translated into functional enzymes, the cell regulates its metabolic activities. While some genes are constantly "turned on," others are expressed only in response to the immediate needs of the cell. It is advantageous for a cell to regulate gene expression so that it can conserve its resources. This is important to conserve the supply of energy, as well as to utilize sparingly the limited pool of metabolic intermediates. By regulating gene expression the organism modifies its phenotype to adapt to its environment. For example, the cell does not produce enzymes needed to catabolize lactose unless lactose is available. Also, the cell does not produce enzymes needed for the synthesis of the amino acid tryptophan when tryptophan is available.

Some regions of DNA are specifically involved in regulating transcription. These regulatory genes can control the synthesis of specific enzymes. Sometimes gene expression is not subject to specific genetic regulatory control. In these cases, the enzymes coded for by such regions of the DNA are **constitutive,** that is, they are continuously synthesized. In contrast to constitutive enzymes, some enzymes are synthesized only when the cell requires them. Some such enzymes are **inducible,** that is, made only in response to a specific inducer substance. Others are **repressible,** that is, made unless stopped by the presence of a specific repressor substance.

OPERONS

In 1961 Francois Jacob and Jacques Monod put forth a hypothesis that induction and repression were under the control of specific proteins. Such proteins would be coded for by regulatory genes. They proposed that regulatory genes were closely associated with the structural genes that code for the enzymes in specific metabolic pathways. Often, several enzymes that have related functions are controlled by the same regulatory gene. Called the **operon model,** the mechanism proposed by Jacob and Monod explains how cells are able to coordinate the expression of genes with related functions.

An **operon** is a cluster of adjacent genes on the chromosome that is controlled by one promoter site. Transcription starting at that promoter site results in the formation of an mRNA coding for several polypeptides. Such an mRNA is said to be **polycistronic,** meaning that it codes for more than one

polypeptide. An operator gene within the operon acts like a switch, turning on and off the transcription of structural genes. Either all or none of the genes of the operon are expressed. This is achieved at the level of transcription by controlling the production of the polycistronic mRNA. Induction and repression of genes in an operon are based on whether or not a regulatory repressor protein binds at a regulatory gene of the DNA, called the *operator.* If the repressor protein binds to the operator, it blocks transcription of the succeeding structural genes.

Some operons are regulated by positive control, which involves the binding of a regulator protein to DNA and the stimulation of gene expression. Others are regulated by negative control, which involves binding of a regulator protein to DNA and the shutting down of gene expression.

Regulating the Metabolism of Lactose—the *lac* Operon

The *lac* operon coordinates the expression of three enzymes that are specifically synthesized by *Escherichia coli* for the metabolism of lactose. These enzymes are: β-galactosidase, galactoside permease, and transacetylase. β-galactosidase cleaves the disaccharide lactose into the monosaccharides galactose and glucose. Galactoside permease is required for the transport of lactose across the bacterial plasma membrane. The role of transacetylase is not yet established. The structural genes that code for the production of these three enzymes occur in a contiguous segment of DNA.

The operon for lactose metabolism is called the **lac operon** (FIG. 7-23). The *lac* operon includes a promoter region where RNA polymerase binds, an operator region where the repressor protein attaches, and three structural genes that code for three proteins that are involved in lactose metabolism. In addition, there is a regulatory gene at another location that codes for the synthesis of a repressor protein. In the absence of lactose, this repressor protein binds to the operator region of the DNA. The operator region occurs between the promoter and the three structural genes. The binding of the repressor protein at the operator region blocks the transcription of the structural genes. This means that in the absence of lactose, the three structural *lac* genes are not transcribed.

The operator region is adjacent to or overlaps the promoter region. The binding of the repressor protein at the operator region interferes with the binding of RNA polymerase at the promoter region. The inducer binds to the repressor protein so that it is unable to bind at the operator region. Thus in the presence of an inducer that binds with the repressor pro-

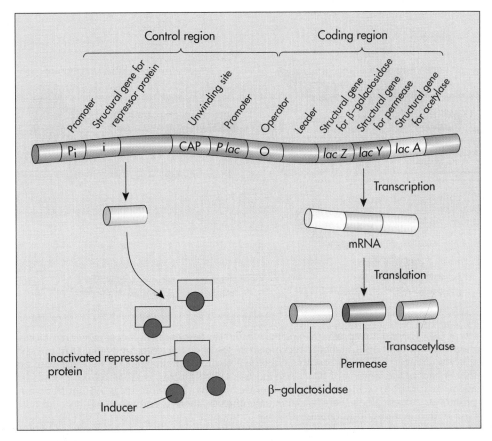

Control region   Coding region

Promoter | Structural gene for repressor protein | Unwinding site | Promoter | Operator | Leader | Structural gene for β-galactosidase | Structural gene for permease | Structural gene for acetylase

Pᵢ | i | CAP | P lac | O | lac Z | lac Y | lac A

Transcription

mRNA

Translation

Inactivated repressor protein

Inducer

β-galactosidase   Permease   Transacetylase

FIG. 7-23   The *lac* operon controls the utilization of lactose. Three structural genes under the control of the *lac* promoter (P *lac*) code for the synthesis of the enzymes needed for lactose utilization. These enzymes are made only when lactose is present.

tein, transcription of the *lac* operon is not blocked and the synthesis of the three structural proteins needed for lactose metabolism proceeds. The *lac* operon is typical of operons that control catabolic pathways; only in the presence of an appropriate inducer is the system turned on.

CATABOLITE REPRESSION

When more than one carbon source such as glucose and lactose is available at the same time, the cell will use the simpler substance first. Thus glucose is used before lactose. The cell turns on the genes for glucose metabolism and does not turn on (represses) the genes for lactose utilization. This type of repression is called **catabolite repression.** It regulates the expression of multiple genes that are under the control of different promoters. Only some genes are controlled by catabolite repression.

Catabolite repression acts via the promoter region of DNA. This is the region where RNA polymerase binds to initiate transcription (FIG. 7-24). To efficiently bind to the promoter region, RNA polymerase requires a protein called the *catabolite activator protein.* The catabolite activator protein, in turn, can-

not bind to the promoter region unless it is bound to cyclic adenosine monophosphate (cAMP).

In the absence of glucose, cAMP is synthesized from ATP by enzymatic action. This maintains an adequate supply of cAMP to permit the binding of RNA polymerase to the promoter region. Thus, when glucose levels are low, cAMP stimulates the initiation of many inducible enzymes.

In the presence of glucose, cAMP levels are greatly reduced. Thus, when glucose is being metabolized, there is not enough cAMP for the catabolite activator protein to bind to promoter region. Consequently, RNA polymerase does not bind to the promoters, and transcription at a number of regulated structural genes ceases in a coordinated manner. Thus, in the presence of an adequate concentration of glucose, a number of metabolic pathways involved in the breakdown of carbohydrates are simultaneously shut off. For example, when glucose is available for catabolism in the glycolytic pathway, disaccharides and polysaccharides are not metabolized because of catabolite repression.

By regulating the metabolism of more complex carbohydrates, the cell conserves its metabolic resources.

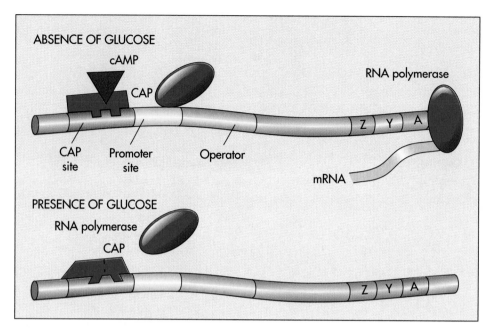

FIG. 7-24 Catabolite repression explains why, in the presence of glucose, several catabolic pathways are shut off. Catabolite repression is based on the need for cyclic AMP (cAMP) to form an activated complex with catabolite activator protein (CAP) at the promoter site that enhances the binding of RNA polymerase. When glucose is metabolized, there is inadequate cAMP to facilitate RNA polymerase binding. Therefore transcription at several promoters ceases. When there is inadequate glucose, there is enough cAMP to bind with CAP and thus transcription occurs at those promoters.

SUMMARY

Molecular Basis of Heredity (pp. 192-194)
- Frederick Griffith, Oswald Avery, Alfred Hershey, and Martha Chase made important contributions to the discovery that the genetic information of a cell is stored within its DNA macromolecules.

Structure of DNA (pp. 194-198)
Nucleotides—Building Blocks of the Genetic Code (p. 194)
- DNA is composed of nucleotides that are linked together. A nucleotide consists of a nucleic acid base, a deoxyribose sugar, and a phosphate group.
- Four nucleic acid bases occur in DNA: cytosine, guanine, adenine, and thymine. The nucleotides are linked by strong covalent bonds. Nucleotides are linked by 3'-5' phosphodiester linkages. At the ends of the DNA strand, there are no linkages and free hydroxyl groups are present. One end has a free hydroxyl group at the 3-carbon position of the monosaccharide (3'-OH end); the other end of the strand has a free phosphate group at the 5-carbon position of the monosaccharide (5'-P free end). This gives DNA directionality.
Chains of Nucleotides—Directionality of DNA (pp. 194-196)
- DNA is a double helix molecule composed of two polynucleotide chains. The chains are held together by hydrogen bonding between complementary nucleotide bases.

DNA Double Helix—Complementarity (pp. 196-198)
- The complementary base pairs are adenine and thymine, which are held together by two hydrogen bonds, and guanine and cytosine, which are held together by three hydrogen bonds. This complementarity establishes the basis for the double helix and the accurate replication of DNA.

Replication of DNA (pp. 198-202)
- Replication of the hereditary information involves synthesizing new DNA molecules that have the same nucleotide sequences as those of the parent organism. The two chains of the DNA double helix are complementary and the nucleotide sequence in one chain specifies the sequence in the other.
- DNA chains are complementary and antiparallel; one chain has the 3'-OH free end and its complementary chain has the 5'-P free end.
Semiconservative DNA Replication (pp. 198-199)
- DNA replication is semiconservative, that is, a parent chain remains intact and a new complementary chain is assembled for each one. Thus each new DNA macromolecule is half old and half new.
Steps in DNA Replication (pp. 199-201)
- DNA replication begins when the double helix unwinds to form a replication fork, separating the chains to serve as templates.
- The parental DNA is pulled apart at the replication fork, providing space for free nucleotides to align op-

posite their complementary bases for the synthesis of new chains.
- DNA replication begins at only one site in bacteria. DNA gyrase twists the DNA as the replication fork moves around the bacterial chromosome.

Formation of a New Chain of Nucleotides—DNA Polymerase (pp. 201-202)

- DNA polymerase links the nucleotides by forming phosphodiester bonds after the nucleotides are aligned by base pairing. DNA polymerase adds nucleotides to the free 3'-OH end of an existing nucleotide chain.
- The continuous, or leading, chain of DNA is the DNA chain that can be continuously synthesized because it runs in the appropriate direction for the continuous addition of free nucleotides to the free 3'-OH end. The lagging, or discontinuous, strand of DNA cannot be synthesized continuously because initiation of its replication can begin only after the double helix has already unwound somewhat. Short DNA fragments are synthesized in the direction opposite the direction in which the parent DNA unwinds. These fragments are joined together by ligases.

Mutations (pp. 203-205)

Types of Mutations (p. 203)
- A mutation is a change (addition, deletion, or substitution) in the nucleotide sequences of DNA. A lethal mutation results in the death of a microorganism or in its inability to reproduce; a conditionally lethal mutation exerts its effect only under certain environmental conditions; an unconditionally lethal mutation is lethal regardless of environmental conditions. Temperature-sensitive mutations alter the range of temperatures over which the microorganisms may grow. Nutritional mutations alter the nutritional requirements for the progeny; nutritional mutants (auxotrophs) require growth factors not needed by the parental (prototrophic) strain.

Factors Affecting Rates of Mutation (p. 205)
- Mutagens are chemicals that increase the rate of mutation. High-energy ionizing radiation causes mutation and can be used for sterilizing objects. Ultraviolet light can cause mutations by producing thymine dimers.

Expression of Genetic Information (pp. 205-213)

- The genotype represents the total informational capacity of the cell. It is mediated by DNA. The phenotype, the discernible characteristics of an organism, including the functional appearance and activities of the cell, is mediated by proteins.
- The sequence of nucleotides within the DNA determines the sequence of amino acids in the protein molecules of the cell.

Genes (pp. 206-207)
- A gene is a segment of the genetic material that has a specific function. Structural genes code for the synthesis of RNA and proteins, respectively determining the sequences of nucleotides and amino acids in these

macromolecules. Regulatory genes control cell activity by specifying when particular structural genes are actually expressed.
- Prokaryotic cells have a single chromosome and therefore are haploid. Eukaryotic cells generally have pairs of matching chromosomes, making them diploid. In homozygous cells the genes at a locus are identical copies; in heterozygous cells the genes differ.

RNA Synthesis (pp. 207-210)
- Protein synthesis involves two stages: transcription to form RNA and translation of the RNA to form a polypeptide chain.
- RNA contains ribose, phosphate, adenine, uracil, cytosine, and guanine. There are three types of RNA. Ribosomal RNA is a structural component of ribosomes. Messenger RNA carries the information from the DNA to the ribosome. Transfer RNA helps align amino acids during protein synthesis in the order specified by mRNA.
- In transcription, the information in the DNA is transferred to RNA. During transcription, DNA serves as a template that determines the order of the bases in the RNA. The RNA that is formed by transcription is complementary to the DNA. RNA polymerase links the bases, forming 3'-5' phosphodiester bonds. The template strand is the DNA chain that codes for the synthesis of RNA.
- Transcription begins at specific promoter regions where RNA polymerase binds.
- The sequence of nucleotides in prokaryotic mRNA corresponds exactly with the sequence of nucleotides in DNA. Eukaryotic genes are split genes, that is, the sequence of nucleotide bases in the mRNA is not complementary to the specific contiguous linear sequence of bases in the DNA. Eukaryotic RNA (heterogeneous nuclear RNA) must be extensively modified after transcription from DNA to form mRNA.

Protein Synthesis—Translation of the Genetic Code (pp. 210-213)
- In translation, mRNA is used to establish the sequence of amino acids that make up the protein. Translation occurs at the ribosomes.
- Translation is a directional process. mRNA is read in a 5'-P to 3'-OH direction. Polypeptides are synthesized from the amino terminal to the carboxyl terminal end.
- The genetic code has 64 possible codons; each codon is a triplet containing three nucleotides. There is more than one codon for most amino acids, and different codons can specify the same amino acid.
- Nonsense codons are ones for which there are no amino acids; the nonsense codons signal termination of synthesis of a polypeptide chain.
- The ribosome moves along the mRNA, exposing one codon at a time. As each triplet is exposed by the ribosome, a transfer RNA (tRNA) brings the specified amino acid to the ribosome; the tRNA has an anticodon region that is complementary to the codon and is responsible for bringing the correct amino acid specified by the codon. The ribosome moves to the next triplet and the process is repeated.

- Translocation is the movement of mRNA, tRNA, and the polypeptide chain along the ribosome.

Regulation of Gene Expression (pp. 214-216)

- The expression of genetic information can be regulated at the level of transcription.
- Constitutive enzymes are continuously synthesized at a constant rate and are not regulated. Inducible enzymes are made only at appropriate times, e.g., when synthesis is induced by appropriate factors.

Operons (pp. 214-215)

- The operon model of gene control explains the basis of control of transcription. An operon consists of structural genes that contain the code for making proteins; an operator region, which is the site where repressor protein binds and prevents RNA transcription; and a promoter region, which is the site where RNA polymerase binds. It is also controlled by a regulatory gene, which codes for the repressor protein.

- The *lac* operon regulates the utilization of lactose. In the presence of lactose an inducer binds to a repressor protein, preventing it from binding to the operator region of the operon; this results in derepression of *lac* operon, and structural genes needed for the utilization of lactose are transcribed until the lactose has been broken down.

Catabolite Repression (pp. 215-216)

- Catabolite repression is a generalized type of repression. Catabolite repression supersedes the control exerted by the operator region. Catabolite repression acts via promoter region of DNA by blocking the normal attachment of RNA polymerase; a catabolite activator protein is needed to bind RNA polymerase to promoter region and cAMP is required for efficient binding to occur. In the presence of glucose, the amount of cAMP is reduced; therefore the catabolite activator protein cannot bind to promoter, and transcription is unable to occur.

CHAPTER REVIEW

REVIEW QUESTIONS

1. Explain the difference between a gene and a chromosome.
2. What is the difference between genotype and phenotype?
3. What is the relationship between DNA and heredity?
4. What is the genetic code?
5. What is a mutagen?
6. What is a mutation?
7. How is DNA replicated in bacterial cells?
8. Describe the process of protein synthesis.
9. Define induction and explain how it regulates gene expression in bacteria.
10. Define catabolite repression and explain how it regulates gene expression in bacteria.
11. What are the different types of mutations?

12. Describe how the Ames test is used to detect carcinogens.
13. Compare and contrast the storage of genetic information in a prokaryotic and a eukaryotic cell.
14. Compare and contrast the expression of genetic information in a prokaryotic and a eukaryotic cell.
15. What is DNA gyrase and what role does it play in DNA replication?
16. How could DNA gyrase be used as a target for an antimicrobial agent for the treatment of disease?
17. How would you go about increasing the rate of mutations?
18. How could you recognize the occurrence of a mutant?
19. How could you design an experiment to select mutants?

CRITICAL THINKING QUESTIONS

1. How does the structure of DNA relate to the ability of this molecule to serve as the universal hereditary molecule of all living organisms? Why is fidelity essential for replication of DNA? How does a bacterial cell replicate its DNA and make very few errors in the process? What is the consequence of making an error during DNA replication?

2. Why is it so important for the bacterial cell to regulate the expression of its genes? What are the advantages and disadvantages of bacterial genes being organized into operons? Why are operons for catabolic pathways normally inducible (turned on by an inducer) and those for biosynthetic pathways normally repressible (turned off by a repressor).

3. Why can eukaryotic cells have split genes? What roles might introns play in the eukaryotic cell?

4. DNA has the sugar deoxyribose and RNA has the sugar ribose. Compared to deoxyribose, ribose has an extra hydroxyl group. The extra hydroxyl group tends to help break phosphate bonds. How would this affect the relative stability of RNA and DNA in a cell? How would this difference in stability be related to the different functions of RNA and DNA in a cell? What are the essential functions of DNA and RNA in a cell?

5. Do all substances that cause mutations in bacteria cause cancer in humans? How could you go about determining whether foods you eat contain substances that might be mutagenic or carcinogenic?

READINGS

Alberts B, D Bray, J Lewis, M Raff, K Roberts, JD Watson: 1989. *Molecular Biology of the Cell*, New York, Garland Press.

Comprehensive text covering the molecular basis of heredity.

Bishop JE and M Waddles: 1990. *Genome: the Story of the Most Astonishing Scientific Adventure of our Time—the Attempt to Map all the Genes in the Human Body*, New York, Simon and Schuster.

Discusses the government-financed program to map every gene in human DNA, the medical, ethical, and scientific questions this effort raises, and searches for the genes that cause specific diseases.

Brock TD: 1990. *The Emergence of Bacterial Genetics*, Cold Spring Harbor, New York; Cold Spring Harbor Laboratory Press.

Definitive history of the development of bacterial genetics as we know it today.

Cairns J, G Stent, J Watson (eds.): 1966. *Phage and the Origins of Molecular Biology*, Cold Spring Harbor, New York; Cold Spring Harbor Laboratory Press.

Collection of essays by the founders of and converts to molecular genetics. Gives a sense of history in the making—the emergence of insights, the wit, the humility, the personalities of the individuals involved.

Dahlberg AE: 1989. The functional role of ribosomal RNA in protein synthesis, *Cell* 57:525-529.

Thorough discussion of the role of RNA in transferring information from DNA to proteins.

Darnell J: 1985. RNA, *Scientific American* 253(4):68-78.

This article suggests that while RNA now functions in an informational capacity, it may have been the original genetic material.

Darnell J, H Lodish, D Baltimore: 1990. *Molecular Cell Biology*, New York, Scientific American Books.

Well-illustrated discussion of the molecular biology of the cell.

Dubos R: 1976. *The Professor, the Institute, and DNA: Oswald T. Avery, his Life and Scientific Achievements*, New York, Rockefeller University Press.

The stories of the discoverer that hereditary characteristics are transmitted by molecules by DNA and his institution, the Rockefeller University.

Feldman M and L Eisenbach: 1988. What makes a tumor cell metastatic? *Scientific American* 259(5):60-85.

Discusses the molecular basis for transformation to cancerous cells.

Felsenfeld G: 1985. DNA, *Scientific American* 253(4): 58-78.

It is the double-helical structure of DNA that allows it to interact with regulatory proteins and other molecules to transfer its hereditary message.

Genetics: Readings from Scientific American: 1981. San Francisco, W. H. Freeman.

Collection of articles on molecular biology.

Grunstein M: 1992. Histones as regulators of genes, *Scientific American* 267(4):68-74B.

A discussion of the role played by histones in regulating gene expression through repression and activation of genes.

Hawkins JD: 1991. *Gene Structure and Expression*, Cambridge, England; Cambridge University Press.

Presents recent ideas and techniques in molecular biology as related to genetics so that students will be able to understand further advances as they occur.

Innis MA, DH Gelfand, JJ Sninsky, TJ White (eds.): 1990. *PCR Protocols: A Guide to Methods and Applications*, San Diego, Academic Press.

Authoritative book describing ways in which PCR can be used, including medical, environmental, and forensic applications.

Maloy SR, JE Cronan Jr, D Freifelde: 1994, *Microbial Genetics*, ed. 2, Boston, Jones and Bartlett.

Comprehensive text covering classical and molecular genetics of microorganisms.

McCarty M: 1987. *The Transforming Principle: Discovering that Genes are Made of DNA*, New York, W. W. Norton.

Describes the discovery of genetic transformation.

McKnight SL: 1991. Molecular zippers in gene regulation, *Scientific American* 264(4):54-64.

Recurring copies of the amino acid leucine in proteins can serve as teeth that zip protein molecules together. This zipper may play a role in turning genes on and off.

Mullis KB: 1990. The unusual origin of the polymerase chain reaction, *Scientific American* 262(4):56-61, 64-65.

The description of the polymerase chain reaction and its discovery by its discoverer and winner of the 1993 Nobel Prize in Chemistry.

Olby R: 1974. *The Path to the Double Helix*, Seattle, University of Washington Press.

The early history of molecular biology is reviewed.

Parker J: 1989. Errors and alternatives in reading the universal genetic code, *Microbiological Reviews* 53(3): 273-298.

Concerns the types of errors and alternate readings, the frequencies with which they occur, and some of the factors that influence these frequencies.

Prescott D: 1988. *Cells*, Boston, Jones & Bartlett.

Chapter 8 contains an excellent introduction to protein synthesis.

Ptashne M: 1989. How gene activators work, *Scientific American* 260(1): 41-47.

Molecular biologists are using what they have learned about turning bacterial genes on and off to study genetic regulation in higher organisms.

Radman M and R Wagner: 1988. The high fidelity of DNA duplication, *Scientific American* 259(2):40-46.

Reviews DNA replication and the systems that ensure accuracy of information duplication.

Venitt S and JM Parry (eds.): 1985. *Mutagenicity Testing*, Oxford University Press.

Reviews the use of microbial test systems identifying mutagens.

Watson J: 1978. *The Double Helix*, New York, Atheneum.

Highly personal view of scientists and their methods, interwoven into an exciting account of how DNA structure was discovered.

Watson JD, N Hopkins, J Roberts, J Steitz, A Weiner: 1987. *Molecular Biology of the Gene*, Menlo Park, CA, Benjamin/Cummings.

A comprehensive review of molecular biology.

Weintraub HM: 1990. Antisense RNA and DNA, *Scientific American* 262(1):40-46.

A discussion of a fascinating mechanism of regulating gene expression.

Zubay G: 1987. *Genetics*. Menlo Park, California, Benjamin/Cummings.

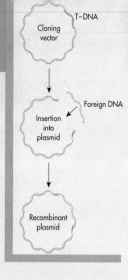

CHAPTER 8

Genetic Recombination and Recombinant DNA Technology

PREVIEW TO CHAPTER 8

In this chapter we will:
- Learn that genes can be transferred from one DNA macromolecule to another, recombining to form new genetic combinations.
- Examine the recombinational processes by which genes are exchanged.
- Study the natural ways by which DNA is transferred from one cell to another.
- Discover how scientists are able to manipulate DNA to form new combinations of genes through recombinant DNA technology.
- See examples of how recombinant DNA technology has been used to genetically engineer new organisms.
- Learn the following key terms and names:

alleles	lysogenic conversion
cloning	R (resistance) plasmids
cloning vector	*rec* (recombination) genes
complementary DNA	recombination
(cDNA)	restriction enzyme
conjugation	reverse transcriptase
conjugative plasmids	specialized transduction
defective phages	temperate phages
donor strain	Ti (tumor-inducing)
endonuclease	plasmid
generalized transduction	transduction
genetic engineering	transformation
homologous	transgenic organisms
recombination	transposase
human gene therapy	transposons

Genetic recombination occurs when genes move from one location to another. Recombination can occur between DNA molecules in the same cell or between DNA molecules from different cells. In some cases, recombination involves the movement of genes to new locations within the DNA macromolecule. Such changes in the locations of genes can alter which regulatory genes control particular genes. In other cases, recombination involves exchanging and/or combining genetic information. In these cases, DNA from two different sources comes together in a new DNA macromolecule. Such recombinational events result in an exchange of different forms of genes that can produce new combinations of genes.

Recombination provides a mechanism for the redistribution of the changes that occur in DNA as a result of mutation. Like mutation, recombination is a source of *genetic diversity*. Recombination is more likely to produce beneficial genetic combinations than mutations because it is less likely to destroy the function of a gene. It also may produce combinations of genes that can provide the organism with a valuable new function. Recombination thus provides a mechanism for generating diversity within the gene pool of a microbial population.

Recombination can produce numerous new combinations of genes, greatly altering the inherited genetic information.

HOMOLOGOUS RECOMBINATION

Homologous recombination occurs when the DNA that is exchanged comes from corresponding genes. An exchange of allelic forms of genes is the result (FIG. 8-1). **Alleles** are alternative forms of genes concerned with the same trait or characteristic. Alleles occur as multiple forms of a gene located at the same *locus* (location) of *homologous chromosomes* (corresponding pairs of chromosomes). Recombination can produce new combinations of alleles.

Homologous recombination, an exchange of segments of DNA with similar nucleotide sequences, produces new combinations of alleles.

Pairs of chromosomes exchange corresponding portions of the DNA in a process called **crossing over** (FIG. 8-2). This is the process that occurs in meiosis when sex cells—such as sperm and egg cells—are formed for reproduction. When chromosomes cross over, a bridge is enzymatically formed between the two DNA strands. The chromosomes then rotate so that the two strands no longer cross each other. At this point they are still held together by covalent linkages. An endonuclease cleaves the DNA strands. A **restriction endonuclease**, or **restriction enzyme**, is an endonuclease enzyme that cleaves a DNA macromolecule at a specific site by breaking bonds within

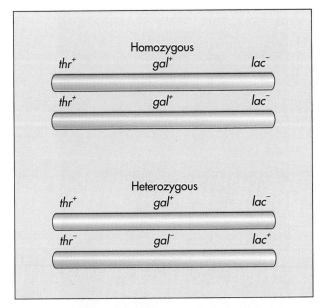

FIG. 8-1 Diploid cells have two sets of genes that may be identical (homozygous alleles) or different (heterozygous alleles).

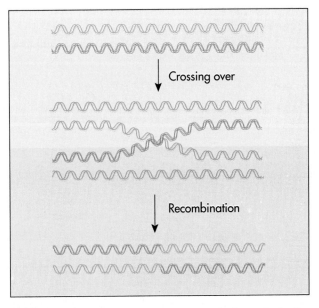

FIG. 8-2 Chromosomal crossing over, which results in the recombination of genes, is a classic example of homologous recombination.

221

the molecule. Two independent chromosomes are formed. This results in the formation of a new combination of genes, that is, a *recombinant DNA macromolecule.*

Portions of a bacterial chromosome can also be transferred from a donor to a recipient bacterium by homologous recombination. In bacteria, specific recombination enzymes and single-stranded regions of DNA are involved in such recombination. An endonuclease can nick one DNA strand, forming a free 3'-OH end within a DNA macromolecule. The free 3'-OH end acts as a primer for DNA synthesis so that one strand of the DNA is copied. The single strand of DNA that is synthesized pairs with the corresponding region of the homologous chromosome. This establishes a union called a **heteroduplex** (FIG. 8-3). The term *heteroduplex* indicates that the two strands of DNA are not fully complementary so that there is a mixture of single- and double-stranded regions of DNA. The heteroduplex forms because the two paired strands of DNA are not exactly complementary. The nucleotide sequences that are complementary form a duplex (a double-stranded complementary segment), while the noncomplementary regions remain unpaired and single stranded. The formation of the heteroduplex is catalyzed by enzymes coded for by *rec* (recombination) genes.

Homologous recombination is catalyzed by recombination (*rec*) gene products.

NONHOMOLOGOUS RECOMBINATION

Nonhomologous recombination does not involve rec enzymes. It permits the joining of DNA molecules from different sources that have very little similarity of nucleotide sequences (little homology). Through this process, segments of DNA can move from one location to another, permitting the insertion of plasmids (extrachromosomal genetic elements) into the bacterial chromosome and the transfer of genes (such as antibiotic resistance genes) from one plasmid location to another (FIG. 8-4). Nonhomologous recombination also permits the incorporation of viral DNA into the DNA of a host bacterial or eukaryotic cell (FIG. 8-5). The process of incorporating viral DNA into host cell bacterial DNA is called **lysogenic conversion.** When this occurs, certain viral genes can be expressed by the bacterial host cells, resulting in bacterial production of proteins coded for by the viral genes.

Nonhomologous recombination can occur between dissimilar segments of DNA.

Insertion Sequences

There are several types of genetic elements that can move from site to site within a bacterial chromosome and undergo nonhomologous recombination. These are called **transposable genetic elements. Insertion sequences** (ISs) are small transposable genetic elements containing about 1,000 nucleotides (FIG. 8-6). The nucleotide bases in the IS regions do not appear

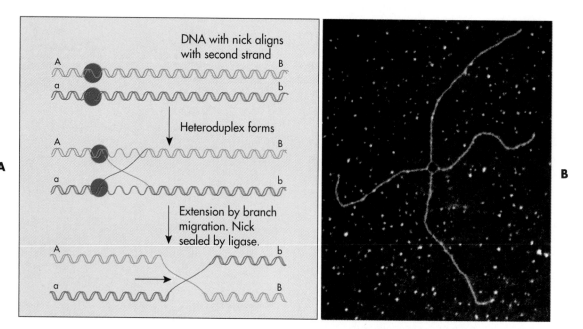

FIG. 8-3 **A,** Illustration of a heteroduplex formation during homologous recombination. **B,** Micrograph of the chi form of a heteroduplex during homologous recombination.

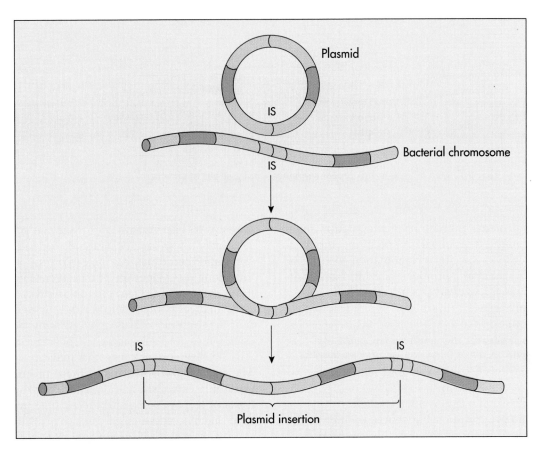

FIG. 8-4 Plasmids can insert into a bacterial chromosome by recombination.

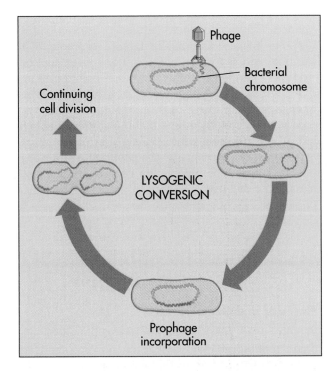

FIG. 8-5 Lysogenic conversion occurs when a temperate phage transfers bacterial DNA that it has acquired, and that DNA recombines with the DNA of the host cell. The bacterial cell then replicates the phage DNA along with the bacterial chromosome DNA.

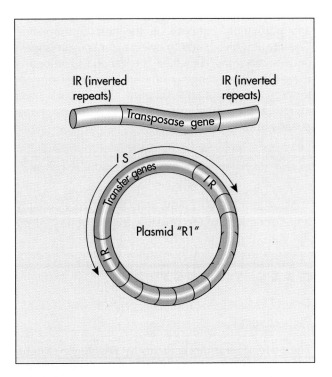

FIG. 8-6 An insertion sequence (IS) has inverted repeats that facilitate its nonreciprocal recombination; an IS may also code for a transposase.

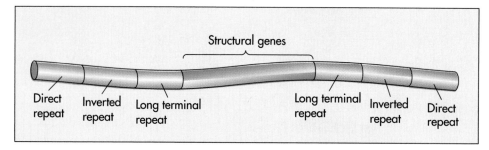

FIG. 8-7   A transposon has repeat sequences and a series of structural genes.

to code for proteins but may have a regulatory function. They may be involved in specifying the locations at which site-specific recombination occurs. ISs, for example, can alter a promoter site and thereby affect the expression of nearby genes. ISs can also insert into structural genes.

> Insertion sequences are small transposable elements that can move from one location to another on the chromosome.

Transposons

Transposons are transposable genetic elements that contain genetic information for the production of structural proteins (FIG. 8-7). They are larger than IS elements. Transposons encode several genes, including an enzyme that brings about transposon insertion called a **transposase.** Transposase does not recognize any particular sequence on the host chromosome. The site at which it acts to insert a transposon is more or less random. Therefore the genetic locations of these elements change at random, jumping from one location to another. Barbara McClintock was awarded the Nobel prize for her work on transposons in corn plants.

> Transposons contain insertion sequences and structural genes; they can move genes more or less at random.

More and more examples are being discovered of novel functions made possible by the ability of organisms to use transposition to move genes to new locations. Many transposons code for antibiotic resistance and their transposition can be important for determining the properties of pathogens and the drugs that can be used to treat microbial infections.

A medically important instance of gene mobilization by transposons is exhibited by African trypanosomes and *Plasmodium.* These protozoa cause African sleeping sickness and malaria, respectively. The surface glycoproteins of these protozoa are important in establishing their infective properties. The hosts of trypanosome and *Plasmodium* infections defend against these organisms by producing substances specifically directed against these surface glycoproteins. Trypanosomes and *Plasmodium* overcome this defense mechanism by transposition. The trypanosome DNA, for example, contains a cluster of several thousand different glycoprotein-encoding sequences. These sequences cannot be transcribed as they are because the cluster lacks a promoter. The promoter site is located within a transposon that periodically jumps randomly from one position to another within the cluster. Changes in the cluster sequence result in the appearance of new surface glycoproteins before the host's defense system is able to kill all the infecting protozoa. By the time new defense substances are made, the protozoa again transpose the genes for production of surface glycoproteins. Host defenses again are foiled. The result is a persistent infection in which the protozoa are able to defeat the human defense systems.

GENE TRANSFER

Gene Transfer and Recombination in Prokaryotes

There are several mechanisms by which genes naturally move from one bacterial cell to another. These methods of gene exchange differ in the way the DNA is transferred between a donor and a recipient cell.

Plasmids and Gene Transfer

Plasmids are small extrachromosomal genetic elements that permit microorganisms to store genetic information in addition to that contained in the bacterial chromosome (FIG. 8-8). Plasmids do not normally contain the genetic information for the essential

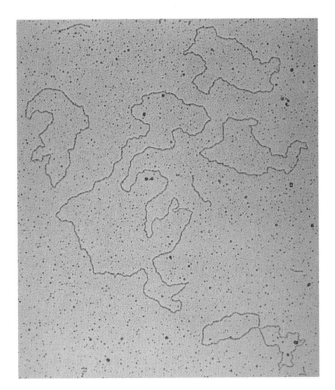

FIG. 8-8 Colorized transmission electron micrograph of plasmid DNA *(green)* from two bacteria. Micrograph of plasmids pBF4 *(larger molecules)* from *Bacillus fragilis* and pSC101 *(smaller molecules)* from *Escherichia coli.*

metabolic activities of the microorganism. They generally contain several genes that code for specialized features, such as antibiotic resistance. Plasmid transfer allows the hereditary information encoded within the plasmid DNA to move from one bacterial population to another. The acquisition or loss of a plasmid alters the genotype of a bacterial strain. Individuals in a bacterial population that possess plasmids contain genetic information different from individuals in the population that lack plasmids.

> **Plasmids, which are extrachromosomal genetic elements that store additional genetic information, are capable of self-replication and can transfer genes from one bacterium to another.**

Although plasmids contain only a very small portion of the bacterial genetic information, they are important. Plasmids can contain (1) genetic information that determines the ability of bacteria to mate and whether a bacterial strain acts as the donor during mating, (2) information that codes for resistance to antibiotics and other chemicals, such as heavy metals, that are normally toxic to bacteria, (3) genetic information for the degradation of various complex organic compounds, such as the aromatic hydrocarbons found in petroleum, and (4) genetic information

for toxin production that renders some bacteria pathogenic to humans.

Bacterial cells can contain more than one plasmid. Certain pairs of plasmids cannot be stably replicated within the same bacterial cell. Incompatible plasmids that cannot exist in the same cell are said to belong to the same incompatibility group (*Inc* group). The genes that prevent co-existence are encoded within the incompatible plasmids themselves.

There are several types of plasmids that serve different functions (FIG. 8-9). The ability of plasmids to transfer genes from one cell to another was discovered in the late 1950s by Joshua Lederberg and Edward Tatum in studies on a particular plasmid of *Escherichia coli,* called the **F plasmid,** for fertility factor. Only cells containing the genes of the F plasmid act as plasmid donors. F plasmids are **conjugative plasmids,** meaning they contain several genes that promote their transfer to other cells. Other nonconjugative plasmids lack such genes and do not readily transfer from one cell to another. In particular, the F plasmids contain a site at which DNA polymerase binds and initiates DNA so that, once transferred, F plasmids can replicate within the recipient cell. The F plasmid can replicate in Gram-negative enteric bacteria—such as *E. coli.* Other conjugative plasmids are capable of moving into numerous other bacterial species.

> **F plasmids are conjugal transfer plasmids that permit transfer of genes from donor to recipient bacterial cells.**

The **colicinogenic plasmids** of *E. coli* carry the genes for a protein, called a colicin. Colicin is toxic to only closely related bacteria. A strain of *E. coli* with a colicinogenic plasmid produces colicins that allow it to fight off competition by other *E. coli.*

Of greater significance are the **R (resistance) plasmids** that carry genes that code for antibiotic resistance. R plasmids can be passed not only from one strain to another but also from one bacterial species to another, such as from *Escherichia coli* to pathogenic strains of *Shigella* or *Salmonella.* Antibiotic-resistant strains of bacteria have become a serious health problem because R plasmids can occur in pathogenic bacteria and the treatment of human bacterial diseases is complicated by the occurrence of these pathogens that are resistant to multiple antibiotics.

> **R plasmids transfer genes for antibiotic resistance.**

Transformation

In **transformation,** a free (naked) DNA molecule is transferred from a donor to a recipient bacterium, fol-

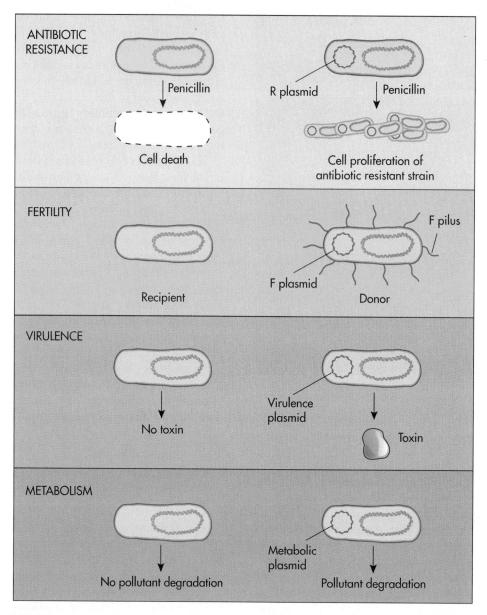

FIG. 8-9 Plasmids have different functions, including resistance to multiple antibiotics.

lowed by recombination within the recipient bacterial cell (FIG. 8-10). The donor bacterium leaks its DNA, generally as a result of lysis and death of the bacterium. The recipient bacterium takes up the leaked free DNA and incorporates it by recombination into its DNA. The classic example of transformation, involving the change of nonpathogenic *Streptococcus pneumoniae* to pathogenic *S. pneumoniae*, was critical in the discovery that DNA is the hereditary macromolecule of cells (see Chapter 7). This discovery of transformation also showed the importance of transformation, since nonpathogenic *S. pneumoniae* were transformed into deadly pathogens.

Transformation involves the transfer of naked DNA, followed by recombination.

For free DNA to be picked up by a recipient cell, the recipient cell must be **competent.** Competency means that the cell must be capable of transporting the donor DNA across its plasma membrane or have specific receptor sites on its surface. Competence probably depends on properties of the cell wall that permit the binding of DNA and its transport into the recipient cell. The recipient cell also must be **compatible,** meaning that once inside the donor, the DNA must not be enzymatically destroyed by the endonucleases of the recipient cell. The highest frequencies of transformation occur when donor and recipient are closely related. If recombination occurs, the progeny are called *recombinants* and are said to be *transformed.* Such recombinants typically have properties different from the original recipient cell.

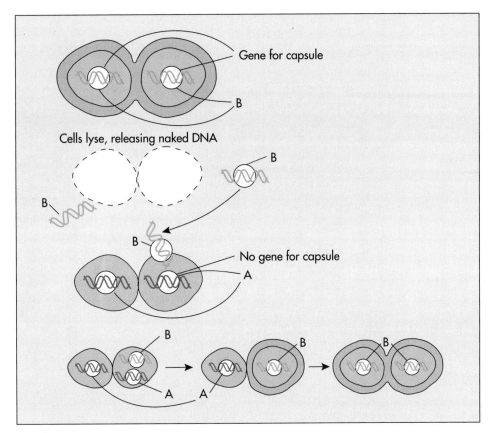

FIG. 8-10 The transformation of *Streptococcus pneumoniae* is the classic example of change in the properties of a bacterial strain due to the transfer of naked DNA. The gene for capsule production (designated B) is released from cells that lyse and taken up by nonencapsulated bacterial cells that have an inactive gene (designated A). Recombination results in the replacement of the inactive A form of the gene with the active B form so that the recombinant cells acquire the ability to produce capsules and, in this case, induces virulence.

Only a few genera of bacteria exhibit natural transformation. These include *Acinetobacter, Bacillus, Haemophilus, Neisseria, Rhizobium, Staphylococcus,* and *Streptococcus.* Some of the genes that are transferred among strains of these bacteria effect pathogenicity and the ability to treat diseases they cause. As examples, genes for capsule production, which make strains more resistant to host defenses and hence increase pathogenicity, and genes for antibiotic resistance, which make infections difficult to treat, are transferred by transformation.

Transduction

In **transduction,** DNA is transferred from a donor to a recipient cell by a viral carrier, followed by recombination within the recipient bacterial cell. Recombination within the recipient cell follows (FIG. 8-11). Viruses normally replicate within a host cell. During this process, the host cell DNA is enzymatically cleaved into pieces, and occasionally some DNA fragments can be accidentally packaged within viral protein coats. Thus a virus sometimes acquires a por-

tion of the DNA of the host cell in which it replicates. When the virus leaves the host cell it carries a segment of the host cell DNA that is protected within the protein coat of the virus. When the virus infects a new host cell, the DNA that was acquired from the earlier host cell is carried along and can recombine with the DNA of the new host cell.

Transduction involves recombination of bacterial DNA after transfer by a virus.

Phages are viruses that replicate within bacterial cells. Phages that acquire bacterial DNA during replication within a host cell retain their ability to infect new host cells. However, they often lose their ability to complete phage replication because some bacterial genes have replaced essential phage genes. These phages are called **defective phages** because they have lost the ability to kill the host cell. The bacterial DNA they acquire replaces some phage genes that are essential for completion of phage replication. Defective phages can carry DNA from a donor bacterial cell and inject it into a recipient bacterial cell. They

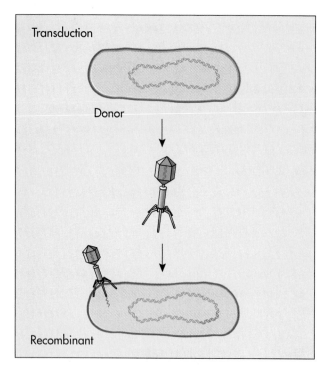

FIG. 8-11 Transduction transfers bacterial DNA via a phage. Bacterial DNA is packaged together with phage DNA during phage replication within a bacterial host cell. This occurs when fragments of bacterial DNA are joined with phage DNA in the phage capsid during phage replication. The phage carries that bacterial DNA along with phage genes to another host cell, where recombination (transduction) occurs.

can randomly carry any host cell genes. If the genes that are transferred by these phages are homologous to ones in the newly infected cell, homologous recombination can occur. This process, called **generalized transduction,** can result in the exchange of any homologous alleles. If recombination occurs, the recombinants are said to be *transduced*. Bacterial genes can be transferred to newly infected cells at low frequency by generalized transduction (FIG. 8-12).

Generalized transduction results in transfer of alleles by homologous recombination.

Other phage are called **temperate phages** because they don't always cause the lysis of the host cell. Instead, the DNA of a temperate phage may become integrated into the bacterial chromosome at specific sites. Such sites are where phage DNA can undergo nonhomologous recombination with host bacterial cell DNA. Sometimes the phage DNA comes out of integration and carries with it bacterial genes that flanked the integration site so that they acquire and transfer only certain host cell genes that are located near a specific site (FIG. 8-13). Because only a limited number of specific genes can be transferred by these phages, the process is called **specialized transduction.**

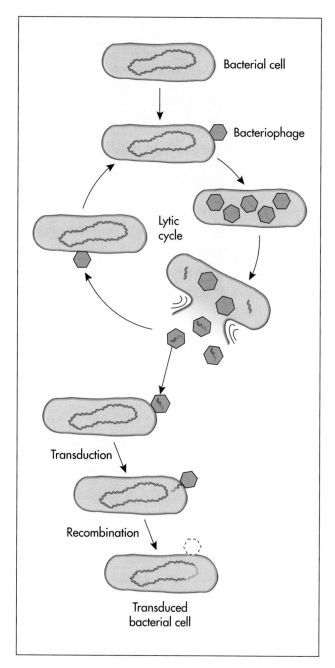

FIG. 8-12 In generalized transduction, various genes are transferred from a donor to a recipient cell via a bacteriophage. Homologous recombination occurs after the exchange.

Specialized transduction results in the transfer of specific genes by nonhomologous recombination.

Conjugation

Conjugation, or **mating,** involves the transfer of DNA from a donor to a recipient by cell-cell contact between the donor and recipient bacterial cells, followed by recombination within the recipient bacterial cell. Preventing physical contact precludes transfer of DNA by conjugation. The necessary physical contact between mating bacteria is established

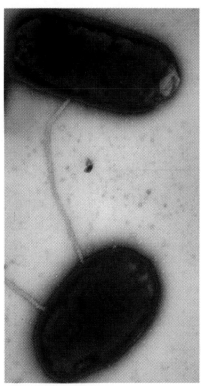

FIG. 8-14 Micrograph of mating cells of *Escherichia coli.* The cells are joined by the F pilus *(blue)* of the donor strain.

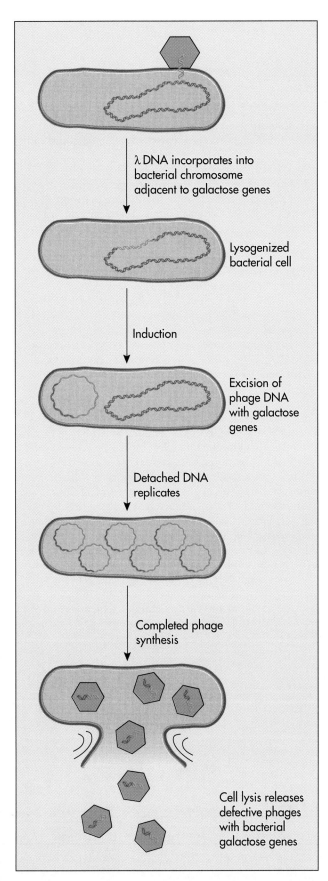

λ DNA incorporates into bacterial chromosome adjacent to galactose genes

Lysogenized bacterial cell

Induction

Excision of phage DNA with galactose genes

Detached DNA replicates

Completed phage synthesis

Cell lysis releases defective phages with bacterial galactose genes

FIG. 8-13 Specialized transduction by lambda (λ) phage results in the transfer of a limited number of specific genes by nonhomologous recombination.

through the **F (fertility) pilus** (FIG. 8-14). As discussed in Chapter 5, pili are filamentous appendages of Gram-negative bacteria that project from the cell's surface and are involved in attachment processes. F pili specifically join mating bacteria. When an F pilus establishes a bridge between two bacterial cells, there is a change in plasma membrane permeability so that DNA can move from one cell to another.

In conjugation, physical contact between bacterial cells established by an F pilus is needed for DNA transfer from donor to recipient cell.

Bacterial strains that produce F pili act as donors during conjugation. Donor strains are designated **F⁺ strains** if the F plasmid, which codes for F pilus production, is independent, that is, free in the cytoplasm. Strains are **Hfr (high frequency recombinant)** if the F plasmid DNA is incorporated into the bacterial chromosome. Strains lacking F pili are recipient strains and are designated **F⁻ strains.**

During bacterial mating, a single strand of donor DNA is replicated. The single-stranded copy of the DNA is transferred to the recipient where the complementary strand is synthesized. The precise portion of the DNA that is transferred depends on the time of mating, that is, how long the F pilus maintains contact between the mating cells.

GENETIC MAPPING

Mating of different strains of bacteria that have different allelic forms of multiple genes can be used for genetic mapping, that is, for determining the relative positions of those genes in the bacterial chromosome. The occurrence of recombinants that result from mating often is used to map the order and, thus, to determine the relative locations (loci) of genes.

In the case of *E. coli,* which has been well studied, mating between Hfr and F⁻ strains has been used to map large sections of the bacterial chromosome. By vibrating a culture of mating bacteria, the cell-to-cell contact is broken and further transfer of DNA ceases. Such interruption of mating can be done at various times after conjugational cell-cell contact begins. The order of genes on the bacterial chromosome can be determined by examining the times at which recombinants for given genes are found.

In mating experiments aimed at mapping the order of genes, the recovery of recombinants of marker genes is normally used as a reference point for establishing the fine structure of the genome. If a gene of unknown location shows a high frequency of recombination along with the marker gene, it is likely that the marker and unknown genes are closely associated in the chromosome. If, however, the genes are far apart, it is unlikely that recombinants of both the marker gene and the gene of unknown location will occur in the progeny.

Let us consider an example of how the locations of genes are determined by the transfer times for recombination as determined by interrupted mating (FIG. *A*). Suppose the genetic markers are threonine biosynthesis *(thr)*, leucine biosynthesis *(leu)*, azide sensitivity *(azi)*, phage T1 sensitivity *(ton)*, lactose utilization *(lac)*, galactose utilization *(gal)*, and streptomycin sensitivity *(str)*. In this example an Hfr that is thr^+, leu^+, azi^s, ton^s, lac^+, gal^+, str^s is mated with an F⁻ strain that is thr^-, leu^-, azi^r, ton^r, lac^-, gal^-, str^r. The

superscript $^+$ indicates that the organism has the genes for biosynthesis or utilization, whereas the superscript $^-$ indicates that the organism lacks functional copies of these genes; the superscript ^{r} indicates resistance, whereas the superscript ^{s} indicates sensitivity.

Mating of the Hfr and F⁻ strains is initiated by mixing the two cultures at time t=0. After mating for 10, 15, 20, 25, 30, 40, 50, and 60 minutes, a portion of the mixed culture is removed and agitated in a blender to interrupt mating and the cells are then plated on a medium containing glucose and streptomycin. On this medium, recombinants that are thr^+ leu^+ str^r will grow because they can synthesize threonine and leucine and are resistant to streptomycin. Recombinants that are thr^+ leu^+ str^r are selected in this manner. The original strains do not have this combination of genes and will not grow. Azide sensitivity *(azi)*, phage T1 sensitivity *(ton)*, lactose utilization *(lac)*, and galactose utilization *(gal)* are unselected markers because this medium does not specifically detect the different alleles of these genes. The thr^+ leu^+ str^r recombinants that form colonies are scored for the alleles of the unselected markers that are present in the selected recombinants by replica plating on media that individually select for *azi, ton, lac,* and *gal.* The frequencies of unselected markers among thr^+ leu^+ str^r selected recombinants are plotted as a function of time until mating is physically interrupted (FIG. *B*). Extrapolation of the frequency of each unselected marker to zero indicates the earliest time at which markers become available for recombination with the chromosome of the F⁻ cell. These times permit the ordering of genes, that is, construction of a genetic map, with the assignment of distances between genes based on the time (in minutes) elapsed from the initiation of conjugation until the earliest time at which a marker from the Hfr strain is detected as a recombinant with the F⁻ strain (FIG. *C*).

A

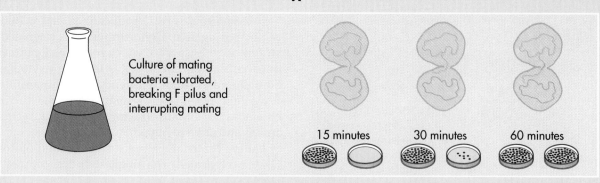

Culture of mating bacteria vibrated, breaking F pilus and interrupting mating

15 minutes 30 minutes 60 minutes

The relative positions of the genes can be established by mating bacteria for varying times and interrupting the mating to halt further gene transfer. The number of recombinants increase with time. Each experiment is assessed by comparison of the number of colonies on a complete medium and a medium restricted to growth of an auxotrophic mutant.

B

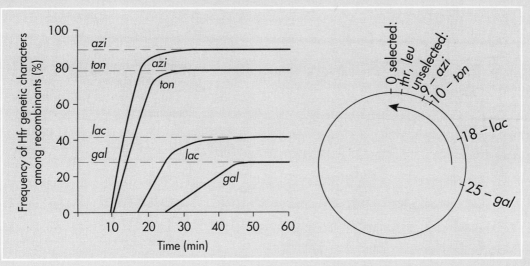

The transfer times of genes and the frequency of recombination indicates the relative positions of genes. Graphs like this are used for gene mapping.

C

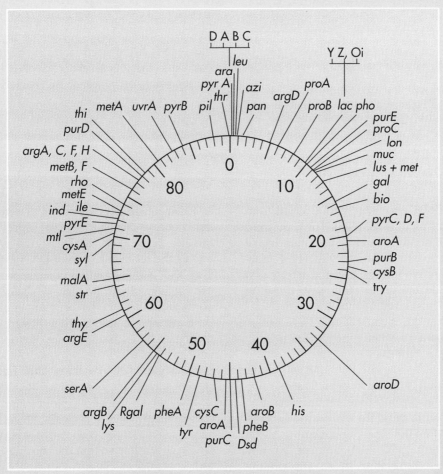

A genetic map is based on recombination frequencies.

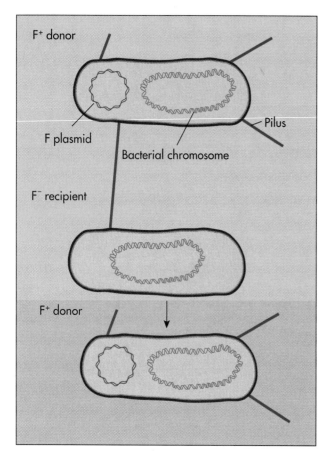

FIG. 8-15 Donor bacteria that have the fertility gene (F gene) produce F pili, which establishes contact for DNA transfer. In F⁺ strains the F gene is on a plasmid. The mating of a donor F⁺ strain with a recipient F⁻ strain *(top)* results in transfer of the F plasmid and the production of F⁺ progeny; there is relatively little recombination.

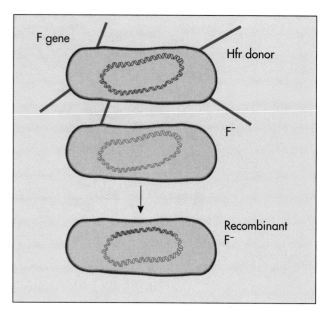

FIG. 8-16 The F gene can be incorporated into the bacterial chromosome to produce donor strains, designated Hfr (high frequency recombination strains). The mating of an Hfr strain with an F⁻ strain *(bottom)* results in transfer of many genes from the bacterial chromosome and a high frequency of recombination; the F gene is not transferred, so that the progeny are F⁻ recipient strains.

When an F⁺ cell is mated with an F⁻ cell, the F plasmid DNA is copied and usually transferred from the donor to the recipient (FIG. 8-15). The F plasmid also stays with the donor. Therefore the offspring of such a mating are mostly donor strains. The F plasmid confers the genetic information for acting as a donor strain. Few if any genes other than those on the F plasmid are transferred in such matings.

> **Mating of an F⁺ strain with an F⁻ strain results in the production of F⁺ strains.**

When an Hfr strain is mated with an F⁻ strain, the genes of the bacterial chromosome are transferred to the recipient cell before those for F pilus production. The F plasmid is often not near the beginning of the DNA that is transferred (FIG. 8-16). Since only rarely is there sufficient mating time to accomplish the complete transfer of the complete bacterial chromosome, the recipient cell normally remains F⁻. How-

ever, a relatively large portion of the bacterial chromosome is transferred from the donor to the recipient. Thus there is a relatively high frequency of recombination of genes of the bacterial chromosome when Hfr strains are mated with F⁻ strains.

> **Mating of an Hfr strain with an F⁻ strain results in the production of F⁻ strains and a high frequency of recombination.**

GENETIC EXCHANGE IN EUKARYOTES

In eukaryotic microorganisms, genetic exchange normally occurs during the sexual reproduction phase of the life cycle (FIG. 8-17). Sexual reproduction involves the recombination of DNA from two parents. The DNA comes together when specialized sexual reproductive cells unite. The *vegetative cells* (growing nonreproductive cells) of many eukaryotic organisms are *diploid,* that is, they have two sets of chromosomes. To exchange genetic information these organisms normally form specialized reproduction haploid gametes that have only one copy of each chromosome.

The conversion of a diploid to a haploid state occurs in the process of **meiosis.** Meiosis is also known as *reduction division* (FIG. 8-18). Meiosis begins after DNA replication, so that the starting cell actually has four copies of each gene. During this tetraploid state the chromosomes are aligned side by side. Recombination can occur by crossing over between homolo-

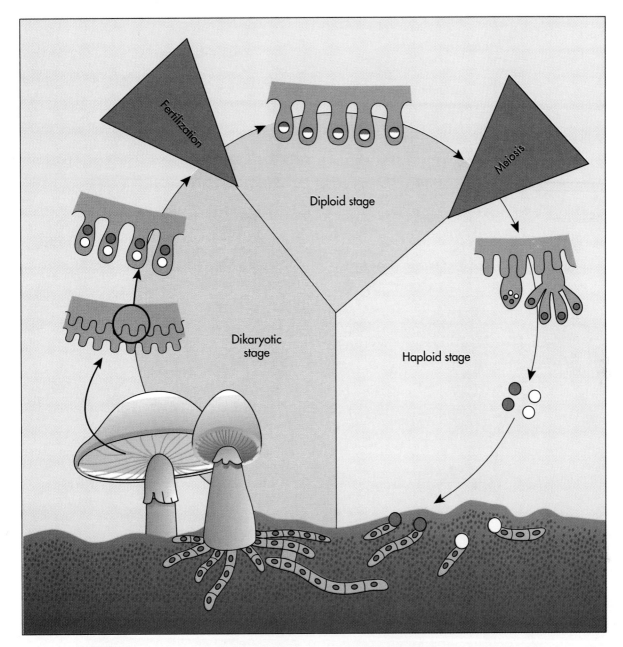

FIG. 8-17 In sexual reproduction, gametes from two parental cells join to form a progeny zygote. The sexual basidiospores (*brown* and *white*) germinate to form mycelia that are dikaryotic. Subsequent fusion of mycelia in basidiomycetes gives rise to the diploid stage.

gous chromosomes. The chromosomes are pulled apart by spindle fibers. A second meiotic division then occurs without further DNA replication, forming four nuclei. Each nucleus contains a haploid number of chromosomes.

> **Homologous recombination occurs during meiosis, the process that results in the reduction of the number of chromosomes and the conversion of a diploid cell into a haploid cell.**

The haploid nuclei of these reproductive cells can later fuse with the nuclei of reproductive cells. The union of the haploid nuclei during the fusion of reproductive cells **(syngamy)** reestablishes the diploid state (FIG. 8-19). One of the haploid sets of chromosomes comes from a **donor strain** (male) and the other haploid set comes from a **recipient strain** (female). The process of gene exchange establishes a life cycle with haploid, dikaryotic, and diploid stages.

Prophase I

Metaphase I

Anaphase I

Telophase I

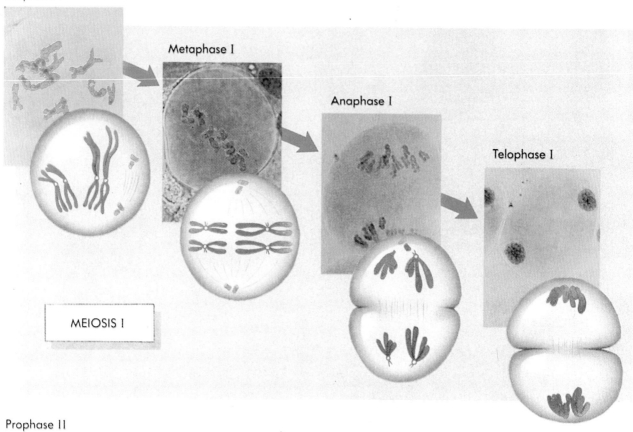

MEIOSIS I

Prophase II

Metaphase II

Anaphase II

Telophase II

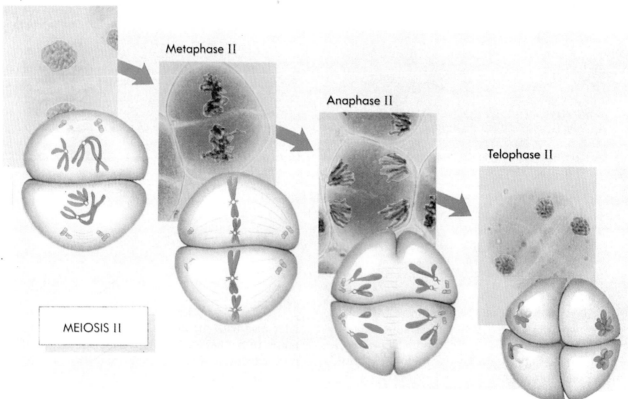

MEIOSIS II

FIG. 8-18 During meiosis, a diploid parent cell (cell with pairs of chromosomes) undergoes reductive division to produce four haploid daughter cells (cells with individual sets of chromosomes). In the first stage of meiosis (prophase), chromosome cross-over occurs between the pairs of homologous chromosomes. Then during metaphase, microtubular spindles form and align the pairs of chromosomes. Next, during anaphase, the chromosomes are pulled apart to opposite poles so that one set of each of the chromosomes is located at each pole at telophase. Cytokinesis occurs so that two cells are formed. These phases are repeated, resulting in the formation of four haploid cells in the second stage of meiosis.

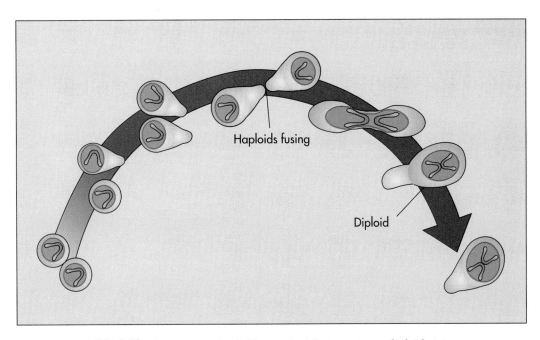

FIG. 8-19 In syngamy haploid gametes join to restore a diploid state.

RECOMBINANT DNA TECHNOLOGY

FORMATION OF RECOMBINANT DNA

Based on an understanding of the natural recombination processes, scientists have learned how to recombine DNA from different sources *in vitro*, that is, in the laboratory outside of living cells. This is the foundation of recombinant DNA technology, the deliberate union of specific genes by scientists to make recombinant DNA macromolecules. Recombinant DNA technology promises to be a powerful tool for understanding basic genetic processes and has tremendous potential industrial applications. DNA macromolecules formed in this manner have a combination of genes from a donor source and from a recipient. The source DNA can be human, plant, bacterial, viral, or even chemically synthesized DNA. The recipient can be the DNA of any organism. Organisms can thus be formed that contain genes from more than one species. Such organisms are called **transgenic organisms.**

> **Recombinant DNA technology is the purposeful union of DNA from different sources in the laboratory outside of living cells.**

The enzymes involved in the normal recombination and replication of DNA are frequently used to join DNA from different sources. Restriction endonucleases are used to cut DNA into specific fragments. Some endonucleases cut the DNA at a **palindromic sequence.** A palindrome is a symmetrical sequence of nucleotide bases that can be read identically in one direction on one strand and in the other direction on the opposite strand. For example, $\frac{-CCTAGG-}{-GGATCC-}$ is a palindromic sequence. When an endonuclease cuts a

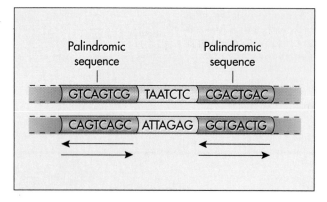

FIG. 8-20 A palindromic sequence has the same genetic information regardless of the direction of reading.

palindromic sequence, DNA with staggered single-stranded ends is thus produced (FIG. 8-20). These ends of the cut DNA can act as cohesive or sticky ends during recombination. This makes them amenable for splicing with segments of DNA from a different source that has been excised by using the same endonuclease. Ligases are used to splice the pieces of DNA. For example, a fragment of DNA

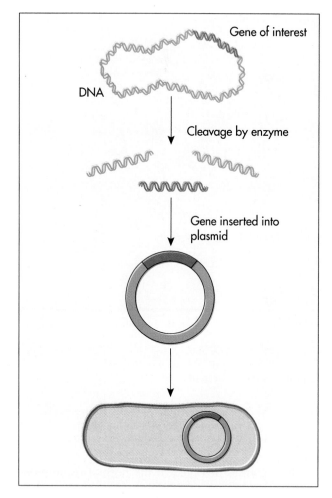

FIG. 8-21 An endonuclease cuts DNA at specific sites within the double helix. There are many different endonucleases that recognize different nucleotide sequences. Ligases join fragments of DNA. Using endonucleases and ligases, specific fragments of DNA can be cut from a chromosome or plasmid and joined together (ligated) to form a recombinant DNA molecule. This is the key to genetic engineering.

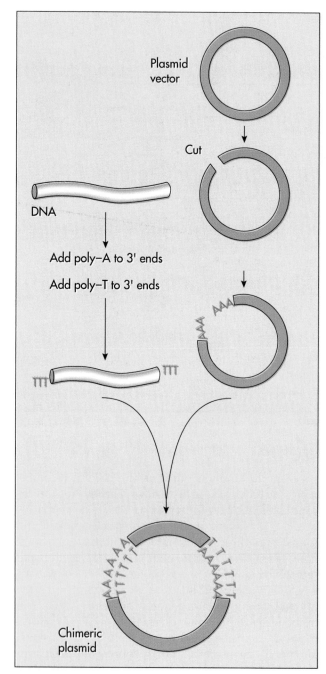

FIG. 8-22 Poly-A and poly-T tails can be added to fragments of DNA to establish homology for recombination.

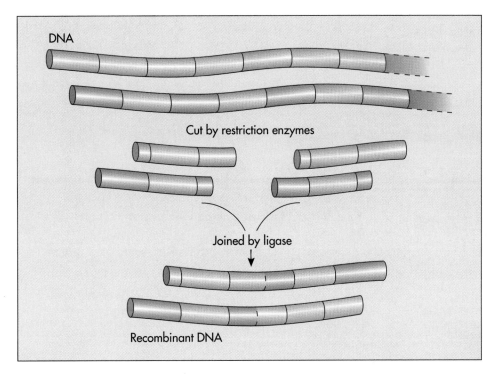

DNA

Cut by restriction enzymes

Joined by ligase

Recombinant DNA

FIG. 8-23 Having homologous ends allows DNA fragments to join. A ligase can join DNA when the ends are homologous.

produced by using endonucleases can be added to a plasmid carrier (FIG. 8-21).

If two different endonucleases are used, one to cut open the recipient DNA and another one to excise a segment of donor DNA, it often is necessary to establish an artificial homology at the terminal ends of the donor and recipient DNA macromolecules (FIG. 8-22). This can be accomplished by using an enzyme that adds adenine to the plasmid DNA to create a polyA (poly-adenine) tail and a different enzyme that adds thymidine to the donor DNA to form a polyT (poly-thymine) tail. Pairing occurs between homologous regions of complementary bases. Ligase enzymes are used to seal the circular plasmid. The tails left by the action of the endonuclease are cleaved *in vitro*, using exonuclease enzymes. Virtually any source of DNA can be used as a donor, including human DNA. By adding a polyT tail to the donor DNA after its excision with an endonuclease, the donor DNA can be made complementary to the polyA tails of the recipient DNA, permitting the formation of a recombinant DNA macromolecule.

If the same endonuclease is used to cut both the donor and recipient plasmid DNA, the strands will have homologous ends. Then it is not necessary to add polyA and polyT tails. The ends of the DNA molecules are sealed by ligases, thus creating a re-

combinant DNA macromolecule that contains a foreign segment of DNA (FIG. 8-23).

Short nucleotide sequences can also be synthesized chemically. This artificial synthesis of a short DNA segment can be accomplished by using an automated DNA synthesizer. The DNA that is produced in this manner can be enzymatically joined to DNA from other sources to create recombinant DNA.

The problem of the discontinuity of the split genes of eukaryotic cells can be overcome by using an mRNA molecule and a **reverse transcriptase.** Reverse transcriptase is an enzyme that uses an RNA template to synthesize a DNA macromolecule. This will produce a DNA macromolecule with a contiguous sequence of nucleotide bases containing the complete functional gene (FIG. 8-24). The single-stranded DNA molecule formed in this procedure is complementary to the complete mRNA molecule and hence is called **complementary DNA** (cDNA). The RNA can be removed by using a nuclease. A complementary strand of DNA can then be synthesized opposite the cDNA strand. The double-stranded DNA derived from a eukaryotic cell can be joined with DNA obtained from a bacterial cell. In this way recombinant DNA can be made that contains eukaryotic and prokaryotic genes.

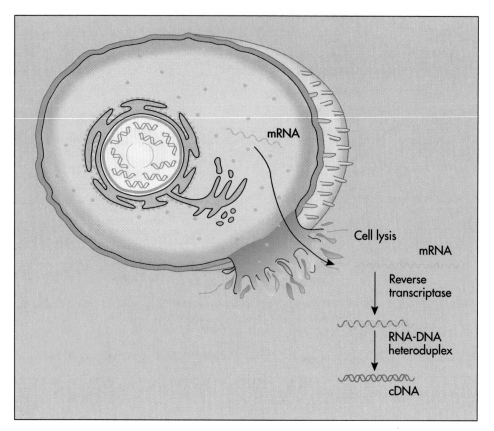

FIG. 8-24 cDNA can be produced from an RNA template using reverse transcriptase. This is useful for genetic engineering of eukaryotic genes because it eliminates introns (intervening sequences) that are present in eukaryotic genes. The cDNA can then be cloned and expressed in prokaryotic cells.

CLONING RECOMBINANT DNA

Once a recombinant DNA molecule is formed it can be transferred to living cells where it can be replicated. When the cells divide, the recombinant DNA is passed to the progeny cells. Genes contained in recombinant DNA can be **cloned** (asexually reproduced to form copies that are identical with the original) in this manner. Various methods are used to add recombinant DNA to cells for cloning. Transformation and transduction are used to facilitate the transfer of recombinant DNA into living cells for cloning. Although most bacterial cells, including those of *E. coli,* are not naturally competent for transformation, scientists can treat cells in the laboratory with agents such as detergents so that they will take up DNA. Thus *E. coli* and most other bacteria can be transformed with recombinant DNA.

Recombinant DNA can be transferred to living cells where it can be replicated and passed onto progeny cells.

Plasmids often are used as vectors (carriers) for the cloning of recombinant DNA. Plasmids can be isolated, and the plasmid DNA can be cut open by using a site specific endonuclease, commonly called a **restriction enzyme.** This creates sites where foreign donor DNA can be inserted. Once the plasmid containing the desired DNA segments is formed, it can be added to a culture of a suitable recipient bacterium that will incorporate the plasmid. Because plasmids will transfer DNA into cells, recombinant plasmids are extremely useful for cloning recombinant DNA. Plasmid DNA, including the foreign DNA segments, can be replicated and passed from one generation to another. Bacteria containing recombinant plasmids act as factories to produce multiple copies of identical cloned genes.

Recombinant plasmids are used for cloning recombinant DNA. Bacteria containing recombinant plasmids produce multiple copies of identical cloned genes.

Plasmid pBR322 was created specifically for use in cloning DNA. It is frequently used for cloning because it has unique sites where restriction enzymes will cut so that new sequences of DNA can be introduced and because it can replicate in *E. coli* (FIG. 8-25). Such a plasmid that is used for the replication of recombinant genes in a host cell (in this case in *E. coli*) is called a **cloning vector.** The use of plasmid pBR322 as a cloning vector is advantageous because it also contains genes for ampicillin and tetracycline resistance and has multiple specific restriction enzyme sites. The presence of antibiotic resistance markers permits its detection using selective media containing these antibiotics. Plasmid pBR322 contains single restriction sites for several endonucleases, including sites within the genes coding for antibiotic resistance. Fragments of foreign DNA can, therefore, be inserted into specific sites. If the insertion occurs at the antibiotic resistance site, resistance is lost. The nucleotide sequence of the antibiotic resistance gene is disrupted. This *insertional inactivation* is useful for detecting the presence of foreign DNA within a plasmid.

Insertional inactivation can detect the presence of foreign DNA in a plasmid.

Protoplast fusion can also be used to transfer DNA from one cell to another so that recombination can occur (FIG. 8-26). Protoplasts are cells that have had their walls removed by enzymatic and/or detergent treatment. They are protected against lysis due to osmotic shock by suspension in a buffer containing a high concentration of a solute, such as sucrose. Protoplast fusion is a particularly useful technique for achieving gene transfer and genetic recombination in organisms with no efficient natural gene transfer mechanism. Interestingly, more than two strains can be combined in one fusion. Recombinants that have inherited genes from all parents in the fusion are generated. The basic procedure involves polyethylene glycol-induced fusion of protoplasts followed by the regeneration of normal cells. An important feature of bacterial protoplast fusion is that it enables establishment of a transient quasi-diploid state during fusion. This permits recombination between complete bacterial chromosomes, as opposed to fragments of the donor bacterial chromosome and the recipient bacterial chromosome.

DNA can be transferred by protoplast fusion.

Regardless of the mechanism of gene transfer, for the information encoded in the cloned DNA sequence to be expressed, it must be transcribed and

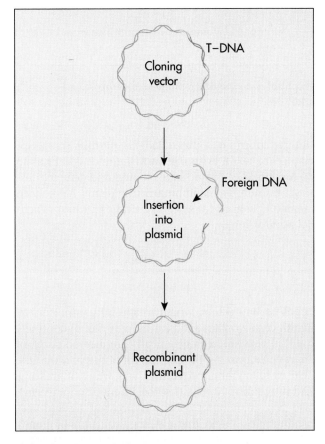

FIG. 8-25 Plasmid pBR322 is often used as a cloning vector because it has a number of unique sites for different endonucleases. This permits the insertion of foreign DNA at specific sites.

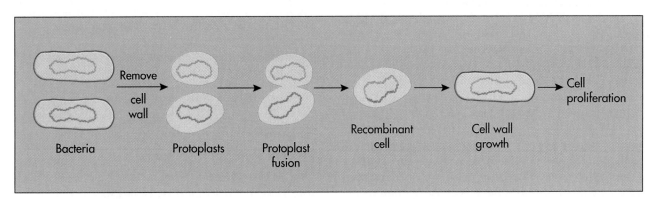

FIG. 8-26 Cell walls of Gram-positive bacteria can be removed to form protoplasts. The fusion of protoplasts permits gene transfer and recombination.

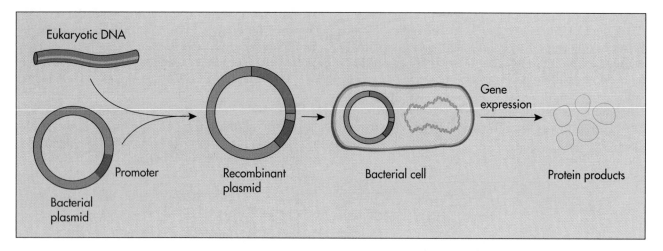

FIG. 8-27   An expression vector has genes under control of a specific promoter so that foreign genes added through recombination can be expressed.

translated to form an active protein molecule. The expression of the foreign genetic information requires that the appropriate reading frame be established and that the transcriptional and translational control mechanisms be turned on to permit the expression of the DNA. In particular, the cloned genes must be under the control of a promoter that permits transcription (FIG. 8-27). With the knowledge of what has to be accomplished to achieve cloning and gene expression, numerous recombinant organisms have been created and used for scientific studies and for practical applications.

GENETIC ENGINEERING

Because the underlying genetic code—specified by the sequences of nucleotides in DNA—is almost always the same in the cells of all organisms, scientists can create cells with entirely new properties using recombinant DNA technology. The use of recombinant DNA technology for this purpose is called **genetic engineering.** Although recombinant DNA technology theoretically can be used to genetically engineer the evolution of entirely new organisms, it is used to add only one or a few genes to an organism.

> **The use of recombinant DNA technology permits the purposeful manipulation of DNA to create organisms with new or different characteristics through genetic engineering.**

Through genetic engineering, scientists can artificially direct recombinant processes for beneficial purposes. Until the development of recombinant DNA technology, bacteria could produce only substances from their own bacterial genes. Now, however, bacteria that have had plant and animal genes added to their DNA by genetic engineers can produce plant and animal proteins such as insulin and human growth hormone (Table 8-1). New vaccines for preventing human diseases have been developed using recombinant DNA technology. Genetically engineered yeast cells produce a vaccine used to prevent hepatitis B infection, which is especially important to health care workers who are at high risk of contracting this disease. Such genetically engineered bacteria are revolutionizing the economics of the pharmaceutical industry and providing new ways of treating and preventing human diseases. Genetic engineering also has great potential for improving agricultural productivity. Gene therapy, the insertion of genes into human chromosomes, may be useful in treating some human diseases. Experiments are underway at the National Institutes of Health with human cells that have added genes that code for a factor that causes shrinkage of tumors. The aim of these experiments is to determine whether genetically engineered human cells can be used to treat cancer.

> **Genetic engineering has opened up many new possibilities for employing microorganisms to produce substances of economic importance and for genetically modifying plants and animals.**

RECOMBINANT PLANTS WITH BACTERIAL GENES

Recombinant Nitrogen Fixers

One of the greatest benefits that may be realized through genetic engineering is the introduction of the capacity to fix nitrogen into plants, such as wheat and rice, that are not able to utilize atmospheric ni-

TABLE 8-1

Some Human Proteins Produced by Recombinant Microorganisms

PROTEIN	PRODUCT NAME	FUNCTION AND USE
Insulin	Humulin, Novolin	Hormone that regulates sugar levels in blood; used in treatment of diabetes
Human growth hormone	Protropin, Humatrope	Hormone that stimulates growth of human body; used in treatment of dwarfism
Bone growth factor	—	Stimulates growth of bone cell; used in treatment of osteoporosis
Interferon Alpha	Berofor, IntronA, Wellferon, Roferon-A , human recombinant alpha interferon	Used in treatment of cancer and viral diseases
Interferon Beta	Frone, Betaseron, human recombinant beta interferon	Used in treatment of cancer and viral diseases
Interferon Gamma	Actimmune	Used in treatment of cancer and viral diseases
Interleukin-2	Proleukin, human recombinant interleukin-2	Used in treatment of immunodeficiencies and cancer
Tumor necrosis factor (TNF)	—	Used in treatment of cancer
Tissue plasminogen activator (TPA)	Actilyse	Dissolves blood clots; used in treatment of heart disease and during heart surgery
Food clotting factor VIII	Recombinate	Stimulates blood clot formation; used in treatment of hemophiliacs
Epidermal growth factor	—	Regulates calcium levels and stimulates growth of epidermal cells; used in treatment of wounds to stimulate healing
Granulocyte colony stimulating factor	Filgrastin, Neupogen	Regulates production of neutrophils in bone marrow; used to prevent infections in cancer patients
Erythropoietin (EPO)	Procrit Epogen	Stimulates red blood cell production; used in treatment of anemia in dialysis patients

trogen. For years scientists have been exploring the relationships between *Rhizobium* and the plants with which this nitrogen-fixing bacterium can establish symbiotic (mutually beneficial) relationships. Microbiologists are studying the genetics and biochemistry of infection by *Rhizobium* with the aim of employing recombinant DNA techniques to genetically engineer plants containing the bacterial genes for nitrogen fixation (*nif* genes).

In one series of studies, the genes for nitrogen fixation were inserted into the genome of a eukaryotic cell, a yeast (FIG. 8-28). Plasmids from *E. coli* and a yeast were cleaved and then fused to form a single hybrid plasmid. This plasmid could be recognized by the yeast cell and integrated into its chromosomal DNA. The genes to be introduced into the yeast were then isolated from the chromosome of *Klebsiella pneumoniae*, a nitrogen fixer. The *nif* genes code for some 17 proteins. Another *E. coli* plasmid was cleaved and the isolated *nif* genes were introduced to form a second hybrid plasmid. Because bacterial DNA had pre-

viously been inserted into one of the yeast chromosomes, the yeast cell recognized the hybrid *E. coli* plasmid. The second plasmid was then integrated into the yeast chromosome. Although the insertion of the prokaryotic *nif* genes into the eukaryotic yeast cell demonstrates that genetic material can be transferred between different biological systems, unfortunately the nitrogen-fixing genes were not expressed in the yeast. More studies are needed to elucidate the factors controlling expression of the *nif* genes before success is obtained.

It is increasingly apparent that the ability to engineer organisms depends on developing a thorough understanding of the molecular biology of gene expression. Once we understand the mechanisms of gene regulation in eukaryotes, we will be able to apply this knowledge through genetic engineering to create crop plants that are able to fix their own nitrogen. This will greatly enhance the ability to produce the world's food supply without the environmental problems caused by the use of chemical fertilizers.

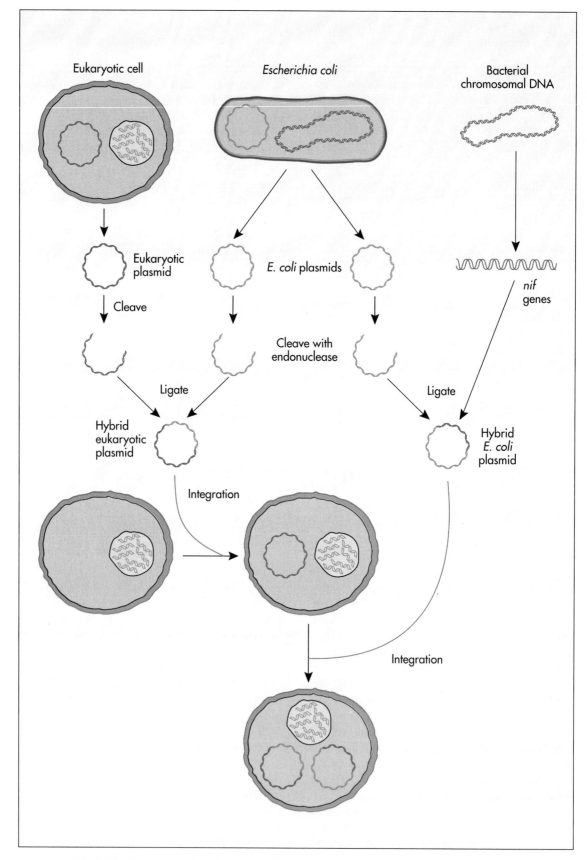

FIG. 8-28   The genes of prokaryotic cells for nitrogen fixation (*nif* genes) can be moved through recombinant DNA technology into eukaryotic DNA, for example, of a yeast. If this could also be accomplished for corn, wheat, and rice, greater food production without chemical fertilizers could be achieved.

First Demonstration of Genetic Engineering

Stanley Cohen and Herbert Boyer in 1973 were the first to demonstrate that eukaryotic genes could be genetically engineered into a bacterial cell (see Figure). Cohen and Boyer used a restriction endonuclease—called *EcoR1* (*Escherichia coli* restriction enzyme number 1)—to cut a plasmid of *Escherichia coli.* They isolated a DNA fragment 9,000 nucleotides long that contained both the nucleotide sequence necessary for replicating the plasmid and a gene that conferred resistance to the antibiotic tetracycline. In the presence of a ligase, this DNA fragment could form a new circular plasmid, which Cohen named pSC101.

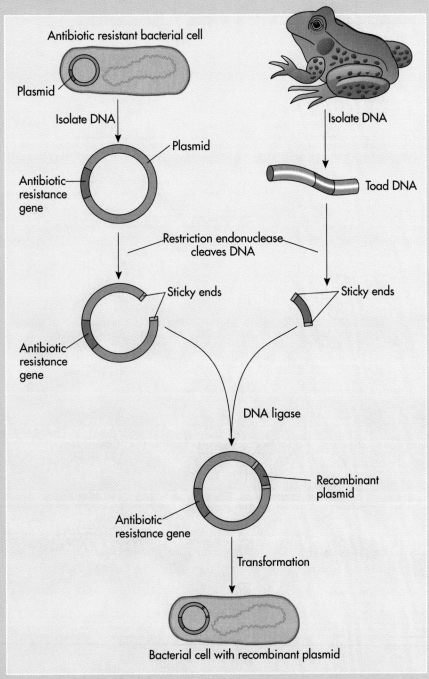

Cohen and Boyer pioneered recombinant DNA technology by moving toad DNA into a bacterial chromosome.

Continued.

FIRST DEMONSTRATION OF GENETIC ENGINEERING—CONT'D

Cohen and Boyer then used *EcoR1* to cut up DNA isolated from a toad cell. They then mixed the toad-DNA fragments with opened-circle molecules of pSC101. They used a ligase to join the bacterial pSC101 DNA with the toad DNA. They then allowed cells of a tetracycline sensitive strain of *E. coli* to take up DNA from the mixture. They plated the cells on a medium containing tetracycline so that only cells that had become resistant to tetracycline would grow. From among these pSC101-containing cells they were able to isolate ones containing the toad DNA. These bacteria had the toad gene spliced into the bacterial pSC101 plasmid.

The pSC101 containing the toad gene is the first product of recombinant DNA technology. It does not exist naturally. It is a form of recombinant DNA, a DNA macromolecule created in the laboratory by molecular geneticists who joined eukaryotic and bacterial DNA. Thus the first recombinant genome produced by genetic engineering was a bacterial plasmid into which a toad gene was inserted.

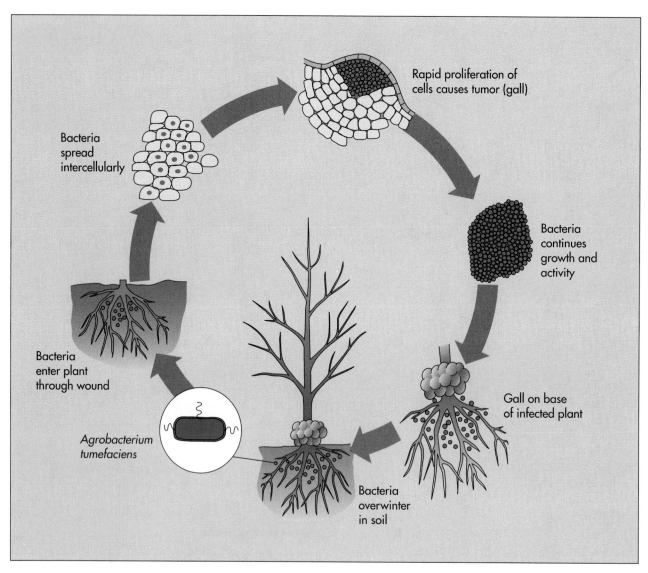

FIG. 8-29 The Ti plasmid of *Agrobacterium tumefaciens* can carry foreign genes and bring about their recombination into plant cells when they are infected with this bacterium.

Gene Transfer Using *Agrobacterium tumefaciens* Ti Plasmid

One approach for improving agricultural crops that holds great promise is to use bacterial plasmids as vectors. The vectors move genetic information from one plant to another. The **Ti (tumor-inducing) plasmid** of *Agrobacterium tumefaciens* appears to be suitable for such a process (FIG. 8-29). *A. tumefaciens* causes crown gall tumors in most dicotyledonous plants. The Ti plasmid of *A. tumefaciens* induces infected plant cells to synthesize nitrogen compounds called opines. During the normal infectious process, a section of the plasmid combines with chromosomal DNA in the nucleus of the plant cell. It may therefore be possible to employ the Ti plasmid as a vector for inserting foreign DNA into the DNA of plant cells. To do so the plasmid would be cut open at a site within the Ti plasmid DNA and the foreign gene spliced into it. When tumor cells are grown in tissue culture, they would continue to carry and replicate Ti-DNA during the normal divisional process. If foreign genes inserted into Ti-DNA are also transmitted to plant progeny, new plant strains can be genetically engineered.

RECOMBINANT BACTERIA WITH HUMAN GENES

Recombinant Human Insulin

Human insulin, a protein produced in the pancreas, is necessary for the regulation of carbohydrate metabolism. The active form of insulin consists of two polypeptides connected by disulfide bridges. The polypeptides are coded by separated parts of an insulin gene, which also codes for additional products. Diabetes, a disease characterized by a shortage of in

sulin, is treated by insulin injections. The source of this insulin traditionally has been commercially produced insulin isolated from beef or pork pancreas. The insulin of most mammals is similar in structure. However, the immune systems of some diabetics can recognize beef or pork insulin as a foreign (nonhuman) protein and destroy the insulin. For this reason, it is preferable to treat diabetics with human insulin.

Human insulin can be produced by using recombinant *E. coli* (FIG. 8-30). The two polypeptides of insulin are made separately and then linked chemically. A nucleotide triplet coding for methionine is

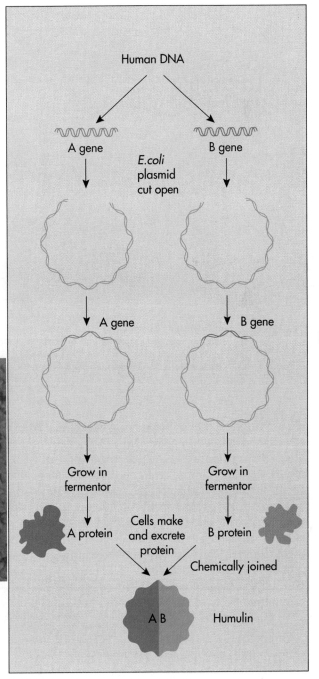

FIG. 8-30 **A,** Crystals of humulin. **B,** Human insulin (humulin) is produced by recombinant strains of *Escherichia coli.* One recombinant strain is genetically engineered to produce the A protein and a second recombinant strain produces the B protein. The two proteins are then chemically combined to produce commercial humulin.

FIRST ENVIRONMENTAL RELEASE OF GENETICALLY ENGINEERED BACTERIA

In 1987 Steven Lindow, a professor at the University of California, released a genetically engineered bacterium into the environment in an experiment (FIG. *A*). The release came after years of debate among scientists about the potential ecological consequences of such releases, extensive governmental oversight, and a court battle with environmentalists. The bacterium that Lindow released is a genetically engineered strain of *Pseudomonas syringae*, designed to reduce frost damage to agricultural crops.

A

Ice-minus bacteria have been formed by recombinant DNA technology and applied to crops to protect them against frost damage. Here a field test is underway in which the ice-minus bacteria are applied at planting to potato plants. The naturally occurring bacteria on plant surfaces contribute to plant damage because they produce surface proteins that catalyze the formation of ice crystals. The ice-minus bacteria lack the surface proteins that initiate ice crystal formation leading to frost damage.

placed between the *lac* promoter genes of *E. coli* and a human insulin polypeptide gene. A ligase is used to attach the nucleotide sequence to plasmid pBR322, which was cloned in *E. coli*. The *E. coli* expresses the genes in the recombinant DNA and produces the polypeptide of human insulin. Different strains are used to produce the two polypeptides needed for insulin, and chemical treatment is then used to link them together. The final product—**humulin**—is identical to insulin purified from the human pancreas. Humulin is commercially produced by the Eli Lily Company.

Through genetic engineering, recombinant strains of *E. coli* have been created that produce human insulin.

Recombinant Human Growth Hormone

Through a series of novel procedures involving a combination of chemical synthesis and isolation of the natural molecules, a gene for a human growth hormone has been constructed. It was placed into *E. coli*. Human growth hormone is a polypeptide that is 191 amino acid units long elaborated by tissues of the pituitary gland. Its absence leads to a form of dwarfism but it can be cured by administration of the hormone. The segment of the gene that codes for the first 34 amino acids of the peptide was constructed chemically from nucleotides. Such short nucleotide sequences are often easy to make using automated DNA synthesizers.

The remainder of the gene was constructed enzymatically. Because the genes that code for human

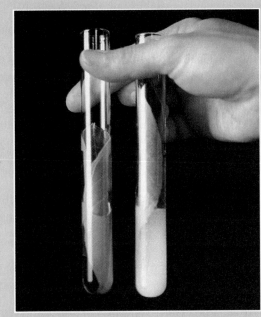

B

The ice-minus bacteria on the plant surface prevent ice crystal formation at temperatures where ice crystals otherwise would form and damage the plant. The bean leaf with the wild-type bacteria on its surface freezes (*right*). The bean leaf with the ice-minus bacteria does not freeze at the same temperature (*left*).

Normally, bacteria living on the surfaces of plants produce proteins that serve as nuclei around which ice condenses. This allows the formation of frost. The genetically engineered strain of *P. syringae*, created by using recombinant DNA methods to delete a specific segment of nucleotides, lacks the gene for producing the protein that initiates ice crystal formation. It is therefore called an ice-minus strain of *P. syringae*. If the normal microbial populations on the plant surface can be replaced by the genetically engineered ice-minus bacteria, the temperature would have to be several degrees lower before ice crystals form and damage the plant (FIG. *B*). This would provide the necessary margin for the survival of many crops. The field test conducted by Lindow demonstrated that application of the recombinant *P. syringae* can prevent ice crystal formation and hence protect plants from frost damage. Application of ice-minus *P. syringae* to plants can potentially save millions of tons of crops that are destroyed annually by frost because genetically engineered ice-minus *P. syringae* can protect plants against frost damage.

growth hormone production in eukaryotic cells are split, reverse transcriptase was employed to copy the gene for the hormone from mRNA obtained from human pituitary tissues. The use of reverse transcriptase simplified the job of cutting and splicing because the DNA produced in this way is colinear with the sequence of nucleotides in the mRNA. Restriction endonucleases cut out the needed fragment. DNA ligase was then used to join the natural and synthetic fragments. The complete gene produced in this manner has been inserted into a modified version of plasmid pBR322 incorporating the *lac* operon. The hormone could therefore be produced independently in bacterial cells. Clinical trials have shown that children who receive injections of bacterially produced human growth hormone for a few months approach normal height for their age group, with no significant side effects.

Recombinant bacteria are used to produce human growth hormone.

HUMAN GENE THERAPY

Recombinant DNA technology can be used to move genes into human cells to cure a disease condition. This potential application of genetic engineering—known as **human gene therapy**—has raised many scientific and ethical questions. On one hand, medical science attempts to alleviate disease conditions, increasing longevity and improving the health and well-being of individuals. On the other hand, science should not try to genetically engineer a new human

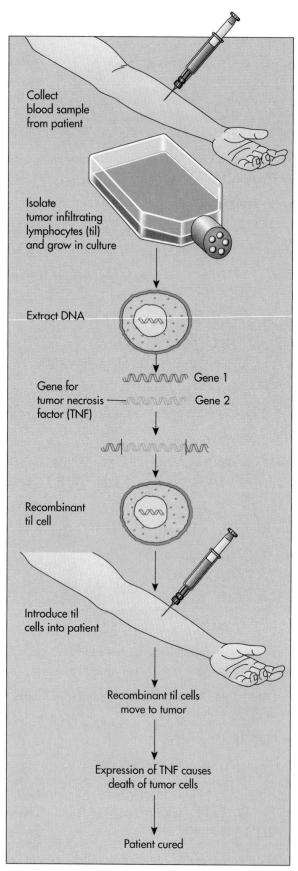

FIG. 8-31 Genes can be added to tumor infiltrating lymphocytes (til cells) through recombinant DNA technology that may help remove malignant tumors.

race. Scientists do not fully understand the regulation of human genes and how specific genes relate to disease conditions, and more needs to be learned.

Human gene therapy is the use of recombinant DNA technology to treat disease.

To help understand the nature of human chromosomes, the Department of Energy and the National Institutes of Health (NIH) have undertaken a major scientific investigation project, called the **Human Genome Project,** to determine the nucleotide sequences of all human genes. This project is aimed at developing an understanding of how human geneotype relates to human phenotype. This research project will provide important insights into how gene therapy might be applied to correct certain disease conditions. At present, the policy under the guidelines of the Recombinant Advisory Committee of NIH is that no human gene therapy can be considered if it allows the use of recombinant genes that can be passed on to the next generation. Therefore gene therapy may not be applied to reproductive cells involved in the passage of heredity information. Gene therapy, however, may be considered on a case-by-case basis when it is to be applied to the somatic cells of an individual.

In the first approved "gene therapy" experiment, Stephen Rosenberg, Michael Blaese, and French Anderson at the NIH inserted the bacterial gene for tetracycline resistance into the DNA of certain human blood cells called tumor infiltrating lymphocytes (til cells) (FIG. 8-31). They did so to track the fate of til cells introduced into the body as part of a therapy for treating advanced forms of cancer. Specifically, these scientists developed a cancer treatment in which they isolate til cells from a cancer patient, grow the til cells in the laboratory, and reinject large numbers of til cells into the patient. In some cases, the til cells attack the tumors (destroying the malignant tumors), but, in other cases, til cell treatment fails. To determine why the treatment succeeds in some cases and fails in others, they used recombinant til cells containing the bacterial tetracycline resistant gene to monitor their fate.

Rosenberg, Blaese, and Anderson next proposed inserting the gene for tumor necrosis factor into the DNA of til cells. This gene codes for a substance that causes tumors to shrink. They inject the special recombinant til cells containing the tumor necrosis factor gene into patients with tumors in an attempt to increase the effectiveness of their cancer treatment. These experiments represent the first attempts to use recombinant DNA technology to directly alter the genetic instructions of human cells within the body.

Recombinant tumor infiltrating lymphocytes have been used to develop new methods for cancer treatment.

SAFETY OF GENETIC ENGINEERING

The potential of genetic engineering to create new gene combinations that have not evolved naturally raises ethical and safety questions. The scientists who discovered that natural restrictions on the transfer of genes between different species could be bypassed by recombinant DNA technology were uncertain about the safety of such procedures, that is, whether recombinant DNA technology would produce organisms dangerous to human health or to the environment. The public feared that scientists would create an Andromeda strain, a fictional but powerful microorganism that would cause widespread disease and wreak havoc on human health.

The capability to create genetically engineered microorganisms to be released into the environment has sparked concern among the public and scientific communities.

In February 1975, scientists gathered at a conference in Asilomar, California, to consider the safety issues related to the construction of genetically engineered microorganisms. Initially it was thought that engineered microorganisms should be contained within laboratories. Methods, therefore, were sought to ensure that microorganisms could not escape containment. Scientists declared a moratorium on the construction of such organisms until a fail-safe strain could be constructed that could not survive outside of special laboratory conditions. Only after scientists had constructed a strain *E. coli* K12 that was so demanding in its nutritional requirements that it could

not live except in a specialized growth medium did scientists resume their work. Also, special facilities were constructed with elaborate air filtration and water treatment systems to ensure that the genetically engineered microorganisms could not escape.

Scientists designed methods for containing genetically engineered microorganisms and for ensuring that if one escaped laboratory containment it could not survive in the environment.

Additionally, governmental guidelines and regulations for preventing accidental release of genetically engineered microorganisms into the environment were established. The NIH formed the Recombinant Advisory Committee (RAC) to oversee the safe use of genetic engineering technology. Despite the extensive controversy about the safety of genetically engineered organisms, experience has shown that the risks are low and manageable and that the benefits are great.

While the initial thrust was to try to contain genetically engineered microorganisms in the laboratory, it soon became clear that great benefits could be realized in constructing genetically engineered organisms to deliberately be released into the environment. Microorganisms to degrade toxic pollutants, to be used as vaccines in humans and other animals, and to enhance agricultural productivity would have to be released into the environment to realize the beneficial uses of genetic engineering for environmental applications. Additional governmental regulations are aimed at ensuring the safety of environmental applications of genetically engineered organisms.

HIGHLIGHT

TRACKING AND CONTAINING GENETICALLY ENGINEERED MICROORGANISMS

Among the major concerns about releasing genetically engineered microorganisms into the environment are the need to track the movement of such organisms and to contain microorganisms that may accidentally cause undesirable side effects. Recognizing these concerns, I initiated two lines of research in my laboratory, one to develop sensitive methods for monitoring genetically engineered microorganisms and the other to develop containment systems that could be used to mitigate any undesirable impact that a deliberate release of such organisms might have.

After several years of work, Robert Steffan, working in my laboratory, developed methods for extracting DNA from environmental samples and using the polymerase chain reaction and gene probes for detecting genetically engineered microorganisms. The combination of PCR and gene probes provides the necessary sensitivity and specificity for tracking genetically engineered

microorganisms in the environment. As few as one genetically engineered microorganism per gram of soil can be detected and positively identified. Thus our work provides an essential tool for tracking the spread of genetically engineered microorganisms with sufficient precision to allay fear that unseen microorganisms were spreading beyond control.

With regard to containment of genetically engineered microorganisms, I decided to use recombinant DNA technology to create suicide vectors, that is, genetic elements that could be placed into genetically engineered microorganisms that would cause such organisms to kill themselves under specific environmental conditions. Soren Molin in Denmark had shown that some naturally occurring genes code for suicide functions and potentially could be used to contain genetically engineered microorganisms. In the studies conducted in my laboratory, Asim Bej, with the collabora-

Continued.

tion of Michael Perlin, Soren Molin, and the late Stephen Cuskey created suicidal microorganisms in which the *lac* operon was used to control expression of a gene called *hok* that codes for a polypeptide that destroys the essential functioning of the cell's plasma membrane. By chemically signalling the *lac* operon to turn on or off, we were able to show that, if necessary, we could cause a genetically engineered microorganism with the suicide vector to kill itself (see Figure). Thus we have engineered a way of containing a genetically engineered microorganism and preventing harmful effects that it might cause.

The use of suicide vectors is appealing to environmentalists and the industrial companies that may pro-duce genetically engineered microorganisms. Industrial companies can use suicide vectors to ensure limited environmental survival and hence the sale of new genetically engineered microorganisms for reuse at the same time that they offer a means for safeguarding against untoward environmental effects. Environmental groups such as the National Audubon Society see the advantage of containment for ensuring environmental safety.

Thus the work undertaken in my laboratory and in Soren Molin's laboratory has gone a long way in developing the methods necessary for reducing risk associated with the deliberate release of genetically engineered microorganisms into the environment.

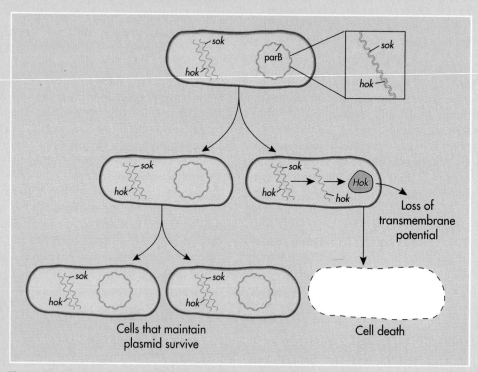

The parB locus of a plasmid in *Escherichia coli* acts as a suicide gene so that the plasmid must be maintained for the cells to survive. One strand of the DNA has the *hok* gene that encodes a polypeptide that kills the cell; if *hok* is expressed the cells die. The other strand of the DNA has the *sok* gene, which produces an antisense mRNA. If the *sok* RNA binds to the *hok* mRNA, the Hok polypeptide is not produced. If *sok* RNA is not there, *hok* is translated. Because the *sok* RNA has a shorter-half life than the *hok* mRNA, if the plasmid with parB is lost, Hok polypeptide is produced and the cell dies. As long as the cell maintains the plasmid with parB, the cell production of *sok* RNA blocks the translation of *hok* mRNA.

SUMMARY

Recombination (pp. 221-224)

- Genetic recombination occurs whenever genes move from one location to another. Movement can be to a new location within a single DNA macromolecule or exchanging or combining DNA from two different sources. Such changes can result in the alteration in the expression of genes or new combinations of genes. Recombination redistributes the changes that result from mutation.

Homologous Recombination (pp. 221-222)

- Homologous recombination occurs when the exchanged DNA comes from corresponding genes and produces new combinations of alleles. Pairs of chromosomes contain the same gene loci pair and exchange corresponding portions of the same chromosomes.

Nonhomolgous Recombination (pp. 222-224)

- Nonhomologous recombination can occur between different DNA segments, even from different species. It permits the insertion of plasmids into bacterial chromosomes, the transfer of genes, and the incorporation of viral DNA into prokaryotic and eukaryotic cells.

Insertion Sequences (pp. 222-224)

- Insertion sequences are small transposable genetic elements composed of about 1,000 nucleotides that can move from site to site within a bacterial chromosome. ISs move by nonhomologous recombination.

Transposons (p. 224)

- Transposons are genetic elements that code for the production of structural proteins. They encode transposase that catalyzes nonhomologous recombination. Tranposase inserts transposons at random within a chromosome. The protozoans that cause African sleeping sickness and malaria have their genes mobilized by transposons, producing random changes in the surface polysaccharides that the host defense system cannot defeat.

Gene Transfer (pp. 224-235)

Gene Transfer and Recombination in Prokaryotes (pp. 224-232)

Plasmids and Gene Transfer (pp. 224-225)

- Plasmids are small extrachromosomal genetic elements that store genetic information that code for specialized functions. Plasmids can replicate and can transfer genes among bacteria that can have more than one plasmid.

- F plasmids are conjugative plasmids that contain special genes that promote their transfer to other cells, where they can replicate because they carry a DNA replication origin. Colicinogenic plasmids code for a protein that is toxic to closely related bacteria. R plasmids carry genes that code for resistance to antibiotics and can be passed not only among strains but also among bacterial species.

Transformation (pp. 225-227)

- Free DNA is transferred from a donor to a recipient bacterium and then is recombined during transformation. The recipient cell must be competent and compatible.

Transduction (pp. 227-228)

- In transduction, DNA is transferred from a donor to a recipient cell by a viral carrier, followed by recombination within the recipient cell. During viral replication within a host cell, some host DNA may break off and be packaged in viral protein coats and passed along as the virus infects a new cell.

- Temperate phage have acquired bacterial DNA and do not kill their host cells because the bacterial DNA replaces genes that code for the completion of phage replication.

- Some temperate phage can carry almost any host cell gene. Generalized transduction results in the transfer

of alleles by homologous recombination. Some temperate phage acquire and transfer only certain host cell genes from specific sites. This process is called specialized transduction. Specialized transduction results in the transfer of genes by nonhomologous recombination.

Conjugation (pp. 228-232)

- During conjugation, DNA is transferred from a donor to a recipient cell via an F pilus. The F pilus is coded for by an F plasmid. If the F plasmid is free in the cytoplasm the donor is an F^+ strain. If it is incorporated into the bacterial chromosome, the donor is a high frequency recombinant (Hfr) strain. Recipient strains do not have F plasmids and do not have F pili.

- During bacterial conjugation, a single strand of donor DNA is replicated and transferred to the recipient where the complementary DNA strand is synthesized. The amount of DNA that is transferred, and hence the number of genes transferred, depends on how long the F pilus maintains contact between the mating cells.

- When F^+ and F^- strains mate, the resultant strains are F^+. When Hfr and F^- strains mate, the recipient strains remain F^- but there is a high frequency of recombination of genes of the bacterial chromosome.

Genetic Exchange in Eukaryotes (pp. 232-235)

- Genetic exchange in eukaryotes usually occurs during sexual reproduction and involves the recombination of DNA from two parents. Diploid eukaryotic vegetative cells form haploid gametes or haploid spores to exchange genetic information. Conversion from diploid to haploid occurs during meiosis. Haploid nuclei of reproductive cells later fuse with nuclei of reproductive cells of an appropriate mating type in syngamy, reestablishing the diploid state.

Recombinant DNA Technology (pp. 235-240)

Formation of Recombinant DNA (pp. 235-237)

- Recombinant DNA technology is the deliberate union of genes to make recombinant DNA macromolecules. Both donor and recipient DNA can be from any source.

- Restriction endonucleases cut the DNA at a palindromic sequence of bases, producing DNA with staggered single-stranded ends that can be spliced with DNA from other sources if excised with the same endonuclease. Ligases splice the pieces of DNA. If different endonucleases cut the DNAs, an artificial homology must be established at the terminal ends of the donor and recipient DNAs by adding a polyA tail to the plasmid and a polyT tail to the donor DNA.

- Reverse transcriptase and an mRNA can produce a DNA macromolecule with a contiguous sequence of nucleotide bases containing complete functional genes to overcome the problem of discontinuity of eukaryotic split genes. Therefore recombinant DNA can be made containing both eukaryotic and prokaryotic genes.

Cloning Recombinant DNA (pp. 238-240)

- Cloning is the asexual reproduction of genes contained in recombinant DNA. Plasmids are often used as carriers for such cloning. Bacteria containing recombinant plasmids produce multiple copies of identical cloned genes.

- Insertional inactivation can detect the presence of foreign DNA within a plasmid. It is caused by the disruption of the nucleotide sequence of a gene by the insertion of foreign DNA.
- Protoplast fusion can transfer DNA from one cell to another.

Genetic Engineering (pp. 240-250)

Recombinant Plants with Bacterial Genes (pp. 240-245)
Recombinant Nitrogen Fixers (pp. 240-244)

- The genetics and biochemistry of infection with the nitrogen-fixing *Rhizobium bacterium* are being studied to try to use recombinant DNA techniques to genetically engineer plants that will fix their own nitrogen.

First Demonstration of Genetic Engineering (pp. 243-244)

- Genetic engineering is the use of recombinant DNA technology for the creation of cells with entirely new properties. In 1973, Cohen and Boyer proved that eukaryotic genes could be genetically engineered into a bacterial cell by inserting a toad gene into a bacterial plasmid.

Gene Transfer Using Agrobacterium tumefaciens *Ti Plasmid (p. 245)*

- The tumor-inducing plasmid of *Agrobacterium tumefaciens* can be used as a vector for moving genetic information from one plant to another.

Recombinant Bacteria with Human Genes (pp. 245-247)
Recombinant Human Insulin (pp. 245-246)

- Human insulin can be produced by using recombinant *E. coli.*

First Environmental Release of Genetically Engineered Bacteria (pp. 246-247)

- The first genetically engineered bacteria to be released into the environment was an ice-minus strain of *Pseudomonas syringae* designed to protect crops from frost damage.

Recombinant Human Growth Hormone (pp. 246-247)

- A gene coding for the production of human growth hormone has been created and placed into *E. coli.* Injections of bacterially produced human growth hormone are successful in allowing children suffering from a form of dwarfism to grow.

Human Gene Therapy (pp. 247-248)

- Human gene therapy is the use of recombinant DNA technology to move genes into human cells to cure disease. The human genome project is designed to determine the nucleotide sequences of all human genes in order to understand how human genotype relates to human phenotype.

Safety of Genetic Engineering (p. 249)

- Scientists have developed methods for safely handling genetically engineered microorganisms. Government regulations have been established to oversee the development and applications of genetic engineering.

CHAPTER REVIEW

REVIEW QUESTIONS

1. Define genetic recombination.
2. By what mechanisms are genes transferred between bacteria?
3. What is a plasmid and what is its function?
4. What is a transposon and what is its function?
5. How is recombinant DNA produced *in vitro*?
6. Compare general and specialized transduction.
7. How could you distinguish between transformation, transduction, and conjugation?
8. What is a gene probe?
9. What are the biological functions of reverse transcriptase, DNA ligase, transposase?
10. What is the difference between transposons and insertion sequences?
11. How can recombination alter the ability to use specific antibiotics in treating disease?
12. How can a bacterium be genetically engineered to produce a human protein?

CRITICAL THINKING QUESTIONS

1. Why is the general public frightened of genetically engineered microorganisms? Could a scientist really create the "Andromeda strain" or the dinosaurs of Jurassic Park? Why has the scientific community lowered its level of concern about the release of genetically engineered microorganisms into the environment? What safeguards have been developed to protect the public and the environment from "potential new pathogens"? What safeguards are needed?

2. How can recombination lead to the evolution of new organisms? How could you go about creating a photosynthetic fungus? How could you improve the agricultural productivity through genetic engineering of plants and microorganisms?

3. What are some of the potential uses for recombinant DNA? How will recombinant DNA technology alter the practice of medicine in the next decade? How can recombinant DNA technology be used to improve environmental quality?

READINGS

Alberts B, D Bray, J Lewis, M Raff, K Roberts, JD Watson: 1989. *Molecular Biology of the Cell*, New York, Garland Publishing.

Excellent coverage of DNA recombination and genetic engineering.

Anderson WF: 1985. Human gene therapy: scientific and ethical considerations, *Journal of Medicine and Philosophy* 10:274-291.

Discusses the application of recombinant DNA technology to human medical treatment.

Bishop JE and M Waldholz: 1990. *Genome: The Story of the Most Astonishing Adventure of our Time—The Attempt to Map All the Genes in the Human Body*, New York, Simon and Schuster.

Typical of many new books explaining the genome project to nonscientists, this one was written by two Wall Street Journal *reporters.*

Brill W: 1985. Safety concerns and genetic engineering in agriculture, *Science* 227:381-384.

Discusses the applicability of genetic engineering to plants.

Culliton BJ: 1990. Gene therapy: into the home stretch, *Science* 249:974-976.

Reviews the work of the Recombinant DNA Advisory Committee of the National Institutes of Health and its role in the control of the future of gene therapy.

Doyle J: 1986. *Altered Harvest: Agriculture, Genetics, and the Fate of the World's Food Supply*, New York, Penguin Books.

Discusses the role of government policy in the genetic engineering industry as applied to agriculture and the food and drug industries.

Fowler JR (ed.): 1989. *Application of Biotechnology: Environmental and Policy Issues*, Boulder, CO; Westview Press.

The results of the AAAS national meeting on the health risk assessment of the environmental aspects of genetic engineering and biotechnology.

Frontiers in recombinant DNA: 1987. *Science* 236:1157, 1223-1268.

A special issue on genetic research.

Hall SS: 1987. *Invisible Frontiers: the Race to Synthesize a Human Gene*, New York, Atlantic Monthly Press.

An account of the race on the cutting edge of science from Spring 1976 to Fall 1978 to synthesize a human insulin gene and to genetically engineer the production of insulin.

Henig RM: 1991. Dr. Anderson's gene machine, *New York Times Magazine*, March 31:30-35.

Dr. W. French Anderson is a pioneer in gene therapy at the National Institutes of Health involved in a gene transfer experiment for the treatment of cancer and an inherited enzyme deficiency.

Krimsky S: 1986. *Genetic Alchemy: the Social History of the Recombinant DNA Controversy*, Cambridge, MA; MIT Press.

Discusses the social aspects of genetic engineering and recombinant DNA technology in the United States.

Kucherlapati R and GR Smith (eds.): 1988. *Genetic Recombination*, Washington, D.C.; American Society for Microbiology.

Presents molecular, biochemical, and biological perspectives on recombination, especially homologous and site-specific recombination. The process of recombination is described for bacteria, bacteriophage, yeasts, Drosophila, and cultured mammalian cells.

Marx JE (ed.): 1989. *A Revolution in Biotechnology*. Cambridge, England; Cambridge University Press.

Begins with an introduction to genetics and molecular biology and continues with chapters on specific areas in biotechnology, including

chemical and pharmaceutical production, single-cell proteins, metal leaching, plant genetic engineering, monoclonal and site-specific antibodies, and gene therapy.

O'Brien TF, M del Pilar Pla, KH Mayer: 1985. Intercontinental spread of a new antibiotic resistance gene on an epidemic plasmid, *Science* 230:87-88.

Widespread dissemination of a newly observed resistance gene on a plasmid suggests that these new genes may be carried to other centers and other plasmids. Such resistance genes might be contained if detected early.

Old RW and SP Primrose: 1987. *Principles of Gene Manipulation*, Oxford, England; Blackwell Scientific.

Explains basic genetic engineering procedures.

Peters P: 1993. *Biotechnology: A Guide to Genetic Engineering*, Dubuque, Iowa; William C. Brown.

A thorough work on the applications of recombinant DNA technology.

Pimental D, MS Hunter, JA La Gro: 1989. Benefits and risks of genetic engineering in agriculture, *BioScience* 39:606-614.

Discusses the environmental aspects of research in plant genetics.

Rodriguez RL and RC Tait: 1984. *Recombinant DNA Techniques: An Introduction*, Redwood City, California; Benjamin Cummings.

A guide to techniques used in genetic engineering.

Shimke RT: 1980. Gene amplification and drug resistance, *Scientific American* 243(5):60-69.

Experiments on the development of drug resistance in mammalian cell cultures that have served as laboratory models for the evolution of duplicate genes.

Stahl FW: 1987. Genetic recombination, *Scientific American* 256(2):90-101.

Review of DNA, chromosomes, and research on how chromosomes are reshuffled during reproduction.

Stewart GJ and CA Carlson: 1986. The biology of natural transformation, *Annual Review of Microbiology* 40:211-235.

Reviews the comparative physiology of transformation by chromosomal fragments, cell-to-cell transformation, and transformation by plasmid DNA.

Suzuki DT and P Knudtson: 1989. *Genetics: the Clash Between the New Genetics and Human Values*, Cambridge, Massachusetts; Harvard University Press.

Begins with a primer on principles of modern genetics that provides a background for a discussion of the ethical issues and need for moral guidelines to make responsible decisions related to genetic engineering.

Verma IM: 1990. Gene therapy, *Scientific American* 263(5):68-72.

The most challenging issue in the treatment of inherited human diseases by inserting a healthy gene is to assure that the therapeutic genes are expressed adequately and persistently in the body.

Watson JD, M Gilman, J Witkowski, M Zoller: 1992. *Recombinant DNA*, ed 2, San Francisco, W. H. Freeman.

Describes procedures for the isolation, identification, and transfer of genes in eukaryotic and prokaryotic cells.

White R and J Lalouel: 1988. Chromosome mapping with DNA markers, *Scientific American* 258(2):40-48.

Variable sequences in the DNA of human chromosomes act as genetic markers that may help trace defective genes and provide elements of the human chromosome map.

UNIT THREE

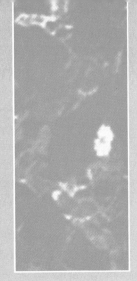

Microbial Growth

and Its Control

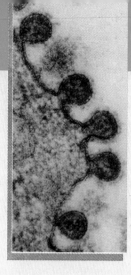

CHAPTER 9

Viral Replication

PREVIEW TO CHAPTER 9

In this chapter we will:

• Learn how viruses replicate within host cells, using their nucleic acids (RNA or DNA) to direct the formation of viral progeny.
• Examine the general replication strategy used by all viruses for their replication and the specific steps used by different bacterial, plant, and animal viruses to accomplish this task.
• See why viruses have tremendous replication potential.
• Learn that viral DNA can be incorporated into host cells, changing the properties of the host cells.
• Learn the following key terms and names:

adsorption
animal viruses
assembly
bacteriophages
budding
burst size
capsid
early protein synthesis
eclipse period
envelope
late protein synthesis
latent period
lysogeny
lytic phages
nucleocapsid

oncogenes
one-step growth curve
penetration
phages
plant viruses
plus RNA strand
prophage
release
reverse transcriptase
reverse transcription
temperate phage
uncoating
viral maturation
virions

Viruses are acellular, nonliving microorganisms that are totally dependent on host cells for replication. They are incapable of independent metabolism, growth, or reproduction. Some microbiologists view viruses as parasites of the host cells within which they replicate. Others consider viruses to be genetic extensions of their host cells. Supportive of both viewpoints is the high degree of specificity necessary between a particular virus and the specific host cell within which that virus replicates. Specificity between virus and host cell is based on the ability of the virus to physically attach to the host cell and the ability of the viral nucleic acid to direct viral replication within that host cell. For a virus to replicate within a host cell, the host cell must be permissive, meaning that the host cell must allow entry and must not degrade the viral nucleic acid genome. The virus also must be compatible with the host cell, meaning that the virus must be capable of using and directing the metabolism of the host cell for viral replication.

> **To replicate, a virus must be able to enter a compatible host cell and, using the host cell's metabolism, the viral nucleic acid must be able to direct viral replication within the host cell.**

Viruses have two essential components: a nucleic acid and a wall-like structure—called the **capsid.** The nucleic acid is either RNA or DNA and carries the hereditary information. The capsid is composed of protein and surrounds the nucleic acid. Together the capsid and enclosed nucleic acid are called the **nucleocapsid.** The replication of a virus involves making a copy of the nucleic acid and capsid protein to construct a new nucleocapsid. Some viruses also have an **envelope** that is acquired mainly from the host cell and composed largely of nuclear or cytoplasmic membrane. In some cases, viral proteins are added to the envelope.

Within a host cell, the viral nucleic acid genome directs the formation of new viruses. It uses the structures and metabolism of the cell to make viral protein and nucleic acid. In many cases, the viral nucleic acid genome actually codes for the shutdown of the metabolic activities normally involved in the host cell's reproduction. The virus then uses the host's biochemical components and anatomical structures for the production of new viruses. In particular, the virus employs the host cell's ribosomes for producing viral proteins and ATP for carrying out biosynthesis. Viral replication results in changes within the host cell, often causing its death.

> **Viral replication occurs only within compatible host cells.**

> **Viral replication requires that a virus gain entry to a host cell and take control of that cell's metabolism.**

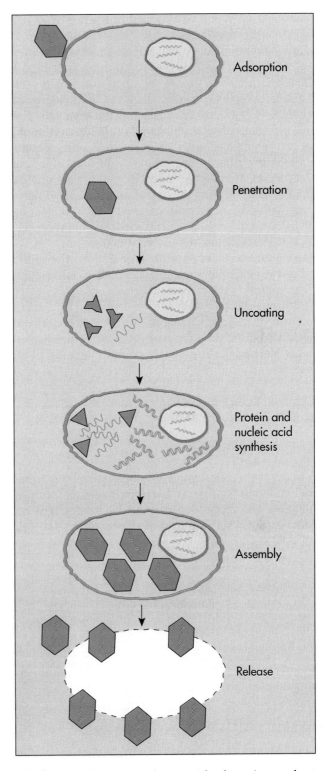

FIG. 9-1 Viral replication begins with adsorption to a host cell. Most animal and plant viruses then penetrate the cell and the viral coat (capsid) is rapidly degraded so that the viral nucleic acid is uncoated and is free within the host cell. Within the host cell the viral nucleic acid directs the synthesis of viral proteins and nucleic acids. Viral progeny (complete virions) are assembled from the synthesized viral nucleic acid and capsid components and subsequently released from the host cell.

STAGES OF VIRAL REPLICATION

The replication of a virus occurs in a series of stages. Viral replication begins with **adsorption** when a virus attaches to the outer surface of a suitable host cell (FIG. 9-1). Generally, adsorption of a virus to a host cell involves specific binding sites on the cell surface, which explains in part the high degree of specificity between the virus and the host cell. The second stage of viral replication, called **penetration,** occurs when the virus or its nucleic acid genome crosses the plasma membrane and enters the host cell. In the third stage, **uncoating,** the viral nucleic acid is released from the capsid. In viruses that replicate within bacterial cells, penetration and uncoating usually occur simultaneously so that the viral nucleic acid genome enters the bacterial host cell. In viruses that replicate in plant and animal cells, the entire virus normally penetrates the host cell and uncoating then occurs in the cytoplasm or nucleus.

Within the host cell, the viral nucleic acid takes control of the cell's metabolism. First, the viral nucleic acid directs the synthesis of proteins **(early protein synthesis).** These early proteins include polymerases that are needed to make copies of the viral nucleic acid. Then multiple new copies of the viral nucleic acid genome are produced **(nucleic acid synthesis).** New proteins needed for capsid production are made last **(late protein synthesis).**

The nucleic acid genome is then packaged into the capsid. This process is called **assembly** or **viral maturation.** During the assembly process the capsid can first be constructed and then the viral nucleic acid packed into the capsid. Alternatively the capsid may be built around the viral nucleic acid to form the nucleocapsid.

Finally, **release** occurs when the assembled viruses leave the host cell. Often, many viruses are released simultaneously and the host cell is killed in the process. The released viruses, called **virions,** are infectious and when they encounter a suitable living host cell the entire process of viral replication begins anew.

> The stages of viral replication are: adsorption to a host cell, penetration, uncoating, early protein synthesis, nucleic acid replication, late protein synthesis, assembly, and release.

REPLICATION OF BACTERIOPHAGES

LYTIC PHAGE REPLICATION

Viruses that replicate only within specific host bacterial cells are known as **bacteriophages** or, simply, **phages.** The replication of most bacteriophages results in the lysis (rupture) of the host bacterial cell when new phages are released. Therefore these bacteriophages are referred to as **lytic phages.** Because release is by lysis, the replication of lytic phages always results in the death of the host cell.

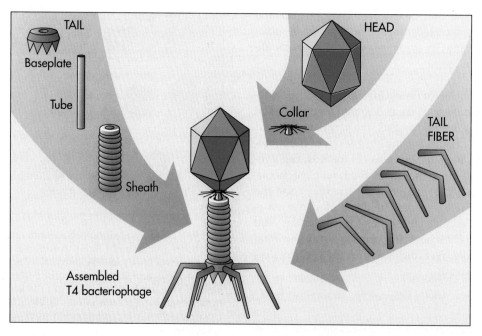

FIG. 9-2 The capsid of a T-even phage has a head that contains the nucleic acid and a complex tail that attaches to the host cell and accomplishes the injection of the nucleic acid.

T-even Phage Replication

One type of bacteriophage, the T-even phage (phage with tails), have a capsid that has distinct head and tail structures (FIG. 9-2). These phage are designated by even numbers, for example, T2, T4, and T6. The DNA genome is contained within the head structure and the tail is involved in adsorption to the host cell. The genomes of T-even phages are double-stranded DNA—like that of cellular organisms.

Replication of a T-even phage begins with the adsorption of a T-even phage tail to a bacterial host cell (FIG. 9-3). There are specific receptor sites on the bacterial cell surface where the phage tail may attach. The bacteriophage's tail releases lysozyme, which breaks down a portion of the bacterial cell wall. This allows the phage tail to penetrate the cell wall but not the plasma membrane. The phage tail then contracts, forcing the phage DNA into the periplasmic space (the region between the outer membrane of the bacterial cell wall and plasma membrane of a Gram-negative bacterial cell). A pore probably is formed in the plasma membrane, allowing the phage DNA to enter the cell. Penetration and uncoating, thus, occur simultaneously. The phage DNA subsequently migrates across the plasma membrane and into the cell.

The entire T-even phage does not penetrate the bacterial cell, but rather, the phage injects only its DNA into the bacterium.

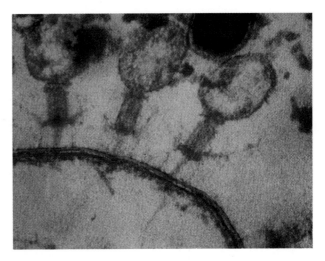

FIG. 9-3 Colorized micrograph showing the attachment of T-even phage to a cell of *Escherichia coli* and the extension of the tail tube as an injection needle.

When the phage DNA enters a compatible bacterial cell, it is not degraded by the nucleases that otherwise would destroy it. This is because the phage DNA is resistant to the nucleases of a compatible cell. The viral DNA is transcribed within the host cell to make viral mRNA. The mRNA is subsequently translated at the ribosomes of the cell to make phage proteins (early protein synthesis). Early phage proteins cause stoppage of normal host cell biosynthesis, that

METHODOLOGY

Assaying for Lytic Phage

Numbers of bacteriophage can be determined by inoculating a suspension of host bacterial cells with the phage. The suspension of host bacterial cells is spread over a solid nutrient growth medium and various dilution of phage are then spread over the same surface. In the absence of lytic bacteriophage, the bacteria form a confluent lawn of growth. Lysis by bacteriophage is indicated by the formation of a zone of clearing or plaque within the lawn of bacteria (see Figure). Each plaque corresponds to the site where a single bacteriophage acted as an infectious unit and initiated its lytic replication cycle. The spread of infectious phage from the initially infected bacterial cell to the surrounding cells results in the lysis of the bacteria in the vicinity of the initial phage particle and hence this zone of clearing. Plaques do not continue to spread indefinitely. With T4 phage, for example, the plaque size is limited because heavy reinfection of a host cell before the time of normal lysis extends the period of synthesis of viral protein and nu-

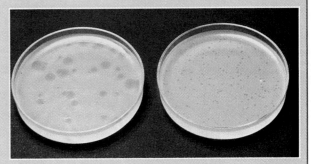

Replication of lytic bacteriophage causes the formation of plaques in a lawn of bacterial cells growing on an agar surface.

cleic acid, thereby preventing completion of the lytic replication cycle. This phenomenon is known as lysis inhibition. The number of plaques that develop and the appropriate dilution factors can be used to calculate the number of bacteriophages in a sample.

is, the bacterial cell ceases making structures for new bacterial cells. Under the direction of the viral nucleic acid genome, the bacterial cell begins to synthesize proteins involved in the replication of phage nucleic acid (nucleic acid synthesis) and then the various proteins that make up the capsid of the phage (late protein synthesis). The head and tail structures of the phage are made up of proteins coded for by different phage genes, with at least 32 genes involved in the formation of the tail structure and at least 55 genes involved in the formation of the head structure of the phage.

After the production of the individual components of the virus, the virus is assembled by packing the nucleic acid genome into the protein capsid (assembly). The assembly of the T-even phage capsid is a complex process. Assembly of the head and tail structures requires several enzymes that are coded for by the phage nucleic acid. The head and tail units of the T-even phage capsid are assembled separately and later combined. The phage DNA is packed into the head structure, and when the head structure is

completely filled with DNA, any extra DNA is cleaved by a nuclease.

Release of the assembled T-even phage occurs because one of the late proteins coded for by the phage is lysozyme. Lysozyme catalyzes the breakdown of the bacterial peptidoglycan wall structure. Lysozyme causes sufficient damage to the cell wall so that the wall is unable to protect the cell against osmotic shock. This results in the lysis of the bacterial cell and the release of the phage into the surrounding medium.

Lysozyme, a product of late protein synthesis, damages the cell wall, leaving it susceptible to osmotic shock and lysis. Thus new phage are released.

Replication of RNA Bacteriophage

There are some RNA phage, such as Qβ and f2, that carry out a lytic replication cycle (FIG. 9-4). The hereditary information for these phage is contained in a single strand of RNA. This RNA, designated as a **plus RNA strand,** contains the viral hereditary infor-

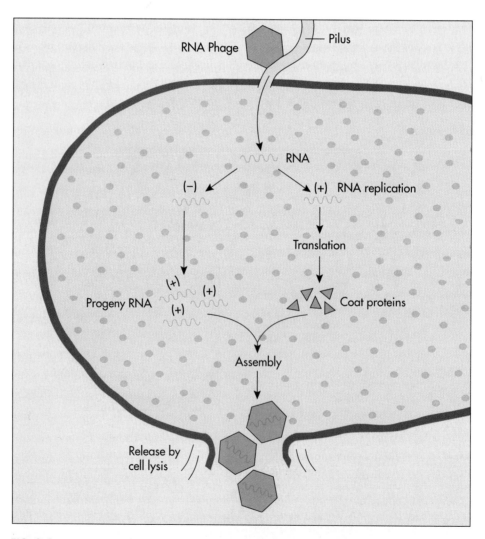

FIG. 9-4   The replication of an RNA phage uses RNA as the hereditary molecule to direct the synthesis of protein for capsid and new RNA genomes.

mation and acts as the nucleic acid genome of the phage. The designation of the RNA genome as a plus strand indicates that it can also serve as an mRNA molecule. The plus strand RNA can be translated at the ribosomes of the bacterial cell to produce the proteins for making the viral capsid.

The RNA also must be replicated so that it can serve as the hereditary macromolecule of the virus. To form new copies of this RNA for viral progeny, the plus strand is used as a template for forming a double-stranded RNA replicate that has both plus and complementary minus RNA strands. To make this replication form, RNA nucleotides in association with RNA polymerase pair with the complementary nucleotides in the plus strand and RNA polymerase links them together. Then the minus RNA strand is used as a template for transcription to produce new plus RNA copies. A copy of plus RNA is then packaged into a capsid and a new phage is assembled. After assembly, the host cell is lysed by an enzyme coded for by the phage. This releases the progeny phage.

The replication of single-stranded RNA phage involves the formation of a double-stranded RNA macromolecule that serves as a template for transcription of new plus RNA.

BACTERIOPHAGE GROWTH CURVE

Lysis of a bacterial cell releases a large number of phage simultaneously. Consequently, the lytic replication cycle exhibits a **one-step growth curve** (FIG. 9-5). The growth curve for lytic bacteriophage

begins with an **eclipse period,** which is the time period from penetration until assembly. During the eclipse period there are no complete infective phage particles because once the phage nucleic acid genome is uncoated it is unable to infect another cell to initiate a new replication cycle. Only complete phage with a capsid can adsorb to host cells and initiate infection of another cell. Until new phage are assembled, therefore, there are no infective phage within the host cell. The eclipse period ends only when an average of one completely assembled infectious phage has been produced for each host cell.

The eclipse period is the time between entry of the phage nucleic acid into a host cell and the formation of a completely assembled phage within that cell.

The **latent period** is longer than the eclipse period, beginning when the phage injects its nucleic acid into a host cell but not ending until the first assembled virus appears outside the host cell. The latent period, thus, starts with penetration and ends after release. The latent period for a T-even phage typically is about 15 minutes. During the time between the end of the eclipse period and the end of the latent period, assembled phage accumulate within the bacterial cell.

The latent period is the time between entry of the viral nucleic acid and the release of viral progeny.

Completely assembled phage continue to accumulate within the bacterial cell until they reach a number known as the **burst size** (FIG. 9-6). The burst size,

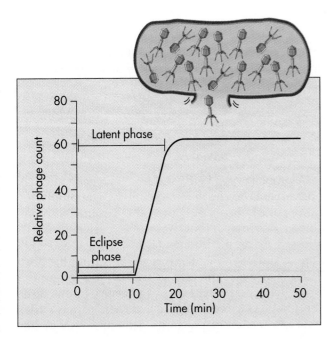

FIG. 9-5 Because many viruses are released simultaneously when a host cell lyses, lytic viruses exhibit a one-step growth curve.

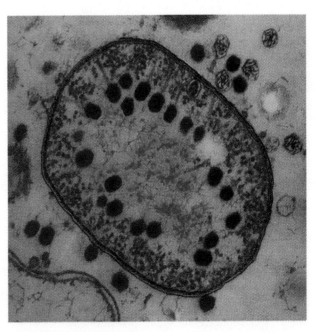

FIG. 9-6 Colorized micrograph of *Escherichia coli* with assembled bacteriophage *(blue)* inside the bacterial host cell.

which varies from cell to cell, represents the average number of infectious viral units that are present when a particular type of host cell lyses or bursts. A typical burst size for a T-even phage is 200. When the cell lyses, the viruses are released into the extracellular fluid. As a result of the simultaneous release of a number of infective phage, the number of phage that can initiate a lytic replication cycle increases greatly in a single step. The entire lytic growth cycle for some T-even phage can occur in less than 20 minutes under optimal conditions so that phage have a tremendous replicative capacity.

> **The burst size is the number of phage released with lysis of the host cell.**

TEMPERATE PHAGE—LYSOGENY

Some DNA bacteriophage are called **temperate phage** because they can infect host cells without replicating the entire phage and without causing lysis of the bacterial host cell (FIG. 9-7). Temperate phage, such as the well-studied phage lambda (λ), do not always cause lysis of the host cell. Rather, these temperate phage often establish a state of **lysogeny** in which a portion of the viral nucleic acid is incorporated into the bacterial chromosome or a bacterial plasmid. This occurs when the phage DNA, having entered a host cell, undergoes recombination with the host cell DNA. Once incorporated into the bacterial DNA, the viral DNA is referred to as a **prophage.** The prophage is replicated with the bacterial DNA during normal host cell DNA replication and the host cell survives and reproduces.

> **Temperate phage can incorporate their DNA into the DNA of the bacterial host cell to establish a state of lysogeny.**

> **The incorporated phage DNA, called a prophage, is replicated with host cell DNA.**

The presence of a prophage prevents reinfection of the host cell by the same phage. The host cell is thus protected against lysis due to phage infection. Many of the genes of the integrated viral DNA are repressed. Regulatory genes prevent transcription of most of the phage genes and therefore few viral proteins are made. Complete phage are not made and the phage genes for cell lysis are not expressed so that the host cell is not killed.

Some phage genes, however, may be expressed. The proteins that are made can greatly alter the properties of the host cell, sometimes accounting for the ability of the cell to cause human disease. For example, cells of the bacterium *Corynebacterium diphtheriae* produce a protein, called diphtheria toxin, when they are in a state of lysogeny because one of the prophage genes that is expressed codes for this protein toxin. It is the production of diphtheria toxin during an infection with *C. diphtheriae* that causes the disease diphtheria. Strains of *C. diphtheriae* that do not carry the prophage are harmless nonpathogens. Similarly the protein toxin produced by *Clostridium botulinum* (botulinum toxin) causes botulism when ingested; the protein toxin produced by *Streptococcus pyogenes* that causes symptoms of scarlet fever and some of the protein toxins produced by *Staphylococcus aureus* that cause toxic shock syndrome are all coded for by prophage genes. Only strains of these bacteria carrying prophage produce the protein toxins responsible for these diseases.

> **The presence of a prophage can alter the properties of a host cell.**

> **Protein toxins coded for by prophage are responsible for some human diseases.**

A prophage can be passed from one generation of host bacteria to the next indefinitely. The prophage, however, can be excised from the bacterial chromosome. This may occur when the bacterial cell is stressed by an environmental factor, such as exposure to ultraviolet light. Once excised from the host cell DNA, the regulatory genes may no longer repress the expression of all phage genes. In this case, the phage can reestablish a lytic replication cycle, with production of phage progeny and lysis of the host cell.

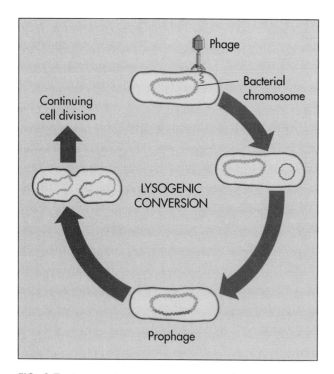

FIG. 9-7 Lysogenic conversion occurs when a temperate phage transfers bacterial DNA that it has acquired and that DNA recombines with the DNA of the host cell. The bacterial cell then replicates the phage DNA along with the bacterial chromosome DNA.

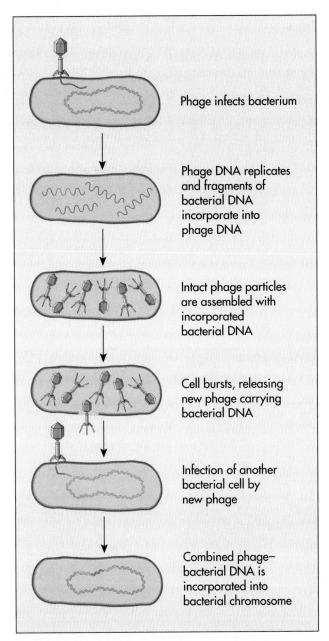

FIG. 9-8 Some viruses can carry bacterial genes. This forms the basis for transduction.

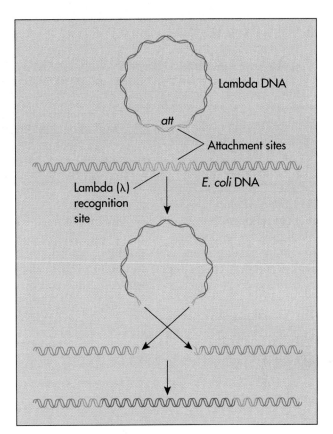

FIG. 9-9 The genome of bacteriophage lambda (λ) can integrate into the bacterial chromosome of *Escherichia coli* at specific sites. The insertion of lambda DNA always occurs at the same site of the *E. coli* genome, between the genes for galactose utilization and biotin synthesis. The lambda phage DNA has a site specific attachment gene *(att)*, and a site specific attachment enzyme (integrase) is involved.

In some cases, an excised prophage carries with it some bacterial DNA (FIG. 9-8). Thus bacterial genes coded for by regions of the bacterial DNA adjacent to the site where the prophage had been inserted into the host cell DNA can be incorporated into phage progeny. When these phage infect new host cells they carry the bacterial genes they acquired. Thus these phage transfer bacterial genes from a donor cell to a recipient cell.

Such phage-mediated transfer of bacterial DNA is a form of transduction called specialized transduction, discussed in Chapter 8. Specialized transduction is so named because only certain specific genes that are located adjacent to the specific site where prophage DNA is incorporated into a host bacterial cell can be transferred. For example, bacteriophage lambda (λ) can establish a state of lysogeny in *Escherichia coli* (FIG. 9-9). Lambda phage have a gene that codes for an integrase. This enzyme recognizes specific nucleotide sequences of the bacterial chromosome and inserts the lambda DNA at sites where those nucleotide sequences occur. The prophage is incorporated into the bacterial chromosome of *E. coli* at the site where the genes for biotin synthesis and galactose utilization are located. When a lambda prophage is excised it can carry with it the adjacent bacterial genes, that is, the *gal* gene for galactose utilization and/or the *bio* gene for biotin synthesis. Such lambda phage can transfer these bacterial genes, changing strains of *E. coli* that require biotin for growth into ones that do not and strains of *E. coli* that cannot utilize galactose into ones that can.

REPLICATION OF PLANT VIRUSES

Specialized transduction transfers certain specific host cell genes located adjacent to the site of the prophage to a recipient cell.

Viruses that replicate within plant cells, called **plant viruses,** follow the same basic steps described earlier for the general replication of viruses. Entry of a plant virus into a susceptible plant often involves abrasions or insect bites. The virus then adsorbs to a susceptible host cell. Both the viral capsid and the viral nucleic acid genome cross the plasma membrane of a plant cell by endocytosis if the plant cell plasma membrane engulfs the viral particle and brings it within the cell. Uncoating of the plant viral nucleic acid then occurs within the plant cell. The viral nucleic acid takes control of the synthetic activities of the host cell, directing the production of viral proteins (early protein synthesis), viral nucleic acid genomes (nucleic acid synthesis) and capsid (late protein synthesis). New viruses are assembled and the assembled viruses are released with the lysis of the host plant cell.

Plant viruses cause various diseases of plants and the viruses are often named after the disease they cause (Table 9-1). Tobacco mosaic virus (TMV), for example, causes a disease of tobacco plants characterized by the occurrence of patchy regions where the plant cells have died due to viral infection. Replicated tobacco mosaic viral particles form crystalline cytoplasmic inclusions within infected plant cells

TABLE 9-1

Some Examples of Plant Viruses and the Diseases They Cause

PLANT VIRUS	DESCRIPTION OF VIRUS	PLANT DISEASE
Cauliflower mosaic virus	Double-stranded DNA virus; reproduces in cytoplasm	Mosaic disease of cauliflower; characterized by localized patches of yellow, dying cells (due to loss of chlorophyll-containing functional photosynthetic cells) or black, dead cells
Cucumber mosaic virus	Naked, icosahedral, RNA virus	Mosaic disease of cucumber; characterized by localized patches of yellow, dying cells (due to loss of chlorophyll-containing functional photosynthetic cells) or black, dead cells
Barley yellow dwarf virus	Small isometric, RNA virus	Dwarfism of barley; characterized by lack of stem development
Tobacco ringspot virus	Polyhedral, RNA virus; transmitted by nematodes	Ringspots on tobacco; characterized by circular patches where plant cells are dying or have died
Tobacco necrosis virus	Isometric RNA virus	Tobacco rot; characterized by softening of plant structures in the regions of viral infection and by blackening of those regions due to death of plant cells
Tobacco mosaic virus	Rod-shaped, helically symmetrical, single-stranded RNA virus	Mosaic disease of tobacco; characterized by localized patches of yellow, dying cells (due to loss of chlorophyll-containing functional photosynthetic cells) or black, dead cells
Tomato bushy stunt virus	Small, cubically symmetrical, RNA virus; resistant to elevated temperatures and organic solvents	Stunting of tomato; characterized by limited growth
Turnip yellow mosaic virus	Icosahedral RNA virus; transmitted by flea beetles	Mosaic disease of turnip, characterized by localized patches of yellow, dying cells (due to loss of chlorophyll-containing functional photosynthetic cells) or black, dead cells
Watermelon mosaic virus	Flexible, rod-shaped, RNA virus; 700-950 nm in size	Mosaic disease of watermelon; characterized by localized patches of yellow, dying cells (due to loss of chlorophyll-containing functional photosynthetic cells) or black, dead cells

HISTORICAL PERSPECTIVE

Plant Viruses Show RNA Can Serve as the Hereditary Macromolecule

In 1935, Wendell Stanley purified and partially crystallized tobacco mosaic virus (TMV). Although it was first thought that TMV crystals were pure protein, later studies showed that they also contained RNA. Experiments with TMV were very important in establishing nucleic acids as the informational molecules in viruses and that RNA could act as the viral hereditary macromolecule .

In 1957, Heinz Fraenkel-Conrat and co-workers isolated TMV from tobacco leaves. From a common weed, they isolated a second, rather similar kind of virus, Holmes ribgrass virus (HRV). Both TMV and HRV consist of protein and a single strand of RNA. Both are plant RNA viruses that infect tobacco plants, causing lesions on the leaves. The two viruses produce different kinds of lesions, however, so that the source of a particular infection can be identified.

In Fraenkel-Conrat's experiment, TMV and HRV were each separated into their protein and RNA components. Hybrid viruses were produced by mixing the HRV RNA and the TMV protein and allowing virus particles to form by self-assembly from the RNA and protein. To determine if viral protein or viral RNA was the hereditary substance, Fraenkel-Conrat now infected healthy tobacco plants with a hybrid virus composed of TMV protein capsids and HRV RNA (see Figure).

When the reconstituted virus particles were added to tobacco leaves, the lesions that developed were of the HRV type. Normal HRV viruses were found in high numbers in the lesions. Thus the RNA from the HRV and not the protein from the TMV contained the information necessary to specify the production of the viruses. Clearly, the hereditary properties of the virus were determined by the RNA and not the protein.

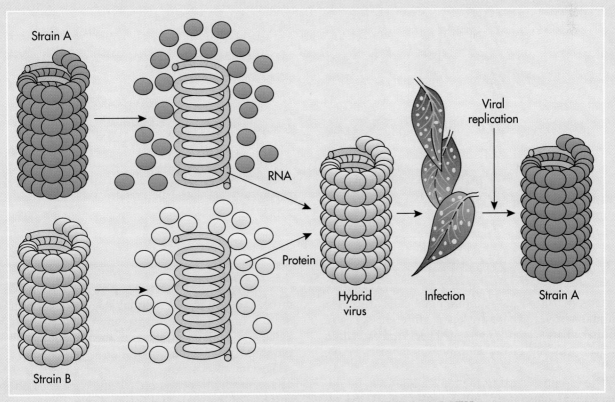

Diagram of Fraenkel-Conrat's experiment (*Strain A*, TMV; *Strain B*, HRV).

(FIG. 9-10). The chloroplast of a tobacco mosaic virus infected leaf becomes chlorotic (yellow due to loss of chlorophyll), leading to the death of the plant cell because it can no longer carry out photosynthesis to supply its cellular energy. The death of the plant cell releases completely assembled TMV progeny and viral nucleic acid that has not been packaged with the protein capsids. Within plants, both completely as-

sembled viral particles and viral RNA can move from one cell to another, establishing new sites of infection. As a consequence of the replication of the viruses within the plant cells, the plant develops characteristic disease symptoms, which include the appearance of a mosaic pattern of chlorotic spots on the leaves that gives the disease and the virus their names.

A

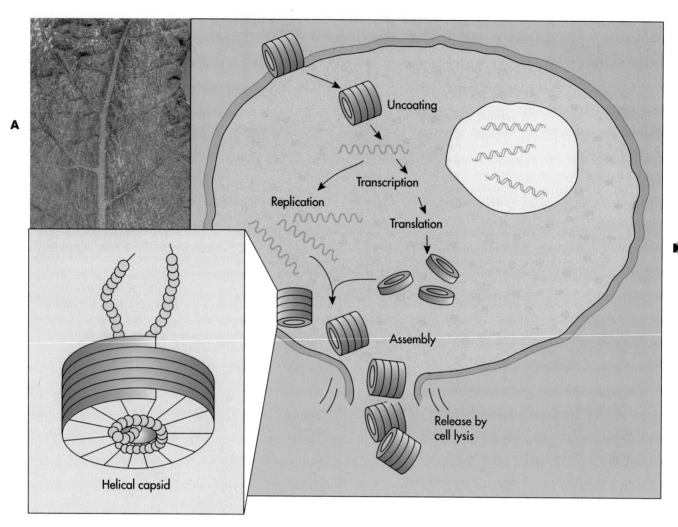

FIG. 9-10 **A,** Leaf infected with tobacco mosaic virus. **B,** The assembly of tobacco mosaic virus involves sequential addition of protein discs to surround the single-stranded viral RNA genome (the ends are shown at the top of the insert).

REPLICATION OF ANIMAL VIRUSES

Viruses that replicate within animal cells are called **animal viruses.** There are many different types of animal viruses; replication of these viruses within human cells can result in numerous diseases (Table 9-2).

The essential steps in the replication cycle of animal viruses, like those of other viruses are: (1) the adsorption of the virus to the surface of the animal cell, (2) penetration so that the intact virus or the viral nucleic acid enters the host cell, (3) uncoating so that the viral nucleic acid is released from the capsid within the host cell, (4) early protein synthesis involving transcription to form viral mRNA and translation using viral-coded messenger RNA to form proteins, (5) replication of viral nucleic acid to form copies of viral nucleic acid genome, (6) late protein synthesis to form late proteins needed for capsids, (7) assembly of complete viral particles, and (8) release of new

viruses. While these essential stages for replication are the same for all viruses, animal viruses are diverse and exhibit variations in the specific details of the stages involved in replication.

The initial adsorption of animal viruses to host cells typically depends on specific chemicals on the surfaces of the virus and the host cell. These surface receptors often help establish the specificity between virus and host cell. Some viruses, such as adenoviruses—which cause upper respiratory infections—have protein spikes that project from the surface of the capsid. These spikes act as the binding sites that permit the virus to adsorb to the surface of a compatible host cell. In this way, adenoviruses attach to the cells lining the upper respiratory tract.

Adsorption of animal viruses to host cells depends on chemicals on the viral surface.

Some Examples of Human Viruses and the Diseases They Cause

GROUP AND TYPE OF VIRUS	DESCRIPTION OF VIRUS	DESCRIPTION OF DISEASE
DOUBLE-STRANDED DNA VIRUSES		
Papilloma virus	Small virus that causes tumors	Replication of the virus within epithelial cells causes warts, which are rough, elevated benign tumors on the skin, urinary tract, or genitals
Herpesvirus	Medium to large size enveloped virus	
Herpes simplex virus (human herpes virus—HHV)		Replication of the virus within epithelial cells causes fever blisters, which are painful or itchy recurrent vesicular lesions usually of the lips or genitals; the virus goes into a dormant stage by entering nerve cells so that the body retains the virus; recurrent outbreaks of symptoms occur; genital herpes is a sexually transmitted disease
Varicella-zoster virus (VZV)		Replication of the virus causes chickenpox, which is characterized by an itchy rash on the scalp or trunk that spreads to the face and leads to vesicle formation and scabs; shingles can occur years later and is characterized by painful, nodular or vesicular lesions that appear in patches on the skin
Epstein-Barr virus (EBV)		Replication of the virus causes infectious mononucleosis, which is characterized by sore throat, swollen lymph nodes, temperature and fatigue; the virus often is transmitted through exchange of saliva
Cytomegalovirus (CMV)		Replication of the virus causes generalized salivary gland disease, which is characterized by sore throat, swollen lymph nodes, fever, and fatigue
Poxvirus	Very large, enveloped, brick-shaped viruses	
Variola virus		Replication of the virus causes smallpox, a once deadly disease that has been eliminated through an effective program of vaccination
SINGLE-STRANDED (PLUS-STRAND) RNA VIRUSES		
Picornavirus	Very small, nonenveloped virus that infects the respiratory tract or gastrointestinal tract	
Poliovirus		Replication of the virus can occur in various body tissues; replication within the nervous system can cause poliomyelitis (commonly called polio or infantile paralysis) that is characterized in some cases by loss of motor function (paralysis); in some cases, viral replication within the spinal column causes meningitis; the virus is often transmitted via contaminated food or water
Rhinovirus		Replication of the virus within the respiratory tract causes the common cold, bronchitis, and croup, which usually are characterized by congestion, coughing, sneezing, and a mild fever
Hepatitis A virus (HAV)		Replication of the virus within the liver causes infectious hepatitis (hepatitis A), which is characterized by a high fever and jaundice (yellowing of the skin due to loss of liver function); the virus is often transmitted via contaminated food or water
Coxsackievirus		Replication of the virus within the respiratory tract causes the common cold, and replication within the oral cavity causes herpangina, which is characterized by lesions at the back of the mouth

Continued.

TABLE 9-2—cont'd		
Some Examples of Human Viruses and the Diseases They Cause		
GROUP AND TYPE OF VIRUS	**DESCRIPTION OF VIRUS**	**DESCRIPTION OF DISEASE**
SINGLE-STRANDED (PLUS-STRAND) RNA VIRUSES—cont'd		
Togavirus	Small, enveloped virus; some transmitted by insects	
St. Louis encephalitis		Replication of the virus within brain cells causes encephalitis, which is characterized by coma and often is fatal; the virus is transmitted via mosquitoes
Rubella virus		Replication of the virus causes German measles, which is characterized by the eruption of a skin rash
Hepatitis C virus (HCV)		Replication of this virus within the liver causes hepatitis C, which is also called non-A, non-B hepatitis; hepatitis C is characterized by a high fever and jaundice (yellowing of the skin due to loss of liver function); the disease is transmitted via blood
Flavivirus Yellow fever virus*	Small, enveloped viruses	Replication of the virus within the liver causes yellow fever, which is characterized by a high fever and jaundice (yellowing of the skin due to loss of liver function); the virus is transmitted via mosquitoes
SINGLE-STRANDED (MINUS-STRAND) RNA VIRUSES		
Orthomyxovirus	Medium to large size, enveloped virus	
Influenza A, B, and C viruses		Replication of these viruses within the respiratory tract causes influenza, which is commonly called flu; influenza is characterized by a high fever, cough, malaise, and body ache; the virus is transmitted through the air; secondary infections can cause fatalities due to pneumonia; susceptible high-risk individuals (elderly and immunocompromised individuals) should be vaccinated to prevent this disease
Paramyxovirus	Medium to large size, enveloped virus	
Measles virus		Replication of this virus causes measles, which is characterized by a red skin rash, high fever, cough, and malaise
Mumps virus		Replication of the virus causes mumps, which is characterized by swelling of one or both salivary glands
Rhabdovirus	Medium-size, enveloped, bullet-shaped virus	
Rabies virus		Replication of the virus within the nervous system causes rabies, which is initially characterized by sensitivity to stimuli such as light and noise, difficulty in swallowing, a fear of water (hydrophobia), followed by delirium, coma, and death; the disease is usually transmitted via the bites of infected animals or in some cases by inhalation and is prevented by vaccination
DOUBLE-STRANDED RNA VIRUSES		
Reovirus Rotavirus	Small, naked virus	Replication of the virus within the gastrointestinal tract causes acute infantile gastroenteritis characterized by severe diarrhea that is a common disease in children
Reovirus		Replication of the virus in humans does not cause a recognized disease

*Yellow fever virus is a flavivirus, a separate group of viruses (flavi means yellow).

GROUP AND TYPE OF VIRUS	DESCRIPTION OF VIRUS	DESCRIPTION OF DISEASE
TABLE 9-2—cont'd		
Some Examples of Human Viruses and the Diseases They Cause		
DOUBLE-STRANDED RNA VIRUSES CONTAINING REVERSE TRANSCRIPTION		
Retroviruses	Medium-size, enveloped virus	
Human immunodeficiency virus (HIV)		Replication of the virus within T lymphocytes of the immune defense system causes acquired immunodeficiency syndrome (AIDS), which is characterized by a loss of immune defenses against infections, leading to opportunistic infections by various pathogens and eventual death
Human T cell leukemia virus (HTLV-I and HTLV-II)		Replication of the virus within T lymphocytes causes leukemia, which is characterized by the malignant growth of infected cells

Some animal viruses are uncoated outside of the host cell so that only the viral nucleic acid enters the host cell. This occurs in enteroviruses, such as polioviruses, which are single-stranded RNA viruses. Most animal viruses, however, enter the host cell by endocytosis and uncoating occurs within the host cell (FIG. 9-11). The plasma membrane of the host cell surrounds the adsorbed virus, forming a membrane-bound vesicle. This vesicle is released within the cell. In some cases the vesicle containing the virus fuses

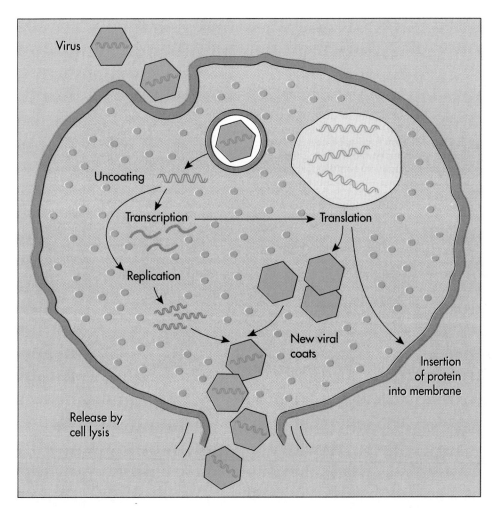

FIG. 9-11 Viruses can enter host cells in several ways. Some animal and plant viruses enter the host cell via endocytosis and subsequently are released into the cytoplasm.

with a lysosome. A lysosome is an organelle that contains digestive enzymes. Lysosomal enzymes degrade the capsid, releasing the viral nucleic acid. In other cases, such as for herpesvirus, the virus is released within the cytoplasm where uncoating occurs when enzymes attack the capsid.

Some animal viruses contain enzymes as well as nucleic acid within the capsid. These viral enzymes, which are released during uncoating, can initiate the synthesis of viral nucleic acid and proteins. The subsequent process of protein synthesis and nucleic acid replication varies greatly, depending on the nature of the nucleic acid acting as the hereditary macromolecule. Some animal viruses are single-stranded DNA, others double-stranded DNA, others single-stranded RNA, and yet others double-stranded RNA. Animal viruses are further distinguished based on the relationship between the nucleic acid of the virus and the viral mRNA (FIG. 9-12). A viral mRNA is designated as a plus strand and its complementary sequence that cannot be used as an mRNA is called a minus strand. Similarly, a DNA strand complementary to a viral mRNA is designated as a minus strand. Thus it is the minus strand of DNA or RNA that serves as the template for the formation of viral mRNA.

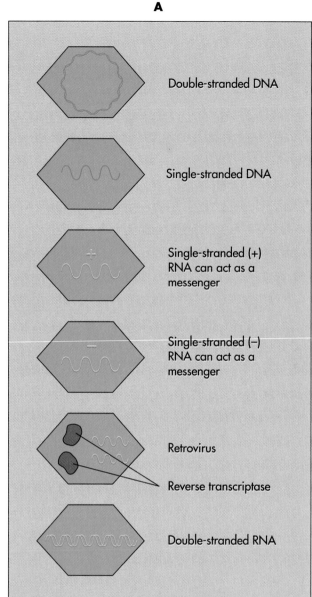

A

Double-stranded DNA

Single-stranded DNA

Single-stranded (+) RNA can act as a messenger

Single-stranded (−) RNA can act as a messenger

Retrovirus

Reverse transcriptase

Double-stranded RNA

FIG. 9-12 **A,** Viruses have a central nucleic acid core, which may be RNA *(gold)* or DNA *(green)* surrounded by a protein coat called a capsid *(blue).* **B,** Some animal viruses, such as adenovirus, have isometric symmetry and a DNA genome. **C,** Other viruses, such as coronavirus, have complex capsids and an envelope with protruding proteins surrounding an RNA genome.

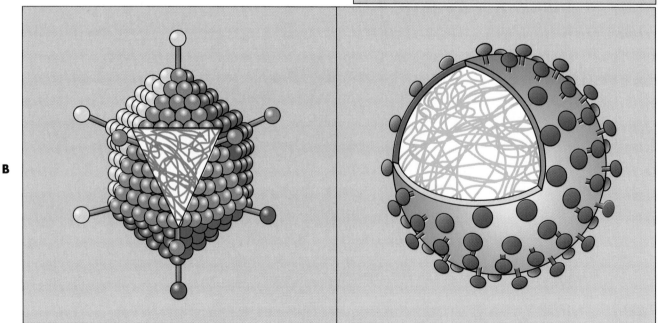

B

C

METHODOLOGY

LABORATORY CULTURE OF ANIMAL VIRUSES

Because viruses only replicate within host cells, the laboratory culture of animal viruses requires that living animal cells be used. In some cases, whole live animals such as mice, rabbits, guinea pigs, and monkeys are used. Some animal viruses can be cultured only in such living animals.

In other cases, animal viruses are cultured in embryonated eggs (fertilized eggs in which an embryo is developing) (FIG. *A*). Most often, embryonated chicken and duck eggs are used. A hole is made through the egg shell and a suspension containing the virions is injected. The cultured viruses can later be harvested and used for

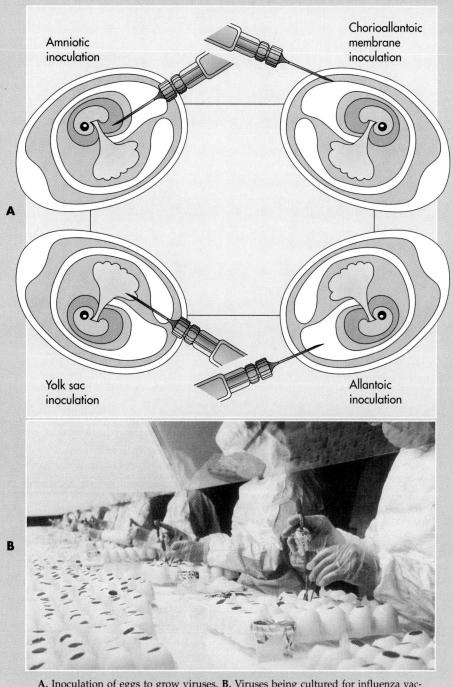

A, Inoculation of eggs to grow viruses. **B,** Viruses being cultured for influenza vaccine production.

Continued.

LABORATORY CULTURE OF ANIMAL VIRUSES—CONT'D

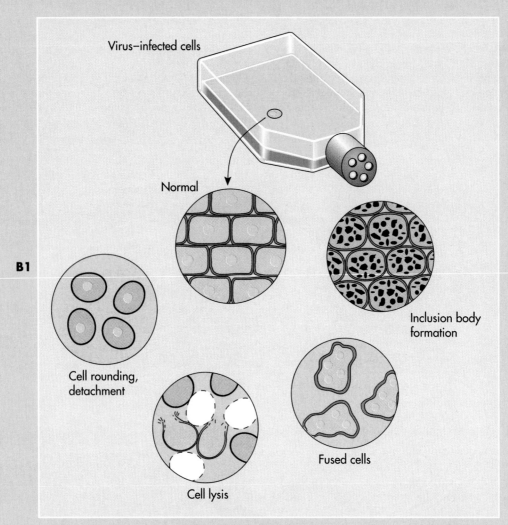

Virus–infected cells

Normal

Inclusion body formation

Cell rounding, detachment

Fused cells

Cell lysis

B1, Infections of animal cells can result in various abnormalities known as cytopathic effects.

scientific studies or for other uses, such as vaccine production.

It is possible to grow animal cells in a fluid broth, called tissue cultures. Viruses can then replicate within the laboratory grown animal cells. Cultivation of viruses in tissue cultures has largely replaced culture in embryonated eggs. For example, rabies vaccine, which used to be made by culturing rabies viruses in embryonated duck eggs, is now made by cultivating rabies viruses in tissue cultures of human fibroblast cells.

Cultivation of animal viruses in tissue culture can be used to determine the numbers of viruses in a suspension, in a method analogous to the plaque assay for enumeration of bacteriophage (FIG. *B*). In a typical procedure, a tissue culture monolayer of animal cells (single layer of animal cells) growing on a plate surface is inoculated with dilutions of a viral suspension and incubated for various periods of time. Viral infection of the animal tissue culture cells may result in plaque formation, indicative of localized death of animal cells, which can be observed microscopically, or, more commonly, with the naked eye. The number of plaques that form and the dilution factors are employed to determine the concentration of viruses in the sample.

Additionally, virus-infected animal cells often develop abnormally. Such abnormalities are visible as a change in appearance, known as the cytopathic effect (CPE). For example, inclusion bodies may form within a cell infected with a virus, the size or shape of the nucleus may change in a cell infected with a virus, or other visible changes may occur. It is possible to observe CPE in animal cell cultures and to determine the number of viruses by counting the number of cells exhibiting the characteristic morphological changes.

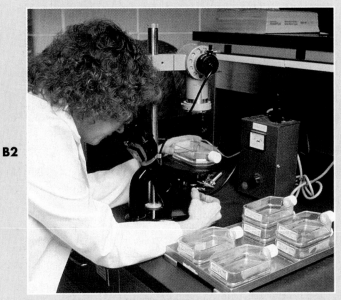

B2, Viruses being grown in tissue culture and observed with an inverted microscope to detect cytopathic effects (CPEs) in the cell culture.

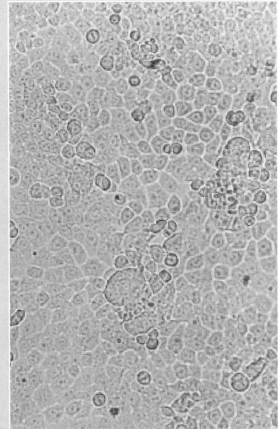

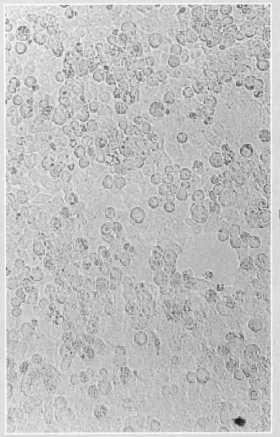

B3, Light micrograph showing the cytopathic effect on HEp-2 cells grown in tissue culture by an infection with adenovirus.

B4, Light micrograph showing the cytopathic effect on HEp-2 cells grown in tissue culture by an infection with respiratory syncytial virus.

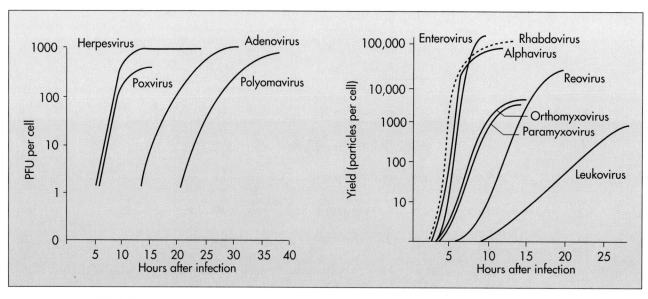

FIG. 9-13 Replication of animal viruses often takes hours, as shown as the time needed to increase the plaque forming units (PFU), or yield, of viral particles.

The DNA of animal viruses generally enters the nucleus. There it is replicated, whereas RNA animal viruses need only enter the cytoplasm of the animal cell to be replicated. In some cases, viral assembly occurs within the cytoplasm, whereas other animal viruses are formed within the nucleus of the host cell.

Once assembled, the virions can be released in one of two ways. Some viruses, such as the rhinoviruses that cause the common cold, are released by lysis and hence death of the host cell. Such viruses exhibit a replication cycle, which closely resembles that of lytic bacteriophage. In such instances there is a step-wise growth curve, with a burst of a large number of viruses released simultaneously. Unlike bacteriophage, however, the single-step growth curve for animal viruses occurs in hours rather than minutes (FIG. 9-13).

Other viruses are released gradually from living host cells by a process called **budding** (FIG. 9-14). Viruses released in this manner include the AIDS-causing human immunodeficiency virus (HIV) and the herpesviruses. Budding is a form of exocytosis in

NEWSBREAK

CONTROVERSY OVER PLANS TO DESTROY REMAINING SMALLPOX VIRUS

Although the disease smallpox has been eliminated through an extensive worldwide immunization program, a few stock cultures of the smallpox virus are maintained in the United States and Russia for scientific studies. A bilateral agreement to destroy these last remaining smallpox viruses, after their DNA sequences have been determined, has sparked controversy within the scientific community. Some scientists feel that biodiversity must be preserved and the stocks of smallpox virus should be maintained. These scientists point to the fact that the stock cultures were deposited in culture collections with the expectation that they would be maintained forever. They argue that as long as such cultures remain in secure collections there is no public health danger. Other scientists feel that smallpox viruses should be further studied as a model of how viruses cause disease. These scientists argue that alternate animal hosts should be developed to permit such studies, even though it raises the risk of reestablishing smallpox as an infectious disease. Public health officials argue that scientific curiosity doesn't warrant the health risk. They point out that the last case of smallpox resulted from an accidental release of the smallpox virus from a laboratory in England where the virus was being studied. The American Society for Microbiology supports destruction of the smallpox virus, to be considered as a special case in the interest of protecting human health. A panel of scientists is expected to meet to read a final decision on the fate of the last remaining smallpox viruses.

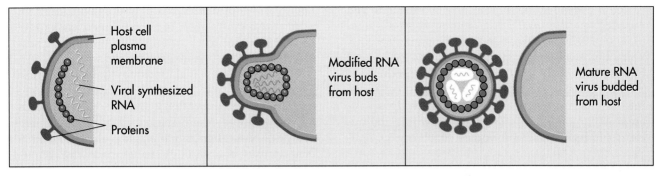

FIG. 9-14   Some enveloped viruses are released from the host cell by budding.

which the plasma membrane of the host cell engulfs an assembled virus, forming a vesicle that is transported outside of the cell. Thus the viruses released are enclosed within a membrane that comes from the host cell and are said to be enveloped. Budding release generally results in a protracted infection in which the host cell may be debilitated.

Budding releases enveloped viruses without killing the host cell.

REPLICATION OF DNA ANIMAL VIRUSES

Double-stranded DNA Viruses

The replication of **adenoviruses,** which often causes respiratory tract infections, is representative of the replication of the double-stranded DNA animal viruses. Following penetration of the virus, the host cell continues its normal metabolic activities for a short period of time (FIG. 9-15). It may take several hours before uncoating of the virus occurs. The uncoated viral nucleic acid genome enters the nucleus

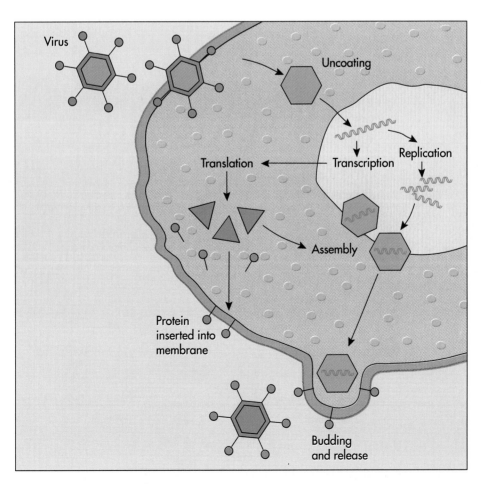

FIG. 9-15   The replication cycle of adenoviruses.

where it codes for the inhibition of normal host cell metabolism. The viral DNA genome acts as a template for its own replication, directing enzymes normally involved in cellular metabolism to make viral nucleic acids and proteins. Viral proteins produced at the ribosomes in the cytoplasm of the host cell move into the nucleus where assembly of adenoviruses occurs. Accumulation of adenoviruses within the nucleus produces inclusion bodies consisting of crystalline arrays of densely packed adenoviruses (FIG. 9-16). The accumulation of adenoviruses leads to lysis and death of the host cell. With lysis of the host cell, numerous adenoviruses are released.

Herpesviruses, which include those that cause genital herpes and infectious mononucleosis, are similarly assembled within the nuclei of the host cells in which they replicate. These viruses acquire an envelope of host cell membrane lipids when the assembled viruses bud through the nuclear membrane. Thus herpesviruses are enveloped viruses that are surrounded by a portion of host cell nuclear membrane. Enveloped herpesviruses are released slowly from the host cell by exocytosis. Once a herpesvirus infection occurs, the host cells remain infected and a latent herpesviral infection persists for the entire life of the host. During latency, the viral nucleic acid persists within the neurons of the nerve ganglia. There is no release of viruses during this period.

Poxviruses, such as the smallpox virus, use a somewhat different replication strategy (FIG. 9-17). Poxviruses are large double-stranded DNA viruses that carry their own RNA polymerase protein, which they use to make mRNAs. These viruses are able to replicate within the cytoplasm of an infected host cell. They are released when there are sufficient assembled viruses to initiate host cell lyses. These released viruses are enveloped; they acquire the envelope from the nuclear membrane.

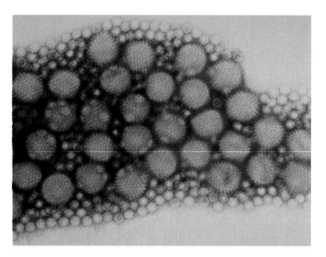

FIG. 9-16 Colorized micrograph of densely packed adenoviruses. (120,000 ×).

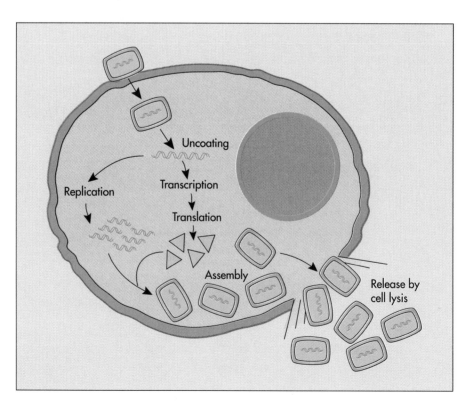

FIG. 9-17 Poxviruses are DNA viruses that replicate within the cytoplasm of a host cell.

Single-stranded DNA Viruses

Parvoviruses contain a single strand of DNA. Many pet owners worry about parvovirus infections that can cause lethal infections of dogs. Some populations of parvoviruses contain only a minus strand or a plus strand of DNA, whereas other populations have roughly equal numbers of viruses with plus and minus strands of DNA. Within a host cell the single strand of DNA of a parvovirus is copied to make a double-stranded DNA macromolecule. Replication of this double-stranded DNA can produce plus and minus strands of DNA for incorporation into the virions.

REPLICATION OF RNA ANIMAL VIRUSES

Double-stranded RNA Viruses

Reoviruses (respiratory enteric orphan viruses) are double-stranded RNA viruses that carry an RNA polymerase that is used for the synthesis of new viral RNA molecules. Reoviruses, such as rotavirus that causes diarrhea in children, contain several different double-stranded RNA molecules. Each of the RNA molecules codes for the production of a different protein. The proteins are assembled into the viral capsid and the viral RNA produced during replication is in-serted into the capsid before release of the completed reoviruses (FIG. 9-18).

Single-stranded RNA Viruses

The single-stranded RNA **picornaviruses,** such as poliovirus, are among the smallest viruses (pico means small). Because poliovirus is a plus strand virus, the RNA genome can function as an mRNA. The poliovirus RNA is translated into proteins that inhibit the host cell's synthesis of RNA and protein. The RNA genome also codes for the production of a large polypeptide that is subsequently cleaved to form several different proteins, including an RNA polymerase and the proteins used to make the viral capsid (FIG. 9-19). The RNA polymerase is used to produce a complementary **replicative minus RNA strand,** which then can serve as a template for the synthesis of new plus strand viral RNA genomes. The assembly of the capsid and insertion of the RNA is followed by the release of a large number of viral particles. Release of the poliovirus occurs because blockage of cellular protein synthesis by the poliovirus leads to breakdown of lysosomes. Lysosomal digestive enzymes lyse the host cell.

In the case of **influenza viruses,** the individual single strands of viral RNA are minus strands that

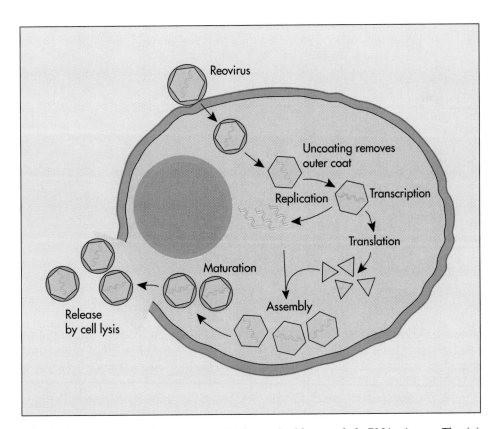

FIG. 9-18   Replication of reoviruses, which are double-stranded RNA viruses. The (+) RNA molecules are placed into capsids, and integral capsid-RNA replicase uses this as a template to form the (−) strand of the double-stranded RNA genome.

HOW NEW VIRAL DISEASES ORIGINATE

Periodically, new infectious viral diseases emerge, challenging scientists and physicians to find their origins. In some cases, these diseases may have existed for some time but have gone undetected. In other cases, genetic changes due to mutation or recombination may have led to new strains with altered virulence. Periodic outbreaks of influenza are due to such genetic changes that result in the evolution of new strains of influenza viruses. Other theories are needed to explain the emergence of seemingly new pathogenic viruses, such as the

human immunodeficiency virus that causes AIDS in the late 1970s and the hantavirus that killed over 20 people in the summer of 1993 (mostly Navajos) in the southwest United States.

In recent years, scientists have found evidence that changing environments is a major cause of emerging infectious diseases. Construction of roadways through jungles and rain forests may allow pathogens to spread rapidly to huge numbers of people. One of the most dramatic indications that humans were disrupting the bal-

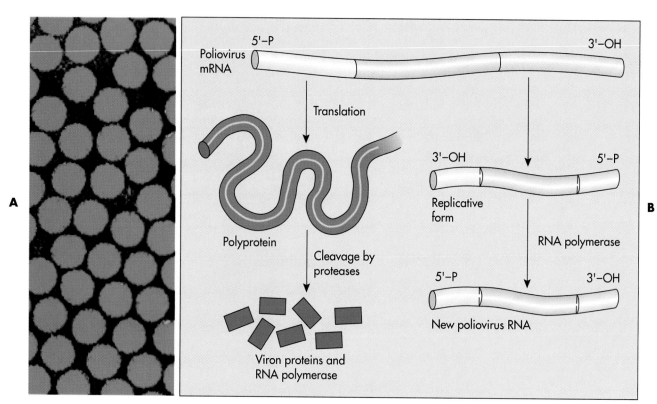

A

B

FIG. 9-19   **A,** Colorized electron micrograph of polioviruses. **B,** Poliovirus RNA serves as a messenger for production of a polyprotein that is then cleaved to form capsid proteins and RNA polymerase. The RNA also serves as the template for producing a replicative form that, in turn, is the template for new poliovirus genomic RNA.

serve as templates for the transcription of mRNAs. Since it is a minus strand it does not directly act as the mRNA. An influenza virus has eight different RNA molecules, each of which codes for a different mRNA. One of these RNA molecules codes for the RNA polymerase required for the production of mRNA. Replication of the minus strand of viral RNA involves the production of a complementary RNA strand that then serves as a template for the synthe-

sis of new minus RNA. Release of influenza virions occurs by budding and the virions are thus enveloped in host cell plasma membrane. One of the proteins on the surface of the virus, neuramidase (N), protrudes through the surrounding membrane that becomes the viral envelope and facilitates the release of the influenza virus. Another protein, hemagglutinin (H), also projects from the surface of an influenza virus through the envelope. Hemagglutinin

ance between pathogens and humans came when Brazil built a highway deep in Amazonian jungle to its new capital, Brasilia. Soon after construction of the highway in the 1950s, viruses, some of which were unknown, were found in the blood of highway workers. One of these viruses, the Oropouche virus, also was found in the blood of a sloth dead at the side of the highway. Oropouche virus was not known to be responsible for epidemics in humans or animals before 1960. In 1961, the Oropouche virus was identified as the cause of a flu-like epidemic in Brazil that afflicted 11,000 people.

While it was clear that Oropouche was to blame for the epidemic, it was not clear how a virus never seen in human beings before had emerged to cause a new dis-

ease. Finding the answer took scientists almost two decades. In 1980, the Oropouche virus was isolated from biting midges. During construction of the highway through the Amazonian jungle, the midges had undergone a population explosion. This led to a huge increase in vectors carrying the Oropouche virus.

Similar environmental changes may underlie the emergence of the new viruses that cause AIDS, Ebola hemorrhagic fever, Marburg hemorrhagic fever, and yellow fever—where the viruses probably initially occurred in monkeys; Rift Valley fever—where the viruses probably initially occurred in cattle, sheep, and mosquitoes; and Hantaan virus—where the viruses probably initially occurred in rodents.

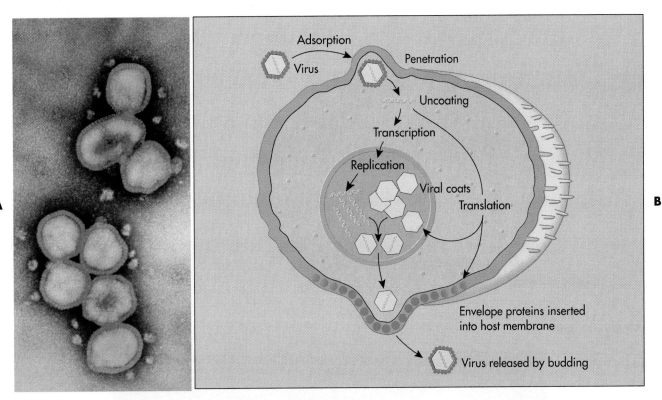

FIG. 9-20   **A**, Colorized electron micrograph of influenza viruses (72,000 ×). **B**, Sequencce of events during influenza virus replication.

is important in the ability of the released viruses to adsorb and enter into new host cells (FIG. 9-20). There are several chemically different H and N proteins and different influenza viruses are designated by these proteins, for example, as H_2N_3. Each year as the flu season approaches, the Centers for Disease Control release information on the strains of influenza viruses that are anticipated to cause major outbreaks of influenza. These strains are designated by the H and N proteins and these designations are generally reported by the local news media.

Retroviruses

Retroviruses are RNA viruses that use **reverse transcriptase** to produce a DNA molecule within the host cell (FIG. 9-21). Reverse transcriptase is an RNA-directed DNA polymerase. The process of making DNA using an RNA template is called **reverse transcription.** The retroviruses use their RNA as a template for producing a complementary DNA that is then integrated into the host genome. The information in the DNA molecule is used to direct the synthesis of RNA, which is accomplished by transcrip-

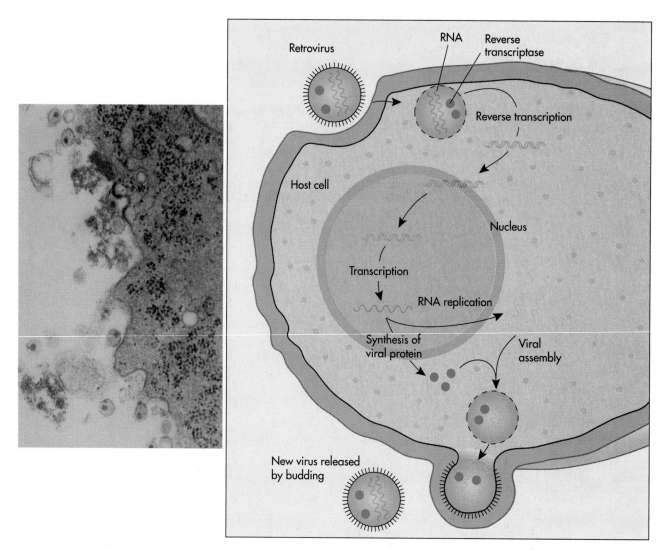

FIG. 9-21 Retroviruses replicate using reverse transcriptase to form DNA that is used to produce viral proteins and RNA genomes for viral progeny. The electron micrograph (*left*) shows the viruses (*blue*) at the surface of a cell.

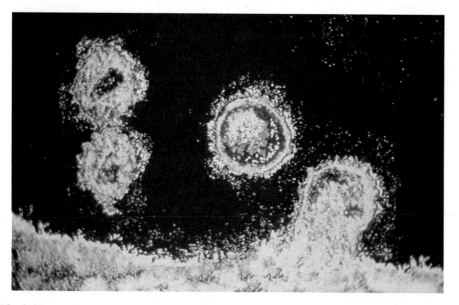

FIG. 9-22 Colorized electron micrograph of the human immunodeficiency virus (HIV) that causes AIDS budding from the host cell that produced it.

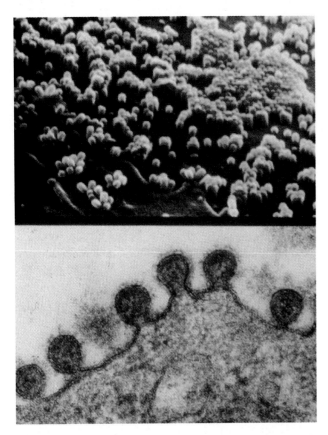

FIG. 9-23 A, Colorized scanning electron micrograph showing budding release of human immunodeficiency virus. **B,** Colorized transmission electron micrograph showing budding release of human immunodeficiency virus.

tion. Some of the RNA acts as mRNA for synthesizing viral proteins, and some of the RNA is put into the RNA of the viral progeny.

Retroviruses carry out reverse transcription, using reverse transcriptase to make DNA from an RNA template.

Retroviruses are released from host cells by budding. This replication does not result in cell lysis and immediate cell death, therefore these viruses can be released slowly and continuously from infected host cells. Eventually, however, the siphoning of cellular resources for viral replication can cause the death of the host cell. The AIDS-causing human immunodeficiency virus (HIV) is a retrovirus that replicates in this manner (FIG. 9-22). Infections with HIV are persistent because of the budding release of assembled viruses from the host cells (FIG. 9-23). The replication of HIV within T lymphocytes, which are essential cellular components of the body's immune defense system, causes a great decrease in the ability of the body to defend itself against microbial infections. Individuals with AIDS thus are susceptible to various opportunistic infections and eventually succumb to the onslaught of such infections.

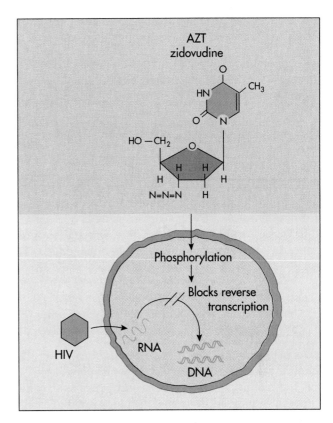

FIG. 9-24 Illustration of action of AZT in blocking HIV replication.

The progress of AIDS can be slowed by using the drug azidothymidine (AZT), also called zidovudine. AZT is incorporated instead of thymidine nucleotides when DNA is made by reverse transcription (FIG. 9-24). The presence of AZT can block replication of HIV and therefore is useful in treating individuals with HIV infections. It does not, however, eliminate all infected cells and is not a cure for AIDS.

TRANSFORMATION OF ANIMAL CELLS

In some cases, infection with an animal virus does not lead to viral replication. Rather, the viral DNA is incorporated into the chromosomal DNA of the host cell. The DNA produced by reverse transcription during the replication of retroviruses, as well as the DNA of some other viruses such as herpes and papillomaviruses, can be incorporated into the host cell's chromosomes. This situation is analogous to that of temperate phage, where the phage DNA is incorporated into the DNA of a host bacterial cell as a prophage. In animal cells the integrated viral DNA is called a **provirus.** Like the prophage, such incorporated viral DNA can be passed from one generation of animal cells to another. It is therefore possible for animals to inherit viral genes. Within the chromosomes of the host cell the viral DNA can be transcribed, resulting in the production of viral-specific

RNA and viral proteins. Animal cells carrying viral DNA may have different properties than uninfected cells.

A provirus is viral DNA that is incorporated into the chromosomal DNA of the host cell.

In particular, the presence of viral-derived DNA within the host cell can transform the animal cell into a malignant (cancerous) cell. Such transformed cells have altered surface properties and continue to grow even when they contact a neighboring cell. This results in the formation of a tumor. Viruses that transform cells and cause cancerous growth are called **oncogenic viruses.** The genes that actually induce cancerous transformations are called **oncogenes.**

Oncogenic viruses are viruses that transform cells so as to cause cancerous growth.

Oncogenes are genes that induce such cancerous transformations.

Viral oncogenes are very similar in nucleotide sequence to oncogenes that occur naturally in animal cells. Animal cells have proto-oncogenes that code for genes involved in normal cellular growth. If activated, these proto-oncogenes act to cause transformation of the cell so that it begins to grow malignantly and rapidly and no longer stops growing when it contacts a neighboring cell (loss of contact inhibition). A viral oncogene that becomes integrated into the DNA of an animal host cell causes transformation only when it is expressed. If a viral oncogene is inserted next to an active promoter (site where RNA synthesis begins), a high level of transcription of the viral oncogene ensues and mRNA is produced. Translation of these mRNAs forms proteins that appear to be involved in transforming a normal cell into a malignant one.

Many oncogene-coded proteins are kinases, which are enzymes that are involved in the transfer of phosphate from ATP to various organic compounds. One normal function of a kinase is to add phosphate to either the amino acid serine or threonine. In normal (noncancerous) cells, phosphate is not added to tyrosine, but in transformed (cancerous) cells, phosphate is added to tyrosine by a kinase. The presence of phosphorylated tyrosine within a protein alters the function of that protein. Apparently the normal regulatory functions of the cell are thus altered. These proteins often become concentrated in the cell nucleus, suggesting that they act as regulator proteins to alter gene expression. Some proteins coded for by oncogenes are known to enhance transcription of a number of genes. Such genes may be central in cell growth and division.

The activation of multiple oncogenes may be necessary to actually cause cancer. In many cases such

activation does not involve viruses. Such activity can be the result of exposure to mutagenic agents that alter gene regulation so that the oncogenes are expressed. **Carcinogens** are agents that cause gene mutations that lead to cancer. Known carcinogens include numerous natural and synthetic compounds such as asbestos, benzene, substances in cigarette smoke, X-rays, gamma rays, and ultraviolet radiation.

Carcinogens are natural or synthetic compounds that cause genetic mutations that lead to cancer.

Additionally, various viral infections may lead to specific forms of cancer. Papillomaviruses, which cause genital warts, frequently are found in cancerous cervical cells. This suggests that these viruses have oncogenes that can cause human cervical cancer. The Epstein-Barr virus, a herpesvirus that causes infectious mononucleosis, appears to be the cause of Burkitt's lymphoma and nasopharyngeal carcinoma, a cancer of the nose and throat. Burkitt's lymphoma is a rare cancer of the lymphatic system that mostly affects children in Africa. Among the RNA viruses, only the retroviruses form DNA during their replication that can be integrated into the host cell's chromosomes. Hence, it is not surprising that the only RNA viruses that are oncogenic are retroviruses. Human T cell leukemia viruses (HTLV 1 and HTLV 2) are examples of such retroviruses that have been shown to cause some types of human leukemia. Leukemias are cancers affecting the white blood cells, and the human T cell leukemia viruses specifically cause transformation of one type of white blood cell: the T cells that are involved in protecting the body against infections.

General Aspects of Viral Replication (pp. 257-258)

- Viruses are composed of a nucleic acid, either RNA or DNA, and a capsid made of protein that surrounds the nucleic acid.
- Within host cells, viral nucleic acid directs the formation of new viruses, using the host cell's ribosomes for producing viral proteins and ATP for carrying out synthesis of viruses.
- A high degree of specificity is required between a virus and the host cell within which it can replicate.

Stages of Viral Replication (p. 257)

- The stages of viral replication generally include adsorption of the virus to specific binding sites on the host cell's surface, penetration of the cell's cytoplasmic membrane by viral nucleic acid, control of the cell's metabolic activities by the viral nucleic acid, use of the host's biochemical components and anatomical structures for production of viral replicates, and release of new viruses.

Replication of Bacteriophages (pp. 258-263)

Lytic Phage Replication (pp. 258-261)

- Bacteriophages are viruses that replicate only within specific host bacterial cells. They are called lytic phages if their replication results in the lysis of the host bacterial cell when the new phages are released.

T-even Phage Replication (pp. 259-260)

- T-even phages are double-stranded DNA viruses whose capsids have distinct head and tail structures. The tail attaches to specific receptor sites on the host and releases lysozyme that breaks down the cell wall, allowing the tail to penetrate. Contraction of the tail forces the phage DNA from the head into the periplasmic space. Normal host cell biosynthesis ceases as the synthesis of proteins for new phage nucleic acids and capsids is directed by the viral nucleic acid. The head and tail units are assembled separately and later combined. Phage DNA is packaged in the head structure. The phage codes for lysozyme, which helps break down the bacterial peptidoglycan wall structure. Lysis of the cell wall releases new phages.

Replication of RNA Bacteriophage (pp. 260-261)

- In RNA phages, a plus RNA strand contains the viral hereditary information and is used as a template for a double-stranded RNA replicate that has both plus and complementary minus RNA strands during its lytic replication cycle. The new plus RNA serves as the hereditary molecule for incorporation into new viruses.

Bacteriophage Growth Curve (pp. 261-262)

- Lytic replication cycles have a one-step growth curve that begins with an eclipse period, which is part of the latent period, the time between the entry of the viral nucleic acid and the release of viral progeny. New phages are released simultaneously when their number reaches the burst size and the cell lyses.

Temperate Phages—Lysogeny (pp. 262-263)

- Temperate phages establish a state of lysogeny in which a portion of the viral nucleic acid is incorporated into a bacterial chromosome or plasmid. The incorporated prophage is replicated with the host cell DNA and passed to later generations. The phage genes usually are not expressed, but when they are, they can greatly alter the properties of the host cell.
- Specialized transduction is the phage-mediated transfer of bacterial DNA. Specific genes located adjacent to the specific site where prophage DNA is incorporated into the host cell can be transferred to a recipient cell.

Replication of Plant Viruses (pp. 264-266)

- Replication of the viruses that infect plant cells is similar to the lytic replication cycle of bacteriophages. It begins with the adsorption of the virus onto a susceptible plant cell; next the viral nucleic acid penetrates into the plant cell; then the viral DNA assumes control of the host cell's biosynthesis activities; the viral nucleic acid codes for synthesis of viral nucleic acid and capsid components; and the viral particles are assembled and released by lysis of the host cell.

Replication of Animal Viruses (pp. 266-282)

- The steps in the replication of animal viruses are: attachment (adsorption of the virus to the surface of the animal cell), penetration (entry of the intact virus or viral nucleic acid into the host cell), uncoating (release of viral nucleic acid from the capsid), transcription (formation of viral mRNA), translation of early proteins (using viral-coded mRNA), replication of viral nucleic acid, translation of late proteins, assembly of complete viral particles, and release of new viruses.
- Newly formed animal viruses can be released by lysis or budding.

Replication of DNA Animal Viruses (pp. 275-277)

Double-stranded DNA Viruses (pp. 275-276)

- Uncoating of a double-stranded DNA virus takes several hours, during which the host cell continues its normal metabolic activities. Once within the nucleus the viral DNA codes for the inhibition of normal host biosynthesis. The viral DNA acts as a template for its own replication. Viral mRNA produced by transcription is translated at host cell ribosomes, producing proteins for assembly of the viral capsid. The viruses are assembled in the nucleus and released by lysis.

Single-stranded DNA Viruses (p. 277)

- Single-stranded DNA viruses copy the single strand within a host cell to make a double-stranded DNA macromolecule. mRNA is produced by transcription to code for the production of capsid proteins. Single-stranded DNA is also produced for viral genomes.

Replication of RNA Animal Viruses (pp. 277-281)

Double-stranded RNA Viruses (p. 277)

- Some double-stranded RNA viruses carry RNA polymerase to synthesize new viral nucleic acid molecules.

Single-stranded RNA Viruses (pp. 277-279)

- In the single-stranded RNA polioviruses, the viral RNA acts as an mRNA and codes for the production

of a single large polypeptide when it enters the host cell. The polypeptide is cleaved, forming RNA polymerase and the proteins used to make the viral capsid. The RNA polymerase is used to produce a complementary replicative minus RNA strand that acts as a template for the synthesis of new plus strand viral RNA.
- In some single-stranded RNA viruses, the RNA within the virion can act as mRNA and therefore is a plus strand. In others, the single strand of viral RNA is a minus strand that serves as a template for the transcription of mRNAs rather than acting as the mRNA.

Retroviruses (pp. 279-281)
- Reverse transcription is the process of making DNA using an RNA template. Single-stranded RNA retroviruses use reverse transcriptase to do this.

Transformation of Animal Cells (pp. 281-282)
- Under certain conditions, animal viral DNA is incorporated as a provirus into the chromosomal DNA of the host cell and is passed on with animal cell reproduction. A provirus can transform the animal cell into a malignant cell and cause the formation of a tumor.
- Viruses that transform cells and cause cancerous growth are called oncogenic viruses. Oncogenes are genes that induce these cancerous transformations. Carcinogens are agents that cause gene mutations that lead to cancer.
- An integrated viral oncogene only causes host cell transformation when it is expressed. Many oncogenes code for the production of kinases, which in transformed cells add phosphate to the amino acid tyrosine rather than serine or threonine. Phosphorylated tyrosine alters the regulatory functions of the cell.

CHAPTER REVIEW

REVIEW QUESTIONS

1. Describe a typical virus.
2. Describe the similarities and differences between the replication of bacteriophage, animal viruses, and plant viruses.
3. Describe the steps in the lytic replication of a T-even bacteriophage.
4. What are the differences between the replication of DNA and RNA animal viruses?
5. What is lysogeny?
6. What is a transformed cell?
7. What is an oncogene?

8. What is a retrovirus and how does its replication differ from other types of viruses?
9. How does viral replication differ from the reproduction of cellular organisms?
10. Compare the budding and lytic modes of animal virus replication.
11. What is the burst size?
12. Describe the "growth curve" for a lytic bacteriophage.
13. What is meant by virus host cell specificity?
14. What advantages does a temperate bacteriophage have over a lytic bacteriophage?

CRITICAL THINKING QUESTIONS

1. Why are viruses obligate intracellular parasites? Why do they need host cells? Why do viruses normally replicate only within certain host cells? What would the consequence be if viruses did not need specific receptors to attach to host cells?
2. Since the smallpox virus only replicates within human host cells, why has it been possible to eliminate smallpox? There are some vials containing smallpox viruses stored in the United States and Russia. Some scientists want these stocks of smallpox virus destroyed. Others want them preserved. Should the virus be destroyed or maintained?
3. Some scientists who argue that the smallpox virus should be preserved want to develop an alternate animal host system so they can study how the smallpox virus causes disease. What would be the benefits and risks of developing an alternate host system?

4. One hypothesis about the origins of the human immunodeficiency virus (HIV) that causes AIDS is that it originated from another virus in African green monkeys. How could a virus that originally infected only monkeys have become infectious for humans? Could other viruses evolve the ability to cause human diseases?
5. What is the significance in finding that several viruses have oncogenic potential? How are these viruses related to cancer?
6. Why are infections with viruses that replicate with budding release, such as herpes viruses and HIV, persistent?
7. How does the replication of HIV differ from the replication of the common cold virus? Why is the body able to eliminate the common cold virus but not HIV? Why are there no useful drugs for controlling or eliminating the common cold? Why can physicians use drugs such as AZT to treat, but not to cure, AIDS?

READINGS

Braun MM, WL Heyward, JW Curran: 1990. The global epidemiology of HIV infection and AIDS, *Annual Review of Microbiology* 44:555-77.

An authoritative review of the global spread of HIV and AIDS

Casjens S (ed.): 1985. *Virus Structure and Assembly*, Boston, Jones and Bartlett.

Reviews how viruses are replicated within host cells.

Compans RW and A Helenius: 1988. *Cell Biology of Virus Entry, Replication, and Pathogenesis*, New York, John Wiley and Sons.

Thorough review of viral replication.

Fields BN, DM Knipe, RM Chanock, MS Hirsch, JL Melnick, TP Monath, B Roizman (eds.): 1990. *Fields Virology*, ed. 2, New York, Raven Press.

Each chapter on all aspects of this topic is written by a different expert in the field.

Fraenkel-Conrat J and PC Kimball: 1982. *Virology*, Englewood Cliffs, New Jersey, Prentice-Hall.

A comprehensive text on virology.

Henig RM: 1988. Viruses revisited, *The New York Times Magazine*, November 13:70.

Discusses viral genetics.

Matthews REF: 1991. *Plant Virology*, ed. 3, N Y, Academic Press.

A text on the viruses that infect plants.

Miller JH: 1989. Evolution set fast-forward, *BioScience* 39:512-513.

Discusses the role of viral genetics in evolution.

Prusiner SB: 1984. Prions, *Scientific American* 251(4):50-60.

Explores the nature of the genome of a prion.

Varmus H: 1987. Oncogenes and transcriptional control, *Science* 238:1337-1339.

Discusses the role of genetic transcription in oncogenic viruses on the genetics of cancer.

Varmus H: 1988. Retroviruses, *Science* 240:1427-1435.

Discusses reverse transcription and the replication of retroviruses.

Webster RG, WJ Bean, OT Gorman, TM Chambers, Y Kawaoka: 1992. Evolution and ecology of influenza A viruses, *Microbiological Reviews* 56(1):152-79.

An in-depth look at influenza viruses.

CHAPTER 10

Bacterial Reproduction and Growth of Microorganisms

PREVIEW TO CHAPTER 10

In this chapter we will:
• Study the reproduction of bacteria.
• See that bacterial reproduction results in a characteristic growth curve.
• Learn that a consequence of bacterial reproduction by binary fission is a high reproductive capacity.
• Examine the factors that influence bacterial growth rates.
• Learn the following key terms and names:

acidophiles
alkalophiles
barophiles
barotolerant
batch culture
binary fission
budding
chemostat
colony forming units
  (CFUs)
continuous culture
cysts
death phase
direct counting
  procedures
doubling time
exponential phase
facultative anaerobes

generation time
growth curve
halophiles
lag phase
log phase
mesophiles
microaerophiles
most probable number
  enumeration procedure
obligate aerobes
obligate anaerobes
optimal growth
  temperature
osmophilic
osmotolerant
psychrophiles
salt tolerant
stationary growth phase

BINARY FISSION

Most bacteria reproduce by **binary fission.** Each bacterial cell divides exactly in half to form two equal-size progeny (daughter) cells (FIG. 10-1). Since bacteria typically are single celled, the reproduction of a single cell accomplishes the reproduction of the entire organism. Binary fission is an asexual process—meaning that a single cell divides to form genetically identical progeny and that genetic recombination does not occur in the process.

During the reproduction of a bacterial cell, the parent cell elongates and the cell wall grows inward, dividing the cell in half. This establishes two progeny cells, each surrounded by a cell wall and a plasma membrane. Each of the progeny cells receives a complete set of hereditary information. Replication of the bacterial chromosome is a prerequisite for reproduction of a duplicate bacterial cell.

Binary fission is the most common means of bacterial reproduction.

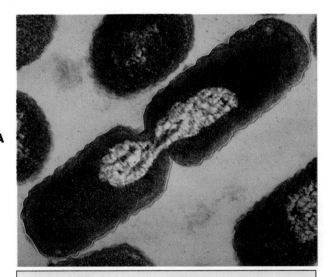

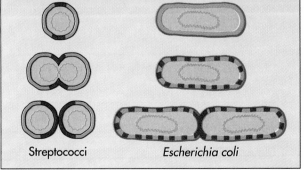

Streptococci · *Escherichia coli*

FIG. 10-1 **A,** Colorized micrograph of *Escherichia coli* dividing by binary fission. **B,** Cell growth occurs at specific sites so that the cell elongates prior to division. The cell wall and plasma membrane are growing inward to separate the cells and the replicated bacterial chromosomes.

During cell division the bacterial chromosome appears to be attached to the plasma membrane and cell wall. Formation of a crosswall or **septum** by the inwardly moving cell wall and plasma membrane physically separates the bacterial chromosomes and distributes them to the two daughter cells. Septum formation pinches off and separates the two complete bacterial chromosomes, providing each progeny cell with a bacterial chromosome (genome) containing a complete set of genetic information. This process requires active protein synthesis to move the bacterial chromosomes to the proper positions. On completion of the crosswall there are two equal-size cells that can separate. Repeating the process results in the multiplication of the bacterial population.

ALTERNATE MEANS OF BACTERIAL REPRODUCTION

Binary fission is the most common mode of bacterial reproduction. Some bacteria multiply by other means. The various modes of replication differ in how the cellular material is apportioned between the daughter cells and whether the cells separate or remain together as part of a multicellular aggregation. For example, bacteria in the genus *Hyphomicrobium* attach to solid surfaces in fresh and saltwater environments and reproduce by budding. **Budding** is a type of division characterized by an unequal division of cellular material. Similarly, *Caulobacter* cell division is unequal; cell division is by elongation of a stalked cell, followed by fission (FIG. 10-2). The daughter cell develops when a crosswall forms, segregating a small portion of the cytoplasm containing

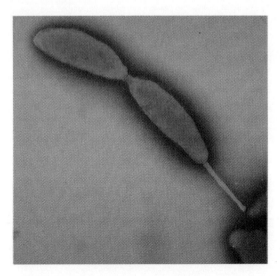

FIG. 10-2 Colorized electron micrograph of *Caulobacter crescentus* showing mother cell with stalk *(pink)* and daughter cell forming by fission.

a duplicate genome. In the case of the actinomycetes, such as *Streptomyces,* reproduction involves the formation of hyphae. In this mode of reproduction the cell elongates, forming a relatively long and generally branched filament, or **hypha**. Regardless of the mode of reproduction, bacterial multiplication requires replication of the bacterial chromosome and synthesis of new boundary layers, including cell wall and plasma membrane structures. All these modes of reproduction are asexual, like binary fission.

Bacterial Spore Formation

Spores are specialized cells produced by some bacteria that are involved in survival or reproduction. The production of spores represents an interesting deviation from vegetative cell reproduction. Some types of spores, including **endospores** (heat resistant spores formed within the cell) and **cysts** (resting or dormant cells sometimes enclosed in a sheath) are not repro-

ductive structures and their production does not increase the number of living cells. In contrast, **arthrospores** (spores formed by the fragmentation of hyphae) are produced by different bacteria as part of their reproductive cycles. The fragmentation of hyphae to produce arthrospores forms numerous progeny cells. Additionally, myxobacteria form reproductive structures, called fruiting bodies, within which numerous spores, called **myxospores** (resting cells of the myxobacteria formed within a fruiting body), are formed. Myxospores are the progeny that result from reproduction of myxobacteria; they are able to survive transport through the air and increase the survival capacity of myxobacteria by permitting dissemination to areas with adequate supplies of nutrients to support bacterial growth and reproduction.

> **Spores are specialized resistant resting cells produced by bacteria. Endospores are involved in survival, but other types of spores, such as arthrospores, are involved with reproduction.**

BACTERIAL GROWTH

Generation Time

Bacterial growth is synonymous with bacterial cell reproduction. Growing bacterial cells increase in biomass through their metabolism in which they convert compounds containing carbon, nitrogen, phosphorus, and other elements into the components of the cell. Most then divide into progeny of equal biomass through binary fission. By its very nature, bacterial reproduction by binary fission results in doubling of the

number of viable bacterial cells. Therefore, during active bacterial growth, the size of the microbial population is continuously doubling. Once cell division begins, it proceeds exponentially as long as growth conditions permit. One cell divides to form two, each of these cells divides so that four cells form, and so forth in a geometric progression. The time required to achieve a doubling of the population size, known as the **generation time** or **doubling time,** is the unit of measure of microbial growth rate (FIG. 10-3).

METHODOLOGY

Logarithms

To conveniently represent large numbers, particularly when the numbers may range over many orders of magnitude (multiples of 10)—as is the case for numbers of bacterial cells that may occur as a few or millions of cells—scientists use a mathematical transformation called the logarithmic transformation. The logarithm to the base 10 (log_{10}) of a number is the exponent indicating the power of 10 to which a number must be raised to produce a given number. Thus the $log_{10}10 = 1$ because the exponent to which 10 must be raised to equal 10 is 1 (10^1). Similarly the $log_{10}100 = 2$ because $10^2 = 100$, and thus the exponent of 10 needed to equal 100 = 2. In a similar fashion the $log_{10}1,000 = 3$, $log_{10}10,000 = 4$, $log_{10}100,000 = 5$, $log_{10}1,000,000 = 6$, and so forth. Obviously in plotting the growth of a bacterial cell over the range of generations that takes 1 cell ($log_{10}1 = 0$) to 1 billion cells ($log_{10}1,000,000,000 = 9$), it is far easier to

plot the logarithm of the cell number using a scale of 0 to 9 than it would be to try to plot the cell numbers on an arithmetic scale of 1 to 1,000,000,000 (see Figure).

GENERATION	GENERATION NUMBER	NUMBER OF CELLS	LOGARITHM OF NUMBER OF CELLS
0		1	0
5	2^5	32	1.51
10	2^{10}	1,024	3.01
15	2^{15}	32,768	4.52
20	2^{20}	1,048,576	6.02
25	2^{25}	33,554,432	7.53
30	2^{30}	1,073,741,824	9.03

Comparison of arithmetic number of cells and logarithmic increase in numbers during bacterial growth.

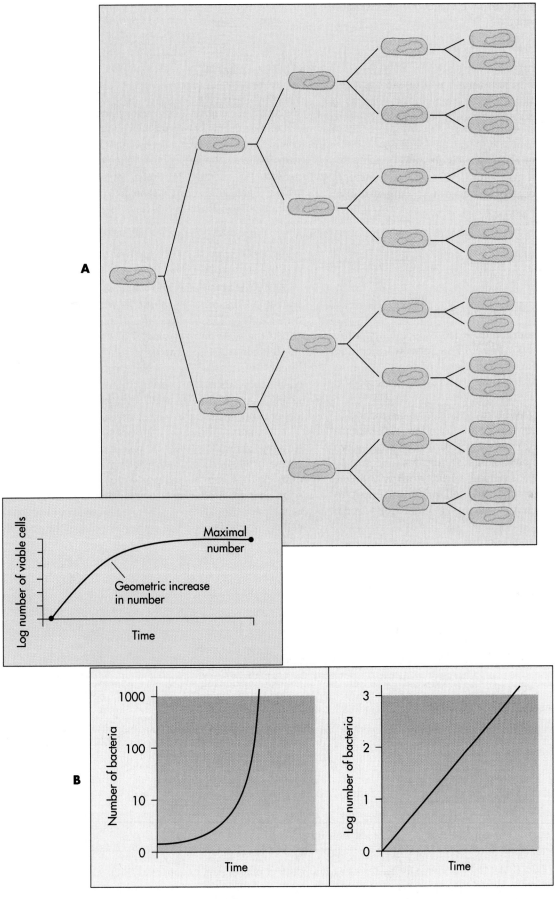

FIG. 10-3 A, During exponential growth the number of cells doubles each generation time. **B,** A graph of the log number of bacteria versus time *(right)* compared with number of bacteria versus time *(left)*.

For a growing bacterial culture, the generation time can be expressed as:

$$g = \frac{t}{n}$$

where g is the generation time, t is time, and n is the number of generations. The number of cells of a growing culture are expressed by the equation:

$$N_t = N_0 \times 2^n$$

where N_0 is the number of cells at time 0, N_t is the number of cells at any time (t) after time 0, and n is the number of generations. Rearranging this formula:

$$n = \frac{\log_{10}N_t - \log_{10} N_0}{\log_{10}2}$$

Since $1/\log_{10}2$ is equal to 3.3, the original equation for the generation time can be written as:

$$g = \frac{t}{3.3 \, (\log_{10} N_t \, 2 \log_{10} N_0)}$$

where g is the generation time, $\log N_t$ is the logarithm to the base 10 of the number of bacteria at time t, $\log N_0$ is the logarithm to the base 10 of the number of bacteria at the starting time, and t is the time period of growth.

By determining cell numbers during the period of active cell division, the generation time can be estimated. A bacterium such as *Escherichia coli* can have a generation time as short as 20 minutes under optimal conditions. Considering a bacterium with a 20-minute generation time, one cell would multiply to 1,000 cells in 3.3 hours and to 1,000,000 cells in 6.6 hours.

Generation time or doubling time is the unit of measure of bacterial growth; it is the time it takes for the size of a bacterial population to double.

Bacterial Growth Curve

When a bacterium is inoculated into a new culture medium, it exhibits a characteristic pattern or change in cell numbers. This pattern is called a **growth curve** (FIG. 10-4). The normal growth curve of bacteria has four phases, the lag phase, the log or exponential growth phase, the stationary phase, and the death phase. During the **lag phase** there is no increase in cell numbers. Rather, the lag phase is a period of adaptation during which bacteria are preparing for reproduction: synthesizing DNA, RNA, other structural macromolecules, and the various enzymes needed for cell division.

After the lag phase, the bacteria begin to multiply by binary fission, doubling in number every time they divide. This is the **log phase** of growth, also called the **exponential phase.** It is so named because the logarithm of the bacterial cell numbers increases linearly with time. During this phase, bacterial reproduction occurs at a maximal rate for the specific set of growth conditions. It is during this period that the generation time of the bacterium is determined.

After some period of exponential growth, the **stationary growth phase** is reached. The stationary phase often occurs when the maximum population density that can be supported by the available resources is reached. Once the stationary growth phase is reached, there is no further net increase in bacterial cell numbers. During the stationary phase the growth

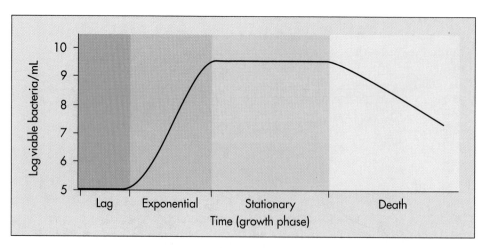

FIG. 10-4 Growth curve for bacteria has four distinct phases: lag, exponential (log), stationary, and death.

rate is exactly equal to the death rate, and cell numbers therefore remain constant. A bacterial population may reach stationary growth when a required nutrient is exhausted, when inhibitory end products accumulate, or when physical conditions do not permit a further increase in population size. The duration of the stationary phase varies, with some bacteria exhibiting a very long stationary phase.

Eventually the number of viable bacterial cells begins to decline. This signals the onset of the **death phase.** During the death phase the number of living bacteria decreases because the rate of cell death exceeds the rate of new cell formation.

> A bacterial growth curve has four phases: (1) lag phase during which bacteria prepare to divide, (2) log or exponential growth phase during which cell numbers increase with regular doublings of viable cells, (3) stationary phase during which cell numbers remain constant, and (4) death phase during which viable cell numbers decline.

BATCH AND CONTINUOUS GROWTH

The normal bacterial growth curve is characteristic of bacteria in **batch culture.** In batch culture, growth occurs in a closed system with fresh sterile medium simply inoculated with a bacterium to which new materials are not added. A flask containing a liquid nutrient medium inoculated with the bacterium *E. coli* is an example of such a batch culture. In batch culture, growth nutrients are expended and metabolic products accumulate in the closed environment. The batch culture models situations such as occur when a canned food product is contaminated with a bacterium.

Bacteria may also be grown in **continuous culture.** In continuous culture, nutrients are supplied and end products continuously removed so that the exponential growth phase is maintained. Because end products do not accumulate and nutrients are not completely expended, the bacteria never reach the stationary phase. A **chemostat** is a continuous culture device in which a liquid medium is continuously fed into the bacterial culture (FIG. 10-5). The liquid medium contains some nutrient in growth-limiting concentration, and the concentration of the limiting nutrient in the growth medium determines the rate of bacterial growth. Even though bacteria are continuously reproducing, a number of bacterial cells are continuously being washed out and removed from the culture vessel. Thus cell numbers in a chemostat reach a plateau.

BACTERIAL GROWTH ON SOLID MEDIA

The development of bacterial colonies on solid growth media follows the basic normal growth curve. The dividing cells do not disperse and the population is densely packed. Under these conditions, nutrients rapidly become limiting at the center of the colony. Microorganisms in this area rapidly reach stationary phase. At the periphery of the colony, cells can continue to grow exponentially even while those at the center of the colony are in the death phase. Bacterial colonies generally do not extend indefinitely across the surface of the media but have a well-defined edge. Therefore individual well-isolated colonies can develop from the growth of individual bacterial cells. The fact that the bacteria have reproduced asexually by binary fission means that all the bacteria in the well-isolated colony should be genetically identical; that is, each colony should contain a clone of identical cells derived from a single parental cell.

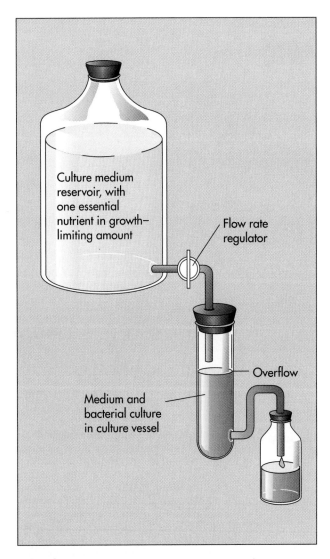

FIG. 10-5 A chemostat continuously provides nutrients with a growth rate limiting factor to a flow-through culture chamber in which bacteria grow.

To assess rates of bacterial reproduction, it is necessary to determine numbers of bacteria. Various methods can be employed for enumerating bacteria. These include viable plate count, direct count, and most probable number (MPN) determinations.

VIABLE COUNT PROCEDURES

The **viable plate count** method is one of the most common procedures for the enumeration of bacteria. In this procedure, serial dilutions of a suspension of bacteria are plated onto a suitable solid growth medium and after a period of incubation (during which single cells multiply to form visible colonies) the number of colonies are counted or enumerated (FIG. 10-6).

Frequently, the suspension is spread over the surface of an agar plate containing growth nutrients **(surface spread technique)** (FIG. 10-7). Alternatively, it can be mixed with the agar while it is still in a liquid state and poured into the plate **(pour plate technique)** (FIG. 10-8). The plates are incubated to allow the bacteria to grow and form colonies. The formation of visible colonies generally takes 16 to 24 hours.

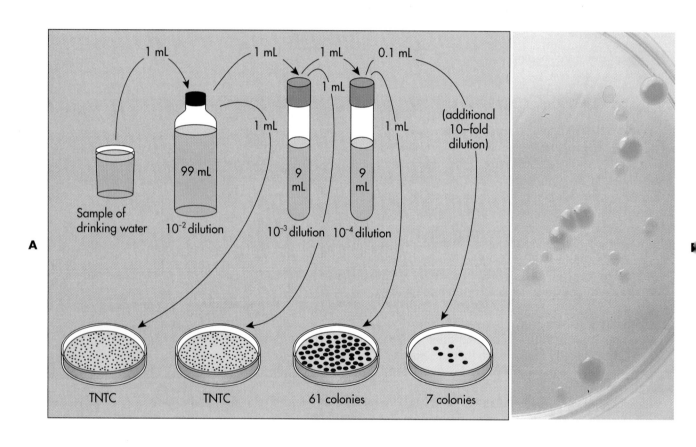

FIG. 10-6 **A,** The plate count procedure is used to determine the viable population in a sample containing bacteria. Dilutions are achieved by adding an aliquot of the specimen to a sterile water dilution tube. If 1 mL of a sample is added to 99 mL of sterile water, the dilution is 1:100 (10^{-2}). (The same dilution could also have been achieved by adding an 0.1 mL sample to 9.9 mL of sterile water). Greater dilutions are achieved by sequentially diluting the sample in series. Adding 1 mL from the first dilution to 9 mL of sterile water achieves an additional tenfold dilution so that the total dilution is 1:1000 (10^{-3}). Adding 1 mL from that second dilution to 9 mL of sterile water achieves a further tenfold dilution so that the total dilution is 1:10000 (10^{-4}). Transferring 1 mL samples from each tube to agar media maintains these dilution factors. Transferring 0.1 mL samples increases the dilution by a factor of 10. After incubation the number of colonies are counted. Counts on the plates in the range of 30 to 300 colonies are used to calculate the concentration of bacteria. The standard notation "TNTC" means too numerous to count (greater than 300 colonies). In this example the plate with 61 colonies would be used to calculate the number of bacteria in the original water sample. Because these colonies developed on a plate in which 1 mL from a 1:10000 dilution was added, the number of bacteria per mL in the original sample is calculated as 6.1 $\times$ 10^5 (61 $\times$ 10^4). **B,** Colonies of lactose fermenting bacteria growing on MacConkey agar.

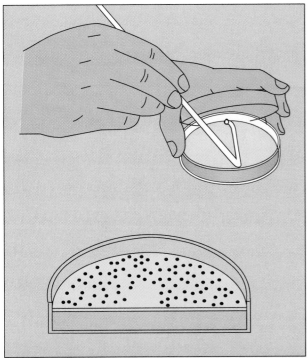

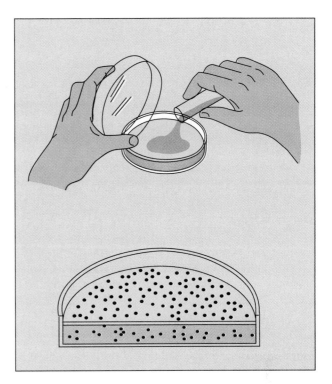

FIG. 10-8 The pour plate technique for isolating and enumerating microorganisms.

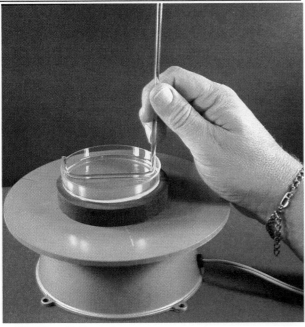

FIG. 10-7 The spread plate technique for isolating and enumerating microorganisms.

It is assumed that each colony arises from an individual bacterial cell. By counting the number of colonies that develop, the **colony forming units** (CFUs), and by taking into account the dilution factors, the concentration of bacteria in the original sample can be determined. Preferably two or three plates are counted to determine numbers of bacteria in a sample. Each plate counted should have 30 to 300

colonies. If bacterial numbers in a sample are low, it is sometimes necessary to filter the suspension to concentrate the bacterial cells by collecting the cells on a membrane filter. The typical membrane filter for collecting bacterial cells is made of nitrocellulose or cellulose acetate and has a pore size of 0.2 to 0.45 μm, which is small enough to trap most bacterial cells. The membrane filter with the trapped bacteria is then placed onto a suitable medium so that bacterial reproduction can occur, and the colonies that develop on the filter are counted. Countable plates are those having between 30 and 300 colonies. Less than 30 colonies is not acceptable for statistical reasons and more than 300 colonies on a plate are likely to produce colonies too close to distinguish as individual CFUs. Such samples are noted as TNTC (too numerous to count).

A major limitation of the viable plate count procedure is that it is selective. There is no single combination of incubation conditions and medium composition that permits the growth of all bacterial types. The nature of the growth medium and the incubation conditions determine which bacteria can grow and thus be counted. Viable counting measures only cells that are capable of growth on the given plating medium under the set of incubation conditions that are used. Sometimes cells are viable but nonculturable unless rigorous steps are taken to acclimate the microorganisms to laboratory culture conditions. The viable plate

UNSAFE ICE CREAM GOES UNDETECTED

An ice cream manufacturing plant in New York City in the early 1970s, in compliance with the required testing procedures to ensure the microbiological safety of its food product, routinely sent samples of the ice cream to a local quality control microbiology laboratory. The laboratory performed viable plate counts to detect coliform bacteria. The presence of coliform bacteria indicates contamination with human fecal matter, making the ice cream unsafe for consumption. The plates were overgrown with coliform bacteria, and the technician at the testing laboratory recorded TNTC, the standard notation for too numerous to count. Records were compiled indicating unsafe levels of contamination but no action was taken. This was because the Board of Health inspector who examined the records did not recognize the abbreviation TNTC and was looking for a number greater than 10 per 100 mL to signal a contamination problem. The inspector did not inquire as to the meaning of TNTC, and it was not until another inspector visited the facility and viewed the records that the problem was detected. The underlying problem with the ice cream was in the plumbing of the building, which had connected the effluent from the restrooms directly to the influent for water used in the manufacture of the ice cream.

count relies on the reproduction of individual bacterial cells to form visible colonies, which are counted to enumerate numbers of bacteria in a sample. Another problem and source of possible error associated with this technique is in the enumeration of bacteria that grow in chains or clumps that are hard to disperse. For example, a chain containing ten attached cells will grow into one colony instead of ten. Therefore using the viable plate count method to measure numbers of bacteria that tend to remain attached to one another can lead to erroneously low values.

The viable plate count procedure is selective because no one combination of incubation conditions and media allows all types of bacteria to grow.

DIRECT COUNT PROCEDURES

Bacteria can also be enumerated by **direct counting procedures.** In this procedure, counting is done without the need to first grow the cells in culture. In one direct count procedure, dilutions of samples are observed under a microscope, and the numbers of bacterial cells in a given volume of sample are counted. These numbers are used to calculate the concentration of bacteria in the original sample (FIG. 10-9). Special counting chambers, such as a hemocytometer or Petroff-Hausser chamber, are sometimes employed to determine the number of bacteria. These chambers are ruled with squares of a known area and are so constructed that a film of liquid of known

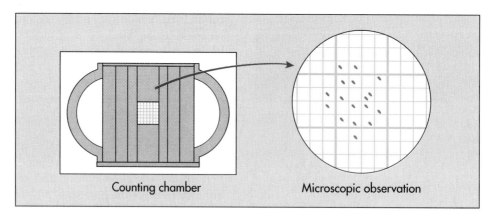

Counting chamber   Microscopic observation

FIG. 10-9 The direct counting procedure using a Petroff-Hauser counting chamber. The sample is added to a counting chamber of known volume. The slide is viewed and the number of cells determined in an area delimited by a grid. In the counting chamber shown, the entire grid has 25 large squares for a total area of 1 mm^2 and a total volume of 0.02 mm^3, formed by the spacing of an overlying coverslip. There are 12 cells within the single large grid (composed of 16 smaller boxes) in this example. Assuming the number of cells in this single grid is representative of all the grids, the number of cells within the total area under the grid is 12 cells. The concentration of cells is therefore 300/0.02 mm^3.

MOST PROBABLE NUMBER (MPN) PROCEDURES

depth can be introduced between the slide and the cover slip. Consequently, the volume of the sample overlying each square is known.

It is often desirable to stain the cells. This helps in visualizing bacterial cells. Alternatively, a known volume of a sample containing a suspension of bacteria is poured through a filter, such as a nitrocellulose 0.2 μm pore size filter. The bacteria are stained on this filter, often using a fluorescent stain, and counted under a microscope. Many fluorescent dyes, such as acridine orange, stain all cells, making it impossible to differentiate living from dead bacteria. The difficulty in establishing the metabolic status of the observed bacteria is a major limitation of this procedure.

Bacterial cells can be enumerated by direct microscopic count procedures.

Instruments also are available for direct counting of bacterial cells. Particle counters, such as a Coulter particle counter, permit the discrimination of particles based on size so that particles the size of bacteria are counted automatically. As long as there are no nonliving interfering particles in the same size range of bacteria, this is a rapid counting method.

Another approach to bacterial enumeration is the determination of the most probable number (MPN). MPN is a statistical method based on probability theory. In a **most probable number enumeration procedure,** multiple serial dilutions are performed to reach a point of extinction. The point of extinction is the dilution level at which not even a single cell is deposited into one or more multiple tubes (FIG. 10-10). A criterion is established for indicating whether a particular dilution tube contains bacteria. Such a criteria can be development of cloudiness or turbidity in a liquid growth medium or gas production (detected by its accumulation in an inverted tube). The pattern of positive and negative test results are then used to estimate the concentration of bacteria in the original sample, that is, the most probable number of bacteria, by comparing the observed pattern of results with a table of statistical probabilities for obtaining those results (Table 10-1). Often a 95% confidence limit is given in MPN tables.

The most probable number procedure is a statistical method based on probability theory used to determine bacterial populations by diluting bacterial samples to the point of extinction.

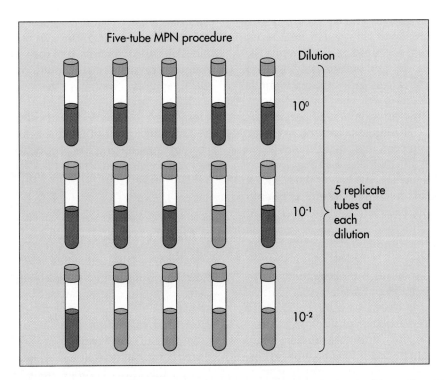

FIG. 10-10 The most probable number (MPN) procedure involves inoculation of multiple tubes with replicate samples of dilutions. The pattern of tubes that show growth *(brown)* and those that do not *(orange)* are compared with a statistical table to calculate the MPN of bacteria in the original sample. In this example all 5 tubes at the 10⁰ show growth, 4 of 5 tubes at the 10⁻¹ dilution show growth, and only 1 of 5 tubes at the 10⁻² dilution show growth. Therefore the MPN bacteria in the original sample is 170 per 100 mL (see Table 10-1).

TABLE 10-1

Table of Most Probable Numbers (MPN)

NUMBER OF POSITIVE TUBES AT THE STATED DILUTION				NUMBER OF POSITIVE TUBES AT THE STATED DILUTION			
10^0	10^{-1}	10^{-2}	MPN/100mL	10^0	10^{-1}	10^{-2}	MPN/100mL
0	1	0	0.18	5	0	0	2.3
1	0	0	0.20	5	0	1	3.1
1	1	0	0.40	5	1	0	3.3
2	0	0	0.45	5	1	1	4.6
2	0	1	0.68	5	2	0	4.9
2	1	0	0.68	5	2	1	7.0
2	2	0	0.93	5	2	2	9.5
3	0	0	0.78	5	3	0	7.9
3	0	1	1.1	5	3	1	11.0
3	1	0	1.1	5	3	2	14.0
3	2	0	1.4	5	4	0	13.0
4	0	0	1.3	5	4	1	17.0
4	0	1	1.7	5	4	2	22.0
4	1	0	1.7	5	4	3	28.0
4	1	1	2.1	5	5	0	24.0
4	2	0	2.2	5	5	1	35.0
4	2	1	2.6	5	5	2	54.0
4	3	0	2.7	5	5	3	92.0
				5	5	4	160.0

FACTORS INFLUENCING BACTERIAL GROWTH

The rates of bacterial growth and death are greatly influenced by environmental parameters. Some environmental conditions favor rapid bacterial reproduction, whereas others do not permit any bacterial growth. Not all bacteria can grow under identical conditions. Each bacterial species has a specific tolerance range for specific environmental parameters. Outside the range of environmental conditions under which a given bacterium can reproduce, it may either survive in a relatively dormant state or may lose viability. Loss of viability means the loss of the ability to reproduce, which leads to death. The effects of environmental factors on bacterial growth and death rates can be seen as differences in rates of reproduction or death of a culture under varying environmental conditions.

It is a particular environmental parameter or interaction of environmental parameters that controls the rate of growth or death of a given bacterial species. Bacteria have particular physiological properties that determine the conditions under which they can grow. In nature, conditions cannot be controlled and many species co-exist; fluctuating environmental conditions favor population shifts because of the varying growth rates of individual microbial populations within the community of a given location. In the laboratory it is possible to adjust conditions to achieve optimal growth rates for a given microorganism. Many laboratory and industrial applications use pure cultures of microorganisms. This facilitates the adjustment of the growth conditions so that they favor optimal growth of the particular bacterial species. Similarly, in industrial fermentors, which are large growth chambers (often thousands of liters) used to grow bacterial cultures, conditions can be adjusted to optimize bacterial growth rates, thereby maximizing the production of desired bacterial metabolic products.

TEMPERATURE

Temperature is one of the most important factors affecting rates of microbial growth. The temperatures at which specific enzymes and cellular structures function varies from one microbial species to another, depending on the specific chemical compositions specified by the genome of the organism. The minimum and maximum temperatures at which a microorganism can grow establish the **temperature growth range** for that microorganism (FIG. 10-11). Within this growth range there will be an optimal growth temperature at which the highest rate of reproduction occurs.

The temperature growth range is defined by the minimum and maximum temperatures at which a microorganism can grow.

ENRICHING FOR SPECIFIC BACTERIA

By taking into account the physiological characteristics of specific bacterial species or types, it is possible to design conditions that favor the growth of those bacteria. This is the basis of enrichment culture technique, a method used to isolate specific groups of bacteria based on designing the culture medium and incubation conditions to preferentially support the growth of a particular bacterial type. Liquid enrichment media tend to select the bacteria that are able to grow best among all the bacteria introduced into the media. For example, to isolate bacteria capable of metabolizing petroleum hydrocarbons, one can design a culture medium containing a hydrocarbon as the sole source of carbon and energy (see Figure). By doing so, one establishes conditions whereby only bacteria that are capable of metabolizing hydrocarbons can grow. Because other bacteria cannot reproduce in this medium, hydrocarbon-utilizing bacteria are thereby selected, resulting in enrichment (increased proportions)

of the selected bacteria. Similarly, a culture medium that favors the growth of autotrophic microorganisms could be designed by providing ammonium ions and carbonate as the sole source of carbon in the medium.

The design of an enrichment procedure takes into consideration the composition of the medium and also environmental factors, such as temperature, aeration, pH, and so forth. For example, temperature can be adjusted to 5° C to favor the growth of microorganisms that live at low refrigerator temperatures or to 37° C to enrich for microorganisms that are capable of growth at the temperature of the human body. Cultures may be aerated by shaking or by bubbling with air to favor the growth of aerobes, or oxygen may be totally excluded to enrich for anaerobes. The enrichment culture technique mimics many natural situations in which the growth of a particular microbial population is favored by the chemical composition of the system and by environmental conditions.

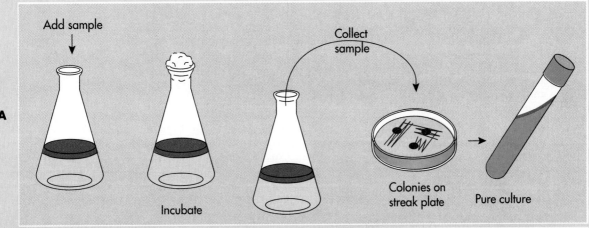

A, To establish an enrichment, a medium is inoculated with a sample, for example, soil or water, that may contain microorganisms with specific characteristics. The medium and the incubation conditions are designed to favor the growth of the microorganisms, for example, microorganisms capable of degrading petroleum hydrocarbons. The desired microorganisms should be able to outcompete others in the sample and increase in number so they then can be isolated and pure cultures established. **B,** Enrichment cultures are designed to selectively support the growth of specific microorganisms. In a medium with petroleum hydrocarbon as the sole source of carbon and energy, hydrocarbon-degrading microorganisms are selectively enriched *(left flask)*. Control showing oil slick and lack of enrichment for hydrocarbon degraders *(right flask)*. Growth of the hydrocarbon-degrading microorganisms emulsifies the oil so that it disperses through the medium in the flask. A pure culture of the hydrocarbon degrader can be isolated from the enrichment culture.

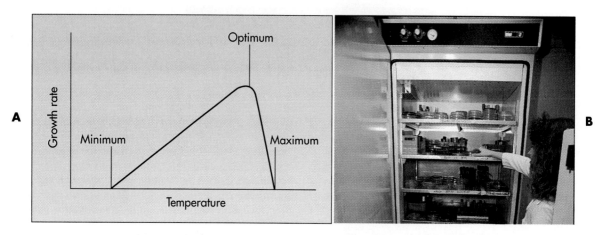

FIG. 10-11   **A,** Microorganisms exhibit specific temperature growth ranges. There is a minimal and a maximal temperature for growth. **B,** An incubator is used to control temperature for culturing microorganisms.

The **optimal growth temperature** is defined as the temperature at which the maximal growth rate occurs. This is the temperature that corresponds to the shortest generation time (Table 10-2). The ability of a microbial species to compete for survival in a given system is favored when temperatures are near its optimal growth temperature. It is not surprising that the optimal growth temperature for most human pathogens is 37° C, the temperature of the human body.

> **Optimal growth temperature is the temperature at which the generation time is shortest and therefore at which the maximal growth rate occurs.**

Different microorganisms have different optimal growth temperatures (FIG. 10-12). Some microorgan-

TABLE 10-2		
Growth Rates for Some Representative Bacteria Under Optimal Conditions		
ORGANISM	**TEMPERATURE (° C)**	**GENERATION TIME (MIN)**
Thermophile		
Bacillus stearo-thermophilus	60	11
Mesophile		
Escherichia coli	37	20
Bacillus subtilis	37	27
Staphylococcus aureus	37	28
Streptococcus lactis	37	30
Pseudomonas putida	30	45
Psychrophile		
Vibrio marinus	15	80

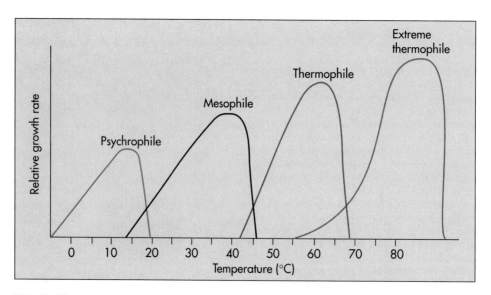

FIG. 10-12   Temperature growth ranges for mesophiles, psychrophiles, and thermophiles.

HISTORICAL PERSPECTIVE

DEEP SEA THERMAL VENT BACTERIA

The work of Holger Jannasch established that microbial activities occur at very low rates in the deep oceans; deep ocean sediments, in effect, are biological deserts because of their low temperatures, high pressures, and low inputs of organic matter. These rates are so low that bologna sandwiches accidentally submerged inside the submersible Alvin were not decomposed during several months of exposure to microorganisms of the deep sea.

What a surprise when investigators found an area of extremely high biological productivity at a depth of 2,550 m in a region of thermal vents (subsea volcanoes) off the Galapagos Islands. The thermal vents warmed the waters, but what was the source of food supporting the growth of worms several feet long and clams several feet across? There was no light to support photosynthesis, and transport of organic matter from the surface was unlikely. The most likely explanation was that chemolithotrophic metabolism by autotrophic bacteria based on oxidation of hydrogen sulfide from the vents was providing the organic matter to support the growth of other organisms (see Figure). Establishing that bacterial chemolithotrophy was the source of organic matter would be difficult; to reach the vents in the Alvin would take hours, time on the bottom to carry out experiments would be extremely limited, and working Alvin's mechanical arms would be difficult. Nevertheless, this was the task undertaken by Holger Jannasch and Carl Wirsen of the Woods Hole Oceanographic Institution.

Jannasch and his associates were able to collect samples using specialized pressurized chambers and return living bacteria from the thermal vents to the laboratory for study. These investigators found that all surfaces intermittently exposed to H_2S-containing hydrothermal fluid were covered with mats composed of layers of prokaryotic, Gram-negative cells interspersed with amorphous manganese-iron metal deposits. Enrichment cultures using thiosulfate as the energy source made from mat material resulted in isolations of different types of sulfur-oxidizing bacteria, including the obligately chemolithotrophic genus *Thiomicrospira*. These studies established that chemolithotrophic bacteria supported the productivity of the thermal rift region.

Jannasch and other scientists then asked about the maximal temperature at which bacteria in thermal vents could grow. Bacteria were observed in waters coming from the vents with temperatures well in excess of 100° C. Could bacteria actually grow there or had the bacteria grown elsewhere at lower temperatures? What was the upper temperature limit at which bacteria can reproduce? Some scientists hypothesized that, since there was liquid water because of the high pressures, bacteria could grow at temperatures of even 500° C.

Experiments were conducted by John Baross and Jody Deming who incubated bacterial samples from the thermal vents in chambers under very high pressures at temperatures of 250° C. Because the chambers had to remain sealed under pressure to maintain the tempera-

A, Photograph of the deep sea submarine ALVIN.

Continued.

DEEP SEA THERMAL VENT BACTERIA—CONT'D

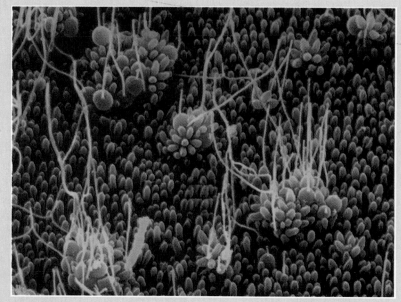

B, Colorized micrograph of deep sea thermal vent bacterial community; the filaments of *Beggiatoa* are abundant.

ture and prevent water from turning to steam, it was impossible to sample the chambers and culture bacteria. Deming and Baross therefore measured protein and nucleic acid content at the end of the experiment, both of which appeared to increase. Based on these observations they reported that bacterial growth occurred in the chambers incubated at 250° C. Their results were immediately questioned by many scientists. Holger Jannasch could repeatedly grow some of the bacteria from the thermal vents at temperatures of 100° to 110° C, but not at higher temperatures. No one was able to repeat the

experiments that purportedly demonstrated bacterial growth at 250° C. Independent confirmation is critical in science. Eventually it was shown by Art Yayanos at the Scripps Institute of Oceanography that the results reported by Deming and Baross could be explained by abiotic changes that occur at high temperature and pressure. Bacterial growth apparently had not occurred at 250° C. The initial report had not met the essential test of the scientific method—that of repeatability by others. The upper demonstrated growth temperature remains about 110° C.

isms grow best at low temperatures. Such organisms, known as **psychrophiles,** have optimal growth temperatures of under 20° C. As long as liquid water is available, some psychrophilic microorganisms are capable of growing below 0° C. Psychrophilic microorganisms are commonly found in the world's oceans and are also capable of growing in a household refrigerator, where they are important agents of food spoilage.

Mesophiles are microorganisms that have optimal growth temperatures in the middle temperature range between 20° and 45° C. Most of the bacteria grown in introductory microbiology laboratory courses are mesophilic. Many mesophiles have an optimal temperature of about 37° C. Many of the normal resident microorganisms of the human body, such as *Eschericia coli,* are mesophiles. Similarly, most human pathogens are mesophiles and thus grow rapidly and establish an infection within the human body.

Thermophilic microorganisms are organisms with high optimal growth temperatures. Thermophiles, such as *Bacillus stearothermophilus,* grow at relatively high temperatures, often growing only above 40° C. The upper growth temperature for extreme thermophilic microorganisms, such as those found in deep thermal rift regions of the areas where volcanic activity heats the ocean water under very high pressure, is about 110° C. Water will remain in a liquid state at temperatures above 100° C when it is subjected to high pressure. Thermophiles have optimal growth temperatures above 45° C and many thermophilic microorganisms have optimal growth temperatures of about 55° to 60° C. One finds thermophilic microorganisms in such exotic places as hot springs and effluents from laundromats. However, many thermophiles can survive very low temperatures, and viable thermophilic bacteria are routinely found in frozen antarctic soils.

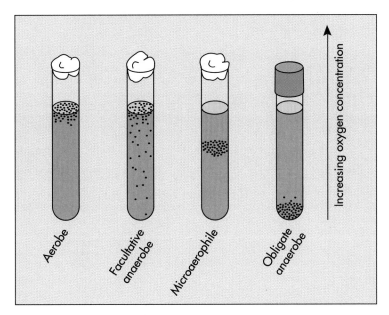

FIG. 10-13 Drawing showing oxygen growth relationships for aerobic, anaerobic, and facultative bacteria in tubes of nutrient media. The growth of aerotolerant anaerobes (obligately fermentative) would be the same as that of facultative anaerobes (organisms capable of fermentation and respiration).

Psychrophiles have optimal growth temperatures of under 20° C; mesophiles grow best between 20° and 45° C; and thermophiles grow best at higher temperatures above 45° C, with 55° to 60° C often being optimal.

OXYGEN

Another factor that greatly influences bacterial growth rates is the concentration of molecular oxygen. Bacteria are classified as aerobes, anaerobes, facultative anaerobes, or microaerophiles, based on their oxygen requirements and tolerances (FIG. 10-13). **Aerobic bacteria (obligate aerobes)** grow only when oxygen is available to support their respiratory metabolism. In laboratory cultures and industrial batch cultures, oxygen is often supplied by forced aeration or mixing (for example, on a rotary shaker) to support the growth of aerobes. **Anaerobic bacteria (obligate anaerobes)** grow in the absence of molecular oxygen. Anaerobic bacteria may carry out fermentation or anaerobic respiration to generate ATP. Some anaerobes have very high death rates in the presence of oxygen, and such organisms are termed **strict anaerobes.** Even the briefest exposure to air can kill strict anaerobes. Other obligately anaerobic bacteria, although unable to grow, have low death rates in the presence of oxygen.

While obligate anaerobes grow only in the absence of molecular oxygen, **facultative anaerobes** such as *E. coli* can grow with or without oxygen. Many facultative anaerobes are capable of both fermentative and respiratory metabolism. Some are capable of both aerobic and anaerobic respiration.

Aerobes need oxygen to support their respiratory metabolism and anaerobes grow in the absence of molecular oxygen, carrying out fermentation or anaerobic respiration; facultative anaerobes grow aerobically or anaerobically.

Although oxygen is required for the growth of many microorganisms, it can also be toxic. Some microorganisms grow only over a very narrow range of oxygen concentrations. Such microorganisms are known as **microaerophiles.** Microaerophiles require oxygen but exhibit maximal growth rates at reduced oxygen concentrations because higher oxygen concentrations are toxic to these organisms.

Oxygen can exist in several energetic states, some of which are more toxic than others. One of these energetic states, called singlet oxygen, is a chemically reactive form that is extremely toxic to living organisms. Phospholipids in bacterial plasma membranes can be oxidized by singlet oxygen, leading to a disruption of membrane function and the death of bacterial cells. Peroxidases in saliva and phagocyte cells (blood cells involved in the defense mechanism of the human body against invading microorganisms) generate singlet oxygen, accounting in part for the antibacterial activity of saliva and the ability of phagocytic blood cells to kill invading microorganisms.

Singlet oxygen is chemically reactive and extremely toxic to living organisms.

The conversion of oxygen to water occurs when oxygen serves as a terminal electron acceptor in respiration pathways. This involves the formation of an

METHODOLOGY

GROWING CULTURES OF AEROBIC AND ANAEROBIC BACTERIA

Under controlled laboratory conditions, it is possible to adjust oxygen concentrations to maximize the growth rate of a particular bacterial species. Because oxygen diffuses only slowly into liquid, the concentration of oxygen frequently limits the growth rate of aerobic and facultatively anaerobic bacteria in liquid culture. To supply oxygen for the growth of aerobic microorganisms and overcome the growth rate limitations caused by low oxygen concentrations, liquid cultures can be agitated at high speed on a shaker table or by an impeller within the culture vessel, or oxygen can be supplied to the culture vessel through forced aeration (FIG. *A*). Interrupting the supply of oxygen to an actively growing culture for even a brief period of time can lead to anaerobic conditions, in some cases causing a rapid die-off of the bacteria. Some microbial populations can lose viability if a rotary shaker is turned off for only a few minutes, such as may occur when changing flasks on the shaker table.

Whereas aeration enhances the rates of aerobic growth, oxygen must be excluded from the growth medium to permit the growth of obligate and strict anaerobes. This can be accomplished by adding chemicals that react with and remove molecular oxygen from the growth medium. For example, sodium thioglycollate is frequently added to liquid culture media for the growth of anaerobes because it reacts with molecular oxygen, removing free oxygen from solution.

Similarly, the amino acid cysteine and other compounds containing sulfhydryl groups can also be used to scavenge molecular oxygen from a growth medium. For liquid cultures, nitrogen may be bubbled through the medium to remove air and traces of oxygen, and then the culture vessel is sealed tightly to prevent oxygen from reentering.

There are many types of anaerobic culture chambers that can be employed to exclude oxygen from the atmosphere (FIG. *B*). Common forms of anaerobic chambers, such as the Gas Pak system, generate hydrogen, which reacts with the oxygen as a catalyst within the chamber to produce water. Carbon dioxide is also generated in this system to replace the volume of gas depleted by the conversion of oxygen to water. It is also possible to combine several approaches to ensure absolute anaerobic conditions. In the Hungate roll tube method, after sterilization of a prereduced medium (a medium from which oxygen is excluded by the incorporation of a chemical that scavenges the free oxygen) within a sealed test tube, the medium is rolled during cooling so that the medium covers the inside of the test tube; the medium is then inoculated with a microorganism under a stream of carbon dioxide or nitrogen and tightly sealed with butyl rubber stoppers to keep oxygen out; the development of microbial colonies can be seen on the tube surface, and individual cultures can be observed without disturbing other cultures.

A

B

A rotary shaker is used to maintain aerobic conditions in liquid cultures.

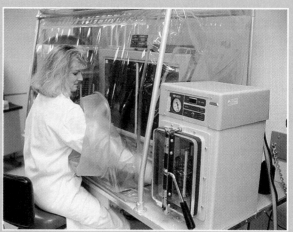

An anaerobic glove box like this may be used to culture anaerobes.

FIG. 10-14 The action of catalase is readily visualized when hydrogen peroxide is added to cells that have produced this enzyme.

intermediary form of oxygen known as the superoxide anion (O_2^-), in addition to forming singlet oxygen. The superoxide anion is converted to hydrogen peroxide and oxygen by the action of the enzyme **superoxide dismutase,** which is produced by most aerobic and facultatively anaerobic bacteria. Superoxide dismutase removes the toxic superoxide anion but forms hydrogen peroxide (H_2O_2), which is also toxic. The reaction that describes the action of superoxide dismutase is:

$$2O_2^- + 2H^+ \longrightarrow H_2O_2 + O_2$$

Hydrogen peroxide is frequently used to kill bacteria, for example, when hydrogen peroxide is applied to a cut to prevent infection. Some bacteria produce enzymes that destroy hydrogen peroxide. **Catalase** converts hydrogen peroxide to water and oxygen. The reaction that describes the action of catalase is:

$$2H_2O_2 \longrightarrow 2H_2O + O_2$$

Bacterial production of catalase can be demonstrated by adding a loopful of a microbial culture to a 3% solution of hydrogen peroxide. The evolution of gas bubbles, oxygen, is evidence of the action of the catalase (FIG. 10-14).

TABLE 10-3		
Bacterial Enzymes that Protect the Cell Against Toxic Forms of Oxygen		
MICROORGANISM	**CATALASE**	**SUPEROXIDE DISMUTASE**
Aerobe	+	+
Facultative anaerobe	+	+
Microaerophile	−	+
Obligate anaerobe	−	−

Obligate aerobes and facultative anaerobes usually produce both catalase and superoxide dismutase (Table 10-3). These enzymes permit such microorganisms to grow without accumulating toxic forms of oxygen. In contrast, obligate anaerobes, such as *Clostridium* species, generally lack these enzymes. The inability of these organisms enzymatically to remove toxic forms of oxygen probably accounts for the fact that they are obligately anaerobic and sensitive to oxygen.

SALINITY

Halophiles are bacteria that specifically require sodium chloride for growth (FIG. 10-15). Moderate halophiles, which include many marine bacteria, grow best at salt (NaCl) concentrations of about 3% NaCl. Extreme halophiles exhibit maximal growth rates in saturated brine solutions. These organisms grow quite well in salt concentrations of greater than 15% NaCl and can grow in places like salt lakes and pickle barrels (FIG. 10-16). High salt concentrations normally disrupt membrane transport systems and denature proteins. Extreme halophiles must possess physiological mechanisms for tolerating high salt concentrations. For example, *Halobacterium*, possesses an unusual plasma membrane and many unusual enzymes that require a high salt concentration for activity.

Halophiles require a high salt concentration for growth.

Most bacteria, however, do not possess these physiological adaptations and cannot tolerate high

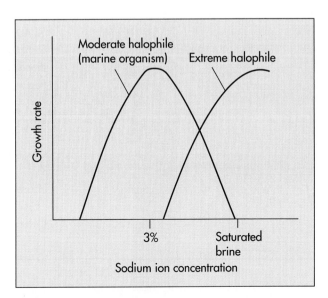

FIG. 10-15 The halophiles require sodium chloride (NaCl) for growth. Marine bacteria typically grow best near 3% NaCl. Some extreme halophiles grow best near 15% NaCl.

FIG. 10-16 Halophiles growing within salt lakes often turn the water pink; this sometimes occurs in Great Salt Lake, Utah.

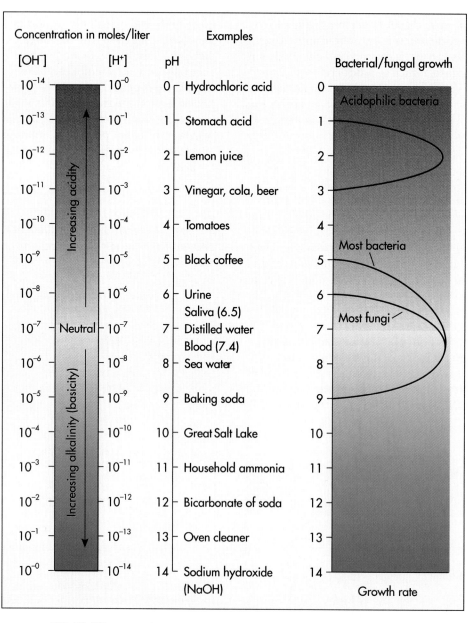

FIG. 10-17 pH scale showing pH values of some common substances.

salt concentrations. The degree of sensitivity to salt varies for different bacterial species. Many bacteria will not grow at a salt concentration of 3%. Some strains of *Staphylococcus,* however, are **salt tolerant** and grow at concentrations greater than 10% NaCl. This physiological adaptation in *Staphylococcus* is important because some members of this genus grow on skin surfaces where salt concentrations can be relatively high.

ACIDITY AND pH

The pH of a solution describes the **hydrogen ion concentration** ([H$^+$]). When bacteria are cultured in the laboratory, they produce acids that can interfere with their own growth. To neutralize the acids and maintain the proper pH, chemicals called **buffers** are included in the growth medium. Peptones and amino acids in some media act as buffers. Many media also contain phosphate salts, which exhibit their buffering effect in the pH growth range of most bacteria. They are also nontoxic and even provide phosphorus, an essential nutrient element.

> **Buffers neutralize acids and maintain the proper pH.**

Microorganisms vary in their pH tolerance ranges (FIG. 10-17). The pH is equal to log [H$^+$] or 1/log [H$^+$]. A neutral solution has a pH of 7.0; acidic solutions have pH values less than 7; and alkaline or basic solutions have pH values greater than 7. Most micro-organisms grow best at near neutral pH. These microorganisms are called **neutralophiles or neutrophiles.** Fungi generally exhibit a wider pH range, growing well over a pH range of 5 to 9, compared to most bacteria, which grow well over a pH range of 6 to 9 (Table 10-4).

Although most bacteria are unable to grow at low pH, there are some exceptional cases. Some bacteria tolerate pH values as low as 0.8. There are even some bacteria, called **acidophiles,** that are restricted to growth at low pH values. Some members of the genus *Thiobacillus* are acidophilic and grow only at pH values near 2; they can grow in sulfuric acid.

Alkalophiles are microorganisms that prefer to grow at high pH values. These microorganisms are found in alkaline environments that are high in sodium such as salt lakes and soils rich in sodium carbonate. The alkalophile *Bacillus alkalophilus* grows best at pH 10.5.

PRESSURE

The solute concentration of a solution affects the osmotic pressure that is exerted across the plasma membrane. The cell walls of bacteria make them relatively resistant to changes in osmotic pressure, but

TABLE 10-4

Table of pH Tolerances of Various Bacteria

ORGANISM	MINIMUM pH	OPTIMUM pH	MAXIMUM pH
ACIDOPHILE			
Thiobacillus thiooxidans	1.0	2.0-2.8	4.0-6.0
Lactobacillus acidophilus	4.0-4.6	5.8-6.6	6.8
NEUTRALOPHILE			
Escherichia coli	4.4	6.0-7.0	9.0
Clostridium sporogenes	5.0-5.8	6.0-7.6	8.5-9.0
Pseudomonas aeruginosa	5.6	6.6-7.0	8.0
Erwinia carotovora	5.6	7.1	9.3
ALKALOPHILE			
Bacillus alcalophilus	8.0-8.5	10.5	11.0

extreme osmotic pressures can result in the death of bacterial cells. In **hypertonic** solutions, bacterial cells may shrink and become desiccated; in **hypotonic** solutions, the cell may burst (FIG. 10-18). Bacteria that can grow in solutions with high solute concentrations are called **osmotolerant.** These bacteria can withstand high osmotic pressures. Some fungi, such as *Xeromyces*, are actually **osmophilic.**

Hydrostatic pressure is another type of pressure that can influence bacterial growth rates. **Hydrostatic pressure** refers to the pressure exerted by a water column as a result of the weight of the water column. Each 10 meters of water depth is equivalent to approximately 1 atmosphere pressure. Most bacteria are relatively tolerant to the hydrostatic pressures in most natural systems but cannot tolerate the extremely high hydrostatic pressures that characterize deep ocean regions. High hydrostatic pressures of greater than 200 atmospheres generally inactivate enzymes and disrupt membrane transport processes. However, some bacteria—referred to as **barotolerant**—can grow at high hydrostatic pressures, and there even appear to be some bacteria—referred to as **barophiles**—that grow best at high hydrostatic pressures.

LIGHT RADIATION

Photosynthetic microorganisms require light in the visible spectrum to carry out photosynthesis. The rate of photosynthesis is a function of light intensity. At some light intensities, rates of photosynthesis

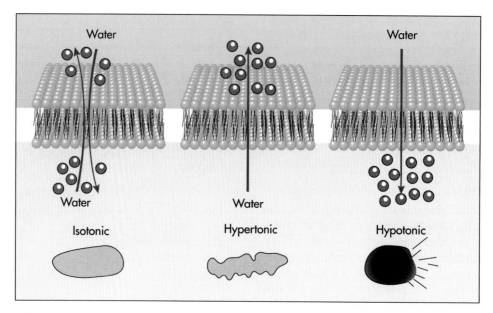

FIG. 10-18 Cells respond to osmotic pressure. In hypotonic solutions, cells may burst.

reach a maximum. Although light intensities above this level do not result in further increases in the rates of photosynthesis, light intensities below the optimal level result in lower rates of photosynthesis.

The wavelength of light also has a marked effect on the rates of photosynthesis. Different photosynthetic microorganisms use light of different wavelengths. For example, anaerobic photosynthetic bacteria use light of longer wavelengths than eukaryotic algae are capable of using. Many photosynthetic microorganisms have accessory pigments that enable them to use light of wavelengths other than the absorption wavelength for the primary photosynthetic pigments. The distribution of photosynthetic microorganisms in nature reflects the variations in the ability to use light of different wavelengths and the differential penetration of different colors of light into aquatic habitats.

The rate of photosynthesis is a function of light intensity and wavelength.

Exposure to visible light can also cause the death of bacteria (FIG. 10-19). Exposure to visible light can lead to the formation of singlet oxygen, which can result in the death of bacterial cells. Some bacteria produce pigments that protect them against the lethal effects of exposure to light. For example, yellow, orange, or red carotenoid pigments interfere with the formation and action of singlet oxygen, preventing its lethal action. Bacteria possessing carotenoid pigments can tolerate much higher levels of exposure to

sunlight than nonpigmented microorganisms. Pigmented bacteria often grow on surfaces that are exposed to direct sunlight, such as on leaves of trees. Many viable bacteria found in the air produce colored pigments.

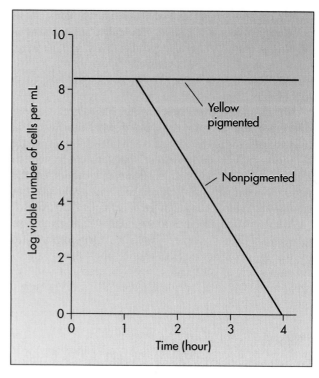

FIG. 10-19 Pigmentation is important in the ability of bacteria to survive exposure to light.

SUMMARY

Bacterial Reproduction (pp. 287-288)

Binary Fission (p. 287)

- Binary fission is the normal form of asexual bacterial reproduction and results in the production of two equal-size daughter cells. Replication of the bacterial chromosome is required to give each daughter cell a complete genome.
- A septum or crosswall is formed by the inward movement of the plasma membrane, separating the two complete bacterial chromosomes in an active protein-requiring process, physically cutting thechromosomes apart and distributing them to the two daughter cells. Cell division is synchronized with chromosome replication.

Alternate Means of Bacterial Reproduction (pp. 287-288)

- Other modes of bacterial reproduction are predominantly asexual, differing in how the cellular material is apportioned between the daughter cells and whether the cells separate or remain together as part of a multicellular aggregation. Budding is characterized by an unequal division of cellular material. In hyphae formation the cell elongates, forming relatively long, generally branched filaments; crosswall formation results in individual cells containing complete genomes.

Bacterial Spore Formation (p. 288)

- Sporulation results in the formation of specialized resistant resting cells or reproductive cells called spores. Endospores are heat-resistant nonreproductive spores that are formed within the cells of a few bacterial genera; cysts are dormant nonreproductive cells sometimes enclosed in a sheath. Myxobacteria form fruiting bodies within which they produce progeny myxospores that can survive transport through the air. Arthrospores are spores formed by hyphae fragmentation that permit reproduction (increase in cell number) of some bacteria.

Bacterial Growth (pp. 288-291)

- Binary fission leads to a doubling of a bacterial population at regular intervals. The generation or doubling time is a measure of the growth phase.

Generation Time (pp. 288-290)

- The time required to achieve a doubling of the population size is known as the generation time or doubling time.
- The generation time is a measure of bacterial growth rate.
- The generation time of a bacterial culture can be expressed as:

$$g = \frac{t}{3.3 \, (\log_{10} N_t - \log_{10} N_0)}$$

Bacterial Growth Curve (pp. 291-292)

- The normal growth curve for bacteria has four phases: lag, log or exponential, stationary, and death. During the lag phase, bacteria are preparing for reproduction, synthesizing DNA and enzymes for cell division; there is no increase in cell numbers. During the log phase the logarithm of bacterial biomass increases linearly with time; this phase determines the generation or doubling time. Bacteria reach a stationary phase if they are not transferred to new medium and nutrients are not added; during this phase there is no further net increase in bacterial cell numbers and the growth rate is equal to the death rate. The death phase begins when the number of viable bacterial cells begins to decline.

Batch and Continuous Growth (p. 291)

- In batch cultures, bacteria grow in a closed system to which new materials are not added. In continuous culture, fresh nutrients are added and end products removed so that the exponential growth phase is maintained.

Bacterial Growth on Solid Media (p. 291)

- On solid media, bacteria do not disperse and so nutrients become limiting at the center of the colony; bacterial colonies have a well-defined edge. Each colony is a clone of identical cells derived from a single parental cell.

Enumeration of Bacteria (pp. 292-296)

Viable Count Procedures (pp. 292-294)

- Numbers of bacteria are determined by viable plate count, direct count, and most probable number determinations. In viable plate counts, serial dilutions of bacterial suspensions are plated on solid growth medium by the pour plate or surface spread technique, incubated, and counted. Since each colony comes from a single bacterial cell, counting the colony forming units, taking into account the dilution factors, can determine the original bacterial concentration.

Direct Count Procedures (pp. 294-295)

- Bacteria enumerated by direct counting procedures do not have to be grown first in culture or stained. Special counting chambers are often used.

Most Probable Number (MPN) Procedures (pp. 295-296)

- In most probable number enumeration procedures, multiple serial dilutions are performed to the point of extinction. Cloudiness or turbidity can be the criterion for existence of bacteria at any dilution level; gas production or other physiological characteristics can be used to determine the presence of specific types of bacteria.

Factors Influencing Bacterial Growth (pp. 296-306)

- Environmental conditions influence bacterial growth and death rates. Each bacterial species has a specific tolerance range for specific environmental parameters. Changing environmental conditions cause population shifts. Laboratory conditions can be manipulated to achieve optimal growth rates for specific organisms.

Temperature (pp. 296-301)

- There are maximum and minimum temperatures at which microorganisms can grow; these extremes of temperature at which growth occurs establish the temperature growth range.

- Several categories of bacteria are defined based on optimal growth temperatures: psychrophiles have optimal growth temperatures of under 20° C; mesophiles have optimal growth temperatures in the middle range (20° to 45° C); and thermophiles grow optimally at higher temperatures, above 45° C.

Oxygen (pp. 301-303)

- Aerobic microorganisms grow only when oxygen is available (respiratory metabolism). Anaerobic microorganisms grow in the absence of molecular oxygen by fermentation or anaerobic respiration. Obligate anaerobes grow only in the absence of molecular oxygen. Facultative anaerobes can grow with or without oxygen and are usually capable of both fermentative and respiratory metabolism. Microaerophiles grow only over a very narrow range of oxygen concentrations; they require oxygen, but high concentrations are toxic.
- Microorganisms possess enzyme systems for detoxifying various forms of oxygen; catalase is involved in the destruction of hydrogen peroxide; superoxide dismutase destroys the toxic superoxide radical.

Salinity (pp. 303-305)

- Most microorganisms cannot tolerate high salt concentrations, but some salt-tolerant bacteria, such as *Staphylococcus,* will grow at high salt concentrations. Halophiles require sodium chloride for growth and extreme halophiles can grow at very high salt concentrations.

Acidity and pH (p. 305)

- The pH of a solution describes its hydrogen ion concentration. Microorganisms vary in their pH tolerance ranges, with fungi generally exhibiting a wider pH range (5 to 9) than bacteria (6 to 9).
- Neutralophiles grow best at near neutral pH. Acidophiles are restricted to growth at low pH values. Some acidophiles grow only at pH 1-2. Alkalophiles grow best at high pH values.

Pressure (p. 305)

- Extreme osmotic pressures can result in microbial death because cells shrink and become desiccated in hypertonic solutions; in hypotonic solutions, cells may burst. Osmotolerant microorganisms can grow in solutions with high solute concentrations. Osmophilic microorganisms require high solute concentrations.
- Hydrostatic pressure is the pressure exerted by a column of water as a result of the weight of the water column (10 meters water = 1 atmosphere of pressure). Most microorganisms are relatively tolerant to hydrostatic pressures in most natural systems, except deep ocean regions.

Light Radiation (pp. 305-306)

- Exposure to visible light can cause death of some microorganisms; some microorganisms produce pigments (often yellow-orange) that protect them against the lethal action of light radiation. Photosynthetic microorganisms require visible light to carry out metabolism and the rate of photosynthesis is a function of light intensity.

CHAPTER REVIEW

REVIEW QUESTIONS

1. Define bacterial growth.
2. What is binary fission?
3. How are microorganisms classified based on optimal growth temperature?
4. Define pH and explain its relation to microbial growth.
5. Define osmotic pressure and explain its relation to microbial growth.
6. How are microorganisms classified based on oxygen requirements?
7. How can oxygen be toxic to cells? How do cells protect themselves from these toxic molecules?
8. What are the phases of microbial growth?
9. What is generation time?
10. Describe some direct and indirect measures of microbial growth.
11. What are the similarities and differences between bacteria growing in the environment and those in a continuous culture?
12. What are the advantages and disadvantages of the viable plate count method to assess bacterial numbers?
13. What are the advantages and disadvantages of the direct microscopic count method to assess bacterial numbers?
14. What special requirements do bacteria need to survive in very hot environments? In very cold environments?
15. Describe the different parts of the bacterial growth cycle. What is happening in the cell and in the population of cells in each phase?
16. What does exponential growth mean? What is happening during the exponential growth phase?
17. How long would it take a single bacterial cell to form 1,000,000 cells if it had a generation time of 30 minutes?

CRITICAL THINKING QUESTIONS

1. Suppose you wanted to isolate a microorganism that was a mesophilic, degraded cellulose and was microaerophilic. What conditions would you have to provide to isolate such a microorganism in the laboratory? Where would you obtain the inoculum for establishing the culture?

2. Some bacteria that live in deep ocean waters are obligate barophiles that tend to lyse or rupture when brought to normal atmospheric pressures. What special requirements would these bacteria need to survive in their high pressure environment? Why can't they survive at the ocean surface? How can they be cultured in the laboratory?

3. Why would you want to distinguish between the numbers of live bacteria and dead bacteria in a population? How would you go about doing this? How would you deal with viable nonculturable bacteria?

4. It takes about 60 minutes to replicate the bacterial chromosome. Given that every daughter cell formed by binary fission must have a complete bacterial chromosome, how can some bacteria reproduce every 30 minutes?

5. Why does the clinical microbiology laboratory employ so many different methods for isolating and identifying pathogenic microorganisms? Why can't one set of standardized conditions be employed?

READINGS

Atlas RM and R Bartha: 1993. *Microbial Ecology: Fundamentals and Applications*, ed. 3, Menlo Park, California; Benjamin/Cummings.

  A text describing the ecology of microorganisms that includes chapters on the effects of environmental conditions on the growth of microorganisms.

Brock TD (ed.): 1986. *Thermophiles: General, Molecular, and Applied Microbiology*, New York, John Wiley.

  Complete coverage of the thermophilic bacteria by an outstanding researcher in the field.

DeLong EF and AA Yayanos: 1985. Adaptation of the membrane lipids of a deep-sea bacterium to changes in hydrostatic pressure, *Science* 228:1101-1103.

  Discusses the physiological effects of hydrostatic pressure on marine bacteria and the role of their membranes on their ability to adapt to this environment.

Dworkin M: 1985. *Developmental Biology of the Bacteria*, Menlo Park, CA; Benjamin/Cummings.

  Describes the special features of bacterial growth.

Ingraham JL, O Maaloe, FC Neidhardt: 1983. *Growth of the Bacterial Cell*, Sunderland, MA; Sinauer Associates.

  Explains biological principles and molecular aspects of bacterial growth.

Jannasch HW and MJ Mottl: 1985. Geomicrobiology of deep-sea hydrothermal vents, *Science* 216:1315-1317.

  A fascinating report on the microorganisms growing in deep-sea thermal vents and how they support the surrounding biological community.

Postgate JR: 1994. *The Outer Reaches of Life*, New York; Cambridge University Press.

  Describes the fascinating adaptations of microorganisms that permit survival under extreme environmental conditions.

Slater JH, R Whittenbury, JWT Wimpenny: 1983. *Microbes in Their Natural Environments*, Thirty-Fourth Symposium of the Society for General Microbiology, England, Cambridge University Press.

  A series of papers on the growth of microorganisms in various natural habitats.

CHAPTER 11

Control of Microbial Growth and Death

PREVIEW TO CHAPTER 11

In this chapter we will:
- Learn that microbial populations can be controlled by limiting growth or increasing death rates.
- Examine the factors that control rates of microbial growth and death.
- See how physical environmental conditions can be modified to control microbial populations.
- Study the chemical approaches for killing or preventing microbial growth.
- Review various types of chemicals used to control microorganisms, including pathogens.
- Learn the following key terms and names:

algicides	high temperature-short
antibiotics	time or HTST process
antimicrobial agents	incineration
antiseptics	infrared radiation
autoclave	low temperature-hold
bactericides	(LTH process)
bacteriostatic	ozonation
chloramination	pasteurization
decimal reduction time (D	preservative
value)	quaternary ammonium
desiccation	compounds (quats)
disinfectant	sanitizer
dry heat sterilization	shelf life
ethylene oxide	sporicidal
sterilization	sterilization
fungicides	thermal death point (TDP)
fungistatic	ultra high temperature
germicides	process (UHT process)
high efficiency particulate	virucides
air filters (HEPA	
filters)	

Rates of microbial growth and death are greatly influenced by several environmental factors. Some environmental conditions favor rapid microbial reproduction; others preclude microbial growth or even result in microbial death. Each microorganism has a certain tolerance range for specific environmental parameters. Outside the range of environmental conditions under which a given microorganism can reproduce, it may either survive in a relatively dormant state or may lose viability. Loss of viability means that it will lose the ability to reproduce and consequently die.

By adjusting environmental conditions, one can increase the death rate of microorganisms. This is an important consideration when trying to kill microorganisms. The ability to kill microorganisms is very important in many instances, such as when trying to reduce the numbers of microorganisms in foods so that they do not spoil, and when it is necessary to totally eliminate microorganisms from pharmaceuticals and medical instruments to make them sterile (free of living organisms) and safe for use with patients. Microbial populations also may be physically removed or excluded so as to limit the numbers of microorganisms that can multiply.

Microbial populations can be controlled by modifying environmental conditions.

It is also possible to alter environmental conditions so that microorganisms do not die but also do not reproduce. This method is used for the preservation of microorganisms, such as in culture collections and food preservation, and for preventing spoilage. Many times the conditions needed to heat sterilize a product alter the texture and color of the desired product. It is for this reason that we use freezing to preserve many foods whose taste and textural qualities are destroyed if sterilized at high temperatures.

There are many factors that determine the effectiveness of a particular agent in controlling microorganisms. These factors include the type of microorganism, the amount and type of material to be treated, the duration of the treatment, the concentration or intensity of the agent, and environmental factors such as pH, temperature, and water availability.

PHYSICAL EXCLUSION OR REMOVAL OF MICROORGANISMS

An effective method for controlling microorganisms is by physically excluding them. **Filtration** can be used to remove microorganisms from liquids and gases. Generally, filtration is accomplished by passage of the substance through a filter with 0.2 to 0.45 μm diameter pores. Many pharmaceuticals, such as solutions that are administered intravenously to patients, are sterilized by passage through such filters. Bacteria and other living organisms are eliminated from the solution by trapping them on the filter, but viruses and some very small bacteria may pass through the filter.

Microorganisms can also be removed from air by passage through **high efficiency particulate air filters** (HEPA filters), which remove particulate material larger than about 0.3 μm. Clean rooms, such as operating theaters and rooms where drugs are packaged, often employ HEPA filters. Many microbiology laboratories also have laminar flow hoods in which air that is filtered through a HEPA filter is blown across the work area to prevent contamination during culturing of microorganisms. While not nearly as effective as a HEPA filter, wearing a face mask helps decrease the exchange of microorganisms between people. Surgical staff wear face masks to prevent exhaling microorganisms into the open surgical wound. Staff and visitors wear masks when they are with patients who have infections that may be transmitted through the air. These precautions generally are adequate. However, greater precaution may be warranted for contact with patients with tuberculosis, and it is now required in many situations that a respirator with a HEPA filter be worn rather than a simple surgical mask.

Regardless of whether filtration or other methods, discussed below, are used to eliminate microorgan-

NEWSBREAK

BRINGING SANITARY CONDITIONS TO HOSPITALS

During the Crimean War in the 1850s, poor sanitary conditions led to the rapid spread of typhus. More soldiers were dying of disease than as a direct result of warfare. Florence Nightingale was called on to help in the military hospitals. She attempted to create sanitary conditions, believing in the healing power of pure air. Her reforms in hospital sanitation markedly improved the survival rate of patients and limited the further spread of disease. Although Nightingale had only contempt for the germ theory of disease, the sanitary conditions she fostered worked because she was stemming the spread of disease-causing microorganisms.

isms, many items that are free of microorganisms are kept sterile by wrapping them in paper, plastic, or foil. Such wrappings are impermeable to microorganisms and prevent recontamination. For example, foods are packaged in cans, jars, or other containers that prevent contamination with microorganisms that could contaminate the foods, causing them to spoil and rendering them a threat to human health. Similarly, syringes, needles, scalpels, and other medical instruments are sealed in sterile packages, as are Petri dishes, pipettes, and many other items used in the microbiology laboratory. Surgical gloves and gowns help prevent microorganisms from passing from one person to another, especially during surgical procedures or where a hospital worker comes in contact with a patient that has a contagious disease.

HIGH TEMPERATURES

Temperature is one of the most important environmental factors affecting the rates of microbial growth and death. Temperature influences the rates of chemical reactions by altering the three-dimensional shapes of proteins, thereby affecting the rates of enzymatic activities. Heat can kill microorganisms by denaturing their enzymes. Consequently, high temperatures can be used to kill microorganisms in order to control their proliferation. At temperatures exceeding the maximal growth temperature, the death rate exceeds the growth rate. The higher the temperature above the maximal growth temperature, the higher the death rate for that microorganism.

> Temperature influences the rates of chemical reactions by affecting the three-dimensional configuration of proteins and thus enzymatic activity; high temperatures kill microorganisms by denaturing their enzymes.

High temperatures can be used to reduce the numbers of microorganisms or to eliminate all viable microorganisms. When using heat to kill microorganisms, the degree of the microorganism's heat resistance must be considered so that the correct exposure times and temperatures are used. Heat resistance varies among different microorganisms; these differences can be expressed through the concept of thermal death point. **Thermal death point** (TDP) is the lowest temperature required to kill all of the microorganisms in a liquid suspension in 10 minutes. Heat is the most widely applicable and effective agent for killing microorganisms and also the most economical and easily controlled.

The heat killing of microorganisms can be described by the **decimal reduction time (D value)** (FIG. 11-1). D is defined as the time required for a tenfold reduction in the number of viable cells at a given

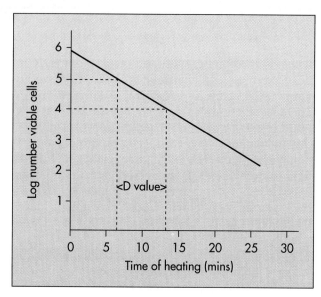

FIG. 11-1   The D value is the time in minutes needed to reduce the number of viable microorganisms by a factor of ten (one log unit).

temperature, that is, the time required for a log reduction in the number of microorganisms. This is the time required to bring about a 90% reduction in the numbers of viable microorganisms. As the temperature is increased above the maximal growth temperature for a microorganism, the decimal reduction time is shortened. The decimal reduction time varies for different microorganisms. In the food industry, the decimal reduction time is important in establishing appropriate processing times for sterilizing food products.

> The **D** value describes the rate of death at a given temperature.

Pasteurization

Pasteurization is a process that uses relatively brief exposures to moderately high temperatures to reduce the numbers of viable microorganisms and to eliminate human pathogens (FIG. 11-2). Such procedures prolong **shelf life** and ensure safety of the food as long as human pathogens are eliminated. A pasteurized food, however, retains viable microorganisms, which means that additional preservation methods are needed to extend the shelf life of the product. These other preservation methods, such as refrigeration, are used to reduce the growth rates of the surviving microorganisms. Pasteurization of milk, for example, is required by law in the U.S. to eliminate pathogenic bacteria, namely *Brucella* sp., *Coxiella burnetii*, and *Mycobacterium tuberculosis*. These bacteria are associated with the transmission of disease via contaminated milk. They are relatively sensitive to elevated temperatures, and pasteuriza-

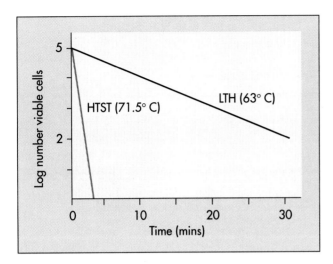

FIG. 11-2 Pasteurization reduces the numbers of viable microorganisms and eliminates human pathogens. The same reduction of viable microorganisms is achieved by heating to 63° C for 30 minutes as is accomplished by exposure to 71.5° C for only 0.33 minutes. The long time hold (LTH) pasteurization process uses 63° C and the high temperature-short time (HTST) pasteurization process uses 71.5° C.

tion is, therefore, normally achieved by exposure of milk to 62.8° C for 30 minutes (**low temperature-hold** or **LTH process**) or 71.7° C for 15 seconds (**high temperature-short time** or **HTST process**). Milk produced in the United States is normally preserved by pasteurization and requires refrigeration to extend its shelf life.

> Pasteurization uses brief exposures to moderately high temperatures to reduce numbers of viable microorganisms and eliminate human pathogens but does not eliminate all viable microorganisms.

Sterilization

Sterilization is the complete elimination of living organisms. In many microbiological procedures, high temperatures are used to kill all viable microorganisms, that is, to sterilize materials. As long as there are no endospore-forming bacteria, boiling at 100° C for 10 minutes is adequate to eliminate microorganisms from water. When the potential for the contamination of water supplies with enteric pathogens exists, boiling ensures the bacteriological safety of the water. In many cases, higher temperatures are needed to ensure that all microorganisms, including endospore producers, are killed.

> Sterilization eliminates all living organisms.

Excessive heat destroys the quality of milk, for example, boiling alters the smell and taste of milk, but brief exposure to high temperatures kills all the mi-

croorganisms in milk without destroying its quality. In the **ultra high temperature** or **UHT process**, exposure to 141° C for 2 seconds is used to sterilize milk. Sterilized milk is currently marketed in several European countries and has been introduced in the United States. It has an indefinite shelf life, provided the container remains sealed.

The heat used in sterilization can be moist or dry heat. Dry heat kills by oxidation. This is, for example, what happens when paper slowly chars in a heated oven when the temperature is below the ignition point of paper. Moist heat kills microorganisms more quickly because the water hastens the breaking of hydrogen bonds that hold proteins in their three-dimensional structure.

Dry heat sterilization requires higher temperatures for much longer exposure periods to kill all the microorganisms in a sample. Exposure in an oven for 2 hours at 170° C (328° F) is generally used for the dry heat sterilization of glassware and other items. The longer period of time and higher temperature (relative to moist heat) are required because heat in water is more readily transferred to a cool body than heat in air. Heat sterilization is very important in medical microbiology where sterile instruments are required for surgical procedures and where contaminated materials must be sterilized before they can be reused or safely discarded.

We routinely sterilize transfer loops by flaming them red hot before aseptically transferring a culture from one site to another. In this case we use a very high temperature for a short time. A similar principle is used in **incineration,** an effective way to sterilize and dispose of contaminated paper cups, bags, and dressings. Medical wastes are safely disposed by using incineration to kill any viable microorganisms.

> Dry heat sterilization uses higher temperatures and often takes longer than moist heat sterilization.

Moist heat is far more penetrating than dry heat and, hence, more effective for killing microorganisms. Steam under pressure is frequently used in sterilization procedures. Steam at 121° C for 15 minutes is at least as effective as a dry oven treatment at 170° C for 2 hours and kills all microorganisms, including endospores (FIG. 11-3). Culture media in bacteriological laboratories are normally prepared by heat sterilization in an **autoclave,** an instrument that permits exposure to steam under pressure. The autoclave is basically a chamber that can withstand pressures of greater than two atmospheres. The materials to be sterilized are placed in a chamber, and the chamber is sealed. Steam then is transferred from a jacket into the chamber, forcing out all the air. The steam is held in the chamber for the necessary time and then vented from the chamber. In the normal heat steril-

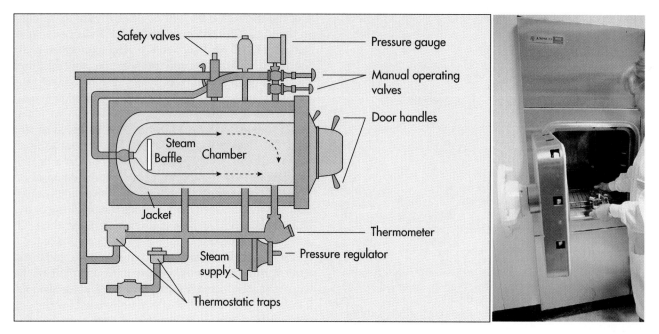

FIG. 11-3 An autoclave is used to sterilize materials by exposure to steam under pressure.

METHODOLOGY

STERILITY TESTING

Testing of sterility is an important quality control procedure. Autoclaves have pressure gauges and thermometers that monitor the sterilization operation. These monitors are supplemented with internal monitors. Chemicals that darken when exposed to specific heat treatments are sometimes included with each sterilization batch. These chemicals are often impregnated into a tape or wrapping material so that, after adequate exposure to elevated temperature, the darkened chemical spells out the word STERILE or AUTOCLAVED (see Figure). In this way the proper operation of the sterilizer is monitored and, simultaneously, the sterilized material is properly labelled.

Another method for monitoring the operation of the sterilizer is to use a biological indicator. This is the best method for assuring adequacy of the sterilization procedure because the method actually monitors the death of heat-resistant endospores. Typically, spores of *Bacillus stearothermophilus* are used for monitoring the effectiveness of steam sterilizers. The indicator consists of a strip of filter paper impregnated with *B. stearothermophilus* spores. The test strip is exposed to the sterilization procedure and viability is tested by placing the strip in a nutrient solution. After incubation for 24 hours the tube containing the spore strip and nutrient solution is examined for growth. The lack of growth indicates successful sterilization; the occurrence of any growth indicates a failure of the sterilization system.

Photograph showing sterility check with special heat-sensitive tape. Left flask is before sterilization and right flask is after autoclaving. The heat sterilization causes the letters on the tape to appear, indicating that the media has been autoclaved.

ization process, the medium is exposed to steam at a temperature of 121° C (which corresponds to 15 pounds per square inch pressure) for 15 minutes in an autoclave.

> An autoclave is an apparatus in which objects or materials may be sterilized by steam under pressure at temperatures over 100° C.

Canning

Canning is the use of heat followed by maintenance of anaerobic conditions (FIG. 11-4). In this preserva-

FIG. 11-4 Canning is widely used to preserve foods and keep them free of pathogenic microorganisms. Canning involves heat killing of microorganisms and hermetic sealing under anaerobic conditions to prevent recontamination and spoilage.

tion method the high temperature exposure kills all the microorganisms in the product, the can or jar acts as a physical barrier to prevent recontamination of the product, and the anaerobic conditions prevent oxidation of the chemicals in the food. Exposure to 115° C for at least 15 minutes is generally considered necessary in home canning to ensure killing of endospore formers in moderate acid (pH 4.5-5.3) to low acid (pH 5.3-7.0) foods (Table 11-1).

> Canning is a food preservation method in which suitably prepared foods are placed in glass or metal containers that are heated, exhausted, and hermetically sealed.

Particular concern is given in canning to ensuring sterility of such foods because of the possible growth of *Clostridium botulinum* and the seriousness of the disease botulism. Somewhat lower temperatures, for example, exposure to 100° C for 10 minutes, are often employed in the home canning of acidic (pH less than 4.5) foods. This can be done in part because of the lowered thermal resistance of microorganisms under acidic conditions and because of the fact that *C. botulinum* is unable to grow at low pH values.

The canning industry typically heats food at 121° C for 2.52 minutes. This is 12 times the D value for the endospores of *C. botulinum* and therefore reduces the probability of the survival of *C. botulinum* endospores to 10^{-12}. Thus, if there were one spore in every can, the probability of contamination remaining after processing should be reduced to one in every trillion cans. Heating at 121° C for 2.52 minutes therefore should ensure the safety of canned foods with respect to possible contamination with *C. botulinum*. Several commercial canning operations that have not adhered to the necessary standards and whose operations resulted in outbreaks of botulism have been put out of business. Most problems occur with home canning, where a lack of knowledge or care can result in insufficient heating.

TABLE 11-1				
Classification of Canned Foods and Their Processing Requirements				
ACIDITY CLASS	**pH**	**REPRESENTATIVE FOODS**	**SPOILAGE AGENTS**	**PROCESSING**
Nonacid-low acid	6-7	Beans, peas, carrots, beets, asparagus, potatoes, poultry, beef, fish, low acid tomatoes	Mesophilic, *Clostridium* spp.	High temperature (121° C)
Acidic-moderately acid	4-5	Tomatoes, pears, peaches, oranges, apricots, apples, pineapples, strawberries, sauerkraut	Aciduric bacteria, fungi	Boiling water (100° C)
Highly acid	3	Pickles, relish, vinegar	Yeasts and other fungi	Boiling water (100° C)

HISTORICAL PERSPECTIVE

DEVELOPMENT OF CANNING

Canning to preserve foods has its origin in 1795 when the French government offered a prize of 12,000 francs for the development of a practical method of food preservation. In 1809 Francois (Nicolas) Appert succeeded in preserving meats in glass bottles that had been kept in boiling water for varying periods of time. Appert was issued a patent for his process in 1810. As early as 1820 the commercial production of canned foods was begun in the United States by W. Underwood and T. Kensett. Appert's discovery that foods could be preserved for prolonged periods of time when they were heated and stored in the absence of oxygen came at the same time that the questions of spontaneous generation and the role of microorganisms in fermentation and putrefaction were being debated by the premier sci-

entists of the day. Spallanzani in 1765 showed that beef broth that had been boiled for an hour and sealed did not spoil. Appert applied to foods the results of Spallanzani's experiments. Not being a scientist, Appert probably did not understand why his method worked or its long-range significance. It was not until almost a half century later that Pasteur, in disproving the theory of spontaneous generation, provided the scientific basis for understanding why Appert's canning method works. Pasteur pointed out that Appert's method, even when modified by using temperatures below 100° C and relatively short incubation times, was a practical method for preventing undesirable ferments. Today, the method of canning, begun almost two centuries ago, is a widely used method for preserving foods.

LOW TEMPERATURES

Low temperatures limit the rates of microbial reproduction and thus can be used to prevent or to limit microbial growth (FIG. 11-5). Refrigeration and freezing are widely used to restrict microbial growth. Samples collected in hospital wards for microbiological analysis sometimes are chilled to prevent microbial growth before reaching the clinical microbiologist. Many foods are kept in the refrigerator or freezer to prevent microbial spoilage. Most mesophilic microorganisms grow extremely slowly at refrigerator temperatures (5° C). Although most pathogenic microorganisms are unable to grow in refrigerated

FIG. 11-5 Refrigeration is used to limit microbial growth. Many foods and products, including microbiological media, shown in this photograph, are preserved by refrigeration.

foods, *C. botulinum* type E will grow and produce toxin. Psychrophilic microorganisms are also able to grow slowly at 5° C. Thus, although refrigeration extends the shelf life of a product, it does not do so indefinitely.

Freezing at temperatures of −20° C or lower precludes microbial growth entirely. Freezing does not kill most microorganisms, although some microbial death may occur during freezing and during thawing as a result of ice crystal damage to microbial membranes. In fact, freezing at extremely low temperatures is routinely used for preserving microorganisms in type culture collections. Therefore, when food is thawed, the microorganisms associated with that food can grow, leading to food spoilage and potential accumulation of microbial pathogens and toxins if the food is not promptly prepared or consumed. Not all food products can be preserved by freezing because of the damage that may occur to the food as a result of ice formation. Desiccation of frozen foods (freezer burn), although not a microbial spoilage process, causes serious quality defects.

> Low temperatures limit rates of microbial reproduction. Temperatures below −20° C preclude microbial growth entirely.

Once thawed, it is generally not advisable to refreeze food products. Freezing, thawing, and refreezing disrupts the texture of the food and in addition permits invasion of the food by microorganisms that are normally restricted to the food surfaces. When thawed a second time, refrozen food products are more prone to microbial spoilage than foods that are allowed to thaw only once.

REFRIGERATION FAILS TO PROTECT AGAINST *YERSINIA ENTEROCOLITICA*

In 1978, containers of chocolate milk were delivered to a school in upstate New York. The milk was refrigerated over a holiday weekend, but it was contaminated with *Yersinia enterocolitica*, a disease-causing bacterium that can grow at refrigerator temperatures. Incubation over the weekend meant that each container of chocolate milk was a culture of *Yersinia*. Fortunately, the ability of a human pathogen to grow at refrigerated temperatures is limited to few species like *Yersinia enterocolitica*. Otherwise, refrigeration would not be useful for preserving foods. Unfortunately, the children who drank the milk containing *Yersinia enterocolitica* developed abdominal pain and other symptoms that normally are characteristic of appendicitis. In all, 200 children were afflicted. One by one they reported to physicians, some of whom diagnosed these as cases of appendicitis, rather than the actual cause—a *Yersinia* infection. Appendicitis is not an infectious disease and would not afflict 200 children simultaneously. Ten appendectomies were performed before the cause of the disease was recognized.

REMOVAL OF WATER—DESICCATION

Many foods are preserved by **desiccation** (removal of water). Water is required for microbial growth. By eliminating water or keeping surfaces dry, microbial growth can be prevented. Exposed wood surfaces are often painted to keep the wood dry enough to preclude microbial growth. Canvas and other textiles are preserved in temperate zones by the lack of water in the air, but in tropical zones these same materials are subject to biodeterioration because the humidity is sufficiently high to permit microbial growth.

Because water is required for microbial growth, many foods can be preserved by desiccation.

Although lack of available water prevents microbial growth, it does not necessarily accelerate the death rate of microorganisms. Some microorganisms, therefore, can be preserved by drying. Active dried yeast is used for baking purposes. After the addition of water, the yeasts begin to carry out active metabolism. **Freeze-drying** (lyophilization) is a common means of removing water that can be used for preserving microbial cultures (FIG. 11-6). During freeze-

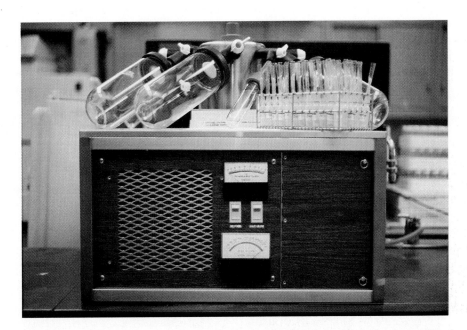

FIG. 11-6 Lyophilization, or freeze-drying, is used to preserve microbial cultures. The instrument used for this process uses a high vacuum and low temperature so that water sublimes (goes from the solid frozen state directly to a gas). This removes water from the specimen without disrupting cellular structures, allowing viability to be maintained.

drying, water is removed by sublimation, that is, water is converted directly from the solid to the gas phase. This process generally eliminates damage to microbial cells from the expansion of ice crystals.

Lack of water prevents microbial growth but does not accelerate the microbial death rate, making possible preservation by drying.

Whereas some microorganisms are relatively resistant to drying, other microorganisms are unable to survive desiccating conditions for even a short period of time. The ability to withstand drying can have important pathogenic implications. *Mycobacterium tuberculosis* is a classic example of an organism capable of withstanding severe desiccation and still remaining infective. In contrast, *Treponema pallidum,* the bacterium that causes syphilis, is extremely sensitive to drying and dies almost instantly in the air or on a dry surface.

Some microorganisms produce specialized spores that can withstand the desiccating conditions of the atmosphere. Such spores generally have thick walls that retain moisture within the cell. Many fungal spores can be transmitted over long distances through the atmosphere; some spores even travel from one continent to another. The transmission of fungal spores through the air is a serious problem in agriculture because it permits the spread of fungal diseases of plants from one field to another.

RADIATION

High-energy, short-wavelength radiation disrupts DNA molecules. Exposure to such radiation may cause mutations, many of which are lethal. Exposure to **gamma radiation** (short wavelengths of 10^{-3} to 10^{-1} nanometers), **X-radiation** (wavelengths of 10^{-3} to 10^{-2} nanometers), and **ultraviolet radiation** (ultraviolet light with wavelengths of 100 to 400 nanometers) increases the death rate of microorganisms and is used in various sterilization procedures to kill microorganisms. Gamma and X-radiation have high penetrating power and are able to kill microorganisms by inducing or forming toxic free radicals (ions). Therefore gamma and X-radiations are referred to as ionizing radiations. Free radicals are highly reactive chemical species that can lead to polymerization and other chemical reactions disruptive to the biochemical organization of microorganisms. Viruses and other microorganisms are inactivated by exposure to ionizing radiation (FIG. 11-7).

Exposure to radiation may cause microbial mutations, induces the formation of toxic free radicals, increases microbial death rates, and is used as a sterilization method.

Sensitivities to ionizing radiation vary. Nonreproducing (dormant) stages of microorganisms tend to be more resistant to radiation than growing organisms. For example, endospores are more resistant than the vegetative cells of many bacterial species.

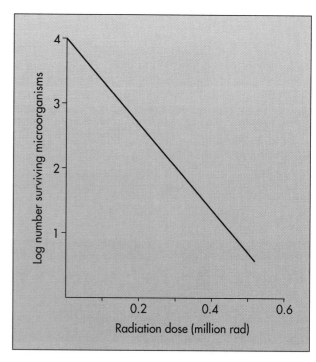

FIG. 11-7 Ionizing radiation effectively kills microorganisms, including viruses.

HIGHLIGHT

SAFETY OF IRRADIATED FOOD

Research on irradiation as a food preservation technology began after World War II when the U.S. Army began a series of experiments irradiating fresh foods for troops in the field. Since 1963, the United States Food and Drug Administration (FDA) has passed rules permitting irradiation to curb insects in food and microorganisms in spices, control parasites in pork, and retard spoilage in fruits and vegetables. On May 2, 1990, the FDA approved the use of irradiation as a safe and effective means to control a major source of foodborne illness—*Salmonella* and other foodborne bacteria in raw chicken, turkey, and other poultry. Food safety experts believe that up to 60% of all poultry sold in the United States is contaminated with *Salmonella* and that perhaps all chicken may be contaminated with *Campylobacter* organisms. Eating poultry contaminated with these organisms may cause disease, with symptoms ranging from a simple stomach ache to incapacitating stomach and intestinal disorders, occasionally resulting in death.

Although the FDA has concluded that irradiation of food is safe, the public remains frightened by any use of radiation. They are fearful that irradiated foods may be contaminated and carry dangerous radioactivity. They associate radiation with atomic bombs and nuclear reactor accidents like Chernobyl and Three Mile Island. Action groups have formed to block the distribution of foods sterilized by irradiation. Three states (Maine, New York, and New Jersey) have banned or issued moratoriums on the sale of irradiated foods. Irradiation

opponents charge that the FDA, the World Health Organization, and the nuclear power industry are conspiring to promote the technique as a way to dispose of nuclear waste.

To counter fears about radiation, the FDA points out that irradiation does not make food radioactive. The specified exposure times and energy levels of radiation sources approved for food cannot induce radioactivity in the food. Food irradiation does not leave a residue that is harmful to people. It removes potentially harmful pathogens and food spoilage microorganisms. During the irradiation process, the genetic material of bacteria is damaged such that they can no longer survive or multiply. No radioactive material is ever added to the product. The same technique is used to sterilize many disposable medical devices.

The FDA requires that irradiated foods be labeled as such so that consumers know what they are buying. A mandatory logo was added in 1986. It consists of a solid circle, representing an energy source, above two petals, which represent food. Like the FDA, the World Health Organization concludes that irradiation can substantially reduce food poisoning. It sees the use of irradiation as a means of reducing food cost because it can reduce food spoilage. The first major food irradiation plants are currently under construction. The success of these plants will depend on consumer acceptance of the products. You can expect to see foods labeled with the FDA-required labels soon on the shelves of your food market.

Exposure to 0.3 to 0.4 Mrads (million units of radiation) is necessary to cause a tenfold reduction in the number of viable bacterial endospores. An exception is the bacterium *Micrococcus radiodurans,* which is particularly resistant to exposure to ionizing radiation. Vegetative cells of *M. radiodurans* tolerate as much as 1 Mrad of exposure to ionizing radiation with no loss of viability. It appears that efficient DNA repair mechanisms are responsible for the high degree of resistance to radiation exhibited by this bacterium.

Ionizing radiation is used to pasteurize or sterilize some products. Most commercially produced plastic Petri plates are sterilized by exposure to gamma radiation. Foods also sometimes are sterilized in this manner. Most such sterilization procedures employ gamma radiation from ^{60}Co or ^{137}Ce. Bacon, for example, can be sterilized by using radiation doses of 4.5 to 5.6 Mrads.

Unlike gamma radiation, **ultraviolet light** (UV) does not have high penetrating power. It is useful for

killing microorganisms only on or near the surface of clear solutions. The strong germicidal wavelength of 260 nanometers coincides with the absorption maximum of DNA, suggesting that the principal mechanism by which ultraviolet light exerts its lethal effect is through the disruption of the DNA. Microorganisms have several mechanisms that can repair the alterations in the DNA that are caused by exposure to ultraviolet light, limiting the effectiveness of using UV exposure to control microbial populations. Exposure to ultraviolet light sometimes is used to maintain the sterility of some surfaces. In some hospitals, benchtops are maintained bacteria-free when not in use by using an ultraviolet lamp. The dangers involved in human exposure to excess ultraviolet radiation include blindness if light is viewed directly.

Ultraviolet light kills microorganisms only on or near the surface of clear solutions by disrupting their DNA.

Long wavelengths of **infrared radiation** (wavelengths of 10^3 to 10^5 nanometers) and **microwave radiations** (wavelengths greater than 10^6 nanometers) also have poor penetrating power. Infrared and microwave radiations do not appear to kill microorganisms directly. Absorption of such long wavelength radiation, however, results in increased temperature. Exposure to infrared or microwave radiations can thus indirectly kill microorganisms by exposing them to temperatures that are higher than their maximal growth temperatures. Because microwaves generally do not kill microorganisms directly, there is some concern in the food industry that cooking with microwave ovens may not adequately kill microorganisms that contaminate food products.

CONTROL OF MICROBIAL GROWTH BY ANTIMICROBIAL AGENTS

Chemical inhibitors are widely used to prevent the spread of disease-causing microorganisms. They are also used to preclude the growth of microorganisms that cause spoilage of foods or biodeterioration of industrial products. Chemicals that kill microorganisms or prevent the growth of microorganisms are called **antimicrobial agents.**

Antimicrobial agents are chemicals that kill or prevent the growth of microorganisms.

Various types of antimicrobial agents are employed to control microbial growth. Many antimicrobial agents are designed to block active metabolism and prevent the organism from generating the macromolecular constituents needed for reproduction. Because resting stages are metabolically dormant and are not reproducing, they are not affected by such antimicrobial agents. Hence, growing microorganisms are more sensitive than dormant stages, such as spores. Similarly, viruses are more resistant than other microorganisms to antimicrobial agents because they are metabolically dormant outside host cells.

Antimicrobial agents are classified according to their application and spectrum of action (Table 11-2).

Microorganisms vary in their sensitivity to particular antimicrobial agents. The suffix -cidal is generally added to indicate the ability of that particular agent to kill microorganisms. A sporicidal agent, for example, kills bacterial spores, a bactericidal agent kills bacteria, and so forth. Other chemical agents are tagged with the suffix -static, indicating that growth of the microorganism is stopped (but the cells are not necessarily killed). A bacteriostatic agent inhibits the growth of bacteria; a fungistatic agent prevents the growth of fungi.

Antimicrobial chemicals are also classified according to their intended use (Table 11-3, p. 322). In particular, antimicrobial agents are differentiated based on whether they can be safely ingested or applied to body tissues. The term *disinfectant* refers to antimicrobial substances that kill microorganisms on inanimate objects. Generally, disinfectants are too harsh to be used on living tissues such as skin. Under appropriate conditions, a disinfectant may produce complete sterilization, that is, disinfectants may kill all microorganisms. In such cases they are termed *sterilizing agents.* However, disinfectants may not inactivate spores. A **sanitizer** represents a particular kind of disinfectant that is used to reduce numbers of bacteria to levels judged safe by public health officials.

TABLE 11-2

Terms Used to Describe Actions of Antimicrobial Agents		
TERM	**ACTION**	**EXAMPLES**
Algicide	Agent that kills algae	Copper sulfate
Bactericide	Agent that kills bacteria	Chlorhexidine, ethanol
Biocide	Agent that kills living organisms	Hypochlorite (bleach)
Fungicide	Agent that kills fungi	Ethanol, zinc pyrithione
Germicide	Chemical agent that specifically kills pathogenic microorganisms	Formaldehyde, silver, mercury
Sporicide	Agent that kills bacterial endospores	Glutaraldehyde
Virucide	Inactivates viruses so that they lose the ability to replicate	Cationic detergents (Cepacol, Zephriam)
Bacteriostatic	Inhibits the growth and reproduction of bacteria	Sorbate, benzoate
Fungistatic	Inhibits the growth and reproduction of fungi	Zinc oxide, calcium propionate

DISCOVERY OF ANTISEPTICS

The use of antiseptics to prevent infections was introduced by Joseph Lister, an English Quaker and physician, who revolutionized surgical practice in 1867 by introducing antiseptic principles (see Figure). The discovery in the early 1850s of anesthesia and its administration to patients made surgery much easier but, of course, did nothing to reduce the incidence of postsurgical disease, which was often as high as 90%, especially in military hospitals.

As a surgeon, Lister investigated various problems relating to inflammation and pus formation and the coagulation of blood in wound healing. Lister knew that in the 1840s, Ignaz Semmelweis, a Hungarian physician who worked in maternity wards in Vienna, showed that physicians who went from one patient to another without washing their hands were responsible for transmitting childbed fever (puerperal fever). In an attempt to minimize postoperative infections, Lister tried to have his wards kept as clean as possible and to perform his surgeries in an environment that was also as clean as possible, but his patients still contracted hospital bacterial infections at an alarming rate.

Lister was also aware that Pasteur in the 1860s demonstrated that microorganisms in the air caused fermentation and putrefaction and reasoned that these same microorganisms could also cause wound infections. Therefore, if wounds could be kept free of microorganisms, he thought, they would not become infected. Lister experimented with solutions of carbolic acid (phenol) as an antiseptic during surgery to kill the microorganisms. He first used bandages soaked in carbolic acid to dress wounds due to compound fractures to diminish the likelihood of infection. Later he used a carbolic acid spray in addition to direct application of this compound during surgical procedures. Lister quickly achieved amazing results and in 1867 was able to report marked decreases in rates of infection and deaths due to infection. He eventually discarded the practice of spraying after 17 years of trials as unnecessary, but he retained the use of direct application. Lister's frequent modifications of his system and the fact that many physicians and surgeons would not accept the germ theory of disease caused many years of delay before Lister's innovations received widespread application. The evolution of aseptic surgery, preventing access by germs to the operative site by the use of sterilization for gowns, drapes, and instruments was simply the logical extension of Lister's work, making his achievement one of the first great triumphs of applied bacteriology in medicine.

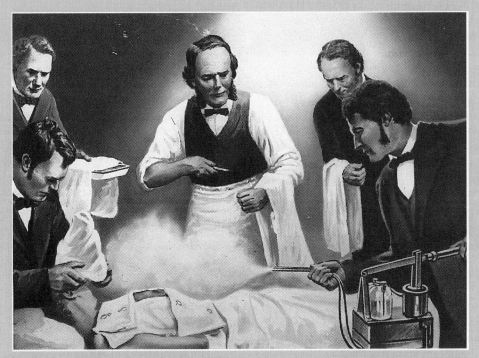

Joseph Lister (1827-1912) recognized the importance of preventing the contamination of wounds to curtail the development of infection. He developed antiseptic methods for preventing infection using carbolic acid (phenol) to treat wounds.

	TABLE 11-3	
Terms Used to Describe Antimicrobial Agents Based on Their Application		
TERM	**DESCRIPTION**	**EXAMPLES**
Antibiotic	Agent produced by microorganisms that inhibits or kills other microorganisms	Penicillin, erythromycin, tetracycline, cephalosporin
Antiseptic	Agent that kills or prevents the growth of microorganisms on living tissues	Mercurochrome, gentian violet, hydrogen peroxide, tincture of iodine, phenolics, ethanol
Disinfectant	Agent that kills microorganisms on inanimate objects	Hypochlorite (bleach), formaldehyde, glutaraldehyde
Sanitizer	A disinfectant that is used to reduce numbers of bacteria to levels judged safe by public health officials	Ethanol

Antiseptics kill or prevent the growth of microorganisms on living tissues. The term *antiseptic* literally means a substance that opposes sepsis, that is, a substance that works against putrefaction or decay. Such agents have minimal toxicity to human tissues and can safely be applied to body surfaces. They may be toxic, however, if ingested. Antiseptics are widely used to prevent disease by eliminating viable microorganisms from surface body tissues. Their use in preventing disease is discussed in detail later in this chapter.

Antibiotics, which are substances produced by microorganisms that inhibit or kill other microorganisms, are used to treat infections within the body. They were first discovered by Alexander Fleming who observed that the bacterium *Staphylococcus aureus* would not grow in the vicinity of the fungus *Penicillium* because this fungus produces the antibiotic penicillin. Since the introduction of penicillin in medicine about 50 years ago, antibiotics have revolutionized medical practice. Today many antimicrobials are produced synthetically and the term *antimicrobic* is often used instead of *antibiotic*. The role of antimicrobics in medicine is discussed in Chapter 17.

> **Disinfectants are antimicrobial substances that kill microorganisms on inanimate objects. Sanitizers reduce bacterial numbers to safe levels. Antiseptics kill or prevent the growth of microorganisms on living tissues. Antibiotics and other antimicrobics are used to treat infections within the body.**

FOOD PRESERVATIVES

Food preservatives are antimicrobial agents that prevent the growth of microorganisms in food products. They preserve food products against spoilage or biodeterioration. They are important for increasing the shelf life (preventing spoilage) of the food and for preventing the growth of disease-causing microorganisms within the food. The use of food preservatives is widespread and represents an important means of preserving foods and other products. A variety of food preservatives are used to protect different food products against specific microorganisms (Table 11-4). There is great public concern over the addition of any chemicals to foods because of the finding that some chemicals that were used as food additives, such as red dye number 2, are potential carcinogens. Still it must be remembered that the effective preservation of food prevents spoilage and the transmission of foodborne diseases. In the United States the FDA is responsible for determining and certifying the safety of food additives and must approve any chemicals that are added to foods as preservatives.

> **Food preservatives are antimicrobial agents that protect a food against microbial spoilage or the growth of disease-causing microorganisms.**

Salt and Sugar

The addition of salt or sugar to a food reduces the amount of available water and alters the osmotic pressure. High salt concentrations, such as exist in saturated brine solutions, are bacteriostatic—that is, they prevent the growth of bacteria. Salting is effectively used for the preservation of fish, meat, and other foods. However, because of the association of high levels of salt in the diet with high blood pressure and heart disease, there is currently great interest in lowering the salt content of foods. Sugars, such as sucrose, also act as preservatives and are effective in preserving fruits, candies, condensed milk, and other foods. Some foods, including maple syrup and honey, are preserved naturally by their high sugar content.

Acids

Various carboxylic acids are inhibitors of microbial growth. Lactic, acetic, propionic, citric, benzoic, and sorbic acids or their salts are effective food preservatives. An examination of the lists of food additives in the various foods in your pantry will rapidly con-

TABLE 11-4

Some Representative Chemical Food Preservatives

PRESERVATIVES	MAXIMUM	TARGET ORGANISMS	FOODS
Propionic acid and propionates	0.32%	Fungi	Bread, cakes, some cheeses
Sorbic acid and sorbates	0.2%	Fungi	Cheeses, syrups, jellies, cakes
Benzoic acid and benzoates	0.1%	Fungi	Margarine, cider, relishes, soft drinks, catsup
Sulfur dioxide, sulfites, bisulfites, metabisulfites	200—300 ppm	Microorganisms	Dried fruits, grapes, molasses
Ethylene and propylene oxides	700 ppm	Fungi	Spices
Sodium diacetate	0.132%	Fungi	Bread
Sodium nitrite	200 ppm	Bacteria	Cured meats, fish
NaCl	None	Microorganisms	Meats, fish
Sugar	None	Microorganisms	Preserves, jellies
Wood smoke	None	Microorganisms	Meats, fish

vince you of the wide use of organic acids as preservatives. The effectiveness of a particular organic acid depends on the pH of the food. For example, at the same pH, citric acid is less effective than lactic acid, which in turn is less effective than acetic acid.

Propionates are primarily effective against filamentous fungi. The calcium and sodium salts of propionic acid are used as preservatives in bread, cake, and various cheeses. Lactic and acetic acids are effective preservatives that form naturally in some food products. Cheeses, pickles, and sauerkraut contain concentrations of lactic acid that normally protect the food against spoilage. Vinegar is dilute acetic acid, which is an effective inhibitor of bacterial and fungal growth. Acetic acid is used to pickle meat products and is added as a preservative to various other products, including mayonnaise and catsup. Both of these preservatives, however, will prevent surface fungal growth on a food only if molecular oxygen is excluded.

Benzoates, including sodium benzoate, methyl *p*-hydroxybenzoate (methylparaben), and propyl-hydroxybenzoate (propylparaben), are extensively used as food preservatives in such products as fruit juices, jams, jellies, soft drinks, salad dressings, fruit salads, relishes, tomato catsup, and margarine. They are also used as preservatives in many pharmaceutical preparations.

Sorbic acid, used primarily as calcium, sodium, or potassium salts (for example, sodium sorbate) is more effective as a preservative at pH 4-6 than the benzoates. Sorbates inhibit fungi and bacteria, such as *Salmonella*, *Staphylococcus*, and *Streptococcus* species. Sorbates are frequently added as preservatives to cheeses, baked goods, soft drinks, fruit juices,

syrups, jellies, jams, dried fruits, margarine, and various other products.

Boric acid is used as a preservative in eyewash and other products. The limited toxicity of boric acid to human tissues makes it suitable for such applications. Boric acid is also used in urine collection jars to prevent bacterial growth between the time of collection and analysis.

Nitrates and Nitrites

Nitrates and nitrites are added to cured meats to preserve the red meat color and protect against the growth of food spoilage and poisoning microorganisms. Nitrates are effective inhibitors of *Clostridium botulinum* in meat products such as bacon, ham, and sausages. Recently, however, there is great concern over the addition of nitrates and nitrites to meats because these salts can react with secondary and tertiary amines to form nitrosamines, which are highly carcinogenic.

DISINFECTANTS AND ANTISEPTICS

Various chemicals are antimicrobial agents that are used as disinfectants and antiseptics (Table 11-5). Disinfectants are used for reducing the numbers of microorganisms on the surfaces of inanimate objects such as floors and walls. Many household cleaning agents contain disinfectants. Disinfectants are also used to limit microbial populations within liquids; for example in swimming pool water. Disinfectants, however, are not considered safe for application to human tissues or for internal consumption. If they are used to destroy microorganisms from a consumable product, such as drinking water, their concen-

TABLE 11-5
Summary of Chemical Agents Used as Disinfectants and Antiseptics

ANTIMICROBIAL AGENT	DESCRIPTION
Phenolics	Phenol is no longer used as a disinfectant or antiseptic because of its toxicity to tissues. Derivatives of phenol such as *o*-phenylphenol, hexylresorcinol, and hexachlorophene are used as disinfectants and antiseptics.
Halogens	Chlorination is extensively used to disinfect water; drinking water, swimming pools, and waste treatment plant effluent are disinfected by chlorination. Organobromine compounds are used to disinfect spas, swimming pools, and cooling towers. Iodine is an effective antiseptic; iodophors are used as disinfectants and antiseptics; the soaps used for surgical scrubs often contain iodophors.
Alcohols	Alcohols are bactericidal and fungicidal but are not effective against endospores and some viruses; ethanol and isopropanol are commonly used as disinfectants and antiseptics. Thermometers and other instruments are disinfected with alcohol, and swabbing of the skin with alcohol is done before injections.
Aldehydes	Formaldehyde is used as a preservative and disinfectant; glutaraldehyde is used to sterilize some surgical equipment.
Heavy metals	Heavy metals such as silver, copper, mercury, and zinc have antimicrobial properties and are used in disinfectant and antiseptic formulations. Silver nitrate was used to prevent gonococcal eye infections. Mercurochrome and Merthiolate are applied to skin after minor wounds. Zinc is used in antifungal antiseptics. Copper sulfate is used as an algicide.
Dyes	Several dyes, such as gentian violet, inhibit microorganisms and are used as antiseptics for treating minor wounds.
Surface-active agents	Soaps and detergents are used to remove microorganisms mechanically from the skin surface. Anionic detergents (laundry powders) remove microorganisms mechanically; cationic detergents, which include quaternary ammonium compounds, have antimicrobial activities. Quaternary compounds (quats) are used as disinfectants and antiseptics.
Oxidizing agents	Ethylene oxide is an excellent sterilizing agent, especially for objects that would be destroyed by heat; ethylene oxide sterilizers are used for the disinfection of plastics and linens. Ozone is a powerful oxidizing agent; ozonation may replace chlorination for the disinfection of drinking water; hydrogen peroxide is a mild antiseptic that is effective against anaerobic bacteria.
Acids	Organic acids can control microbial growth and are frequently used as preservatives. Sorbic, benzoic, lactic, and propionic acids are used to preserve foods and pharmaceuticals. Benzoic, salicylic, and undecylenic acids are used to control fungi that cause diseases such as athlete's foot.

tration must be reduced before the product is consumed.

Concentration and **contact time** are critical factors that determine the effectiveness of disinfectants and antiseptic agents against a particular microorganism. Microorganisms vary in their sensitivity to particular antimicrobial agents. Generally, growing microorganisms are more sensitive than organisms in dormant stages, such as spores. Many antimicrobial agents are aimed at blocking active metabolism and preventing the organism from generating the macromolecular constituents needed for reproduction. Because resting stages are metabolically dormant and are not reproducing, they are not affected by such antimicrobial agents. Similarly, viruses are more resistant than other microorganisms to antimicrobial agents because they are metabolically dormant outside host cells.

Antimicrobial agents are chemicals that kill or prevent the growth of microorganisms.

Disinfectants obviously should have high germicidal activity. They should rapidly kill a wide range of microorganisms, including spores. The agent should be chemically stable and effective in the presence of organic compounds and metals. The ability to penetrate into crevices is desirable. It is essential that a disinfectant not destroy the materials to which it is applied. Furthermore, it should be inexpensive and aesthetically acceptable.

Effective disinfectants kill a wide range of microorganisms, including spores; are chemically stable; work in the presence of organic compounds and metals; penetrate crevices; are safe on surfaces; and are inexpensive and aesthetically pleasing.

Two factors must be evaluated in determining the effectiveness of antiseptics: the agent must produce effective antimicrobial activity and must not be toxic to living tissues. A particularly meaningful approach for comparing antiseptics that encompasses the antimicrobial activity and the toxicity to tissues is the generation of a toxicity index. In the tissue toxicity test, germicides are tested for their ability to kill bacteria and their toxicity to chick-heart tissue cells. The toxicity index is defined as the ratio of the greatest dilution of the product that can kill the animal cells in 10 minutes to the dilution that can kill the bacterial cells in the same period of time and under identical conditions. For example, if a substance is toxic to chick-heart tissue at a dilution of 1:1,000 and bactericidal for *Staphylococcus aureus* at a dilution of 1:10,000, the toxicity index would be 1,000/10,000 or 0.1. Typical toxicity values for iodine solution and Merthiolate are 0.2 and 3.3, respectively. Ideally, an antiseptic should have a toxicity index of less than 1.0, that is, it should be more toxic to bacteria than to human tissue.

> Antimicrobial activity and lack of toxicity to living tissue must be determined for antiseptics by generating a toxicity index.

Halogens—Chlorine, Bromine and Iodine

The halogens—chlorine, bromine, and iodine—are effective microbicidal elements that are widely used as disinfectants. Chlorine kills microorganisms by disrupting membranes and inactivating enzymes (FIG. 11-8). Various inorganic and organic forms of chlorine are used for disinfection purposes. Hypochlorite solutions are commonly used in disinfecting and deodorizing procedures. Sodium hypochlorite (Clorox bleach) is widely used as a household disinfectant. Food processing plants and restaurants also use calcium and sodium hypochlorite solutions to disinfect utensils. In some hospitals, hypochlorite is used to disinfect rooms, surfaces, and nonsurgical instruments.

> **Chlorine kills microorganisms by disrupting membranes and inactivating enzymes.**

The germicidal action of chlorine is based on the formation of hypochlorous acid when it is added to water. Hypochlorous acid releases an active form of oxygen that reacts with cellular biochemicals. Chlorine gas condensed into liquid form is widely used for such disinfection. **Chlorination** (treatment with chlorine) or **chloramination** (treatment with chloramines) is the standard treatment for disinfecting drinking water in most communities. It is also used to disinfect effluents from sewage treatment plants to minimize the spread of pathogenic microorganisms. Because most forms of chlorine are inactivated in the presence of organic matter, this disinfection process does not completely eliminate microorganisms from the water.

The disinfection of water is very important because the same water supplies that are used as drinking water often are also used for human waste disposal. The outfall from one city's sewage treatment plant often flows downstream to the drinking water intake of another city. When the disinfection of drinking water is inadequate, other measures, such as boiling before use, must be employed to ensure water safety. For example, when flooding washes untreated sewage directly into the water supply, such alternative disinfection procedures become necessary. Outbreaks of diseases such as cholera and typhoid fever often occur if drinking water supplies are not completely disinfected.

Campers often use organic compounds containing chlorine to disinfect water. Their campsite water supply may be contaminated with fecal matter and associated human pathogens. Halazone (parasulfone dichloramidobenzoic acid) and succinchlorimide are examples of such chlorine-containing compounds. These organic chlorides, which are quite stable in tablet form, become active when placed in water. A halazone concentration of 4 to 8 milligrams per liter (mg/L) safely disinfects water containing *Salmonella typhi*, the bacterium that causes typhoid fever, within 30 minutes. Succinchlorimide at a concentration of 12 mg/L will disinfect water within 20 minutes. NASA uses iodine in space vehicles to treat potable water.

Chlorine is also used to disinfect swimming pools. Liquid chlorine and hypochlorite solutions are frequently used for such purposes. A residual chlorine level of 0.5 mg/L will achieve control of microbial populations and prevent the multiplication of pathogens in swimming pools. Such levels of chlorine are relatively harmless to human tissues, al-

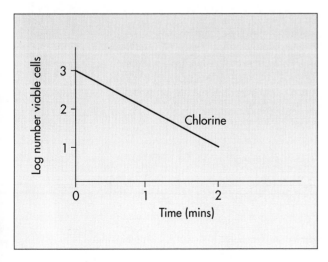

FIG. 11-8 Chlorine rapidly kills microorganisms and is widely used for the disinfection of water.

EVALUATION OF THE EFFECTIVENESS OF DISINFECTANTS

Concentration and contact time are critical factors that determine the effectiveness of an antimicrobial agent against a particular microorganism. Several standardized test procedures have been employed for evaluating the effectiveness of disinfectants. The classic test procedure used until a few decades ago is the phenol coefficient (FIG. *A*). Phenol, which is also known as carbolic acid, was the antimicrobial chemical used by Joseph Lister for preventing infections following compound fractures and for antiseptic surgical practice. The phenol coefficient test compares the activity of a given product with the killing power of phenol under the same test conditions.

To determine the phenol coefficient, dilutions of phenol and the test product are added separately to test cultures of *Staphylococcus aureus* or *Salmonella typhi*. The tests are run in liquid culture. After exposure for 5, 10,

and 15 minutes, a sample from each tube is collected and transferred to a nutrient broth medium. After incubation for 2 days, the tubes from the different disinfectant dilutions are examined for visible evidence of growth. The phenol coefficient is defined as the ratio of the highest dilution of a test germicide that kills the test bacteria in 10 minutes but not in 5 minutes to the dilution of phenol that has the same killing effect. For example, if the greatest dilution of a test disinfectant producing a killing effect was 1:100 and the greatest dilution of phenol showing the same result was 1:50, the phenol coefficient would be 100/50 or 2.0.

The phenol coefficient indicates the relative toxicity of various disinfectants but does not establish the appropriate concentration that should be used for disinfecting surfaces. The Association of Official Analytical Chemists (AOAC) use-dilution method, which

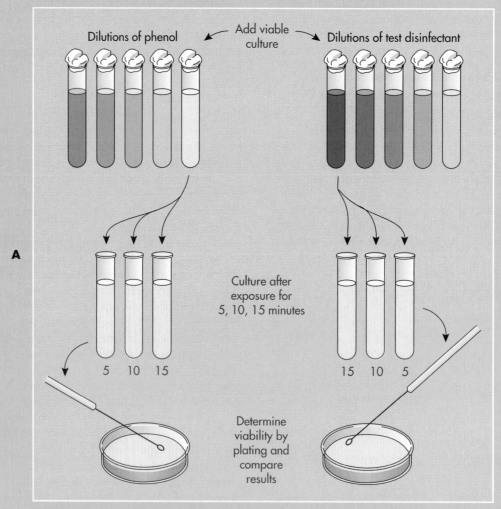

The phenol coefficient determines the relative effectiveness of a disinfectant to that of phenol for killing microorganisms.

has replaced the phenol coefficient as the standard method for evaluating the effectiveness of disinfectants, establishes appropriate dilutions of a germicide for actual conditions (FIG. *B*). In this procedure, disinfectants are tested against *Staphylococcus aureus* strain ATCC 6538, *Salmonella choleraesuis* strain ATCC 10708, and *Pseudomonas aeruginosa* strain ATCC 15442. Small stainless steel cylinders are contaminated with specified numbers of the test bacteria. After drying, the cylinders are placed into a series of specified dilutions of the test disinfectant. At least 10 replicates of each organism at the test dilutions of the disinfectant are used. The cylinders are *(1)* exposed to the disinfectant for 10 minutes, *(2)* allowed to drain, *(3)* transferred to appropriate culture media, and *(4)* incubated for 2 days. After incubation the tubes are examined for growth of the test bacteria. No growth occurs if the disinfectant is effective at the test concentration. An acceptable use-dilution is one that kills all test organisms at least 95% of the time.

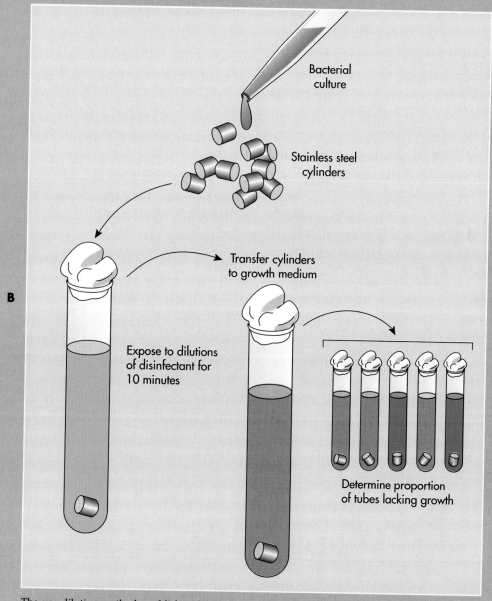

B

Bacterial culture

Stainless steel cylinders

Transfer cylinders to growth medium

Expose to dilutions of disinfectant for 10 minutes

Determine proportion of tubes lacking growth

The use dilution method establishes the appropriate concentration of a disinfectant that should be used.

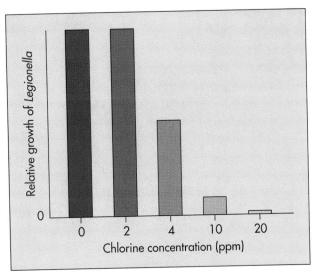

FIG. 11-9 Chlorination is used to control *Legionella* in cooling towers; chlorine effectively kills *Legionella*.

though prolonged exposure can cause irritation to the eyes and will bleach swimsuits.

Chlorination of water in air-conditioning cooling towers is important to control populations of *Legionella pneumophila*, the bacterium that causes Legionnaire's disease (FIG. 11-9). Outbreaks of Legionnaire's disease have frequently been traced to aerosols released from cooling towers that are then dispersed through the air. It has also been suggested that disinfection of home water heaters may be necessary to prevent the multiplication of *L. pneumophila* and its subsequent release in aerosols produced by shower heads. When excessive levels of *L. pneumophila* are detected, shock treatment with calcium hypochlorite at a dose of 50 mg/L per day will lower the concentration of this disease-causing bacterium to acceptable levels.

Generally, bromine is too toxic to be used safely around people. However, organic-bromine complexes are used to disinfect the water in swimming pools, hot tubs, and whirlpools.

Iodine is a very effective antiseptic agent because it is bactericidal and sporicidal so that it kills all types of bacteria, including endospores. It is also effective against many protozoa. Iodine is used in alcohol solution (tincture of iodine) and in combination with organic molecules (an iodophor). Iodine combined with polymers such as polyvinylpyrrolidone is a particularly effective antiseptic because the iodine is released slowly. Iodine is frequently applied to minor wounds to kill microorganisms on the skin, thereby preventing infection of the wound. Normally it does not seriously harm human tissues but tincture of iodine stains tissue and may cause local skin irritation and occasional allergic reactions.

Iodophors are less irritating than tincture of iodine and do not stain the skin surface. Antiseptic iodophors are used routinely for preoperative skin cleansing and disinfection. Wescodyne and Betadine are frequently used for preoperative disinfection of skin and laboratory paraphernalia. In surgical procedures, the surgical staff often scrubs with soap impregnated with iodine, and the patient's skin in the area of the incision is normally treated with iodine before beginning the surgical procedure. Betadine and Isodine are frequently used for this purpose. A standard surgical scrub with a 10% solution (1% available iodine) decreases the cutaneous bacterial population by 85% and is particularly effective against Gram-negative bacteria.

Phenolics

Phenol (carbolic acid) is probably the oldest recognized disinfectant. Its use as a germicide in operating rooms was introduced by Joseph Lister in 1867. Phenol and its chemical derivatives (phenolics) disrupt plasma membranes, inactivate enzymes, and denature proteins, thereby exerting antimicrobial activities. Phenolics are particularly useful as disinfectants because they are very stable when heated or dried. They also retain their activity in the presence of organic material. Phenolics are commonly used for the disinfection of hospital floors and walls.

> **Phenolics disrupt plasma membranes, inactivate enzymes, and denature proteins. They are stable when heated or dried and remain active in the presence of organic material.**

Cresols, which are phenolic derivatives of coal tars, are good disinfectants. Although cresols are only moderately effective against bacterial spores and are disruptive to human tissues, they are useful for disinfecting inanimate objects. The active ingredient in Lysol, a commonly used household disinfectant, is the cresol *o*-phenylphenol. In hospitals the commercial phenolics One-Stroke Vesthene and Wexide are used to disinfect emergency rooms and operating rooms (FIG. 11-10). The distinctive aroma of these phenolics gives many hospitals their characteristic smell. In addition to their use for disinfecting floors and walls, phenolics are incorporated into telephone poles, railroad ties, and other wood products to prevent microbial deterioration of the wood.

Several phenolic compounds are widely used antiseptics. Resorcinol (meta-dihydroxybenzene) is only about one third as active as phenol, but it is both bactericidal and fungicidal. Resorcinol is used in the treatment of acne, ringworm, eczema, psoriasis, dermatitis, and other cutaneous lesions. It is usually applied as a 10% ointment or lotion. Hexylresorcinol is commonly used in mouthwashes and in over-the-counter drugs used for treating sore throats. Thymol is used in vaginal deodorants at a concentration of

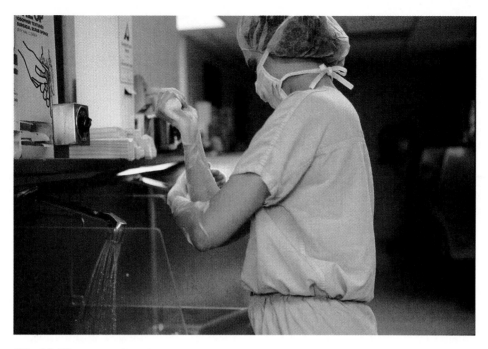

FIG. 11-10 Hospital staff scrub to disinfect skin and reduce the chances of tranferring pathogens.

1% because of its antibacterial and antifungal activities.

Hexachlorophene is one of the most useful of the phenol derivatives. Combined with a soap, it is a highly effective skin disinfectant. Unlike most phenolic compounds, hexachlorophene has no irritating odor and has a high residual action. Hexachlorophene is more effective against Gram-positive than against Gram-negative bacteria. A 3% solution of hexachlorophene will kill *Staphylococcus* within 30 seconds, but up to 24 hours may be required to kill Gram-negative bacteria. Because most bacteria on the skin are Gram-positive, hexachlorophene, the active ingredient in pHisoHex, once was commonly used by surgeons, physicians, and other health care workers. It was also used in the 1960s for daily bathing of newborns to prevent fatal *Staphylococcus* infections. However, it was found that frequent bathing of infants with hexachlorophene could lead to neurological damage and therefore this practice was largely discontinued. Over-the-counter preparations of hexachlorophene were banned in the United States by the FDA, but hexachlorophene is still used for limited purposes in hospitals.

The use of pHisoHex as a surgical scrub has largely been replaced by the use of povidone iodine (scrub soap form of Betadine) and Hibiclens (chlorhexidine gluconate). These two scrubs are effective against many microorganisms commonly encountered in hospitals. The antimicrobial action of iodine has already been discussed. Hibiclens is chemically different from other antimicrobials that are commonly used in the United States. It maintains a high level of antimicrobial activity in the presence of organic matter, such as blood, and does not irritate or dry the skin. **Chlorhexidine** is not a phenol, although its structure and application resemble hexachlorophene. It is frequently used for disinfection of skin and mucous membranes as an alternative to hexachlorophene. It is combined with a detergent or alcohol for surgical hand scrubs and preoperative skin preparation in patients. In such applications, it works more rapidly than hexachlorophene and is equally persistent on the skin. The skin does not absorb it, and no toxicity has been reported. Its killing effect is related to damage to the plasma membrane. It is effective against most vegetative bacteria, but not against spores.

Detergents

Detergents also are effective for removing microorganisms from floors and walls. One end of a detergent molecule is hydrophilic and mixes well with water. The other end is hydrophobic and is attracted to nonpolar organic molecules. If the detergents are electrically charged, they are termed *ionic*. Anionic (negatively charged) detergents are only mildly bactericidal. Anionic detergents are used as laundry detergents to remove soil and debris. They also reduce numbers of microorganisms associated with the item being washed. Cationic (positively charged) detergents are highly bactericidal, that is, they kill bacteria. In particular, cationic detergents are effective against *Staphylococcus* and various viruses. This quality makes them excellent candidates for disinfecting agents for hospital use.

The most widely used cationic detergents are **quaternary ammonium compounds (quats).** These compounds have four organic groups bonded to a nitrogen atom. Some examples of commonly used quats are Ceepryn (cetylpyridinium chloride), Phemerol (benzethonium chloride), and Zephiran (benzalkonium chloride). These chemicals are commonly used in antiseptic scrubs and mouthwashes. They are effective against fungi and Gram-positive bacteria. Their bactericidal action appears to be based on the disruption of plasma membrane and enzyme function. The quaternary ammonium compounds that are effective antimicrobial agents are used in concentrations that are not irritating to human tissues. Several quaternary ammonium cationic detergents are used as antiseptic agents. These compounds are relatively nonirritating to human tissues at concentrations that are inhibitory to microorganisms. However, they act slowly and are inactivated by soaps. They are also adsorbed by cotton and other porous materials, which can severely interfere with their effectiveness as antiseptics. Many mouthwash formulations contain quats, as do storage solutions for contact lenses. Many hospitals also use quats for disinfecting floors and walls, as antiseptics, and for surgical scrubs.

> **Quaternary ammonium compounds have four organic groups bonded to a nitrogen atom. Their bactericidal action is based on disrupting plasma membranes and enzymes.**

There are, however, some problems associated with the use of quats as disinfectants. Their antimicrobial activity is lowered if they are absorbed by porous or fibrous materials such as gauze bandages. Hard water containing calcium or magnesium ions interferes with their action. Also, they can cause metal objects to rust. More importantly, rather than killing *Pseudomonas* species, quats actually support the growth of these bacteria. As such, quats are not used in operating theaters because of the danger that they will permit *Pseudomonas* to survive and infect surgical wounds.

Alcohols

Alcohols are among the most effective and heavily relied on agents for disinfection (FIG. 11-11). **Alcohols** denature proteins, disrupt membrane structure, and act as a dehydrating agent, all of which contribute to their effectiveness as an antiseptic. Even viruses are inactivated by alcohol. Methanol, ethanol, and isopropanol are commonly used for disinfection. Of the three, isopropyl alcohol has the highest bactericidal activity and therefore is the most widely used.

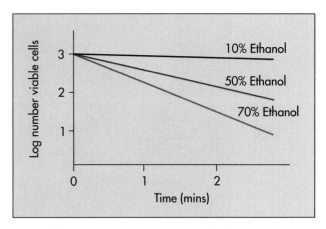

FIG. 11-11   Alcohol is a widely used antiseptic for killing microorganisms on skin. Its effectiveness is concentration dependent. The most effective concentration is about 70% (higher concentrations are less effective).

In practice, a solution of 70% to 80% alcohol in water is generally employed, although isopropyl alcohol is effective in solutions of up to 99%. On the skin, 70% ethanol kills nearly 90% of the cutaneous bacterial population within 2 minutes. Before puncturing the skin with a hypodermic syringe, the area is generally wiped with alcohol. Cabinet surfaces are frequently disinfected by wiping with alcohol. Some medical instruments are left soaking in alcohol to maintain their sterility. Oral thermometers are often wiped with alcohol to kill disease-causing microorganisms that otherwise might be transmitted from one patient to another. Even though brief exposure to alcohol is not sufficient to achieve sterility, it reduces the numbers of microorganisms to levels that make infection unlikely.

> **Alcohols denature proteins and disrupt membrane structure.**

Aldehydes

Formaldehyde and glutaraldehyde are useful for disinfecting medical instruments. These **aldehydes** kill microorganisms by denaturing proteins. Instruments can be sterilized by placing them into a 20% solution of formaldehyde in 70% alcohol for 18 hours. Formaldehyde, however, leaves a residue and it is necessary to rinse the instruments in sterile water before use. A solution of glutaraldehyde at pH 7.5 kills *Staphylococcus* within 5 minutes and *Mycobacterium tuberculosis* within 10 minutes, but endospores may survive for up to 12 hours. The use of glutaraldehyde is limited by its expense. It is used for sterilizing specialized medical instruments such as bronchoscopes.

> **Aldehydes denature proteins, thus killing microorganisms.**

Acids

Several acids are used as antiseptics. Acetic acid at a concentration of 5% is bactericidal and at lower concentrations is bacteriostatic. It is occasionally used at a concentration of 1% in surgical dressings. *Pseudomonas aeruginosa* is particularly susceptible to acetic acid, and this acid may be employed in burn therapy. It is used in vaginal douches to suppress fungal and protozoan infections of the vaginal tract.

Undecylenic acid is active against various fungi, including fungi that cause superficial mycoses. It is usually compounded with zinc but may also be used alone. Compounded undecylenic acid contains 2% to 5% undecylenic acid and 20% zinc undecylenate. This antiseptic agent is very useful for the treatment of ringworm. Undecylenic acid is the active ingredient in Desenex, which is used to treat athelete's foot and other fungus infections of the skin. Benzoic acid and salicylic acid used in combination also inhibit fungal growth. Whitfield's ointment contains benzoic acid and salicylic acid in a ratio of 2:1. This ointment is used to prevent fungal growth on the feet, as occurs in athlete's foot.

Ethylene Oxide

Ethylene oxide has several applications as a sterilizing agent. The ethylene portion of the molecule reacts with proteins and nucleic acids. **Ethylene oxide** kills all microorganisms and endospores. It is toxic and explosive in its pure form, so it is usually mixed with a nonflammable gas such as carbon dioxide or nitrogen.

A special autoclave-type sterilizer is used for ethylene oxide sterilization. Several hours of exposure to 12% ethylene oxide at 60° C is used for sterilization. Its remarkable penetrating power is one reason why ethylene oxide was chosen to sterilize spacecraft sent to land on the moon and planets—using heat to sterilize the electronic gear on these vehicles was not practical.

Because of their ability to sterilize without heat, gases like ethylene oxide are also widely used on medical supplies and equipment that cannot withstand steam sterilization. Examples include disposable sterile plasticware such as syringes and Petri plates, linens, sutures, lensed instruments, artificial heart valves, heart-lung machines, and mattresses. Many large hospitals have ethylene oxide chambers, some large enough to sterilize mattresses, as part of their sterilizing equipment. Additionally, some foods such as nuts and spices are sterilized by exposure to ethylene oxide.

Ethylene oxide sterilizes without heat, making it useful for sterilizing materials that cannot withstand high temperatures.

Hydrogen Peroxide

Hydrogen peroxide (H_2O_2) is an effective antiseptic. The molecule is unstable and degrades into water and oxygen. Anaerobic bacteria are particularly sensitive to peroxides because they do not have catalase, an enzyme that degrades peroxides. Hydrogen peroxide concentrations of 0.3% to 6.0% are used as antiseptics. A 3% solution is often used to cleanse and disinfect wounds. Although its germicidal action is brief, the effectiveness of hydrogen peroxide against anaerobic bacteria is important because several deadly anaerobic bacteria often are associated with soils that may contaminate wounds. Higher concentrations of 6.0% to 25.0% can be used in sterilization. Such treatment is useful for surgical implants and contact lenses because it leaves no residual toxicity after a few minutes of exposure. H_2O_2 is not effective against Gram-positive bacteria, including staphylococci found on skin.

Ozone

Ozone is a strong oxidizing agent that kills microorganisms by oxidizing cellular biochemicals. In some communities, ozone replaces chlorination as the primary means of disinfecting drinking water. Unlike chlorination, **ozonation** leaves no residue. The current cost of ozonation limits its widespread introduction for disinfection purposes. However, ozonation costs may eventually come down, particularly with the advantage of leaving no residue.

Dyes

Various dyes that are bacteriostatic or bactericidal are widely used as antiseptics. For example, crystal violet, which is also called gentian violet, is a potent bacteriostatic agent for Gram-positive bacteria. It is bactericidal at concentrations of less than 1:10,000. The mechanism of action of this compound against Gram-positive bacteria appears to be very similar to that of penicillin. It blocks a final step in the synthesis of cell wall material. Crystal violet has been used for the treatment of vaginal tract infections because the protozoan *Trichomonas* and the fungus *Candida albicans,* two common etiologic agents of vaginitis, are very sensitive to this dye.

Heavy Metals

Microorganisms are inhibited by heavy metals such as mercury, silver, and copper. Mercuric chloride was once used as a disinfectant solution, but because it is inactivated by organic matter, it is not widely used anymore. Copper sulfate is effective as an algicide. This compound is frequently added to swimming pools and aquaria to control algal growth.

Heavy metals are also used in antiseptic formulations. Mercury, zinc, and silver are examples of heavy

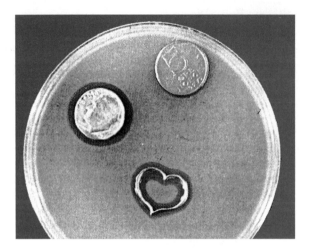

FIG. 11-12   Photograph of oligodynamic action of heavy metals.

metals used to kill microorganisms. The inhibitory effect of heavy metals is termed **oligodynamic action** (FIG. 11-12). Heavy metal ions can react with sulfhydral groups (—SH) on proteins, which inactivates them.

Silver nitrate has classically been applied to the eyes of newborn human infants to kill possible microbial contaminants. It is particularly important in the prevention of the transmission of gonococcal infections from mother to newborn. Silver combines with proteins and disrupts bacterial surface structures. Silver nitrate and silver sulfadiazine are used to treat severe burns. Preventing infections of tissues exposed by burns is critical to the recovery of the patient.

Organic mercury compounds are effective antiseptics for the treatment of minor wounds and as pre-

TABLE 11-6

Some Representative Antibiotics

ANTIBIOTIC	MECHANISM OF ACTION	TARGET ORGANISMS	SOME USES
Penicillin G	Inhibits bacterial cell wall synthesis	Gram-positive bacteria	Streptococcal sore throat, gonorrhea, syphilis
Ampicillin	Inhibits bacterial cell wall synthesis	Gram-positive and Gram-negative bacteria	Middle ear infections, urinary tract infections caused by *Enterococcus faecalis*, infections caused by some strains of *Escherichia coli*
Methicillin	Inhibits bacterial cell wall synthesis	Gram-positive and Gram-negative bacteria	Penicillinase-producing *Staphylococcus aureus* infections
Cephalosporins	Inhibits bacterial cell wall synthesis	Gram-positive and Gram-negative bacteria	Urinary tract and other infections caused by *Escherichia coli*, middle ear infections and meningitis caused by *Haemophilus influenza*
Streptomycin	Inhibits bacterial protein synthesis	Gram-negative bacteria	Bubonic plague, tularemia
Neomycin	Inhibits bacterial protein synthesis	Gram-negative bacteria	Sometimes used as a topical ointment for general cuts and abrasions of the skin
Chloramphenicol	Inhibits bacterial protein synthesis	Gram-positive and Gram-negative bacteria	Meningitis caused by *Haemophilus influenza* or *Neisseria meningitidis*, typhoid fever
Tetracycline	Inhibits bacterial protein synthesis	Gram-positive and Gram-negative bacteria	Pneumonia caused by *Mycoplasma*, non-gonococcal urinary tract infections
Bacitracin	Inhibits bacterial cell wall synthesis	Gram-positive bacteria	Topical ointment for general cuts and abrasions of the skin
Erythromycin	Inhibits bacterial protein synthesis	Gram-positive and Gram-negative bacteria	Whooping cough, diphtheria, diarrhea caused by *Campylobacter* and pneumonia caused by *Legionella* or *Mycoplasma*
Rifampicin	Inhibits bacterial RNA synthesis	Gram-positive bacteria and some Gram-negative bacteria	Tuberculosis and Hansen disease (leprosy)
Nystatin	Damages plasma membrane	Yeast	*Candida albicans* infections of skin and vagina
Griseofulvin	Inhibits mitosis	Fungi	Tinea (ringworm) of hair and nails
Amphotericin B	Damages plasma membrane	Fungi	Histoplasmosis, cryptococcal meningitis

servatives in serums and vaccines. The organic mercurials are bacteriostatic and relatively nontoxic. Mercurochrome (merbromin) was the first organic mercurial antiseptic to be introduced. Mercurochrome actually has limited bacteriostatic action and the lowest therapeutic index of the commercial mercurial antiseptics. Metaphen and Merthiolate are more effective.

Salts of zinc are used as mild antiseptics. Zinc pyrithione is used as an antidandruff agent. Calamine lotion, for example, contains zinc oxide. Calamine lotion is used in the treatment of ringworm, impetigo, and various other cutaneous diseases. White lotion, which contains zinc sulfate at a concentration of 4%, is also used to treat skin diseases and infections. A mixture of a long-chain fatty acid and the zinc salt of the acid is commonly used as an antifungal powder or ointment. It is particularly effective in the treatment of athlete's foot. The zinc salt also acts as an astringent and aids in healing superficial lesions, as does zinc oxide paste. Zinc oxide paste also is commonly recommended for treating diaper rash and its concurrent bacterial or fungal infections.

ANTIBIOTICS

Antibiotics are antimicrobial substances that were originally produced by other microorganisms. Many antibiotics today are synthetically manufactured by pharmaceutical companies. However, the original parent compound was isolated from a microorganism. For example, penicillin, the first antibiotic to be discovered, was isolated from a fungus, *Penicillium chrysogenum*. Streptomycin was first isolated from the bacterium *Streptomyces griseus*.

Antibiotics are used clinically to prevent or treat human diseases caused by microorganisms (Table 11-6). A more complete description of antibiotics and their use in combatting human infections is presented in Chapter 18. Antibiotics are also used in controlling microorganisms in nonhuman animals. They are routinely added to animal feeds to protect swine, cattle, poultry, and other farm animals from microorganisms. The addition of antibiotics to animal feeds stimulates the growth of the animals by controlling infections, especially in an animal's intestinal tract. This probably contributes to faster weight gain by the animals, which in turn reduces the amount and cost of feeding them. They are especially useful in preventing outbreaks of disease when large numbers of animals are housed together, such as in chicken coops. There is concern, however, that the widespread use of antibiotics in animal husbandry will lead to greater abundances of antibiotic-resistant microorganisms and that residues of antibiotics will cause problems in individuals who consume these food products.

S U M M A R Y

Control of Microorganisms by Physical Environmental Factors (pp. 311-320)
- Environmental factors influence rates of microbial growth and death. Some environmental conditions increase microbial growth and others decrease or end microbial growth. Microbial populations can be controlled by modifying the physical conditions under which they live. Each microbial species has a specific tolerance range for specific environmental conditions.

Physical Exclusion or Removal of Microorganisms (pp. 311-312)
- Filtration is an effective method for achieving sterility of liquids and gases.

High Temperatures (pp. 312-315)
- High temperatures can be used to kill all microorganisms in a sample. This is heat sterilization. Decimal reduction time describes the heat killing of microorganisms. It is the time required for a tenfold reduction in the number of viable cells at a given temperature. The D value decreases as the temperature is raised above maximal. The D value for heat-resistant, endospore-forming microorganisms is used for processing in the canning industry.
- Pasteurization uses relatively brief exposures to moderately high temperatures to reduce the number of viable microorganisms and to eliminate human pathogens. Pasteurization reduces the number of microorganisms but does not sterilize.
- Sterilization uses high temperatures to kill all viable microorganisms. Autoclaving, using steam under pressure, is used to sterilize many materials by exposure at 121° C at 15 pounds per square inch pressure for 15 minutes. Dry heat sterilization requires higher temperatures and longer exposure times.
- Canning uses elevated heat followed by conditions that ensure the maintenance of aerobic conditions to preserve foods.

Low Temperatures (p. 316)
- Low temperatures limit the rates of microbial reproduction and can be used to prevent or limit microbial growth. Freezing at −20° C or lower precludes microbial growth entirely. Refrigeration at 5° C limits the rate of microbial growth.

Removal of Water—Dessication (pp. 317-318)
- Desiccation removes the water required for microbial growth; it does not necessarily increase the death rate. Some microorganisms can be preserved by drying. Water can be removed by sublimation in freeze-drying (lyophilization).
- Microorganisms are unable to grow at low water activity. Adding salt limits microbial growth and increases the shelf life of some foods. Drying is used to

preserve many foods, such as powdered milk, dried fruits, and cereals.

Radiation (pp. 318-320)

- Microorganisms can be killed by exposure to certain forms of radiation. Gamma and X-radiation have high penetrating power and kill microorganisms. Ultraviolet light does not have high penetrating power but kills microorganisms on surfaces. Infrared and microwave radiations have poor penetrating power and do not appear to kill microorganisms directly. The absorption of long wavelength radiation results in increased temperature, killing microorganisms.

Control of Microbial Growth by Antimicrobial Agents (pp. 320-333)

- Chemical inhibitors are used to prevent the spread of disease-causing microorganisms and to preclude the growth of microorganisms that cause the spoilage of foods or biodeterioration of industrial products.

Types of Antimicrobial Agents (pp. 320-322)

- Antimicrobial agents are classified according to their application and spectrum of action. Biocides are agents that kill living organisms or inactivate viruses. Germicides are chemical agents that kill microorganisms. Virucides inactivate viruses; bactericides kill bacteria; algicides kill algae; and fungicides kill fungi.

Food Preservatives (pp. 322-323)

- Chemical additives are used in the preservation of food and other products. The addition of salt or sugar to food reduces the amount of available water and alters the osmotic pressure, creating bacteriostatic conditions.
- Various low molecular weight carboxylic acids are effective inhibitors of microbial growth. Propionates are effective against filamentous fungi and are used in milk and bread dough products. Lactic and acetic acids are effective, naturally occurring preservatives in such food products as cheeses, pickles, and sauerkraut. Benzoates are used as preservatives in fruit juices, jams, jellies, soft drinks, salad dressings, catsup, margarine, and pharmaceuticals. Sorbic acid inhibits fungi and bacteria at pH 4-6. Boric acid is used as a preservative in eyewash and other products.
- Nitrites and nitrates are added to cured meats to preserve red meat color and protect against the growth of food spoilage and poisoning microorganisms.

Disinfectants and Antiseptics (pp. 323-333)

- Disinfectants are antimicrobial substances that kill or prevent the growth of microorganisms and are used on inanimate objects. Disinfectants must have high germicidal activity; rapidly kill a wide range of microorganisms, including spores; be chemically stable and effective in the presence of organic compounds and metals; be able to penetrate crevices; and be inexpensive and aesthetically acceptable.
- Concentration and contact time are critical factors in determining effectiveness of antimicrobial agents; standardized tests evaluate the effectiveness of disinfectants.
- Antiseptics are antimicrobial agents with relatively low toxicities to human tissues so that they can be applied to the skin.

- Antiseptics are used for surface applications to biological tissue but are not necessarily safe for consumption. A toxicity index is the ratio of the greatest dilution of the product that can kill animal cells in 10 minutes to the dilution that can kill bacterial cells in the same period of time and under identical conditions. Antiseptics should have a toxicity index of less than 1.0.
- Chlorine kills microorganisms by disrupting membranes and inactivating enzymes. Chlorination is the standard treatment for disinfecting drinking water and effluents from sewage treatment plants.
- Iodine is bactericidal and sporicidal and is also effective against protozoa. Iodophors are used for preoperative skin cleansing and disinfection.
- Phenolics disrupt plasma membranes, inactivate enzymes, and denature proteins; they are very stable when heated or dried and retain activity in the presence of organic material; they are in common household disinfectants and are used in hospital wards and operating theaters. Phenolics include resorcinol and hexachlorophene. Resorcinol is bactericidal and fungicidal and is used in the treatment of acne, ringworm and other skin infections. Hexachlorophene combined with soap is a highly effective skin disinfectant and deodorant.
- Detergents, particularly quaternary ammonium compounds, are effective disinfectants used for removing microorganisms from floors and walls. Anionic detergents are used as laundry detergents to remove soil and thus lower the numbers of associated microorganisms. Quaternary ammonium cationic detergents are used as antiseptic agents.
- Alcohols are the most effective and most used agents for sterilization and disinfection; they denature proteins and disrupt membrane function.
- Aldehydes, such as formaldehyde and glutaraldehyde, are used for disinfecting and sterilizing medical instruments.
- Acids used as antiseptic agents include acetic acid and undecylenic acid.
- Peroxides, in the form of hydrogen peroxide, are effective nontoxic antiseptics because they are unstable and degrade into a reactive form of oxygen that is toxic to microorganisms. They are used with surgical implants and contact lenses.
- Dyes, such as crystal violet, block the final step in the synthesis of cell wall material.
- Ethylene oxide sterilization is used in hospitals to disinfect materials that cannot withstand steam sterilization.
- Ozone is a strong oxidizing agent that kills microorganisms by oxidizing cellular biochemicals. Ozonation is used to disinfect drinking water.
- Heavy metals, —mercury, silver, and copper— inhibit microorganisms. Heavy metals are inhibitory to microbial growth because of their oligodynamic action.

Antibiotics (p. 333)

- Antibiotics are antimicrobial substances produced by microorganisms.
- Antibiotics are used to treat disease in humans and nonhuman animals.

CHAPTER REVIEW

REVIEW QUESTIONS

1. Define sterilization.
2. Define disinfection.
3. Explain how you could disinfect different types of materials.
4. Describe how you would sterilize several different types of materials.
5. Define antisepsis and explain its relation to the elimination or suppression of microbial growth.
6. What is a germicide and what is its relation to the elimination or suppression of microbial growth?
7. How is microbial growth affected by the type of microorganism and environmental conditions?
8. What physical methods can be employed to control microbial growth?
9. What factors determine the effectiveness of a disinfection agent?
10. How are disinfectants evaluated?
11. What is pasteurization?
12. How does refrigeration increase the shelf life of milk?
13. Why don't you have to refrigerate sugar-coated cereal?
14. What is a phenol coefficient?
15. Which heavy metals are used as disinfectants or antiseptics?
16. How are the halogens—chlorine, bromine, and iodine—used as disinfectants?

CRITICAL THINKING QUESTIONS

1. How can you determine if a disinfectant or antiseptic is bacteriostatic or bactericidal? How can you determine the appropriate disinfectant to use for cleaning the floor of a hospital? How can you determine the appropriate antiseptic for treating minor abrasions?
2. What are the possible consequences of irradiation of food? Is it safe to eat perishable food that has been irradiated to extend its shelf life? What precautions, if any, should be taken with irradiated food?
3. What factors would you have to take into account in search for a new antiseptic? A new disinfectant? A new food preservative?
4. If your local water treatment plant shuts down because of a natural disaster, how could you treat the water at home to ensure that it is safe to drink?
5. Many health food stores market foods that are free of preservatives. What special precautions must be used with such foods? Are such foods safe for consumption?
6. What special precautions are taken during surgical procedures? Why does the surgical staff scrub if they are going to wear gloves? Are the gloves, gowns, and masks worn by the surgical staff aimed at protecting the patient or the physicians and nurses?

READINGS

Baumler E: 1984. *Paul Ehrlich, Scientist for Life*, New York, Holmes and Meier.
   Uses unpublished documents to examine Ehrlich's scientific contributions. Explores the connection between Ehrlich and the German chemical industry in the development of antimicrobial chemicals.
Block SS: 1991. *Disinfection, Sterilization and Preservation*, ed. 4, Malvern, Pennsylvania; Lea and Febiger.
   An extensive work covering various methods for killing and preventing the growth of microorganisms.
Castle M: 1980. *Hospital Infection Control*, NY, John Wiley & Sons.
   A practical discussion of the maintenance of sterility and asepsis in the hospital environment.
Favero MS: 1980. Sterilization, disinfection, and antisepsis in the hospital. In Lennette FH et al., eds: *Manual of Clinical Microbiology*, American Society for Microbiology, Washington, D.C.
   A review of the methods used to control microorganisms in hospitals.
Graham JC: 1981. The French connection in the early history of canning, *Journal of the Royal Society of Medicine* 74:374-381.
   The story of the work of Nicolas Appert.

Nadolny MD: 1980. Infection control in hospitals. What does the infection control nurse do? *Am J Nurs* 80:430-440.
   A practical view of controlling microorganisms in the hospital setting.
Russell A: 1982. *Destruction of Bacterial Spores*, Orlando, Academic Press.
   Factors affecting the activity of sporicidal agents and their possible practical uses are the main focus of this book.
Russell AD, WB Hugo, GAJ Ayliffe: 1982. *Principles and Practice of Disinfection, Preservation, and Sterilization*, Oxford, England; Blackwell Scientific.
   Describes the application of antimicrobial agents, including the bases for their uses in controlling microorganisms.
Witkop B: 1982. Paul Ehrlich: his ideas and his legacy, In Bernhard CG, E Crawford, P Sorborn, eds: *Science, Technology, and Society in the Time of Alfred Nobel*, pp. 146-166. Oxford, Pergamon Press.
   Portrays Ehrlich's role and contribution to developments in the history and character of modern science.

UNIT FOUR

Microorganisms and

Human diseases

CHAPTER 12

Microorganisms and Human Diseases

PREVIEW TO CHAPTER 12

In this chapter we will:

• Examine the interactions between microorganisms and humans, learning why those interactions sometimes result in infections and human disease.
• See how Koch's postulates establish the scientific basis for determining that a specific microorganism causes a specific human disease.
• View the special properties of pathogenic microorganisms—called virulence factors—that permit them to overcome body defenses and to cause human diseases.
• Study the portals of entry through which microorganisms enter the body to establish an infection and the routes via which disease-causing microorganisms are transmitted.
• See that by knowing routes by which infections are transmitted we can take steps to control infectious diseases.
• Examine the methods that are used to reduce exposure to pathogens, including the use of hygienic--sanitary procedures.
• Learn the following key terms and words:

Even before the existence of microorganisms had been shown, Girolamo Fracastoro, a sixteenth-century Italian scholar, proposed a disease theory. He said that disease occurs when seeds too small to be seen pass from one thing to another. However, it was not until the latter half of the nineteenth century that **Robert Koch** used the scientific method to demonstrate that specific microorganisms cause specific diseases (FIG. 12-1). Koch was a German country physician who began his microbiological studies isolated from the scientific community. He worked alone with primitive tools and materials. One of the professors with whom Koch studied, Jacob Henle, was a strong advocate of the **germ theory of disease.** This theory said that disease was due to microorganisms that could be transmitted through the air or by contact and that could multiply in the body.

Koch hypothesized that some diseases were caused by microorganisms and that specific microorganisms cause specific diseases. He reasoned that to establish that the cause of a specific disease was a specific microorganism, the same microorganism

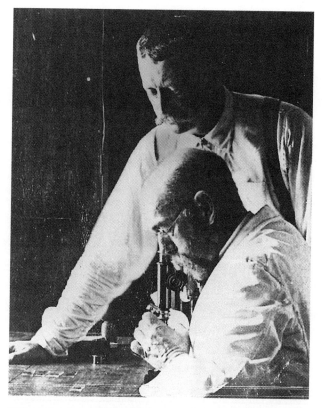

FIG. 12-1 Robert Koch (1843-1910) pioneered studies in medical microbiology and developed many of the basic methods essential for the study of microbiology. Koch's postulates for establishing the etiology of infectious diseases and the methodological techniques he developed are still used today in scientific investigations.

would have to be found regularly in diseased animals, that the microorganism would have to be isolated, and that the isolated microorganism would have to be shown to be capable of producing the disease in a previously healthy animal.

As a result of his medical practice, Koch questioned whether microorganisms caused the diseases that were prevalent in his time and place, such as tuberculosis, anthrax, and cholera. From 1873 to 1876, Koch studied the cattle disease anthrax. He conducted controlled experiments to show that microorganisms grown outside an animal could cause disease and that specific microorganisms are the cause of specific diseases. Koch examined the blood of animals that died of anthrax. He consistently observed countless colorless, nonmoving, rod-shaped bacteria in their blood. He hypothesized that these rod-shaped bacteria specifically caused anthrax. Koch demonstrated the validity of this hypothesis. He collected a small amount of blood containing the rod-shaped bacteria from an animal with anthrax and injected it into healthy animals. He reasoned that *if the* bacteria in the blood were the cause of anthrax, *then* the animals he injected would develop the disease and die. He observed that, in fact, the animals he injected with the bacteria-containing blood all died of anthrax. He collected blood samples from these dead animals and observed that they all contained numerous rod-shaped bacteria, just like those seen in the original blood samples. He did not observe such bacteria in the blood of healthy animals. This strongly suggested that these rod-shaped bacteria were the cause of anthrax but did not prove it to Koch. It was not a true proof because, besides the bacteria, all the other elements of blood were also injected in the experiment.

Koch developed a method for growing the bacteria outside the animal to prove that the rod-shaped bacteria alone cause anthrax. He grew the rod-shaped bacteria in beef broth and on jellied beef soup free of any other cells. Then he injected the pure bacteria into healthy animals. He observed that soon after the injections with pure bacterial cells, the animals developed anthrax with the same symptoms of naturally occurring anthrax. Healthy animals that were not injected or that were injected with solutions not containing the rod-shaped bacteria did not develop anthrax. These observations demonstrated that the rod-shaped bacteria, which Koch called anthrax bacilli, are the unique cause of the disease anthrax. Koch went on to determine the causative organisms for several other diseases, including tuberculosis and cholera. His experiments changed the study of disease from philosophy to science.

Robert Koch demonstrated that the scientific method could be used to study the relationships between microorganisms and human diseases, showing that specific human diseases are caused by specific microorganisms.

Koch was able to show that *if* a specific microorganism infects an animal, *then* a specific disease will occur. He also showed that *if* that specific microorganism is absent, *then* that specific disease does not occur. Koch's experiments established the steps necessary for identifying the causative agent of a disease. "To obtain a complete proof of a causal relationship, rather than mere co-existence of a disease and a parasite, a complete sequence of proofs is necessary. This can only be accomplished by removing the parasites [pathogenic microorganisms] from the host, freeing them of all tissue elements to which a disease-inducing effect could be ascribed, and by introducing these isolated parasites [pathogens] into a healthy animal with the resulting reproduction of the disease with all its characteristic features."

The steps necessary to establish that a specific microorganism causes a specific disease have become known as **Koch's Postulates** (FIG. 12-2). These postulates can be used to establish the cause-and-effect relationship between a specific microorganism and a particular disease, even diseases of plants and animals. Koch's Postulates are:

1. The microorganism must be present in all animals suffering from the disease and absent from all healthy animals.

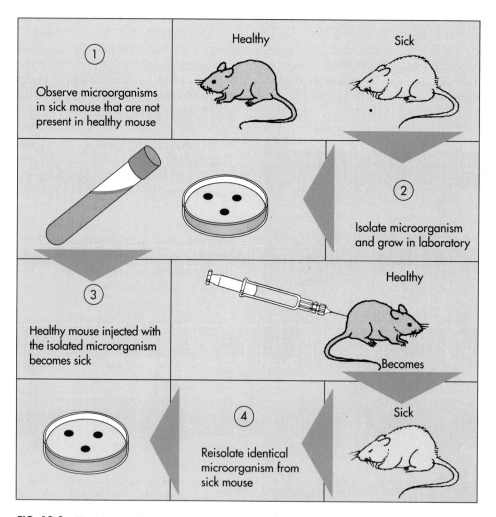

FIG. 12-2 Koch's postulates are used to define a causal relationship between a pathogenic microorganism and a specific disease. There are four postulates that make up the four steps needed to prove that a particular microorganism causes that disease: 1, The organism should be present and observed in all sick animals with the disease and absent from all healthy animals. 2, The organism must be isolated from the sick animals and grown in pure culture in the laboratory. 3, When the pure culture is inoculated into a healthy animal, that animal must become sick and develop symptoms of the disease. 4, The organism must be reisolated from the experimentally infected animal that had become sick.

2. The microorganism must be grown outside the diseased animal host as a pure culture, that is, free of other organisms.
3. When such a pure culture of the microorganism is inoculated into a healthy susceptible animal, that animal must develop the symptoms of the disease.

4. The microorganism must be reisolated from the experimentally infected animal and shown to be identical to the original microorganism.

Koch's Postulates form the basis for determining that a particular disease is caused by a given microorganism.

HISTORICAL PERSPECTIVE

APPLICATION OF KOCH'S POSTULATES TO LEGIONNAIRE'S DISEASE

In the summer of 1976, the American Legion was holding its annual convention in Philadelphia. The meetings were held at an elegant old hotel where many of the Legionnaires were also housed. A large number of people who had been in the hotel suddenly became ill, suffering from high fever and symptoms of pneumonia (weakness, headache, chills, cough). Some scientists hypothesized that the disease was caused by a poison that had entered the hotel's air-conditioning system. Some news reports speculated that the disease was the result of terrorist activity against the American Legion. Other scientists hypothesized that the disease was caused by a microorganism and began to apply the first of Koch's postulates in search of any specific microorganism that appeared in all specimens obtained from the afflicted individuals. Initially they could find none. They isolated many microorganisms but none with the necessary consistency to establish a cause-and-effect relationship between its presence and the observed disease. Within a few weeks, 29 of the 182 people stricken with the disease died. The cause remained a mystery.

Scientists systematically continued their analyses, chemically examining specimens for possible poisons and searching for evidence of any potential disease-causing microorganisms. After 90,000 hours of investigations that cost over $2 million and employed virtually all conventional procedures for isolating microorganisms, the breakthrough came when a group of scientists inoculated guinea pigs with tissues from the lungs of six individuals who died of the disease and consistently found a rod-shaped bacterium. They realized that the disease-causing microorganism was probably present in the lung tissue in far larger numbers than were being demonstrated by conventional microscopy. They, therefore, treated the specimens that they had collected with a silver-impregnated dye, which resulted in the clear observation of apparently identical bacteria in the lung tissues of all those afflicted with the disease. Such bacteria were not seen in tissues from healthy individuals or those who had died of other diseases. This observation fulfilled Koch's first postulate.

Although the bacterium could not initially be grown in pure culture, that is, free of any other living organisms, it eventually was grown on a medium supplemented with the amino acid cysteine. Growth in pure culture fulfilled Koch's second postulate that the organism be grown outside its host animal. The third of Koch's postulates was fulfilled by inoculating animal models rather than humans, with the finding that the inoculated animals became ill. Finally, the same bacteria were again isolated from the sick animals, fulfilling Koch's fourth postulate and completing the proof that the observed bacterium, which has been named *Legionella pneumophila*, causes Legionnaires' disease.

VIRULENCE FACTORS OF PATHOGENIC MICROORGANISMS

What is it about some microorganisms that allows them to establish **infections** (growth of microorganisms within the body) that cause **human diseases** (disruption of normal body functions)? Disease-causing microorganisms **(pathogens)** possess special properties, referred to as **virulence factors.** Virulence factors are intrinsic properties of pathogenic microorganisms that allow them to overcome the host defenses that generally protect the human body and so establish an infection. The **virulence** of pathogenic microorganisms, that is, the degree of their ability to induce human disease, depends in large part on these virulence factors, which may permit growth of the pathogen within the body or may disrupt body functions.

Virulence factors are special properties that enhance the ability of pathogenic microorganisms to cause disease.

COLONIZATION AND ABILITY TO GROW WITHIN THE BODY

Adhesion Factors

Specific virulence factors that enhance the ability of a microorganism to attach to the surfaces of mammalian cells are termed *adhesins.* Adhesins are proteins or polysaccharides on the surface of the microorganism that bind to specific receptor sites on the surface of other cells. The production of adhesins is an important factor in determining the ability of potential pathogens to colonize body tissues. Many pathogenic bacteria must adhere to mucous membranes to establish infections. In the absence of specific attachment mechanisms these would-be invaders of the body could not adhere, or stick, to establish an infection. Capsules and slime layers contribute to the ability of bacteria to attach or adhere to particular host cells or tissues. The fimbriae (pili) of several pathogenic bacteria and their associated adhesins appear to have a key role in permitting these bacteria to adhere to host cells and establish infections. For example, enteropathogenic strains of *Escherichia coli* (strains that invade the gastrointestinal tract) have particular adhesins associated with their fimbriae that permit them to bind to the mucosal lining of the intestine. Similarly, *Vibrio cholerae* is able to adhere to the mucosal cells lining the intestine by its fimbriae, allowing the establishment of an infection. The fimbriae of *Neisseria gonorrhoeae* permit this pathogen to colonize the genital tract, leading to infection and a case of gonorrhea.

> Adhesins allow microorganisms to attach to the surfaces of host cells.

Likewise, the adsorption of viruses onto specific receptor sites of human cells establishes the necessary prerequisite for the uptake of the viruses by those cells. This leads to viral replication, disruption of normal host cell function, and the production of disease. Some viruses, such as adenoviruses, have external spikes that aid in their attachment to host cells.

Some protozoa possess specialized structures that allow their attachment to host tissues. *Giardia intestinalis* is a protozoan that invades the gastrointestinal tract. It attaches to the cells lining the small intestine by a specialized adhesive disc. Peristaltic contractions do not dislodge *Giardia* because of its strong attachment. This protozoan then burrows into the tissue and uses its flagella to expel tissue fluids.

> The ability of pathogenic microorganisms to attach to particular cells and tissues establishes specific tissue affinities for pathogenic microorganisms.

Invasiveness

Invasiveness refers to the ability of microorganisms to invade human tissues. It also includes the ability to multiply on or within the cells and tissues of the human body. Microorganisms that possess invasive properties are able to establish and to spread infections within host cells and tissues. Some enzymes produced by microorganisms contribute to the invasiveness of microbial pathogens. They enable the microorganisms to destroy body tissues and cells (Table 12-1). For example, *Clostridium perfringens* produces lecithinase, an enzyme that hydrolyzes lecithin (FIG. 12-3). Lecithin is a lipid component of eukaryotic plasma membranes. Production of lecithinase by *C. perfringens* destroys the integrity of the membranes of many cells. It is the primary cause of gas gangrene and the extensive tissue damage associated with this disease. Various phospholipases produced by microorganisms can also destroy animal cell plasma membranes.

> Invasiveness is the ability of microorganisms to invade tissues and to reproduce within the body.

> Enzymes enhance invasiveness by enabling the microorganisms to destroy body tissues and cells.

TABLE 12-1

Some Extracellular Enzymes Involved in Microbial Virulence

ENZYME	ACTION	EXAMPLES OF BACTERIA THAT PRODUCE THESE ENZYMES
Hyaluronidase (spreading factor)	Breaks down hyaluronic acid	*Streptococcus pyogenes*
Coagulase	Blood clots, coagulation of plasma	*Staphylococcus aureus*
Phospholipase	Lyses red blood cells	*Staphylococcus aureus*
Lecithinase	Destroys red blood cells and other tissue cells	*Clostridium perfringens*
Collagenase	Breaks down collagen (connective tissue fiber)	*Clostridium perfringens*
Fibrinolysin (kinase)	Dissolves blood clots	*Streptococcus pyogenes*

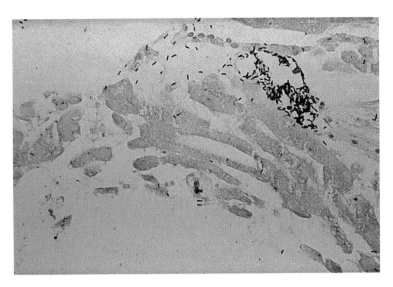

FIG. 12-3 Light micrograph showing *Clostridium perfringens (purple rod-shaped cells)* in necrotic tissue in a case of gas gangrene. This bacterium destroys surrounding areas of its growth.

Some *Staphylococcus* and *Streptococcus* species produce fibrinolysin. Fibrinolysin catalyzes the breakup of fibrin clots. The formation of fibrin clots during infection is part of the host defense mechanism to wall off the invading microorganisms. Fibrinolysin enhances the invasiveness of pathogenic strains of *Staphylococcus* and *Streptococcus* by allowing these pathogens to spread to surrounding tissues. Additionally, some *Staphylococcus* species, such as *Staphylococcus aureus*, produce coagulase. Coagulase converts fibrinogen to fibrin. The deposition of fibrin around staphylococcal cells protects them against phagocytosis by white blood cells. Phagocytosis is a host defense mechanism in which certain white blood cells engulf and destroy invading bacteria.

Hyaluronidase is produced by various species of *Staphylococcus, Streptococcus,* and *Clostridium.* It breaks down hyaluronic acid, the substance that holds together the cells of connective tissues. Pathogens that produce hyaluronidase spread through body tissues. Hyaluronidase is, therefore, referred to as the "spreading factor."

Some *Clostridium* species also produce **collagenase.** Collagenase is an enzyme that breaks down the proteins of collagen tissues, a gelatinous substance found in connective tissue, bone, and cartilage. The production of collagenase enhances the invasiveness of pathogens.

> The breakdown of fibrous tissues enhances the invasiveness of pathogenic microorganisms.

Growth and Survival Enhancing Factors

To establish an infection, bacteria that enter the body must be able to grow there. Environmental conditions at most body sites limit the growth of microorganisms. Pathogens often have virulence factors that enable them to overcome host defenses. Some microorganisms are able to grow under conditions that prevail in the human body.

The ability of bacteria to grow within blood is limited by the lack of available iron (FIG. 12-4). The binding of iron to specific chemicals produced by the body, namely transferrin and lactoferrin, is a host defense mechanism that prevents most microorganisms from establishing an infection in the blood. Some pathogens, however, are able to overcome this limitation and take the iron they require from blood. To supply themselves with iron needed for reproduction, some pathogens produce low molecular weight compounds involved in iron transport, called **siderophores.** Siderophores bind iron tightly. They remove iron normally bound to transferrin or other iron-binding compounds found in blood. *Neisseria meningitidis* and *Mycobacterium tuberculosis* species, which cause the diseases meningitis and tuberculosis, respectively, sequester iron without producing siderophores. They synthesize an outer membrane protein that removes iron directly from transferrin.

Some pathogens have structures that protect them against destruction by body defenses. Possession of such structures enhances their survival within the body. Capsules protect some bacteria against phagocytosis (FIG. 12-5). Capsules surrounding the cells of strains of *Streptococcus pneumoniae,* for example, permit these bacteria to evade the normal defense mechanisms of the host respiratory tract. These bacteria reproduce in the lungs and cause pneumonia. The virulence of other bacteria, including *Haemophilus influenzae* and *Klebsiella pneumoniae,* is also enhanced by capsule production.

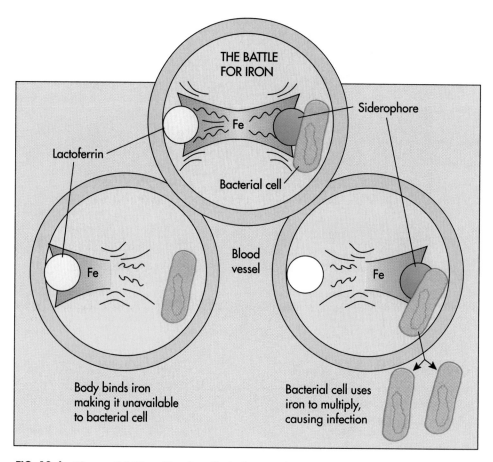

FIG. 12-4 The availability of free iron limits bacterial growth in blood. The body produces substances that withhold iron. Some pathogens produce siderophores that enable them to obtain the necessary iron for growth.

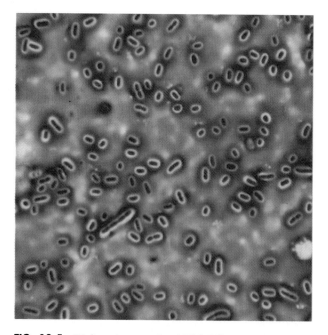

FIG. 12-5 Light micrograph of *Klebsiella pneumoniae* surrounded by capsules that interfere with phagocytosis. (2,300×).

TOXIGENICITY

Pathogenic microorganisms produce various toxins that cause discernible damage to human host systems. Some can cause death. **Toxins** are biological poisons. They disrupt the normal functions of cells and are generally destructive to human cells and tissues. **Toxigenicity** refers to the ability of a microorganism to produce such substances.

Endotoxin

Bacterial endotoxin is equated with the lipopolysaccharide (LPS) component of the Gram-negative bacterial cell wall. All Gram-negative bacteria have LPS in their cell walls. LPS is, however, toxic when it is released from the cell. When Gram-negative bacteria die, their cell walls disintegrate and LPS endotoxin is released. Some growing Gram-negative bacteria can also release LPS. The physiological effects of LPS endotoxins include fever, circulatory changes, and other general symptoms, such as weakness and nonlocalized aches (FIG. 12-6). Endotoxin released by *Salmonella* and *Shigella* species is responsible in part

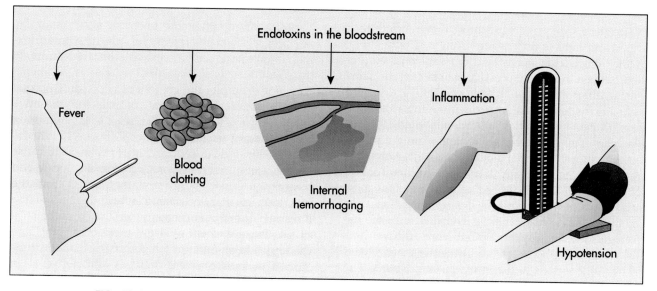

FIG. 12-6 Endotoxin produced by gram-negative bacteria has various physiological effects on the body.

for diseases, such as gastroenteritis, caused by these pathogens.

Exotoxins

Various Gram-negative and Gram-positive pathogenic bacteria produce proteins that act as toxins. Protein toxins are called **exotoxins** because they are typically released by growing cells. Exotoxins are specific to the microorganism producing the toxin and cause specific disease conditions because of their modes of action (Table 12-2). Most bacterial protein toxins are composed of a receptor protein component and a toxic component. The receptor component attaches to a target cell. The toxic component enters the cell and disrupts normal cell activity. Protein toxins are generally more potent than lipopolysaccharide toxins. Far less protein toxin is needed to produce serious disease symptoms. About 30 grams of diphtheria toxin—a protein exotoxin—could kill 10 million people and 1 gram of botulinum exotoxin could kill everyone in the United States (over 225 million people).

Toxigenicity is the ability of microorganisms to produce biological poisons called toxins.

Protein toxins are called exotoxins and lipopolysaccharide toxins are called endotoxins.

TABLE 12-2			
Some Protein Exotoxins Produced by Microorganisms That Cause Disease in Humans			
MICROORGANISM	**TOXIN**	**DISEASE**	**ACTION**
Clostridium botulinum	Several neurotoxins	Botulism	Paralysis; blocks neural transmission
Clostridium tetani	Neurotoxin	Tetanus	Spastic paralysis; interferes with motor neurons
Corynebacterium diphtheriae	Cytotoxin	Diphtheria	Blocks protein synthesis
Bordetella pertussis	Pertussis toxin	Whooping cough	Blocks G proteins that are involved in regulation of cell pathways
Streptococcus pyogenes	Hemolysin	Scarlet fever	Lysis of blood cells
Staphylococcus aureus	Enterotoxin	Food poisoning	Intestinal inflammation
Aspergillus flavus	Cytotoxin	Aflatoxicosis	Blocks transcription of DNA, thereby stopping protein synthesis
Amanita phalloides	Cytotoxin	Mushroom food poisoning	Blocks transcription of DNA, thereby stopping protein synthesis

Neurotoxins

Neurotoxins are protein exotoxins that interfere with the functioning of the nervous system. They usually work by blocking nerve cell transmissions. Neurotoxins, even those produced by members of the same genus, differ markedly in their modes of action. The neurotoxins responsible for the symptoms of botulism, caused by *Clostridium botulinum,* bind to nerve synapses. This blocks the ability to transmit impulses through motor neurons. Blocking essential functions, such as movement of the diaphragm needed for respiration, results in death. The neurotoxin **tetanospasmin** produced by *Clostridium tetani,* the causative agent for the disease tetanus, interferes with the peripheral nerves of the spinal cord (FIG. 12-7). Tetanospasmin blocks the ability of these nerve cells to transmit signals to the muscle cells properly. The symptomatic spastic paralysis of tetanus results. These spasms are characteristic of tetanus, which, like botulism, is often fatal.

Neurotoxins affect the nervous system.

Enterotoxins

Various enteropathogenic bacteria, such as *Salmonella, Shigella,* and *Vibrio* species, produce **enterotoxins.** Enterotoxins are protein exotoxins that stimulate the cells of the gastrointestinal tract in an abnormal way. For example, the cholera enterotoxin (choleragen) is produced by *Vibrio cholerae,* the causative agent for the disease cholera (FIG. 12-8). Choleragen blocks the conversion of cyclic AMP to ATP. The resulting elevated concentrations of cyclic AMP cause the release of inorganic ions, including chloride and bicarbonate ions. These ions are released from the mucosal cells that line the intestine and enter into the intestinal lumen. The change in the ionic balance causes the movement of large amounts of water into the lumen in an attempt to balance the osmotic pressure. This causes severe diarrhea and dehydration that sometimes results in the death of infected individuals.

Enterotoxins cause an inflammation of the tissues of the gastrointestinal tract.

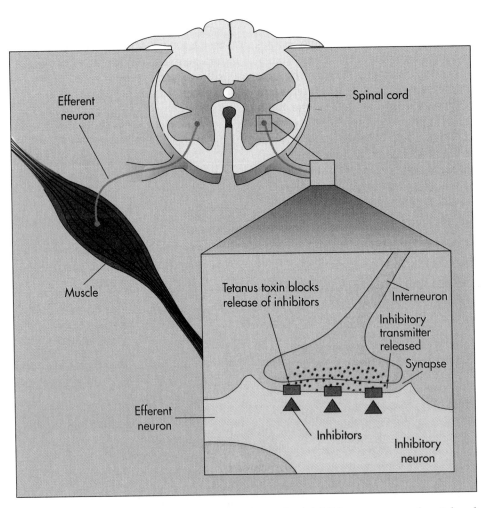

FIG. 12-7 Tetanus toxin blocks release of inhibitors from inhibitory neurons of peripheral nerves of the spinal cord. This results in spastic paralysis.

Cytotoxins

Cytotoxins are protein exotoxins that kill cells by enzymatic attack or by blocking essential cellular metabolism. Some cytotoxins interfere with protein synthesis. For example, **diphtheria toxin,** produced by *Corynebacterium diphtheriae*, inhibits protein synthesis in mammalian cells. It preferentially affects cardiac and renal tissues. This toxin blocks the translation of mRNA. It prevents the addition of amino acids and thus the elongation of the peptide chain during protein synthesis.

Cytotoxins interfere with cellular functions.

Other cytotoxins cause the lysis of blood cells, such as human erythrocytes. Such cytotoxins are called **hemolysins** because their action results in the release of hemoglobin from red blood cells. Hemolysins are produced by pathogenic *Streptococcus*, *Staphylococcus*, and *Clostridium* species. When bacteria that produce hemolysins are grown on blood agar plates, zones of clearing form around the colonies. A zone of clearing around a bacterial colony growing on a blood agar plate results from the complete lysis of red blood cells and is referred to as **beta hemolysis** (β hemolysis). The partial clearing or greening of the blood agar plate around the bacterial colony results from the partial lysis of red blood cells and is referred to as **alpha hemolysis** (α hemolysis).

In addition to red blood cells, white blood cells are killed by some microbial cytotoxins. For example, leukocidin produced by *Staphylococcus aureus* causes lysis of leukocytes, contributing to the pathogenicity of this organism. Leukocytes are white blood cells (WBCs) that take part in the human defense against infection.

Bacteria are not the only microorganisms that produce cytotoxins. Dinoflagellates, which are algae, produce highly potent toxins that cause paralytic shellfish poisoning. Many mushrooms are highly poisonous because of the potency of the **mycotoxins** (fungal toxins) they produce. In the case of the most famous of the poisonous mushrooms, the "death an-

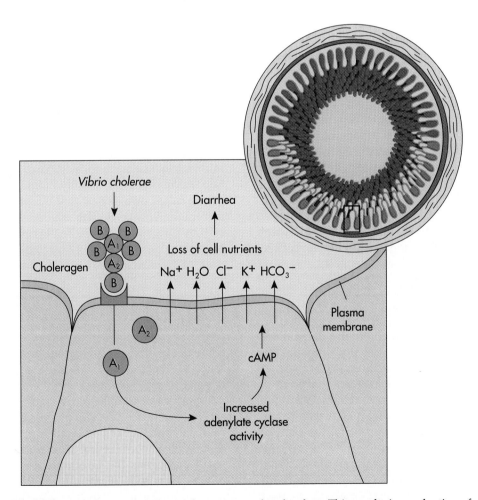

FIG. 12-8 Cholera toxin activates the enzyme adenylcyclase. This results in production of increased levels of cyclic AMP. High levels of cyclic AMP alter transport of ions across the gut lumen. Sodium ion transport is blocked and chloride ions and water move into the intestine from the blood. This causes severe diarrhea and water loss.

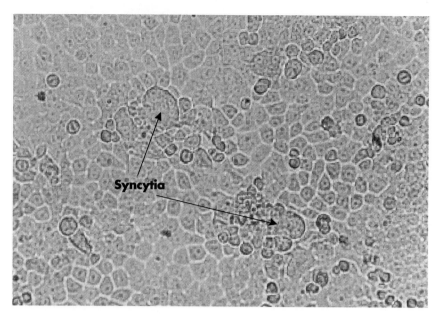

FIG. 12-9 Light micrograph showing the cytopathic effect on HEp-2 cells grown in tissue culture by an infection with respiratory syncytial virus. Syncytia are formed by fusion of virus-infected cells.

gel" *Amanita,* the toxin alpha-amanitin blocks transcription of DNA by interfering with RNA polymerase. Similarly, fungal *Aspergillus* species produce aflatoxins. The aflatoxins bind to DNA and prevent transcription of genetic information, resulting in various adverse effects—ranging from malignancy to death—in humans and other animals.

Viruses also have virulence factors that contribute to their ability to cause disease. As discussed in Chapter 9, when viruses replicate within host cells they often produce substances that can destroy or interfere with the normal functioning of cells. The observable changes in the appearance of cells infected with viruses are collectively known as **cytopathic effects** (CPE) (FIG. 12-9). In some cases, human cells infected with viruses die. For example, polio viruses kill the human cells they infect. In other cases, infected cells develop nonlethal abnormalities. Inclusions sometimes occur within the nucleus or cytoplasm of infected cells. Cells infected with rabies virus develop such inclusions only within the cytoplasm. Cells infected with adenovirus develop inclusions within the nucleus. Some viruses, such as measles virus, cause infected cells to fuse. Giant multinuclear cells are formed. Such cells do not function normally. Additionally, some viruses possess genes, called **oncogenes,** that transform normal cells into malignant cells. The transformation of animal cells by viruses appears to be the cause of some forms of cancer.

INFECTION AND DISEASE

The relationships between microorganisms, humans, and disease are dynamic. Usually humans are healthy. Economic, medical, and social factors, genetics, or individual lifestyle choices can contribute to tipping the balance for or against microbial initiation of disease in a particular individual. Virtually any microorganism can cause disease under the right set of conditions. The common adage "when you are tired and run-down you are more prone to infection," though an obvious oversimplification, has much validity. This is because an infectious disease is as much the result of the failure of the human defense system as it is the result of the special properties of pathogenic microorganisms. Put another way, diseases caused by microorganisms occur when the inherent properties of that microorganism enable it to overcome host defenses and/or because there is a breakdown in the human host defenses against disease-causing microorganisms.

A sufficient number of microorganisms is necessary to initiate an infective process that results in disease. The number of pathogens needed to establish a disease is known as the **infectious dose** (ID). For some pathogens the infectious dose is one, but for others, hundreds of thousands of microorganisms may be necessary to overwhelm the host defenses. Various factors influence the infectious dose that is required to initiate a disease, including the route of entry of the pathogen and the state of the host defenses.

In many cases, diminished host defenses permit relatively low numbers of potential pathogens to establish an infection. Malnutrition, for example, results in lowered amounts of antimicrobial body fluids and inadequate host defenses to protect against infective microorganisms. A person whose resistance to infection is weakened by disease, chemotherapy, or burns constitutes a **compromised host.** Broken skin or mucous membranes, a suppressed immune system, and/or impaired cell activity can compromise a host.

Predisposing factors affect the occurrence of disease, making an individual or population more susceptible to a certain disease. These factors sometimes alter the course of the disease. In some diseases, urinary tract infections, for example, gender is a predisposing factor, as women have a higher incidence than men. Genetic background may also be a factor. People with the genetically inherited sickle-cell trait, for example, are more resistant to malaria than others. Other predisposing factors may include environment, climate, weather, nutrition, lifestyle, age, fatigue, occupation, pre-existing illnesses or conditions, and medication.

The establishment of a microbially caused disease is a function of the virulence of the particular microorganism, the dosage (numbers) of that microorganism, and the resistance of the host individual.

The simple presence of an organism is not synonymous with a disease condition. A disease condition is a state in which the body does not function normally. Many microorganisms live on the surfaces of body tissues without causing disease. Such microorganisms constitute the so-called normal micro-

biota. Many of the surfaces of the body, particularly of the gastrointestinal tract, are covered by dense populations of these microorganisms. The normal microbiota actually contribute to the body's defenses by competing with and warding off potentially invasive pathogens.

The progression of the effects of microoganisms on their hosts can be expressed as contamination, colonization, infection, and disease (FIG. 12-10). **Contamination** simply indicates the presence of microorgan-

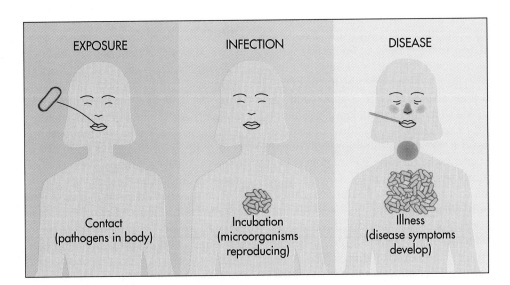

FIG. 12-10 There is a regular progression from infection, to exposure, to disease. Exposure results in contact and contamination. Infection occurs when microorganisms reproduce in the body. Illness occurs when disease symptoms develop.

isms not usually present in the body. If these microorganisms survive within the body, they may begin to grow at the site of contamination. This process is called **colonization.** The continued growth of microorganisms results in an **infection,** that is, the multiplication of microorganisms within the host organism whether or not disease results from that growth. For protozoa and larger parasites, such as helminthic worms, the term *infestation* rather than infection often is used.

The invasion or infection of the body by a microorganism, even by a pathogen that typically causes disease, does not always result in disease. A disease condition occurs only when the potential of the microorganism to disrupt normal functions is fully expressed. Many infections with potentially pathogenic microorganisms do not lead to disease. Sometimes healthy individuals become carriers of potentially pathogenic microorganisms. A person can be infected with the microorganisms but have not yet or will not develop a disease condition as a result of the infection. **Subclinical infections** are infections that do not produce symptoms in the host. In such cases, too few microorganisms are present or the host's defense mechanisms effectively control the invading pathogens.

> **Infection is the growth of microorganisms within the body.**

> **Disease is the disruption of normal body functions.**

It is important to recognize that diseases caused by pathogenic microorganisms account for nearly half of all human illness. The other half can be caused by various conditions such as nutritional deficiencies. Scurvy and rickets are caused by a shortage of vitamins C and D, respectively. Some disease conditions are present at birth as a result of a condition that developed during formation of the fetus or during embryological development. Such congenital diseases include cleft palate and heart defects. Some diseases, like Down syndrome and Tay-Sachs disease, are inherited. Metabolic diseases, such as some forms of diabetes mellitus, may also be inherited and result in abnormalities in the body's biochemistry. Degenerative diseases result from the wearing down of part of the body and result in the loss of functional abilities. Cirrhosis and emphysema are examples of degenerative diseases. Neoplasms or tumors are new growth of cells or tissues that may or may not be cancerous. Disease can result when one's immune systems attack one's own cells. Such "autoimmune diseases" in which the body attacks itself include systemic lupus erythematosus and rheumatoid arthritis. The final category of disease is idiopathic—disease for which there is no known cause.

> **Only about one half of all human disease is caused by pathogenic microorganisms.**

> **Other causes of disease include nutritional deficiencies, birth defects, genetic abnormalities, metabolic imbalances, degeneration, and autoimmune and hypersensitivity reactions.**

PATTERN OF DISEASE

Signs and Symptoms

When disease occurs, it is characterized by a specific set of signs and symptoms. A **sign** is a characteristic of a disease that can be observed by examining the patient. A **symptom** is a disease characteristic that can be observed or felt only by the patient. Signs of disease may include swelling, redness, coughing, runny nose, fever, vomiting, and diarrhea (Table 12-3). Symptoms of disease may include pain, shortness of breath, nausea, and malaise. Malaise is the general state of not feeling well; it implies a feeling of weakness. The signs and symptoms of a disease are usually related to the tissue damage being done by the pathogen.

> **A sign of a disease is observable, while a symptom is observed or felt only by the patient.**

The combination of signs and symptoms that occur together are known as the disease **syndrome.** Specific diseases have characteristic syndromes. It is this combination of signs and symptoms that physicians often use in the absence of clinical analyses to diagnose a disease and to determine the appropriate course of action. The syndrome that characterizes a disease can be mild to severe and reversible or irreversible. Some diseases leave aftereffects, such as paralysis after infection with the poliovirus. Aftereffects of a disease are called **sequelae.**

Stages of Disease

The course of an infectious disease generally follows a regular pattern of discrete stages (FIG. 12-11). The progress of any infectious disease in a given patient from the time of contamination with a pathogen until the disease process is completed can be divided into the following stages: incubation, prodromal, period of illness (also called acute phase), period of decline (also called progression of acute phase), convalescence, and full recovery.

The **incubation period** occurs after the pathogen enters the body and before any signs or symptoms appear. During the incubation period the microorganism has invaded the host and is typically migrating to various tissues. It has not yet begun to increase to sufficient numbers or to produce enough toxins to cause discomfort, nor to cause the individual to be in-

TABLE 12-3

Descriptions of Signs and Symptoms Associated with Infectious Diseases

MANIFESTATION OF DISEASE	DESCRIPTION
SYMPTOMS	
Abdominal pain	Symptoms of local pain between the bottom of the rib cage and the groin
Anorexia	Loss of appetite
Breathing difficulty	Breathlessness or tightness in the chest
Dizziness	A sense of being dazed and unsteady accompanied by a spinning sensation
Earache	Pain in one or both ears, either sharp and stabbing or dull and throbbing
Fainting	Sudden feeling of weakness and unsteadiness that may result in a brief loss of consciousness
Fatigue	Weakness or weariness
Headache	Pain in the head that may range from mild to severe and incapacitating
Itching	General or local sensation that makes one want to scratch
Malaise	Vague, generalized feeling of not being well
Nausea	Stomach distress with distaste for food and an urge to vomit
Numbness	Loss of feeling
Pain	An unpleasant sensation occurring in varying degrees of severity as a result of injury or disease
Painful and/or stiff neck	Pain or discomfort that may or may not be accompanied by a slight headache
Painful urination	Discomfort when urinating
Palpitations	Increased awareness of heartbeat that occurs when heart tries to compensate for shortages in the blood by pumping blood faster than normal
Sore throat	Any rough or raw feeling in the back of the throat that causes discomfort, especially when swallowing
Tingling sensation	Prickling sensation (pins and needles)
Toothache	Pain in a tooth, or in the teeth and gums generally, felt either as a dull throb or a sharp twinge
Irritation	Itching or soreness in or around an area of the body
SIGNS	
Abnormal-looking bowel movement	Red blood in bowel movement; off color, odorous, or watery bowel movement
Abnormal-looking urine	Urine that differs from the usual straw color, that is cloudy, blood-tinged, dark yellow or orange
Abnormal vaginal discharge	Discharge that differs in color, consistency, and quantity from what is usual between menstrual periods
Convulsion	Abnormal violent and involuntary contraction or series of contractions of the muscles
Coughing	A noisy expulsion of air from the lungs that may produce phlegm or be dry
Coughing up blood	Coughing that produces bright red, rusty brown, or pink and frothy colored phlegm
Diarrhea	Frequent passing of unusually loose and runny bowel movements
Fever	Temperature of 100° F or above
Hearing loss	Deterioration in the ability to hear some or all sounds in one or both ears
Hoarseness (loss of voice)	Any abnormal huskiness in the voice; may be so severe that little or no sound can be made
Inflammation	Local response to cellular injury marked by swelling, redness, heat, and pain
Jaundice	Yellowish pigmentation of the skin and body fluids caused by the deposition of bile pigments
Loss of weight	Loss of 10 lbs. or more over a period of 10 weeks or less without a deliberate change in eating and/or exercise patterns
Pallor	Extreme or unnatural paleness
Rashes	Discolored (often red-pink) and/or raised areas of itchy skin
Runny nose	Completely or partially blocked nose with liquid discharge
Swellings under the skin	Lumps or raised spots that can be seen or felt under the skin; may be inflamed or same color as surrounding skin
Swollen abdomen	Generalized swelling over the whole abdomen between the bottom of the ribcage and the groin
Vomiting	Throwing up the contents of the stomach; may be preceded by nausea
Wheezing	Noisy, difficult breathing, particularly when breathing out

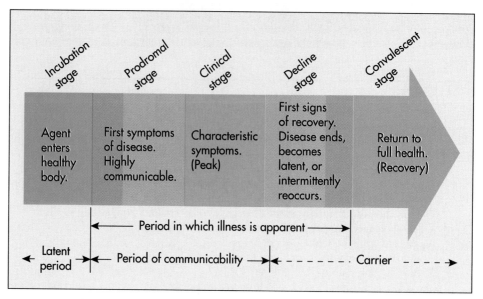

FIG. 12-11 Diseases progress through characteristic stages, from the initial prodromal stage following infection, during which the pathogen multiplies in the body, through the acute phase, when there are disease symptoms, to the recovery stage, during which the pathogen and disease symptoms are eliminated.

fective. This incubation period varies for different diseases, from a few hours to months, or even years, as is the case for AIDS and Hansen disease (Table 12-4). Incubation and periods of communicability are not always the same. HIV-positive individuals can transmit the virus during the prolonged period prior to the onset of AIDS. In other diseases with short incubation periods, the disease generally is not communicable until after the incubation period. While hepatitis A has an incubation period of 3 to 4 weeks, infected individuals can spread the virus only 1 or 2 weeks before onset of the disease.

The onset of symptoms marks the start of the **prodromal stage.** Now the patient is aware of discomfort but does not have adequately precise symptoms or signs to permit a diagnosis, that is, for the physician to determine the identity of the disease. However, sufficient numbers of the pathogen may be present to make the patient contagious to others. Moreover, the immune defenses have detected the infection and have become operative.

The **period of illness** occurs in the next stage. The various signs and symptoms that characterize the particular disease occur in this period. This is when the disease is most severe. During the period of illness **(acute stage)**, the patient often is sufficiently ill to alter his or her normal work or school activities. This phase of the disease progresses toward death or convalescence. Recovery depends on whether the body's defense systems or medical treatments are adequate. Assuming the disease is not fatal, the signs and symptoms begin to disappear during the **convalescent period (period of decline).** Convalescence progresses to a carrier stage or to freedom

from the pathogen. In some cases, the body's defense system may protect the person from recurrence of the infection for several months, several years, or life. Full **recovery** marks the end of the disease syndrome.

The stages of an infectious disease are: incubation, prodromal, period of illness, period of decline, convalescence, and recovery.

Types of Infections

A **primary infection** is an initial infection in a previously healthy host. Such an infection sometimes permits the invasion of the body by other microorganisms and thus the establishment of a **secondary infection.** Often, a secondary infection occurs because the primary infection depletes the host defenses so that other microorganisms can grow within the body. Bacterial pneumonia is a frequent secondary infection following influenza virus infection. While physicians aim to cure disease, medical treatment of a primary infection may actually contribute to the establishment of a secondary infection. A **superinfection** is a secondary infection that results from the destruction or debilitation of the microorganisms that normally inhabit the body. For example, frequently following broad-spectrum antibiotic use, the balance of microorganims in the gastrointestinal tract is disrupted and superinfection by *Clostridium difficile* occurs. The result of superinfection with *C. difficile* (antibiotic associated pseudomembranous colitis) can cause mild symptoms of diarrhea and gastrointestinal upset and may lead to more severe symptoms and death.

TABLE 12-4
Incubation Periods of Diseases

DISEASE	CAUSATIVE AGENT	INCUBATION PERIOD
VIRAL DISEASES		
Acquired immunodeficiency syndrome (AIDS)	Human immunodeficiency virus (HIV)	21 days-10+ years
Chickenpox	Varicella-zoster virus	7-21 days
Common cold	A variety of viruses, including rhinoviruses, enteroviruses, adenoviruses, orthomyxoviruses, paramyxoviruses, and coronaviruses	12 hours-3 days
German measles	Rubella virus	14-21 days
Hepatitis A	Hepatitis A virus	14-42 days
Hepatitis B	Hepatitis B virus	60-90 days
Hepatitis C (nonA-nonB Hepatitis)	Hepatitis C virus	15-64 days
Influenza	Influenza virus	1-3 days
Measles	Measles virus	7-14 days
Mumps	Mumps virus	14-28 days
Poliomyelitis	Poliovirus	3-35 days
Rabies	Rabies virus	2-8 weeks
BACTERIAL DISEASES		
Cholera	*Vibrio cholerae*	Few hours-5 days
Diphtheria	*Corynebacterium diphtheriae*	2-5 days
Epidemic typhus	*Rickettsia prowazekii*	7-14 days
Gonorrhea	*Neisseria gonorrhoeae*	2-9 days
Hansen disease	*Mycobacterium leprae*	7 months-5 years
Legionellosis	*Legionella pneumophila*	2-10 days
Meningococcal meningitis	*Neisseria meningitidis*	2-10 days
Primary atypical pneumonia	*Mycoplasma pneumoniae*	8-21 days
Scarlet fever	*Streptococcus pyogenes* (Group A *Streptococcus*)	1-3 days
Shigellosis	*Shigella dysenteriae*	1-7 days
Staphylococcal food poisoning	*Staphylococcus aureus*	1-6 hours
Syphilis	*Treponema pallidum*	10 days-10 weeks
Tetanus	*Clostridium tetani*	4 days-3 weeks
Tuberculosis	*Mycobacterium tuberculosis*	4-12 weeks
Tularemia	*Francisella tularensis*	2-10 days
Typhoid fever	*Salmonella typhi*	1-3 weeks
Whooping cough	*Bordetella pertussis*	7-14 days
FUNGAL DISEASES		
Histoplasmosis	*Histoplasma capsulatum*	5-18 days
San Joaquin Valley fever (coccidioidomycosis)	*Coccidioides immitis*	1-4 weeks
PROTOZOAN DISEASES		
Amebic dysentery	*Entamoeba histolytica*	Few days-several months
Giardiasis	*Giardia lamblia*	2-7 days
Malaria	*Plasmodium falciparum*	12 days
	Plasmodium vivax	14 days
	Plasmodium malariae	30 days

Types of Disease

A disease is characterized as **acute** if it develops rapidly and runs its course quickly. **Chronic diseases** develop more slowly, are usually less severe than acute diseases, and generally persist for a long, indeterminate period of time. Measles and influenza are acute diseases, whereas tuberculosis and Hansen disease (leprosy) are chronic diseases. An intermediate condition between an acute and a chronic disease is called **subacute.** Gum disease is a subacute disease.

Acute diseases develop and complete their courses rapidly. Chronic diseases develop slowly, are less severe, and last longer.

Latent diseases are characterized by periods that are sign and symptom free. Syphilis is characterized by periods of latency. **Localized infections,** such as boils, are confined to specific areas of the body. **Systemic** or **generalized infections,** such as septicemia (growth of bacteria throughout the circulatory system) affect the entire body, with the pathogens widely distributed in many tissues.

TRANSMISSON OF INFECTIOUS AGENTS

Infectious diseases that can be spread from one host to another are said to be **communicable** or **contagious.** The term *communicable disease* implies direct transmission from one person to another. Preventing such diseases often is accomplished by avoiding contact with infected individuals. Measures such as quarantine were devised to avoid exposure. Measles, German measles, influenza, gonorrhea, and genital herpes are all highly communicable. This means that the pathogens causing these diseases are readily transmitted with high frequency from an infected individual to a susceptible host.

Some diseases are not caused by agents that are communicable from one human to another. Tetanus, rabies, and Lyme disease are examples of noncommunicable infectious diseases. This means that they are acquired from the environment and are not spread directly from one person to another. Some noncommunicable diseases are caused not by the effects of invading microorganisms on host tissues, but rather by the ingestion of toxins made by the invading microorganisms. Such diseases are called **intoxications** rather than infections. For example, staphylococcal food poisoning is an intoxication that results from the ingestion of enterotoxin rather than the growth of *Staphylococcus* bacteria in the body.

The source of an infectious agent is known as the **reservoir.** Humans are the principal reservoirs for microorganisms that cause human diseases. Individuals infected with a pathogen act as the source of infection for others. The pathogens that cause contagious diseases move from one infected individual to the next. People who come in contact with someone suffering from a contagious disease are at risk of contracting that disease unless they are immune. If they are immune, their host defenses protect them against that particular pathogen.

In some cases, infected individuals do not develop disease symptoms. Such individuals are called asymptomatic carriers or simply **carriers.** Although they do not become sick, carriers are important reservoirs of infectious agents. The classic case of disease transmission by such a carrier occurred in the early 1900s when a cook, Mary Mallon, known as "Typhoid Mary," spread typhoid fever from one community to another.

Some diseases can be transmitted to humans by direct contact with infected animals, by ingesting contaminated meat, or, more frequently, by vectors. **Vectors** are organisms that carry the disease agent to the host. The vector need not develop disease. It only transmits the disease agent from a reservoir to a susceptible individual. Arthropods, such as mosquitoes, are frequently the vectors of human disease.

Pathogens also can be transmitted from an infected mother to her fetus or infant. Syphilis and rubella can be transmitted across the placenta. Hepatitis, gonorrhea, and chlamydial infections can be acquired as the newborn passes through the birth canal. Transmission between individuals can also be by sexual intercourse, touching, breathing aerosols (airborne, minute droplets of water that contain microorganisms), blood transfusions, or contaminated hypodermic needles.

The reservoirs of human pathogens can be nonliving sources such as soil and water. For example, tetanus is generally acquired when spores of *Clostridium tetani,* which are widely distributed in soil, contaminate a wound. Often diseases acquired from such sources are noncommunicable. Such diseases are singular events and are not normally transmitted from one infected individual to the next.

A reservoir is a source of an infectious agent, which may be air, water, soil, animals, or people.

ROUTES OF DISEASE TRANSMISSION

There are various modes by which pathogens are transmitted from a source to a susceptible individual (FIG. 12-12). Pathogenic microorganisms gain access to the body through a limited number of routes. These specific routes are known as **portals of entry.** The routes of entry are the respiratory tract, gastroin-

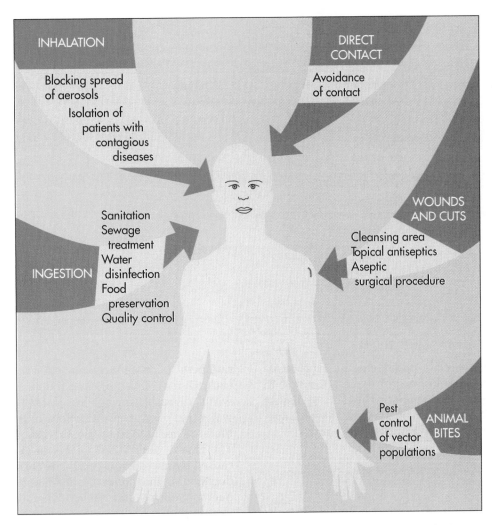

FIG. 12-12 Routes of transmission from an infected to a susceptible individual.

testinal tract, genitourinary tract, skin, and wounds. The invasive properties of specific pathogens permit them to penetrate the body's defense mechanisms through a specific portal of entry. Most pathogenic microorganisms will cause disease only if they enter the body via this specific route (Table 12-5, p. 356). For example, depositing *Clostridium tetani* on the intact skin surface does not result in disease, while deposition of *C. tetani* into deep wounds results in the deadly disease tetanus. The reasons that pathogens are restricted in where they can establish an infection are based on the host defenses associated with different body tissues and the inherent properties of the microorganism. In many cases, pathogenic microorganisms establish localized infections in the region of the portal of entry. In other cases, pathogens are able to spread systemically and establish infections involving other body tissues.

The sites at which microorganisms leave the body are called **portals of exit.** Pathogens of the genitourinary and gastrointestinal systems generally exit the body with body fluids or feces. Pathogens of the respiratory system exit through the nose or mouth in fluids expelled during coughing, sneezing, and speaking. Some pathogens, such as the bacteria that cause tuberculosis, are resistant to desiccation and can remain viable in the air for a long while. These organisms can be transmitted via the air, and inhalation of contaminated air can lead to infection. Similarly, gastrointestinal pathogens that exit infected individuals in fecal matter may contaminate food and water when local sanitation conditions are poor. Exudates from erupting lesions on the skin, as in chickenpox, may also be portals of exit. Direct contact with such exudates can lead to transmission of the disease.

A portal of entry is the site where a pathogen enters the body.

A portal of exit is the route via which pathogens leave the body.

TABLE 12-5			
Portals of Entry for Some Specific Disease-causing Microorganisms			
PORTAL OF ENTRY	**MICROORGANISM**	**TYPE OF MICROORGANISM**	**DISEASE**
Skin	*Staphylococcus aureus*	Bacterium	Impetigo
	Papilloma virus	Virus	Warts
	Trichophyton and *Epidermophyton* species	Fungus	Athlete's foot; tinea
Gastrointestinal tract	*Salmonella typhi*	Bacterium	Typhoid fever
	Poliovirus	Virus	Poliomyelitis
	Giardia lamblia	Protozoan	Giardiasis
Genitourinary tract	*Treponema pallidum*	Bacterium	Syphilis
	Herpes simplex virus	Virus	Genital herpes
Respiratory tract	*Bordetella pertussis*	Bacterium	Whooping cough
	Influenza virus	Virus	Influenza
	Histoplasma capsulatum	Fungus	Histoplasmosis
Wound	*Clostridium perfringens*	Bacterium	Gas gangrene
	Rabies virus	Virus	Rabies
	Sporotrix schenckii	Fungus	Rose-handler's disease

For disease to occur, pathogens must be transferred from the portal of exit of the infected host to the portal of entry of a new host. The route of disease transmission is established by the ways in which specific pathogens can enter and leave the body.

The transmission of infectious agents involves the movement of pathogens from a source to the appropriate portal of entry.

Person-to-person body contact is required for **direct contact transmission.** Shaking hands, kissing, contact with sores, sexual contact, and poor hygienic personal habits can spread pathogens in this manner. The direct fecal-oral route of disease transmission, for example, occurs when pathogens are spread from unwashed hands into the mouth. **Indirect contact transmission** occurs via fomites. **Fomites** are contaminated nonliving objects such as clothing, eating and cooking utensils, bedding, toys, and money. Ingestion of water contaminated with human fecal matter represents indirect fecal-oral transmission. This type of transmission frequently occurs in regions of the world with inadequate water treatment facilities for sewage treatment and disinfection of drinking water.

The transmission of pathogens occurs via restricted routes. For this reason it is possible to control our interactions with microbial populations in ways that reduce the probability of contracting infectious diseases. The methods employed for preventing exposure to specific disease-causing microorganisms vary, depending on the route of transmission. Many modern sanitary practices are aimed at reducing the incidence of diseases by preventing the spread of pathogenic microorganisms or by reducing their

numbers to concentrations that are insufficient to cause disease. Mosquito and rodent control, sanitary waste disposal, sewage treatment, chlorination of water supplies, pasteurization, and various other methods are used to restrict the spread of pathogens. The greatly diminished incidence of many diseases caused by microorganisms is the consequence of understanding the modes of transmission of pathogenic microorganisms and preventive measures that reduce exposure to disease-causing microorganisms.

NOSOCOMIAL (HOSPITAL-ACQUIRED) INFECTIONS

Medical procedures are designed to cure diseases. Some procedures used in the treatment of disease, however, can inadvertently introduce pathogenic microorganisms into the body and initiate an infectious process. Even a puncture wound with a sterile hypodermic syringe can pick up microorganisms from the vicinity of the puncture and carry them through the skin surface. The routine cleansing of wounds and use of topical antiseptics after minor skin punctures and abrasions are accepted prophylactic measures to prevent the establishment of infections.

Patients in hospitals often are in a debilitated state of health. Their body defenses are weak. They are, therefore, susceptible to various infectious diseases. The term *nosocomial infections* is used to describe a hospital-acquired infection (FIG. 12-13). Nosocomial infections include pneumonia acquired in hospitals, urinary tract infections that develop as a result of the insertion of a catheter, and infections of the genital tract that develop from gynecological procedures. Nosocomial infections affect approximately 2 million patients hospitalized annually in the United States

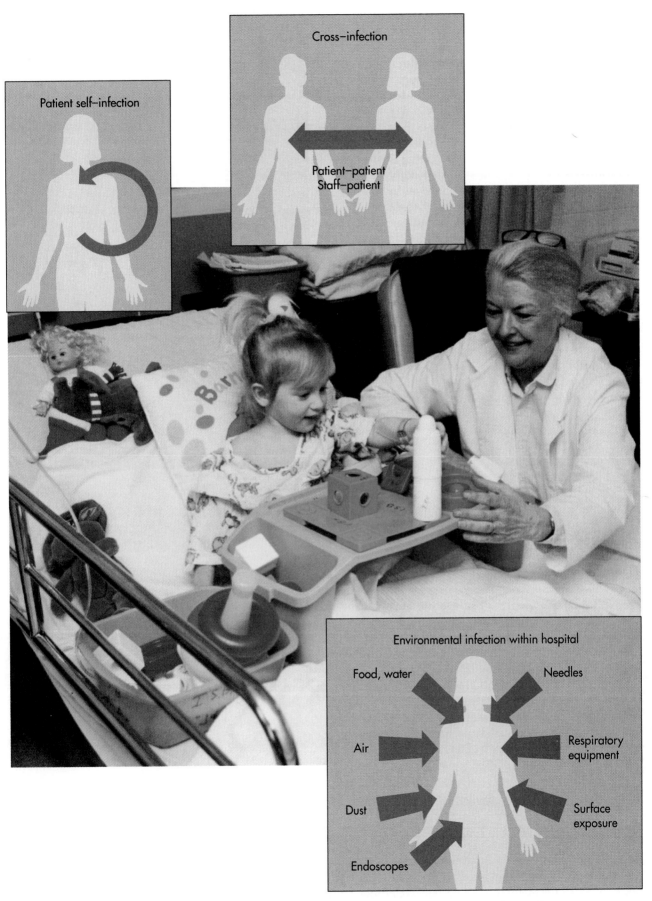

FIG. 12-13 Nosocomial infections occur when pathogens spread through a hospital.

alone. The numbers of nosocomial infections have been reduced in the United States during the past decade. This has been accomplished by increasing epidemiological standards and procedures such as educational seminars, the infection control nurse, and hospital committees designed to identify and break the routes of transmission of pathogens to patients.

Surgical procedures often expose deep body tissues to potentially pathogenic microorganisms. A surgical incision circumvents normal body defense mechanisms. Great care is taken in modern surgical practices, therefore, to minimize microbial contamination of exposed tissues. These practices include the use of clean operating rooms with minimal numbers of airborne microorganisms, sterile instruments, masks, and gowns. All of these prevent the spread of microorganisms from the surgical staff to the patient. The application of topical antiseptics before making incisions also prevents accidental contamination of the wound with the indigenous skin microbiota of the patient (FIG. 12-14). After many surgical procedures, antibiotics are given for several days as a prophylactic measure.

Despite all of these precautions of maintaining aseptic practices, infections still sometimes occur after surgery. Infections after surgery can be serious because the patient is already in a debilitated state. The onset of such infections is generally marked by fever. A purulent lesion may develop around the wound. Serious complications may follow open heart surgery if the patient develops endocarditis, caused by *Staphylococcus* or *Streptococcus* species. In surgical procedures involving cutting the intestines, the normal gut microbiota may contaminate other body tissues unless great care is taken to minimize such contamination. Antibiotics are also used in such cases to prevent microbial growth. The specific microorganisms causing infections of surgical wounds and the specific tissues that may be involved depend on the nature of the surgery and the tissues that are exposed to potential contamination with pathogens.

Surgical practices use elaborate aseptic procedures to minimize potential infection, but nevertheless, infections sometimes occur after surgery, attesting to the vulnerability of the body to microbial infection when the skin barrier is disrupted and host defense mechanisms are impaired.

RESPIRATORY TRACT AND AIRBORNE TRANSMISSION

We inhale 10,000 to 20,000 liters of air per day. This volume of air usually contains between 10,000 and 1,000,000 microorganisms. It should not be surprising, therefore, that the respiratory tract provides a portal of entry for many human pathogens. Potential pathogens freely enter the respiratory tract through

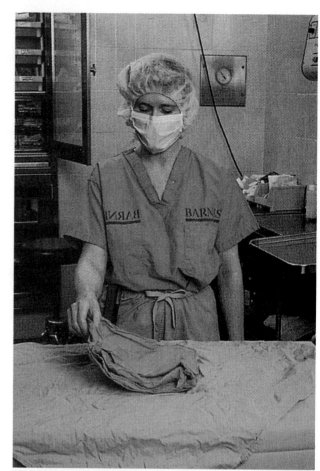

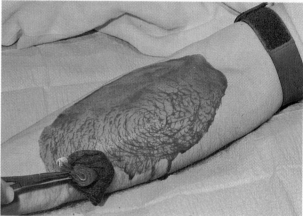

FIG. 12-14 **A,** Medical staff use gowns, masks, sterile instruments, and clean rooms to reduce the risk of infection, especially during surgical procedures. **B,** Iodine being applied prior to a surgical procedure.

the normal inhalation of air. Various viruses, bacteria, and fungi are able to multiply within the tissues of the respiratory tract. Sometimes they cause localized infections. At other times they enter the circulatory system through the numerous blood vessels associated with the respiratory tract and spread through the bloodstream to other sites in the body

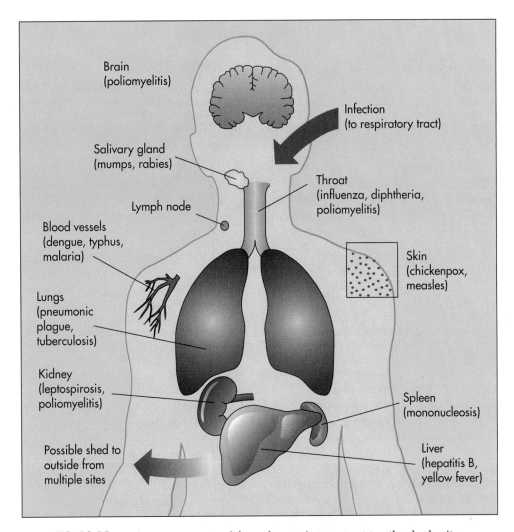

FIG. 12-15 Pathogens can spread from the respiratory tract to other body sites.

(FIG. 12-15). To establish an infection via the respiratory tract, a pathogen must overcome the natural defense mechanisms that are particularly extensive in the lower respiratory tract. There are numerous phagocytic cells in this area. While the potential for respiratory infection is great, fortunately, the actual rate of disease is low.

Airborne transmission occurs when pathogenic microorganisms are transferred from an infected to a susceptible individual via the air. Droplets regularly become airborne during normal breathing, but the coughing and sneezing associated with respiratory tract infections are primarily responsible for the spread of pathogens in aerosols and thus the airborne transmission of disease. Airborne pathogens often become suspended in aerosols. Aerosols are clouds of tiny water droplets suspended in air. The incidence of these diseases can be reduced by covering one's nose and mouth while coughing and sneezing and avoiding contact with contagious individuals (FIG. 12-16). These are practices we are taught to follow at an early age.

FIG. 12-16 Sneezing propels aerosols containing microorganisms. In this manner pathogens are transmitted through the air. Use of a handkerchief can block the spread of aerosols containing pathogens.

Transmission through the air is undoubtedly the main route of transmission of pathogens that enter via the respiratory tract. In spite of conditions of dryness, extreme temperatures, and ultraviolet radiation that characterize the air and which prevent them from growing in that environment, microorganisms still reach new hosts through the air. Some bacteria, particularly Gram-positive bacteria, can survive for several months in dust particles. Bacterial and fungal spores and naked viruses can live even longer. The incidence of airborne infections has increased in recent years because so many new buildings are sealed and have self-contained recirculating air systems for temperature control.

Potential pathogens enter the respiratory tract through the inhalation of airborne pathogens.

GASTROINTESTINAL TRACT—WATER AND FOODBORNE TRANSMISSION

Microorganisms routinely enter the gastrointestinal tract in association with ingested food and water. Waterborne and foodborne pathogens can infect the digestive system and cause gastrointestinal symptoms. The large resident microbiota that develops in the human intestinal tract after birth is important for the maintenance of good health. This population is usually not involved in disease processes and is normally noninvasive. Waterborne and foodborne transmission generally involves transmission of pathogens that enter via the mouth and exit via the anus. Generally, the establishment of infection through the gastrointestinal tract requires a relatively large infectious dose. This means that a relatively large number of pathogenic microorganisms are required to successfully overcome the inherent defense mechanisms of the gastrointestinal tract. High infectious doses often are encountered in waters contaminated by sewage or other sources of human fecal matter.

GENITOURINARY TRACT—SEXUAL TRANSMISSION

The genitourinary tract provides the portal of entry for pathogens that are directly transmitted during sexual intercourse. Such infections are known as **venereal** or **sexually transmitted diseases.** The physiological properties of the pathogens causing these diseases restrict their transmission, for the most part, to direct physical contact. They have very limited natural survival times outside infected tissues. The overall control of sexually transmitted diseases rests with breaking the network of transmission. This necessitates public health practices that seek to identify and treat all sexual partners of anyone diagnosed as having one of the sexually transmitted diseases.

NEWSBREAK

TRANSMISSION OF RABIES

Several years ago, frequent requests appeared in the media for help in finding a potentially rabid dog that bit a child. The media would give a description of the dog with a plea that unless the dog was located and tested, the child would have to undergo a series of painful vaccine injections. Often, as a result of this search, the dog was located, and the child was spared the pain of vaccine administration. Today, we no longer hear such pleas for finding a potentially rabid dog because dogs are rarely carriers of the rabies virus. Vaccination of pets has virtually eliminated rabies from the domestic dog population. Today rabies is transmitted via bites of raccoons, prairie dogs, and other wild animals (see Figure). It is not reasonable to ask for help in locating in the forest a raccoon that bites someone. Prudence requires vaccine administration in such cases without the search for the animal. Fortunately, the shift in rabies transmission from domestic to wild animals occurred at the same time that a new and far less painful vaccine became available.

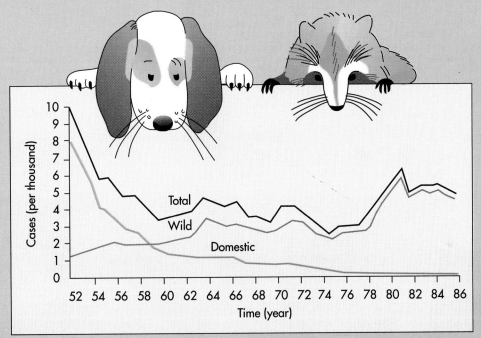

Mandatory vaccination of domestic animals has resulted in the near elimination of rabies from dogs and other pets. The rabies virus remains endemic to many wild animals that act as reservoirs and vectors of this disease.

SUPERFICIAL BODY TISSUES—DIRECT CONTACT TRANSMISSION

In some cases the deposition of pathogenic microorganisms on the skin surface can lead to an infectious disease. Since they require direct contact between skin and microorganisms for transmission to occur, these diseases are called **contact diseases**. Some diseases transmitted in this manner are superficial skin infections. On others, the pathogens are able to enter the body and spread systemically. Relatively few microorganisms possess the enzymatic capability to establish infections through the skin surface. Some microorganisms, however, are able to enter the subcutaneous layers through the channels provided by hair follicles. Transmission of some contact diseases may follow minor abrasions that allow the pathogens to circumvent the normal skin barrier.

PARENTERAL ROUTE

Punctures, injections, bites, cuts, wounds, surgical incisions, and cracking skin due to swelling or drying establish portals of entry to a host for a potential pathogen. Such access is called the parenteral route (from Greek *para* [beside] and *enterik* [intestinal tract]. Microorganisms thus gain entry to the body by being deposited directly into the tissues beneath the skin or into the mucous membrane.

Animal Bites and Disease Transmission

Many animals are carriers of microorganisms and transmit pathogens to humans through bites. Animals that transmit pathogens are called **vectors**. In some cases the nonhuman animal also suffers disease, but in many cases such animals are only carriers. Animal bites simultaneously disrupt the skin barrier and inoculate the wound with microorganisms whose pathogens potentially may be life threatening. Arthropods, particularly insects, commonly act as vectors of some very dangerous human pathogens. Their bites can establish serious infections.

Many human infections that are transmitted via animal bites involve animals that act as reservoirs for the pathogens. The animal populations that maintain the pathogen and act as reservoirs may themselves

suffer from a disease caused by that pathogen. These diseases, termed *zoonoses*, are defined as infectious diseases of nonhuman animals transmissible to humans (FIG. 12-17). The animal and human populations act as alternate hosts for the proliferation of the pathogens. In considering the transmission of such diseases it is important to examine the reservoirs and vectors of the infectious agents, as well as the nature of the specific etiologic agent of the disease. Prevention of infectious diseases that enter the body through animal bites often involves maintaining control of infected reservoir and carrier animal populations. For example, mosquito control programs are employed to prevent the spread of many arthropod-borne diseases, such as yellow fever and malaria. It should be remembered, however, that even though

FIG. 12-17 Zoonoses occur when pathogens spread from nonhuman animals to humans.

animals have a critical role in the transmission of such diseases, it is the viruses, bacteria, or protozoa that are the actual etiologic agents.

> **Animal bites that disrupt the protective skin layers provide the main portals of entry by which pathogenic microorganisms penetrate the skin and gain entry to the body.**

Wounds

If skin tissues are mechanically interrupted by a wound, a portal of entry is opened through which pathogenic microorganisms may enter the body. Breaking the skin surface not only provides a portal of entry for potentially pathogenic microorganisms but also often inoculates microorganisms directly into the circulatory system and inner body tissues. For example, when a child scrapes a hand or knee on the ground, the wound is frequently contaminated with dirt and associated microorganisms.

Wounds disrupt the protective barrier of the skin. They provide a portal of entry through wich microorganisms can enter the circulatory system and deep body tissues. Microorganisms on the skin surface can readily pass through the opening of a wound. As a result, many infections associated with wounds are caused by opportunistic pathogens derived from the normal microbiota of the skin. An opportunistic pathogen is one that does not harm the healthy host but can take advantage or opportunity to harm an unhealthy one. To avoid entry of bacteria into a wound, the area is usually covered with gauze to protect against contamination.

Staphylococcus aureus is the most frequent cause of wound infections. *Streptococcus pyogenes* also is often associated with infections of wounds. In cases of severe wounds, where the integrity of the gastrointestinal tract is disrupted, enteric bacteria are frequently the causative agents of wound infections. In many cases, wound infections are localized at the site of the wound. Such infections can spread systemically and may involve many body tissues and organs. Infections with *S. aureus*, established through skin wounds, can spread and form abscesses in bone marrow (osteomyelitis) and other body tissues, including the spine and brain. Superficial wounds can generally be treated with topical antiseptics or antibiotics to prevent the establishment of infections. Serious deep wounds, however, may require the prophylactic use of systemic antibiotics to prevent the onset of serious infections. Infections of deep wounds may involve anaerobic bacteria of the genera *Clostridium,* *Bacteroides,* and *Fusobacterium,* as well as *Staphylococcus* and *Streptococcus* species.

> **Minimizing contamination of wounds with foreign matter and associated microorganisms is important in controlling infections that may occur when the skin barrier is broken.**

Burns

Burns remove the protective skin layer, exposing the body to numerous potential pathogens. Microbial infection after extensive burns, where a large portion of the skin is damaged, is a very serious complication that often results in the death of the patient. *Staphylococcus aureus, Streptococcus pyogenes, Pseudomonas aeruginosa, Clostridium tetani,* and various fungi often cause infections in burn victims. It is important to avoid contamination of the burn area that can introduce opportunistic pathogens into exposed tissue. To ensure that infections are detected in time to permit treatment with antimicrobial agents, microbiological tests are frequently performed on burn victims. If infections develop, prognosis depends on the size of the burn, the extent of infection, and the physiological state of the patient.

PREVENTION OF DISEASES BY AVOIDING EXPOSURE TO PATHOGENIC MICROORGANISMS

EPIDEMIOLOGY—DETERMINING SOURCES AND ROUTES OF DISEASE TRANSMISSION

The development of a basic understanding of the interrelationships between humans and microorganisms has led to practices that prevent or diminish the incidence of human diseases caused by microorganisms. Of particular importance is an understanding of the routes of transmission of specific pathogens. Such understanding lets us avoid exposure to pathogens.

Epidemiologists are scientists who study the factors and mechanisms that govern the spread of disease within a population. They consider the causes, or **etiology,** of a disease and the factors involved in the transmission of infectious agents. With this information the epidemiologist evaluates the statistical probability that a susceptible individual will be exposed to a particular pathogen and that such exposure will result in disease transmission. The likelihood of disease transmission depends on the concentration and virulence of the pathogen, the distribution of susceptible individuals, and the potential sources of exposure to the pathogenic microorganisms.

Epidemiology is the study of the factors and mechanisms that govern the spread of disease within a population.

Epidemiologists often act as detectives to locate the origin of a disease outbreak. In some cases, they search for a source of tainted food, in others, direct contact with infected individuals, and so forth. The number of cases reported each day and the locations of disease occurrences enables epidemiologists to distinguish between a **common source outbreak** and a **person-to-person (propagated) epidemic.** Common source outbreaks are characterized by a sharp rise and rapid decline in the number of cases. Person-to-person transmission is characterized by a relatively slow and prolonged rise and decline in numbers of cases (FIG. 12-18). In the United States, the Centers for Disease Control (CDC) in Atlanta, Georgia, compiles the statistics necessary for such determinations.

If the number of cases of a disease is constant in a particular geographic area, but the number of cases and the severity of the disease are insufficient to be a public health problem, the disease is classified as **endemic.** For example, plague is endemic to the southwestern United States because the causative bacterium, *Yersinia pestis,* is present in prairie dogs. **Epidemics** occur when a disease has a very high incidence in a population and when the death rate

and/or the degree of potential harm is high enough to pose a public health problem. There are annual outbreaks of cholera in southeastern Asia, especially after monsoon flooding washes *Vibrio cholerae* into the drinking water supply. When an epidemic spreads worldwide it is called a **pandemic.** The last great pandemic of influenza occurred in 1918-1919 when over 20 million people in the United States alone died from this disease.

The incidence of a disease is the number of new cases over a specific period of time. The prevalence of a disease is the number of people infected at any given time. The **morbidity rate** of a disease is the number of cases usually expressed per 100,000 people per year. The **mortality rate** is the number of deaths due to that disease in relation to the total population, also expressed per 100,000 people per year. These statistics are kept and studied in the United States at the Centers for Disease Control in Atlanta and are reported in the *Morbidity and Mortality Weekly Report.*

A disease outbreak can often be traced to a single source of exposure. Epidemiologists act as detectives searching for the source of tainted food or the direct contact with infected individuals to locate the origin of a disease outbreak. They use various methods to identify and to track the sources of disease, including conventional methods for culturing bacteria and modern molecular methods such as the use of gene probes. They know that disease transmission occurs by air, eating contaminated food or water, and/or direct contact with infected individuals or contaminated objects (fomites). By determining where and what people have eaten, where they have been, and with whom they have been in contact, epidemiologists establish patterns of disease transmission. This information then can be used to interrupt the further spread of the pathogens causing the disease.

The epidemiologist also is able to determine how a disease outbreak in a population can be effectively controlled or prevented. Indeed, it is the aim of the epidemiologist to identify the sources of disease outbreaks and to advise public health officials on the steps that should be taken to prevent such disease outbreaks. Essentially the epidemiologist is able to advise public health officials about what sources of pathogens should be treated or avoided, or what steps should be taken to lower the probability that individuals will contract a disease if exposed to the particular pathogens.

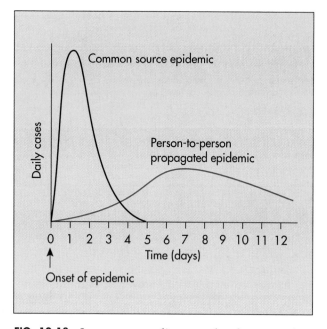

FIG. 12-18 In some cases, disease outbreaks occur when numerous individuals are exposed to a common source of the pathogen. Such outbreaks are characterized by a sudden rapid rise in the number of cases. Person-to-person transmission of disease results in epidemic outbreaks of disease that are characterized by a slower rise in the number of cases over a more prolonged time period.

A greatly diminished incidence of many diseases caused by microorganisms is the consequence of an understanding of the modes of transmission of pathogenic microorganisms and using preventive measures to reduce exposure to disease-causing microorganisms.

HIGHLIGHT

CONTACT TRACING

The goal of a contact tracing or partner reporting program is to control the spread of an infectious disease. When an individual has an infectious disease it may be possible to identify from whom he or she caught it and to whom the disease may have been passed. If these contacts can be persuaded to be examined and treated, it may be possible to limit the spread of the disease. Not all contacts are infected and some infected individuals are asymptomatic (not having symptoms of disease).

Such programs have long been carried out for many sexually transmitted diseases, including syphilis and gonorrhea. They have also been used for other infectious diseases, such as tuberculosis, meningitis, and some imported tropical diseases. They are not undertaken for less serious infectious diseases or for those diseases for which the transmission mechanism is indirect or not understood.

Trained public health workers interview patients to explain the nature of the disease, its mode of transmission, and the possible complications if it is not treated. In strictest confidence, patients are asked for the names of the persons with whom they have been in contact in ways relevant to the mode of transmission of their disease. Investigators want to know the identities of primary and secondary contacts. A primary contact is one from whom an infection may have been contracted and a secondary contact is one to whom the disease may have been transmitted.

When many people are infected with a disease, mass education directed at the general population can be used to prevent the further spread of the disease. When few individuals are infected, attempts must be made to locate the infected individuals and educate them individually because of the potential public health consequences of the actions of an uninformed infected individual.

Patients cannot be compelled to give the names of such contacts. Neither can the contacts be compelled to see a physician. Both must be persuaded of the public health benefit, the personal necessity, and the personal and public responsibility they have to do so. A major assumption of any contact tracing program is that an individual does not want to infect others unknowingly. This is particularly important when the individuals named are women of childbearing age who might transmit an infection to an unborn child.

A contact notification program for AIDS prevention to try to limit the spread of AIDS has been established. The program tries to find individuals in the heterosexual or homosexual communities who have been exposed and do not know it and offer them risk reduction education and serological testing. In San Francisco, where 50% of all gay men are infected with the AIDS virus, there has been an active program of mass education that has had a great impact in decreasing the level of unsafe sexual practices that facilitate transmission of the HIV. Among the general heterosexual population, HIV infection in 1987 was much less than that in the homosexual population. Therefore the homosexual population was considered to be a high-risk group. However, all infected individuals of either population may carry the virus and unknowingly infect others because the incubation period for AIDS can be 10 years or longer. Heterosexuals do not feel the immediacy of the epidemic, since they are not members of the classic high-risk groups.

While some states have made contact tracing a legal requirement, they still cannot force patients to comply. In those states, anyone who has been diagnosed as having AIDS is asked for the names of their sexual partners or individuals with whom they have shared needles so that they can be contacted and the situation explained to them. Investigators usually receive a high level of cooperation from the contact when they handle the interview with sensitivity, confidentiality, and a depth of knowledge about the subject on which the contact can rely.

One of the most controversial questions about contact tracing and notification is whether physicians and other health care workers should undergo mandatory HIV testing and whether they should be required to notify their patients. Several bills have been considered by the U.S. Congress that would make HIV testing and patient notification mandatory, with severe penalties for health care workers who do not comply; none of these bills has passed. The movement to make patient notification a requirement for any HIV-infected health care worker arises from the case of a dentist who apparently infected some of his patients by contaminating instruments used in the dental procedures. As a result there were public demands for increased protection from such exposures. In several well-publicized cases, hospitals and dental clinics have notified large numbers of patients who may have been exposed to a health care worker with AIDS. The American Medical Association has proposed voluntary testing of physicians and notification of patients by HIV-positive physicians performing certain procedures, such as surgery, where blood contamination may inadvertently occur. Physicians feel they are at greater risk of contamination by patients with AIDS than are the patients. By identifying those infected with HIV and notifying their contacts it is hoped that appropriate measures can be taken to minimize the spread of AIDS.

HYGIENIC AND ASEPTIC PRACTICES

Soon after the germ theory of disease was accepted, hygienic and aseptic practices were instituted that have greatly reduced the incidence of disease. Physicians and all health care workers now take great precautions regarding hygienic practices to prevent infections. Childbirth, for example, used to carry with it a very high risk of infection to the mother from the midwife's and the physician's contaminated hands. Today's midwives and obstetricians thoroughly wash their hands, wear sterile surgical gloves, and use sterile equipment, thereby reducing the risk of infection during delivery. Indeed, limiting exposure to pathogens is fundamental to a healthful society. The hygienic methods employed for preventing exposure to specific disease-causing microorganisms vary, depending on the particular route of transmission.

Human-microorganism contacts can be controlled in ways that reduce the incidence and spread of infectious diseases.

Avoiding sexual contact with individuals suffering from sexually transmitted diseases, such as syphilis and gonorrhea, interrupts the transmission of the pathogens that cause these diseases. Avoidance of sexual contact with infected individuals and proper use of prophylactic condoms, the main methods for controlling the spread of sexually transmitted diseases, are essential.

Many modern sanitary practices are aimed at reducing the incidence of diseases that spread through water by reducing concentrations of pathogens. Concentrations are reduced to levels insufficient to cause disease. Proper sewage treatment and drinking water disinfection programs reduce the likelihood of contracting a disease through contaminated water. Failure to maintain water quality often results in outbreaks of disease. For example, cholera outbreaks often occur when sewage is allowed to mix with drinking water supplies. This frequently occurs in the Far East when monsoon rains cause flooding that results in contamination of drinking water supplies.

Recognition of the fact that many serious diseases, such as typhoid, are transmitted through water contaminated with fecal material is the basis for enforcement of strict water quality control. Chloramination of municipal water supplies is widely used to prevent exposure to the pathogenic microorganisms that occur in water supplies and thus to ensure the safety of drinking water. Chloramination is treatment with chloramines, which are organic compounds that are toxic to microorganisms. Before the twentieth century, rivers in Europe were open sewers, carrying typhoid and other diseases to cities along their paths of flow. The sanitary practices applied to water supplies may be the single largest factor in reducing the incidence of infectious diseases.

Control measures are also applied throughout the food industry to prevent the transmission of disease-causing microorganisms through food products. Extensive quality control testing is required in most countries to prevent outbreaks of diseases associated with food supplies. Pasteurization of milk is a good example of a process designed to reduce exposure to pathogenic microorganisms that occur and proliferate in untreated milk. Restaurant workers are required to wear plastic gloves and to wash their hands frequently to avoid the accidental contamination of food they handle with soil or other substances that may harbor populations of disease-causing microorganisms.

Steps can be taken to reduce the probability of encountering pathogenic microorganisms, thus reducing the incidence and spread of infectious diseases.

ISOLATION AND QUARANTINE

We are continuously exposed to microorganisms carried through the air, in water and foods, and on the surfaces of virtually all objects that we contact. Only in the rarest of cases, when the immune system is totally nonfunctional, is absolute avoidance of contact with microorganisms ever attempted. However, we often isolate individuals with infectious diseases to prevent their serving as a source of infection for others (FIG. 12-19).

Quarantine is an ancient procedure that was introduced to physically contain in one area humans potentially carrying pathogenic microorganisms, thereby preventing the spread of dread diseases such as plague. Classically, people were quarantined for 40 days, and individuals often were prevented from entering walled cities for this period of time to prevent the entry of disease-carrying individuals. Historically, the isolation of leprosy (Hansen disease) patients in remote colonies is an example of the extreme steps taken to prevent contact of such individuals with the general population. Such practices decrease the probability of exposure to pathogenic organisms and prevent, or at least reduce, the transmission of disease. Today, this extreme practice is not needed, even for leprosy, because of the use of antimicrobial agents. We have also recognized that this disease is not as infectious as once thought. In fact, a leper colony was disbanded in southwestern Louisiana in 1990 (although a leper colony still exists in Hawaii).

Today, individuals who are particularly susceptible to infections are isolated from the general public. Special precautions are employed to minimize their exposure to microorganisms. For example, because of their susceptibility to infections, newborns are isolated from the general public—an important precaution for preventing the spread of infectious diseases. Burn victims are similarly isolated and visitors are re-

FIG. 12-19   Patients with infectious diseases are sometimes isolated to prevent the spread of pathogens.

quired to use surgical masks and gowns to reduce the likelihood of transmitting disease-causing microorganisms to exposed tissues. Preventing the exposure of individuals whose immunological defense mechanisms are compromised as a result of such conditions as treatment for cancer or organ transplant to airborne pathogenic microorganisms is an important aspect of patient management practice.

Isolation of individuals with contagious microbial diseases, in which the infectious agent is airborne, is often practiced. For example, children with measles, chickenpox, or mumps are often kept away (isolated) from other children who are not immune to these diseases. Surgical masks are worn in the presence of patients with tuberculosis or other pulmonary diseases. Other precautions, such as maintaining a safe distance from the infected individual, are also taken to minimize exposure. Such practices decrease the probability of exposure to pathogenic organisms and prevent, or at least reduce, the transmission of disease.

CONTROL OF VECTORS

Practices such as the use of insecticides are normally employed to control insect and other animal populations that act as **vectors** (carriers) of pathogenic microorganisms. Vectors acquire pathogens from a **reservoir**, which may be an animal population or nonbiological source of the pathogen. The most notable vectors of pathogenic microorganisms are mosquitoes, lice, ticks, and fleas. For example, fleas—the vector of plague—pick up *Yersinia pestis* from rats—

the reservoir. Some public health measures, such as mosquito control programs, are aimed at reducing the sizes of these vector populations. This would lower the probability of exposure to the pathogenic microorganisms capable of causing diseases such as

NEWSBREAK

BIOTECHNOLOGICAL VECTOR CONTROL

Imaginative biotechnological approaches for vector control are being developed. In one approach, sterile females are developed and released into the vector population. Since the females are sterile, reproduction cannot occur and the size of the vector population diminishes. This approach is being used to control specific pathogen-carrying mosquito populations. In another approach, genetic engineering is being used to create a strain of *Anopheles* mosquito that will not act as a vector for malaria-causing protozoa; the hope is that introduction of this "anti-malaria mosquito" will displace the wild type mosquitoes that act as vectors for malaria. Attempts to control tse tse flies that act as vectors for the protozoa that cause African sleeping sickness are based on putting genes for resisting the protozoa into bacteria and then getting the bacteria to grow in the intestines of tse tse flies. Tse tse flies with the genetically engineered bacteria in their intestines cannot act as vectors.

plague, typhus fever, yellow fever, malaria, and various other diseases transmitted by insect vectors. The famous story of the Pied Piper of Hamlin leading the rats with their fleas to the sea is a fairy-tale description of vector and reservoir control. Sometimes it is very difficult to control vectors. They may be distributed over vast areas of land so that pesticide application is difficult. Also, some insect vectors have developed resistance to certain pesticides.

It was possible to eliminate smallpox because humans were the only reservoirs of the pathogens. It is not, however, possible to eliminate bubonic plague because various wild animal populations serve as reservoirs for the bacterial causative agent of plague, *Yersinia pestis*, that is transmitted to humans by flea vectors. It is possible, though, to avoid being bitten by vectors. Clothing that covers the body surfaces, such as long-sleeved shirts, high socks, and hats, protects against tick bites. Since ticks carry the bacterium that causes Lyme disease, such protection is important in lessening the risk of tick bites, thus reducing the likelihood of contracting this disease. Mosquito netting is essential to protect against malaria-carrying mosquitoes in parts of Africa and the Far East.

SUMMARY

Koch's Postulates—Proving Specific Microorganisms Cause Specific Diseases (pp. 339-341)

- Robert Koch believed the germ theory of disease: that some diseases are caused by microorganisms and that specific diseases are caused by specific microorganisms. He established the series of proofs necessary to prove the cause of a specific disease.
- Koch's postulates state that the causative agent of a disease has to be found in all subjects with that disease and that it should be absent from healthy individuals, the agent has to be isolated, the isolated agent has to be shown to be able to cause the disease by inoculation into healthy individuals who become ill after inoculation, and that the same organism must be reisolated.

Virulence Factors of Pathogenic Microorganisms (pp. 341-348)

- Virulence factors are special properties of pathogenic microorganisms that enable them to cause disease. Adhesins enable pathogens to adhere to host cell surfaces to enable them to establish an infection. Capsules, slime layers, surface carbohydrates and proteins, flagella, and pili also aid in attachment.

Colonization and Ability to Grow Within the Body (pp. 342-344)

Invasiveness (pp. 342-344)

- Invasiveness is the ability of microorganisms to invade human tissues and grow and reproduce within host cells and tissues. Some pathogens produce enzymes that increase their invasive capacity by destroying body tissues and cells.
- Siderophores are low molecular weight compounds involved in iron transport produced by pathogens to bind needed iron taken from the blood, enabling the pathogens to grow.

Toxigenicity (pp. 344-348)

- Pathogens produce toxins that disrupt normal cell functions. Endotoxins come from the lipopolysaccharide component of Gram-negative cell walls. Exotoxins are protein toxins. Neurotoxins are toxins that affect the nervous system. Enterotoxins inflame the tissues of the gastrointestinal tract. Cytotoxins kill cells by enzymatic attack or blocking cellular metabolism.

- Cytopathic effects are the observable changes in cells infected with viruses. Some of these changes are lethal.

Infection and Disease (pp. 348-354)

- Contamination is the presence of foreign microorganisms in a host. These microorganisms may colonize the site of contamination and cause an infection. The infection may cause changes in normal body functions, a disease.
- The number of pathogens needed to establish a disease varies, depending on the nature of the pathogen and the state of the host's defenses. A compromised host is more susceptible to infection.
- In addition to pathogenic microorganisms, disease can also be caused by factors such as genetic defects, nutritional deficiencies, metabolic imbalances, degeneration, and autoimmune and hypersensitivity reactions.

Pattern of Disease (pp. 350-354)

- Diseases are characterized by specific sets of characteristic signs and symptoms.
- Diseases follow a regular pattern of stages: incubation, prodromal, period of illness, period of decline, convalescence, and recovery.
- Diseases are acute or chronic and localized or systemic.

Transmission of Infectious Agents (pp. 354-363)

- Infectious diseases that can spread from infected host to susceptible individual are communicable. Intoxications are diseases caused by the ingestion of toxins.
- A reservoir is a source of an infectious agent, such as humans, soil, and water. Diseases can be spread by direct or indirect contact transmission. Direct contact transmission is person-to-person and indirect transmission contact is with contaminated nonliving objects called fomites. Carriers are infected individuals who have no disease symptoms.
- Public health efforts and modern sanitary practices attempt to reduce the incidence of disease by controlling the spread or numbers of potential pathogens through controlling the conditions in the routes of transmission.

- A human pathogen usually must enter the body through a specific portal of entry to establish an infection.

Routes of Disease Transmission (pp. 354-363)

- Disease pathogens are transmitted by air to the respiratory tract, by food and water to the gastrointestinal tract, by sexual contact to the genitourinary tract, and by direct contact to superficial body tissues.
- Parenteral routes of disease transmission include punctures, injections, bites, cuts, surgical incisions, and cracked skin. All disrupt the protective layers of the skin and provide a portal of entry for potential pathogens. Burns also remove the skin and the burn area is easily contaminated.
- Zoonoses are infectious diseases of nonhuman animals that are transmissible to humans.
- Nosocomial infections are infectious diseases acquired while a patient is in a hospital.
- Aseptic procedures and prophylactic antibiotics are used to prevent their occurrence.

Prevention of Diseases by Avoiding Exposure to Pathogenic Microorganisms (pp. 363-368)

Epidemiology—Determining Sources and Routes of Disease Transmission (pp. 363-365)

- Epidemiology is based on the statistical probability that exposure of a susceptible individual to a particular pathogen will result in disease transmission. By determining where and what people have eaten, where they have been, and with whom they have been in contact, the epidemiologist can establish a pattern of disease transmission.
- A common source outbreak is characterized by a sharp rise and rapid decline in the number of cases. A person-to-person epidemic is characterized by a relatively slow rise and decline in the number of cases.
- Because the transmission of pathogens occurs via restricted routes it is possible to control our interactions with microbial populations in ways that reduce the probability of contracting infectious diseases.

- Etiology is the cause of disease and is studied by epidemiologists, with factors involved in the transmission of infectious agents, to determine how to control disease outbreaks.
- Morbidity is the incidence rate of disease. Mortality is the death rate. The morbidity rate of common source outbreaks rise and fall rapidly, while person-to-person epidemics start and end slowly.
- Diseases are endemic if the number of cases remains constant. Diseases are epidemic if the incidence and death rates are high. Pandemics are international outbreaks of disease.

Hygienic and Aseptic Practices (p. 366)

- Avoiding sexual contact with infected individuals and exercising appropriate prophylactic measures controls the spread of sexually transmitted diseases.
- Pathogens can infect deep body tissues and the circulatory system when the intact skin is broken as a result of cuts, wounds, bites, punctures, surgical incisions, or injections. To prevent such infections, surgical rooms, instruments, garments, gloves, and masks are kept sterile and wounds are cleansed and treated with antiseptics.
- Fecal contamination of food and water is associated with disease transmission via water. Proper sewage treatment and drinking water disinfection programs reduce the likelihood of the spread of disease.
- Sanitary methods and quality control measures in the food industry prevent transmission of disease-causing microorganisms through food products. Food preservation methods attempt to eliminate microorganisms from food or limit their growth rates.

Isolation and Quarantine (pp. 366-367)

- Disease transmission can be interrupted by avoiding physical contact with pathogenic microorganisms. Quarantine removes potential carriers of disease from the general population.

Control of Vectors (pp. 367-368)

- Vectors are carriers of disease agents. Reservoirs are constant sources of infection agents found in nature.

CHAPTER REVIEW

REVIEW QUESTIONS

1. What structures contribute to the virulence of pathogens?
2. What is the difference between toxigenicity and invasiveness?
3. What is a toxin? What is the difference between an endotoxin and an exotoxin?
4. What are the differences between the toxins produced by *Clostridium botulinum* and *C. tetani*?
5. What are parenteral routes of disease transmission?
6. What are potential reservoirs of agents of infectious human diseases?
7. What are carriers?
8. How do carriers contribute to disease transmission?

9. What are portals of entry? What are different portals of entry in the human body?
10. What are portals of exit? How do microorganisms escape from an individual?
11. What are vectors? How do they contribute to the transmission of disease?
12. Why would some strains of a bacterium be pathogenic and not others?
13. How can you distinguish a common source outbreak from person-to-person transmission?
14. How is a knowledge of the routes of disease transmission used to control the spread of disease in a hospital?
15. How is quarantine used to control infectious diseases?
16. How does vector control limit the spread of disease?

READINGS

Ayoub EM, GH Cassell, WC Branche, TJ Henry (eds.): 1990. *Microbial Determinants of Virulence and Host Response*, Washington, D.C.; American Society for Microbiology.

Summary and review of associated advances in bacteriology, mycology, immunology, and host-parasite interactions as they affect microbial virulence.

Bruce-Chwatt LJ and J DeZulueta: 1980. *The Rise and Fall of Malaria in Europe: A Historico-Epidemiological Study*, Oxford, Oxford University Press.

History of malaria and malaria eradication in different countries and regions of Europe.

Clark VC and PM Bavoil (eds.): 1994. *Bacterial Pathogenesis: Part A, Identification and Regulation of Virulence Factors*, San Diego, Academic Press.

An in-depth discussion of the factors involved in microbial virulence showing how specific intrinsic properties of pathogens contribute to disease.

Clark VC and PM Bavoil (eds.): 1994. *Bacterial Pathogenesis: Part B, Interaction of Pathogenic Bacteria with Host Cells,* San Diego, Academic Press.

An in-depth discussion of the factors involved in microbial virulence showing how the interactions of pathogens with host cells contribute to disease.

Coleman W: 1987. *Yellow Fever in the North: the Methods of Early Epidemiology*, Madison, Wisconsin; University of Wisconsin Press.

The author investigates the methods used in epidemiology before the germ theory of disease. He uses reports of outbreaks of yellow fever in France in 1861, Wales in 1865, and Gibraltar in 1828 to demonstrate the impact of epidemiological studies on French and British medical theory.

Collier R: 1974. *The Plague of the Spanish Lady: the Influenza Pandemic of 1918-1919*, London, Macmillan.

Describes the progress of the disease during a 17-week period as it swept through the world, telling the stories of victims, healers, exploiters, and scientists who worked to prevent its recurrence.

Cotran RS, V Kumar, SL Robbins: 1994. *Robbins Pathologic Basis of Disease*, Philadelphia, Saunders.

Classic comprehensive reference book on infectious diseases.

Davis JM and GT Shires (eds.): 1990. *Principles and Management of Surgical Infection*, Philadelphia, Lippincott.

How to prevent and control the bacterial infections that can follow surgical procedures.

Ewald PW: 1993. The evolution of virulence, *Scientific American*, 268(4):86-93.

A thought-provoking article on how microorganisms develop properties that enable them to cause disease.

Fischetti VA: 1991. Streptococcal M protein, *Scientific American* 264(6):58-65.

A description of an important protein that contributes to the virulence of streptococci.

Gasser RA, AJ Magill, CN Oster, EC Tramont: 1991. The threat of infectious disease in Americans returning from Operation Desert Storm, *New England Journal of Medicine* 324:859-864.

Discusses the infectious diseases that American service personnel may bring back from the Middle East, including some that are uncommon and rapidly fatal, some that may spread to contacts, some that may cause chronic conditions, and some that may not emerge for many years.

Harrison G: 1978. *Mosquitoes, Malaria, and Man: A History of the Hostilities since 1880*, New York, Dutton.

Account of the first organized war against an insect written by a military historian who concludes that we cannot expect to eradicate this disease; we can only work to decrease the number of deaths that it causes.

Howard BJ: 1994. *Clinical and Pathogenic Microbiology*, ed. 2, St. Louis, Mosby.

An advanced general textbook of medical microbiology.

Kluger MJ: 1979. *Fever: Its Biology, Evolution and Function*, Princeton, New Jersey, Princeton University Press.

Demonstrates how fever is beneficial to an infected host. Discusses regulation of body temperature, laboratory findings on the evolution of the febrile response, and the role of fever in disease.

Last JM (ed.): 1983. *Dictionary of Epidemiology*, Oxford, England; Oxford University Press.

Reference book that defines the terminology used in the field of epidemiology.

Lennox J: 1985. Those deceptively simple postulates of Professor Koch, *American Biology Teacher* 47:216-221.

An interesting and simple presentation of Koch's postulates.

Lillienfeld AM (ed.): 1980. *Times, Places, and Persons: Aspects of the History of Epidemiology*, Baltimore, Johns Hopkins University Press.

Combines the history of the origin and development of numerical studies and statistical approaches to epidemiology with the stories of smallpox, yellow fever, and pellegra.

Lilienfeld DE, PD Stolley, AM Lilienfeld: 1994. *Foundations of Epidemiology*, ed. 3, New York, Oxford University Press.

Describes the basis for epidemiology and gives cases of how epidemiologists approach the study of disease spread.

Marks G and WK Beatty: 1976. *Epidemics*, New York, Scribner's.

The authors use primary source material, including eyewitness accounts, to tell the stories of various forms of contagion from ancient times to the present. The emphasis is on the growth of the epidemics, rather than their treatment or impact.

Matossian MAK: 1989. *Poisons of the Past: Molds, Epidemics, and History*, New Haven, Yale University Press.

Explores the effects of various diseases, including food poisoning and other mycotoxicoses, on historical events.

McNeill WH: 1976. *Plagues and Peoples*, Garden City, New York, Doubleday.

This description of the impact of infectious diseases on the rise and fall of civilizations was designed to provoke discussion and research.

Murray PR: 1994, *Medical Microbiology*, ed. 2, St. Louis, Mosby.

An advanced general textbook of medical microbiology.

Nutton V: 1983. The seeds of disease: an explanation of contagion and infection from the Greeks to the Renaissance, *Medical History* 27:1-34.

Explores the development of theories beginning with the ancient Greeks of how and why diseases occur and spread.

Rietschel ET and H Brade: 1992. Bacterial endotoxins, *Scientific American* 267(2):54-61.

A thorough discussion of bacterial endotoxins and their importance.

Rosen G: 1960. *History of Public Health*, New York, MD Publications.

The story of western community health action from the ancient Greeks to the present. Describes the social, political, and technical developments affecting the development of public health.

Roth JA (ed.): 1988. *Virulence Mechanisms of Bacterial Pathogens*, Washington, D.C., American Society for Microbiology.

> *Bacterial virulence as examined by different disciplines, including medicine, veterinary science, genetics, biochemistry, immunology, and microbiology.*

Roueche B: 1980. *The Medical Detectives*, New York, Times Books.

> *Articles reprinted from* New Yorker *magazine from 1947 to 1980, each presenting a medical mystery about the diagnosis or source of an ailment or both. The work these stories describe is a combination of scientific deduction and detective legwork.*

Salyers AA and DD Whitt: 1994. *Bacterial Pathogenesis: A Molecular Approach*, Washington, ASM Press.

> *A comprehensive and authoritative volume on the molecular basis of pathogenesis..*

Savage DC and M Fletcher (eds.): 1985. *Bacterial Adhesion: Mechanisms and Physiological Significance*, New York, Plenum Press.

> *The surfaces, mechanisms, and consequences of bacterial attachment to both animate and inanimate objects are studied in marine science, chemistry, medicine, agriculture, and engineering. This book helps investigators in all areas understand the phenomenon.*

Schaechter M, G Medoff, D Schlessinger: 1989. *Mechanisms of Microbial Disease*, Baltimore, Williams & Wilkins.

> *Comprehensive volume of microbiology and the physiopathology of communicable diseases caused by bacteria, fungi, and viruses.*

Schlossberg D: 1994. *Infections of Leisure*, New York, Springer-Verlag.

> *Discusses the association of infections with leisure activities such as fishing, camping, hiking, and swimming; also covers infections associated with pets, including dogs, cats, and birds.*

Schlossberg D (ed.): 1990. *Infections of the Nervous System*, New York, Springer-Verlag.

> *Covers bacterial infections and viral diseases, especially meningitis, that affect the central nervous system.*

Soule BM and E Larson: 1994. *Infections and Nursing Practice*, St. Louis, Mosby.

> *Takes a holistic approach to the prevention and control of infections in various health care settings.*

Spink WW: 1978. *Infectious Diseases: Prevention and Treatment in the 19th and 20th Century*, Minneapolis, University of Minnesota Press.

> *A history of the control of infectious diseases, emphasizing twentieth century treatment of specific diseases, particularly the discovery of the sulfonamides and other chemical and antibiotic agents.*

Thomas G and M Morgan-Witts: 1982. *Anatomy of an Epidemic*, Garden City, New York, Doubleday.

> *The story of Legionnaire's disease.*

Vinken PJ, GW Bruyn, HL Klawans (eds.): 1988. *Microbial Disease*, Amsterdam, Elsevier.

> *Reviews the etiology and diagnosis of the bacterial, viral, and parasitic diseases that affect the nervous system.*

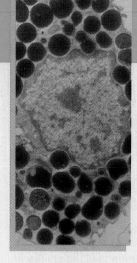

C H A P T E R   13

Nonspecific Host Defenses Against Microbial Infections: The Immune Response

PREVIEW TO CHAPTER 13

In this chapter we will:
- See why most microorganisms cannot infect the human body, causing infections and disease.
- Examine several nonspecific lines of defense that protect the human body against microbial infections.
- Study the physical, chemical, and cellular lines of defense that contribute to the protection of the body against microbial invasion.
- Learn the following key terms and names:

agranulocytes
alternate pathway
alveolar macrophages
basophils
classical pathway
complement
eosinophils
fever
granulocytes
immune adherence
inflammatory exudate
inflammatory response
interferons
keratin
Kupffer cells
lactoferrin
leukocytes
lymph
lymphocytes
lysozyme
macrophages

membrane attack complex
  (MAC)
microglia
monocytes
mucociliary escalator
  system
mucous membrane
mucus
neutrophils
nonspecific defenses
normal microbiota
opsonization
phagocytes
phagocytosis
polymorphonuclear
  neutrophils (PMNs)
pyrogens
skin
transferrin
wandering macrophages

Each of us is continuously exposed to microorganisms, many of which have the potential for growing within the human body and disrupting body functions. Every drop of water that strikes our body when we shower contains hundreds or thousands of bacteria. When we brush our teeth, thousands of microorganisms are driven through the gums into the circulatory system. Our foods are laden with microorganisms, as is the air we breathe. Why then are we not continuously ill? Why can't most microorganisms invade and infect the human body?

The reason lies in a complex network of interactive, overlapping defense systems that protect us against potentially pathogenic microorganisms. Collectively these defense systems are called *immunity*. The totality of this system determines whether we are resistant to disease-causing pathogens or susceptible to infection and disease. We may view infection and disease as the result of failed defense systems. Fortunately, given the magnitude of our exposure to microorganisms, we rarely suffer infections. Because of the body's integrated defense systems, most microorganisms are prevented from establishing infections.

The body defenses include surveillance and protective responses. These defenses are comprised of various molecules and cells, distributed through the body, that defend against invasion by microorganisms and foreign substances. The first lines of defense against invasion of the body by microorganisms are a set of physical and chemical barriers known as the *nonspecific defenses*. The nonspecific defenses of the body represent a generalized system that guards against the wide variety of microbial pathogens. Most nonspecific defenses are innate and offer protection from the moment of birth. Being already in place, they often are able to prevent the replication and spread of pathogens in the body. These nonspecific defenses are supplemented by the specific immune response, which is discussed in Chapter 14. Together the nonspecific and specific defense systems protect the body from infection and facilitate recovery from disease when infections do occur (FIG. 13-1).

The nonspecific defenses are especially important in preventing infections of the body. Nonspecific barriers to microbial invasion of the body, which include the skin, phagocytic cells, and various antimicrobial chemicals, block the entry of microorganisms into the

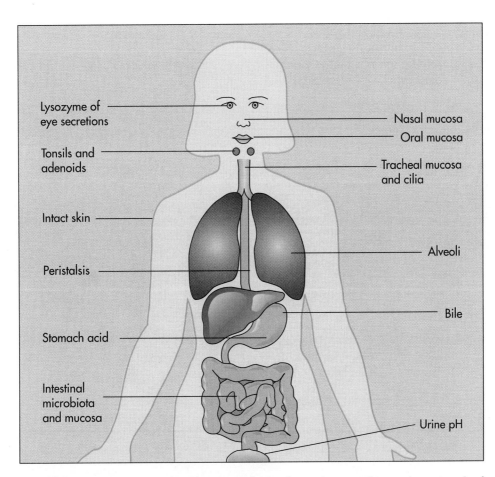

FIG. 13-1 The body is protected against infection by an integrated extensive network of nonspecific and specific immune responses. Body surfaces represent the first line of protection against microbial infection.

body. Nonspecific defenses also seek out and destroy most potential disease-causing microorganisms that enter the body before they replicate and establish an infection. The nonspecific defenses continuously patrol the bloodstream for the presence of foreign cells or molecules. When such cells or substances are detected the system responds with multiple attacks to eliminate them.

Nonspecific defenses, consisting of various physical, chemical, and cellular barriers, represent the first lines of defense that protect the body against microbial invasion.

PHYSICAL BARRIERS

The first line of defense against microbial infections are physical barriers that block the ability of most microorganisms to enter the body. Intact body surfaces represent the first line of defense against microorganisms. They physically block the entry of pathogens into the body. Preventing microorganisms from entering the body blocks infection and is an effective means of disease prevention.

SKIN

Most microorganisms are noninvasive and so do not penetrate the skin. The skin is composed of a thin outer epidermis composed of layers of epithelial cells and a thicker underlying dermis layer composed of connective tissue (FIG. 13-2). The outer surface of the skin layer, the epidermis, is designed to block the entry of microorganisms. The cells of the epidermis are tightly packed so that there are not many channels through which microorganisms could move. Several continuous layers of these tightly packed epithelial cells make the epidermis a formidable physical barrier. To add to its resistance toward microbial penetration, the surface layer of the epidermis contains *keratin*, a substance that is not readily degraded enzymatically by microorganisms. Keratin is a fibrous and insoluble protein that resists penetration of water. Since the body is frequently exposed to microorganisms in aqueous environments, as in aerosols or suspensions in liquids, this impermeable keratinized layer provides a formidable external barrier to microorganisms. Additionally, the outer layer of skin consists predominantly of dead cells that are continuously being sloughed off or shed so that the complete epidermis is replaced once or twice a month. This prevents infection by viruses, which require live cells for their replication.

The skin blocks the entry of many pathogens into the body.

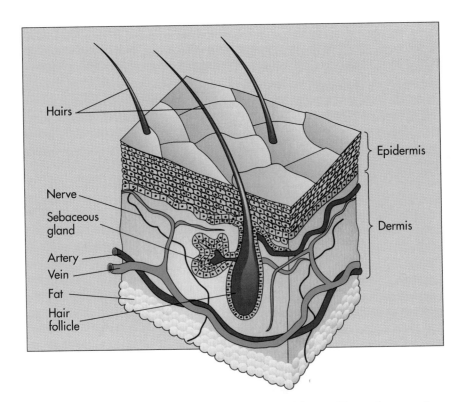

FIG. 13-2   The skin structure has epithelial and dermal layers. There also are sebaceous glands that secrete sebum onto the skin surface.

The importance of the skin as a protective barrier can readily be seen when breaks in the intact skin occur. Cuts and wounds that break the continuity of skin exposes the body to numerous microorganisms that can then establish subcutaneous (below the skin) infections. Often these subcutaneous infections are caused by staphylococci that normally live on skin and hair surfaces. Disrupting the protective barrier of the skin also allows microorganisms to enter the circulatory system and deep body tissues. This results in infection unless precautionary actions are taken. To avoid entry of bacteria and other microorganisms into a wound, the area is cleansed and is usually covered with gauze to protect against contamination. Washing and antimicrobial agents are used to lower the probability of infection following wounds and burns. Care is taken in surgical procedures to prevent the entry of microorganisms into exposed tissues.

MUCOUS MEMBRANES

Whereas the skin covers the outside of the body, many internal body surfaces are covered by a lining called the **mucous membrane.** Like the skin, mucous membranes consist of two layers: a surface epithelial layer and an underlying connective tissue layer. Mucous membranes line the surfaces of the respiratory tract, gastrointestinal tract, and genitourinary tract. These are important and effective barriers to invasion by microorganisms. Mucus, which is secreted by goblet cells and subepithelial glands, accumulates on the surface of the epithelial layer of the mucous membrane. The sticky mucus accumulates on the surface of the cells of the mucous membrane, where it traps microorganisms, preventing most potential pathogens from penetrating into the body.

Mucous membranes protect against the invasion of pathogenic microorganisms by making it difficult for them to attach to and penetrate the linings of the respiratory, gastrointestinal, and genitourinary tracts.

The respiratory tract, for example, is protected in part against the invasion of pathogenic microorganisms because it has a mucous membrane and cilia (FIG. 13-3). The nose has a mucous membrane and mucus-covered hairs that filter inhaled air so as to trap many microorganisms and prevent them from moving further into the respiratory tract. Protection of the respiratory tract is critical because the air we breathe contains numerous microorganisms. The contact of the air in the lungs with the blood that brings oxygen to the body would allow airborne microorganisms to invade the body if it were not for these defenses. The ciliated epithelial cells that line the upper respiratory tract have cilia on the exposed side. The cilia move with an upward wave-like motion that resembles the movement of an escalator.

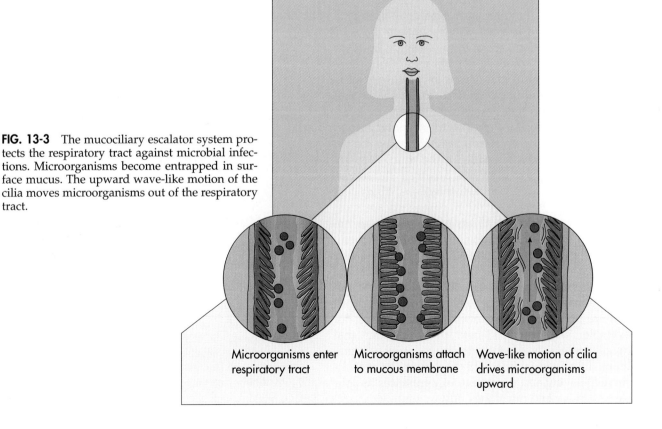

FIG. 13-3 The mucociliary escalator system protects the respiratory tract against microbial infections. Microorganisms become entrapped in surface mucus. The upward wave-like motion of the cilia moves microorganisms out of the respiratory tract.

Microorganisms enter respiratory tract

Microorganisms attach to mucous membrane

Wave-like motion of cilia drives microorganisms upward

Some of the mucus and microorganisms are swept out of the body through the oral and nasal cavities by this wave-like action. This system, called the *mucociliary escalator system,* effectively acts as a filter to prevent potential pathogens from penetrating the surface tissues of the respiratory tract. Chronic smokers are more prone to infection because the smoke damages ciliated epithelial cells and weakens the defense of the mucociliary escalator system.

The mucociliary escalator system protects the respiratory tract from infection with pathogens.

Sneezing and coughing also tend to remove many of these microorganisms from the respiratory tract. The gag reflex helps remove postnasal drip and mucus swept up by the ciliated epithelium of the bronchi, with its associated microorganisms. Additionally, many mucus-covered microorganisms are carried out of the respiratory tract and move into the digestive tract where they are swept up the trachea to the epiglottis and over to the esophagus to be swallowed and digested.

Like the respiratory tract, the digestive tract also is lined by a mucous membrane that makes it difficult for pathogenic microorganisms to attach and to penetrate into the body. Here again, the mucous membrane helps prevent the invasion of the body by microorganisms through the lining of the gastrointestinal tract. The genitourinary tract similarly is protected by mucous membranes that make it difficult for microorganisms to penetrate.

Although the mucous membrane is generally protective of the respiratory, gastrointestinal, and genitourinary tracts, it is more penetrable than the skin, and some pathogens are able to enter the body by moving through mucous membranes. Conditions that remove water from mucus, such as the dry air in many heated buildings, tend to make the mucus layer thinner and less protective. Many more infections of the respiratory tract occur in winter in part due to such drying of the mucus that lessens the defenses of the body. The cells of the mucous membrane are alive, and if viruses penetrate the mucus they may be able to establish infections. High concentrations of viruses or other pathogens are most likely to overwhelm the ability of the mucous membrane to protect against infection. The mucous membrane is inadequate for protecting the body against influenza viruses, for example, when high numbers of these viruses are inhaled. Some pathogenic microorganisms tend to stick to the mucous membrane and are protected by the mucus against drying. For example, *Treponema pallidum,* the bacterium that causes syphilis, is able to penetrate the mucous membrane of the genital tract. This bacterium is very sensitive to drying and mucus actually enhances its ability to survive and infect the body.

FLUID FLOW

Just as you wash to remove dirt containing microorganisms from the skin, some body tissues are protected against accumulations of microorganisms by the flow of fluids across their surfaces. Movement of fluids washes the surfaces of various body tissues. For example, microorganisms that do not adhere to surfaces in the oral cavity are washed into the stomach by the fluid flow of saliva. Urine, which generally is a sterile body fluid, flushes microorganisms from the surfaces of the urinary tract. The urea in

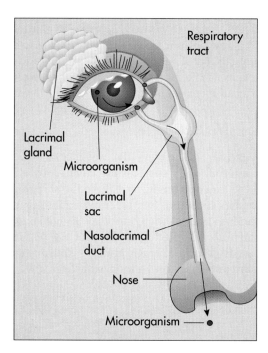

FIG. 13-4 Tears continuously wash the eye surface, cleansing it of microorganisms. Lysozyme kills many bacteria. Tears drain into the nose, carrying surviving microorganisms into the respiratory tract.

NEWSBREAK

DANGERS OF CATHETERIZATION

The insertion of a catheter to drain urine from the bladder sometimes is a necessary medical procedure. However, it also can be the cause of urinary tract infections. The catheter may carry microorganisms into the bladder, inoculating it with infecting microorganisms. Even when proper technique is used to prevent contamination of the catheter from the hands of the medical staff, microorganisms from the external regions of the urinary tract may contaminate the catheter and be carried into the bladder. The lack of urine flow through the urethra when urine is draining through the catheter also increases the likelihood of urinary tract infection.

urine is toxic to many microorganisms so that washing with urine also disinfects the surfaces of the urinary tract. Tears, produced and removed by the lacrimal apparatus, remove microorganisms from the eyes. Tears move from the lacrimal glands across the surface of the eye under the eyelid to the corner of the eye near the nose where they pass through the lacrimal canals into the nose (FIG. 13-4). Blinking spreads the tears over the entire eye surface. Normally, tears are removed as they are formed so that new microorganism-free fluid continuously washes the eyes. If a foreign substance dirt carrying microorganisms enters the eye, the irritation stimulates greater tear production so that there is increased cleansing.

MICROBIAL BARRIERS—NORMAL MICROBIOTA

Although the surfaces of the body are protected against invasion by microorganisms, some microorganisms are able to grow on them. Most body surfaces are covered with microorganisms that do not invade other parts of the body. These microorganisms are not harmful. They even contribute to the protection and healthful state of the body. The average adult human has 10^{13} eukaryotic animal cells (human cells) and 10^{14} associated prokaryotic and eukaryotic cells of microorganisms; stated another way, the normal human being is composed of just over 10^{14} cells, only 10% are human and the remaining 90% are microbial. Most of these are bacteria associated with the gastrointestinal tract.

Microorganisms associated with particular body tissues are called **normal microbiota** or **normal microflora**. Although the term *microflora* is used extensively, *microbiota* is preferred because it avoids any inference that microorganisms are little plants. The normal microbiota are indigenous (naturally occurring) and establish a dynamic and mutually beneficial association with body tissues (FIG. 13-5).

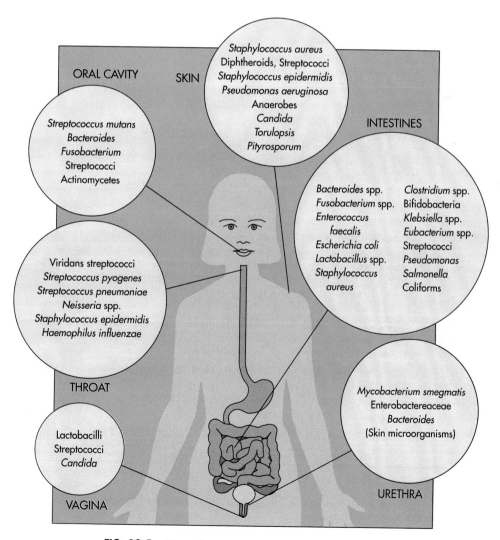

FIG. 13-5 Normal microbiota at various body sites.

A specific characteristic microbiota is associated with most body surfaces, with distinct populations inhabiting the surface tissues of the skin, oral cavity, respiratory tract, gastrointestinal tract, and genitourinary tract (Table 13-1).

The normal microbiota inhabits most body surfaces, with specific microorganisms characteristically associated with specific body locations.

Not all body tissues, though, provide suitable habitats for the growth of microorganisms. For example, most of the urinary tract lacks a resident microbiota. Only the distal end of the urinary tract has a resident microbiota. Urine that has not yet con-

tacted this extremity (that is, urine in the kidney, ureter, and bladder) is considered a sterile body fluid. Similarly, blood is considered a sterile body fluid because the circulatory system does not possess a resident microbiota. In reality, various microorganisms frequently enter the bloodstream but normally do not establish growing populations within the circulatory system. For example, a segment of the circulatory system associated with the liver, the hepatic portal system, normally contains low numbers of bacteria that pass through the intestinal wall as a result of abrasions in the lining of the intestinal tract caused by food particles. These bacteria are routinely eliminated from the circulatory system by specialized

TABLE 13-1		
Normal Microbiota of Various Body Sites		
BODY SITE	**RESIDENT MICROBIOTA**	**FACTORS INFLUENCING MICROBIAL COMMUNITY COMPOSITION**
Skin	Gram-positive bacteria *Staphylococcus* and *Micrococcus* most abundant; Gram-positive *Corynebacterium*, *Brevibacterium*, and *Propionibacterium* also occur; few fungi and few Gram-negative bacteria except in moist regions	Low water activity and fatty acids produced from sebum limit numbers and types of microorganisms on the skin.
Oral cavity	*Streptococcus* species, such as *S. mutans* on teeth and *S. sanguis* on saliva-coated surfaces, are abundant, as are obligate anaerobes; Gram-negative coccoid members of the genus *Veillonella* and Gram-positive species of *Bacteroides*, *Fusobacterium*, and *Peptostreptococcus*	Polysaccharide production by resident microbiota that forms plaque and allows adherence to surfaces in the oral cavity. Scavenging of molecular oxygen by facultative anaerobes allows growth of obligate anaerobes.
Gastrointestinal tract	Obligate and facultative anaerobes of the genera *Lactobacillus*, *Streptococcus*, *Clostridium*, *Veillonella*, *Bacteroides*, *Fusobacterium*, *Escherichia*, *Proteus*, *Klebsiella*, and *Enterobacter*	Abundance of substrates for growth of abundant resident microbiota. Scavenging of molecular oxygen by facultative anaerobes allows growth of obligate anaerobes.
Upper respiratory tract (nasal cavity and nasopharynx)	*Streptococcus*, *Staphylococcus*, *Moraxella*, *Neisseria*, *Haemophilus*, *Bacteroides*, and *Fusobacterium*	Ability to resist nonspecific defenses.
Lower respiratory tract	None	Phagocytic cells prevent colonization by a resident microbiota.
Upper urinary tract (kidneys and bladder)	None	Filtration and outward fluid flow prevent establishment of resident microbiota.
Vaginal tract	*Streptococcus*, *Lactobacillus*, *Bacteroides*, and *Clostridium*; coliforms; spirochetes; yeasts, including members of the genus *Candida*	Large surface area and secretions of nutrients permits growth of abundant microbiota. Acidity limits species within resident microbial community.

phagocytic blood cells that occur in blood vessels of the liver.

Usually the normal microbiota of the human body are nonpathogenic, that is, they do not cause disease. They grow on body surfaces and do not invade the body's tissues. By colonizing body tissues, the normal microbiota preempt the colonization of those tissues by other microorganisms that may cause disease. Many of the normal microbiota produce antimicrobial substances that act to prevent the establishment of infection by pathogenic microorganisms. Other mechanisms of antagonism by the normal microbiota against would-be pathogens also enhance host resistance to disease. These additional mechanisms include reduction of the oxygen concentration and competition for available nutrients so that pathogens do not find favorable growth conditions.

The normal microbiota contribute to host defense against infection by pathogens.

The normal microbiota contribute to the nutrition of the host by synthesizing essential nutrients. For example, germ-free animals require vitamin K in their food because it is not synthesized by the resident microbiota of the gastrointestinal tract as it is in animals with a normal microbiota. The microbiota of the gastrointestinal tract also synthesize biotin, riboflavin, and other vitamins that they supply to the animal host. Thus the maintenance of a "healthy" indigenous microbiota is essential to the maintenance of a healthy individual.

The acquisition of the normal microbiota by humans begins during birth and continues thereafter. Although a few microorganisms and parasites can migrate through the placenta, the human fetus is normally sterile. The acquisition of the normal microbiota occurs in stages and therefore is termed a *successional process*. The growth of microorganisms on body tissue surfaces alters the local environmental conditions, leading to the successional changes in the populations of microorganisms associated with the tissues until a relatively stable, normal microbiota is established.

The microbiota that live on body surfaces form a carefully balanced complex microbial community. When the normal microbiota are adversely affected an imbalance may occur that may lead to the development of disease. In this case the usually nonpathogenic microbiota can cause disease. In such cases these microorganisms are called *opportunistic pathogens*. For example, the use of antibiotics sometimes disrupts the balance of the microbial community of the gastrointestinal tract, permitting the growth of *Clostridium difficile*. *C. difficile* is normally found in this location, but when its growth is uncontrolled, it causes a severe and sometimes fatal gastrointestinal tract infection (antibiotic-associated pseudomembranous enterocolitis). Similarly, women taking antibiotics sometimes develop vaginitis due to an overgrowth of the fungus *Candida albicans*, which is normally held in check by the indigenous bacteria of the vaginal tract.

METHODOLOGY

USE OF GERM-FREE ANIMALS TO DETERMINE THE ROLE OF NORMAL MICROBIOTA IN HOST DEFENSE SYSTEMS

Germ-free (gnotobiotic) animals are good experimental models for investigating the interactions of animals and microorganisms. To determine the role of the normal microbiota, animals can be delivered by aseptic Caesarean section (the surgical removal of the fetus from the uterus via the abdomen) so they will not be contaminated by the normal microbiota of the vagina and birth canal during vaginal delivery. Germ-free animals can then be raised in the absence of microorganisms by being kept in a sterile environment. They are fed sterile food and water and given sterile air to breathe. Comparing animals possessing normal associated microbiota with germ-free animals permits the exploration of the complex relationships between microorganisms and host animals.

Germ-free animals develop abnormalities of the gastrointestinal tract. They are more susceptible to disease than animals with normal associated microbiota. Germ-free animals are more susceptible to bacterial infection. Organisms such as *Bacillus subtilis* and *Micrococcus luteus*, which are harmless to other animals, cause disease in germ-free animals. More exotic pathogenic microorganisms such as *Vibrio cholerae* and *Shigella dysenteriae* are far more readily able to establish infections where there are no normal microbiota. They do not have to compete for survival within the intestinal tract. At the same time, though, germ-free animals are resistant to *Entamoeba histolytica*, the causative organism of amebic dysentery. This is because this protozoan requires normal intestinal bacteria as a food source. Likewise, tooth decay is no problem to germ-free animals, even those on high sugar diets, because they do not have lactic acid bacteria—the bacteria that cause tooth decay—in their oral cavities.

CHEMICAL DEFENSES

Some of the fluids that wash body tissues also contain antimicrobial chemicals. Additionally, blood and lymph contain several chemical factors that defend against microbial infections. These chemicals limit the abilities of microorganisms to infect the body. Antimicrobial chemicals may inhibit the growth of microorganisms or may kill potential pathogens.

ACIDITY

Acids kill or prevent the growth of most microorganisms. Various body tissues are protected by the low pH environment created by acid production. The skin, for example, is bathed by secretions from the sebaceous (oil) glands that deposit a substance called sebum on the outer surface. Sebum, which is rich in lipids, prevents drying of the skin and hairs so they do not become brittle. The indigenous microorganisms on the surface of the skin can break down these lipids into free fatty acids. This contributes to the acidity of the skin and inhibits the growth of other microorganisms.

> **Acidity of various body fluids prevents the growth of many pathogens, contributing to the defenses of the genitourinary tract and gastrointestinal tract.**

The normal vaginal pH in postpubescent and premenopausal women (those producing estrogen) is maintained at about pH 4 (typically pH 4.4 to 4.6). Low pH is generally inhibitory to the growth of most microorganisms. The low vaginal pH is partially due to the presence of *Streptococcus* and *Lactobacillus* species (Döderlein's bacillus, which are nonpathogenic and acid-tolerant bacteria) that produce lactic acid from their fermentation of glycogen. The low pH is generally inhibitory for the growth of pathogens, including the bacterium *Neisseria gonorrhoeae* that causes gonorrhea.

The acid of the stomach provides a chemical barrier that acts to prevent microbial invasion of the body. Gastric acid, which is produced by glands in the stomach, is a mixture of hydrochloric acid, enzymes, and mucus. Most microorganisms entering the digestive tract are unable to tolerate the low pH (normally 1 to 2) of the stomach. Thus the number of viable microorganisms is greatly reduced during passage through the stomach. Bile and digestive enzymes in the intestines further reduce the numbers of surviving microorganisms. Microorganisms indigenous to the lower intestinal tract protect the host against invasion by pathogens by producing acidic metabolic fermentation products such as lactic acid and acetic acid; the natural microbiota of the gastrointestinal tract form antagonistic relationships

with nonindigenous microorganisms. Some microorganisms produce bacteriocins, which are substances toxic to the same or similar species. As a result, most nonindigenous microorganisms entering the intestinal tract are degraded during passage through it or are removed, along with large numbers of indigenous microorganisms, in the passage of fecal material from the body.

> **Many pathogens cannot survive in the pH environments that characterize much of the human body.**

LYSOZYME

Lysozyme, an enzyme that degrades the cell walls of bacteria, is found in several body fluids, including saliva, mucus, and colostrum. Lysozyme confers antimicrobial activity on these body fluids. Bacteria bathed in lysozyme die from osmotic shock because they no longer are protected by rigid cell walls. Lysozyme is especially effective in defending against infections caused by Gram-positive bacteria. Eggs, which are essential for reproduction, are surrounded by lysozyme and bathed in mucus secretions, protecting them from infection. The continuous washing of the eye with tears containing lysozyme generally prevents the growth of microorganisms on the tissues of the eye. In a similar way, since sweat, saliva, and mucus contain lysozyme, swallowing, coughing, and sneezing expose bacteria to lysozyme-containing body fluids, thus reducing the number of potential pathogens.

> **Lysozyme confers antibacterial activity on various body fluids, including tears, because it degrades bacterial cell walls.**

IRON-BINDING PROTEINS

Some chemicals within the body bind iron, thereby withholding this essential growth element from pathogenic microorganisms. By limiting the amount of available iron, these compounds limit the growth of pathogens. *Lactoferrin* and *transferrin* are examples of such iron-binding compounds. Lactoferrin is present in tears, semen, breast milk, bile, and nasopharyngeal, bronchial, cervical, and intestinal mucosal secretions. Transferrin is present in serum and the intercellular spaces of many tissues and organs. Transferrin transports iron from the small intestine, where the iron is absorbed, to the tissues, where the iron is used.

Transferrin and lactoferrin bind iron, limiting the growth of pathogens in the blood.

Since iron is stored intracellularly in a form that is tightly bound, it is not readily available to support microbial growth within the body's tissues. The concentration of free iron in the blood and other tissues is normally over a billion times lower than iron concentration required for growth by most microorganisms. Systemic bacterial infections are precluded in large part by the lack of free iron in the blood. Conversely, when the iron supply is more abundant, infection is more likely; for example, during menstruation there is more free iron in the vaginal tract and during this time a woman is more likely to contract the sexually transmissible disease, gonorrhea.

INTERFERONS

The body is protected in part against viral infections by the production of interferons, which block viral replication by rendering host cells nonpermissive. **Interferons** are a family of inducible glycoproteins produced by human cells in response to viral infections and other microbial pathogens that reproduce or replicate within host cells. Interferon is produced by infected tissue cells (α and β interferons) and by certain lymphocyte blood cells (γ interferon) that are part of the body's defense system. Alpha interferon (IFN-α) is made predominantly by mononuclear phagocytes; beta interferon (IFN-β) predominantly by fibroblasts; and gamma or immune interferon (IFN-γ) by T lymphocytes.

These interferons limit the abilities of viruses to replicate within the cells of the body because they make cells unsuitable hosts for viral replication (FIG. 13-6). They prevent the replication of viral pathogens within protected cells. Interferons work by inducing the production of proteins that block the replication of viruses. The proteins induced by interferon block translation of viral mRNAs by interfering with initiation of translation and by degrading viral mRNAs.

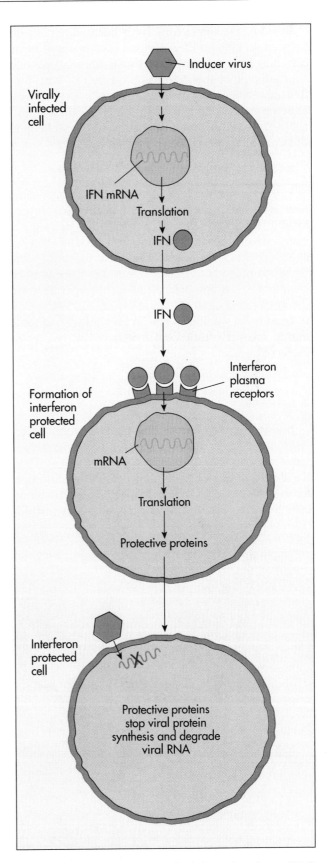

FIG. 13-6 Virally infected cells release interferon (IFN) that protects neighboring uninfected cells from viral infections. The mechanisms of interferon action are very complex and result in degradation of viral RNA and blockage of viral protein synthesis.

In addition, interferons have been shown to inhibit rapid cell proliferation because of these same activities that interfere with translation and protein synthesis. Since interferon is produced in very limited quantities, only neighboring cells are immediately protected. Interferons do not block the entry of the virus into a cell but rather prevent the replication of viral pathogens within protected cells.

Interferons help defend against viral infections.

Interferon production is considered a nonspecific resistance factor because interferon proteins do not exhibit specificity toward a particular pathogenic virus, which means that interferon produced in response to one virus is also effective in preventing the replication of other viruses. Interferons appear to be an important component of the elaborate integrated defense system against viral infections, and their production has a significant role in preventing and facilitating recovery from viral infections like the common cold.

Interferon induces the production of an antiviral protein that blocks viral replication within human host cells.

Besides its role in protecting against viruses, interferon acts as a regulator of the complex defense network that protects the body against infections and the development of malignant cells. As such, interferon is involved in the control of phagocytic blood cells that engulf and kill various pathogens (including bacteria) and abnormal or foreign mammalian cells (including cancer cells).

Because of the importance of interferon in controlling viral infections and the proliferation of malignant cells, its commercial production is being developed with the expectation that interferon administration will prove useful in the treatment of certain diseases. The human genes coding for the production of interferon have been cloned into *Escherichia coli*, creating by genetic engineering a bacterial strain that produces this human protein. Such genetically engineered bacteria are able to produce sufficient quantities of interferon for therapeutic uses. Interferon produced by genetically engineered bacteria is currently used in various experimental medical protocols.

COMPLEMENT

Besides interferon, blood contains a family of glycoprotein molecules, collectively called **complement,** that have a role in the removal of invading pathogens. Complement is especially important in preventing and limiting bacterial infections. Complement glycoproteins are designated C1, C2, C3, and so forth, with the numbers assigned based on the order of their discovery. As the name implies, complement augments or complements other defenses that protect the body against microbial infections. Complement glycoproteins work together in an integrated fashion so that as one component becomes activated, it in turn activates another complement component.

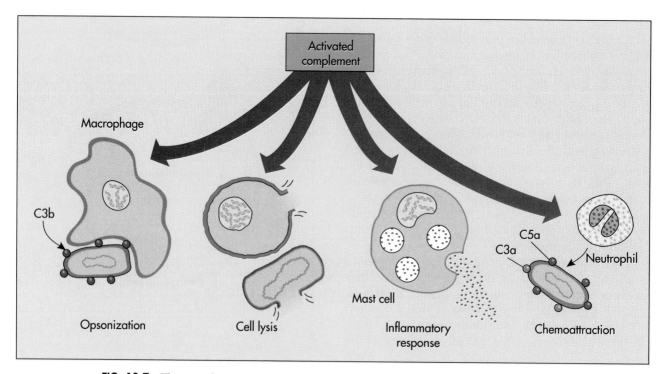

FIG. 13-7 The complement system is a multicomponent mechanism that destroys invading microorganisms. Complement molecules lead to enhanced phagocytosis (opsonization), cell lysis, inflammation, and chemoattraction of PMNs.

The complement system attacks and destroys invading microorganisms (FIG. 13-7). The result of complement activation leads to various nonspecific defense responses in the host.

Complement is a family of glycoproteins that help protect against bacterial and microbial infections.

Several factors can initiate the complement system. Endotoxin (LPS) in the Gram-negative cell wall triggers the complement system. This nonspecific initiation of the complement system is referred to as the *alternative pathway* (FIG. 13-8). The activation of this alternative pathway enables complement to prevent infections of the circulatory system by Gram-negative

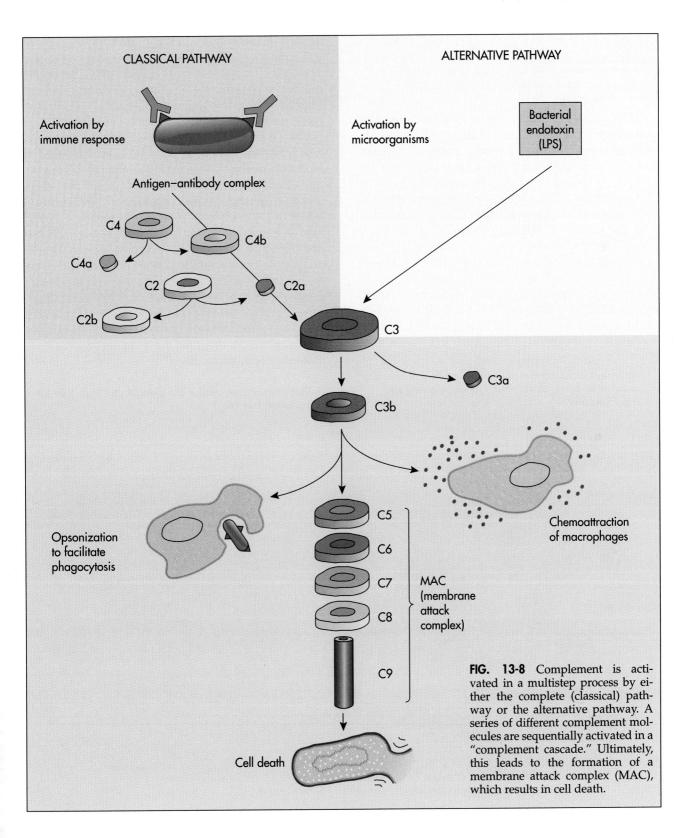

FIG. 13-8 Complement is activated in a multistep process by either the complete (classical) pathway or the alternative pathway. A series of different complement molecules are sequentially activated in a "complement cascade." Ultimately, this leads to the formation of a membrane attack complex (MAC), which results in cell death.

bacteria. The complement cascade can also be activated or triggered by specific antigen-antibody complexes, discussed in Chapter 14. This is called the *classical pathway.* When antibodies, which are substances made by the body, combine with an antigen on a cell surface, a specific complement molecule can bind to a region of the antibody molecule; this forms an antigen-antibody–complement complex that activates the classical pathway.

> **The complement system can be activated by the alternate pathway by endotoxin or the classical pathway by antigen-antibody complex.**

When activated, complement molecules attach to the surface of an invading cell in a cascade fashion, one complement molecule adding after another in an ordered sequence. Complement proteins activate one another, usually by cleaving the next protein in the order. The cleaved fragments have new enzymatic or physiological functions. Some of these fragments attack the invader's plasma membrane by joining to form a *membrane attack complex (MAC)* (FIG. 13-9). The MAC penetrates the plasma membrane, forming a pore that leads to osmotic lysis of the bacterial cell. Complement molecules of the MAC insert themselves into the membrane and produce circular lesions through which the cell's contents leak. Cell lysis results in the death of the cell.

> **Activated complement molecules act in a sequential manner to bring about the lysis of invading microorganisms.**

In addition to bacterial cell lysis by the MAC formed in the classical or alternative pathways, initiation of the complement cascades leads to other results. Some of the cleaved complement fragments attach to the surface of bacterial or fungal cells, which enhances their being phagocytized by leukocytes (see Phagocytosis). In addition, other cleaved complement fragments chemically attract other phagocytic cells (macrophages) to the site of an infection. One of the complement molecules (complement C3a) acts as a chemotactic stimulus for certain leukocytes called *neutrophils.* Another (complement C3b) attaches to cell surfaces and leads to enhanced binding to neutrophils and macrophages, which then are able to carry out phagocytosis.

The coating of a bacterial cell with complement, which leads to enhanced phagocytosis, is referred to as **opsonization.** Enhanced phagocytosis occurs because both neutrophils and macrophages have receptors on their surfaces for complement C3b. The binding of this complement molecule to a pathogen permits the establishment of a bridge between the pathogen and a neutrophil or macrophage so that the phagocytic cell remains in contact with the pathogen. Individuals lacking this specific complement have inadequate phagocytic activities and are particularly susceptible to bacterial infections. Bacterial cells that have been opsonized have a 1,000-fold greater chance of being engulfed by phagocytic cells than non-opsonized bacteria.

> **Complement enhances the phagocytic killing of bacterial pathogens.**

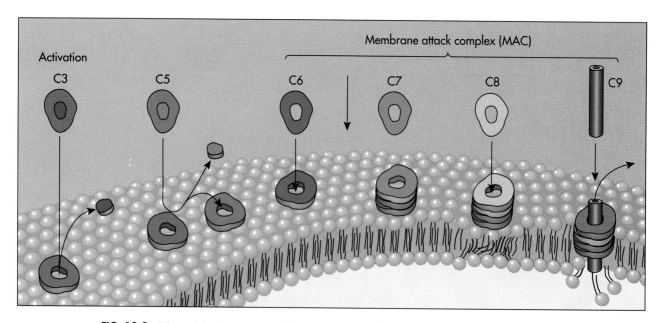

FIG. 13-9 The result of activation of complement is a cascade of complement molecules to a cell that produces damage to the plasma membrane and cell leakage (death). Plasma membranes are disrupted by the membrane attack complex (MAC).

PHAGOCYTOSIS

Phagocytosis (from the Greek *phago,* referring to eating, and *cyte,* referring to cell) involves the engulfment and ingestion of foreign cells, generally followed by the destruction of the engulfed cells. This is a highly efficient host defense mechanism against the invasion of microorganisms. Microorganisms that enter the circulatory system are subject to phagocytosis by various cells of the blood, the mononuclear phagocyte system, and the lymphatic system. Various cells at fixed body sites are also capable of phagocytosis. Cells involved in the phagocytic capture and destruction of microorganisms are called **phagocytes.**

> **Phagocytes engulf and digest foreign substances, including pathogenic microorganisms that invade the body.**

As discussed above, complement molecules have a role in activating phagocytes. In addition, antibody can have a role in activating phagocytes. This phenomenon is known as **immune adherence.** Immune adherence occurs when complement molecules or antibody molecules bind to the surface of a microorganism. They then interact with receptors on the surface of a phagocytic blood cell. This enhances the efficiency by which the blood cell engulfs and destroys the invading microorganism.

> **Phagocytosis is a highly efficient host defense mechanism against the invasion of microorganisms.**

PHAGOCYTIC CELLS

Several types of **leukocytes** (white blood cells) are involved in nonspecific phagocytic defenses against pathogenic microorganisms (Table 13-2). All blood cells, including white blood cells that are involved in the defense of the body and red blood cells that carry oxygen and carbon dioxide throughout the body, and platelets are derived from a common ancestor called a *stem cell* that is in the bone marrow. Some leukocytes, called *granulocytes,* contain cytoplasmic granules. These granules contain different types of substances that contribute to the differential staining of granulocytes (Table 13-3). Some granulocytes, **basophils,** contain basic substances such as histamine that allow them to be stained with basic dyes, such as methylene blue. Release of the basic granules enhances phagocytic activity by other cells. **Eosinophils** are granulocytes that react with acidic dyes and become red when stained with the dye eosin. The release of the acidic granular substances is important in defending against parasites, such as worms, that are too large to be engulfed by leukocytes and destroyed by phagocytosis. **Neutrophils** (also called polymorphonuclear neutrophils, polymorphs, or PMNs) contain granules that exhibit no preferential staining—that is, they are stained by neutral, acid, and basic dyes. The leukocytes that do not contain granular inclusions (*agranulocytes*) include the *monocytes* and *lymphocytes.* Monocytes are important in the nonspecific immune response, and lymphocytes are especially important in the specific immune response.

> **Leukocytes are white blood cells that are important in nonspecific defense against pathogenic microorganisms.**

> **The different types of leukocytes include granulocytes (basophils, eosinophils, and neutrophils) and agranulocytes (monocytes and lymphocytes).**

Polymorphonuclear neutrophils (PMNs), the most abundant phagocytic cells in blood, are pro-

TABLE 13-2		
Normal Cellular Composition of Adult Human Blood		
CELL TYPE	**NUMBER PER mL**	**FUNCTION**
Leukocytes (white cells)	$4.5\text{-}9.0 \times 10^6$	
Granulocytes		
Neutrophils	$3.0\text{-}6.8 \times 10^6$	Phagocytosis
Basophils	$2.5\text{-}9.0 \times 10^4$	Contributes to the inflammatory response
Eosinophils	$1.0\text{-}3.6 \times 10^5$	Defense against certain parasites
Mononuclear cells		
Lymphocytes	$1.0\text{-}2.7 \times 10^6$	The specific immune response—antibody production (B lymphocytes) and cell-mediated immunity (T lymphocytes)
Monocytes	$1.5\text{-}1.7 \times 10^5$	Phagocytosis
Platelets	$1.4\text{-}3.8 \times 10^8$	Blood clotting
Erythrocytes (red cells)	$3.6\text{-}5.4 \times 10^9$	Carries O_2 to and CO_2 away from tissues

TABLE 13-3

Leukocyte Granules and Some of Their Contents

CELL TYPE	GRANULE	CONTENTS OF GRANULE	FUNCTION
Neutrophil	Lysosomes (acidic primary granules)	Acid hydrolases	Degrades proteins and other molecules
		Lysozyme	
	Basic secondary granules	Lactoferrin	Removes iron
		Lysozyme	Degrades peptidoglycan
		Peroxidase	Degrades peroxides
Basophil (and mast cell derived from basophil)	Basophil granule	Histamine	Constricts smooth muscle, increases permeability of blood vessels
		Heparin	Inhibits blood clotting
		Serotonin	Increases permeability of blood vessels
Eosinophil	Eosinophil granule	Histaminase	Inactivates histamine
		Aryl sulfatase	Inactivates slow reactive substance of anaphylaxis (SRS-A)
		Major basic protein	Immunity to parasites

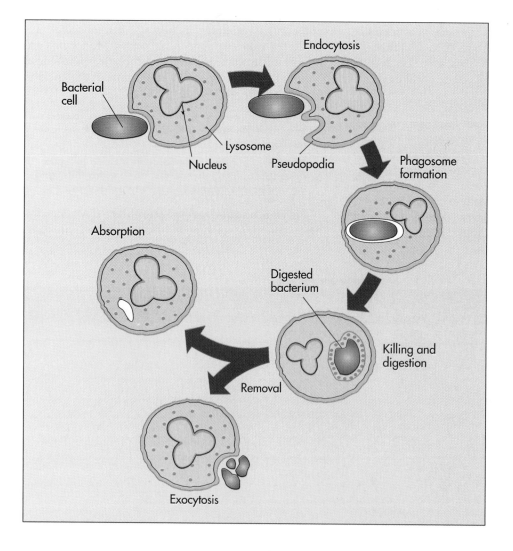

FIG. 13-10 Neutrophils phagocytize and degrade microorganisms that invade the body. The bacterial cell that has been captured by endocytosis is enclosed within the cell in a phagosome. Lysosomes fuse with the phagosome to form a phagolysosome. The lysosomal enzymes digest the bacterial cell and the cellular debris of the degraded bacterial cell are absorbed or released from the neutrophil cell.

CHARACTERISTIC	POLYMORPHONUCLEAR NEUTROPHIL (PMN)	MACROPHAGE
Origin	Bone marrow	Bone marrow and tissues
Length of time in bone marrow	14 days	2.25 days
Length of time in blood	7-10 hours	20-40 hours
Length of time in body	4 days	Months-years
Number of cells in blood	2.5×10^6 to 7.5×10^6/mL	0.2×10^6 to 0.8×10^6/mL
Number of cells in marrow	2.5×10^7 to 7.5×10^7/mL	—
Number of cells in tissues	—	2.5×10^8 to 7.5×10^8/mL
Molecules secreted	Lysozyme	Lysozyme, tumor necrosis factor, interleukin-1, complement, and over 80 other molecules

TABLE 13-4
Comparison of Neutrophils and Macrophages

duced in the bone marrow. They are continuously present in circulating blood, affording protection against the entry of foreign materials. These leukocytes exhibit chemotaxis and are attracted to foreign substances, including invading microorganisms. PMNs engulf, kill, and digest microorganisms along with particulate matter that may be present, such as cell debris (FIG. 13-10). These neutrophils live for only a few days in the body but are replenished from the bone marrow in high numbers (Table 13-4).

Monocytes are mononuclear phagocytic cells. They are larger than neutrophils. They are the precursors of macrophages and are able to move out of the blood to tissues that are infected with invading microorganisms. Outside the blood, monocytes become enlarged, forming phagocytic **macrophages.** Macrophages are long-lived in the body, persisting in tissues for weeks or months (see Table 13-4). Once these differentiated macrophages are formed, they are capable of reproducing to form additional macrophages. This is in contrast to neutrophils, which are terminal cells that are short-lived in the body; neutrophils are nonreproductive and must be replenished from the bone marrow. Macrophages, like neutrophils, are able to engulf, kill, and digest microorganisms. As discussed in Chapter 14, they also have an important role as antigen-presenting cells in the immune response.

Polymorphonuclear neutrophils and macrophages are the main phagocytic cellular defenders of the body against microbial infections.

Macrophages are important in the elimination of viral infections. Macrophages secrete soluble proteins, called cytokines, when they encounter a virus. This signals the beginning of the immune response.

Among the different kinds of cytokines produced are gamma-interferon and interleukin-1. Gamma-interferon activates other monocytes to mature into macrophages. Interleukin-1 also activates the system responsible for the fever associated with infection.

Some of the lysosomal enzymes of macrophages are different from those of neutrophils. Some microorganisms are resistant to the enzymatic activities of neutrophils and/or macrophages. For example, *Mycobacterium tuberculosis* survives and even multiplies within macrophages. As a result, some microorganisms survive, continuing to grow and later to cause infection because of the failure of these phagocytic cells to kill the invading pathogens. This is one of the reasons that tuberculosis is a persistent disease and is difficult to treat.

Macrophages are distributed throughout the body, including fixed sites within the mononuclear phagocyte system (FIG. 13-11). The *mononuclear phagocyte system*, formerly called the *reticuloendothelial system*, refers to a systemic network of phagocytic macrophage cells distributed through a network of loose connective tissue and the endothelial lining of the capillaries and sinuses of the human body. The phagocytic cells associated with the lining of the blood vessels in bone marrow, liver, spleen, lymph nodes, and sinuses constitute this host defense system.

Macrophages are mononuclear phagocytes in the spleen, liver, lymph nodes, and blood.

Macrophages are able to phagocytize infecting microorganisms because fluid circulating between tissues of the body drains into the vessels of the lymphatic system, where the fluid is called *lymph.* Infecting microorganisms also can enter the lymph and get

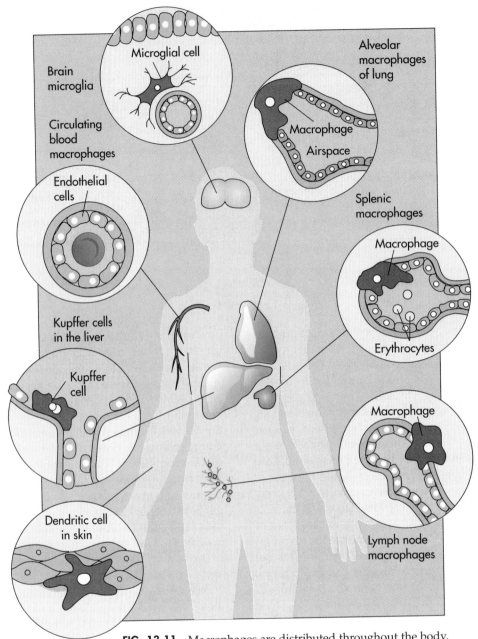

FIG. 13-11 Macrophages are distributed throughout the body, protecting different body sites from infection.

ing microorganisms also can enter the lymph and get carried to the lymph nodes where phagocytic macrophages can capture and digest them. In this way, infecting microorganisms can be removed from the lymph. The lymph drains into the lymphatic or thoracic ducts and returns to the blood at the subclavian veins, which are located near the heart.

Some of the macrophages in the mononuclear phagocyte system occur at fixed sites and are designated with particular names. For example, *microglia* are macrophages of the central nervous system; *Kupffer cells* are phagocytic cells that line the blood vessels of the liver; *alveolar macrophages* are macrophages

fixed in the alveolar lining of the lungs; and *histiocytes* are fixed macrophages in connective tissues. Other macrophages of the mononuclear phagocytic system are called *wandering cells* because they move freely into tissues where foreign substances have entered. Wandering macrophages are attracted to these tissues through chemotaxis by chemical stimuli elicited by the foreign material. Wandering macrophages occur in the peritoneal lining of the abdomen and the alveolar lining of the lung, as well as in other tissues. The presence of relatively high numbers of macrophages in the respiratory tract is important in preventing the establishment of pathogens and a nor-

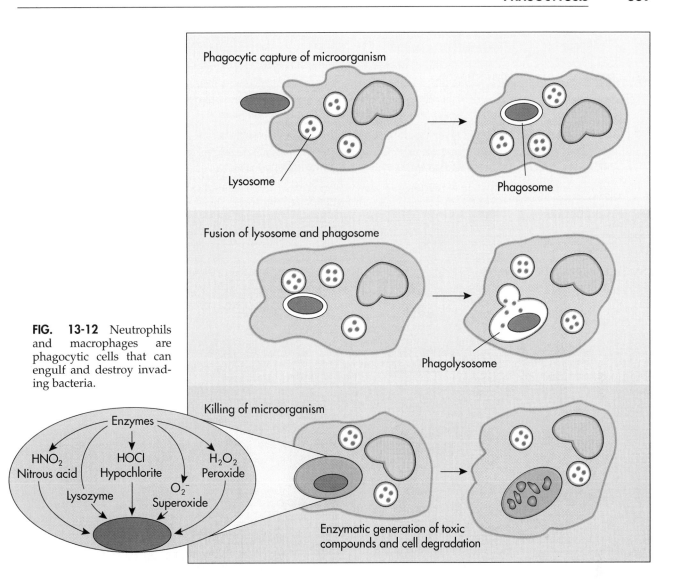

Phagocytic capture of microorganism

Lysosome

Phagosome

Fusion of lysosome and phagosome

Phagolysosome

Killing of microorganism

Enzymes

HNO$_2$
Nitrous acid

HOCl
Hypochlorite

H$_2$O$_2$
Peroxide

Lysozyme

O$_2^-$
Superoxide

Enzymatic generation of toxic
compounds and cell degradation

FIG. 13-12 Neutrophils and macrophages are phagocytic cells that can engulf and destroy invading bacteria.

mal indigenous microbiota within the tissues of the lower respiratory tract.

Monocytes, macrophages, and PMNs are the main phagocytic cells, that is, "professional phagocytes" of the host defense system. Other cells in the body can be phagocytic but are not as efficient as these professional phagocytes in engulfment and destruction of particles. The phagocytic cells engulf and destroy most bacteria that attempt to invade the body. They are like a defending army consisting of 2 trillion cells, approximately 2% of all the body's cells.

MECHANISM OF PHAGOCYTIC KILLING

Phagocytic blood cells can have numerous lysosomes that contain hydrolytic enzymes capable of digesting microorganisms (FIG. 13-12). During phagocytosis, the microorganism is engulfed by the pseudopods (cytoplasmic extensions) of the phagocytic cell and is transported by endocytosis across the plasma mem-

brane, where it is contained within a vacuole called a *phagosome*. The phagosome migrates to and fuses with a lysosome, producing a *phagolysosome*. Within the phagolysosome, an engulfed microorganism is exposed to enzymes and chemicals, including degradative enzymes and enzymes that can kill the microorganism. Phagocytic cells kill or degrade engulfed microorganisms in at least three different ways (Table 13-5).

Phagocytes can kill microorganisms they capture by several mechanisms.

During phagocytosis there is an increase in oxygen consumption by the phagocytic cells associated with elevated rates of metabolic activities (FIG. 13-13). This phenomenon is called the respiratory burst and results from the cell's requirement for ATP to power phagocytosis. Oxygen-dependent enzymes in the lysosome (and phagolysosome) form toxic deriv-

TABLE 13-5	
Mechanisms by Which Phagocytes Kill Microorganisms	
MECHANISM	**DESCRIPTION**
Oxygen-dependent	Conversion of molecular oxygen to the superoxide anion, hydrogen peroxide, singlet oxygen, and hydroxyl radicals that are toxic to microorganisms
Oxygen-independent	Degradative enzymes, including lysozyme, phospholipases, proteases, RNase, and DNase that destroy the macromolecules of microorganisms; defensins damage bacterial and fungal plasma membranes and also disrupt enveloped viruses
Nitrogen-dependent	Formation of reactive nitrogen intermediates (RNIs), including nitric oxide (NO), nitrite (NO_2^-) and nitrate (NO_3^-) that kill microorganisms

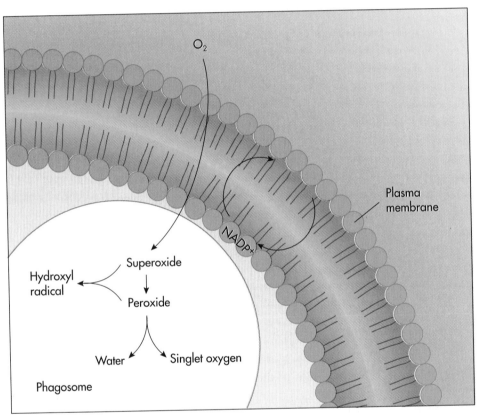

FIG. 13-13 There is a respiratory burst during phagocytic killing that produces toxic forms of oxygen (superoxide, peroxide, hydroxyl radical, and singlet oxygen).

atives from oxygen, converting molecular oxygen to the superoxide anion, hydrogen peroxide, singlet oxygen, and hydroxyl radicals, all of which are toxic to microorganisms.

Phagocytosis also involves a shift in metabolism from a respiratory to a fermentative process, with the consequent production of lactic acid. This leads to a decrease in pH, which enhances the activity of many lysosomal enzymes. Oxygen-independent mechanisms of killing by phagocytes include various degradative enzymes that are associated with lysosomes. These enzymes include lysozyme, phospholipases, proteases, RNase, and DNase that contribute to the destruction of the ingested microorganism.

Polymorphonuclear neutrophils also synthesize *defensins*, which are a family of peptides with antimicrobial activity. Defensins are stored in cytoplasmic granules and delivered to phagocytic vacuoles. They can increase the permeability of bacterial and fungal plasma membranes, contributing to their death. They also affect enveloped viruses (but not nonenveloped viruses), inactivating them so that they cannot replicate.

Once the engulfed microorganisms have been killed and degraded, the remaining material is transported to the plasma membrane of the human cell within a vacuole; it is removed from the phagocytic cells by exocytosis or is consumed within the phagocytic cell.

METCHNIKOFF AND THE DISCOVERY OF PHAGOCYTOSIS

Eli Metchnikoff was a moody, temperamental Russian who did his most creative scientific work in a manic state (see Figure). His studies were highly detailed and characterized by detached observation of his subject. He meticulously reported his microscopic observations in 1884 of what happens to a microorganism when it invades an organism. He began by examining starfish larvae stuck with thorns. He went on to examine microbial infections of *Daphnia*, a tiny freshwater organism. He infected *Daphnia* with spores of the fungus *Monospora*. This was an ideal model system because *Daphnia* is simple and transparent and *Monospora* is large and easily seen without staining. Metchnikoff was able to observe *Monospora* in the abdominal cavity of the infected *Daphnia*. He saw the fungus penetrating the *Daphnia's* intestinal wall as a result of peristalsis. Immediately, blood corpuscles began to surround and attach themselves to the fungal spores. He observed the blood corpuscles as circulating, colorless, phagocytic cells adapted to the uptake of solid particles. The mobile cells, which he called phagocytes, migrated to the area of infection, where they engulfed and digested the mi-

croorganisms. In this way the phagocytes are protected against the infecting microorganisms.

Metchnikoff was able to observe and describe the changes the fungal spores underwent until they were destroyed and separated into irregular grains. He reasoned that the same processes explained how the bacterial cells that he saw inside white blood cells in the human body got there and how the blood cells destroyed such bacteria. This pioneering work established the role of cellular components of the blood in destroying disease-causing microorganisms. Metchnikoff went on to study phagocytosis by leukocytes in human blood. Because he found this process to be even more active in animals recovering from an infection, he concluded that phagocytosis was the main line of defense against infection. He recognized two types of phagocytic leukocytes: polymorphonuclear neutrophils, which he called microphage, and larger cells, which he called macrophage. Today, we recognize the phagocytic activity of human white blood cells as one of the key primary lines of defense against invasion of the body by pathogenic microorganisms.

Eli Metchnikoff (1845-1916), shown at work in his laboratory, first proposed the cellular theory of immunity, opposing the accepted humoral theory, which said that soluble substances in body fluids were responsible for immunity.

FEVER

Humans are *homeothermic animals,* meaning that they maintain a body temperature within a fairly constant range. Usually during a 24-hour period, the body temperature of a healthy individual fluctuates about 1° C to 1.5° C. Although "normal" body temperature is considered to be 37° C, many people's body temperature may fluctuate around 36° C or 38° C. Part of the brain called the hypothalamus regulates body temperature.

Fever is an abnormal increase in body temperature. Fever enhances the body's natural defense mechanisms by stimulating phagocytosis, increasing the rate of enzymatic reactions that lead to degradation of microorganisms and tissue repair, intensifying the action of interferons, and causing a reduction in blood iron concentrations—iron is required by many bacteria for growth.

Many microorganisms produce substances that enter the bloodstream and result in fever by directly or indirectly stimulating the hypothalamus. Chemicals that cause fever are called **pyrogens** (Greek *pyr* + *genes,* meaning fire or heat producing). Some examples of pyrogens are: (1) the lipopolysaccharide (endotoxin) molecules of Gram-negative bacteria, (2) fragments of the peptidoglycan molecule in the cell walls of Gram-negative and Gram-positive bacteria, and (3) specific fever-inducing exotoxins, such as toxic-shock syndrome toxins produced by some *Staphylococcus aureus* strains and a red blood cell–destroying toxin produced by *Streptococcus pyogenes.*

> **Fever contributes to the defense of the body by stimulating phagocytes and enhancing the rates of enzyme activities involved in the destruction of invading pathogens.**

Bacterial endotoxins and peptidoglycan cause phagocytic cells that have ingested these substances to release interleukin-1 (IL-1), also called *endogenous pyrogen.* Bacterial exotoxins that are produced during infections enter the bloodstream and also stimulate macrophages and monocytes to release IL-1, which in turn stimulates the hypothalamus to release prostaglandins. Prostaglandins cause the hypothalamus to readjust its thermostat to a higher temperature, thus causing fever. The body responds to signals from the hypothalamus by constriction of blood vessels, increased rate of metabolism, and muscular contractions (shivering). This initial condition is called a chill. It reflects the fact that even though the body temperature is rising the skin remains cold. The body temperature rises to the new "thermostat setting" of the hypothalamus and remains in this state of fever (typically up to 39° C) until the IL-1 is depleted from the blood. The body then responds with dilation of blood vessels and sweating to attempt to lose heat and return to normal temperature. This heats the skin and results in sweating. This phase of the fever response is called *crisis* and indicates a period of recovery when the body temperature is returning to normal.

Some fevers are continuous; that is, the body temperature remains elevated during the progression of an infection. Typhoid fever is an example of a continuous fever. An intermittent or spiking fever is one in which the temperature is elevated but fluctuates widely (>1° C). A remittent fever is one that abates (returns to normal) for a short period or intervals. Malaria and many other bacterial infections result in remittent fevers. Relapsing fever, caused by *Borrelia recurrentis,* is manifested by recurring episodes of fever and normal temperatures.

INFLAMMATORY RESPONSE

The **inflammatory response** represents a generalized response to infection or tissue damage. It is designed to localize invading microorganisms and arrest the spread of the infection (FIG. 13-14). The inflammatory response is characterized by four symptoms: reddening of the localized area, swelling, pain, and elevated temperature.

Redness results from capillary dilation that allows more blood to flow. Dilation occurs because basophils and mast cells release vasodilators, such as histamine, during the inflammatory response. The term *dilation* is a misnomer because many of the capillaries remain constricted; during "dilation" there are simply fewer constricted capillaries, permitting

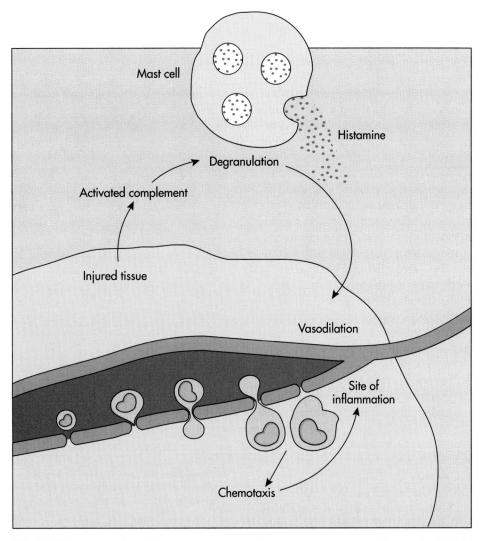

FIG. 13-14   During the inflammatory response there is an increase in polymorphs (PMNs) along with fluid accumulation during the acute stage. Later there is an increase in macrophages during the chronic phase. The release of activated complement leads to mast cell degranulation with release of histamine that results in vasodilation.

increased blood circulation through more open (dilated) capillaries. The *elevated temperature,* which is a localized phenomenon, also occurs because capillary dilation permits increased blood flow through these vessels with the associated high metabolic activities of neutrophils and macrophage. The increased blood flow brings heat to the site of an infection. Increased blood flow also is important because it brings more PMNs and monocytes to the site of inflammation. The transition of monocytes to macrophages that move into tissue spaces enhances and prolongs the inflammatory response. The dilation of blood vessels is accompanied by "increased capillary permeability," causing *swelling* as fluids accumulate in the bases surrounding tissue cells. Actually, the swelling is due to increased permeability of the venules, but the term *increased capillary permeability* is entrenched in the clinical terminology used to describe this phenomenon.

The inflammatory response is characterized by redness, swelling, pain, and elevated temperature, which are due to increased blood flow that results from capillary dilation.

Pain, in the case of inflammation, is due to lysis of blood cells that triggers the production of bradykinin and prostaglandins. These are substances produced by human cells that alter the threshold and intensity of the nervous system response to pain. Bradykinin decreases the firing threshold for pain nerve fibers, and the prostaglandins, PGE1 and PGE2, intensify this effect. Aspirin, which is often used to decrease pain, antagonizes prostaglandin formation but has little or no effect on bradykinin formation. Thus aspirin can decrease, but not eliminate, the pain associated with the inflammatory response.

The lysis of blood cells and the release of bradykinin and prostaglandin cause the pain associated with inflammation.

The dilation of blood vessels in the area of the inflammation increases blood circulation, allowing increased numbers of phagocytic blood cells to reach the affected area. It focuses the nonspecific defenses and the specific immune response (discussed below) at a specific body site. *Kinins,* which are substances in blood plasma, become activated during the inflammatory response. These cause vasodilation and attract phagocytes, particularly polymorphonuclear neutrophils. Prostaglandins, which are released from damaged cells, and leukotrienes, which are produced by basophils and mast cells, also cause vasodilation. This permits increased numbers of phagocytic blood cells to reach the area of inflammation. PMNs are initially most abundant. In the later stages of inflammation, monocytes and macrophages of the mononuclear phagocyte system predominate. The phagocytic cells are able to kill many of the ingested microorganisms. Phagocytic blood cells migrate to the affected tissues, passing between the endothelial cells of the blood vessel by a process known as *diapedesis.* The death of phagocytic blood cells involved in combatting the infection results in the release of histamine, prostaglandins, and bradykinins, which, in addition to their other effects, are vasodilators, that is, substances that increase the internal diameter of blood vessels.

Additionally, specialized cells that line connective tissues, called *mast cells,* react with complement. Mast cells are basophils that have become fixed within tissues. The reaction of a mast cell with complement leads to the release of large amounts of histamine contained within these cells (FIG. 13-15). These substances are biochemical mediators that alter the circulation during the inflammatory response. Thus the death of some phagocytic cells and the release of certain biochemicals enhance the inflammatory response.

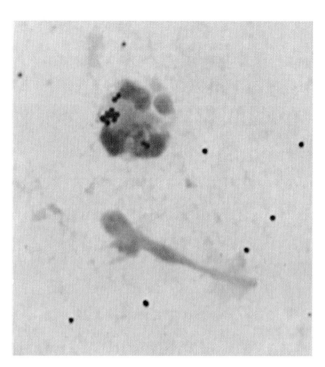

FIG. 13-16 Pus from an infected wound consists of leukocytes and bacterial cells. In this light micrograph, typical clusters of *Staphylococcus aureus (purple cells)* are seen overlying a white blood cell in pus from an abscess.

The inflammatory response increases the flow of blood to the site of infection and focuses the nonspecific defense systems on that localized region.

The area of the inflammation also becomes walled off as the result of the development of fibrinous clots. The deposition of fibrin isolates the inflamed area, cutting off normal circulation. The fluid that forms in the inflamed area is known as the *inflammatory exudate,* commonly called *pus* (FIG. 13-16). This exudate contains dead microorganisms, dead phagocytic cells, debris, and body fluids. After the removal of the exudate, the inflammation may terminate and the tissues may return to their normal function. This is why physicians often lance boils to remove the exudate. The final stage of the inflammatory response involves the repair of the affected tissue. New cells are formed and added to replace damaged tissue. In some cases only cells that comprise the functional cells of the tissue are formed. This leads to the restoration of normal tissue. In other cases, cells of the supportive connecting tissue are also produced. This produces scar tissue.

The complex reactions of the inflammatory response work in an integrated fashion to contain and eliminate infecting pathogenic microorganisms and to restore normal or near normal tissue.

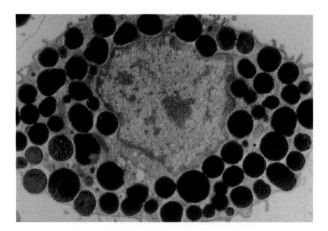

FIG. 13-15 Colorized electron micrograph showing a mast cell with numerous dense granules of histamine.

SUMMARY

Nonspecific Host Defenses Against Microbial Infections

- The body's resistance to infection is determined by the state of its defensive surveillance and protective response systems.
- The defense systems consist of the innate immune system, which provides nonspecific immunity, and the adaptive immune system, which is an acquired response to specific disease-causing substances.

Physical Barriers (pp. 374-377)

- Physical barriers prevent most microorganisms from entering the body.
- Keratin prevents microorganisms from colonizing or penetrating the skin.
- The mucous membrane secretes mucus, which traps microorganisms.
- The mucociliary escalator system protects the respiratory tract from potential pathogens.

Microbial Barriers—Normal Microbiota (pp. 377-379)

- Normal microbiota consists of microorganisms that are usually associated with particular body tissues.
- The normal microbiota prevent other microorganisms from colonizing the areas where they grow by producing antimicrobial substances, reducing oxygen concentrations, and competing for available nutrients.
- Normal microbiota synthesize essential nutrients.

Chemical Defenses (pp. 380-384)

- The pH level of many body systems and organs limits the species of microorganisms that can survive and reproduce in the body.
- Enzymes produced in the body kill invading pathogens. Lysozyme is present in tears, sweat, saliva, and mucus.
- Some chemicals produced by the body bind to iron, withholding this essential growth element from foreign microorganisms. Lactoferrin and transferrin are iron binding compounds.
- Some blood cells produce interferon, which makes host cells unsuitable for viral replication.
- The complement system is a nonspecific defense mechanism of the body. Complement is a group of proteins in blood plasma. Complement fragments cause the lysis of foreign microorganisms.

Phagocytosis (pp. 385-391)

- Various leukocytes (white blood cells) protect the body against infection because they are phagocytic.
- Granulocytes are leukocytes that contain cytoplasmic granules. Eosinophils, basophils, and neutrophils are types of granulocytes.
- Agranulocytic leukocytes are monocytes and lymphocytes.
- Leukocytes (white blood cells) that engulf and digest microorganisms are called phagocytes. An engulfed microorganism is transported into the phagocyte, where it is contained in a phagosome. The phagosome fuses with a lysosome, which contains degradative enzymes. Within the phagolysosome, microorganisms are killed by several mechanisms, including exposure to toxic forms of oxygen.
- Neutrophils and macrophages are the major types of phagocytic cells.
- Complement molecules bind to the surfaces of foreign microorganisms. This is known as immune adherence. Complement attracts phagocytes.
- Monocytes enlarge and form macrophages outside the blood. Macrophages engulf and digest microorganisms. The mononuclear phagocyte system is a network of macrophages that is distributed throughout the body's connective tissue and capillary and sinus linings.
- Macrophages secrete cytokines, which have anti-viral properties.

Fever (p. 392)

- Fever is a response to the presence of pyrogens. The defenses of the body are more efficient at elevated temperatures so that the rise in temperature helps eliminate infecting microorganisms.
- Interleukin 1 acts as a mediator of the fever response. It triggers the release of prostaglandins by the hypothalamus to reset the "body's thermostat" at a higher temperature, causing fever. When the body temperature rises, chills occur because the skin is still cold. As the fever diminishes, heat is released to the skin and sweating occurs during the period known as crisis.

Inflammatory Response (pp. 392-394)

- The inflammatory response is characterized by reddening, swelling, pain, and fever.
- Vasodilators increase the internal diameter of blood vessels. Vasodilators come from mast cells.
- Capillary dilation causes redness.
- Increased capillary permeability causes swelling.
- The death of phagocytes involved in the inflammatory response results in the release of histamine, prostaglandins, and bradykinins. Bradykinin and prostaglandins cause pain.
- Inflammatory exudate (pus) is the fluid that forms in the inflamed area.

CHAPTER REVIEW

REVIEW QUESTIONS

1. What physical factors contribute to the resistance of the body to invasion by microorganisms?
2. What is the normal microbiota? How does it contribute to natural immunity?
3. What are interferons? How do they contribute to host immunity?
4. How does the body withhold iron that could allow the growth of pathogens in the blood?
5. How does lysozyme protect the body against infections? What body tissues are protected by lysozyme?
6. What is complement? What are the pathways for complement activation? How does complement contribute to the defense of the body against bacterial invasion?
7. What is phagocytosis and how does it contribute to our resistance to pathogenic microorganisms?
8. How do phagocytes eliminate infecting bacteria from the body?
9. What blood cells are involved in the nonspecific defense of the body against microbial infection? What function does each play?
10. What is an inflammatory response and how does inflammation act to prevent the spread of pathogens throughout the body?
11. What mediates the fever response? How does fever help limit infections?
12. Why do chills and sweating occur during the fever response? What does each indicate in terms of progression of the infection in relation to the body's fever response?
13. How is the respiratory tract protected against invasion by pathogenic microorganisms?
14. How is the gastrointesinal tract protected against invasion by pathogenic microorganisms?
15. How is the genitourinary tract protected against invasion by pathogenic microorganisms?

CRITICAL THINKING QUESTIONS

1. Given the nonspecific defenses of the body how could an individual have contracted a common cold? A case of bacterial pneumonia? A *Salmonella* infection of the gastrointestinal tract? A bacterial infection of the urinary tract bladder?
2. A few years ago experiments in Britain demonstrated that the virus that causes the common cold can be transmitted when one individual covers his or her nose and mouth while sneezing, shakes another individual's hand, and that second individual then rubs his or her eyes. How could this occur?
3. Sometimes one infection leads to another. For example a bacterial infection of the lungs causing pneumonia sometimes is a sequel to influenza. Why does this happen?
4. Antibiotics often are prescribed for treating bacterial infections. However, the extended use of antibiotics sometimes leads to other microbial infections. Why do such infections occur?

READINGS

Gallen JI and AS Fauci (eds.): 1982. *Phagocytic Cells*, New York, Raven Press.

Volume 1 in the series Advances in Host Defense Mechanisms. Topics discussed include monocyte and macrophage development, pyrogens, neutrophils, and eosinophils.

Jaret P: 1986. Our immune system: The wars within, *National Geographic*, June, pp. 702-736.

A very readable account of current progress in the study of the human immune system; includes striking photographs by Lennart Nilsson.

McNabb PC and TB Tomasi: 1981. Host defense mechanisms at mucosal surfaces, *Annual Review of Microbiology* 35:477-496.

Discusses how mucous membranes act as effective barriers against microbial infections, including the mechanisms by which they function.

Pestka S: 1983. The purification and manufacture of human interferons, *Scientific American* 249(2):36-43.

Describes the production of human interferon by recombinant DNA bacteria.

Rother K and GO Till (eds): 1988. *The Complement System*, London, Springer-Verlag.

Review of complement and complement reaction, biological function of complement, and in vivo manipulation of the complement system.

Sharon N and H Lis: 1993. Carbohydrates in cell recognition, *Scientific American* 268(1):82-89.

This article describes how sugars on the surfaces of cells enable cells to identify and interact with one another and how drugs that target such surface recognition carbohydrates can be used to stop infection and inflammation.

Watson RR: 1984. *Nutrition, Disease Resistance, and Immune Function*, New York, Marcel Dekker.

Immunological measures can assess nutritional stress. Describes interactions of disease organisms and nutritional stresses with changes in the host's physiology and immunology.

Weller PF: 1991. The immunobiology of eosinophils, *New England Journal of Medicine* 324:1110-1118.

Discusses the current state of knowledge about the role of eosinophils in tissues and in disease processes.

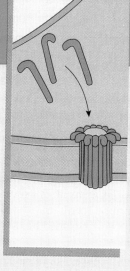

CHAPTER 14

Specific Host Defenses Against Microbial Infections: The Immune Response

PREVIEW TO CHAPTER 14

In this chapter we will:
- Introduce the components of the immune response.
- Examine the properties of the immune response that confer specificity and memory, and the acquired (learned) ability to detect foreign substances.
- Explore the properties of antigens.
- Describe the lymphocytes that are the underpinnings of the immune response.
- Discuss antibody-mediated immunity and show how the body recognizes and responds to foreign substances.
- Learn about the various immunoglobulins and their functions in antibody-mediated immunity.
- Examine cell-mediated immunity and show how various cells are involved in the recognition and elimination of foreign and abnormal cells.
- Discover how the body acquires immunity and develops a memory system.
- Learn the following key terms and names:

acquired immunity
active immunity
agglutination
agglutinins
anamnestic response
antibodies
antibody-mediated
  immune response
antigenic determinant
antigen-presenting cells
  (APCs)
antigens
artificial immunity
B lymphocytes (B cells)
CD (cluster of
  differentiation) antigen
cell-mediated immune
  response
complement fixation
cytokines
cytotoxic T cells (T_C)
delayed hypersensitivity T
  helper cells (T_{DTH})
epitope
Fab fragments
Fc fragment
gamma (γ)-interferon
haptens
helper T cells (T_H)
hemagglutination
hybridomas

IgA (immunoglobulin A)
IgD (immunoglobulin D)
IgE (immunoglobulin E)
IgG (immunoglobulin G)
IgM (immunoglobulin M)
immune interferon (IFN-γ)
immune response
immunogenicity
immunoglobulins
immunological tolerance
killer cells (K cells)
lymphokines
major histocompatibility
  complex (MHC)
memory response
monoclonal antibodies
natural immunity
natural killer cells (NK
  cells)
neutralization
opsonization
passive immunity
plasma cells
primary immune response
secondary immune
  response
specific immune system
suppressor T cells (T_S)
T cell receptors (TCRs)
T lymphocytes (T cells)
toxoids

The nonspecific defenses of the human body are augmented by a second line of defenses called the **specific immune system** or simply the **immune response**. This is a learned response that recognizes specific substances that are foreign to the body, including specific strains and species of pathogens. Failure to recognize and respond to foreign substances can render one susceptible to infectious disease. After detecting foreign substances that are not part of the body, the immune response system acts to eliminate the specific foreign substances. It can recognize and attack microorganisms by responding to the specific chemicals produced by the microorganisms. This physiological response is especially important as a defense against infection and for protection against disease.

The immune response is essential and failures of the system, such as acquired immunodeficiency syndrome (AIDS) and severe combined immunodeficiency (SCID), result in mortality. Specific immunity is an acquired response through which the body specifically detects and destroys specific substances. The adaptive immune response is characterized by specificity, memory, and the acquired (learned) ability to detect foreign substances. As a consequence of acquired immunity, we usually suffer from many diseases only once. For example, exposure to the pathogens that cause the diseases measles and diphtheria produce lifelong immunity to these infections.

The specific immune response is characterized by specificity, memory, and the acquired ability to detect and to eliminate foreign substances.

The specific immune response recognizes substances that are different in some way from the body's normal (self) substances. The ability to differentiate "self" from "nonself" at the molecular level is necessary for the development of the specific immune response. The human immune response is able to recognize substances that differ from the normal macromolecules of the body. The specificity of the immune response permits the recognition of even very slight biochemical differences between molecules. Consequently, the macromolecules of one microbial strain can elicit a different response from those of even a very closely related strain of the same species.

The immune response system is the body's mechanism that responds to the invasion by pathogenic microorganisms and other foreign substances.

The immune response reacts to particular chemicals, which are called **antigens**. Antigens activate the immune response and react with the cells and chemicals of the immune response system. Once a response to an antigen has occurred, a memory system

is established that permits a rapid and specific secondary response on reexposure to the same substance (FIG. 14-1). Thus the body acquires immunity only after exposure to an antigen, and the specific immune response is therefore adaptive. The ability to recognize and respond rapidly to pathogenic microorganisms establishes a state of immunity that prevents infection with those specific pathogens. The ability to recognize the microorganisms that previously elicited an immune response forms the basis for acquiring or developing immunity to specific diseases.

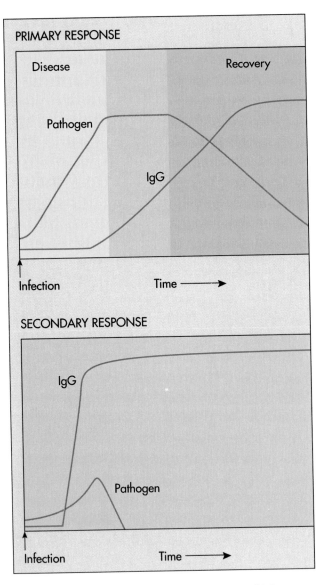

FIG. 14-1 The primary immune response, which occurs after the first exposure to an antigen, is characterized by a lag period of about a week or longer and a slow rise in antibody production. This permits the pathogen to reach concentrations that result in disease. The secondary response, which occurs with subsequent exposure to an antigen that the body "remembers," is characterized by a short lag and rapid rise in antibody concentration to high levels. The rapid production of antibody helps eliminate the pathogen and prevents disease.

Acquiring immunity protects us against infections by pathogens that our body has learned to recognize as foreign. It is activated by and produces a specific reaction to each infectious agent, normally eradicating that agent from the body. The ability to recognize a microorganism that has previously elicited an immune response is the basis for acquiring immunity to specific diseases. As a consequence of such *acquired immunity,* we usually suffer from many diseases, such as chicken pox, only once. We can also intentionally expose ourselves to specific foreign macromolecules, through the use of vaccines, to artificially establish a state of immunity. The use of vaccination to prevent disease is discussed in Chapter 17—after the basis for its use is established in this chapter.

ANTIGENS

An **antigen** is any macromolecule that activates the immune system and that subsequently reacts with the cells or substances (effector molecules) of the immune response that are specific for that antigen (FIG. 14-2). Various macromolecules can act as antigens, including all proteins, most polysaccharides (especially large polysaccharides), nucleoproteins, lipids (glycolipids and lipoproteins), and various small chemicals if they are attached to proteins or polypeptides. The two essential properties of an antigen are its *immuno-* *genicity* (ability to stimulate the immune system) and its *specific reactivity* (ability to react with the specific effector molecules of the immune system).

> **Antigens are macromolecules that activate the immune response system and react with the cells and effector substances of that system.**

For example, an antigen can elicit the formation of an antibody (an effector molecule of the immune response system) and then can subsequently react with that antibody to form an antigen-antibody complex or immune complex. The antibody molecule does not react with the entire antigen molecule. Rather, the antigen molecule consists of a reactive portion, the **epitope** or **antigenic determinant,** that reacts chemically with the antibody molecule to form an antigen-antibody complex (FIG. 14-3).

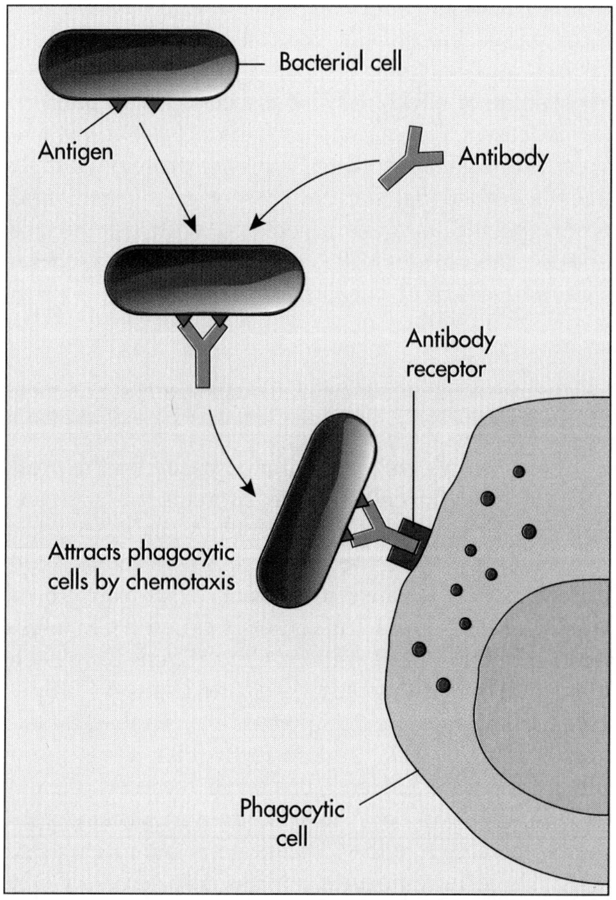

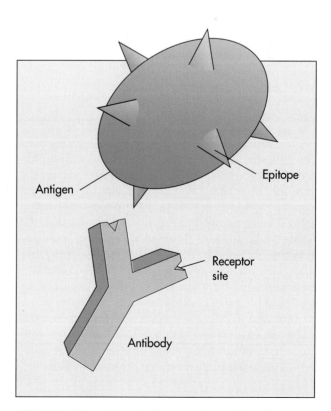

FIG. 14-2 Antigens activate and react with effector cells and molecules of the immune response. The binding of antibody to antigen of a bacterial cell surface enhances the capture of that bacterium by a phagocytic cell.

FIG. 14-3 The epitope is the part of an antigen that reacts with antibody.

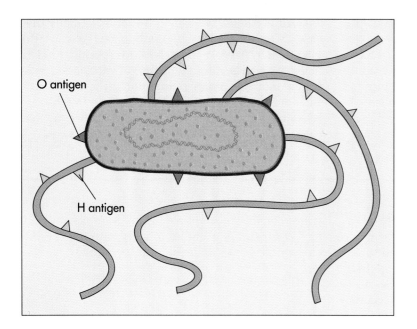

FIG. 14-4   A bacterial cell can have numerous antigens on its surface. The antigens on the cell of *Salmonella* are called O antigens and those on the flagella are called H antigens.

An epitope or antigenic determinant is the part of the antigen molecule that interacts with an antibody.

Some molecules, called **haptens,** have antigenic determinants but are too small to elicit the formation of antibodies by themselves. Penicillin is an example of a hapten that by itself does not initiate the formation of antibody. A hapten, however, can complex with a larger molecule, a carrier, and thereby become an antigen. A hapten can react with antibody molecules that already have been formed, but it is unable to elicit the formation of antibody. A complete antigen molecule both reacts and elicits the production of specific antibodies.

In many cases, antigens are associated with cell surfaces and are therefore called *surface antigens.* Human cells have specific surface antigens; for example, human blood types are determined by the presence or absence of antigens designated A and B on the surfaces of red blood cells. Microorganisms, including viruses and bacteria, also have many surface antigens, some of which may be associated with particular structures. For example, strains of *Salmonella* have specific antigens associated with the proteins of their flagella, called *flagellar antigens* or *H antigens,* and other specific antigens associated with the surface lipopolysaccharides (LPS) of the cell wall, called *somatic antigens* or *O antigens* (FIG. 14-4).

LYMPHOCYTES

The immune defense system depends on lymphocytes that can detect foreign antigens and initiate the immune response. Lymphocytes are differentiated into B lymphocytes (B cells), T lymphocytes (T cells), natural killer cells (NK cells), and killer cells (K cells) (Table 14-1); all have different roles in the immune defenses of the body. B and T lymphocytes originate from bone marrow stem cells and become differentiated during maturation (FIG. 14-5). B cells are responsible for *antibody-mediated immunity (humoral immunity)* and T cells are responsible for *cell-mediated immunity.* Antibody-mediated immunity and cell-mediated immunity represent two distinct but interconnected parts of the immune response system.

There are two interconnected arms of the immune response: antibody-mediated immunity that depends on B cells and cell-mediated immunity that depends on T cells.

Lymphocytes are differentiated based on the presence of specific cell surface proteins or antigens bound to their plasma membranes. A surface marker that identifies a specific line of cells or a stage of cell differentiation because it interacts with a group or cluster of individual antibodies is called a CD (cluster of differentiation) antigen. CD antigens have been identified on various blood cells and tissue cells. Specific CD antigens are designated by a number, for example as CD4, CD8, and so forth (FIG. 14-6). Lymphocytes with CD4 antigens have received a great deal of attention because HIV, the virus that causes AIDS, can attach to the CD4 antigens and infect these cells. These lymphocytes additionally have a CD26 by which the HIV gains access into the cell.

Lymphocytes are differentiated by the presence of specific markers on their surfaces.

TABLE 14-1

Types of Lymphocytes and Their Functions

LYMPHOCYTES	FUNCTION
T helper cells (T_H)	Help or assist other T cells and B cells to express their immune functions
Delayed-type hypersensitivity T cells (T_{DTH})	Participate in activation of macrophages, which contributes to delayed-type hypersensitivity reactions
Cytotoxic T cells (T_C)	Kill and lyse target cells that express foreign antigens (cells containing obligate intracellular parasites and tumor cells)
T suppressor cells (T_S)	Suppress or inhibit the immune function of other lymphocytes
T memory cells	Long-lived cells that recognize previously encountered T-dependent antigens
B lymphocytes	Differentiate into antibody-producing plasma cells and B memory cells in response to an antigen
Plasma cells	Actively secrete antibody
B memory cells	Long-lived cells that recognize a previously encountered antigen
Natural killer (NK) or null cells	Kill and lyse target cells that express foreign antigens
Killer cells (K)	Kill and lyse target cells that express foreign antigens

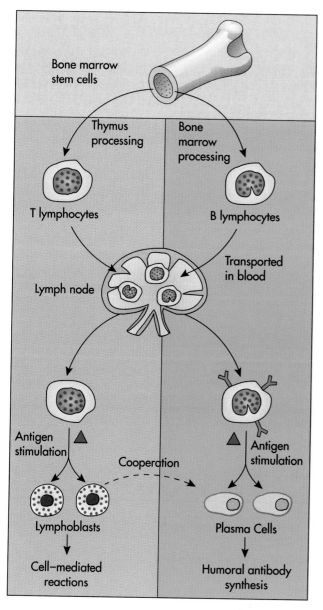

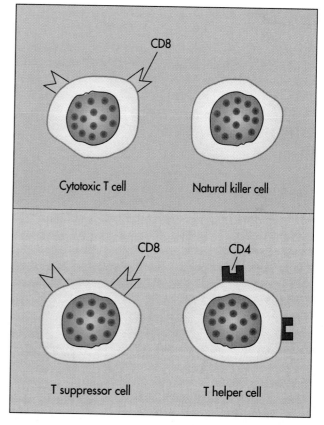

FIG. 14-6 T cells and natural killer cells are differentiated based on the presence or absence of CD4 and CD8 receptors.

FIG. 14-5 B and T cells are differentiated at different locations in the body, within the bone marrow and thymus respectively. They are then transported to secondary sites, such as lymph nodes. Subsequently, they are activated by exposure to antigen. The T cells give rise to cell-mediated reactions and B cells produce antibodies.

B Cells

The term *B cell* or *B lymphocyte* actually refers to bursa-dependent lymphocytes, so named because these lymphocytes are differentiated in chickens and other birds in a lymphoid organ known as the *bursa of Fabricius*. Even though humans do not possess a bursa of Fabricius, the designation B lymphocyte is applied to lymphocytes that can differentiate into antibody-synthesizing cells in the human body. Human B lymphocytes appear to mature in the bone marrow. They come out of the bone marrow and are found predominantly in lymphatic tissues, including the spleen, tonsils, and lymph nodes. Within T cell independent regions of the lymphoid tissues, B cells undergo secondary processing. In response to antigenic stimulation, they then give rise to antibody-secreting plasma cells or long-lived memory B cells (FIG. 14-7). B cells thus form the basis for antibody-mediated immunity.

B cells are involved in antibody-mediated immunity.

T Cells

The term *T cell* or *T lymphocyte* refers to thymus-dependent lymphocytes, so named because these lymphocytes are differentiated in the thymus gland. The precursors of T cells originate in the bone marrow and pass through the liver and spleen before reaching the thymus gland, where they are processed. Within the thymus these cells are called *thymocytes*. After they leave the thymus they are called *T cells*. Although differentiated in the thymus gland, T cells are inactive until they mature later in other lymphoid tissues. T cells are processed within specialized T-cell domains of lymphoid tissues. The thymus-dependent differentiation of T cells or thymocytes occurs during childhood, and by puberty the secondary lymphoid organs of the body generally contain their full complement of T cells. The T cells then generally circulate throughout the body. They are critical in cell-mediated immunity, a part of the overall immune response that recognizes and attempts to eliminate abnormal cells, including cells infected with viruses and other pathogenic microorganisms, malignant cells, and cells from another individual, as in the case of donated organs and transplanted tissues.

T cells are involved in cell-mediated immunity.

There are several types of T cells that have different functions, including regulation of the overall im-

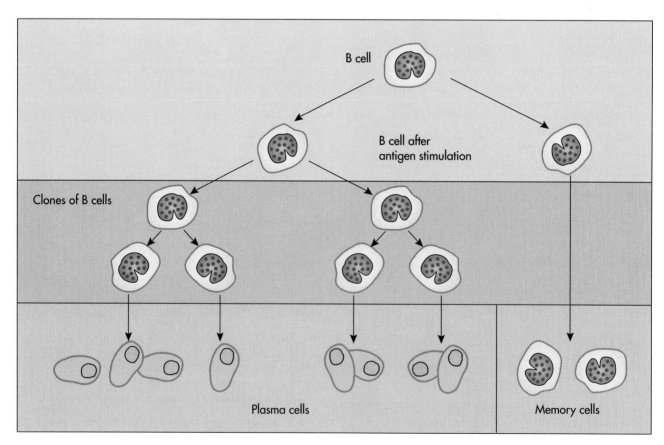

FIG. 14-7   When stimulated by an antigen a B cell differentiates, forming B memory cells and antibody-secreting plasma cells.

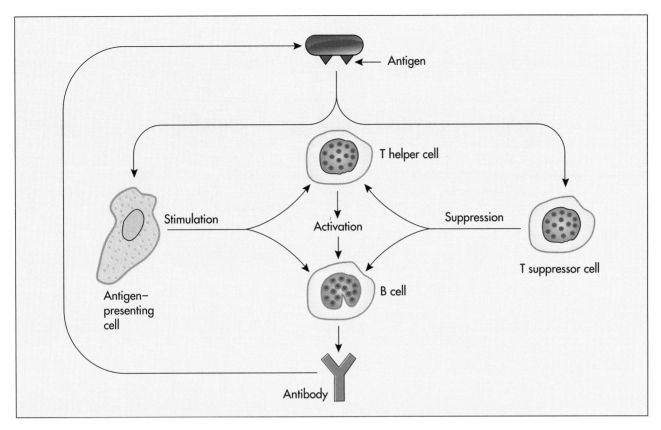

FIG. 14-8 Various T cells help regulate the immune response. T helper cells activate the immune response, j whereas T suppressor cells tend to turn the system off. The T helper cells are stimulated by antigen-presenting cells and, in turn, activate B and T cells.

mune response (FIG. 14-8). Each of the different types of T cells has distinguishing surface CD antigens that enable them to recognize and interact with other cells. In addition, after exposure to a particular antigen, each type of T cell can produce a subset of T memory cells, which are long-lived and responsible for the secondary T cell response.

T Helper Cells

T helper cells (T_H) have CD4 surface antigens. They interact with B lymphocytes and are required for B cells to produce antibody (FIG. 14-9). T_H cells also activate other types of T lymphocytes. Therefore T_H cells generally activate the immune response and are critical for its overall regulation. Both the antibody-mediated and cell-mediated immune responses usually depend on the activation of T_H lymphocytes. For T and B lymphocytes to function in response to a particular protein antigen, they must not only interact with the antigen but also require the presence and interaction with T_H lymphocytes. T_H cells are stimulated and activated by interaction with macrophages and dendritic cells of the spleen and lymph nodes; they also can be activated by B lymphocytes. Some types of T_H cells are also involved in delayed-type hypersensitivity—these are called **delayed hyper-**

sensitivity T helper cells (T_{DTH}). These T_{DTH} cells are responsible for immunological responses such as: (1) the skin rash of poison ivy, (2) the response to certain bacteria such as *Mycobacterium tuberculosis,* and (3) rejection of transplanted tissues.

T Suppressor Cells

T suppressor cells (T_S) are also important in the regulation of the immune response. T_S cells have CD8 surface antigens. They produce substances that are inhibitory to other T lymphocytes. T_S cells tend to turn off the immune response, thereby preventing excessive reactions that could destroy normal body cells.

Cytotoxic T Cells

Cytotoxic T cells (T_C), like suppressor T cells, have CD8 surface antigens. T_C cells are responsible for eliminating abnormal body cells that have been invaded by intracellular parasites such as viruses and some microorganisms; they also recognize tumor cells (FIG. 14-10). On recognition of these "different" body cells, cytotoxic T lymphocytes have the ability to lyse and destroy them. This provides an important mechanism in the defense against viral infection and cancer. Cytotoxic T cells appear to kill target cells by

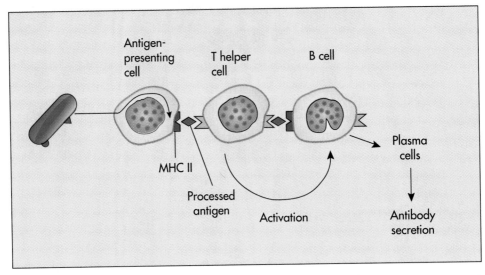

FIG. 14-9   T helper cells modulate the response of B cells to an antigen. The interaction of T helper cells with B cells is necessary for achieving proper antibody production.

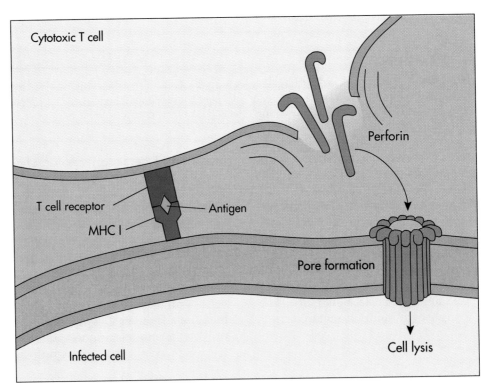

FIG. 14-10   The activation of cytotoxic T cells leads to the elimination of abnormal body cells infected with parasites or viruses. There is an interaction between a T cell receptor and a protein complex (MHC I) that bears processed antigen from an infected cell. This results in the release of perforins, which form pores in the infected cell, resulting in cell lysis.

at least two different mechanisms. Activated T_C cells produce cytoplasmic granules that contain a membrane pore-forming protein called *perforin* and toxic proteins. These proteins are directed to the point of contact between the T_C cell and target cell and lead to the osmotic lysis of the target cell. In addition, cytotoxic T cells secrete a protein toxin that causes fragmentation of nuclear DNA and produce IFN-γ (gamma interferon), which has antiviral activity and is an immune system activator.

There are various types of T cells that have different functions within the immune response.

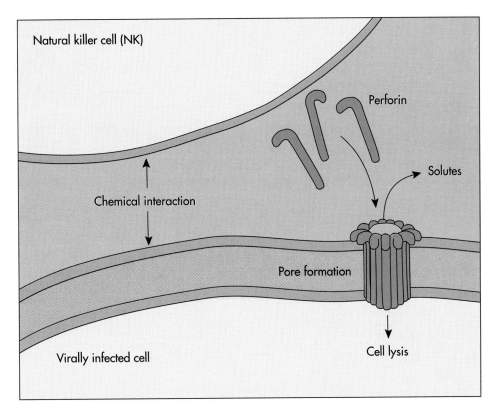

FIG. 14-11 Natural killer cells (NK cells) are cytotoxic to virally infected cells. A chemical interaction triggers release of perforins from the NK cell that results in killing of the infected cell.

NATURAL KILLER CELLS

Natural killer cells (NK cells) are large granular lymphocytes that can recognize cell surface changes that occur on some virally infected cells and some tumor cells. They are neither T cells nor B cells; they lack surface immunoglobulins and CD4 and CD8 markers on their surfaces. This is why they are sometimes called *null cells*. NK cells pierce holes in the cells they attack, causing them to lyse (FIG. 14-11). This kind of reaction, in which a lymphocyte kills a target cell, is called *cytotoxicity*. The activity of NK cells is stimulated by interferons.

> **Natural killer or null cells are a special subset of lymphocytes that are neither T nor B cells.**

KILLER CELLS

Killer cells (K cells) are large granular lymphocytes that have antibody-binding receptors on their surfaces. They are capable of binding to cells that have been coated with antibodies. They then lyse these cells. This function of K cells is called antibody-dependent cell-mediated cytotoxicity (ADCC).

> **Killer cells are lymphocytes that are responsible for antibody-dependent cell-mediated cytotoxicity.**

INTERACTIONS WITH ANTIGEN-PRESENTING CELLS

Antigen-presenting cells (APCs), which include macrophages and other specialized cells (dendritic cells) in skin and lymphoid organs, help activate the immune response. B cells respond more efficiently when antigens are processed by APCs than they do to free antigen. T cells only respond to antigens that have been processed and presented by APCs (FIG. 14-12). APCs bind antigen on their surfaces in association with a special protein called the **major histocompatibility complex** (MHC). MHC proteins are found on almost all cells in the body. They were first identified as the main antigenic determinants of tissue or graft rejection when tissue from one individual is transplanted into another individual. There are two classes of MHC proteins that serve different functions. MHC class I molecules bind antigens produced within cells, such as proteins made during viral replication. MHC class II molecules bind antigens made outside of human cells, such as proteins produced by bacteria growing within the body. In this manner, APCs present antigen to T cells in association with MHC molecules.

> **T cells respond to antigens associated with MHC on antigen-presenting cells.**

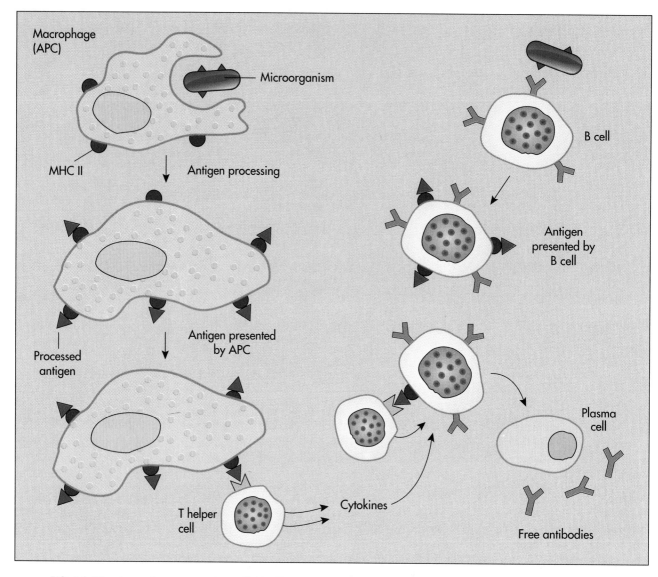

FIG. 14-12   An antigen-presenting cell (APC) reacts with an antigen in a way that enhances the response of T and B cells. B cells can also function as antigen-presenting cells. APCs process antigen and display it in association with MHC class II proteins. The antigen-MHC class II complex is recognized by T helper cells that are then stimulated to release cytokines. Sone cytokines stimulate T cells while others stimulate B cells to become antibody-secreting plasma cells.

ANTIBODY-MEDIATED IMMUNITY

In the **antibody-mediated immune response,** specific glycoproteins called **antibodies** or **immunoglobulins** are made when foreign antigens activate B cells. Antibodies are found in body secretions and serum, which is the fluid portion of coagulated (clotted) blood that is free of blood cells and the clotted material. In old medical terminology, blood and other vital body fluids were considered as "humors," after the Latin word for fluids. Thus antibody-mediated immunity is sometimes referred to as *humoral immunity* because the antibody molecules flow extracellularly through the body fluids. The key to antibody immunity is the ability of antibodies to react specifically with antigens. Several types of reactions can occur when an antibody combines with an antigen to form an antigen-antibody complex (Table 14-2).

Foreign antigens enter the body in various ways. These antigens may be carried from the tissues into lymph fluid, which then drains through the lymph nodes (FIG. 14-13). Alternatively, antigens that gain access to the blood are filtered in the spleen. Foreign antigens thereby come into contact with B cells that are residents of the lymph nodes and spleen. Other antigens may be phagocytized and carried to lymphoid tissues by macrophages.

The formation of antibody begins when a B lymphocyte comes in contact with an antigen. B cells react with antigens because they have specific antigen

TABLE 14-2		
Functions of Antibodies		
TYPE OF ANTIBODY	**FUNCTION**	**DESCRIPTION**
Agglutinin	Agglutination	Antibodies that react with antigens on the surfaces of the cells or particles to form antigen-antibody complexes that cause clumping together of cells or particles
Precipitin	Precipitation	Antibodies that react with soluble antigens to form insoluble antigen-antibody complexes that precipitate
Antitoxin	Neutralization	Antibodies (antitoxins) that react with antigens (toxins) to form antigen-antibody complexes that inactivate (neutralize) the toxins
Complement-fixing	Activation of complement	Antibodies that react with antigens to form antigen-antibody complexes that trigger the complement cascade
Neutralizing	Inactivation of viruses	Antibodies that react with antigens on virus surfaces that are necessary for adsorption of the viruses to host cells; the formation of antigen-antibody complexes blocks viral attachment to host cells and prevents viral replication; because the virus cannot initiate replication it is inactivated
Opsonizing	Opsonization	Antibodies that react with antigens on the surfaces of bacterial or fungal cells, coating the cell and thereby enhancing phagocytosis of that cell

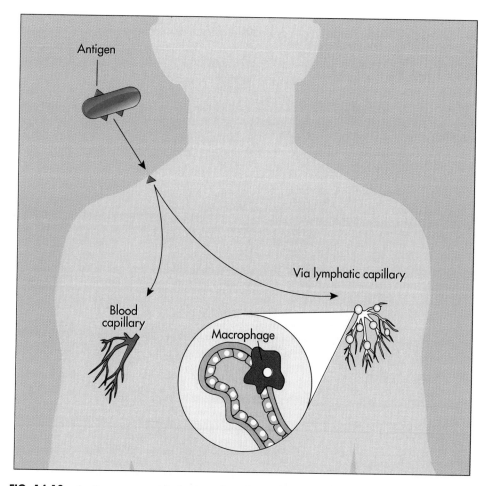

FIG. 14-13 Antigens can enter the lymph and circulate through the body, coming in contact with cells of the immune response system. Some antigens may be carried to lymphoid tissues (lymph nodes and spleen) by macrophages.

receptors on their surfaces. The antigen receptor is an immunoglobulin molecule that will react only with a specific antigen. Each individual B lymphocyte contains the genetic information for initiating an immune response to a single specific antigen. The immunoglobulin antigen receptor is located on the plasma membrane of the differentiated B lymphocyte. There are at least 10^6 different types of B cells in the body, each with the ability to recognize a different antigenic determinant. The activation of a particular B cell leads to *clonal selection and expansion* (FIG. 14-14), that is, cell division of the specific B lymphocyte. This cell division creates a population or clone of similar B lymphocytes that produces antibodies with the same specificity.

B cells that react with antigens divide and form many additional B cells.

The B lymphocyte also becomes an antigen-presenting cell (APC) as a result of its processing of an antigen. The antigen is modified within the B cell and the processed antigen is placed on the surface of the B cell. The B lymphocyte displays the processed antigen in association with its class II MHC proteins. The activated B lymphocyte, in turn, activates T_H cells, and activated T_H cells then release substances that specifically stimulate the B lymphocyte into cell growth, differentiation, and proliferation.

B cells that have processed antigens become antigen-presenting cells that activate the immune response.

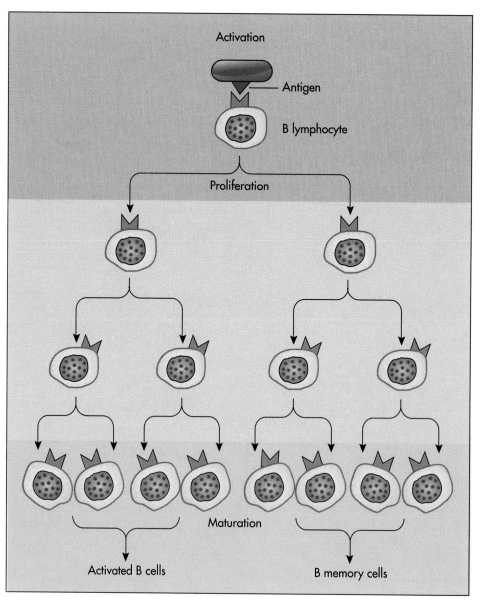

FIG. 14-14 Activation of a B cell leads to formation of B memory cells with the same surface receptors, that is, clonal selection and expansion.

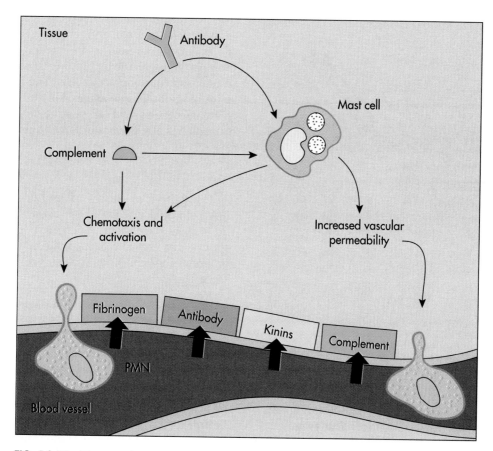

FIG. 14-15 The specific immune response and nonspecific defenses act together to eliminate pathogens and to repair body tissues. The immune reactions involving antigen, antibody, and complement result in degranulation of mast cells. The substances released from the mast cells cause increased permeability of blood vessels so that PMNs can move into the surrounding tissue. This enhances the inflammatory response. Proteins such as fibrinogen are released, aiding in tissue repair.

As a result of activation and maturation, some B lymphocytes differentiate into **plasma cells** that secrete large amounts of antibody specific for the stimulating antigen. These cells are the major producers of antibody molecules that are secreted into the blood or lymph. Other B cells differentiate into B memory cells. These resting cells are very long-lived—perhaps as long as 20 to 30 years—and are responsible for the secondary immune response, which is discussed later. The activation of B cells also links the antibody-mediated immune response to the nonspecific defense system so that they act together to eliminate infecting microorganisms and to repair body tissues (FIG. 14-15).

B cells give rise to plasma cells that synthesize and secrete antibodies and memory cells.

ANTIBODIES (IMMUNOGLOBULINS)

Antibodies are glycoproteins that are made in response to specific antigens and react with those anti-

gens. The reaction of an antibody with an antigen forms an antigen-antibody complex (immune complex). They are secreted into lymph and then drain into the blood. The major concentrations of antibodies in the body are in the serum fraction of blood. Antibodies are **immunoglobulins,** so named because of their role in the immune response and because they are globular (spherically shaped).

Antibodies are immunoglobulins.

All immunoglobulin molecules have the same basic molecular structure consisting of four peptide chains: two identical heavy chains and two identical light chains (FIG. 14-16). The terms *heavy* and *light* refer to the relative molecular weights of the polypeptide chains. There are more amino acids in the heavy chain; hence, it has a greater molecular weight than the light chain. The chains are joined by disulfide bridges that link the chains together.

Immunoglobulins have two light and two heavy chains that are linked together.

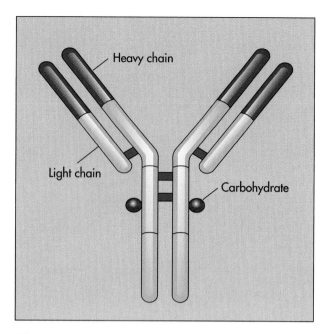

FIG. 14-16 An immunoglobulin has two heavy and two light chains joined by disulfide bonds.

but not forming Fc fragments. It is the Fab fragment that actually binds to antigen molecules, whereas the Fc fragment augments the action of the immunoglobulin molecule by binding to complement molecules and/or phagocytic cells. The Fc portion of the immunoglobulin molecule will only bind complement molecules and/or phagocytic cells if the Fab portion has already bound an antigen.

The Fab portion of an immunoglobulin reacts with specific antigens.

There are five classes of immunoglobulins: IgG, IgA, IgM, IgD, and IgE. Each class of immunoglobulin serves a different function in the immune response. The characteristics of these five major classes of immunoglobulin molecules are summarized in Table 14-3.

There are five classes of immunoglobulins: IgG is the most common; IgE is the least common; IgM is made early in the course of an infection; IgA is important in protecting surface tissues; and the precise role of IgD is probably a B cell antigen receptor.

IgG

IgG (Immunoglobulin G) is the largest immunoglobulin fraction, generally comprising approximately 80% of the body's immunoglobulins. It is the predominant circulating antibody and readily passes through the walls of small vessels (venules) into body tissues, where it reacts with antigen and stimulates the attraction of phagocytic cells to invading microorganisms. Reactions of IgG with surface antigens on bacteria activate the complement system and at-

The heavy and light peptide chains can be broken apart by enzymes. Papain cleaves an immunoglobulin molecule to form two identical **Fab fragments** that contain the antigen-combining site and an additional **Fc fragment,** which is crystallizable (FIG. 14-17). The Fc portion may contain amino acid sequences that anchor the immunoglobulin molecule to the plasma membranes of cells. Pepsin cleaves the immunoglobulin molecule at a different location than papain, forming an antibody-binding fragment

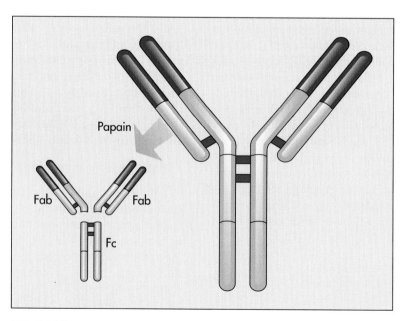

FIG. 14-17 Papain cleaves IgG into differing fragments, including Fab and Fc fragments; the Fab fragment binds to antigen.

TABLE 14-3

Properties of the Five Classes of Immunoglobulins

PROPERTY	IMMUNOGLOBULIN				
	IgG	IgA	IgM	IgD	IgE
Molecular weight and dimer	150,000	160,000	900,000	185,000	200,000
Number of basic four-peptide units	1	1, 2	5	1	1
Heavy chains	γ	α	μ	δ	ϵ
Light chains	$\kappa + \lambda$	$\kappa + \lambda$	$\kappa + \lambda$	$\kappa + \lambda$	$\kappa + \lambda$
Number of antigen binding sites	2	2, 4	10	2	2
Concentration range in normal serum	8-16 mg/mL	1.4-4 mg/mL	0.5-2 mg/mL	0-0.4 mg/mL	17-450 μg/mL
Percentage of total imunoglobulin	80	13	6	0-1	0.002
Complement fixation					
Classical	+	−	++	−	−
Alternative	−	±	−	−	−
Crosses the placenta	+	−	−	−	−
Fixes to homologous mast cells and basophils	−	−	−	−	+
Binds to macrophages and neutrophils	+	±	−	−	−
Major characteristics	Most abundant Ig of body fluids; combats infecting bacteria and toxins	Major Ig in seromucous secretions; protects external body surfaces	Effective agglutinator produced early in immune response	Mostly present in lymphocyte surface	Protects external body surfaces; responsible for atopic allergies; response to helminthic infestations
Structure	IgG (monomer)	IgA (dimer)	IgM (pentamer)	IgD (monomer)	IgE (monomer)

ELECTROPHORETIC SEPARATION OF IMMUNOGLOBULINS

Immunoglobulins can be separated by using electrophoresis (see Figure). Electrophoresis is a method that uses an electronic field to separate molecules such as immunoglobulins on the basis of their charge, size, and shape. A protein sample is usually placed on a solid support medium, such as agarose or polyacrylamide. The support medium in turn is placed between two electrodes, a positively charged anode and a negatively charged cathode. When electric current is applied, protein migration occurs. The rate of migration is characteristic for each protein. This rate is determined by the protein's net charge, which is a function of pH, and molecular weight. The classes of immunoglobulins form relatively broad electrophoretic bands. This broad band is indicative of the heterogeneity of molecules within each class. Such analyses are useful in clinical medicine in diagnosing certain diseases, such as those associated with deficiencies of the immune response.

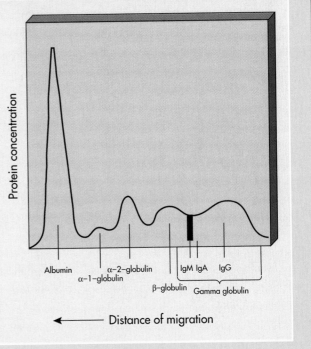

Immunoglobulins can be separated by electrophoresis into various globulins, including IgA and IgG.

---

tract additional neutrophils to the site of the infection. IgG is the only antibody that can cross the placenta and confers immunity on the fetus that lasts for up to a few months after birth. IgG can combine with toxins and neutralize or inactivate them. It also has a major role in preventing the systemic spread of infection through the body and in facilitating recovery from many infectious diseases.

IgA

IgA (Immunoglobulin A) occurs in mucus, in secretions such as saliva, tears, and sweat, and in blood. It also is found in human milk. It is important in the respiratory, gastrointestinal, and genitourinary tracts, where it protects surface tissues against invasion by pathogenic microorganisms (FIG. 14-18). IgA also is secreted into human breast milk and colostrum and has a role in protecting nursing newborns against infectious diseases.

The IgA molecules bind with surface antigens of microorganisms, preventing the adherence of such antibody-coated microorganisms to the mucosal cells lining the respiratory, gastrointestinal, and genitourinary tracts. IgA molecules do not initiate the classical complement pathway but can activate the alternative complement pathway. Plasma contains relatively high concentrations of monomeric IgA molecules,

but it is the dimers of IgA that bind to receptors on the surface of secretory cells, leading to their secretion into body fluids. The IgA picks up an additional secretory protein that protects it against proteases and promotes its secretion. In mucus secretions, IgA is the major immunoglobulin molecule involved in the immune response that protects external body surfaces.

IgM

IgM (Immunoglobulin M) is the highest molecular weight immunoglobulin, occurring as a pentamer; that is, IgM contains five monomeric units of the basic four-peptide chain immunoglobulin molecule. IgM molecules are formed before IgG molecules in response to exposure to an antigen (FIG. 14-19). Because of its high number of antigen-binding sites, the IgM molecule is effective in attaching to multiple cells that have the same surface antigens. As such, it is important in the initial response to a bacterial infection and in the activation of complement. During the later stages of infection, IgG molecules are more important. IgM molecules occur primarily in the blood serum and lymph. Together with IgG molecules, IgM molecules are important in preventing the circulation of infectious microorganisms through the circulatory system.

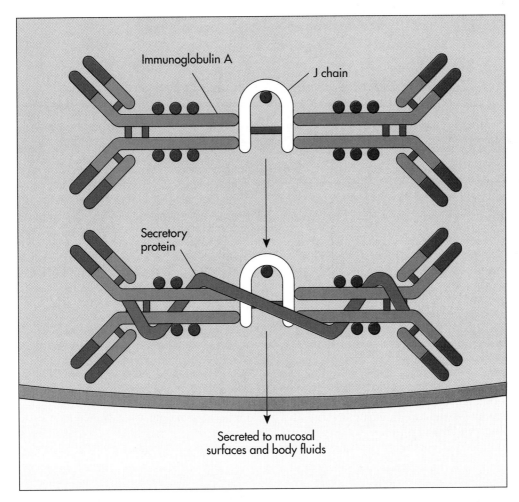

FIG. 14-18 IgA is a dimer that is secreted from plasma cells. A secretory protein binds to the IgA, allowing it to be secreted. The secreted IgA binds to mucosal membranes, protecting body surfaces from infection.

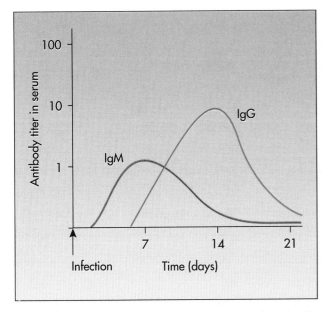

FIG. 14-19 During an infection, IgM is produced relatively early. IgM is produced before IgG.

IgD

IgD (Immunoglobulin D) antibody molecules, together with IgM, are present on the surface of some lymphocyte cells. Although the precise role of IgD remains to be fully defined, it appears to have a role as an antigen receptor in lymphocyte activation and suppression. Within blood plasma, IgD molecules are short-lived, being particularly susceptible to proteolytic degradation.

IgE

IgE (Immunoglobulin E) molecules are normally present in the blood serum as a very low proportion of the total immunoglobulins. The ratio of IgG to IgE is normally 50,000:1. IgE serum levels, though, are elevated in individuals with allergic reactions such as hay fever and in some persons with chronic parasitic infections. The main role of IgE appears to be the protection of external mucosal surfaces by mediating the attraction of phagocytic cells and the initiation of the

GENETIC BASIS FOR IMMUNOGLOBULIN DIVERSITY

The human immune system has vast capacity to generate different antibodies that recognize and bind to millions of potential antigens. The genetic variability coding for this diversity of antibody molecules provides the basis for the immune response to many different antigens. Before the recognition that eukaryotic cells had split genes, it was difficult to understand how the genome encoded the information for the enormous diversity of antibodies.

One early theory of immunoglobulin diversity proposed that each human cell contains a separate gene encoding each antibody chain that an individual is capable of synthesizing. However, there is not enough DNA in human cells to accomplish this and still have genes that code for other functions; the human genome contains perhaps several hundred thousand genes, and only a small fraction of these can specify antibodies. Another proposal was that each cell contains a small number of genes for encoding antibody chains but that these genes are so susceptible to mutation that multiple mutations accumulating in mature B cells confer on the organism the ability to produce various antibodies. This theory also cannot adequately account for the diversity of antibodies. Thus the theory, which was held for many years, that there was a one-to-one correspondence between genes and polypeptides, cannot account for the diversity of antibodies; another theoretical explanation was necessary.

The essence of the explanation of how a limited number of genes can generate the diversity of antibodies is that the genes ultimately specifying the structure of each antibody are not present as such in germ cells (the male sperm and the female egg) or in the cells of the early embryo. The currently accepted theory is that there are variable and constant regions of antibodies that are encoded by separate groups of genes. Polypeptides making up the antibodies can be synthesized from information contained in several gene fragments scattered over the genome. Recombination permits shuffling and joining of the components so that billions of different combinations can be generated (see Figure). The reshuffling results in numerous combinations of varieties of the light and heavy chains that make up the complete immunoglobulin molecules. There are three clusters of genes, one for heavy chain synthesis and two for light chain synthesis. Within each cluster, there are a number of gene fragments that contain the information for the complete immunoglobulin molecules.

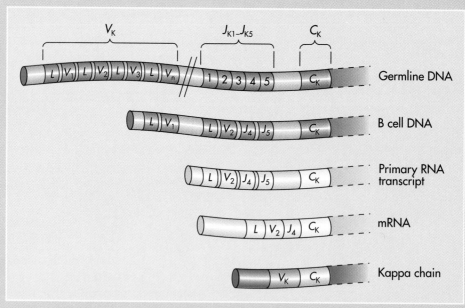

Immunoglobulin genes that code for the light chains are split. In B cells, rearrangements result in joining of some V and J genes. Transcription followed by elimination of introns produces a functional mRNA. The translation results in one of the chains of the immunoglobulin.

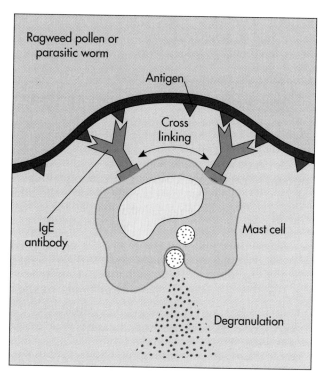

FIG. 14-20 IgE is important in the activation of the inflammatory response.

inflammatory response (FIG. 14-20). IgE molecules are important because they bind to mast cells and basophils, which can lead to the initiation of the inflammatory response if the IgE reacts with an antigen. IgE molecules are attached to specific receptors on the surfaces of mast cells and basophils by their Fc region so that the antigen-combining sites protrude from these cells. When specific antigens, such as ragweed pollen or those associated with parasitic worms, bind to these surface IgE molecules, it causes mast cells and basophils to degranulate (release the contents of their granules into the blood and tissues). Degranulation, unfortunately, also produces undesirable reactions such as hay fever, asthma, and other allergic responses.

ANTIGEN-ANTIBODY REACTIONS

Antibody-mediated immunity is important in preventing and eliminating microbial infections. The bases of these reactions depends on the reactions of antigen with antibody molecules. The reactions of IgA molecules with bacteria and viruses in the fluids surrounding surface tissues, for example, prevent the adsorption of many potential pathogens onto these surface barriers. In this way, IgA antibody molecules prevent the establishment of infections. IgG antibody molecules, acting as antitoxins, are able to neutralize toxin molecules by combining with them, thus blocking their reactions and preventing the onset of disease symptomatology. Even poisonous cobra venom

can be neutralized by reaction with appropriate antibody molecules. The reaction of IgG with antigen on a cell surface also permits the addition of complement. The complement attaches to the Fc portion of the antibody, activating the classical complement pathway. This leads to the death of the cell and is an important way of eliminating pathogens from the body. The two main methods used by the immune system to defend against bacterial infections are complement-mediated lysis, which antibody reactions trigger via the classical pathway, and phagocytic killing, which is enhanced through opsonization when antibodies react with antigens on a cell's surface.

Opsonization

The interactions of antibody with surface antigens of bacterial cells also render many pathogenic bacteria more susceptible to phagocytosis. In fact, the ingestive phagocytic attack on most bacteria requires an initial antigen-antibody reaction before phagocytic blood cells can engulf the invading bacteria. Both polymorphonuclear neutrophils and macrophages express receptors for the Fc portion of antibody on their surface. The increased phagocytosis associated with antibody-bound antigen, called **opsonization,** is important in the destruction of pathogenic bacteria (FIG. 14-21). Phagocytic killing of bacteria is largely due to the activities of neutrophils and mononuclear phagocytes (monocytes and macrophages). Fixed macrophages in the spleen and liver are important in clearing bacteria from the blood. Antigen-antibody reactions are required to overcome infections by bacteria, such as *Haemophilus influenzae,* that are inherently resistant to phagocytosis. The major mechanism for fungal clearance is by activated macrophages and phagocytosis. These *in vivo* reactions

NEWSBREAK

A PLAYWRIGHT'S VIEW OF OPSONIZATION

In the preface to his play *The Doctor's Dilemma,* the British playwright George Bernard Shaw gave a laymen's view of the 1920 state-of-the-art understanding of opsonization. He likened opsonization to buttering bread, stating that no Englishman would eat dry toast, but that buttered toast was a delight to eat. Buttering the toast made it palatable and easier to ingest. This he said was analogous to macrophage ingesting antibody-coated bacteria. In Shaw's view, antibody to a macrophage is like butter to an Englishman.

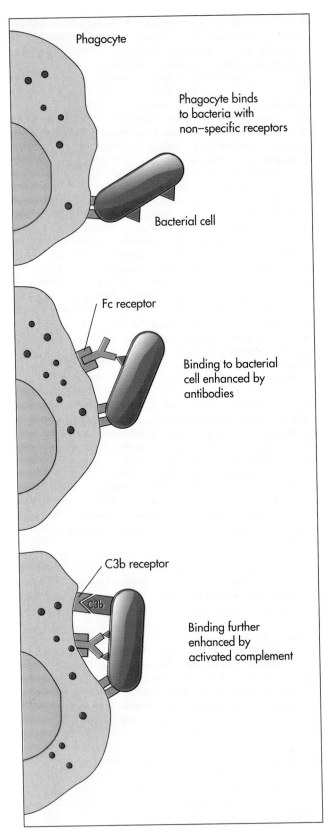

FIG. 14-21 Opsonization involves binding of a bacterial cell to the surface of a phagocyte via a number of different mechanisms, including immunoglobulin and complement receptors. By binding to multiple receptors the bacterial cell becomes firmly attached and engulfed by the phagocytic cell. It is subsequently ingested and digested.

between antigen and antibody molecules constitute a major line of defense against invading bacteria and other microorganisms.

Complement Fixation

Following the reaction of antigen and antibody, complement can bind to the antigen-antibody complex. This reaction is called **complement fixation.** Only free complement can bind to antigen-antibody complexes and so complement fixation reduces the amount of complement available to continue this reaction. The fixation of complement activates the complement system by the classical pathway that was discussed in Chapter 13. This leads to the formation of membrane attack complexes (MACs). These MACs assemble in the outer membrane of Gram-negative bacteria or the cytoplasmic membrane of Gram-positive bacteria. They form pores in these membranes that result in lysis of the cell. Individuals with an inherited deficiency in the late components of the complement cascade that form MACs have a greater chance of developing infections due to *Neisseria* bacteria. Complement-mediated lysis of fungal cells is less efficient than that of bacterial cells because fungi have very thick cell walls. These walls keep the complement proteins from reaching the fungal plasma membrane. Complement activation has a limited role in the elimination of viruses from the body. Complement may be responsible for the destruction of viruses with lipid envelopes.

Neutralization

One of the reasons that the immune response is effective in preventing disease is that the toxins contributing to the virulence of pathogenic microorganisms usually have antigenic properties. Antibodies produced against toxins are referred to as *antitoxins*. Bacterial LPS toxins are antigenic and are responsible for the initiation of antibody production against many Gram-negative bacteria. Most protein toxins (exotoxins) are highly antigenic, eliciting the synthesis of high concentrations (titers) of antibody. Reaction with the toxin eliminates its ability to act as a poison that disrupts normal body functions. The formation of the antigen(toxin)-antibody(antitoxin) complex *neutralizes* the toxin (FIG. 14-22).

Antitoxins are antibodies produced against toxins.

Neutralization is the result of the formation of an antigen(toxin)-antibody(antitoxin) complex.

Even denatured proteins often retain their antigenic properties. Of particular importance are denatured proteins called **toxoids.** Toxoids are commonly formed by treatment of toxins with glutaraldehyde or formaldehyde. Toxoids retain their antigenic properties but do not cause the onset of disease symp-

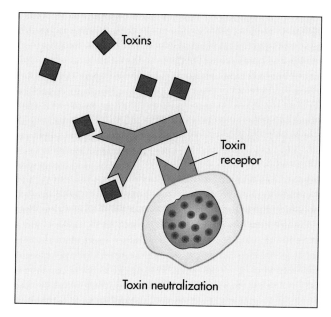

FIG. 14-22 The physiological effects of a toxin can be neutralized by reaction with an antibody. Such antibodies are called antitoxins. The binding prevents the toxin from reacting with its target site.

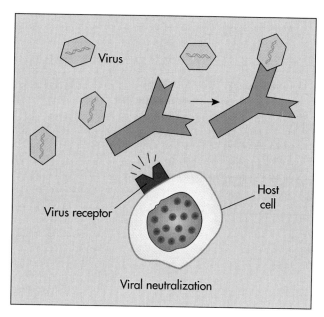

FIG. 14-23 Neutralizing antibodies can inhibit viral replication by binding to the virus so that the virus cannot adsorb to a host cell. This prevents viral infection.

toms. They are antigenically active macromolecules but are no longer toxins. Toxoids are useful for eliciting antibody-mediated immune responses. They are used as vaccines for protecting individuals against diphtheria, tetanus, and various other diseases caused by pathogenic microorganisms that produce protein toxins.

Toxoids are denatured protein toxins that have lost their toxicity but retain antigenicity.

Similarly, antibodies formed against viral proteins combine with these proteins and thereby prevent viruses from attaching to host cells. This effectively inhibits the viral replication cycle. These antiviral antibodies are called *neutralizing antibodies* (FIG. 14-23). Early in the course of a viral infection, the action of specific antiviral antibodies is the most important host defense mechanism. Neutralizing antibodies bind to virus particles and inhibit their attachment to specific receptors on host cells. Opsonizing antibodies probably enhance the clearance of viruses by phagocytic cells. The formation of specific antibodies to viruses is the basis for the development of vaccines to a variety of virally caused human diseases.

Serological Reactions

Serological reactions are antigen-antibody reactions run *in vitro* (from the Latin, meaning in glass); these reactions are performed in laboratory vessels such as test tubes or plastic dishes. The identification of antigen-antibody complexes is one of the hallmarks of

serology. There are several types of serological reactions that are widely used for identifying microorganisms (Table 14-4). Discussed in Chapter 16, these serological tests are widely used in the clinical laboratory for helping diagnose the causes of disease.

Serological reactions, which are *in vitro* antigen-antibody reactions, are used for clinical diagnoses.

Reaction of antibody molecules with antigens that are located on the surface of cells or other particles are not soluble and can lead to the aggregation or clumping of the cells or particles; such reaction is called **agglutination** (FIG. 14-24). Antibodies that combine with antigens on cell or particle surfaces are called **agglutinating antibodies** or **agglutinins.** If the antigens are located on the surfaces of red blood cells and the addition of antibody leads to clumping of the cells, this is called **hemagglutination.** Hemagglutination is the basis for blood typing and distinguishing the presence of A type antigen or B type antigen on the surface of human red blood cells (FIG. 14-25). Individuals with type A blood have erythrocytes with the A antigen on their surface and will be agglutinated by anti-A antibodies. Individuals with type B blood have erythrocytes with the B antigen on their surface and will be agglutinated by anti-B antibodies. Type O erythrocytes have neither the A nor B antigen and will not be agglutinated by either anti-A serum or anti-B serum. Finally, type AB blood has erythrocytes with both markers and will be agglutinated by both anti-A and anti-B sera.

TABLE 14-4

Descriptions of Serological Reactions

REACTION	DESCRIPTION	USE
Agglutination	Clumping of cells or particles because of reactions of antigens on the surfaces of the cells or particles with antibodies to form antigen-antibody complexes	Detection of cells with specific surface antigens
Hemagglutination	Clumping of red blood cells because of reactions of antigens on their surfaces with antibodies to form antigen-antibody complexes	Blood typing; detection of specific microorganisms
Passive agglutination	Clumping of particles because of reactions of antigens that have been added to them with antibodies to form antigen-antibody complexes or because of antibodies that have been added to them with antigens to form antigen-antibody complexes	Detection of specific microorganisms, such as *Streptococcus* species in cases of strep throat
Complement fixation	Reactions of complement with an antigen-antibody complex so that the complement is bound (fixed) and no longer free; fixed complement can not react further with antigen-antibody complexes and thus cannot contribute to reactions such as cell lysis	Indirect detection of infecting microorganisms, such as the classical detection of *Treponema pallidum* in cases of syphilis
ELISA (enzyme-linked immunosorbent assay)	Reaction of an antibody that has been tagged with an enzyme or the substrate of an enzyme so that the reaction can be detected based on the products of the enzymatic reaction	Sensitive detection of antigen-antibody reactions
Direct ELISA	Reaction of an antigen with an antibody that has been tagged with an enzyme or the substrate of an enzyme so that the reaction can be detected based on the products of the enzymatic reaction	Detection of viral or bacterial proteins, hormones, drugs, and other antigens in a specimen
Indirect ELISA	Reaction of the antibody of an antigen-antibody complex with an antibody that has been tagged with an enzyme or the substrate of an enzyme so that the reaction can be detected based on the products of the enzymatic reaction	Detection of the presence of antibodies to HIV virus, rubella virus (German measles), *Mycoplasma pneumoniae*, *Helicobacter pylori*, and *Borrelia burgdorferi* (Lyme disease)
RIA (Radioimmunoassay)	Reaction of an antibody that has been tagged with a radioisotope so that the reaction with an antigen to form an antigen-antibody complex can be detected	Detection of hormones such as human growth hormone, insulin and thyroxine; detection of steroids such as corticosterone and progesterone; detection of prostaglandins and leukotrienes; detection of drugs such as digoxin, morphine, and phenobarbital
Immunofluorescence	Reaction of an antibody tagged with a fluorescent dye that can be visualized	
Direct immunofluorescence	Reaction of an antibody tagged with a fluorescent dye directly with an antigen so that the reaction can be visualized	Detection of *Legionella pneumophila*
Indirect immunofluorescence	Reaction of an antibody tagged with a fluorescent dye with the antibody of an antigen-antibody complex so that the reaction can be visualized	Detection of *Treponema pallidum*

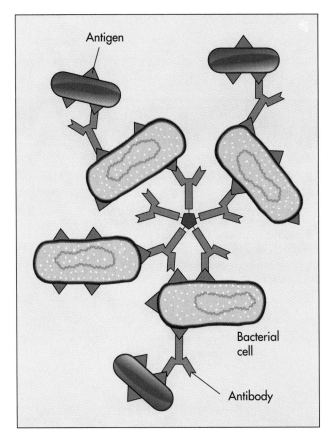

FIG. 14-24 Reaction of cell surface antigens with immunoglobulins can result in clumping (agglutination) of cells.

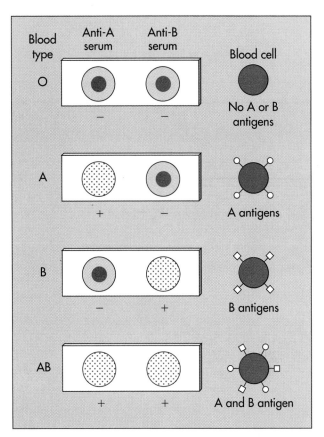

FIG. 14-25 Agglutination reactions are used to determine the major blood types. This is based on separate reactions with anti-A antibodies *(left side of slide)* and anti-B antibodies *(right side of slide)* to determine the presence or absence of A and B antigens on the red blood cell surface. Agglutination is shown here as the formation of clumps of cells *(small dotted pattern)* and lack of agglutination by the presence of a pellet of unreacted cells in the center; this is how the reaction would appear in the wells of a microtiter plate where many blood typing reactions are run.

The **enzyme-linked immunosorbent assay** (ELISA) uses labels attached to antibody molecules to identify them. It is far more sensitive than agglutination reactions so that antigen-antibody reactions can be detected. ELISA can be used for detecting the presence of antigen in a sample (FIG. 14-26, *A*). Antibody specific for the antigen is bound to a well in a plastic microtiter plate. A sample that may or may not contain antigen is added to the well. If antigen is present, it will bind to the antibody. Unbound antigen is washed out of the well. A second antibody solution is added to the well, but these antibodies have an enzyme molecule bound to them. Frequently used are horseradish peroxidase, glucose oxidase, and alkaline phosphatase. Excess enzyme-bound antibody is washed out of the well. Then, the substrate for the enzyme is added. The reaction of the enzyme forms a colored product that can be measured spectrophotometrically. The amount of colored product is proportional to the amount of antigen bound in the mi-

crotiter well. The direct ELISA (sandwich ELISA) can be used for detecting viral or bacterial proteins, hormones, drugs, and other antigens in a specimen.

Indirect ELISAs (FIG. 14-26, *B*) are used for detecting the presence of antibody in serum. In the indirect method, antigen is bound to the microtiter well and serum (containing antibody) is added to it. Excess and unbound antibody is washed out of the well. If antibody specific to the antigen is present in the serum, it will bind to the antigen in the microtiter well. A second antibody is then added; this antibody is animal (usually goat or rabbit) anti-human IgG that has been conjugated with enzyme. If the primary antibody has bound to antigen, the secondary antibody will bind to it and can be detected by the addition of the enzyme substrate. Indirect ELISAs are used in clinical laboratories to detect the presence of antibodies to HIV virus, rubella virus (German measles), *Mycoplasma pneumoniae*, *Helicobacter pylori*, and *Borrelia burgdorferi* (Lyme disease).

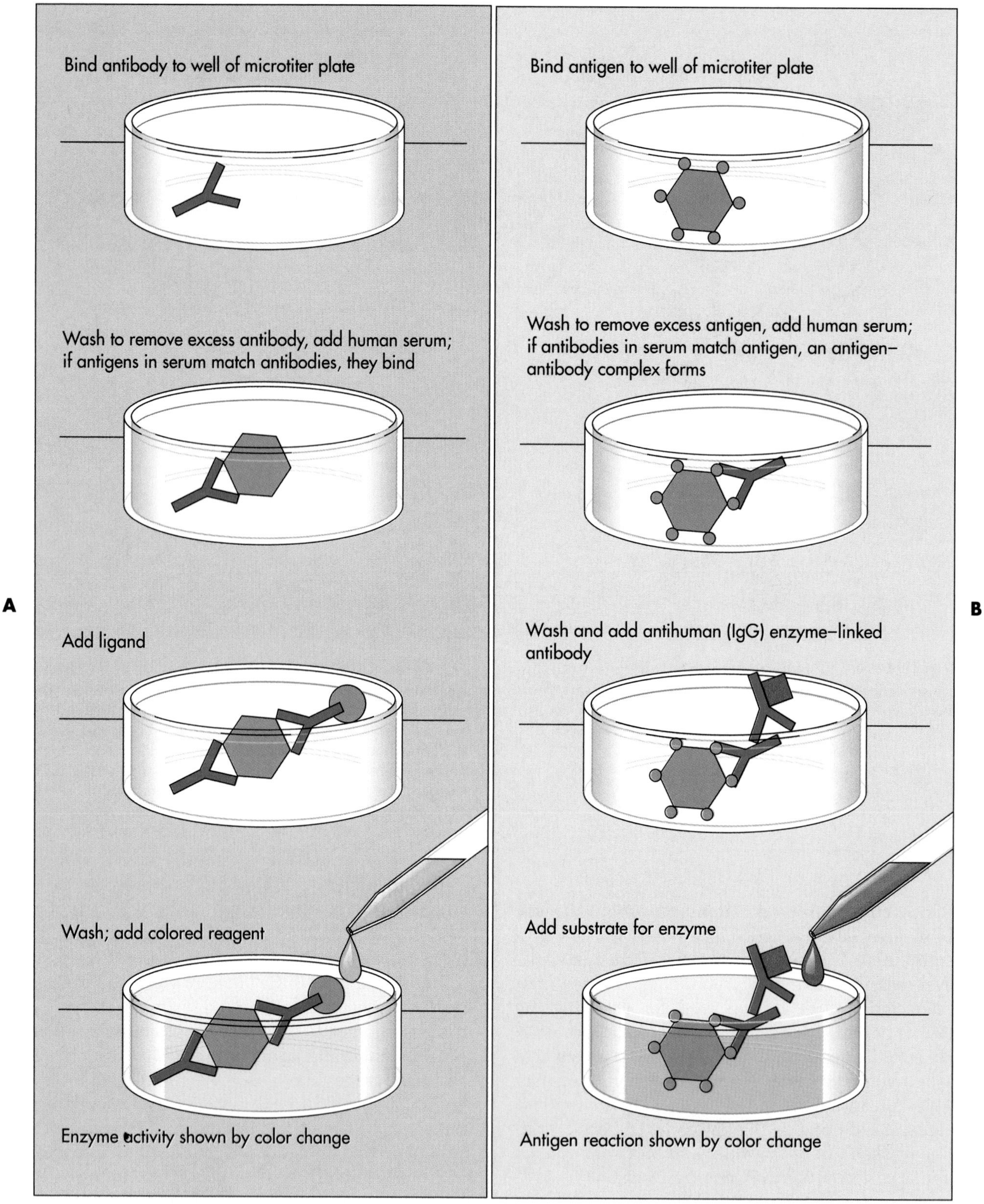

Bind antibody to well of microtiter plate

Bind antigen to well of microtiter plate

Wash to remove excess antibody, add human serum; if antigens in serum match antibodies, they bind

Wash to remove excess antigen, add human serum; if antibodies in serum match antigen, an antigen–antibody complex forms

A

B

Add ligand

Wash and add antihuman (IgG) enzyme–linked antibody

Wash; add colored reagent

Add substrate for enzyme

Enzyme activity shown by color change

Antigen reaction shown by color change

FIG. 14-26 **A,** The direct ELISA procedure uses antibody bound to the walls of a microtiter plate to trap antigen. A second antibody molecule with an attached ligand, typically a substrate for an enzymatic reaction, is added. When the enzyme is added, activity is shown by a color change indicating the reaction of the enzyme with its substrate. **B,** The indirect ELISA procedure uses antigen bound to the walls of a microtiter plate to trap human antibody if it is present in serum. A second antihuman-antibody IgG molecule with an attached enzyme is added. When the enzyme substrate is added, activity is shown by a color change indicating the reaction of the enzyme with its substrate.

METHODOLOGY

CULTURE OF HYBRIDOMAS AND MONOCLONAL ANTIBODIES

The fact that each specific antibody is synthesized by a different cell line of B lymphocytes and their derived plasma cells makes it difficult to study antibody structure and antibody-antigen interactions. However, certain cells called *myelomas* can be used to produce large quantities of one type of antibody; *myeloma* cells can be cultured indefinitely in tissue culture techniques (normal cells tend to die off after a specific number of transfers in tissue culture media). Most myelomas produce antibody of unknown specificity. However, a technique was developed in 1975 by Cesar Milstein and Georges Köhler in England that fused B lymphocytes of known antibody production with myeloma cells. This cell fusion created immortalized hybrid cells, called *hybridomas*, that produce large amounts of *monoclonal antibody* of a particular specificity. A monoclonal antibody reacts with a single type of antigenic determinant. The hybridomas are formed, screened, and selected for the production of antibody specific for a particular antigen. The clones of these hybridoma cells produce, in tissue culture, large amounts of specific monoclonal antibody (see Figure).

These highly specific monoclonal antibodies are useful in clinical procedures and may prove useful in the treatment of some diseases. If, for example, monoclonal antibodies could be made from antigens that are unique to particular pathogens, these antibodies could be used to treat specific diseases. Monoclonal antibodies are already used for the diagnosis of allergies and infectious diseases, such as hepatitis, rabies, and some venereal diseases. Early stages of cancer may also be detectable with monoclonal antibodies because certain types of cancer cells have surface antigens that differ from those of normal cells. It is possible that, in the future, monoclonal antibodies will be used to treat cancer and infectious diseases. Monoclonal antibodies, used alone or chemically attached to drugs, could locate and destroy a cancer cell or pathogenic microorganism without damaging the healthy tissue surrounding it.

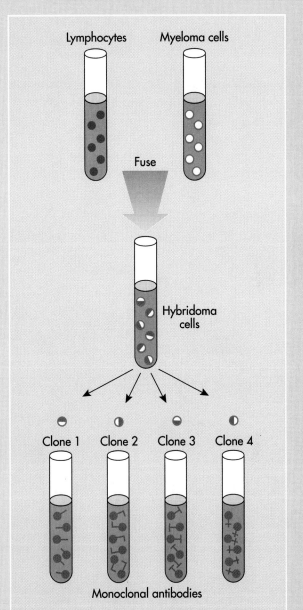

Monoclonal antibodies are produced by fusing lymphocytes with specific IgG encoding genes with myeloma cells to form hybridoma cells. Clones of these hybridoma cells reproduce rapidly and express single antibodies.

CELL-MEDIATED IMMUNITY

Whereas the antibody-mediated immune response system recognizes substances that are outside host cells, the **cell-mediated immune response** is effective in recognizing modified host cells (FIG. 14-27). Cell-

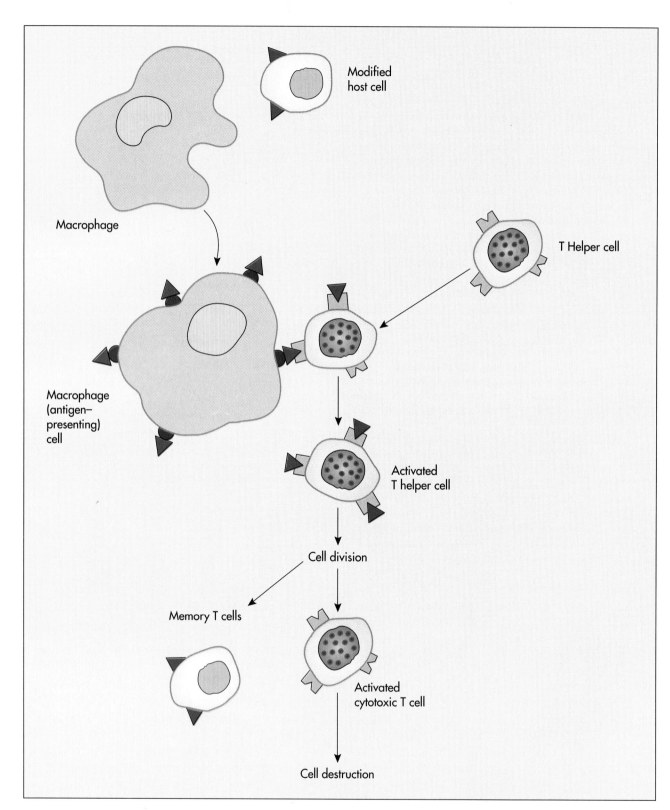

FIG. 14-27 The cell-mediated response recognizes modified cells and helps eliminate these from the body.

mediated immunity is important in controlling infections where the pathogens are able to reproduce within human cells. These include infections caused by viruses, some bacteria such as rickettsias and chlamydias, and some parasitic protozoa such as trypanosomes. Cell-mediated immune responses include: (1) delayed-type hypersensitivity in response to intracellular bacteria such as *Listeria monocytogenes* and *Mycobacterium tuberculosis* and fungal infections such as *Histoplasma capsulatum* and *Cryptococcus neoformans*; (2) cytotoxic T lymphocyte response to virally infected cells, tissue transplants, and tumors; and (3) response to tumor cells and tissue grafts by natural killer (NK) cells. In addition, the cell-mediated immune response may be important in surveillance for and destruction of naturally occurring tumor cells.

In a number of intracellular viral infections, antibodies are ineffective. In such cases, cell-mediated immunity augments the antibody-mediated immune response. Although antibody molecules can neutralize free viruses, antibodies are unable to penetrate and attack viruses multiplying within host cells. It is the cell-mediated immune response that has the capability of eliminating cells infected with viruses. The principal specific defense mechanism against well-established viral infections is provided by cytotoxic T lymphocytes (T_C). These cells recognize viral antigens associated with class I MHC molecules on the surfaces of infected cells. T_C lyse virally infected cells. In some viral infections the resultant tissue damage is often the product of the activities of these T_C rather than the virus itself.

ACTIVATION OF T CELLS

The cell-mediated immune response is based on a variety of lymphocytes, including the T_H cells, T_C cells, T_{DTH}, and NK cells described earlier in this chapter. When T cells are activated they are able to recognize foreign or abnormal cells. T helper cells have a central role in the activation of this system. They, like other T cells, have specific surface receptors that mediate their interaction with other cells. Activation of T helper cells results in the secretion of cytokines or lymphokines, which are the effector molecules of the cell-mediated immune response.

There are various results from the activation of T lymphocytes and other cells of the cell-mediated immune response. T_H cells are needed for B cells to synthesize antibody to protein antigens. T_C cells can lyse virally infected cells by coming in contact with those cells and secreting perforins and toxic proteins. NK cells produce substances that lyse tumor cells and virally infected cells and other substances that regulate the production of blood cells in the bone marrow. T_{DTH} cells secrete cytokines that work

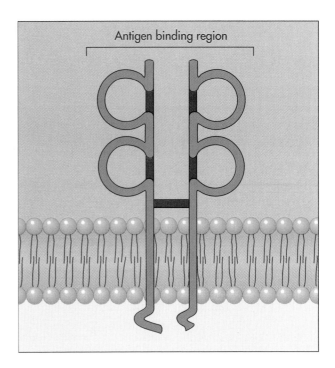

FIG. 14-28 Structure of the T cell receptor found on the surface of T lymphocytes.

on macrophages and cause them to become activated.

Different T cells are specific for one kind of antigen. Whereas B cells recognize antigen by their surface immunoglobulin molecules, T cells recognize antigens by specific surface molecules called **T cell receptors** (TCRs) (FIG. 14-28). These TCRs are very similar in structure to immunoglobulin molecules. TCR cannot recognize antigens in a free, soluble form. It can only recognize antigens that are complexed with MHC molecules on antigen-presenting cells. The TCR actually recognizes both the antigen and the MHC molecule on the antigen-presenting cell. When antigens are bound to an APC, the APC secretes interleukin-1 (IL-1), which in turn activates the T cell.

CYTOKINES

Activated T lymphocytes and other leukocytes respond to antigen recognition by secreting small effector molecules called **cytokines** (FIG. 14-29). These effector molecules are called *lymphokines* if they are produced by lymphocytes, or *monokines* if they are produced by monocytes or macrophages. In addition to activated T lymphocytes, macrophages and fibroblasts can secrete specific cytokines. Cytokines are generally produced during the effector phase of natural or specific immunity and have stimulatory or in-

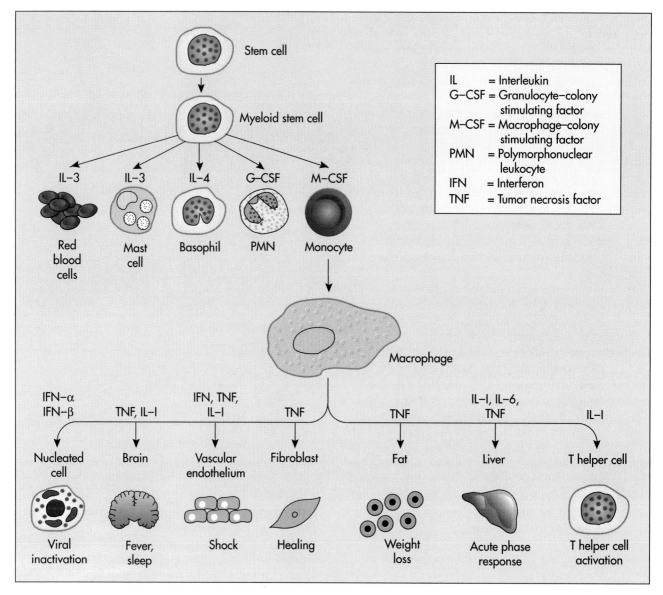

FIG. 14-29 Cytokines produced by activated macrophages help regulate the cell-mediated immune response. Other cytokines (not shown) produced by lymphocytes (lymphokines) also regulate the cell-mediated response.

hibitory effects on other cells of the immune system. Cytokines serve as communication and regulatory functions between different immune cells.

T cells produce cytokines when they are stimulated.

Various cytokines are produced by different T cells (Table 14-5). These include: (1) interleukin-2 (IL-2), which stimulates reproduction of T cells; (2) macrophage chemotactic factor (MCF), which causes the attraction and accumulation of macrophages to the site of lymphokine release; (3) migration inhibition factor (MIF), which inhibits macrophages from migrating once they have reached the site of lymphokine attraction; (4) macrophage activating factor (MAF), which results in an alteration of macrophage

cells that increases their lysosomal activities and thus their ability to kill and ingest organisms; (5) skin reactive factor, which enhances capillary permeability and thus the movement of monocytes across the vascular spaces; (6) immune interferon (IFN-γ), which activates antiviral proteins, preventing further intracellular multiplication of viruses; and (7) macrophage colony stimulating factor (CSF), which regulates the production of macrophages.

Activated T lymphocytes secrete **interleukin-2** (IL-2); IL-2 is secreted and can be reabsorbed by the same cell. IL-2 is, thus, an autocrine or self-stimulator. As a result of T cell activation and cytokine production, T cell metabolism is stimulated, and enhanced mitosis leads to cell division and expansion of T cell clones.

TABLE 14-5	
Effects of Cytokines on Various Cells	
LYMPHOKINE	**MODE OF ACTION**
Macrophage chemotactic factor	Attracts macrophages to site of lymphokine release, typically causing an accumulation of macrophages at a site of inflammation
Migration inhibition factor (MIF)	Inhibits migration of macrophages away from the site of inflammation
Macrophage activating factor (MAF)	Stimulates phagocytic activity of macrophages (γ-interferon) by enhancing their lysosomal activities
Immune interferon (IFN-γ)	Activates antiviral proteins, T cells, and macrophages
Interleukin-1 (IL-1)	Stimulates activities of T cells, B cells, and macrophages
Interleukin-2 (IL-2)	Stimulates antigen activated T helper cells
Interleukin-3 (IL-3)	Stimulates growth and differentiation of various blood cells
Interleukin-4 (IL-4)(B cell growth factor)	Stimulates production of B cells
Interleukin-5 (IL-5)	Activates eosinophils and activates and stimulates growth of B cells
Interleukin-7 (IL-7)	Stimulates growth and differentiation of B cells
Leukocyte inhibitory factor	Inhibits migration of phagocytic neutrophils away from the site of infection
Leukocyte chemotactic factor	Attracts phagocytic neutrophils to the site of infection
Skin reactive factor	Enhances capillary permeability and movement of monocytes across vascular spaces
Platelet-activating factor	Activates platelets to aggregate
Eosinophil chemotactic factor	Attracts phagocytic neutrophils to the site of parasitic infection

Type I interferons (IFN-α and IFN-β) are produced by various human cells in response to viruses. These cytokines produce an anti-viral state in noninfected neighboring cells. In addition, natural killer cells are capable of killing and lysing a variety of cells that contain viruses. One of the specific cytokines, **immune** or **gamma** (γ-interferon [IFN-γ]) is secreted by lymphocytes in response to a specific antigen to which they have been sensitized or stimulated to divide. In some of its physiological effects, this immune interferon is different from other interferon molecules and may kill tumor cells. Like other interferons,

immune interferon molecules have antiviral activities, stimulating the synthesis of antiviral proteins, including 2,5 adenylate polymerase. This polymerase, when bound to double-stranded RNA, a viral replicative intermediate, activates an endonuclease that can cleave viral RNA. The primary function of IFN-γ, though, appears to be different from that of α and β interferons. Immune interferon may regulate the proliferation of the lymphoid cells that are stimulated to divide in response to interactions with antigenic biochemicals. Immune interferon may also enhance phagocytosis by macrophages as well.

IMMUNOLOGICAL TOLERANCE

While the immune system is essential for protection against "foreign substances," it is critical that this system not attack the body's own cells. The immune system is designed to ignore *self-antigens*, that is, antigens associated with one's own cells.

Unresponsiveness to self-antigens is called **immunological tolerance**. Tolerance of T helper lymphocytes is a central mediator of overall immunologi-

cal tolerance because T_H cells are required for activation of other T lymphocytes, as well as B lymphocytes. Tolerance in T_H and T_C lymphocytes occurs during the maturation of these cells in the thymus. Many self-reacting clones of T cells are systematically recognized and destroyed in the thymus gland and therefore never get the opportunity to circulate in the blood or be deposited in peripheral lymphoid tissues such as the spleen. This type of tolerance is called *clonal deletion.*

Alternatively, tolerance may be induced by the development of T cells that are *anergic* or unresponsive to self-antigens. **Anergy** means *without working.* Anergy shuts off self-reactive cells as a result of the way in which the antigens are presented. CD4+ T helper cells require activation by antigen bound to MHC class II molecules, as well as by other stimulatory factors. Some T_H cells may recognize antigen-presenting cells that carry self-antigens but do not respond to that antigen because of a lack of accessory stimulator molecules. Anergy induction occurs when a T cell is presented with an antigen-specific signal but without a second cosignal. Such T cells fail to produce IL-2 and do not divide. They may produce other cytokines as well as inhibitory proteins.

The development of immunological tolerance in B lymphocyte clones occurs by a similar mechanism as seen in T lymphocytes. Since their maturation occurs in the bone marrow, it is likely that specific recognition and deletion of self-antigen specific B cell clones occurs in that site. Also, clonal unresponsiveness may arise in B cell populations, forming *anergic B cell clones,* which are B cells that do not respond to antigenic stimulation.

MEMORY RESPONSE—BASIS OF ACQUIRED IMMUNITY

Because relatively few B and T cells with the appropriate receptors are present at the first exposure to a given antigen, the primary antibody-mediated immune response is characteristically slow, producing relatively low yields of antibody and cell-mediated immune responses. There is a long lag period in the **primary immune response,** during which selection, differentiation, and cloning of appropriate B and T cell lines must occur (FIG. 14-30). After an antigen elicits an immune response, there is an increase in the number of B and/or T lymphocytes capable of reacting with that antigen. Because antigen binds to and selects for cells having receptors of the highest affinity, this process results in an increase in the number of lymphocyte cells with receptors of high affinity for the particular antigenic molecule. The cloning of these cells establishes a bank of secondary lymphocytes that act as memory cells. Memory cells are long-lived resting cells. They do not secrete antibody.

On subsequent exposure to the same antigen, perhaps years later, the memory cells are activated and rapidly divide into a clone. The clone of B or T cells produces a large population of cells that can initiate the **secondary immune response** rapidly and efficiently (see FIG. 14-30). The secondary immune response is usually faster, longer lasting, and of greater intensity (more extensive production of immunoglobulin by B cells or activation of T cells) than the primary immune response. The secondary immune response is also called a **memory response** or **anamnestic response.**

When an individual's immune system is active in responding to the foreign antigen, the immune response is called **active immunity** (FIG. 14-31). Active immunity may occur naturally as the result of an infection or may be artificial as the result of vaccination. The term *active-acquired immunity* indicates production of antibody within the individual's body as a result of the learned (acquired) ability to recognize antigens. Such acquisition may be natural as a result,

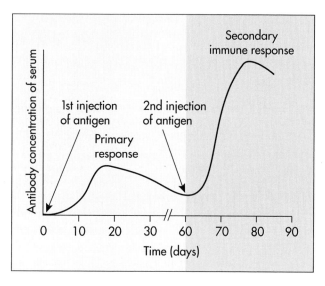

FIG. 14-30 The primary immune response is characterized by a long lag period and a slow production of antibodies. The secondary immune response has a short lag period and rapid production of antibodies.

for example, of an infection—in which case, it is called *natural immunity. Artificial immunity,* in contrast, indicates medical treatment as opposed to natural means of acquiring immunity. Vaccination to prevent disease is discussed in Chapter 16.

> **Active-acquired immunity involves the development of a memory system and the ability of the body to respond rapidly to specific antigens.**

Immunity also may be conferred on an individual by the transfer of serum, secretions such as milk and colostrum, or immune cells from another individual (see FIG. 14-31). The recipient of such a transfer becomes immune to specific foreign antigens without

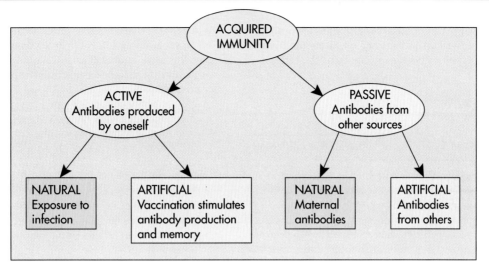

FIG. 14-31   In active immunity the body produces its own immunoglobulins. In passive immunity the immunoglobulins are obtained from another source.

his or her immune system being activated. This form is called **passive immunity.** The individual does not produce his or her own antibodies and does not acquire a memory response in this case. *Natural passive immunity* occurs when a fetus acquires antibodies from its mother. An infant does not initially produce its own antibodies but relies initially on the maternal antibodies that crossed the placenta. Within a few months these will have been eliminated from the infant's circulatory system unless they are replenished, for example, from human breast milk. Transfer of antibodies from mother to child during breast-feeding is an example of natural passive immunity. Passive immunity may also be artificial through the medical administration of antibodies. The administration of antitoxin for treating snake bites is an example of such artificial passive immunity. Preventing disease through the administration of antibodies is discussed in Chapter 15.

Passive immunity occurs when antibody production has occurred outside the individual's body.

VIRTUES OF BREAST-FEEDING

Breast-fed infants tend to be healthier than those given formula. The reason is that human milk and colostrum (the milk produced a few days before and after childbirth before production of normal human milk) contains IgA and macrophages that help defend against infection. Over 80% of the cells in colostrum are macrophage that can phagocytize pathogenic microorganisms. Because a breast-feeding woman is exposed to many of the same antigens as the infant, she is likely to make immunoglobulins that specifically protect the infant. The gastrointestinal tract of the infant is especially prone to infection because the normal microbiota have not yet established a stable and protective community. IgA helps protect the surface of the gastrointestinal tract of the infant against infection. Bacterial and viral infections that cause infant diarrhea are prevented by the IgA in colostrum.

SUMMARY

Specific Host Defenses Against Microbial Infections—the Immune Response (pp. 398-399)

- The specific immune response is characterized by specificity, memory, and the acquired ability to detect and to eliminate foreign substances.
- The immune response detects substances that are foreign to the body.
- The immune response comprises antibody-mediated and cell-mediated immunity.

Antigens (pp. 399-400)

- Antigens activate the immune response and react with the cells and chemicals of the immune response system.
- An epitope or antigenic determinant is the part of the antigen molecule that interacts with an antibody.
- Some molecules called haptens have antigenic determinants but are too small to elicit the formation of antibodies by themselves.

- Antigens exhibit immunogenicity (ability to stimulate the immune system) and specific reactivity (ability to react with specific effector molecules of the immune system).

Lymphocytes (pp. 400-406)

- Lymphocytes are differentiated into B lymphocytes (B cells), T lymphocytes (T cells), natural killer cells (NK cells), and killer cells (K cells); all have different roles in the immune defenses of the body.
- Lymphocytes are differentiated based on the presence of specific cell surface proteins or antigens bound to their plasma membranes.
- A surface marker that identifies a specific line of cells or a stage of cell differentiation because it interacts with a group or cluster of individual antibodies is called a CD (cluster of differentiation) antigen.
- B lymphocytes (B cells), which are differentiated in bone marrow, give rise to plasma cells when stimulated by antigen.
- Plasma cells produce and release antibodies. Each clone of a plasma cell line secretes a single specific antibody.
- T lymphocytes are differentiated in the thymus.
- Cytotoxic T cells ($CD8^+$ cells) lyse abnormal cells.
- Helper T cells ($CD4^+$ cells) activate the immune response.
- Suppressor T cells ($CD8^+$ cells) regulate the immune response.
- Natural killer or null cells are a special subset of lymphocytes that are neither T nor B cells.
- Antigen-presenting cells (APCs) include macrophages and other specialized cells (dendritic cells) in skin and lymphoid organs. APCs help activate the immune response. APCs bind antigen on their surfaces in association with a special protein called the major histocompatibility complex (MHC).
- B cells that have processed antigens become antigen-presenting cells that activate the immune response.
- T cells respond to antigens associated with MHC on antigen-presenting cells.

Antibody-mediated Immunity (pp. 407-421)

- Antibody-mediated immunity is important in preventing and eliminating microbial infections.
- In the antibody-mediated immune response, specific glycoproteins called antibodies or immunoglobulins are made when foreign antigens activate B cells.
- The key to antibody immunity is the ability of antibodies to react specifically with antigens.
- The activation of a particular B cell leads to clonal selection and expansion of B cells specific to an individual antigen.

Antibodies (Immunoglobulins) (pp. 409-415)

- Antibodies (immunoglobulins) are glycoproteins that are made in response to specific antigens and react with those antigens.
- Antibodies (immunoglobulins) recognize and bind to specific antigens.
- Immunoglobulins have two light and two heavy chains that are linked together.
- The Fab portion of an immunoglobulin reacts with specific antigens.

- There are five classes of immunoglobulins: IgG is the most common; IgE is the least common; IgM is made early in the course of an infection; IgA is important in protecting surface tissues; and the precise role of IgD is unknown.
- Monoclonal antibodies are the result of fusing myeloma cells with lymphocyte cells to produce hybridomas. Hybridomas synthesize specific antibodies. Clones continue to produce the same antibody characteristic of the parent cell.

Antigen-antibody Reactions (pp. 415-421)

- Increased phagocytosis associated with antibody-bound antigen is called opsonization.
- Antitoxins are antibodies produced against toxins.
- Neutralization is the result of the formation of an antigen(toxin)-antibody(antitoxin) complex.
- Toxoids are denatured protein toxins that have lost their toxicity but retain antigenicity.
- Serological reactions are *in vitro* antigen-antibody reactions; they are used for clinical diagnoses.
- Aggregation or clumping of the cells, microorganisms, or particles due to antigen-antibody reaction is called agglutination.
- Antibodies that combine with antigens on cell or particle surfaces are called agglutinating antibodies or agglutinins.
- Hemagglutination is agglutination of red blood cells.
- Enzyme-linked immunosorbent assay (ELISA) uses labels attached to antibody molecules to identify them.

Cell-mediated Immunity (pp. 422-425)

- The cell-mediated immune response depends on natural killer cells and various T lymphocytes.
- T cells recognize antigens by specific surface molecules called T cell receptors (TCRs).
- T cells produce cytokines when they are stimulated.
- Cytotoxic and aggressor T cells detect abnormal cells by antigens on tissue cell surface. These antigens are called major histocompatibility complex antigens. These MHC antigens are important in the distinction between self- and nonself tissues.

Immunological Tolerance (pp. 425-426)

- Unresponsiveness to self-antigens is called immunological tolerance.
- Clonal selection results in populations of B and T cells that specifically respond to foreign and not self-antigens.

Memory Response—Basis of Acquired Immunity (pp. 426-427)

- Acquired immunity protects the body against pathogens the immune system has learned to recognize as foreign.
- There is a long lag period in the primary immune response, during which selection, differentiation, and cloning of appropriate B and T cell lines must occur.
- The cloning of B lymphocytes capable of reacting with a specific antigen creates a bank of memory cells that can initiate a secondary immune response whenever that antigen reappears in the body.
- The secondary immune response is also called a

memory response or anamnestic response; it is characterized by a short lag period and rapid response.
- When an individual's immune system is active in responding to the foreign antigen, the immune response is called active immunity.

- Active-acquired immunity involves the development of a memory system and the ability of the body to respond rapidly to specific antigens.
- Passive immunity occurs when antibody production has occurred outside the individual's body.

CHAPTER REVIEW

REVIEW QUESTIONS

1. What is an antigen?
2. What is an antigenic determinant?
3. What is the difference between an antigen and a hapten?
4. What characteristics are similar for all lymphocytes? What characteristics are different for different subclasses of lymphocytes?
5. What roles do B cells have in the immune defense?
6. What roles do different T cells have in the immune defense?
7. How do helper T cells regulate immune reactions?
8. What are antigen-presenting cells? Why are they important in the immune reaction?
9. What is an antibody? How does the reaction of antibody with antigen defend the body against disease?
10. Describe the structure of an immunoglobulin.
11. What are the five classes of immunoglobulins? Compare the role of each immunoglobulin class in the immune response.
12. What types of reactions occur between antigen and antibody? How are these important in the natural defense of the body against disease? How are they used in clinical diagnosis?
13. What is a cytokine (lymphokine)?
14. What functions do cytokines have in the immune response?
15. Discuss the differences between a primary and a secondary (memory) immune response.
16. What is the difference between active and passive immunity? Between natural and artificial immunity?

CRITICAL THINKING QUESTIONS

1. How does the body learn to differentiate self-antigens from foreign antigens? How could one develop autoimmunity in which the body's immune response mistakenly responds to self-antigens?
2. What is the basis of memory and specificity that characterize the immune response? How do each develop? What happens when the body detects a foreign antigen?
3. What immune reactions occur when a virus infects the body? What immune reactions occur when a bacterium infects the body?
4. Discuss the differences between the antibody-mediated and cell-mediated immune response systems. Why do we need two elaborate defense systems?

5. If one develops active immunity against a pathogen, can one ever contract the disease caused by that pathogen? If so, how?
6. How can pathogens evade the immune response? How can some pathogens, such as the protozoan that causes malaria (Plasmodium) persist in the body for prolonged periods? How can some individuals become carriers of disease, for example, of typhoid?
7. What are the likely consequences of an increasing population of immunocompromised individuals who lack an effective immune response? (The Centers for Disease Control estimates that over the next decade up to 10% of the American population will be immunocompromised.)

READINGS

Abbas AK, AH Lichtman, JS Pober: 1991. *Cellular and Molecular Immunology,* Philadelphia, W.B. Saunders.
   A text that emphasizes the molecular basis of the immune response.
Baron S, F Dianzani, GJ Stanton, WR Fleischmann (eds.): 1987. *The Interferon System,* Austin, University of Texas Press.
   The third in a series of reviews about the interferon system does not repeat information in earlier reviews. Covers induction, large-scale production, purification and characterization, cellular, physiologic, molecular mechanisms, and clinical uses of interferon.
Engleman EG, SKH Foung, J Larrich, A Raubitschek (eds.): 1985. *Human Hybridomas and Monoclonal Antibodies,* NY, Plenum Press.
   Presents details on the methodology of human hybridoma production for investigators using these techniques.

Feedman M, J Lamb, MJ Owen (eds.): 1989. *T Cells,* New York, John Wiley and Sons.
   Reviews the normal physiology of T lymphocytes and the role of T lymphocytes in disease, with emphasis on the molecular explanation of T cell function.
Jaret P: 1986. Our immune system: The wars within, *National Geographic,* June, pp. 702-736.
   A very readable account of current progress in the study of the human immune system, with striking photographs by Lennart Nilsson.
Johnson HM, JK Russell, CH Pontzer: 1992. Superantigens in human disease, *Scientific American* 266(4):92-101.
   A discussion of antigens that can stimulate multiple cells of the immune response and their involvement in human disease.

Leder P: 1982. The genetics of antibody diversity, *Scientific American*, May, pp. 102-116.

An account of how a few hundred genes are shuffled to make millions of antibody combinations, by the man who first worked it out.

Marrack P and J Kappler: 1986. The T cell and its receptor, *Scientific American*, February, pp. 36-45.

An account of the discovery of the T cell receptor and what is known so far of its structure and function.

Milstein C: 1980. Monoclonal antibodies, *Scientific American* 243(4):66-74.

The fusion of particular antibody-producing cells with cancer cells produces clones of cells that secrete antibody directed at only one antigen.

Schwartz RH: 1993. T cell anergy, *Scientific American* 269(2):62-71.

An interesting discussion of the development of immunotolerance.

Sharon N and H Lis: 1993. Carbohydrates in cell recognition, *Scientific American* 268(1):82-89.

This article describes how sugars on the surfaces of cells enable cells to identify and interact with one another; it also shows how drugs that target such surface recognition sugars can be used to stop infection and inflammation.

Smith KA: 1990. Interleukin-2, *Scientific American* 262(3):50-57.

The article discusses the role of the cytokine interleukin-2 and regulation of the immune response and its possible role in treating disease.

Tonegawa S: 1985. The molecules of the immune system, *Scientific American* 253(4):122-131.

An account of what is currently known of the structure of the B and T receptors.

Verma IM: 1990. Gene therapy, *Scientific American* 263(5):68-84.

A discussion of the use of human gene therapy to treat genetic diseases in order to form a functional immune system.

Watson RR: 1984. *Nutrition, Disease Resistance and Immune Function*, New York, Marcel Dekker.

Immunological measures can assess nutritional stress. Describes interactions of disease organisms and nutritional stresses with changes in the host's physiology and immunology.

Young JD-D and ZA Cohn: 1988. How killer cells kill, *Scientific American* 258(1):38-47.

Killer lymphocytes attack tumor cells and cells infected by viruses. They kill by secreting proteins that link to form pores in target cells that then leak to death.

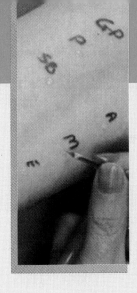

CHAPTER 15

Immune Response and Human Disease

PREVIEW TO CHAPTER 15

In this chapter we will:
- Discover how the body's immune defense system can be used to protect against infection and/or disease.
- Examine some limitations of the immune response.
- Review various diseases that result from immunodeficiencies.
- See that the failure to recognize self-antigens results in autoimmunity.
- Observe that excessive immune reactions cause hypersensitivity and allergy.
- Consider that it is sometimes necessary to suppress the immune system.
- Learn the following key terms and words:

acquired immunodeficiency syndrome (AIDS)
adjuvants
allergen
allergy
anaphylactic hypersensitivity (anaphylaxis) (type I hypersensitivity)
anaphylactic shock
antibody-dependent cytotoxic hypersensitivity reactions (type II hypersensitivity)
Arthus immune-complex reactions
asthma
attenuated vaccine
autoimmune diseases
autoimmunity
booster vaccinations
Bruton congenital agammaglobulinemia
cell-mediated or delayed hypersensitivity (type IV hypersensitivity) reactions
contact dermatitis
DiGeorge syndrome
erythroblastosis fetalis
gammaglobulin
generalized anaphylaxis
glomerulonephritis
graft-versus-host (GVH) disease

granuloma
haptens
hemolytic disease of the newborn
hypersensitivity
immediate hypersensitivity
immune complex-mediated hypersensitivity (type III hypersensitivity)
immunization
immunodeficiency
inactivated vaccine
inhalation
late-onset hypogammaglobulinemia
myasthenia gravis (MG)
orally
Rh incompatibility
rheumatoid arthritis
sensitization
serum sickness
severe combined immunodeficiency (SCID)
systemic lupus erythematosus
tissue rejection
toxoids
universal donors
universal recipients
vaccination
vaccines
vector vaccines

PREVENTION OF DISEASES USING THE BODY'S IMMUNE RESPONSE

Of the many diseases that plagued societies for centuries, smallpox was among the most serious. Numerous individuals contracted and died of this viral disease. Those who survived, however, no longer seemed to be susceptible. They had become resistant to infection with the smallpox virus. Even without any knowledge of the immune response, some individuals reasoned that it was possible to acquire immunity (resistance to disease). In China, where many herbs and other substances were long used to treat disease, children inhaled dried scabs from smallpox victims to protect them against serious smallpox infections. Those that developed mild cases of smallpox and survived were subsequently resistant (immune) to this disease. This practice was carried out as early as the thirteenth century and, by the early eighteenth century, individuals throughout the Far East were exposing themselves to smallpox viruses to develop immunity.

In Turkey, elderly women collected material from the sores (pustules) of mild cases of smallpox and placed it into a walnut shell. A small amount of the material was then injected into a vein of an individual to protect her or him from smallpox. The injected individual would become ill within a week, developing fever, sores, and pustules. But in another week, the sores and pustules would heal. There generally was no permanent scarring and after recovery the individual was resistant to smallpox.

This practice of ingrafting was introduced in England in 1718 by Lady Mary Montagu, whose husband had been the British ambassador to Turkey. Lady Mary was a striking English beauty until the age of 26 when she contracted smallpox. Although she survived, she was permanently scarred and bemoaned that "my beauty is no more." It was perhaps because of her experience with smallpox that Lady Mary became interested in ingrafting when she was living in Constantinople. In a letter to her family she wrote "The Small Pox, so fatal and general amongst us, is here entirely harmless by the invention of ingrafting." When she returned to England, Lady Mary used her considerable influence in the court of King George I to gain publicity for the increased use of ingrafting. She even arranged a test of her idea on prisoners and orphans, then a common practice. Today such testing of humans is viewed as unethical. Despite her efforts, ingrafting was not accepted by the scientists and physicians of the time as a useful practice for preventing disease. Too often, ingrafting resulted in scarring and in some cases it produced fatal cases of smallpox.

It was not until a report by Edward Jenner to the Royal Society of London, 80 years after Lady Mary tried to introduce ingrafting in England, that credence was given in Europe to the practice of immunization to protect against smallpox (FIG. 15-1). Jenner was a middle-class country doctor, whose inter-

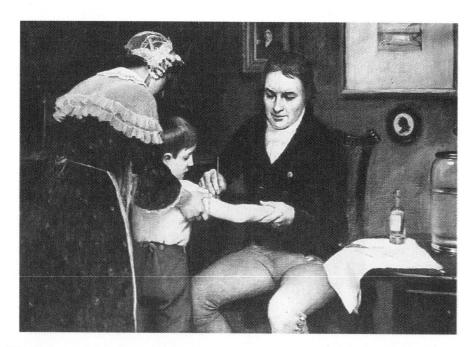

FIG. 15-1 Edward Jenner (1749-1823) vaccinated James Phipps (in about 1800) with cowpox material, resulting in the development of resistance to smallpox infection by the boy and thereby establishing the scientific credibility of vaccination to prevent disease.

est in science was typical of his position: scholarly but amateurish. Although it is unclear when he began to develop his ideas about vaccination, Jenner apparently did not develop them from the reports of Lady Mary Montagu about ingrafting in Turkey. He clearly knew of the dairy country's folk belief that cowpox, which is caused by vaccinia virus, protected its victims from subsequent infections of smallpox. In particular, milkmaids were known for their immunity to smallpox. Jenner also knew that Benjamin Jesty, a Dorset farmer, had vaccinated his wife and two children with cowpox material taken from a sore on the udder of an infected cow in 1774. Although smallpox was ravaging the population in the vicinity, it did not affect the Jesty family.

Jenner performed his first inoculation in 1796 on a child with material from the cowpox lesions of a dairymaid. Six weeks later, Jenner inoculated the boy with smallpox virus. The boy did not develop smallpox, indicating that he was resistant to the disease. In this way Jenner experimentally demonstrated the validity of his hypothesis that exposure to cowpox viruses made one resistant to infections with smallpox viruses. Jenner continued his experimental studies, inoculating other children with cowpox viruses. Although friends and other physicians advised him against damaging his reputation by publishing results that were at such variance with accepted knowledge, Jenner decided to present his findings to the Royal Academy. His June 1798 report, *An Inquiry into the Causes and Effects of the Variolae Vaccine,* describing the value of vaccination with cowpox as a means of protecting against smallpox established the basis for the immunological prevention of disease. His work met the criteria of the scientific method: his hypothesis had been experimentally tested by observations of control and experimental groups, and it was also repeatable by others. Immunization had gained scientific credibility; medical practice and the quest to eliminate smallpox had taken a giant step forward. The work begun with Jenner's discovery of the effectiveness of vaccination in preventing smallpox culminated in the 1970s with the eradication of smallpox from the face of the Earth. Today, vaccines are employed for preventing many diseases, such as tetanus, diphtheria, measles, and so forth. Research continues to develop new vaccines for preventing many other diseases.

IMMUNIZATION (VACCINATION)

The usefulness of immunization rests with its ability to render individuals resistant to a disease without actually producing the disease. This is accomplished by exposing the individual to antigens associated with a pathogen in a form that does not cause disease. This medical process of intentional exposure to antigens is called **immunization** or **vaccination.** Implementation of immunization programs has drastically reduced the incidence of several diseases and has greatly increased life expectancies (FIG. 15-2). Many once widespread deadly diseases such as whooping cough and diphtheria are rare today because of immunization programs.

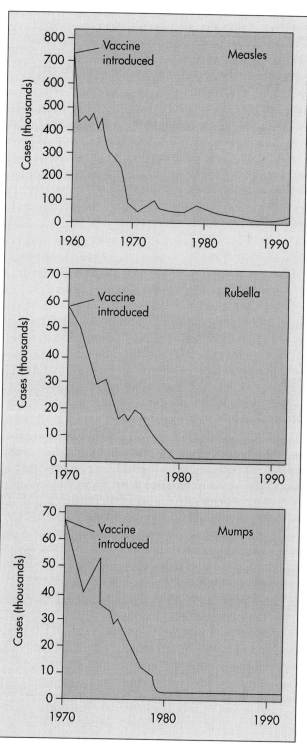

FIG. 15-2 Immunization has resulted in a great decline in the incidence of several diseases.

SHOULD THE GOVERNMENT MANDATE IMMUNIZATION?

During the last few years there have been debates in Congress and in the media as to whether the government should mandate immunization. Public health officials stress the importance of herd immunity in protecting the overall population against the spread of pathogens. They particularly emphasize the need to protect children against diseases such as polio, diphtheria, whooping cough (pertussis), and tetanus that can be fatal. Most municipalities have accepted this view and require proof of immunization against several diseases before a child can attend school. Some parents argue that because there are occasional adverse side effects from immunization, the government should not require the vaccination of children. Problems with the polio and pertussis vaccines that cause a low rate of side effects, but include some fatalities, resulted in lawsuits, and sizeable damage awards led some companies to cease making vaccines. Although a vaccine effective against chicken pox has been developed, it is not widely used to prevent this disease. Chicken pox normally is a relatively mild disease in childhood with few fatalities, but is more severe in adults. Booster vaccines would be needed to protect adults if, as children, they were vaccinated against chicken pox, so government agencies are not yet recommending its use.

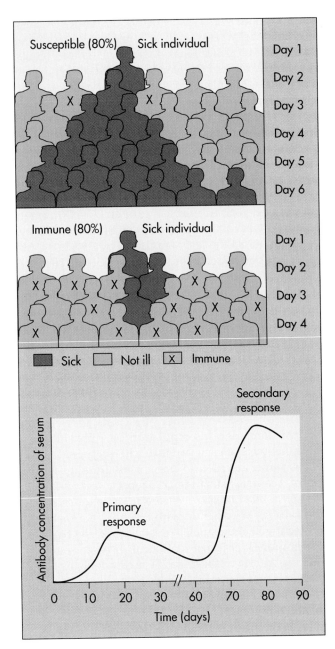

Sick ☐ **Not ill** ☒ **Immune**

FIG. 15-3 When only a few individuals are immune a pathogen can spread rapidly through a population, resulting in an epidemic. When over 70% of the individuals in a population are immune the propagation from individual to individual is not sustained and epidemics do not occur. This is because a sufficient number of individuals have developed a secondary response.

Scientific Basis of Immunization

There are several scientific principles underlying the use of immunization to prevent individuals from contracting specific diseases and for preventing epidemic outbreaks of diseases (FIG. 15-3):

1. Any macromolecule associated with a pathogen can be an antigen—an antigen is not the entire pathogen (see discussion of antigens in Chapter 14). Hence, one can use specific target antigens associated with pathogens to elicit the immune response without causing disease.

2. After exposure to an antigen the body may develop an anamnestic (memory) response (see discussion of anamnestic response in Chapter 14). Subsequent exposure to the same antigen can then bring about a rapid and enhanced immune response that can prevent replication of the infecting microorganism and/or the effects of toxins it produces so that disease does not occur. In this manner, intentional exposure to an antigen through vaccination can establish an anamnestic response that renders the individual resistant to disease.

3. When a sufficiently high proportion of a population is immune to a disease, epidemics do not occur. This is because individuals who are immune are no longer susceptible and thus no longer participate in the chain of disease transmission. When approximately 70% of a population is immune, the entire population generally is protected, a concept known as *herd immunity*.

Herd immunity can be established by artificially stimulating the immune response system through the use of vaccines, rendering more individuals insusceptible to a particular disease and thereby protecting the entire population.

Vaccines

Vaccines are preparations of antigens whose administration artificially establishes a state of immunity without causing disease. Vaccines are designed to stimulate the normal primary immune response. This results in a proliferation of memory cells and the ability to exhibit a secondary memory or anamnestic response on subsequent exposure to the same antigens. The antigens within the vaccine need not be associated with active virulent pathogens. The antigens in the vaccine need only elicit an immune response with the production of antibodies or cytokines. Antibodies and/or cytokine-producing T cells possess the ability to react with the critical antigens associated with the pathogens against which the vaccine is designed to confer protection.

> **Vaccines are preparations of antigens that stimulate the primary immune response, producing memory cells and the ability to exhibit an anamnestic response to a subsequent exposure to the same antigens without causing disease.**

Vaccines may contain antigens prepared by killing or inactivating pathogenic microorganisms; vaccines also may use attenuated or weakened strains that are unable to cause the onset of severe disease symptoms (Table 15-1). Some of the vaccines that are useful in preventing diseases caused by various microorganisms are listed in Table 15-2. Most vaccines are administered to children (Table 15-3), some are administered to adults (Table 15-4), and some are used for special purposes. Travellers receive vaccines against pathogens prevalent in the regions they are visiting that do not occur in their home regions (Table 15-5). In each of these cases the use of the vaccine is pro-

phylactic and aimed at preventing diseases caused by pathogens to which the individual may be exposed.

Although vaccines are normally administered before exposure to antigens associated with pathogenic microorganisms, some vaccines are administered after suspected exposure to a given infectious microorganism. In these cases the purpose of vaccination is to elicit an immune response before the onset of disease symptoms. For example, tetanus vaccine is administered after puncture wounds may have introduced *Clostridium tetani* into deep tissues, and rabies vaccine is administered after animal bites may have introduced rabies virus. The effectiveness of vaccines administered after the introduction of the pathogenic microorganisms depends on the relatively slow development of the infecting pathogen before the onset of disease symptoms. It also depends on the ability of the vaccine to initiate antibody production before active toxins are produced and released to the site where they can cause serious disease symptoms.

TABLE 15-2

Descriptions of Widely Used Vaccines

DISEASE	VACCINE
ANTIVIRAL VACCINES	
Smallpox	Attenuated live virus
Yellow fever	Attenuated live virus
Hepatitis B	Recombinant
Measles	Attenuated live virus
Mumps	Attenuated live virus
Rubella	Attenuated live virus
Polio	Attenuated live virus (Sabin)
Polio	Inactivated virus (Salk)
Influenza	Inactivated virus
Rabies	Inactivated virus
ANTIBACTERIAL VACCINES	
Diphtheria	Toxoid
Tetanus	Toxoid
Pertussis	Acellular extract from *Bordetella pertussis*
Meningococcal meningitis	Capsular material from 4 strains of *Neisseria meningitidis*
Haemophilus influenzae type b (Hib) infection	Capsular material from *Haemophilus influenzae* type b conjugated to diphtheria protein
Cholera	Killed *Vibrio cholerae*
Plague	Killed *Yersinia pestis*
Typhoid fever	Killed *Salmonella typhi*
Pneumococcal pneumonia	Capsular material from 23 strains of *Streptococcus pneumoniae*

TABLE 15-1

Comparison of Attenuated and Inactivated Vaccines

FACTOR	ATTENUATED/ LIVE	INACTIVATED/ NONLIVING
Route of administration	Natural route, e.g., orally	Injection
Doses	Single	Multiple
Adjuvant	Not required	Usually needed
Duration of immunity	Years to life	Months to years
Immune response	IgG, IgA, IgM, cell mediated	IgG, little or no cell mediated

TABLE 15-3

Recommended Vaccination Schedule for Normal Children

VACCINE	ADMINISTRATION	RECOMMENDED AGE	BOOSTER DOSES
Diphtheria, pertussis, tetanus (DPT)	Intramuscular injection	2, 4, 6, and 15 months	One intramuscular booster at 4-6 years; tetanus and diphtheria booster at 14-16 years
Measles, mumps, rubella (MMR)	Subcutaneous injection	15 months	One subcutaneous booster of MMR or just the measles portion at 4-6 years
Haemophilus influenzae type b (Hib) conjugate	Intramuscular injection	2, 4, 6, and 15 months	None
Hepatitis B	Intramuscular injection	2, 6, and 18 months	None
Polio (Sabin)	Oral	2, 4 , and 15 months (also 6 months for children in high-risk areas)	One oral booster at 4-6 years

TABLE 15-4

Recommended Vaccination Schedule for Adults

VACCINE	ADMINISTRATION	RECOMMENDATIONS
Tetanus, diphtheria (Td)	Intramuscular Td injection	Repeated every 10 years throughout life
Adenovirus types 4 and 7	Intramuscular	For military population only
Influenza	Intramuscular	For individuals over 65 years old; individuals with chronic respiratory or cardiovascular disease
Pneumococcal	Intramuscular or subcutaneous	For individuals over 50 years old, especially those with chronic diseases
Staphylococcal	Subcutaneous, aerosol inhalation, oral	For treatment of infections caused by *Staphylococcus*

TABLE 15-5

Recommended Vaccinations for Travellers

VACCINE	ADMINISTRATION	RECOMMENDATIONS
Cholera	Intradermal, subcutaneous, or intramuscular	For individuals travelling to or residing in countries where cholera is endemic
Plague	Intramuscular	Only for individuals at high risk of exposure to plague
Typhoid	Oral; (booster, intradermal)	For individuals travelling to or residing in countries where typhoid is endemic; booster is recommended every 3 years
Yellow fever	Subcutaneously	For individuals travelling to or residing in countries where yellow fever is endemic; a booster is recommended every 10 years

Attenuated Vaccines

Some vaccines consist of living strains of microorganisms that do not cause disease. Such strains of pathogens are said to be **attenuated** because they have weakened virulence. Pathogens can be attenuated, that is, changed into nondisease-causing strains, by various procedures, including moderate use of heat, chemicals, desiccation, and growth in tissues other than the normal host. Vaccines containing viable attenuated strains require relatively low amounts of the antigens because the microorganism is able to replicate after administration of the vaccine, resulting in a large increase in the amount of antigen available within the host to trigger the immune response mechanism. The principle disadvantage of living attenuated vaccines is the possible reversion to virulence through mutation or recombination. Also, even strains may cause disease in individuals who lack adequate immune responses, such as those with AIDS.

Vaccines can be made using live attenuated microbial strains.

Most attenuated vaccines are for viral diseases. These vaccines have been developed by screening for mutant viruses that have diminished to negligible virulence. These vaccine strains often are developed by culture at low temperature and by repeated passage (replication cycles) through cells cultured *in vitro* (animal cell tissue culture) (FIG. 15-4). The Sabin polio vaccine, for example, uses viable polioviruses attenuated by growth in tissue culture. Three anti-

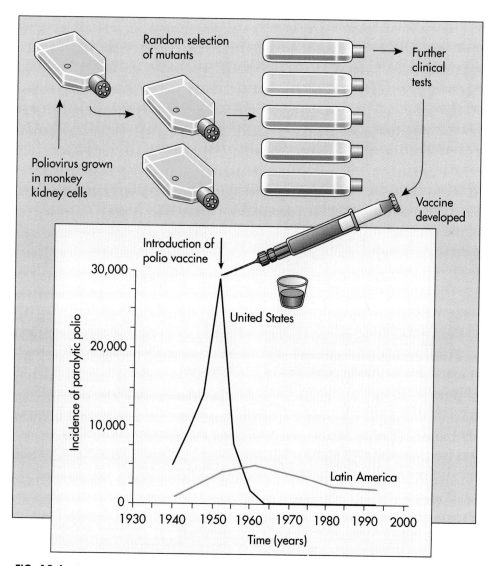

FIG. 15-4 Attenuated vaccines can be produced by multiple passage through animal cell tissue culture. Poliovirus for the oral vaccine was developed by passage through monkey kidney cells. Following clinical trials the use of the vaccine greatly reduced the incidence of polio in the United States. Elsewhere, where the vaccine is not used, the incidence of polio remains high.

PASTEUR DEVELOPS AN ATTENUATED VACCINE FOR RABIES

Louis Pasteur furthered the development of vaccines when in 1880 he reported that attenuated microorganisms could be used to develop vaccines against chicken cholera. The production of these vaccines depended on prolonging the time between transfers of the cultures. This fact was accidentally discovered through an error by Charles Chamberland, who used an old culture during one of the experiments he was conducting with Pasteur. The old culture contained attenuated microorganisms, that is, weakened or altered microorganisms that were less virulent. Following his work on chicken cholera, Pasteur directed his attention to the study of anthrax. Because he enjoyed being the center of attention and controversy, Pasteur staged a dramatic public demonstration to test the effectiveness of his anthrax vaccine. Witnesses were amazed to see that the 24 sheep, 1 goat, and 6 cows that had received the attenuated vaccine were in good health, whereas all of the animals in this experiment that had not been vaccinated were dead of anthrax.

In 1885, Pasteur announced to the French Academy of Sciences that he had developed a vaccine for preventing another dread disease, rabies (see Figure). Although he did not understand the nature of the causative organism, Pasteur developed a vaccine that worked. Pasteur's motto was "Seek the microbe," but the microorganism responsible for rabies is a virus, which could not be seen under the microscopes of the 1880s. Pasteur, none the less, was able to weaken the rabies virus by drying the spinal cords of infected rabbits and allowing oxygen to penetrate the cords. Thirteen inoculations of successively more virulent pieces of rabbit spinal cord were injected over a period of 2 weeks during the summer of 1885 into Joseph Meister, a 9-year-old boy who had been bitten by a rabid dog.

"Since the death of the child was almost certain, I decided in spite of my deep concern to try on Joseph Meister the method which had served me so well with dogs... I decided to give a total of 13 inoculations in 10 days. Fewer inoculations would have been sufficient, but one will understand that I was extremely cautious in this first case. Joseph Meister escaped not only the rabies that he might have received from his bites, but also the rabies which I inoculated into him."

The development of the rabies vaccine crowned Pasteur's distinguished career.

genically distinguishable strains of polioviruses are used in the Sabin vaccine. These viruses are capable of multiplication within the digestive tract and salivary glands but are unable to invade nerve tissues and thus do not produce the symptoms of the disease polio. The vaccines for measles, mumps, rubella, and yellow fever similarly use viable but attenuated viral strains. Attenuated strains of rabies viruses can be prepared by desiccating the virus after growth in the central nervous system tissues of a rabbit or following growth in a chick or duck embryo.

The BCG (bacille Calmette-Guerin) vaccine is an example of an attenuated bacterial vaccine. This vaccine is administered in Britain to children 10 to 14 years old to protect against tuberculosis. It is used in the United States only for high-risk individuals. This mycobacterial strain was developed from a case of bovine tuberculosis. It was cultured for over 10 years in the laboratory on a medium containing glycerol, bile, and potatoes. During that time it accumulated mutations so that it no longer was a virulent pathogen. In over 70 years of laboratory culture the BCG mycobacterial strain has not reverted to a virulent form.

Killed/Inactivated Vaccines

Some vaccines are prepared by killing or inactivating microorganisms so they cannot reproduce or replicate within the body and are not capable of causing disease. When microorganisms are killed or inactivated by treatment with chemicals, radiation, or heat, the antigenic properties of the pathogen are retained. Killed/inactivated vaccines generally can be used without the risk of causing the onset of the disease associated with the virulent live pathogens. The vaccines used for the prevention of whooping cough (pertussis) and influenza are representative of the preparations containing antigens that are prepared by inactivating pathogenic microorganisms.

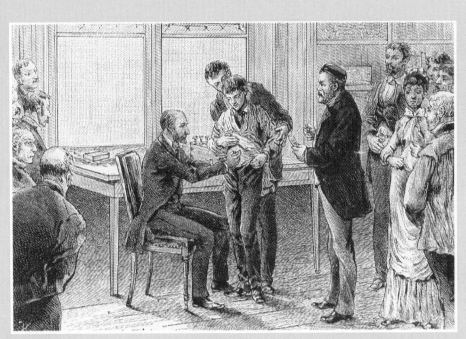

Pasteur administering rabies to Joseph Meister, a 9-year-old boy, who had been bitten by a rabid dog. The vaccine saved the boy's life.

Even when the vaccines are killed cells, problems can occur in some cases. A small percentage of children, for example, have allergic-type reactions to the pertussis component of the standard DPT (diphtheria-pertussis-tetanus) vaccine. Some people are now questioning the wisdom of government-mandated administration of this vaccine. Most manufacturers of this vaccine ceased its production rather than face liability lawsuits associated with such reactions. The supply of DPT vaccine is currently dangerously short. Enhanced quality control programs by the major remaining producer and the development of a new form of the vaccine promise to reduce the incidence of adverse reactions.

There have been several problems with inadequate inactivation of vaccines, leading to disease outbreaks when the vaccines were administered. In 1976 there was a scare about an impending outbreak of swine flu. Some people given swine flu vaccine actually contracted flu because of the inadequate inacti-

vation of the viruses in hastily prepared vaccines. Others developed a neurological disorder called Guillain-Barré syndrome after vaccination against swine flu. In the 1950s several tragic cases of polio occurred in children given the Salk polio vaccine, an inactivated vaccine prepared from a very virulent strain of poliovirus. This incident occurred because of the failure to fully inactivate some batches of the vaccine prepared by treatment of polioviruses with formaldehyde. Because the Salk vaccine is prepared from a particularly virulent strain of poliovirus, replication of the virus in individuals inoculated with the improperly prepared vaccines caused paralytic polio.

The failure of the quality control program for the Salk vaccine, in part, led to the general switch to the "live" attenuated Sabin polio vaccine. The Sabin vaccine is prepared with attenuated viral strains. These strains are not particularly virulent and do not invade the nervous system, causing paralysis. The

ALBERT SABIN AND THE ORAL POLIO VACCINE

About 2 years before his death in March, 1993, I had dinner with Albert Sabin during which he recounted his efforts to introduce his oral polio vaccine. Sabin clearly felt that attenuated viruses were superior, in part because of the natural route of introduction that made them more likely to be used worldwide. He extolled with pride the efforts of Rotary Club International to distribute the oral polio vaccine in developing nations and described the extraordinary efforts of some governments to ensure complete vaccination of the populace by declaring national vaccination days. Suddenly pounding the table with passion, he expressed bitterness over his rivalry with Jonas Salk, blaming Salk for haste in introducing his polio vaccine that resulted in some children being injected with viruses that had not been inactivated. He viewed the rivalry as political, feeling that politics rather than science had been the key

to the initial decision to introduce the inactivated Salk vaccine in the United States. Sabin described his problems attaining permission to conduct clinical trials of his vaccine and how he had decided to initially test it on his own children. If he couldn't trust it for his own children, he argued, why should anyone else? He told how out of frustration he had decided to conduct tests of his vaccine in the former Soviet Union and how this resulted in its being labelled the "red vaccine" that should not be used in America. Even 30 years later, he had difficulty accepting that he had been persecuted by Senator Joe McCarthy during the anticommunist era of the 1950s when anyone and anything associated with the Soviet Union was defamed. It was evident that Sabin still could not accept that political views had been used to prevent children from receiving the medical benefits of vaccination with his oral polio vaccine.

Sabin vaccine is administered orally and the virus multiplies within the gastrointestinal tract. Although the virus is attenuated, mutations and recombinations are still possible during replication. Some recent cases of polio have been reported with the Sabin vaccine, causing the reevaluation of the relative merits of the Salk versus the Sabin vaccine.

In some cases the toxins responsible for a disease are inactivated and used for vaccination. Some vaccines, for example, are prepared by denaturing microbial exotoxins. The denatured proteins produced are called **toxoids.** Protein exotoxins, such as those involved in the diseases tetanus and diphtheria, are suitable for toxoid preparation. The vaccines for preventing these diseases employ toxins inactivated by treatment with formaldehyde. These toxoids retain the antigenicity of the protein molecules. This means that the toxoids elicit the formation of antibody and are reactive with antibody molecules but, because the proteins are denatured, they are unable to initiate the reactions associated with the active toxins that cause disease.

Toxoids are denatured proteins that are used as vaccines.

Individual Microbiological Components

Individual components of microorganisms can be used as antigens for immunization. For example, the polysaccharide capsule from *Streptococcus pneumoniae* is used to make a vaccine against pneumococcus pneumonia. This vaccine is used in high-risk pa-

tients, particularly individuals over 50 years old who have chronic diseases, such as emphysema. Another vaccine has been produced from the capsular polysaccharide of *Haemophilus influenzae* type b, a bacterium that frequently causes meningitis in children 2 to 5 years old. The Hib vaccine, as it is called, is being widely administered to children in the United States. This vaccine is not always effective in establishing protection in children under 2 years old. It is administered to children between 18 and 24 months old who attend day care centers because they have a greater risk of contracting *H. influenzae* infections.

The first vaccine to provide active immunization against hepatitis B (Heptavax-B) was prepared from hepatitis B surface antigen (HBsAg). This antigen was purified from the serum of patients with chronic hepatitis B. Immunization with Heptavax-B is about 85% to 95% effective in preventing hepatitis B infection. It was administered predominantly to individuals in high-risk categories such as health care workers. It has been replaced by a newer recombinant vaccine, Recombivax HB. To produce Recombivax HB, a part of the hepatitis B virus gene that codes for HBsAg was cloned into yeast. The vaccine is derived from HBsAg that has been produced in yeast cells by recombinant DNA technology.

Attempts were made to make a vaccine against gonorrhea using pili from *Neisseria gonorrhoeae*, the bacterium that causes this disease. The vaccine produced, however, was not successful because long-lasting immunity against *N. gonorrhoeae* does not develop. The military, though, has used this vaccine to achieve short-term immunity. Synthetic proteins are also being considered as potential antigens for protection against various diseases.

> **Individual components of a microorganism can be used in a vaccine to elicit an immune response.**

Vector Vaccines

Recombinant DNA technology is being used to create vaccines containing the genes for the surface antigens for various pathogens. Such **vector vaccines** act as carriers for antigens associated with pathogens other than the one from which the vaccine was derived. The attenuated virus used to eliminate smallpox is a likely vector for simultaneously introducing multiple antigens associated with different pathogens, such as the chicken pox virus. Several prototype vaccines using the smallpox vaccine as a vector have been made (FIG. 15-5).

Booster Vaccines

Multiple exposures to antigens are sometimes needed to ensure the establishment and continuance of a memory response. Several administrations of the Sabin vaccine are needed during childhood to establish immunity against poliomyelitis. A second vaccination is necessary to ensure immunity against measles. Only a single vaccination, though, is needed to establish permanent immunity against mumps and rubella.

In some cases, vaccines must be administered every few years to maintain the anamnestic response capability. Periodic **booster vaccinations** are necessary, for example, to maintain immunity against tetanus. A booster vaccine for tetanus is recommended every 10 years.

Adjuvants

Some chemicals, known as **adjuvants,** greatly enhance the antigenicity of other chemicals (FIG. 15-6). The inclusion of adjuvants in vaccines therefore can greatly increase the effectiveness of the vaccine. When protein antigens are mixed with aluminum compounds, for example, a precipitate is formed that is more useful for establishing immunity than the proteins alone. Alum-precipitated antigens are released slowly in the human body, enhancing stimulation of the immune response. The use of adjuvants can eliminate the need for repeated booster doses of

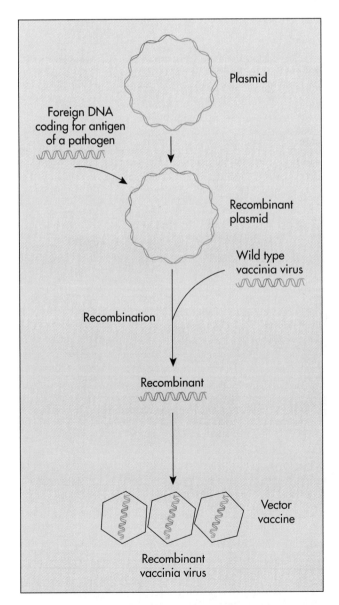

FIG. 15-5 New vaccines can be formed by using recombinant DNA technology to form vector vaccines; for example, using vaccinia virus as a carrier.

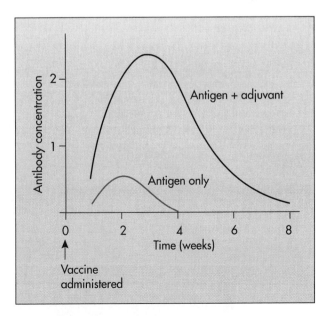

FIG. 15-6 Adjuvants enhance antigenicity and can greatly improve the effectiveness of a vaccine.

HIGHLIGHT

ELIMINATION OF SMALLPOX

The greatest success in preventing disease through the use of vaccines can be seen in the case of smallpox (see Figure). The vaccine for preventing smallpox contained a live strain of pox virus prepared from scrapings of lesions from cows or sheep. The scrapings were treated with 1% formaldehyde to kill bacterial contaminants and 40% glycerol to stabilize the viral antigens. These viral antigens are quite labile (easily destroyed), which is why attenuated viral preparations were required for successful vaccination to achieve a state of immunity. Various viral strains have been used for the production of commercial vaccines. Although the commercial strains used for vaccine production were presumed to have been derived from cowpox virus, it now appears, based on its antigenic properties, that an attenuated strain of smallpox virus may have been inadvertently substituted for the cowpox virus. Because of the length of time that this virus has been cultivated, it is difficult to identify positively its original source, but certainly the pox virus used for vaccine preparation differs from the cowpox viruses found in nature.

Regardless of the origins of the viral strain used in its vaccines, smallpox, a once dreaded disease, has been completely eliminated through an extensive worldwide immunization program conducted under the auspices of the World Health Organization (WHO). The success of the WHO program depended on the use of lyophilized vaccines to overcome the problem of inactivation of the viral antigens in hot climates. The program was not without risks because the virus used for vaccination was virulent enough to cause a fatality rate of 1 in 1 million vaccinations. By immunizing a sufficient portion of the world's population against smallpox, though, it was possible to interrupt the normal transmission of smallpox virus from infected individuals to susceptible hosts. A consequence of the success of this immunization program is that it is no longer necessary to vaccinate against smallpox. The successful elimination of smallpox through a vaccination program was dependent on the fact that humans are the only known host for the smallpox virus and that the virus has a relatively short survival time outside human host tissues. Smallpox presumably is eliminated permanently and as such is the only infectious human disease known to have been eliminated through human intervention, ingenuity, and cooperation.

Smallpox was eliminated through a massive worldwide vaccination program.

the antigen, which increases the intracellular exposure to antigens to establish immunity. It also permits the use of smaller doses of the antigen in the vaccine.

Some bacterial cells are effective adjuvants. The killed cells of *Bordetella pertussis,* used in the DPT vaccine, are adjuvants for the tetanus and diphtheria toxoids used in this vaccine. Similarly, mycobacteria are effective adjuvants. Freund's adjuvant, which consists of mycobacteria emulsified in oil and water, is especially effective in enhancing cell-mediated immune responses. This adjuvant, however, can induce tissue damage and is not used for that reason.

> **Chemical adjuvants are used in vaccines to increase the antigenicity of other chemical components and hence the effectiveness of the vaccine. They can eliminate the need for booster doses of the antigen.**

Routes of Vaccination

The effectiveness of vaccines depends on how they are introduced into the body. Antigens in a vaccine may be given via a number of routes: *intradermally* (into the skin), *subcutaneously* (under the skin), *intramuscularly* (into a muscle), *intravenously* (into the bloodstream), and into the mucosal cells lining the respiratory tract through *inhalation,* or *orally* into the gastrointestinal tracts. Killed/inactivated vaccines normally must be injected into the body, whereas attenuated vaccines often can be administered orally or via inhalation. The effectiveness of a given vaccine depends in part on the normal route of entry for the particular pathogen. For example, polioviruses normally enter via the mucosal cells of the upper respiratory or gastrointestinal tracts. The Sabin polio vaccine, therefore, is administered orally, enabling the attenuated viruses to enter the mucosal cells of the gastrointestinal tract directly. It is likely that vaccines administered in this way stimulate secretory antibodies of the IgA class in addition to other immunoglobulins. Intramuscular administration of vaccines, like the Salk polio vaccine, is more likely to stimulate IgM and IgG production. IgG is particularly effective in halting the spread of pathogenic microorganisms and toxins produced by such organisms through the circulatory system.

ARTIFICIAL PASSIVE IMMUNITY

Passive immunity can be used to prevent diseases when there is not sufficient time to develop an acquired immune response through vaccination. The administration of pooled gamma globulin that contains various antibodies, specific immunoglobulins, or specific antitoxins provides immediate protection (Table 15-6). Before the development of antibiotics, passive immunization—often using horse sera—was widely practiced. Unfortunately, precipitation from extensive antigen-antibody complex formation caused kidney damage when horse sera was routinely administered. Today the use of passive immunity to treat disease is limited to cases of immunodeficiencies and to specific reactions to block the adverse effects of pathogens and toxins.

Various antitoxins (antibodies that neutralize toxins) can be used to prevent toxins of microbial or other origin from causing disease symptoms. The administration of antitoxins establishes passive artificial immunity. Antitoxins are used to neutralize the

TABLE 15-6	
Substances Used for Passive Immunization	
SUBSTANCE	**USE**
Gamma globulin (human)	Prophylaxis against various infections for high-risk individuals, such as those with immunodeficiencies; lessening intensity of diseases, such as hepatitis after known exposure
Hepatitis B immune globulin	To prevent infection with hepatitis B virus after exposure, such as via blood contaminated needles
Rabies immune globulin	Used in conjunction with rabies vaccine to prevent rabies after a bite from a rabid animal; used around wound to block entry of virus
Tetanus immune globulin	Used in conjunction with tetanus booster vaccine to prevent tetanus after a serious wound; used around wound to block entry of virus
Rh immune globulin (Rhogam)	To prevent an Rh-negative woman from developing an anamnestic response to the Rh antigen of an Rh-positive fetus; administered during third trimester or after birth
Antitoxin (various)	To block the action of various toxins, such as those in snake venom, those from spiders, and those produced by microorganisms, including diphtheria toxin and botulinum toxin

NEWSBREAK

COMBATTING CANCER

Cancer is the result of the uncontrolled growth of malignant cells. These cells reproduce more rapidly than normal cells, form abnormal shapes, and fail to stop growing when they contact other cells. They lose their adhesiveness so that they may break off (metastasize) and move to other body sites where new malignant growths then form. Malignant cells also invade connective tissues and various organs of the body. The formation of malignant tumors disrupts body functions and will cause death.

Numerous treatments are used in the attempt to control malignant growths, including surgical removal, chemical inhibition (chemotherapy), exposure to radiation (radiation therapy), and administration of various cytokines (immunotherapy). In some cases, when the cancer spreads throughout the body, such as in cases of leukemia (malignant growth of white blood cells), massive doses of chemotherapy and/or radiation therapy can be employed to kill the malignant cells. Such treatments also damage healthy body cells and tissues. If the lymphocytes of the immune system are eliminated along with the malignant leukocytes, the individual could not survive due to a lack of an immune response system. To prevent this, bone marrow is sometimes surgically removed. This bone marrow, which contains the cells that give rise to the body's blood cells, including lymphocytes, is frozen. After chemotherapy and/or radiation therapy, the bone marrow cells are thawed and reinjected into the body. This therapy can be tried as long as the malignant cells have not invaded the bone marrow. If possible, a compatible donor can be used as a source of bone marrow cells. The donor must have the same or nearly the same antigens on his or her blood cells as the patient. Otherwise, as the immune function is restored, the donor bone marrow cells will be attacked and rejected. Typically six HLA antigens (human leukocyte antigens) on the bone marrow cells must match for it to be compatible. Similar matching is necessary for any organ or tissue donation used for transplantation. An international donor program helps locate and match donors and recipients.

Various immunological therapies can be used to augment chemotherapy and radiation therapy. These include the administration of cytokines, such as tumor necrosis factor, interleukin, and interferon. In some cases these immunotherapies produce dramatic results, even complete remission of metastasized tumors. The administration of interleukin-2, for example, stimulates the natural activities of killer cells that will attack the malignant growth.

HIGHLIGHT

FRUSTRATING EFFORTS TO FIND A VACCINE FOR AIDS

It is not always easy to find antigens associated with pathogens that confer long-term active immunity. Desperate efforts are now underway to formulate a vaccine that will prevent AIDS. Years of research, however, have failed to produce vaccines against other sexually transmitted diseases such as syphilis, as well as other prevalent diseases, including malaria and tooth decay.

Producing an AIDS vaccine will be very difficult because of several properties of HIV. Being a virus, it replicates intracellularly and is released by budding. Hence a method will have to be found for preventing the initial adsorption of the virus to host cells or for detecting and eliminating host cells within which the virus is replicated. Also, the DNA made during replication of this retrovirus is incorporated into the host cell chromosomes. Use of an attenuated HIV that still permits incorporation of DNA into the chromosomes of the host cell could lead to malignancy. To complicate matters, HIV exhibits variable surface antigens so that multiple antigens may have to be employed to ensure that the body's immune system recognizes infecting HIV. So far, the most promising approaches are aimed at blocking the sites on the virus that bind to host cell receptors and are involved in entry of HIV into the host cells. Some research groups are targeting the CD4 binding protein of HIV; others are exploring CD26 as a site for blocking the ability of the virus to enter and to infect human host cells.

toxins in snake venom, saving the victims of snake bites. The toxins in poisonous mushrooms can also be neutralized by administration of appropriate antitoxins. The administration of antitoxins and immunoglobulins to prevent disease occurs after exposure to a toxin and/or an infectious microorganism.

Antitoxins are antibodies that neutralize toxins and can be used to prevent toxins from causing disease symptoms.

It is also possible to establish passive immunity by the administration of gamma globulin, which contains mainly IgG and some IgM and IgA. This is a widely used treatment in Africa for many diseases. It is important that the gamma globulin used for establishing passive immunity is pooled in order to combine the immune functions from many people. Passive immunity lasts for a limited period of time because IgG molecules have a finite lifetime in the body. The administration of IgG does not establish an anamnestic response capability. The administration of IgG is also particularly useful therapeutically in preventing disease in persons with immunodeficiencies and other high-risk individuals.

IMMUNODEFICIENCIES

Failures of the immune response can compromise the ability of the human body to resist infection. Such failures may be due to an inadequate or an inappropriate immune response. If an individual has an inadequate immune response (immunodeficiency), he or she will not be protected against many infectious diseases. Individuals with immunodeficiencies are subject to numerous infections with opportunistic pathogens (Table 15-7). Immunodeficiencies may be congenital, that is, the result of an inherited genetic abnormality or acquired from external causes at some time during the life of the individual.

SEVERE COMBINED IMMUNODEFICIENCY

The most devastating type of congenital immunodeficiency is **severe combined immunodeficiency** (SCID). Individuals with severe combined deficiency have neither functional B nor T lymphocytes. Such individuals are incapable of any immunological response. Any exposure of such individuals to microorganisms can result in the unchecked growth of the microorganisms within the body. This results in certain death.

Individuals suffering from severe combined deficiency can be kept alive in sterile environments. They must be protected from any exposure to microorganisms. In a well-publicized case, a boy named David was kept alive in a sterile bubble chamber for 14 years (FIG. 15-7). Everything entering the chamber—air, water, food—was sterilized. As long as he was not exposed to microorganisms, he was able to survive. Tragically, he died as a result of an attempt to cure his immunodeficiency. He was given a bone

TABLE 15-7	
Common Infections in Immunocompromised Individuals	
DEFICIENCY	**INFECTING AGENT**
Damaged tissues (burns, wounds, trauma)	*Aspergillus* species, *Candida* species, *Pseudomonas aeruginosa*, *Staphylococcus aureus*, *Staphylococcus epidermidis*, *Streptococcus pyogenes*
T lymphocytes	Cytomegaloviruses, herpes simplex viruses, varicella-zoster virus, *Listeria monocytogenes*, *Mycobacterium* species, *Nocardia* species, *Aspergillus* species, *Candida* species, *Cryptococcus neoformans*, *Histoplasma capsulatum*, *Pneumocystis carinii*, *Strongyloides stercoralis*
B lymphocytes	*Staphylococcus aureus*, *Streptococcus* species, *Haemophilus influenzae*, *Neisseria meningitidis*, *Escherichia coli*, *Giardia lamblia*, *Pneumocystis carinii*
Severe combined immunodeficiency (SCID)	Same infecting agents for T and B lymphocyte deficiencies
Phagocytic cells (PMNs and macrophages)	*Aspergillus* species, *Candida* species, *Nocardia* species, *Staphylococcus aureus*, *Streptococcus pyogenes*, *Haemophilus influenzae*, *Escherichia coli*, *Klebsiella* species, and *Pseudomonas aeruginosa*
Complement	*Staphylococcus aureus*, *Streptococcus pneumoniae*, *Pseudomonas*, *Proteus*, *Neisseria* species

FIG. 15-7 In a widely publicized case of severe combined immunodeficiency (SCID) a boy named David was kept alive by isolating him in a sterile chamber. He was known as the bubble baby. He was delivered by caesarian section under aseptic conditions and kept in a sterile environment. He died after an attempt to infuse his body with lymphocytes to establish an immune response. Today gene therapy and other treatments are being used to cure or treat SCID.

marrow graft from a sibling with compatible bone marrow in an attempt to establish functional lymphocytes in his body. He developed an adverse reaction that proved fatal. In other cases, such bone marrow transplants have been effective, including some performed within weeks of the unsuccessful treatment of David.

A new treatment for some cases of SCID is the administration of the enzyme adenosine deaminase (ADA). Accumulation of adenosine compounds is toxic to lymphocytes and ADA is needed to prevent toxicity. In the absence of ADA, B and T lymphocytes die. Approximately 35% of the cases of SCID are due to ADA deficiency. Administering ADA can be therapeutic as long as it is not detected as a foreign antigen. To block its recognition as an antigen that would trigger an adverse immune reaction, the ADA is chemically linked to polyethylene glycol (PEG). PEG coats the ADA and blocks its recognition as an antigen. PEG-ADA treatment is being used effectively to treat some cases of SCID.

The inability to produce ADA in an individual with SCID is due to a defective gene. Gene therapy is also being tried in an attempt to cure this condition (FIG. 15-8). Cells can be obtained from a patient and a functional gene for ADA production inserted into the cells by genetic engineering. The recombinant cells can then be introduced into the patient. Early clinical trials have shown significant improvements in the immune responses of children treated with this gene therapy.

People with no immunological response system have severe combined immunodeficiency (SCID).

Bone marrow transplants and PEG-ADA are used to treat SCID.

DiGeorge Syndrome

DiGeorge syndrome results from a failure of the thymus to develop correctly. It is probably caused by an abnormal fetal development that interferes with the

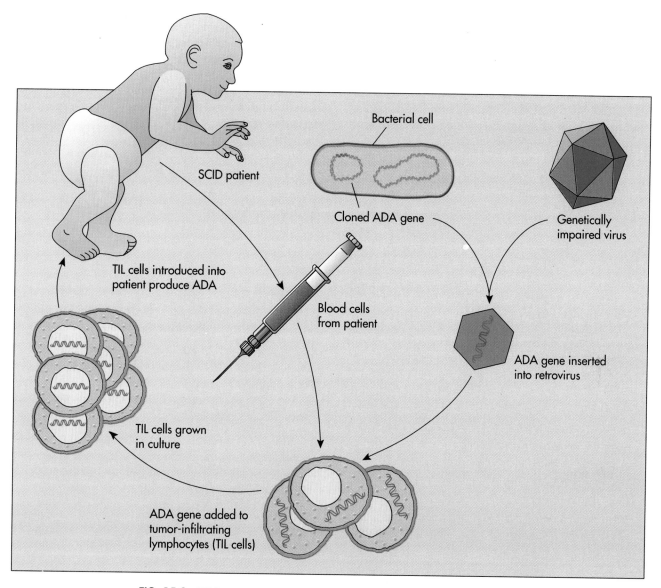

FIG. 15-8 PEG-ADA gene therapy can be used in the treatment of SCID.

proper formation of the thymus. T lymphocytes in individuals with this disease do not become properly differentiated. Signs and symptoms of this disease often are apparent at birth. They include deformities such as low-set ears, fish-shaped mouth, undersized jaw, and wide-set eyes. Elevated serum phosphate and low serum calcium also are characteristic of DiGeorge syndrome. A low phosphate diet and calcium supplements are used to achieve acceptable levels of phosphate and calcium in the blood. Individuals suffering from this condition do not exhibit cell-mediated immunity. Therefore they are prone to viral and other intracellular infections. Avoidance of infecting agents is important in the management of patients with DiGeorge syndrome.

Children with DiGeorge syndrome that contract measles do not show the characteristic skin rash associated with this disease (the rash is due to activated T cells in the skin). Because T helper cells are involved in enhancing antibody production by B cells, the antibody-mediated or humoral response is also depressed in individuals suffering from DiGeorge syndrome. The complete absence of the thymus is rare. Partial DiGeorge syndrome—in which some T cells are produced, although in lower numbers than in individuals with fully functional thymus glands—is more common.

In DiGeorge syndrome there is a shortage of T cells.

BRUTON CONGENITAL AGAMMAGLOBULINEMIA

Bruton congenital agammaglobulinemia results in the failure of B cells to differentiate and produce antibodies. Individuals with Bruton disease have a normal cell-mediated response. This immunodeficiency disease affects only males. Boys with Bruton agammaglobulinemia are particularly subject to bacterial infections, including those by pyrogenic (fever-inducing) bacteria such as *Staphylococcus aureus, Streptococcus pyogenes, S. pneumoniae, Neisseria meningitidis,* and *Haemophilus influenzae.* The treatment of this disease involves the repeated administration of pooled gamma globulin to maintain adequate levels of antibody in the circulatory system.

In Bruton disease, B cells do not differentiate and produce antibodies.

LATE-ONSET HYPOGAMMAGLOBULINEMIA

The most common form of immunodeficiency is known as **late-onset hypogammaglobulinemia.** In this condition, there is a deficiency of circulating B cells and/or B cells with IgG surface receptors. Such individuals are unable to respond adequately to antigen through the normal differentiation of B cells into antibody-secreting plasma cells. Other immunodeficiencies may affect the synthesis of specific classes of antibodies. For example, some individuals exhibit IgA deficiencies, producing depressed levels of IgA

METHODOLOGY

GENE THERAPY WITH TUMOR INFILTRATING LYMPHOCYTES

The cellular immune defense system recognizes abnormal cells when they arise in the body. It attempts to eliminate such cells. In this manner the immune system is able to hold in check most malignant (cancer-forming) cells when they occur. T cells detect abnormal antigens on the surfaces of malignant cells and attack those cells. Sometimes the cellular immune response is adequate and malignant tumors do not develop. In other cases the proliferation of malignant cells leads to the growth of cancerous tumors.

Recognizing that T cells have the capacity to attack malignant cells in the body, Stephen Rosenberg and colleagues at the National Institutes of Health postulated that they could develop a method for cancer treatment based on the body's own immune response. They sought to isolate T cells that could recognize specific types of malignancies. They then developed methods for culturing this specialized class of T cells, which they called tumor infiltrating lymphocytes (TIL cells). They hypothesized that if they could culture large numbers of TIL cells from a patient and could reinject the cultured TIL cells into that same patient, those TIL cells would then attack the developing malignant tumors. The patient would be receiving his or her own genetically modified cells.

In a number of cases where they carried out this procedure, there was remarkable regression of the tumors. Some patients responded dramatically and the cancer went into total remission. In other cases, however, the procedure failed to check the growth of the tumors and the patients died of cancer.

It was not clear why the treatment worked in some cases and failed in others. Did the injected TIL cells survive in the body? Did they reach the sites of tumors? Would the injection of lymphokines, such as interleukin, enhance the abilities of TIL cells to destroy malignant tumors? To answer these questions Rosenberg needed a method for tracking the fate of the TIL cells that he had cultured and introduced back into the patient's body. Rosenberg, with Michael Blaise and French Anderson, proposed to genetically label the TIL cells. They obtained the gene for tagging the TIL cells from a bacterium. It is a gene that codes for neomycin resistance. This gene does not occur naturally in humans. Anywhere the neomycin resistance gene would be found in the patient could be directly tied to the introduced TIL cells.

The proposal to introduce genetically altered TIL cells into human subjects was reviewed by the Recombinant Advisory Committee of the National Institutes of Health. The persons serving on that committee recognized the profound significance of the proposed experiments. Not only could these experiments lead to improved cancer treatment, they also would pioneer the field of gene therapy. It was clear that the next step in development would be to use genetic engineering to alter human cells to perform different functions. Cells could be modified genetically and introduced into the body to cure disease. After many long debates about the safety and scientific validity of the experiments, the Recombinant Advisory Committee approved the experimental plan of Rosenberg, Blaise, and Anderson.

Shortly thereafter, TIL cells obtained from several patients were marked with the neomycin resistance gene and introduced back into those patients. The researchers were able to follow the specific movement of the TIL cells that they had introduced. They were able to improve the treatment regime so as to enhance survival of the introduced TIL cells.

Rosenberg, Blaise, and Anderson next proposed to genetically alter TIL cells by introducing the gene for tumor necrosis factor. Lymphocytes that produce tumor necrosis factor are able to cause the shrinkage of malignant tumors. Again after extensive debates, the Recombinant Advisory Committee of the National Institutes of Health recommended that clinical trials of such recombinant cells be permitted. These trials represented the first true attempts at gene therapy. A new era in modern medicine based on recombinant DNA technology had begun. A new treatment was added in the continuing battle against cancer.

antibodies. Such individuals are prone to infections of the respiratory tract and body surfaces normally protected by mucosal cells that secrete IgA.

> **In late-onset hypogammaglobulinemia there is a shortage of B cells.**

COMPLEMENT AND CELLULAR DEFICIENCIES

Immunodeficiencies may also affect the complement system. People who fail to produce sufficient amounts of a specific type of complement, called C3 complement, are unable to respond properly to bacterial infections. The lack of an active complement system limits the inflammatory response and the killing of pathogenic bacteria.

Immunodeficiencies may also result from inadequate functioning of monocytes, neutrophils, and macrophages. Phagocytic cells lacking enzymes that produce hydrogen peroxide and other antimicrobial forms of oxygen do not have proper lysosomal functions that kill bacteria. Pathogenic bacteria are able to multiply within such metabolically deficient phagocytes. Antibiotics can be used to protect individuals who are deficient in both complement and active phagocytic cells against invading pathogenic bacteria.

MALIGNANT CELL DEVELOPMENT

The development of malignant (cancer) cells can also be viewed as a failure of the immune response. In this case the failure to recognize and to respond properly to inappropriate cells within the body allows malignant cells to proliferate in an uncontrolled manner. Kaposi sarcoma develops in many individuals with AIDS because the failing immune response is unable to prevent malignancies (FIG. 15-9).

ACQUIRED IMMUNODEFICIENCY SYNDROME (AIDS)

Acquired immunodeficiency syndrome or **AIDS** is caused by an infection with the human immunodeficiency virus (HIV) (FIG. 15-10). This virus is able to adsorb specifically to lymphocytes through the CD4 surface receptor. Most of the lymphocytes with the CD4 molecule are T_H cells. HIV enters the T_H cell via CD26. Replication of HIV within T_H cells leads to death of some of the infected cells, lowering numbers of T helper cells. Nearly all the T_H cells in blood, lymph nodes, and spleen are destroyed. In addition, macrophages and microglial cells in the brain may become infected with HIV, causing dementia. Individuals with AIDS are subject to infection by a wide variety of disease-causing microorganisms and to the development of a form of cancer known as Kaposi sarcoma.

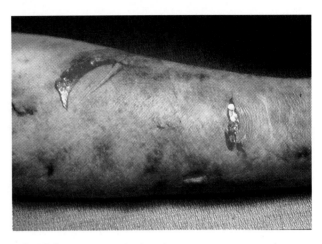

FIG. 15-9   An individual with Kaposi sarcoma on the arm.

> **Individuals with AIDS exhibit immunosuppression because HIV infection kills T helper cells.**

As the T_H cells are killed, the ratio of T_H to T_S cells decreases, from about 2.0 to less than 0.5. This is because the T_H cells decline from over 100,000/mL in healthy individuals to less than 50,000/mL as a consequence of HIV infection. Because individuals with AIDS have more T suppressor than T_H cells, the immune response does not work efficiently. This preponderance of T suppressor cells depresses other immune functions. When numbers of T_H cells are decreased, B cells are not stimulated to produce sufficient numbers of antibodies to combat infections. Amounts of lymphokines produced are insufficient to activate macrophages and cytotoxic T cells. Infected T_H cells release soluble suppressor factor, which inhibits certain immune responses. The T helper cells that survive do not have surface receptors for antigens. They are incapable of recognizing antigens. Thus the first step in the immune response is blocked.

> **T suppressor cells predominate over T helper cells and depress immune functions in individuals with AIDS.**

Because HIV reduces the effectiveness of the immune system, the body is unable to rid itself of HIV once the infection is established. The key to controlling this disease rests with prevention. HIV is transmitted by direct sexual contact, by exchange of blood, and from mother to fetus. Casual contact does not result in transmission of the virus. Sexually promiscuous individuals and intravenous drug abusers who share contaminated needles are at high risk of contracting this disease. Steps have been taken to protect the blood supply used for transfusions. Blood is routinely tested for the presence of HIV. Tissues for transplantation are also tested. Health care

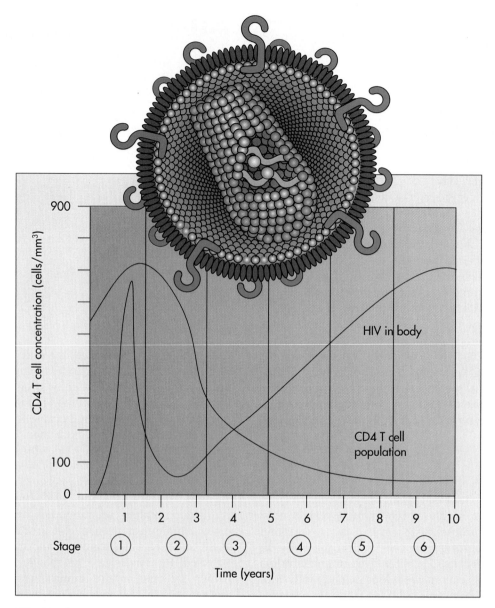

FIG. 15-10 The human immunodeficiency virus (HIV) causes AIDS. HIV has an RNA genome and reverse transcriptase contained within a capsid surrounded by an envelope. The envelope has surface proteins that protrude and are involved in adsorption to host cells.

workers must take special precautions to avoid infection due to exposure to HIV-containing blood. Health care workers with AIDS have a special obligation to ensure that they do not transmit HIV to their patients.

Some drugs have proven effective in prolonging the life expectancies of AIDS patients by retarding the replication of HIV. These drugs interfere with the replication of HIV. HIV is a retrovirus. Retroviruses carry out reverse transcription during replication; they copy their RNA into DNA using a viral enzyme called reverse transcriptase. Azidothymidine (AZT) has been approved for treatment of AIDS. At least 40% of individuals treated with AZT develop intoler-

ance and must cease taking the drug. Another drug, dideoxyinosine (ddI), may be used as an alternate to AZT. Both AZT and ddI block the formation of functional DNA during reverse transcription. AZT and ddI do not eliminate HIV. They only slow down the rate of HIV replication and resultant destruction of T cells.

There currently is no cure for AIDS. There is no vaccine for its prevention. Reducing the likelihood of exposure, such as by using condoms, is necessary to limit the spread of this disease. As the disease progresses the immune system becomes less and less capable of defending the body against infection. Eventually the disease is fatal.

THE CASE OF PATIENT ZERO

Even before AIDS was recognized as a specific disease, epidemiologists in California in the late 1970s were seeking the source of the unusual cases of immunodeficiency among homosexual men. The Los Angeles Cluster Study was conducted to investigate the origins of a disease then called GRID, Gay-Related Immune Disease. This study tried to determine the links between individuals diagnosed as having GRID (AIDS) based on interviews with individuals afflicted with the disease. Among the individuals interviewed was a 28-year-old French-Canadian airline steward, Gaetan Dugas.

Dugas, a homosexual, was exceptionally popular. He had determinedly made his way out of a small town provincial life, where he was regularly beaten because he was gay. He found his own niche as a star of the homosexual jet set. His looks, physique, and personal charm made him the object of attraction of many gay men. He travelled frequently between New York, San Francisco, Los Angeles, Vancouver, and Toronto. He was sexually active in all these cities. Based on his own estimates, Dugas had approximately 250 sexual contacts per year, and after an active sex life of 10 years, as many as 2,500 different sexual partners.

In 1980, Dugas discovered purplish spots near his ear. These spots were diagnosed as Kaposi sarcoma, a rare type of skin cancer. The unprecedented frequency of Kaposi sarcoma among gay men in the late 1970s was among the first clues that led to the recognition of AIDS. Dugas continued to be sexually active even after beginning chemotherapy treatment. After an interview with public health officials, he was identified as one of the sexual partners of a known victim of GRID. This was the first time that two victims of what seemed to be a new epidemic were linked.

Gaetan Dugas was a member of a group of gay men who were together in New York in the summer of 1976. All of the early cases of GRID were associated with members of that group. Of the first 19 cases of GRID in the Los Angeles area, 8 had sex with Dugas or with one of his sex partners. Of the first 248 gay men diagnosed with GRID in the United States, 40 had sex with Dugas or with someone who had. Statisticians figured that the odds that it could be sheer coincidence that 40 of these 248 men might all have had sex with the same man or with men sexually linked to him were zero. And so Gaetan Dugas became known as "Patient Zero."

On April 1, 1982, Dugas was told by a physician at the hospital at University of California-San Francisco that he should stop having sex. He refused, saying that it hadn't been proven to him that his disease could be spread. In May 1982, Dugas went to the Centers for Disease Control in Atlanta to give plasma for lab research and blood from which to isolate viruses. He complained bitterly that he was treated like a lab rat there. He had skin cancer for 2 years and was sick of being a guinea pig for physicians who couldn't help him.

Although Dugas was aware of all the available information on AIDS, nothing he read and no advice from physicians could persuade him to stop having sex. He continued to endanger the lives of others by having sex with them. When he was feeling well, he continued working for Air Canada and continued to have sexual relations in Canada and the United States.

Dugas suffered from bouts of illness, especially *Pneumocystis* pneumonia. His health began to fail in late 1983, over 3 years after his initial diagnosis with Kaposi sarcoma. While recovering from his fourth bout with *Pneumocystis* pneumonia, his kidneys failed after the strain of so many years of infection. He died on March 30, 1984.

It can never be known with absolute certainty whether Gaetan Dugas was the person who brought the AIDS virus to North America. Clearly, though, the first cases of AIDS in New York and Los Angeles can be linked to Dugas, who was one of the first 10 patients on the continent. There can be no doubt of the important role Gaetan Dugas played in spreading HIV and AIDS.

AUTOIMMUNITY

In some individuals the immune response fails to recognize self-antigens. In such cases the immune system attacks one's own body, a condition known as **autoimmunity.** The inability to recognize self-antigens results in reactions that kill some of one's own cells. There are a number of **autoimmune diseases** that result from the failure of the immune response to recognize self-antigens. Such autoimmune diseases often result in the progressive degeneration of tissues.

Autoimmunity occurs when the body's immune system attacks the body's own cells.

Some autoimmune diseases affect single sites within the body. Graves disease, for example, is an autoimmune disease that affects the thyroid. In Graves disease the body produces an antibody that reacts with the receptor for thyroid-stimulating hormone. In contrast, some autoimmune diseases affect sites throughout the body. In systemic lupus erythe-

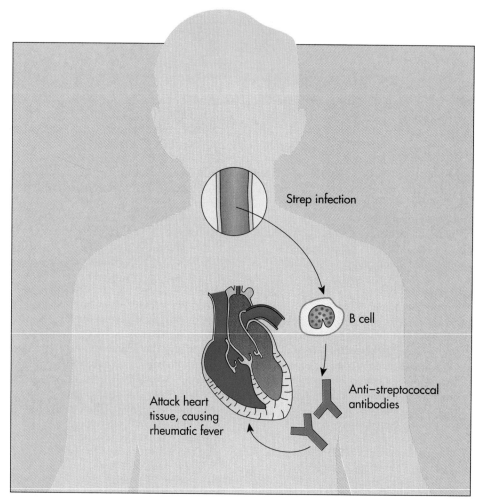

FIG. 15-11   An autoimmune response can occur following an infection with *Streptococcus pyogenes.* The normal immune response produces antibodies against streptococcal antigens. However, these anti-streptococcal antibodies can cross-react with heart tissue and cause damage that may result in later heart failure.

matosus, numerous autoantibodies are produced that react with self-antigens. They attack blood cells and cells at multiple body sites. Antigen-antibody complexes circulate and settle in the glomeruli of the kidney. In cases of myasthenia gravis, antibodies react with nerve-muscle junctions. In autoimmune hemolytic anemia, antibodies react with red blood cells, causing anemia. Immunosuppressive substances are available to prevent the self-destruction of body tissues by the body's own immune response.

Various other disease conditions reflect the failure of the immune system to recognize self-antigens. These self-antigens are similar to antigens associated with pathogenic microorganisms. For example, rheumatic fever is an autoimmune disease that results following an infection with group A streptococci (*Streptococcus pyogenes*) (FIG. 15-11). Some antibodies made in response to group A streptococcal antigens can also react with myosin of the heart muscle tissue. After a strep throat, therefore, antibodies

made against the group A streptococci cross react with myosin in some individuals, causing tissue damage to the heart. These individuals develop rheumatic fever. Damaged heart valves may cause heart failure years later. The immune complexes between antibody and myosin or related antigens may also cause arthritis and kidney failure.

When the immune system cannot distinguish between self and nonself cells, autoimmune diseases result.

Rheumatoid Arthritis

Rheumatoid arthritis is a commonly occuring disease. Although rheumatoid arthritis is usually associated with older people, it often develops early in life. It is a chronic inflammation of the joints, especially the hands and feet. It can lead to crippling disabilities. This form of arthritis often begins with a joint inflammation from an infection that causes phagocytic cells to release lysozymes. These degradative enzymes attack and alter certain antigens. B cells make IgM antibodies in response to the antigens and cause more inflammation in the joints.

Treatment of rheumatoid arthritis is designed to relieve the symptoms. There is no cure. Hydrocortisone lessens inflammation and reduces joint damage. Aspirin is also used because hydrocortisone produces side effects. Aspirin also reduces inflammation and pain.

MYASTHENIA GRAVIS

Myasthenia gravis (MG) is an autoimmune disease that affects the neuromuscular system. It is characterized by weakness and rapid fatigue of the skeletal muscles. It affects muscles in the limbs and the muscles used in eye movement, speech, and swallowing. Patients with MG have a high incidence of thyroid abnormality, reduced levels of complement, and antiskeletal muscle antibody. This disease is rare. It affects 3 persons in every 100,000. Twice as many women as men are affected. The disease usually appears in late childhood to middle age.

Normal muscle contraction requires that pores in the membranes of neurons that stimulate muscles be open. It appears that antibodies that react with self-antigens may be blocking these pores in people with myasthenia gravis. When the pores are blocked, the neurons do not release acetylcholine. Acetylcholine initiates muscle cell contraction. Myasthenia gravis is treated with drugs that inhibit the enzyme that breaks down acetylcholine. The slowing of acetylcholine breakdown allows each muscle longer time to act. This compensates for the decreased amount of acetylcholine.

SYSTEMIC LUPUS ERYTHEMATOSUS

Systemic lupus erythematosus is a widely disseminated, systemic autoimmune disease. Erythematose means red and lupus means wolf. The name of the disease comes from a butterfly-shaped rash that appears on the nose and cheeks. It was thought that the rash looked like a wolf bite. This disease occurs four times as often in women as in men, usually during the reproductive years. Patients have reduced complement levels and high levels of immune complexes in their serum and glomeruli.

In this disease autoantibodies are made primarily against components of chromatin (DNA, RNA, and proteins). Immune complexes are deposited between the dermis and epidermis and in blood vessels, joints, glomeruli of the kidneys, and central nervous system. They cause inflammation and interfere with normal functions wherever they are. The symptoms depend on where the antigen-antibody complexes most interfere with function. Usually there is inflammation of the blood vessels, heart valves, and joints. A skin rash appears. Many victims die from kidney failure as glomeruli fail to remove wastes from the blood.

GRAVES DISEASE

Patients with Graves disease, which include former President and Mrs. George Bush, suffer from overproduction of hormones produced by the thyroid. Normally the pituitary secretes thyroid-stimulating hormone, which controls the amount of thyroid hormone released. Antibodies to thyroid-stimulating hormone receptors are produced in Graves disease patients. These antibodies trigger thyroid cells to produce the hormones. The antibodies are not subject to hormonal feedback control and so the thyroid continues to produce, and overproduce, hormones. Treatment of this disease involves destruction of part of the thyroid. This is often accomplished using the radioisotope ^{131}I, which is concentrated in the thyroid gland and subsequently kills thyroid cells.

MULTIPLE SCLEROSIS

Multiple sclerosis (MS) occurs in people 20 to 50 years of age. Common signs are sensory and visual motor dysfunction. The etiology of this disease is unknown. It is generally believed, however, that MS is a T-cell-mediated autoimmune disease. Macroscopic lesions called plaques are found in the central nervous systems of MS patients. The lesions contain macrophages and lymphocytes. The term *multiple sclerosis* was originally used to describe the wide distribution of these lesions. There is also breakdown in the myelin sheath that surrounds nervous tissue.

HYPERSENSITIVITY REACTIONS

In some individuals there is an excessive immune reaction that causes physiologically adverse reactions. An excessive immunological response to an antigen can result in tissue damage and a physiological state known as **hypersensitivity.** In a hypersensitivity reaction, the immune system inappropriately overreacts to a specific antigen it usually ignores. This results in tissue damage within the body. The term *allergy* (from the Greek *allos* and *ergia,* meaning altered reaction) is used in layman terms as a synonym for a hypersensitivity reaction. Allergies are the result of physiological changes caused by certain types of immune responses to various substances (such as in foods, dust, pollen, animal dander, or drugs) that do not normally activate the immune system. A person who has such a hypersensitivity is said to be allergic, and the specific antigen that elicits the hypersensitivity response is called an **allergen.** Individuals suffering from allergies know too well that immune responses may occasionally be dysfunctional.

> **A hypersensitivity or allergic reaction is an inappropriate excessive immune response to an antigen (allergen) that the body normally ignores.**

Individuals are not born with allergies. After exposure to an allergen, the individual may become sensitized to it, meaning that the allergen elicits an inappropriate immune response that results in memory. Subsequent contacts with the allergen result in an elevated immune response. Hypersensitivity reactions may be immediate, occurring shortly after exposure to the antigen, or delayed, occurring a day or more afterward. These reactions are divided into four major categories, depending on their mechanism of action and the amount of time it takes to see the response after exposure to the antigen (Table 15-8).

ANAPHYLACTIC HYPERSENSITIVITY

Anaphylactic hypersensitivity (anaphylaxis) (type I hypersensitivity) occurs when antigens react with antibodies bound to mast cells in the tissues or basophils in the blood (FIG. 15-12). This condition is also known as **immediate hypersensitivity** because it occurs rapidly (within minutes) after exposure to the antigen that triggers this response. Type I hyper-

sensitivity reactions may be localized or systemic. The type of response often depends on the nature of the allergen, the amount of allergen, and how it is introduced into the body.

Individuals who develop anaphylactic hypersensitivities produce IgE antibodies as a result of exposure to a specific allergen. These IgE antibodies bind via their Fc regions to the surfaces of mast cells and basophils. The surface of a mast cell can be covered with as many as 500,000 IgE receptors. Mast cells and basophils also contain granules in their cytoplasm. When an allergen reacts with several IgE molecules on a sensitized mast or basophil cell, the cell degranulates, that is, it releases the contents of its granules into the surrounding blood or tissues. The release of the contents of basophil or mast cell granules establishes the basis for several physiological responses. One of the most important substances released is histamine, which causes blood vessels to dilate (vasodilation) and become more permeable. Blood flow is increased and the escape of fluid and cells from the blood vessels is also increased This results in tissue swelling and redness. In addition, histamine causes increased secretion of mucus, contraction of smooth muscle, and constriction of bronchial air passageways. Other substances that are newly synthesized and released when mast cells and basophils are stimulated by antigen include prostaglandins and leukotrienes. Some leukotrienes are called slow-reacting substances of anaphylaxis (SRS-A). These molecules also contribute to vasodilation and smooth muscle contraction. Leukotrienes and prostaglandins also stimulate nerve endings to cause pain and itching.

		TABLE 15-8		
		Hypersensitivity Reactions		
TYPE	**NAME**	**TIME AFTER EXPOSURE TO ANTIGEN**	**DESCRIPTION**	**EXAMPLE**
Type I	Anaphylactic hypersensitivity	5-30 min	Antigen binds to IgE on mast cells and basophils, resulting in degranulation and release of substances such as histamine, heparin, and serotonin	Allergy to drugs, food, plant pollens, insect venoms; hay fever; asthma
Type II	Antibody-dependent cytotoxic hypersensitivity	4-12 hours	IgM or IgG antibodies react with antigens on the body's own cells	Transfusions with incompatible blood types; hemolytic disease of the newborn
Type III	Immune-complex hypersensitivity	4-10 hours	Antigen-antibody complexes activate complement and trigger an inflammatory response	Arthus reaction (localized), serum sickness (spread throughout body)
Type IV	Delayed-type hypersensitivity	24-72 hours	Cell-mediated due to T_{DTH} cells	Contact dermatitis to soaps, metals, poison ivy; tuberculin reaction

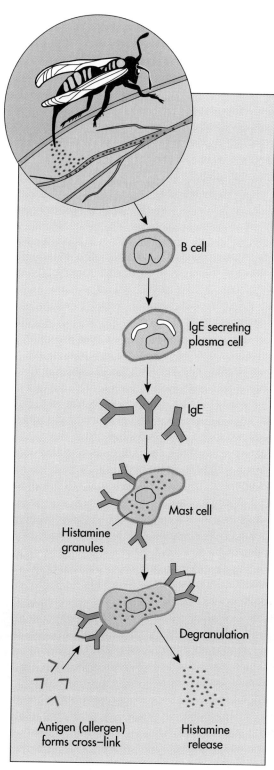

Initially, when hypersensitive individuals come in contact with an allergen for the first time, clones of B lymphocytes are activated by the allergen. These B lymphocytes differentiate into plasma cells that make IgE antibodies against the allergen. The reason that IgE is made preferentially is not yet known. Many individuals inherit the genetic trait for producing high levels of IgE and are prone to develop type I hypersensitivities. When specific IgE antibodies are synthesized in response to an allergen, they move through the bloodstream to mast cells in connective tissue where they become firmly fixed to the receptors. This is the process of *sensitization*. The next time the individual is exposed to the same allergen, that allergen can react directly with the IgE fixed to mast cells and basophils. It does not cause B lymphocytes to initiate antibody synthesis. These cells are now sensitized to the antigen. The immune response is now primed to respond to attack the allergen. The initial sensitizing dose of the allergen must be fairly high. Subsequent eliciting doses can be quite small.

Sensitization is the initial recognition of a foreign substance against which B cells make IgE antibodies.

Generalized Anaphylaxis

A sudden release of large amounts of potent vasodilators, such as histamine and leukotrienes into the bloodstream, can cause extensive dilation of blood vessels. If this occurs, there is a rapid loss of blood pressure, which can produce a condition called *anaphylactic shock* or *generalized anaphylaxis*. Respiratory or cardiac failure may be the outcome of anaphylactic shock and this is, therefore, a potentially life-threatening condition. Injected antigens such as those from insect stings and antibiotics more commonly trigger such systemic anaphylactic hypersensitivity reactions in some people than antigens that are introduced into the body by other means. Prompt intravenous administration of adrenaline (epinephrine) is used to counter anaphylactic hypersensitivity reactions. Epinephrine raises the blood pressure. The action of the vasodilators released as a result of the hypersensitivity reaction is thus reversed.

Localized Anaphylaxis

Localized anaphylaxis includes atopic (out of place) allergies at specific body sites. Some examples of localized anaphylaxis include hay fever, asthma, hives, and gastrointestinal upset. Localized anaphylaxis produces different effects, depending on the site of the body. Typically, these reactions occur in individuals who are hypersensitive to dust and dust mites, foods, plant pollens, or animal dander. For example, a person who is allergic to grass pollen inhales the allergen (pollen grains) and experiences a predomi-

FIG. 15-12 In allergic individuals, exposure to an antigen (allergen) that reacts with IgE leads to degranulation of mast cells. Degranulation is triggered when the allergen forms a cross-linkage between the two adjacent IgE molecules on the surface of a mast cell. The release of histamine causes vasodilation. During systemic anaphylaxis, such as occurs following allergic reactions to drugs and bee stings, blood pressure drops due to vasodilation and breathing and heart irregularities develop. If untreated the person dies as breathing and heart beating cease. Epinephrine is administered to restore respiratory and heart function.

HIGHLIGHT

DETECTING AND TREATING ALLERGIES

Allergies can be diagnosed by skin tests (FIG. *A*). Subcutaneous or intradermal injection of antigens results in a localized inflammation reaction if the individual is allergic to that antigen. The symptoms of atopic allergies can be controlled, at least in part, by avoiding the identified allergens and by using antihistamines. Antihistamines combine with histamine receptors so that the action of the principal chemical mediator of the allergic reaction, histamine, is thus blocked. Swelling and redness are reduced.

In addition to treating the immediate symptoms of an allergic reaction, attempts can be made to desensitize the individual (FIG. *B*). This procedure generally is time consuming and costly. Desensitization usually is achieved by identifying and then administering in-

creasingly strong doses of the allergen. The mechanism by which desensitization works may consist of directing immunoglobulin production to the production of IgG antibodies, called blocking antibodies. They are not involved in the allergic response. IgE, a critical mediator of atopic allergies, is produced in less quantity. IgG antibodies complex with the allergen before it can react with IgE. Mast cells thus do not release the substances that cause allergic symptoms. Suppressor T cells sensitized to the allergen also increase during this process. Once made, IgG molecules can react with the allergen molecules before they reach IgE. This prevents the degranulation of mast cells and basophils and blocks the allergic reaction. Over a period of time the allergic response can be reduced or eliminated.

A

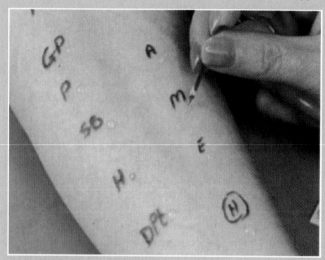

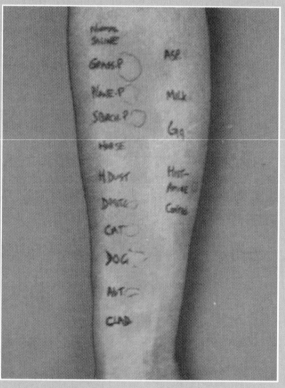

Skin tests are used to diagnose allergies. Antigens are placed below the skin. Development of a red, raised area is indicative of an allergic response.

nantly upper respiratory tract response, including watery eyes, runny nose, itching, coughing, and sneezing. Since these effects are mainly caused by the release of histamine, antihistamines are drugs that can be used to block its action.

In some cases, the hypersensitivity reaction primarily affects the lower respiratory tract. This pro-

duces the condition known as asthma. **Asthma** is characterized by shortness of breath and wheezing. These symptoms occur because the allergic reaction causes a constriction of the bronchial tubes, producing spasms. The primary mediator of asthma is not histamine but SRS-A. Therefore, antihistamines cannot be used in treating this condition. The treatment

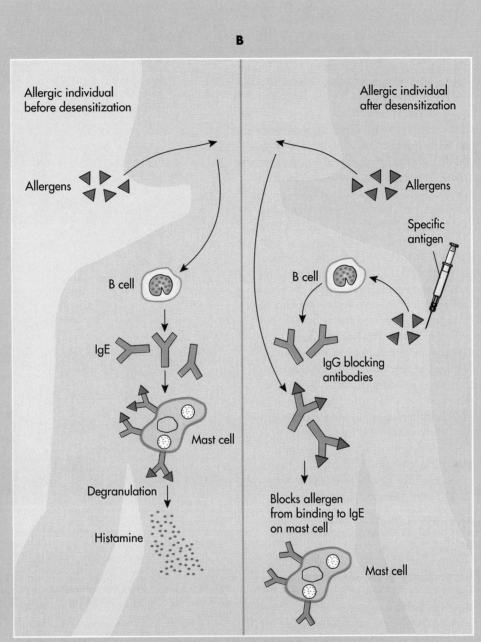

Desensitization can be used for the treatment of allergies. Allergy shots can be an effective way of desensitizing an individual to a particular allergen. Desensitization introduces low doses of the allergen into the skin that elicits an IgG-mediated immune response. This blocking IgG antibody prevents the allergen on subsequent exposure from reacting with IgE and thus prevents the allergic response.

of asthma generally involves administration of epinephrine or aminophylline. Corticosteroid hormones can also be used to control severe asthma because they inhibit activation of mast cells and basophils. Asthmatic attacks can also be prevented by inhalation of sodium cromolyn, which also makes mast cells less susceptible to activation.

Individuals who are hypersensitive to certain foods ingest the allergen that then upsets the gastrointestinal tract or is spread throughout the body in a more systemic response. Food allergies can result in hives or rashes on the skin, nausea, vomiting, or diarrhea. Common foods that cause allergies are eggs, wheat, milk, fish or shellfish, and nuts.

ANTIBODY-DEPENDENT CYTOTOXIC HYPERSENSITIVITY

Antibody-dependent cytotoxic hypersensitivity reactions (type II hypersensitivity) occur by a different mechanism than anaphylactic hypersensitivity (FIG. 15-13). In type II hypersensitivity reactions, an antigen on the surface of the cell combines with an antibody. This stimulates phagocytic attack or initiates the sequence of the complement pathway that results in cell lysis and death.

Antibody-dependent cytotoxic response occurs after transfusions with incompatible blood types. Blood serum contains antibody to any antigens that do not occur in the plasma membranes of the red blood cells of that individual. A person with type A blood has antigen A on red blood cell surfaces and circulating anti-B antibody. If that person were given a transfusion with type B blood that has antigen B on blood cell surfaces and anti-A antibody in the serum, the circulating antibodies in the recipient would react with the surface antigens of the donor cells. Formation of antigen-antibody complexes on the surface of the red blood cells would activate the complement system, resulting in the lysis of the donated cells. It is therefore essential that blood transfusions be made with compatible blood types. A transfusion reaction can occur when matching antigens and antibodies are present in the blood at the same time. This is a type II hypersensitivity reaction against a foreign antigen. Foreign erythrocytes in the transfused blood are agglutinated (clumped), complement is activated, and red blood cells undergo hemolysis (rupture). The patient's symptoms include fever, low blood pressure, back and chest pains, nausea, and vomiting.

> **Antibody-dependent cytotoxic hypersensitivity (type II) reactions are caused by transfusions with incompatible blood types.**

Persons with type O blood are sometimes called **universal donors** and individuals with type AB blood are sometimes called **universal recipients.** The reason for this is that type O red blood cells lack both A and B antigens on their surfaces. They lack the antigens generally associated with transfusion incompatibility. Regardless of the circulating antibodies in the recipient, the donated blood cells do not have antigens with which to react. The anti-A and anti-B antibodies in the donated blood are rapidly diluted when introduced into the larger volume of blood in the recipient. Similarly, persons with type AB blood do not have circulating antibodies against either A or B antigens. They lack antibodies that react with the A and B antigens on blood cells that are introduced regardless of cell type. However, the concepts of the universal donor and the universal recipient refer only to the major A and B antigens. There are various other antigens on blood cell surfaces, in-

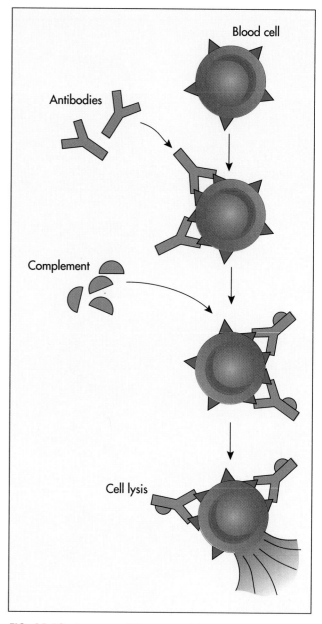

FIG. 15-13 In a type II hypersensitivity reaction, antigen on a cell, such as on a red blood surface, combines with antibody.

cluding the Rh antigen, that can cause incompatibility reactions. Therefore, except in emergencies, transfusions are given only after adequate analysis of cell antigens and only with matching blood types.

Rh incompatibility between mother and fetus is another example of type II hypersensitivity (FIG. 15-14). Approximately 85% of the human population has an antigen on their red blood cells known as the Rh factor. These people are Rh positive. Antibodies that react with Rh antigens are not found in the serum of Rh-negative people. Exposure to the Rh antigen will sensitize Rh-negative individuals to produce anti-Rh antibodies. Rh incompatibility occurs

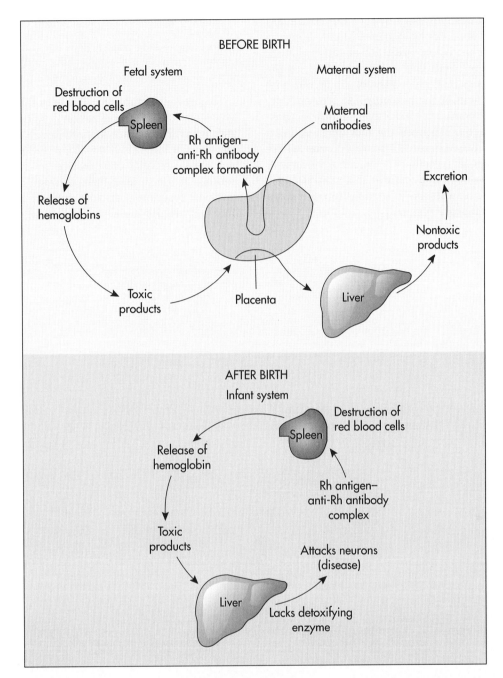

FIG. 15-14 Before birth the products formed by the attack of anti-Rh antibodies pass across the placenta and are detoxified by the maternal liver. After birth the liver of the newborn is unable to detoxify these products, which attack the nerves as well as red blood cells, causing disease.

when the father is Rh positive, the mother is Rh negative, and the fetus is Rh positive. In this case, the mother develops Rh antibodies in response to exposure to the Rh antigens of the fetus. Generally, the mother is only exposed to fetal Rh antigens at the time of birth. She does not develop an immune response until after the birth of the first child. In subsequent pregnancies, however, the anti-Rh antibodies (IgG) circulating through the mother's body can cross the placenta and attack the cells of a subsequent Rh-positive fetus. This causes anemia. During development of the fetus, fetal blood is purified because molecules pass from the fetus across the placenta and then through the mother's liver. After birth, the fetal blood is no longer purified by the maternal circulatory system and the infant develops jaundice. This disease is called **hemolytic disease of the newborn** (previously called **erythroblastosis fetalis**). It is characterized by an enlarged liver and spleen. If untreated, the mortality rate is about 10%.

Hemolytic disease of the newborn can be treated by removal of the fetal Rh-positive blood. Blood is replaced by transfusion with Rh-negative blood. Cells in the new blood will not be attacked by the anti-Rh antibodies that crossed the placenta and now are circulating within the newborn. At a later time, when the anti-Rh antibodies passively acquired (passive natural immunity) from the mother have been diluted and eliminated, these transfused cells are replaced by Rh-positive cells produced by the infant. Cell destruction does not occur at this time because the maternal antibodies have been eliminated from the infant's circulatory system.

Hemolytic disease of the newborn is a type II hypersensitivity reaction that occurs in Rh-positive infants whose mothers are Rh negative and whose fathers are Rh positive.

To prevent hemolytic disease of the newborn, passive artificial immunization of the Rh-negative mother with Rhogam (anti-Rh antibodies) is done at the time of birth (FIG. 15-15). Anti-Rh antibodies react with the fetal Rh-positive cells that enter the mother at the time of birth through traumatized tissue. The reaction of anti-Rh antibodies with Rh-positive cells limits the development of an immune re-

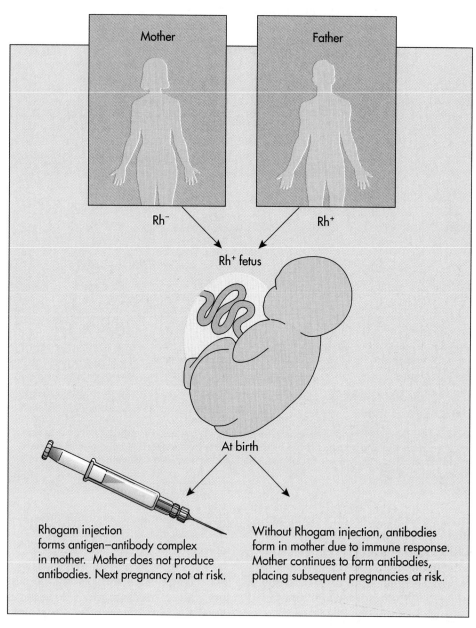

FIG. 15-15 Hemolytic disease of the newborn occurs when an Rh-negative mother becomes sensitized and produces antibodies that attack an Rh-positive fetus. Administration of Rhogam prevents the mother from developing an immune response that would produce anti-Rh antibodies that would attack the fetus.

sponse in the mother. Anti-Rh antibodies bind to the Rh antigens that may have been introduced from the baby to the mother at the time of birth.

Their recognition by the immune system of the mother is prevented. Thus, artificial passive immunization is used to prevent the development of active natural immunity. This treatment is repeated at each birth when the baby is Rh positive and the mother is Rh negative. As a result of this treatment, a serious antibody-dependent cytotoxic hypersensitivity reaction can be prevented.

IMMUNE COMPLEX-MEDIATED HYPERSENSITIVITY

Immune complex-mediated hypersensitivity (type III hypersensitivity) reactions occur when the formation of antibody-antigen complexes triggers the onset of an inflammatory response (FIG. 15-16). Such an inflammatory response is part of the normal immune response. If there are large excesses of antigen, the antigen-antibody–complement complexes may circulate and become deposited in various tissues. Inflammatory reactions from such deposition of immune complexes can cause localized physiological damage to kidneys, joints, and skin. These reactions are called type III hypersensitivity or **Arthus immune-complex reactions.**

> The deposition of antibody-antigen–complement complexes in tissues causes inflammatory reactions.

In the Arthus reaction, the site becomes infiltrated with neutrophils (FIG. 15-17). This leads to extensive injury to the walls of the local blood vessels because lysosomal enzymes and vasoactive substances (substances that cause dilation of blood vessels) are released from these cells. An Arthus reaction sometimes occurs in the lungs because of repeated exposure to antigens on the surfaces of inhaled particulate matter. Symptoms generally include cough, fever, and difficulty in breathing. These symptoms typically develop over a period of 4 to 6 hours. The attack usually subsides within a few days after the removal of the source of the antigen. Persons in particular occupations have a high risk of developing this condition. For example, farmers often develop this reaction because of repeated exposure to the airborne spores of actinomycetes growing on hay. Sugar cane workers, mushroom growers, cheesemakers, and pigeon fanciers are also prone to this condition because of exposure to airborne antigens associated with their activities.

Serum sickness is another type of immune complex disorder. This disease results when patients are given large doses of foreign sera. Serum sickness was more prevalent before the use of antibiotics. Using horse serum antitoxins to provide passive immunization against such infectious diseases as tetanus and

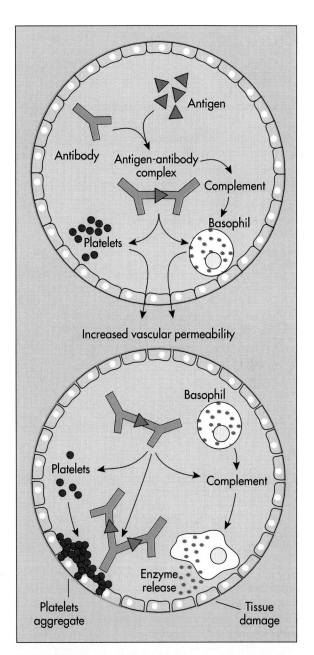

FIG. 15-16 Type 3 hypersensitivity reactions result in inflammation and deposition of immune complexes in blood vessel walls. The immune response that forms antigen-antibody–complement complex triggers degranulation of mast cells and basophils. This causes increased vascular permeability. Enzymes released in the process lead to tissue damage and platelet aggregation. The platelets that aggregate block oxygen transfer and cause further damage.

diphtheria was once practiced extensively before the discovery and use of antibiotics. Antilymphocyte serum for immunosuppression is used to protect against rejection of transplanted tissues.

The antigens in these foreign sera stimulate an immune response. Because large doses of serum are given, these antigens have not been degraded and cleared from the body by the time circulating antibodies appear. Immune complexes form between the

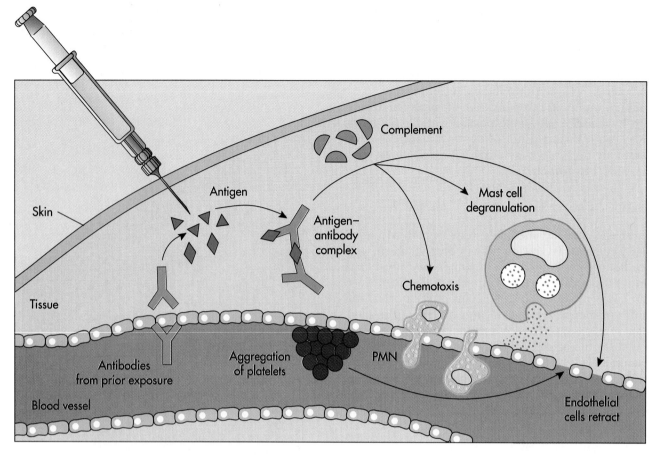

FIG. 15-17 In the Arthus reaction, antigens enter the tissues where they react with antibody and form antigen-antibody complexes. These complexes activate complement and lead to mast cell degranulation and PMN chemotaxis. Enzymes released lead to an inflammatory response and tissue damage.

residual antigens and the circulating antibodies. These antigen-antibody complexes are deposited at various body sites, including joints, kidneys, and blood vessel walls. Symptoms of serum sickness generally appear 7 to 10 days after injection of the foreign serum. Symptoms include fever, nausea, vomiting, malaise, hives, enlarged lymph nodes, and pain in muscles and joints. Numbers of circulating leukocytes are decreased. In many cases of serum sickness, the immune complexes are carried to the kidneys and cause nephritis (inflammatory disease of the kidneys).

> Serum sickness is a type III hypersensitivity reaction caused by the presence of antigens in patients given large doses of foreign serums.

This condition of **glomerulonephritis** can also be brought about by persistent infections (FIG. 15-18). As a result of such infections, antigen-antibody complexes are formed that are deposited within the glomeruli of the kidneys. Immune complexes formed by antibody reactions with antigens produced by

Streptococcus pyogenes (the causative agent of strep throat), hepatitis B virus (the cause of serum hepatitis), *Plasmodium* species (protozoa that cause malaria), and *Schistosoma* spp. (helminthic worms that cause schistosomiasis) may lead to this condition. The persistence of these infections provides a continuing supply of antigen to react with circulating antibodies produced by the infected individual. The immune complex that forms accumulates in the kidneys. Eventually nephritis due to complex-mediated hypersensitivity results.

CELL-MEDIATED (DELAYED) HYPERSENSITIVITY

Cell-mediated or **delayed hypersensitivity (type IV hypersensitivity) reactions** involve sensitized T_{DTH} lymphocytes. As the name implies, these reactions occur only after a prolonged delay after exposure to the antigen. Such reactions often reach maximal intensity 24 to 72 hours after initial exposure. Delayed hypersensitivity reactions occur as allergies to various microorganisms and chemicals. **Contact dermatitis** results from exposure of the skin to various

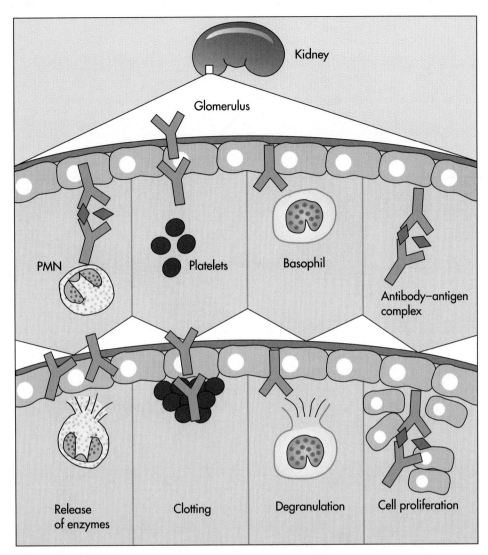

FIG. 15-18 Multiple reactions triggered by antigen-antibody complexes occur as a result of serum sickness. As a result of immune complex deposition, PMNs degranulate and release destructive enzymes, platelets clot, basophils degranulate, and some cells (mesangial cells) proliferate, blocking the glomerulus. These reactions cause kidney damage and the disease glomerulonephritis.

chemicals. Skin rashes are typical of delayed hypersensitivity reactions.

> **Type IV hypersensitivity reactions occur 24 to 72 hours after exposure to the antigen.**

> **Contact dermatitis is a delayed hypersensitivity reaction.**

Poison ivy is one of the best-known examples of contact dermatitis (FIG. 15-19). Contact with catechols in the leaves of the poison ivy plant leads to the development of a characteristic rash with itching, swelling, and blistering. Catechols appear to act as **haptens.** They react with skin proteins to form active immunogens. Lipids in the skin retain the catechols. Catechols combine with skin proteins to bring about a cell-mediated response involving T_{DTH} cells. No dermatitis (skin rash) occurs on primary exposure to

poison ivy. Subsequent exposure of sensitized individuals to the oils of the poison ivy plant results in dermatitis after an initial delay of several days. Various other agents, including metals, soaps, cosmetics, and biological materials, can also cause contact dermatitis. The treatment of contact dermatitis often involves the administration of corticosteroids, which depress cell-mediated immune reactions.

Intracellular bacterial parasites, such as *Mycobacterium tuberculosis* and *Listeria monocytogenes,* also elicit a delayed-type hypersensitivity response. In these situations, activated T_{DTH} cells recognize foreign antigens on the surfaces of infected cells and activate macrophages. Activated macrophages then eliminate the bacteria. Some bacteria may evade being killed or degraded by the macrophages. In such cases, additional macrophages proliferate and aggre-

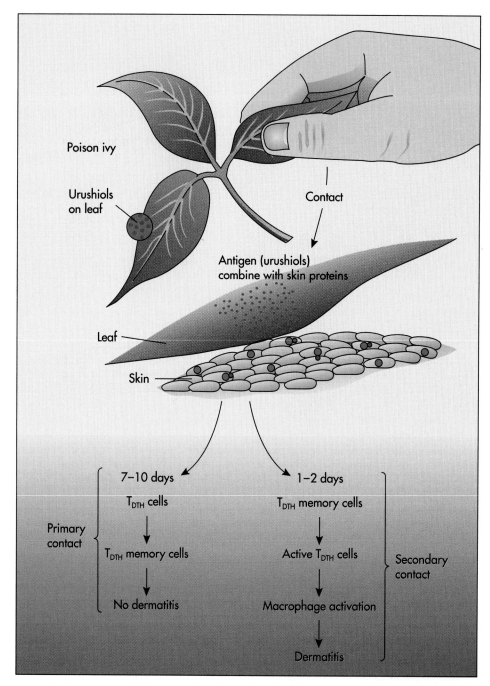

FIG. 15-19   Contact dermatitis, such as poison ivy, involves reactions of T_{DTH} cells that activate macrophages.

gate, leading to the formation of a characteristic nodular or granular mass called a *granuloma* in the tissues surrounding the original infection. The presence of granulomas may be indicative of a chronic cell-mediated response to a microorganism. By surrounding the microorganism, they attempt to prevent its spread to other parts of the body.

The immune response to invasion by fungal cells is very similar to that of bacteria. In addition, many pathogenic fungi can survive inside macrophages and elicit a delayed-type hypersensitivity response. Thus chronic infections with *Histoplasma capsulatum, Coccidioides immitis,* and *Sporothrix schenckii* can lead to granulomatous (granuloma formation) and inflammatory responses. Some parasitic infections may invoke a delayed-type hypersensitivity response. Chronic parasitic infections can lead to granuloma formation and inflammation in the tissues.

The immune response can be undesirable in some cases, such as when tissues are transplanted from one individual to another. In this case the immunological recognition of and response to the foreign antigens of a donor results in *tissue rejection*. In a sense, the normal immune defense mechanism is dysfunctional in transplantation and grafting because in these cases it does not serve the desired or useful function. The ability to transplant major organs such as kidneys and hearts is dependent on the ability to control the normal immune response and prevent rejection of the transplanted tissues (FIG. 15-20).

Tissues contain surface antigens that are coded for by the major histocompatibility complex. Reactions to these antigens generally prevent successful transplantation of tissues unless the transplant tissues come from a compatible donor. A compatible donor is one with an identical histocompatibility complex. Finding a compatible donor is virtually impossible. Even siblings (except identical twins) do not share many histocompatibility antigens. Suppression of the immune response (immunosuppression), therefore, is usually necessary in organ transplant cases.

Immunosuppression is used to permit transplantation of tissues from one individual to another.

Several drugs, including cyclosporin and prednisone, may be used to suppress the normal immune response. The most widely used immunosuppressant for blocking rejection of major organ transplants is cyclosporin. This drug is produced by microorganisms. It blocks rejection of transplanted tissue by suppressing the cell-mediated immune system. Cyclosporin, however, is extremely expensive. Currently it costs over $800 per month to prevent rejection of a transplanted heart. This medication continues to be a requirement for the survival of the successful transplant patient.

Cyclosporin and prednisone are used as immunosuppressants in organ transplant patients.

Immunosuppressed individuals are especially prone to infections and must take prudent precautions, including vaccinations and prophylactic use of antibiotics. Transplant patients also have an increased risk for developing cancer within 5 years af-

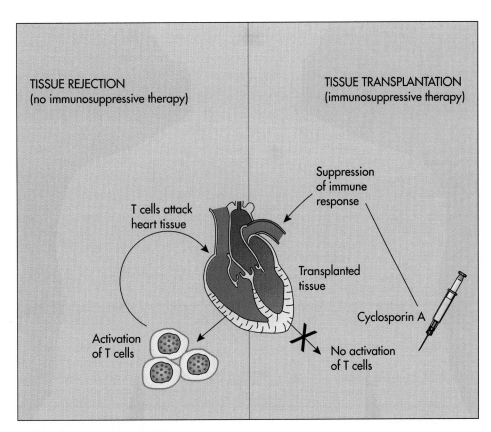

TISSUE REJECTION
(no immunosuppressive therapy)

TISSUE TRANSPLANTATION
(immunosuppressive therapy)

T cells attack heart tissue

Suppression of immune response

Transplanted tissue

Cyclosporin A

Activation of T cells

No activation of T cells

FIG. 15-20 The immune system can be suppressed to prevent tissue rejection in cases of transplantation. Cyclosporin A is commonly used to block T cell activation. In the absence of such immunosuppressive therapy, T cells are activated and attack the transplanted tissue, leading to rejection.

ter the transplant because the cell-mediated immune response is important for the control of malignancy. Suppression of the immune response renders the individual susceptible to pathogens or even opportunists that are normally excluded by the host's immune defense mechanisms. Consequently, extraordinary measures must be practiced to protect such individuals. Extensive antibiotic therapy and hospitalization in wards supplied with HEPA (high efficiency particulate) filtered air are used with these patients.

Immunosuppression in organ transplant cases can cause **graft-versus-host (GVH) disease** (FIG. 15-21).

This occurs when the transplanted (grafted) tissue contains lymphocytes that respond to the MHC antigens of the tissues of the recipient. Nearly all body tissues have class I MHC surface antigens. The immune system of the recipient is unable to control the lymphocytes in the transplanted tissue that begin to attack the recipient's tissues. This commonly is a problem in bone marrow transplants. Bone marrow contains large numbers of B and T cells that can initiate an immune response against the immunosuppressed recipient. In some cases the result is fatal.

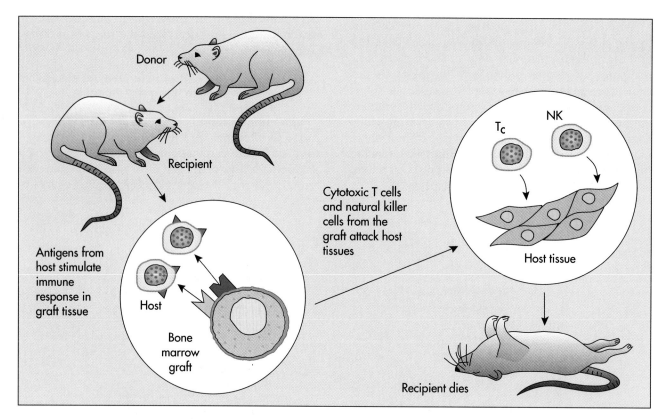

FIG. 15-21 In graft-versus-host disease the graft (usually bone marrow) contains T_C cells and NK cells, which initiate an immune response against host cell antigens. This immune response causes destruction of host cell tissues especially in the skin, liver, and gastrointestinal tract. These reactions may be fatal.

S U M M A R Y

Prevention of Diseases Using the Body's Immune Response (pp. 432-445)

Immunization (Vaccination) (pp. 433-443)

- Epidemics cannot be transmitted from one person to another when a sufficiently high proportion of the population is immune to a disease.
- Vaccines are preparations of antigens designed to stimulate the normal primary immune response. Exposure to the antigens in a vaccine results in a proliferation of memory cells and the ability to exhibit a

secondary memory or anamnestic response on subsequent exposure to the same antigens.
- Vaccines may contain antigens prepared by killing or inactivating pathogenic microorganisms or may use attenuated strains that are unable to cause the onset of severe disease symptoms.
- Some vaccines are prepared by denaturing microbial exotoxins to produce toxoids; these toxoids can elicit the formation of antibody and are reactive with antibody molecules but are unable to initiate the bio-

chemical reactions associated with the active toxins that cause disease conditions.

- Microorganisms used for preparing inactivated vaccines are killed by treatment with chemicals, radiation, or heat. They retain their antigenic properties but cannot cause the onset of the disease caused by the live virulent pathogen.
- Living but attenuated strains of microorganisms used in vaccines are weakened by moderate use of heat, chemicals, desiccation, and growth in tissues other than the normal host. Such vaccines can be administered in relatively low doses because the microorganism is still able to replicate and increase the amount of antigen available to trigger the immune response mechanism.
- Microbial components, such as capsules and pili, can be used to make vaccines.
- Recombinant DNA technology can be used to create vector vaccines that contain the genes for the surface antigens of various pathogens.
- Adjuvants are chemicals that enhance the antigenicity of other biochemicals. They are used in vaccines, often eliminating the need for repeated booster doses and permitting the use of smaller doses of the antigen.
- Vaccine antigens may be introduced into the body intradermally, subcutaneously, intramuscularly, intravenously, or into the mucosal cells lining the respiratory and gastrointestinal tracts.
- Some vaccines are administered after suspected exposure to a given infectious microorganism to elicit an immune response before the onset of disease symptomatology. In such cases the development of the infecting pathogen must be slow and the vaccine must be able to initiate antibody production quickly before active toxins are produced and released.

Artificial Passive Immunity (pp. 443-445)

- Passively acquired immunity comes about when antibodies produced in another organism are introduced into the body. Artificially acquired passive immunity is derived from the injection of antibodies into an individual to provide immediate protection against a pathogen or toxin.

Immunodeficiencies (pp. 445-451)

- An immunodeficiency is the result of an inadequate immune response.
- It can be inherited or acquired.

Severe Combined Immunodeficiency (pp. 445-446)

- People with severe combined immunodeficiency disease (SCID) have neither functional B or T lymphocytes. They have no immunological response. Any infection can be fatal.
- Adenosine deaminase linked to polyethylene glycol (PEG-ADA) is used to treat SCID. Bone marrow transplants are also done.

DiGeorge Syndrome (pp. 446-447)

- If the thymus does not develop properly, T cells are not differentiated and DiGeorge syndrome results.

Bruton Congenital Agammaglobulinemia (pp. 447-448)

- If B cells do not differentiate and produce antibodies, Bruton congenital agammaglobulinemia results. This condition affects only males.

- It is treated with IgG to maintain antibody levels in the circulatory system.

Late-onset Hypogammaglobulinemia (pp. 448-449)

- Late-onset hypogammaglobulinemia is the most common immunodeficiency. Individuals with this condition are deficient in circulating B cells and/or B cells with IgG surface receptors.

Complement and Cellular Deficiencies (p. 449)

- If C3 complement is not produced, the body cannot defend against bacterial infections.
- Defective monocytes, neutrophils, and macrophages may cause immunodeficiencies.

Malignant Cell Development (p. 449)

- If the immune response system does not recognize and respond to the presence of abnormal cells, malignant cells can continue to grow.

Acquired Immunodeficiency Syndrome (AIDS) (pp. 449-451)

- AIDS is caused by the human immunodeficiency virus (HIV). HIV destroys T helper cells. The proportion of T suppressor cells increases, which depresses immune functions. B cells do not produce sufficient antibodies. Amounts of lymphokines are lowered.
- There is no cure for AIDS. Prevention is the only way to control it.
- Azidothymidine (AZT) and dideoxyinosine (ddI) slow the rate of HIV replication.

Autoimmunity (pp. 451-453)

- Autoimmunity is the result of the inability of the body to properly recognize self-antigens. The immune system kills the body's own cells.

Rheumatoid Arthritis (pp. 452-453)

- Rheumatoid arthritis is chronic joint inflammation. Lysozymes attack antigens and B cells make IgM in response, causing inflammation. There is no cure.

Myasthenia Gravis (p. 453)

- Myasthenia gravis affects the neuromuscular system. The pores in neuron membranes are blocked. These neurons stimulate muscles. Blocked neurons do not release acetylcholine, which initiates muscle cell contraction. Treatment is with drugs that inhibit the enzyme that breaks down acetylcholine.

Systemic Lupus Erythematosus (p. 453)

- Systemic lupus erythematosus is a systemic autoimmune disease. Autoantibodies are made against DNA components. The deposition of immune complexes causes inflammation.

Graves Disease (p. 453)

- Graves disease is an autoimmune disease of the thyroid in which antibodies to thyroid-stimulating hormone receptors are produced, allowing the overproduction of hormones.

Multiple Sclerosis (p. 453)

- Multiple sclerosis is believed to be a T-cell-mediated autoimmune disease. Lesions containing macrophages and lymphocytes are found in the central nervous system of people with MS.

Hypersensitivity Reactions (pp. 453-464)

- Hypersensitivity is an excessive immunological response to an antigen. It is an inappropriate immunological response.

Anaphylactic Hypersensitivity (pp. 454-457)
- Type I hypersensitivity reactions can be life threatening. They occur when antigens react with antibodies bound to mast or to basophil blood cells.
- Antigens that initiate anaphylactic hypersensitivity reactions are called allergens. The reaction of an allergen and a mast or basophil cell causes the cell to rupture and release vasoactive substances. Vasoactive substances can cause anaphylactic shock.
- Allergies are immunological hypersensitivity reactions that produce exaggerated responses in sensitized individuals.
- Fc fragments of IgE bind to mast and basophil cells. IgE antibodies attach to receptors on mast cells in connective tissue.
- Sensitization is the initial recognition of a foreign substance against which B cells are stimulated to make IgE antibodies.
- The next time the allergen appears, it reacts directly with the IgE fixed to the mast cells and basophils.
- Hay fever is an atopic allergy caused by the interaction of allergens with cell-bound IgE on the mucosal membranes of the upper respiratory tract and conjunctival tissues.
- Asthma is a hypersensitivity reaction of the lower respiratory tract that causes constriction of the bronchial tubes.
- Allergies are treated by avoiding allergens, using antihistamines and corticosteroids, and desensitization.

Antibody-dependent Cytotoxic
Hypersensitivity (pp. 458-461)
- In type II hypersensitivity reactions, an antigen on a cell surface combines with an antibody, stimulating phagocytic attack or the initiation of the complement pathway.
- Antibody-dependent cytotoxic hypersensitivity reactions occur after transfusions with incompatible blood types because antibodies against all nonself antigens circulate in the blood. Donated cells are lysed.
- Type O blood types are universal donors and type AB blood types are universal recipients. Type O blood cells have neither A nor B antigens.
- Hemolytic disease of the newborn occurs when an Rh-negative mother and an Rh-positive father have an Rh-positive child. The mother develops Rh antibodies in response to the fetus' Rh antigens.
- During subsequent pregnancies, these anti-Rh antibodies can cross the placenta and cause anemia in the fetus. Such mothers are passively immunized with anti-Rh antibodies to limit the development of the bank of memory immune response cells.

Immune Complex-mediated Hypersensitivity (pp. 461-462)
- In type III hypersensitivity reactions the formation of antibody-antigen–complement complexes triggers the onset of an inflammatory response. Neutrophils come to where these complexes are deposited.
- Serum sickness is caused by the presence of antigens in large amounts of foreign serum. Glomerulonephritis is the result of the deposition of antibody-antigen complexes in the glomeruli of the kidneys.

Cell-mediated (Delayed)
Hypersensitivity (pp. 462-464)
- Type IV hypersensitivity reactions involve T lymphocytes.
- Contact dermatitis is a delayed hypersensitivity reaction, usually occurring 24 to 72 hours after contact with the allergen.

Transplantation and Immunosuppression (pp. 465-466)
- Immunosuppression, the minimization of immune reactions, is necessary to prevent rejection of transplanted or grafted tissues.
- Cyclosporin and prednisone suppress the normal immune response.
- Immunosuppression can cause graft-versus-host disease. The recipient's immune system cannot control the lymphocytes in the transplanted tissue.

CHAPTER REVIEW

REVIEW QUESTIONS

1. What substances can be used to prepare vaccines?
2. What are the differences between attenuated and inactivated vaccines?
3. Compare the Salk and Sabin polio vaccines.
4. How was immunization used to eliminate smallpox? Why can't this be done for the common cold?
5. What are the consequences of a child being born with severe combined immunodeficiency?
6. What are the differences between immediate hypersensitivity and delayed-type hypersensitivity?
7. What is an autoimmune disease? Name some specific autoimmune diseases.
8. What is allergy? How can it be diagnosed?
9. How does desensitization work to control allergy?
10. Who should be immunized against influenza?
11. What are the consequences of deficiencies in the complement system?
12. What are the consequences of deficiencies in the mononuclear phagocyte system?
13. How are immunodeficiency diseases related to the cells of the immune system?
14. What are the consequences of using immunosuppressive drugs to treat a patient?
15. Compare the four types of hypersensitivity reactions.
16. How is hemolytic disease of the newborn prevented?
17. What are the problems in the transplantation of organs from one individual to another? How are these overcome?

CRITICAL THINKING QUESTIONS

1. What characteristics of some microorganisms make vaccination against them almost impossible? Why is it unlikely that we will see a vaccine against the common cold in the near future?
2. Why doesn't the immune system protect humans from developing AIDS?
3. What role should animals have in the development of vaccines?
4. Should animal tissues be used for transplantation into humans?
5. How can recombinant DNA technology be used to develop new vaccines?

READINGS

Baxby D: 1981. *Jenner's Smallpox Vaccine: The Riddle of the Vaccine Virus and Its Origin*, London, Heinemann Educational.

    Discusses the historical nature of the vaccinia virus, speculates on its origin, and contemplates its present status in the medical world.

Buisseret P: 1982. Allergy, *Scientific American*, August, pp. 86-95.

    An up-to-date account of what is known about this common problem, stressing that allergy is a disorder of the immune system.

Chanock RM et al. (eds.): 1987. *Vaccines 87: Modern Approaches to New Vaccines: Prevention of AIDS and Other Viral, Bacterial, and Parasitic Diseases*, Cold Spring Harbor, New York, Cold Spring Harbor Laboratory.

    Proceedings of a conference on the state-of-the-art in vaccine technology.

Chase A: 1982. *Magic Shots: A Human and Scientific Account of the Long and Continuing Struggle to Eradicate Infectious Diseases by Vaccination*, New York, Morrow.

    Describes the work of Jenner on smallpox, Pasteur on rabies, and Theiler on yellow fever, and looks at the possibilities for vaccines against colds and cancers.

Coulter HL and BL Fisher: 1985. *DPT, A Shot in the Dark*, San Diego, Harcourt, Brace Jovanovich.

    Discusses the possible complications and sequelae of the use of the DPT vaccine as the preventative inoculation against whooping cough.

Dardick KR and HH Neumann: 1990. Foreign Travel & Immunization Guide, Ordell, NJ, Medical Economic Books.

    Gives the recommended vaccines by country for travel to various regions of the world.

Findlay S: 1987. Three new vaccines for junior, *U.S. News and World Report* 103(November 2):14.

    Discusses progress on new vaccines for children against chicken pox, meningitis, and pertussis.

Klein A: 1972. *Trial by Fury: The Polio Vaccine Controversy*, New York, Scribner's.

    This account of the development of the polio vaccines illustrates what science can do, how science works, and how public opinion influences what is done.

Lawrence J: 1985. The immune system in AIDS, *Scientific American*, June, pp. 702-736.

    An excellent overview of how T cells function in the immune response and how AIDS thwarts that response.

Mills J and H Masur: 1990. AIDS-related infections, *Scientific American* 263(2):50-57.

    A discussion of the opportunistic infections that occur when the body's immune system no longer is protective because of an HIV infection and AIDS.

Neustaedter R: 1990. *The Immunization Decision: Guide for Parents*, Berkeley, CA, North Atlantic Press.

    A useful guide regarding immunizations that should be given to children.

Old LJ: 1988. Tumor necrosis factor, *Scientific American* 258(5):59-75.

    The author discovered that infection stimulates production of tumor necrosis factor. TNF has an anticancer effect that is now being tested as an anticancer drug.

Raffel S: 1982. Fifty years of immunology, *Annual Reviews of Microbiology* 36:1-26.

    Review of the status and evolution of the study of immunology beginning with the author's first class in 1930 and following the field and his part in it for the next 50 years.

Rennie J: 1990. The body against itself, *Scientific American* 263(6):106-115.

    A discussion of how the body's immune defenses can go awry and cause damage to the body.

Sanders P: 1982. *Edward Jenner, the Cheltenham Years, 1795-1823: Being a Chronicle of the Vaccination Campaign*, Hanover, NH, University Press of New England.

Smith KA: 1990. Interleukin-2, *Scientific American* 262(3):50-57.

    A discussion of one of the very important regulators of the body's immune defense system.

von Boehmer H and P Kisielow: 1991. How the immune system learns about self, *Scientific American* 265(4):74-81.

    A discussion of the molecular basis of how the immune defense system distinguishes self from foreign antigens.

Welch WJ: 1993. How Cells Respond to Stress, *Scientific American* 268(5):56-64.

    An examination of how cells cope with infection, autoimmune disease, and cancer.

Woodrow GC and MM Levine: 1990. *New Generation Vaccines*, New York, Marcel Dekker.

    Brings together multidisciplinary approaches for creating and improving vaccines, demonstrating the use of recent biotechnological advances, including recombinant DNA and monoclonal antibodies.

CHAPTER 16

Diagnosis of Human Diseases

PREVIEW TO CHAPTER 16

In this chapter we will:
• See the approaches used for diagnosing infectious diseases.
• Learn how different specimens are handled to determine the cause of a disease.
• Examine the applications of serological techniques with respect to identification of pathogens.
• Examine the applications of gene probes with respect to identification of pathogens.
• Find out how computers are being used to assist the diagnostician.
• Discover how the clinical microbiologist aids the physician.
• Learn the following key terms and names:

alpha hemolysis	lumbar puncture
anemia	Mantoux test
antigenic differences	metabolic products
API-E system	Micro-ID system
band cells	Minitek system
beta hemolysis	nasopharyngeal swabs
cerebrospinal fluid (CSF)	nucleic acid hybridization
chronic diseases	oxidase activity
counter	passive agglutination
immunoelectrophoresis	polymerase chain reaction
(CIE)	(PCR)
cytopathic effects (CPE)	radioimmunoassays (RIA)
differential blood count	selective media
differential media	serological test procedures
direct fluorescent	signs
antibody staining	skin testing
(FAB)	stab cell
disease syndrome	symptoms
Enterotube system	systemic bacterial
enzyme-linked	infection
immunosorbent assay	throat swabs
(ELISA)	transtracheal aspiration
erythema (reddening)	tuberculin reaction
gene probes	Weil-Felix test
heterophile antibodies	
indirect	
immunofluorescence	
tests	

Despite preventive measures and host defenses, infectious diseases occur. Diagnosis of these diseases is important for instituting appropriate treatment. The reproduction of pathogenic microorganisms within the body often produces specific diseases that are associated with characteristic signs and symptoms. These signs and symptoms aid the physician in diagnosing the disease and in determining the proper course of treatment. **Signs** are objective changes such as a rash or fever that a physician can observe. **Symptoms** are subjective changes in body function, such as pain or loss of appetite, that are experienced by the patient. A characteristic group of signs and symptoms constitutes a **disease syndrome.** Often the physician is able to diagnose a disease based exclusively on the symptoms reported by the patient and the signs observed. In other cases, more elaborate laboratory tests are necessary to identify the cause of the disease.

> A disease syndrome is a characteristic group of signs and symptoms.
>
> Signs are objective, observable changes produced by disease. Symptoms are subjective body changes.

In **acute diseases** the symptoms and signs develop rapidly, reaching a height of intensity, and end fairly quickly. Measles, cholera, and influenza are all examples of acute diseases. In **chronic diseases** the symptoms persist for a prolonged period of time. The persistent cough associated with chronic bronchitis is typical of the long-term signs associated with a chronic disease.

The physician observing a patient with a red sore throat and a fever assumes that these signs are the result of a microbial infection (FIG. 16-1). In many such cases, when the presumptive evidence strongly indicates pharyngitis (infection of the pharynx), treatment is usually administered without rigorous clinical diagnosis and confirmation of the cause even though it would be appropriate to identify the etiologic agent by laboratory testing.

DIFFERENTIAL BLOOD COUNTS

When the identification of a microbial infection is not clear-cut, additional presumptive evidence of an infectious process is needed. It can be obtained by performing a **differential blood count** in which the relative concentrations of the different types of blood cells are determined. This clinical procedure can provide a general indication of the nature of the infecting agent, that is, if the disease is mediated by a virus, bacterium, fungus, or protozoan. Changes in the

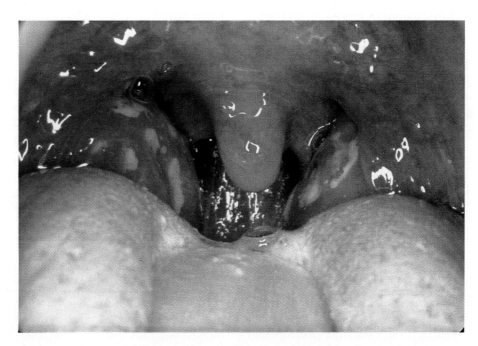

FIG. 16-1 Appearance of the mouth and throat in streptococcal tonsillitis. Streptococcal infection of the throat often causes readness and the appearance of grey-white patches.

	TABLE 16-1				
	Differential Blood Counts for Representative Bacterial Infections				
	NUMBER (PER µL)				
CELL TYPE	**NORMAL**	**SCARLET FEVER**	**APPENDICITIS**	**STAPHYLOCOCCAL SEPTICEMIA**	**TULAREMIA**
Leukocytes, total	7,500	16,680	13,800	34,950	19,550
Basophils	0-1	2	0	0	1
Eosinophils	2-4	0	0	0	0
Immature neutrophils (myelocytes + stabs)	0-6	100	70	40	60
Mature neutrophils (segmented)	58-66	58	20	46	23
Lymphocytes	21-30	18	10	8	12
Monocytes	4-8	7	0	3	5

composition of the blood usually occur as a consequence of a microbial infection. Such changes generally result from the immune response. The changes are reflected in shifts in the relative quantities and types of white blood cells.

A differential blood count determines the relative concentrations of different types of blood cells to provide a general indication of the nature of an infecting agent.

An elevated white blood cell count is characteristic of many systemic infections. A systemic bacterial infection, for example, is normally characterized by a progressive increase in neutrophils (Table 16-1); neutrophils are phagocytic white blood cells. Young neutrophils known as **stab** or **band cells** particularly increase. As compared to mature neutrophils, stab cells have a U-shaped nucleus that is slightly indented but not segmented. The increase in stab cells is known as a shift to the left. The shift refers to a blood cell classification system in which immature blood cells are positioned on the left side of a standard reference chart and mature blood cells are placed on the right. The recovery phase of an infection is characterized by a reduction in fever, a decrease in the total number of leukocytes, and an increase in the number of monocytes. Gradually, the relative numbers of the various white blood cells return to their respective normal ranges.

Systemic bacterial infections are generally characterized by a progressive increase in the number of neutrophils.

Some localized infections, like abdominal abscesses, also result in increased neutrophils. The diagnosis of appendicitis is aided by differential white blood cell counts that show increased neutrophils (FIG. 16-2). Not all bacterial infections, though, show this characteristic increased white blood cell count.

Most bacterial infections result in increased numbers of neutrophils. Many viral infections similarly result in lowered numbers of white blood cells. A general indication of whether a disease is of bacterial or viral origin, therefore, may be obtained by performing a white blood cell count and determining whether there is a significant shift in the quantity of neutrophils.

Changes in quantities of eosinophils may also indicate the nature of the infection. An eosinophil is a type of white blood cell that is involved in nonspecific resistance against pathogenic microorganisms. The number of eosinophil cells generally declines during systemic bacterial infections. Increased numbers of eosinophils are signs of allergic diseases and parasitic infections, including those caused by protozoans. Thus the observation of elevated numbers of eosinophils is useful in the preliminary diagnosis of such diseases.

During bacterial infections the numbers of eosinophils generally decline, but increase during parasitic infections and allergic diseases.

In some diseases, such as infectious mononucleosis, there are characteristic changes in the white blood cells (FIG. 16-3). In this disease, there is a transient increase in B lymphocytes, which characteristically are enlarged—making them appear like monocytes—and show obvious changes in the nucleus, including the shape, size, and density of the nuclear region. These changes are useful in the diagnosis of this disease.

Some microbial infections, malaria for example, result in decreased numbers of red blood cells **(anemia)**. Thus a simple examination of the blood often gives a preliminary indication of the etiology of a disease condition, establishing the direction of additional test procedures for positively identifying the causative agent of a disease.

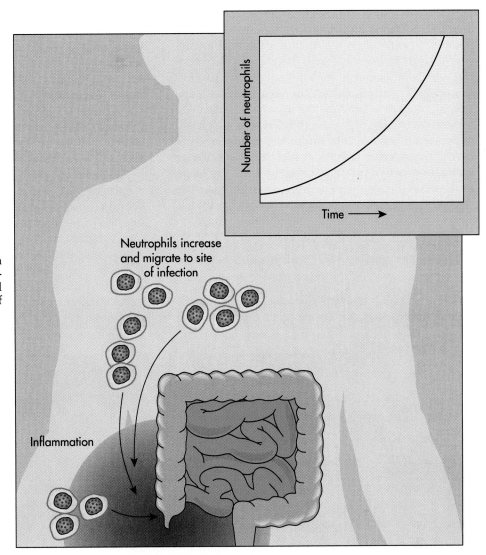

FIG. 16-2 The concentration of white blood cells (neutrophils) in circulating blood increases greatly in cases of appendicitis.

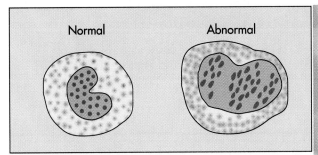

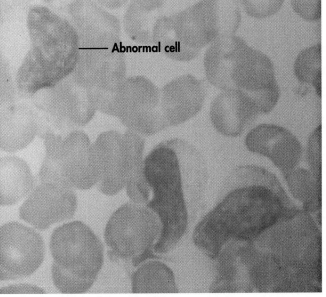

FIG. 16-3 A major characteristic of infectious mononucleosis is the presence of numerous atypical mononuclear cells in the peripheral blood. The cells vary greatly in size and shape. The nucleus may be round, bean-shaped, or lobulated; the cytoplasm is vacuolated and more basophilic than usual. Dividing cells are found in the peripheral blood, and mitotic activity is greatly enhanced, but cellular structure is not fundamentally deranged.

SKIN TESTING

Skin testing based on delayed hypersensitivity reactions can be another useful procedure in the presumptive diagnosis of several diseases. In **skin testing**, sterile antigens from a test organism or substance are injected below the skin surface. The development of induration that may be accompanied by redness in 24 to 72 hours is evidence for a delayed hypersensitivity reaction, indicating previous exposure to that specific antigen. A positive skin test may indicate an active infection caused by the organism from which the antigens are derived but could just as well reflect an earlier exposure to that organism.

> The development of induration 72 hours after an intradermal injection of a test organism is evidence of a delayed hypersensitivity reaction, which indicates previous exposure to an antigen.

The classic skin test for a microbial infection is the **tuberculin reaction** for detecting probable cases of tuberculosis (FIG. 16-4). A purified protein derivative extract (PPD) from *Mycobacterium tuberculosis* is injected subcutaneously and the area around the injection is observed for evidence of a delayed hypersensitivity reaction. The **Mantoux test** is commonly used in the United States. An appropriate dilution of PPD is injected intradermally into the superficial layers of the skin of the forearm. A positive test results in **erythema** (reddening) and **induration** (hardening) of the skin, with the peak reaction occurring between 48 to 72 hours.

Similar skin tests are available for the diagnosis of coccidioidomycosis, using coccidioidin, an antigen derived from *Coccidioides immitis*; histoplasmosis, using histoplasmin, a crude filtrate from *Histoplasma capsulatum*; leprosy, using lepromin derived from *Mycobacterium leprae*; brucellosis, using brucellergen obtained from a *Brucella* species; and lymphogranuloma venereum, using lygranum from *Chlamydia* species.

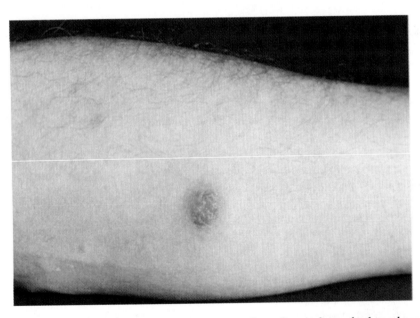

FIG. 16-4 A positive test for hypersensitivity to tuberculin. A dose of tuberculoprotein was pricked into the skin. After 3 days, the site is raised into a firm plaque. Vesicles that formed at the sites of injection have ruptured, and necrosis is beginning in the central zone.

CULTURAL APPROACHES FOR IDENTIFYING PATHOGENS

SCREENING AND ISOLATION PROCEDURES

Disease symptoms, changes in blood composition, and skin testing can serve as indicators of microbial infection and possibly the nature of the microorganisms causing the disease. Many other nonmicrobiological factors may produce similar symptoms and clinical findings. The positive diagnosis of an infectious disease, therefore, requires the isolation and identification of the pathogenic microorganism or the identification of antigens specifically associated with a given microbial pathogen. Various procedures are employed for the isolation of pathogenic microorganisms from different tissues (FIG. 16-5; Table 16-2, pp. 476-477). Many identification tests can only be performed using pure cultures consisting of a single variety of isolated pathogen. Accuracy, reliability, and speed are important factors governing the selection of clinical identification protocols. Obtaining

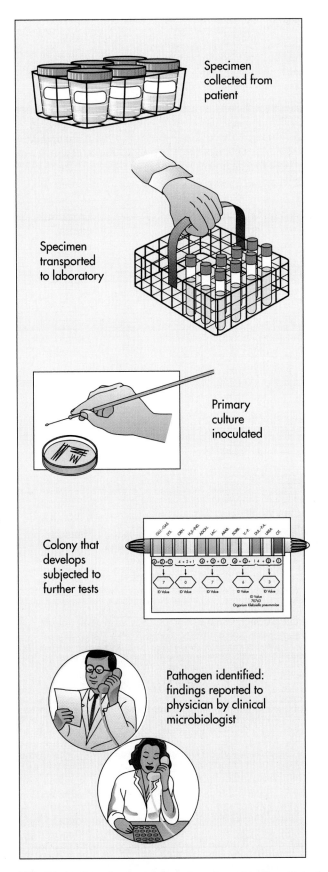

reliable clinical specimens is sometimes difficult and must be done with care. Specimens should be cultured as soon as possible after collection for accuracy. In some cases specimens must be placed immediately into transport media, which facilitate the viability of frail pathogens between the time of collection and culturing and prevent the overgrowth by more robust organisms that are members of the normal microbiota.

> An infectious disease cannot be positively diagnosed until the microbial pathogen or its associated antigens are isolated and identified.

Very different procedures are required for the isolation of different types of microorganisms. For example, procedures used for the isolation of pathogenic bacteria are not applicable for the isolation of viruses, and procedures designed for the recovery of aerobic microorganisms preclude the isolation of obligate anaerobes. Most clinical procedures are designed to screen and facilitate the recovery of the etiologic agents of disease that predominate within specific tissues. When the symptoms suggest that the disease may be caused by a rare pathogen and/or routine screening fails to detect a probable causative microorganism, additional specialized isolation procedures may be required. In the following pages, we will examine some of the routine procedures employed for the isolation of pathogens from various sites of the human body so that the etiologic agents of infectious diseases can be identified (FIG. 16-6).

Upper Respiratory Tract Cultures

For the isolation of pathogens from the upper respiratory tract, throat and nasopharyngeal cultures are collected using sterile swabs. The swabs are placed in

FIG. 16-5 Culturing of microorganisms is a mainstay of clinical identification by culturing. The cultures usually are identified by morphological and metabolic characteristics determined from the cultures. The cultural methods are supplemented by immunological and gene probe tests.

FIG. 16-6 Samples are collected and sent to the clinical microbiology laboratory where they are analyzed. Numerous cultures are examined and identified. The results of tests are reported to the physician, who uses the information to guide the selection of appropriate treatments.

TABLE 16-2

Some Procedures Used for the Diagnosis of Various Diseases, Indicating the Collection Method, Culture Medium, and Bacteria Detected

BODY PART	COLLECTION METHOD	CULTURE MEDIUM	ORGANISM	RESULT	DISEASE
Upper respiratory tract: throat and nasopharyngeal cultures	Sterile cotton swabs	Blood agar	*Streptococcus pyogenes*	Beta-hemolysis	Pharyngitis, rheumatic fever
		Chocolate agar	*Haemophilus influenzae* *Neisseria meningitidis*		Epiglottitis Meningitis
		Bordet-Gengou agar	*Bordetella pertussis*		Whooping cough
		Tellurite serum agar	*Corynebacterium diphtheriae*	Smooth, glistening gray-black colonies	Diphtheria
Lower respiratory tract	Transtracheal aspiration of sputum	Blood agar	*Streptococcus pyogenes, Staphylococcus aureus*		Pneumonia
		Chocolate agar	*Streptococcus pneumoniae*		
		MacConkey agar	*Klebsiella pneumoniae, Haemophilus influenzae*		
		Sabouraud agar	*Coccidioides immitis, Candida albicans*		
		Lowenstein-Jensen agar	*Mycobacterium tuberculosis*	Brown-cream colonies; acid-fast, red cells	Tuberculosis
Central nervous system	Lumbar puncture for cerebrospinal fluid	Liquid enrichment media	*Streptococcus pneumoniae*		Meningitis
		Blood agar	*Neisseria meningitidis*		
		Chocolate agar	*Haemophilus influenzae*		

BODY PART	COLLECTION METHOD	CULTURE MEDIUM	ORGANISM	RESULT	DISEASE
Circulatory tract blood	Venipuncture	Radiolabeled glucose medium Roll-tube streak anaerobic culture	Various	Radiolabeled gas production	Septicemia
Urinary tract	Midstream catch of voided urine	Blood agar Cysteine lactose electrolyte-deficient agar MacConkey agar, Eosin Methylene Blue agar	*Escherichia coli* *Klebsiella* sp., *Proteus* sp., *Pseudomonas* sp., *Salmonella* sp. *Serratia* sp., *E. coli*, and other Gram-negative rods	Less than 10^5 bacteria/mL	Urinary tract infections
Genital tract	Urethral exudate (males); swabs from cervix, vagina, and rectum (females)	Thayer-Martin medium and chocolate agar	*Neisseria gonorrhoeae*	Gram-negative kidney bean shaped cells	Gonorrhea
Intestinal tract	Stool samples	Hektoen enteric media, xylose-lysine-desoxycholate media, brilliant green, EMB, Endo, and MacConkey agars	*Salmonella* spp., *Shigella* spp.		Gastroenteritis
Eyes and ears	Fluids	Blood, chocolate, MacConkey agar	Various	Growth	Bacterial infection
Skin	Swabs, aspirates, or washings from lesions	Anaerobic culture techniques	*Clostridium tetani, C. perfringens*	Growth	Tetanus Gas gangrene

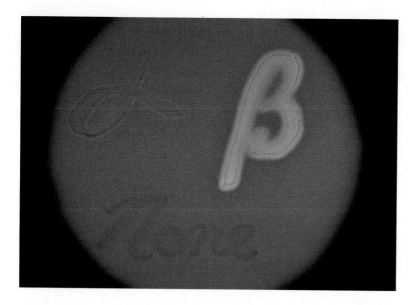

FIG. 16-7 Sheep blood agar plate exhibiting all three types of hemolysis: complete hemolysis (β hemolysis), partial hemolysis (α hemolysis), and no hemolysis (γ hemolysis).

sterile transport media to prevent desiccation during transit to the laboratory. These cultures are streaked onto blood agar plates—prepared by using defibrinated sheep red blood cells in the United States or defibrinated horse red blood cells in the United Kingdom—and incubated in an atmosphere of 5% to 10% CO_2 for isolation of microorganisms. Human red blood cells are not used in the preparation of blood agar because the natural antibodies inhibit the recovery of bacteria, particularly *Streptococcus* species. Blood agar plates permit the detection of **alpha hemolysis** (greening of the blood around the colony, indicating partial lysis of red blood cells) or **beta hemolysis** (zones of clearing around the colony, indicating complete lysis of red blood cells) (FIG. 16-7). Some *Streptococcus* species show no change in the blood agar around the colony; this indicates a lack of hemolysis. *S. pyogenes* forms relatively small colonies and demonstrates beta hemolysis on blood agar. It is the predominant pathogen detected by using throat swabs and blood agar plates. The detection of β-hemolytic streptococci is important because these organisms can cause rheumatic fever. β-hemolytic *Staphylococcus aureus* cultures, as well as many other hemolytic bacteria, may also be detected by using this procedure.

Haemophilus influenzae, Neisseria meningitidis, and *N. gonorrhoeae* can also be detected by using throat swabs and plating on various media. These organisms grow better on chocolate agar than on plain blood agar (FIG. 16-8). Chocolate agar is a medium prepared by heating blood agar until it turns a characteristic brown color. Thayer-Martin medium is preferred for isolation of *N. gonorrhoeae* because growth of normal throat microbiota is inhibited by antibiotics (colistin sulfate, trimethoprim, vancomycin, and nystatin) that have been added. Infection of the upper

respiratory tract with these pathogens can lead to serious diseases, such as bacterial meningitis, if the infection spreads. Early diagnosis is very important in such cases. Also, *Haemophilus influenzae* may cause acute epiglottitis, the rapid diagnosis of which is important because death can result within 24 hours.

Nasopharyngeal swabs may also be used for determining the presence of *Bordetella pertussis. B. pertussis* is the causative organism of whooping cough.

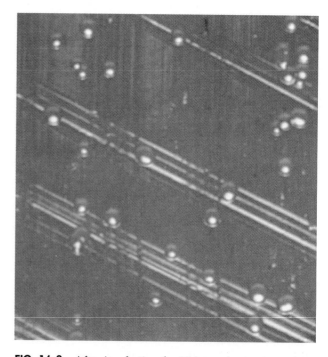

FIG. 16-8 After incubation for 24 hours in an atmosphere enriched with carbon dioxide, colonies of *Neisseria gonorrhoeae* are small, discrete, transparent, and hemispherical on chocolate agar (blood agar in which the blood has been heated).

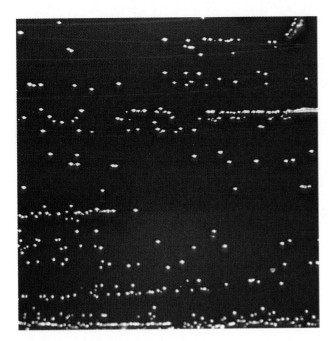

FIG. 16-9 Colonies of *Bordetella pertussis* on Bordet-Gengou potato medium. The dome-shaped colonies have a smooth, reflective metallic surface.

The isolation of *B. pertussis* requires plating on a special medium, such as Bordet-Gengou potato medium (FIG. 16-9). This medium contains potato, glycerol, and blood that neutralize growth-inhibiting substances that may be present in the agar or peptone (meat digest). Cultures of *B. pertussis* also may be obtained by having the patient cough directly onto a plate of Bordet-Gengou medium.

When diphtheria is suspected, additional special procedures must be carried out. The observation of bacteria demonstrating typical snapping division indicates the possible presence of *Corynebacterium diphtheriae*. These bacteria divide by an unusual type of binary fission, which forms daughter cells that remain partially attached to one another at an angle. As a result of this snapping division, *C. diphtheriae* cells appear like Chinese characters, or sometimes in palisades where cells are lined up parallel to one another like a picket fence. However, this is not a positive diagnosis because other bacteria with similar morphology may be present. Usually, several media are employed in the culture of *C. diphtheriae*. Colonies of *C. diphtheriae* on tellurite serum agar, a medium used for the culture of *Corynebacterium* species, appear smooth, glistening, and gray-black. Gram stains prepared on colonies that develop on tellurite agar can confirm the presence of morphologically typical *C. diphtheriae*. This gives an early indication of the presence of this bacterium before more rigorous identification procedures can be performed.

In addition to culturing for bacterial pathogens, throat swabs can be used to collect viral pathogens. The laboratory growth of viruses employs tissue cultures rather than bacteriological media. Rhesus monkey kidney cells are used most frequently for viral tissue culture. Antibiotics are added to the tissue culture to prevent bacterial and fungal growth. Viruses are washed from throat swabs by using appropriate synthetic medium such as Hanks' or Earls' basal salt solutions. The solution is then added to tissue culture tubes or plates.

Viral infection of tissue culture cells normally produces morphological changes, known as **cytopathic effects** (CPE). These changes can be observed readily by microscopic observation (FIG. 16-10, p. 480). Some viruses exhibit a characteristic CPE that can be used in the identification of the virus, but in other cases, serological procedures and/or electron microscopy are necessary to identify viral isolates.

Sterile cotton swabs are used to collect throat and nasopharyngeal cultures, which are streaked onto blood agar plates or placed into tissue cultures for isolation and identification of bacterial and viral pathogens from the upper respiratory tract.

Lower Respiratory Tract Cultures

Isolating microbial pathogens from the lower respiratory tract is a more formidable task than culturing organisms from the upper respiratory tract. Sputum, an exudate containing material from the lower respiratory tract, is frequently used for the culture of lower respiratory tract pathogens. Unfortunately, sputum samples vary greatly in quality and should therefore be examined microscopically before screening. The microscopic examination is carried out to determine whether the samples are suitable for the culture of lower respiratory tract organisms. Acceptable sputum samples should have a high number of neutrophils, should show the presence of mucus, and should have a low number of squamous epithelial cells. A large number of epithelial cells generally indicates contamination with oropharyngeal secretions, and such samples are not suitable.

To ensure the quality of lower respiratory tract specimens, transtracheal aspiration may be employed. In this technique a needle is passed through the neck, a catheter is extended into the trachea, and samples are then collected through the catheter tube. Collection of sputum by using the transtracheal procedure avoids contact with oropharyngeal microorganisms so that the clinician is certain of the source of any isolates that are obtained. It is important to know where the isolated strains originate because some microorganisms are not likely to be associated with disease when they occur among the normal microbiota of the upper respiratory tract, but if these same or-

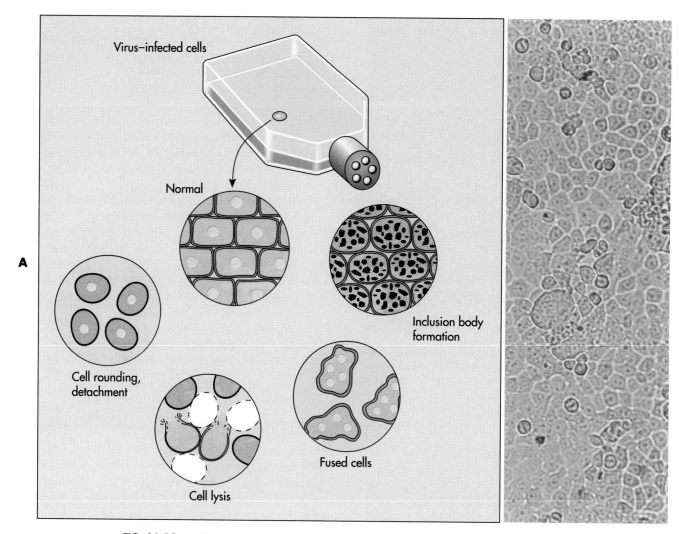

A

Virus–infected cells

Normal

Inclusion body formation

Cell rounding, detachment

Cell lysis

Fused cells

FIG. 16-10   **A,** Infections of animal cells can result in various abnormalities known as cytopathic effects. **B,** Micrograph showing the cytopathic effect on HEp-2 cells grown in tissue culture by an infection with respiratory syncytial virus.

ganisms are found in the lower respiratory tract they are prime candidates for the etiologic agents of disease.

Sputum, often collected by transtracheal aspiration, is used for the culture of pathogens from the lower respiratory tract.

The routine examination of sputum involves plating on appropriate selective, differential, and enriched media, for example, on blood agar, chocolate agar, and MacConkey agar (a medium used for the isolation of Gram-negative enteric bacteria). The bacterial pathogens detected by using this technique include *Streptococcus pneumoniae, Staphylococcus aureus, Streptococcus pyogenes, Klebsiella pneumoniae,* and *Haemophilus influenzae,* among the other pathogens that cause lower respiratory tract infections.

In cases of suspected tuberculosis and diseases caused by fungi, additional procedures are necessary. Examination of stained smears of sputum are useful in detecting such infections. Yeast-phase cells of *Histoplasma capsulatum, Coccidioides immitis,* and *Candida albicans* may be observable in such smears. In cases where these organisms appear to be present, fungal culture media, such as Sabouraud agar supplemented with antibacterial antibiotics, should be employed for culturing the suspected fungal pathogens. For the diagnosis of tuberculosis, the acid-fast stain procedure can reveal the presence of *Mycobacterium tuberculosis* in sputum samples (FIG. 16-11). Members of the genus *Mycobacterium* are acid-fast and appear red when stained by this procedure. Sputum showing presumptive evidence of the presence of *M. tuberculosis* should be cultured by using Lowenstein-Jensen medium—or other suitable

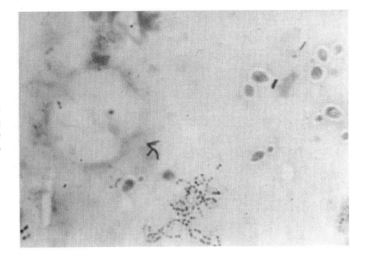

FIG. 16-11 Light micrograph of smear of tuberculous sputum after acid-fast staining. In this field there is a small group of tubercle bacilli—a pair and a single tubercle bacillus (red). Yeasts and streptococci are also present (blue).

medium that support the growth of *M. tuberculosis*—for the positive diagnosis of this organism.

Cerebrospinal Fluid Cultures

In cases of suspected infection of the central nervous system, cerebrospinal fluid (CSF) can be obtained by performing a lumbar puncture. It is important to determine rapidly whether there is a microbial infection of the cerebrospinal fluid because such infections (bacterial, fungal, or viral meningitis) can be fatal if not rapidly and properly treated. Several chemical tests can be performed immediately to determine whether a CSF infection is of probable bacterial, fungal, or viral origin. Most bacterial and fungal infections of the cerebrospinal fluid greatly reduce the level of glucose, whereas viral infections do not alter the glucose level; much of the glucose is metabolized to lactic acid. This difference can be determined rapidly by measuring glucose and lactic acid concentrations in the cerebrospinal fluid.

> **Cerebrospinal fluid obtained by lumbar puncture is used to isolate pathogens of the central nervous system.**

Cerebrospinal fluid can be screened further for possible bacterial infection by observing Gram-stained slides and by culture techniques. The growth of bacteria in a liquid enrichment medium can easily and rapidly be detected as an increase in turbidity. Cultures can also be obtained from CSF by plating on blood and chocolate agar. To provide a sufficient inoculum, the CSF is routinely centrifuged to concentrate the bacteria and the sediment is used for inoculation. Bacteria commonly associated with cases of bacterial meningitis include *Streptococcus pneumoniae, Neisseria meningitidis, Haemophilus influenzae, Streptococcus pyogenes, Staphylococcus aureus, Escherichia coli, Klebsiella pneumoniae,* and *Pseudomonas aeruginosa.*

Streptococcus pneumoniae and *Neisseria meningitidis* probably are the most frequent etiological agents of bacterial meningitis in adults, whereas *Haemophilus influenzae* is the most frequent in children but rarely occurs in cases of adult bacterial meningitis.

It should be noted that various other microorganisms can cause meningitis, including viruses, anaerobic bacteria, fungi, and protozoa. Anaerobic bacteria can be cultured in thioglycollate broth or other media in the absence of free oxygen. Fungi and protozoa can be observed microscopically and identified by using both cultural and serological test procedures.

> **In cases of suspected infections of the central nervous system, rapid and accurate diagnosis of the infecting agent is crucial for determining the appropriate modes of treatment.**

Blood Cultures

The detection of bacteria in blood is important in diagnosing many diseases, including septicemia. Septicemia is a systemic bacterial infection of the blood stream (blood poisoning). Blood for culturing bacteria typically is collected by venipuncture, using aseptic technique. The numbers of infecting bacteria in the blood are often low and may vary with time. It is therefore necessary to collect and examine blood samples at various time intervals. For example, in respiratory infections the number of bacteria that enter the blood fluctuates over a period of time. In endocarditis the numbers of bacteria in the blood generally remain relatively constant. Both aerobic and anaerobic culture techniques are needed to ensure the growth and detection of any bacteria in the blood. There are several effective methods employed in clinical laboratories for isolating anaerobes, ensuring that exposure to air is avoided. The roll-tube method may be required for the culture of very strict anaerobic bacteria.

FIG. 16-12 Detector module for the BACTEC NR infrared blood culture system (Becton-Dickinson). The needles are inserted through each bottle's septum to withdraw headspace gas and analyze it for the presence of CO_2. Fresh gas is then injected into the bottles and the needles are heat-sterilized before the next bottle is moved automatically into position.

Blood is collected by venipuncture, using aseptic technique, for culturing bacteria.

Another rapid screening procedure for the presence of bacteria in blood employs a medium containing radiolabeled glucose (FIG. 16-12). The conversion of glucose to radiolabeled carbon dioxide is rapidly determined with great sensitivity by using automated instrumentation.

Initial screening of the blood for bacterial contaminants can be accomplished in liquid media, aerobically and anaerobically. Increased turbidity is used as the index. Liquid media can also serve as an enrichment culture before plating on such solid media as blood agar, chocolate agar, and MacConkey agar. The blood sample should be diluted (blood:broth ratio of 1:10 or greater) and an inhibitor of coagulation and phagocytosis added to remove residual bactericidal factors in the blood. Additionally, if there was previous antibiotic treatment, blood specimens may require further dilution, or an appropriate antibiotic inactivator may be added to permit growth of bacterial pathogens in the patient's blood.

Urine Cultures

To detect urinary tract infections a urine sample is collected and cultured. Precautions are normally taken to avoid contamination with exogenous bacteria during voiding of the urine sample. This can be accomplished, for instance, by washing the area around the opening of the urinary tract and using a midstream catch of voided urine. Urine, though, normally becomes contaminated with bacteria during discharge through the urethra, particularly in females. Therefore culture of the urine should be performed qualitatively and quantitatively. The urine often is centrifuged and the sedimented cells are inoculated onto culture media. High numbers of a given microorganism are indicative of infection rather than contamination of the urine during discharge (FIG. 16-13).

Urine samples are taken from a midstream catch of voided urine after the area around the opening of the urinary tract is washed to minimize contamination with exogenous bacteria.

In general, greater than 10^5 bacteria/mL indicates a urinary tract infection. Plating should be performed on a general medium, like blood agar, and on selective media, such as cysteine lactose electrolyte deficient agar (CLED), MacConkey agar, or Eosine–Methylene Blue (EMB) agar. Bacteria of clinical significance that may be found in the urine include *Escherichia coli, Klebsiella, Proteus, Pseudomonas, Sal-*

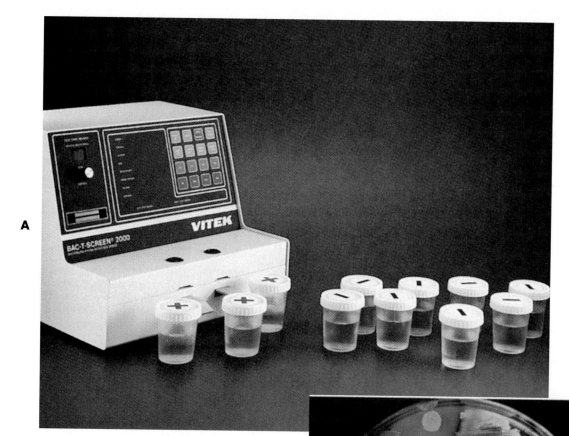

FIG. 16-13 **A,** Bac-T-Screen instrument for detection of bacterial urinary tract infection. The instrument detects an increase in turbidity (growth). **B,** Half-plate with antibiotic discs for the identification of urinary tract pathogens. Blood agar was poured into half of the plate, and when this set, MacConkey agar was added to the other half. The plate was sown with a continuous zigzag movement across the two media and then heavily inoculated in a narrow strip near the junction of the two media. On the strip were placed disc of four antibiotics: nitrofurantoin (NI), naladixic acid (NA), ampicillin (AP) and sulfonamide (SF). This provides a prompt identification of the causative agent of the urinary tract abnormality, as well as suggesting an appropriate treatment. *Escherichia coli* grows on both media. In the thin part of the MacConkey wedge it has exhausted the lactose and the pH has reverted to alkalinity. The organism is sensitive to all four antibiotics.

monella, Serratia, Streptococcus, and *Staphylococcus* species. Gram-negative enteric bacteria are the most frequent etiologic agents of urinary tract infections.

Urethral and Vaginal Exudate Cultures

Examination of urethral and vaginal exudates centers on the detection of microorganisms that cause sexually transmitted diseases. Most notable among these organisms are *Neisseria gonorrhoeae, Chlamydia* sp., and *Treponema pallidum.* In males, one symptom of gonorrhea is the painful release of a urethral exudate. Gram-stained slides of the exudate are made. If Gram-negative, kidney bean-shaped diplococci are present, a diagnosis of gonorrhea is suggested. In females, gonorrhea is more difficult to detect because of the high numbers and various normal microbiota associated with the vaginal tract. Culture techniques using inoculation with swabs collected from the cervix, vagina, and rectum can be employed for detecting *N. gonorrhoeae* in females.

Urethral exudates from males and swabs collected from female cervices, vaginas, and rectums are cultured to identify pathogens causing sexually transmitted diseases.

Thayer-Martin medium and chocolate agar, incubated under an atmosphere of 5% to 10% CO_2, are employed for the culture of *N. gonorrhoeae*. Screening for pathogens infecting the genital tract such as *Haemophilus ducreyi, Streptococcus pyogenes, Staphylococcus aereus,* and *Candida albicans* is accomplished by plating on blood agar, chocolate agar, MacConkey agar, and Sabouraud agar.

Chlamydia can often be identified by direct microscopic examination of Giemsa-stained or iodine-stained specimens. Alternatively, these bacteria can be detected by fluorescent antibody staining. These methods detect inclusion bodies with multiple cells of *Chlamydia. Treponema pallidum,* the causative organism of syphilis, cannot be cultured on laboratory media. Exudates from primary or secondary lesions can be examined by using dark-field microscopy and fluorescent antibody staining for the detection of *T. pallidum*. The characteristic morphology and movement of spirochetes can be readily detected by dark-field microscopic observation. Since nonpathogenic spirochetes may also be present in the specimen, serological tests that are specific for *T. pallidum* are also performed.

Fecal Cultures

Stool specimens are normally used for the isolation of microorganisms that cause intestinal tract infections. The common enteric bacterial pathogens are *Salmonella* spp., *Shigella* spp., enteropathogenic *Escherichia coli, Vibrio cholerae, Campylobacter* spp., and *Yersinia enterocolitica*. Fecal matter contains numerous nonpathogenic microorganisms. It is therefore necessary to employ selective and differential media for the isolation of intestinal tract pathogens. **Selective media** contain components that select for the growth of particular microorganisms. **Differential media** contain indicators that permit the recognition of microorganisms with particular metabolic activities.

> The causative organisms of intestinal tract infections are identified from samples collected from stool specimens that are plated onto selective and differential media.

Common selective and differential media that are employed for the isolation of intestinal tract pathogens include *Salmonella-Shigella* agar (SS), hektoen-enteric agar (HE), xylose-lysine-desoxycholate agar (XLD), brilliant green medium, Eosin–Methylene Blue agar (EMB), endo medium, and MacConkey medium. Two different differential and selective media, such as MacConkey agar and hektoen enteric agar, are often used for the isolation of intestinal tract pathogens.

It is often necessary to carry out an enrichment culture before *Salmonella* and *Shigella* species can be

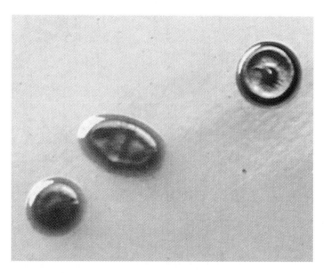

FIG. 16-14 *Salmonella* colonies on Wilson and Blair medium. The strain of *S. paratyphi B* grows first as mucoid colonies, which eventually flatten. The sequence is illustrated from left to right in this photograph. The colonies are becoming progressively more collapsed until the one at the right has a crater with a central knob.

isolated by using differential or selective solid media (FIG. 16-14). For example, in cases of suspected typhoid fever it may be necessary to carry out an enrichment in appropriate selective medium, such as GN (Gram-negative) broth or selenite F broth, before isolation of *Salmonella typhi* can be achieved. In cases of suspected viral infections, tissue cultures can be inoculated with fecal matter for the culture of enteric viral pathogens. Characteristic cytopathic effects and serological procedures can then be employed in the identification of viral pathogens. Additionally, electron microscopy can be used for the direct detection of viruses, such as rotavirus, that cause viral gastroenteritis (but this is rarely done). For the detection of intestinal parasites, direct microscopic examination of fecal matter and observation of protozoan cysts and trophozoites is used.

Eye and Ear Cultures

Fluids collected from eye and ear tissues can be inoculated onto blood agar, chocolate agar, and MacConkey agar, or other defined media, to culture bacterial pathogens commonly found in these tissues. Additionally, a Gram-stain slide can be prepared and observed to identify the presence of bacteria. Microscopic observation of stained slides can indicate whether the infection is due to bacterial pathogens. In the case of eye infections, it is particularly important to differentiate between bacterial and viral infections to determine the appropriate treatment.

Skin Lesion Cultures

Material from skin lesions, including wounds and boils, can be collected for culture purposes with swabs, aspirates, or washings. Such material, though, often is contaminated with endogenous bacteria. Various bacteria can infect wounds and cause localized skin infections. Both aerobic and anaerobic culture techniques are required for the screening of wounds for potential pathogens. Particular concern must be given to the possible presence of *Clostridium tetani* and *C. perfringens* because these anaerobes cause serious diseases. Various fungi may also be involved in skin infections. Appropriate fungal culture media are required for the isolation of these organisms. Dermatophytic fungi and actinomycetes that cause skin infections can also be detected by direct microscopic examination of skin tissues. The characteristic morphological appearance of filamentous fungi and bacteria in stained microscopic preparations often permits rapid presumptive diagnosis of the disease. There are sharp morphological differences between bacteria and fungi that can form the basis for differentiating one from another.

Swabs, aspirates, and washings are used to collect culture material from skin lesions.

CONVENTIONAL TESTS FOR DIAGNOSING PATHOGENIC MICROORGANISMS

The identification of *Staphylococcus* species is based on the observation of typical grape-like clusters, catalase activity, coagulase production, and mannitol fermentation. Members of the genus *Staphylococcus* are strongly catalase positive, whereas *Streptococcus* species are catalase negative. **Catalase activity** can be readily observed by adding a drop of hydrogen peroxide to a portion of a colony and observing the production of gas bubbles. The culture must be growing on a nonblood medium because blood contains catalase and will give a false positive result. Pathogenic strains of *S. aureus* are generally coagulase positive and produce acid from mannitol. Strains of *S. epidermidis*, in contrast, are negative for both coagulase production and mannitol fermentation. Coagulase activity, the ability to clump or clot blood plasma, is associated with virulent strains of *S. aureus* and other staphylococci (FIG. 16-15).

Gram-negative bacteria are usually differentiated initially based on **oxidase activity.** Oxidase refers to the presence of cytochrome oxidase, which is an important component of the electron transport chains of many bacteria. This cytochrome can transfer an electron to a colorless test reagent, thus turning it pink-purple. In the oxidase test, a portion of a colony of cells is mixed with the test reagent. If the reagent turns pink-purple, the cells are considered to be oxi-

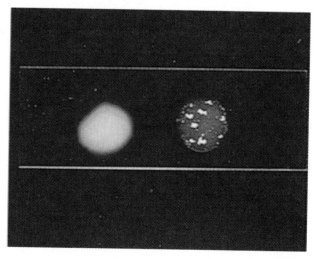

FIG. 16-15 Slide coagulase test which is useful for the diagnosis of *Staphylococcus.* Loopfuls of *S. epidermidis (left)* and of *S. aureus (right)* were emulsified in water to form thick, even suspensions on a slide. To each a small loopful of undiluted human plasma was added. When the *S. aureus* preparation was stirred with the loop, within 5 seconds the cells were clumped into large aggregates that adhered to the slide; this is a positive reaction. The *S. epidermidis* preparation shows a negative reaction. A false-positive reaction might occur if the plasma contained sufficient antibody; the resulting agglutination would take longer than 20 seconds to develop and would not show the tendency to be stuck to the slide by fibrin.

dase positive. If the reagent remains colorless, the cells are considered to be oxidase negative.

Streptococcus species are separated into serological groups based on **antigenic differences.** Biochemical test procedures are also routinely employed to distinguish species and groups of streptococci. Presumptive identification of the particular streptococcal group can be made based on the type of blood hemolysis, sensitivity to the antimicrobics bacitracin and optochin, metabolism of hippurate and esculin, and ability to grow in the presence of 6.5% NaCl.

A novel approach to the identification of obligate anaerobes used in clinical laboratories involves the gas liquid chromatographic (GLC) detection of **metabolic products** (FIG. 16-16). The anaerobes are grown in a suitable medium and the short-chain, volatile fatty acids produced are extracted in ether. Fatty acids detected in this procedure include acetic, propionic, isobutyric, butyric, isovaleric, valeric, isocaproic, and caproic acids. The pattern of fatty acid production can be used to differentiate and identify various anaerobes. When coupled with observations of colony and cell morphology (staining reaction, such as with Gram stain and cell shape) and a limited number of biochemical tests, common anaerobes isolated from clinical specimens can be identified.

MINIATURIZED COMMERCIAL IDENTIFICATION SYSTEMS

Several commercial systems have been developed for the rapid identification of pathogenic bacteria (FIG. 16-17). These systems are widely used in clinical microbiology laboratories because of the high frequency of isolation of Gram-negative rods that are indistinguishable except for characteristics determined by detailed physiological and/or serological testing. These systems typically run many tests simultaneously. The pattern of test results is compared with a data bank containing test results from known microorganisms. Statistical tests are used to determine the best match of characteristics between the clinical isolate and those in the data bank. Therefore the identification is based on statistical probability. Using computers greatly speeds up the identification process. Often the results of computer analysis indicate several potential identifications, giving the likelihood for each. Systems commonly used in clinical laboratories include the Enterotube, API 20-E, Minitek, Micro-ID, Enteric Tek, and r/b enteric systems. The pattern of test results obtained in these systems is converted to a numerical code that can be used to calculate the identity of an isolate. The numerical code describing the test results obtained for a clinical isolate is compared with results in a data bank describing test reactions of known organisms.

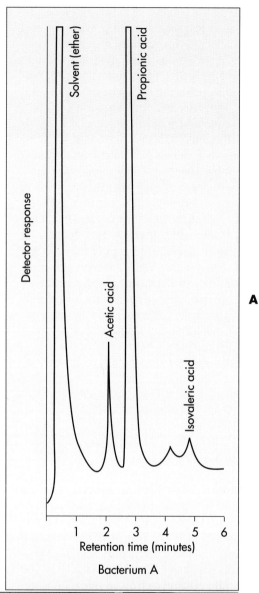

A

FIG. 16-16 A, Characteristic fatty acid profile used to identify anaerobic pathogens. **B,** Gas chromatograph for detecting diagnostic fatty acids in bacterial identification.

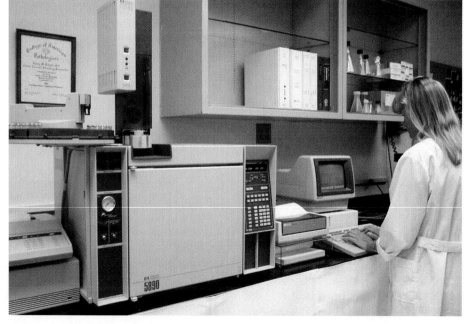

B

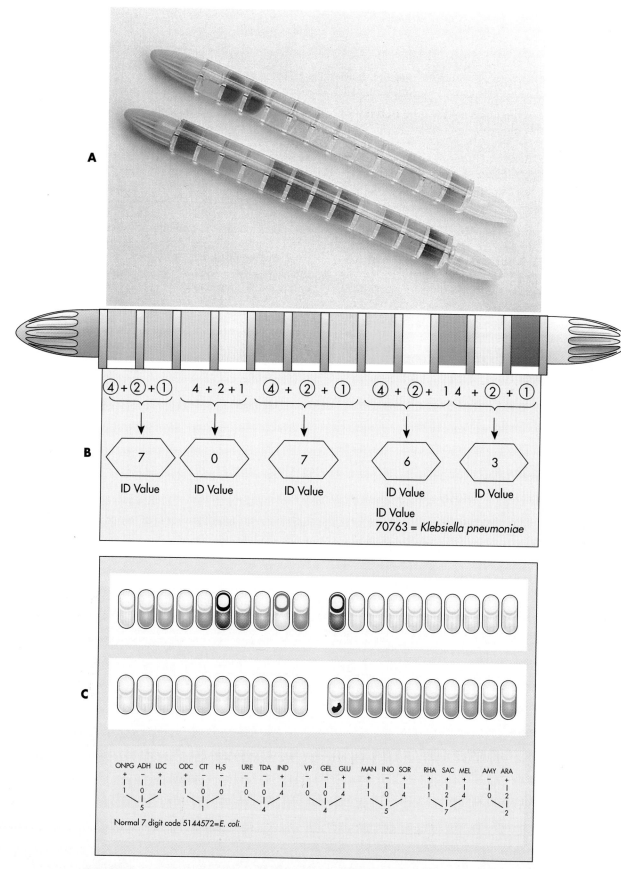

FIG. 16-17 **A,** Enterotube for the identification of clinical isolates. **B,** API 20 system for the identification of clinical isolates. **C,** Reactions of the API 20 system for the identification of pathogens.

COMPUTER-AIDED MICROBIAL IDENTIFICATION

Computers greatly facilitate the identification of microorganisms. When computers are used, the data gathered on an unknown microorganism can rapidly be compared to a data bank containing information on the characteristics of defined taxa. The classical approach to microbial identification uses a diagnostic identification key, consisting of a series of questions that lead through a classification system to the determination of the identity of the organism. Generally a series of yes-no questions leads one to the identification of a bacterium (FIG. A). Such a diagnostic key can readily be programmed for computer-assisted identification of unknown isolates.

Because computers rapidly perform large numbers of calculations and comparisons, computerized identification systems have also been developed to assess the statistical probability of correctly identifying a microorganism. In these methods the results of a series of characteristics of an unidentified microorganism are scored and compared to the test results of known microorganisms. Unlike keys, where individual tests are critical in reaching proper diagnostic identification, these identification systems assess the statistical likelihood of obtaining a particular pattern of test results.

Many commercial identification systems used in clinical laboratories are based on such statistical identification methods. Several commercial systems simplify the process of identification by calculating a numerical profile to describe unambiguously the pattern of test results (FIG. B). The numerical profile of an unknown organism can be compared with the test pattern of a defined species (a particularly easy operation using a computer) to determine the probability that the test results could represent a member of that species. Because of the critical nature of making correct identifications in medical microbiology, a positive identification in a clinical identification system generally requires that the unknown organism be far more similar to one species to which it is identified as belonging than to any other species. For example, in some identification systems an unknown microorganism must be a thousand times more similar to one group than to all other groups in the system to establish a positive identification. Computerized systems such as the Vitec system are used routinely in major clinical laboratories (FIG. C).

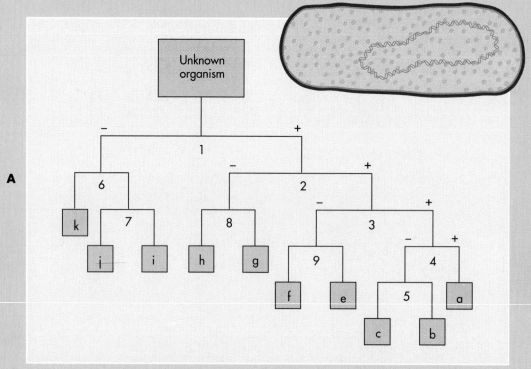

A dichotomous key approach often is used for identifying bacteria.

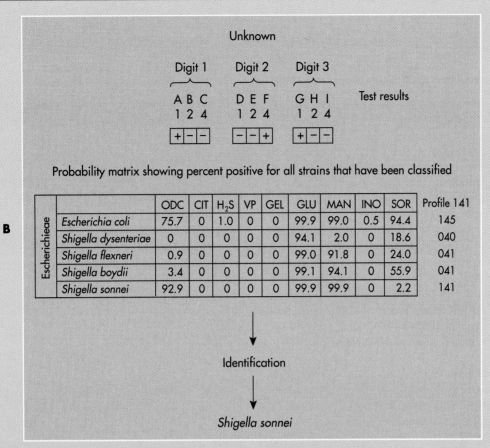

Unknown

Digit 1	Digit 2	Digit 3	Test results
A B C	D E F	G H I	
1 2 4	1 2 4	1 2 4	
+ − −	− − +	+ − −	

Probability matrix showing percent positive for all strains that have been classified

B

	ODC	CIT	H$_2$S	VP	GEL	GLU	MAN	INO	SOR	Profile 141
Escherichia coli	75.7	0	1.0	0	0	99.9	99.0	0.5	94.4	145
Shigella dysenteriae	0	0	0	0	0	94.1	2.0	0	18.6	040
Shigella flexneri	0.9	0	0	0	0	99.0	91.8	0	24.0	041
Shigella boydii	3.4	0	0	0	0	99.1	94.1	0	55.9	041
Shigella sonnei	92.9	0	0	0	0	99.9	99.9	0	2.2	141

(Escherichieae)

↓

Identification

↓

Shigella sonnei

A numerical profile can be created to identify an unknown organism. This approach is employed in several widely used commercial systems for the identification of clinical isolates.

C2

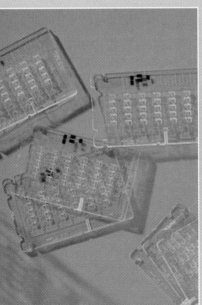

The Vitek system is one of several computerized identification systems widely used in clinical microbiology laboratories. The system provides rapid and accurate identification. **C1**, Plastic cards containing dried substrate in tiny wells for use in Vitek system. The wells are filled with urine or a suspension of an organism by vacuum suction. **C2**, Filling module and incubator modules, computer for data entry and overview, and printer components of the Vitek System.

Some of the commercial systems list a series of possible identifications indicating the statistical probability that a given organism (biotype) could yield the observed test results.

Miniaturized commercial identification systems produce a pattern of test results that is converted to a numerical code that is compared with a data bank of test results for known organisms.

All of the commercial systems employ miniaturized reaction vessels. Some are designed for automated reading and computerized processing of test results. The systems differ in how many and which specific biochemical tests are included. They also differ in whether they are restricted to identifying members of the family Enterobacteriaceae or whether they can be utilized for identifying other Gram-negative rod-shaped bacteria. The test results obtained with all these systems show excellent correlation with conventional test procedures. These package systems yield reliable identifications as long as the isolate is one of the organisms the system is designed to identify.

The **Enterotube system** contains eight solid media from which several different physiological tests can be run (Table 16-3). This system has a self-contained inoculating needle that is touched to a colony on the isolation plate and drawn through the tube. The characteristics determined in the Enterotube are used to generate a description of the test results. The **API 20-E system,** as the name implies, uses 20 miniature

TABLE 16-4

Metabolic Tests Used in the API 20-E System

| TEST | REACTIONS | |
	POSITIVE	NEGATIVE
ONPG (β-galactosidase)	Yellow	Colorless
Arginine dihydrolase	Red-orange	Yellow
Lysine decarboxylase	Red-orange	Yellow
Ornithine decarboxylase	Red-orange	Yellow
Citrate utilization	Dark blue	Light green
Hydrogen sulfide production	Blackening	Colorless
Urease	Cherry red	Yellow
Tryptophan deaminase	Red-brown	Yellow
Indole production	Red ring	Yellow
Voges-Proskauer	Red	Colorless
Gelatin liquefaction	Pigment diffusion	No pigment diffusion
Glucose fermentation	Yellow	Blue-green
Mannitol fermentation	Yellow	Blue-green
Inositol fermentation	Yellow	Blue-green
Sorbitol fermentation	Yellow	Blue-green
Rhamnose fermentation	Yellow	Blue-green
Sucrose fermentation	Yellow	Blue-green
Melibiose fermentation	Yellow	Blue-green
Amygdalin fermentation	Yellow	Blue-green
Arabinose fermentation	Yellow	Blue-green

TABLE 16-3

Metabolic Tests Used in the Enterotube System

| TEST | VISUAL REACTIONS | |
	POSITIVE	NEGATIVE
Glucose fermentation	Yellow	Red
Gas from glucose	Bubbles	No bubbles
Lysine decarboxylase	Purple-blue	Yellow
Ornithine decarboxylase	Purple-blue	Yellow
Hydrogen sulfide production	Black media	No blackening
Indole production	Red ring	No red ring
Lactose fermentation	Yellow	Red
Phenylalanine deaminase	Brown	Light green
Dulcitol fermentation	Yellow	Light green
Urease	Red	Light yellow
Citrate utilization	Deep blue	Light green

capsule reaction chambers (Table 16-4), from which over 100 taxa of Gram-negative rod-shaped bacteria can be identified.

The **Minitek system** consists of reagent-impregnated paper discs to which a broth suspension is added for the determination of characteristic reactions. The number of discs and tests employed is variable. Generally, 17 tests at a time are used to differentiate clinical isolates. The profile of test results obtained with the Minitek system is useful in identifying obligate anaerobes, as well as facultative enteric bacteria.

The **Micro-ID system,** in contrast, is based on constitutive enzymes (enzymes that are always synthesized and present in a microorganism). It is designed primarily for identifying members of the Enterobacteriaceae, although it can also be used for biotyping *Haemophilus influenzae* and *H. parainfluenzae*. In the Micro-ID system, only oxidase-negative strains are tested. Therefore bacteria must be screened for oxidase activity. This test determines if a microorganism contains cytochrome oxidase, which is a component of electron transport chains. All commercial ID systems actually recommend performing the oxidase test before attempting an identification. The Micro-ID system lists possible identifications and probabilities based on the results of the 15 biochemical test reactions. Identifications with the Micro-ID system can be accomplished in as little as 4 hours.

There are several especially rapid systems that automatically read test results. For example, the microscan system has various panels that can be used to test the metabolic activities of clinical isolates. The incubations are in microtiter plates, which are like miniaturized test tubes. Many tests are run simulta-neously, and a spectrophotometer automatically reads the results. For the identification of Gram-positive bacteria, 34 tests are included in the diagnostic panel. Within a few hours, the results of the tests are being analyzed by the computer data acquisition system. The rapid identification can be forwarded to the clinical microbiologist and also the physician and attending staff. This system and other systems, such as the Vitek system, are used in many clinical microbiology laboratories because they can handle large numbers of specimens efficiently and provide the physician with rapid and accurate results.

IMMUNOLOGICAL APPROACHES FOR IDENTIFYING PATHOGENS

In addition to using growth and biochemical characteristics for the identification of pathogenic microorganisms, various **serological test procedures** are employed for identifying disease-causing microorganisms. Serological tests are immunological reactions run *in vitro*. Some specific examples will be discussed here to illustrate the application of these serological procedures for clinical diagnostic purposes. Serological tests are particularly useful in identifying pathogens that are difficult or impossible to isolate on conventional media. They are also useful for identifying pathogenic strains of microorganisms that are not easily distinguished by biochemical testing. For example, over 2,000 serotypes in the genus *Salmonella* are defined by the O (somatic cell) and H antigens (flagella), with each serotype defined by a constellation of O and H antigens. The identification of path-ogenic viruses and nonculturable bacteria, such as *Treponema pallidum,* generally depend on serological testing.

> Serological tests determine *in vitro* immunological reactions to identify pathogens hard to isolate on conventional media and to distinguish between varieties of microbial strains.

AGGLUTINATION

Several diseases can be rapidly diagnosed by employing agglutination tests. Agglutination is the formation of insoluble aggregates of antibody-antigen complexes where the antigen is particulate (such as a protein on a cell). For example, infectious mononucleosis, caused by the Epstein-Barr virus, is routinely diagnosed with the aid of agglutination tests. The blood serum of individuals infected with this virus contains antibodies that will agglutinate sheep red blood cells. Such antibodies are called **heterophile antibodies** because they cross-react with antigens other than the ones that elicited their formation. Heterophile antibodies may be present in normal sera, but in individuals with infectious mononucleosis, the concentration of such antibodies is greatly elevated. By testing the ability of serial dilutions of blood serum to cause agglutination of sheep red blood cells, the titer (concentration) of heterophile antibodies can be determined. In performing these tests the patient's serum is heated to 56° C for 30 minutes to destroy complement that could result in lysis rather than ag-glutination. A titer of 1:256 or greater is considered as presumptive evidence of infectious mononucleosis.

> Heterophile antibodies are found in the blood serum of individuals infected with the Epstein-Barr and other viruses; they cross-react with antigens other than the ones that elicited their formation.

NEWSBREAK

TORCH TESTS DURING PREGNANCY

Because some microorganisms can cross the placenta and lead to congenital diseases, a series of serological tests have been developed called the TORCH series. These tests are aimed at determining if the mother is infected during pregnancy with any of four different microorganisms—the protozoan *Toxoplasma gondii* (TO), rubella virus (R), cytomegalovirus (C), and herpes virus (H). Each of these microorganisms can produce congenital diseases that damage the newborn, causing blindness, brain damage, and various deformities. The TORCH test is based on detecting circulating antibodies to these microorganisms.

Additional differential tests for infectious mono-nucleosis can be carried out to confirm the diagnosis. In these tests the patient's serum is mixed with guinea pig kidney tissue in one tube and beef ery-throcyte antigen in another. Guinea pig tissue does not absorb heterophile antibodies produced in re-sponse to the Epstein-Barr virus. Neither does it re-duce the ability of the serum to agglutinate sheep cells. Beef erythrocyte antigens, on the other hand, do adsorb these heterophile antibodies, lowering the ability of the serum to agglutinate sheep red blood cells.

The detection of heterophile antibodies is also use-ful in diagnosing several other diseases. Heterophile antibodies are cross-reactive. For example, the **Weil-Felix test** is used to diagnose some diseases caused by *Rickettsiae* species. In these tests, antibodies pro-duced in response to a particular rickettsial infection agglutinate strains of *Proteus*, designated OX-19, OX-2, and OX-K. The results of the Weil-Felix test are considered to give presumptive diagnoses of rick-ettsial diseases. Febrile agglutinins, which are anti-bodies produced in response to various fever-pro-ducing bacteria, can similarly be detected by aggluti-nation tests.

The agglutination test procedures discussed so far depend on the occurrence of antigens on cell sur-faces. In cases where the antigen to be detected is sol-uble and not associated with a cell or other particle, **passive agglutination** tests can be employed (FIG. 16-18). In such tests the antigen is attached to a parti-cle surface before running an agglutination test. The particle may be a cell or, more commonly, a latex bead. Very small quantities of antigen-coated beads are required for these tests.

> **Passive agglutination tests are used when the anti-gen to be identified is soluble and not associated with a cell or other particle.**
>
> **Antigen-coated latex beads are used in these tests.**

Passive agglutination using latex beads is rapidly replacing various older test procedures. It is now em-ployed for detecting *Haemophilus influenzae, Strepto-coccus pneumoniae, Neisseria meningitidis, Staphylococ-cus aureus,* and rubella virus. Latex beads may also be coated with soluble antibodies and used in aggluti-nation tests.

Pneumonia caused by *Streptococcus pneumoniae* can be detected by passive agglutination. Blood serum from patients with pneumococcal pneumonia possesses a protein called C reactive protein (CRP). The CRP reacts with the C polysaccharide of the cap-sular material of *S. pneumoniae.* Anti-CRP attached to latex beads will react with C reactive protein, causing readily observed agglutination. A positive agglutina-

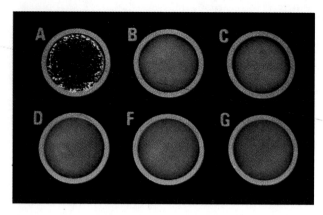

FIG. 16-18 Photograph showing passive agglutination (streptex) for diagnosis. The reaction in well *A* shows ag-glutination and identifies the presence of group A strepto-cocci that have reacted with latex particles coated with anti-group A antibodies. The reactions in the other wells are negative.

tion test of the patient's serum with anti-CRP–coated latex beads gives presumptive evidence of pneumo-coccal pneumonia.

Passive agglutination can also be used to detect *Treponema pallidum* in the diagnosis of syphilis. The antigens from *T. pallidum* can be conjugated with red blood cells in passive agglutination tests run to detect the presence of antibodies against *T. pallidum* in a pa-tient's serum. Another test used for the diagnosis of syphilis is designed to detect the presence of an IgE antibody, known as reagin, produced in individuals infected with *T. pallidum.* This IgE antibody reacts with cardiolipin, as well as with an antigen, on the surface of *T. pallidum.* The presence of reaginic anti-body can be determined by using cardiolipin ex-tracted from beef hearts by alcohol. The reaction of reagin with cardiolipin results in an agglutination or flocculation (clumping) that can be visualized by us-ing low power microscopy.

COUNTER IMMUNOELECTROPHORESIS

Counter immunoelectrophoresis (CIE) is also used in several diagnostic procedures. CIE can be useful in detecting the presence of various microorganisms in body fluids such as cerebrospinal fluid, urine, and blood. This procedure depends on a precipitin reac-tion to identify homology between antigen and anti-body. A precipitin (precipitation) reaction is the for-mation of insoluble aggregates of antibody-antigen complexes where the antigen is soluble (such as an extracellular protein). It relies on electrophoresis to move the antigen and antibody through a gel toward each other. If the antigen and antibody can form ap-propriate complex, a precipitation reaction occurs.

CIE is used to detect both antigens and antibodies in body fluids and antigens from microbial cultures. *Streptococcus pneumoniae, Haemophilus influenzae, Neisseria meningitidis, Pseudomonas aeruginosa, Klebsiella pneumoniae, Escherichia coli, Staphylococcus aureus, Mycoplasma* spp., and *Legionella pneumophila* can all be detected clinically using counter immunoelectrophoresis.

ENZYME-LINKED IMMUNOSORBENT ASSAY

Still another approach to serological diagnosis of disease-causing microorganisms is the **enzyme-linked immunosorbent assay** (ELISA) (see FIG. 14-26). This method permits the detection of antibody by combining antigen-antibody with an enzyme that gives a color-producing reaction that can be read spectrophotometrically. The ELISA assays can be used for the detection of antibodies such as those produced in response to infections with *Salmonella, Yersinia, Brucella, Rickettsia, Treponema, Legionella, Mycobacterium,* and *Streptococcus* species. The ELISA assay can also be used to detect the toxins (antigens) produced by *Vibrio cholerae, Escherichia coli,* and *Staphylococcus aureus.*

RADIOIMMUNOASSAYS

Radioimmunoassays (RIA) are also applicable to the detection of pathogenic bacteria (FIG. 16-19). In this approach a radioisotopic label, usually ^{125}I, rather than a color reaction, is used to assay the extent of antigen-antibody reaction. The RIA assays are extremely sensitive. RIA assays can be carried out for the detection of *Staphylococcus aureus* enterotoxins, *Clostridium botulinum* toxin, *Neisseria meningitidis, Haemophilus influenzae, Pseudomonas aeruginosa,* and *Escherichia coli.*

IMMUNOFLUORESCENCE

Direct fluorescent antibody staining (FAB) uses a defined antibody conjugated with a fluorescent dye to stain unknown microorganisms. In this procedure, microorganisms are stained only if the antibody reacts with the surface antigens of the microorganisms on the slide. Viewing with a fluorescence microscope permits the specific detection of microorganisms that react with the fluorescent antibody stain. In this procedure, specific pathogenic microorganisms can be identified even in the presence of other microorganisms. The fluorescence antibody staining method is useful for the identification of clinical isolates and for detecting pathogens, such as *Treponema pallidum* or *Legionella pneumophila,* within exudates from infected tissues.

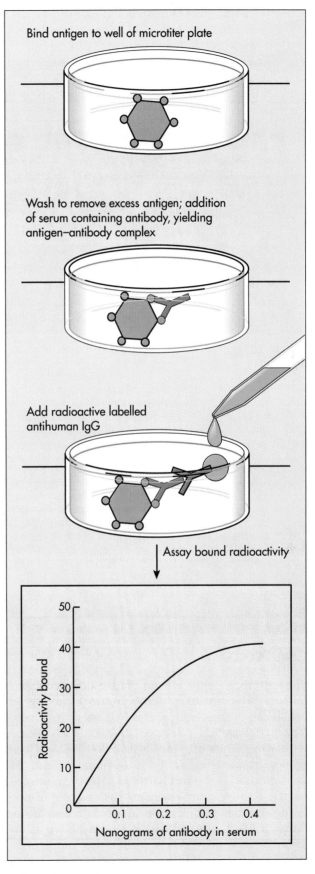

FIG. 16-19 The RIA procedure detects the ability of antibody to bind a radiolabeled antigen.

Indirect immunofluorescence tests are also employed in identifying bacteria, such as *Treponema pallidum* (FIG. 16-20). In the fluorescent test for treponemal antibodies (FTA), dead *T. pallidum* cells are fixed to a slide and a patient's serum is added. The slide is washed and a fluorescent anti-immunoglobulin is added. If the antibodies in the patient's serum react with the *T. pallidum* on the slide, the bacteria will be stained (resulting in a positive test). If no reaction occurs, the fluorescent antibody is washed off the slide, and no fluorescing bacteria are visible (resulting in a negative test).

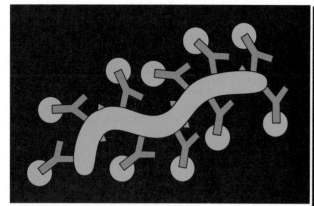

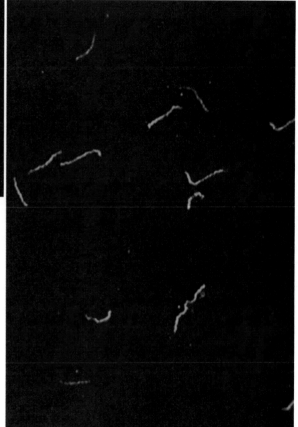

FIG. 16-20 **A,** Fluorescent antibody staining is used for detection of specific pathogens. Some microorganisms that are difficult to culture are identified in this manner from direct smears. **B,** Detection of syphilitic antibody in a patient's serum using fluorescent antiglobulin. A smear of *Treponema pallidum* was exposed to the serum of a patient with syphilis. Later, the excess serum was washed off and the preparation was similarly treated with fluorescein-conjugated antihuman-globulin. The smear was then examined by ultraviolet fluorescence microscopy. The spirochetes can be seen in this positive reaction because the antibody was bound to them and to the fluorescent antiglobulin. Since human serum may contain cross-reacting antibody against commensal treponemes, the patient's serum was first absorbed with an extract of the Reiter treponeme. The test is therefore called the fluorescent treponemal antibody absorption (FTA-ABA) test.

MOLECULAR METHODS FOR IDENTIFYING PATHOGENS

GENE PROBES

The most recent approach for diagnosing pathogenic microorganisms employs **gene probes.** Gene probes are short nucleotide sequences that are complementary to nucleotide sequences that uniquely occur within the DNA of a pathogen (FIG. 16-21). Gene probes use **nucleic acid hybridization.** Nucleic acid hybridization is the formation of a double-stranded nucleic acid by complementary base pairing of two single-stranded nucleic acids. In hybridization procedures, a DNA solution that has been heated is allowed to cool slowly. This enables the complementary strands to reassociate—a process called **reannealing.** Reannealing occurs only if the base sequences of the two strands are complementary. Nucleic acid hybridization provides a powerful tool for clinical identification of pathogens because of the specificity of nucleotide sequences in organisms. Specific genes associated with species of pathogens can be targeted for hybridization with gene probes. If reannealing between the gene probe and DNA or RNA from a clinical specimen occurs, it indicates that the gene probe specifically hybridized with the complementary target sequence, and thus that the target organism was present in the specimen from the patient.

> **Gene probes are single-stranded DNA or RNA molecules that are complementary to target nucleotide sequences.**

Detection of nucleic acid hybridization is usually done with membrane filters made of cellulose nitrate or nylon. The first step in filter hybridization is the attachment of single-stranded target nucleic acids to the filter surface. Next, the filters are prehybridized to block nonspecific nucleic acid binding sites. Labelled probe DNA is added to the filters and the probe is allowed to hybridize by annealing with the complementary target sequences to form double-stranded DNA. After hybridization, excess unbound labelled probe is washed off and the hybrid (target:probe) sequences are detected. In some cases radiolabeled gene probes are used and detection is accomplished using X-ray film. In most clinical procedures nonradioactive probes are employed. Biotin commonly is used to label the probe. Biotin can be detected colorimetrically following reaction with strepavidin and enzymatic development of the color reaction. Fluorescent and chemiluminescent labels are also employed for gene probes. The use of gene probes in the clinical laboratory provides a basis for the specific diagnosis of pathogens such as the human immunodeficiency virus that causes AIDS.

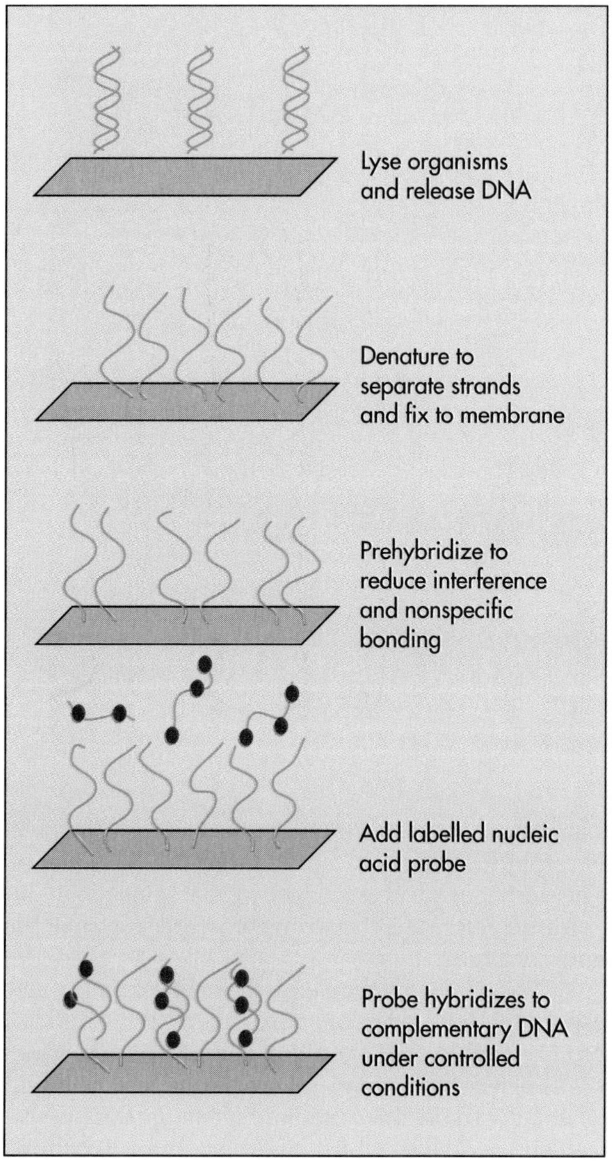

POLYMERASE CHAIN REACTION

The polymerase chain reaction (PCR) is an *in vitro* method for amplification of DNA sequences (FIG. 16-22). PCR can be applied to DNA recovered from clinical specimens and cultures. Detection of specific gene sequences can be used to diagnose the presence of specific microbial populations and the presence of specific functional genes. The target region for amplification is defined by two unique oligonucleotide primers flanking a DNA segment. The oligonucle-

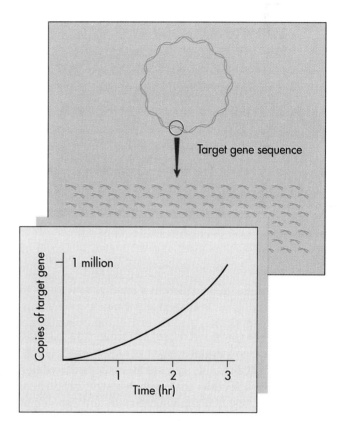

FIG. 16-21 In DNA hybridization procedures for gene probe detection, cells are lysed to release double-stranded DNA. The DNA is denatured to convert it to single-stranded target DNA. The single-stranded DNA is affixed to a membrane. A prehybridization solution is used to prevent nonspecific binding to the membrane. A labelled nucleic acid probe (gene probe) is added. (The label may be a dye or a radioactive element.) The labelled probe hybridizes to complementary regions (if any) of the target DNA so that the reaction can be detected.

FIG. 16-22 PCR is used to increase the concentration of a target gene sequence. A million copies of the target gene sequence can be made within a few hours. PCR involves three stages. First the DNA is denatured. This is accomplished by heating to convert the double helical DNA to single-stranded DNA. Then primers complementary to the nucleotide sequences flanking target region are annealed. This is accomplished by lowering the temperature to permit primer annealing. Then the DNA is extended by the action of *taq* polymerase. Because *taq* polymerase is heat stable, the process can be repeated to increase the number of copies of the target sequence.

METHODOLOGY

GENE PROBES FOR *LEGIONELLA*

Legionella pneumophila, which causes Legionnaire's disease, is widely distributed in environmental waters. It is transmitted to humans via aerosols, most commonly from air-conditioning cooling towers but also from shower heads and other sources. To detect this pathogen, researchers in my laboratory developed a gene probe-PCR method that specifically detects *L. pneumophila* (see Figure).

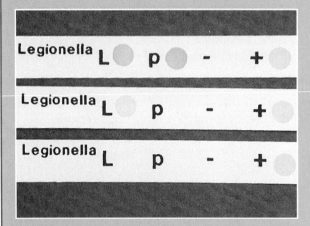

Nucleic acid hybridization for detection of *Legionella*. The blue dots indicate where hybridization has occurred. The dots at the + are the positive control; these must be blue to read the test result. The dots at the − are the negative control; these must remain white to read the test result. A blue dot at the *L* indicates the presence of *Legionella* species. A blue dot at *p* indicates the presence of *Legionella pneumophila*.

We chose a region of the macrophage infectivity potentiator *(mip)* gene as a target for the gene probe. This gene enhances the ability of *L. pneumophila* to infect macrophages. Our choice was based on the fact that the *mip* gene was reported to occur in pathogenic *L. pneumophila* and the nucleotide sequence for this gene had been determined. We were able to compare the reported nucleotide sequence for the *mip* gene with the nucleotide sequences from other genes using a computerized genetic data bank. By doing so, we identified several candidate regions with 25 to 50 nucleotides. We then synthesized short chains of nucleotide (oligonucleotides) with nucleotide sequences complementary to the nucleotide sequences in the *mip* gene. Next, we tested our candidate gene probes against hundreds of known bacterial strains to determine whether the gene probe was really specific for *L. pneumophila* and whether it could detect all strains of this pathogen. Our results indicated that several of the gene probes had the necessary specificity.

To increase the sensitivity of this gene probe system we coupled gene probe detection with the polymerase chain reaction (PCR) to amplify the target *mip* gene. Biotin was chemically attached to some of the nucleotides so that it was incorporated into the DNA made during PCR. The biotin was detected by an assay that produces a blue color so that the gene probe could be detected by the intensity of color (a nonradioactive assay). Using this method we were able to detect single cells of *L. pneumophila* with great precision. The system with minor modifications is being commercially produced by Roche Diagnostics Corporation.

otide primers are designed to hybridize to regions of DNA flanking a desired target gene sequence, annealing to opposite strands of the target sequence. The primers are then extended across the target sequence using a heat stable DNA polymerase (frequently *taq* DNA polymerase) in the presence of free deoxyribonucleotide triphosphates. The product of each PCR cycle is complementary to and capable of binding primers, so that the amount of DNA synthesized is doubled in each successive cycle. The essential reagents for PCR are a thermostable DNA polymerase, oligonucleotide primers, deoxyribonucleotides, template (target) DNA, and magnesium ions. An automated temperature cycler is used to cycle between the temperatures for DNA melting, primer annealing, and primer extension. The heat stable DNA polymerase retains activity through sufficient PCR cycles to achieve greater than one mil-

lion-fold amplification of the target DNA sequence. The primer sequences and the primer annealing temperature determine the stringency (specificity) of the amplification.

PCR is useful for the rapid detection of pathogens, especially those whose *in vitro* cultivation is difficult or not possible. PCR detection of *Chlamydia trachomatis,* a bacterium that only grows within host cells and is sexually transmitted, and of *Mycobacterium tuberculosis* and *M. avium,* which take a long time to culture, are important new diagnostic tests being introduced into clinical microbiology laboratories. The list of pathogens that can be diagnosed by PCR is increasing rapidly (Table 16-5).

There are several advantages to using PCR for diagnosis of pathogens. PCR can be applied to fixed tissues (frozen or formalin fixed), reducing the potential dangers involved in transport and handling of specimens with live virulent pathogens. Diseases can be diagnosed in hours that might take days by other methods. Specimens can be examined directly without culturing pathogens so that diagnoses can be accomplished rapidly and safely.

PCR is very sensitive so that the presence of pathogens can be detected very early in an infection. Only a few cells of the pathogen in a specimen are necessary to yield a positive diagnosis. For this reason PCR is very useful for screening blood supplies to detect HIV and other pathogens. Although blood interferes with PCR, the sensitivity of PCR permits dilution of samples so that interference can be eliminated.

PCR also has great specificity. Using PCR, pathogens with specific virulence factors can be specifically diagnosed. Both invasive factors and toxigenicity can be detected. PCR permits the simultaneous screening for multiple pathogens that might be causing a disease condition. In cases of multiple agent infections, all the pathogens causing the disease can be detected in a single reaction.

Because of the amplification power of PCR, it is critical to avoid even a trace of contamination with DNA containing the target sequence. Contamination of the PCR amplification reaction with products of a previous PCR reaction, that is, product carryover, cross-contamination between samples, contamination with nucleic acids from the laboratory environment, and even from the skin of the laboratory personnel, can create a false positive result. Fear of obtaining false positives has limited the acceptance of PCR in clinical laboratories. As general preventive measures, precautions such as pre-aliquoting reagents, using pipettes exclusively for specific steps of the reaction, using positive displacement pipettes, or using tips with barriers to prevent contamination of the pipette barrel, and physical separation of the amplification reaction preparation from the area of amplified product analysis can minimize the possibility of such contamination. One approach to eliminate the danger of contamination causing false positive results is to make contaminating PCR products susceptible to degradation. This method involves substituting uracil for thymine in the PCR reaction mixture so that the DNA made by PCR contains uracil. Because normal DNA from a pathogen would contain only thymine, the DNA made by PCR can be selectively destroyed by treating with an enzyme that will degrade any DNA containing uracil. The degraded DNA can not be amplified by PCR and in this way possible contaminating DNA is eliminated. This overcomes a major concern of using PCR in the clinical diagnostic laboratory.

TABLE 16-5

Diagnostic Tests That Utilize PCR

MICROORGANISM	SOURCE
VIRUSES	
Coxsackievirus	Heart tissue
Cytomegalovirus	Urine
Herpes simplex virus	Cerebrospinal fluid
Hepatitis C virus	Serum
Measles virus	Various tissues
Papillomavirus	Specimens from cervix
Rotavirus	Fecal material
Rubella virus	Placenta
BACTERIA	
Chlamydia trachomatis	Specimens from urethra, cervix, or conjunctiva
Helicobacter pylori	Stomach tissue
Legionella pneumophila	Environmental waters and fluid from lung
Mycobacterium tuberculosis	Cerebrospinal fluid and sputum
Mycobacterium avium-intracellularae	Fluid from lung
Mycoplasma pneumoniae	Throat swab
Rickettsia rickettsii	Blood
FUNGI	
Candida albicans	Blood, urine, and sputum
Pneumocystis carinii	Fluid or tissue from lung
PROTOZOA	
Toxoplasma gondii	Cerebrospinal fluid
Trypanosoma cruzi	Blood

SUMMARY

Indicators That Diseases are of Microbial Etiology (pp. 471-474)

- Pathogenic microorganisms produce specific diseases that are associated with characteristic signs (objective changes) and symptoms (subjective changes) in body functions. A disease syndrome is a characteristic group of signs and symptoms.

Differential Blood Counts (pp. 471-473)

- The relative concentrations of the different types of blood cells are determined by performing a differential blood count to identify what kind of microorganism is causing the infection. Systemic bacterial infections are characterized by increases in stab or band cells and decreases in numbers of eosinophils. Increases in eosinophils indicate allergic diseases and parasitic infections. Many viral infections result in lowered numbers of white blood cells. Malaria results in anemia, a decreased number of red blood cells.

Skin Testing (p. 474)

- In skin testing based on delayed hypersensitivity re-

actions, antigens from a test organism are injected below the skin surface. A positive reaction, erythema
(reddening) and/or induration (hardening) of the
skin, indicates an active or earlier infection caused by
the organism from which the test antigen was derived.

Cultural Approaches for Identifying Pathogens (pp. 474-491)

Screening and Isolation Procedures (pp. 474-485)

- Positive diagnosis of an infectious disease requires the
isolation and identification of the pathogenic microorganism or the identification of the antigens specifically associated with a given microbial pathogen.

- Upper respiratory tract pathogens are isolated by collecting throat and nasopharyngeal cultures using
sterile swabs that are streaked onto blood agar plates
and incubated in an atmosphere of 5% to 10% CO_2.
Alpha hemolysis is greening of the blood around the
colony. Beta hemolysis is a zone of clearing around
the colony, indicating the presence of *Streptococcus
pyogenes*. Some bacterial colonies show no clearing of
blood.

- *Haemophilus influenzae, Neisseria meningitidis,* and *N.
gonorrhoeae* grow better on chocolate agar. Thayer-
Martin medium is preferred for isolation of *N. gonorrhoeae* because growth of normal throat microbiota is
inhibited by antibiotics (colistin sulfate, trimethoprim, vancomycin, and nystatin) that have been
added. Bordet-Gengou potato medium is used for the
isolation of *Bordetella pertussis,* the causative organism
of whooping cough. This medium contains potato,
glycerol, and blood that neutralize growth-inhibiting
substances that may be present in the agar or peptone
(meat digest).

- Laboratory growth of viruses employs tissue cultures, usually from rhesus monkey kidney cells; antibiotics are added to prevent bacterial and fungal
growth. Viral infection of tissue culture cells produces
morphological changes (cytopathic effects) observable by microscopic examination.

- Sputum samples, exudate containing material from
the lower respiratory tract, vary in quality and should
be screened microscopically to determine suitability.
Acceptable samples contain high numbers of neutrophils and mucus, and low numbers of squamous
epithelial cells. Transtracheal aspiration ensures a better quality specimen. Sputum samples are plated on
selective, differential, and enriched media. Stained
smears of sputum can detect fungal infections.

- Cerebrospinal fluid is obtained by lumbar puncture
when infection of the central nervous system is suspected. The glucose level indicates whether the infection is bacterial (greatly reduced) or viral (the same).
Growth of bacteria in liquid enrichment media causes
an increase in turbidity.

- Blood cultures are collected at various intervals by
venous puncture using aseptic technique. Aerobic
and anaerobic cultures are performed. An automated
rapid screening procedure for the presence of bacteria
in blood detects the conversion of radiolabeled glucose to carbon dioxide.

- Urine samples are collected by a midstream catch of
voided urine to minimize contamination with normal
genitourinary microbiota. Greater than 10^5 bacteria
per milliliter of a given microorganism indicates an
infection. Samples are plated on general and selective
media. Gram-negative enteric bacteria are the most
frequent etiologic agents of urinary tract infections.

- Urethral and vaginal exudate cultures are collected to
detect the presence of microorganisms that cause sexually transmitted diseases, such as *Neisseria gonorrhoeae, Chlamydia trachomatis,* and *Treponema pallidum.*
Thayer-Martin medium and chocolate agar and incubation under an atmosphere of 5% to 10% CO_2 are
used to culture *N. gonorrhoeae. T. pallidum* cannot be
cultured on standard laboratory media; dark-field
microscopy and fluorescent antibody staining are
necessary.

- Fecal cultures are collected to isolate the causative organisms of intestinal tract infections. Because fecal
matter contains a large number of nonpathogens, selective and differential media are used in culturing.
Selective media contain components that select for
the growth of a particular microorganism. Differential media contain indicators that permit recognition
of microorganisms with particular metabolic activities.

- Eye and ear fluids are cultured on defined media and
Gram stains are determined. It is important to differentiate between bacterial and viral infections of the
eye to determine appropriate treatment.

- Skin lesion culture material can be collected with
swabs, aspirates, or washings but it is often contaminated with exogenous bacteria. Samples are grown
under anaerobic and aerobic conditions. Dermatophytic fungi and actinomycetes can be detected by direct microscopic examination of skin tissues.

Conventional Tests for Diagnosing Pathogenic
Microorganisms (p. 485)

- Conventional tests can be used for the identification
of Gram-positive bacteria. *Staphylococcus* species are
identified by the observation of typical grape-like
clusters, strongly positive catalase activity, coagulase
production, and mannitol fermentation.

Miniaturized Commercial Identification
Systems (pp. 486-491)

- Miniaturized commercial identification systems produce a pattern of test results that is converted to a numerical code that is compared with results in a data
bank describing the test reactions of known organisms. Some systems are designed for automated reading and computerized processing of test results. Systems differ in how many and which specific
biochemical tests are included–whether they are restricted to identifying members of the family Enterobacteriaceae or can be used for identifying other
Gram-negative rods.

Immunological Approaches for Identifying Pathogens (pp. 491-494)

Agglutination (pp. 491-492)

- Serological test procedures utilize immunological reactions *in vitro*. Agglutination is the clumping reac-

tion of surface antigen and antibody. The presence of heterophile antibodies is determined by agglutination tests in which the patient's serum is mixed with a known antibody that cross-reacts with antigens other than the one that elicited its formation. Passive agglutination tests are used when the antigen to be detected is soluble and not associated with a cell or other particle. The antigen is attached to a latex bead.

Counter Immunoelectrophoresis (pp. 492-493)

- Counter immunoelectrophoresis is useful in detecting the presence of various microorganisms in body fluids. It is based on a precipitin reaction to identify homology between antigen and antibody and relies on immunodiffusion with electrophoresis driving the antigen and antibody toward each other.

Enzyme-linked Immunosorbent Assay (p. 493)

- Enzyme-linked immunosorbant assay permits detection of antibody by combining antigen–antibody with an enzyme conjugant that produces a colored reaction that can be read spectrophotometrically. ELISA can also be used to detect some toxins.

Radioimmunoassays (p. 493)

- Radioimmunoassays use radioisotope labels to assay the extent of antigen–antibody reactions.

Immunofluorescence (pp. 493-494)

- Direct fluorescent antibody staining uses defined antibody conjugated with fluorescent dye to stain an unknown organism. Microorganisms are stained only if antibody reacts with the organism's surface antigens. The slides are viewed with a fluorescence microscope.

- Indirect immunofluorescence tests are used for identifying *Treponema pallidum*. Dead *T. pallidum* cells are fixed to a slide to which the patient's serum and fluorescent anti-immunoglobulin are added. If the serum antibodies react with the specimen on the slide, the bacteria will be stained and the test will be positive.

Molecular Methods for Identifying Pathogens (pp. 494-497)

Gene Probes (pp. 494-495)

- Gene probes are short nucleotide sequences that are complementary to nucleotide sequences that occur uniquely within the DNA of a pathogen. Gene probes use nucleic acid hybridization. When the gene probe and DNA or RNA from a clinical specimen reanneal, the target organism is present in the patient's specimen.

Polymerase Chain Reaction (pp. 495-497)

- The polymerase chain reaction (PCR) is an *in vitro* method for amplification of DNA sequences.
- PCR is useful for the rapid detection of pathogens.

CHAPTER REVIEW

REVIEW QUESTIONS

1. How are differential blood counts used to determine if a disease is caused by a microorganism?
2. How are microorganisms isolated from the lower respiratory tract to identify pathogens? What precautions must be taken to ensure the origin of the culture?
3. How are miniaturized identification systems used in the clinical microbiology laboratory? Describe the tests used in three different systems. What are the advantages of miniaturized identification systems in a clinical microbiological laboratory?
4. Why is it important to quickly and accurately identify cultures sent to the clinical microbiology laboratory?
5. What are serological tests? How are they used in the identification of pathogens?
6. How is immunofluorescence used to identify *Trepo-nema pallidum*? Why are serological methods critical for identifying this pathogenic bacterium?
7. What is a cytopathic effect? How is it used in the diagnosis of infectious mononucleosis?
8. What are the differences between signs and symptoms of an illness?
9. Why are pure cultures essential for the identification of clinically important pathogens?
10. Why are transport media used?
11. What roles do selective and differential media have in the identification of pathogens?
12. What is the importance of rapidly culturing specimens as soon as they enter the clinical laboratory?
13. What events occur from the time a patient develops an infectious disease until the disease is diagnosed?

CRITICAL THINKING QUESTIONS

1. How would you go about identifying the causative organism in a case of bacteremia (systemic bacterial infection of the blood)? How could you tell if the pathogen was *Streptococcus pyogenes*? *Staphylococcus aureus*?
2. What criteria do you have to consider to design a gene probe for use in the specific identification of a microorganism? What are the advantages and disadvantages of gene probes compared to conventional identification methods?
3. Why hasn't PCR been more extensively used in the clinical microbiology laboratory?
4. It is predicted that modern developments in serology and molecular biology will make the current clinical microbiology laboratory obsolete in the next 10 years. Is this likely to happen? Will all future diagnoses be made at bedside?

READINGS

Balows A, WJ Hausler, EH Lennette (eds.): 1988. *Laboratory Diagnosis of Infectious Diseases: Principles and Practices*, New York, Springer-Verlag.

Volume 1 covers bacterial, mycotic, and parasitic diseases and volume 2 reviews the viral, rickettsial, and chlamydial diseases.

Balows A, WJ Hausler, KL Herrmann, HD Isenberg, HJ Shadomy (eds.): 1991. *Manual of Clinical Microbiology*, Washington, D.C.; American Society for Microbiology.

The "must have" reference book for clinical microbiologists, infectious disease specialists, pathologists, medical technologists, clinicians, and all others involved in clinical and diagnostic microbiology, public health, and epidemiology.

Baron EJ, LR Peterson, SM Finegold: 1994, *Bailey and Scott's Diagnostic Microbiology*, ed. 9, St. Louis; Mosby.

This classic reference is a comprehensive reference to the full range of diagnostic microbiology, including bacteriology, virology, mycology, and parasitology.

Bondi A (ed.): 1984. *Urogenital Infections: New Developments in Laboratory Diagnosis and Treatment*, New York, Plenum Press.

Reviews the methods used in the laboratory for clinical diagnosis and how the choice of therapy for the treatment of urinary tract infections is made.

Garcia LS and DA Bruckner: 1993. *Diagnostic Medical Parasitology*, ed. 2, Washington, D.C.; American Society for Microbiology.

An excellent diagnostic parasitology manual.

Gardner P: 1984. *Manual of Acute Bacterial Infections: Early Diagnosis and Treatment*, Boston, Little, Brown.

A handbook for the diagnosis and treatment of bacterial infections.

Gerba CP and SM Goyal (eds.): 1982. *Methods in Environmental Virology*, New York, Marcel Dekker.

Explains techniques used in virology for the identification of viruses and the diagnosis of viral diseases.

Gutierrez Y: 1990. *Diagnostic Pathology of Parasitic Infections With Clinical Correlations*, Philadelphia, Lea & Febiger.

Reviews the parasites, the physiopathology, and the diagnosis of parasitic diseases.

Isenberg HD: 1993. *Clinical Microbiology Procedures Handbook*, Washington, D.C.; American Society for Microbiology.

An authoritative guide to over 300 procedures used in the clinical microbiology laboratory.

Koneman EW, SD Allen, WM Janda, PC Schreckenberger, WC Winn: 1994. *Introduction to Diagnostic Microbiology*. Philadelphia; JB Lippincott.

Text on clinical microbiology aimed at students in laboratory technician programs.

Lennette EH (ed.): 1985. *Laboratory Diagnosis of Viral Infections*, New York, Marcel Dekker.

A handbook of procedures used in the clinical laboratory to diagnose diseases caused by viruses.

Matthews PM, DL Arnold et al: 1991. *Diagnostic Tests in Neurology*, New York, Churchill Livingstone.

Handbook reviewing diagnostic procedures for diseases of the nervous system.

Milgromm F, CJ Abeyounis, K Kano (eds.): 1981. *Principles of Immunological Diagnosis in Medicine*, Philadelphia, Lea & Febiger.

Explains immunologic techniques and serodiagnosis in the identification of bacterial, viral, and immunologic diseases.

Mulvihill ML: 1991. *Human Diseases: A Systemic Approach*, Norwalk, CT; Appleton & Lange.

Discusses the etiology and diagnosis of disease for students entering a health career. Covers general mechanisms of diseases and commonly occurring infectious and noninfectious diseases of each system.

Persing DH, TF Smith, FC Tenover, TJ White (eds.): 1993. *Diagnostic Molecular Microbiology Principles and Applications*, Washington, D.C.; American Society for Microbiology.

Shows how PCR can be used in the clinical microbiology laboratory.

Resnick MI, AA Caldamone, JP Spirnak: 1991. *Decision Making in Urology*, Philadelphia, B. C. Dekker.

Reviews the pathology of the urogenital system and the diagnosis of urologic diseases, including male and female genital diseases, adrenal gland diseases, and urogenital neoplasms.

Rose NR, H Friedman, JL Fahey (eds.): 1986. *Manual of Clinical Laboratory Immunology*, Washington, D.C.; American Society for Microbiology.

Guide to immunological methods and their applications in the diagnosis of microbial infections and also diseases such as autoimmunity, allergy, and cancer.

Spector S (ed.): *Cumulative Techniques and Procedures in Clinical Microbiology*, Washington, D.C.; American Society for Microbiology.

Each series booklet focuses on a specific diagnostic concern, laboratory technique, or infectious agent and consolidates, in brief and practical style, the latest information on optimal procedures.

Stevens RW: 1986. *Diagnostic Devices Manual and Directory: Immunology and Microbiology*, New York, Marcel Dekker.

Van Regenmortel MHV and AR Neurath: 1985. *Immunochemistry of Viruses: The Basis for Serodiagnosis and Vaccines*. Amsterdam, Elsevier Science Publishing.

Discusses the immunological aspects of viral diseases and the role of immunochemistry in the use of serum in diagnostic procedures and the creation of viral vaccines.

Vinken PJ, GW Bruyn, HL Klawans (eds.): 1988. *Microbial Disease*, Amsterdam, Elsevier Science Publishing.

Reviews the etiology and diagnosis of bacterial, viral, and parasitic diseases, particularly those of the nervous system.

Wentworth BB (ed.): 1987. *Diagnostic Procedures for Bacterial Infections*, Washington, D.C.; American Public Health Association.

Explains procedures approved by the American Public Health Association for the laboratory diagnosis of bacterial infections.

CHAPTER 17

Treatment of Infectious Diseases

PREVIEW TO CHAPTER 17

In this chapter we will:
- Study the ways in which physicians treat infectious diseases.
- Learn how antimicrobial drugs are used to treat diseases.
- Examine the various modes of action of different antimicrobics.
- Discover why bacterial infections are easier to treat with antimicrobics than viral, fungal, and protozoan infections.
- Learn the following key terms and names:

acyclovir	minimal lethal
aminoglycosides	concentration (MLC)
amphotericin B	minimum inhibitory
antibiotics	concentration (MIC)
antimalarials	multiple antibiotic
antimicrobial agents	resistance
antimicrobial	narrow spectrum
susceptibility testing	antibiotic
antimicrobics	nystatin
azidothymidine (AZT)	penicillins
bacitracin	pentamidine
broad spectrum	polyene antibiotics
antimicrobic	polymyxins
cephalosporins	quinolones
chemotherapeutic agent	rifampicin (rifampin)
chloramphenicol	Schlicter test
combined drug therapy	selective toxicity
dideoxycytidine (ddC)	serum killing power
dideoxyinosine (ddI)	streptomycin
erythromycin	synergism
isoniazid	tetracyclines
Kirby-Bauer method	trimethoprim
metronidazole	vancomycin
microbicidal	zone of inhibition
microbiostatic	
minimal bactericidal	
concentration (MBC)	

When the body's normal defenses cannot prevent or overcome a disease, the infected individual becomes ill. The physician then uses various treatments to alleviate the patient's symptoms and to cure the patient of the disease. Often the physician uses drugs to treat the disease. The term **chemotherapeutic agent** applies to any drug used for any medical condition, whether it is a simple headache, a "strep" throat, malaria, high blood pressure, or cancer. For example, aspirin, which is used to alleviate symptoms associated with the body's inflammatory response, and penicillin, which is used to eliminate infecting bacteria, are chemotherapeutic agents.

For many diseases there are no treatments that eliminate the infecting agents. In these instances the physician uses supportive therapies that rely on the body's immune defenses to fight off the infection. The simplest of the supportive therapies is encompassed in the sage advice "rest, drink plenty of fluids, and stay warm." By following this advice, the body is best able to mount physiologically an immune response against the infection. In other cases the physician selects drugs to maintain the comfort of the patient, such as aspirin and other drugs to limit pain.

In some cases supportive therapies that maintain essential body functions are necessary. They permit the body's immune defense system sufficient time to respond to the infection. The patient must necessarily remain alive for the immune system to function. An infection may result in the severe loss of water that may upset the body's electrolyte balance. This may cause death due to shock and cardiac arrest. In such cases, administration of an intravenous solution to maintain the necessary electrolyte balance is a needed supportive therapy (FIG. 17-1). If an infection blocks normal gas exchange, oxygen may be used to temporarily maintain respiratory function.

Supportive therapies do not directly attack the infecting agents but rather permit the body's immune defenses sufficient time to respond to the infection.

NEWSBREAK

SHOULD SCIENTISTS SEEK MEDICAL CURES FOR DISEASE?

While scientists and physicians search for ways of preventing and treating disease, some have begun to argue fervently against finding new cures for disease. In some cases the arguments are on religious grounds—some religious sects shun medical treatment and rely on their faith in cases of illness. Jeremy Rifkin, an ardent foe of biotechnology, and representatives of the handicapped argue that because the United States government forbids discrimination against individuals with handicaps, scientists should not be allowed to develop new treatments for disease conditions, such as blindness; they are fearful that the government may force individuals with handicaps to use the treatments to cure the handicap condition. In other cases the arguments against developing new medical treatments are economic; some stress the high cost of health care and the economic inability to support health-demanding populations with increased life expectancies. Pharmaceutical companies are hard pressed to invest in the development of new medical treatments because of the high cost of bringing new drugs and treatments to the marketplace. Governments are reevaluating how much to invest in medical research.

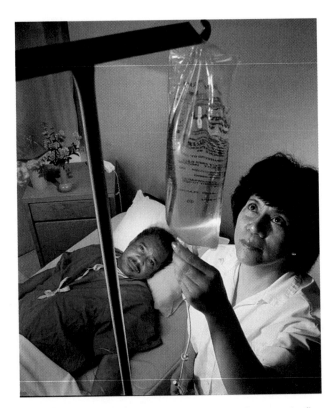

FIG. 17-1 In many cases an IV drip is used to supply fluids and salt solutions to maintain proper electrolyte balance so as to prevent shock. This supportive treatment is provided while the body's own immune system acts to eliminate a microbial infection that can not be eliminated through the use of antibiotics.

In contrast to supportive therapies, antimicrobial drugs are used to eliminate infecting agents. **Antimicrobial drugs** are the class of chemotherapeutic agents used to treat infectious diseases. They act by killing microorganisms or by interfering with the growth of microorganisms. Such antimicrobial agents can eliminate an infecting agent and thereby cure a disease. There are numerous antimicrobial agents available for treating diseases caused by microorganisms. Such drugs have become an essential part of modern medical practice.

The **antimicrobial agents** used in medical practice are aimed at eliminating infecting microorganisms. They can also act to prevent the establishment of an infection. **Antibiotics,** which are defined as antimicrobial substances produced by microorganisms, have been used in medicine only since the mid-1940s (FIG. 17-2).

Many of the antimicrobial compounds that are used today are in fact produced by microorganisms and therefore are properly called antibiotics. Some antimicrobial compounds, however, are produced partly or entirely by chemical synthesis. Even though not of microbial origin, similar compounds synthesized by organic chemists usually also are called antibiotics. To avoid problems in terminology the all-inclusive term **antimicrobic** often is used. In this section the terms antibiotic and antimicrobic will be used interchangeably. These chemicals should not be confused with the large number of drugs used in medical practice for alleviating the symptoms of disease or for treating diseases that are not caused by microorganisms.

> Antibiotics are antimicrobial agents produced by microorganisms. Antimicrobics may be "true" antibiotics or they may be chemically synthesized.

> Many infectious diseases are treated with antimicrobial agents that eliminate infecting microorganisms or prevent the establishment of an infection.

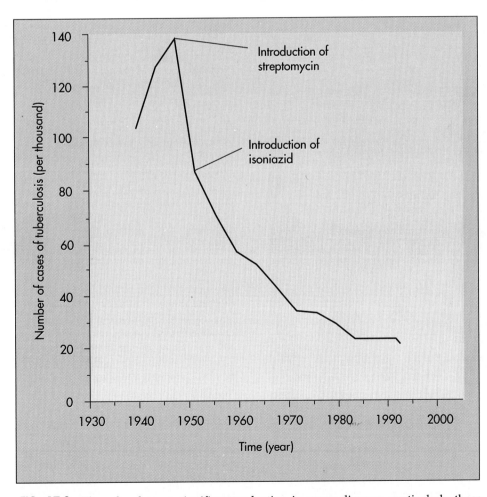

FIG. 17-2 There has been a significant reduction in many diseases, particularly those caused by bacterial infections, as a result of the introduction of antibiotics into medical practice after 1946. For example, the introduction of streptomycin and isoniazid into medical practice has greatly reduced the incidence of tuberculosis.

HISTORICAL PERSPECTIVE

DEVELOPMENT OF ANTIMICROBIAL THERAPY

Humans have long sought to find chemical remedies for their ills. Various plants were used as herbal remedies by ancient European and Far Eastern cultures. Some remedies could kill infecting microorganisms, but many were also highly toxic to humans.

Modern chemotherapy began at the end of the nineteenth century with the work of Paul Ehrlich in Germany (FIG. *A*). From 1780 to 1796, Ehrlich worked in the laboratory of Robert Koch, and in 1796 he became director of the first of his own institutes, which he dedicated to finding "substances which have their origin in the chemist's retort," that is, substances produced by chemical synthesis, to cure infectious diseases. Ehrlich hypothesized the existence of a "magic bullet"—some chemical agent that would selectively find and destroy pathogens while not harming the host. He discovered that chemical derivatives of atoxyl, an arsenical, was capable of antimicrobial activities. The arsenical compound 606, salvarsan, proved to be effective in treating syphilis. The Catholic church denounced Ehrlich for this discovery, claiming that he was interfering with God's punishment for sin.

A

Paul Ehrlich, working in the laboratory of Robert Koch, was the first to demonstrate that chemical agents could specifically be used to treat microbial infections. He pioneered antimicrobial treatment of disease.

Sahachiro Hata, a Japanese expert on spirochetes, found that the related arsenical compound 914, neosalvarsan, was an effective cure for syphilis and relapsing fever, both caused by spirochetes. The use of neosalvarsan in 1912 represents the first widespread use of synthetic drugs; these drugs became known as "magic bullets" and were portrayed as being able to find and kill disease-causing germs (FIG. *B*). These arsenic-containing drugs were also quite toxic to the patient. Prolonged usage or using too high a dose could kill the patient, but judicious use could control the diseases caused by spirochetes.

A major breakthrough in chemotherapy occurred in 1929 when the Scottish bacteriologist Alexander Fleming, working in a London teaching hospital, reported on the antibacterial action of cultures of a *Penicillium* species (FIG. *C*). Fleming observed that the mold *Penicillium notatum* killed his cultures of the bacterium *Staphylococcus aureus* when the fungus accidentally contaminated the culture dishes. It is likely that the fungal contaminant of Fleming's cultures, which was to bring medical practice into the modern era of drug therapy, blew into his laboratory from the floor below, where an Irish mycologist was working with strains of *Penicillium*. Such a serendipitous event can change history, but in science it takes a special individual like Fleming to recognize the significance of the observation. As Pasteur said, "Chance favors the prepared mind." Fleming's chance discovery of the effect of mold contamination on a bacterial culture plate was possible only because Fleming's background and knowledge had been enriched by the growth of general scientific awareness such that what he saw made sense to him and fitted an historical pattern of scientific investigation.

After growing the fungus in a liquid medium and separating the fluid from the cells, Fleming discovered that the cell-free liquid was an inhibitor of many bacterial species. His publication on the active ingredient, which he called penicillin, was the first report of the production of an antibiotic. Fleming himself pointed out that the pioneering work on antibiotics was thwarted because of the failure of microbiologists to pursue the chemical investigations necessary to separate and purify the active agents in their extracts. By the 1930s, however, microbiology had become far more chemistry oriented, and this approach culminated in the preparation of solid penicillin. In 1940, 10 years after Fleming's initial report, Howard Florey and Ernst Chain at Oxford University successfully isolated and purified penicillin. Other scientists established the therapeutic value of penicillin, and this antibiotic remains the cornerstone of the modern medical treatment of many infectious diseases. Productive *Penicillium* strains were isolated for use in mass production of the antibiotic. The most fa-

B

Antimicrobial agents were originally depicted as magic bullets that could specifically target and destroy invisible pathogens.

mous of these strains was originally isolated from a cantaloupe bought at a market in Peoria, Illinois.

Major advances in the development of chemotherapeutic agents continued to be made in the 1930s. Gerhard Domagk was a German physician employed as a chemist by a dye works company, the forerunner of modern pharmaceutical companies. Domagk developed prontosil, the first sulfa drug, effective in the treatment of streptococcal infections. At the Pasteur Institute, a husband-and-wife research team, the Trefouels, discovered the active constituent of prontosil. This active constiuent was sulfanilamide, the first real wonder drug, so called because of its amazing ability to cure serious diseases, such as pneumonia.

In the early 1940s, the Russian immigrant soil microbiologist, Selman Waksman, and co-workers at Rutgers University in New Jersey found that various bacteria of the actinomycetes group produced antibacterial agents. Streptomycin, produced by *Streptomyces griseus*, became the best known of the new antibiotic wonder drugs. The antibiotics produced by actinomycetes generally have a broader spectrum of action than penicillin and thus can be used to treat a number of diseases for which penicillin is ineffective. Nearly 85% of antibiotics in current use are produced by actinomycetes. The importance of penicillin and other antibiotics in treating diseases of microbial origin cannot be overestimated.

C

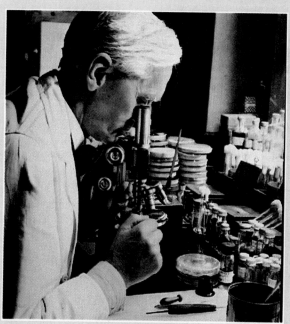

Sir Alexander Fleming (1881-1955) discovered the antibiotic penicillin. He had the insight to recognize the significance of the inhibition of bacterial growth in the vicinity of a fungal contaminant when most other scientists probably would have simply discarded the contaminated plates.

An antimicrobial agent must exhibit selective toxicity to be of therapeutic use. A therapeutically useful antimicrobial agent must inhibit infecting microorganisms and exhibit greater toxicity to the infecting pathogens than to the host organism. A drug that kills the patient is of no use in treating infectious disease, whether or not it also kills the pathogens. Even selective, therapeutically useful antimicrobics, though, can produce side effects. As a rule, antimicrobics are of most use in medicine when the mode of action of the antimicrobial chemical involves biochemical features of the invading pathogens not possessed by normal host cells.

Some antimicrobial agents are **microbicidal,** they kill microorganisms. Others are **microbiostatic,** they inhibit the growth of microorganisms but do not actually kill them (FIG. 17-3). Microbiostatic agents can prevent the proliferation of infecting microorganisms. They hold populations of pathogens in check until the host's normal immune defense mechanisms can eliminate the invading pathogens. By their very nature, antibiotics and synthetic antimicrobics must exhibit selective toxicity because they are produced by one microorganism and exert varying degrees of toxicity against others. The discovery and use of antibiotics have revolutionized medical practice by the twentieth century.

The control of microorganisms pathogenic to humans is fundamental to the practice of modern medicine.

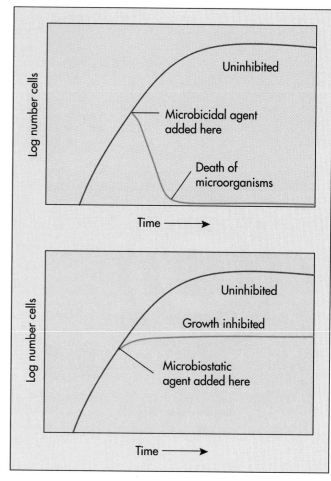

FIG. 17-3 Microcidal agents kill microorganisms, whereas microbiostatic agents only inhibit their growth.

SELECTION OF ANTIMICROBIAL AGENTS IN CHEMOTHERAPY

The selection of a particular antimicrobial agent for treating a given disease depends on several factors. The sensitivity of the infecting microorganism to the particular antimicrobial agent is particularly important. The side effects of the antimicrobial agent, with regard to direct toxicity to mammalian cells and to the microbiota normally associated with human tissues must be considered. Biotransformations of the particular antimicrobial agent can occur *in vivo* that affect whether the antimicrobial agent will remain in its active form for a sufficient period of time to be selectively toxic to the infecting pathogens. The chemical properties of the antimicrobial agent determine its distribution within the body and whether adequate concentrations of the active antimicrobial chemical will be able to reach the site of infection in order to inhibit or kill the pathogenic microorganisms causing the infection.

Selection of an antimicrobial agent depends on microbial sensitivity, possible side effects, possible *in vivo* biotransformations, and distribution within the body.

PHARMACOLOGICAL PROPERTIES

The pharmacological properties of a potential antimicrobial agent must be considered by the physician in deciding whether to treat an infection with a particular drug. It is important that the drug have a high degree of **selective toxicity.** It is useless to consider using drugs that are effective against a pathogen if the drug will also kill the patient. Generally the mode of action of therapeutically useful agents exploits differences between human cells and those of the pathogen. Ideally, the antimicrobial agent should target a property of the pathogen that does not occur in the host organism. This is easier to achieve for bacterial infections because of the differences between prokaryotic bacterial cells and eukaryotic human cells than it is when the infection is caused by viruses, fungi, or protozoa. Even drugs that have a high degree of selective toxicity may produce adverse side effects in humans (Table 17-1).

Physicians must guard against adverse reactions, monitoring patients' responses and adjusting the dosage of an antimicrobial agent or selecting another drug when necessary.

To be used in treating an infection, it is critical that the properties of the drug permit it to reach the site of that infection. The selection of a particular antimicrobial agent for treating a specific infection depends on the ability to get the drug to the site where it is needed, not just on the ability of the drug to inhibit or kill the pathogen (Table 17-2). It is important to consider the level of the antibiotic that may be reached at a given body site (Table 17-3). Many antibiotics, for example, are toxic to the pathogenic bacteria that cause urinary tract infections, but relatively few can reach and be concentrated in the tissues of the urinary tract in their active form. Therefore only a limited number of drugs are effective in treating these infections.

It is also important to consider the potential interactions between drugs that may be employed at the same time to treat a patient. Some infections must be treated with more than one antimicrobial agent at the same time. This is called *combined drug therapy*. When more than one antimicrobial agent is used simultaneously, one antimicrobial agent can influence the effects of another antimicrobial agent. In some cases treatment with two drugs enhances the effectiveness of treatment. This is called drug *synergism*. In other cases one drug can interfere with the inhibitory effects of a second antimicrobial agent.

TABLE 17-1

Major Toxicities of Selected Antimicrobics

ANTI-MICROBIC AGENT	MECHANISM	SIGNS
Aminoglycosides	Binds hair cells of organ of Corti	Deafness
	Binds vestibular cells	Vertigo
	Competitive neuromuscular blockage	Respiratory paralysis
	Tubular necrosis	Nephrotoxicity
Amphotericin	Distal tubular damage	Nephrotoxicity
	Renal tubular acidosis	Nephrotoxicity
Carbenicillin	Inhibition of platelet aggregation	Bleeding
Cephalosporins	Cortical stimulation	Myoclonic seizures
Cephaloridine	Proximal tubular damage	Nephrotoxicity
Chloramphenicol	Damages stem cell	Aplastic anemia
	Inhibits protein synthesis	Reversible anemia
Clindamycin	Proliferation of *Clostridium difficile*	Diarrhea
Emetine	Permeability changes	Hypotension
Isoniazid	Liver cell damage	Hepatitis
Neomycin	Villus damage	Malabsorption
Penicillins	Cortical stimulation	Myoclonic seizures
Polymyxins	Noncompetitive neuromuscular blockage	Respiratory paralysis
	Tubular necrosis	Nephrotoxicity
Rifampin	Liver cell damage	Hepatitis
Sulfonamides	Glucose 6-phosphate deficiency	Hemolytic anemia
	Collecting duct obstruction	Nephrotoxicity
Tetracyclines	Liver cell damage	Hepatitis
	Degradation products	Fanconi syndrome

NEWSBREAK

ROLES OF THE PHYSICIAN AND CLINICAL MICROBIOLOGIST

Clinical microbiologists and physicians have different roles in determining the appropriate drugs to use in treating infectious diseases. Sometimes there are territorial conflicts. Clinical microbiologists generally are Ph.D.s who are not licensed to practice medicine. They cannot prescribe antibiotics and cannot tell or even advise physicians which drugs to use for treating an infectious disease. The clinical microbiologist is restricted to advising about test results in the laboratory that can aid the physician in determining the appropriate therapies. If the physician needs assistance in interpreting the test results from the laboratory, he or she must consult an infectious disease specialist with an M.D. degree.

Distribution of Antimicrobics to Specific Body Areas

BODY SITE	USEFUL ANTIMICROBICS
Bone	Penicillins, tetracyclines, cephalosporins, lincomycin, and clindamycin antimicrobics penetrate bone and bone marrow; levels are higher in infected bone than in normal bone.
Central nervous system	Various penicillins achieve adequate concentrations in the brain to treat brain abscess. Penicillin G and ampicillin can achieve adequate cerebrospinal fluid (CSF) levels; oxacillin, naficillin, and methicillin can be used to treat staphylococcal meningitis. CSF levels of chloramphenicol are adequate to treat *Streptococcus, Neisseria,* and *Haemophilus* spp. but not most Gram-negative bacteria. Cefoxamine, moxalactam, and cefoperazone enter CSF in the presence of inflammation in concentrations that are adequate to treat *Streptococcus* spp., *Neisseria* spp., *Haemophilus* spp., *Klebsiella* spp., and *Escherichia coli* infections.
Ears and sinuses	Most of the penicillins reach levels in the middle ear fluid in sufficient concentrations for the treatment of otitis media. Concentrations of sulfonamides and trimethoprim in sinuses are adequate to treat infections.
Eyes	Few antimicrobics penetrate the eye well. Levels of penicillins and cephalosporins in the aqueous humor are less than 10% of the peak serum levels and inhibit only highly sensitive bacteria.

Achievable Levels of Some Common Antibiotics in Various Body Fluids

ANTIBIOTIC	ACHIEVABLE PEAK BLOOD LEVELS (μg/mL)	ACHIEVABLE URINE LEVELS (μg/mL)	DOSE
Clindamycin	1-4	>20	Oral 150-300 mg
	6-10	>60	IV 300-600 mg
Erythromycin	1-2		Oral 250-500 mg
	10-20		IV 300 mg
Penicillin	2-3	>300	Oral 500 mg
	6-8	>300	IM 500 mg
	4-7	>300	IV 500 mg
Ampicillin	1-3	>50	Oral 250-500 mg
	2-6	>20	IM 250-500 mg
	10-25	>100	IV 1,000-1,500 mg
Cephalothin	3-18	>100	Oral 250-500 mg
	9-24	>300	IM 500-1,000 mg
	30-85	>1,000	IV 1,000-2,000 mg
Gentamicin	2-10	>20	IV/IV 1-2 mg
Tetracycline	1-2	>200	Oral 250-500 mg
	10-20	>200	IV 500 mg
Chloramphenicol	10-12	>100	Oral 1,000 mg
	20-30	>200	IV 1,000 mg
Nitrofurantoin	<1	>100	Oral 50-100 mg

IV, Intravenous. *IM,* Intramuscular.

BODY SITE	USEFUL ANTIMICROBICS
Pleural and pericardial fluids	Most of the penicillins, cephalosporins, sulfonamides, macrolides, clindamycin, chloramphenicol, and antituberculosis drugs diffuse into serus cavities.
Pulmonary	Concentrations of most antibiotics within the lung are satisfactory, provided there is sufficient blood flow. Penicillins and tetracyclines show variable sputum concentrations. Antituberculosis agents, such as isoniazid and rifampin, achieve appreciable levels in pulmonary tissue.
Skin	Tetracyclines and clindamycin concentrate in skin tissue and are effective in treatment of acne.
Synovial fluid	Most antibiotics used in the treatment of joint infection reach inflamed joints in adequate concentrations.
Urinary tract	Treatment of kidney and other urinary tract infections depends largely on the concentrations in the urine rather than on serum levels. Nalidixic acid and nitrofurantoins are effective in treating urinary tract infections.

ANTIMICROBIAL SUSCEPTIBILITY TESTING

Antimicrobial susceptibility testing determines the effectiveness of specific drugs against specific pathogens. Determination of the antimicrobial susceptibility of a pathogen is important in aiding the clinician to select the most appropriate agent for treating that disease. It is pointless to prescribe an antibiotic that is ineffective against the microorganism causing the disease.

NEWSBREAK

METHICILLIN RESISTANT STAPHYLOCOCCI

Penicillins are frequently used to treat staphylococcal infections. Staphylococci, however, sometimes develop resistance to these drugs. Consequently, physicians have greatly restricted the use of methicillin for treating infections with staphylococci. In Australia, a government board must approve the use of methicillin by any physician. This is done on a case by case basis. Despite the restricted use of methicillin, some strains of *Staphylococcus aureus* are resistant to methicillin. These strains are called MRSA for methicillin resistant *Staphylococcus aureus*. Strains of MRSA are difficult to treat, particularly when they cause heart infections. Identifying such strains and determining their susceptibility to alternate antibiotics is essential.

The clinical microbiology laboratory provides information with regard to the activities of antimicrobial agents. It tests agents to be used against microorganisms that have been isolated and identified as the probable causative agents of disease. The laboratory uses standardized *in vitro* testing. Antibiotic susceptibility testing relies on the observation of antibiotics inhibiting the growth and/or killing cultures of microorganisms *in vitro*. It provides the physician with the information needed to prescribe the proper antibiotics for treating infectious diseases.

Physicians select appropriate antimicrobics. They must avoid indiscriminate administration of antibiotics because the selective pressures of excessive antibiotic usage can and have led to the evolution of antibiotic-resistant strains of pathogens. These antibiotic-resistant pathogens become problems because they cause infections that do not respond to the antibiotics routinely used to treat specific diseases.

Concern is mounting about the overuse of antibiotics. An undesired side effect of such excessive antimicrobic usage is that it gives a competitive advantage to strains that are resistant to antibiotics, thereby selecting for disease-causing antibiotic resistant strains. The reason for concern about how we use antibiotics is that numerous bacterial strains have acquired the ability to resist the effects of some antibiotics. The use of antibiotics favors the natural selection of strains of bacteria that are resistant to that particular antibiotic. These strains are naturally occurring antibiotic resistant mutants that are selected for in the presence of the drug. (FIG. 17-4)

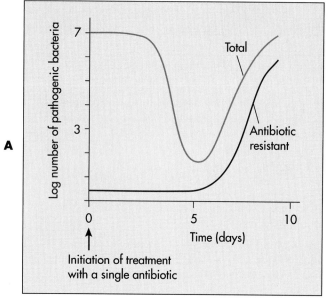

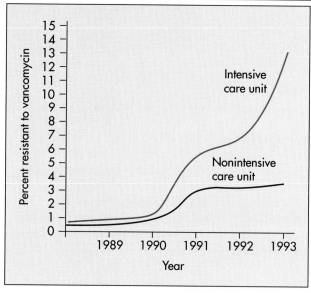

FIG 17-4   A, The incidence of antibiotic resistance has been increasing, making it more difficult for physicians to treat some infections. **B,** Antibiotic resistance is seen within a hospital as the periodic occurrence of bacterial isolates containing multiple antibiotic resistant plasmids. **C,** There has been an increased incidence of vancomycin-resistant enterococci in hospitals in the 1990s. This is seen as a major emerging problem in control of infectious diseases, especially if vancomycin-resistance incidence increases in staphylococci.

There are three major mechanisms that lead to antibiotic resistance: inactivation of the antibiotic by a microbial enzyme, prevention of the antibiotic from reaching its target cell structure, and alteration of the target cell structure so that it is no longer affected by the antibiotic. Some bacterial strains, generally those containing R plasmids, even have **multiple antibiotic resistance.** The American Medical Association has alerted its member physicians to the danger of overusing antibiotics. It is now considered proper medical practice to perform culture and sensitivity studies to determine the proper antibiotic for treating a patient. Only in cases of life-threatening infections should antibiotics be used without such testing to avoid the selective pressure for the development of antibiotic resistant pathogens.

In vitro **testing in the clinical microbiology laboratory provides information on the activity of antimicrobial agents against probable disease-causing microorganisms.**

Agar Diffusion Antimicrobial Susceptibility Testing

The **Kirby-Bauer method** for testing the effectiveness of antimicrobial agents was developed in the early 1960s. This method employs antibiotic-impregnated filter-paper discs that are placed onto agar plates inoculated with a test organism (FIG. 17-5). The antibiotic diffuses from the paper disc onto the agar. A clear region around the disc occurs when the antimicrobic agent inhibits microbial growth. This clear region is called a **zone of inhibition.** The diameter of

TABLE 17-4

Interpretation of Zones of Inhibition for Kirby-Bauer Antibiotic Susceptibility Testing

ANTIBIOTIC	DISC CONC.	INHIBITION ZONE DIAMETER (mm)		
		RESISTANT	INTERMEDIATE	SUSCEPTIBLE
Amikacin	0.01 mg	13 or less	12-13	14 or more
Ampicillin	0.01 mg	11 or less	12-13	14 or more
Bacitracin	10 units	8 or less	9-11	13 or more
Cephalothin	0.03 mg	14 or less	15-17	18 or more
Chloramphenicol	0.03 mg	12 or less	13-17	18 or more
Erythromycin	0.015 mg	13 or less	14-17	18 or more
Gentamicin	0.01 mg			13 or more
Kanamycin	0.03 mg	13 or less	14-17	18 or more
Lincomycin	0.002 mg	9 or less	10-14	15 or more
Methicillin	0.005 mg	9 or less	10-13	14 or more
Nalidixic acid	0.03 mg	13 or less	14-18	19 or more
Neomycin	0.03 mg	12 or less	13-16	17 or more
Nitrofurantoin	0.3 mg	14 or less	15-16	17 or more
Penicillin G—staphylococci	10 units	20 or less	21-28	29 or more
Penicillin—other organisms	10 units	11 or less	12-21	22 or more
Polymyxin	300 units	8 or less	9-11	12 or more
Streptomycin	0.01 mg	11 or less	12-14	15 or more
Sulfonamides	0.3 mg	12 or less	13-16	17 or more
Tetracycline	0.03 mg	14 or less	15-18	19 or more
Vancomycin	0.03 mg	9 or less	10-11	12 or more

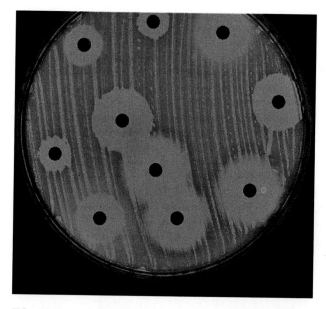

FIG. 17-5 The Bauer-Kirby test (also called the Kirby-Bauer test) is a standardized procedure for determining antimicrobic susceptibility. The diameter of the zone of inhibition (clear area around antimicrobic impregnated discs) indicates the sensitivity of the microorganism to that antimicrobic.

the zone of inhibition reflects the solubility properties of the particular antibiotic. It depends on the concentration gradient established by diffusion of the antibiotic into the agar and the sensitivity of the given microorganism to the specific antibiotic. Standardized zones for each antibiotic disc have been es-

tablished to determine whether the microorganism is susceptible (S), intermediately sensitive (I), or resistant (R) to the particular antibiotic. The results of Kirby-Bauer testing indicate whether a particular antibiotic has the potential for effective control of an infection caused by a particular pathogen (Table 17-4).

The qualitative susceptibility of microorganisms to antimicrobial agents can be determined on agar plates by using filter-paper discs impregnated with antimicrobial agents in the Kirby-Bauer test.

The Kirby-Bauer agar diffusion test procedure is designed for use with rapidly growing bacteria. This method is not directly applicable for use with filamentous fungi, anaerobes, or slow-growing bacteria. Modifications of the media composition and incubation conditions, however, can be made for testing the antibiotic susceptibility of such microorganisms.

Automated Liquid Diffusion Methods

Many clinical laboratories use automated systems for determining the antibiotic susceptibility of a pathogen. In such systems, the antibiotics are added to a broth rather than a solid agar. The system uses a spectrophotometer to measure light scattering as an index of microbial growth. The concentrations of the antibiotics and the density of growth are used for assessing antibiotic sensitivities. Various automated systems are available for performing this procedure, including the Autobac, Microscan, BBL sceptre, Vitek AMS, and Abbott MSII. These automated systems simplify and enhance the reliability of antimicrobial susceptibility testing. This asset makes it likely that these tests will be used more frequently, thereby reducing excessive use of inappropriate antimicrobics by some physicians.

Minimum Inhibitory Concentration (MIC)

Another approach to antimicrobial susceptibility testing is the determination of the **minimum inhibitory concentration** (MIC) that will prevent microbial growth (FIG. 17-6). This procedure determines the concentration of an antibiotic that is effective in preventing the growth of the pathogen and gives an indication of the dosage that should be effective in controlling the infection in the patient. A standardized microbial inoculum is added to tubes containing serial dilutions of an antibiotic. Growth of

the microorganism is monitored as a change in turbidity. In this way, the break point or minimum inhibitory concentration of the antibiotic, that is, the lowest concentration of the antimicrobic that prevents growth of the microorganism *in vitro*, can be determined.

The minimum inhibitory concentration indicates the minimal concentration of the antibiotic that must be achieved at the site of infection to inhibit the growth of the microorganism being tested. By knowing the MIC and the theoretical levels of the antibiotic that may be achieved in body fluids, such as blood and urine, the physician can select the appropriate antibiotic, the dosage schedule, and the route of administration. Generally, a margin of safety of 10 times the MIC is desirable to ensure successful treatment of the disease.

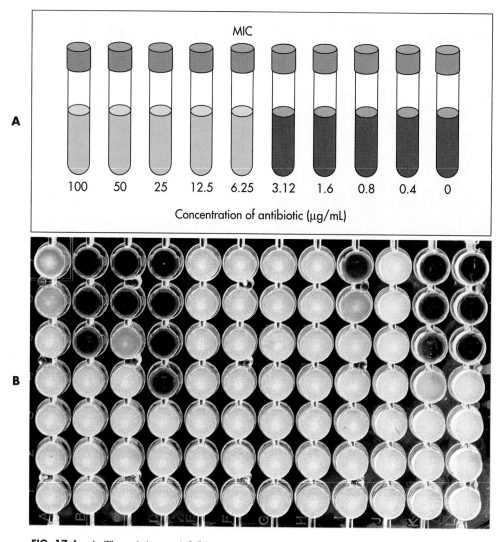

FIG. 17-6 **A,** The minimum inhibitory concentration indicates the lowest concentration of an antimicrobic that prevents growth. In this example the MIC is 6.25 mg/mL; it is the lowest concentration that precludes growth (*clear orange*). Growth occurs at lower concentrations (*cloudy brown*). **B,** A microtiter plate showing the determination of the MICs for several different antimicrobics.

Minimum inhibitory concentration testing determines the concentration of an antibiotic effective against a pathogen and indicates an effective dosage.

The MIC is the lowest concentration of antimicrobic that prevents the growth of a microorganism *in vitro*.

MIC determinations can be used for determining the antibiotic sensitivity of aerobic and anaerobic microorganisms. The use of microtiter plates, which require only a few hundred microliters per sample well, and automated inoculation and reading systems make the determination of MICs feasible for use in the clinical laboratory. MICs can even be performed on normally sterile body fluids without isolating and identifying the pathogenic microorganisms. For example, blood or cerebrospinal fluid containing an infecting microorganism can be added to tubes containing various dilutions of an antibiotic

and a suitable growth medium. An increase in turbidity would indicate the growth of microorganisms and the fact that the antibiotic at that concentration was ineffective in inhibiting microbial growth. A lack of growth would indicate that the pathogenic microorganisms were susceptible to the antibiotic at the given concentration. By determining the minimum inhibitory concentration, the appropriate dosage of the right antibiotic can be selected for treating an infectious disease.

Minimal Bactericidal Concentration (MBC)

MIC determinations are used for establishing the concentration of an antibiotic that will inhibit growth. They are not designed to determine whether the antibiotic is microbicidal. It is, however, also possible to determine the **minimal bactericidal concentration** (MBC). The MBC is also known as the **minimal lethal concentration** (MLC) (FIG. 17-7). The

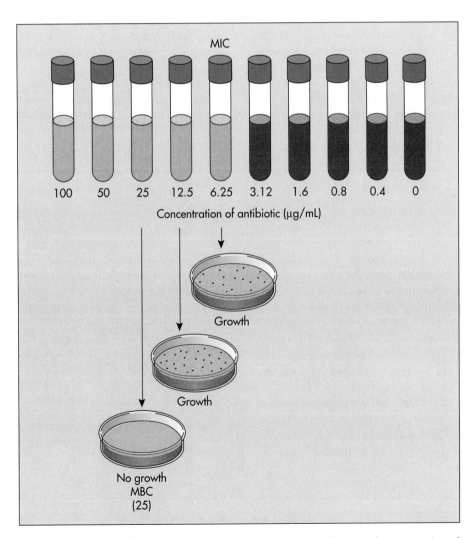

FIG. 17-7 The minimum bactericidal concentration (MBC) of an antibiotic requires the demonstration that microorganisms have lost the ability to reproduce. In this example, although cell growth is inhibited at concentrations of 6.25 and 12.5, viable cells remain, which is shown by the formation of colonies (growth) on an agar plate lacking the antimicrobic. No viable cells are detected (no growth on agar plates) at 25 μg/mL, which therefore is the MBC.

MBC refers to the lowest concentration of an antibiotic that will kill a defined proportion of viable organisms in a bacterial suspension during a specified exposure period. Generally, a 99.9% kill of bacteria at an initial concentration of 10^5–10^6 cells/mL during a 17- to 24-hour exposure period is used to define the MBC.

> **The minimal bactericidal concentration is the lowest concentration of an antibiotic that will kill a defined proportion of viable organisms in a bacterial suspension during a specified period of exposure.**

To determine the minimal bactericidal concentration, it is necessary to plate the tube suspensions showing no growth in tube dilution (MIC) tests onto an agar growth medium. This is done to determine whether the bacteria are indeed killed or whether they survive exposure to the antibiotic at the concentration being tested. Determination of the MIC is adequate for establishing the appropriate concentration of an antibiotic that should be administered for controlling the infection in patients with normal immune response levels. Determination of the MBC is essential for patients with endocarditis (inflammation of the endocardium or lining to the heart) because the patient's immune response cannot be relied on to remove the infecting microorganisms. It is particularly useful in determining the appropriate concentration of an antibiotic for use in treating patients with lowered immune defense responses, such as may occur in patients receiving chemotherapy treatment for cancer.

Serum Killing Power

Another approach to determining the effectiveness of an antibiotic is to measure **serum killing power**. Serum killing power is a measure of the effectiveness of an antimicrobic in a patient's blood. The **Schlichter test** determines serum killing power. Instead of adding dilutions of an antibiotic to suspensions of bacteria in a growth medium, a bacterial suspension is added to dilutions of the patient's serum. The ability of the bacteria to grow in the patient's blood is assessed by measuring changes in turbidity. If the patient is being treated with an antibiotic, no bacterial growth should occur. The break point in the dilution where bacterial growth occurs reflects the concentration of the antibiotic in the patient's blood and the *in vivo* effectiveness of the antibiotic in controlling the infection. Inhibition at dilutions of the patient's serum of greater than or equal to 1:8 is considered an acceptable level.

> **Serum killing power is a measure of the effectiveness of an antimicrobic in a patient's blood as measured by changes in turbidity.**

ANTIBACTERIAL AGENTS

The choice of a particular antibiotic for treating bacterial infections depends in part on the chemical properties of the specific infecting bacterial strain. The differences in the cell structures of bacterial cells and eukaryotic human cells form the basis for the effective use of antibiotics against bacterial infections. The unique peptidoglycan of the bacterial cell wall and the 70S ribosome of the bacterial cell represent two major sites against which antibacterial agents may be targeted. Most of the common antibiotics used in medicine for treating bacterial infections are, in fact, inhibitors of cell wall or protein synthesis.

In many cases, physicians make an educated guess as to which antibiotic is appropriate for treating a particular infection. The selection of the antibiotic is based on the most likely pathogen causing the given disease symptomatology and the antibiotics generally known to be effective against such pathogens. Some antibiotics are more selective than others with respect to the bacterial species that they can inhibit. A **narrow spectrum antibiotic** may be targeted at a particular bacterial pathogen, for instance, at Gram-positive cocci, or at a particular bacterial species. In contrast, some antibiotics have a **broad spectrum of action.** These antibiotics inhibit a relatively wide range of bacterial species, including both Gram-positive and Gram-negative types. Many times a physician will select a broad spectrum antibiotic using a shotgun approach of treatment. Only in special cases, such as when a patient fails to respond to a particular antibiotic and the infection persists, is an attempt normally made to isolate the pathogenic bacterium and to determine the range of specific antibiotic sensitivity of that organism.

> **Narrow spectrum antibiotics are targeted against specific bacterial pathogens, broad spectrum antibiotics inhibit a wide range of bacterial species.**

> **Overuse of antibiotics is a problem because of the development of disease-causing antibiotic resistant strains. Physicians should request cultures and sensitivities before prescribing antibiotic therapy.**

CELL-WALL INHIBITORS

Several antibiotics act by inhibiting the synthesis of the bacterial cell wall. Most target peptidoglycan. Because of its unique chemistry and the fact that it only occurs in bacteria, peptidoglycan is an excellent target for achieving selective toxicity. Hence antibiotics such as penicillins and cephalosporins that are cell-wall inhibitors are often the antibiotics of choice in treating infections (Table 17-5).

TABLE 17-5

Some Diseases and Their Causative Organisms for Which Penicillins and Cephalosporins Are Recommended

CAUSATIVE ORGANISM	DISEASE	DRUG OF CHOICE
GRAM-POSITIVE COCCI		
Staphylococcus aureus	Abscesses Bacteremia Endocarditis Pneumonia Meningitis Osteomyelitis Cellulitis	Penicillin G A penicillinase-resistant penicillin
Streptococcus pyogenes	Pharyngitis Scarlet fever Otitis media, sinusitis Cellulitis Erysipelas Pneumonia Bacteremia Other systemic infections	Penicillin G Penicillin V
Streptococcus (viridans group)	Endocarditis Bacteremia	Penicillin G
Streptococcus faecalis (Enterococcus faecalis)	Endocarditis Urinary tract infection Bacteremia	Penicillin G Ampicillin Penicillin G
Streptococcus bovis	Endocarditis Urinary tract infection Bacteremia	Penicillin G
Streptococcus (anaerobic species)	Bacteremia Endocarditis Brain and other abscesses Sinusitis	Penicillin G
Streptococcus pneumoniae (pneumococcus)	Pneumonia Meningitis Endocarditis Arthritis Sinusitis Otitis	Penicillin G
GRAM-NEGATIVE COCCI		
Neisseria gonorrhoeae (gonococcus)	Genital infections Arthritis-dermatitis syndrome	Ampicillin or amoxicillin Penicillin G Ampicillin or amoxicillin Penicillin G
Neisseria meningitidis (meningococcus)	Meningitis Bacteremia	Penicillin G
GRAM-POSITIVE RODS		
Bacillus anthracis	"Malignant pustule" pneumonia	Penicillin G
Corynebacterium diphtheriae	Pharyngitis Laryngotracheitis Pneumonia Other local lesions	Penicillin G
Erysipelothrix rhusiopathiae	Erysipeloid	Penicillin G
Clostridium perfringens	Gas gangrene	Penicillin G
Clostridium tetani	Tetanus	Penicillin G

TABLE 17-5—cont'd

Some Diseases and Their Causative Organisms for Which Penicillins and Cephalosporins Are Recommended

CAUSATIVE ORGANISM	DISEASE	DRUG OF CHOICE
GRAM-NEGATIVE RODS		
Haemophilus influenzae	Otitis Sinusitis Bronchitis Epiglottitis	Amoxicillin Ampicillin
Enterobacter aerogenes	Urinary tract infection	Cephamandole
Klebsiella pneumoniae	Urinary tract infection Pneumonia	Cephalosporin
Pasteurella multocida	Wound infection Abscesses Bacteremia Meningitis	Penicillin G
Bacteroides spp.	Oral disease Sinusitis Brain abscess Lung abscess	Penicillin G
Fusobacterium nucleatum	Ulcerative pharyngitis Lung abscess Genital infections Gingivitis	Penicillin G
Streptobacillus moniliformis	Bacteremia Arthritis Endocarditis Abscesses	Penicillin G
SPIROCHETES		
Treponema pallidum	Syphilis	Penicillin G
Treponema pertenue	Yaws	Penicillin G
Leptospira interrogans	Weil disease Meningitis	Penicillin G
ACTINOMYCETES		
Actinomyces israelii	Cervical, facial, abdominal, thoracic, and other lesions	Penicillin G

Penicillins

The **penicillins** are widely used antibiotics that inhibit the formation of bacterial cell walls. Penicillins are synthesized by strains of the fungus *Penicillium*. The penicillins contain a β-lactam ring that is a strained four membered ring structure (FIG. 17-8). Penicillins bind to several enzymes involved in cell wall peptide formation, blocking normal wall synthesis. These enzymes are located in the plasma membrane of bacteria and are called penicillin binding proteins. Bacterial cell walls lacking the normal cross-linked peptide chains are subject to attack by autolysins. Autolysins are enzymes that degrade the cell's own cell-wall structures. The result is that, in the presence of penicillins, growing bacterial cells are subject to lysis because without functional cell wall structures the bacterial cell is not protected against osmotic shock.

> Penicillins inhibit the formation of peptide cross-linkages within the peptidoglycan backbone of the cell wall.

Many of the penicillins, such as penicillin G, have a relatively narrow spectrum of activity. They are most effective against Gram-positive cocci, including *Staphylococcus* species. Ampicillin, an amino-substituted penicillin, is a broad spectrum antibiotic that is active against many Gram-negative rods, including *Escherichia coli*, *Haemophilus influenzae*, *Shigella* sp., and *Proteus* sp.. The effectiveness of ampicillin against Gram-negative bacteria is based on its ability

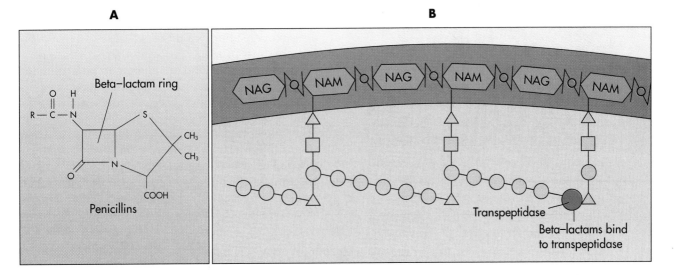

FIG. 17-8   A, The penicillins have beta lactam rings. **B,** Penicillins, which are widely used antimicrobics, block synthesis of bacterial cell walls.

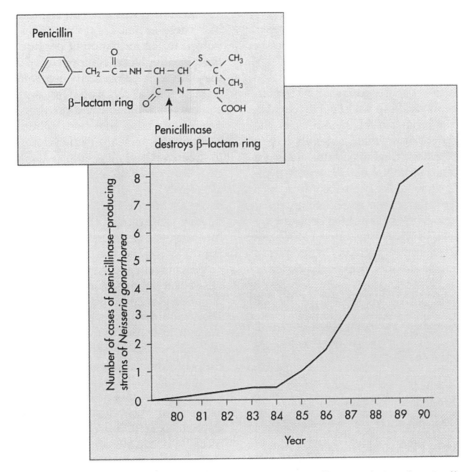

FIG. 17-9   Penicillinase destroys the β-lactam ring of penicillins, rendering them ineffective. Bacteria that produce penicillinases are penicillin resistant.

to pass through the outer membrane and to reach the cell wall peptidoglycan layer.

Penicillins are subject to inactivation by penicillinases (β-lactamases). Penicillinase-producing bacterial strains are able to degrade the β-lactam ring structures of many penicillins. This renders them ineffective in treating such bacterial strains (FIG. 17-9). For example, penicillin G is normally effective against *Neisseria gonorrhoeae*, a Gram-negative diplococcus that causes gonorrhea, but some penicillinase-pro-

ALLERGIES TO PENICILLIN

Some individuals develop allergies to drugs such as penicillin. Physicians must use alternate antibiotics in treating such individuals. It is important that physicians be aware of individuals with allergies to drugs. Many individuals wear identification tags to alert physicians about such allergies. This is critical, since allergic reactions to drugs can cause anaphylactic shock and death.

ducing strains of *N. gonorrhoeae* have now been found. The use of antibiotics other than penicillin G in the treatment of cases of gonorrhea caused by these strains is now required.

Some bacteria produce penicillinases that degrade the β-lactam rings of many penicillins.

Cephalosporins

Like the pencillins, the cephalosporins act by inhibiting bacterial cell-wall synthesis and are bactericidal. Cephalosporins are produced by members of the fungal genus *Cephalosporium*, which generally have a broad spectrum of action. Cephalosporins contain a β-lactam ring in their structure similar to the penicillins (FIG. 17-10). Many of the cephalosporins, such as cefoxitin and cephalothin, are also relatively resistant to β-lactamases. As such, the cephalosporins are useful in treating various infections caused by Gram-positive and Gram-negative bacteria. Many physicians are now using broad spectrum cephalosporins

where the use of narrow range and more specifically directed penicillins would be adequate. Cephalosporins are most prudently used as alternatives to penicillins for patients who are allergic to penicillin and for pathogens that are not penicillin sensitive.

Cephalothin is often the antibiotic of choice for treating severe staphylococcal infections, such as endocarditis. Its use avoids complications in cases where the infecting *Staphylococcus* species produces β-lactamases. Cefamandole is widely used in treating pneumonia because it is active against *Haemophilus influenzae*, *Staphylococcus aureus*, and *Klebsiella pneumoniae*. These are frequently the causative agents of respiratory tract infections resulting in pneumonia. Cephalosporins may also be used in place of penicillins for the prophylaxis of infection by Gram-positive cocci following surgical procedures.

Vancomycin and Bacitracin

Several other antibiotics also inhibit cell-wall synthesis, including vancomycin, bacitracin, and cycloserine. These antibiotics do not block the enzymes involved in the formation of peptide cross-linkages in the peptidoglycan component of the wall. Rather, they block other reactions involved in the synthesis of the bacterial cell wall. Cycloserine is a structural analogue of D-alanine and can prevent the incorporation of D-alanine into the peptide units of the cell wall. In the presence of D-cycloserine, the enzymes that convert L-alanine into D-alanine and then link two D-alanine molecules together are inhibited. Therefore the D-alanine–D-alanine subunits that are necessary for cell-wall synthesis cannot be adequately synthesized. The therapeutic use of cycloserine, however, is limited by its toxic reactions involving the central nervous system.

Vancomycin and bacitracin prevent the linkage of the *N*-acetylglucosamine and *N*-acetylmuramic acid moieties that compose the peptidoglycan molecule. These antibiotics are bactericidal. The use of bacitracin is restricted to topical application because this antibiotic causes severe toxic reactions and kidney damage. Vancomycin is especially effective against strains of *Staphyloccocus aureus*. Vancomycin is a very toxic drug but it is probably the most effective drug in clinical use against penicillinase-producing staphylococci. It is used in the treatment of streptococcal endocarditis and staphylococcal infections of such devices as prosthetic heart valves.

INHIBITORS OF PROTEIN SYNTHESIS

All living organisms depend on the ability to synthesize proteins. Proteins are needed for incorporation into cell structures and for use as enzymes. Blocking protein synthesis prevents microbial growth and reproduction and leads to the death of the microorgan-

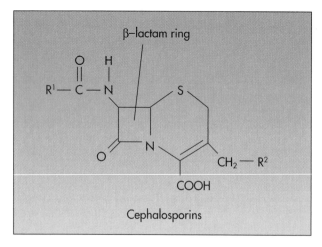

FIG. 17-10 Cephalosporins have a beta lactam ring structure and act in the same manner as penicillins.

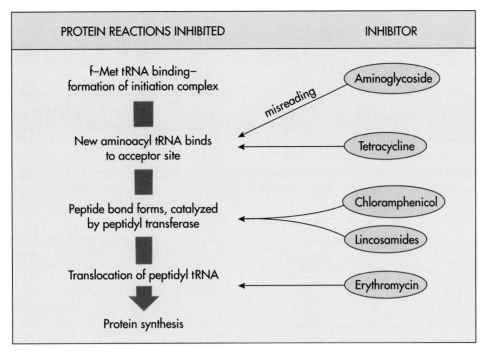

PROTEIN REACTIONS INHIBITED	INHIBITOR

f–Met tRNA binding–
formation of initiation complex

Aminoglycoside

misreading

New aminoacyl tRNA binds
to acceptor site

Tetracycline

Peptide bond forms, catalyzed
by peptidyl transferase

Chloramphenicol

Lincosamides

Translocation of peptidyl tRNA

Erythromycin

Protein synthesis

FIG. 17-11 Aminoglycosides, such as streptomycin and gentamycin, inhibit bacterial protein synthesis. These antibiotics also have relatively high toxicities to humans and are used only when necessary. They sometimes are administered by IV drip to control their concentrations accurately.

ism. Several inhibitors of protein synthesis are widely used in medicine. These inhibitors act at different stages of protein synthesis (FIG. 17-11).

Aminoglycosides

The antibiotics streptomycin, gentamicin, neomycin, kanamycin, tobramycin, and amikacin are inhibitors of bacterial protein synthesis. They are grouped together as **aminoglycosides** because they have similar structures. The aminoglycoside antibiotics contain amino sugars (sugars with amine groups) that are connected to inositol or to other amino sugars by glycosidic bonds.

The aminoglycosides are used almost exclusively in the treatment of infections caused by Gram-negative bacteria (Table 17-6). These bactericidal antibiotics are relatively ineffective against anaerobic bacteria and facultative anaerobes growing under anaerobic conditions. Their action against Gram-positive bacteria is also limited. Aminoglycosides are pro-

TABLE 17-6		
Some Diseases and Their Causative Organisms for Which Aminoglycoside Antibiotics Are Recommended		
CAUSATIVE ORGANISM	**DISEASE**	**DRUG OF CHOICE**
GRAM-NEGATIVE RODS		
Enterobacter aerogenes	Urinary tract, other infections	Gentamicin, tobramycin
Proteus spp.	Urinary tract, other infections	Gentamicin, tobramycin
Pseudomonas aeruginosa	Bacteremia	Gentamicin, tobramycin
Acinetobacter spp.	Various nosocomial infections, bacteremia	Gentamicin
Yersinia pestis	Plague	Streptomycin ± tetracycline
Serratia marcescens	Various nosocomial and opportunistic infections	Gentamicin
GRAM-POSITIVE RODS		
Mycobacterium tuberculosis	Tuberculosis	Streptomycin + other antibiotics

duced by actinomycetes. For example, **streptomycin** is produced by *Streptomyces griseus,* neomycin by *S. fradiae,* kanamycin by *S. kanamyceticus,* and gentamicin by *Micromonospora purpurea.* Amikacin is a semisynthetic derivative of kanamycin.

Aminoglycosides bind to the 30S ribosomal subunit of the 70S prokaryotic ribosome and do not affect eukaryotic 80S ribosome function. They block bacterial protein synthesis and decrease the fidelity of translation of the genetic code (FIG. 17-12). Aminoglycosides disrupt the normal functioning of the ribosomes by interfering with the initial step of protein synthesis that occurs during translation. Interference in protein synthesis results in the death of the bacterium. Various mutations, though, can occur that reduce the effect of misreading some mRNA molecules. In some cases this can even lead to a dependence on a streptomycin-induced misreading of the genetic information.

Aminoglycosides inhibit bacterial protein synthesis and are used against Gram-negative bacterial infections.

Sensitive bacteria transport aminoglycosides across the membrane, where these antibiotics accumulate intracellularly. Resistant strains may lack a mechanism for aminoglycoside transport across the membrane. Resistant strains may also produce enzymes that degrade or transform the aminoglycoside molecules. For example, various enzymes associated with the plasma membranes of some bacterial strains are capable of chemically modifying aminoglycoside antibiotics. Additionally, mutations can occur that alter the site at which the aminoglycosides normally bind to the bacterial ribosomes. Some *Pseudomonas aerugi-nosa* strains, for example, possess ribosomes to which streptomycin is unable to bind.

Aminoglycosides are useful in treating many diseases, but some have limitations. Streptomycin has a serious side effect. It affects the eighth cranial nerve and can cause deafness with prolonged use. For this reason streptomycin is used in the treatment of only a limited number of bacterial infections. It is sometimes used in the treatment of brucellosis, tularemia, endocarditis, plague, and tuberculosis. Gentamicin is effective in treating urinary tract infections, pneumonia, and meningitis. Gentamicin is, however, extremely toxic and thus is used only in the cases of severe infections that may prove lethal if unchecked. Its use is important when the infecting bacteria are not sufficiently sensitive to other less toxic antibiotics. Tobramycin has properties similar to gentamicin. *Pseudomonas aeruginosa,* however, is particularly sensitive to tobramycin and thus this antibiotic is sometimes used for the treatment of pneumonia and other infections when caused by *Pseudomonas* species. Neomycin is active against a broad spectrum of Gram-negative bacteria. It is primarily used in topical application for various infections of the skin and mucous membranes. Kanamycin is a narrow spectrum antibiotic. It is frequently used by pediatricians for infections due to *Klebsiella* spp., *Enterobacter* spp., *Proteus* spp., and *Escherichia coli.* Amikacin has the broadest spectrum of activity of the aminoglycosides. It is the antibiotic of choice for treating serious infections caused by Gram-negative rods acquired in hospitals because such nosocomial (hospital acquired) infections are often caused by bacterial strains that are resistant to multiple antibiotics, including other aminoglycosides.

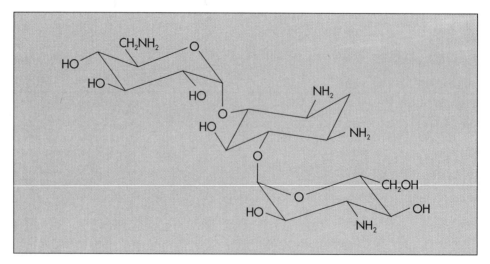

FIG. 17-12 Structure of an aminoglycoside antimicrobic. These antimicrobics block protein synthesis at the ribosomes of bacterial cells.

TABLE 17-7
Some Therapeutic Uses of Tetracyclines

CAUSATIVE ORGANISM	DISEASE
Haemophilus ducreyi	Chancroid
Brucella spp.	Brucellosis
Vibrio cholerae	Cholera
Pseudomonas mallei	Glanders
Pseudomonas pseudomallei	Melioidosis
Borrelia recurrentis	Relapsing fever
Rickettsia spp.	Murine typhus
	Brill disease
	Rocky Mountain spotted fever
Chlamydia trachomatis	Trachoma
	Inclusion conjunctivitis
	Nonspecific urethritis

TABLE 17-8
Some Therapeutic Uses of Chloramphenicol

CAUSATIVE ORGANISM	DISEASE
Salmonella spp.	Typhoid fever
	Paratyphoid fever
	Bacteremia
Haemophilus influenzae	Pneumonia
	Meningitis
Pseudomonas pseudomallei	Melioidosis
Campylobacter fetus	Enteritis
Bacterioides fragilis	Brain abscess
	Lung abscess
	Intra-abdominal abscess
	Bacteremia
	Endocarditis
Rickettsia spp.	Typhus fever

Tetracyclines

The **tetracyclines,** like the aminoglycosides, bind specifically to the 30S ribosomal subunit. This prevents the addition of amino acids to a growing peptide chain. Tetracycline action is bacteriostatic. Different tetracycline antibiotics have the same basic 4-ring structure with some variations in the substituents. The tetracyclines are produced by various *Streptomyces* species. Chlortetracycline (aureomycin) is produced by *S. aureofaciens,* oxytetracycline by *S. rimosus,* and demeclocycline by *S. aerofaciens;* methacycline, doxycycline, minocycline, and tetracycline are all semisynthetic derivatives.

The tetracyclines are effective against various pathogenic bacteria, including *Rickettsia* and *Chlamydia* species (Table 17-7). Tetracyclines, for example, are used therapeutically in treating the rickettsial infections of Rocky Mountain spotted fever, typhus fever, and Q fever and the chlamydial diseases of lymphogranuloma venereum, psittacosis, inclusion conjunctivitis, and trachoma. Tetracyclines are useful in treating many other bacterial infections, including pneumonia caused by *Mycoplasma pneumoniae,* brucellosis, tularemia, and cholera. Tetracyclines can bind to calcium in bone and teeth. This may result in yellowing or discoloration of the tooth enamel surface. Therefore tetracycline is not recommended for use during pregnancy or for infants and children. Tetracycline also causes liver and kidney damage in some individuals with prolonged use.

Tetracyclines are effective against *Rickettsia* and *Chlamydia* species.

Chloramphenicol

Unlike the antibiotics discussed so far that inhibit bacterial protein synthesis, **chloramphenicol** acts primarily by binding to the 50S ribosomal subunit. Consequently, peptide bonds are not formed when chloramphenicol is present. Chloramphenicol, which is produced by *Streptomyces venezuelae,* is a fairly broad-spectrum, bacteriostatic antibiotic, active against many species of Gram-negative bacteria (Table 17-8). Resistance to chloramphenicol is generally associated with the presence of an R plasmid that codes for enzymes able to transform the chloramphenicol molecule.

Chloramphenicol, a broad spectrum antibiotic effective against many Gram-negative bacteria, binds to the 50S ribosomal subunit, preventing protein synthesis.

Chloramphenicol has some toxic effects, including aplastic anemia (decrease in red blood cell number due to damaged or destroyed stem cells in the bone marrow), that limit its therapeutic uses to those where the benefits outweigh the dangers associated with toxic reactions. Chloramphenicol is used for treating typhoid fever, as well as various other infections caused by *Salmonella* species. It is also effective against anaerobic pathogens and can be used effectively in treating diseases such as brain abscesses that are normally caused by anaerobic bacteria.

Erythromycin

Erythromycin also acts by binding to 50S ribosomal subunits, blocking protein synthesis. Erythromycin is bacteriostatic and most effective against Gram-positive cocci, such as *Streptococcus pyogenes.* Erythromycin is not active against most aerobic Gram-negative rods but does exhibit antibacterial activity against some Gram-negative organisms, such as *Pasteurella multocida, Bordetella pertussis,* and *Legionella pneumophila* (Table 17-9). Therapeutically, erythromy-

TABLE 17-9	
Some Therapeutic Uses of Erythromycin	
CAUSATIVE ORGANISM	**DISEASE**
Flavobacterium meningosep-ticium	Meningitis
Campylobacter fetus	Enteritis
Legionella pneumophila	Legionnaire's disease
Mycoplasma pneumoniae	Atypical pneumonia

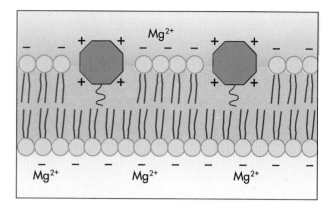

FIG. 17-13 Polymyxin B disrupts the cytoplasmic membrane, causing leakage and cell death.

cin is recommended for the treatment of Legionnaire's disease and is also effective in treating diphtheria, whooping cough, and the type of pneumonia caused by *Mycoplasma pneumoniae*. Erythromycin may also be used as an alternative to penicillin in treating staphylococcal infections, streptococcal infections, tetanus, syphilis, and gonorrhea.

> **Erythromycin acts by binding to 50S ribosomal subunits, blocking protein synthesis, and is effective against diseases caused by Gram-positive cocci, Legionnaire's disease, diphtheria, whooping cough, and mycoplasmal pneumonia.**

Rifampicin

Rifampicin (rifampin) blocks protein synthesis at the level of transcription. This antimicrobic inhibits RNA polymerases and thus can inhibit transcription. This antibiotic is more effective against bacterial RNA polymerases than mammalian RNA polymerases and therefore can be used therapeutically in treating some bacterial diseases. Also, because the half lives of bacterial messenger RNAs are very short (1 to 2 minutes), blocking transcription rapidly leads to stoppage of translation so that protein synthesis is inhibited. Rifampicin is effective against several Gram-positive organisms and *Neisseria* spp. Rifampicin is used in combination with other antibiotics in the treatment of mycobacterial diseases such as tuberculosis and leprosy. Rifampicin can penetrate tissues and reach therapeutic levels in cerebrospinal fluid and abscesses. This characteristic is probably an important factor in its antitubercular activity because the tuberculosis pathogen is usually located intracellularly in macrophages. An unusual side effect of rifampicin is the appearance of orange-red urine, feces, saliva, sweat, and even tears.

> **Rifampicin is a protein synthesis inhibitor that can penetrate tissues and reach therapeutic levels in cerebrospinal fluid and abscesses, making it effective against some Gram-positive bacteria, *Neisseria* spp., and especially tuberculosis pathogens.**

INHIBITORS OF MEMBRANE TRANSPORT

The plasma membrane is the site of action of some antibacterial agents. The **polymyxins** interact with the plasma membrane, causing changes in the structure of the bacterial cell membrane and leakage of cell contents (FIG. 17-13). Polymyxin B is bactericidal, but its effectiveness is restricted to Gram-negative bacteria. The action of polymyxin B is related to the phospholipid content of the cell wall and membrane complex. The principle use of polymyxin B and colistin (polymyxin E) is in the treatment of infections caused by *Pseudomonas* species and other Gram-negative bacteria that are resistant to penicillins and the aminoglycosides. Because polymyxins are toxic to human cells, they are mainly used topically. However, polymyxin B and colistin are sometimes useful in treating severe urinary tract infections caused by Gram-negative bacteria that are resistant to other antibiotics. Polymyxin B is also a component (with bacitracin and neomycin) of nonprescription ointments used to treat superficial skin abrasions and cuts.

> **Polymyxins damage the cytoplasmic membranes of Gram-negative bacteria, causing changes in the structure of the bacterial cell membrane and leakage of the cell's contents.**

DNA REPLICATION INHIBITORS

Some antimicrobial agents act by blocking bacterial DNA replication. In particular the **quinolones** interfere with DNA gyrase, the enzyme involved in unwinding DNA during replication of the bacterial chromosome (FIG. 17-14). Quinolones prevent replication of DNA needed for bacterial multiplication. Bacteria exposed to quinolones elongate rather than divide normally. The quinolones include nalidixic acid, ciprofloxacin, norfloxacin, amifloxacin, and enoxacin. These antimicrobics are effective against a broad range of Gram-positive and Gram-negative bacteria. They are also effective against mycobacteria. Nalidixic acid is effective against most Gram-negative bacteria that cause urinary tract infections.

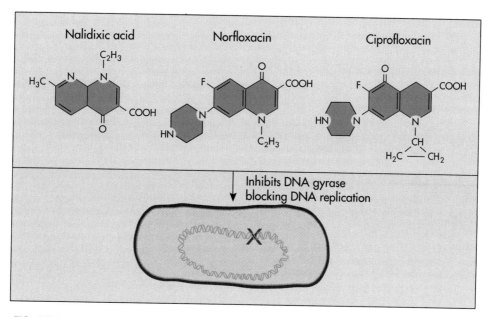

FIG. 17-14 Quinolones inhibit DNA gyrase, blocking bacterial chromosome replication.

Ciprofloxacin is a broad spectrum antimicrobic that has relatively low toxicity to human cells. It can be administered orally for treating infections, including those caused by pseudomonads. Before the discovery of "cipro," such infections required treatment with aminoglycosides administered intravenously.

Quinolones block bacterial DNA replication by interfering with DNA gyrase.

INHIBITORS WITH OTHER MODES OF ACTION

There are several other antibiotics that act by different mechanisms. Some of these antibiotics are useful in treating specific infections, such as tuberculosis. Others are particularly useful in treating infections of particular tissues, such as urinary tract infections. For example, the sulfonamides, sulfones, and *p*-aminosalicylic acid are useful antibacterial agents. The sulfones are useful in treating leprosy.

Sulfonamides, sulfones, and *p*-aminosalicylic acid are structural analogs of the vitamin para-aminobenzoic acid. This makes them useful antibacterial agents (FIG. 17-15). A cell mistakenly using an analog, such as sulfonamide, in place of the normal substance, para-aminobenzoic acid in this case, produces molecules that are unable to perform their essential metabolic functions. In this case there is a failure of critical coenzyme functions. Folic acid is an essential coenzyme composed in part of para-aminobenzoic acid. Mammalian cells are unable to synthesize folic acid. They require an intake of folic acid as part of their diet and must be able to take it into their cells via an active transport system. Bacterial cells, in contrast, normally synthesize their required folic acid and are unable to transport folic acid across their

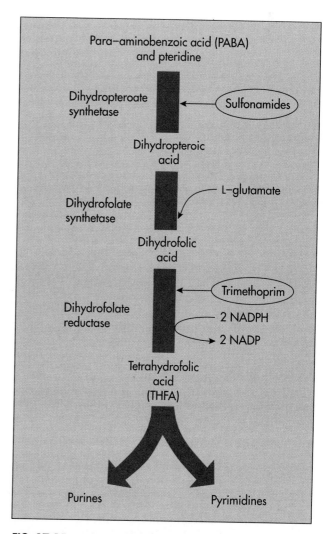

FIG. 17-15 Sulfonamides and trimethoprim inhibit, in series, the steps in the synthesis of tetrahydrofolic acid by interacting with key enzymes in the pathway. These antimicrobics often are used in combination.

plasma membranes. The analogs of para-amino-benzoic acid are effective competitors with the natural substrate for the enzymes involved in the synthesis of folic acid. As such they are able to inhibit the formation of this required coenzyme, causing a bacteriostatic effect.

> **A cell mistakenly using an analog in place of the normal substance forms molecules that are unable to perform their essential metabolic functions.**

Trimethoprim is an inhibitor of dihydrofolate reductase, especially in bacteria. Dihydrofolic acid is a coenzyme required for the synthesis of thymidine and purines. Trimethoprim is effective in blocking bacterial growth by preventing the formation of the active form of the required coenzyme. Trimethoprim is a broad spectrum antibacterial agent and is effective in the treatment of many urinary and intestinal tract infections. It is used primarily for the treatment of urinary infections due to *Escherichia coli*, *Proteus* spp., *Klebsiella* spp., and *Enterobacter*. The effectiveness of trimethoprim is enhanced when antibiotic therapy is coupled with sulfamethoxazole. The most widely used sulfa today is a combination of trimethoprim and sulfamethoxazole (TMP-SMZ). This combination is an excellent example of drug synergism, reducing the concentration needed to one tenth of what is used when each is used alone. This combination has a broader spectrum of action and reduces the emergence of resistant strains.

> **Trimethoprim blocks bacterial growth by preventing the formation of the active form of coenzyme required for the synthesis of thymidine and purines.**

Isoniazid is not particularly effective against *Mycobacterium tuberculosis* when used alone. It is gener-

NEWSBREAK

MULTIPLE DRUG RESISTANT TUBERCULOSIS

Outbreaks of multiple drug resistant strains of tuberculosis causing mycobacteria, known as MDR-TB, were first noted in New York and Miami in 1990. Twenty-nine patients in one hospital were diagnosed with multiple drug resistant tuberculosis. Of these, 27 were infected with HIV, 23 had AIDS, and 21 died within 2 months of being diagnosed with MDR-TB. Eight cases of TB were also reported among health care workers caring for MDR-TB patients. Five of the eight were HIV-positive. Laboratory tests performed on mycobacteria isolates from these patients showed that the majority of the strains were resistant to multiple drugs, including isoniazid, streptomycin, rifampicin, and ethambutol. Since then, many individuals with AIDS have died from tuberculosis caused by infections with *Mycobacterium avium*. The disease is called MATS for *M. avium tuberculosis*. Strains causing MATS normally infect birds but are able to invade the bodies of immunocompromised individuals. They have proven to be multiple drug resistant, making them difficult to control.

ally used in association with other antibiotics in treating tuberculosis. The specific mode of action of isoniazid is not known, but its primary action appears to involve the inhibition of mycolic acid biosynthesis. Mycolic acids are unique components of the cell walls of mycobacteria, and blockage of the biosynthesis of these compounds could specifically inhibit mycobacteria.

ANTIFUNGAL AND ANTIPROTOZOAN AGENTS

Bacterial infections can be treated with relative ease. The fact that fungi, protozoa, and humans all have eukaryotic cells limits the sites against which antimicrobial agents can be selectively directed to control eukaryotic pathogens of human beings. Consequently, there are relatively few antimicrobial agents of therapeutic value effective against infections caused by eukaryotic microorganisms, with particularly few effective for treating systemic infections.

ANTIFUNGAL AGENTS

There are sufficient differences between a fungal cell and human cell so that some therapeutically useful compounds with antifungal activity have been discovered (Table 17-10). The **polyene antibiotics,** amphotericin B and nystatin, are used in treating vari-

ous fungal diseases. Polyene antibiotics act by altering the permeability of the plasma membrane, leading to the death of the affected cells. Interactions of polyenes with the sterols in the plasma membranes of eukaryotic cells appear to form channels or pores in the membrane, allowing leakage through the membrane. Differences in the sensitivity of various organisms is determined by the concentrations of sterols in the membrane. Because mammalian cells, like fungi, contain sterols in their plasma membranes, it is not surprising that polyene antibiotics also cause alterations in the membrane permeability of mammalian cells and toxicity to mammalian tissue, as well as the death of fungal pathogens.

> **Polyene antibiotics alter the permeability properties of plasma membranes, causing the death of cells.**

TABLE 17-10

Some Therapeutic Uses of Antifungal Agents

CAUSATIVE ORGANISM	DISEASE	DRUG OF CHOICE
Candida albicans	Skin and superficial mucous membrane lesions	Amphotericin B Nystatin
Cryptococcus neoformans	Meningitis	Amphotericin B + flucytosine
Candida albicans	Pneumonia	Amphotericin B
Aspergillus spp.	Meningitis	—
Mucor spp.	Skin lesions	—
Histoplasma capsulatum	Lung lesions Histoplasmosis	Amphotericin B
Coccidioides immitis	Coccidioidomycosis	Amphotericin B
Blastomyces dermatitidis	Blastomycosis	Amphotericin B

TABLE 17-11

Some Drugs Used in the Treatment of Diseases Caused by Protozoan Pathogens

INFECTING ORGANISM AND DISEASE	DRUG OF CHOICE
Entamoeba histolytica	Diiodohydroxyquin
Asymptomatic cyst passer	
Mild intestinal disease	Metronidazole
Severe intestinal disease	Metronidazole
Hepatic abscess	Metronidazole
Giardia lamblia (giardiasis)	Quinacrine hydrochloride
Balantidium coli (balantidiosis)	Oxytetracycline
Trichomonas vaginalis (vaginitis)	Metronidazole
Pneumocystis carinii (pneumocystis pneumonia)	Trimethoprim-sulfamethoxazole
Toxoplasma gondii (toxoplasmosis)	Pyrimenthamine + trisulapyrimidines
Leishmania donovani (kala azar, visceral leishmaniasis)	Sodium stibogluconate
Leishmania tropica (oriental sore, cutaneous leishmaniasis)	Sodium stibogluconate
Leishmania braziliensis (American mucocutaneous leishmaniasis)	Sodium stibogluconate
Trypanosoma gambiense (African trypanosomiasis)	Pentamidine
Trypanosoma rhodesiense (African trypanosomiasis)	Suramin
T. gambiense or T. rhodesiense in late disease with central nervous system involvement	Malarsoprol
Trypanosoma cruzi (South American trypanosomiasis; Chagas disease)	Nifurtimox

Nystatin is primarily used in the treatment of topical infections by members of the fungal genus *Candida*. Vaginitis and thrush, caused by *Candida albicans*, are effectively treated by using nystatin.

Amphotericin B has a relatively broad spectrum of activity. It is used in the treatment of systemic fungal infections. Amphotericin B is the most effective therapeutic agent for treating systemic infections due to fungi. The potential toxic side effects of amphotericin B usage, however, such as kidney damage, require careful supervision of its administration. Patients receiving amphotericin B almost invariably exhibit some toxic side effects, but without the administration of this drug, systemic fungal infections are almost invariably fatal. Amphotericin B is used in the treatment of cryptococcosis, histoplasmosis, coccidioidomycosis, blastomycosis, sporotrichosis, and candidiasis. Patients requiring administration of amphotericin B must be hospitalized so that the initial reaction to the therapy can be carefully supervised.

In addition to the polyene antibiotics, such imidazole derivatives as miconazole and clotrimazole have a broad spectrum of antifungal activities. They are used in the topical treatment of superficial mycotic infections. These two antimicrobial agents appear to alter membrane permeability, leading to the inhibition and/or death of selected fungal species.

ANTIPROTOZOAN AGENTS

Treatment of human protozoan diseases with antimicrobial agents presents a special problem because many of the pathogenic protozoa exhibit a complex life cycle. These life cycles often include stages that develop intracellularly within mammalian cells. Different antimicrobial agents are generally needed for use against different forms of the same pathogenic protozoan (Table 17-11). The agent of choice depends on the stage of the life cycle and the involved tissues. For example, the protozoan species of the genus *Plasmodium* that cause malaria exhibit complex life cycles, part of which are carried out in the liver and blood of human beings. The erythrocytic stage of the *Plasmodium* life cycle that occurs within human blood cells is the most sensitive to antimalarial drugs. The life stages that occur within the liver are difficult to treat and the sporozoites injected into the bloodstream by mosquitoes are not affected by antimalarial drugs.

Pathogenic protozoa exhibit complex life cycles and different antimicrobial agents are needed against different forms of the same protozoan, depending on the stage of its life cycle and the tissues involved.

The antimalarials effective against the erythrocytic forms of the protozoan include chloroquine and amodiaquine. Neither of these is effective against the stages of the *Plasmodium* spp. that occur in the liver. These antimalarial agents appear to interfere with DNA replication. This is the rapid interruption of schizogony or multiple division—the form of protozoan reproduction—that occurs within red blood cells. The sensitivity of malarial protozoa to these drugs depends on the active transport of these compounds into the protozoa and the selective accumulation of the drugs intracellularly.

Chloroguanide is also used in the suppression of malaria. This drug is transformed within the body to a triazine derivative that inhibits the enzyme dihydrofolate reductase. It interferes with the essential metabolic reactions involving this coenzyme that are required for the proliferation of the malaria protozoa. Chloroguanide binds more strongly to the plasmodial enzyme than to the comparable mammalian dihydrofolate reductase. This accounts for its selective inhibition. For the radical cure of malaria, that is, the eradication of the erythrocytic and liver stages of the protozoan, primaquine is normally used. The precise mode of action of primaquine has not been elucidated.

Several other drugs are used in the treatment of various other protozoan infections. As in the case of malaria, the life cycle of the particular protozoan determines which agents will be effective in controlling the infection. Quinacrine hydrochloride is used to treat cases of *Giardia lamblia* infections, a protozoan disease spread through contaminated water that has become a major problem in the United States.

Metronidazole is also used in the treatment of *Giardia lamblia* infections. Another use is in cases of dysentery caused by the protozoan *Entamoeba histolytica*. Metronidazole (Flagyl) is a widely used antiprotozoan drug. It is active not only against parasitic protozoa; it is active also against obligate anaerobic bacteria. As an antiprotozoan agent, it is the drug of choice for vaginitis caused by *Trichomonas vaginalis* and is also effective when vaginitis is caused in anaerobic conditions by the catalase-negative bacterium *Gardnerella vaginalis.* It is also used in the treatment of giardiasis and amebic dysentery. Metronidazole interferes with hydrogen transfer reactions, specifically inhibiting the growth of anaerobic microorganisms, including anaerobic protozoa.

Pentamidine isothionate is used in the early stages of African trypanosomiasis. It is now also used in the treatment of *Pneumocystis carinii* pneumonia, a common complication in immunocompromised persons and victims of AIDS. The drug's mode of action is unknown but it appears to bind to DNA. Nifurtimox is a member of the synthetic nitrofuran drugs and is effective against the trypanosome-caused Chagas disease. Treatment may produce the side effects of nausea and convulsions.

ANTIVIRAL AGENTS

The search for antiviral drugs comparable to the antibiotics used to control bacterial infections has been fruitless for the most part. There are no broad spectrum antiviral agents currently in clinical use. Most viral infections cannot be treated effectively by using antiviral chemicals. The integral role of the mammalian host cell in the process of viral replication complicates the difficulty of finding compounds that specifically inhibit viral replication. Most compounds that prevent the reproduction of viruses also interfere with mammalian cell metabolism. This results in such adverse effects on human cells so as to preclude the therapeutic use of such agents. Very few antiviral agents have been found with clinical applicability, and these generally have a narrow spectrum of antiviral activity (Table 17-12).

> **No broad spectrum antiviral agents are currently in clinical use. Most viral infections cannot be effectively treated with antiviral chemicals.**

The best "treatment" for many viral diseases is prevention through the appropriate use of vaccines. Controlling the vectors that act as carriers for viruses is also important in the prevention of viral diseases. Generally, recovery from a viral infection depends on the natural immune defense response of the body. For most viral infections, treatment is aimed at maintaining a physiological state in which an effective immune response can be ensured, by following the sage advice, "rest and drink plenty of fluids."

The discovery of the role of interferons in the natural immune response to viruses holds promise for the future. This depends on whether interferons can be produced commercially and if their administration proves to be of therapeutic value for specific disease conditions. Genetic engineering seems to provide the greatest hope for the commercial production of interferons.

One of the few promising antiviral agents that has been discovered is acyclovir. Acyclovir (9-[2-hydroxyethoxymethyl] guanine) has proven to be the best antiherpes drug so far discovered (FIG. 17-16). Acyclovir is effective in treating cold sores and genital herpes caused by herpes simplex. It also may be of value in treating chickenpox and shingles. Acyclovir is a nucleoside analog. It is converted *in vivo* to an

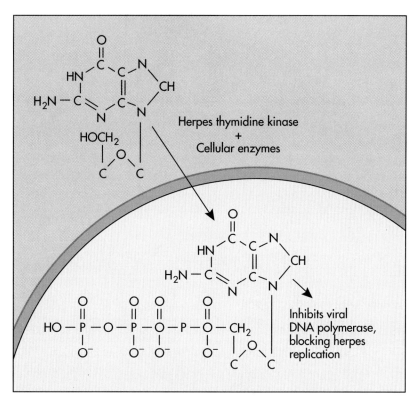

FIG. 17-16 The activity of an antiviral agent against different herpesviruses is correlated with their ability to induce a thymidine kinase, hence acyclovir is most active against HSV, least active against CMV.

TABLE 17-12

Some Antiviral Agents and Their Therapeutic Uses

CAUSATIVE ORGANISM	DISEASE	DRUG OF CHOICE
Herpes simplex virus	Keratoconjunctivitis	Acyclovir, Vidarabine
	Encephalitis	Acyclovir
	Cold sores	Acyclovir
	Genital herpes	Acyclovir
Varicella-zoster virus	Shingles	Acyclovir
	Chickenpox	Acyclovir
Influenza virus A	Influenza	Amantadine
HIV	AIDS	Zidovudine (AZT)
		Didanosine (ddI)
		Zalcitabine (ddC)

acylguanosine triphosphate that inhibits herpes simplex viral DNA polymerase, thus blocking viral DNA replication. The activation of acyclovir is initiated by a viral-directed thymidine kinase enzyme that converts this compound to an acycloguanosine monophosphate, which is subsequently converted to acycloguanosine diphosphate and triphosphate. In an uninfected cell there is limited conversion of acyclovir to the toxic forms of the chemical. Because an enzyme coded for by the herpesvirus is required to activate acyclovir, this compound exhibits selective antiviral activity, making it therapeutically valuable.

Azidothymidine (AZT) (also called Zidovudine and Retrovir), dideoxyinosine (ddI) (also called didanosine and VIDEX), and dideoxycytidine (ddC) (also called zalcitabine and HIVID) have been found to be effective in the treatment of acquired immunodeficiency syndrome (AIDS) (FIG. 17-17). HIV, which causes AIDS, is a retrovirus and contains reverse transcriptase that is needed for the successful replication of the virus (see Chapter 21 for a discussion of AIDS). AZT, ddI, and ddC are DNA nucleotide analogs that prevent the formation of viral-directed DNA by retroviruses.

AZT was first synthesized as a possible anticancer drug but proved ineffective against tumors. However, it was found to be effective in blocking the ability of the reverse transcriptase of this virus to form the viral-directed DNA that is needed for the successful replication of the virus (see Chapter 9 for a discussion of HIV replication). The AZT is sequentially converted in the body by cellular enzymes into monophosphate, diphosphate, and triphosphate forms. The triphosphate of AZT is an analog of the

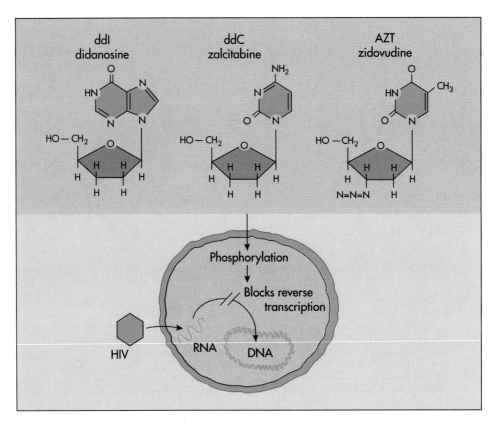

FIG. 17-17 DNA nucleotide analogs, such as AZT, ddI, and ddC, block reverse transcription by retroviruses. HIV reverse transcriptase is 100 times more sensitive to zidovudine triphosphate (AZT) than is host cell DNA polymerase, but toxic effects are not uncommon.

DNA base thymidine. The AZT triphosphate inhibits the viral reverse transcriptase and terminates DNA chain elongation prematurely. The United States Food and Drug Administration has approved use of this drug for the treatment of most individuals with AIDS. Unfortunately, the cost of administration of this drug is extremely expensive. This drug also can cause anemia and decrease the number of granulocytes in the blood.

> **AZT is used in the treatment of AIDS. It is an analog of the DNA base thymidine and blocks HIV reverse transcriptase from forming viral-directed DNA needed for successful HIV replication.**

The other chemotherapeutic agents, ddI and ddC, that are used to treat HIV infection are also nucleotide analogs of adenosine and cytidine, respectively. These dideoxynucleotides also act as DNA chain elongation terminators. The drugs ddI and ddC, like AZT, are toxic to human cells and have serious side effects with prolonged use. Because of their side effects and conflicting results in clinical trials, there is controversy over when best to initiate their use in treating HIV-infected individuals. AZT, ddI, and ddC are effective in limiting viral replication and delaying the onset of AIDS, but they are not cures for this deadly disease.

SUMMARY

Therapies That Treat Symptoms and Support the Body's Immune Response (p. 502)

- Chemotherapeutic agents include all drugs used for all diseases. Supportive therapies include rest, fluids, and staying warm and dry. Supportive therapies help maintain essential body functions while allowing the body's immune defense system time to respond to infection.

Treating Diseases With Antimicrobial Agents (pp. 503-506)

- The use of antimicrobial agents in modern medical practice is aimed at eliminating microorganisms or preventing the establishment of an infection. To be of therapeutic value an antimicrobial agent must be selective, exhibiting greater toxicity against the infecting pathogen.

Selection of Antimicrobial Agents in Chemotherapy (pp. 506-514)

- Antimicrobial agents are chosen for treating specific infectious disease based on sensitivity of the infectious agent to the antimicrobic; possible side effects, including direct toxicity to mammalian cells and associated microbiota; possible *in vivo* biotransformations; and the chemical properties of the antimicrobic that may affect its distribution within the body.

Pharmological Properties (pp. 506-509)

- Drugs must exhibit toxicity to be pharmacologically useful.
- Drugs must reach the site of infection in toxic concentrations for treatment of disease.

Antimicrobial Susceptibility Testing (pp. 509-514)

- Culture and sensitivity studies should be done to determine proper antibiotic therapy. The Kirby-Bauer agar diffusion test is used to determine if a bacterial isolate is sensitive or resistant to specific antibiotics. Automated liquid diffusion tests are based on light scattering in broth.
- Minimum inhibitory concentration procedures test the lowest concentration of antibiotic that will inhibit microbial growth to determine the most effective antibiotic and the appropriate dosage. The minimal bactericidal concentration is the lowest concentration of an antibiotic that will kill a defined proportion of organisms in a bacterial suspension during a specified exposure period.
- The Schlicter test, in which bacterial suspensions are added to dilutions of the patient's serum, determines serum killing power, a measure of the effectiveness of an antimicrobic.

Antibacterial Agents (pp. 514-524)

- Antibiotics are biochemicals produced by microorganisms that inhibit the growth of, or kill, other microorganisms. They also may be chemically synthesized or may be chemically modified forms of microbial biosynthetic products. Narrow spectrum antibiotics are targeted at specific pathogens; broad spectrum antibiotics inhibit a wider range of bacterial species.
- There is rising concern about the overuse of antibiotics because of the development of antibiotic-resistant bacterial strains. Some bacterial strains, particularly those with R plasmids, have multiple antibiotic resistance.

Modes of Action of Antibacterial Agents (pp. 514-524)

- The mode of action of most antimicrobial agents involves biochemical features of the pathogen that are not possessed by normal host cells. The effectiveness of most of the common antibiotics used in treating bacterial infections is based on the inhibition of cell wall or protein synthesis.
- Penicillins, from the *Penicillium* fungi, and cephalosporins, from the *Cephalosporium* fungus, inhibit formation of peptide cross-linkages within the peptidoglycan backbone of the cell wall, leaving the bacteria subject to attack by autolysins and rendering them vulnerable to osmotic shock. They do not remove intact cell walls and so are ineffective against resting or dormant cells. They are cell-wall inhibitors.
- Cycloserine, bacitracin, and vancomycin also inhibit bacterial cell-wall synthesis but have side effects that limit their therapeutic usefulness.
- Aminoglycoside antibiotics are inhibitors of protein synthesis and are produced by actinomycetes. They include streptomycin, gentamicin, neomycin, kanamycin, tobramycin, and amikacin and are used against Gram-negative bacteria. To be effective they must be transported across the plasma membrane. They bind to the 30S bacterial cell ribosomal subunit, blocking protein synthesis and decreasing fidelity to genetic code transmission.
- Tetracyclines bind specifically to the 30S ribosomal subunit, preventing the addition of amino acids to the growing peptide chain.
- Chloramphenicol acts by binding to the 50S ribosomal subunit. It is a broad spectrum antibiotic, active against many Gram-negative species.
- Erythromycin also binds to 50S ribosomal subunits, blocking protein synthesis, and is most effective against Gram-positive cocci.
- Rifampin blocks protein synthesis at transcription and is effective against mycobacterial diseases because it can penetrate tissues and reach therapeutic levels in cerebrospinal fluid and abscesses.
- The polymyxins inhibit membrane transport by interacting with the plasma membrane and causing changes in the structure of the bacterial cell, causing the cell to leak. Polymyxin B is bactericidal and effective against Gram-negative bacteria. Polymyxins B and E (colistin) are used in the treatment of infections caused by *Pseudomonas* species and other Gram-negative bacteria resistant to penicillin and aminoglycoside antibiotics.
- Quinolones act as antimicrobial agents by interfering with DNA gyrase and thus blocking bacterial DNA replication. They are effective against a broad spectrum of Gram-positive and Gram-negative bacteria and also mycobacteria.
- Sulfonamides, sulfones, and para-aminosalicylic acid interrupt normal bacterial metabolism.
- Trimethoprim blocks bacterial growth by preventing the formation of a coenzyme required for metabolism. It is a broad spectrum antibiotic used primarily for treatment of urinary tract infections caused by *Escherichia coli*, *Proteus* spp., *Klebsiella*, and *Enterobacter*.
- Isoniazid is effective against *Mycobacterium tuberculosis*. Its action appears to involve inhibition of biosynthesis of mycolic acid, a component of mycobacterial cell walls.

Antifungal and Antiprotozoan Agents (pp. 524-526)

Antifungal Agents (pp. 524-525)

- Polyene antibiotics are antifungal agents that act by altering the permeability properties of the plasma membrane. Nystatin is used against topical infections caused by *Candida* species. Amphotericin B, which is produced from *Streptomyces nodosus*, is a broad spectrum antibiotic used in the treatment of systemic fungal diseases.

- Miconazole and clotrimazole are broad spectrum antifungal agents used in the topical treatment of superficial infections because they alter membrane permeability.

Antiprotozoan Agents (pp. 525-526)

- Diseases caused by protozoans are difficult to treat because many of the pathogenic protozoa exhibit a complex life style and different antiprotozoan agents are needed for different life stages of the same pathogens.
- Antimalarial agents include chloroquine and amodiaquine, which appear to interfere with DNA replication and are effective against erythrocytic forms of *Plasmodium*. Chloroguanide interferes with metabolic reactions involving a coenzyme needed for reproduction of the malaria protozoa.
- Metronidazole, an antimicrobial is a widely used antiprotozoan agent.

Antiviral Agents (pp. 526-528)

- There are no broad spectrum antiviral agents because the essential role of mammalian cells in viral replication complicates the difficulty in finding compounds that specifically inhibit viral replication. Viral disease should be prevented by the use of vaccines and by controlling viral vectors.
- Interferons are involved in the natural immune response to viruses. Genetic engineering may have a role in their commercial production.
- Acyclovir is used in treating herpetic ocular disease, herpes encephalitis, cold sores, and genital herpes. It is a nucleoside analog that inhibits herpes simplex viral DNA polymerase, thus blocking viral DNA replication.
- Azidothymidine (AZT) (also called Zidovudine and Retrovir), dideoxyinosine (ddI) (also called didanosine and VIDEX), and dideoxycytidine (ddC) are used in the treatment of AIDS because they block the formation of DNA needed for viral replication.

CHAPTER REVIEW

REVIEW QUESTIONS

1. Explain what is meant by selective toxicity with respect to antibiotics.
2. What factors must a physician consider when determining the appropriateness of antibiotics for therapeutic use?
3. What information is supplied by the clinical microbiology laboratory?
4. Discuss the differences between broad and narrow spectrum antimicrobics.
5. Why is it essential to perform antimicrobial susceptibility testing on pathogenic isolates?
6. What is the Kirby-Bauer test?
7. What is an MIC? Why is this an increasingly common test in clinical microbiology laboratories?

8. Should penicillin be prescribed for treating the common cold? Explain.
9. Why is it easier to find antibacterial agents than to discover useful antifungal agents?
10. Why is it so difficult to find antimicrobial agents for treating viral diseases?
11. What causes drug resistance in microorganisms?
12. Discuss the mode of action of penicillin.
13. Why is penicillin ineffective against bacteria that produce β-lactamases?
14. Why is an inhibitor of transcription not useful in treating bacterial infections of humans?
15. Discuss the mode of action of streptomycin.
16. Discuss the mode of action of acyclovir.

CRITICAL THINKING QUESTIONS

1. How does a physician select an antibiotic for treating an infectious disease? Describe several approaches used to determine the sensitivity of a pathogen to antibiotics. Is antimicrobial sensitivity the sole criterion for selecting an antibiotic? Discuss.
2. What antibiotics should a physician prescribe for each of the following conditions?
   a. urinary tract infection

   b. upper respiratory tract bacterial infection
   c. fungal infection of the vaginal tract
   d. herpes encephalitis
   e. malaria
3. Why must antibiotics be administered by prescription? What problems are associated with indiscriminate use of antibiotics, or if they were made nonprescription drugs?

READINGS

Abramowicz M (ed.): 1980. *Handbook of Antimicrobial Therapy*, New Rochelle, New York; The Medical Letter, Inc.
Discusses the therapeutic use and possible adverse effects of drugs, antibiotics, and other anti-infective agents in the treatment of respiratory tract diseases.

Baldry P: 1976. *The Battle Against Bacteria: A Fresh Look*, Cambridge, England; Cambridge University Press.
A history of the fight against bacterial diseases, with special emphasis on the development of antibacterial drugs.

Bryan LE: 1982. *Bacterial Resistance and Susceptibility to Chemotherapeutic Agents*, Cambridge, England; Cambridge University Press.
Written by a physician/microbiologist/researcher to give an integrated overview of antibiotic activity taking into account drug distribution and elimination, host tissue antagonism, and mechanisms of action.

Dowling HF: 1977. *Fighting Infection: Conquests of the 20th Century*, Cambridge, Massachusetts; Harvard University Press.
A medical historian describes advances made in this century in the control of infectious diseases by prophylactic and therapeutic measures.

Gallo RC and F Wong-Staal: 1990. *Retrovirus Biology and Human Disease*, New York, Marcel Dekker.

Greenwood D and G O'Grady (eds.): 1985. *The Scientific Basis of Antimicrobial Chemotherapy*, New York, Cambridge University Press.
Discusses the scientific bases for the use of drugs as anti-infective agents in the treatment of communicable diseases.

Hooper DC and JS Wolfson: 1993. Quinolone Antimicrobial Agents, ed. 2, Washington, D.C.; American Society for Microbiology.
Provides complete discussion of quinolones and their expanding clinical applications.

Hooper DC and JS Wolfson: 1991. Fluroquinolone antimicrobial agents, *New England Journal of Medicine* 324:384-394.
Evaluates the current status of the quinolones, considers the mechanisms of action and resistance, activity in vitro, *pharmacokinetics, clinical efficacy, and adverse effects.*

Jacoby GA and GL Archer: 1991. New mechanisms of bacterial resistance to antimicrobial agents, *New England Journal of Medicine* 234:601-612.
New and novel mechanisms of bacterial resistance to antimicrobial agents have appeared along with alterations of the old mechanisms. This article focuses on the response to the newer, broad spectrum antibiotics and the increased and often indiscriminate use of the older agents.

Jeljaszewicz J and G Pulverer: 1986. *Antimicrobial Agents and Immunity*, London, Academic Press.
Explains the role of anti-infective agents in immunology and allergies.

Kagan BM: 1980. Antimicrobial Therapy, Philadelphia, Saunders.
Seventy-one authorities contributed to this volume on the therapeutic use of drugs, antibiotics, and other anti-infective agents in the treatment of disease.

Nelson JD and C Grassi: 1980. *Current Chemotherapy and Infectious Disease*, Washington, D.C.; American Society for Microbiology.
Proceedings of a conference that discussed the use of drug therapy, anti-infective agents, and antineoplastic agents in the treatment of communicable diseases.

Newman MG and KS Kornman: 1990. *Antibiotic/Antimicrobial Use in Dental Practice*, Chicago, Quintessence Publishing.
A handbook of drug therapy for diseases of the mouth and teeth, explaining the therapeutic use and adverse effects of antibiotic treatment.

Nsanze H: 1986. Drug resistance, *World Health*, November:24-25.
Presents the problem of drug resistance in the treatment of sexually transmitted diseases.

Poupard, JA, LR Walsh, B Kelger (eds.): 1994. *Antimicrobial Susceptibility Testing: Critical Issue for the 90s*, New York, Plenum Press.
Discusses the clinical methods used for determining appropriate antibiotics that may be used in medical practice.

Root RK and MA Sande: 1984. *New Dimensions in Antimicrobial Therapy*, New York, Churchill Livingstone.
This volume in the series "Contemporary Issues in Infectious Diseases" is about the pharmacological basis of antibiotic therapy.

Siporin C, CL Heifetz, JM Domagala (eds.): 1990. *The New Generation of Quinolones*, New York, Marcel Dekker.
Covers major aspects of quinolone science from the initial chemical discovery through current clinical applications.

Stout S and B Goodner: 1994. *The Infection Control Book*, St. Louis, Mosby.
Based on OSHA and CDC recommendations, this book presents a basic approach to infection control in all health care settings.

Williams JD (ed.): 1979. *Antibiotic Interactions*, London, Academic Press.
Topics in this book as they relate to antibiotic interactions include laboratory methods, role of the cell envelope, biochemical and pharmacokinetic basis of these interactions, and possible uses of interactive chemotherapy.

Wolfson JS and DC Hooper (eds.): 1989. *Quinolone Antimicrobial Agents*, Washington, D.C.; American Society for Microbiology.
Consolidates existing information on this new class of synthetic drug, including general information and applications for specific bacterial infections.

UNIT
FIVE

Infectious Diseases

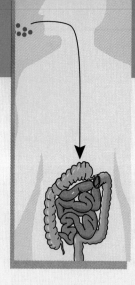

CHAPTER 18

Diseases Caused By Viruses: Diseases of the Respiratory, Gastrointestinal, and Genital Tracts

PREVIEW TO CHAPTER 18

In this chapter we will:
- Learn about the human diseases of the respiratory, gastrointestinal, and genital tracts caused by viruses.
- Examine the properties of viruses that enable them to cause human disease at these body sites.
- Review the specific causative agents, routes of transmission, characteristic symptoms, and treatments for each of these major viral diseases of humans.
- Learn the following key terms and names:

antigenic drift
antigenic shift
common cold
cytomegalovirus (CMV)
cytomegalovirus inclusion
  disease
gastroenteritis
genital herpes
genital warts
hantavirus pulmonary
  syndrome
hemagglutinin (H)
hepatitis
hepatitis A
hepatitis B
hepatitis C
herpes simplex type virus
human viruses

influenza
influenza viruses
intestinal flu
mumps
mumps virus
neuramidase (N)
nonA-nonB hepatitis
Norwalk virus (Norwalk
  agent)
papilloma viruses
respiratory syncytial virus
  (RSV)
Rotavirus
serum hepatitis
viral pneumonia
yellow fever
yellow fever virus

Viruses are distinguished from other organisms by their lack of cellular structure (see Chapter 3 for a review of the structure of viruses). Viruses lack the capacity for independent existence. Outside of susceptible cells, viruses are inert particles—subject to destruction by exposure to heat, ultraviolet light, and other factors. Within infected host cells, viruses have great replication potential—using the metabolism and structures of the cells to support formation of many new viruses (see Chapter 9 for a review of viral replication). Because viruses replicate only within the living cells of susceptible organisms, the transmission of disease-causing viruses depends on the movement of viruses from an infected cell to a susceptible cell. A viral infection occurs only when a virus actually enters a cell and is able to take control of cellular activities. Each of the different types of viruses exhibits great specificity for the cells in which it can replicate, which accounts for the fact that viral infections typically affect particular body tissues.

To establish an infection, a virus must invade the body through a portal of entry. A portal of entry, such as the respiratory tract, brings it into contact with susceptible cells. Further, it must not be destroyed by the body's numerous defense mechanisms before entering a cell in which it can replicate. Factors that promote transmission from one infected individual to another include the release of high numbers of viruses from the infected individual, proximity to a susceptible individual, environmental routes of transmission that limit dilution and/or destruction of viruses, and genetic properties of the virus that permit it to resist destruction by the body's immune defense systems.

Once an infection is established, it can be propagated by spreading from cell to cell. This is accomplished by direct transfer between adjoining cells or by transport via body fluids to other susceptible cells. Infection need not result in the clinical development of symptoms that characterize disease. Sometimes the magnitude of tissue damage is not sufficient to produce clinical symptoms. The factors that determine whether a clinical illness will develop in a person already infected depend in part on the dosage, virulence, and portal of entry of the agent, but more importantly, they depend on certain intrinsic properties of the host, such as the host's physiological/immunological state. The elderly, the malnourished, and those who are physiologically stressed are particularly prone to viral infections.

Whether or not infection and disease result from a particular virus depends on the nature of the virus and the physiological status of the body.

Although disease-causing bacteria and other pathogens produce toxins, viruses do not. Rather, it is the process of viral replication that alters infected cells (Table 18-1). Many viruses cause damage directly by redirecting the host cells' biosynthetic activities into viral replication and preventing normal cell function. Viral replication may cause damage to lysosomes, and the resulting release of lysosomal enzymes into the cytoplasm can kill an infected cell. Often host cells are killed as the final step in viral replication

---

TABLE 18-1

Mechanisms of Viral Cellular Pathogenesis

MECHANISM	REPRESENTATIVE VIRUSES
Inhibition of protein synthesis	Polioviruses, herpes simplex virus, togaviruses, poxviruses
Inhibition and degradation of cellular DNA	Herpes simplex virus
Changes in structure of cell membrane	
Insertion of glycoproteins	All enveloped viruses, reoviruses
Syncytia formation	Herpes simplex virus, varicella-zoster, paramyxoviruses, human immunodeficiency virus
Disruption of cytoskeleton	Herpes simplex virus
Changes in permeability	Togaviruses, herpesviruses
Inclusion bodies	
Negri bodies (cytoplasmic)	Rabies
Owl's eye (nuclear)	Cytomegalovirus
Cowdry's type A (nuclear)	Herpes simplex virus, measles virus
Nuclear basophilic inclusion bodies	Adenoviruses
Cytoplasmic acidophilic inclusion bodies	Poxviruses
Perinuclear acidophilic inclusion bodies	Reoviruses
Toxicity of components of the virion	Adenovirus fibers

that results in the release of the newly formed viruses. This may result in lesions in the affected tissues and organs and, hence, disease. Viral infections may also cause the body's own immune defense mechanisms to attack infected cells, damaging the body's own tissues in an effort to remove the viruses that are harbored within the infected cells. Almost all body tissues are subject to damage as a result of viral infections. For each virus, there is a characteristic incubation period before there is sufficient damage to body cells and tissues to cause disease (Table 18-2).

	TABLE 18-2				
	Incubation Times of Some Viral Infections				
VIRUS	**DISEASE**	**INCUBATION TIME (DAYS)**	**VIRUS**	**DISEASE**	**INCUBATION TIME (DAYS)**
Influenza virus	Influenza	1-2	Varicella-zoster virus	Chickenpox	13-17
Rhinovirus	Common cold	1-3	Mumps virus	Mumps	16-20
Enterovirus			Rubella virus	German measles	17-20
Adenovirus			Epstein-Barr virus	Infectious mononucleosis	30-50
Myxovirus					
Coronavirus			Hepatitis A virus	Infectious hepatitis	15-40
Parainfluenza virus	Croup	3-5			
Respiratory syncytial virus	Pneumonia	3-5	Hepatitis B virus	Serum hepatitis	50-150
Dengue virus	Dengue fever	5-8	Rabies virus	Rabies	30-100
Herpes simplex virus	Cold sores	5-8	Papilloma virus	Warts	50-150
Poliovirus	Poliomyelitis	5-20	Human immunodeficiency virus	AIDS	365-3,650
Measles virus	Measles	9-12			
Smallpox virus	Smallpox	12-14			

CLASSIFICATION OF HUMAN VIRUSES

Relatively few viruses are capable of replicating within human cells. These are classified as **human viruses.** Like other viruses, human viruses are classified principally based on their structure, for example, the shape of the capsid (the protein coat surrounding the virus), the symmetry of the capsid (many human viruses have icosahedral symmetry [icosahedron, from the Greek *ikosi,* meaning twenty, and *edron,* meaning faces]), the number of capsomers (subunits that make up the capsid), whether the virus is naked or enveloped, the nature of the nucleic acid (DNA or RNA), and whether the nucleic acid is single- or double-stranded. About a dozen taxonomic groups of viruses are recognized based on various combinations of these factors (Table 18-3).

Various types of viruses cause human diseases.

VIRAL DISEASES OF THE RESPIRATORY TRACT

Several viruses can infect and cause diseases of the respiratory tract (Table 18-4). Perhaps 90% of respiratory infections are caused by viruses. Airborne transmission from person to person is probably the most important route of spread for most viruses causing respiratory tract diseases. Viruses released into the air from an infected individual are transported as aerosols, such as those formed by sneezing and coughing. When virus-containing aerosols are inhaled, the respiratory tract is exposed to numerous viruses. Even though macrophages and the mucociliary escalator system generally protect the respiratory tract from viral infections, some viruses can penetrate this defense system and establish an infection (FIG. 18-1, p. 538). When this occurs, viruses may enter the circulatory system and spread to other sites or establish an infection within the respiratory tract.

Some viruses are able to adhere specifically to the cells of the respiratory tract. Adherence is the essential first step for the entry and intracellular replication of viruses within host cells of the respiratory tract. Specific cell receptors for individual viruses

TABLE 18-3

Characteristics of Some Viruses That Cause Human Diseases

VIRAL GROUP	GENOME	CAPSID SYMMETRY	OTHER
Adenoviruses	Double-stranded DNA	Icosahedral	Medium sized; produce an array of crystalline particles within host cells
Arenaviruses	Single-stranded RNA	Helical	Enveloped
Bunyaviruses	Single-stranded RNA	Helical	Enveloped
Calciviruses	Single-stranded RNA	Helical	
Coronaviruses	Single-stranded RNA	Helical	Enveloped
Filoviruses	Single-stranded RNA	Helical	
Flaviviruses	Single-stranded RNA	Icosahedral	Enveloped
Hepadnaviruses	Double-stranded DNA	Icosahedral	Enveloped
Herpesviruses	Double-stranded DNA	Icosahedral	Medium; enveloped, establish latent infections
Orthomyxoviruses	Single-stranded RNA	Helical	Enveloped; surface projections or spikes
Papovaviruses	Double-stranded circular DNA	Icosahedral	Some, such as papillomavirus, cause tumors
Paramyxoviruses	Single-stranded RNA	Helical	Enveloped; larger than orthomyxoviruses
Parvoviruses	Single-stranded DNA	Icosahedral	Small; resistant to heat inactivation; can withstand exposure to 60° C for 30 minutes; includes "defective" viruses that cannot multiply alone but can reproduce with the aid of another virus, such as adeno-associated satellite viruses
Picornaviruses	Single-stranded RNA	Icosahedral	Includes rhinoviruses, enteroviruses, and hepatitis A viruses
Poxviruses	Double-stranded DNA	Complex; brick shaped	Large; complex internal structure covered by two membranes—an inner membrane surrounding the genome core, an outer membrane that also surrounds the genome core, and two elliptical lateral bodies; surface of the outer membrane studded with tubes; some have an additional outer envelope; may include enzymes, such as RNA polymerase, within the viral particle
Reoviruses	Double-stranded RNA	Icosahedral	Known as arboviruses because they are transmitted by arthropod vectors
Retroviruses (Lentiviruses)	Single-stranded RNA	Helical	Carry out reverse transcription; carry a reverse transcriptase and tRNAs
Rhabdoviruses	Single-stranded RNA	Helical	Enveloped; bullet shaped
Togaviruses	Single-stranded RNA	Icosahedral	Known as arboviruses because they are transmitted by arthropod vectors; enveloped

TABLE 18-4

Some Viruses That Cause Human Diseases of the Respiratory Tract

VIRAL GROUP	VIRUS	DISEASE	VIRAL GROUP	VIRUS	DISEASE
Adenoviruses	Adenovirus	Common cold, pharyngitis, bronchitis, pneumonia	Paramyxoviruses	Parainfluenza virus	Common cold, pharyngitis, pneumonia
				Measles virus	Pneumonia
Bunyavirus	Hantavirus	Hantavirus pulmonary syndrome		Respiratory syncytial virus	Pneumonia
Coronaviruses	Coronavirus	Common cold	Parvoviruses	Adeno-associated satellite virus	Common cold
Herpesviruses	Herpes simplex	Pharyngitis	Picornaviruses	Rhinovirus	Common cold, pharyngitis
	Epstein-Barr	Pharyngitis			
	Cytomegalovirus	Pneumonia		Coxsackievirus	Common cold
	Varicella-zoster	Pneumonia		Echovirus	Common cold
Orthomyxoviruses	Influenza virus	Influenza and pneumonia		Enterovirus	Pneumonia
				Chickenpox virus	Pneumonia

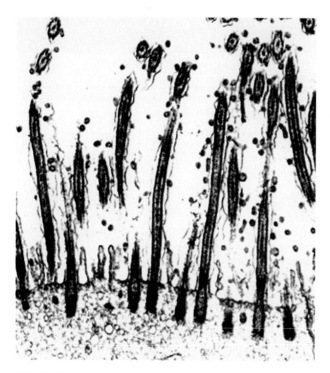

TABLE 18-5

Viruses Causing Common Colds

VIRUS	TYPES INVOLVED
Rhinovirus	Over 100 types with several at any given time often causing disease within a community
Coxsackievirus A	Several types, especially type A21
Influenza virus	Several types
Parainfluenza virus	Four types
Respiratory syncytial virus (RSV)	One type
Coronaviruses	Several types
Adenovirus	Ten types
Echoviruses	Several types

FIG. 18-1 Colorized micrograph of influenza viruses *(blue)* penetrating the mucociliary escalatory system *(brown)* of the respiratory tract.

have a crucial role in susceptibility. The viruses that cause respiratory tract infections, such as the viruses that cause influenza, rhinoviruses that cause the common cold, and respiratory syncytial viruses that cause pneumonia, are generally limited to the respiratory tract. Symptoms develop because of the breakdown of the constituents of host cells killed by viruses that are absorbed into the bloodstream. Fever, for example, is produced by the liberation of pyrogens when polymorphonuclear leukocytes are killed fighting a viral infection. Tissue damage from cell lysis (bursting that causes death) and tissue necrosis (decay) often occur with peeling of the surface layer of respiratory membranous tissue as a result of localized viral replication.

Respiratory diseases most often are caused by viruses. In most cases the viruses establish localized infections that cause tissue damage with resultant disease symptoms.

The Common Cold

The **common cold** is the name given to a cluster of diseases characterized by similar symptoms. Colds occur when certain viruses infect the cells lining the nasal passages and pharynx. Such viral infections of the upper respiratory tract produce an inflammatory response and tissue damage in the infected regions. Many different viruses can cause this disease (Table 18-5). Some 90 immunologically distinct rhinoviruses

are responsible for 25% of colds in adults and 10% of colds in children. Other viruses that cause colds include adenoviruses and coronaviruses. Hence, it is not surprising that immunity does not offer continuous protection against all of these different viruses.

A variety of viruses, including various rhinoviruses, cause the common cold.

As with many other respiratory diseases, colds occur primarily during the winter months. This may in part be due to the physiological stress posed by exposure to cold temperatures and the excessive drying of mucous membranes in heated buildings with low humidity. It may also be a consequence of increased contact between individuals during indoor winter activities that permits transmission of viruses through the air from an infected individual to a susceptible one. In the United States alone, more than 200 million work and school days are lost each year because of colds.

The symptoms of the common cold are the result of a localized inflammation in the upper respiratory tract (FIG. 18-2). This inflammation causes the release of mucous secretions and is generally accompanied by sneezing and sometimes coughing (FIG. 18-3). Symptoms during the course of the common cold include nasal stuffiness, sneezing, coughing, headache, *malaise* (a vague feeling of discomfort), sore throat, and sometimes a slight fever. There are various cold remedies that can alleviate the symptoms of a cold, but there is no specific cure for a cold. Fortunately, the common cold is a self-limiting syndrome, with recovery usually occurring within 1 week as a result of the natural immune defense response.

Many individuals experience the symptoms of the common cold, particularly sneezing and coughing.

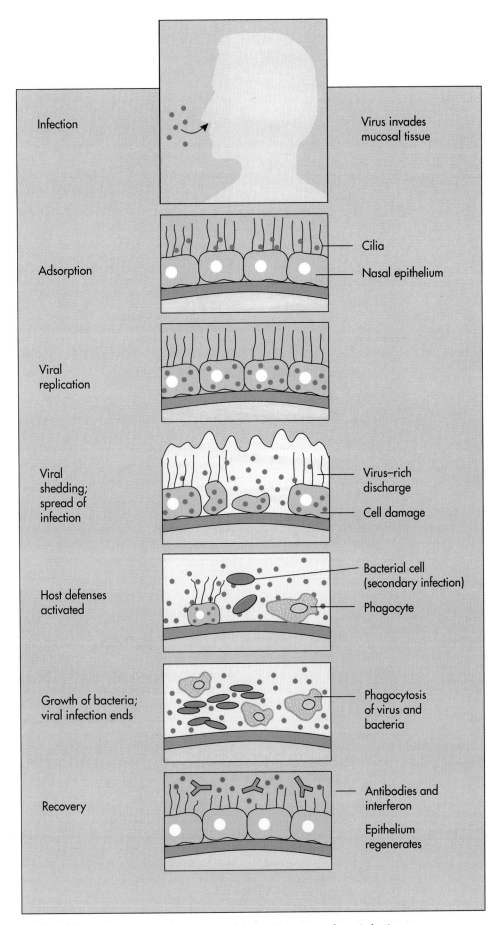

FIG. 18-2 Pathogenesis of common cold showing stages from infection to recovery.

FIG. 18-3 Coughing and sneezing from a common cold spread aerosols containing the viruses.

HIGHLIGHT

TRANSMISSION OF THE COMMON COLD

For decades it has been accepted that the viruses causing the common cold are transmitted via the air and enter the body through the respiratory tract, but scientific proof was lacking. Therefore researchers in England in the late 1970s set out to demonstrate the suspected method of transmission. To their surprise they were unable to demonstrate it experimentally. Their experiments consisted of intentionally exposing healthy individuals to the airborne droplets generated by the sneezing and coughing of people with colds. Very few of these experimentally exposed individuals developed colds. Having failed to demonstrate the airborne transmission of colds, these scientists asked whether there could be another route of entry.

Among the possibilities they considered was that cold viruses are transmitted by direct contact and enter the body through the eyes. They asked volunteers who were healthy to shake hands with individuals who had colds and then to rub their eyes. In most cases these healthy volunteers soon developed colds. The results of these experiments suggested that the eyes rather than the respiratory tract are the initial portal of entry for rhinoviruses that cause the common cold. After entering through the eyes, the viruses then migrate to the tissues of the respiratory tract. This occurs via the lacrimal drainage system of the eyes.

Disinfectant companies were quick to respond to the results of these experiments, advertising the advantages of spraying disinfectant on bathroom counter tops to prevent the spread of cold-causing rhinoviruses that may be picked up and rubbed into the eyes.

Although the evidence gathered in these studies indicates that cold-causing viruses could enter the body through the eyes, it did not eliminate the possibility that they also can enter via the respiratory tract. The respiratory tract defense systems of the healthy individuals used in the first series of experiments may have been sufficiently strong to prevent infection. Had their systems been weakened, aerial transmission of cold-causing viruses could have occurred.

In fact, recent studies identified a receptor site on surfaces of the cells lining the nasal passage that allows rhinoviruses to infect those cells. The nasal passage cells appear to be primary sites for initial infections with rhinoviruses that cause the common cold. Other cold-causing viruses may have other binding sites. Interestingly, the nasal rhinovirus binding sites are identical with the cell surface receptors, called intercellular adhesion molecules, involved in the inflammatory response. During the inflammatory response the numbers of intercellular adhesion molecules on the surfaces of mucosal cells increases greatly—to as many as 350 million per cell. This means that the stress that causes inflammation of the nasal passages increases the opportunities for rhinovirus infection and makes available many new sites for spread to other cells. The identification of the rhinovirus receptor site in the nasal passage reaffirms the idea that cold-causing viruses can be transmitted through air and enter the body through the respiratory tract. Hence, covering one's nose and mouth when coughing and sneezing remains a prudent measure for limiting the spread of the common cold.

INFLUENZA

The **influenza viruses,** which are members of the or-thomyxoviruses, are commonly referred to as the **flu viruses.** They cause a disease of the respiratory tract called **influenza** or **flu** (FIG. 18-4). Influenza is characterized by the sudden onset of a fever, with temperatures abruptly reaching 102° to 104° F approximately 1 to 3 days after actual exposure and onset of infection. The development of this disease is further characterized by malaise, headache, and muscle ache. In uncomplicated cases of influenza, the viral infection is self-limiting and recovery occurs within a week. However, infection with influenza virus can be a serious and debilitating disease. Complications such as bacterial pneumonia are prevalent among the elderly and individuals with compromised host defense responses following influenza infections. Such individuals should be immunized against the prevalent strains of influenza virus before the outbreak of influenza epidemics because these complications can result in death.

> Influenza, which can be a serious infection of the respiratory tract, is caused by influenza viruses.

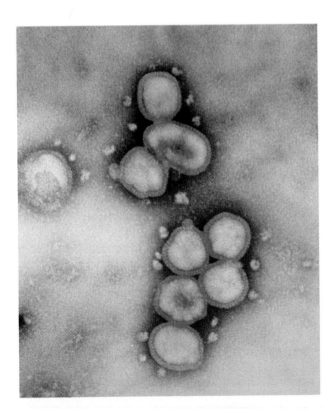

FIG. 18-4 Colorized electron micrograph of influenza virus. (84,600×). Each type of influenza virus is characterized by its specific nucleoprotein (soluble) antigen. Antigens present in the envelope (V-antigens) give subtype or strain-specific reactions. The outer membrane has rodlike projections containing hemagglutinin, and neuraminidase is also present on the surface.

Influenza is transmitted by inhalation of droplets containing flu viruses. Droplets are released into the air by sneezes and coughs from the respiratory tracts of infected individuals. Outbreaks of influenza spread worldwide via airborne transmission from the site of an initial outbreak with a new strain, and it is possible to track the disease as it spreads from one area to another. Each year epidemiologists make predictions about the severity of influenza outbreaks, and public health officials take the necessary steps to immunize high-risk individuals and warn the general public about the dangers of this disease. There have been several major, or pandemic, outbreaks of influenza during the nineteenth and twentieth centuries that have occurred when a sufficient proportion of the population was susceptible to a particular strain of influenza virus.

Major outbreaks of influenza are associated with the evolution of new strains of influenza viruses (Table 18-6). There are three major types of influenza viruses, designated influenza A, influenza B, and influenza C viruses. The types are differentiated according to the antigens associated with their nucleoproteins. The genomes of types A and B influenza viruses consist of 8 RNA segments; the type C influenza viruses have 7 RNA segments. The viral envelope contains protein spikes that project from the surface of the virus (FIG. 18-5). There are two kinds of spikes: H (hemagglutinin) and N (neuraminidase) spikes. The H spikes cause clumping of red blood cells (hemagglutination). Presumably the H spikes are important for the ability of the influenza virus to attach to cells in the body and are also a valuable aid in the serological identification of the particular

TABLE 18-6				
Human Influenza Viruses				
TYPE	SUBTYPE	YEARS	CLINICAL SEVERITY
A	H3N2	1889-1917	Moderate
	H1N1 (swine)	1918-1928	Severe
	H0N1	1929-1946	Moderate
	H1N1	1947-1956	Mild
	H1N1	1977-present	Mild
	H2N2 (Asian)	1957-1967	Severe
	H3N2 (Hong Kong)*	1968-present	Moderate
	H1N1 (Beijing)	1993-present	Moderate-severe
B	—	1940	Moderate
C	—	1947	Very mild

*Amino acid and base sequence analysis suggests that recombination between H3N8 (from ducks) and H2N2 gave rise to H3N2.

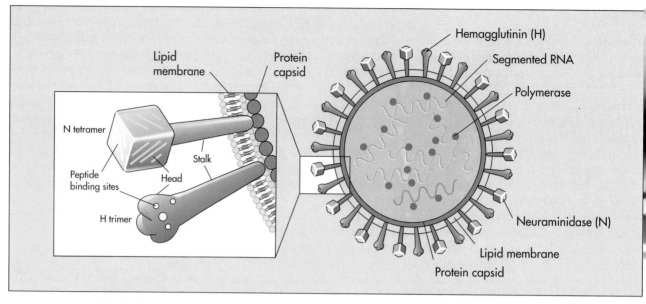

FIG. 18-5 Influenza viruses have envelope proteins called H and N spikes that protrude from the surface.

strain of influenza virus. Antibodies against the H spikes are very important in the resistance against infection by that particular strain of influenza virus. These antibodies prevent attachment of influenza viruses to surface receptors. The N spikes contain neuraminidase activity. These are less important in developing resistance to influenza infections than the H spikes. The N spikes are involved in the release of viruses from infected cells following viral replication.

There are several antigenically distinguished types of influenza viruses designated type A, B, and C; the influenza viruses have H and N protein spikes on their surfaces.

Among the type A influenza viruses there are 13 varieties of H spikes, although most of them occur in influenza viruses that infect birds, such as ducks, geese, chickens and turkeys. Type A influenza viruses are also present in swine, horses, and other mammals. Of these various H spikes, five are important in causing human disease—H0, H1, H2, H3, and Hswine. There are nine varieties of N spikes, but only two, N1 and N2, are important in causing human disease. The nomenclature of influenza viruses often indicates the type of virus, the location from which it was isolated, the year, and the variety of H and N spikes that are present. For example, A/Hong Kong/68/H3N2 and A/Japan/57/H2N2 are two strains of influenza virus.

Specific strains of influenza virus are designated by variations in the protein composition of their H and N spikes. For example, the Hong Kong strain of influenza virus, first seen in the Orient in 1968, is a type A influenza virus designated as H3N2. The antigenic designation is important because it indicates to epidemiologists whether there has been a substantial change in the antigenic properties of the virus and thus whether a sufficient proportion of the population will be susceptible to that strain so that an epidemic outbreak is likely.

HIGHLIGHT

WHY NEW INFLUENZA VIRUSES ORIGINATE IN THE FAR EAST

Many new strains of influenza viruses originate in the Far East where the most common animal hosts for influenza viruses (ducks, chickens, and pigs) live in close proximity to each other and to humans. Often, infections exist with multiple strains, including animal and human types. This creates conditions that favor mixing of gene pools and recombination to form new strains of influenza viruses. Within the intestines of an infected animal, recombination produces the new strains that contain some human and some duck, swine, or other animal influenza virus genes. This permits antigenic shifts so that the new strains of influenza viruses are not recognized by the body's immune system. This leads to epidemic or even pandemic outbreaks of influenza, such as occurred in 1918 when the most devastating epidemic in the history of humankind resulted in 20 to 40 million deaths.

Influenza viruses can exhibit genetic variability and undergo evolution. New strains result from mutation, gene reassortment, and RNA recombination when simultaneous infection of a host by an animal and a human influenza virus occurs. Minor mutations that occur in the genes coding for the H and N spikes constantly lead to new varieties of these antigens. This change is called **antigenic drift.** Less frequently, there are major changes in the H or N spikes that occur when RNA from different strains of influenza viruses recombine or reassort. This change, called **antigenic shift,** produces vastly different varieties of influenza viruses that spread through the human population. As a result of antigenic shift, many individuals do not have protective antibodies to these new forms of viral antigens, and epidemics of influenza can occur. Antigenic drift also contributes to enhanced infectivity of new viral strains that arise between episodes of major antigenic shift. Since there are large populations of geese, ducks, and swine in eastern Asia that serve as reservoirs for human influenza viruses, genetic recombination can lead to variant strains from these areas—for example, Hong Kong flu and Beijing flu.

> **The evolution of new strains of influenza viruses means that major outbreaks of flu regularly occur when a sufficient proportion of the population is susceptible to permit frequent person-to-person transmission.**

Major outbreaks caused by type A influenza virus occur every 2 to 4 years, those caused by type B influenza virus every 4 to 6 years; type C influenza virus normally infects animals—it rarely infects humans and causes mild cases of influenza (FIG. 18-6). The 1918-1920 influenza pandemic (Spanish flu) that resulted in 20 million deaths was associated with a type A influenza virus. The determination that two deaths at Fort Dix, New Jersey, in 1976 were caused by a swine flu virus aroused sufficient fear to initiate a major nationwide immunization program. Unfortunately, some individuals given the swine flu vaccine developed a neurological disorder called Guillain-Barré syndrome.

Clinical diagnosis of influenza depends on the isolation of an influenza virus. Tissue culture, with a cell line such as monkey kidney cells, is the isolation technique used. Increased titers of antibody in the patient's serum that are reactive with flu viral antigens must then be detected serologically. As with most other viral diseases, treatment of uncomplicated cases of influenza centers on treating the symptoms, with recovery from the disease dependent on the immune response of the infected individual.

Primary control of influenza is achieved by vaccinating individuals who are prone to the complications resulting from this disease, leaving others unprotected to suffer periodically from influenza. Vaccination against one type of influenza virus does not make one resistant to infection with an antigenically different strain of this virus. Vaccination is recommended for anyone over 65, those with chronic illnesses such as heart disease, lung disease, and diabetes, and for healthy individuals who perform essential services, such as medical personnel. Vaccination is important because influenza causes a significant number of deaths. For example, the death rate in the United States due to influenza in 1980 (a nonepidemic year) was 0.3 deaths per 100,000 population.

> **A vaccine against influenza is used to protect high-risk individuals.**

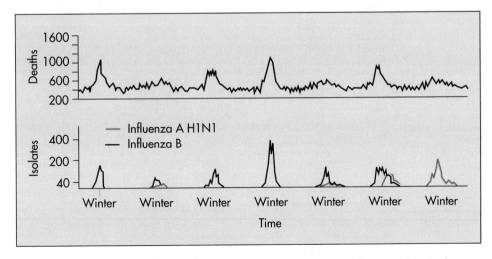

FIG. 18-6 Influenza outbreaks show regular cyclic fluctuations due to antigenic changes in the influenza viruses and associated changes in the susceptibilities of individuals in the population. The highest incidence of influenza occurs during the winter.

HANTAVIRUS INFECTIONS STRIKE THE SOUTHWESTERN UNITED STATES

A hantavirus (Muerto Canyon virus), related to the one that causes Korean hemorrhagic fever, has been identified as the cause of a disease outbreak in the southwestern United States in the summer of 1993 that killed several Navajo Indians. The disease, called hantavirus pulmonary syndrome, initially resembles influenza, but as the infection progresses, blood plasma fills the lungs. The disease has a high mortality rate (75%). Hantaviruses are commonly found in rodents in the United States, and individuals living in rodent-infested neighborhoods often show evidence of infection. Fortunately, the symptoms of these infections have been mild, and acute infections causing severe bleeding and kidney failure, which are characteristic of most hantavirus infections, are virtually unknown in the United States. Hantavirus pulmonary syndrome is unique among hantavirus diseases in that it primarily affects the lungs. The route of transmission from rats to humans is not yet known; scientists hypothesize that in the southwestern United States the hantavirus is spread from deer mice through contaminated urine, feces, and saliva. It appears that the virus is a new genetic variant of one that has been present in the United States for a long time, one that is far more virulent than its predecessors.

VIRAL PNEUMONIA

Various viruses, including influenza, chickenpox, and enteric viruses, cause **viral pneumonia** (Table 18-7). Viral pneumonia is an inflammation of the lungs due to a viral infection. It is a serious disease that often is life-threatening because the accumulation of fluids that results from the inflammation can block essential gas exchange in the lungs. The specific virus causing a particular case of viral pneumo-

nia is rarely isolated and identified because most clinical laboratories are not equipped to test for viruses.

The **respiratory syncytial virus** (RSV) is the most common cause of viral pneumonia in infants. RSV is a member of the parainfluenza class. This virus is usually transmitted via droplet inhalation. The infection proceeds downward along the respiratory mucosa from one infected cell to the next. The respira-

TABLE 18-7

Causes of Viral Pneumonia

VIRUS	CLINICAL CONDITION	COMMENTS
Adenovirus	Pharyngoconjunctival fever, pharyngitis, atypical pneumonia (military recruits)	No antivirals; vaccines not generally available
Cytomegalovirus (CMV)	Interstitial pneumonia	In immunodeficient patients (e.g., AIDS)
Influenza A or B	Primary viral pneumonia or pneumonia associated with secondary bacterial infection	Pandemics (type A) and epidemics (type A or B); increased susceptibility in elderly or in certain chronic diseases
Measles	Primary viral (giant cell) pneumonia in immunodeficient individuals; commonly leads to secondary bacterial pneumonia	Adult infection rare but severe; King and Queen of Hawaii both died of measles when they visited London in 1824
Parainfluenza (types 1-4)	Croup, pneumonia in children (under 5 yrs); upper respiratory illness (often subclinical) in older children and adults	Antivirals and vaccines not available
Respiratory syncytial virus (RSV)	Pneumonitis pneumonia (infants); common cold syndrome (adults)	Peak mortality in 3-4-month-old infants; secondary bacterial infection rare
Varicella-zoster virus (VZV)	Pneumonia in young adults suffering primary infection	Uncommon; recognized 1-6 days after rash; lung lesions may eventually calcify

tory syncytial virus causes cell fusion (syncytium formation) of infected cells because the envelope of the virus has a fusion protein that causes the viral envelope to bind to the host cell membrane. The fusion protein permits adsorption of the virus to the host cell and also causes fusion of the membranes of infected cells. The immune system is inadequate to protect against RSV. Infants develop RSV infections when they still have passive immunity from maternal antibodies. A specific immune globulin can be used to protect newborns who are considered to be at high risk of RSV infections. RSV can also cause a mild upper respiratory tract infection in older children and adults.

The respiratory syncytial virus is the major cause of viral pneumonia in infants.

Approximately 55,000 infants are hospitalized and 2,000 die each year in the United States because of respiratory syncytial viral infections (RSV). Serious epidemics caused by RSV sometimes occur in hospital nurseries. Flu-like symptoms, coughing, and wheezing, which typically last for more than a week, characterize RSV infections. Aerosol administration of the antiviral drug ribavirin is sometimes used to diminish the severity of symptoms of RSV infections in infants and young children. Ribavirin is believed to block transcription.

VIRAL DISEASES OF THE GASTROINTESTINAL TRACT

The gastrointestinal tract is an important portal of entry for infection by viruses (Table 18-8). Transmission by the fecal-oral route—in which viruses excreted from the gastrointestinal tract of an infected individual reach other susceptible persons through fecally contaminated hands, food, water, or inanimate objects—is frequently involved in the transmission of disease-causing viruses. Some viruses are resistant to the hydrochloric acid in the stomach and the bile acids in the intestine, enabling them to reach susceptible cells in the intestine. They then can replicate within those cells or spread through the circulatory system to other parts of the body, including the accessory organs of the digestive tract, such as the liver. Replication of viruses within the gastrointestinal tract can disrupt essential life-supporting functions and also can result in losses of water that cause electrolyte imbalances and shock.

VIRAL GASTROENTERITIS

Gastroenteritis is an inflammation of the lining of the gastrointestinal tract. This disease can be caused by various viruses, including enteroviruses (viruses that replicate primarily in the gastrointestinal tract), coxsackieviruses, polioviruses, and members of the echovirus group. The viruses that cause gastroenteritis kill the intestinal epithelial cells in which they replicate. The destruction of cells lining the intestine, which normally absorb electrolytes, results in decreased fluid absorption and a net secretion of fluid that results in diarrhea.

Viruses that cause gastroenteritis replicate within cells lining the gastrointestinal tract. This results in large numbers of viruses released in fecal matter. The fecal-oral route, involving contamination of food with fecal matter containing viruses, is an important mode of transmission of viral gastroenteritis. The **Norwalk virus (Norwalk agent)** is a small RNA virus that is identified as responsible for an outbreak of "winter vomiting disease" that occurred in Norwalk, Ohio, in 1968. Norwalk virus and Norwalk-like viruses are the most widely recognized agents of foodborne and waterborne viral gastroenteritis.

Rotavirus, a large RNA virus, is a very common etiologic agent for diarrhea in infants, particularly in socioeconomically depressed regions of the world. Rotaviruses are members of the reovirus class. Replication of rotaviruses causes severe diarrhea as a result of damage to cells in the small intestine (FIG. 18-7).

TABLE 18-8		
Some Viruses That Cause Human Diseases of the Gastrointestinal Tract		
VIRAL GROUP	**VIRUS**	**DISEASE**
Calciviruses	Norwalk agent	Winter vomiting disease
	Hepatitis E	Hepatitis
Hepadnaviruses	Hepatitis B	Hepatitis
Herpesviruses	Cytomegalovirus	Cytomegalovirus inclusion disease
Paramyxoviruses	Mumps virus	Mumps
Picornaviruses	Coxsackievirus	Gastroenteritis
	Enterovirus	Gastroenteritis
	Poliovirus	Gastroenteritis
	Hepatitis A	Hepatitis
Reoviruses	Reovirus	Infantile diarrhea
	Rotavirus	Infantile diarrhea
Togaviruses	Hepatitis C	Hepatitis
	Yellow fever virus	Yellow fever

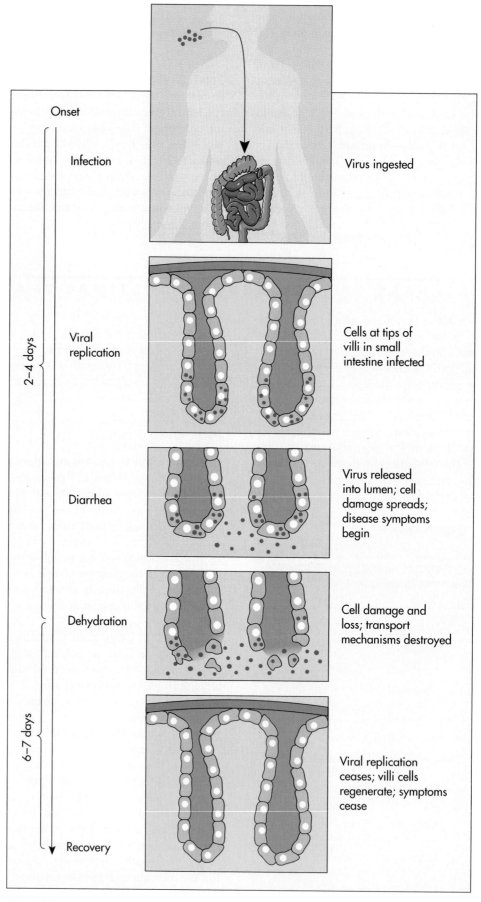

FIG. 18-7 Pathogenesis of rotavirus infection showing stages from infection to recovery.

Rotaviruses are responsible annually for three million cases of diarrhea, 500,000 physician visits, and $200 to $400 million in hospital costs.

The replication of various viruses within the gastrointestinal tract, including the Norwalk agent and rotavirus, can cause an inflammation of the intestines called gastroenteritis.

The characteristic symptoms of viral gastroenteritis include sudden onset of gastrointestinal pain, vomiting, and/or diarrhea (FIG. 18-8). In most cases, viral gastroenteritis is a self-limiting disease, often referred to as the *24-hour flu* or *intestinal flu.* (This disease is not caused by influenza viruses and the reference to gastrointestinal flu is colloquial and not scientifically descriptive). Recovery normally occurs within 12 to 24 hours of the onset of disease symptoms. As a result of the vomiting and diarrhea, however, there can be a severe loss of body fluids and dehydration. The loss of water and resultant imbalances in electrolytes can have serious consequences, particularly in infants, where viral gastroenteritis is sometimes fatal. Treatment generally is supportive and may involve replenishment of body fluids and maintenance of proper electrolyte balance.

HEPATITIS

Hepatitis is a systemic viral infection caused by various diverse **hepatitis viruses** that primarily infect the hepatocyte cells of the liver (Table 18-9). These viruses are taxonomically unrelated but are called hepatitis viruses because they cause similar disease symptoms as a result of infection of the liver. The liver is an accessory organ of the gastrointestinal tract that produces bile. Cell damage, which results in release of the hepatitis viruses into the bloodstream, is not due to the cytopathic properties of the virus itself but rather to the activation of the immune system that kills host cells infected with the viruses.

Hepatitis is caused by hepatitis viruses replicating within the liver.

The major types of hepatitis viruses are designated types A, B, C, D, and E. Type C hepatitis is still referred to as nonA-nonB hepatitis by the Centers for Disease Control. Type E hepatitis is also referred to as enteric nonA-nonB hepatitis. Hepatitis A virus is a member of the picornavirus class, hepatitis B virus is a member of the hepadnavirus class, hepatitis C virus is a member of the togavirus class, and hepatitis E virus is a member of the calcivirus class. Currently, hepatitis D virus has not been classified.

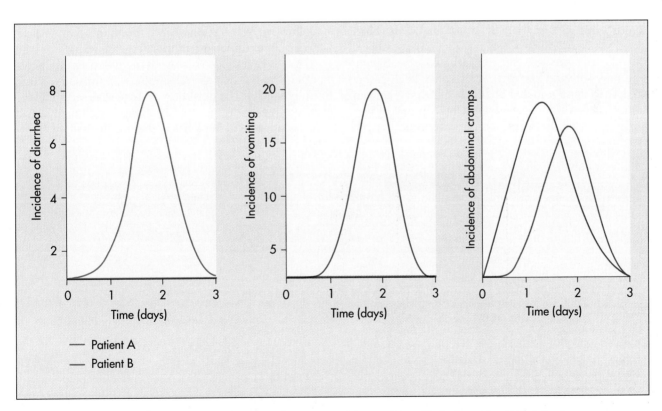

FIG. 18-8 Viruses that cause viral gastroenteritis are transmitted via contaminated food and water. Large numbers of the virus are shed in feces from infected individuals. Two volunteers were orally administered a stool filtrate from an individual infected with Norwalk agent. Both individuals developed viral gastroenteritis characterized by nausea and abdominal cramps; one individual vomited repeatedly and the other had severe diarrhea.

TABLE 18-9				
Comparison of Hepatitis Caused by Different Viruses				
VIRUS	**GROUP**	**GENOME**	**INCUBATION PERIOD**	**TRANSMISSION**
Hepatitis A virus (HAV)	Picornavirus (enterovirus)	Single-stranded RNA	2-4 weeks	Fecal-oral
Hepatitis B virus (HBV)	Hepadnavirus	Double-stranded DNA	4-12 weeks	Sexually transmitted, exchange of blood
Hepatitis C virus (HCV) (nonA-nonB)	Togavirus	Single-stranded RNA	8 weeks	Exchange of blood
Hepatitis D virus (delta-agent)	—	Single-stranded RNA	2-12 weeks	Exchange of blood
Hepatitis E virus (HEV) (enteric nonA-nonB)	Calcivirus	Single-stranded RNA	6-8 weeks	Fecal-oral

Type A Hepatitis

Type A hepatitis virus normally enters the body via the gastrointestinal tract and causes **type A hepatitis** (formerly called infectious hepatitis). Nearly 25,000 cases of hepatitis A are reported each year in the United States. It is usually transmitted by the fecal-oral route and is prevalent in areas with inadequate sewage treatment. The virus is shed in the feces of infected individuals and can contaminate water used for drinking. Outbreaks of type A hepatitis occur when untreated rural water sources become contaminated with fecal matter. Even in areas with chlorinated water supplies, outbreaks can occur because of the relative resistance of the virus to chlorine. Several outbreaks of viral hepatitis A also have been associated with contaminated shellfish that have concentrated viruses from sewage effluents. Spread of hepatitis A can occur in restaurants where food is pre-

pared by infected personnel. Other outbreaks of hepatitis A have occurred among children in day care centers.

Hepatitis type A virus is transmitted through contaminated food and water.

The incubation period for hepatitis A virus is several weeks before disease symptoms appear (FIG. 18-9). The initial symptoms of hepatitis A include fever, abdominal pain, and nausea, followed by jaundice, yellowing of the sclerae, and, often, the skin, indicative of liver impairment caused by the virus. Damage to the liver cells also results in increased serum levels of enzymes, such as transaminases, normally active in liver cells. The detection of increased serum levels of these enzymes is used in diagnosing this disease. In most cases of hepatitis A there is no specific treat-

NEWSBREAK

OUTBREAKS OF HEPATITIS A IN KENTUCKY

Three separate outbreaks of hepatitis A occurred in Kentucky within a 3-year period, all due to different sources. The first outbreak occurred simultaneously in several rural regions of the Commonwealth. At first there appeared to be no connection. The only thing these outbreaks had in common was that the individuals afflicted lived in homes that had water delivered and stored in cisterns. The water supply companies were different and each initially claimed to have obtained the water it supplied from different sources. Further investigation disclosed that all the water companies were clandestinely obtaining the water from a common waterfall. A family living above the waterfall had

hepatitis and their untreated sewage was released into the stream that fed the waterfall.

In the second incident, all those who contracted hepatitis had eaten raw oysters at the same restaurant chain. The problem was traced to a Gulf Coast supplier of oysters who had obtained the oysters from a coastal area contaminated with sewage.

The third outbreak was traced to restaurants that served green salads that were obtained from a common supplier. The lettuce had been contaminated with human feces by a farmer with hepatitis who had applied the feces as fertilizer. One individual died from liver failure in this latter outbreak.

ment, the infection is usually self-limiting, and recovery normally occurs within 4 months. Its incidence, however, can be reduced by using passive immunization with immunoglobulin G, which is used when there is a high probability of exposure to the hepatitis A virus.

Hepatitis B and NonA-nonB Hepatitis

Hepatitis caused by hepatitis B virus and by hepatitis C virus is typically transmitted via serum and hence often is called **serum hepatitis.** Clinical manifestations of **hepatitis B viral infections** generally are very similar to those described for hepatitis A in-

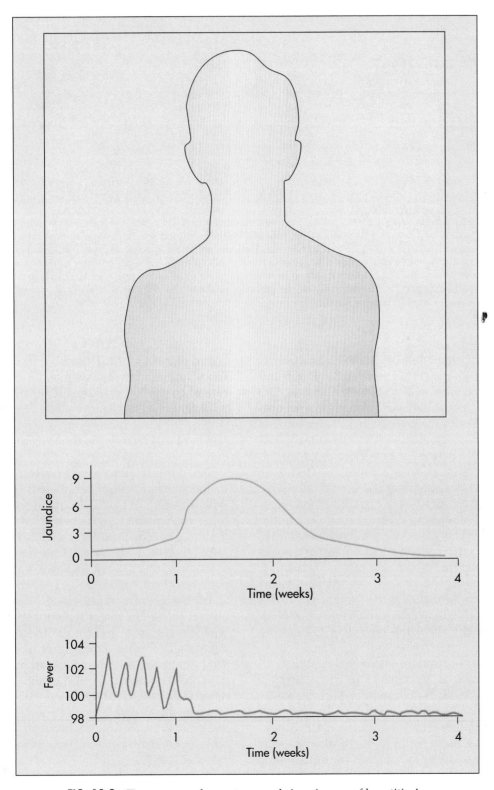

FIG. 18-9 Time course of symptoms and signs in case of hepatitis A.

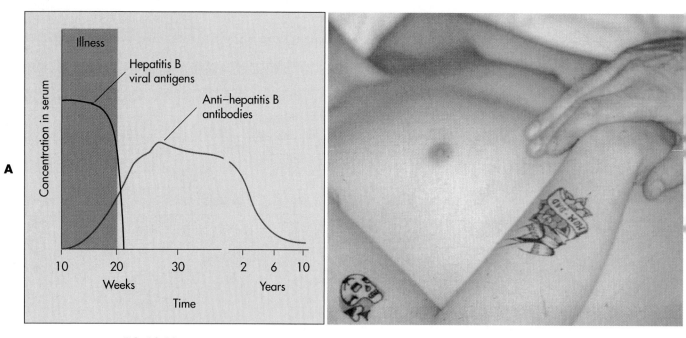

FIG. 18-10   **A,** Virus B hepatitis. Long-incubation (virus B) hepatitis is generally transmitted by blood or blood products. As little as 0.004 ml of blood serum containing hepatitis B virus has been known to cause infection. Patients and staff in renal dialysis units are particularly vulnerable to this disease, and there is also a high incidence in drug addicts and homosexuals. **B,** The patient shown here was one of a group of young men infected from a tattooing needle. The intensity of the jaundice is shown by the contrast with the normal color of the physician's hand.

fections. The duration of serum hepatitis, however, is much more prolonged (FIG. 18-10). Also, types B and C hepatitis infections tend to be more serious. The mortality rate from hepatitis B normally is low, usually less than 1% for the population at large, but it is higher for certain groups, such as the elderly, and during particular outbreaks. Five thousand deaths in the United States each year occur resulting from liver damage caused by hepatitis B and C viral infections. Approximately 25,000 cases of hepatitis B and hepatitis C viral infections are reported each year in the United States. Many cases of hepatitis, however, are unreported and the United States Centers for Disease Control estimate that 300,000 new cases of hepatitis B and C may actually occur each year in the United States alone. The incidence is much higher in Africa and Asia. Worldwide, approximately 200 million people chronically carry the type B or C hepatitis viruses.

The normal route of transmission of hepatitis B, C, D, and E viruses is by transfer of virus-carrying blood from one individual to another. These hepatitis viruses may also be transmitted via secretions such as saliva, sweat, breast milk, and semen. All of these viruses can enter the body through a break in the skin or a break in a mucous membrane. Transmission of these viruses can also occur via sexual contact. These viruses can be transmitted to the fetus across the pla-

centa, to a baby at the time of birth, or in early infancy from an infected mother. There also is a high rate of transmission of serum hepatitis among drug addicts, who frequently share contaminated syringes. The incubation period is about 60 to 180 days, which makes tracing the origin of an outbreak difficult.

Blood transfusions have frequently been involved in the transmission of serum hepatitis. Individuals with a history of jaundice, one of the clinical symptoms of hepatitis, are generally rejected as blood donors. The blood supply is screened for known hepatitis antigens. Tests are available for the direct detection of hepatitis B virus. An indirect test, based on the presence of antibodies can be used to detect hepatitis C, but this is not as reliable as the direct test for hepatitis B virus. The major type of hepatitis associated with blood transfusions in the United States is now hepatitis C. Proper screening of blood for hepatitis viral antigens has reduced the transmission rate of hepatitis B but not of type C hepatitis. This is because detection methods for screening blood for hepatitis B antigens are available, but screening procedures for antigens associated with hepatitis C viruses and other nonA-nonB hepatitis viruses have yet to be developed.

Unfortunately, no method used today is effective in rendering whole blood containing hepatitis viruses totally free of the agent. Plasma can be treated with the viricide betapropiolactone to inactivate any hepatitis viruses that may be present. Hepatitis B virus resists irradiation and desiccation. It represents a significant risk to health care workers who are exposed to blood, including dried blood. Hepatitis B can be prevented by vaccination (see Chapter 15). When the vaccine was first introduced, immunization was restricted to health care workers and others at high risk of contacting contaminated blood, such as intravenous drug abusers. This was because of the limited availability of the vaccine and its high cost. Now hepatitis B vaccination is recommended for all, with three doses of vaccine given by 18 months.

Hepatitis B and C (nonA-nonB) viruses are transmitted through serum, for example, through blood transfusions and contaminated syringe needles.

Hepatitis B virus can be the cause of lifelong chronic liver infection. Perhaps 10% of those infected with hepatitis B become chronic carriers. Chronic hepatitis can lead to cirrhosis (hardening of the liver) and hepatocellular carcinoma (liver cancer). The carrier rate of hepatitis B virus in blood donors in the United States appears to be between 0.5% and 1%, but in other countries the carrier rate may be as high as 5%.

A severe outbreak of hepatitis B in Massachusetts involved a simultaneous infection with a hepatitis B virus and a hepatitis D virus (originally called delta agent). The hepatitis D virus was discovered in 1977 as the cause of an outbreak of severe hepatitis in Italy.

It is a defective virus, so-called because it is unable to cause infections on its own, reproducing only when the host is already infected with hepatitis B virus. The hepatitis D virus is a single-stranded RNA virus that is contained in a hepatitis B capsid. Chronic infection with hepatitis D virus causes an asymptomatic or mild hepatitis B viral infection to become acutely worse. Individuals simultaneously infected with hepatitis B and D hepatitis virus have a higher incidence of severe liver damage and a higher mortality rate than those who only had antibodies against hepatitis B. This hepatitis D virus is responsible for a particularly virulent form of hepatitis (with a reported mortality rate of 17%) affecting natives of the Amazon Basin and countries surrounding the Mediterranean Sea. In Europe and North America, there have also been outbreaks among intravenous drug users.

Some serious cases of hepatitis involve the simultaneous infection with hepatitis B virus and a defective virus, called the hepatitis D virus, that is unable to infect cells on its own.

YELLOW FEVER

Yellow fever is caused by a small RNA togavirus, called the **yellow fever virus.** Like hepatitis, multiplication of the yellow fever virus within the liver causes jaundice and the subsequent yellowing of the skin in people with lighter skin tones, from which this disease derives its name. The yellow fever virus is transmitted by mosquito vectors, predominantly by *Aedes aegypti.* There are two epidemiologic pat-

HIGHLIGHT

HEALTH CARE WORKERS AT RISK FOR HEPATITIS INFECTIONS

Health care workers who come into contact with blood from patients have a considerably higher incidence of serum hepatitis than the population at large. Doctors, for example, have five times, and dentists two or three times, the normal rate of infection. Special precautions are used by health care workers to limit the transmission of serum hepatitis. Whenever possible, disposable equipment is used for routine hospital and clinical procedures and these items are sterilized before being discarded. Nondisposable syringes, needles, tubing, and other types of equipment used for obtaining blood specimens or for the administration of therapeutic agents should be sterilized thoroughly before reuse. The most common means of chemical sterilization are not reliable for destroying hepatitis-causing viruses. Boiling in water for at least 20 minutes, heating in a drying oven at a temperature of 180° C for 1 hour, and autoclaving are procedures that are effective in destroying hepatitis viruses.

When a health care worker or other individual is believed to have been exposed to blood containing hepatitis viruses, passive immunity by injection of immunoglobulin is used as a precautionary measure. Both IgG and hepatitis B immune globulin are used for the prevention of hepatitis B infection. Hepatitis B immune globulin has a high level of immune globulins and often is recommended for health care workers who accidentally stick themselves with a needle or whose mucous membranes are exposed to blood containing hepatitis viruses. Health care workers and other high-risk individuals today can be protected against hepatitis B by vaccination.

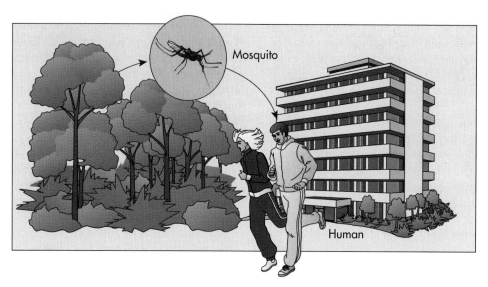

FIG. 18-11   Yellow fever virus is transmitted by the mosquito *Aedes aegypti*. It normally is transmitted to monkeys during the enzootic cycle but also can be transmitted to humans, causing yellow fever.

terns of transmission (FIG. 18-11). In urban transmission of yellow fever, the mosquito *Aedes aegypti* transfers blood from an infected to a susceptible individual. In jungle transmission of yellow fever, the virus is passed normally among monkeys, with occasional transfer via mosquito vectors to human beings. Outbreaks of yellow fever were a major problem in the construction of the Panama Canal, leading to Walter Reed's instrumental work in establishing the relationship between yellow fever and mosquito vectors in 1901.

Yellow fever virus is transmitted by mosquito vectors.

The onset of yellow fever is marked by anorexia (loss of appetite), nausea, vomiting, and fever. Yellow fever viral multiplication results in liver damage, causing jaundice. Symptoms generally last less than 1 week, after which, recovery begins or death occurs. The fatality rate for yellow fever is about 5% but may be as high as 40% during epidemics. There is no effective antiviral drug at present for treating yellow fever but the disease can be prevented by vaccination. Two strains of yellow fever virus are used for vaccination. The 17D vaccine uses an attenuated strain of yellow fever virus produced by culture in chick embryos and is administered subcutaneously. The Dakar strain is grown in mouse brain tissue and is introduced by scratching into the skin.

Today, yellow fever occurs primarily in remote tropical regions: Central America, South America, the Caribbean, and Africa. The urban form of yellow fever transmission has been largely controlled by effective mosquito eradication programs, but the jungle form of transmission cannot easily be interrupted

NEWSBREAK

DEMONSTRATING VECTORS TRANSMIT DISEASE

A Cuban physician, Dr. Carlos Finlay, not the American Army doctor Walter Reed, first suggested that mosquitoes were the injectors of the unknown causal agent of yellow fever. Reed designed highly controlled experiments with paid human volunteer subjects who were bitten by mosquitoes who had previously bitten patients infected with yellow fever. Many volunteer subjects died. Reed and his team discovered that the specific agent of yellow fever had passed through a filter along with the filtrate with which subjects were inoculated.

because of the large natural reservoir of yellow fever viruses maintained in monkey populations.

MUMPS

Salivary glands are accessory glands of the digestive system. They are the sites for the replication of an enveloped RNA virus called the **mumps virus.** The mumps virus is a member of the paramyxovirus class. Replication of the mumps virus causes the disease **mumps,** which is characterized by enlargement of the infected salivary glands (FIG. 18-12). Swelling on both sides (bilateral parotitis) occurs in about 75% of individuals with mumps, with swelling in one gland usually preceding swelling on the other side by about 5 days. This disease primarily occurs during childhood.

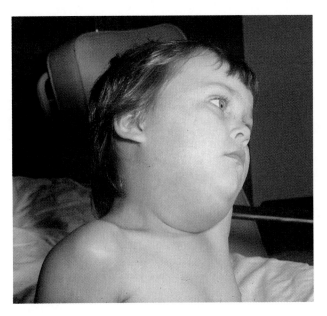

FIG. 18-12 Mumps is a generalized infection with a wide range of clinical manifestations. The illness begins with fever and malaise, quickly followed by trismus (spasm of the jaw muscles) and pain behind the angle of the jaw. Within 24 hours the parotid gland begins to swell. Both parotid glands are involved in 70% of patients with parotitis.

The mumps virus is transmitted via contaminated droplets of saliva. The initial infection appears to occur in the upper respiratory tract, with subsequent dissemination to the salivary glands and other organs. The average incubation period for mumps is 18 days, and the swollen salivary glands generally persist for less than 2 weeks. The mumps virus may spread to various body sites, and although the effects of the disease are normally not long lasting, there can be several complications (Table 18-10). For example, mumps is a major cause of deafness in childhood. In males past puberty, the mumps virus can cause or-

chitis (inflammation of the testes), but although it is commonly believed, mumps rarely results in male sterility. There is no specific treatment for mumps but the disease can be prevented by vaccination with MMR (measles-mumps-rubella) vaccine. The mumps portion of this vaccine is prepared with attenuated mumps virus; a single vaccination appears to confer lifelong immunity.

> Mumps is a common childhood disease that is a major cause of deafness; it can be prevented by the MMR vaccine.

CYTOMEGALOVIRUS INCLUSION DISEASE

Cytomegalovirus (CMV) is the cause of the severe and often fatal disease of newborns known as **cytomegalovirus inclusion disease.** CMV is a member of the herpesvirus class. It enters the body through a break in the skin or break in a mucous membrane. Transmission can occur congenitally, via transfusion of contaminated blood, sexual contact, or transplantation of an infected organ, or via aerosols entering the respiratory tract. From the point of entry the virus travels to the blood and on to the various organs. CMV is excreted in body fluids such as saliva, blood, semen, urine, and milk. Reproduction of CMV causes abnormalities of the infected human cells, characterized by an increase in size (cytomegaly) and the development of inclusion bodies within the nucleus (FIG. 18-13). The virus may remain inactive within the body or may multiply within various organs. CMV can be transmitted from mother to child across the placenta or during the process of birth. Congenital transmission of this virus can cause various abnormalities. Between 10% and 20% of stillborn infants show abnormal cells characteristic of cytomegalovirus inclusion disease. In newborns, CMV

TABLE 18-10		
Pathogenesis of Mumps		
SITE OF GROWTH	**DISEASE CONDITION**	**COMMENT**
Salivary glands	Inflammation; parotitis (swelling of salivary glands)	Unilateral or bilateral; virus shed in saliva (from 3 days before to 6 days after symptoms)
Meninges	Meningitis	Common (10% cases)
Brain	Encephalitis	Complete recovery usual; deafness rare complication
Kidney	None	Virus present in urine
Testis	Orchitis	Common (20% in adult males); not a significant cause of sterility
Pancreas	Pancreatitis	Rare (possible role in juvenile diabetes)
Mammary gland	Mastitis	Virus detectable in milk; 10% post-pubertal females
Thyroid	Thyroiditis	Rare
Myocardium	Myocarditis	Rare
Joints	Arthritis	Rare

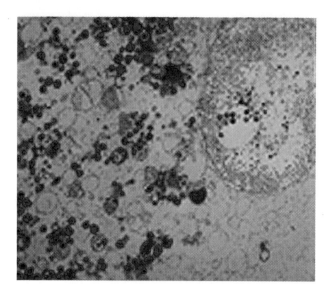

FIG. 18-13 Colorized electron micrograph of cytomegalovirus *(blue)* within an infected cell.

generally affects the salivary glands, liver, kidneys, and lungs.

> **Cytomegalovirus inclusion disease, which can be a fatal infection in newborns, is caused by cytomegalovirus (CMV).**

Up to 80% of adults over 35 years of age show antibodies that indicate prior or active infection with CMV. In most cases the virus is inactive, producing a latent infection, with the viruses probably residing within lymphocytes. CMV can be transmitted to a new host via a latently-infected blood transfusion or latently-infected transplant organ. The viral infection can be activated when the immune system is suppressed, such as when immunosuppressive drugs are

NEWSBREAK

HEART DISEASE AND CYTOMEGALOVIRUS INFECTIONS

The discovery that heart disease may be linked to cytomegalovirus (CMV) infections has cardiologists contemplating new approaches to treating heart disease. In many cases, heart disease occurs when the arteries supplying blood to heart muscle become blocked by a buildup of plaque. It has been thought that the plaque is due to a buildup from exogenous sources; it now appears that cytomegalovirus infections cause cells that line the inner walls of the arteries to reproduce excessively, leading to the accumulations of plaque cells. This fits the observation that the process of artery closure begins in the blood vessel wall as cells proliferate and form a tumor-like growth. Cytomegalovirus interferes with p53 proteins that normally cause human cells to slow their growth. By doing so they allow cells to reproduce in a manner similar to cancerous cells. This could explain why 40% to 50% of patients treated with balloon angioplasty to remove plaque redevelop clogging of the heart arteries within 6 months. Now the challenge is to find a way of stopping CMV from causing heart disease.

used after transplants. Alternatively, endogenous latent CMV may be reactivated if the host becomes pregnant or the host's immune system becomes suppressed due to drug therapy or other causes—such as AIDS.

VIRAL DISEASES OF THE GENITAL TRACT

Some viruses establish infections in the genital tract, causing diseases such as genital herpes and genital warts (Table 18-11). These viruses generally have poor survival potential outside infected cells and hence are transmitted almost exclusively by direct sexual contact. Replication of these viruses within the cells of the genital tract causes lesions. The lesions contain viruses that can be transmitted by direct contact. There has been an alarming rise in sexually transmitted diseases, including viral diseases that directly affect the genital tract and those where the virus enters the body via the genital tract and cause disease elsewhere in the body. The increase in the rate of incidence of sexually transmitted diseases may not be due only to contemporary sexual behavioral patterns but may also reflect changes in reporting and recording cases of these diseases.

GENITAL HERPES

Genital herpes is caused by **herpes simplex type 2 virus.** It is most frequently transmitted by sexual contact and causes infection of the genitalia. (Herpes simplex type 1 viral infections generally are not sexually transmitted and most frequently occur above the waist in contrast to type 2 genital infections.) It is estimated that 20 million Americans now have genital herpes and that there will be at least a half a million new cases each year unless effective means are found for controlling this disease (FIG. 18-14). The herpesvirus does not survive outside the body for long periods of time, so that contact with fomites (toilet seats, bedding, etc.) is not likely to transmit the virus. Additionally, condoms are effective in preventing transmission of the virus.

TABLE 18-11		
Some Viruses That Cause Human Diseases of the Genital Tract		
VIRAL GROUP	VIRUS	DISEASE
Herpesviruses	Herpes simplex type 2	Genital herpes
Papovaviruses	Papillomavirus	Genital warts, cervical cancer

In women the primary site of herpes infection is the cervix. The infection may also involve the vulva and vagina. In men the herpes simplex virus frequently infects the penis, where the infection causes genital soreness and ulcers in the infected areas (FIG. 18-15). The virus and manifestations of infection may be transmitted to other areas of the body, most notably the mouth and anus. Genital herpes may have particularly serious repercussions for pregnant women because the virus can be transmitted to the infant during vaginal delivery and, in rare cases, may cross the placenta, causing damage to the infant's central nervous system and/or eyes. Herpes is lethal in up to 60% of infected newborns, and for surviving babies there is a 50% risk of blindness or neurological damage. Caesarean deliveries are often performed to prevent transmission to the newborn when the mother has an active outbreak of herpes. A high percentage of women who have cervical cancer also have antibodies to herpes simplex type 2. However, the exact relationship between this virus and cervical carcinoma has not been elucidated.

Ulcers produced from herpes simplex type 2 infections generally heal spontaneously in 10 to 14 days. However, because of the budding mode of reproduc-

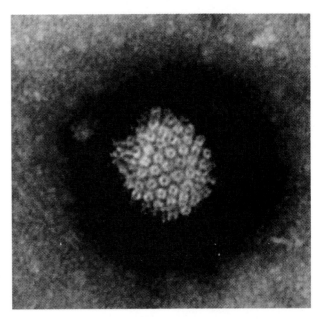

FIG. 18-14 Colorized electron micrograph of the capsid of a herpesvirus with no envelope present. (272,000×).

tion of herpes viruses, the infection is not eliminated when the ulcers heal; rather, a reservoir of infected cells remains in the body in a latent state within the ganglia of sensory neurons. Subsequently, multiplication of the viruses can produce secondary ulcers, even in the absence of additional sexual activity. It is not known exactly what initiates subsequent attacks of herpes, but such recurrences may be triggered by sunlight, sexual activity, menstruation, and/or stress. Generally the secondary ulcers heal more rapidly than the primary lesion. The disease remains trans-

FIG. 18-15 A, Genital herpes lesions on the penis of an adult male. **B,** Genital herpes lesions on the labia of a female. Infection is commonly transmitted by sexual intercourse but not invariably. The prevalence is much greater in women.

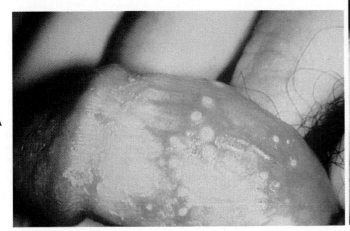

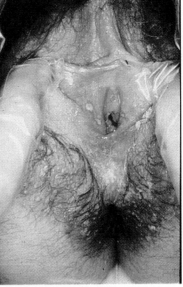

missible, interfering with establishing stable sexual relationships. Herpes simplex type 2 can be transmitted even when the carrier displays no visible viral lesions—usually a few days before, during, and a few days after visible lesions.

> **Herpesviruses replicate by budding so that, even when the lesions that occur as a result of herpes infections heal, a reservoir of infected cells remains within the body.**

Several of the newly developed antiviral drugs are useful in the treatment of herpes viral infections. In particular, acyclovir and interferon show promise of reducing the severity of the symptoms; neither of these drugs, though, promises a cure for the disease.

GENITAL WARTS

Warts, including **genital warts,** are benign tumors caused by infections with **papilloma viruses.** These viruses are transmitted by direct contact, normally infecting the skin and mucous membranes. Warts can

develop on the genitals and direct sexual contact usually is the source of the infecting papilloma viruses when this occurs (FIG. 18-16). Warts generally do not appear for several weeks after infection. Chemical treatments—for example, acid application—and/or physical methods—for example, freezing—can be used for the removal of warts, but these benign tumors also will disappear without treatment. Some papilloma viruses cause malignancies—cervical cancer in women and urinary tract cancer in men. Adolescent females who have had extensive sexual contacts and have developed genital warts infections face an increased probability of contracting cervical cancer. Therefore, frequent pap smears are suggested for women diagnosed with genital warts.

> **Genital warts are caused by papillomaviruses and can lead to cervical cancer.**

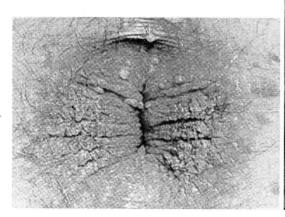

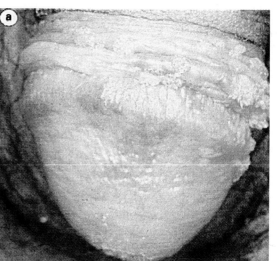

FIG. 18-16   **A,** Genital warts in the vulvo-perineal area of a female. **B,** Genital warts on the male penis are usually multiple and on the shaft are often flat.

SUMMARY

Viruses and Human Diseases (pp. 535-536)

- Various human diseases are caused by viral infections as a result of physiological damage to host cells during viral replication.
- Typically, when viruses are released from host cells the host cells lyse; when sufficient numbers of cells are killed, there can be tissue damage.
- The body's immune defenses may also attack and destroy viral infected cells, causing tissue damage.

Classification of Human Viruses (p. 536)

- There are about a dozen taxonomic groups of viruses. These groups are classified based on their structure, including capsid shape, number of capsomers, whether the virus is naked or enveloped, whether it

has DNA or RNA, and whether its nucleic acid is single- or double-stranded.

Viral Diseases of the Respiratory Tract (pp. 536-545)

- The respiratory route is the most common site of viral infections. Viruses typically enter the respiratory tract in aerosols.

The Common Cold (pp. 538-540)

- The common cold is a cluster of related diseases with symptoms of nasal stuffiness, sneezing, coughing, headache, malaise, sore throat, and slight fever.
- There is no treatment because the common cold is a self-limiting syndrome.
- Rhinoviruses are the most frequent causative agent.

Influenza (pp. 541-543)

- Influenza is caused by influenza viruses (flu viruses). There are three major types of influenza viruses, designated A, B, and C. There also are numerous antigenic subtypes that are responsible for different outbreaks of the disease.
- Genetic changes produce new strains of flu viruses.
- Flu is transmitted by inhalation of droplets containing flu viruses released into the air from the respiratory tracts of infected individuals; symptoms are sudden fever, malaise, headache, and muscle ache.
- Vaccination is advised for individuals prone to the complications resulting from this disease.

Viral Pneumonia (pp. 544-545)
- Viral pneumonia often occurs as a complication of influenza and can be caused by various viruses.
- The respiratory syncytial virus causes pneumonia in infants, who cough and wheeze.

Viral Diseases of the Gastrointestinal Tract (pp. 545-554)
- Viruses causing gastrointestinal tract infections typically are transmitted via the fecal-oral route and enter the body through contaminated food or water.

Viral Gastroenteritis (pp. 545-547)
- Viral gastroenteritis can be caused by various viruses, including enteroviruses, coxsackieviruses, polioviruses, and members of the echovirus group.
- Contamination of food with fecal material is an important mode of transmission.
- Characteristic symptoms include gastrointestinal pain, vomiting, and/or diarrhea, with recovery normally within 12 to 24 hours.

Hepatitis (pp. 547-551)
- Hepatitis type A virus is usually transmitted by the fecal-oral route and primarily infects the liver. Symptoms include fever, abdominal pain, nausea, and jaundice.
- There is no treatment. The infection is self-limiting.
- Hepatitis type B and C infections (serum hepatitis) are more serious and last longer than type A infec-

tions. The type B and C viruses are transferred by blood, saliva, sweat, breast milk, and semen.
- A vaccine is available for hepatitis B virus but not for hepatitis type C.

Yellow Fever (pp. 551-552)
- Yellow fever is caused by an RNA virus transmitted by mosquito vectors, producing symptoms of anorexia, nausea, vomiting, fever, and jaundice. There is a vaccine available.

Mumps (pp. 552-553)
- Mumps is caused by the mumps virus carried by contaminated droplets of saliva, producing swelling of the salivary glands.
- There is no effective treatment for mumps but a vaccine is available.

Cytomegalovirus Inclusion Disease (pp. 553-554)
- Cytomegalovirus inclusion disease is transmitted across the placenta from mother to child during birth and via blood transfusions.
- The disease causes abnormalities of the infected human cells and is characterized by an increase in size and the development of inclusion bodies within the nucleus.
- There is no treatment.

Viral Diseases of the Genital Tract (pp. 554-556)

Genital Herpes (pp. 554-556)
- Genital herpes, a sexually transmitted disease of the genital tract, is caused by herpes simplex type 2 virus. It produces genital soreness and ulcers in infected areas.
- Acyclovir and interferon may reduce the severity of the symptoms.

Genital Warts (p. 556)
- Genital warts are caused by a papillomavirus. This virus can also cause cancer. These warts are sexually transmitted and characterized by the development of benign tumors. They can be chemically or physically removed or left alone.
- There is no drug that eliminates papillomaviruses.

CHAPTER REVIEW

REVIEW QUESTIONS

1. How are viruses classified, including those that cause human diseases?
2. What viral infections cause diseases of the respiratory tract?
3. How are viral diseases of the respiratory tract transmitted? How can transmission be halted?
4. Why isn't it possible to cure or to prevent the common cold by immunization?
5. Who should be immunized against influenza? Why?
6. How do genetic changes influence outbreaks of influenza? Why are there periodic epidemics of influenza?
7. What viral infections cause diseases of the gastrointestinal tract?
8. How are viral diseases of the gastrointestinal tract transmitted? How can transmission be halted?

9. What viruses cause hepatitis? Should these all be called hepatitis viruses?
10. What are the differences and similarities between hepatitis A, B, and C?
11. Why was there a major outbreak of yellow fever during the building of the Panama Canal? What was learned about the transmission of yellow fever virus at that time? How can yellow fever be controlled?
12. What are the normal signs and symptoms of mumps? What are some possible complications?
13. What viral infections cause diseases of the genital tract?
14. How are viral diseases of the genital tract transmitted? How can transmission of these diseases be interrupted?
15. Why is genital herpes a persistent infection?
16. What are the possible consequences of papillomavirus infections?

CRITICAL THINKING QUESTIONS

1. Is it possible to carry out a global immunization program that would eliminate influenza? What are the similarities and differences between influenza and smallpox that affect your answer?

2. Why are viral infections of the gastrointestinal tract more serious in newborns than in adults?

3. Why must health care workers take special precautions to avoid contracting hepatitis? What precautions must health care workers employ?

CASE STUDY

HISTORY AND ASSESSMENT

Shannon, a 28-year-old female college student, came to the student health clinic with a chief complaint of flu-like symptoms—fatigue and achiness. She hoped to get a flu shot or other medication because she needed energy to study for upcoming midterm exams. Shannon explained to the nurse that she began to feel ill several days ago, when she lost her appetite, became nauseous, and vomited several times. She complained that she lost all her energy and was having a difficult time studying. The "flu" has been so bad that Shannon quit smoking in the past week. (Shannon usually smokes one pack of cigarettes per day.) *Cessation of smoking can be a sign of significant anorexia (loss of appetite).*

On initial assessment, the nurse noticed that the sclerae (white part of the eye) of Shannon's eyes were slightly icteric (yellowed). *The liver processes blood bilirubin, a yellow-to-brown pigmented molecule derived from recycling of red blood cells. Damage to the liver is thus often reflected by elevated blood bilirubin, which can deposit in body tissues such as the skin and the sclerae—this condition is called jaundice. From your reading, what are some causes of liver damage?*

Her vital signs were normal, except for a temperature of 102° F.

PHYSICAL EXAMINATION

The clinic physician examined Shannon, noting during the EENT (eye, ear, nose, and throat) examination that Shannon's eyes were slightly icteric. During the abdominal examination, the physician noticed that the liver was slightly enlarged; Shannon said that it hurt when the physician palpated it. The skin appeared somewhat jaundiced, although this was difficult to assess because Shannon said she had been using the tanning booth to prepare for a trip to Miami on spring break. Shannon admitted that several of her friends were similarly ill.

Shannon told the physician that she thought her urine was darker than usual but attributed that to the fact that she had not been drinking much water and was somewhat dehydrated. *What else might the darkened urine indicate?*

The physician questioned Shannon about her use of alcohol. Shannon admitted to being only a "social drinker" and doesn't really like the taste of alcohol.

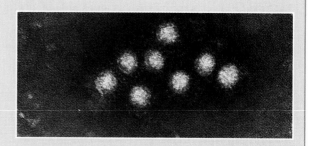

PROCEDURES AND LABORATORY TESTS

Results of Shannon's blood chemistry test were significant for the following values: SGOT 814 IU/liter and SGPT 1382 IU/liter. *SGOT and SGPT are also known as AST and ALT, respectively. They are enzymes present in many body cells, where they function for normal cell metabolism. Hepatocytes (liver cells) contain unusually large amounts of these enzymes. Damage to liver cells will result in diffusion of these (and other) cytoplasmic enzymes into nearby capillaries, causing blood concentrations of these enzymes to increase. Normal levels of these two enzymes in the blood are less than 40 IU/liter. Viral hepatitis usually leads to AST and ALT levels of 500 to 2000 IU/liter, with AST levels less than ALT levels. Alcoholic hepatitis rarely elevates the AST and/or ALT levels above 300 IU/liter.*

Urine dipstick screening showed that an abnormal amount of bile was present in the urine sample.

Based on the physical findings and laboratory results, the physician is certain that Shannon has a form of hepatitis (inflammation of the liver). The cause of the inflammation must now be determined. Alcoholic hepatitis was ruled out, based on the patient history and the fact that the SGOT and SGPT levels were so high. Many of Shannon's presenting symptoms are compatible with an infection of the Epstein-Barr virus (causing infectious mononucleosis), but again, the SGOT and SGPT levels are above what is normally seen with this condition. The physician ordered a "hepatitis screen," a blood test to determine if there are any antigens or antibodies to the various strains of hepatitis viruses that are present in Shannon's blood.

The results of the hepatitis screen showed the presence of IgM antibody against the hepatitis A virus (anti-HAV IgM). No other results of the screen were significant.

DIAGNOSIS

The results of the hepatitis screen confirm the diagnosis of a hepatitis A infection. *The hepatitis A virus is a small, icosahedral, single-stranded RNA virus (see Figure). Infection occurs via the digestive tract. The incubation period is 2 to 6 weeks, during which time the virus infects the liver and is excreted in the feces. The patient is infectious until approximately the time that jaundice is evident.*

The physician notified the local health department of the diagnosis, which, in turn, notified the Centers for Disease Control. An epidemiological investigation was initiated. How could Shannon have contracted hepatitis A virus? How would the epidemiologists determine the source of the infection?

TREATMENT AND COURSE

No specific treatment was prescribed for Shannon. She was told to lighten her work load until she felt better. Shannon rescheduled her midterms and left campus to spend spring break recovering at her parents house instead of in Miami. Two weeks later, she returned to school feeling fine, took her midterms, and continued her studies.

It was also recommended that Shannon's friends experiencing the same symptoms be seen at the clinic. *Including Shannon, eight students were diagnosed with hepatitis A that semester. This "mini-epidemic" was traced to a food service worker at the dormitory who had been sick a month earlier. Signs in restaurant restrooms reminding employees to wash their hands before returning to work are an attempt to reduce the risk of passing hepatitis to patrons should any workers contract hepatitis. Hepatitis A is spread by a fecal-oral transmission route.*

The physician's final recommendation was for Shannon's roommate and anyone with whom she had intimate contact to receive a serum immunoglobulin injection. *Serum immunoglobulin contains antibodies against hepatitis A and can lessen the severity of acute hepatitis if given within two weeks of exposure.* Why did the physician recommend prophylatic use of immune serum?

READINGS

Aral SO and KK Holmes: 1991. Sexually transmitted diseases in the AIDS era, *Scientific American* 264(2):62-69.
Gonorrhea, syphilis, and chancroid have nearly disappeared in most industrialized nations except the U.S., where drug-resistant strains are ravaging urban minority populations beset with poverty, social disintegration, prostitution, and drug addiction.

Baker S: 1986. What killed the Greeks? *Omni* 8(6):36-37.
Presents the views of Alexander Langmuir on Thucydides' description of ancient plagues and epidemics, which may have included toxic shock syndrome.

Beveridge WIB: 1977. *Influenza, the Last Great Plague: An Unfinished Story of Discovery,* London, Heinemann.
Describes early investigations on the cause of influenza, attempting to show how the disease affected the individual and the group. The author, a veterinarian, also addresses the problem of flu in animals and the development of types, subtypes, and variants of the virus.

Came PE and LA Caliguiri (eds.): 1982. *Chemotherapy of Viral Infections,* Berlin, Springer-Verlag.
Discussions by various contributors on the pathogenicity of viruses, drug therapy for viral diseases, and the pharmacology and therapeutic use of antiviral agents.

Castleman M: 1988. Cures for the common cold: a consumer guide, *Utne Reader,* January/February: 80-82.
Reviews the possibilities for cold remedies.

Evans AS (ed.): 1989. *Viral Infections of Humans: Epidemiology and Control,* New York, Plenum Medical Book Co.
Reviews the epidemiology, prevention, and control of viral diseases.

Fincher J: 1989. America's deadly rendezvous with the Spanish lady, *Smithsonian* 19(1):130-132.
Reviews the role of the influenza epidemic in the history of WWI.

Fraenkel-Conrat H and RR Wagner (eds.): 1980. *Virus-Host Interactions: Viral Invasion, Persistence, and Disease,* New York, Plenum Press.
Explores the metabolism of viruses and the role it plays in viral disease.

Haskell M: 1990. *Love and Other Infectious Diseases: A Memoir,* New York, Morrow.
The wife of movie critic Andrew Sarris tells the story of his battle with a cytomegalovirus infection.

Hull R, F Brown, C Payne: 1989. *Virology: Directory and Dictionary of Animal, Bacterial and Plant Viruses,* New York, M. Stockton Press.
Gives descriptions of the major viruses.

Kleinmann L: 1988. Yuppie flu: do you have it? *Health* 20(4):64-66.
A review of chronic fatigue syndrome and the role the Epstein-Barr virus may play in it.

Miller JH: 1989. Diseases for our future, *BioScience* 39(8):509-517.
Global ecological changes in the near future may allow viruses to expand their ranges and diversify into new epidemic strains.

Neustadt RE and HV Fineberg: 1983. *The Epidemic That Never Was: Policy-Making and the Swine Flu Scare,* New York, Vintage Books.
Discusses how medical policy is made in the United States today with special reference to the attempt at preventive inoculation against the prospective swine flu epidemic in 1976.

Olson LC: 1982. *Virus Infections: Modern Concepts and Status,* New York, Marcel Dekker.
Discusses viruses in relation to current human diseases.

Osborn JE (ed.): 1977. *History, Science and Politics: Influenza in America 1919-1976,* New York, Prodist.
The pandemic of 1918-1920, the scientific decision-making process leading the 1976 national swine flu immunization campaign, and the costs and benefits of that campaign are reviewed.

Pyle GF and KD Patterson: 1987. The geography of influenza, *Focus* 37(Fall):16-23.
Discusses medical geography in relation to influenza.

Radetsky P: 1989. The ultimate parasite, *Discover* 10:20-21.
Discusses the impact of delta viruses in hepatitis.

Tiollais P and MA Buendia: 1991. Hepatitis B virus, *Scientific American* 264(4):116-123.
An article describing hepatitis and the virus that causes spread of this disease via contaminated body fluids.

Tsao J et al: 1991. The three-dimensional structure of canine parvovirus and its functional implications, *Science* 251:1456-1464.
The three-dimensional model of single-stranded DNA canine parvovirus helps our understanding of how viruses attack cells and use cells to duplicate themselves.

Vinken PJ, GW Bruyn, HL Klawans (eds.): 1989. *Viral Disease,* Amsterdam, Elsevier Science Publishing.
Reviews the human diseases that are caused by viruses.

Weisberger BA: 1984. Epidemic, *American Heritage,* 35(October/November):57-64.
Public health measures in the 1900's in Memphis, Tennessee, to combat yellow fever are examined.

Zuckerman AJ: 1987. Viral hepatitis, *World Health,* December:24-25.
Review of the worldwide programs to prevent hepatitis.

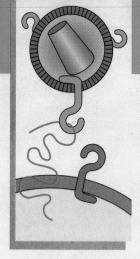

CHAPTER 19

Viral Diseases of the Central Nervous, Cardiovascular, and Lymphatic Systems, Skin, and Eye

Some viruses that enter the circulatory system are able to migrate via the blood and enter the central nervous system, causing diseases of the nervous system (Table 19-1, p. 562). Viruses that infect the central nervous system can enter the body via the respiratory, gastrointestinal, or genital tracts or, more commonly, may be injected directly into the circulatory system through the bite of an infected animal carrier. Mosquitoes and other biting arthropods can be vectors of viruses that invade the central nervous system. To enter the central nervous system, viruses must cross the blood-brain barrier or the blood-cerebrospinal fluid barrier (FIG. 19-1) The blood-brain barrier consists of tightly joined endothelial cells surrounded by glial cells (astrocytes). The blood-cerebrospinal fluid barrier consists of tightly joined choroid plexus epithelial cells. These barriers block the entry of most microorganisms into the central nervous system. Some viruses can cross the barriers within leukocytes by infecting cells that make up the barrier. Within the brain, the replication of viruses can cause an inflammation known as *encephalitis.* From the cerebrospinal fluid, viruses can invade the meninges and cause meningitis. Viruses that invade the central nervous system may cause lesions due to viral replication or may induce damage as a result of the immune response to the viral infection. In either case, cell damage and neurological dysfunction may result.

RABIES

Rabies is principally a disease of warm-blooded animals other than humans. The rabies virus, a bullet-shaped, single-stranded RNA virus, can be transmitted to people through the bite of an infected animal

(FIG. 19-2). Rabies occurs in two forms—"furious" and "dumb." Animals with the "furious" type of rabies snap and bite; animals with the "dumb" form of rabies exhibit paralysis. In urban settings, dogs and cats are most frequently involved in the transmission of rabies to humans. Many municipalities require that pets be immunized against rabies and require a certificate of rabies vaccination to obtain a pet license. Wild animals involved in the transmission of rabies include foxes, skunks, jackals, mongooses, squirrels, raccoons, coyotes, badgers, and bats.

Rabies viruses multiply within the salivary glands of infected animals. The viruses normally enter humans in an animal's saliva through a portal of entry established by an animal's bite. The rabies viruses are not able to penetrate the skin by themselves, and deposition of infected saliva on intact skin does not necessarily result in transmission of the disease. Transmission from bats can also occur via the respiratory tract because of the aerosols formed in the atmosphere around dense populations of infected bats. The rabies viruses can also cross intact mucous membranes, including those lining the eyes. One case of rabies transmission occurred as the result of a corneal transplant from an infected individual. Veterinarians and animal handlers need to be aware of possible transmission from animals they are treating, and vaccination is sometimes recommended for those likely to be exposed to the virus to avoid the risk of contracting rabies.

When rabies viruses enter the human body through an animal bite, they are normally deposited within muscle tissues. The viruses can multiply in skeletal muscles and connective tissue. It may remain localized there for days to months, generally in num-

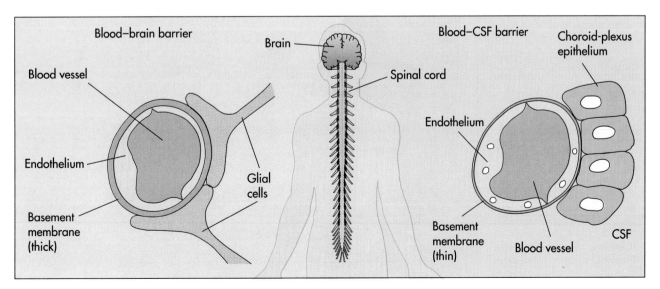

FIG. 19-1 The blood-brain barrier and blood-cerebrospinal fluid (CSF) barrier prevent most microorganisms from entering the central nervous system from the circulatory system.

TABLE 19-1

Some Viruses That Cause Human Diseases of the Central Nervous System

VIRAL GROUP	VIRUS	DISEASE
Arenaviruses	Lymphocytic choriomeningitis virus	Aseptic meningitis
Bunyaviruses	California encephalitis virus	Encephalitis
Herpesviruses	Herpes simplex virus	Encephalitis
	Cytomegalovirus	Aseptic meningitis, encephalitis
	Varicella-zoster virus	Aseptic meningitis
Paramyxoviruses	Mumps virus	Aseptic meningitis, encephalitis
	Measles virus	Subacute sclerosing panencephalitis
Picornaviruses	Polioviruses	Poliomyelitis, aseptic meningitis
	Coxsackieviruses	Aseptic meningitis
	Echoviruses	Aseptic meningitis
Retroviruses (Lentiviruses)	HIV	Subacute encephalitis, dementia
Rhabdoviruses	Rabies virus	Rabies
Togaviruses	Japanese encephalitis virus	Aseptic meningitis, encephalitis
	Eastern equine encephalitis virus	Aseptic meningitis, encephalitis
	Western equine encephalitis virus	Aseptic meningitis, encephalitis
	Louping ill virus	Aseptic meningitis, encephalitis
	Rubella virus	Aseptic meningitis, encephalitis
	Ilheus virus	Encephalitis
	St. Louis encephalitis virus	Encephalitis
	Venezuelan encephalitis virus	Encephalitis
	Murray Valley encephalitis virus	Encephalitis

bers too low to evoke an immune response. From there the viruses move into the peripheral nerve endings and migrate to the central nervous system. They replicate in nerve cells and move to the central nervous system via the axoplasm. Once within nerve

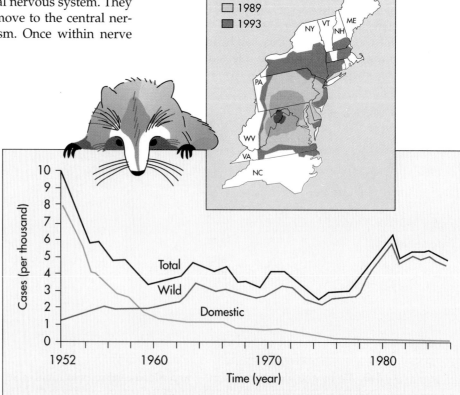

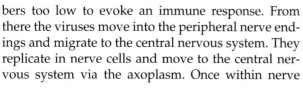

FIG. 19-2 Various animals transmit rabies. Domestic animals, which used to be major sources of disease, have been replaced by wild animals as the source of most cases of rabies. The map indicates the gradual spread of rabies among raccoons in the northeastern United States from 1979 to 1993.

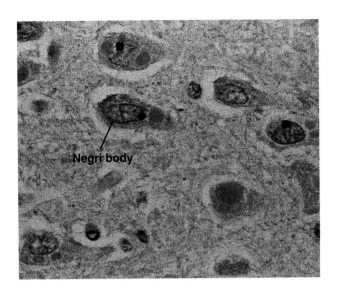

FIG. 19-3 Multiple cytoplasmic Negri bodies in pyramidal neurons of the hippocampus in rabies.

cells, the viruses are protected against the body's immune response. Viral multiplication in the brain causes encephalitis (inflammation of the brain). Multiplication of rabies viruses within the brain causes a number of abnormalities, manifested as the symptoms of this disease. Cytoplasmic inclusion bodies, known as **Negri bodies,** develop within the nerve cells of the brain (FIG. 19-3). The initial symptoms of rabies include anxiety, irritability, depression, and sensitivity to light and sound. These symptoms are followed by difficulty in swallowing and development of **hydrophobia** (fear of water). This fear of water really is a fear of swallowing because of the occurrence of a convulsive reflex that can result in choking. The sight of water can trigger spasms. As the infection progresses, there is paralysis, coma, and death because of extensive damage to the cells of the spinal column and brain.

> Rabies viruses are transmitted by infected animals. Replication of rabies viruses within the central nervous system causes neurological abnormalities.

Once the clinical symptoms of rabies begin, the disease is considered to be almost invariably fatal. Death often occurs within a few days of the manifestation of neurological signs of rabies. Treatment, therefore, requires vaccination and the use of passive immunity by immunoglobulin injection before the symptoms become manifest. The vaccination is used to prevent replication of viruses that have entered the body. Passive immunoglobulin injection is used to inactivate any viruses remaining at the site of the bite. A few individuals have survived symptomatic cases of rabies, but such cases are extraordinarily unusual. Fortunately, the symptomatic disease is very rare in the United States, with less than 10 deaths per year. The use of immune treatment to protect individuals

who may have been infected with the rabies virus keeps the rate low.

Anyone bitten by an animal that may be infected with rabies virus must be vaccinated to prevent death. If the animal can be located it is possible to use a fluorescent antibody test to detect viral antigens in the saliva, spinal fluid, or brain tissue of the animal. Unfortunately, because most cases are now associated with wild animals, rather than pet dogs and cats, locating and positively identifying any animal that bit a human is almost impossible. Therefore vaccination is employed in cases of all such animal bites, especially when the bite was unprovoked.

The rabies vaccine stimulates antibody synthesis. This prevents proliferation of the virus before it can cause irreversible damage to the central nervous system. If someone is bitten by an animal with rabies, antibodies are applied to the wound area to inactivate the virus and vaccination is initiated to establish active immunity against rabies. For many years the vaccine used was prepared from rabies virus propagated in duck eggs and inactivated by β-propiolactone. The treatment involved 21 daily injections followed by booster inoculations 10 and 20 days later. These injections were administered through the abdomen and were very painful. Recent improvements in the vaccine have reduced the number of injections required to establish immunity. The new rabies vaccine, prepared by growing the virus in human cell tissue culture, has a higher concentration of the necessary antigens for eliciting an immune response and generally only five injections over a 28-day period are required. These injections are administered intrauscularly into the thigh. Only if the animal is conclusively shown not to be infected with rabies can the

NEWSBREAK

VACCINATING PETS AND OTHER ANIMALS

Pet dogs and cats are vaccinated against rabies and several other diseases. Dogs are vaccinated against rabies and parvovirus and cats are vaccinated against feline leukemia virus. Municipalities generally require proof of vaccination against rabies before issuing a pet license. Through effective vaccination, rabies has almost been eliminated from domestic pets so that dogs and cats no longer are a significant vector in the transmission of rabies to humans. Vaccinating raccoons, ground squirrels, and other wild animals against rabies is far more challenging. A genetically engineered vaccine has been developed that can be added to a bait (food source) and dropped into forests and other areas. When the animals consume the bait, they are vaccinated.

immunization procedure be safely omitted for the bitten individual.

POLIOMYELITIS

Polioviruses may enter the body through the gastrointestinal or respiratory tracts. Polioviruses are able to multiply within the tissues of the intestines. The fecal-oral route via contaminated water supplies is probably the most common route of transmission. Initially polioviruses replicate within the gastrointestinal tract but subsequently move into the bloodstream and are disseminated to other tissues, including lymphatic tissues, where additional viral replication occurs. Polioviruses move through the blood to the brain where they continue to multiply within neural tissues and cause varying degrees of damage to the nervous system. They replicate within the

nerve cells of the gray matter of both the brain and spinal cord, destroying those nerve cells when viral replication causes cell lysis.

Poliovirus causes **poliomyelitis,** which is commonly referred to as **polio.** The initial symptoms of polio include headache, vomiting, constipation, and sore throat. In many cases, these early symptoms are followed by obvious neural involvement, including paralysis due to the destruction of nerve cells in the spinal column. The pathogenesis of poliomyelitis can result in paralysis, meningitis, and encephalitis (FIG. 19-4). Although the paralysis can affect any motor function, in over half of the cases of paralytic poliomyelitis, the arms and/or legs are involved. Fortunately, the paralytic symptoms occur in only 1% to 2% of polio cases—98% to 99% of polio cases are nonparalytic. In many cases, polio virus infections fail to show any evidence of clinical symptoms at all.

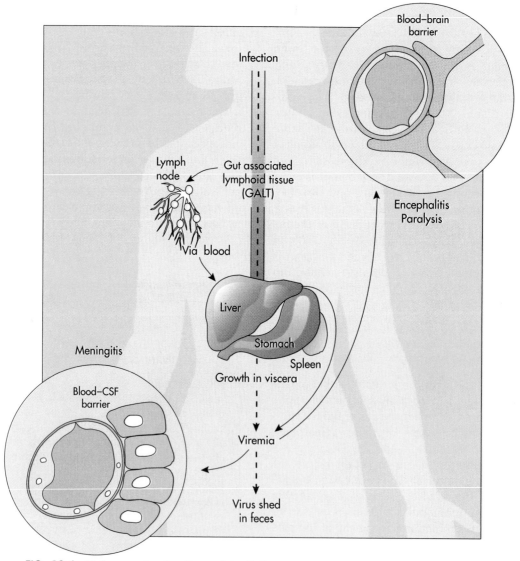

FIG. 19-4 Pathogenesis of poliomyelitis. Polioviruses enter the body through the gastrointestinal tract and spread elsewhere via blood and lymph. The virus can spread into the central nervous system (CNS) either via the blood-brain or blood-CSF barriers. Paralysis results from multiplication of polioviruses within the CNS.

Polioviruses cross the blood-brain barrier and replicate within the central nervous system where they cause lesions that can result in paralysis.

During polio epidemics the incidence of infection is particularly high among children. Because of this fact the disease is also called **infantile paralysis** (FIG. 19-5). Polio also strikes adults, and in fact, the fatality rate in adults is much higher than in children. The use of the **Salk** and **Sabin polio vaccines** has dramatically reduced the incidence of this disease. (The use and modes of action of these vaccines are discussed in Chapter 15.) Before the development of these vaccines there were about 21,000 paralytic cases of polio annually in the United States. The number of cases in the United States is now near zero (FIG. 19-6). It remains important, however, that preschool children continue to be immunized. Although the World Health Organization projects that polio could be gone by the year 2000, the virus is not yet eliminated.

Polio can be prevented by vaccination.

Major outbreaks of poliomyelitis have traditionally been associated with transmission among children in close contact in a schoolroom. Many school systems now require evidence of polio vaccination

FIG. 19-5 Polio may cause paralysis. Children are stricken with polio more often than adults, leading to the name infantile paralysis.

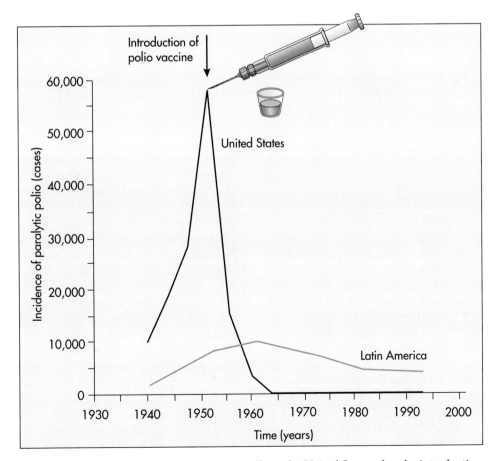

FIG. 19-6 Paralytic polio declined dramatically in the United States after the introduction of the Salk inactivated and the Sabin oral vaccines. The incidence of this disease remains high in Latin America where vaccination against polio has not been as widely practiced.

HIGHLIGHT

RETURN OF POLIO AS A NEW DISEASE SYNDROME

After several decades of mass inoculations of infants and children in a classic public health triumph that seemed to have made the images of crutches and iron lungs vestiges of a long forgotten era, polio is back. It is not appearing as the crippling and wasting scourge of children, but as new symptoms in adults who once had the disease and now—though perhaps handicapped—thought of themselves as recovered.

There were about 300,000 living survivors of polio in 1985. An estimated 40,000, approximately 15%, of these people may be developing post-poliomyelitis muscular atrophy (PPMA). PPMA seems to occur some 30 years after the original illness. The symptoms include extreme fatigue, severe muscle pain, and a muscular weakness that may be slowly progressive over many years. Some sufferers feel they are losing muscle mass. The muscle fatigue and weakness brought on by PPMA can make simple tasks, such as climbing stairs or opening jars, difficult. Patients face loss of muscle strength because PPMA seems to affect the same muscles that were damaged years ago.

It is impossible to predict which polio survivors will be afflicted. It is also impossible to predict the course of the symptoms once they begin. The severity of the fa-

tigue, muscle weakness, and pain may level off and remain stable for several years or the symptoms may worsen every 2 or 3 years. The cause is not known, and neither is prevention, treatment, or cure.

There are several theories about what may be producing these symptoms. One is based on the fact that as all people grow older they lose nerve cells in the spinal cord. The loss is usually compensated for by other cells, but polio victims have already lost some of the cells that control muscle movement. Another theory is that PPMA may be caused by a reactivated dormant poliovirus. This would be similar to the way in which a dormant chicken pox virus can cause shingles many years after the initial infection. Other theorists believe it may be related to a deficiency in the patient's immune system. Whatever its cause, PPMA is not life-threatening.

Medication such as aspirin to control joint and muscle pain is working for many patients. Doctors also recommend good nutrition, good health practices, and no overexertion. Post-polio patients are learning to accept this added burden to their already burdened bodies and are adjusting to these changed conditions in their lives just as they made bigger and more drastic adjustments when polio first entered their lives so many years ago.

---

before a child may be enrolled. As many as 50% of urban poor children may currently remain unprotected against polio. The lower rate of occurrence of poliomyelitis means that there are fewer paralyzed individuals to serve as visible reminders of the seriousness of this disease. Constant efforts to reinforce parental awareness of the importance and success of vaccination against poliomyelitis are worthwhile.

Currently in the United States the Sabin oral vaccine is used. This vaccine mimics the natural oral infection route. Occasionally and inexplicably, however, this attenuated living vaccine strain undergoes a genetic transformation after it is in the body. Reverting to its original virulent form, the vaccine strain poliovirus causes the disease it was designed to prevent. About 10 children per year develop polio as a result of genetic transformation of the vaccine strain of poliovirus. Hence some groups have been urging a return to using the Salk vaccine, which was widely used in the early 1950s. The Salk vaccine uses an inactivated strain that cannot replicate. There is resistance to using the Salk vaccine, however, because it must be injected, and children dislike such painful medical procedures. Also, there were some major outbreaks of polio that resulted from improperly inactivated vaccine batches when the Salk vaccine was first introduced. Thus the choice of

which polio vaccine to use remains a dilemma, but for the moment the Sabin vaccine remains the one of choice.

ASEPTIC MENINGITIS

Invasion of the central nervous system by various viruses can cause an inflammation of the **meninges** (membranes surrounding the brain and spinal column). Such a condition is known as **aseptic meningitis.** The term *aseptic meningitis* indicates that no microorganisms are observed in smears and no bacteria or fungi are recovered in cultures. Examination of spinal fluid using serological tests can confirm the viral origin of this disease. These tests reveal the presence of antibodies to particular viruses known to cause aseptic meningitis. Various viruses can cause meningitis (Table 19-2). Often echoviruses and coxsackieviruses are the causative agents. In addition, viral meningitis can also be caused by other viruses such as poliovirus, measles virus, mumps virus, Epstein-Barr virus, varicella-zoster virus, and arthropod-borne encephalitis viruses. There are about 10,000 cases of aseptic meningitis in the United States each year. Frequently, outbreaks of aseptic meningitis affect several members of a community or family, which helps epidemiologists identify the source of the viruses causing this disease.

TABLE 19-2		
Viruses Causing Meningitis		
VIRAL GROUP	**VIRUS**	**COMMENTS**
Arenaviruses	Lymphocytic choriomeningitis virus	Uncommon infection from urine of mice, hamsters, and other rodents carrying the virus
Herpesviruses	Herpes simplex virus	Uncommon
Paramyxoviruses	Mumps virus	A quite common complication
Picornaviruses	Poliovirus Coxsackievirus Echovirus	Sometimes a complication Most common in immunocompromised individuals A rare complication to a gastrointestinal infection
Togavirus	Eastern equine encephalitis virus Western equine encephalitis virus Louping ill virus Japanese encephalitis virus	Occurs in eastern U.S. Occurs in western U.S. Occurs in Scotland Occurs in Japan, India, and Southeast Asia

Aseptic meningitis is an inflammation of the meninges caused by viral infection; the name is derived from the fact that no microorganisms are seen during diagnostic observation of cerebrospinal fluid.

Echoviruses and coxsackieviruses are enteroviruses that are transmitted via the fecal-oral route. Viral particles that are ingested pass through the stomach intact and then can replicate in the intestinal tract. From the intestinal tract, these viruses enter the circulatory system. They can cross the blood-brain and blood-cerebrospinal fluid barriers and infect the meninges. The symptoms of aseptic meningitis include fever, stiffness of the neck, fatigue, and irritability. Treatment is supportive and involves maintenance of essential body functions and reliance on natural body defenses for recovery. The infection tends to be of short duration, with most patients recovering without complication in 3 to 14 days.

TABLE 19-3				
Encephalitis in Humans				
DISEASE	**VECTOR**	**RESERVOIR**	**GEOGRAPHIC DISTRIBUTION**	**COMMENT**
Eastern equine encephalitis	*Aedes* mosquito	Birds	Eastern United States, Canada, Brazil, Cuba, Panama, Dominican Republic, Trinidad, Philippines	Effective vaccine for equines, 50% fatality
Venezuelan equine encephalitis	Mosquito	Rodents, horses	Brazil, Colombia, Ecuador, Trinidad, Venezuela, Mexico, Florida, Texas	Rare, less than 10% in adults
California encephalitis	*Aedes* mosquito	Small mammals	California, Midwest	Rarely fatal, endemic to a few locations
Western equine encephalitis	*Culex* mosquito	Birds	Western United States, Argentina, Canada, Mexico, Guyana, Brazil	Effective vaccine for equines, 2% fatality
St. Louis encephalitis	*Culex* mosquito	Birds	United States, Trinidad, Panama	Large urban outbreaks, 10% fatality
Japanese B encephalitis	*Culex* mosquito	Birds, pigs	Japan, Guam, Eastern Asian mainland, India, Malaya	Predominantly rural, 8% fatality
Murray Valley encephalitis	*Culex* mosquito	Birds	Australia, New Guinea	Up to 70% fatality
Ilheus	Mosquito	Birds	Brazil, Guatemala, Honduras, Trinidad	—
Tick-borne group (Russian spring-summer encephalitis group)	Tick	Birds, mammals	Central and Eastern Europe, Russia, Canada, Malaya, United States, Finland, Japan, India, Great Britain	Up to 10% fatality

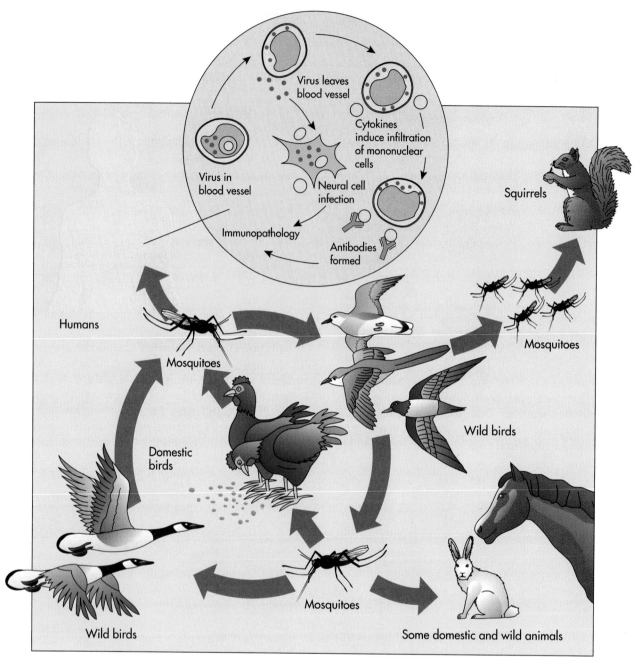

FIG. 19-7 Transmission routes for various types of encephalitis. Often birds, especially chickens, are involved in the transmission cycle. Mosquitos transmit the viruses to humans. The viruses reach neural cells of the central nervous system and initiate immune reactions that result in encephalitis (see inset). Death can occur from encephalitis.

VIRAL ENCEPHALITIS

Encephalitis is a disease defined by an inflammation of the brain. It can be caused by various viruses (Table 19-3). Postinfection encephalitis sometimes occurs after measles, rubella, influenza, and other viral infections. Many of the viruses capable of causing encephalitis in humans are maintained in populations of various vertebrates, particularly birds and rodents, as well as populations of arthropods (FIG. 19-7). Transmission to human beings, via an arthropod vector in which the virus has multiplied, generally repre-

sents a dead end in the transmission cycle. Viral encephalitis is not transmitted from one human to another. Viruses typically accumulate in the saliva of the arthropod. This facilitates transfer to a person bitten by the arthropod. The specific viral etiologic agent, arthropod vector, and geographic distribution are different for each of these forms of encephalitis (see Table 19-3). Outbreaks of viral encephalitis exhibit seasonal cycles, with increased numbers of cases occurring during summer when the arthropod vector populations are at their peak.

There are several types of viral encephalitis caused by different viruses. Each type has a different mode of transmission, generally involving a different arthropod vector.

Infections with encephalitis-causing viruses begin with **viremia** (viral infection of the bloodstream). Localization of the viral infection follows within the central nervous system, where lesions develop. The locations of the lesions within the brain are characteristic for each type. With the exception of St. Louis encephalitis, in which kidney damage also occurs, the pathologic changes in cases of encephalitis are normally restricted to the central nervous system.

Encephalitis symptoms begin with fever, headache, and vomiting. Stiffness and then paralysis, convulsions, psychoses, and coma follow. Different forms of viral encephalitis have different outcomes. For example, in symptomatic cases of eastern equine encephalitis the mortality rate is near 50%; in western equine encephalitis, the fatality rate is under 5%.

No antiviral drug has been developed for the treatment of encephalitis caused by arthropod-borne viruses. However, vaccines have been developed that are effective in establishing immunity against the specific viruses that cause encephalitis. As a rule, control of arthropod-borne viral encephalitis depends on control of vector populations. Insecticides have been widely used in public health programs to reduce the population levels of the mosquitoes and ticks that are the vectors of viral encephalitis. Mosquito eradication programs are often intensified when positive diagnoses of encephalitis raise the possibility of widespread outbreaks.

Not all forms of encephalitis are arthropod-borne. For example, herpes simplex virus, which is not normally transmitted through vectors, can cause meningitis and encephalitis (herpes encephalitis). Acyclovir has been demonstrated to be effective in treating herpes simplex virus encephalitis and approved for use in treating this disease, but this drug has not been proven effective against other encephalitis-causing viruses.

REYE SYNDROME

Reye syndrome is an acute pathological condition affecting the central nervous system. It is a serious complication associated with outbreaks of influenza and chickenpox. This disease syndrome also occurs after infections with other viruses, and the specific relationship to influenza and chickenpox viruses is not clear. Reye syndrome principally is associated with children. It was thought largely restricted to those under 16 years of age but is now known to affect those up to 19 years old. Fever and severe vomiting often occur at the onset of Reye syndrome, followed by the accumulation of fluids in the brain (cerebral edema), coma, and death due to intracranial pressure.

There is a high incidence of Reye syndrome when aspirin is used for treating the symptoms of a viral infection. The reasons for this are not yet clear. Consequently, pediatricians and aspirin manufacturers have warned against the use of aspirin for children with influenza and other viral infections of the respiratory tract. Studies indicate that children with chickenpox or flu who are given aspirin are 25 times more likely to develop Reye syndrome than those who are not treated with aspirin.

To lower the risk of Reye syndrome, children with viral infections should not be given aspirin.

GUILLAIN-BARRÉ SYNDROME

Another condition involving the central nervous system associated with influenza infections, Epstein-Barr virus infections, and herpes zoster infections is **Guillain-Barré syndrome.** This disease normally is rare, but in 1976 there was an increased incidence of the syndrome associated with an active immunization program against a predicted outbreak of swine flu. The predicted epidemic did not occur. Swine flu is caused by an influenza virus that normally occurs in pigs but can be transmitted to humans, with fatal consequences. The outbreak of Guillain-Barré syndrome appears to have resulted from contamination of some batches of vaccine with influenza viruses that had not been adequately inactivated. This disease involves a rapidly developing inflammation of several nerves that results in spreading muscular weakness of the extremities and possible paralysis. Treatment is supportive.

CREUTZFELDT-JAKOB DISEASE AND RELATED DISEASES

Creutzfeldt-Jakob disease is a rare disorder that results in the loss of intellectual capacity **(dementia).** The worldwide incidence of Creutzfeldt-Jakob disease is about one in a million. Interest in this disease centers on the finding that it is caused by a **prion.** Prions are infectious proteins that do not appear to contain any nucleic acids, which distinguishes prions from all other infectious agents. The replication of these proteins within the body, however, can cause degeneration within the central nervous system. It is not yet understood how prions replicate, but it is clear that when prions infect mammalian cells, new prions are produced and abnormalities develop in the mammalian system. The natural mechanism for the transmission of Creutzfeldt-Jakob disease is not known. Artificial transmission to humans has occurred via transplantation of infected tissues such as corneal grafts and human growth hormone extracted from infected pituitary tissues. The agent that causes Creutzfeldt-Jakob disease has not been isolated from body surfaces, secretions, or excretions. It is transmissible by exposure to blood, cerebrospinal fluid, and brain, and other tissues. Care is required in handling these fluids or tissues from infected patients.

Diseases caused by prions develop slowly and were originally thought to be caused by viruses; therefore they have been termed *slow viral diseases.* These slow infections are characterized by prolonged incubation periods of months, years, or possibly decades, during which there are no symptoms. Once the illness begins, however, it progresses steadily and generally leads to death.

Creutzfeldt-Jakob disease is one of several apparently related diseases that are characterized by similar clinical and pathological symptoms. **Scrapie,** a disease of sheep and goats, has similar symptoms and is known to be caused by prions. Bovine spongioform encephalitis, or "mad cow" disease is another related disease. Prions have been hypothesized to cause Alzheimer's disease, which is the commonest form of human senile dementia and the fourth leading cause of death in the United States, but this has not yet been proven. Additional studies are needed to positively identify the cause of Alzheimer's disease and to find ways of curing or preventing degenerative diseases of the nervous system.

Other apparently related human diseases are **Gerstmann-Sträussler syndrome** and **kuru.** The occurrence of kuru is very restricted because this disease is transmitted through the ingestion of human brain tissue during ritual cannibalism in New Guinea. Such tribal customs are no longer practiced and the incidence of kuru has declined.

Prions are infectious proteins that have been implicated as the causes of various degenerative nervous disorders.

The initial symptoms of scrapie, kuru, and Gerstmann-Sträussler syndrome are difficulty in walking and loss of coordination. These symptoms indicate involvement of the cerebellum. A late symptom of kuru is dementia. Creutzfeldt-Jakob disease sometimes shows early symptoms that resemble kuru but usually begins with loss of memory. Pathological changes in these diseases are restricted to the central nervous system.

VIRAL DISEASES OF THE CARDIOVASCULAR AND LYMPHATIC SYSTEMS

Viruses that enter the circulatory system often are eliminated by the various immune defense systems in the blood or are simply transported to other tissues of the body. In some cases, however, viruses infect and replicate within the cells of the circulatory system—causing disease (Table 19-4). Diseases of the cardiovascular and lymphatic systems include those, such as AIDS, that affect the immune system. Because of the central role of the immune system in protecting the body against infections, such suppression of the immune system is life-threatening. Likewise, viral infections that impair heart function, such as myocarditis, often cause death. Treatment of these diseases is supportive.

TABLE 19-4

Some Viruses That Cause Human Diseases of the Cardiovascular and Lymphatic Systems

VIRAL GROUP	VIRUS	DISEASE
Arenaviruses	Lassa fever virus Bolivian hemorrhagic fever virus	Lassa fever Bolivian hemorrhagic fevers
Bunyavirus	Rift Valley virus Congo-Crimean virus Hantavirus	Rift Valley fever Congo-Crimean hemorrhagic fever Korean hemorrhagic fever
Flavivirus	Kyasanur Forest virus	Kyasanur Forest hemorrhagic fever
Herpesviruses	Epstein-Barr virus Cytomegalovirus	Infectious mononucleosis Mononucleosis-like syndrome
Parvoviruses	Human parvovirus	Erythema infectiosum
Reoviruses	Colorado tick fever virus	Colorado tick fever; hemorrhage
Retroviruses	HTLV-I	T cell leukemia, lymphoma
Oncornaviruses	HTLV-2	Hairy cell leukemia
Lentiviruses	HIV	AIDS
Togaviruses	Dengue virus Yellow fever virus	Dengue fever Hemorrhagic fever

ACQUIRED IMMUNODEFICIENCY SYNDROME (AIDS)

In 1979, a new disease—**acquired immunodeficiency syndrome** or **AIDS**—started to appear. Within a decade this disease would reach epidemic proportions, dominating health news, taxing health care delivery systems, forcing a redirection of medical research, and most importantly, adversely affecting the lives of millions. The AIDS epidemic is the "plague" that threatens modern society. Since its discovery, a great deal has been learned about AIDS. However, despite intensive scientific investigation, for the near future neither a cure nor a vaccine for preventing this disease is likely. The World Health Organization estimates that the number of cases of HIV infection will rise from 15 million in 1995 to 40 million by the year 2000.

Characteristics of HIV

AIDS is caused by a retrovirus named **HIV (human immunodeficiency virus)** (FIG. 19-8). A retrovirus is an RNA virus that uses viral reverse transcriptase to make DNA and then uses that DNA for making new viruses. The virus is composed of a central core that contains two identical single strands of RNA, each associated with one reverse transcriptase and several core proteins. The core is surrounded by a matrix protein that in turn is surrounded by a lipid envelope. The envelope contains two important surface glycoproteins: gp41 and gp120. The gp120 glycoprotein functions in the attachment of the virus to host cells. The gp41 glycoprotein is a fusion protein that is involved in the joining of infected host cells with adjacent noninfected cells; gp41 also helps anchor the gp120 glycoprotein to the envelope.

Since its discovery, much has been learned about the molecular biology of HIV that may help lead to a method for controlling AIDS. The RNA genome of HIV contains three genes that are found in all retroviruses. The *gag* gene (group-specific *a*ntigen) codes for the core proteins, the *pol* gene codes for the reverse transcriptase, and the *env* gene codes for the two major envelope glycoproteins. There are several other genes in the genome that have regulatory functions. Two of these genes are *tat*, which increases viral gene expression, and *rev*, which codes for a protein that transports the viral mRNA molecules out of the nucleus.

The ability of HIV to infect cells depends on adsorption to specific receptors. HIV preferentially binds to the CD4 receptor and infects cells bearing that receptor, including T helper (T_H) lymphocytes, monocytes, and dendritic cells. The CD4 receptor is used by T_H cells to recognize antigens in association

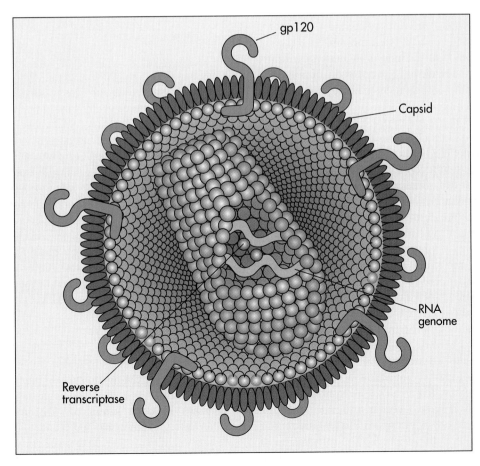

FIG. 19-8 The structure of HIV.

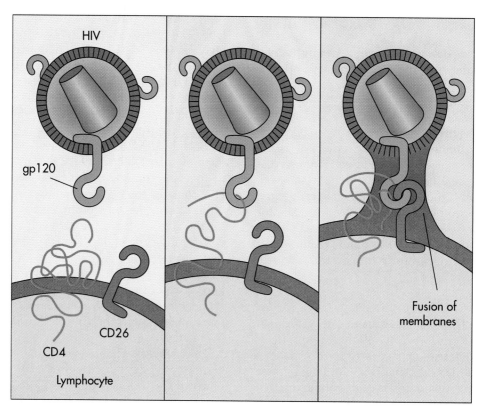

FIG. 19-9 The entry of the human immunodeficiency virus into T helper cells involves HIV gp120 binding to CD4 receptor of the T cell with CD26 assistance for entry.

with the major histocompatibility complex. In 1993, it was learned that an additional antigen on T_H cells, CD26, acts as a coreceptor. The CD4 receptor initially binds the gp120 envelope glycoprotein of the HIV, and then the CD26 antigen assists the entry of the virus into the cell (FIG. 19-9).

After entry of the virus into the T_H cell, the T_H cell must become activated for replication of the virus and destruction of the host cell to occur. HIV-infected T lymphocytes may become activated in response to HIV antigens or to other microbial antigens. Viral genomes are replicated with a rather high rate of mutation. This leads to an increase in viral diversity not only in the population but also within a given individual. This form of genetic variation in HIV may have a part in the virus evading the host defense mechanism. It may also account for the emergence of more virulent strains of HIV in the population.

After new viral proteins and RNA genomes are synthesized and assembled in the cytoplasm, virions exit the cell by budding and go on to infect other cells. The *tat* gene product is necessary for host cell death, but it is unclear how HIV actually kills T_H cells. In some cases, the viral DNA synthesized by the reverse transcriptase becomes integrated into the host DNA and the virus goes into a latent stage and does not replicate within the cell. The integration of viral DNA into host DNA is accomplished by an integrase.

AIDS is caused by the human immunodeficiency virus (HIV). The virus predominantly infects and kills T helper cells and other cells with CD4 receptors.

Monocytes, macrophages, and dendritic cells are similarly infected but are generally not destroyed by the HIV. However, these cells carry the gp41 fusion protein that allows infected cells to fuse with other infected or noninfected cells. As a result, the virus can spread throughout cells in the body. In addition, multinucleated cells are formed, especially in the brain. Mononuclear and dendritic cells also act as a reservoir for HIV, making it almost impossible to completely rid the body of this virus. Because some macrophages move through the body, macrophages infected with HIV carry the AIDS-causing virus to various parts of the body, including the brain. Some brain cells are killed when HIV is carried into the central nervous system.

Body Response to HIV Infection

HIV replication leads to a decline in T_H cells. Since the T_H cell is often at the center of activating T and B lymphocyte responses (see Chapter 13), functional changes in both kinds of lymphocytes are seen. Macrophages are not killed by HIV, but their normal functions, which include destruction of invading bacteria and assisting lymphocytes in protecting the body against infection, are disrupted. Antigen presentation by macrophages to T and B lymphocytes is impaired. The overall effect of HIV infection is a depressed immune response.

HIV infection leads to lowered numbers of T_H cells and a depressed immune response.

Initial infection with HIV is sometimes accompanied by mild flu-like symptoms such as fever, sore throat, and fatigue. Thereafter, HIV-infected individuals may be symptom free for a long period (5 to 7 years on the average), although T_H cell counts progressively declines. Ultimately, individuals with AIDS exhibit immunosuppression. As a result, they are subject to infection by numerous opportunistic pathogens, as well as true pathogens, and to development of several forms of cancer (Table 19-5). Generally one infection follows another in people with this disease until death occurs. In 1993, the case definition for AIDS included individuals who: (1) tested HIV positive (ELISA assay for HIV p25 viral core antigen), (2) had less than 200 CD4 T lymphocytes/mm^3 of blood—instead of 800-1300 CD4 T lymphocytes/mm^3 in healthy individuals, or (3) greater than 200 CD4 T lymphocytes/mm^3 of blood and opportunistic infection or cancer.

About 21% of homosexual or bisexual males with AIDS develop an unusual cancer of blood vessel walls called **Kaposi sarcoma.** This type of cancer is visible as purplish patches on the skin (FIG. 19-10) but it occurs internally as well. Another form of cancer that occurs particularly in African children with AIDS is Burkitt lymphoma. This form of cancer may

TABLE 19-5	
Principal Opportunistic Pathogens in AIDS Patients	
ORGANISM	**TREATMENT**
BACTERIA	
Legionella pneumophila	Erythromycin
Listeria monocytogenes	Ampicillin, penicillin G, gentamicin
Nocardia asteroides	Sulfonamide, trimethoprim-sulfamethoxazole
Mycobacterium tuberculosis	Isoniazid + rifampin + pyrazinamide + ethambutol
Mycobacterium avium-intracellulare	Clarithromycin + ethambutol + rifampin
Salmonella spp.	Ampicillin, chloramphenicol
FUNGI	
Candida spp.	Amphotericin B, fluconazole
Coccidioides immitis	Amphotericin B, itraconazole
Cryptococcus neoformans	Amphotericin B + flucytocine; fluconazole
Histoplasma capsulatum	Amphotericin B, itraconazole
Pneumocystis carinii	Pentamidine, trimethoprim-sulfamethoxazole
PROTOZOA	
Toxoplasma gondii	Triple sulfonamides + pyrimethamine
VIRUSES	
Herpes simplex	Acyclovir, foscarnet
Cytomegalovirus	Ganciclovir, foscarnet
Varicella-zoster	Acyclovir, foscarnet
Measles	Supportive
JC virus	Supportive
Adenovirus	Supportive

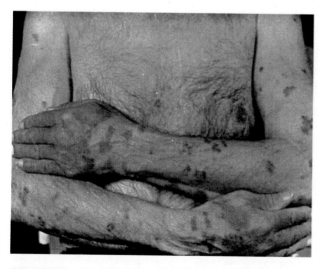

FIG. 19-10 Kaposi sarcoma skin lesion. Patients with acquired immune deficiency syndrome (AIDS) may present to a physician a wide range of opportunistic infections and a few rare malignancies. The skin lesions in Kaposi sarcoma, seen in this photograph, usually appear first on the legs as discrete, bluish-red macules or papules, which later become pigmented.

be related to reactivation of latent Epstein-Barr virus (discussed later in this chapter).

Macrophages that are infected with HIV can cross the blood-brain barrier and spread the infection to the central nervous system. In particular, microglial cells of the brain or other infiltrating macrophages in the brain become infected with HIV. Multinuclear giant cells form due to cell fusion, as do nodules of inflammatory cells. As a result, HIV infection in the brain leads to subacute encephalitis, dementia, or other neurological problems.

Transmission of HIV

The AIDS-causing human immunodeficiency virus is transmissible primarily by sexual contact, both homosexual and heterosexual (Table 19-6). Contaminated body fluids, including semen and blood, carry the virus. In the United States the homosexual male population is the most widely infected group, but in other countries heterosexual transmission appears to be a significant route of spread and both genders are equally infected. In addition to sexual transmission, the virus is spread through contaminated blood and blood products and can cross the placenta from mother to fetus (FIG. 19-11). Steps have been instituted to screen the blood supply used for transfu-

	TABLE 19-6			
	AIDS Patients in the United States			
	TOTAL NUMBER OF CASES BY YEAR AND PERCENT OF TOTAL CASES FOR EACH YEAR			
PATIENT GROUP	**1981-1983**	**1986**	**1990**	**1993**
Homosexual/bisexual	562 (69)	8,322 (67)	24,053 (56)	36,000 (52)
IV drug user	98 (12)	1,674 (13)	10,161 (24)	17,000 (25)
Both homosexual and IV drug user	74 (9)	925 (8)	2,445 (5)	2,800 (4)
Hemophilia-coagulation disorder	7 (1)	119 (1)	369 (1)	500 (1)
Heterosexual contacts	57 (7)	470 (4)	2,799 (7)	7,000 (10)
Transfusion recipients	3 (<1)	275 (2)	884 (2)	1,500 (2)
Undetermined	17 (2)	510 (5)	1,948 (5)	4,000 (6)
TOTAL	818	12,295	42,659	68,800

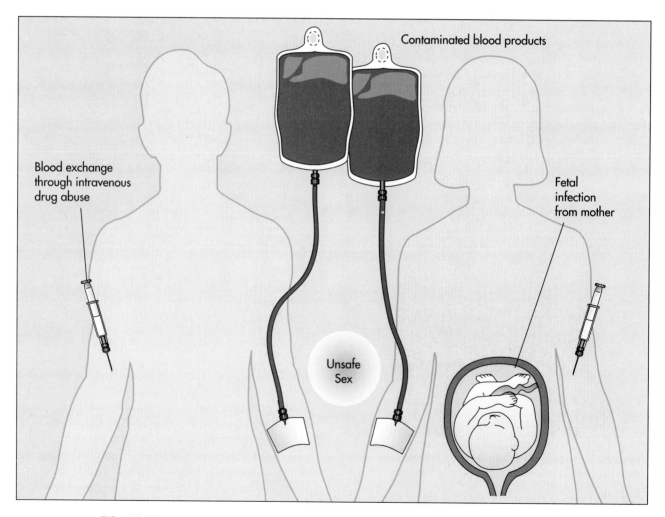

FIG. 19-11 Routes of AIDS transmission. The disease is spread (1) congenitally, (2) through sexual contact, and (3) through contaminated blood.

HIGHLIGHT

PRECAUTIONS FOR HEALTH CARE WORKERS TO PREVENT HIV TRANSMISSION

Given that HIV can be transmitted through blood, health care workers must take precautions to protect themselves against contracting AIDS. Physicians, nurses, medical technologists, dentists, dental hygienists, emergency medical technicians, and other health care workers have become extremely cautious about possible contamination with HIV-containing blood. They use extreme care when working around "sharps," such as needles and scalpels. Gloves are now routinely worn for even the most noninvasive procedures. Some surgical gloves have been reinforced to minimize the possibility of being cut by a sharp medical instrument. Not only health care workers but many others, from law enforcement officers to boxing referees, now wear gloves as a precaution. Athletes who are injured and bleeding must be removed from the game, their bleeding stopped, and the contaminated clothing disinfected before being allowed to reenter the competition.

The Centers for Disease Control (CDC), which develops isolation precautions annually, issued recommendations for prevention of HIV transmission in health-care settings in 1987. These recommendations emphasized the need to treat all blood and body fluids from all patients as potentially infective. In 1989, guidelines were issued to cover the activities of all public safety workers, including firefighters, emergency medical technicians, paramedics, and law enforcement officers. Universal blood and body fluid precautions as recommended by the CDC should be followed consistently in the care of all patients because the infective status of a patient usually can be neither ascertained nor relied on. These universal precautions protect against the transmission of HIV and other blood-borne pathogens such as hepatitis B virus. They include using barrier protections (gloves, masks, eyewear, and gowns); washing hands and skin after contamination or removing gloves, and before seeing another patient; preventing needle-sticks and scalpel cuts; minimizing mouth-to-

mouth resuscitation and using protective devices; refraining from direct patient care when lesions are weeping; being extra cautious when pregnant; and receiving hepatitis B vaccine.

Besides concern over the possible transmission of HIV to health care workers, individuals seeking medical care are concerned about contracting an HIV infection from the physician, dentist, or other health care workers. A well-publicized case of several individuals being infected by a Florida dentist with AIDS has caused alarm. Although the manner of disease transmission in these cases remains a mystery, the most likely explanation is that a contaminated dental drill was the source of infection. The public demands reassurance that they are not put at risk when seeking health care. Clinics have notified patients when a doctor or dentist who had treated them died of AIDS and offer free HIV testing. So far the transmission of HIV from health care worker to patients during medical procedures appears to be limited to the one Florida dentist.

Nevertheless, there have been demands for mandatory testing of health care workers, but this has not been done. Physicians who are HIV positive are asked to voluntarily notify their patients and to refrain from procedures that would put the patient at risk of infection. The CDC has made recommendations to minimize the risk of the transmission of an infectious agent from a health care worker to a patient. These guidelines include compliance with universal precautions; guidelines for disinfection and sterilization; identification of exposure-prone medical and dental invasive procedures; personal knowledge of HIV status by health care workers who perform invasive procedures, that is, voluntary testing; review by an expert panel of infected health care workers before performing exposure-prone, invasive procedures; and informing patients of their health care workers' HIV status.

---

sions in the United States to ensure its safety. This has not been done so thoroughly in other countries. Intravenous drug abusers also are at high risk of contracting this disease through contaminated syringes. Intravenous drug abuse is increasingly the cause of HIV transmission. Although the virus is found in saliva of infected individuals, there is no evidence for transmission by even prolonged casual contact; only direct sexual contact, exchange of blood (including via contaminated syringe needles), and transplacental movement of HIV from an infected mother to the fetus (cause of congenital AIDS) are known to cause HIV transmission. Blood-sucking insects, such as mosquitoes, have not been found to be vectors in the transmission of HIV.

HIV is transmitted through sexual contact and contaminated blood.

Diagnosis of AIDS

Frequently used assays for HIV are ELISAs that detect the presence in serum of the viral core antigen, p24, or antibodies to the HIV. Individuals that have reactive serum after repeated tests are considered to be HIV positive. This indicates that the individual may be infected with HIV. Further confirmation of diagnosis may be made using an indirect immunofluorescence assay (IFA) to detect virus or Western immunoblot assay to detect HIV-specific antibodies in tissues and fluids. The most sensitive method of detecting HIV is a polymerase chain reaction (PCR) assay.

Treatment of AIDS

AIDS is incurable at this time but treatment with azidothymidine (AZT) and other drugs limits replication of the HIV virus. AZT is a helpful treatment; it blocks the reverse transcription needed by this retrovirus for replication. AZT is useful in limiting replication of the virus and alleviates some of the effects of infection with HIV. Additional drugs, such as ddI and ddC, have also shown promise in slowing the progression of infection with HIV that leads to AIDS. Patients with this disease can be treated with passive immunization and antibiotics to protect them against life-threatening secondary infections.

Impact of AIDS

Clearly AIDS is a serious and frightening disease. It has produced some excessive public reactions. In some hospitals, staff members have refused to treat patients with AIDS. Establishments in San Francisco catering to the homosexual community have been closed on the grounds that they constitute a public health hazard. These reactions are reminiscent of those in our dark past when people with Hansen disease were labeled lepers and driven from society. Until we fully understand how to treat and prevent the spread of AIDS, unfounded public reactions are likely to continue.

By 1994, there had been at least 84,000 cases of AIDS in the United States (FIG. 19-12). According to the World Health Organization (WHO) there were many more cases elsewhere in the world. These are clearly underestimates of the true incidence of AIDS. The actual number of cases of AIDS probably represents 10% of the individuals who are infected with

METHODOLOGY

APPLICATION OF KOCH'S POSTULATES TO AIDS

When the deadly disease, acquired immunodeficiency syndrome (AIDS), was first recognized in 1981, scientists reasoned that there must be an underlying cause for the frequent occurrence of opportunistic infections and that the afflicted individuals must have acquired a disease that impaired the normal functioning of the body's immune defense system that protects against these diseases. They began to apply Koch's Postulates to establish the specific microorganism responsible for the disease.

They found that there was a deficiency in the immune systems of these individuals that left them subject to numerous other infections with microorganisms normally found in healthy individuals. The fact that these microorganisms, called opportunistic pathogens, which are normally associated with healthy animals, can cause disease when the immune defense system is weakened is at variance with Koch's first postulate, which states that disease-causing microorganisms should be absent from healthy animals and present only in diseased animals. The widespread occurrence of unusual and opportunistic infections in patients with AIDS, which eventually leads to their death, pointed to the underlying deficiency in the immune response but did not explain why that immune deficiency occurred.

Some scientists hypothesized that it was the result of a microbial infection and began to hunt for a microorganism that occurred only in individuals with AIDS and not in healthy individuals. Almost simultaneously in 1984, Robert Gallo and his colleagues at the National Institutes of Health in Washington, D. C. and Luc Montagnier and his research team at the Pasteur Institute in Paris isolated a virus from patients with AIDS. The virus that they found is now called the human immun-odeficiency virus (HIV). HIV was found in a very few healthy individuals, but HIV was found regularly in patients with AIDS and in individuals who had sexual relations or had shared intravenous drug apparatus with individuals with AIDS, and who would, in fact, likely later develop the disease. This finding fulfilled Koch's first postulate.

The virus was grown in the laboratory, outside of the infected individuals. However, because a virus can only replicate within host cells, the virus could not be grown totally free of animal cells as specified by Koch. The virus was grown in a tissue culture of animal cells, that is, in animal cells growing in a culture vessel, meeting the intent of Koch's second postulate.

Researchers were then faced with a problem. Fulfilling Koch's third postulate required that they inject individuals with a pure culture of the virus and observe that all those injected develop AIDS. To inject humans would be unethical because, assuming that the scientists were right and HIV caused AIDS, the infection would cause the deaths of the test subjects. Therefore the scientists injected the pure virus into chimpanzees, which they used as animal models to represent what would occur in a human. Some of the chimpanzees developed some of the symptoms of AIDS, such as swollen lymph nodes. However, none of the chimpanzees developed a clear case of AIDS and hence Koch's third postulate was not precisely fulfilled. Nevertheless, recognizing that a microorganism can produce different symptoms in different animals, the evidence strongly pointed to the fact that AIDS is most likely caused by HIV infections. These experiments also indicated that some degree of flexibility is necessary when trying to fulfill Koch's postulates.

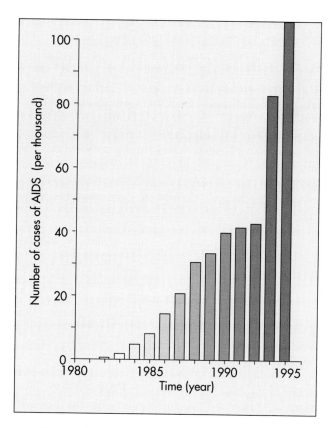

FIG. 19-12 AIDS in the United States is increasing in epidemic proportions. The large increase in 1993 is due, in part, to a change in how the Centers for Disease Control defined cases of AIDS.

HIV. By the year 2000 it is estimated that one in ten Americans will be infected with HIV. The rate in other regions of the world is already higher. By 1995, the World Health Organization estimates eight million individuals will be infected with HIV in Africa. In central Africa one in four individuals appears to be infected with HIV. Blood often is not screened and frequently is used in place of other less available medicines, 10% to 20% of the daily transfusions in Africa will be contaminated with HIV. In central Africa, 10% to 20% of all infants will be born to infected mothers. About 30% of these infants are expected to have congenital AIDS. A similar rate of infection of newborns has been found in some sites in the United States, where there is a very high incidence of AIDS.

The devastating impact of AIDS around the world, both in economic and human terms, has led to major public health efforts to educate the public about AIDS and to identify and to publicize ways to halt the spread of this disease. Part of the problem in the development of vaccines against HIV has been the lack of suitable host animals in which to study viral pathogenesis and immune protection. Individuals who have been infected with HIV make antibodies to the viral antigens; however, these do not appear to protect the patient. Also the virus can hide within macrophages and dendritic cells and escape immune surveillance.

NEWSBREAK

THE FACES OF AIDS

AIDS has struck all parts of society, claiming as victims many well-known celebrities. Each case of infection with HIV and each person who has died of AIDS represents a tragic story. Ryan White, a child who contracted AIDS through a blood product used to treat his hemophilia, struggled to continue attending school and carry on a normal life. He was met by prejudice and fear. He and his family fought to bring the message that HIV is not casually transmitted and that those with HIV infections should not be shunned. Kimberly Bergalis was infected by her dentist, who also infected several other of his patients. She pleaded before the U.S. Congress to make HIV testing mandatory and to protect the public against infections from health care workers. Several artists and entertainers have died of AIDS that they contracted through homosexual activities, including romantic movie star Rock Hudson, the premier ballet

dancer of the twentieth century Rudolf Nureyev, artist Keith Haring, fashion designer Perry Ellis, actor Anthony Perkins, and movie director Tony Richardson. Ray Sharkey, an actor, contracted HIV through contaminated intravenous drug needles. Arthur Ashe, one of America's best tennis players contracted HIV through a blood transfusion given during a surgical procedure. His death emphasizes the risks of blood transmission of HIV and the need to screen the blood supply. His death did not prevent, however, the use of blood in France and Germany that was known to be contaminated with HIV. Magic Johnson, one of the greatest basketball players, contracted HIV through extensive heterosexual activities. Despite the infection, he played on the 1992 winning Olympic basketball team, but soon after retired from professional basketball. Each case of AIDS represents an individual tragedy and a loss to society.

Curbing the Spread of AIDS

Blood screening has been instituted for certain groups, such as military personnel and prison inmates, to identify individuals who have been exposed to HIV. Some communities distribute clean syringes to drug abusers to prevent the spread of HIV through contaminated needles. Some individuals now stockpile their own blood for use in later surgical procedures rather than risk exposure to blood that may contain HIV. Former Surgeon General of the United States, C. Everett Koop recommended the institution of an extensive program of AIDS education in the public schools and that condoms be widely used to block the sexual spread of HIV. Advertisements now appear on television and in the lay media with the message that the proper use of condoms, that is, the practice of "safe sex," is an important way to limit the spread of HIV. Although the only certain way of avoiding sexual transmission of HIV is abstinence and monogamy, latex condoms will block the transmission of HIV when properly used. Scientists will undoubtedly eventually find methods for preventing and better ways of treating AIDS, as they have for most other infectious diseases. At this time avoiding exposure and taking all prudent precautions to avoid infection are essential.

> In the absence of a vaccine that offers protection against HIV infection, a drastic change to limit sexual activities and intravenous drug abuse is essential for halting the epidemic rise in deaths attributable to AIDS.

LEUKEMIA AND OTHER FORMS OF CANCER

Most forms of human cancer have not been shown to be associated with viruses. A few types, however, are apparently caused by viruses (Table 19-7). Cancer is the result of uncontrolled reproduction of malignant cells. When the malignancy occurs in white blood cells, the disease called **leukemia** results in excessive white blood cell proliferation. Several viruses have been found that can transform normal human white blood cells into malignant cells. The particular viruses that have been demonstrated to cause human leukemias are designated **human T-cell leukemia viruses** (**HTLV-I** and **HTLV-II**). These human T-cell leukemia viruses are retroviruses that, like HIV, reproduce by initially reversing the transcription process using reverse transcriptase to form a DNA molecule, which then acts as a template for the production of new viruses. In several cases the segments of DNA produced by reverse transcriptase have been found to be virtually identical to **oncogenes** (cancer-causing genes); that is, they are identical to the naturally occurring animal cell genes that code for the transformation of the cell into a cancer cell line.

> Oncogenic viruses cause the transformation of infected cells that results in malignancies.

In addition to the human T-cell leukemia retroviruses, some other viruses cause malignancies in humans and lower animals. Considerable evidence links the Epstein-Barr virus to two cancers: **Burkitt lymphoma** and **nasopharyngeal cancer.** Nasopharyngeal cancer due to infections with the Epstein-Barr virus is common in southern China and other parts of Asia. The incidence of Burkitt lymphoma is high in East Africa and New Guinea, particularly in males 6 to 14 years old; it is a very rare disease in the United States and Europe. The reason for the restricted geographical occurrences of these diseases is unknown.

In Burkitt lymphoma the Epstein-Barr virus infects immature B cells. This leads to malignancy and the excessive production of lymphocytes, which become localized in tissues of the lymphatic system. Most commonly, Burkitt lymphoma develops into a tumor of the lower jaw; other organs, including the thyroid, liver, and kidneys, may also be sites of tumor development. In cell cultures, Epstein-Barr virus will infect human lymphoid cells, transforming some of

TABLE 19-7		
Viruses Causing Human Cancer		
VIRUS	**CANCER**	**COMMENT**
Epstein-Barr virus	Burkitt lymphoma, nasopharyngeal carcinoma	May also cause Hodgkin disease; malaria may be a cofactor in Burkitt lymphoma
Human papillomavirus	Cervical cancer, skin cancer	Association between genital warts and cervical cancer
Hepatitis B virus	Liver cancer	Aflatoxin may be a cofactor
HTLV-I	T cell leukemia	Retrovirus
HTLV-2	T cell leukemia	Retrovirus
Herpes simplex virus 2	Cervical cancer	May be a cofactor with papillomavirus

the cells into Epstein-Barr virus-carrying cells that have malignant characteristics. When inoculated into monkeys, the cells produce fatal lymphomas.

INFECTIOUS MONONUCLEOSIS

In addition to its association with Burkitt lymphoma, the **Epstein-Barr (EB) virus** causes **infectious mononucleosis.** EB virus is classified as a herpesvirus. Burkitt lymphoma and infectious mononucleosis are very different manifestations of infections with EB virus, and there does not appear to be any increase in the incidence of malignancy in individuals who contract infectious mononucleosis. The EB virus occurs in oropharyngeal secretions of infected individuals and appears to be transmitted primarily by direct and indirect exchange of oropharyngeal secretions.

Infectious mononucleosis most commonly occurs in young adults 15 to 25 years of age. This fact is probably due to the exchange of saliva during kissing, a prevalent activity often involving more partners for this age group than others. From the mouth, B cells and monocytes in the tonsils or adenoids are infected. The virus is carried from there via circulating B cells and monocytes to other lymphoid tissues such as the cervical lymph nodes and spleen. Once it establishes an infection, the EB virus normally remains within the body indefinitely. Up to 90% of the people in the United States have antibodies, indicating a prior infection with the EB virus. In the course of infectious mononucleosis, mononuclear white blood cells are affected, leading to characteristic changes in the white blood cells—including a change in the appearance of the nucleus—that are diagnostic of this disease.

The symptoms of infectious mononucleosis include a sore throat, low-grade fever that generally peaks in the early evening, enlarged and tender lymph nodes, general tiredness, and weakness. The liver and spleen may also be affected by this condition. In most cases of infectious mononucleosis the symptoms are relatively mild and the acute stage of the illness lasts less than 3 weeks. In young children, for example, infectious mononucleosis often is very mild or asymptomatic. Treatment is supportive for the symptoms of sore throat, low-grade fever, enlarged and tender lymph nodes, fatigue, and weakness.

Infectious mononucleosis, caused by the Epstein-Barr virus, is characterized by swollen lymph nodes, low-grade fever, and fatigue.

The EB virus is also suspected as the cause of some cases of **chronic fatigue syndrome** in adults. People with this condition often have difficulty performing routine mental tasks and become so debilitated that they are unable to work. There have recently been several identifiable outbreaks of this condition, and many persons suffering from it have much higher than normal antibody levels for the EB virus. The disorder, called **chronic Epstein-Barr virus syndrome,** has been dubbed the "Yuppie plague" because it is often diagnosed in professional women in their 20s and 30s. Scientists think some patients never completely recover from Epstein-Barr viral infections but instead develop long-standing symptoms of chronic fatigue.

The symptoms of chronic Epstein-Barr virus syndrome include fever and persistent fatigue that seems unconnected to any specific illness. Because chronic fatigue may represent an abnormal response to various different infections, many scientists are skeptical about the validity of attributing chronic fatigue to an Epstein-Barr viral infection. The proper diagnosis of a case of chronic fatigue syndrome requires the identification of multiple symptoms—particularly fatigue, low-grade fever, sore throat, and swollen lymph nodes—that persist or recur over at least a 6-month period.

MYOCARDITIS

The myocardium or heart muscle is subject to the inflammatory disease **myocarditis.** Although this disease may also be of bacterial, fungal, or protozoan origin, it is usually the result of a viral infection. The viruses that most frequently cause myocarditis belong to the coxsackievirus (enterovirus) group. After an initial infection of the respiratory or gastrointestinal tracts, the virus reaches the heart via the blood or lymph. Once there, the infecting virus damages the myocardium, leaving scar tissue.

DENGUE FEVER

Dengue fever is caused by an RNA togavirus transmitted by the mosquito *Aedes aegypti.* More than 100 million cases of dengue fever are estimated to occur worldwide annually. Outbreaks of dengue fever occur in the Caribbean, South Pacific, and Southeast Asia where the mosquito carrier is abundant. Few cases of dengue fever have occurred in the United States but the mosquito that carries the dengue virus is now abundant east of the Mississippi River and future outbreaks within the United States are anticipated. Avoidance of mosquitoes, for example through the use of insect repellants, is useful in preventing this disease. There is no vaccine and no antiviral drug treatment.

The dengue virus replicates within the circulatory system. It causes **viremia** (viral infection of the bloodstream) that persists for 1 to 3 days during the febrile period (characterized by fever). Multiplication of the dengue virus within the circulatory system causes vascular damage and a characteristic rash. It has been postulated that immune reactions contribute to the formation of complexes that initiate internal bleeding. Previous exposure to dengue virus and the presence of cross-reacting antibody seem to be important in determining the severity of disease symptoms. Antibodies enhance the ability of the dengue virus to infect white blood cells. Damage to circulatory vessels appears to occur when antigen-

antibody complexes activate the complement system with the release of vasoactive compounds (FIG. 19-13). Antibody-enhanced infection can lead to a potentially fatal syndrome involving internal bleeding, severe dehydration, and shock. Treatment is supportive, maintaining electrolyte balance, and controlling blood loss.

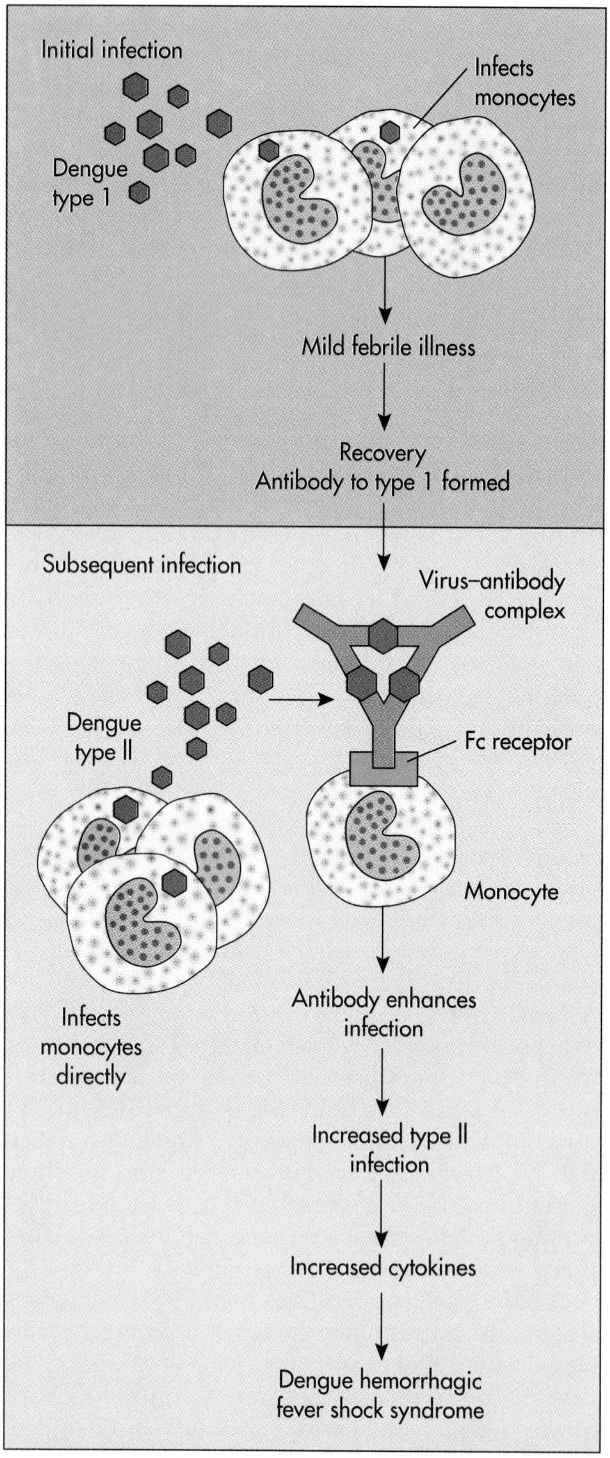

VIRAL HEMORRHAGIC FEVERS

There are several viral infections that result in hemorrhaging (bleeding) (Table 19-8). Such infections are known as the **viral hemorrhagic fevers.** One of these, **Lassa fever,** was discovered in 1969 when three nurses working in missionary hospitals in Nigeria died from an unknown disease and laboratory technicians performing virologic studies of serum from a survivor contracted the same disease. The agent causing this disease is an arenavirus called the Lassa fever virus. Lassa fever is a major health problem in western and central Africa.

In its early stages the symptoms of Lassa fever are similar to many other viral infections, but as the infection progresses some patients develop severe symptoms including extensive vomiting, bleeding, and high fever. The disease can be fatal when blood loss and electrolyte balances cannot be maintained.

> **Some viral infections result in severe hemorrhaging that often is fatal.**

Another viral hemorrhagic fever is caused by the **Marburg virus.** The Marburg virus was discovered in 1967, when 31 persons were infected by African monkeys imported into Europe. Other hemorrhagic fevers of viral origin, such as **Crimean-Congo hemorrhagic fever,** have since been reported. The viruses causing hemorrhagic fevers are transmitted very rapidly from person to person by contact with bodily fluids. Nosocomial transmission via hypodermic needles accounted for one half of the cases of **Ebola hemorrhagic fever** in Zaire in 1976. The fatality rates of these viral hemorrhagic fevers are very high for an infectious disease. Patients suffer headache, muscle pain, skin or gastrointestinal hemorrhaging, and shock.

Hemorrhagic fever, with renal syndrome (formerly called **Korean hemorrhagic fever,** or epidemic fever) is caused by a hantavirus. This hantavirus is named for the Hantaan River in Korea. The disease begins with a fever and flushed face. Disease symptoms progress to include bleeding, nausea, and kidney failure, and eventually the alterations of the electrolyte concentrations result in high blood pressure, shock, and death. Many soldiers developed this disease during the Korean War. Over 200,000 cases of hantavirus infection cause hemorrhagic fever in Asia annually—mostly in China. There are 4,000 to 20,000 deaths each year from this disease.

FIG. 19-13 Pathogenesis of dengue.

TABLE 19-8

Arboviruses That Can Cause Fevers and Hemorrhagic Diseases

VIRUSES	DISEASE	GEOGRAPHICAL DISTRIBUTION	VECTOR	ANIMAL RESERVOIR
Yellow fever	Fever, hepatitis	Africa, Central and South America	Mosquito *Aedes* sp.	Monkeys
Dengue (4 serotypes) (flavivirus)	Fever, rash (hemorrhagic shock syndrome)	India, southeast Asia, South Pacific, South America, Caribbean	Mosquito	None
Kyasanur Forest (flavivirus)	Hemorrhagic fever	India	Tick	Monkeys, rodents
Ross River (alphavirus)	Fever, arthralgia arthritis	Australia	Mosquito	Birds
Rift Valley fever (bunyavirus)	Fever, sometimes hemorrhage	Africa	Mosquito	Sheep, cattle, camels
Sandfly fever (phlebovirus)	Fever (mild disease)	Asia, South America, Mediterranean	Sandflies	Gerbils
Congo-Crimean hemorrhagic fever (bunyavirus)	Fever, hemorrhage	Asia, Africa	Tick	Rodents
Colorado tick fever (reovirus)	Fever, myalgia	USA (Rocky Mountains)	Tick	Rodents

VIRAL DISEASES OF THE SKIN

Some viruses circulate throughout the body. They can cause damage to various body sites, including circulatory vessels and skin tissues (Table 19-9). Replication of these viruses typically results in characteristic skin rashes, such as the red rash of measles infections. Viruses that cause diseases affecting the skin may be transmitted by direct contact, as in the case of smallpox, but more commonly enter the body by other routes, as in the cases of measles and German measles where the causative viruses enter the body through the respiratory tract and reach the skin via the circulatory system.

TABLE 19-9

Some Viruses That Cause Human Diseases of the Skin

VIRAL GROUP	VIRUS	DISEASE
Herpesviruses	Herpes simplex virus	Cold sores, fever blisters, genital herpes
	Varicella-zoster virus	Chickenpox, shingles
	Human herpes virus type 6	Roseola infantum
Papovaviruses	Papillomavirus	Common warts, genital warts, plantar warts
Paramyxoviruses	Measles virus	Measles maculopapular rash
Parvoviruses	Human parvovirus	Erythema infectiosum (facial rash)
Picornaviruses	Coxsackie virus A	Herpangina; hand, foot, and mouth disease
Poxviruses	Molluscum contagiosum	Fleshy papular eruption
Togaviruses	Rubella virus	German measles rash
	Dengue virus	Dengue fever rash

SMALLPOX

Smallpox is the only human disease that has been totally eliminated (FIG. 19-14). This disease is caused by variola virus, a member of the poxvirus class. Elimination of smallpox is a remarkable event considering that in the Middle Ages every individual had an 80% chance of contracting smallpox. The last naturally occurring case of smallpox occurred in Somalia, Africa, in 1977. One later case in 1978 was the result of a laboratory accident in England. The successful elimination of smallpox was the result of an extensive worldwide vaccination program, coordinated by the World Health Organization, and the facts that humans are the only known host for the smallpox (variola) virus and that there are no asymptomatic carriers of the disease. As a result of the elimination of susceptible hosts, the virus and the dread disease were eliminated. Consequently, it is no longer necessary to vaccinate individuals against this disease. (The only exceptions are certain workers at laboratories, such as the Centers for Disease Control, where smallpox viruses are still maintained for research purposes.)

> **Smallpox, once the major plague of humans, has been eliminated by a worldwide vaccination program.**

When it existed, the major route of transmission of smallpox was via aerosols entering the respiratory

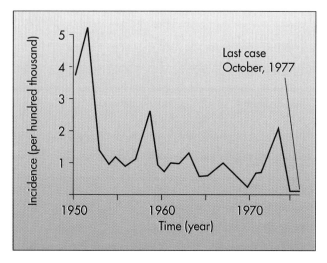

FIG. 19-14 Smallpox was eliminated through a global vaccination program.

tract. The virus moved through the circulatory system, reaching the skin, where it caused the noticeable symptoms of this disease (FIG. 19-15). The growth of the virus in the epidermal layers of the skin caused the development of a rash that eventually formed pustules (raised blisters containing pus). These lesions developed over the entire body. Lesions also occurred internally, where the virus infected mucous membranes. In addition to the obvious lesions over

HISTORICAL PERSPECTIVE

SMALLPOX ERADICATION

By the early 1900s, vaccination programs begun in the 1700s by Edward Jenner had been largely successful in eliminating smallpox from North America and Europe. This vaccination program used an attenuated strain of vaccinia virus. In 1967, the World Health Organization began a campaign to eliminate smallpox from the world, focusing on South America, Africa, India, and Indonesia. The program, which cost over 150 billion dollars, employed vaccination, surveillance, and containment of cases. The massive campaign was successful. In 1974 there were 218,000 cases of smallpox in the world; by 1978 there were none. The last case of naturally occurring smallpox occurred in Somalia in October 1977. The eradication of smallpox was possible because all cases could be readily identified so that surveillance and containment were possible (no carrier states of smallpox develop and there are no subclinical cases), humans are the only host (no animal reservoir exists), and an effective vaccine was available.

With the eradication of smallpox, the only smallpox viruses that remained were in laboratories where they were studied. A tragic accidental release of smallpox viruses from a laboratory in England resulted in a fatal case of smallpox, the last recorded death due to this disease. Russia and the United States agreed that the remaining stocks of smallpox virus would be destroyed by December 21, 1993, after the genomes had been completely sequenced. This agreement was very controversial in the scientific community. Many scientists argued that studies on the virus should continue and that alternate animal hosts should be created so that the pathogenesis of the virus could be studied further. Some scientists argued that all forms of biological diversity should be preserved. Other scientists held that the public health risk of accidental release from the laboratory make the elimination of the remaining smallpox viruses a special case.

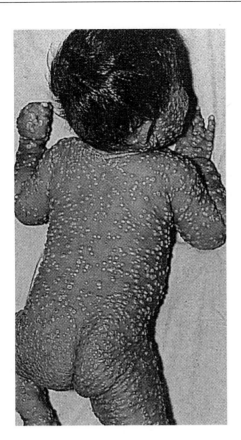

FIG. 19-15 Lesions, such as those shown were common prior to the elimination of smallpox.

the entire body, the symptoms of smallpox included fever, malaise, and pain in the head, back, and stomach. In one form of this disease, variola major, the mortality rate was greater than 20%, but in another form, variola minor, less than 1% of the individuals infected with the poxvirus died. Individuals who recovered from smallpox were immune to the disease.

MEASLES (RUBEOLA)

Measles, or **rubeola,** is a highly contagious disease occurring on a worldwide basis almost exclusively in children. The **measles virus,** a paramyxovirus, is readily transmitted from an infected child to a susceptible host, as illustrated by the fact that there is greater than a 90% incidence of an acute infection after exposure to measles virus by susceptible children. In one dramatic outbreak, over 4,000 susceptible inhabitants of Greenland contracted measles during a 6-week period in 1951 (FIG. 19-16).

The virus initially multiplies in the mucosal lining of the upper respiratory tract. The measles viruses then appear to be disseminated to lymphoid tissues, where further multiplication occurs. Before the onset

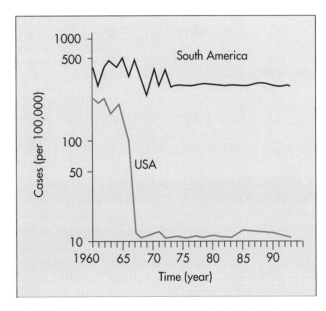

FIG. 19-16 Measles has been nearly eliminated in the United States through vaccination but the incidence of this disease remains high in other regions.

TABLE 19-10		
Pathogenesis of Measles		
SITE OF VIRUS GROWTH	CONDITION	COMMENT
Lung	Pneumonia	Life-threatening pneumonia in malnourished child or poor medical care
Ear	Otitis media quite common	Common complication
Oral mucosa	Koplik spots	Severe ulcerating lesions in malnourished child
Conjunctiva	Conjunctivitis	Can lead to secondary bacterial infection and blindness
Skin	Maculopapular rash	Blotchy red rash; hemorrhagic rashes may occur ("black measles") in malnourished child
Intestinal tract	Gastroenteritis	Diarrhea
Urinary tract	None	Virus detectable in urine

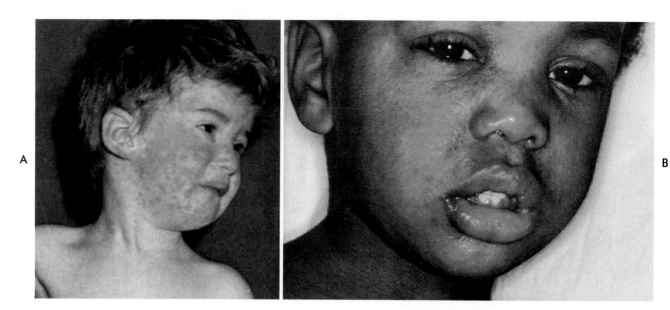

FIG. 19-17 A, Measles rash on first day in a caucasian child. A transient erythematous rash during the prodromal period may be confused with scarlet fever, but careful inspection of the mouth will usually disclose Koplik spots. The true rash appears behind the ears and along the hairline, quickly affects the face, and spreads progressively from above downward. On the first day of the rash the face is heavily covered but elsewhere the spots are scanty. **B,** Measles in an African-American child. Measles may be difficult to diagnose in a dark-skinned patient. Koplik spots may be found during the prodromal period.

of symptoms, large numbers of measles viruses are shed in secretions of the respiratory tract, the eyes, and in urine, promoting the rapid epidemic spread of this disease. Infection with measles viruses can involve a number of organs (Table 19-10). There is a high rate of mortality associated with measles in regions of the world where malnutrition predominates and medical treatment facilities are limited. When measles is fatal, the virus generally invades the central nervous system, causing encephalitis.

Measles is characterized by the eruption of a blotchy red skin rash. The rash appears approximately 14 days after exposure to the measles virus (FIG. 19-17). It generally appears initially behind the ears, spreading rapidly to other areas of the body during the next 3 days. Disease symptoms often begin a few days before the onset of the rash. These initial symptoms include high fever, coughing, sensitivity to light, and the appearance of **Koplik spots.** These are lesions on mucous membranes seen as red spots with a white dot in the center that occur in the oral cavity, generally first appearing on the inner lip (FIG. 19-18). These symptoms are due to the cell damage that occurs when measles viruses replicate and the host immune system responds.

Treatment is normally supportive, including rest and the intake of sufficient fluids. In uncomplicated cases, the fever disappears within 2 days, and the in-

dividual returns to normal activities a few days later. If the fever persists for more than 2 days after the eruption of the rash, it is likely that a complication, such as bronchitis or pneumonia, has developed. In these cases additional treatment is needed to cure the secondary infection. Prolonged or latent measles infections may also cause immune complex reactions. Such immune complex hypersensitivity reactions may be responsible for neurological complications of measles infections, including the condition subacute sclerosing panencephalitis (SSPE). This is a slow viral disease that may take 5 to 10 years to develop. Often it effects children or young adults who were infected with the measles virus before age one. SSPE results in a sudden rapid degeneration of the nerve cells of the brain that is fatal. With the decline of measles in the United States this condition has disappeared here, but still occurs elsewhere in the world.

Measles can be prevented by childhood immunization. After the introduction of the measles vaccine in 1963, the rate of measles infection in the United States declined (FIG. 19-19). The number of cases in the United States was greater than 500,000 per year before the introduction of the measles vaccine. By 1978 the number of cases had dropped to 27,000. Epidemiologists were confidently predicting that measles would be eradicated from the United States by 1982. That confidence was based in part on the fact that, like smallpox, the measles virus infects

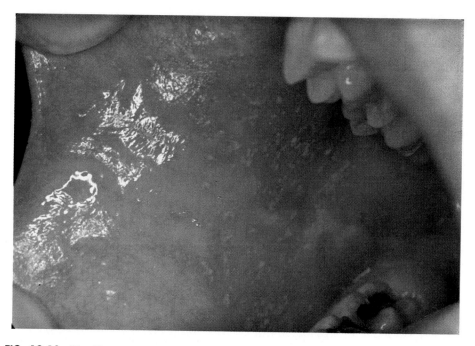

FIG. 19-18   Koplik spots are found on mucous membranes during the prodromal stage and are easily detected on the mucosa of the cheeks opposite the molar teeth, where they resemble coarse grains of salt on the surface of the inflamed membrane. Histologically the spots consist of small necrotic patches in the basal layers of the mucosa.

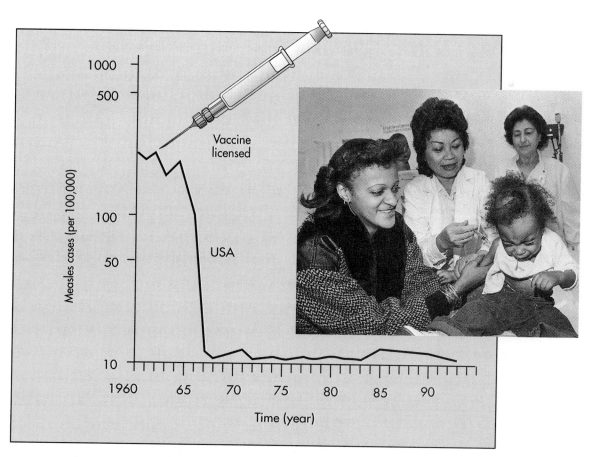

FIG. 19-19   The incidence of measles in the United States has declined greatly since 1963 when a measles vaccine was introduced.

only human cells. Thus by breaking the person-to-person transmission chain the disease could be eliminated.

However, in 1983 there were still almost 1,500 cases of measles in the United States, where several major measles outbreaks occurred in the years following. Some outbreaks involved elementary school age inner city children who had not been vaccinated. Most of the cases of measles, though, occurred among high school and college students who, as children, received a vaccine that had been inactivated by exposure to excessive heat or light. It was assumed that a single vaccination would confer lifelong immunity. This proved to be wrong. A booster vaccine is required to ensure immunity and two doses rather than one of vaccine are now used to ensure the effectiveness of immunization. This improved vaccination program has been effective. By 1994 there were fewer than 1,000 cases of measles in the United States and by 1996 the goal is to eliminate the disease.

While the United States has achieved control of measles and effectively eliminated this disease, measles remains a significant disease elsewhere in the world. The worldwide incidence of measles has not declined because there has been no effective global immunization program. There are still 1.5 million deaths per year worldwide due to measles.

Measles can be prevented by vaccination; it has been virtually eliminated in the United States, but not enough people have been vaccinated worldwide so outbreaks of measles continue to occur.

GERMAN MEASLES

Like measles, transmission of **rubella virus,** the causative agent of **German measles (rubella),** appears to be via droplet spread. Rubella virus is classified as a togavirus. The initial infection occurs in the upper respiratory tract. In contrast to the measles virus, however, the rubella virus exhibits a relatively low rate of infectivity, and thus relatively prolonged exposure appears to be needed for the establishment of an infection (FIG. 19-20). After multiplication in

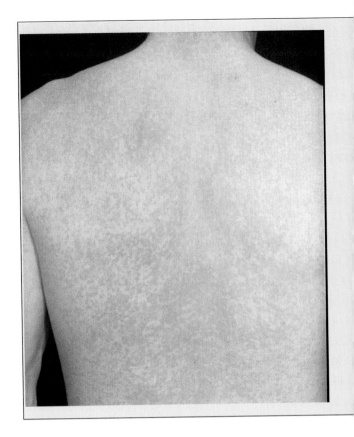

the mucosal cells of the upper respiratory tract, rubella viruses appear to be disseminated systemically through the blood. Rubella viruses can infect several body sites (Table 19-11). Approximately 18 days after initiation of the infection by the rubella virus, a characteristic rash, appearing as flat pink spots, occurs on the face and subsequently spreads to other parts of the body. Enlarged and tender lymph nodes and a low-grade fever characteristically precede the occurrence of the German measles rash.

In children and adolescents, rubella is usually a mild disease. If it is acquired during pregnancy, however, the fetus can become infected with the rubella virus, resulting in **congenital rubella syndrome.** This

TABLE 19-11		
Pathogenesis of Rubella		
SITE OF VIRUS GROWTH	CONDITION	COMMENT
Respiratory tract	Mild sore throat, coryza, cough	Viral shedding so that patient is infectious 5 days before to 3 days after symptoms
Skin	Rash	—
Lymph nodes	Lymphadenopathy	Swelling of lymph nodes
Joints	Arthritis	Due to circulating immune complexes
Placenta/fetus	Fetal damage	Congenital rubella

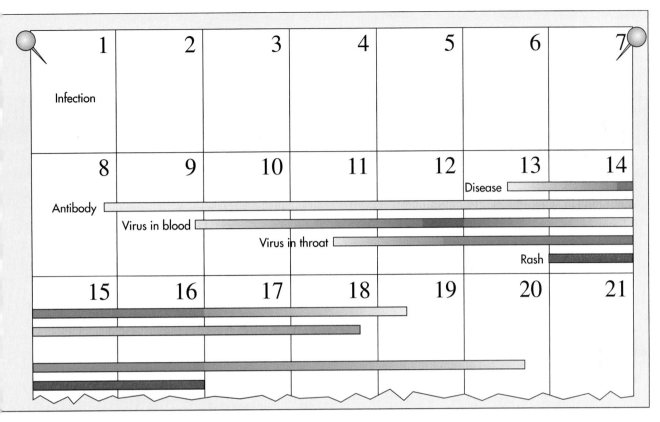

FIG. 19-20 Pathogenesis of rubella over time, as illustrated in this calendar, shows the sequence of disease development. The rash of rubella consists initially of discrete, delicate pink macules (*left*) but sometimes maculopapular and hemorrhagic elements may be found. The severity of the rash varies considerably and is easily missed when lesions are sparse.

syndrome is characterized by the development of multiple abnormalities in the infant, including mental retardation and deafness. Generally the earlier in pregnancy the German measles occurs, the more serious are the effects. There is a very high rate of mortality, exceeding 25%, in cases of congenital rubella syndrome. Vaccination has greatly reduced the incidence of rubella in children and is also used to confer immunity on women of childbearing age who had not contracted the disease at an earlier age.

Congenital rubella syndrome, which results in multiple abnormalities in the infant, occurs when a woman contracts German measles during pregnancy.

CHICKENPOX AND SHINGLES

Chickenpox (varicella) is caused by the **varicella-zoster virus.** This virus is a member of the herpesvirus group, which also causes shingles (zoster). Ninety percent of all cases of chickenpox occur in children under 9 years of age (FIG. 19-21). In children, chickenpox is generally a relatively mild dis-

ease, but when this disease occurs in adults the symptoms are more frequently severe. Chickenpox is a highly contagious disease that is transmitted probably via contaminated droplets and by direct contact with vesicle fluid containing varicella-zoster viruses.

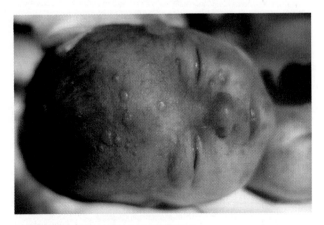

FIG. 19-21 Distribution of rash in a case of chickenpox. The rash in chickenpox occurs all over the body. It is heaviest on the trunk and diminishes in intensity toward the periphery.

The initial site of viral replication has not been positively established but appears to be in the upper respiratory tract. Local vesicular lesions occur in the skin after dissemination of the virus through the blood. These skin lesions become encrusted, and the crusts fall off in about 1 week. Vesicles also occur on mucous membranes, especially in the mouth.

In some cases the varicella-zoster virus spreads to the lower respiratory tract, resulting in pneumonia. It is in this way that several other tissues, including the central nervous system, can also be involved in complicated cases of chickenpox. Approximately 100 deaths occur each year in the United States due to encephalitis caused by the spread of chickenpox viruses to the central nervous system. Complications from a case of chickenpox are a particular danger in immunocompromised individuals. Gamma-2-immune globulin is used as a prophylactic measure when such individuals develop a case of chickenpox. Ad-

ditionally, there is an elevated incidence of Reye syndrome associated with occurrences of chickenpox, especially if aspirin is used. Unlike the other childhood viral diseases, vaccination against chickenpox has not yet been introduced, and outbreaks of chickenpox continue to show regular seasonal cycles of the same magnitude.

Chickenpox is not yet prevented by vaccination.

In adults, **shingles** is the principal disease resulting from reactivation of infections with the varicella-zoster virus. The varicella-zoster virus acquired in childhood can remain in a latent state within the body, perhaps as a provirus in which the viral DNA can be incorporated into human chromosomes in a dormant state of lysogeny. The viruses are maintained in sensory ganglia of spinal or cranial nerves. Reactivation of the latent virus leads to inflammation of these nerves. This reaction is the cause of shingles

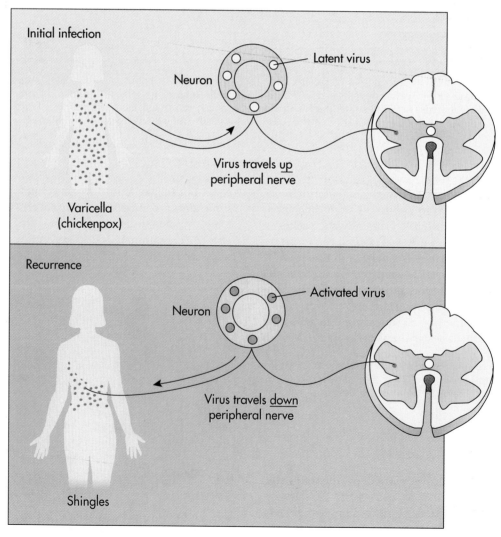

FIG. 19-22 Pathogenesis of shingles, showing the relationship between childhood chickenpox and the development of shingles in adults.

SHOULD CHILDREN BE IMMUNIZED AGAINST CHICKENPOX?

A vaccine against chickenpox is one of the newest live attenuated vaccines. The vaccine called Varivax uses the Oka strain of varicella virus. It is licensed for general use in Japan and Korea. If approved for use in the United States it most likely will be given along with the measles-mumps-rubella vaccine. It is predicted that introduction of the vaccine in the United States could rapidly reduce the number of cases of chickenpox, from nearly four million to less than a quarter of a million cases. The vaccine is expensive; it would cost $157 million per year to carry out an effective immunization program in the United States. Although the vaccine has been shown to be safe and efficacious, its potential use in the United States and elsewhere is controversial. The point of controversy is whether to institute a costly vaccination program to protect children against chickenpox that will increase the possibility that adults will contract this disease. Adults would require two doses of the vaccine to develop immunity. Chickenpox in children almost always is a mild disease but can be serious if it occurs in adults, sometimes causing brain damage. What do you think?

(FIG. 19-22). There is usually pain and tenderness along the affected sensory nerves. A rash develops on the skin in areas supplied by affected sensory nerves. The rash usually lasts for 2 to 4 weeks, but the pain may last for weeks or months. The skin lesions contain chickenpox viruses, and susceptible individuals who come in contact with these lesions can acquire chickenpox. Usually the symptoms of shingles are restricted to one side of the body at a time because the disease follows the branches of the cutaneous sensory nerves.

Shingles is the result of the reactivation of an infection with varicella-zoster virus that remained dormant in the body as a provirus.

WARTS

Warts are benign tumors of the skin caused by **papillomaviruses.** Papillomaviruses are small DNA viruses. Transmission appears to occur primarily by direct contact of the skin with wart viruses from an infected individual. Indirect transfer also may occur through fomites (inanimate objects). Human papillomaviruses appear to infect only humans and no other animals. Children develop warts more frequently than adults. These viruses infect cells at the basal layers of the epidermis layer of the skin. When these cells migrate to the skin surface they carry the wart viruses along with them. The viral infection becomes manifest in the outer layers of the skin because the wart-virus-infected cells proliferate and form a thick layer (FIG. 19-23).

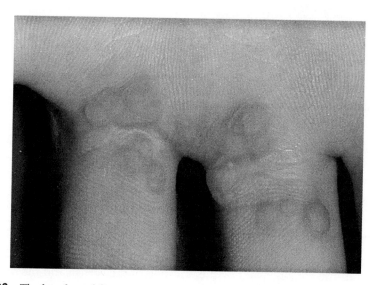

FIG. 19-23 The hands and fingers are a common site for the formation of warts due to infection with papillomaviruses.

The development of warts can occur on any body surface. The appearance of the warts varies, depending on location. At present there is no effective antiviral treatment for human warts, and therapy often involves destruction of infected tissues by applying acid or freezing. In general, human warts are self-limiting, and recovery can be expected without treatment within 2 years.

Warts are benign tumors caused by papillomaviruses.

HERPES SIMPLEX INFECTIONS

Herpes simplex 1 virus is most frequently involved in nongenital herpes infections. Infections with herpes simplex 1 virus most commonly involve the skin above the waist, usually the lips and mouth. The virus appears to be transmitted through direct contact of the surface epithelial tissues with the virus. A focal infection develops around the site of inoculation, and dissemination of the virus occurs from the primary focal lesion. In most cases the lesions are limited to the epidermis and surface mucous membranes.

The development of herpes simplex viruses in the mouth and on the lips is seen as the development of lesions known as **cold sores** or **fever blisters** (FIG. 19-24). This is a very common occurrence. Similar lesions also develop in infected regions of the skin. The lesions generally heal within several days of their appearance. The virus, however, is not cleared from the body. Herpes simplex viruses enter the trigeminal nerve that innervates the infected skin region and go into a latent stage in which viral replication does not occur. The virus is reactivated by stressful stimuli such as ultraviolet light, hormonal changes, and fever. Herpes simplex virus comes out of the nerve

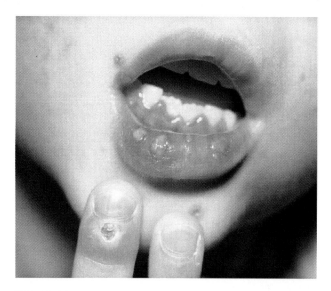

FIG. 19-24 Herpes infections of the oral cavity are common and result in recurring lesions.

and reinfects areas of the skin near the original infection. New lesions develop recurrently, indicative of the persistence of herpes simplex. Recurrent outbreaks are usually slower in duration. The infection is lifelong. Active lesions contain infectious herpes simplex viruses and direct contact should be avoided. Dental and other health care workers must be aware of the possible spread of infection to their own fingers. Several of the newly developed antiviral drugs, such as acyclovir, are effective against herpes viruses and may prove useful in treating herpes infections.

Cold sores or fever blisters are caused by infections with herpes simplex 1 viruses.

VIRAL DISEASES OF THE EYE

The eyes normally are only rarely infected with viruses. In some cases, however, viruses directly contaminate eye tissues and initiate an infection that causes disease of the eye tissue (Table 19-12).

EPIDEMIC KERATOCONJUNCTIVITIS (SHIPYARD EYE)

Several viruses cause **conjunctivitis** (inflammation of the mucous membranes, or conjunctivae, of the

eyes). Adenoviruses are the etiologic agents of **epidemic keratoconjunctivitis** or **shipyard eye**. The disease derives its name from the fact that outbreaks have occurred in shipyards and industrial plants where trauma to the eye may occur as an occupational hazard. Adenoviruses also may occur in swimming pools, and transmission of the disease can be waterborne. Shipyard eye can result in the inflammation of the cornea and the conjunctivae.

TABLE 19-12					
Some Viruses That Cause Human Diseases of the Eye					
VIRAL GROUP	**VIRUS**	**DISEASE**	**VIRAL GROUP**	**VIRUS**	**DISEASE**
Adenoviruses	Adenovirus types 3, 7, 8, 19	Conjunctivitis	Paramyx- oviruses	Measles virus	Conjunctivitis
Herpesviruses	Herpes simplex virus	Corneal lesions, keratitis	Picornaviruses	Enterovirus 70	Acute hemor- rhagic con- junctivitis
	Varicella-zoster virus	Conjunctival le- sions		Coxsackievirus A	Conjunctivitis
			Togaviruses	Rubella virus	Cataracts

HERPETIC KERATOCONJUNCTIVITIS

Herpes simplex virus type 1 can cause an ocular in-fection called **herpes corneales.** This disease, which occurs in children and adults, affects the cornea. Ul-cerations of the cornea caused by infection with her-pes simplex virus can cause blindness in some cases.

Herpes simplex viruses cause corneal lesions and ker-atitis. The use of antiviral drugs, such as iododeoxy-uridine and acyclovir, is effective in controlling her-pes simplex viral infections of the eye. However, her-pes infections persist within the body and recurrences can occur despite treatment with antiviral drugs.

SUMMARY

Viral Diseases of the Central Nervous System (pp. 561-570)

Rabies (pp. 561-564)
- Rabies, a disease of the central nervous system, is caused by the rabies virus, which is transmitted via the deposition of saliva through the bite of an infected animal. It can also be spread via the respiratory tract through aerosols in the atmosphere around dense populations of infected animals. Symptoms include anxiety, irritability, depression, sensitivity to light and sound, hydrophobia, difficulty in swallowing, paral-ysis, and coma. There is a vaccine.

Poliomyelitis (pp. 564-566)
- Poliomyelitis, another disease of the central nervous system, is caused by the poliovirus. Its symptoms in-clude headache, constipation, vomiting, and sore throat, followed by neural involvement, especially paralysis. Vaccination is standard.

Aseptic Meningitis (pp. 566-567)
- Aseptic meningitis is caused by various viruses that inflame the meninges, often echoviruses and coxsack-ieviruses. Fever, stiff neck, fatigue, and irritability are the symptoms. Treatment is supportive.

Viral Encephalitis (pp. 568-569)
- Viral encephalitis can be caused by various viruses transmitted by the bite of an infected arthropod vec-tor. The symptoms are often subclinical but may in-clude fever, headache, and vomiting, followed by stiffness, paralysis, convulsions, psychoses, and coma. Treatment is supportive, depending on the symptoms. Vaccines are available and the disease can be controlled by vector eradication programs.

Reye Syndrome (p. 569)
- Reye syndrome is a disease of the central nervous system associated with outbreaks of influenza and other viral infections among children. Sudden fever, nausea, vomiting, delirium, and coma are symptoms. Treatment is symptomatic. Limiting the use of aspirin in treating children with viral infections seems to pre-vent this disease.

Guillain-Barré Syndrome (p. 569)
- Guillain-Barré syndrome is associated with influenza. The 1976 outbreak of this disease appears to have re-sulted from the contamination of vaccines with viable influenza viruses prepared for an expected outbreak of swine flu. Treatment is supportive.

Creutzfeldt-Jakob Disease and Related Diseases (pp. 569-570)
- Creutzfeldt-Jakob disease, Gerstmann-Sträussler syn-drome, and kuru are diseases of the central nervous system caused by prions, which are infectious pro-teins that do not appear to contain any nucleic acids. These diseases are characterized by long periods of incubation during which there are no symptoms. Dif-ficulty in walking, loss of coordination and memory, and dementia are followed by death. Treatment is supportive.

Viral Diseases of the Cardiovascular and Lymphatic Sys-tems (pp. 570-581)

Acquired Immunodeficiency Syndrome (AIDS) (pp. 571-578)
- Acquired immunodeficiency syndrome (AIDS) is a sexually transmitted disease of the cardiovascular

and lymphatic systems caused by the HIV-3 virus. It is also transmitted through blood and across the placenta. It suppresses the immune response, producing malaise, swollen lymph nodes, and various secondary infections such as *Pneumocystis carinii* pneumonia and Kaposi sarcoma.

Leukemia and Other Forms of Cancer (pp. 578-579)

- Oncogenic viruses cause malignancies. Human T-cell leukemia viruses, RNA viruses that reproduce using reverse transcription to form a DNA molecule that then acts as the template for the production of new viruses, cause leukemia. They are retroviruses.
- Burkitt lymphoma is a form of cancer caused by the Epstein-Barr virus, producing excessive amounts of lymphocytes that localize in the lymphatic system.

Infectious Mononucleosis (p. 579)

- Infectious mononucleosis is also caused by the Epstein-Barr virus transmitted via aerosols and by direct exchange of oropharyngeal secretions containing the virus. Treatment is supportive for the symptoms of sore throat, low-grade fever, enlarged and tender lymph nodes, fatigue, and weakness.

Myocarditis (p. 579)

- Myocarditis, the inflammation of the heart muscle, can be caused by bacteria, fungi, or protozoas, but is usually caused by a virus, especially a coxsackievirus. The virus reaches the heart via the blood or lymph after infecting the respiratory or gastrointestinal tract.

Dengue Fever (pp. 579-580)

- Dengue fever is caused by an RNA virus transmitted by the mosquito *Aedes aegypti,* producing a rash and fever.

Viral Hemorrhagic Fevers (pp. 580-581)

- Viral hemorrhagic fevers are viral infections that result in bleeding. They include Lassa fever, Marburg virus fever, Crimean-Congo hemorrhagic fever, Ebola fever, and Korean hemorrhagic fever. These diseases produce a variety of symptoms.

Viral Diseases of the Skin (pp. 581-590)

Smallpox (pp. 582-583)

- Smallpox has been eliminated by worldwide vaccination. It was caused by the smallpox (variola) virus

and transmitted by aerosols, producing symptoms of rash, fever, malaise, and pain in the head, back, and stomach.

Measles (Rubeola) (pp. 583-586)

- Measles is caused by the measles virus. It is transmitted from an infected individual to a susceptible host, producing skin rash, high fever, coughing, sensitivity to light, and Koplik spots. Bed rest and fluids are prescribed; there is a vaccine.

German Measles (pp. 586-587)

- German measles is caused by the rubella virus transmitted by droplet spread, producing characteristic spots, enlarged and tender lymph nodes, and a low-grade fever. Vaccination is recommended.

Chickenpox and Shingles (pp. 587-589)

- Chickenpox and shingles are caused by varicella-zoster virus transmitted via contaminated droplets or direct contact with vesicle fluid containing the virus. Skin lesions are characteristic of chickenpox. Symptoms of shingles are inflammation of the ganglia of the spinal or cranial nerves, pain and tenderness, and a rash along the affected nerves.

Warts (pp. 589-590)

- Warts are skin diseases caused by papillomaviruses transmitted by direct contact. Destruction of the infected tissue is the only treatment.

Herpes Simplex Infections (p. 590)

- Herpes simplex infections of the oral cavity are caused by the herpes simplex 1 virus, transmitted by direct contact of the surface epithelial tissues with the virus, producing cold sores and fever blisters. It can be treated with acyclovir.

Viral Diseases of the Eye (pp. 590-591)

Epidemic Keratoconjunctivitis (Shipyard Eye) (p. 590)

- Epidemic keratoconjunctivitis (shipyard eye) is caused by adenoviruses and is associated with trauma to the eye. The cornea is keratinized and the conjunctivae are inflamed.

Herpetic Keratoconjunctivitis (p. 591)

- Herpetic keratoconjunctivitis is caused by herpes simplex type 1 virus and affects the cornea. It is treated with iododeoxyuridine and acyclovir.

C H A P T E R   R E V I E W

REVIEW QUESTIONS

1. What viral infections cause diseases of the central nervous system?
2. How are viral diseases of the central nervous system transmitted? How can transmission of these diseases be interrupted?
3. How has the transmission and treatment of rabies changed in the last decade? Why is it essential to vaccinate anyone suspected of having been infected with the rabies virus? Why can the rabies vaccine be adminis-

tered after someone has been bitten by an animal carrying the rabies virus?

4. What are the similarities and differences between meningitis and encephalitis?
5. How are vectors involved in the transmission of viral encephalitis.
6. What viral infections cause diseases of the cardiovascular and lymphatic systems?
7. How are viral diseases of the cardiovascular and lym-

phatic systems transmitted? How can transmission of these diseases be interrupted?

8. Which viruses can cause human cancers?

9. Describe HIV and explain how the properties of this virus causes AIDS.

10. Describe AIDS, including the pathogenesis of this disease. Why are individuals with AIDS susceptible to opportunistic infections?

11. How are viral diseases of the skin transmitted? How can transmission of these diseases be interrupted?

12. What viral infections cause diseases of the skin?

13. What are the similarities and differences between chickenpox and shingles? How can these diseases be caused by the same virus?

14. Why are herpes infections persistent? Why do lesions from herpes viral infections recur?

15. What viral infections cause diseases of the eye?

16. How are viral diseases of the eye transmitted? How can transmission of these diseases be interrupted?

17. How are viral infections diagnosed?

CRITICAL THINKING QUESTIONS

1. Why is immunization used against some viral diseases and not against others? Should immunization be used to prevent chickenpox?

2. Why was it possible to eliminate smallpox through worldwide immunization? Could the same be done for AIDS?

3. What can be done to halt the AIDS epidemic? Why is it proving difficult to develop a vaccine against AIDS?

4. Could vaccines be used to eliminate rabies? How?

5. Why are there so few drugs that are effective against viral infections?

6. How is the immune system used to treat viral diseases?

CASE STUDY

HISTORY AND ASSESSMENT

Kristie, a 5-year-old child, was brought by her father to their HMO (Health Maintenance Organization). Her father explained to the nurse that he and a friend had taken Kristie camping over the weekend. They were forced to shorten the camping trip when Kristie became very irritable and started to develop a rash. The nurse disrobed Kristie and carefully placed her on an examining table. The nurse observed a rash over Kristie's trunk, scalp, arms, and face. *At this point, the health care provider must consider a wide differential diagnosis, including an allergy to a new food or poisonous plant on the camping trip, a viral infection of measles or chickenpox, or a bacterial or fungal infection of the skin.*

The rash appeared to be more severe over the trunk and scalp. The nurse obtained the following vital signs: Pulse rate 110, Respiratory rate 24, Temperature 104° F. *Do any of the above findings appear abnormal to you? If so, what could be causing the abnormality?*

PHYSICAL EXAMINATION

The physician examined Kristie and verified the rash, most prominent on the trunk and scalp. The individual lesions were fluid-filled vesicles 1 to 4 mm in size, containing a clear fluid. Some of the vesicles had been ruptured and these lesions appeared dry and crusted on a reddened base. The doctor also noted that several vesicles contained yellow pus, and were larger than other lesions. *Why might some lesions contain yellow pus while others do not? Based on the physical examination and the ap-*

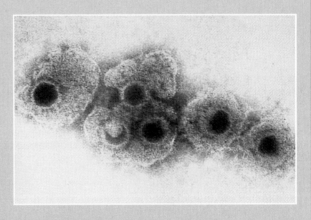

pearance of the rash, the doctor has a good idea of what is causing Kristie's illness. Based on your reading of this chapter, what do you think it is?

PROCEDURES AND LABORATORY TESTS

The physician ordered blood drawn for a chemistry profile (to assess the level of electrolytes in the blood) and a CBC (complete blood count). The only abnormality found was an elevated white blood cell count that is found in cases of infection.

A chest x-ray was also taken but was within normal limits. *Why did the physician order a chest x-ray?*

Continued.

DIAGNOSIS

Putting all of the facts together, the physician determined that Kristie had an infection of the varicella-zoster virus (see Figure), commonly known as chickenpox. *Chickenpox is a very common childhood disease, with approximately two thirds of all cases appearing in the 5 to 9 year age group. The virus is spread through an airborne route and enters the child through the respiratory tract. A characteristic rash is the predominant sign on physical examination. The individual lesions are tiny vesicles of 1 to 4 mm, containing a clear fluid. If these vesicles rupture, they form dry crusted lesions. If these vesicles do not rupture, the fluid becomes cloudy.*

The physician remembers that some of the lesions had pus-filled centers, not a clear or cloudy fluid. This can be a sign of a *superinfection*—a bacterial infection and a viral infection. Scratching of the broken lesions can lead to contamination by bacteria and subsequent bacterial infection. *Superinfections are often caused by the normal flora of the skin, such as* Staphylococcus epidermidis, *when there is a break in the skin disrupting its function as a protective barrier for the body.*

Occasionally the virus will cause respiratory complications resulting in pneumonia. This occurs more often in adults. With Kristie's normal chest x-ray, this complication has been ruled out.

TREATMENT

Ordinarily, chickenpox does not require any specific therapy. Kristie's father was told to apply wet compresses to the lesions to help reduce the itching and to bathe Kristie often with soap and water. He was told to put mittens on Kristie if she began scratching the lesions and to put clean clothes on her each day. *All of the advice given to Kristie's father is designed to decrease the chances of superinfection. Chickenpox lesions typically clear up without scarring in 1 to 2 weeks. Lesions that are disturbed by scratching and become infected have a greater chance of scarring.*

COURSE

Kristie went home that day and was put to bed by her father. He monitored her temperature, which returned to normal the next day *(fever usually only lasts for the first 1 to 2 days after the appearance of the rash)* and kept her comfortable. Kristie began to feel better over the next few days, and a week later, Kristie was brought back to the HMO for a follow-up visit. The rash was disappearing and her vital signs were normal. Kristie was well on the road to recovery. *The varicella-zoster virus may lie dormant in nerve cell bodies for a number of years following a bout of chickenpox. It can become active later in life and travel to the nerve endings, causing another skin eruption. Based on your reading of this chapter, what is this second skin disease called? What is unique about the distribution of the rash compared to that of the chickenpox rash?*

READINGS

Adamson PB: 1977. The spread of rabies in Europe and the probable origin of this disease in antiquity, *Journal Royal Asiatic Society* 1977:140-144.

Interesting discussion of history of rabies.

Anderson RM and RM May: 1992. Understanding the AIDS pandemic, *Scientific American* 266(5):58-61, 64-66.

An in-depth view of the spread of AIDS.

Anonymous: 1987. Smallpox: never again! *World Health*, August/September:1-25.

A special issue devoted to the eradication of smallpox and the role of the World Health Organization in this endeavor.

Anonymous: 1989. An image of the past? *World Health*, December 1989:2-29.

A special issue devoted to the problems involved with polio vaccines and polio vaccination programs.

Beardsley T: 1990. Oravske kuru, *Scientific American* 263(2):24+.

Discusses public health efforts in Czechoslovakia to deal with Creutzfeldt-Jakob disease and the virus that causes it.

Becker Y: 1983. *Molecular Virology: Molecular and Medical Aspects of Disease-causing Viruses of Man and Animals*, The Hague, Matinus Nijhoff Publishers.

A review of the molecular virology of viruses.

Brookmeyer R and MH Gail: 1994. *AIDS Epidemiology: A Quantitative Approach.* New York, Oxford University Press.

Describes the spread of AIDS and shows the basis for the AIDS epidemic.

Cohen PT, MA Sande, P Volberding (eds.): 1994. *The AIDS Knowledge Base: A Textbook on HIV Disease from the University of California, San Francisco, and the San Francisco General Hospital*, Boston, Little, Brown.

Provides a critical perspective to AIDS based on the experiences in San Francisco.

Cassidy J: 1990. Rabies, *Current Health 2* 16:27-29.

Discusses the current status of rabies.

Diamond JM: 1990. A pox on our genes: smallpox vanished twelve years ago, but its genetic legacy may still linger within us, *Natural History:* February, pp. 26-30.

Discusses the role of smallpox on natural selection, immunogenetics, and blood groups.

Dimock NJ, PD Griffiths, CR Madeley: 1990. *Control of Virus Diseases.* Cambridge, England; Cambridge University Press.

Discusses vaccines and other methods used for preventing viral diseases.

Dimock NJ, and PD Minor (eds.): 1989. *Immune Responses, Virus Infections, and Disease*, Oxford, England; Oxford University Press.

A publication of the Society for General Microbiology in England based on the proceedings of a meeting of the society.

Dixon B: 1986. Decade of the killer brain infection, *Science 86* 7:62-64+.

The story of the 1916-1926 encephalitis lethargica epidemic.

Evans AS (ed.): 1989. *Viral Infections of Humans*, New York, Plenum Medical Book Co.

Reviews the epidemiology, prevention, and control of viral diseases.

Farrar WE: 1994. *Areas of Infections of the Nervous System*, St. Louis, Mosby.

Covers all common infections of the nervous systems with color photographs of clinical and histological appearances and radiological images.

Fenner L: 1985. Shingles: painful ghost of chickenpox, *FDA Consumer* 18:10-13.

Explains the symptoms and cause of shingles and describes the research currently being done to alleviate the symptoms and prevent the disease.

Fraenkel-Conrat H and RR Wagner (eds.): 1983. *Virus-host Interactions: Receptors, Persistence, and Neurological Diseases,* New York, Plenum Press.

Surface receptors on viruses affect the physiopathology of viral diseases.

Friedman-Kien AE (ed.): 1994. *Color Atlas of AIDS,* ed. 2, Philadelphia, Saunders.

Gives a pictorial view of the development of AIDS.

Galasso GJ, TC Merigan, RA Buchanan (eds.): 1984. *Antiviral Agents and Viral Diseases of Man,* New York, Raven Press.

Presents the agents used in antiviral therapy and the concepts that dictate their ability to manage disease.

Grimes DE and RM Grines: 1994. *AIDS and HIV Infection,* St. Louis, Mosby.

Covers the pathology and HIV infections leading to AIDS.

Grmek MD: 1990. *History of AIDS: Emergence and Origin of a Modern Pandemic,* Princeton, NJ; Princeton University Press.

Analyzes the recent history of the disease and speculates on its prehistory. Includes a history of virology and the discovery of oncogenes, retroviruses, and slow viruses and how, therefore, virologists were ready in 1980 to identify the causal organism of AIDS.

Halstead LS and G Grumby: 1994. *Post-Polio Syndrome,* St. Louis, Mosby.

Reviews the late emerging effects of polio and their treatments and discusses new methods of respiratory management and new theories of immunologic aspects.

Halstead SB: 1988. Pathogenesis of dengue: challenges to molecular biology, *Science* 239:476-481.

Suggests that dengue antibody has a role in mediating infection of phagocytic target cells. Presents a system with which to study the function and structure of the dengue virus.

Horowitz J: 1985. Polio's painful legacy, *New York Times Magazine,* July 7:16-19+.

Discusses the problem of post-polio syndrome, a condition affecting many poliomyelitis victims years after the original disease incident.

Hsiao K, Z Meiner, E Kahana et al: 1991. Mutation of the prion protein in Libyan Jews with Creutzfeldt-Jakob disease, *New England Journal of Medicine* 324:1091-1097.

Relates how a consistent mutation was found in the prion-protein gene among 11 Libyan Jews with Creutzfeldt-Jakob disease. This strongly suggests a genetic role in the onset of this disease.

Kaslow RA and DP Francis (eds.): 1989. *The Epidemiology of AIDS: Expression, Occurrence, and Control of Human Immunodeficiency Virus Type 1 Infection,* New York, Oxford University Press.

Series of articles on AIDS

Kurstak E (ed.): 1992. *Control of Virus Diseases,* ed. 2, New York, Marcel Dekker.

Includes up-to-the-minute reports on the global attempt to control AIDS and hepatitis—with contributions from worldwide experts.

Lahart CJ: 1994. *Q & A in HIV/AIDS,* St. Louis, Mosby.

Uses a question and answer format to address current diagnostic and management strategies and controversies.

Leibowitch J: 1985. *A Strange Virus of Unknown Origin,* New York, Ballantine Books.

The story of this French researcher's investigations of the etiological agent that causes AIDS.

Mills J and L Corey (eds.): 1989. *Antiviral Chemotherapy: New Directions for Clinical Application and Research,* New York, Elsevier.

Discusses the current state of drug therapy and the antiviral agents used in the treatment of viral diseases.

Minerbrook S and FL Kritz: 1990. Return of a childhood killer, *U.S. News and World Report* 109:63-64.

Discusses the rise in cases of measles along with the decrease in the incidence of vaccination against this disease.

Munsat TL: 1991. Poliomyelitis—new problems with an old disease, *New England Journal of Medicine* 324:1206-1207.

Explains the history of post-polio syndrome and reviews current research in the area.

Olsen LF and WM Schaffer: 1990. Chaos vs. noisy periodicity: alternative hypotheses for childhood epidemics, *Science* 249:499-504.

Discusses measles and chickenpox in relation to epidemiology and the chaos theory.

Oppenheim IA: 1987. The Epstein-Barr virus: beyond mono, *Current Health 2* 13:18-19.

Discusses the consequences of Epstein-Barr viral infections.

Oppenheim IA: 1987. The spotted story about measles after childhood, *Current Health 2* 13:16-17.

Discusses the effects of measles in young adults.

Palca J: 1991. The sobering geography of AIDS, *Science* 252:372-373.

Relates how the rate at which the epidemic is spreading in North America may be slowing, but suggests it is poised to take off in Asia and Latin America and has not abated in Africa.

Radetsky P: 1989. Forgotten but not gone, *Discover* 10:22+.

Discusses the occurrence of polio especially as a public health problem in developing countries and efforts at vaccination programs.

Russell WC and JW Almond (eds.): 1987. *Molecular Basis of Virus Disease,* Cambridge, England; Cambridge University Press.

Reviews the microbiology of viral diseases.

Schlossberg D (ed.): 1990. *Infections of the Nervous System,* New York, Springer-Verlag.

Discusses diseases of the central nervous system, including viral diseases such as meningitis.

Schupbach J: 1989. *Human Retrovirology: Facts and Concepts,* Berlin, Springer-Verlag.

Explores the Retroviridae species, including HTLV viruses and the infections they cause.

Sears C: 1989. The heart part, *American Health* 8:44.

Discusses the failure of many dentists to follow appropriate procedures before dental treatments for heart patients to prevent endocarditis.

Silverstein AM: 1981. *Pure Politics and Impure Science: the Swine Flu Affair,* Baltimore, Johns Hopkins University Press.

An immunologist studies the American political system's response to a potential health crisis, which did not materialize because the virus did not spread as predicted. The author agrees with and defends the fired public health officials who recommended the immunization program.

Smith JS: 1990. *Patenting the Sun: Polio, the Salk Vaccine and the Children of the Baby Boom,* New York, Morrow.

Follows the social and political history of the poliomyelitis epidemics and the development of the poliomyelitis vaccine and implementation of the vaccination program in the United States. A personal essay on the interplay among medicine, politics, psychology, culture, and collective memory.

Stanley SR and V Carty: 1994. *Compendium of HIV/AIDS Positions, Policies, And Documents,* Washington, American Nurses Association.

A compilation of the guidelines for nurses and other medical staff for the safe treatment of individuals with AIDS.

Winkler WG and K Bogel: 1992. Control of rabies in wildlife, *Scientific American* 266(6):86-92.

A description of a fascinating use of vaccines for controlling rabies in wild animal populations.

Zamula E: 1990. Reye syndrome: the decline of a disease, *FDA Consumer* 24:20-23.

Follows the decline in the incidence of Reye syndrome with the publicity program against the use of aspirin for the control of fevers in childhood diseases.

Zigas V: 1990. *Laughing Death: the Untold Story of Kuru,* Clifton, NJ; Humana Press.

Relates the history of kuru in Papua, New Guinea.

Zuckerman AJ, JE Banatvala, JR Pattison (eds.): 1987. *Principles and Practices of Clinical Virology,* Chichester, Eng.; John Wiley & Sons.

Includes chapters on herpesvirus infections, diarrheal diseases, respiratory tract diseases, measles, rubella, mumps, enteroviruses, and other viruses that cause human diseases.

CHAPTER 20

Diseases Caused by Bacteria: Diseases of the Respiratory and Gastrointestinal Tracts

PREVIEW TO CHAPTER 20

In this chapter we will:
- Learn about the diseases of humans caused by bacteria.
- Examine the properties of pathogenic bacteria.
- Review the major diseases of humans caused by
  bacteria that affect the respiratory and gastrointestinal
  tracts, examining the specific etiologic agent, the route
  of transmission, the major symptoms, and the
  treatments of each disease.
- Learn the following key terms and names:

Several factors are important in establishing the virulence of bacterial pathogens. Virulence is the ability of bacteria to establish infections and to cause human diseases. These factors are the ability of the bacteria to invade the human body, to avoid the immune defenses that are designed to destroy invading bacteria, and to disrupt body functions in a way that causes disease.

The surface layers surrounding a pathogenic microorganism can interfere with the action of phagocytic blood cells. Phagocytes engulf and destroy pathogens when they invade the human body. Capsules surrounding the cells of *Streptococcus pneumoniae, Haemophilus influenzae,* and *Klebsiella pneumoniae* permit these bacteria to avoid destruction by phagocytic blood cells, allowing them to reproduce and to cause pneumonia.

The pili of several pathogenic bacteria have a key role in permitting these bacteria to adhere to host cells. Once adherence occurs, these bacteria can establish an infection. For example, the pili of *Neisseria gonorrhoeae* are important in their ability to adhere to the lining of the genitourinary tract and initiate an infection that causes gonorrhea. Certain strains of *Escherichia coli* and *Vibrio cholerae* have pili that permit them to bind to the mucosal lining of the intestine.

Some bacteria produce toxins that enable them to cause human diseases (FIG. 20-1). **Toxins** disrupt the normal functions or are generally disruptive to target human cells and tissues. Bacterial **endotoxin,** which refers to the lipopolysaccharide (LPS) portion of the cell wall of Gram-negative bacteria, causes various physiological effects when it is released into the body. These effects include fever, circulatory changes that produce shock, and other general symptoms, such as weakness and nonlocalized aches. Protein toxins **(exotoxins)** cause distinctive clinical symptoms. Often, protein toxins are referred to by the dis-

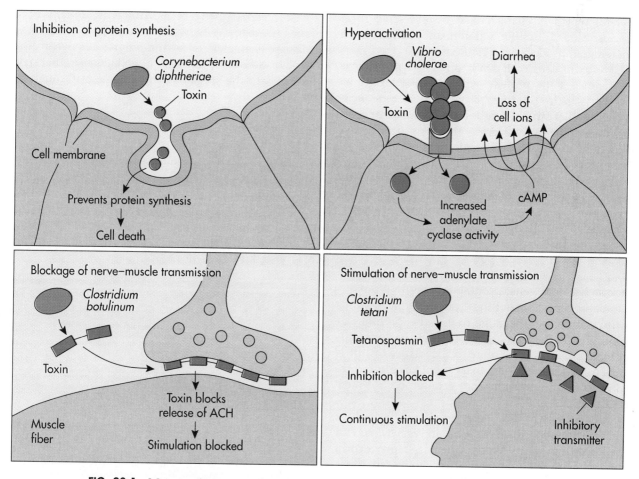

FIG. 20-1   Many pathogens produce toxins that disrupt body functions, resulting in disease. Diptheria toxin from *Corynebacterium diptheriae* blocks protein synthesis. Cholera toxin from *Vibrio cholerae* activates adenylcyclase, producing cyclic AMP; sodium ion transport is blocked and chloride ions and water move into the intestine from the blood. This causes severe diarrhea and water loss. Botulinum toxin from *Clostridium botulinum* blocks release of acetylcholine from nerve cells, causing paralysis due to blockage of motor neuron transmission. Tetanus caused by *Clostridium tetani* is characterized by arching of the back and neck and "lockjaw" due to convulsive spastic paralysis.

ease they cause, such as diphtheria toxin or botulinum toxin. Toxins are also described according to the cells and associated body systems they affect, such as neurotoxins that affect the nervous system, enterotoxins that affect the gastrointestinal tract, and cytotoxins that interfere with cellular functions.

Neurotoxins interfere with nerve transmissions. The neurotoxins produced by *Clostridium botulinum* block the release of acetylcholine from nerve cells of the central nervous system, causing the loss of motor function *(flaccid paralysis)*. The neurotoxin tetanospasmin produced by *Clostridium tetani* interferes with the peripheral nerves of the spinal cord. Tetanospasmin blocks the ability of nerve cells properly to transmit signals to muscle cells, by blocking the relaxing or inhibitory nerve signals; this causes the symptomatic spasmodic paralysis of tetanus *(spastic paralysis)* and other signs such as "lockjaw."

Enterotoxins cause an inflammation of the gastrointestinal tract. Typically, enterotoxins cause excessive secretions of fluid and electrolytes from the lining of the gastrointestinal tract. For example, cholera, which is characterized by excessive fluid loss, is caused by the enterotoxin cholera toxin (choleragen) produced by *Vibrio cholerae*.

Cytotoxins cause cell death, often by lysis and/or interference with protein synthesis. For example, diphtheria toxin, produced by *Corynebacterium diphtheriae*, inhibits protein synthesis in mammalian cells. The toxin blocks transferase reactions during the translation of mRNA by specifically blocking the reaction that is necessary for the elongation of a polypeptide during protein synthesis. This prevents subsequent addition of amino acids and thus the elongation of the peptide chain. Cytotoxins that cause the lysis of human erythrocytes are termed *hemolysins* because their action results in the release of hemoglobin from these red blood cells. *Streptococcus* species produce various hemolysins, including streptolysin O, an oxygen-labile and heat stable protein, and streptolysin S, an acid-sensitive and heat-labile protein. The hemolytic action of these cytotoxins produces zones of clearing when these bacteria are grown on blood agar plates. A complete zone of clearing due to hemolysis around a bacterial colony growing on a blood agar plate is referred to as β-hemolysis and the partial hemolysis is referred to as α-hemolysis.

Several bacteria produce enzymes that destroy body tissues. For example, hyaluronidase breaks down hyaluronic acid, the substance that holds together the cells of some tissues. Bacterial pathogens that produce hyaluronidases spread through body tissues and, therefore, hyaluronidase is called a *spreading factor*. Various species of *Staphylococcus*, *Streptococcus*, and *Clostridium* produce hyaluronidases. Some *Clostridium* species also produce collagenase, an enzyme that breaks down the proteins of tissues. *C. perfringens*, which causes gas gangrene—a disease characterized by the progressive decay of infected tissues—produces a collagenase that contributes to the spread of this organism through the human body.

Some bacteria produce toxins that disrupt normal cell activity.

Some toxins cause clinical symptoms.

Different toxins affect cells in different ways, causing a variety of damages.

CLASSIFICATION OF BACTERIAL PATHOGENS

Various bacterial genera contain species that are human pathogens. Some of these are the typical Gram-positive and Gram-negative rods and cocci that can be cultured in the laboratory on routine bacteriological media. Others have more exotic shapes or nutritional requirements. Some that cannot be cultured require serological procedures or gene probes for their identification.

There are two medically important genera of Gram-positive cocci: *Streptococcus* and *Staphylococcus*. **Streptococci** grow in chains, whereas **staphylococci** produce grape-like clusters. Streptococci can cause many diseases in humans (Table 20-1). β-hemolytic streptococci, which cause most streptococcal infections, cause lysis of red blood cells; this can readily be seen as a zone of clearing around colonies of these bacteria growing on blood agar plates (FIG. 20-2). Some of the most virulent β-hemolytic streptococci

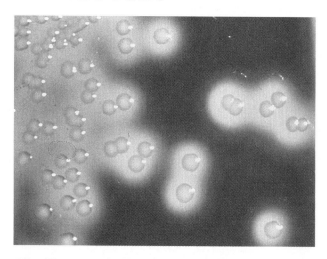

FIG. 20-2 Blood agar plate showing beta hemolysis (zones of clearing due to complete hemolysis of red blood cells) around colonies of *Streptococcus pyogenes*.

TABLE 20-1

Some Human Infections Caused by Streptococci and Enterococci

SPECIES	GROUP	DISEASE
Streptococcus agalactiae	Group B	Meningitis in newborns, wound infections
Streptococcus anginosus	Group F, G, and L	Endocarditis, pharyngitis
Streptococcus bovis	Group D	Endocarditis
Streptococcus equisimilis	Group C	Endocarditis, pharyngitis
Enterococcus faecalis, Enterococcus faecium	Group D	Urinary tract infections, endocarditis
Streptococcus milleri	Group F	Endocarditis, tissue abscesses
Streptococcus mitis	Group O, M	Endocarditis, tooth abscesses
Streptococcus mutans	Not grouped	Caries
Streptococcus salivarius	Group K	Endocarditis
Streptococcus sanguis	Group H	Endocarditis, caries
Streptococcus pneumoniae	Not grouped	Pneumonia
Streptococcus pyogenes	Group A	Skin infections, puerperal fever, pharyngitis, scarlet fever

are the group A streptococci, so designated because of a particular polysaccharide surface antigen that occurs in the cell walls of these bacteria. Streptococci cause mild infections of the upper respiratory tract, "strep throat," and more serious infections such as rheumatic fever. Staphylococci occur at many sites of the body, especially the skin, without causing disease. Like the streptococci, however, some staphylococci secrete extracellular enzymes and toxins that can cause human diseases. One of these, coagulase, clots plasma and camouflages the bacteria from the immune system, which makes it a virulence factor (FIG. 20-3). Coagulase production is useful in classification, since only the more pathogenic species,

Staphylococcus aureus, make this enzyme. *S. aureus* causes mild infections of the upper respiratory tract, boils, pimples, and serious infections in deep tissues, such as osteomyelitis (infection of the bone marrow) and endocarditis (infection of the heart valves).

Streptococcus and Staphylococcus are medically important genera of Gram-positive coccal bacteria.

The principal Gram-negative coccus of medical importance is the diplococcus *Neisseria.* Like other Gram-negative bacteria, *Neisseria* has an outer membrane that contains endotoxin (lipopolysaccharide) that can cause disease symptoms. Species of *Neisseria* cause gonorrhea and meningitis.

FIG. 20-3 The production of coagulase distinguishes *Staphylococcus aureus* from *Staphylococcus epidermidis.* **A,** The slide test demonstrates bound coagulase by its ability to cause staphylococcal cells to clump within a few seconds when mixed with plasma. **B,** The tube test detects free coagulase by its ability to cause plasma to clot after incubation at 37° C for about 4 hours.

Both anaerobic and aerobic Gram-positive rods cause various human diseases. Several *Clostridium* species are medically important (Table 20-2). Clostridia are anaerobic endospore formers (FIG. 20-4). *C. botulinum* causes botulism, *C. tetani* causes tetanus, *C. perfringens* causes gas gangrene. The symptoms of these diseases are produced by the toxins formed by these bacteria. Another important pathogen among the Gram-positive rods is *Listeria;* this anaerobe occasionally causes serious infections in infants and adults who are immunocompromised. Also *Corynebacterium diphtheriae,* which causes diphtheria, is a falcultative Gram-positive rod.

The Gram-negative rods are a large group of bacteria that includes many important pathogens (Table 20-3). The enteric bacteria (members of the family *En-*

TABLE 20-2

Human Infections Caused by Clostridia

SPECIES	DISEASE
Clostridium botulinum	Botulism
Clostridium difficile	Normal inhabitant of colon; opportunistic pathogen that causes antibiotic-associated colitis
Clostridium histolyticum	Gas gangrene
Clostridium novyi	Gas gangrene
Clostridium perfringens	Gas gangrene (most common cause)
Clostridium septicum	Gas gangrene
Clostridium tetani	Tetanus

TABLE 20-3

Human Infections Caused by Gram-negative Bacteria

SPECIES	DISEASE
Pseudomonas aeruginosa, Pseudomonas cepacia	Opportunistic pathogen associated with burns, wounds, catheters, and cystic fibrosis patients
Brucella abortus, Brucella melitensis	Brucellosis (undulant fever)
Francisella tularensis	Tularemia (rabbit fever)
Bordetella pertussis	Whooping cough
Legionella pneumophila	Legionnaire's disease
Enterobacteriaceae	
Escherichia coli	Pathogenic strains cause diarrhea or urinary tract infection; common cause of meningitis in newborns
E. coli O157:H7	Dysentery
Edwardsiella tarda	Opportunistic pathogen causing gastroenteritis
Citrobacter	Opportunistic secondary infections
Klebsiella pneumoniae	Pneumonia
Enterobacter	Opportunistic secondary infections
Hafnia	Opportunistic secondary infections
Serratia	Opportunistic secondary infections
Proteus	Opportunistic secondary infections
Providencia	Opportunistic secondary infections
Morganella	Opportunistic secondary infections
Salmonella typhi	Salmonellosis, typhoid fever
Salmonella choleraesuis	Septicemia
Salmonella enteritidis	Gastroenteritis
Shigella boydii	Dysentery
Shigella dysenteriae	Dysentery
Shigella flexneri	Dysentery
Shigella sonnei	Dysentery
Yersinia pestis	Plague
Yersinia pseudotuberculosis	Gastroenteritis
Yersinia enterocolitica	Gastroenteritis
Pasteurella multocida	Septicemia, skin abscesses
Haemophilus influenzae	Epiglottitis, meningitis
Gardnerella vaginalis	Vaginitis
Calymmatobacterium granulomatis	Granuloma inguinale (donovanosis)

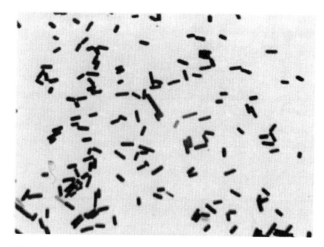

FIG. 20-4 Light micrograph of smear from a culture of *Clostridium welchii*. The morphology of this organism is sufficiently distinctive to enable experienced workers to distinguish typical strains from other clostridia. Such strains consist predominantly of single cells that are stout, relatively short, and have truncated ends and no spores.

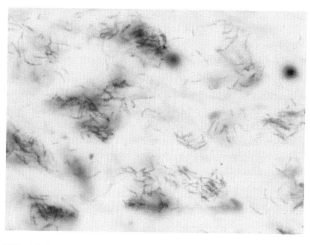

FIG. 20-5 Light micrograph of *Mycobacterium tuberculosis* (*red cells*) in a sputum sample of an individual with tuberculosis. The appearance of red rods after acid-fast staining indicates the presence of mycobacteria and is diagnostic of tuberculosis.

terobacteriaceae) are divided among those that ferment lactose—for example, *Escherichia coli*—and those that do not—for example, *Salmonella* and *Shigella*. *E. coli* is a natural inhabitant of the gastrointestinal tract and contributes to human nutrition. If *E. coli* enters the urinary tract, however, it is an opportunistic pathogen and, in fact, is a major cause of urinary tract infections. Also, some strains of *E. coli* produce toxins that affect the gastrointestinal tract; these enterotoxigenic strains are not the same strains of *E. coli* that inhabit the intestinal tracts of healthy individuals.

Shigella species cause bacterial dysentery. *Salmonella* species cause various diseases, including gastroenteritis and typhoid fever.

Several Gram-negative rods have very specialized physiological requirements for growth. Some of these bacteria are able to grow within specific tissues of the human body and cause disease. These bacteria tend to be small and to have complex nutritional requirements. Examples of such bacteria are *Haemophilus*, which causes pneumonia and meningitis, *Bordetella*, which causes whooping cough, and *Legionella*, which causes an acute respiratory infection responsible for Legionnaire's disease.

Mycobacterium is an acid-fast bacterium, so designated because it retains the primary stain during the acid-fast staining procedure (FIG. 20-5). It does so because of the unusual lipids in its cell wall. *Mycobacterium* has a Gram-positive cell wall that is loaded with lipids (waxes). The lipids protect mycobacteria against the immune defense system, enabling some mycobacteria to cause persistent human diseases (Table 20-4). *M. tuberculosis* causes tuberculosis, and *M. leprae* causes Hansen disease, which used to be known as leprosy.

Besides rods and cocci, some bacteria form curved cells. *Vibrio cholerae*, for example, which causes cholera, is a comma-shaped cell. Spirochete bacteria have helical cells that are coiled around a central filament. *Treponema pallidum*, which causes syphilis, and *Borrelia burgdorferi*, which causes Lyme disease, are examples of spirochetes that cause human diseases.

Some bacteria can only reproduce within animal cells. Chlamydiae are very small obligate intracellular bacteria that only replicate inside host eukaryotic cells. They are Gram-negative cocci. They cause sex-

TABLE 20-4	
Human Infections Caused by Mycobacteria	
SPECIES	**DISEASE**
Mycobacterium avium-intracellulare complex	Opportunistic respiratory infections, especially in AIDS patients
Mycobacterium bovis	Tuberculosis
Mycobacterium fortui-tum-chelonae complex	Wound and postsurgical infections
Mycobacterium kansasii	Opportunistic respiratory infections
Mycobacterium leprae	Leprosy (Hansen disease)
Mycobacterium marinum	Swimming pool granuloma
Mycobacterium scrofulaceum	Scrofula (cervical lymphadenitis)
Mycobacterium tuberculosis	Tuberculosis
Mycobacterium ulcerans	Skin ulcers

ually transmitted diseases and diseases of the eyes. These bacteria obtain their energy and nutrition from the host cells.

Rickettsias are also small, obligate intracellular bacteria that obtain their energy and nutrition from the host cells in which they reproduce. They are Gram-negative rods. *Rickettsia* cause typhus, Rocky Mountain spotted fever, and other diseases. In most cases rickettsia live in animals other than humans. Lice, fleas, and ticks serve as reservoirs and vectors.

Transmission of rickettsias to humans normally is through the bites of these vectors.

The previously described bacteria all have a cell wall that protects the plasma membrane of the bacterial cell. *Mycoplasma*, however, lacks a cell wall. *M. pneumoniae* can grow within the lower respiratory tract where it causes pneumonia. This disease often is called "walking pneumonia." When cultured in the laboratory, the colonies have a "fried egg" appearance.

BACTERIAL DISEASES OF THE RESPIRATORY SYSTEM

Each of us inhales about 1 million bacteria each day. Some of these bacteria are pathogens, capable of causing diseases affecting the respiratory tract (Table 20-5). In particular, aerosols released by coughing and sneezing carry pathogens from one individual to another. To establish an infection these bacteria must overcome the immune defense network that protects the respiratory tract. They must elude the mucociliary-escalator system and if they reach the lower respiratory tract, they must escape numerous phagocytic cells. Bacteria that have capsules tend to be protected against phagocytosis and are the most frequent causes of lower respiratory tract infections.

PHARYNGITIS (SORE THROAT)

The growth of bacteria in the pharynx can cause an inflammation of the upper respiratory tract known as **pharyngitis** or more commonly, as *sore throat*. The common symptoms of this disease are pain in the throat, redness of the lining of the throat (FIG. 20-6), and fever. In some cases the tonsils are also inflamed, in which case the disease is called **tonsillitis.** Before

the widespread use of antibiotics to control bacterial infections of the upper respiratory tract, tonsils often were surgically removed from those who suffered from repeated tonsillitis. The most common causes of pharyngitis and tonsillitis are infections with *Streptococcus pyogenes.* Also called Group A *Streptococcus*, the Gram-positive cocci tend to form chains and produce toxins, including hemolysins that cause beta hemolysis of red blood cells. Other bacteria can cause pharyngitis (Table 20-6).

When the causative agent of pharyngitis is *Streptococcus pyogenes*, the disease is commonly called **strep throat**. Although strep throat is readily treated using penicillin or various other antimicrobics, its diagnosis and treatment are important because the toxins produced by *S. pyogenes* can spread to other parts of the body and cause more serious diseases, notably scarlet fever and rheumatic fever. It is now routine practice in cases of suspected strep throat to streak a throat swab onto blood agar or to perform a serological test. The development of small (pinpoint) white colonies surrounded by a zone of clearing is diagnostic of β-hemolytic streptococci, such as *S. pyogenes*.

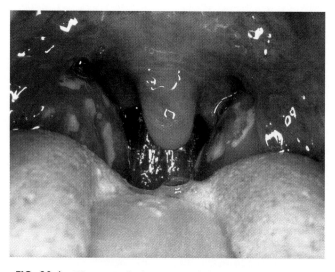

FIG. 20-6 Photograph showing red throat of pharyngitis.

TABLE 20-5

Bacterial Diseases of the Respiratory Tract

DISEASE	ORGANISM	CHARACTERISTICS
UPPER RESPIRATORY TRACT INFECTIONS		
Bronchitis	*Streptococcus pneumoniae, Mycoplasma pneumoniae,* and other bacteria	Inflammation of bronchi and bronchioles
Diphtheria	*Corynebacterium diphtheriae*	Inflammation of pharynx, often with pseudomembrane forming over mucous membranes
Epiglottitis	*Haemophilus influenzae* type b	Inflammation of epiglottis accompanied by pain in swallowing
Laryngitis	*Streptococcus pneumoniae* and *Haemophilus influenzae* type b	Inflammation of larynx, frequently with loss of voice
Otitis media	*Streptococcus pneumoniae, Streptococcus pyogenes* and *Haemophilus influenzae* type b	Infection of middle ear with formation of pus, leading to pressure and pain
Pharyngitis	*Streptococcus pyogenes* and other bacteria	Sore throat
Sinusitis	*Streptococcus pneumoniae, Streptococcus pyogenes, Staphylococcus aureus* and *Haemophilus influenzae* type b	Infection of nasal sinuses
LOWER RESPIRATORY TRACT INFECTIONS		
Whooping cough	*Bordetella pertussis*	Disease is divided into catarrhal stage, paroxysmal stage, and convalescent stage
Pneumonia	*Streptococcus pneumoniae, Klebsiella pneumoniae, Mycoplasma pneumoniae, Staphylococcus pneumoniae*	Inflammation of bronchi and alveoli of lungs
Tuberculosis	*Mycobacterium tuberculosis*	Formation of tubercles in lungs with inflammation
Ornithosis	*Chlamydia psittaci*	Inflammation of bronchi and alveoli; transmitted by birds
Q fever	*Coxiella burnetii*	Atypical pneumonia transmitted by ticks, droplet inhalation or contact with fomites
Nocardiosis	*Nocardia asteroides*	Pneumonia-like disease seen in immunocompromised individuals
Legionnaire's disease	*Legionella pneumophila*	Inflammation of bronchi and alveoli; pneumonia transmitted by inhalation of droplets from contaminated water sources

TABLE 20-6

Causes of Acute Bacterial Pharyngitis

ORGANISM	CHARACTERISTICS
Arcanobacterium haemolyticum	Similar to streptococcal pharyngitis; infects mainly teenagers and young adults
Borrelia vincenti	Vincent's angina; caused by fusospirochetes; most common in adolescents and adults
Corynebacterium diphtheriae	Diphtheria; a thick, patchy, gray-green membrane forms over the mucous membranes of the pharynx, tonsils, soft palate and nose
Haemophilus influenzae type b	Can cause epiglottitis
Mycoplasma pneumoniae	Indistinguishable from streptococcal pharyngitis
Neisseria gonorrhoeae	Often asymptomatic; usually transmitted via orogenital contact
Streptococcus pyogenes (Group A Streptococcus)	Most common cause (about 80%) of acute bacterial pharyngitis in 5-15 year olds; may be associated with a scarlatinal rash (scarlet fever); occurs most frequently from October to April

blood cells in the medium. The test may be supplemented by the inclusion of a disc impregnated with bacitracin because group A strains of *S. pyogenes* are particularly sensitive to this antibiotic and will not grow in the vicinity of the disc. Alternative immunological procedures permit identification of *S. pyogenes* infections without culturing, allowing rapid diagnosis (within minutes) that can aid the physician in selecting the appropriate method of treatment.

> **Gram-positive cocci, including streptococci, cause upper respiratory tract infections that result in inflammation (pharyngitis).**
>
> **The use of antibiotics has decreased the likeliness of spread of such infections.**

PNEUMONIA

Pneumonia is a lower respiratory tract infection. It is an inflammation of the lungs, that can be caused by several bacteria, as well as by various other microorganisms (Table 20-7). More than half of all pneumonia cases are caused by bacteria, and this disease ranks among the top causes of death from infectious diseases (FIG. 20-7).

Often, pneumonia occurs as a complication of another disease condition, such as a viral infection (FIG. 20-8). Such secondary infections typically occur when an individual is run-down and his or her physiological state depresses the effectiveness of the immune response system. This may happen after surgery or during the course of treatment for another disease. Consequently, pneumonia frequently is a **nosocomial infection;** that is, an infection acquired during hospitalization. The lack of movement and deep breathing in postsurgical and other patients reduces the efficiency of the normal defense mechanisms to clear the lungs of mucus and bacteria, and the accumulation of fluids favors the establishment of a microbial infection in the lower respiratory tract. To prevent the development of pneumonia in such patients, inhalation devices—which may be as simple as a tube with a ball that moves when air is blown into the tube—are frequently used by the patients to ensure that they are breathing deeply enough to clear their lungs of fluids.

The bacteria that cause pneumonia most frequently enter the lungs in air. Transport of pathogens to the lungs through the bloodstream can also result in pneumonia. The most frequent etiologic agent of bacterial pneumonia is *Streptococcus pneumoniae* (pneumococcus), a Gram-positive, capsule-forming coccus (FIG. 20-9). Pneumococcal pneumonia is often an *endogenous disease;* that is, the infection originates from the microorganisms normally associated with the individual's own upper respiratory tract. Several other bacteria, including *Staphylococcus aureus, Haemophilus influenzae,* and *Klebsiella pneumoniae,* are also responsible for a significant number of cases of pneumonia. Individuals with AIDS often develop pneumonia, frequently caused by *Mycobacterium avium* and *M. intracellulare.* Group A β-hemolytic streptococci occasionally cause pneumonia; they tend to cause severe pneumonia, as exemplified by the fatal infection of Jim Henson, creator of Miss Piggy, Kermit the Frog, and the other muppets. Henson died within hours of being hospitalized and despite the immediate administration of six different antibiotics.

Klebsiella pneumonia is most common in individuals who are debilitated; malnutrition is a contributing factor. Male alcoholics over age 50, particularly those who also are heavy cigarette smokers, are most susceptible. While *Klebsiella* pneumonia accounts for only about 1% of pneumonias, it has a high fatality rate (85%) unless rapidly diagnosed and treated. This bacterium is not sensitive to penicillin, and treatment with other antibiotics, such as cephalosporins, is necessary.

During the development of pneumonia, bacteria reproduce in the lung tissue. They form lesions in the lung tissue. The resultant host defense response of phagocytic cells and inflammation results in the accumulation of fluid. Coughing causes movement of fluid containing bacteria to the adjacent tissue, spreading the infection. The bacteria then move into the circulatory system and pleural cavity surround-

TABLE 20-7			
Frequency of Major Types of Bacterial Pneumonia			
CLINICAL DESCRIPTION	**CAUSATIVE AGENT**	**CASES (%)**	**TREATMENT**
Pneumococcal lobar pneumonia (classical pneumonia)	*Streptococcus pneumoniae*	Over 90	Penicillin
Primary atypical pneumonia	*Mycoplasma pneumoniae*	5-10	Tetracycline or erythromycin
Klebsiella (Friedlander) pneumonia	*Klebsiella pneumoniae*	1-5	Gentamicin
"Flu" pneumonia	*Haemophilus influenzae* type b	1-5	Ampicillin

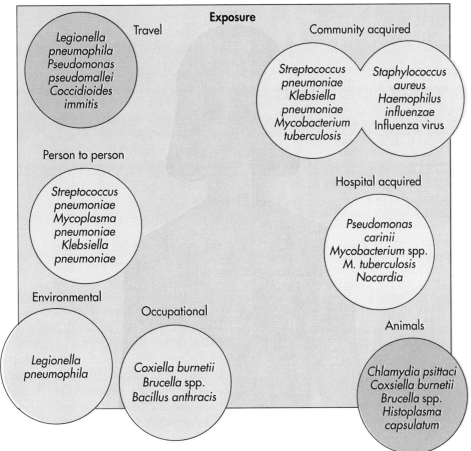

Exposure

Travel

Legionella
pneumophila
Pseudomonas
pseudomallei
Coccidioides
immitis

Community acquired

Streptococcus
pneumoniae
Klebsiella
pneumoniae
Mycobacterium
tuberculosis

Staphylococcus
aureus
Haemophilus
influenzae
Influenza virus

Person to person

Streptococcus
pneumoniae
Mycoplasma
pneumoniae
Klebsiella
pneumoniae

Hospital acquired

Pseudomonas
carinii
Mycobacterium spp.
M. tuberculosis
Nocardia

Environmental

Legionella
pneumophila

Occupational

Coxiella burnetii
Brucella spp.
Bacillus anthracis

Animals

Chlamydia psittaci
Coxsiella burnetii
Brucella spp.
Histoplasma
capsulatum

FIG. 20-7 There are many pathogens capable of causing pneumonia in adults, and the etiology is related to risk factors such as the exposure to pathogens through occupation, travel, and contact with animals. The elderly are more likely to be infected and tend to be more severely ill than young adults. These infections are often reactivating endogenous infections rather than community or hospital acquired.

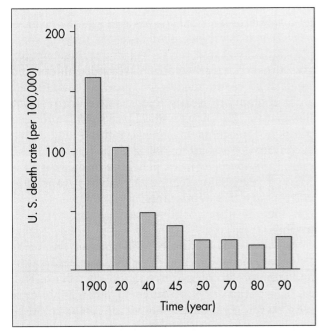

FIG. 20-8 Incidence of pneumonia and death rate due to pneumonia.

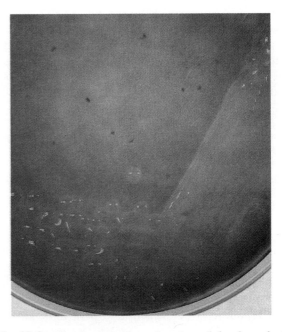

FIG. 20-9 Blood agar plate showing alpha hemolysis (greening due to partial hemolysis of red blood cells) around colonies of *Streptococcus pneumoniae*.

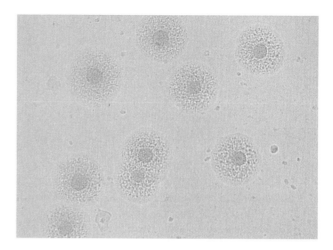

FIG. 20-10 Colonies of *Mycoplasma* with characteristic "fried-egg" appearance.

ing the lungs. The fluid that develops during pneumonia interferes with gas exchange in the lungs, and without treatment the death rate from pneumococcal pneumonia is about 30%. With antibiotic treatment, the fatality rate is reduced to about 1%.

Antibiotic treatment is used to cure bacterial pneumonia. Penicillin is the antibiotic of choice for treating pneumonia caused by *Streptococcus pneumoniae.* A vaccine (Pneumovax) has been developed from the capsular material of 23 common strains of *S. pneumoniae,* which are responsible for 80% of the cases of pneumococcal pneumonia in the United States. The vaccine is effective in preventing pneumonia and is currently used as a preventive measure for susceptible groups such as the elderly and debilitated individuals with impaired defense systems.

Pneumonia, an infection of the lower respiratory tract that often is fatal, usually is caused by indigenous bacteria after another infection or when the natural defense systems are impaired.

PRIMARY ATYPICAL PNEUMONIA

Mycoplasma pneumoniae causes **primary atypical pneumonia.** This name is derived from the facts that conventional culturing procedures produce negative results and the disease does not respond to penicillin treatment (FIG. 20-10). *M. pneumoniae* lacks a cell wall and therefore is not sensitive to penicillin. This organism, however, is sensitive to tetracycline and erythromycin, which block protein synthesis at the ri-

bosomes and can be used effectively in treatment. Unlike other mycoplasmas, *M. pneumoniae* can attach to the epithelial surface of the respiratory tract. This bacterium does not penetrate the epithelial cells nor does it produce a protein toxin, but the hydrogen peroxide released by the bacterium causes cell damage, including loss of the cilia lining the respiratory tract and death of surface endothelial cells.

Penicillin is ineffective in treating primary atypical pneumonia because *Mycoplasma* lacks a cell wall.

LEGIONNAIRE'S DISEASE (LEGIONELLOSIS)

Legionnaire's disease or **Legionellosis** is another atypical form of pneumonia. It is caused by *Legionella pneumophila* and related species in this genus. *L. pneumophila* is a Gram-negative, *fastidious* (having complex nutritional and physiological requirements for growth), rod-shaped organism, whose highly specific nutritional requirements for growth complicated early attempts to isolate the causative organism of this disease. Infections with *L. pneumophila* produce purulent (pus-containing) alveoli in the lungs. This bacterium is resistant to killing by phagocytic blood cells. *L. pneumophila* has a gene, the macrophage infectivity potentiator gene, that enables it to invade macrophage and neutrophils in the lungs. *Legionella* can then multiply within macrophage and neutrophils.

L. pneumophila produces β-lactamases (enzymes that attack penicillins and other β-lactam antibiotics) and therefore is not sensitive to most penicillins and cephalosporins. It is sensitive to other antibiotics, with erythromycin the antibiotic of choice. In addition to the typical symptoms of pneumonia, Legionnaire's disease is often characterized by kidney and liver involvement and by an unusually high incidence of associated gastrointestinal symptoms. The fever associated with this disease starts low but then typically reaches 104° to 105° F. If untreated the fatality rate is about one in six.

Legionnaire's disease received its name because the first detected disease outbreak occurred during a convention of the American Legion in Philadelphia during July 1976. The first 90,000 hours of investigation of the outbreak of this disease cost over $2 million and employed virtually all conventional isolation procedures, yet failed to reveal its causative agent. Finally the breakthrough, revealing that this disease is of bacterial etiology, came by using indirect immunofluorescent staining with antibodies from the blood of affected individuals (FIG. 20-11). Later, it was discovered that the bacterium, subsequently named *Legionella pneumophila* (lung-loving *Legionella*), could be grown on a chocolate agar medium (medium made with heated blood) if iron and cysteine were included as growth factors.

It was also later found by examining stored blood that a 1968 outbreak of a disease in Pontiac, Michigan, the etiology of which had never been identified, had been caused by a different strain of *Legionella*. Various other outbreaks of this disease have since been identified, and it is now clear that there are several different clinical manifestations of respiratory tract infections caused by *Legionella*. Mild infections include **Pontiac fever,** whereas the more severe manifestations occur as Legionnaire's disease.

Species of *Legionella* appear to be natural inhabitants of bodies of water. They grow in large numbers as free living cells and within protozoa. During periods of rapid evaporation, such as occur during summer, the bacteria can become airborne in aerosols, and inhalation of contaminated aerosols can lead to the onset of the illness. In some cases, outbreaks of Legionnaire's disease have been traced to air-conditioning cooling systems. These bacteria multiply in the cooling system waters, which are rapidly evaporated to provide cooling, and inadvertently are permitted to become airborne and circulate through air-conditioning systems. Addition of biocides to air-conditioning systems is used to prevent spread of this disease.

Legionella pneumophila **contaminates aerosols from evaporating water such as water cooling towers.**

When inhaled, *L. pneumophila* **causes Legionnaire's disease, which is treatable with erythromycin.**

NEWSBREAK

CRUISE SHIP EVACUATED AFTER OUTBREAK OF LEGIONNAIRE'S DISEASE

Twelve hundred people who had set off on a pleasure cruise were evacuated from a cruise ship when it reached Bermuda because of potential exposure to *Legionella pneumophila* from the ship's water system. This followed confirmation of several cases of Legionnaire's disease among individuals from the previous three cruises. Passengers were given leaflets informing them that there was a low risk of exposure to the bacteria that cause Legionnaire's disease when they boarded the cruise ship in New York. Before the ship sailed, the water temperature was raised to 60° C and the whirlpools were drained as precautionary measures. As the ship made its way to Bermuda, multiple cases were confirmed among previous passengers. This pointed directly to the ship as the source of infection. With more than a dozen confirmed cases and more than three dozen more suspected cases the cruise ship line ordered evacuation of the ship when it reached port in Bermuda. The passengers were given the option of transferring to another ship or flying home. Epidemiologists from the Center for Disease Control boarded the ship and found evidence of *Legionella* in the water system—especially in the hot tubs. Water systems that form aerosols, particularly those used for cooling, are the most common sources of *Legionella pneumophila*. Chlorination is effective in controlling this bacterial pathogen. The ship set out to sea, and chlorine (50 ppm) was added to the water lines. After disinfection of the water system the ship will again board passengers.

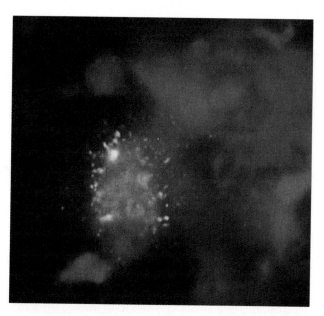

FIG. 20-11   Micrograph after fluorescent antibody staining for the specific detection of *Legionella pneumophila (green)*.

PSITTACOSIS

Another type of atypical pneumonia, **psittacosis,** is caused by *Chlamydia psittaci. C. psittaci* is an obligate intracellular bacterial pathogen, that is, one that grows exclusively within infected animal cells. Psittacosis is also known as **ornithosis** or **parrot fever** because birds act as a reservoir for *C. psittaci.*

Parakeets, canaries, other pet birds, and domestic fowl are frequently the sources of human infection. The primary route of transmission of psittacosis is from birds to human beings via aerosol dispersal of droplets and contaminated dust particles.

C. psittaci also causes a systemic infection throughout the body, multiplying in macrophage and monocytes before systemic dissemination through the bloodstream. The symptoms of psittacosis, which generally are mild, include fever, headache, malaise, and coughing. In uncomplicated cases, recovery normally occurs within 1 week, aided by the use of tetracyclines.

Birds and fowl are the reservoir for *Chlamydia psittaci,* which causes psittacosis.

BRONCHITIS

Bronchitis is an inflammatory disease of the bronchial tubes. It can be caused by several viruses and bacteria, including *Streptococcus* sp., *Staphylococcus* sp., *Haemophilus influenzae,* and *Mycoplasma pneumoniae.* It is difficult to define the specific etiologic agent because bronchitis almost always occurs as a complication of another disease condition, such as pharyngitis. Bronchitis often appears as a secondary infection when bacteria endogenous to the upper respiratory tract move to the lower respiratory tract. The onset of bronchitis is marked by the development of a cough that eventually yields mucopurulent sputum, which contains mucus and pus, reflective of the development of bronchial congestion. Acute bronchitis caused by bacteria can be effectively treated with antibiotics such as penicillin and tetracycline.

The development of chronic bronchitis is not due to microbial infection alone. Irritation of the bronchi by repeated microbial infections and/or the inhalation of irritants such as cigarette smoke also may be involved. These irritations compromise the normal secretory and ciliary function of the bronchial mucosa, and the resulting excessive mucus secretion in the bronchi favors bacterial growth and the establishment of infection. The mucus secretions result in a chronic cough that characterizes bronchitis.

Bronchitis is caused by microbial infection that causes an inflammation of the bronchi.

ACUTE EPIGLOTTITIS

When *Haemophilus influenzae* infects the epiglottis it can cause **acute epiglottitis,** which can result in death within 24 hours. The epiglottis is the structure that covers the larynx (voice box) and prevents the flow of fluids, intended to move into the gastrointestinal tract, from entering the lower respiratory tract. When the epiglottis is infected with *H. influenzae* it becomes severely inflamed and no longer properly opens and closes the entrance to the lower respiratory tract. Blockage of the entrance to the larynx can result in asphyxiation. Often the onset of acute epiglottitis is sudden and a surgical incision into the trachea, a tracheotomy, must be performed immediately to maintain an airway and prevent death.

DIPHTHERIA

Diphtheria is caused by the bacterium *Corynebacterium diphtheriae. C. diphtheriae* is a club-shaped bacterium that stains irregularly and shows incomplete separation of cells after division (FIG. 20-12). Actually only strains of *C. diphtheriae* harboring a temperate phage produce the toxin that causes diphtheria because the gene encoding the toxin is in the viral DNA. Diphtheria toxin is a potent protein exotoxin exhibiting toxicity against almost all mammalian cells.

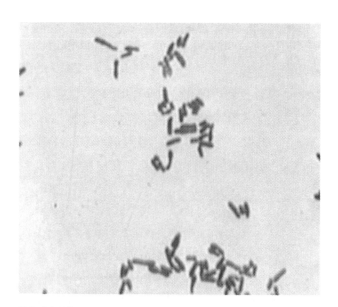

FIG. 20-12 Light micrograph of smear from a culture of a diphtheria bacillus. The club shape (thick at one end and tapering toward the other), the arrangement in so-called Chinese letter patterns, and the slight curvature of the rods are all typical of the corynebacteria. Gram-stained smears are of little value in the identification of *Corynebacterium diphtheriae;* methylene blue and special stains for metachromatic granules are more useful.

C. diphtheriae is normally transmitted via droplets from an infected individual to a susceptible host. In the new host it establishes a localized infection on the surface of the mucosal lining of the upper respiratory tract. The bacteria generally do not invade the tissues of the respiratory tract. Rather the exotoxin produced by the bacteria on the surface of the respiratory tract disseminates through the body; the toxin inhibits protein synthesis at the 80S ribosomes of eukaryotic cells, and this causes the severe symptoms of this disease. In severe infections with *C. diphtheriae*, symptoms include low-grade fever, cough, sore throat, swelling of the lymph glands, difficulty in swallowing due to swelling of the throat, and the formation of grey-white pseudomembranes on the tonsils and the back of the throat (FIG. 20-13). Death can result from physical blockage of respiratory gas exchange. Disseminated diphtheria toxin also can destroy cardiac, kidney, and nervous tissues, causing death. Treatment of diphtheria involves the use of antitoxin to block the cytopathic effects of diphtheria toxin. The use of antitoxin prevents the occurrence of serious symptoms associated with diphtheria toxin.

Corynebacterium diphtheriae produces an exotoxin that kills human cells.

Antitoxin is used to block the cytopathic effects of diphtheria toxin.

The extensive use of vaccines to prevent diphtheria has greatly reduced the incidence of this disease but has not altered the case fatality ratio (FIG. 20-14). In immunized individuals, infection with toxin-producing strains of *C. diphtheriae* is generally restricted to a localized pharyngitis with no serious complications. Diphtheria, however, remains a serious problem in socioeconomically depressed regions of the world where extensive immunization is not practiced.

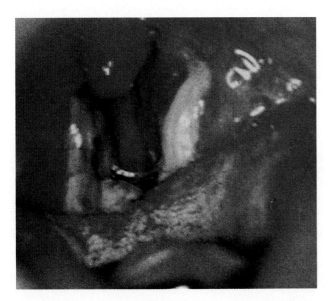

FIG. 20-13 A grey pseudomembrane composed of fibrin, necrotic epithelial cells, PMNs, other blood cells, and cells of *Corynebacterium diphtheriae* forms during diphtheria. In severe diphtheria the membrane may be thin and transparent, especially at the spreading edge. The older parts of the membrane are usually greyish-yellow but if there has been bleeding into the membrane the color may alter to green or black. The membrane is firmly adherent and forcible removal causes slight bleeding. The underlying mucosa is not ulcerated and the membrane forms again after 24 hours.

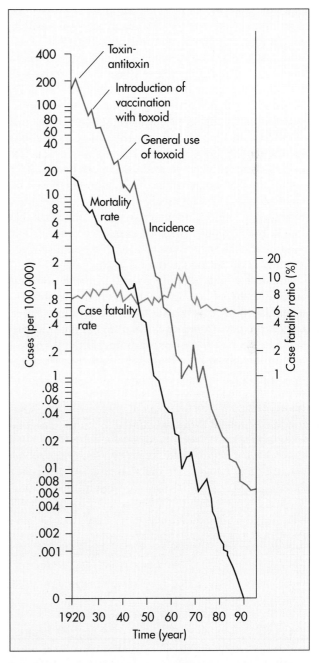

FIG. 20-14 Incidence and mortality due to diphtheria has dramatically declined after the introduction of antitoxin for treatment and toxoid for vaccination to prevent this disease.

WHOOPING COUGH (PERTUSSIS)

Whooping cough or **pertussis** derives its name from the distinctive symptomatic cough associated with this disease. These short coughs produce copious amounts of mucus, followed by a gasp for air that produces a characteristic "whoop" sound. Other symptoms of pertussis resemble those of the common cold, although vomiting often occurs after severe coughing episodes. Strangulation on aspirated vomitus is often the cause of death for the few fatalities associated with this disease. Whooping cough is caused by *Bordetella pertussis*, a Gram-negative coccobacillus, that has fastidious nutritional requirements and produces several toxins (FIG. 20-15). Pili and other surface components, such as a hemagglutinin, are important in the attachment of the bacterium to the epithelium of the respiratory tract. The organism does not penetrate the respiratory tract. *B. pertussis* is capable of reproducing within the respiratory tract, and high numbers of *B. pertussis* are found on the surface tissues of the bronchi and trachea in cases of pertussis. It is transmitted from one individual to another via aerosols. The symptoms of the disease generally are restricted to the respiratory tract. Symptoms begin 1 to 3 weeks after infection. The blockage of gas exchange can cause central nervous system anoxia. Secondary pneumonia may also occur as a result of damage to the respiratory tract.

Treatment of whooping cough primarily involves maintenance of an adequate oxygen supply. The administration of pertussis vaccine has greatly reduced the occurrence of whooping cough, and prevention of the disease is accomplished by routine immunization of infants. This vaccine is part of the DPT vaccine. There may be side effects from the pertussis portion of the vaccine, including pain at the site of vaccine administration, fever, and malaise (while not serious, this occurs in up to 20% of immunized infants), convulsions (up to 0.5% of immunized infants), and permanent neurological damage (<0.001% of immunized infants).

> **Whooping cough (pertussis) is caused by *Bordetella pertussis* and is characterized by a distinctive cough.**

TUBERCULOSIS

Tuberculosis is caused by *Mycobacterium tuberculosis* and related mycobacterial species. Mycobacteria are slender, acid-fast rods that have a lipid-rich cell wall. Tuberculosis is primarily transmitted via droplets from an infected to a susceptible individual. The mycobacteria causing this disease can also be transmitted through the ingestion of contaminated food, and before the extensive use of pasteurization, milk contaminated with *M. tuberculosis* or *M. bovis* was associated with outbreaks of this disease. About 90% of infections with *M. tuberculosis* produce asymptomatic disease conditions. In the remaining 10% of cases there usually is pulmonary disease.

The frequency of tuberculosis has been greatly reduced from the early to mid-1900s when this disease affected so many Americans that most cities were forced to construct separate hospitals for its treatment (FIG. 20-16). Immunization with the live attenuated BCG vaccine (*Bacillus* Calmette-Guerin [named

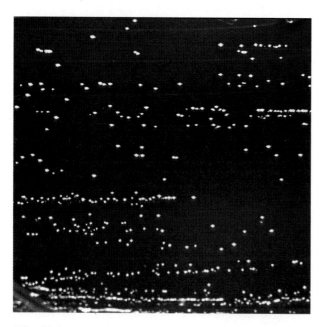

FIG. 20-15 *Bordetella pertussis* growing on a differential medium, Bordet-Gengou agar, after 48 hours of incubation.

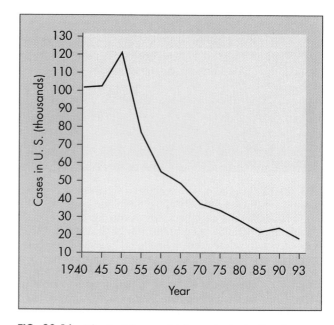

FIG. 20-16 The incidence of tuberculosis declined, starting in the 1950s, with the introduction of antimicrobics such as streptomycin.

after the discoverers]) can be used when tuberculosis is prevalent. The vaccine does not prevent infection but initiates the immune response so that proliferation of the mycobacteria in the body is limited. Antibiotics are effective in controlling most infections with *M. tuberculosis* and other tuberculosis-causing mycobacteria. There is an increased incidence, however, of antibiotic resistant strains that pose a serious new health threat.

Nevertheless, tuberculosis remains a common disease, with about 25,000 cases occurring annually in the United States. The number of cases is increasing, especially among individuals with AIDS. The global occurrence of tuberculosis is estimated to be 150 million persons infected with tubercule bacilli, 20 million sputum-positive individuals capable of disseminating the disease, 3 to 5 million new cases each year, and 600,000 deaths per year. Other estimates are even higher, suggesting that up to 20% of the world's population may by infected with tuberculosis-causing mycobacteria. Many of these strains are resistant to multiple antibiotics. Immunocompromised individuals are especially at risk from such strains.

> **There has been a significant increase in tuberculosis in immunocompromised individuals, such as individuals with HIV infections.**

Many of the new cases are caused by antibiotic resistant strains of mycobacteria. Besides *M. tuberculosis,* many of the cases of tuberculosis in immunocompromised individuals are caused by *M. avium* and *M. intracellulare* (*M. avium* complex), species normally associated with birds. New public health programs are needed to cope with the increased prevalence of tuberculosis among individuals with AIDS, intravenous drug users, and those living in poverty, including the homeless.

In the pulmonary form of tuberculosis, *M. tuberculosis* multiplies within the lower respiratory tract. This leads to a wasted appearance that accounts for the old-fashioned name of the disease—*consumption.* There is inflammation and lesions of lung tissue. When lung tissue is infected with *M. tuberculosis,* exudative lesions initially form. These lesions contain the mycobacteria, phagocytic leukocytes, and exudate fluid. This reaction appears as an area of nonspecific inflammation that resembles pneumonia. *M. tuberculosis* escapes the phagocytic activity of macrophages and does not elicit a strong antibody-mediated immune response. Infection with *M. tuberculosis* elicits a cellular immune response because bacteria are able to reproduce within phagocytic cells, and a delayed hypersensitivity reaction is typical. The delayed hypersensitivity reaction acts to contain the infection, but it also is responsible for the damage to body tissues that occurs from tuberculosis. The macrophages responding to the infection undergo a major modification, becoming concentrically

arranged in the form of elongated cells to create "tubercules," which are characteristic of this disease. (The name of the disease is based on the formation of these tubercules.) Small granulomas form, consisting of epithelioid cells and giant cells. This granulomatous inflammation occurs 2 to 4 weeks after the formation of an initial lesion and only after a cell-mediated immune response occurs. The granulomas later become necrotic and cheesy in composition. They may heal and become calcified, persisting this way for life.

> *Mycobacterium tuberculosis* is able to evade the immune defense mechanisms and grow within the pulmonary system, producing inflammation and lesions of lung tissue.

Dormant mycobacteria can remain viable for indefinite periods within the tubercules, and the infectious process can be reactivated at a later time, when physiological factors reactivate the disease. Secondary tuberculosis due to reactivation of dormant mycobacteria usually is a consequence of impaired immune function, such as due to AIDS. Malnutrition and stress are important factors relating to the susceptibility to tuberculosis. They also affect the course of the disease. In some cases the infection is restricted to the area of primary lesions, but in others it spreads into various other tissues. Mycobacteria can spread into blood and cause disseminated disease (*miliary tuberculosis*). Tuberculosis can occur in bone and central nervous system tissues. Disease symptoms, including fatigue, weight loss, and fever, generally do not appear until there are extensive lesions in the lung tissues. In some cases the vertebrae disintegrate, rendering the victims hunchbacks. As a result of the slow growth rate of *M. tuberculosis* and the ineffectiveness of phagocytic cells in killing this bacterial species, tuberculosis is generally a persistent and progressive infection. Without treatment, the disease is often fatal.

Diagnosis of tuberculosis is based on chest X-rays and skin testing to establish potential tuberculosis infections (FIG. 20-17). Direct microscopy (acid-fast stain) and cultural examination of sputum samples verify the presence of acid-fast rods. Confirmation of a positive, active case of tuberculosis is difficult and time consuming. Culturing of mycobacteria from sputum samples takes weeks because of the slow growth of this bacterium.

Skin testing often is performed when there is reason to suspect that an individual, for example, a family member or health care worker, may have been exposed to someone with an active case of tuberculosis. The skin test involves the introduction of tuberculin proteins into the skin and observation for a delayed hypersensitivity reaction. The Tine and Mantoux tests are used for this purpose. In the **Mantoux test** a standard amount of purified protein derivative

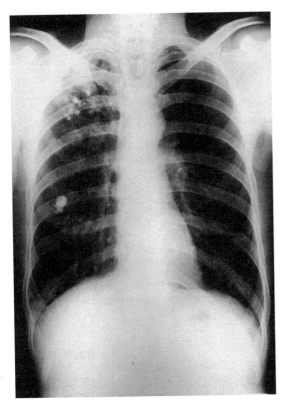

FIG. 20-17   X-rays reveal calcified areas (tubercles) where mycobacteria infect the lung in cases of tuberculosis, as shown here as a clear zone in the right lung (*left side of X-ray*).

(PPD) from *M. tuberculosis* is introduced intradermally in the forearm. The test area is examined 48 hours later and, if negative or doubtful, is reread at 72 hours. The test is read by measuring the extent of induration (hardening) and reddening around the site of the injection. A positive test indicates previous exposure to tuberculosis-causing mycobacteria but does not necessarily indicate an active infection. Such a test result can occur if an individual has been immunized against tuberculosis or has been treated and recovered from a case of tuberculosis.

When tuberculosis is diagnosed, it can be treated successfully with antibiotics. Treatment generally is for a long time period. Multiple antibiotics, such as streptomycin, rifampin, isoniazid, and ethambutol, are needed because during the prolonged treatment (over 6 months) of this persistent infection, the mycobacteria can mutate and become antibiotic resistant. Because it is highly unlikely that the mycobacteria will undergo multiple mutations during the treatment period that would make them resistant to two or more different antibiotics, using multiple antibiotics provides the necessary margin of safety for treatment.

Q FEVER

Q fever is the only rickettsial disease that does not manifest as a rash and also the only one normally transmitted via the respiratory tract. All other rickettsial diseases are transmitted by arthropod vectors and have characteristic rashes. Q fever is caused by the rickettsia *Coxiella burnetii*. *C. burnetii* occurs within nonhuman animal populations, such as cattle and sheep, where it is transmitted via tick vectors. The bacteria can become airborne on fomites (inanimate objects, such as hair and dust particles) and establish human infections by invading the lower respiratory tract, leading to a systemic infection. Possible sources of human infection include dried animal feces and urine. *C. burnetii* can also be transmitted to humans through contaminated milk; it is killed by pasteurization of milk. The symptoms of Q fever often include fever, headache, chest pain, nausea, and vomiting. Infection with *C. burnetii* can also affect the heart (endocarditis) and liver (hepatitis). Tetracyclines and chloramphenicol are normally used in treating this disease.

Coxiella burnetii **is the only rickettsial bacterium causing human disease that is not transmitted by arthropod vectors.**

BACTERIAL DISEASES OF THE GASTROINTESTINAL TRACT

Each day 27,000 Americans become ill due to spoiled or improperly handled food. In many cases, foods are contaminated with fecal matter containing disease-causing bacteria; often the disease is characterized by diarrhea (loose watery stools) (Table 20-8). Release of bacteria in the stools leads to the transmission of disease by the fecal-oral route. Foodborne infections develop after ingestion of the contaminated food. Additionally, there is a second type of foodborne disease that occurs when toxin-producing bacteria grow in food. The ingestion of toxin in contaminated food results in food poisoning. Various foods can serve as

the source of foodborne toxins and infections (Table 20-9). The growth of toxin-producing bacteria in the food can cause disease from the ingestion of the toxin, even in the absence of an infection within the body of the individual. Alternately, the growth of bacteria within food increases the size of the inoculum that is ingested, increasing the likelihood that the bacteria will evade or overcome natural body defenses. Within an ingested food, bacteria may be protected and escape destruction by the acids and enzymes of the gastrointestinal tract that protect it against invasion by bacteria. To prevent such infec-

TABLE 20-8
Characteristics of Bacterial Gastrointestinal Tract Infections

PATHOGEN	INCUBATION PERIOD (DAYS)	DURATION (DAYS)	SYMPTOMS			
			DIARRHEA	VOMITING	PAIN (CRAMPS)	FEVER
Bacillus cereus	0.3-0.5	0.5-1	Moderate	None	Moderate	None
Campylobacter	2-10	3-21	Severe	None	Moderate	Moderate
Clostridium perfringens	0.3-1	0.5-1	Moderate	None	Moderate	None
Escherichia coli	1-3	2-3	Severe	Slight	Slight	Slight
Salmonella	0.25-2	2-7	Moderate	Slight	None	Slight
Shigella	1-4	2-3	Severe	None	Slight	Slight
Vibrio cholerae	2-3	2-7	Severe	Slight	None	None
Vibrio parahaemolyticus	0.3-2	2-3	Moderate	Slight	Slight	Slight
Yersinia enterocolitica	4-7	7-14	Moderate	None	Moderate	Slight

TABLE 20-9
Causes of Food Poisoning and Foodborne Infections

BACTERIAL SPECIES	FOODS	CASES (%)
Staphylococcus aureus	Meat dishes, desserts, salads with mayonnaise	38
Salmonella spp.	Chicken, other meats, milk and cream, eggs	31
Clostridium perfringens	Cooked and reheated meats and meat products	14
Campylobacter jejuni	Chicken, milk	7
Shigella sp.	Water	4.5
Yersinia enterocolitica	Pork, milk	4
Bacillus cereus	Rice and other starchy foods	1
Clostridium botulinum	Home-canned vegetables (especially beans and corn), smoked fish	0.3
Vibrio parahaemolyticus	Seafoods	0.2

NEWSBREAK

DIARRHEAL DISEASES DURING OPERATION DESERT STORM

Soldiers returning from the 1990 Gulf War, Operation Desert Storm in Kuwait, brought back various infections. Such is frequently the case in overseas military operations. During this war, acute gastroenteritis was the most common problem. Up to 10% of American soldiers were afflicted with bacterial gastrointestinal infections, most commonly caused by enteropathogenic *Escherichia coli* and *Shigella sonnei*, with lesser infections with *Salmonella* and *Campylobacter* species. Throughout American military history, diarrheal diseases have been the most common infection among soldiers.

tions, foods are preserved by various means, such as heating and removing oxygen (canning), adding salt (pickling), lowering pH (acidification), and lowering temperature (refrigeration and freezing). These means of preservation are designed to prevent microbial spoilage of food and, most importantly, to prevent toxin accumulation that would cause disease and to preclude the growth of sufficient numbers of bacteria to act as an infection-initiating inoculum.

STAPHYLOCOCCAL FOOD POISONING

Strains of *Staphylococcus aureus* cause food poisoning because they produce toxins. *Staphylococcus* species are the most frequent source of bacterial foodborne illness in the United States, followed by *Salmonella* species. Staphylococci often enter foods from the skin surfaces of people handling food. They grow in the food and produce toxin that accumulates. The toxin is heat stable and may remain active even in cooked foods. The prevention of staphylococcal food poisoning depends on proper handling and preservation of food products to prevent contamination and subsequent growth of enterotoxin-producing strains of staphylococci. *S. aureus* is able to reproduce within many different types of food products. Custard-filled bakery goods, dairy products, processed meats, potato salad, and various canned foods are frequently the source of the toxin in cases of this type of food poisoning.

Staphylococcal toxins do not exert a direct local effect. They are absorbed through the bloodstream and circulate back to the digestive tract to initiate a **staphylococcal food poisoning syndrome.** The toxin does not exert its effect directly on the digestive tract. The toxin is a neurotoxin that acts on the vagus nerve (vomit center of the brain), causing vomiting. The symptoms of staphylococcal food poisoning occur relatively rapidly after ingestion of toxin-containing food, usually within 1 to 6 hours. The greater the dose of toxin, the sooner the onset of disease symptoms and generally the more severe the disease. Symptoms generally include nausea, vomiting, and abdominal pain. Fever does not occur. Within 8 hours of their onset, symptoms typically subside, and complete recovery usually occurs within a day or two.

> *Staphylococcus aureus* **can grow in food and produce toxins that cause food poisoning, characterized by vomiting, shortly after ingestion.**

SALMONELLOSIS

Salmonella are Gram-negative, facultatively anaerobic rods that are unable to use lactose as a growth substrate. Many different *Salmonella* species, especially the numerous serotypes of *Salmonella enteritidis,* can infect the gastrointestinal tract, causing a form of **gastroenteritis** (inflammation of the intestines) called **salmonellosis.** Birds and domestic fowl, particularly ducks, turkeys, and chickens, including their eggs, are commonly identified as the sources of *Salmonella* infections. Studies of chicken in supermarkets in the United States have found that up to 80% contain *Salmonella.* Inadequate cooking of large turkeys and the ingestion of raw eggs contribute to a significant number of cases of salmonellosis. Some cases are associated with the handling of pet turtles and pet chicks (Easter time novelties), which are often contaminated with high numbers of *Salmonella,* followed by ingestion of the bacteria from contaminated hands. During the 1980s, there was a major increase in the number of cases of salmonellosis in the northeastern United States that was associated with contaminated eggs. Bacterial cells, including those of *Salmonella* species, can pass through an eggshell and contaminate the egg. Currently 1 in 50 consumers of eggs in the United States is exposed to a *Salmonella* contaminated egg every year. If thoroughly cooked, *Salmonella* is

HIGHLIGHT

AN OUTBREAK OF *SALMONELLA* INFECTION ASSOCIATED WITH THE CONSUMPTION OF RAW EGGS

In 1991 several individuals in a city developed symptoms of gastrointestinal infections that included diarrhea, fever, abdominal cramping, nausea, and chills. The pathogenic bacterium *Salmonella enteritidis* was isolated from the stools of 15 sick individuals, confirming that an outbreak of an infectious disease had occurred. Interviews with the 15 individuals revealed that they had eaten at the same restaurant within a 9-day period. This pointed to the restaurant as the likely source of infection. Twenty-three employees of the restaurant reported that they had similar disease symptoms during the same period, supporting the view that the restaurant was the source of the outbreak.

Epidemiologists hypothesized that consumption of a particular food was responsible for the *Salmonella* infections. Fourteen of the fifteen sick patrons of the restaurant reported that, among other foods, they had eaten Caesar salad at the restaurant. Eleven individuals who had eaten at the restaurant during the same 9-day period and had not become ill reported that they had not

eaten Caesar salad. Stool specimens of restaurant employees who had eaten the Caesar salad and become ill were positive for *Salmonella enteritidis.* Not all restaurant employees who were positive for *Salmonella enteritidis,* however, reported eating Caesar salad. Those who had not eaten Caesar salad had eaten other foods with raw eggs.

The Caesar salad dressing was prepared by combining 36 egg yolks with olive oil, anchovies, garlic, and warm water. The salad dressing generally was prepared each morning except for a 3-day period when a single batch was prepared and stored in a refrigerator that was found later to be at 15.6° C when it should have been at 5° C. Tracing the source of the eggs during the outbreak of the disease indicated that they all had come from a single flock of chickens.

The information gained in this study points to the consumption of raw eggs as the source of *Salmonella enteritidis* infection. It highlights the risks of infection associated with the consumption of uncooked eggs.

killed, but if eggs are eaten raw or are inadequately cooked there is a risk of *Salmonella* infection.

Like many other bacteria that cause gastrointestinal tract infections, *Salmonella* species have pili that enable them to adhere to the lining of the gastrointestinal tract. Most *Salmonella* species grow within the gastrointestinal tract without penetrating the mucosal lining. Such *Salmonella* infections are normally characterized by abdominal pain, fever, and diarrhea that lasts for 3 to 5 days. The onset of disease symptoms normally occurs 8 to 30 hours after ingestion of contaminated food. Nausea and vomiting may be initial symptoms but usually do not persist after pain and diarrhea begin. The feces may contain mucus and blood. During acute salmonellosis the feces may contain one billion *Salmonella* cells per gram, leading to fecal contamination of water and food supplies that contributes to the transmission of this disease. Generally, salmonellosis is self-limiting, with recovery occurring without antibiotic treatment within 1 week, although a carrier state sometimes is established. A carrier shows no symptoms of disease, although there is an active infection, and the individual remains a source of bacteria that can infect others. Use of antibiotics generally makes the disease condition worse and is used only when a systemic infection occurs. In some cases the strains of *Salmonella* are resistant to multiple antibiotics. The occurrence of such multiple antibiotic resistant strains has increased because of the use of antibiotics in feeds for poultry to try to control disease in densely reared populations of poultry.

Salmonella **species have pili that attach to the lining of the gastrointestinal tract and initiate infection. Poultry is a major source of** *Salmonella* **infections.**

Shigellosis

Shigellosis is also known as **bacterial dysentery.** It is characterized by frequent watery stools that contain blood. This disease is an acute inflammation of the intestinal tract caused by species of *Shigella* (Gram-negative nonmotile rods), including *S. flexneri, S. sonnei,* and *S. dysenteriae.* The transmission of *Shigella* species occurs by the fecal-oral route with contaminated water and food often involved in outbreaks. *Shigella* species are able to penetrate the cells lining the large intestine and multiply in those cells. Growth leads to large numbers of *Shigella* in the feces, which are a source of spreading infection. Areas of intense inflammation develop around the multiplying bacteria. Microabscesses form and spread within the gastrointestinal tract, leading to bleeding ulceration. Enterotoxin generally is produced by *Shigella* species but its role in the pathogenesis of this disease is unclear because strains that do not produce toxin cause the same disease.

The symptoms of shigellosis include abdominal pain, fever, and diarrhea, with mucus and blood in the feces. Although recovery typically occurs 2 to 7 days after onset, the severe dehydration associated with this disease can cause shock and lead to death in children, among whom the incidence of bacterial dysentery is highest. Because *Shigella* strains contain plasmids coding for antibiotic resistance, combined therapy with tetracycline, ampicillin, and nalidixic acid is often employed for treating bacterial dysentery in children.

Shigella **causes bacterial dysentery. Contaminated water often is the source of** *Shigella* **infections.**

Campylobacter Gastroenteritis

Campylobacter fetus var. *jejuni* has been found to be the causative agent of many cases of gastroenteritis in infants (FIG. 20-18). This Gram-negative, motile, curved bacterium is transmitted via contaminated food. There are large animal reservoirs of *Campylobacter* species in cattle, sheep, rodents, and birds. Infections are from consumption of contaminated food and water. Domestic pets (dogs and cats) can become infected and subsequently transmit infection to humans. Infants usually are infected from contaminated water or unpasteurized milk. Young children, the elderly, and immunocompromised individuals frequently are infected from undercooked chicken, raw mushrooms, or water. Disease symptoms range from mild to severe and include abdominal pain, diarrhea,

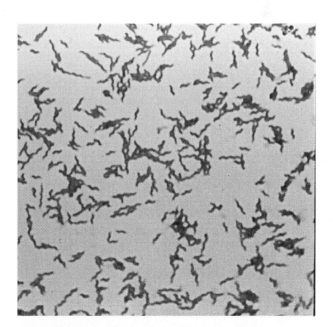

FIG. 20-18 Light micrograph showing organisms of the genus *Campylobacter*, which are now recognized as an important cause of gastroenteritis. They are Gram-negative rods; many are curved or show a "seagull" configuration.

fever, and bloody stools. Treatment may include administration of antibiotics, such as erythromycin and tetracyclines, and maintenance of proper fluid and electrolyte balance.

Campylobacter fetus **often is the cause of gastroenteritis in infants.**

TRAVELER'S DIARRHEA

Enterotoxin-producing strains of *Escherichia coli* cause mild and severe forms of gastroenteritis. *E. coli* are Gram-negative, lactose-fermenting rods that in most cases do not invade the body through the gastrointestinal tract. The toxin released by cells growing on the surface lining of the gastrointestinal tract causes diarrhea. Travelers from the United States to Mexico often suffer severe diarrhea, "Montezuma's Revenge," as a result of ingestion of strains of *E. coli* foreign to their own microbiota in fecal-contaminated water and, therefore, should generally avoid drinking the water.

Traveler's diarrhea is caused by the ingestion of strains of toxin-producing strains of *E. coli*, generally from fecal-contaminated water.

In some cases, pathogenic strains of *E. coli* invade the body through the mucosa of the large intestine and cause a serious form of dysentery. Invasive strains of *E. coli* are primarily associated with contaminated food and water in Southeast Asia and South America. The ability to invade the mucosa of the large intestine depends on the presence of a specific K antigen in enteropathogenic serotypes of *E. coli*. The enterotoxins produced by *E. coli* cause a loss of fluids from intestinal tissues. With proper replacement of body fluids and maintenance of the essential electrolyte balance, infections with enterotoxic *E. coli* normally are not fatal. Without such medical treatment, however, *E. coli* infections can be fatal. Enteropathogenic strains of *E. coli* are a major cause of infant mortality in developing nations and in rural areas of the United States.

A particular strain of *E. coli* has been responsible for a number of deaths in the United States and is an emerging pathogen of concern. This strain, *E. coli* O157:H7 produces verotoxins that cause severe diarrhea. (This toxin is so named because of its toxicity to tissue cultures of "vero" cells.) The mucosa lining of the gastrointestinal tract can be destroyed and hemorrhage may result. Verotoxin may also attack the epithelium of the kidney, causing kidney failure.

VIBRIO PARAHAEMOLYTICUS GASTROENTERITIS

Vibrio parahaemolyticus is responsible for a large number of cases of gastroenteritis in Japan and perhaps in the United States. *V. parahaemolyticus* occurs in marine environments. Ingestion of contaminated seafood, particularly raw fish and shellfish, is the main route of transmission. Gastroenteritis caused by *V. parahaemolyticus* requires establishment of an infection within the gastrointestinal tract, rather than simple ingestion of an enterotoxin. The symptoms of gastroenteritis generally appear 12 hours after ingestion of contaminated food and include abdominal pain, diarrhea, nausea, and vomiting. Recovery from this form of gastroenteritis normally occurs in 2 to 5 days, and the mortality rate is very low.

CHOLERA

Cholera is caused by the Gram-negative, curved rod, *Vibrio cholerae*, serotypes cholerae and El Tor (FIG. 20-19). This disease is primarily transmitted via water supplies contaminated with fecal matter from infected individuals. Cholera is a particular problem in socioeconomically depressed countries, where there is poor sanitation and inadequate sewage treatment and where medical facilities have a limited capacity to deal with outbreaks of this disease. *V. cholerae* is also a natural inhabitant of estuaries, and some cases of cholera along the Gulf Coast of the United States have been traced to contaminated shellfish.

Cholera is endemic in the Ganges delta. There are annual epidemic outbreaks of cholera in India and Bangladesh associated with the monsoon floods, which wash fecal matter into drinking water supplies

NEWSBREAK

OUTBREAK OF *ESCHERICHIA COLI* O157:H7 TRACED TO FAST FOOD CHAIN HAMBURGERS

An outbreak of foodborne infections in 1993 in the northwestern United States was traced to the consumption of hamburgers at the *Jack in the Box* fast food restaurant chain. The outbreak occurred in several states. Hundreds of individuals developed infections that resulted in bloody diarrhea, and some developed hemolytic uremic syndrome (a disease syndrome that produces anemia and renal failure). Several children died of the infection. The causative agent was *Escherichia coli* O157:H7. This bacterium had contaminated the hamburger meat. The cooking temperature used in these restaurants was inadequate to kill it. The Food and Drug Administration has issued warnings to thoroughly cook hamburgers and has recommended cooking temperatures for hamburgers be raised to 86.1° C. Rare hamburgers are not considered safe and no longer will be available at fast food chain restaurants.

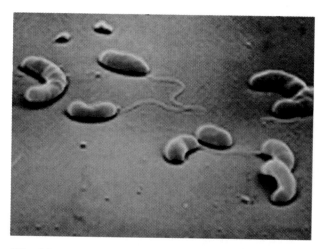

FIG. 20-19 Colorized scanning electron micrograph of *Vibrio cholerae* showing comma-shaped rods and single polar flagellum.

NEWSBREAK

DANGERS OF CONSUMING RAW OYSTERS

Although considered a food delicacy by many, the consumption of raw oysters is responsible for the transmission of several diseases. Oysters are filter feeders and concentrate particles, including bacteria and viruses, from the waters in which they live. *Vibrio cholerae* (the causative agent of cholera) and *V. vulnificus* (the causative agent of gastroenteritis and septicemia that is especially serious and an often fatal illness in immunocompromised individuals and those with pre-existing liver disease) are among the pathogens that frequently become concentrated in oysters. Both of these bacteria naturally occur in coastal waters where oysters grow and are commercially harvested. The problem in the United States is greatest along the Gulf Coast where oysters are commercially collected. Almost 90% of fatal cases of septicemia caused by *V. vulnificus* are associated with the consumption of Gulf Coast raw oysters. The Food and Drug Administration recommends that immunocompromised individuals and those with liver disease avoid the consumption of raw oysters and only consume shellfish that is fully cooked.

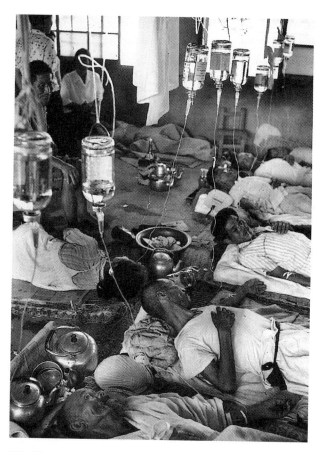

FIG. 20-20 During a severe outbreak of cholera, as occurs in the Far East during monsoon season, many individuals are stricken. Supportive treatment consisting of intravenous (IV) administration of electrolyte solutions is given to stricken individuals.

(FIG. 20-20). In endemic areas of Asia the death rate is normally 5% to 15% of those affected, but during the seasonal epidemic outbreaks, the mortality rate may reach 75%. The high mortality rate reflects the inadequacy of health care systems to provide supportive therapy, such as intravenous solutions to maintain electrolyte balance, during massive outbreaks of cholera. A major cholera epidemic killed hundreds of thousands of individuals who fled the civil war in Rwanda. The water supplies at the refugee camp at Goma Zaire became contaminated due to inadequate sanitation. Twenty thousand individuals were dying per day at the height of the epidemic in the summer of 1994, making it the worst outbreak of cholera in over 200 years. The disease is also now endemic to South America. *V. cholerae* has contaminated water in Peru and surrounding countries. Fish, which is eaten raw in some of these countries, is responsible for a significant number of cases of cholera in South America.

V. cholerae multiplies within the small intestine and produces the enterotoxin responsible for the symptoms of cholera (FIG. 15-21). These symptoms occur suddenly, and include nausea, vomiting, abdominal pain, profuse diarrhea with large amounts of mucus in stools (rice water stools), and severe dehydration, followed by collapse, shock, and in many cases death. *V. cholerae* itself does not invade the body and is not disseminated to other tissues. Rather, the enterotoxin produced by *V. cholerae* binds to the epithelial cells of the small intestine, causing changes in the membrane permeability of the mucosal cells, which initiates secretion of water and electrolytes

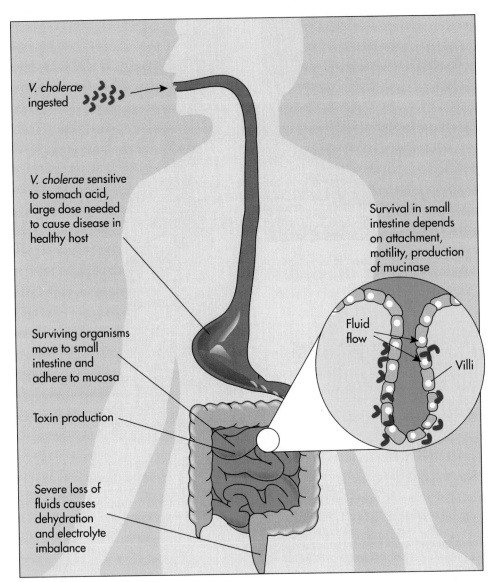

V. cholerae ingested

V. cholerae sensitive to stomach acid, large dose needed to cause disease in healthy host

Survival in small intestine depends on attachment, motility, production of mucinase

Surviving organisms move to small intestine and adhere to mucosa

Fluid flow

Villi

Toxin production

Severe loss of fluids causes dehydration and electrolyte imbalance

FIG. 20-21 Pathogenesis of cholera. *Vibrio cholerae* grows on the surfaces of the intestines. This leads to the flow of fluids from the villi and results in severe fluid loss due to diarrhea. The electrolyte imbalance can cause death.

into the lumen of the small intestine. Excessive fluid loss can occur within 24 hours. The rapid loss of fluid from the cells of the gastrointestinal tract associated with this disease often produces shock, and if untreated there is a high mortality rate. The treatment of cholera centers on replacing fluids and maintaining the electrolyte balance to combat shock. There is a vaccine against cholera but it is not very effective and only works for 6 months.

> **Cholera is caused by *Vibrio cholerae*. This disease is endemic to the Far East where major epidemics occur when monsoon rains wash fecal matter into drinking water supplies.**

YERSINIOSIS

Yersinia enterocolitica is a Gram-negative short rod that produces a severe form of gastroenteritis called **yersiniosis.** *Y. enterocolitica* has been found in water, milk, fruits, vegetables, seafoods, and poultry. This organism is psychrotrophic and thus is able to reproduce within refrigerated foods, where it can multiply and reach the level of an infectious dose. In fact, *Y. enterocolitica* grows better at 25° C than at 37° C.

Outbreaks of yersiniosis are most common in western Europe but have also been confirmed in the United States. The symptoms of an infection with *Y. enterocolitica* resemble appendicitis and include abdominal pain, fever, diarrhea, vomiting, and leukocytosis. Often an appendectomy is performed before this disease is properly diagnosed. In an outbreak of yersiniosis in New York involving over 200 school children, the infection was traced to a common source of contaminated chocolate milk. Ten children underwent unnecessary appendectomies before the true etiology of the disease was established.

> **Yersiniosis can be confused with appendicitis, causing performance of unnecessary appendectomies. *Yersinia enterocolitica* can grow at body and refrigerator temperatures.**

CLOSTRIDIUM PERFRINGENS GASTROENTERITIS

Some cases of gastroenteritis are caused by infections with *Clostridium perfringens*. *C. perfringens* is a large, Gram-positive, endospore-forming, obligately anaerobic rod that also causes gas gangrene. Outbreaks of this form of gastroenteritis are associated with meat and meat-containing stews. It is especially a problem with undercooked sausage meats. Meat provides the amino acids required by clostridia and cooking lowers the oxygen level so that these anaerobes can grow. The spores of *C. perfringens* can survive the temperatures used in cooking many meats, and if incubated in a warm gravy, there is sufficient time for the spores to germinate and the growing bacteria to produce sufficient toxin to cause this disease.

This organism grows in the intestinal tract where it produces a toxin that causes the typical symptoms of abdominal pain and diarrhea. The toxin alters the permeability of the intestinal wall, resulting in the loss of water and electrolytes and producing the symptoms of diarrhea from 8 to 12 hours after ingestion. Most cases are mild and self-limiting and probably are never clinically diagnosed.

BACILLUS CEREUS GASTROENTERITIS

Bacillus cereus is a commonly occurring, usually harmless, large, Gram-positive, endospore-forming organism. *B. cereus* grows in various foods, particularly those with high starch contents. It can cause outbreaks of foodborne illness, often resembling *C. perfringens* infections and causing diarrhea or nausea and vomiting. Several toxins may be involved in producing the differing symptoms. The disease is generally self-limiting.

APPENDICITIS

There are more than 200,000 cases of **appendicitis** per year in the United States. Appendicitis is an inflammation of the appendix occurring when there is an obstruction of the lumen of the appendix. It is caused by a mixture of bacterial populations that constitute

the normal microbiota of the intestinal tract, and virtually all bacterial members of the normal microbiota of the gastrointestinal tract can contribute to appendicitis. The symptoms of appendicitis normally include abdominal pain, localized tenderness, fever, nausea, vomiting, and leukocytosis. Treatment usually involves surgical removal of the appendix.

Serious complications arise if the infection is permitted to progress and the appendix ruptures, releasing bacteria into the abdominal cavity. In such cases, **peritonitis** (infection of the peritoneal cavity) and **systemic bacteremia** (bacteria in the blood system) occur, and there is a high rate of mortality. Antibiotics are not effective in treating acute appendicitis and are unable to stop the development of infection before rupture of the appendix can occur. However, in cases where appendicitis is complicated by rupture of the appendix, antibiotic treatment is essential in controlling the spread of infection to other tissues.

> **Appendicitis occurs when there is a physical blockage of the appendix that leads to an overgrowth by the normal indigenous microorganisms. The appendix can rupture, leading to peritonitis and systemic bacteremia.**

PEPTIC ULCERS

The spiral-shaped bacterium *Helicobacter pylori* can cause ulcer formation (FIG. 20-22). This organism produces urease, an enzyme that breaks down ureas into ammonia. Ammonia is basic and neutralizes gastric acids, permitting survival of *H. pylori* in the stomach and the upper region of the small intestine (duodenum). Growth of *H. pylori*, which produces cytotoxins, may be involved in the pathogenesis of peptic ulcers (lesions of the gastric mucosal membrane). It is likely that killing of cells in the mucosal

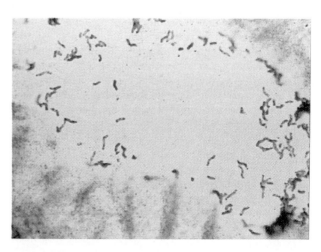

FIG. 20-22   Light micrograph showing the growth of *Helicobacter pylori (purple curved cells)* in the stomach. Growth of *H. pylori* causes gastric ulcers.

lining removes the protection against gastric acid and that exposure to hydrochloric acid and pepsin in the gastric fluid causes the ulcer formation. *H. pylori* is difficult to culture but can be identified by serological and gene probe methods. It is found in almost all tissues where ulcers form. Treatment with bismuth salts and antibiotics is successful, but infections generally recur. The most effective treatment for eradicating *H. pylori* is the use of bismuth subsalicylate, tetracycline, and metronidazole for a 2-week period. Such treatment reduces the risk of recurrent disease

to below 10% for the first year after treatment. The epidemiology of *H. pylori* infections has not yet been established. Ulcers that form in the duodenum are called *duodenal ulcers;* ulcers that form in the stomach are called *gastric ulcers.* Duodenal ulcers account for about 80% of all ulcers. They most commonly occur in men between ages 20 and 50. Gastric ulcers are more common in middle-aged and elderly men, especially those who are malnourished, alcoholics, and chronic users of aspirin. There often is a sequence of remission and recurrence of ulcers.

SUMMARY

Bacteria and Human Diseases (pp. 597-598)

- Bacterial pathogens typically have virulence factors that enable them to invade the body, avoid destruction by the body's immune system, and disrupt body functions.
- Bacteria cause disease by producing toxins that disrupt the normal functioning of human cells and tissues.

Classification of Bacterial Pathogens (pp. 598-602)

- Both Gram-negative and Gram-positive bacteria cause human diseases. *Streptococcus* and *Staphylococcus* are medically important Gram-positive cocci. *Neisseria* is a medically important Gram-negative diplococcus. Many pathogens are Gram-negative rods, some of which have highly specialized physiological requirements for growth.

Bacterial Diseases of the Respiratory System (pp. 602-612)

- Pharyngitis or sore throat is a respiratory disease caused by *Staphylococcus aureus, Streptococcus pyogenes,* and other bacteria that are spread by aerosols containing respiratory secretions. It causes sudden high fever, colored purulent sputum, and congestion. It is treated with penicillin and erythromycin.
- Pneumonia is an inflammation of the lungs, often occurring as a complication of another disease. It is caused by *Streptococcus pneumoniae* and is treated with penicillin.
- Primary atypical pneumonia is caused by *Mycoplasma pneumoniae* that is spread by aerosols containing respiratory secretions. It is treated with tetracycline and erythromycin.
- Legionnaire's disease is caused by *Legionella pneumophila* and is spread by aerosols from evaporating water. It produces typical pneumonia symptoms and may lead to liver and kidney involvement and associated gastrointestinal symptoms. It is treated with erythromycin.
- Psittacosis, caused by *Chlamydia psittaci,* is spread from birds to humans via aerosol dispersal of droppings and contaminated dust particles. Its symptoms are fever, headache, malaise, and coughing. It is treated with tetracyclines.

- Bronchitis is almost always a complication of another disease and is marked by the development of a cough that eventually yields mucopurulent sputum. It is treated with penicillin and tetracycline.
- Acute epiglottitis is caused by *Haemophilus influenzae* infecting the epiglottis and can result in asphyxiation. A tracheotomy will surgically maintain the open airway.
- Diphtheria is a disease of the respiratory tract caused by *Corynebacterium diphtheriae.* It is spread by aerosols containing respiratory secretions, producing symptoms of pharyngitis, low-grade fever, cough, difficulty in swallowing, and swollen lymph glands. It is treated with antitoxins and antibiotics, including penicillin, tetracycline, and chloramphenicol. There is a vaccine.
- Whooping cough or pertussis is caused by *Bordetella pertussis.* It is spread by aerosols containing respiratory secretions and is much like the common cold but characterized by the distinctive characteristic cough. Although a vaccine has decreased the incidence of whooping cough, treatment is with erythromycin and tetracyclines while maintaining an adequate oxygen supply.
- Tuberculosis is a respiratory disease caused by *Mycobacterium tuberculosis* and other *Mycobacterium* species transmitted by droplets from an infected person to a susceptible one or by eating contaminated food, producing symptoms of fatigue, weight loss, fever, and lesions in the lungs. It is treated with antibiotics. A vaccine is available.
- Q fever is caused by *Coxiella burnetii* and is spread by aerosols with dust contaminated by infected animals. Symptoms include fever, headache, chest pain, nausea, and vomiting. Tetracyclines and chloramphenicol are used in treating this disease of the respiratory tract.

Bacterial Diseases of the Gastrointestinal Tract (pp. 613-620)

- Staphylococcal food poisoning is caused by eating food contaminated with *Staphylococcus* toxin produced by *Staphylococcus aureus.* It produces symptoms of nausea, vomiting, abdominal pain, and diarrhea. There is no treatment.

- Salmonellosis, a gastroenteritis caused by different *Salmonella* species, is caused by the ingestion of contaminated food, usually birds or eggs, or water contaminated with fecal material. There is no treatment for the symptoms of abdominal pain, fever, and diarrhea.
- Shigellosis or bacterial dysentery is caused by species of *Shigella* ingested in contaminated food or water. Symptoms include abdominal pain, fever, and diarrhea with mucus and blood. Antibiotics are used in the treatment of children with this disease, in whom it is also extremely important to prevent dehydration.
- Infant gastroenteritis is often a *Campylobacter* infection transmitted by contaminated food or water. Severe cases include symptoms of abdominal pain, diarrhea, fever, and bloody stools. Treatment includes antibiotics while maintaining proper fluid and electrolyte balance.
- Traveler's diarrhea is gastroenteritis caused by enterotoxin-producing strains of *Escherichia coli*, producing diarrhea and abdominal cramping. There is no treatment beyond replacement of body fluids and electrolytes.
- *Vibrio parahaemolyticus* gastroenteritis is caused by *Vibrio parahaemolyticus* that is transmitted via contaminated shellfish, causing abdominal pain, diarrhea, nausea, and vomiting.
- Cholera is a gastrointestinal disease caused by *Vibrio cholerae* serotypes cholerae and El Tor that are transmitted via water supplies contaminated with fecal matter from infected individuals. The symptoms include abdominal pain, diarrhea with "rice water" stools, nausea, and vomiting, followed by collapse, shock, and death. Treatment with tetracycline is accompanied by fluid replacement and maintenance of electrolyte balance.
- Yersiniosis is caused by *Yersinia enterocolitica* that is ingested with contaminated water, milk, fruit, vegetables, or seafood, producing symptoms of abdominal pain, fever, diarrhea, vomiting, and leukocytosis.
- *Clostridium perfringens* gastroenteritis is a form of food poisoning caused by eating food containing toxin produced by *Clostridium perfringens*. There is no treatment for the abdominal pain and diarrhea it causes.
- *Bacillus cereus* produces different toxins that when ingested produce different symptoms of gastroenteritis, including diarrhea, nausea, and vomiting.
- Appendicitis is caused by a mixture of bacterial populations from the normal microbiota of the intestinal tract. Surgical removal of the appendix relieves the symptoms of abdominal pain, localized tenderness, fever, nausea, vomiting, and leukocytosis.
- Peptic ulcers are formed in the stomach and duodenum because of the growth of *Helicobacter pylori*, which damages the mucosal lining of the gastrointestinal tract.

CHAPTER REVIEW

REVIEW QUESTIONS

1. What are the differences between food poisoning and foodborne infections?
2. How do different bacterial toxins contribute to disease?
3. How are foodborne infections transmitted from person to person?
4. Which bacteria can cause food poisonings?
5. What organism causes traveler's diarrhea? How can the traveler avoid such an infection?
6. Which bacteria can cause pneumonia? What are the characteristics of these pathogens?
7. What are the characteristics of diphtheria and the organism that causes this disease? How is this disease controlled?
8. What are the characteristics of pertussis and the organism that causes this disease? How is this disease controlled?
9. How is Legionnaire's disease transmitted? What is the causative organism? How is the disease treated?
10. Why is tuberculosis so difficult to treat? What are some of the commonly used treatments for tuberculosis?
11. Why is pneumonia caused by *Mycoplasma pneumoniae* considered to be atypical?
12. What are the characteristics of the disease cholera? Why does the incidence of this disease often increase after natural disasters?
13. How are poultry and poultry products involved in the transmission of salmonellosis?
14. What bacteria are common causes of gastroenteritis in newborns and infants? Why are such diseases often life threatening in these individuals?
15. How does appendicitis differ from other gastrointestinal infections?

CRITICAL THINKING QUESTIONS

1. How would you determine if an outbreak of food poisoning was due to *Staphylococcus aureus* from a particular food source or food handler?
2. Why is tuberculosis reemerging as a significant worldwide disease?
3. Why are waterborne and foodborne diseases so prevalent in the world? How can waterborne and foodborne pathogens be controlled worldwide?
4. Why is cholera spreading throughout South and Central America? Is it likely to spread into the United States?

CASE STUDY

HISTORY AND ASSESSMENT

Kyle, a 42-year-old man, was admitted to the emergency room at 7 am on July 15 with a chief complaint of severe coughing and shortness of breath. Because it was hard for Kyle to talk, his wife provided the history. His wife said that Kyle had been having a persistent, dry cough for several months and it was getting worse. Recently, he began to cough up sputum that had some blood in it. *Dry coughs tend to be caused by viruses or fungi, not bacteria. Based on what you know so far, what type of microorganism would you expect is causing Kyle's cough? Could there be a combination of microorganisms?*

Kyle's wife explained that Kyle awakened at 6:30 am to get ready for work, began coughing, and suddenly became very short of breath and experienced a sharp pain in his chest. They immediately came to the emergency room. *Why do you think Kyle's coughing is worse in the morning? (HINT: Consider how gravity affects the clearing of secretions.)*

Sudden chest pain and subsequent shortness of breath suggests a rupture of the membranes enclosing the lungs, called a pneumothorax. Do you think that coughing alone was the cause of Kyle's pneumothorax?

Kyle's wife stated that he complained in the past few months of not feeling well and of having less energy to go to work. *Why does he have less energy? (HINT: Physiological energy relies on adequate oxygenation of blood by the lungs.)*

The nurse briefly examined Kyle, noting that he was in good physical condition but was in obvious respiratory distress. She called the doctor immediately and obtained the following vital signs: Pulse rate 122, Respiratory rate 32 and shallow, Blood pressure 150/94, and Temperature 101. *The elevated pulse rate and blood pressure may indicate anxiety over his condition. The rapid, shallow breathing indicates respiratory distress. The temperature is elevated, indicating an infection.*

At this point in the exam, what organs or systems would you like to see physically examined? Explain.

PHYSICAL EXAMINATION

Notable physical examination findings included the observance of labored breathing and diminished lung sounds in the right lung fields on auscultation of the lung. Diminished lung sounds are consistent with functional or anatomic loss of lung tissue, such as might occur with the presence of fluid in a section of the lung tissue or with compression of lung tissue from a pneumothorax or hemothorax.

PROCEDURES AND LABORATORY TESTS

Several chest X-rays (CXRs) were ordered from different angles. (These X-rays were performed with a portable X-ray machine in the emergency room with the doctor in attendance because Kyle was too unstable to go to the X-ray department.) Venous blood was drawn for blood chemistry and a complete blood count. Arter-

ial blood was drawn to assess the oxygenation level and blood pH of Kyle's blood. A sputum sample was obtained and sent for culture following a subsequent coughing attack. Blood was noted in the sputum.

The chest X-rays showed a spontaneous pneumothorax of the right lung. In addition, there were multinodular infiltrates in the apices of both lungs (consistent with the presence of microorganisms in the upper parts of the lungs and an immune reaction to these microorganisms). As expected, the arterial blood gases showed a decreased level of oxygen in the blood.

TREATMENT AND DIAGNOSIS

The physician's first objective was to stabilize the patient by treating the pneumothorax. A chest tube was inserted between two ribs into the space between the right lung and the chest wall. This tube was connected to a suction device that helps inflate the lung. Kyle became more comfortable within 45 minutes of the procedure and was transferred to the medical intensive care unit (MICU).

What caused the pneumothorax and what were the infiltrates on the chest X-ray? By putting all of the clues together, the attending physician had a pretty good idea of the diagnosis. The physician went to the laboratory to make a slide of Kyle's sputum sample. By using a special acid-fast staining technique and examining the specimen under a microscope, the physician identified acid-fast bacilli, consistent with *Mycobacterium tuberculosis* (see Figure). The culture report confirmed this a few days later.

The physician prescribed for Kyle a specific regimen of isoniazid (INH), rifampin, and ethambutol for the treatment of tuberculosis. *Why were three antimicrobics prescribed?*

COURSE

Despite the antibiotic treatment, Kyle's condition worsened, blood oxygenation fell, blood carbon dioxide levels increased, and Kyle died.

Kyle's wife was given a tuberculin skin test to determine if she had been infected with tuberculosis. Even though the skin test was negative, she was begun on a preventive regimen of INH for 1 year due to her close association with Kyle during his illness. *Why was prolonged administration of INH required?*

READINGS

Bartlett CLR, AD Macrae, JT Macfarlane: 1986. *Legionella Infections*, London, Edward Arnold.

Describes the present knowledge of the chemistry, pathology, treatment, epidemiology, sources of infection, and methods of control of the disease, and the characteristics and habits of the family Legionellaceae.

Bloom BR (ed.): 1994. *Tuberculosis: Pathogenesis, Protection, and Control*, Washington, ASM Press.

A series of authoritative chapters on all aspects of tuberculosis in the 1990s, including the emergence of new species of mycobacteria causing this disease, the relationship of emerging infections to AIDS, and the rise of antimicrobic resistant strains of mycobacteria causing tuberculosis.

Butler T: 1983. *Plague and Other Yersinia Infections*, New York, Plenum Medical Book Co.

Clear and easy to read book with helpful illustrations that describes diseases caused by plague and nonplague yersiniae, which both have a predeliction for lymph nodes and disseminate via the bloodstream, share certain antigens and toxins, and have plasmids.

Chambers JS: 1938. *The Conquest of Cholera: America's Greatest Scourge*, New York, Macmillan.

A biography of the disease in the New World from its arrival with Irish immigrants in the 1830's through five American epidemics to its conquest on this continent.

Doyle MP: 1989. *Foodborne Bacterial Pathogens*, New York, Marcel Dekker.

A useful reference on bacterial pathogens transmitted via foods.

Edwards DD: 1987. Poker players, pneumonia and cat tales, *Science News* 132:255.

Discusses Q fever and the role cats play in the transmission of the rickettsia that causes this disease.

Evans AS and PS Brachman (eds.): 1989. *Bacterial Infections of Humans: Epidemiology and Control*, NY, Plenum Medical Book Co.

Explores the epidemiology, prevention, and control of bacterial infections.

Farthing MJG and GT Keusch (eds.): 1989. *Enteric Infection: Mechanisms, Manifestations, and Management*, New York, Raven Press.

Review of bacterial diseases of the gastrointestinal system, including enteritis.

Godal T: 1983. Mycobacterial diseases, *World Health*, Nov: 4-5.

Discusses current state of knowledge about diseases caused by mycobacteria, including leprosy and tuberculosis.

Gorbach SL (ed.): 1986. *Infectious Diarrhea*, Boston, Blackwell Scientific Publications.

Explains the microbiological basis of and treatment alternatives for diarrhea.

Hauschild AHW and KL Dodds (eds.): 1992. *Clostridium botulinum: Ecology and Control in Foods*, New York, Marcel Dekker.

This in-depth resource addresses the ecology of Clostridium botulinum, which affects the degree of food contamination, and its control in various foods—summarizing worldwide data on this organism in food and the environment and the principles of its control in specific food commodities and products.

Hui YH (ed.): 1994. *Foodborne Disease Handbook*, New York, M. Dekker.

Discusses foodborne diseases, especially those caused by microorganisms.

Lattimer GL and RA Ormsbee: 1981. *Legionnaires' Disease*, New York, Marcel Dekker.

Describes the salient features of Legionnaire's disease for clinicians, pathologists, microbiologists, immunologists, epidemiologists, and public health workers.

Murphy J: 1986. A comeback for whooping cough, *Time* 127(June 30):78.

Problems with lawsuits and production have led to a decrease in the availability of DPT vaccine and a rise in the incidence of whooping cough.

Pelling M: 1978. *Cholera, Fever and English Medicine, 1825-1865*, New York, Oxford University Press.

The story of public health efforts in nineteenth century England to uncover the epidemiology, prevention, and treatment of cholera.

Pennington JE (ed.): 1989. *Respiratory Infections: Diagnosis and Management*, New York, Raven Press.

Explains the diagnosis and treatment of various types of pneumonia, describing their pathogenesis, clinical occurrence, etiological agents, and therapeutic considerations.

Rosenthal E: 1990. Return of consumption, *Discover* 11:80-83.

Reveals the recent rise in the incidence of tuberculosis.

Rubin RH and L Weinstein: 1977. *Salmonellosis: Microbiologic, Pathologic, and Clinical Features*, New York, Stratton Intercontinental Medical Book Corp.

Explains the clinical features of the wide variety of Salmonella infections with guidelines for treatment.

Ryan F: 1993. *The Forgotten Plague*, Boston, Little Brown and Co.

This account chronicles the efforts to cure tuberculosis and provides insight into today's recurrence of this disease and the efforts to find a cure for AIDS.

Schlossberg D (ed.): 1994. *Tuberculosis*, ed. 3 New York, Springer-Verlag.

Reviews epidemiology and host factors involved in TB, clinical manifestations of the disease, and atypical mycobacterial infections. Complete discussion of tuberculosis, including discussion of the emergence of antimicrobic resistance.

Smith FB: 1988. *The Retreat of Tuberculosis 1850-1950*, London, Croom Helm.

The history of tuberculosis in Great Britain that examines the causes for the disease's decline in recent years, questioning whether it was really due to advances in scientific medicine.

Van Heyningen WE and JR Seal: 1983. *Cholera: the American Scientific Experience, 1947-1980*, Boulder, CO; Westview Press.

Describes the role of the U.S. medical service and army and navy doctors in the war against cholera during World War II when the United States sought to protect its soldiers and allies in tropical theaters of war.

Varnam AH and MG Evans: 1991. *Foodborne Pathogens: An Illustrated Text*, St. Louis; Mosby.

A very well-illustrated book showing micrographs of foodborne pathogens and the pathologies of infected regions of the gastrointestinal tract.

Wachsmuth IK, PA Blake, O Olsvik: 1994. *Vibrio cholerae and Cholera: Molecular To Global Perspectives*, Washington, ASM Press.

Authoritative discussion of cholera and the bacterium that causes this disease, including consideration of contemporary outbreaks of this disease around the world.

Weiss R: 1988. TB troubles: tuberculosis is on the rise again, *Science News* 133:92-93.

The rise in tuberculosis is linked to the increasing incidence of AIDS.

Ziporyn TD: 1988. *Disease in the Popular American Press: the Case of Diphtheria, Typhoid Fever, and Syphilis, 1870-1920*, New York; Greenwood Press.

How the stories of disease have been presented in newspapers in the United States.

CHAPTER 21

Bacterial Diseases of the Central Nervous, Cardiovascular, and Lymphatic Systems

PREVIEW TO CHAPTER 21

In this chapter we will:
• Review the major diseases of humans caused by
  bacteria that affect the central nervous, cardiovascular,
  and lymphatic systems, examining the specific etiologic
  agent, the route of transmission, the major symptoms,
  and the treatments of each disease.
• Learn the following key terms and names:

bacterial meningitis
bartonellosis
black death
blood poisoning
body louse
botulism
brain abscess
brucellosis
buboes
bubonic plague
carditis
childbed fever
endemic typhus fever
epidemic typhus fever
gas gangrene
infant botulism
listeriosis
lockjaw
Lyme disease

meninges
murine typhus
plague
pneumonic plague
puerperal fever
relapsing fever
rheumatic fever
rickettsialpox
Rocky Mountain spotted
  fever
scrub typhus
septicemia
tetanus
tetanus neonatorum
toxic shock syndrome
tularemia
typhoid fever
undulant fever
wound botulism

The tissues of the central nervous system normally are sterile and protected against infection. If bacteria enter the brain or spinal column they can establish infections and life-threatening diseases such as brain abscess and meningitis. Most bacteria that cause central nervous system infections enter the body through the respiratory tract and move through the circulatory system and lymph nodes to reach the central nervous system. Some bacteria grow elsewhere in the body and interfere with central nervous system functions by producing neurotoxins; neurotoxins need not be produced within the central nervous system to cause disease. Tetanus toxin, for example, is produced by *Clostridium tetani* growing at the site of a wound and botulinum toxin typically is produced by *C. botulinum* completely outside of the body in canned food.

BACTERIAL MENINGITIS

The meninges are the membranes surrounding the brain and spinal cord. Meningitis is an inflammation of the meninges that can be caused by various bacteria. It typically occurs after an infection in the respiratory tract, when bacteria are transmitted via the circulatory system to the central nervous system. Usually the presence of high numbers of bacteria in blood or lymph over a prolonged period is necessary before transmission to the meninges will occur. Injuries that expose the central nervous system to bacterial contaminants provide an alternate portal of entry. Once the central nervous system is infected, the symptoms of disease typically occur rapidly. The limited immune response within the central nervous system is unable to block the onset of disease once such infection occurs. Bacterial meningitis is characterized by sudden fever, severe headache, painful rigidity of the neck, nausea, and vomiting, and frequently by convulsions, delirium, and coma. Meningitis can be fatal within hours or days and diagnosis and the initiation of antibiotic therapy must be accomplished quickly to reduce the risk of brain damage, retardation, and death.

Several bacteria can cause meningitis (Tables 21-1 and 21-2). Most of these possess pili for attachment and are coated with surface polysaccharides that in-

TABLE 21-1

Correlation of Age With the Etiologic Agents of Bacterial Meningitis

ETIOLOGIC AGENT	PERCENT OF ISOLATES		
	INFANTS (UNDER 2 MONTHS)	CHILDREN (2 MONTHS TO 15 YEARS)	ADULTS (OVER 15 YEARS)
Neisseria meningitidis	1	25	25
Haemophilus influenzae	2	50	2
Escherichia coli and other Gram-negative rods	50	2	5
Streptococcus agalactiae (Group B *Streptococcus*)	35	3	5
Streptococcus pneumoniae	2	10	50
Staphylococcus aureus	5	2	8
Other	5	8	5

TABLE 21-2

Clinical Features of Major Causes of Bacterial Meningitis

CAUSATIVE ORGANISM	CLINICAL SIGNS AND SYMPTOMS	MORTALITY RATE (% OF CASES)
Neisseria meningitidis	Pharyngitis and fever followed by typical signs and symptoms* of meningitis; sometimes red, macular rash or arthritis; can occur in epidemics	7-10
Haemophilus influenzae	Respiratory symptoms followed by typical signs and symptoms of meningitis	3-10
Escherichia coli and other Enterobacteriaceae	Typical signs and symptoms of meningitis	40-80
Streptococcus agalactiae (Group B *Streptococcus*) and other *Streptococcus* sp.	Typical signs and symptoms of meningitis	20-50
Staphylococcus sp.	Typical signs and symptoms of meningitis	—

*Typical signs and symptoms of bacterial meningitis include fever, chills, headache, vomiting, stiff neck, opisthotonos (a spasm where the back arches), and positive Brudzinski sign (pain and resistance when patient is lying down and neck is raised upward).

terfere with phagocytosis. These bacteria are able to survive passage through the circulatory system, reach the central nervous system, and enter the cerebrospinal fluid. There are relatively few phagocytes and low concentrations of immunoglobulins and complement in cerebrospinal fluid, so that the bacteria can survive. There also are some nutrients, such as glucose, that can support bacterial growth within the cerebrospinal fluid. The diagnosis of the particular etiologic agent of meningitis depends on the isolation and identification of pathogens from the cerebrospinal fluid. *Neisseria meningitidis, Haemophilus influenzae, Streptococcus pneumoniae, S. agalactiae,* and *Escherichia coli* are the most common etiologic agents of bacterial meningitis. *E. coli* often causes meningitis in infants but not in adults. *S. agalactiae* is also a common cause of meningitis in infants. Most cases of streptococcal meningitis occur among children under 12 months of age. Although pneumococcal meningitis accounts for only about 13% of the cases of reported bacterial meningitis in the United States, the mortality rate of 26% is very high.

H. influenzae type b is by far the most common cause of bacterial meningitis in children. There are an estimated 20,000 cases of meningitis each year caused by *H. influenzae*. It is one of the leading causes of mental retardation in children under 5 years of age. While *H. influenzae* type b accounts for most of the cases of reported bacterial meningitis (48%), it has the lowest mortality rate of the common types of bacterial meningitis (6%). It occurs mostly in children under 4 years of age and is especially prevalent at about age 6 months. The antibodies acquired from the mother provide protection that lasts for only about 2 months. The carrier state in the throats of children is quite high, and transmission in day care centers is now an increasing problem. An *H. influenzae* infection begins as an upper respiratory tract infection, a mild pharyngitis. The bacteria enter the circulatory system and can invade monocytes. The monocytes carry these bacteria to the central nervous system, permitting them to enter the cerebrospinal fluid. A vaccine made of antigenic polysaccharide capsule (the Hib vaccine) has been developed; it is conjugated to DPT vaccine and is routinely administered to children, which should reduce the number of cases of meningitis.

Neisseria meningitidis is a Gram-negative diplococcus. It is also known as the meningococcus and is often the causative agent of bacterial meningitis in persons between 5 and 40 years of age (FIG. 21-1). There are about 2,500 cases of meningococcal infections in the United States each year. This organism frequently causes infections in those in close contact with an infectious individual. At particular risk are family members, schoolmates, and medical personnel with intimate contact (for example, those performing car-

NEWSBREAK

HAEMOPHILUS INFLUENZAE— A MISNAMED ORGANISM

During the 1918 influenza pandemic, scientists searching for the causative organism isolated a bacterium from many individuals with the disease. They prematurely concluded that influenza was caused by this bacterium and named it *Haemophilus influenzae*. Applying Koch's postulates showed that this bacterium does not cause this disease. Rather *H. influenzae* causes secondary pneumonia after influenza, which is caused by influenza viruses. Unfortunately, once a bacterium is officially named and published in the *International Journal of Systematic Bacteriology*, the name cannot be changed. Hence, *H. influenzae* remains a misnamed bacterium that causes various human diseases, but not the one implied by its epithet *influenzae*.

diovascular pulmonary resuscitation). The lipopolysaccharides of the cell wall (endotoxin) of *N. meningitidis* trigger an inflammatory response and are responsible for the disease symptoms. The virulence of the organism is largely determined by the presence of a capsule that protects it against phagocytosis.

An infection with *N. meningitidis* can progress so rapidly that death occurs within a few hours of the onset of symptoms. Several antibiotics are used in the treatment of meningitis, and the specific antibiotic of

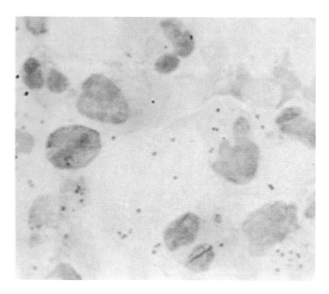

FIG. 21-1 Micrograph showing meningococci in cerebrospinal fluid. The Gram-negative cocci are in pairs; some are intracellular. The long axes of pairs of cocci are parallel and not in line, as with the pneumococcus.

choice is determined by the antibiotic susceptibility of the causative agent of the disease and the antibiotic's ability to cross the blood-brain barrier. In many cases penicillin is the antibiotic of choice. Culture and sensitivity tests traditionally, however, take at least a day so that antibiotic therapy generally must be instituted before a final diagnosis is made. Knowing the most likely etiologic agents for a particular age group aids the physician in selecting an appropriate antibiotic. Serological methods can provide the necessary identification of the etiologic agent within minutes or hours, permitting rapid positive diagnosis and institution of proper therapy for treating bacterial meningitis. Such serological methods, however, require high numbers of bacteria in the cerebrospinal fluid (generally $>10^5$/mL), by which point the disease often is at a crisis stage. More sensitive detection (less than 10^2/mL) can be achieved using the polymerase chain reaction to amplify the diagnostic regions of the DNA of *N. meningitidis* and other bacteria that cause bacterial meningitis.

> Meningitis, an inflammation of the meninges, is caused by various bacteria, notably *Neisseria meningitidis*, *Haemophilus influenzae*, *Streptococcus pneumoniae*, and *Escherichia coli*. Each of these bacteria is most commonly the cause of meningitis for particular age groups.

CLOSTRIDIAL DISEASES

Several diseases of the central nervous system are caused by *Clostridium* species that produce neurotoxins. These clostridia do not grow within the central nervous system but produce exotoxins that disrupt nerve cell transmissions (Table 21-2). In some cases, the toxin is ingested in contaminated foods. In other cases, toxin production occurs at the site of *Clostridium* infection in the tissues and the toxin is spread to the nervous system.

Tetanus

Tetanus is sometimes referred to as **lockjaw** because the muscles of the jaw and neck contract convulsively so that the mouth remains locked closed. It is

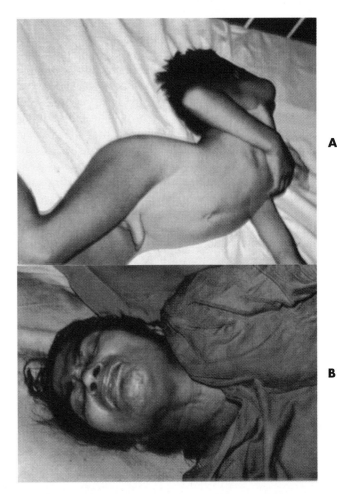

FIG. 21-2 Death due to tetanus is characterized by **(A)** arching of the back and neck and **(B)** "lockjaw" due to convulsive spastic paralysis.

caused by *Clostridium tetani*, a Gram-positive anaerobic rod that forms endospores (FIG. 21-2). This bacterium produces a neurotoxin that inhibits relaxation (inhibition signals) of neurons, which produces severe muscle spasms (spastic paralysis). Transmission to humans normally occurs as a result of a puncture wound that inoculates the body with spores of *C. tetani*, which are widely distributed in soil. Virtually

	TABLE 21-3			
	Neurotoxins of Clostridia			
ORGANISM	TOXIN	PRODUCTION	MODE OF ACTION	SIGNS
Clostridium botulinum	Botulinum toxin (types A-F)	Intracellular; released when bacteria lyse	Blocks release of acetylcholine at nerves	Flaccid paralysis, double vision, difficulty in swallowing, and respiratory paralysis
Clostridium tetani	Tetanus toxin (tetanospasmin)	Secreted	Inhibits antagonists of motor neurons	Spastic paralysis, lockjaw, severe skeletal muscle spasms, respiratory paralysis

any type of wound into which foreign material carrying spores of *C. tetani* is introduced may lead to the development of tetanus. Wounds should be washed, often with sterile water flushing, to try to remove material that may carry *C. tetani* and other infecting bacteria. Attempts should be made to prevent oxygen depletion that favors the growth of clostridia. If anaerobic conditions develop at the site of the wound, endospores of *C. tetani* germinate and multiplying bacteria produce neurotoxin. *C. tetani* is noninvasive and multiplies only at the site of inoculation, but the neurotoxin produced by *C. tetani* spreads systemically, causing the symptoms of this disease (FIG. 21-3).

> While *Clostridium tetani* is noninvasive and multiplies only at the site of inoculation, the neurotoxin it produces spreads systemically, causing the symptoms of tetanus.

Tetanus may occur months after contamination of a wound due to the outgrowth of subcutaneous spores of *C. tetani*. Tales of the association of rusty nails with this disease probably began because farmers often developed tetanus after stepping on nails that were contaminated with soil and the endospores of *C. tetani*, but clearly rusty nails are not the cause of this disease. If untreated, tetanus is frequently fatal, but if recovery occurs, there are no lasting effects. Many newborns in developing countries die of *tetanus neonatorum*, which results from postnatal infection of the degenerating remnant of the severed umbilical cord.

Tetanus can be treated by the administration of tetanus antitoxin to block (neutralize) the action of the neurotoxin. The disease can be prevented by immunization with tetanus toxoid to preclude the development of infection with *C. tetani*. Tetanus booster vaccinations are frequently given after wound injuries to ensure immunity against this disease. Public awareness and the effective use of vaccines have reduced the incidence of tetanus in the United States to fewer than 100 cases per year.

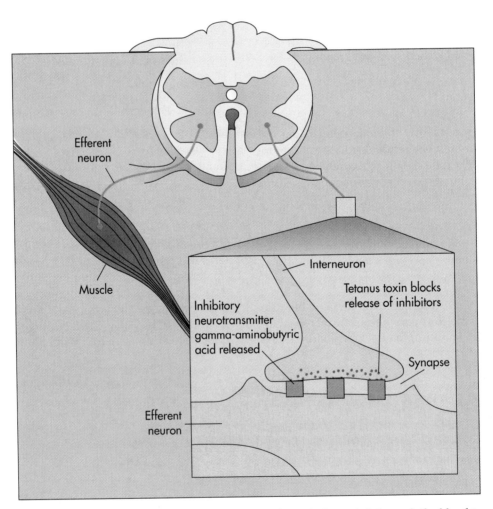

FIG. 21-3 Tetanus toxin causes spastic paralysis. The toxin is carried through the blood to the central nervous system, where it binds with neurons. The release of inhibitory mediators in spinal synapses is blocked by the toxin. This leads to overactivity of motor neurons.

Botulism

Botulism is caused by neurotoxins produced by *Clostridium botulinum*. The toxins are absorbed from the intestinal tract and transported via the circulatory system to motor nerve synapses where their action blocks normal neural transmissions (FIG. 21-4). The botulinum neurotoxin inhibits motor signals, which results in flaccid paralysis. These toxins are produced when *C. botulinum* grows in foods, and hence, botulism is a form of food poisoning. This disease was known in the 1800s as the sausage disease because of the ability of *C. botulinum* to grow in sausage meats, especially when other bacteria were eliminated by boiling for a short period in the preparation of sausage. *Botulus* means sausage in Latin.

There are fewer than 100 cases of botulism in the United States annually. Nevertheless, because this disease has a high fatality rate, great care is taken to prevent it. Over 90% of the cases of botulism involve improperly home-canned food, such as beans and low-acid (yellow) tomatoes. The endospores of *C. botulinum* are heat resistant and can survive prolonged exposure at 100° C. Contaminated canned foods provide an optimal anaerobic environment for the growth of *C. botulinum*. This bacterium, however, cannot grow and produce toxin at low pH (below 4.5) and thus is not a problem in acidic food products. Proper canning and cooking of food can prevent this disease. *C. botulinum* can be eliminated by heating to 121° C for 15 minutes. The toxin it produces also is destroyed by cooking food. Symptoms of botulism can appear 8 to 48 hours after ingestion of the toxin; an early onset of disease symptoms normally indicates that the disease will be severe. These symptoms often begin with nausea and are followed by evidence of neurological impairment, such as double or blurred vision; there is difficulty in swallowing and weakness as the effects of the neurotoxin progress.

Botulism, caused by neurotoxins produced by *Clostridium botulinum*, can cause paralysis of the respiratory muscles and death by asphyxiation.

Toxin accumulates in foods when the anaerobe *C. botulinum* grows.

Type E toxins are associated with the growth of *C. botulinum* in fish or fish products, and most outbreaks of botulism in Japan are caused by type E tox-

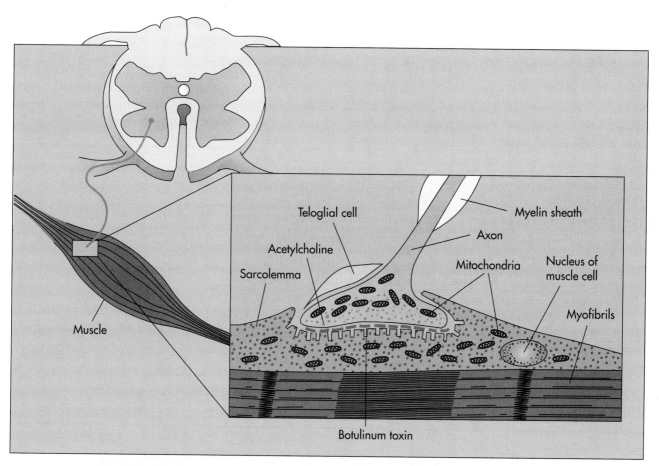

FIG. 21-4 Botulinum toxin blocks neural transmission by preventing the release of the neurotransmitter acetylcholine. Blockage of nerve transmissions to muscle cells results in paralysis.

ins because large amounts of fish are consumed there. Type A is the predominant toxin in cases of botulism in the United States, and type B toxin is most prevalent in Europe. Type A toxin botulism is generally more severe than the disease caused by other types of toxin. In severe cases of botulism, there is paralysis of the respiratory muscles, and despite improved medical treatment, the mortality rate is still about 25%. Antibiotics are not useful because botulism is a food poisoning (food intoxication) and not a foodborne infection. If the disease is diagnosed quickly, antitoxins can be administered to prevent the action of the toxin. This is effective in preventing botulism.

Infant Botulism

C. botulinum is normally not capable of establishing an infection in adults because of the low pH of the stomach and upper end of the small intestine. However, in infants, before the colonization of the intestinal tract by acid-producing bacteria, such as *Lactobacillus* species, *C. botulinum* can reproduce within the tissues of the gastrointestinal tract. When they occur they produce neurotoxin that causes the often fatal **infant botulism.** There are about 100 cases of infant botulism identified in the United States each year, but the actual number is much higher. There is evidence that a few cases of sudden infant death syndrome (SIDS), or crib death, can be attributed to *C. botulinum.* Accordingly, additional concern is being given to the food products, such as honey, that infants may consume with respect to the possible ingestion of *C. botulinum* endospores. Pediatricians now warn mothers against allowing their infants to eat honey, which is used in some regions to entice infants to nurse.

Wound Botulism

Wound botulism is caused when *C. botulinum* spores are inoculated into a wound. In many ways, wound botulism resembles tetanus except that it is caused by a different *Clostridium* species that produces a different neurotoxin. Spores of *C. botulinum* are abundantly distributed in soils, and if inoculated into a deep wound this bacterium can initiate an infection. *C. botulinum* is noninvasive but can produce a potent neurotoxin that spreads throughout the body. The toxin blocks neural transmissions, inhibiting motor signals and causing paralysis that may result in respiratory failure and death. If the disease is diagnosed in time, antitoxins can be used to neutralize the toxins.

Listeriosis

Listeria monocytogenes, a Gram-positive rod that can grow at temperatures from 5° C to 37° C, is the causative organism of **listeriosis.** (The genus name is for Joseph Lister who introduced antiseptic practices in medicine; the species epithet reflects the involvement of monocytes in the disease.) This bacterium is widely distributed in soil and water, coming mostly from animal feces. Listeriosis usually is a mild, often symptomless disease in adult humans. In human outbreaks, the organism is mostly foodborne, especially via dairy products (cheeses) and uncooked foods. Unpasteurized milk may be a source of *L. monocytogenes. Listeria* can grow in the refrigerator at 5° C. This permits it to reach high enough numbers in foods stored there to reach an infectious dose. Within the body, *L. monocytogenes* is not destroyed by phagocytic cells and even proliferates within them. There is a large increase in monocytes during listeriosis. It travels from the gastrointestinal tract, through the circulatory system, to the central nervous system. It can cross the placenta and infect a fetus.

Growth of *L. monocytogenes* in the central nervous system usually results in meningitis. The organism seldom appears in healthy adults, but is the leading cause of meningitis in immunosuppressed patients; it is a common cause of meningitis in individuals with AIDS. It also often causes meningitis in the elderly and very young children. In pregnant women, the growth of *L. monocytogenes* on the placenta leads to a high rate of spontaneous abortion or stillbirth. If the newborn survives, it may have septicemia (a condition called blood poisoning) and meningitis, with mortality about 60%. Penicillin and ampicillin are the antibiotics of choice for treatment.

Brain Abscess

Under some conditions, bacteria are able to invade the brain and cause a **brain abscess.** The development of a brain abscess (localized accumulation of fluid) usually occurs as a complication of an infection at another site in the body, most frequently from bone and ear infections. The bacteria usually found in cases of brain abscesses include species of *Streptococcus, Staphylococcus, Haemophilus, Nocardia, Bacteroides, Peptostreptococcus, Peptococcus, Fusobacterium,* and *Actinomyces,* although anaerobes are more frequently the cause. When these bacteria invade the brain, they initiate an inflammatory response and the formation of pus, that is, a characteristic abscess. Fever, headache, nausea, and various neurological symptoms, such as blurred vision, are characteristic of this disease. Brain abscesses are obviously very serious and potentially life threatening. Diagnosis of the causative agent and prompt initiation of treatment with appropriate antibiotics is critical. Penicillin and chloramphenicol are frequently used for treating brain abscesses.

Bacteria entering the body from any portal of entry can reach the circulatory system and be disseminated throughout the body. In some cases the tissues and organs of the circulatory system themselves become the sites of infection. Some bacteria cause heart disease; others infect various blood cells. Because of the essential life-support functions of the cardiovascular and lymphatic systems such infections can be life threatening.

RHEUMATIC FEVER

Rheumatic fever is a complication that sometimes results from pharyngitis caused by streptococci (FIG. 21-5). It occurs because the systemic spread of *Strepto-coccus pyogenes* toxins causes the body to make antibodies that can mistakenly attack body tissues or form complexes that interfere with body functions. The symptoms of rheumatic fever normally begin a little over 2 weeks after the onset of a sore throat associated with an upper respiratory tract infection with *S. pyogenes* that causes pharyngitis. The most susceptible group is between 5 and 15 years of age. Approximately 0.05% of streptococcal throat infections in the United States results in rheumatic fever; 1% of such cases causes rheumatic fever in tropical countries. The symptoms vary, but characteristically there is a high fever, painful swelling of various body joints, and cardiac involvement, including subsequent development of heart murmurs due to permanent dam-

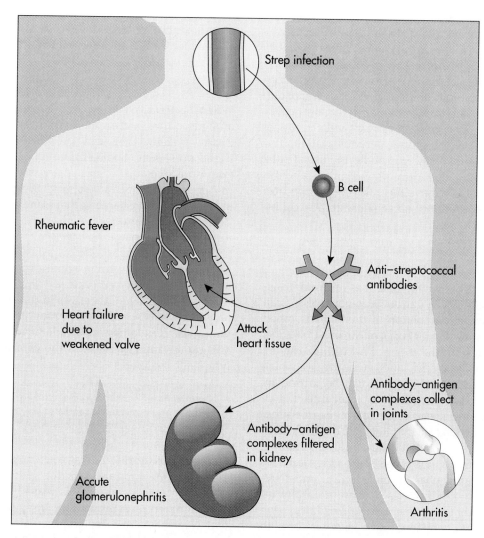

FIG. 21-5 Pathogenesis of rheumatic fever. Antistreptococcal antibodies can cross-react with heart tissue antigens, causing damage to heart valves and potential later heart failure. Antigen-antibody complexes can accumulate in the kidney, causing glomerulonephritis, and in joints, causing arthritis.

631

age to heart valves in affected children. Heart damage occurs because of reactions initiated by antibody made against streptococcal antigens cross-reacting with cardiac tissue. This leads to inflammation that damages the heart. Rheumatic fever can cause arthritis. This occurs when antigen-antibody complexes collect in the synovial fluid of the joints, leading to inflammation. Because of the serious manifestations of rheumatic fever, it is important to diagnose the etiologic agents of sore throats in children, who sometimes are given penicillin as a prophylactic measure to prevent occurrences of rheumatic fever.

Because rheumatic fever can lead to permanent damage to heart valves, it is important to diagnose the etiologic agents of sore throats in children and to treat streptococcal infections effectively.

The causal relationship between *S. pyogenes* and the symptoms of rheumatic fever occurs because antibodies produced in response to a group A streptococcal infection also react with cardiac antigens. This produces an "autoimmune" or immune-complex response that actually results in cardiac damage. Granulomas form in the heart, and myocarditis or pericarditis develops. Heart valves can be damaged so that, years later, heart failure occurs. Subcutaneous and nervous tissues are also damaged. In some cases the antistreptococcal antibodies react with neurons, causing damage to the central nervous system. The treatment of rheumatic fever includes the use of anti-inflammatory drugs to reduce tissue damage and the use of antibiotics to remove the infecting streptococci if there is still evidence of an active streptococcal infection.

CARDITIS

In addition to cardiac involvement as one manifestation of rheumatic fever, various microorganisms can infect the heart. Inflammation of the heart is called **carditis.** There are several types of carditis that affect different tissues of the heart. Pericarditis can be caused by *Staphylococcus,* and myocarditis, which most often is caused by viruses, can be caused by rickettsia. **Infective endocarditis** is an infection of the endocardium, a specialized membrane of epithelial and connective tissue that lines the cardiac chambers and forms much of the heart valve structure. Acute endocarditis occurs when *Staphylococcus aureus* invades the heart; subacute endocarditis most frequently occurs when bacteria that are normally indigenous to body surfaces enter the circulatory system and attach to cardiac tissues. Microbial growth on heart tissues causes serious abnormalities and if untreated is often fatal. Viridans streptococci (a heterogeneous group of alpha hemolytic *Streptococcus* species) are the most frequently identified etiologic agents of endocarditis. Subacute endocarditis is com-

mon among individuals with underlying minor heart defects or heart disease, such as rheumatic fever.

Individuals with prosthetic heart valves have an increased risk of developing subacute endocarditis. Because viridans streptococci are present in high numbers in the oral cavity and dental treatments are often implicated in initiating entry of these bacteria into the bloodstream, individuals with prosthetic heart valves are given prophylactic doses of antibiotics (generally penicillin) before dental treatments.

Microbial growth on heart tissue causes serious abnormalities, and if untreated carditis is often fatal.

Infective endocarditis normally begins with a gradual onset of fever and general tiredness over a period of days or weeks. A heart murmur is detectable in 95% of endocarditis cases. Treatment of bacterial infections of the heart tissues usually employs broad-spectrum antibiotics that are not inactivated by β-lactamases and often requires prolonged administration of high doses of antibiotics to ensure elimination of the infecting bacteria.

SEPTICEMIA

When bacteria overcome the defenses of the body and reproduce within the cardiovascular and lymphatic system, they can produce **septicemia,** or **blood poisoning.** Septicemia can be caused by both Gram-positive and Gram-negative bacteria. *Staphylococcus aureus* and *Streptococcus pyogenes,* for example, are Gram-positive bacteria that sometimes cause septicemia. The Gram-negative bacteria *Escherichia coli, Enterobacter aerogenes, Pseudomonas aeruginosa, Serratia marcescens,* and *Proteus* species are frequently identified as the causative agents of blood poisoning, releasing lipopolysaccharides (endotoxins) from their cell walls when they lyse in the circulatory system. The release of large amounts of endotoxin within the circulatory system damages blood vessels and causes inflammation of the cardiovascular and lymphatic systems.

Symptoms of septicemia are caused by the release of lipopolysaccharides (endotoxins) from the cell walls of lysed infecting Gram-negative bacteria.

Characteristically, red streaks appear under the skin along the arms and legs due to the inflammation of the lymph vessels. There is also a high fever and a drop in blood pressure and the body goes into shock that may be fatal. Treatment must be supportive, maintaining body functions and limiting the effects of shock. Antibiotics are used to eliminate the infecting bacteria but may initially exacerbate the disease by causing the lysis of more bacteria, with the release of additional endotoxin into the circulatory system.

TYPHOID FEVER

Typhoid fever is a systemic infection caused by the Gram-negative bacterium *Salmonella typhi*. Outbreaks are associated with contaminated water supplies and the handling of food products by infected individuals. Improper sanitary habits of food handlers is important in the spread of this disease. Many restaurants post signs in restrooms reminding workers to wash their hands, a message aimed at ensuring proper sanitary habits that will block the transmission of *S. typhi*. In some states, food handlers are required to wear gloves to prevent contamination of food.

Although the portal of entry for *S. typhi* normally is the gastrointestinal tract, infections with this organism do not initially cause gastroenteritis. Rather, the infecting bacteria rapidly enter the lymphatic system and disseminate through the circulatory system. Phagocytosis by neutrophil cells does not kill *S. typhi*, and the bacteria continue to multiply within phagocytic blood cells (FIG. 21-6).

> Although the causative organism of typhoid fever, *Salmonella typhi*, enters the body through the gastrointestinal tract, the disease involves the lymphatic system.

After invasion of the mononuclear phagocyte system, infection with *S. typhi* becomes localized in lymphatic tissues, particularly in Peyer patches of the intestine, where ulcers can develop. Localized infections develop that damage the liver and gallbladder and sometimes also the kidneys, spleen, and lungs. The symptoms of typhoid fever, which develop in a stepwise fashion over a 3-week period, include fever

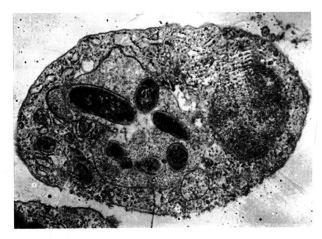

FIG. 21-6 Colorized electron micrograph showing *Salmonella typhi (purple cells)* in a macrophage.

(104° F), headache, apathy, weakness, abdominal pain, and a rash with rose-colored spots. If no complications occur, the fever begins to decline at the end of the third week. There is a high carrier rate because the bacteria can remain and multiply within infected cells. *S. typhi* can continue to grow within phagocytes in the spleen and intestinal tract. It also can grow in the gallbladder, providing a source of bacterial cells that enter the intestine and are shed in feces. This is important in the continued spread of infection from carriers. The mortality rate averages 10% in untreated cases of typhoid fever. Although there is reluctance to use chloramphenicol because it may cause aplastic anemia, its use has reduced the death rate due to typhoid to approximately 1%.

HIGHLIGHT

THE CASE OF TYPHOID MARY

The association between a cook named Mary Mallon and typhoid fever was first brought to the attention of public health authorities in 1906 during an outbreak of typhoid at Oyster Bay, Long Island, New York. Six members of a household of eleven people at which Miss Mallon worked were stricken with the disease. It was locally believed that the water was the source of contamination. The sanitary facilities, water, and food supplies, especially milk and soft clams, within the house were examined and found to be uncontaminated.

Because the first person in the house became sick on August 27, Dr. George A. Soper, who was in charge of the investigation, reasoned that the "infectious matter which produced the epidemic had been taken with food or drink on or before August 20." He learned that a new

cook had been engaged just 3 weeks before the fever began. She came from an employment bureau with excellent references and remained only a short time, leaving about 3 weeks after the outbreak of typhoid.

Dr. Soper suspected the cook was the source of infection. He recognized the importance of an interview with this woman as a source of facts from which the cause of the epidemic could be determined. The cook, Mary Mallon, was located but refused to speak to anyone about herself except to repeat facts that were already well known. She was hostile and indignant, refusing to consent to any sort of examination. Dr. Soper was very disappointed that she was not as interested as he was about learning if there was a connection between typhoid and herself.

Continued.

HIGHLIGHT

THE CASE OF TYPHOID MARY—cont'd

Dr. Soper investigated the cook's background without her cooperation and discovered that in 1904 she had worked for 9 months in another household in Sands Point, Long Island, in which four of the seven servants, but none of the four family members, came down with typhoid. He also discovered that in 1902 Mary Mallon cooked for a New York family during their summer vacation in Maine and that seven of nine members of that household became ill within a short time of her coming. Mary and the head of the family, who had had a previous bout with typhoid, were the only ones unaffected. Dr. Soper learned of at least three other cases in which there was typhoid fever in a household shortly after Mary Mallon had been employed there as a cook. She always left shortly after someone became ill with typhoid.

Mary Mallon was sent to a state hospital on an island in the East River of New York City in March 1907 by the New York City Health Department for examination of her feces and urine. A healthy, robust woman, she struggled to fight off the five policemen assigned to take her away. She was found to have large numbers of typhoid bacilli in her feces. After 16 months of treatment, the numbers of bacilli in her stools were reduced but not eliminated. She was placed in confinement because she was a carrier of typhoid bacteria. It was felt that this was the most effective measure to take because most reasonable people would not want to be responsible for the inadvertent injury of others and would take care of themselves in such a way as to minimize any danger they might present to their communities.

In the spring of 1909, after 2 years of confinement, Mary Mallon, now widely known as "Typhoid Mary," sued for a writ of *habeas corpus*, demanding her release. By then her case was already so highly publicized and her nickname so famous that the *New York Times* devoted an editorial to her. On July 1, 1909, the editors said that Typhoid Mary should in her own best interest submit to examination and, if necessary, to treatment because her problems would never be over until she had some official statement that she was not a carrier of typhoid. Her attorneys argued that she was being deprived of her liberty without ever having been accused of committing a crime, or knowingly having done injury to anyone or anything. They also contended that she was being held without a hearing, apparently under life sentence, which was clearly a violation of her constitutional rights. The New York State Supreme Court, however, ruled that Mary Mallon must remain at River-

side Hospital on North Brother Island in the East River, that releasing her would be dangerous to the health of the community, and the court would not assume the responsibility of releasing her.

It was not until July 1910 that Mary Mallon was released from confinement by the Board of Health. The new commissioner of health stated that she had "been shut up long enough to learn the precautions that she ought to take. As long as she observes them I have little fear that she will be a danger to her neighbors." Chief among the things she was told to do were observe strict personal cleanliness and not be involved in the preparation of food for others. She also had to report to the Department regularly. In 1911, Mary Mallon sued the City of New York and its Health Department for $50,000 in damages for her 3-year confinement and because of the difficulty she was then having trying to earn a living. In her petition she maintained that she had never had typhoid fever or any other diagnosed disease and that she was not the typhoid germ carrier she was claimed to be by city authorities.

Four years after her release, however, public health authorities again were looking for Typhoid Mary as the source of a typhoid epidemic in Newfoundland, New Jersey. Mary Mallon had been a cook there. In the interim she broke her parole and disappeared, assuming different names and leaving little trace of her whereabouts. In 1915, 25 employees of the Sloane Maternity Hospital were stricken with typhoid and two died. Mary Mallon worked in the kitchen there, leaving just before authorities discovered her. She was finally found, working as a cook in a private home, and was sent back to the Riverside Hospital in 1915 where she remained until her death in 1938. The issues generated by this case, including imprisonment for having an infectious disease and forced surgery, were instrumental in the founding of the American Civil Liberties Union.

By the time her story was completed, 51 cases and 3 deaths were attributed to her. What is astonishing about this case is the fact that Typhoid Mary knew the danger she presented to others and how to avoid it and yet seemed to deliberately risk the lives of others and her own freedom by continuing to work as a cook. She even chose to work in a hospital where the chance of detection and severe punishment was great. Mary Mallon was described as intelligent, resourceful, independent, mysterious, noncommunicative, self-reliant, and brave. Her motivation for continuing to act in ways that spread disease will forever remain a mystery.

Toxic Shock Syndrome

Toxic shock syndrome is caused by strains of *Staphylococcus aureus*. Toxins produced by strains of *S. aureus* cause high fever, nausea, vomiting, a sunburn-like rash often followed by shock, and a sudden drop in blood pressure. The association of toxic shock syndrome with the use of tampons received a great deal of publicity in the early 1980s, forcing one major manufacturer to remove its product from the market. The absorption of magnesium by the fibers of the tampon may be a factor in toxic shock syndrome. As the concentration of the metal in the vagina is reduced, the staphylococci grow more slowly but are stimulated to produce toxin. This is similar to the mechanism involving a deficit of iron responsible for the production of the toxin that causes diphtheria. It is also possible that air in the tampons provided oxygen that favored the aerobic growth of *S. aureus* and production of toxin. Toxic shock syndrome has also been observed occasionally in users of contraceptive vaginal sponges and even in men with staphylococcal infections, including postsurgical infections; the first reported case of toxic shock syndrome was in a young boy who had an infected leg wound.

> **Toxic shock syndrome, while not an exclusively female disease, is associated with the use of vaginal tampons.**

Puerperal Fever

Puerperal fever is a systemic bacterial infection. It is also called **childbed fever** because it may be acquired via the genital tract during childbirth or abortion. The most frequent etiologic agents of this disease are β-hemolytic group A and group B *Streptococcus* species, although *Staphylococcus, Pseudomonas, Bacteroides, Peptococcus, Peptostreptococcus,* and *Clostridium* species, as well as other bacteria, can also cause this disease. The source of infection typically comes from the obstetrician, obstetrical instruments, or bedding, but can also come from the individual's own normal microbiota. Before the introduction of aseptic procedures, puerperal fever was often a fatal complication after childbirth, and it remains an important complication following childbirth and abortion procedures. Like other septicemias, this disease is characterized by a high fever and if not treated has a high mortality rate. Penicillin is usually effective in treating postpartum sepsis. Proper obstetric procedures generally prevent this disease.

Plague

Plague is caused by *Yersinia pestis,* a Gram-negative, nonmotile, pleomorphic (irregularly shaped) rod that exhibits bipolar staining (staining at the ends of the

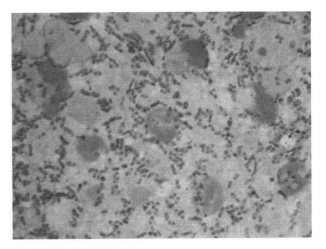

FIG. 21-7 Light micrograph showing smear of *Yersinia pestis* in a liver lesion. The numerous plague bacilli show the bipolar staining characteristic of this species in tissue preparations. Individual cells are shaped rather like safety pins. Since some other Gram-negative rods show bipolar staining, this appearance is not diagnostic of *Y. pestis.*

cell) (FIG. 21-7). It is normally maintained within populations of wild rodents where it is transferred from infected to susceptible rodents by fleas. It can also be transmitted to humans by direct contact with infected animals, animal products, or aerosols containing *Y. pestis.* Within the gut of the flea, reproduction of *Y. pestis* blocks digestion, causing the flea to increase the frequency of its feeding attempts and so to bite more animals and increase the probability of disease transmission. The transmission of plague was extremely widespread during the Middle Ages because of poor sanitary conditions and the abundance of infected rat populations in areas of dense human habitation (FIG. 21-8 p. 636). The development of rat control programs and improved sanitation methods in urban areas has greatly reduced the incidence of this disease. It is not possible, however, to completely eliminate plague in human beings because of the large number of alternate hosts in which *Y. pestis* is maintained. In rural environments, for example, *Y. pestis* is found in ground squirrels, prairie dogs, chipmunks, rabbits, mice, rats, and other animals.

> **Plague is caused by *Yersinia pestis* and transmitted by fleas. The disease cannot be completely eliminated because of the large number of alternate hosts in which *Y. pestis* can live.**

The introduction of *Y. pestis* into humans through flea bites initiates a progressive infection that can involve any organ or tissue of the body. Phagocytosis is effective in killing many of the invading bacteria, but some cells of *Y. pestis* produce a capsule that makes them resistant to phagocytosis. This establishes intra-

FIG. 21-8   A mother waits outside a hospital to admit her child suffering from the highly infectious pneumonic plague in Surat, India, Sept. 23, 1994. Hundreds of people were hospitalized and many died in the epidemic.

NEWSBREAK

OUTBREAK OF PNEUMONIC PLAGUE

For the first time in a century an outbreak of pneumonic plague began in September 1994 in Surat, India. Unlike the typical route of transmission of plague through rat flea vectors, this pneumonic form of plague is transmitted from one individual to another via the air. The incubation period is 2 to 5 days. Pneumonic plague spreads rapidly in densely populated areas, such as those in India where the outbreak occurred. The initial infection occurs in the lungs and if untreated with antibiotics rapidly leads to death. Over 100 deaths were reported in Surat as a result of pneumonic plague. Fear of contracting plague sent tens of thousands fleeing the region. The Indian government imposed a quarantine to try to prevent the spread of the disease. However, enough people had already fled so that there were additional outbreaks of plague in several other cities in India. Other nations issued warnings to travelers and established quarantine stations.

cellular infections and bacteria continue to multiply and to spread through blood and lymph. In **bubonic plague,** *Y. pestis* becomes localized and causes inflammation in the regional lymph nodes, especially about the armpits, neck, groin, and upper legs. Enlarged lymph nodes are called **buboes,** from whence the name of the disease is derived (FIG. 21-9). (*Bubo* is from the Greek, meaning groin.) The symptoms of bubonic plague include malaise, fever, and severe pain in the areas of the infected regional lymph nodes. Severe tissue necrosis (tissue death), caused by exotoxins produced by *Y. pestis,* can occur in various areas of the body. Blackened skin appears because of subcutaneous hemorrhages. It was this symptom that gave the name **"black death"** to the disease in the Middle Ages. As the infection progresses, the symptoms become severe. The disease starts with a fever and later involves septicemia, hemorrhaging, and multiple sites of infection (spleen, liver, lungs, and central nervous system). Complications can involve pneumonia, meningitis, and coagulation of blood. Without treatment the fatality rate is 60% to 100%. Death occurs in 3 to 5 days due to high concentrations of toxins produced by the numerous proliferating bacteria.

In some cases plague can involve the pulmonary system. Transmission of *Y. pestis* then can readily occur through droplet spread, establishing outbreaks of **pneumonic plague.** This was responsible for epidemics of plague during the Middle Ages when mil-

lions of people died of plague. When *Y. pestis* invades the lungs, the disease often progresses rapidly and is manifest by severe prostration, respiratory difficulties, and death within a few hours of onset. Plague can be effectively treated with antibiotics, and streptomycin generally is the drug of choice against

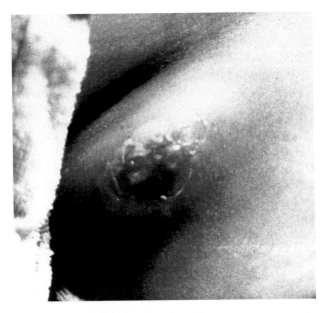

FIG. 21-9   In a case of bubonic plague the lymph node (*bubo*) has undergone suppuration and the lesion has drained spontaneously.

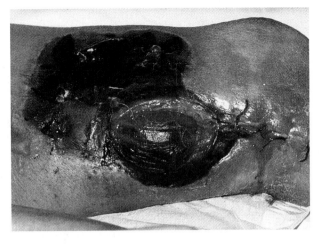

FIG. 21-10   In this case of gas gangrene there is a discharge from the lower end of the wound. The gangrenous area is turning black due to tissue necrosis.

Y. pestis, although other antibiotics, such as tetracycline, can also be used.

GAS GANGRENE (CLOSTRIDIAL MYONECROSIS)

Deep wounds not only provide a portal of entry for microorganisms, but the tissue damage often interrupts circulation to the area and the metabolism of adjacent tissues and contaminating bacteria remove oxygen. This creates conditions that permit the growth of obligately anaerobic bacteria. **Gas gangrene** is a serious infection that may result from the growth of *Clostridium perfringens* and other *Clostridium* species. The name of this disease may be misleading, since gas may not form until late in the infection. Therefore a better name for this disease is **clostridial myonecrosis,** indicating that muscle tissue is destroyed by *Clostridium* species. These clostridia occur in soil and within the normal microbiota of the large intestine. Gas gangrene is often due to a mixed infection of organisms capable of producing a very large number of enzymes that greatly enhance their invasive capabilities. The development of gas gangrene depends on the deposition of endospores of *Clostridium* in wound tissue and the development of anaerobic conditions due to necrosis of local tissues that permit the germination and multiplication of these obligately anaerobic bacteria.

Gas gangrene-causing *Clostridium* species produce toxins. Diffusion of these toxins extends the area of dead and anaerobic tissues. The exotoxins produced by these *Clostridium* species are tissue necrosins and hemolysins that spread from the infected tissues to uninfected tissues, killing them. This allows the infection to spread to adjacent areas as these areas lose circulation and become anaerobic; this accounts in part for the rapid spread of infection (FIG. 21-10). The growing *Clostridium* species produce carbon dioxide and hydrogen gases and the formation of odoriferous low molecular weight metabolic products. The gas that accumulates is primarily hydrogen because it is less soluble than carbon dioxide. The buildup of gas pockets in the tissues causes a crackling or rattling called crepitation. In most cases, the onset of gas gangrene occurs within 72 hours of the occurrence of the wound and, if untreated, the disease is fatal. Even with antimicrobial treatment, there is a high rate of mortality, and therefore, radical surgery—amputation—is often employed to prevent the spread of infection. If treated rapidly enough, localized areas of necrotic tissue can be excised—a process called debridement—and high doses of penicillin, tetracycline, and antitoxin are administered to block the spread of the infection.

The prevention of gas gangrene depends on ensuring that wounds are not suitable environments for the growth of the anaerobic *Clostridium* species. This requires that wounds have adequate drainage to prevent establishment of anaerobic conditions and that foreign material and dead tissue be removed. Hyperbaric oxygen chambers are sometimes used to treat gas gangrene in an effort to kill the oxygen-sensitive *Clostridium* species before they spread to the point where radical surgery is needed to prevent the death of the patient.

> The growth of *Clostridium* species in the anaerobic tissues of serious wounds releases toxins that cause the disease gas gangrene. Oxygenation of the tissues and radical surgery are used to treat this disease.

RELAPSING FEVER

Relapsing fever is caused by *Borrelia recurrentis*. *B. recurrentis* is a Gram-negative spirochete—a bacterium that is helically coiled around a central filament that is attached to the ends of the cell (FIG. 21-11). It is endemic to rodents in many parts of the world, including the western United States. *B. recurrentis* normally is transmitted to humans from rodents through tick bites of the soft tick *Ornithodorus*. This is the endemic form of the disease. In the tick the bacteria are transmitted by transovarial passage, through the ovaries to the eggs so that it is passed congenitally from generation to generation. Under crowded conditions, epidemic outbreaks of relapsing fever can occur, with transmission from human to human occurring via the human body louse. This is the epidemic form of the disease. Lice bite and defecate into the wound, inoculating the body. The irritation causes the bitten person to rub the area, increasing the likelihood of getting bacteria into the wound and initiating an infection. Poor sanitary conditions, such as occur during wars and natural disasters, favor the spread of this disease because it is under these conditions that body lice proliferate.

Body lice are the vectors that carry Borrelia species, the causative organisms of relapsing fever.

The clinical manifestations of relapsing fever occur intermittently. There is normally a sudden onset of fever approaching 105° F that lasts for 3 to 6 days. The fever then falls rapidly and remains normal for 5 to 10 days. During this time the infecting bacteria alter their surface antigens so that they evade the body's immune response. This permits a relapse of the fever. The second onset of fever generally lasts 2 to 3 days. Additional relapses may occur, each with diminishing fever and duration. Generally there are two relapses in the epidemic form and three to four relapses in the endemic form of the disease. During the course of relapsing fever, anatomical abnormalities develop in the spleen. Lesions may also occur in various other body organs. The natural removal of *Borrelia* from the body depends on an antibody-mediated immune response rather than on phagocytosis. Penicillin, tetracycline, and chloramphenicol are effective against *Borrelia* and are used in treating this disease. With proper treatment, over 95% of patients with relapsing fever recover.

BARTONELLOSIS

Bartonellosis is an infection caused by the obligately intracellular, Gram-negative bacterium, *Bartonella bacilliformis*. This disease is transmitted to people through the bites of infected sand flies that occur only in the valleys of the Andes Mountains and thus occurs only in specific regions of Peru, Ecuador, and Colombia. Prevention of bartonellosis depends on avoiding contact with infected sand fly vectors. The sand flies carrying *Bartonella* are nocturnal, and the use of insect repellents and protective clothing after dark are useful for people in the Andes. *B. bacilliformis* proliferates within the endothelial cells of blood vessels after inoculation by the bite of the sand fly vector. The bacteria reenter the circulatory system and infect erythrocytes, generally causing severe anemia accompanied by fever, headache, and delirium. The mortality rate from bartonellosis can be as high as 40%. Chloramphenicol can reduce the fatality rate.

Bartonella bacilliformis, the causative organism of bartonellosis, infects erythrocytes in the blood.

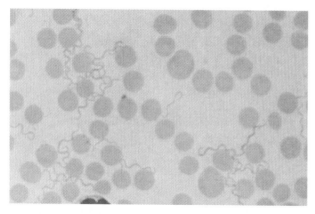

FIG. 21-11 Light micrograph showing *Borrelia recurrentis* (*thin helical cells*) in a smear of mouse blood. The mouse had been inoculated with the blood of a patient suffering from relapsing fever.

TABLE 21-4		
Human Infections Caused by *Rickettsia*		
SPECIES	**DISEASE**	**VECTOR**
Rickettsia prowazekii	Epidemic typhus	Body louse
Rickettsia typhi	Murine typhus	Flea
Rickettsia rickettsii	Rocky Mountain spotted fever	Tick
Rickettsia akari	Rickettsialpox	Mite
Rickettsia tsutsugamushi	Scrub typhus	Chigger
Coxiella burnetii	Q fever	Ticks (also droplet inhalation)
Rochalimaea quintana	Trench fever	Body louse

RICKETTSIAL DISEASES

Rickettsia species are obligate intracellular parasites. They do not have the mechanisms for generating sufficient ATP to support growth and reproduction. They obtain ATP from the animal cells in which they grow. *Rickettsia* species cannot be cultured on agar media but they can be grown using tissue culture techniques. Most *Rickettsia* species that cause human diseases are transmitted by vectors (Table 21-4).

Rocky Mountain Spotted Fever

Rocky Mountain spotted fever is caused by *Rickettsia rickettsii*. This microorganism is transmitted to humans through tick bites (FIG. 21-12). *R. rickettsii* is normally maintained in various tick populations, such as wood and dog ticks, where it multiplies within the midgut and is passed congenitally via the ovaries from one generation to the next. Humans are accidental hosts of *R. rickettsii* as a result of occasional bites by infected ticks (FIG. 21-13).

When injected into human beings, *R. rickettsii* multiply within the endothelial cells lining the blood vessels. Vascular lesions occur and account for the production of the characteristic skin rash associated with this disease. The rash is most prevalent at the extremities, particularly on palms of the hands and soles of the feet. Lesions probably also occur in the meninges, causing severe headaches and a state of

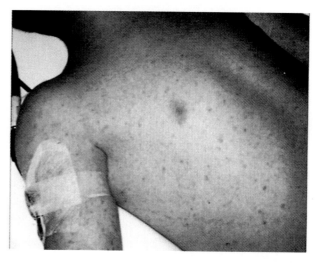

FIG. 21-12   Rocky Mountain spotted fever in a boy characterized by moderately severe eruption of a red skin rash.

mental confusion. If treated with antibiotics, such as chloramphenicol and tetracycline, the disease is rarely fatal; if untreated, the overall mortality rate is greater than 20%. Prevention of Rocky Mountain spotted fever primarily involves control of population levels of infected ticks and the avoidance of tick bites. Clothing that covers the body, insect repellents, and limited time in tick infested areas lower the probability of being bitten by an infected tick.

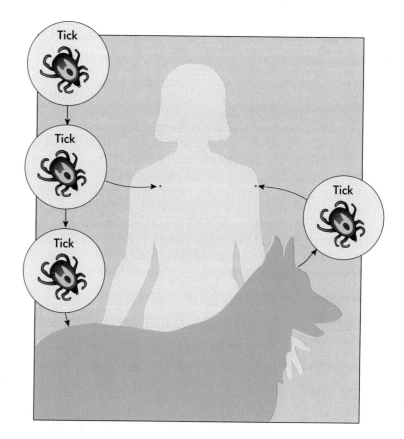

FIG. 21-13   Rocky Mountain spotted fever is transmitted by ticks. Ticks transmit *Rickettsia rickettsii* congenitally from one tick to another and also transmit this bacterium to animals. Humans often acquire infections from wood and dog ticks carrying *R. rickettsii*.

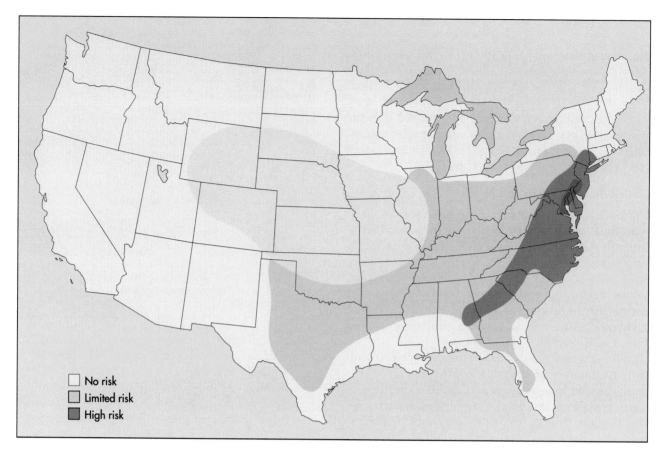

FIG. 21-14 Although initially endemic to the Rocky Mountain region, Rocky Mountain spotted fever today occurs primarily east of the Mississippi River.

The causative organism of Rocky Mountain spotted fever, *Rickettsia rickettsii,* causes vascular lesions that produce the characteristic skin rash. *R. rickettsii* is transmitted by ticks.

Rocky Mountain spotted fever is most common in spring and summer when ticks and people are most likely to come in contact. It occurs in regions of North and South America. During the mid-twentieth century many cases of Rocky Mountain spotted fever occurred in the United States in the region of the Rocky Mountains, but relatively few cases have been reported there in recent years (FIG. 21-14). However, outbreaks of Rocky Mountain spotted fever have risen dramatically in the eastern United States, where over 1,000 cases of this disease now occur each year.

Typhus Fever

There are several types of typhus fever. All types are caused by rickettsias transmitted to humans via biting arthropod vectors (FIG. 21-15). **Epidemic typhus fever** (also called infectious typhus or classical typhus fever) is caused by *Rickettsia prowazekii* and is transmitted to humans via the body louse. Lice contract the disease from infected humans and in turn

pass the rickettsia on to susceptible human hosts. Under crowded conditions with lice infestation, the disease can spread easily and cause large numbers of cases. For example, millions of cases occurred in World War I and in the concentration camps during World War II. From 1918 to 1922, 30 million cases occurred in Eastern Europe and the Soviet Union. The disease still occurs in South America and Africa. It has been eliminated from the United States as a result of delousing campaigns; the last case occurred in 1922.

The different types of typhus fever are all caused by rickettsias transmitted to humans via biting arthropod vectors.

R. prowazekii multiplies within the epithelium of the midgut of the louse. When an infected body louse bites another human it defecates at the same time, depositing feces containing *R. prowazekii,* which enter through the wound created by the bite; rubbing the wound due to the irritation introduces louse feces and associated *R. prowazekii* into the body. The onset of epidemic typhus involves fever, headache, and rash. The heart and kidneys are frequently the site of

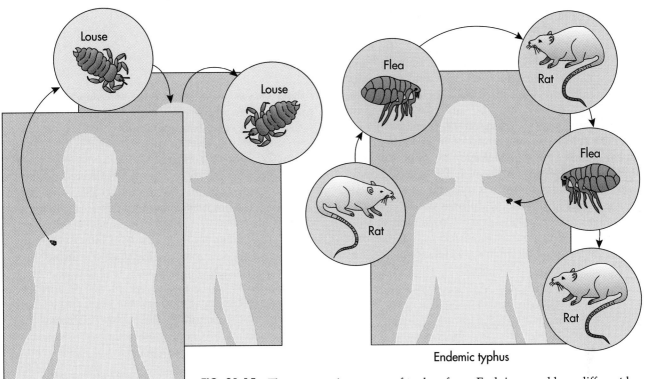

FIG. 21-15 There are various types of typhus fever. Each is caused by a different bacterial species and each has a characteristic route of transmission involving different factors. Epidemic typhus occurs when body lice transmit *Rickettsia prowazekii* from one human to another. Endemic typhus occurs when rat fleas transmit *R. typhi* from infected rats to humans.

vascular lesions. If untreated, the fatality rate in persons 10 to 30 years old is approximately 50%. Survivors often maintain the rickettsia within cells of the body and the infection can be reactivated, even 50 years later, to produce *Brill-Zinsser disease.* Chloramphenicol, tetracycline, and doxycycline are effective in treating epidemic typhus.

Murine typhus, or **endemic typhus fever,** is caused by *Rickettsia typhi.* It is transmitted to humans by rat fleas. Murine typhus is normally maintained endemically in rat populations through transmission by rat fleas. As with louse-borne typhus, when an infected flea bites a human, it deposits pathogenic bacteria in its fecal matter. The symptoms of murine typhus are similar to those of classical typhus fever but are generally milder. Chloramphenicol and tetracycline are effective in treating this disease, and there is a relatively low mortality rate. Prevention of murine typhus depends on limiting rat populations, which also limits the size of the vector rat flea population.

Scrub typhus is almost exclusively restricted to Japan, Southeast Asia, and the western Pacific Islands. During the war in Vietnam, many American soldiers contracted this disease. The disease is caused by *Rickettsia tsutsugamushi,* which is maintained in a reservoir population of rats. It is transmitted to people through the bite of mite vectors (chiggers).

R. tsutsugamushi is normally transmitted congenitally in mite populations and is only accidentally introduced into the human population. In humans, multiplication of *R. tsutsugamushi* occurs at the site of inoculation and is later distributed through the circulatory system; eventually the infection becomes localized in the lymph nodes. The symptoms of scrub typhus include fever, severe headache, and rash. There is a characteristic lesion of the skin, known as the *eschar,* that occurs at the site of initial infection. Chloramphenicol and tetracycline are effective in treating this form of typhus. Controlling vector populations and minimizing the chance of being bitten by an infected mite are means by which this disease can be prevented.

Rickettsialpox

Rickettsialpox is caused by *Rickettsia akari.* This organism is transmitted to humans by the mouse mite. *R. akari* is congenitally maintained in populations of the mouse mite, and in addition to serving as a reservoir for *R. akari,* the mouse mite acts as a vector. Rickettsialpox occurs almost exclusively in urban environments of the United States and the Soviet Union, where there are not enough mice for the mites to feed on. It is under such conditions that infected mites will bite people.

As with other rickettsial diseases, the bacterial infection can spread along the vessels of the circulatory system, causing vascular lesions. Systemic spread of the infection is followed by the onset of the illness, which is manifest by fever, headache, and secondary lesions. The rash associated with this disease may cover any part of the body, but unlike Rocky Mountain spotted fever, the rash rarely develops on the palms or soles. The symptoms of rickettsialpox disappear without treatment beginning approximately 10 days after the onset of disease. The use of tetracyclines and other antibiotics is effective in ending the manifestation of disease symptoms within 1 day of beginning antimicrobial therapy. Prevention of rickettsialpox depends on effective rodent control programs; the elimination of mice from urban settings can effectively end this disease.

Rickettsialpox is a self-limiting disease whose symptoms disappear without treatment about 10 days after onset.

TULAREMIA

Tularemia is caused by *Francisella tularensis,* a Gram-negative fastidious coccobacillus. The bacterium is transmitted by biting arthropods (ticks and deerflies), inhalation of contaminated aerosols, and ingestion of contaminated food and water. It may gain entry to the body directly through the skin, particularly through minute openings, such as at hair follicles, and near the fingernails. Transmission to humans can occur through ingestion of contaminated meat and handling of infected animals. Rabbits are especially dangerous. Tularemia is a particular occupational problem for individuals, such as hunters, butchers, cooks, agricultural workers, and so on, who handle potentially contaminated animals. There is a vaccine available for such high-risk workers.

Tularemia is an occupational hazard for people who work with animals potentially contaminated with *Francisella tularensis.*

Localized ulcers result from the multiplication of *F. tularensis.* A systemic infection with localization within the regional lymph nodes also occurs. Often, finger ulcers are symptomatic of tularemia that is contracted through the handling of infected animals. In most cases of tularemia, there is an elevated fever, development of skin ulceration, and enlargement of the lymph nodes. The death rate in untreated cases of tularemia acquired through the skin is approximately 5%, but higher mortality rates approaching 30% occur when the disease is contracted through inhalation. Streptomycin, tetracycline, and chloramphenicol are effective in controlling this disease. Prevention of tularemia can be achieved by avoiding contaminated animals, and because rabbits have an especially high rate of infection with tularemia, special precautions should be taken in their handling.

BRUCELLOSIS

Brucellosis is caused by several *Brucella* species. *Brucella* are Gram-negative, small, nonmotile, aerobic rods. This disease is an infectious disease of nonhuman animals. *B. abortus* occurs in cattle, *B. suis* in swine, *B. melitensis* in goats, and *B. canis* in dogs. These *Brucella* species can be transmitted to humans through ingestion of contaminated milk and handling infected animals. Pasteurization has greatly reduced the transmission of brucellosis via the gastrointestinal tract, and most human infections today result from direct contact with infected animals. Brucellosis is an occupational hazard in farming, veterinary practice, and meatpacking plants, where *Brucella* species can enter through the skin, particularly through minor abrasions. Most of the 200 cases that occur annually in the United States are associated with handling of swine carcasses.

Pasteurization has greatly reduced the transmission of brucellosis via the gastrointestinal tract.

Once *Brucella* species enter the body, the bacteria spread rapidly through the mononuclear phagocyte system and multiply within phagocytic cells. Infecting *Brucella* species normally become localized in the regional lymph nodes. There may be enlargement of the spleen and liver, and other symptoms include weakness, chills, malaise, headache, backache, and fever. The fever may rise and fall in some cases; thus this disease is often referred to as **undulant fever.** The death rate in untreated cases of brucellosis is about 3%. The use of antibiotics is effective in treating this disease and reduces the death rate to near 0%. Antibiotic treatment, however, requires prolonged administration of combinations of drugs, such as tetracyclines and streptomycin, because *Brucella* can live intracellularly. The prevention of brucellosis involves eliminating the disease in animals such as cattle, sheep, and goats, the reservoirs of *Brucella.* It is important to segregate and treat infected animals to prevent the spread of this disease through animal herds. Vaccines are effective in limiting the spread of brucellosis through some animal populations, such as cattle, but there are no effective vaccines for other animals, such as hogs.

LYME DISEASE

Lyme disease was named for the small Connecticut community in which the disease was first recognized in 1975. It is an inflammatory disorder caused by the spirochete *Borrelia burgdorferi.* The disease is transmitted to humans by *Ixodes* ticks. The white-

HIGHLIGHT

EPIDEMIOLOGY OF LYME DISEASE

In October 1975 the Connecticut Department of Health received independent calls reporting multiple cases of what appeared to be arthritis in children in Lyme and Old Lyme, rural towns in that state. Despite being assured by their physicians that arthritis was not infectious, the callers were not satisfied. An epidemic investigation ensued in which the extent, characteristics, mode of transmission, and etiology of the cluster of cases were studied. As characteristic of many epidemiological investigations, public health officials began by trying to locate all individuals who had sudden onset of swelling and pain in the knee or other large joints lasting a week to several months. An odd, large skin rash, repeated attacks at intervals of a few months, fever, and fatigue were the reported symptoms. State epidemiologists questioned parents and physicians—asking were the cases related, were there other similar cases, was this an infectious form of arthritis, and what forms of arthritis are infectious? Next the epidemiologists determined the time, place, and personal characteristics of these cases. The incidence of onset of disease seemed to cluster in late spring and summer and lasted from a week to a few months. The cases were concentrated in three adjacent towns on the eastern side of the Connecticut River and most patients lived in wooded areas near lakes and streams. Of the 51 cases, 39 were children about evenly split between boys and girls. There were no familial patterns. Epidemiologists created an epidemic curve, listing the cases by the time of onset, and began calling the disease "Lyme arthritis."

The clustering of cases, the fact that most began in late spring or summer and that they were most frequently located in wooded areas along lakes or streams suggested a disease transmitted by an arthropod. A study was undertaken to determine if this was a communicable disease. Cases of the disease were matched with a similar group of control or unaffected persons for age, sex, and other relevant factors. It was found that affected people were more likely to have a household pet than those who were unaffected. Pet owners are more likely to come in contact with ticks that their dogs and cats might pick up in the woods. The importance of this finding was emphasized when combined with the fact that one fourth of the patients reported that their arthritic symptoms were preceded by an unusual skin rash that started as a red spot that spread to a 6-inch ring. A dermatology consultant recalled a similar skin outbreak reported in Switzerland in 1910 that was attributed to tick bites.

This was only suggestive evidence that a tick bite might initiate an infectious disease. The connection between the rash and the disease had to be strengthened. Now public health authorities had to ask if patients with such a rash always progress to develop Lyme arthritis. A prospective study looked for new patients with a rash. Of 32 new cases of the characteristic skin rash, 19 progressed to show signs and symptoms of Lyme disease. The tick connection was strengthened after an entomological study found that adult ticks were 16 times more abundant on the east side of the Connecticut River than the west. This corresponded to the proportion of incidence of the disease on each side of the river. Also, more tick bites were reported by the arthritis sufferers than by their unaffected neighbors. A surveillance network was set up in Connecticut and surrounding states to gather information about other cases. These investigations showed more adult victims than children and also more serious manifestations, including neurological and heart diseases.

The Rocky Mountain Public Health Laboratory in Montana was asked to assist in the investigation because of its expertise in the area of tickborne disease. They found unusual spirochetes in the guts of many of the ticks sent from Connecticut. Spirochetes, which are bacteria with curved cells wound around a central filament, are often difficult to culture and so it would be difficult to prove that these were the causative organisms of Lyme disease. Therefore they first tried to infect laboratory animals with the infected ticks. The rabbits developed rashes resembling those seen in humans. A spirochete was isolated from the ticks and when pure cultures were inoculated into rabbits, the rabbits developed the characteristic rash. The infected rabbits contained antispirochetal antibodies in their serum. The identification was complete when the spirochete was isolated from human cases. The spirochete was classified as a member of the *Borrelia* genus and named *Borrelia burgdorferi* after the entomologist who discovered the organisms in the ticks.

Lyme disease accounted for more than 90% of the vector-borne infectious diseases in the United States in 1992. The distribution of the disease was highly correlated with the distribution of the principal tick vectors. The nearly twentyfold increase in cases since the early 1980s may be a consequence of increased surveillance and improved diagnostic methods, or a real increase in disease prevalence due to increases in deer and tick populations and closer human contact with these animals.

tailed deer and white-tailed mouse are the reservoirs of *B. burgdorferi*. The distribution of the disease is restricted to regions with these animal reservoirs and *Ixodes* ticks. Increases in deer populations and human activities such as clearance of woodland and hiking and camping in wooded areas have increased the likelihood of being bitten by a deer tick. This has resulted in an increased number of cases of Lyme disease. Although the territory over which this disease has occurred is increasing, the majority of the approximately 7,000 reported cases of Lyme disease have occurred in the Northeast, upper Midwest, and California.

Lyme disease usually begins with a distinctive skin lesion (FIG. 21-16). The circularly expanding annular skin lesions are hardened (indurated) with wide borders and central clearing. Although these lesions may reach diameters of 12 inches or more, they are painless. Accompanying symptoms may resemble a mild flu, with some patients experiencing symptoms resembling mild meningitis or encephalitis, hepatitis, musculoskeletal pain, enlarged spleen, and cough. Weeks to months later, the patient often shows signs of the second stage of the disease, developing arthritic joint pain, and sometimes neurologic or cardiac abnormalities. The neurological complications may include visual, emotional, and memory disturbances, temporary paralysis of a facial nerve, and movement difficulties. The third stage, which

NEWSBREAK

DOGS TOO CAN CONTRACT LYME DISEASE

Lyme disease, or Borreliosis, is a tick-borne bacterial disease affecting humans and animals. It was first discovered in the United States in a group of children by Dr. W. Burgdorfer in Lyme, Connecticut, in 1975 and reported in 1985 to also affect dogs. In dogs the symptoms include arthritis, lameness, fever, lethargy, and loss of appetite. The bacterium that causes Lyme disease, *Borrelia burgdorferi*, is carried and spread by the deer tick, *Ixodes dammini*, but can also be spread by dog (wood) ticks, *Dermacentor variabilis*, and other varieties of ticks. More than 30,000 cases in humans have been reported from 47 states. In 1994, Lyme disease is the most commonly occurring tick-borne disease in the United States. It is estimated that the incidence of this disease in dogs is six times more prevalent than in humans. Dogs that run in infested fields can introduce the infected ticks to suburban areas. Dogs travelling with their owners can spread the disease to new locations. The number of cases is rising and the disease is becoming more widespread in the United States and in other countries. A vaccine can be administered to prevent this disease in dogs. The vaccine initially is administered twice (2 to 3 weeks apart) and annual booster vaccines are then recommended to maintain immunity.

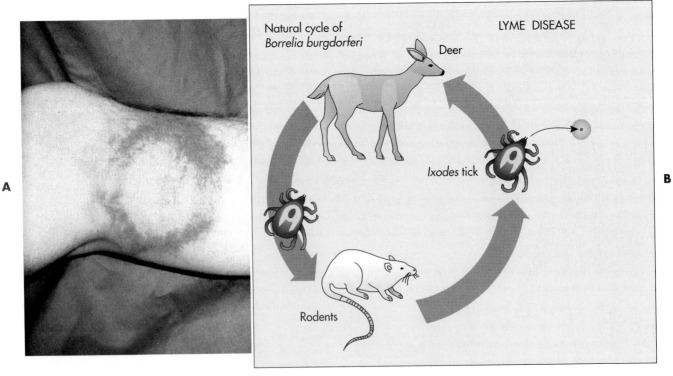

FIG. 21-16 **A,** Characteristic rash of Lyme disease. **B,** Lyme disease is transmitted to humans by ticks, principally deer ticks.

may appear months to years after infection, is characterized by crippling arthritic symptoms in one or more joints, especially in the knees, and severe neurological symptoms that mimic multiple sclerosis. These symptoms are believed to be caused by the body's immune defense system's attempts to fight the infective agent, rather than by the organism itself. The antibody complexes produced in response to the infective agent cause the joints to become inflamed. Treatment with trivalent antibodies neutralizes the toxins produced by *B. burgdorferi.*

Lyme disease is characterized by the development of arthritis and neurological symptoms that result when the body's immune defenses react to infections with the spirochete *Borrelia burgdorferi.*

Recent epidemiologic and laboratory investigations indicate that Lyme disease may be a different manifestation of *erythema chronicum migrans,* a syndrome long recognized in Europe. Both diseases are caused by the spirochete, *Borrelia burgdorferi,* transmitted to humans by the bites of *Ixodes* ticks. Similar spirochetes have been recovered from the blood of Lyme disease patients and cultured from *Ixodes* ticks, and patients with Lyme disease have antibodies to the cultured spirochetes. When this disease is diagnosed, penicillin is an effective treatment. There is also a new vaccine that appears to be effective for preventing Lyme disease.

The occurrence of Lyme disease has been concentrated in the Northeast and other areas where *Ixodes* ticks carry *Borrelia burgdorferi.*

SUMMARY

Bacterial Diseases of the Central Nervous System (pp. 625-630)

- Meningitis is a disease of the central nervous system that can be caused by various bacteria, most commonly *Neisseria meningitidis, Haemophilus influenzae, Streptococcus pneumoniae,* and *Escherichia coli.* It is transmitted by droplet spread, with the initial infection occurring in the respiratory tract followed by transmission via the bloodstream to the meninges. Symptoms include sudden fever, severe headache, painful rigidity of the neck, nausea, and vomiting, and, frequently, convulsions, delirium, and coma. Treatment is with antibiotics, usually penicillin.
- Tetanus is caused by *Clostridium tetani,* which produces a neurotoxin that causes spastic paralysis. *C. tetani* is transmitted via a puncture wound that introduces spores from the soil, producing severe muscle spasms and difficulty in swallowing. There is a vaccine available that prevents thousands of cases of this disease. Treatment involves vaccination and administration of tetanus antitoxin.
- Botulism is caused by the ingestion of foods containing neurotoxins produced by *Clostridium botulinum,* producing paralysis of the respiratory muscles and death by asphyxiation. It is treated with trivalent ABE antibodies.
- *Clostridium botulinum* can reproduce in the higher pH environment of an infant stomach and elaborate neurotoxins. Some cases of crib death may be due to infant botulism. Infant food products must be free of *C. botulinum* endospores.
- Wound botulism is caused by spores of *Clostridium botulinum* inoculated into a deep wound. The neurotoxin produced blocks neural transmissions. Antitoxin can be administered.
- Listeriosis is a foodborne disease of the central nervous system caused by *Listeria monocytogenes* and affects immunosuppressed adults (individuals with AIDS) and pregnant women. It is treated with penicillin and ampicillin.

- Brain abscesses are caused by bacteria, including species of *Streptococcus, Staphylococcus, Haemophilus, Nocardia, Bacteroides, Peptostreptococcus, Peptococcus, Fusobacterium,* and *Actinomyces.* They develop as a complication of an infection at another body site, producing symptoms of fever, headache, nausea, and various neurological symptoms. They are treated with antibiotics, frequently penicillin, chloramphenicol, and metronidazole.

Bacterial Diseases of the Cardiovascular and Lymphatic Systems (pp. 631-645)

- Septicemia or blood poisoning is caused by *Escherichia coli, Enterobacter aerogenes, Pseudomonas aeruginosa, Serratia marcescens,* and *Proteus* sp. It produces red streaks under the skin along the arms and legs, a high fever, low blood pressure, and shock. Antibiotics accompany attempts to maintain body functions and limit the effects of shock.
- Rheumatic fever is an autoimmune disease caused by the cross-reactivity of antibodies made against *Streptococcus pyogenes.* Its symptoms include high fever, painful swelling of body joints, and cardiac involvement. It is treated with penicillin and anti-inflammatory drugs.
- Carditis is usually caused by viridans streptococci and occurs when normally nonpathogenic members of the microbiota that are associated with body surfaces enter the circulatory system and attach to cardiac tissue. It can be treated with broad-spectrum antibiotics.
- Typhoid fever is caused by *Salmonella typhi* that is ingested in contaminated water and food handled by infected individuals. Symptoms are fever, headache, apathy, weakness, abdominal pain, and rash. Chloramphenicol is the treatment of choice.
- Toxic shock syndrome is caused by *Staphylococcus aureus* that enter lesions in the vaginal wall through the use of tampons. It also can enter through surgical wounds. Treatment is strictly supportive.

- Puerperal, or childbed, fever is most frequently caused by beta-hemolytic group A and B *Streptococcus* species. It is spread by the obstetrician, obstetrical instruments, or bedding. It is treated with penicillin.
- Plague is caused by *Yersinia pestis* transmitted to humans via transfer within rodent populations by fleas who then bite people. Symptoms include inflammation and enlargement of lymph nodes, malaise, fever, pain, and tissue necrosis. It is treated with antibiotics, generally streptomycin.
- Gas gangrene is an infection that often appears in the aftermath of a wound. It is caused by *Clostridium perfringens* and other *Clostridium* species when *Clostridium* endospores are deposited in wound tissue where anaerobic conditions exist. The necrotic tissue must be excised; sometimes this means amputation. Treatment is with penicillin.
- Relapsing fever is caused by various species of *Borrelia*. It is transmitted to humans via the bite of the body louse or ticks. Symptoms occur intermittently, a sudden high fever falls rapidly and remains normal, followed by a second onset of fever. It is treated with tetracycline and chloramphenicol.
- Bartonellosis is caused by *Bartonella bacilliformis* transmitted through bites of infected sand flies, causing severe anemia, fever, headache, and delirium. It is treated with chloramphenicol.
- Rocky Mountain spotted fever is caused by *Rickettsia rickettsii* transmitted via tick bites, causing skin rash,

headaches, and mental confusion. It is treated with chloramphenicol.
- Typhus fever is caused by *Rickettsia prowazekii* transmitted via body lice and causes fever, headache, rash, and heart and kidney lesions. It is treated with chloramphenicol, doxycycline, and tetracycline. Other forms of the disease include Brill-Zinsser, murine or endemic typhus, and scrub typhus.
- Rickettsialpox is caused by *Rickettsia akari* transmitted to humans by the mouse mite, producing symptoms of fever, headache, primary lesions at the bite site, and secondary lesions and rash, which are treated with tetracyclines.
- Tularemia is caused by *Francisella tularensis* ingested in contaminated meat and from handling infected animals. It produces ulcers on fingers and skin, elevated fever, and enlarged lymph nodes. It is treated with streptomycin, tetracycline, and chloramphenicol.
- Brucellosis is caused by several species of *Brucella* transmitted via ingestion of contaminated meat and handling infected animals. Symptoms include enlargement of spleen and liver, chills, weakness, headache, backache, and a fever that rises and falls. It is treated with tetracyclines. There is an animal vaccine.
- Lyme disease is caused by a spirochete transmitted by tick bites, producing a rash followed by arthritis of the knee and possible neurological and cardiac abnormalities. It is treated with penicillin.

CHAPTER REVIEW

REVIEW QUESTIONS

1. Which bacterial species cause meningitis? How are these pathogens transmitted? How do they enter the cerebrospinal fluid?
2. How can bacterial meningitis be differentiated from viral meningitis?
3. How are vaccines being used to prevent meningitis?
4. How are vectors involved in the transmission of diseases caused by rickettsias?
5. What are the similarities and differences between botulism, infant botulism, and wound botulism?
6. How can a streptococcal infection cause rheumatic fever?

7. How is plague transmitted? Why was it a major cause of death in the Middle Ages? Why are there still cases of plague today?
8. Describe the signs of Lyme disease. What is the cause of this disease? How is it transmitted? How can it be prevented?
9. What is an asymptomatic carrier? How are carriers involved in the transmission of typhoid fever?
10. How were certain superabsorbant tampons associated with toxic shock syndrome?
11. How is gas gangrene treated?

CRITICAL THINKING QUESTIONS

1. Why is there a high mortality rate in cases of bacterial meningitis despite the fact that there are many antibacterial agents?
2. Why is the incidence of typhus fevers high during times of war? What can be done to reduce the incidence of typhus?

3. Could Lyme disease be an old disease that has only recently been recognized? What could have contributed to an increased incidence of infections with *Borrelia burgdorferi* and the recognition that Lyme disease is an infectious disease?

CASE STUDY

HISTORY AND ASSESSMENT

Denzel, a 2-year-old boy, was brought to the emergency room by his mother at 8 pm. He had been febrile (elevated temperature) and irritable for several hours during the day. At approximately 7 pm, Denzel vomited and had a convulsion. Denzel's mother brought him directly to the emergency room.

On assessing Denzel's condition, the nurse noted that he was lying on the table with his knees slightly bent. He appeared irritable. He resisted her attempts to move him on the stretcher, especially when she tried to move his head. A widespread skin rash was noted. Vital signs were as follows: Pulse rate 96, Respiratory rate 18, Blood pressure 110/70, and Temperature 105° F. *What is your assessment of these vital signs?*

Denzel's mother told the nurse that he had a cold for a day or two preceding this incident.

PHYSICAL EXAMINATION

On EENT (eyes, ear, nose, and throat) examination, the physician noted that Denzel's pupils were of uneven size, the right pupil was larger than the left, and the left eye appeared to deviate somewhat to the left.

What body system appears to be involved with these findings?

Next, the physician noted that Denzel resisted any attempt to straighten his legs from a flexed position. An attempt to flex the child's chin to chest was also met with strong resistance and great pain. *These examination findings suggest meningitis (an inflammation of the meninges). The meninges are the tissues that cover the brain and spinal cord. Among their functions, the meninges create a blood–brain barrier (BBB) that controls what crosses between blood and brain tissue; generally, substances do not easily cross the BBB. Flexion of the neck stretches the inflamed meninges, causing pain. Knee flexion helps release tension on the inflamed meninges.*

Through the physical assessment, the physician determined that Denzel had the clinical symptoms of meningitis. Quickly determining the causative microorganism and beginning appropriate therapy is essential for Denzel's well being.

PROCEDURES AND LABORATORY TESTS

Denzel was transferred to the Medical Intensive Care Unit (MICU) for further diagnosis and treatment. Orders were written for blood chemistry, urinalysis, clotting times, blood cultures (every hour for 3 hours), lumbar puncture, and CSF (cerebrospinal fluid, a fluid that bathes the brain and spinal cord) analysis and culture. *A lumbar puncture is a procedure in which a needle aspiration of CSF is made in the lumbar (lower back) region of the spinal column. CSF analysis, including checking for leukocytes and glucose concentrations, can help determine whether there is an infection of the central nervous system. CNS infection is a very serious condition.*

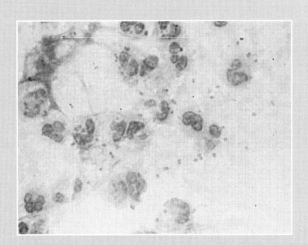

Cultures of the blood and CSF are not available for 12 to 24 hours after the procedures are performed, but the physician should begin effective treatment immediately. *The physician does not know for certain what type of microorganism—virus, bacteria, or fungus—is causing Denzel's symptoms and decides to look at the CSF under a microscope. Bacteria or fungi present in the CSF can be identified by microscopy. The physician also knows that a rash similar to Denzel's is seen in approximately 50% of the cases of meningitis due to* Neisseria meningitidis, *but that a similar rash can also be seen in cases of meningitis caused by viruses and other bacteria.*

The physician examines the CSF under the microscope and sees a prevalence of Gram-negative cocci (see Figure). This is confirmed the next day by the preliminary CSF culture report.

DIAGNOSIS

The physician knows that the three most prevalent organisms causing meningitis in Denzel's age group are Hemophilus influenzae *(Gram-negative rods),* Neisseria meningitidis *(Gram-negative cocci), and* Streptococcus pneumoniae *(Gram-positive cocci). With this information and the results of the doctor's Gram stain, identify the causative microorganism.*

If the physician's investigation showed a Gram-negative rod, what type of meningitis would you suspect?

TREATMENT AND COURSE

The physician began Denzel on a regimen of Ceftriaxone, an antibiotic, to treat his meningitis caused by *Neisseria meningitidis.* Several days later the child's condition improved. Denzel was discharged from the hospital with no further complications. *The mortality rate for* N. meningitidis *meningitis in the United States is 5% to 10%.*

READINGS

Adams EB, DR Lawrence, JWG Smith: 1969. *Tetanus*, Oxford, Blackwell Scientific.

> *Outlines the experimental and laboratory backgrounds of the clinical practice of tetanus prevention.*

Bayliss JH: 1980. The extinction of bubonic plague in Britain, *Endeavour* 4:58-66.

> *An interesting account of how plague was eliminated from Great Britain.*

Bergdoll MS and PJ Chesney: 1991. *Toxic Shock Syndrome*, Boca Raton, FL; CRC Press.

> *Thorough coverage of research on all aspects of TSS, including epidemiology, clinical aspects, pathology, role of antibodies, causative agent, toxins, tampon involvement, and prevention.*

Beveridge WIB: 1977. *Influenza: the Last Great Plague: An Unfinished Story of Discovery*, London, Heinemann.

> *Describes early investigations into the causes of influenza, shows how this disease affects the individual and the group, and looks at the pandemics of the past 200 years.*

Coyle P: 1993. *Lyme Disease*, St. Louis, Mosby.

> *Covers the full scope of Lyme disease, including dermatologic, neurologic, musculoskeletal, cardiologic, and ophthalmologic aspects.*

Farrar WE: 1994. *Areas of Infections of the Nervous System*, St. Louis, Mosby.

> *Covers all common infections of the nervous systems with color photographs of clinical and histological appearances and radiological images.*

Gottfried RS: 1983. *The Black Death: A Natural and Human Disaster in Medieval Europe*, New York, Free Press.

> *The author views plague as an environmental and ecological disaster and his book stresses its social and economic effects, especially its impact on the medical profession.*

Gregg CT: 1985. *Plague: An Ancient Disease in the Twentieth Century*, Albuquerque, NM; University of New Mexico Press.

> *Discusses episodes of outbreaks of this disease in the United States and the epidemiology of each outbreak.*

Harden VA: 1990. *Rocky Mountain Spotted Fever: History of a Twentieth Century Disease*, Baltimore, Johns Hopkins University Press.

> *Reveals important insights into the impact of public policy on medical research and public health in the history of the efforts to diagnose, prevent, and treat spotted fever.*

Jankovic J and MF Brin: 1991. Therapeutic uses of botulinum toxin, *New England Journal of Medicine* 324:1186-1191.

> *One of the most lethal biological toxins has been found to be of therapeutic value in the treatment of various neurological and ophthalmological disorders and has been approved for use by the FDA.*

Kaye D (ed.): 1992. *Infective Endocarditis*, New York, Raven Press.

> *An overview of bacterial infection of the heart.*

Magilligan DJ and EL Quinn: 1986. *Endocarditis: Medical and Surgical Management*, New York, Marcel Dekker.

> *A discussion of the courses and treatment of infection of the heart.*

Mitscherlich E and EH Marth: 1984. *Microbial Survival in the Environment: Bacteria and Rickettsiae Important in Human and Animal Health*, Berlin, Springer-Verlag.

> *A handbook for the study of microbial ecology that concentrates on microbial growth, especially of pathogenic bacteria and rickettsia.*

Rail CD: 1985. *Plague Ecotoxicology: Including Historical Aspects of the Disease in the Americas and the Eastern Hemisphere*, Springfield, IL; C. C. Thomas.

> *Reviews the history and epidemiology of plague, disease susceptibility, chemically induced animal infections, and the adverse effects of environmental pollutants.*

Robertson N: 1982. Toxic shock, *New York Times Magazine*, September 19:30-33+.

> *The personal story of this* New York Times *reporter's bout with toxic shock syndrome.*

Simpson LL (ed.): 1989. *Botulinum Neurotoxin and Tetanus Toxin*, San Diego, Academic Press.

> *Draws together information from protein chemists, microbiologists, virologists, pharmacologists, immunologists, and clinicians emphasizing the similarities between these two toxins.*

Smith LDS and H Sugiyama: 1988. *Botulism: the Organism, its Toxins, the Disease*. Springfield, IL; C. C. Thomas.

> *Covers the history, the cultural and serological characteristics, the structure, and identification of* Clostridium botulinum *and its presence in food, people, animals, fish, and birds.*

Stollerman GH: 1975. *Rheumatic Fever and Streptococcal Infection*, New York, Grune & Stratton.

> *Concerned particularly with the possible complications these diseases may produce.*

Twigg G: 1985. *The Black Death: A Biological Reappraisal*, New York, Schocken Books.

> *A review of the history and etiology of plague and its role in the history of medicine in medieval Europe.*

Young EJ and MJ Corbel: 1989. *Brucellosis: Clinical and Laboratory Aspects*, Boca Raton, FL; CRC Press.

> *The pathogenesis of brucellosis, and the presentation to the physician and clinical microbiologist is discussed in this book.*

Zinsser H: 1935. *Rats, Lice, and History: Being a Study in Biography, which, after Twelve Preliminary Chapters Indispensable for the Preparation of the Lay Reader, Deals with the Life History of Typhus Fever*, New York, Blue Ribbon Books.

> *A uniquely personal approach to the history of endemic and epidemic typhus.*

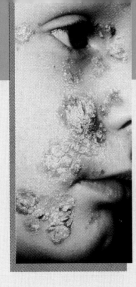

Bacterial Diseases of the Urinary Tract, Genital Tract, Skin, Eyes, Ears, and Oral Cavity

PREVIEW TO CHAPTER 22

In this chapter we will:
• Review the major diseases of humans caused by
  bacteria that affect the urinary tract, genital tract, skin,
  eyes, ears, and oral cavity, examining the specific
  etiologic agent, the route of transmission, the major
  symptoms, and the treatments of each disease.
• Learn the following key terms and names:

acne
acute necrotizing
   ulcerative gingivitis
anthrax
chancroid
conjunctivitis
cystitis
dental caries
Donovan bodies
epidemic conjunctivitis
erysipelas
eschar
folliculitis
Gardnerella vaginitis
gingivitis
glomerulonephritis
gonococcal ophthalmia
gonococcal urethritis
gonococcus
gonorrhea
granuloma inguinale
granuloma venereum
Hansen disease
impetigo
inclusion conjunctivitis
juvenile periodontitis
keratitis
late syphilis
leprosy
leptospirosis
lymphogranuloma
   venereum

necrotizing ulcerative
   gingivitis
nongonococcal urethritis
   (NGU)
nonspecific urethritis
   (NSU)
ophthalmia neonatorum
otitis externa
otitis media
pelvic inflammatory
   disease (PID)
periodontal disease
periodontitis
pinkeye
PPNGs (penicillinase-
   producing *N.
   gonorrhoeae*)
prostatitis
pyelonephritis
salpingitis
scalded skin syndrome
scarlatina
scarlet fever
swimmer's ear
syphilis
trachoma
trench mouth
urethritis
Vincent infection
Weil disease

The internal portions of the urinary tract normally are sterile tissues. Although bacteria colonize the external regions of the urinary tract, including the distal third of the urethra, the outward flow of urine generally prevents bacteria from entering the kidneys and other internal portions of the urinary tract. When some bacteria, such as *Escherichia coli* or *Proteus* opportunistically enter the urinary tract they can grow and cause inflammation. Such infections often are caused by poor hygiene practices that allow fecal matter to enter the urinary tract, as well as by urinary catheterization.

URETHRITIS AND CYSTITIS

Many microorganisms can cause **urinary tract infections** (UTIs). The most common causative agents are Gram-negative bacteria, particularly those normally occurring in the gastrointestinal tract. Accidental contamination of the urinary tract with fecal matter appears to be one of the most important means of transmission of such infections. *E. coli* probably is the most common cause of urinary tract infections, but *Klebsiella, Enterobacter, Serratia, Proteus,* and *Pseudomonas* are also isolated relatively frequently from urinary tract infections (FIG. 22-1). These bacteria are all flagellated, allowing them to effectively "swim upstream." Nosocomial infections with *Serratia* species and *Pseudomonas aeruginosa* most often occur after catheterization procedures because the insertion of a catheter can carry microorganisms from the extremities of the genitourinary tract, contaminating the inner tissues.

The urethra and urinary bladder are most frequently the sites of infections within the urinary tract. The resulting infections are referred to as **urethritis** and **cystitis,** respectively. The symptoms of urethritis normally include pain and a burning sensation during urination. Cystitis is characterized by suprapubic pain and the urge to urinate frequently. Various antibiotics, such as sulfonamides, nitrofurantoin, and nalidixic acid, are used in treating infections of the lower urinary tract. Depending on whether there is an obstruction or another underlying cause involved in the establishment of an infection, the prognosis is normally good.

PYELONEPHRITIS

Infections of the urethra may spread to the kidney and the renal pelvis, causing **pyelonephritis.** Typical signs and symptoms of kidney infection include urinary urgency, burning during urination, blood or pus in the urine, and fever. Pyelonephritis can cause kidney failure and death. *E. coli* is the most frequent etiologic agent of this disease in patients who de-

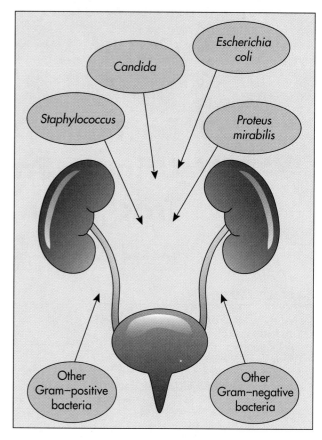

FIG. 22-1 Several microorganisms cause most urinary tract infections (UTIs). When *Escherichia coli*, which inhabits the gastrointestinal tract as a nonpathogen, enters the urinary tract, it causes disease. *E. coli* is the most frequent cause of UTIs.

velop infections outside of hospital settings. *Proteus mirabilis, Enterobacter aerogenes, Klebsiella pneumoniae, Pseudomonas aeruginosa, Streptococcus* sp., and *Staphylococcus* sp. are additional causative agents of this disease. As with other urinary tract infections, the use of antibiotics is generally effective in curing pyelonephritis.

GLOMERULONEPHRITIS

Infections with *Streptococcus pyogenes* can result in **glomerulonephritis.** This is an inflammation of the glomeruli of the kidney. Glomerulonephritis arises from nonurinary tract infections that liberate soluble antigens into the blood. These antigens trigger antibody production, and antigen–antibody complexes form that can stick to the glomeruli. Glomeruli are the capillaries involved in filtering blood as it passes through the kidneys. When they become inflamed, kidney function is impaired. The best evidence now suggests that glomerulonephritis is an immune-complex disease that follows a streptococcal infection and other immune-complex–forming diseases. It is not due to an active kidney infection. Symptoms include fever, elevated blood pressure, and the presence of protein and red blood cells in the urine. Blood and protein in the urine result from the altered permeability of the glomeruli in the kidney. Antibiotics, such as penicillin and chloramphenicol, are useful in treating the initial infection. In most cases there is no chronic damage to the kidneys and recovery from glomerulonephritis is complete.

> **Glomerulonephritis is an immune-complex disease that follows a streptococcal infection and impairs kidney function.**

LEPTOSPIROSIS

Leptospirosis, or **Weil disease,** is principally a disease of nonhuman animals. It is considered to be one of the most important zoonoses (animal disease that can be transmitted to humans). It is caused by the spirochete *Leptospira interrogans,* which is a Gram-negative, aerobic bacterium. Morphologically, *L. interrogans* is a thin filament coiled into a spiral. Infected animals, especially rodents and dogs, excrete leptospires in their urine, and contact with contaminated water harboring viable *Leptospira* species will infect humans (FIG. 22-2). *L. interrogans* can enter the body through the skin, especially through small abrasions. Contact of leptospires with mucous membranes increases the likelihood of infection. Swimming in farm ponds and working in rice paddies also poses a risk of exposure to *Leptospira*. Immunization of dogs against *L. interrogans* is important because they may serve as reservoirs of *L. interrogans*. Penicillin and streptomycin or tetracycline are effective antibiotics used to treat leptospirosis.

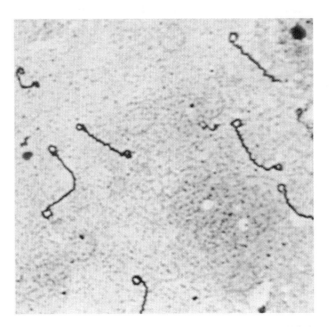

FIG. 22-2 Light micrograph showing a smear from a culture of *Leptospira*. These extremely thin organisms are rendered visible by the precipitation of silver stain on their surfaces.

During the acute phase of this illness, the symptoms normally include a high spiking fever, chills, headache, muscle ache, malaise, abdominal pain, nausea, and vomiting. Various body organs can be involved, including the liver, meninges, and kidney. Most fatal cases result from kidney infection and subsequent renal failure. When severe symptoms develop, the death rate from Weil disease is about 10% to 40%.

> **The spirochete *Leptospira interrogans* causes leptospirosis; the source of infections often is urine-contaminated water.**

BACTERIAL DISEASES OF THE GENITAL TRACT

Most infections of the genital tract are sexually transmitted diseases (STDs). These diseases are caused by various bacteria (Table 22-1) and other microorganisms. Such diseases are contracted by direct sexual contact with an infected individual, generally during sexual intercourse. Previously, these diseases were called venereal diseases (from Roman goddess of love, Venus), which means diseases of love. Since these diseases have more to do with sexual contact than love, the name has been changed to STDs. The physiological properties of the pathogens causing STDs restrict their transmission, for the most part, to direct physical contact because the etiologic agents of sexually transmitted diseases have very limited natural survival times outside infected tissues. Several

sexually transmitted diseases, such as syphilis, herpes simplex, and chancroid, cause lesions in tissues that facilitate the transmission of other sexually transmitted diseases, especially AIDS.

The rise in the rate of incidence of sexually transmitted diseases documents contemporary sexual behavioral patterns but may also in part reflect changes in the reporting and recording of cases of these diseases. At present, outbreaks of some STDs are reaching epidemic proportions, with over 3,000 Americans learning daily that they have contracted an STD. The social implications inherent in the transmission of these illnesses often overshadow the fact that they are infectious diseases. STDs must be treated as medical problems, with the emphasis on curing the pa-

TABLE 22-1

Sexually Transmitted Bacterial Diseases

DISEASE	CAUSATIVE BACTERIUM	CHARACTERISTICS
Chancroid	*Haemophilus ducreyi*	Sharp-edged, flat, painful ulcers with swelling of inguinal (groin) lymph nodes
Gardnerella vaginitis	*Gardnerella vaginalis*	Vaginal discharge with fishy odor
Gonorrhea	*Neisseria gonorrhoeae*	In males: urethritis, purulent discharge; in females: often asymptomatic, urethritis, vaginitis, cervicitis
Granuloma inguinale (donovanosis)	*Calymmatobacterium granulomatis*	Irregular-shaped, painless ulcers on genitals
Lymphogranuloma venereum	*Chlamydia trachomatis*	Swelling of inguinal (groin) lymph nodes
Nongonococcal urethritis	*Chlamydia trachomatis, Ureaplasma urealyticum, Mycoplasma hominis*	Symptoms similar to gonorrhea but milder
Syphilis	*Treponema pallidum*	Primary stage: hard chancre Secondary stage: skin rash Tertiary stage: latency, gummas, cardiovascular and neurological disorders

tient and reducing the incidence of disease by preventing spread of the infectious agents. The overall control of STDs rests with breaking the network of transmission, which necessitates public health practices that seek to identify and treat all sexual partners of anyone diagnosed as having one of the sexually transmitted diseases.

GONORRHEA

Gonorrhea is a sexually transmitted disease caused by the Gram-negative, kidney bean-shaped, diplococcus, *Neisseria gonorrhoeae*, often referred to as **gonococcus** (FIG. 22-3). *N. gonorrhoeae* is a fastidious organism readily killed by drying (desiccation) and exposure to metals. The sensitivity of *N. gonorrhoeae* to desiccation makes negligible the likelihood of transmission of gonorrhea through inanimate objects, such as toilet seats in public restrooms. *N. gonorrhoeae* infects the mucosal cells lining the epithelium, usually of the genital tract, adhering to these cells via its pili. It can be transmitted by anal, oral, or vaginal intercourse. It can also be transmitted to eye tissue. The urethra, cervix, rectum, pharynx, and conjunctivae can be infected.

There are about 1 million cases of gonorrhea reported in the United States each year. That means that every day 2,700 Americans discover that they have gonorrhea. Many more cases undoubtedly occur that are never reported to public health authorities. Following the widespread use of birth control pills in the 1960s there was a large increase in cases of gonorrhea (FIG. 22-4). This occurred because the hor-

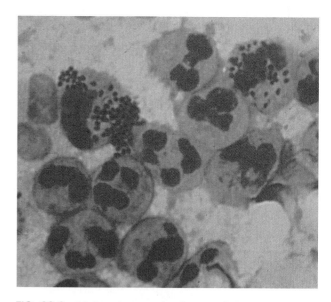

FIG. 22-3 Light micrograph of intracellular, Gram-negative diplococci. The presence of these bacteria in a urethral discharge is diagnostic for gonorrhea and in a vaginal discharge is presumptive for gonorrhea.

mones in oral contraceptives cause increased pH values in the vaginal tract, removing one line of protection against infection by *N. gonorrhoeae*, which is relatively sensitive to acidic conditions. Gonorrhea is readily treated with antibiotics, with penicillin the antibiotic of choice. Unfortunately, the number of penicillin-resistant strains, called PPNGs (penicillinase-producing *N. gonorrhoeae*), is increasing. For treatment of PPNGs, spectinomycin is the antibiotic of choice.

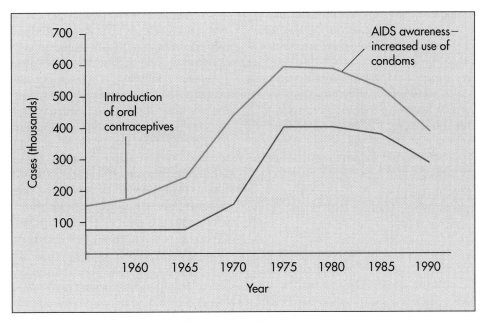

FIG. 22-4 The increase in gonorrhea from 1960 to 1970 coincides with the greater use of oral contraceptives and an increase in sexual activity (men, *blue*; women, *red*). The decrease in this disease after 1980 probably reflects an increase in safe sex practices.

Gonorrhea is caused by the Gram-negative, fastidious, diplococcus *Neisseria gonorrhoeae*. Transmission is by sexual contact.

In men, gonorrhea results in a characteristic painful, purulent urethral discharge (FIG. 22-5). The pus results from the migration of phagocytic leukocytes to the site of infection. Symptoms of gonorrhea in males are usually apparent less than 1 week after infection. If the disease is untreated, obstructions due to scarring of the tubes through which sperm move may produce sterility. Also *N. gonorrhoeae* can infect the prostate gland, causing acute inflammation called **prostatitis.**

Male infections with *Neisseria gonorrhoeae* are painful and can cause serious complications.

In women, infections with *N. gonorrhoeae* can cause various symptoms such as pain and swelling due to inflammation, vaginal discharge, and abnormal menstrual bleeding. Often, the early stages of gonorrhea in women are not associated with any overt symptoms, and many women with gonorrhea in fact remain asymptomatic carriers. If asymptomatic carriers do not seek treatment, they may be a significant factor in increasing the number of cases of this disease. Because many women with gonorrhea are asymptomatic, the eyes of all infants are routinely washed immediately after birth with erythromycin sulfate to prevent infections of the eye (ophthalmia neonatorum—discussed later in this chapter), which could result from the transmission of *N. gonorrhoeae*

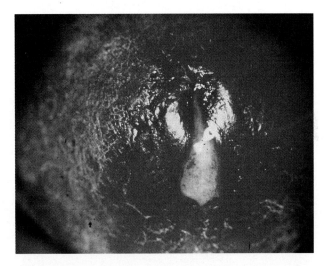

FIG. 22-5 The most common presentation of gonorrhea in men is purulent urethral discharge, which appears within a few days after exposure and is associated with dysuria. Untreated, this will often last for many weeks before clearing spontaneously.

from mother to infant during passage through the vaginal tract.

N. gonorrhoeae may spread, however, to other parts of the female genital tract, including the cervix and uterus. If the infection spreads to the fallopian tubes, it causes a chronic infection called **salpingitis.** This condition typically is characterized by abdominal pain. Salpingitis can cause infertility, fertilization outside the uterus, and failure of the fertilized egg to implant in the uterus, a life-threatening situation called an ectopic pregnancy. A gonococcal infection may spread to the urethra, causing an inflammation called **gonococcal urethritis.**

Gonorrhea also can lead to **pelvic inflammatory disease** (PID), which results from a generalized bacterial infection of the uterus, pelvic organs, uterine tubes, and ovaries (FIG. 22-6). PID may cause infertility without overt symptoms of gonorrhea. Occlusions of the fallopian tubes due to scarring produces sterility in some cases of gonorrhea.

> **Female infections with *N. gonorrhoeae* include gonorrhea, salpingitis, gonococcal urethritis, and pelvic inflammatory disease.**

N. gonorrhoeae may spread from the original site of infection via the blood to cause a disseminated gonococcal disease. This most frequently appears as an arthritis-dermatitis syndrome. The small joints of the hands and feet are most frequently affected by arthritis, and skin lesions usually develop on the wrist, elbows, and ankles.

There are several serotypes of *N. gonorrhoeae* and no long-term immunity exists. Therefore individuals can become infected more than once. There is currently no vaccine available for gonorrhea.

SYPHILIS

Approximately 30,000 cases of the sexually transmitted disease **syphilis** are reported in the United States each year (FIG. 22-7). The causative organism, *Treponema pallidum*, is a bacterial spirochete that is fastidious in its growth requirements. *T. pallidum* is readily killed by drying, heat, and disinfectants such as soap and arsenic and mercury-containing compounds. The fact that *T. pallidum* is unable to survive for very long outside the body makes transmission through inanimate objects (fomites) virtually nonexistent. Transmission depends on direct contact with the infective syphilitic lesions that contain *T. pallidum*. Syphilis may also be transmitted in saliva and therefore poses a threat to dentists or when kissing. Serological tests are employed for the diagnosis of syphilis because *T. pallidum* cannot be cultured using cell-free media; fluorescent antibody staining is useful in viewing *T. pallidum* and positively diagnosing cases of syphilis (FIG. 22-8).

> **Syphilis is caused by *Treponema pallidum*, a sexually transmitted, fastidious bacterium.**

> ***T. pallidum* doesn't survive long outside the human body; therefore, syphilis cannot be transmitted through contact with inanimate objects.**

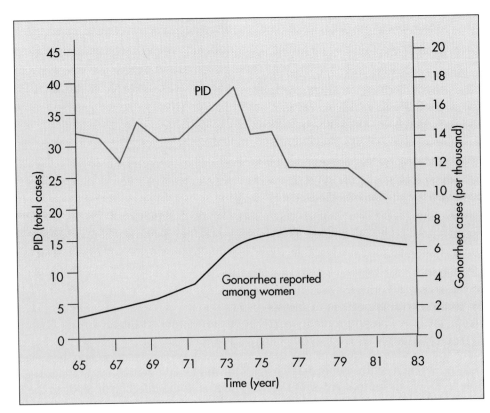

FIG. 22-6   Pelvic inflammatory disease (PID) can be a complication in women from an infection with *Neisseria gonorrhoeae* and other sexually transmitted pathogens.

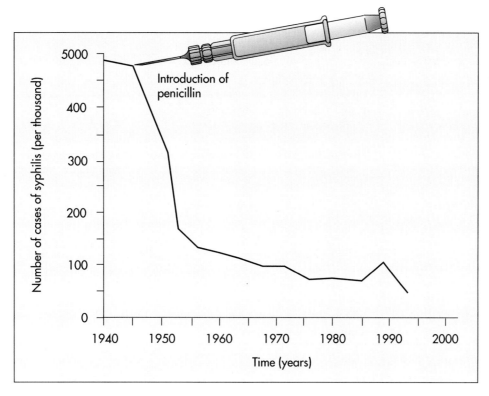

FIG. 22-7 The incidence of syphilis declined after the introduction of penicillin.

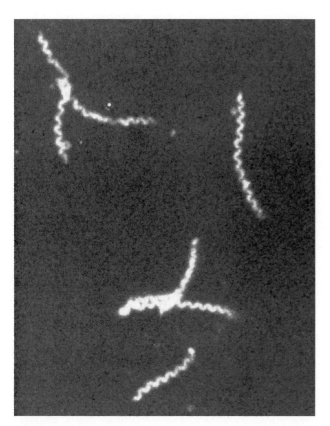

FIG. 22-8 Micrograph showing *Treponema pallidum*, which causes syphilis, is a slender spirochete.

NEWSBREAK

ORIGIN AND HISTORICAL TREATMENT OF SYPHILIS

Syphilis first appeared in Europe at the beginning of the sixteenth century. Even though it was called the "French disease" (except in France), it was believed to have been brought back from America by Christopher Columbus' crew. The name syphilis comes from a poem written by Girolamo Fracastoro, who introduced the concept of the theory that germs cause disease. Fracastoro's poem recounted the legend of the shepherd Syphilus being stricken with the disease.

As the disease became prevalent in Europe, desperate attempts were made to find cures. One treatment was to elevate body temperature; this was effective in some cases since *Treponema pallidum* is sensitive to even 37° C. People bathed in hot spas, which became centers for treatment of syphilis. Later, chemical treatments were developed, including the use of arsenic and arsenic-containing substances. The work of Paul Ehrlich led to the use of salvarsan and neosalvarsan (arsenic-containing compounds) as the first chemotherapeutic agents for treating infectious disease. These drugs were considered to be "magic bullets" that shot the bacteria dead.

TABLE 22-2		
The Pathogenesis of Syphilis		
STAGE OF DISEASE	**SIGNS AND SYMPTOMS**	**PATHOGENESIS**
Primary syphilis 2-10 weeks after initial infection (depends on inoculum size)	Primary chancre at site of infection; enlarged inguinal nodes; spontaneous healing	Proliferation of treponemes in regional lymph nodes
Secondary syphilis 1-3 months after primary syphilis; lasts 2-6 weeks	Flu-like illness; myalgia; fever; headache; mucocutaneous rash; spontaneous resolution	Multiplication of treponemes and production of lesions in lymph nodes, liver, joints, muscles, skin, and mucous membranes
Latent syphilis lasts 3-30 years	None	Treponemes dormant in body
Tertiary syphilis	Neurosyphilis; general paralysis; insanity, unusual walk (tabes dorsalis) Cardiovascular syphilis; aortic lesions, heart failure Gummas in skin, bone, testes	Cell-mediated hypersensitivity and inflammation; treponemes rare or absent

Syphilis manifests in three distinct stages and a latent period (Table 22-2). During the primary stage of syphilis, a chancre develops (FIG. 22-9). A **chancre** is a painless, small, fluid-filled lesion that erodes at the site of *Treponema* inoculation. Chancres have hardened, raised edges with clear bases. Primary lesions generally occur on the genitalia, but in 10% of the cases the primary lesions occur in the oral cavity. The average incubation time for the manifestation of primary syphilis is 21 days after infection. The primary lesions typically heal within 3 to 6 weeks.

The secondary stage of syphilis normally begins 6 to 8 weeks after the appearance of the primary chancre. During this stage there are open lesions on the skin and lesions of the mucous membranes that contain infective *T. pallidum*. Lesions may appear on the lips, tongue, throat, penis, vagina, and numerous other body surfaces. There may be additional symptoms of systemic disease during this stage, such as headache, low-grade fever, enlargement of the lymph nodes, and so on. The secondary stage is the most infectious because of the numerous spirochetes in the multiple lesions.

After the secondary phase, syphilis enters a characteristic latent period during which there is an absence of any clinical symptoms of the disease. The latent phase marks the end of the infectious period of syphilis and may persist for many years. *T. pallidum* may be able to survive in the body because its surface is rich in lipids that are antigenically unreactive, enabling it to evade recognition and elimination by the immune defenses. This is because only dead and dying cells of *T. pallidum* appear to reveal antigens to which the body's immune system responds.

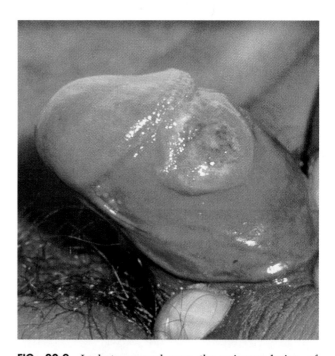

FIG. 22-9 In heterosexual men the primary lesion of syphilis, the chancre, is most commonly found on the glans penis or in the sulcus and less commonly on the penile shaft. The chancre is indurated but is not tender and is often associated with enlarged but painless inguinal lymph nodes. Dark-field preparations are made from serum exuded from the chancre. Classically, the chancre in a female (not shown) appears in the genital region after an incubation period of 21 to 35 days (extremes, 9 to 90 days) as a single lesion in 50% of cases. It evolves rapidly from a macule to a papule, which erodes and forms a round, painless ulcer with a clean surface and surrounding hard induration. It heals within 3 to 10 weeks, leaving a thin atrophic scar in some cases. Vulval lesions may be readily recognized but cervical lesions are commonly overlooked.

NEWSBREAK

HISTORICAL FIGURES, SYPHILIS, AND CIVILIZATION

Although an unproven and controversial hypothesis, syphilis is believed by many to have returned to Europe from the Americas in 1493 with Columbus and his crew. An early name for the disease in Spain was Espanola, indicating the belief that the disease came from sexual contacts in Haiti. Columbus himself may have been infected. In 1496 on his second voyage to the Americas, Columbus began having attacks of fever that were consistent with the second stage of syphilis. On his third voyage in 1498 he developed signs of arthritis consistent with the tertiary stage of syphilis when joints are often attacked. Also on that voyage, he began to hear voices and to view himself as an "ambassador of God," suggestive of dementia of tertiary syphilis. By the time Columbus returned to Spain in 1504, he lacked physical strength to walk, probably indicative of heart damage, and he was totally disoriented, probably due to brain damage, which is characteristic of late syphilis.

King Henry VIII of England, who was known to be sexually promiscuous, also seems to have contracted syphilis. The Shakespearean depiction of Henry VIII indicates the dementia of late syphilis. The first four children fathered by Henry VIII with Catherine of Aragon were stillborn or died shortly after birth, probably due to congenital syphilis. The fifth child, Mary, survived but also appears to have had congenital syphilis; she was disfigured and died suddenly at age 42, probably due to a syphilitic heart aneurism. Charles VIII of France also appears to have had syphilis. His four children all died in infancy, apparently due to congenital syphilis, ending the Valois dynasty. This led to the reign of Francis I, whose reign was contemporary with Henry VIII. He also appears to have suffered from syphilis and French civilization likewise suffered because of his delusions of grandeur and irritable inconsistencies, also characteristics of late syphilis.

The tertiary phase of syphilis, also known as **late syphilis,** usually does not occur until years after the initial infection. If the disease reaches this stage, it no longer is treatable with antibiotics. This stage is generally noninfectious, since treponemes are relatively rare or absent. During tertiary syphilis, damage can occur to any organ of the body, sometimes causing extensive tissue damage. Damage to the body occurs due to the body's inflammatory response and hypersensitivity reactions. Gummas, or granulomatous inflammations, in the skin or connective tissues are common. In many cases lesions develop that are similar to those of tuberculosis. In about 10% of the cases of untreated syphilis, the tertiary phase involves the aorta, and damage to this major blood vessel can result in death. In approximately 8% of the cases of untreated syphilis, there is central nervous system involvement with various neurological manifestations, including personality changes, paralysis, deafness, blindness, and loss of speech. Blindness, insanity, and bone deformations are characteristic of advanced stages of syphilis.

Syphilis manifests in three stages.

The opportune time for treatment is during the primary stage when there are open lesions.

Irreversible damage may occur if tertiary syphilis develops.

The risks of debilitating symptoms and death make syphilis a very serious STD. Fortunately,

syphilis can be treated with penicillin. For individuals allergic to penicillin, various other antibiotics are effective, including tetracycline, doxycycline, and erythromycin. The appropriate time for treatment is during the primary stage. No long-term immunity develops after infections with *T. pallidum,* and individuals who are cured by treatment with antibiotics remain susceptible to contracting syphilis again. There is no effective vaccine available for syphilis. As with other STDs, the overall control of syphilis rests with finding and treating all sexual contacts who may have contracted this disease and may be involved in the further transmission of it.

T. pallidum can be transmitted across the placenta of pregnant women with syphilis. The fetus can be infected and stillbirth or congenital syphilis in the newborn may follow. Stillbirth is likely if pregnancy occurs during the primary or secondary stages of syphilis. Congenital syphilis is most likely during the latent period of the disease, usually resulting in mental retardation and neurological abnormalities in the infant. The probability of survival of such infants depends on the specific nature of the neurological impairment.

GARDNERELLA VAGINITIS

Vaginal infections by the sexually transmitted *Gardnerella vaginalis* are sometimes called *nonspecific vaginitis.* This is because the disease is an interaction between *G. vaginalis* and anaerobic bacteria in the vagina, neither of which will produce the disease alone. *G. vaginalis* is a Gram-negative, anaerobic rod. One third of all cases of vaginitis are of this type and occur when the pH of the vagina, which normally is less than 4, rises above 5. Diagnosis is based on the fishy odor, vaginal discharge, and microscopic observation of sloughed-off vaginal epithelial cells covered with Gram-negative rods (referred to as clue cells in the discharge). Metronidazole, a drug that eradicates the anaerobes essential to continuation of the disease but allows the normal lactobacilli to repopulate the vagina, is the primary treatment.

NONGONOCOCCAL URETHRITIS (NGU)

Nongonococcal urethritis (NGU), also known as **nonspecific urethritis** (NSU), is a sexually transmitted disease. It results in inflammation of the urethra caused by bacteria other than *Neisseria gonorrhoeae.* Compared to gonococcal urethritis, NGU has a longer incubation period. It is estimated that between 4 and 9 million people in the United States have nongonococcal urethritis. Females often are asymptomatic; males usually notice some pain and discharge

during urination. In males the epididymis may become inflamed; in females the cervix or fallopian tubes may become blocked, causing infertility.

Mycoplasma hominis and *Ureaplasma urealyticum,* both of which lack cell walls, are frequent causes of NGU. These bacteria are inhibited by antibiotics such as tetracyclines but not by penicillin.

Most cases of NGU appear to be caused by *Chlamydia trachomatis,* which are small, obligately intracellular bacteria. *C. trachomatis,* like *Neisseria gonorrhoeae,* can be transmitted during birth from mother to the eyes of the newborn. *Chlamydia* infections of the eye can be serious. Because many females are asymptomatic, erythromycin is applied to the eyes of newborns shortly after birth to protect them against *C. trachomatis* and *N. gonorrhoeae.*

LYMPHOGRANULOMA VENEREUM

Lymphogranuloma venereum is a STD caused by *Chlamydia trachomatis* (FIG. 22-10). This disease is more common in the tropics than in the United States where only a few hundred cases each year occur. In lymphogranuloma venereum, *C. trachomatis* enters the body through small abrasions in the genitourinary tract. The chlamydial cells are phagocytized and carried to lymph nodes—the primary sites affected in this disease. Swelling and tenderness usually occur 5 to 21 days after the primary lesions in the area of initial infection heal. In about 25% of cases, there is a bloody, mucopurulent rectal discharge. Prolapsed (falling) rectum is also a serious consequence of lymphogranuloma venereum but this is rare. Although treatment with tetracycline and other antibiotics is usually effective in preventing the onset of late and serious symptoms, it does not seem to shorten the usual time of 4 to 6 weeks required for enlarged

NEWSBREAK

ORAL CONTRACEPTIVES AND SEXUALLY TRANSMITTED DISEASES

The introduction of oral contraceptives in the 1960s was almost immediately followed by a significant increase in the number of cases of gonorrhea. Other sexually transmitted diseases did not increase at that time, suggesting the cause was a specific relationship between oral contraceptives and gonorrhea rather than a change in sexual activities. The relationship was attributed to a slight rise in vaginal pH that occurs as a result of the hormones in oral contraceptives that would lessen the defense against infection with *Neisseria gonorrhoeae.* A similar increase in cases of *Chlamydia* associated with the use of oral contraceptives was later reported. Both findings, however, have been questioned because of possible confounding factors in the statistical analyses that led to these conclusions, such as the sexual activities of the male partners. Some reports indicate lowered cases of symptomatic nongonococcal pelvic inflammatory disease in women using oral contraceptives, suggesting that the effects of infection with *Chlamydia* but not with *N. gonorrhoeae* are inhibited by oral contraceptives. Thus the impact of the use of oral contraceptives on sexually transmitted diseases remains complex and controversial.

FIG. 22-10 Direct immunofluorescence of elementary bodies of *Chlamydia trachomatis,* which appear as bright green dots.

lymph nodes to heal. In some cases, the *C. trachomatis* may go into a latent stage, and these asymptomatic individuals may be reservoirs for this disease.

CHANCROID

Chancroid is a relatively rare STD caused by *Haemophilus ducreyi,* a Gram-negative, rod-shaped bacterium. Chancroid occurs most frequently in the underdeveloped nations of Africa, the Caribbean, and Southeast Asia, but the incidence of this disease is increasing in the United States. Soft chancres develop 3 to 5 days after sexual exposure, and untreated lesions may persist for months. Chancroidal ulcers heal quickly but often leave deep scars. Sulfanilamide and tetracycline are used in the treatment of this disease.

GRANULOMA INGUINALE

Granuloma inguinale is also known as **granuloma venereum** or **donovanosis.** It is caused by a Gram-negative, rod-shaped bacterium, *Calymmatobacterium granulomatis.* It is characterized by ulcerations of the skin or mucosa of the genitals and pubic region. This disease is the rarest of all STDs in North America and Europe. This disease most frequently occurs and is endemic in tropical and subtropical countries, including India, the west coast of Africa, islands of the South Pacific, and some South American countries. There appears to be a very low rate of infection for this STD, and in many cases sexual partners do not contract this disease. Initial lesions occur on the genitalia and the first lesions appear 9 to 50 days after sexual intercourse with an infected individual. *C. granulomatis* can invade and multiply within monocytes so that it evades the immune defenses of the body. Scrapings from the lesions reveal the presence of *C. granulomatis* within large mononuclear cells called **Donovan bodies.** The genitalia develop characteristic ulcers, older portions of which exhibit loss of pigmentation. Chloramphenicol, erythromycin, and tetracycline, as well as other antibiotics, are used in the treatment of this disease.

BACTERIAL DISEASES OF THE SKIN

The skin is colonized by numerous bacteria, especially Gram-positive cocci because of their resistance to desiccation. These include staphylococci that are resistant to high salt concentrations. In some cases excessive growths of these bacteria cause skin infections. Bacteria may also circulate through the blood and release toxins that affect skin tissues. Breaks in the skin surface also permit bacterial infections of subdermal tissues.

STAPHYLOCOCCAL SKIN INFECTIONS

Staphylococcus aureus occurs on the skin where it can grow in part because of its relative resistance to desiccation and high salt concentrations. Pathogenic strains of *S. aureus* produce toxins, some of which can damage skin tissues. For example, some strains produce exfoliative toxin, which is responsible for **scalded skin syndrome,** a disease characterized by the skin on the palms and soles and other parts of the body peeling off in sheets when touched (FIG. 22-11). *S. aureus* can cause infections of hair follicles, or **folliculitis.** These infections often are accompanied by the formation of *pimples* (small inflammatory regions of swelling of the skin) and *boils* (localized regions of pus surrounded by inflamed tissue).

After an incubation period of about 2 to 4 days, *S. aureus* infection in hair follicles leads to an inflammatory response with an influx of neutrophils to the site. In addition, fibrin is deposited as the body attempts to wall off the infection. The result is an abscess that

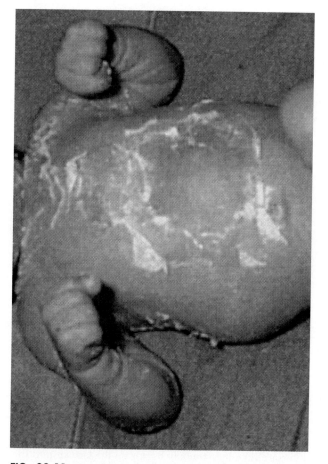

FIG. 22-11 In a case of scalded-skin syndrome the destructive effect of bacterial toxin may result in permanent damage to the skin, or death.

contains pus consisting of plasma, bacteria, and dead white blood cells. As bacterial toxins continue to erode the surrounding tissues, the infection may extend deeper into the subcutaneous layers and become a *carbuncle* (a painful spreading inflammation of the subcutaneous tissue that causes sloughing of the skin) or a *furuncle* (a painful spreading inflammation of the subcutaneous tissue that involves more than one follicle). *S. aureus* also causes *styes*—inflammatory swellings like a small boil on the edge of the eyelid.

S. aureus also is the primary cause of impetigo of the newborn, a serious problem in hospital nurseries. *Streptococcus pyogenes* may also be involved in cases of impetigo, sometimes as a dual infection with *Staphylococcus aureus*. **Impetigo** is a superficial skin infection characterized by isolated *pustules* (small, round elevations containing pus) that become crusted and rupture (FIG. 22-12). The disease is spread largely by contact, with the bacteria penetrating the skin through minor abrasions.

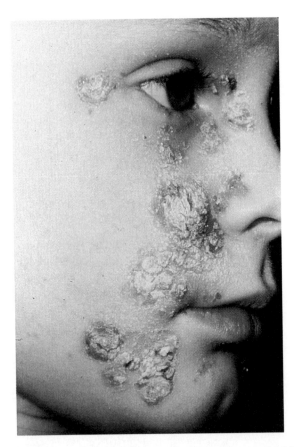

FIG. 22-12 *Impetigo contagiosa* of face. Impetigo is a highly contagious superficial infection of the skin, caused either by streptococci, staphylococci, or both. The disease may affect an apparently normal skin or complicate some underlying skin condition. It commonly begins on the face around the mouth or nose, spreading with alarming rapidity to other parts of the body. In streptococcal impetigo the exudate dries to form a thick crust with a golden-yellow color.

Staphylococcus aureus **can cause skin infections because it can survive the conditions of high salt and dryness of the skin and sometimes produce tissue-damaging toxins.**

STREPTOCOCCAL SKIN INFECTIONS

Group A β-hemolytic streptococci can cause several infections of the skin. These bacteria produce various toxins, including hemolysins, streptokinases (that dissolve fibrin blood clots), hyaluronidase (that breaks down connective tissue), and leukocidins (that kill white blood cells). These toxins allow the bacteria to spread from the initial site of infection. In addition, many strains of streptococci have an M protein on the outer surface of their cell walls. M proteins contribute to the pathogenicity of these bacteria by allowing them to attach to epithelial cells and to evade phagocytosis.

Impetigo can be caused by Group A streptococci and is common in infants and children. Sometimes, impetigo is caused by a combined infection of streptococci and staphylococci. This infection can be treated with penicillin or erythromycin.

Erysipelas occurs when streptococcal infections invade the deeper regions of the dermis, causing the skin to erupt into reddish patches that enlarge and thicken and swell at the margins. The reddening is

---

NEWSBREAK BOX

PUBLIC FRENZIED ABOUT FLESH-EATING BACTERIA

The spring of 1994 brought alarming news when the media reported the emergence of a new flesh-eating strain of *Streptococcus pyogenes* (Group A streptococci). It is possible that this bacterium has been the cause of undiagnosed disease for some time and that the current frenzy about it comes largely from media recognition that ten individuals died from this infection in 1994. Streptococci produce various toxins, some of which destroy skin, muscle, and connective tissue. The current strain of Group A streptococci has been reported to destroy 1 inch of tissue per hour. Death ensues rapidly following infection, often within 24 hours. Despite the rareness of this disease, the public is alarmed because of the gory description of how it kills. The rapid progression of the disease makes treatment difficult. Like some other diseases that severely damage skin and underlying tissues (such as gas gangrene), radical removal of the infected region may be necessary to limit spread. Antibiotics can then be administered in an attempt to prevent death.

caused by toxins produced by the streptococci as they invade new areas. High fever is common. Erysipelas is most likely to occur in the very young and the very old. Penicillin or erythromycin normally are used for treating this disease.

PSEUDOMONAS SKIN INFECTIONS

Pseudomonas aeruginosa is a common cause of skin infections in immunocompromised patients and those with wounds, catheters, or second- or third-degree burns. *P. aeruginosa* is a ubiquitous Gram-negative bacterium that is present in soil, water, and the skin of some individuals. It can be transmitted to patients via contaminated respiratory care equipment, catheters, intravenous fluids, flowers and plants, and even some foods such as lettuce. It can cause folliculitis (inflammation of skin follicles) in individuals exposed to improperly sanitized swimming pools, whirlpools, or hot tubs.

Many strains of *P. aeruginosa* produce a blue-green pigment. Consequently, blue-green pus is a good indication of infection by this organism (FIG. 22-13). *P. aeruginosa* is quite resistant to many conventionally used disinfectants and antibiotics. Treatment of *Pseudomonas* infections can therefore be difficult. New antibiotics, such as carbenicillin, ticarcillin, and cefoperazone, have been developed to specifically deal with *Pseudomonas* infections.

LEPROSY (HANSEN DISEASE)

Leprosy, or **Hansen disease,** is caused by *Mycobacterium leprae.* There is an extremely long incubation period for leprosy, usually 3 to 5 years, before the onset of disease symptoms. Leprosy is a chronic disease and symptoms vary, but the earliest detectable symptoms generally involve skin lesions or loss of sensation in a localized region of the skin (FIG. 22-14). The disease can be diagnosed by observing acid-fast bacterial cells in scrapings from lesions.

Leprosy can occur in two main forms: tuberculoid or neural leprosy and lepromatous or progressive leprosy. The lepromatous, or cutaneous, form of leprosy is characterized by disfiguring nodules in the skin all over the body. This form of the disease is progressive and can produce deformities of various parts of the body such as the hands and nose. Bacteria in the lesions are infective and can be transmitted through the skin or mucous membranes to susceptible individuals. There may be 1 billion viable cells of *M. leprae* per gram of skin in advanced cases of lepromatous leprosy, and direct skin contact appears to be very important in the transmission of this disease.

There also is a tuberculoid or neural form of leprosy. This form of the disease produces lesions around the peripheral nerves, causing a loss of sen-

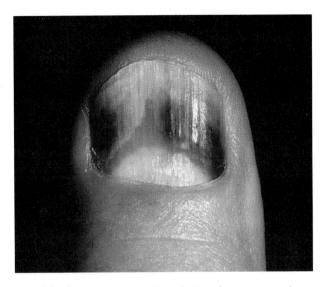

FIG. 22-13 Areas infected with *Pseudomonas aeruginosa* may appear green due to the growth of this pathogen, as has occurred in this nail infection.

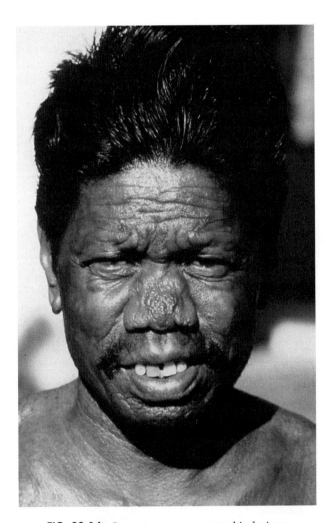

FIG. 22-14 Leprosy causes severe skin lesions.

sation in the regions of the nodules. In tuberculoid leprosy there are fewer lesions and lower numbers of bacteria and, therefore, lower infectivity. Unlike other mycobacterial species, *M. leprae* is able to reproduce within nerve cells (called Schwann cells), so that damage to the nervous system occurs. This leads to anesthesia or loss of sensation especially in peripheral nerves. The local anesthesia renders the individual prone to repeated trauma and secondary bacterial infections.

During the course of this disease, many organs and tissues of the body may be infected in addition to the infection of nerve cells characteristic of all forms of leprosy. In the tuberculoid form, there are relatively few areas of skin involved; but in lepromatous leprosy, multiplication of *M. leprae* is not contained by the immune defense mechanisms, and the bacteria are disseminated through many tissues. Lepromatous leprosy progresses with extensive skin involvement. There is a loss of eyebrows and thickening and enlargement of the nostrils, ears and cheeks, resulting in a leonine (lion-like) facial appearance. The nasal septum is destroyed and the nasal mucosa is loaded with bacteria.

One of the problems associated with studying leprosy has been the difficulty in culturing this bacterium. *M. leprae* has never been cultured *in vitro* on nonliving media. Recently, *M. leprae* has been found associated with nine-banded armadillos—these animals may provide an alternative host in which to study the pathogenesis of leprosy.

Treatment of leprosy can be achieved by using dapsone, which is bacteriostatic, or rifampin, which is bactericidal. Prolonged treatment with antimicrobial agents is needed to maintain control of leprosy infections. Before the use of antibiotics, leprosy was a dread and deadly disease and individuals with this disease were exiled to leper colonies. Today, fortunately for the 3 million cases of leprosy that still occur each year, leprosy is rarely fatal, and complete recovery occurs after treatment in many cases. Since leprosy can now be treated by chemotherapy at home, a leper colony in southern Louisiana was disbanded in 1990. However, a leper colony still exists in the United States in Hawaii.

> **Leprosy or Hansen disease, transmitted by direct contact with *Mycobacterium leprae*, has a very long incubation period and also must be treated with antimicrobial agents for long periods to achieve control.**

ANTHRAX

Anthrax is primarily a disease of herbivorous sheep, cattle, horses, hogs, camels, and goats. It can occasionally be transmitted to people. The disease is caused by *Bacillus anthracis*, a Gram-positive, aerobic, endospore-forming rod. This bacterium produces a toxin that increases vascular permeability, leading to hemorrhage, necrosis, and pulmonary edema. Anthrax can be cutaneous from skin contact, pulmonary from inhalation of endospores, or gastrointestinal from ingestion of *B. anthracis*. Although the pulmonary form is far less common than the cutaneous form, it generally is fatal.

Deposition of spores of *B. anthracis* under the epidermis permits germination. Several exotoxins are produced by the growing bacteria, including one that is appropriately named lethal toxin. The localized accumulation of toxin causes necrosis of the tissue in the formation of a blackened lesion called an **eschar.** The development of cutaneous anthrax can initiate a systemic infection, and untreated cutaneous anthrax has a fatality rate of 10% to 20%. Cutaneous anthrax can be treated with penicillin and other antibiotics, reducing the death rate to under 1%. Pulmonary and gastrointestinal anthrax are usually not diagnosed in time to save the patient and the organism is detected only on postmortem examination. If diagnosis is made while the patient is still alive, massive intravenous administration of penicillin, tetracycline, or erythromycin may be effective.

Contact with animal hair, wool, and hides containing spores of *B. anthracis* is often implicated in transmission of anthrax; therefore it is also known as *woolsorters' disease*. Animal handlers and veterinarians are

NEWSBREAK

ARMADILLOS, HUMANS, AND LEPROSY

There are only two animals in which *Mycobacterium leprae* reproduces: humans and armadillos. About 30% of the armadillos in the wild are infected with this bacterium. The worldwide incidence among humans is approximately 15 million. Five thousand individuals in the United States have leprosy, and fewer than 200 new cases appear each year, most of them among immigrants from areas where the disease is endemic. At one point most individuals with leprosy in the United States were forcibly placed in specialized facilities, such as the leper home in Carville, Louisiana, established in 1894 and maintained for that purpose until the early 1990s. Only one such "leper colony" remains, located in Hawaii. Most individuals with leprosy in the United States are treated today with multiple antibiotics on an outpatient basis. Treatment in some cases also involves the use of thalidomide to reduce the pain of skin lesions that occur in leprosy; this is the same drug that caused thousands of birth defects when used as a sleeping pill in the 1960s.

HIGHLIGHT

BIOLOGICAL WARFARE AGENTS

The capabilities of pathogenic microorganisms to cause debilitating and lethal human diseases makes them potential agents of biological warfare. Anthrax is particularly suitable as a biological agent because endospores of *Bacillus anthracis* can be stored indefinitely and because the disease can be transmitted by aerosols. The former Soviet Union, as well as probably other nations, developed biological warfare programs employing anthrax and other agents. An accidental release of anthrax spores occurred at a test facility in the Soviet Union during the 1980's, resulting in the deaths of nearby animals. Similarly, plague, although not caused by an endospore-producing bacterium, can be transmitted via the air and hence *Yersinia pestis* is a likely candidate as a biological warfare agent. Cholera and diseases caused by toxins, such as botulinum toxin and staphylococcal enterotoxin, likewise, could have devastating effects on military and civilian populations; these agents could be disseminated via food and water.

The fear of biological weapons, their unpredictability, and their unacceptable effects on humankind, has led to an International Convention banning them. The United States took the lead in establishing an international ban on biological weapons. By international convention, signatory nations are to cease all research and development of offensive biological weapons. By declaring that even if attacked with biological weapons the United States would not respond by using such weapons, the United States has taken the lead in attempting to eliminate biological weapons from the arsenal of weapons of mass destruction. Many other nations have taken similar stances against biological warfare.

The fear remains, however, that some nations will not comply with international conventions on banning biological weapons. Unfortunately, verification of compliance with the Biological Weapons Convention is very difficult. Unlike enforcement of treaties limiting nuclear weapons and banning chemical weapons, biological weapons development could easily be confused with natural occurrences of disease outbreaks and with medical efforts to protect humans against pathogens. Development of vaccines against anthrax, tularemia, and other devastating diseases also could serve the dual purpose of developing biological agents that could be used for offensive military purposes. An effective vaccine against a disease can, in some cases, be converted into an offensive biological weapon by simply eliminating a final inactivation step that normally renders the organism harmless. Thus the threat of biological warfare remains.

---

at high risk of exposure to *B. anthracis.* Avoiding contact with infected animals and preventing the development of anthrax in farm animals through the use of anthrax vaccine have effectively reduced the incidence of this disease.

> **Transmission of *Bacillus anthracis*, the causative agent of anthrax, often is associated with contact with infected animal hair, wool, or hide.**

SCARLET FEVER

Scarlet fever is the result of a β-hemolytic *Streptococcus* infection in which the toxin spreads to the skin. Generally this occurs as a result of an infection with an erythrogenic toxin-producing strain of *Streptococcus pyogenes* that begins as an upper respiratory tract infection. Scarlet fever is characterized by the development of a characteristic rash over the body. The generalized rash is due to the toxins produced by certain *Streptococcus* strains that contain temperate phage (FIG. 22-15). Mild cases of this disease generally are called **scarlatina,** although scarlet fever and scarlatina are really different names for the same disease. As with other β-hemolytic streptococcal infec-

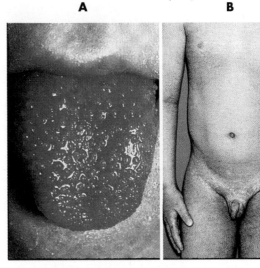

FIG. 22-15 Scarlet fever is caused by an erythrogenic strain of *Streptococcus pyogenes* invading a susceptible host. There is a bright flush on the cheeks and chin, which contrasts vividly with the pallor around the mouth. **A,** The tongue becomes strawberry red. **B,** Elsewhere there is an erythematous background of varying intensity with tiny superimposed red spots or puncta. Over the distal parts of the limbs the rash may condense into discrete macules. Pallor around the mouth is seen in many other conditions, especially lobar pneumonia.

tions, penicillin is the drug of choice and most frequently used for the treatment of scarlet fever. The antibiotic eliminates the streptococcal infection and thus the spread of toxin through the body to the skin. The body develops an immune response so that the disease only occurs once.

> The characteristic rash of scarlet fever is caused by toxins produced by *Streptococcus* species that contain lysogenic phage.

ACNE

Acne is a common problem during adolescence. More than 65% of teenagers develop this problem to some extent (FIG. 22-16). Acne occurs when microbial invasion at the base of hair follicles is associated with excessive secretions by the sebaceous glands. The disease is characterized by inflammatory papules, pustules, and cysts, most commonly on the face, upper back, and chest. Various bacteria, including *Propionibacterium acnes* (a Gram-positive rod that carries out a propionic acid fermentation and also is called *Corynebacterium acnes*) and *Staphylococcus epidermidis*, have been implicated as the etiologic agents of acne. *P. acnes* is able to metabolize sebum, producing fatty acids that elicit an inflammatory response. The inflammation can produce scarring if the lesions are scratched, picked, or even just pressed too hard.

Acne most frequently occurs during puberty because of the hormonal changes that occur during that period. Once adaptation to mature levels of sex hormones occurs, adults are normally not susceptible to this inflammatory disease. The United States Food and Drug Administration has reported that sulfur, benzoyl peroxide, and salicylic acid, alone or in combination with resorcinol, are safe and effective in the treatment of acne. Many nonprescription remedies for acne contain these ingredients.

> Acne is an inflammatory skin disease caused by bacteria, for example, *Propionibacterium acnes*; acne mostly affects teenagers with changing levels of hormones.

FIG. 22-16 Typical lesions of acne often are dark colored because of melanin pigment from infected follicles.

BACTERIAL DISEASES OF THE EYES AND EARS

EYE INFECTIONS

The eyes are continuously bathed in tears that contain lysozyme. Most bacteria, therefore, are unable to cause eye infections. A few bacteria, nevertheless, are able to infect the eyes and cause disease. Contact lenses, if not properly maintained, may help disseminate bacteria to the eyes.

Conjunctivitis

Conjunctivitis is an inflammation of the conjunctiva, a mucous membrane covering much of the anterior surface of the eyeball and the lining of the inner surface of the eyelids, that results from bacterial infections of the eye. Contamination of the eye with foreign material and injuries of the eye tissue favor the establishment of such eye infections. It can be transmitted by the exudate from an infected eye. Bacterial conjunctivitis can be caused by various bacteria, including *Haemophilus* sp., *Moraxella* sp., *Staphylococcus aureus*, *Streptococcus pneumoniae*, *Pseudomonas aeruginosa*, *Corynebacterium* sp., *Chlamydia* sp., and *Neisseria gonorrhoeae*. The symptoms of conjunctivitis generally include swelling and reddening of the eyelids and the formation of purulent discharges. The eye becomes red and itchy because of extensive dilation of the capillaries. There may also be some photophobia and blurring of vision. Many cases of bacterial conjunctivitis are self-limiting and the disease symptoms disappear within 1 week after onset. Antimicrobial agents are useful in limiting the severity of the disease and sulfacetamide, neomycin, polymyxin B, tobramycin, and gentamicin are frequently used in treatment.

> **Bacterial conjunctivitis is an inflammation of the membrane covering the cornea that makes the eyelids swell and redden.**

Pinkeye

Haemophilus aegyptius is a fastidious, Gram-negative, rod-shaped bacterium known as the Koch-Weeks bacillus. It lacks a capsule and may actually be a strain of unencapsulated *H. influenzae*. It is the common etiologic agent of **epidemic conjunctivitis,** or **pinkeye,** among school children. This is a form of conjunctivitis that can be transmitted by contact with exudate from an infected eye. *H. aegyptius* is highly infectious and pinkeye can rapidly spread among school children. Symptoms of pinkeye can include extreme swelling of the eyelids, extensive discharge from the eye, and bleeding within the conjunctiva, as well as the redness and itching associated with conjunctivitis inflammations (FIG. 22-17). Pinkeye can be treated with sulfonamides and tetracyclines.

Ophthalmia Neonatorum

Some forms of conjunctivitis are caused by sexually transmitted pathogens. Adults, for example, can acquire **gonococcal ophthalmia** by rubbing their eyes with contaminated fingers. Newborns are at particular risk for acquiring sexually transmitted pathogens from lesions within the vaginal tracts of their mothers during vaginal births (FIG. 22-18). *Neisseria gonorrhoeae*, the causative agent of **ophthalmia neonatorum** (conjunctivitis of the newborn) can be transmitted in this manner. When ophthalmia neonatorum occurs, it almost always involves both eyes. The symptoms of this disease include discharge of blood and pus and swelling of the eyelids; if untreated, the disease can cause blindness. Erythromycin, penicillin, and tetracyclines are used in treating cases of ophthalmia neonatorum.

Inclusion Conjunctivitis

In addition to *Neisseria gonorrhoeae, Chlamydia trachomatis* can be transmitted from infected mother to the eyes of the newborn during birth. When *C. trachomatis* infects eye tissues it causes **inclusion conjunctivitis.** The symptoms of this disease include reddening of the eyelid and a pussy discharge. In newborns these symptoms usually appear within 36 hours of birth. Inclusion conjunctivitis caused by *C. trachomatis* can also be acquired by adults and children by direct contact with infected individuals and from swimming in nonchlorinated waters contaminated with this bacterial species from genital sources. The disease can be treated with tetracyclines and sulfonamides.

Trachoma

Chlamydia trachomatis also is the causative agent for **trachoma.** Trachoma occurs worldwide but is most prevalent in dry regions of the world, including the southwest United States, where this disease is a particular problem on Indian reservations. The worldwide incidence of trachoma may be as high as 500

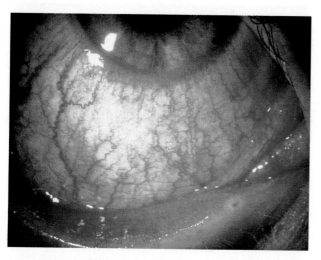

FIG. 22-17 Pinkeye is a common contagious disease among children caused by *Haemophilus aegyptus*.

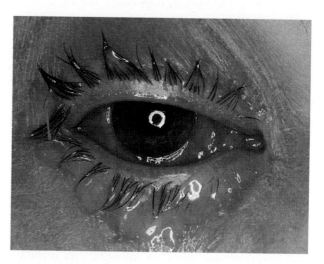

FIG. 22-18 An eye with a *Neisseria gonorrhoeae* infection.

million cases. This disease is a type of keratoconjunctivitis, so named because the conjunctivae become inflamed and the cornea becomes covered with keratin protein (keratinized). Transmission of the disease is through direct or indirect contact of the eye with infectious material.

The development of the chlamydial infection in cases of trachoma initially involves a localized infection of the conjunctivae, but this is followed by spread to other areas of the eye, including the cornea. In *C. trachomatis* infections, vascular papillae and lymphoid follicles are formed, eye tissues can become deformed and scarred, and a chronic infection can lead to partial or total loss of vision. Trachoma is the major cause of preventable blindness in the world. Scarring of the eye often is due to scratching the cornea with lesions that develop on the eye lid. Several antibiotics, including sulfonamides, tetracyclines, and erythromycin, are useful in treating trachoma.

Keratitis

In some cases, infections of the eye cause inflammation of the cornea, generally called **keratitis.** These infections are particularly serious because they can produce scarring of the cornea with permanent vision impairment. Symptoms of infections involving the cornea often include fever, headache, swelling of the tissues around the eye, and discharge of pus. Occasionally, as when the cornea is infected with *Pseudomonas aeruginosa,* perforation may result within 24 hours after onset of infection. *Streptococcus pneumoniae* can cause the formation of cataracts (clouding of the lens of the eye obstructing the passage of light). Other bacterial species, including *Moraxella lacunata, Neisseria meningitidis, Staphylococcus aureus,* and *Streptococcus* spp., may also cause damage of the cornea. Treatment of keratitis can employ various antibiotics, depending on the specific etiologic agent.

EAR INFECTIONS

A few bacteria can infect the ears. These can enter the outer ear from the outside or reach the middle ear from the pharynx (throat) via the auditory canal (eustachian tube).

Otitis Externa

Several microorganisms can cause inflammations of the outer ear, or **otitis externa.** In some cases these infections involve the tympanic membrane. *Escherichia coli, Proteus* sp., *Streptococcus pyogenes,* and *Staphylococcus aureus* are opportunistic pathogens of the outer ear. In many cases outer ear inflammation is the result of a mixed infection involving these organisms. Ampicillin, penicillin, chloramphenicol, erythromy-

cin, and various other antibiotics are used to treat such infections.

One type of outer ear infection, known as **swimmer's ear,** is caused by *Pseudomonas aeruginosa*. This disease is prevalent among swimmers, which accounts for its name. Bathing in *Pseudomonas*-contaminated waters is responsible for the transmission of this disease. *P. aeruginosa* is highly resistant to chlorine, and hence swimming pools and hot spas are easily contaminated with this bacterium. In cases of swimmer's ear there is an inflammation of the lining of the external ear. Generally a broad-spectrum antibiotic is used to treat this condition.

Otitis Media

Otitis media is an inflammatory disease of the mucosal lining of the middle ear. Bacterial infections of the middle ear normally originate from an upper respiratory infection or from the normal microbiota of the upper respiratory tract, with the bacteria entering the ear through the auditory (eustachian) tube, the principal portal of entry to the ear. Certain individuals are more prone to middle ear infections than others. In particular, children are predisposed to middle ear infections if they have adenoids (enlarged masses of lymphoid tissue) in the nasopharynx. In these children, middle ear infections are an extension of their tendency to contract repeated upper respiratory tract infections. Today, plastic tubes are placed into the ears of some children to keep the auditory tubes open and limit the occurrences of ear infections.

Streptococcus pneumoniae is the etiologic agent in over 50% of the cases of otitis media. Ampicillin is effective in treating such infections. *Streptococcus pyo-*

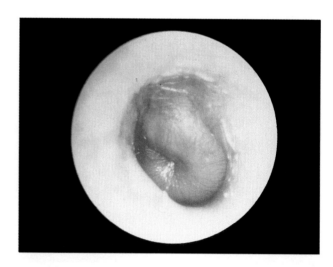

FIG. 22-19 Acute otitis media is characterized by a bulging ear drum and a fiery red tympanic membrane due to inflammation.

genes and *Haemophilus influenzae* are also frequently the causative agents of middle ear infections. Various antibiotics, including tetracyclines, chloramphenicol, and penicillin, are effective in treating middle ear infections caused by these organisms. However, antibiotic resistance is a growing problem with otitis media. Manifestations normally include severe pain and fever and sometimes loss of balance due to the accumulation of fluids. In cases caused by *Streptococcus* species, the tympanic membrane is usually fiery red (FIG. 22-19). There is generally a loss of hearing. Otitis media can become a chronic condition and the pressure on the tympanic membrane can lead to thickening, scarring, or even rupture. Such disruptions of the tympanic membrane cause significant losses of hearing.

BACTERIAL DISEASES OF THE ORAL CAVITY

The oral cavity is the site of abundant bacterial growth. The teeth, tongue, and mucous membranes of the mouth are covered with bacteria. Many of these bacteria only carry out anaerobic metabolism. These bacteria do not cause disease unless they grow excessively.

Excessive growth of microorganisms in the mouth can cause diseases of the tissues of the oral cavity. Too much lactic acid production by streptococci and acid production by other bacteria in the oral cavity causes dental caries. The lactic acid would not cause problems except that it gets trapped against the surfaces of teeth because the streptococci also produce a polysaccharide that results in plaque formation. The excessive growth of other bacteria in the junction between the gum and the tooth causes an inflammation that can lead to periodontal disease, which can result in tooth loss.

DENTAL CARIES

One of the most common human diseases caused by microorganisms is **dental caries**. Caries is initiated at the tooth surface as a result of the growth of *Streptococcus* species that are normal indigenous inhabitants of the oral cavity. These streptococci can initiate caries because they have the following essential properties: (1) they can adhere to the tooth surface because they produce a polymeric substance, which causes them to remain in contact with the tooth surface; and (2) they produce lactic acid as a result of their fermentative metabolism, thereby dissolving the dental enamel surface of the tooth. *S. mutans, S. sanguis,* and *S. salivarius* are implicated as the causative agents of dental caries. Initially *S. mutans* can adhere to proteins in the saliva that coat the tooth surface and form pellicle. These bacteria produce extracellular glucosyl transferases, enzymes that catalyze the formation of extracellular glucans (polymers of glucose) and levans (polymers of fructose) from dietary sucrose. Glucosyl transferase splits the disaccharide sucrose into glucose and fructose and then links the glucose molecules together into long chain polymers called *dextrans*. Some of these dextrans are sticky and help *S. mutans* adhere more firmly to the tooth. Dextrans also contribute to the binding of other bacteria to the tooth surface. This accumulation of a mixed microbial community in a dextran matrix is called **dental plaque** (FIG. 22-20). The accumulation of bacteria in plaque may be 500 cells thick. There is a high degree of structure within plaque, indicative of sequential colonization by different bacterial populations and the different positions of each population within this complex bacterial community.

Dental caries is a common human microbial disease caused by species of *Streptococcus* that produce the enzyme responsible for the accumulation of dental plaque.

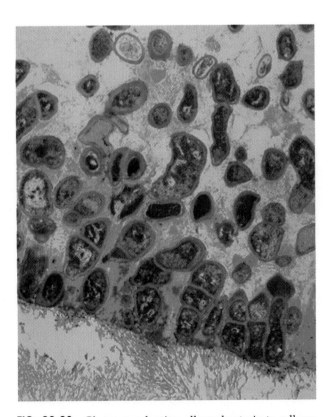

FIG. 22-20 Glucan production allows bacteria to adhere to teeth, forming dental plaque. This colorized micrograph shows bacterial cells and the dextran *(red)* matrix of plaque. (10,000×).

Dental caries can be prevented by limiting dietary sugar substrates from which the bacteria produce acid and plaque. Removing accumulated food particles and dental plaque by periodic brushing and flossing of the teeth is also helpful. Such hygienic practices are particularly effective if performed after eating meals and snacks. Additionally, tooth surfaces can be rendered more resistant to microbial attack by including calcium in the diet, as by drinking milk, and by fluoride treatments of the tooth surface. The administration of fluorides in the diet, such as by consumption of fluoridated water, during the period of tooth formation can reduce dental caries by as much as 50%. The fluoride alters the chemical composition of the tooth so that it is more resistant to acid.

PERIODONTAL DISEASE

In addition to causing dental caries, microorganisms growing in the oral cavity can cause diseases involving the supporting tissues of teeth (**periodontal disease**). Periodontal disease can manifest as **gingivitis** (inflammation of the gingiva [gums]), **necrotizing ulcerative gingivitis** (trench mouth), **periodontitis** (inflammation of the periodontium, also known as pyorrhea), and **juvenile periodontitis** (noninflammatory degeneration of the periodontium, leading to bone regression). Periodontal disease develops when dental plaque accumulates between the tooth and the surrounding tissues.

Although it is unlikely that Koch's postulates can ever be satisfied to identify specific etiologic agents of periodontal disease, it appears that different defined bacterial populations are involved in different stages of development of periodontal disease. Healthy periodontal tissues in humans appear to be associated with relatively few indigenous microorganisms. Most of these organisms are Gram-positive cocci that are located above the gingiva (supragingival) on the tooth surface. Microorganisms commonly associated with these tissues include *Streptococcus mitis*, *Streptococcus sanguis*, *Staphylococcus epidermidis*, *Rothia dentocariosa*, *Actinomyces viscosus*, *Actinomyces naeslundii*, and, occasionally, species of *Neisseria* and *Veillonella*, among others. The onset of gingivitis is marked by characteristic changes in the bacterial populations associated with gingival plaque. Initially, there appears to be an overgrowth of the normal supragingival plaque accompanied by a large increase in the proportion of *Actinomyces* species, such

as *A. viscosus*. These *Actinomyces* species may be important in the attachment of other bacterial populations. As the inflammation of the gingiva progresses, anaerobic Gram-negative bacteria, such as species of *Veillonella*, *Campylobacter*, and *Fusobacterium*, become prevalent, locating primarily on the surface of the plaque in subgingival sites.

Acute necrotizing ulcerative gingivitis (also known as **trench mouth** or **Vincent infection**) is the only form of periodontal disease in which invasion of microorganisms into the tissues occurs. Clinical manifestations of trench mouth include fever, swelling, ulceration, and bleeding of the gums with tissue necrosis, causing ulcers between the teeth. A spirochete infiltrates the tissues and is the likely cause of this disease. *Treponema vincentii*, *Leptotrichia buccalis*, and *Eikenella* spp. are believed to be involved in development of this disease condition. This infection occurs mostly in adolescents and young adults. The name trench mouth is derived from the prevalence of this condition in the trenches of World War I. Fatigue, poor diet, poor oral hygiene, and anxiety are important predisposing factors that have a role in the establishment of this disease. Trench mouth can be treated with penicillin and metronidazole.

> Trench mouth is the only form of periodontal disease in which microorganisms invade the tissues.

The development of advanced forms of periodontal disease, involving tissues beyond the gingiva, is serious and can lead to the loss of bone surrounding the teeth and to tooth loss. Microbial infection of the subgingiva can establish abscesses and pockets between the tooth and supporting tissues (periodontal pocket) in which microorganisms can proliferate. Microorganisms growing in such protected pockets erode the bone tissue and destroy periodontal membranes, causing loosening of the teeth that can lead to tooth loss. The control of periodontal disease can be achieved by frequent removal of accumulations of dental plaque, by daily flossing, and periodic professional dental cleanings. Flossing removes plaque that has an important role in the development of periodontal disease; prophylactic dental treatment also removes calcified plaque (calculus) that may have a role in the advance of periodontal disease.

> Advanced forms of periodontal disease that can lead to tooth loss can be prevented by brushing, flossing, and regular professional dental cleanings.

Bacterial Diseases of the Urinary Tract (pp. 650-651)

- Urethritis and cystitis are bacterial diseases of the urinary tract most commonly caused by *Escherichia coli*, but also can be caused by *Klebsiella, Enterobacter, Serratia, Proteus,* and *Pseudomonas*. The symptoms of urethritis are pain and a burning sensation during urination. The symptoms of cystitis are suprapubic pain and the urge to urinate frequently. Treatment is with various antibiotics, including nitrofurantoin and nalidixic acid.
- Pyelonephritis is caused most commonly by *Escherichia coli,* but also can be caused by *Klebsiella pneumoniae, Proteus mirabilis, Enterobacter aerogenes, Pseudomonas aeruginosa, Streptococcus* sp., and *Staphylococcus* sp. Treatment with antibiotics controls the symptoms of urgent and frequent urination.
- Glomerulonephritis is an immune complex disease that follows infections caused by *Streptococcus pyogenes*. It produces symptoms of high spiking fever, chills, headache, muscle ache, malaise, abdominal pain, nausea, and vomiting. Antibiotics such as chloramphenicol and penicillin are used to treat the original infections and thereby prevent the development of the hypersensitivity reaction that causes this disease.
- Leptospirosis is caused by *Leptospira interrogans* entering through the skin, especially through small abrasions, causing high spiking fever, chills, headache, muscle ache, malaise, abdominal pain, nausea, and vomiting. It is treated with penicillin.

Bacterial Diseases of the Genital Tract (pp. 651-659)

- Sexually transmitted diseases are contracted by direct sexual contact with infected individuals.
- Gonorrhea is caused by *Neisseria gonorrhoeae* transmitted by direct sexual contact. In the male the symptoms are painful purulent urethral discharge and in the female the symptoms are inflammation, pain, and swelling of the cervix, abnormal vaginal discharge, and abnormal menstrual bleeding. Treatment is with antibiotics, usually penicillin.
- Syphilis is caused by *Treponema pallidum* transmitted by direct contact with infected syphilitic lesions containing *T. pallidum,* which enters the body via abrasions of the epithelium and penetrates the mucous membranes. It can also be transmitted across the placenta from an infected mother to her fetus. Symptoms in the primary stage are chancres at the site of inoculation; in the secondary stage, cutaneous lesions and lesions of the mucous membranes on the lips, tongue, throat, penis, vagina, and other body surfaces, also low-grade fever, headache, and enlargement of lymph nodes (latent stage shows no symptoms); and tertiary stage, damage at any organ. Penicillin or other antibiotics are used in treatment.
- *Gardnerella* vaginitis is caused by the bacterium *Gardnerella vaginalis* and produces a light, frothy vaginal discharge with a fishy smell. It is treated with metronidazole.
- Nongonococcal urethritis is caused by *Chlamydia trachomatis* and also *Mycoplasma hominis*. It is a sexually transmitted disease that can also be transmitted at birth from mother to the newborn's eyes. Females are often asymptomatic and males have pain and discharge during urination. It is treated with antibiotics, such as tetracyclines.
- Lymphogranuloma venereum is also a sexually transmitted disease caused by *Chlamydia trachomatis* carried to lymph nodes, producing swollen and tender nodes there and a bloody, nonpurulent rectal discharge. It is treated with antibiotics, such as tetracyclines.
- Chancroid is a sexually transmitted disease caused by *Haemophilus ducreyi* characterized by the development of soft chancres. It is treated with tetracycline and sulfanilamide.
- Granuloma inguinale is a sexually transmitted disease caused by *Calymmatobacterium granulomatis* in which the genitalia develop characteristic ulcers, older portions of which lose their pigmentation. It is treated with tetracycline, chloramphenicol, and erythromycin.

Bacterial Diseases of the Skin (pp. 659-664)

- *Staphylococcus aureus* is a pathogenic bacteria that enters the body through natural openings in the skin, causing styes, boils, carbuncles, impetigo, and folliculitis.
- *Pseudomonas aeruginosa* often is the cause of skin infections.
- Streptococcal skin infections include impetigo and erysipelas, which can be treated with penicillin and erythromycin.
- Hansen disease is caused by *Mycobacterium leprae.* It is transmitted through the skin or mucous membranes via direct contact or droplets, causing skin lesions and damage to nerve tissue. It is treated with dapsone and rifampin.
- Anthrax is caused by *Bacillus anthracis* transmitted by direct contact of skin with its endospores via the respiratory tract through inhalation and via the gastrointestinal tract through ingestion. It causes tissue necrosis and is treated with penicillin.
- Scarlet fever is caused by *Streptococcus pyogenes* that is a secondary complication to pharyngitis. It produces a characteristic rash and is treated with penicillin.
- Acne is a skin infection caused by *Propionibacterium acnes* and *Staphylococcus epidermidis*. The invasion of these microorganisms to the base of hair follicles initiates excessive secretions by the sebaceous glands, producing inflammatory papules, pustules, and cysts on the face, upper back, and chest.

Bacterial Diseases of the Eyes and Ears (pp. 664-667)

- Conjunctivitis, an inflammation of the mucous membrane covering the cornea, is caused by various bacteria, including *Haemophilus* sp., *Moraxella* sp., *Staphylococcus aureus, Streptococcus pneumoniae, Pseudomonas aeruginosa, Corynebacterium* sp., *Chlamydia* sp., and *Neisseria gonorrhoeae*. It produces swelling and reddening of the eyelids, formation of purulent discharges, photophobia, and blurred vision. It is treated with antimicrobial agents, including sulfacetamide, neomycin, and gentamicin.

- Pinkeye is caused by *Haemophilus aegyptius*, making eyelids swell, discharge in the eyes, bleeding within the conjunctiva, redness, and itching. It is treated with sulfonamides and tetracyclines.
- Ophthalmia neonatorum is caused by *Neisseria gonorrhoeae* passed from an infected mother to the eyes of her newborn during childbirth. It produces a discharge of blood and pus and swelling of the eyelid. It is treated with erythromycin, penicillin, and tetracyclines.
- Inclusion conjunctivitis is caused by *Chlamydia trachomatis* passed from an infected mother to the eyes of her newborn during childbirth. It causes reddening of the eyelid and a pussy discharge; it is treated with tetracyclines and sulfonamides.
- Trachoma is caused by *Chlamydia trachomatis*. It is an inflammation of the conjunctivae and covers the cornea with dead keratin protein.
- Keratitis is an inflammation of the cornea caused by *Pseudomonas aeruginosa, Streptococcus pneumoniae, Moraxella lacunata, Neisseria meningitidis, Staphylococcus aureus*, and *Streptococcus* sp. It causes fever,

headache, swelling of the tissues around the eye, pussy discharge, and perforation of the cornea.

- Otitis externa, infections of the outer ear, are caused by *Escherichia coli, Proteus* sp., *Streptococcus pyogenes*, and *Staphylococcus aureus*. Swimmer's ear is an infection caused by swimming in *Pseudomonas*-infected water. It is treated with ampicillin, penicillin, chloramphenicol, erythromycin, and other antibiotics.
- Otitis media, infections of the middle ear, can be caused by *Streptococcus pneumoniae, S. pyogenes*, or *Haemophilus influenzae*. These bacteria are transmitted from an upper respiratory infection entering the ear through the eustachian tube, producing symptoms of severe pain and fever. Antibiotics treatments include tetracyclines, chloramphenicol, and penicillin.

Bacterial Diseases of the Oral Cavity (pp. 667-668)

- Dental caries are caused by *Streptococcus* sp. They can be prevented by limiting dietary sugars and including calcium in the diet, brushing and flossing teeth, and getting fluoride treatments.
- Periodontal disease is caused by *Actinomyces* sp. that may initiate gingivitis, with more and different microorganisms involved as the disease progresses. This disease develops when food particles and dental plaque are allowed to accumulate between the tooth and surrounding tissues and produce inflammation of the gingiva.

CHAPTER REVIEW

REVIEW QUESTIONS

1. What are sexually transmitted diseases? Why are some bacteria restricted to this mode of transmission?
2. What are the stages of syphilis? What is happening in each? When can syphilis be diagnosed?
3. Why are the eyes of newborns washed with erythromycin sulfate?
4. How can pathogens infect the urinary tract? Why are women more likely than men to have urinary tract infections?
5. What is pelvic inflammatory disease (PID)? How is it caused?
6. What causes leptospirosis and how is this disease transmitted?
7. What causes tooth decay? How can dental carries be prevented?

8. How does periodontal disease develop?
9. Which bacteria cause diseases of the skin? How are these treated?
10. Which bacteria cause diseases of the eye? How are these treated?
11. Which bacteria cause diseases of the ears? How are these treated?
12. Why has leprosy historically been such a feared disease? How can this disease be treated?
13. What is the cause of anthrax? How is this disease transmitted?
14. What causes acne?
15. Why is acne associated with puberty?
16. Why is pinkeye commonly spread in day care centers?

CRITICAL THINKING QUESTIONS

1. What can be done to reduce the incidence of sexually transmitted diseases?
2. Why are there periods of increasing incidence of some bacterial sexually transmitted diseases? Why are others decreasing or remaining constant?

3. Can a vaccine be developed to prevent dental caries?
4. How can a vaccine be developed to prevent dental caries?
5. Why aren't vaccines used to prevent sexually transmitted bacterial diseases?

CASE STUDY

HISTORY AND ASSESSMENT

John, a 39-year-old man, made an appointment with his physician because he was concerned about a skin rash that progressed in severity over the past 2 weeks. John denied any itching associated with the rash, which he thought was unusual. He stated that he also "felt bad," with headache and joint pains. John attributes this to stress of his job. John said he had recently been hiking with his girlfriend in the foothills of the Adirondack Mountains in upstate New York. He is worried that he may have contracted some sort of allergy to a plant there. John admits to being allergic to penicillin, but he is not aware of having any other allergies to plants, drugs, or foods. John is very concerned and would like to have this rash cleared up. He worries that it will hamper his relationship with his girlfriend.

The nurse charted John's vital signs as: Pulse rate 76, Respiratory rate 12, Blood pressure 132/72, and Temperature 99.8° F. *Are these vital signs within normal limits?*

The nurse also noted a rash of pink lesions symmetrically distributed on the torso but in greater density on John's palms, forearms, and the soles of his feet. *Now consider a differential diagnosis between a contact dermatitis and an infection with a microorganism. How would you expect the distribution of the rash of a contact dermatitis (for example, poison ivy) to differ from the distribution of the rash caused by a microorganism carried throughout the body by the bloodstream? Which do you think has caused John's rash?*

PHYSICAL EXAMINATION

The physician examined John and verified the nurse's findings, observing that some of the lesions were filled with pus. In addition, the physician observed a generalized lymphadenopathy (swelling of lymph nodes) and several lesions on the patient's palate and pharynx. Several patches of scalp hair appeared thin; this concerned John because he had not noticed any significant hair thinning over the past several months. On examination of John's genitals, the physician found what appeared to be a healing ulceration on the underside of the penis. The doctor questioned John about any recent sexual encounters; John admitted to sexual relations with two women in the past 6 months, his present girlfriend and a woman he met at a friend's party. John denied using a condom during any of these encounters. *The physician is fairly certain that John has acquired a sexually transmitted disease. Based on your reading, what do you think is the cause of John's symptoms.*

PROCEDURES AND LABORATORY TESTS

To confirm the diagnosis, the physician scraped one of the lesions, squeezed some tissue fluid onto a microscope slide, and then examined the fluid under a darkfield microscope. The physician saw small, spiral shaped microorganisms swimming about in the fluid (see Figure). *The physician isolated the causative agent of John's rash—Treponema pallidum.*

The physician also ordered a VDRL blood test for syphilis, which was sent to the lab and returned with a positive result the next day, confirming the diagnosis.

DIAGNOSIS

The physical findings and the results of the darkfield microscopy were enough for the physician to diagnose secondary syphilis in John. Syphilis is a sexually transmitted disease caused by the spirochete *Treponema pallidum. Primary syphilis has a characteristic lesion, called a chancre, at the point of initial infection by the spirochete. The chancre is typically found on or around the genitals. Without treatment, the chancre heals completely within 4 to 6 weeks. Secondary syphilis appears 2 to 8 weeks after the appearance of the chancre and is characterized by viral symptoms, lymphadenopathy, and a rash. Distribution of a rash to the palms of the hands and soles of the feet is highly suggestive of syphilis. If untreated, the symptoms of secondary syphilis will resolve in 2 to 6 weeks. Some infected and untreated patients may develop late syphilis with cardiovascular and central nervous system complications.*

TREATMENT AND COURSE

The treatment of choice for a secondary syphilis infection is an injection of 2.4 million units of benzathine penicillin G. However, the physician chose to treat John with an oral dose of 500 mg of tetracycline hydrochloride four times a day for 15 days. *Review the admission notes of this case study to determine why this alternate therapy was used.*

The physician also recommended that John's sexual partners be treated for exposure to syphilis. *The physician recommends treatment without a confirming blood test of the partners because of confirmed exposure to the microorganism. A blood test may not show a positive result for up to 90 days after exposure.*

The physician also advised John about the risks of unsafe sex today and ways to practice "safer sex."

John's rash cleared up over the next 2 weeks and the chancre healed. John will require additional blood tests over the next year to assess the effectiveness of the treatment.

READINGS

Aral SO and KK Holmes: 1991. Sexually transmitted diseases in the AIDS era, *Scientific American* 264(2):62-69.
> *An important discussion of the increasing incidence of sexually transmissible diseases.*

Bialasiewicz AA and KP Schaal (eds.): 1994. *Infectious Diseases of the Eye*, ed. 5, Boston, Butterworth-Heinemann.
> *Covers microbial infections of the eye.*

Bluestone CD and JO Klein: 1988. *Otitis Media in Infants and Children*, Philadelphia, W. B. Saunders.
> *Provides a thorough explanation of this major problem in clinical pediatrics.*

Chase A: 1983. *The Truth About STD: The Old Ones-Herpes and Other New Ones-The Primary Causes-The Available Cures*, New York, Morrow.
> *Relates the historic and current status of sexually transmitted diseases, describing diagnoses, treatment, and epidemiology. The author believes that societal elements, such as wars, poverty, fear, family collapse, and hopelessness, have triggered and encouraged the spread of these diseases now and in the past.*

Cunliffe WJ: 1989. *Acne*, London, Year Book Medical Publishers.
> *Part of the series "Focal Points in Dermatology" designed for practicing physicians.*

Goodfield J: 1985. *Quest for the Killers*, Boston, Birkhauser.
> *The history of communicable disease control is presented through the stories of the diseases kuru, hepatitis B, schistosomiasis, leprosy, and smallpox.*

Gussow Z: 1989. *Leprosy, Racism, and Public Health: Social Policy in Chronic Disease Control*, Boulder, CO; Westview Press.
> *Presents the social aspects that determined government policy decision-making that controlled the treatment of persons with Hansen disease (leprosy) in the United States and how these decisions were influenced by racism.*

Hobson D and KK Holmes (eds.): 1977. *Nongonococcal Urethritis and Related Infections*, Washington, D.C.; American Society for Microbiology.
> *Proceedings of a symposium that assessed the clinical spectrum of chlamydial and ureaplasmal infections to define from the clinical and epidemiological standpoints the nature and extent of morbidity of NGU and related infections.*

Maibach HI and G Hildick-Smith (eds.): 1965. *Skin Bacteria and their Role in Infection*, New York, McGraw-Hill.
> *A review of the microbiology of the skin, covering the bacteria themselves, the clinical infections they cause, their epidemiology, and prevention.*

Miller CH and CJ Palenik: 1994. *Infection Control and Management of Hazardous Materials for the Dental Team*, St. Louis, Mosby.
> *Describes safety aspects of dental practice with regard to infectious microorganisms.*

Nabers CL and WEH Stalker: 1990. *Periodontal Therapy*, Toronto, B.C. Decker.
> *Reviews the current treatment methods for periodontal diseases.*

Newbrun E: 1989. *Cariology*, Chicago, Quintessence Publishing Co.
> *Explores the causes, effects, and treatment of dental caries.*

Nisengard RJ and MG Newman: 1993. *Oral Microbiology and Immunology*, ed. 2, Philadelphia, W. B. Saunders.
> *Text on microbiology of the oral cavity.*

Pastorek JG and SA Gall: 1994. *Obstetric and Gynecologic Infectious Disease*, New York, Raven Press.
> *Describes the varied infections of the genital tract, including sexually transmitted diseases.*

Peppe JS, GN Holland, K Wilhelmas: 1994. *Ocular Infection and Immunity*, St. Louis, Mosby.
> *Covers the principles of ocular immunology and many viral, bacterial, fungal and parasitic infections and immune-related conditions that affect the eye.*

Sears C: 1989. Where the bad bugs are, *American Health* 8(April):50.
> *Reviews the microbiology of the mouth and explains the role of bacteria in gum disease and tooth decay.*

Smyth GDL: 1980. *Chronic Ear Disease*, New York, Churchill Livingstone.
> *Otitis media and tympanoplasty are the major diseases discussed,*

Spagna VA and RB Prior: 1985. *Sexually Transmitted Diseases*, New York, Marcel Dekker.
> *Informative coverage of sexually transmitted diseases.*

Stiller R: 1974. *The Love Bugs: A Natural History of the VD's*, Nashville, T. Nelson.
> *This history of syphilis and gonorrhea includes all aspects of these venereal diseases—diagnosis, treatment, manifestations, and effects of the infections.*

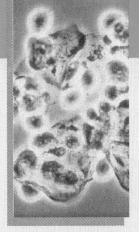

CHAPTER 23

Diseases Caused by Eukaryotic Organisms

PREVIEW TO CHAPTER 23

In this chapter we will:

• Learn about the diseases of humans caused by
eukaryotic microorganisms and some multicellular
parasites.
• Review the major diseases of humans caused by fungi,
algae, protozoa, and some multicellular parasitic
worms—examining the specific etiologic agent, the
route of transmission, the major symptoms, and the
treatments of each disease.
• Learn the following key terms and names:

aflatoxin
African sleeping sickness
African trypanosomiasis
amebiasis
amebic dysentery
American
  trypanosomiasis
aspergillosis
athlete's foot
babesiosis
balantidial dysentery
balantidiasis
blastomycosis
cestodes
Chagas disease
coccidioidomycosis
cryptococcal meningitis
cryptosporidiosis
cysticercosis
dermatomycoses
dermatophytes
elephantiasis
ergotism
flukes
giardiasis
histoplasmosis
hookworm
hydatid cyst
hydatidosis
jock itch

kala-azar disease
leishmaniasis
malaria
merozoites
mushroom poisoning
Naegleria
  microencephalitis
paralytic shellfish
  poisoning
pinworm
Pneumocystis pneumonia
ringworm
roundworm disease of
  humans
San Joaquin fever
schistosomiasis
schizogony
sporotrichosis
tapeworms
thrush
tineas
toxoplasmosis
trematodes
trichinosis
tsetse fly vector
vaginitis
valley fever
vulvovaginitis
whipworm

Of the numerous fungi very few cause human disease (Table 23-1). Most are nonpathogenic or cause disease in plants. The limited number of fungal diseases is both surprising and fortunate. Surprising, because fungi produce spores that can survive in the environment and disseminate over long distances, and thus are readily transmissible to humans. Fortunate, because relatively few antimicrobics are effective in treating fungal infections of humans without causing serious side effects. Diseases caused by fungal growth are called *mycoses*. Fungi that establish infections on external body surfaces are treatable without harmful effects. Some fungi produce toxins when growing in the environment. If ingested these toxins can cause serious disease symptoms and death. Neurotoxins produced by various mushrooms are extremely powerful toxins. Avoiding ingestion of such fungal toxins is key to avoiding these diseases.

FUNGAL DISEASES OF THE RESPIRATORY TRACT

The fungi that cause infections of the respiratory tract come from environmental sources. They are typically limited in geographic distribution. They enter the body as spores carried through the air. Usually only fungal spores can escape the body's immunological defense system and establish infections. Disease-causing fungi have an interesting property called dimorphism, which is the ability to grow sometimes as single yeast-like cells and other times as filamentous forms. In the environment these fungi grow as filamentous forms, producing multicellular mycelia, but within the human body they grow as yeast-like forms, reproducing as unicellular fungi.

Histoplasmosis

Histoplasmosis is caused by *Histoplasma capsulatum*. This fungus grows in a filamentous form in the environment but changes and grows as a yeast-like fungus when it infects the human body (FIG. 23-1). It is a dimorphic fungus, meaning that it grows as single cells (yeast form) and other times as hyphae (filamentous form). The natural habitat for *H. capsulatum* is soil. It is found in tropical regions of the world and within restricted regions of the central United States, sometimes called the "histo belt." Most individuals in the Ohio and Mississippi river valleys show evidence of having been infected with or at least exposed at some time to *H. capsulatum* (FIG. 23-2). This fungus is especially found in soils contaminated with bird and bat droppings. The fungus does not infect birds but the organic matter from the bird wastes

TABLE 23-1
Diseases Caused by Fungi

DISEASE	CAUSAL ORGANISM	COMMENT
ALLERGIES		
Cheese washer's lung	*Penicillium casei*	Cheese
Maltster's lung	*Aspergillus clavatus*	Barley malt
Maple-bark stripper's lung	*Cryptostroma corticale*	Maple tree bark
Sequoiosis	*Aureobasidium pullulans, Graphium*	Redwood sawdust
Suberosis	*Penicillium frequentans*	Cork
Wood-pulp worker's disease	*Alternaria*	Wood pulp
Farmer's lung	*Faenia rectivirgula*	Stored hay
SUPERFICIAL MYCOSES (dermatomycoses)		
Tinea capitis	*Microsporum, Trichophyton*	Scalp
Tinea pedis (athlete's foot)	*Epidermophyton*	Between toes, skin
Jock itch	*Trichophyton, Epidermophyton*	Genital region
SUBCUTANEOUS MYCOSES		
Sporotrichosis	*Sporothrix schenckii*	Arms, hands
Chromoblastomycosis	Several genera	Legs, feet
SYSTEMIC MYCOSES		
Cryptococcosis	*Cryptococcus neoformans*	Lungs, meninges
Coccidioidomycosis	*Coccidioides immitis*	Lungs
Histoplasmosis	*Histoplasma capsulatum*	Lungs
Blastomycosis	*Blastomyces dermatitidis*	Lungs, skin
Candidiasis (opportunistic)	*Candida albicans*	Oral cavity, intestinal tract

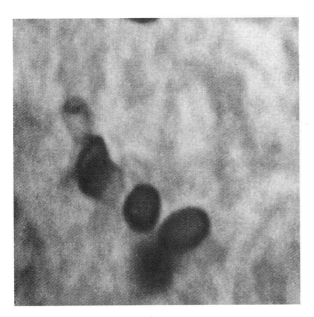

FIG. 23-1 Histological section of the lung showing yeast forms of *Histoplasma capsulatum*.

BIRD KILLS TO PREVENT HISTOPLASMOSIS

The association of *Histoplasma capsulatum* with bird roosts has been used to justify the killing of large flocks of blackbirds. The birds create a nuisance because their noise disrupts sleeping in residential neighborhoods and their wastes can destroy the paint finishes of cars and houses. Noise, such as shooting off canons, and other methods of persuading the birds to leave the area usually fail. Because of the potential threat of histoplasmosis from the roosts, court orders sometimes are issued that permit the killing of tens of thousands of birds in a roost. The killing is done by spraying with the detergent tergitol on a night when the temperature will drop below 0° C. The detergent removes the lipids that protect the birds against the cold and as a result the birds freeze to death.

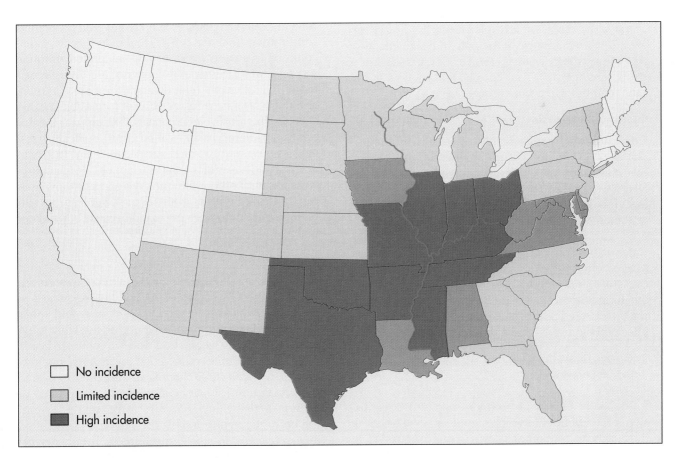

No incidence

Limited incidence

High incidence

FIG. 23-2 Histoplasmosis is endemic to the Ohio and Mississippi river valleys. Individuals in this area show immunological evidence of a high incidence of infection with *Histoplasma capsulatum*. Over 90% of individuals tested in the region show skin reactivity to histoplasmin.

supports the growth of *Histoplasma.* Dust particles released from bird roosts appear to be involved in some outbreaks of this disease. Large-scale kills of blackbirds have been used to limit the spread of *H. capsulatum.* Inhalation of airborne soil and contaminated dust particles can result in human infections. It is a common problem in chicken houses in Arkansas.

Histoplasma capsulatum, **the causative organism of histoplasmosis, is associated with bird droppings and is endemic to the Ohio and Mississippi river valleys.**

Spores of *H. capsulatum* that are deposited in the alveoli of the lungs initiate an infection. The fungus spreads via the lymphatic system to the regional lymph nodes. *H. capsulatum* can grow within phagocytic cells and is able to persist within the lungs and the tissues and cells of the mononuclear phagocyte system. The symptoms of histoplasmosis typically resemble those of a mild cold. In some individuals the infection can cause a chronic progressive disseminated disease that resembles tuberculosis. In immunocompromised individuals, the disseminated disease can occur years after the initial infection. The systemic spread of *H. capsulatum* can be fatal. Amphotericin B is used in the treatment of cases of progressive disseminated histoplasmosis, but mild cases are not treated because of the potential side effects of this antimicrobic. There is about a 50% success rate of using amphotericin B to cure histoplasmosis in immunocompromised individuals.

Coccidioidomycosis

Coccidioidomycosis, which is caused by the fungus *Coccidioides immitis,* occurs frequently in the San Joaquin Valley of California and hence is also referred to as **valley fever** or **San Joaquin Fever.** Many individuals living in this area show a positive skin test, indicating that they have been infected at some time with this fungus. Visitors to the southwestern states of Nevada, California, Utah, Arizona, and New Mexico often develop symptoms of a mild cold because of this infection. Because of the association of the fungal spores with the arid soils of the southwestern United States, soils from this region are disinfected before shipping to other areas.

Coccidioidomycosis occurs in the southwestern United States.

Normally, *C. immitis* occurs in soil, and transmission of coccidioidomycosis involves inhalation of dust particles containing spores of this fungus. When deposited in the bronchi or alveoli, *C. immitis* elicits an inflammatory response. In some cases, *C. immitis* remains localized in the area of the primary lesion, but the organism can be distributed to other parts of the body. The infection is most likely to become systemic in immunocompromised individuals, such as those with AIDS. The fungus can move through the blood from the primary site of infection in the lung, reaching the brain and spinal column where it can cause meningitis.

When the fungus is restricted to the respiratory tract, the symptoms of coccidioidomycosis include chest pain, fever, malaise, and a dry cough. In most such cases, no special treatment is required for the cure of localized coccidioidomycosis, and upon recovery the individual is immune to this disease. However, when there is evidence of systemic dissemination of *C. immitis,* and particularly if meningitis occurs, antibiotics such as amphotericin B are used.

Blastomycosis

Blastomycosis is caused by the fungus *Blastomyces dermatidis.* In addition to humans this fungus infects captive marine mammals, such as dolphins. *B. dermatidis* is a dimorphic fungus that normally inhabits soils, particularly in the southeastern United States, Central America, and Africa. The primary site of infection is the lungs, from which the fungus can be disseminated to many other body tissues. The symptoms of blastomycosis, which are generally mild and self-limiting, include cough, fever, and general discomfort. In severe cases, the cell-mediated immune response produces granulomas (nodules of inflamed tissue). Skin lesions (cutaneous ulcers) are common, and there can be extensive abscess formation and tissue destruction in severe cases. The appearance of nonencapsulated, thick-walled, multinucleate yeast cells in pus, sputum, or tissue sections and their isolation in cultures establish the diagnosis of blastomycosis. As with other systemic mycoses, blastomycosis can be treated with amphotericin B, but hydroxystilbamidine is sometimes preferred because of its lower toxicity.

Aspergillosis

Immunosuppressed hosts or individuals exposed to high numbers of spores of *Aspergillus* are subject to respiratory disease. *Aspergillus fumigatus* and other species of *Aspergillus* cause **aspergillosis.** They grow well in decaying vegetation and so compost heaps are common sources of infection for farmers and gardeners. In some cases there is an allergic response to the inhalation of *Aspergillus,* with symptoms of asthma that include difficulty in breathing. Within the body *Aspergillus* may form a hyphal mass called an *aspergilloma.* Invasive infections of pulmonary aspergillosis can be very dangerous. A large aspergilloma in the lung can block respiratory gas exchange and cause death due to asphyxiation. Predisposing factors include an impaired immune system, cancer, or diabetes. In immunocompromised individuals the fungal infection may become disseminated. Amphotericin B is used in treating aspergillosis.

Pneumocystis Pneumonia

Pneumocystis **pneumonia** is caused by the fungus *Pneumocystis carinii* (FIG. 23-3). This somewhat rare form of pneumonia has received attention recently because it occurs with relatively high frequency in patients with acquired immunodeficiency syndrome (AIDS). It may be the most common opportunistic infection in individuals with AIDS and often is the index defining when an HIV-infected individual has developed AIDS. The occurrence of *Pneumocystis* pneumonia appears to be associated with failures of the immune system, as occurs in cases of AIDS. It also occurs in some premature babies and in the elderly. Patients receiving drugs that suppress their immune system also are prone to this disease. Hospitalized patients whose immune systems are stressed due to disease conditions such as leukemia are also susceptible. The reservoir for *P. carinii* within hospitals has not been identified; in most hospitals, patients with this infection are isolated.

> **AIDS patients are subject to frequent bouts with *Pneumocystis* pneumonia.**

If untreated *Pneumocystis* pneumonia generally is fatal. During the course of this disease, lung alveoli characteristically fill with fluid, thereby preventing gas exchange. The disease can be treated with drugs such as pentamidine. Children at high risk, such as those with acute lymphocytic leukemia, are treated with trimethoprim-sulfamethoxazole. This drug may

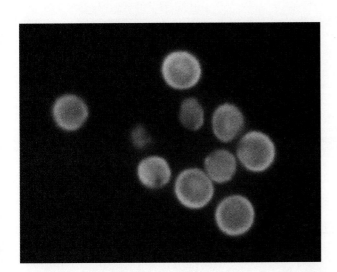

FIG. 23-3 *Pneumocystis carinii* pneumonia diagnosed by fluorescent monoclonal antibody stain of bronchoalveolar washings.

also be effective in use with AIDS patients, although prolonged usage may lead to decreased counts of white and red blood cells and platelets. Trimetrexate, dapsone, and aerosolized pentamidine are used for treatment of *Pneumocystis* pneumonia.

FUNGAL DISEASES OF THE GASTROINTESTINAL TRACT

The disease aspergillosis is an infective process. *Aspergillus* can also produce a toxin while growing outside the body that can cause serious human disease symptoms. When growing on improperly stored grains or peanuts, *Aspergillus* produces **aflatoxin.** Export of aflatoxin-contaminated grains is a problem and as a result peanut flour is no longer exported. Peanut butter may contain aflatoxin and peanut butter that does not contain preservatives may permit growth of *Aspergillus* and accumulation of this toxin. Ingestion of aflatoxin disrupts liver function and may also cause liver cancer. Tests for aflatoxin are performed by the food industry to ensure the safety of foods.

FUNGAL DISEASES OF THE VAGINAL TRACT

Fungal overgrowth of the vaginal tract by *Candida* sometimes occurs as a result of population shifts among the microorganisms that compose the normal vaginal tract microbiota. *Candida* is a yeast-like fungus. The excessive growth of *Candida albicans* causes changes in the mucosal cells lining the vaginal tract and inflammation **(vaginitis).** The symptoms include burning sensation and an increased vaginal discharge, which is thick and cheesy. *Candida* infections are common during pregnancy when the pH of the vaginal tract increases. The use of birth control pills also tends to result in higher pH values in the vagina. The administration of antibacterial antibiotics, such as tetracyclines, which can adversely affect the indigenous bacterial populations, also favors the development of *Candida* infections of the vaginal tract. Nystatin, applied topically, generally is used for *Candida* infections.

> ***Candida* infections of the vagina are associated with increases in the pH of the area.**

FUNGAL DISEASES OF THE CENTRAL NERVOUS SYSTEM

Mushroom Poisoning

Basidiomycetes are mushroom-forming fungi. Some basidiomycetes produce potent neurotoxins that can be absorbed through the gastrointestinal tract. The ingestion of poisonous mushrooms, such as *Amanita phalloides*, is normally fatal. *A. phalloides* is called the death angel mushroom because of its beautiful appearance (FIG. 23-4) and its production of numerous deadly toxins. The toxins produced by *A. phalloides,* and other species of *Amanita,* produce symptoms of food poisoning 8 to 24 hours after they are ingested. Initial symptoms include vomiting and diarrhea. Later, degenerative changes occur in liver and kidney cells, and the patient often becomes comatose. Death may occur within a few days of ingesting as little as 5 to 10 mg of toxin. Because some highly poisonous mushrooms resemble other edible mushrooms, great care must be exercised to identify properly wild mushrooms that are selected for eating. Treatment of mushroom poisoning generally is supportive and not always effective.

Ergotism

Ergotism, or **St. Anthony's Fire,** results from eating grain containing ergot alkaloids (nitrogen-containing organic bases) produced by the fungus *Claviceps purpurea. Claviceps* grows in rye grain that becomes moist because of improper storage, resulting in the accumulation of ergot alkaloids in the grain. Ergot alkaloids cause degeneration of the capillary blood vessels, producing neurological impairment due to inadequate circulation to the nervous system and often death. Symptoms of ergotism may include vomiting, diarrhea, thirst, hallucinations, convulsions, and lesions on the hands and feet. The convulsions and hallucinations are common symptoms reflecting the effects of the ergot alkaloids on the central nervous system. Various outbreaks of mass hallucinations have been traced to contamination of food with ergot alkaloids.

Cryptococcal Meningitis

Cryptococcus neoformans is a yeast-like, capsule-forming fungus that can cause **cryptococcal meningitis.** This disease is also known as cryptococcosis. *C. neoformans* is associated with soil contaminated by pigeon droppings, and transmission of this fungus to humans results from inhalation of dust associated with pigeon droppings. The fungus does not grow within the pigeons but can grow well in the nutrient-rich wastes of the birds. Infection initially affects the lungs. In most cases cryptococcosis is restricted to the respiratory tract, where it causes a disease resembling influenza. In some cases the fungus spreads systemically to invade the brain and meninges, causing meningitis. This often happens to individuals receiving immunosuppressive therapy for other disease conditions. The onset of disease often is slow, occurring over days or weeks. The progression of symptoms typically includes fever, headache, stiff neck, and disorientation. If untreated, cryptococcal meningitis is usually fatal, but effective treatment can be achieved using amphotericin B.

Cryptococcus neoformans, **which is the causative agent of cryptococcal meningitis, is associated with soil and dust that is contaminated with pigeon droppings.**

FIG. 23-4 *Amanita muscaria* is a beautiful but potentially deadly mushroom.

HIGHLIGHT

SALEM WITCH HUNTS: A CASE OF ERGOTISM?

The colonial American village of Salem was the site of witch persecution near the end of the seventeenth century. Attempts to explain why Salem became a center for purported witch activities and witch trials at this particular time and place have been unsatisfactory. Historians have suggested several explanations, including social pressure on adolescents and the need of the group for a scapegoat.

Our knowledge of what happened during this time comes from the court records of the witch trials, because in the late 1600s, the practice of witchcraft was a serious crime. These trials tried to prove that the sick children were under the spell of witches and did not attempt to provide clear, complete accounts of everything that happened. In 1692, 30 children and teenagers in Essex County, Massachusetts, claimed to be victims of bewitchment and accused a slave in the household of two of the victims and two elderly women of ill repute of being the source of their affliction, of having "bewitched" them. The victims suffered from what were believed, according to English folk tradition, to be specific, common symptoms of bewitchment, including convulsions, sensations of being pinched, pricked, or bitten; temporary blindness, deafness, and speechlessness; burning sensations; visions; and various other sensations, such as flying outside of their bodies, and being torn, picked, or pulled apart. Three adults and seven infants and young children died after suffering similar symptoms; three cows also died.

Scientists today recognize many of the symptoms reported in the Salem witch trials as characteristic of ergotism. Ergotism is caused by consumption of ergot alkaloids, which are chemicals produced by the fungus *Claviceps purpurea*. This fungus grows on rye, most likely when rye is grown in low, moist, shaded land and especially on newly cultivated land. The development of ergot alkaloids is encouraged by a severely cold winter followed by a cool, moist growing season. The cold weakens the rye plant, while the wet weather promotes fungal growth.

Ergot alkaloids produce various symptoms. Symptoms of early and mild ergotism are giddiness, headache, fatigue, diarrhea, dizziness, chills, sweating, depression, nausea, vomiting, and pain in the limbs. A sensation of ants under the skin, cold hands and feet, muscle twitches, and spasms of the limbs, tongue, and facial muscles are symptoms of more severe cases. In the most severe cases the victim has epileptic-like convulsions and eats ravenously between fits. Such a sufferer may lie as if dead for 6 to 8 hours and later show paralysis of the lower arms, jerking arms, delirium, and loss of speech. Death may occur on the third day after onset of symptoms.

Ergot alkaloids are powerful hallucinogens like LSD. People who have eaten ergot-contaminated rye and who are under the influence of LSD are highly suggestible and often may think they see images. Young people are more vulnerable to ergot than their elders because they ingest more food per unit of body weight and, therefore, more poison per pound.

The first symptoms of bewitchment appeared in Salem Village in late 1691. The incidence increased the following spring, slowed in June, and peaked between July and September. Ergot can remain chemically stable in storage for up to 18 months and, we know from diaries and inventories from the time, that the rye crop harvested in the summer often remained unthreshed in the barn when other foods were available and might not be used until November or December, when the symptoms began to appear. Tree rings show that the winters of 1690-91 and 1691-92 were especially cold. Rye flourished in such weather, whereas other crops failed, and so rye may have been consumed in greater amounts in those years. Epidemics of convulsions often occurred in years where the tree rings indicated cold weather.

Farmers used swampy, sandy, marginal land because the population of Salem was growing and so was food demand. This type of land, when drained, is suitable for the cultivation of rye, but also the type most likely to be infected with ergot. All 22 households in Salem affected in 1692 were located on or at the edge of moist, sandy, acidic, loamy soil. The possibility that ergotism might have played a role in this interesting incident in American history is intriguing. It may be possible that the witchcraft affair might actually have been an unrecognized public health problem. Microorganisms may have been the cause of hallucinations that led masses of people to believe they had seen the work of witches.

TABLE 23-2			
Epidemiology of Dermatomycoses			
DISEASE	**CAUSATIVE AGENT**	**TRANSMISSION**	**EXAMPLES OF SOURCES**
Tinea capitis (ringworm of the scalp)	*Microsporum* sp. *Trichophyton* sp.	Direct or indirect contact	Lesions, combs, toilet articles, headrests
Tinea corporis (ringworm of the body)	*Epidermophyton* sp. *Microsporum* sp. *Trichophyton* sp.	Direct or indirect contact	Lesions, floors, shower stalls, clothing
Tinea pedis (ringworm of the feet—athlete's foot)	*Epidermophyton* sp. *Trichophyton* sp.	Direct or indirect contact	Lesions, floors, shoes and socks, shower stalls
Tinea unguium (ringworm of the nails)	*Epidermophyton* sp. *Trichophyton* sp.	Direct contact	Lesions
Tinea cruris (ringworm of the groin—jock itch)	*Epidermophyton* sp.	Direct or indirect contact	Lesions, athletic supports

FUNGAL DISEASES OF THE SKIN

Some fungi only infect the skin and its appendages, such as hair and nails (Table 23-2). Such fungi are called **dermatophytes.** These dermatophytic fungi cause infections of the skin called **dermatomycoses.** Contact of the skin with a spore of a dermatophyte can initiate an infection. Dermatophytic fungi then grow filamentously within the dead keratin-containing layers of the skin. Colonization of the skin by dermatophytic fungi initiates a cell-mediated immune response that generally occurs 10 to 35 days after infection. While preventing further lateral spread of the fungus, the cell-mediated immune response also causes inflammatory damage to the skin tissue.

Tinea (Ringworm)

Three genera of fungi are involved in fungal infections of the skin. Such infections are known as *cutaneous mycoses. Trichophyton* can infect hair, skin, or nails; *Microsporum* usually involves only the hair or skin; and *Epidermophyton* affects only the skin and nails. Dermatophytes grow on the keratin present in these locations, causing infections called **tineas** or **ringworms** (FIG. 23-5). The term *ringworm* is based on the development of a raised red welt in cases of tinea corporis (ringworm of the body) that led some to believe that an earthworm had crawled under the skin and coiled up in a ring. There are several diseases known as ringworm or tinea that are distin-

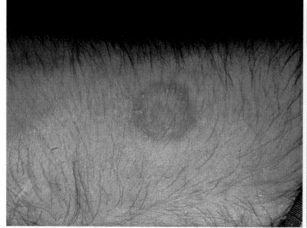
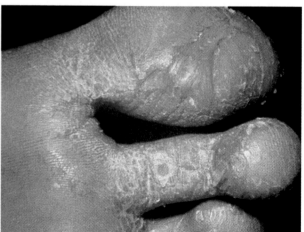

FIG. 23-5 **A,** Classic annular erythematous lesion due to *Microsporum* species showing an advancing active periphery and scaling in the central area in a case of tinea corporis. **B,** Tinea pedis (athlete's foot) showing scaling of skin patches in the characteristic location involving the toes, toe webs, and sole of the foot.

guished based on which regions of the body are infected. These diseases are normally localized and never fatal. Ringworm of the groin, or *jock itch,* is known as tinea cruris; ringworm of the feet, or *athlete's foot,* is known as tinea pedis; ringworm of the beard is called tinea barbae; and ringworm of the body is called tinea corporis. Tinea capitis, ringworm of the scalp, is fairly common among elementary school children, often resulting in bald patches. It is usually transmitted by contact with fomites. Dogs and cats are also frequently infected with fungi that cause ringworm in children.

> **Dermatophytic fungi grow on the keratin in skin and nails.**

Transmission of dermatophytic fungi is enhanced by conditions of high moisture and sweating. Retention of this moisture increases the probability of contracting superficial infections of the skin. The transmission of athlete's foot, for example, is often associated with the high moisture levels and bare feet of athletes in a locker room, although it is now known that this disease is not acquired unless the individual has skin abrasions through which the fungus can enter. Drying feet well and using antifungal agents, however, can reduce the spread of athlete's foot.

It is virtually impossible to protect all body areas against potential infection with superficial dermatophytic fungi. Therefore the incidence of dermatomycoses is high. Most cases of tinea can be treated by topical application of antifungal agents. The topical drug of choice for tinea infections is usually miconazole or clotrimazole. Other topical agents in nonprescription remedies are tolnaftate and zinc undecylenate. Whitefield ointment (a mixture of salicylic and benzoic acids) is often used to treat tinea. An oral antibiotic, griseofulvin, is often useful in these infections because it can localize in skin tissue, but it has side effects such as rashes.

Sporotrichosis

Sporotrichosis is a subcutaneous mycosis caused by the fungus *Sporothrix schenckii.* This infection occurs when this fungus is inoculated into the skin as a result of a minor injury. The disease is sometimes called gardener's disease because horticulturalists often get splinters or cuts that permit infection with *S. schenckii.* The fungal infection causes lesions, which normally occur on the extremities and spread to other parts of the body. The spread of the disease occurs slowly, permitting time for diagnosis and therapy. As a result, the death rate due to sporotrichosis is very low. The cutaneous form of sporotrichosis can be treated with a solution of iodine, but if the fungus is widely disseminated, administration of amphotericin B is necessary.

FUNGAL DISEASES OF THE ORAL CAVITY—THRUSH

Candida albicans is a yeast-like fungus. Candida can grow at various body sites, including the skin (moist regions are most susceptible), nails, vagina (discussed earlier in this chapter), and digestive tract, including the oral cavity (mouth). Growth of *C. albicans* in the mouth can cover the tissues, causing **thrush.** Thrush is characterized by a white frothy ("cheesy") covering of the tongue, filling the oral cavity (FIG. 23-6). This disease occurs commonly in immunosuppressed patients and in newborns. Transmission to newborns is common because *Candida* is a normal inhabitant of the vaginal tract and often grows excessively during pregnancy. Transfer to newborns during birth results in development of superficial infections. The lack of a developed immune system that would provide protective IgA and the fact that the normal microbiota (which compete for nutrients and limit the growth potential of *Candida*) has not yet been established contribute to the susceptibility of newborns to thrush. Nystatin and other antifungal drugs are effective in controlling this disease.

FIG. 23-6   Chronic oral thrush.

Algae are rarely considered as etiologic agents of disease. **Paralytic shellfish poisoning,** however, is caused by toxins produced by dinoflagellate algae, most often *Gonyaulax* (FIG. 23-7 p. 683). Blooms of *Gonyaulax* cause red tides in coastal marine environments, and during such algal blooms, the algae and the toxins they produce can be concentrated in bivalve shellfish, such as clams and oysters (FIG. 23-8). The ingestion of shellfish or fish containing toxins produced by *Gonyaulax* can lead to symptoms that resemble botulism. Shellfishing is banned in areas of *Gonyaulax* blooms to prevent this form of food poisoning. Shellfish taken from waters infected with red tide should not be eaten because they may contain toxins produced by the algae *Gonyaulax*. There is no treatment once the disease occurs.

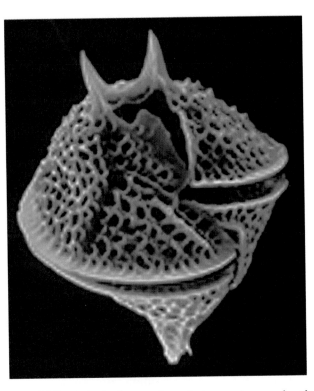

FIG. 23-7 Colorized scanning electron micrograph of *Gonyaulax*, the cause of red tide.

FIG. 23-8 Red tide occurs when there are large blooms of dinoflagellates.

DISEASES CAUSED BY PROTOZOA

Like other eukaryotic microorganisms, few species of protozoa cause human diseases (Table 23-3, p. 683). However, the most frequently occurring microbial infection of humans—malaria—is caused by protozoa. Protozoa, including those that cause malaria, exhibit complex life cycles, alternating between active growing forms and resting stages (cysts and spores). The resting stages often are responsible for the transmission of infections. Cysts can survive long enough in the environment so as to permit transmission. Spores do not last long outside of host organisms; vectors often are responsible for their transmission.

Some protozoa that cause human infections enter the body via contaminated food or water and establish infections of the gastrointestinal tract. Most of the protozoans that cause diseases affecting the cardiovascular, lymphatic, and central nervous systems are carried by vectors. The control of protozoan dis-

TABLE 23-3			
Examples of Medically Important Protozoa			
PHYLUM	**GROUP**	**PATHOGEN**	**DISEASE**
Sarcomastigophora	Amoebae	*Entamoeba histolytica*	Amebiasis, amebic dysentery
		Acanthamoeba spp., *Naegleria fowleri*	Amebic meningoencephalitis
Apicomplexa	Coccidia	*Cryptosporidium* spp.	Cryptosporidiosis
Ciliophora	Ciliates	*Balantidium coli*	Balantidiasis
Sarcomastigophora	Blood and tissue flagellates	*Leishmania tropica*	Cutaneous leishmaniasis
		L. braziliensis	Mucocutaneous leishmaniasis
		L. donovani	Kala-azar (visceral leishmaniasis)
		Trypanosoma cruzi	American trypanosomiasis
		T. brucei gambiense, T. brucei rhodesiense	African sleeping sickness
Sarcomastigophora	Flagellates of digestive tract and genitals	*Giardia lamblia*	Giardiasis
		Trichomonas vaginalis	Trichomoniasis
Apicomplexa	Sporozoa	*Plasmodium falciparum, P. malariae, P. ovale, P. vivax*	Malaria
		Pneumocystis carinii	*Pneumocystis* pneumonia
		Toxoplasma gondii	Toxoplasmosis

eases is generally by avoidance and elimination of these vectors. Mosquito control programs and prophylactic doses of chloroquinines, for example, are used to reduce the incidence of malaria.

PROTOZOAN DISEASES OF THE GASTROINTESTINAL TRACT

Amebic Dysentery

Amebic dysentery, or **amebiasis,** is caused by several protozoa, including the protozoan *Entamoeba histolytica.* This disease occurs worldwide; it is most frequent in tropical and subtropical countries, where the incidence may exceed 50%. The protozoan exhibits a life cycle that involves a trophozoite stage (active stage of the protozoan) that is motile by pseudopod formation (extension of the cytoplasm) and a nonmotile resistant cyst. Amebic dysentery occurs when the cysts of this protozoan contaminate water supplies (FIG. 23-9). This often is a result of inadequate sewage treatment and because the cysts of *E. histolytica* are not killed by the normal chlorination methods used to treat municipal drinking water. Cysts of *E. histolytica* can also be transmitted to the digestive tract through the oral cavity via contaminated fomites and hands.

Cysts of the protozoa *Entamoeba histolytica* are not killed by chlorination, contaminate water supplies, and cause amebic dysentery.

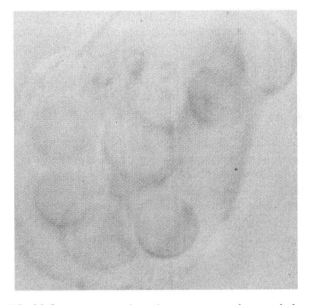

FIG. 23-9 Diagnosis of amebiasis is most often made by finding cysts of *Entamoeba histolytica* in stool.

Within the small intestine the cyst wall can break down so that the cyst opens (excysts) and develops into trophozoites (actively growing protozoa). The trophozoites move into the large intestine where they multiply and periodically form new cysts. Infections with *E. histolytica* may be asymptomatic or may involve mild or severe diarrhea and abdominal pain. The trophozoites may live harmlessly within the large intestine, feeding on bacteria, yeasts, and red blood cells, and reproducing by binary fission. Periodically cysts may be formed that pass out of the body in stools. Asymptomatic carriers can shed numerous cysts, which are the infective form involved in the transmission of disease. Cysts can survive in the environment for extended periods. The main source of cysts is contamination with feces from asymptomatic carriers. Infection occurs through the ingestion of contaminated water or food. Contamination may occur through inadequate sewage treatment; foods may be contaminated by food handlers with poor hygiene practices. The cysts pass through the stomach and exist within the small intestine. Each cyst gives rise to eight trophozoites.

In some cases the trophozoites invade the epithelial cells of the colon, resulting in the formation of ulcers or lesions. The trophozoites spread, cutting off the blood supply to the mucosal lining of the intestine, which leads to sloughing off of mucosal cells and enlargement of these lesions. Diarrhea results after the protozoa invade the intestinal mucosa. Because of the lesions in the mucosa, mucus and blood characteristically appear in the feces. Complications can include perforation of the intestine and invasion of other organs. In some cases *E. histolytica* can spread to the liver, lung, or skin, causing ulcer or abscess formation in these tissues.

Several antiprotozoan drugs, such as metronidazole, are effective in treating amebiasis. The choice of which one to use depends on whether the infection is restricted to the intestinal tract or whether other organs, such as the liver, are involved.

Giardiasis

Giardia lamblia is a flagellated protozoan that causes a gastrointestinal tract infection called **giardiasis** (FIG. 23-10). It was the first intestinal microorganism observed; it was described in the recorded observations of his own stool made by Antonie van Leeuwenhoek in 1681. It is the most common cause of epidemic waterborne diarrheal disease in the world. It is especially common among children in developing nations. In the United States almost 4% of the population appears to be infected by this organism. *G. lamblia* can be transmitted via contaminated water and direct oral-fecal contact. The disease can be endemic in nursery schools, where direct oral-fecal contact often occurs. The organism can also be transmitted during oral-anal sexual activities and giardiasis is common among homosexuals.

G. lamblia can live within the small intestine without causing any symptoms of giardiasis. Excessive growth of the organism, however, can cause disease symptoms that include diarrhea, dehydration, mucus secretion, and flatulence. Diarrhea occurs due to

an inflammatory response to damaged epithelial cells that interferes with the normal absorption of fluids within the intestine. The stools in cases of giardiasis are characteristically loose and foul smelling. Metronidazole or quinacrine hydrochloride are generally used in the treatment of this disease.

G. lamblia has two life-cycle stages: a motile flagellate trophozoite stage and a cyst stage. The cysts are the infective forms. Infection occurs when cysts are ingested, typically in contaminated water. Acid will stimulate excystation of the cysts so that cysts that pass through the stomach break open and give rise to trophozoites in the small intestine. The trophozoites have a ventral disc that acts like a suction cup, permitting them to adhere to the wall of the small intestine. The trophozoites multiply within the upper regions of the small intestine where they divide by binary fission. Sufficient numbers of trophozoites are produced to cover a large portion of the mucosal surface. Cysts are formed from trophozoites at regular intervals. A trophozoite rounds up to form a cyst with a thick resistant wall. Cysts pass out of the body in large numbers in stools.

The cysts can survive in the environment for long periods. This protozoan is relatively resistant to chlorination. It is difficult to detect in water supplies in time to protect the public. Therefore it is not surprising that giardiasis is one of the most common waterborne diseases in the United States. Filtration can be used to help eliminate *Giardia* from potable water supplies. Epidemics of giardiasis have occurred when public potable water supplies have been contaminated. Because waterways are used as receptors

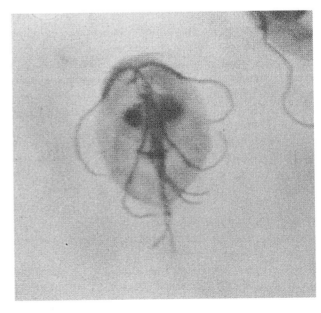

FIG. 23-10 Micrograph of *Giardia lamblia* trophozoite associated with an acute infection.

of sewage and sources of potable waters, such transmission is common. Transmission of *G. lamblia* also occurs from drinking water contaminated by wild animals, particularly beavers. Strains of *Giardia* that infect humans can also multiply within animals such as beavers so that drinking water from a brook is ill advised. Campers can prevent infections by boiling water to kill *Giardia*, including *Giardia* cysts.

Giardiasis is a waterborne disease of the gastrointestinal system caused by Giardia lamblia.

Cryptosporidiosis

Cryptosporidiosis is a disease caused by the protozoan *Cryptosporidium parvum* that is characterized by diarrhea. It is a common infection in individuals with AIDS. Even in immunocompetent individuals the diarrhea can be profuse and last for 3 weeks. In immunocompromised individuals, such as those with AIDS, the diarrhea can be even more prolonged and can be life threatening.

C. parvum is widely distributed in many animals. It has a complex life cycle with both sexual and asexual phases. In the small intestine the cysts give rise to nonmotile sporozoites that invade the epithelial cells. The sporozoites divide, forming many merozoites that further invade epithelial cells. Sexual reproduction produces cysts that are released in stools. Transmission is via ingestion of cysts, which are even more resistant to chlorination than the cysts of *G. lamblia*. Ingestion of food or water containing the cysts results in new cases of cryptosporidiosis. There have been several major outbreaks of cryptosporidiosis from municipal water supplies, for example in Milwaukee, where *C. parvum* has contaminated the water supply system. In this single outbreak, over 400,000 individuals were afflicted. Treatment is necessary only in severe cases and for immunocompromised individuals. The macrolide antibiotic spiramycin has been used with some success.

Cryptosporidiosis is an important waterborne emerging infectious disease.

Balantidiasis

Balantidium coli is the only ciliated protozoan that causes human infections. It causes the disease **balantidiasis,** or **balantidial dysentery.** This disease is relatively rare in the United States but occurs elsewhere in the world. *B. coli* is transmitted via food and water contaminated with feces containing cysts of *B. coli*. Swine may serve as reservoirs for the protozoan. The typical symptoms of balantidial dysentery are abdominal pain, nausea, vomiting, diarrhea, and loss of weight. In most cases the trophozoites of this protozoan reproduce within the large intestine, causing a mild form of the disease. In some cases, however, *B. coli* causes ulcerations of the colon that can be fatal. The disease can be treated effectively with metronidazole.

PROTOZOAN DISEASES OF THE GENITAL TRACT

Trichomonas Vulvovaginitis

T. vaginalis is a common sexually transmitted protozoan. Over 50% of women in some regions of the United States are infected with this protozoan. Infections with *T. vaginalis* often cause negligible symptoms. In women, 20% to 50% show no symptoms and in men up to 90% have no symptoms of infection. Therefore the disease often is untreated in males and promiscuous males can be a source of significant transmission. The incidence of *T. vaginalis* infections is higher in women than in men.

T. vaginalis usually does not grow well at acidic conditions but sometimes grows in the vagina, causing **vulvovaginitis.** It grows best when hormonal changes cause a rise in the vaginal pH. Excessive growth of *Trichomonas vaginalis* causes changes in the mucosal cells lining the vaginal tract (FIG. 23-11). Symptoms of vulvovaginitis, the infection of the

NEWSBREAK

CRYPTOSPORIDIUM CONTAMINATES WATER SUPPLIES

In the summer of 1993 the municipal water supply of Milwaukee was the source of a *Cryptosporidium* outbreak. Many residents of Milwaukee developed diarrhea before the cause of infection was identified. Pharmacies ran out of antidiarrheal drugs and hundreds of people reported to emergency room clinics for treatment. The city quickly issued a warning not to drink the water. *Cryptosporidium* is difficult to detect, generally requiring microscopic analysis of high volumes of water. The identification process usually takes a week, so that the water is consumed long before the pathogenic protozoan is detected. In the case of Milwaukee the source of infection probably originated in farm runoff carrying the protozoa from infected animals. Milwaukee is not the only city to experience severe outbreaks of diarrheal disease due to *Cryptosporidium,* and the Milwaukee case is not isolated. In 1987 13,000 people suddenly developed diarrhea during an outbreak of cryptosporidiosis in Georgia.

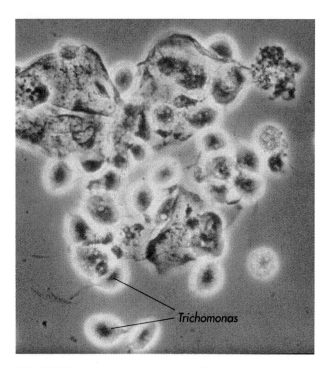

FIG. 23-11 Direct examination of a vaginal secretion shows pus cells interspersed with the pear-shaped nucleated and flagellated causal organisms *Trichomonas vaginalis*.

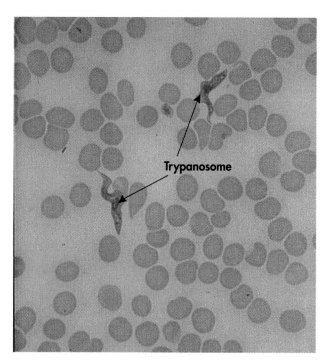

FIG. 23-12 In a case of African trypanosomiasis, trypanosomes can be seen in a thin blood smear.

vulva and vagina, include increased vaginal discharge and burning. In cases in which *T. vaginalis* is the causative agent, which is in about 20% of the cases, there is normally a profuse, greenish, odorous discharge. Treatment with metronidazole is effective in controlling vulvovaginitis caused by *Trichomonas*.

> **Vulvovaginitis, caused by *Trichomonas vaginalis*, is characterized by a heavy, greenish, odorous vaginal discharge.**

Protozoan Diseases of the Central Nervous System

African Sleeping Sickness

African trypanosomiasis is also known as **African sleeping sickness.** It is caused by infections with *Trypanosoma gambiense* and *T. rhodesiense*. These protozoa are transmitted to humans through the tsetse fly vector. Tsetse flies acquire *Trypanosoma* species from various vertebrate animals; cattle are reservoirs of *T. rhodesiense*; humans are the sole mammalian host for *T. gambiense*. In humans, infections with *T. gambiense* or *T. rhodesiense* are disseminated through the mononuclear phagocyte system, and there is evidence of localization within regional lymph nodes. Trypanosomes alter surface antigens as they grow within the body so that the immune system response is inade-

quate; by the time antibodies are made to one type of surface antigen, the protozoan has changed so that it has different surface antigen. Multiplication of the protozoa can cause damage to heart and nerve tissues. Progression through the central nervous system takes from months to years. If untreated, the initially mild symptoms, which include headaches, increase in severity and lead to fatal meningoencephalitis.

> **The tsetse fly is the vector for the spread of *Trypanosoma gambiense* and *T. rhodesiense*, which cause African sleeping sickness.**

Diagnosis is by clinical symptoms and by detection of trypanosome protozoa (trypanosomes) in blood smears, cerebrospinal fluid, or lymph node aspirate (FIG. 23-12). If the disease is diagnosed before there is central nervous system involvement, it can be successfully treated with antiprotozoan agents, such as suramin. If there is central nervous system involvement, melarsoprol, an arsenical, is used for treating the disease. Prevention of African trypanosomiasis involves controlling population levels of the tsetse fly, accomplished by clearing vegetation to destroy the natural habitats of the tsetse fly. Because trypanosomes often express new surface antigens during infections that outpace the immune response, it is unlikely that effective vaccines will be developed to prevent this disease.

PROTOZOAN DISEASES OF THE CARDIOVASCULAR AND LYMPHATIC SYSTEMS

Malaria

On a worldwide basis **malaria** is one of the most common human infectious diseases. Descriptions of this disease (fever, periodic shivering, and enlargement of the spleen) were recorded in Egypt as early as the sixteenth century B.C. The annual incidence today is about 150,000,000 cases, making it perhaps the most common infectious disease (FIG. 23-13). Malaria has been largely eliminated from North America and Europe but remains the most serious infectious disease in tropical and subtropical regions of the world. About one third of the world's population is estimated to be infected with malaria-causing protozoa (*Plasmodium* species).

Malaria is caused by four species of the protozoan *Plasmodium* (Table 23-4, p. 688). *P. vivax* and *P. falciparum* are most frequently involved in human infections. The female *Anopheles* mosquito is the vector responsible for transmitting malaria to human beings. The *Plasmodium* species causing malaria have the most complex life cycle of any microorganism causing human infections (FIG. 23-14). Sexual reproduction of the *Plasmodium* within the mosquito permits genetic recombination. Sporozoites (resting cells) de-

velop within the mosquito vector; it is essential that the mosquito live long enough for this to happen or transmission of the disease does not occur.

After inoculation into the body, which occurs by mosquito bites, the sporozoites of *Plasmodium* begin to reproduce within liver cells. Multiplication of the *Plasmodium* sporozoites occurs by **schizogony** (a form of asexual division in which many new protozoa are produced through multiple fission) by which a single sporozoite can produce as many as 40,000 **merozoites** (feeding stage in the life cycle of this organism). The merozoites attach to receptors on the surfaces of red blood cells. The lack of one red blood cell surface receptor in most West Africans explains their resistance to *P. vivax*. Sickle cell anemia also contributes to resistance to malaria because of the lack of suitable red blood cells in which the merozoites can reproduce; merozoites normally invade red blood cells during the erythrocytic phase of the life cycle of *Plasmodium*.

Invasion of these erythrocytes by merozoites begins the erythrocytic phase of malaria. This results in the development of trophozoites (ring stages), which become segmentors, resulting in merozoite formation. *Plasmodium* merozoites reproduce within red blood cells, killing large numbers of red blood cells

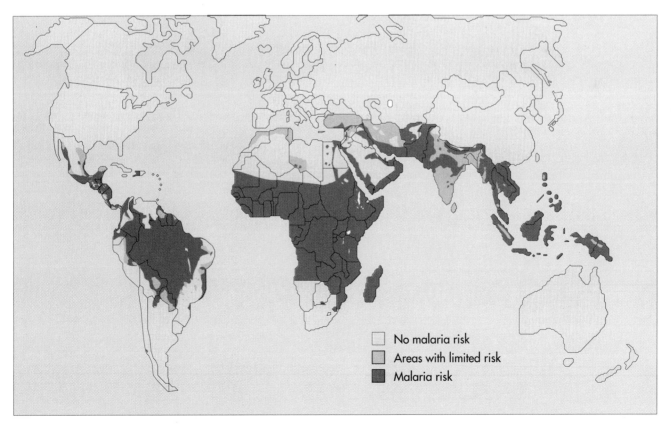

FIG. 23-13 Malaria is one of the most prevalent infectious diseases in the world. It occurs primarily in tropical and subtropical regions.

TABLE 23-4

Summary of Important Characteristics of Human Malarias

ETIOLOGIC AGENT	P. falciparum	P. vivax	P. ovale	P. malariae
Incidence	Common	Common	Uncommon	Uncommon
Primary hepatic schizogony	1-40,000 in 5.5-7 days	1-10,000 in 6-8 days	1-15,000 in 9 days	1-2000 in 13-16 days
Secondary hepatic schizogony	1-8 to 24 (av.16) in 48 hr	1-12 to 24 (av.16) in 48 hr	1-6 to 16 (av. 8) in 48 hr	1-6 to 12 (av. 8) in 72 hr
Incubation period	8-27 days (average 12)	8-27 days (average 14) (rarely months)	9-17 days (average 15)	15-30 days
Duration of un-treated infection	0.5-2.0 yr	1.5-4.0 yr	1.5-4.0 yr	1-30 yr
Erythrocytic cycle	48 hr	48 hr	48 hr	72 hr
Mortality	High in nonimmunes	Uncommon	Rarely fatal	Rarely fatal

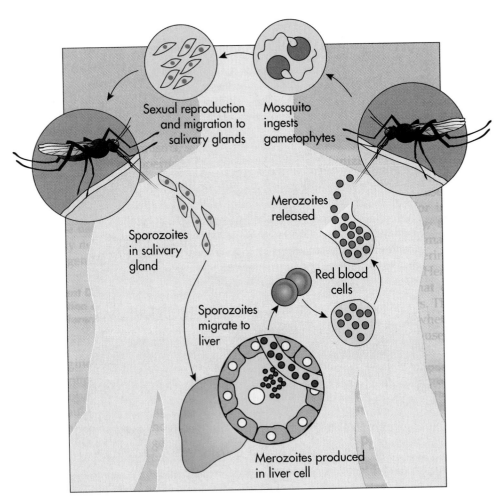

FIG. 23-14 *Plasmodium* species that cause malaria have complex life cycles, portions of which are carried out in the mosquito vector. Within the human body, different stages of the malaria-causing protozoa multiply in blood cells and the liver.

and causing anemia. Multiplication of merozoites that results in killing of red blood cells is responsible for the fever and chills that symptomatically occur in cases of malaria.

Plasmodium **merozoites reproduce within red blood cells, killing them and causing anemia.**

Symptoms of malaria begin approximately 2 weeks after the infection established by the mosquito bite. Symptoms include chills, fever, headache, and muscle ache. Malaria symptoms are periodic, with symptomatic periods generally lasting less than 6 hours. Schizogony occurs every 48 hours for *P. vivax* and *P. ovale*, and every 72 hours for *P. malariae*, resulting in a synchronous rupture of infected erythrocytes. The symptoms of the disease are caused by the asexual erythrocytic cycle, but the periodic onset of disease symptoms coincides with the rupture of infected erythrocytes.

Symptoms of malaria are recurring and coincide with the rupture of infected erythrocytes every 48 to 72 hours.

Malarial infections persist for long periods. Malaria is rarely fatal, except when the disease is caused by *P. falciparum*. The most severe complications occur when there is cerebral involvement, which is characterized by progressive headache, neck stiffness, convulsions, and coma. There is no vaccine, but the disease can be prevented by drug prophylaxis. Individuals traveling to areas with high rates of malaria, such as southeast Asia and Africa, often use antimalarial drugs, such as chloroquine, to avoid contracting this disease. Quinine, the active ingredient of cinchona bark, has been known to be an effective treatment for over 400 years and remains the drug of choice in treating malaria. However, many strains of *Plasmodium* are relatively resistant to antimalarial drugs, so this course of action is not always effective. Also, quinine treatment can have complications, including massive intravascular hemolysis known as blackwater fever. The use of insect netting and other measures to prevent being bitten by an infected mosquito are extremely important in avoiding contracting malaria. In the United States control measures have been effective, but periodic increases in morbidity have occurred after overseas military ventures.

Babesiosis

Babesiosis is a protozoan disease that occurs on the offshore islands near Cape Cod, Massachusetts. It is caused by *Babesia microti*, a protozoan that normally lives within the red blood cells of rodents. *Babesia* can be transmitted to humans by tick vectors. When this occurs, the infection causes a relatively mild, self-limiting febrile disease. The parasites in the red blood cell take on several forms, the most common of which are two to four small, pear-shaped forms. Individuals who have had their spleens removed and elderly individuals who are infected may develop more severe symptoms that resemble malaria. These symptoms include fever, sweating, joint and muscle pain, nausea, and vomiting. Chloroquinine has been used to treat babesiosis.

Leishmaniasis

Leishmaniasis is caused by infections with members of the protozoan genus *Leishmania* (Table 23-5). It is transmitted to human beings by sand fly vectors (FIG. 23-15). *Leishmania* species reproduce in humans and other animals intracellularly as a nonmotile form, the *amastigote*. Leishmaniasis is geographically restricted to regions where sand flies can reproduce and acquire *Leishmania* species from infected canines and rodents. The protozoa reproduce within the sand

TABLE 23-5			
Epidemiology of Leishmaniasis			
ETIOLOGIC AGENT	**DISEASE**	**GEOGRAPHIC DISTRIBUTION**	**SITE OF LESIONS**
Leishmania donovani	Kala-azar, visceral leishmaniasis	Mediterranean, southern Europe, central Asia, China, India, Sudan	Macrophages of the deep viscera (liver, spleen, bone marrow)
Leishmania tropica	Oriental sore, Old World cutaneous leishmaniasis	Mediterranean basin, central Asia, India	Macrophages surrounding skin lesions
Leishmania mexicana	New World and Old World cutaneous leishmaniasis	Mexico, Guatemala, Central America, Amazonia	Macrophages surrounding skin lesions
Leishmania braziliensis	New World cutaneous leishmaniasis	Central America, South America	Macrophages of the nose and pharynx

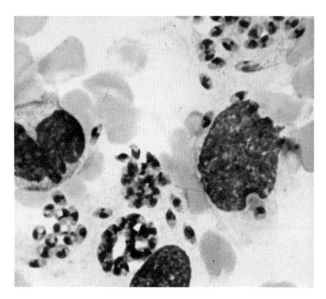

FIG. 23-15 Leishmanial organisms within macrophages in aspirate from lesion of Old World leishmaniasis.

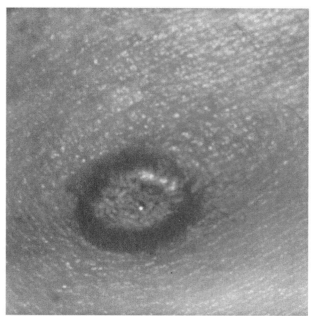

A

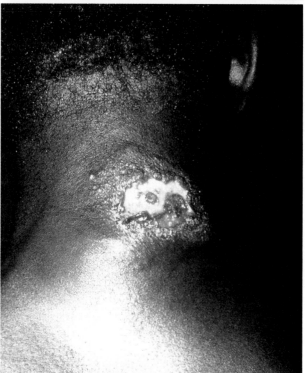

B

flies and are present in the saliva. In sand flies the protozoa exists in a flagellated form, the promastigote. During the Gulf War (Operation Desert Storm), many United States soldiers were bitten by sand flies and infected with *Leishmania.* Many developed leishmaniasis after they returned home.

Four species of *Leishmania* cause leishmaniasis in humans. *L. mexicana* and *L. tropica* cause infections limited to the skin, sometimes referred to as Old World cutaneous leishmaniasis, or oriental sore (FIG. 23-16). *L. braziliensis* causes infections of skin. The most serious form of leishmaniasis is caused by *L. donovani,* which multiplies throughout the mononuclear phagocyte system and causes **kala-azar disease.**

The lesions caused by *Leishmania* infections may be minor or extensive. The formation of extensive ulcers can produce permanent scars. While cutaneous leishmaniasis syndromes are self-limiting, the kala-azar syndrome can be fatal. Untreated cases of kala-azar disease normally last several years and in the terminal stages can involve liver and heart damage and hemorrhages. Cutaneous forms of leishmaniasis can be treated with amphotericin B. Kala-azar syndrome can be treated with sodium stibogluconate, an antimony-containing drug, or with amphotericin B.

Prevention of leishmaniasis involves controlling vector and reservoir populations. Insecticide impregnated bed nets help prevent transmission of this disease. The use of DDT has been effective in some regions in eliminating the sand fly vector. In other cases infected rodent populations have been controlled, greatly reducing the incidence of leishmaniasis. Infections result in long-term immunity, an unusual occurrence for protozoan diseases. In endemic areas

FIG. 23-16 A, Early stage of *Leishmania tropica* infection shows local lesion with erythema at site of sandfly bite. **B,** Cutaneous ulcer on neck caused by *L. braziliensis.*

children are often inoculated with pustular material from mild cases of the disease. This immunization procedure resembles practices in the Far East employed at the time of Jenner to prevent smallpox. There is a strong likelihood that a vaccine will be developed to prevent leishmaniasis.

Sand flies are the vector for *Leishmania* species and can be eliminated by DDT.

Toxoplasmosis

Toxoplasmosis is caused by the protozoan *Toxoplasma gondii,* a member of the sporozoa. Meat containing cysts of *T. gondii* is frequently involved in the transmission of this disease, although *T. gondii* can also be transmitted congenitally (FIG. 23-17). Cats are the definitive hosts for *T. gondii* and are involved in most cases of toxoplasmosis. In a definitive host, the protozoan carries out a complete life cycle, which for *T. gondii* involves asexual and sexual reproduction. The asexual reproduction involves motile trophozoites and nonmotile cysts. Cysts are formed from the trophozoites. The cysts join by sexual reproduction to produce oocysts. This occurs in the epithelial cells of the small intestine. Cysts and oocysts pass out of the body in feces and contaminate soils. Grazing animals, birds, and rodents acquire the organism

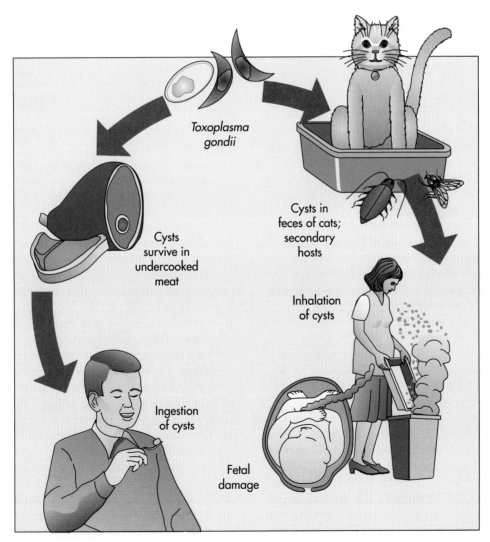

Cysts survive in undercooked meat

Cysts in feces of cats; secondary hosts

Inhalation of cysts

Ingestion of cysts

Fetal damage

Toxoplasma gondii

FIG. 23-17 Transmission of *Toxoplasma* often involves cats, which acquire the protozoa when grazing on contaminated soil. The disease is most likely serious when contracted by a pregnant woman and passed to her fetus.

N E W S B R E A K

DOMESTIC PETS AS SOURCES OF DISEASE

Pet dogs and cats can be reservoirs of infection for their owners. Cat feces are involved in the transmission of toxoplasmosis (a protozoan disease). Feces from dogs are involved in the transmission of toxocariasis (a disease caused by the worm *Toxocara*). Other diseases are transmitted by direct contact such as ringworm caused by dermatophytic fungi. Tick vectors carry *Rickettsia rickettsii* from dogs, causing Rocky Mountain spotted fever. Exotic pets also transmit disease. Pet turtles and chicks are sources of *Salmonella* infections. Birds transmit *Chlamydia psittaci*, which causes parrot fever. Exotic birds may also be reservoirs for viruses that cause human encephalitis.

from the soil by ingestion. Cats acquire the protozoa by ingesting infected rodents or birds.

Cats are hosts for the protozoan Toxoplasma gondii, the causative organism of toxoplasmosis.

The main route of transmission of *T. gondii* to humans is by the fecal-oral route from contact with contaminated hands and fomites, or by ingestion of infected meat that has not been cooked to a temperature high enough to kill the protozoa. The domestication of cats has led to a great increase in human infections with *T. gondii*. Because oocysts are passed in the feces of infected cats, pregnant women are advised not to handle cats or clean cat litter boxes and not to eat undercooked meat because of possible contraction of toxoplasmosis; *T. gondii* can be transmitted congenitally to the fetus.

Humans often are infected from eating undercooked contaminated meat or by direct contact with feces containing Toxoplasma gondii.

The oocysts of *T. gondii* give rise to trophozoites within the intestines of infected individuals. There the trophozoites of *T. gondii* multiply and kill infected cells. They then are disseminated from the gastrointestinal tract via the bloodstream to other organs and tissues. Cell-mediated immunity is involved in containing infections of *T. gondii*. The symptoms and prognosis of toxoplasmosis depend on the virulence of the infecting strain of *T. gondii* and on the immune state of the infected individual. In most cases toxo-

plasmosis is asymptomatic. When symptoms occur, muscle pain and fever are characteristic. Recent immunological surveys suggest that 50% of adults in the United States have been infected. When multiple organs are involved in the infection, the consequences are serious and can be fatal. This is most likely to occur in immunocompromised individuals. Some infections with *T. gondii* have been successfully controlled with antimalarial drugs, such as pyrimethamine. Serious consequences of toxoplasmosis occur when there is central nervous system involvement, which is particularly prevalent in the congenital transmission of this disease.

Trypanosomiasis

American trypanosomiasis, or **Chagas disease,** occurs in Latin America. It is caused by *Trypanosoma cruzi*, which is usually transmitted to humans by infected cone-nosed bugs, commonly referred to as the kissing bug. *T. cruzi* is a flagellate protozoan, but in vertebrate hosts it forms a nonflagellate form, the amastigote. Dogs and cats are reservoirs of *T. cruzi*. The vectors of Chagas disease normally live in the mud and wood houses of South America, and construction of better housing eliminates the habitat for vector populations that brings them into close contact with humans. Cone-nosed bugs feed at night, biting sleeping people on the lips, face, and forearms.

American trypanosomiasis is caused by the protozoan Trypanosoma cruzi, which is transmitted to humans by cone-nosed bugs.

When *T. cruzi* infects human hosts, the protozoa initially multiply within the mononuclear phagocyte system. Later, the myocardium, which is the muscular tissue of the heart, and nervous systems are also invaded. Organisms localize in the heart and interfere with the ability of the myocardial fibers to contract efficiently, sometimes causing them to atrophy altogether. Damage to heart tissue occurs as a result of this infection. In 90% of the cases there is spontaneous remission, but 10% of the patients hospitalized with this disease die during the acute phase because of heart failure. Death due to heart disease resulting from Chagas disease may also occur well after recovery from the acute phase. Chagas disease is the leading cause of cardiovascular death in South America, and the incidence of this disease in Brazil is extraordinarily high. Several antiprotozoan drugs, such as aminoquinoline and nifurtimox, are effective in treating Chagas disease if the symptoms are recognized early, but once the progressive stages have begun, treatment is supportive rather than aimed at eliminating the infecting agent.

PROTOZOAN DISEASES OF THE CENTRAL NERVOUS SYSTEM

Naegleria Microencephalitis

The protozoans *Naegleria fowleri* and *N. gruberi* are known to cause the neurological disease ***Naegleria* microencephalitis.** This protozoan normally reproduces in stagnant freshwater bodies in warm climates. They are common in sediments at bottoms of ponds and in swimming pools. Only a few cases per year are reported in the United States. Children who swim in ponds or streams are the usual victims. The organism initially infects the nasal mucosa. The protozoa migrate along the olfactory tracts and reach the meninges of the central nervous system. They can proliferate in the brain and spinal column, causing meningitis. The fatality rate is nearly 100%.

DISEASES CAUSED BY MULTICELLULAR PARASITES

Clinical microbiologists sometimes locate evidence for diseases caused by multicellular parasites in samples sent to the laboratory for examination. Eggs and cysts of multicellular parasites can be confused with protozoa and other microorganisms. Therefore it is important for the microbiologist to have some knowledge of diseases caused by intestinal worms and other multicellular parasites. The multicellular parasites that cause human infestations usually are Platyhelminthes (flatworms) or Aschelminthes (roundworms). These worms acquire required nutrition by absorption from the host's tissues and body fluids. The sexual reproduction of these animals involves production and fertilization of eggs. The eggs frequently are involved in the transmission of disease to humans.

PARASITIC DISEASES OF THE GASTROINTESTINAL TRACT

Tapeworms

Tapeworms, or **cestodes,** are flatworms (platyhelminths) that infect the intestinal tracts of animals, including humans. These tapeworms can cause human infestations (Table 23-6). Each tapeworm has a host in which the larval stages can develop into adult worms. For example, the larval stages of the beef tapeworm *(Taenia saginata),* the pork tapeworm *(Taenia solium),* and the fish tapeworm *(Diphyllobothrium latum)* develop into adult forms within the human intestinal tract. Eating raw or inadequately cooked beef containing the larval stage of *T. saginata* is responsible for transmission of beef tapeworm to humans.

TABLE 23-6

Human Tapeworm Infections			
SPECIES	**ACQUISITION**	**ANIMAL HOSTS**	**SITE OF INFESTATION**
Diphyllobothrium latum	Larvae in fish	Fish-eating mammals	Intestine
*Dipylidium caninum**	Larvae in fleas	Dogs, cats	Intestine
Echinococcus granulosus (hydatid disease)	Eggs passed by dogs	Sheep	Liver, lung, brain
*Echinococcus multilocularis**	Eggs passed by carnivores	Rodents	Liver
*Hymenolepis diminuta**	Larvae in insects	Rats, mice	Intestine
Hymenolepis nana	Eggs, or larvae in beetles	Rodents	Intestine
Pseudophyllid tapeworms* (sparganosis)	Larvae in other hosts	Many vertebrates	Subcutaneous tissues, eyes
Taenia multiceps	Eggs passed by dogs	Sheep	Brain, eye, subcutaneous tissue
Taenia saginata	Larvae in beef	Cattle	Intestine
Taenia solium (cysticercosis)	Larvae in pork, eggs in food or water contaminated with human feces	Pigs	Intestine, brain, eyes

* Rare infection.

Similarly, eating inadequately cooked pork containing the larval forms of *T. solium* is the source of transmission of pork tapeworm to humans. Eating raw and pickled fish containing the larval forms of *D. latum* transmits the fish tapeworm to humans.

> **Ingestion of raw or inadequately cooked beef, pork, or raw fish containing larval forms of *Taenia saginata*, *Taenia solium*, or *Diphyllobothrium latum* can transmit beef, pork, and fish tapeworms, respectively, to humans.**

The scolex or head of the tapeworm emerges from the larval form within the human intestinal tract. It attaches to the lining of the intestine. The scolex has several suckers and hooks, in most cases, that allow it to attach to the mucosal lining of the small intestine. The tapeworm derives its nutrition from its human host. It matures and grows to a length of several meters. The name tapeworm is based on the resemblance of these parasites to a tape measure. Tapeworms are segmented and the posterior segments (proglottids) become filled with eggs. These egg-bearing segments break off and are shed with human feces.

Contaminated fecal matter is also the source of transmission of these tapeworms to their intermediate hosts. The eggs develop into larval stages in the intermediate host. Once inside the intermediate host, the eggs hatch and develop into larvae. The larvae become disseminated throughout the body of the intermediate host. The larvae of the beef tapeworm can remain viable in the muscle system of the cow for only about 9 months. The larvae of the pork tapeworm, however, can remain viable for years. The passage between humans and the intermediate hosts for the worms perpetuates the life cycles of these tapeworms.

Surprisingly, most infected individuals show few if any symptoms. When symptoms occur they generally include loss of weight, anemia, and abdominal pain. Niclosamide is used to treat intestinal infestations with these tapeworms. Observation of proglottids and eggs in human fecal matter is diagnostic of tapeworm infestations.

The development of tapeworms within humans is not always restricted to the intestines. For example, ingestion of the larvae of the pork tapeworm results in intestinal infestation, but ingestion of the eggs of this tapeworm leads to an infection in which larvae can develop throughout the body. The ingestion of the larvae leads to the adult tapeworm in the intestine. The eggs are released from the gravid segments of the adult and hatch in the host's intestine, becoming larvae that can migrate throughout the body. This condition is called **cysticercosis.** The liver, muscles, heart, and brain are commonly the sites of larval de-

velopment. In cases where the larvae develop within the heart or central nervous system, the infestation is usually fatal. There is no satisfactory treatment for cysticercosis.

> **Cysticercosis occurs when pork tapeworm eggs or larvae develop in parts of the human body other than the intestines, usually the liver, muscles, heart, and brain.**

The tapeworm *Echinococcus granulosus* (canine tapeworm) also causes a human infestation that is not restricted to the intestines. The larval stages of this tapeworm form within a fluid-filled bladder called a **hydatid cyst.** This tapeworm develops in several animals, including dogs.

In dogs, *E. granulosus* grows within the small intestine with no apparent ill effects. Humans become accidental intermediate hosts for this tapeworm by close association with dogs. In humans, the cysts most frequently develop in the liver because that is where most ingested eggs become trapped. Bone, brain, lungs, and kidneys are other sites where the hydatid cysts (cysts with watery centers) may develop. The development of the hydatid cyst causes an inflammatory reaction that isolates the cyst within a walled-off region. Development of the cyst can cause liver impairment. If the cyst ruptures, the release of fluids from the cyst can cause a hypersensitivity reaction and death from anaphylactic shock. Diagnosis of **hydatidosis** before rupture of the cysts is important. The cysts are normally detected by X-rays. Surgery then is performed to remove the intact, walled-off, hydatid cysts.

> **Hydatid cysts can develop in the liver, bone, brain, lungs, and kidneys and can cause a hypersensitivity reaction if they rupture.**

Flukes

Flukes, or **trematodes,** which are members of the class trematoda in the phylum Platyhelminthes, are much smaller than the tapeworms. The largest human parasite trematode is 8 centimeters in length. Humans are alternate hosts for these parasitic worms, most of which have complex life cycles that involve alternation of hosts. Trematodes can cause several human diseases and these diseases can be widespread (Table 23-7, p. 695). For example, in some villages in China, there is a 100% infection rate of the local population with the intestinal fluke *Fasciolopsis buski.* This high infection rate occurs because night soil (human feces) is used as fertilizer and because sewage contaminated by night soil with flukes drains directly into adjacent water supplies. Ingestion of contaminated plants leads to dissemination of this

	TABLE 23-7	
	Human Fluke Infections	
SPECIES	**ACQUISITION**	**SITE OF INFESTATION**
Schistosoma haematobium	Penetration of skin by larval stages released from snails	Blood vessels of bladder
Schistosoma japonicum	Penetration of skin by larval stages released from snails	Blood vessels of intestine
Schistosoma mansoni	Penetration of skin by larval stages released from snails	Blood vessels of intestine
Clonorchis sinensis	Ingesting fish infected with larval stages	Liver
Fasciola hepatica	Ingesting vegetation (watercress) with larval stages	Liver
Paragonimus westermani	Ingesting crabs infected with larval stages	Lungs

parasite. In humans, *F. buski* matures within the small intestine. Once the infection progresses, diarrhea and abdominal pain occur. The loss of body fluids can be fatal, particularly in children. This parasitic disease can be treated with hexylresorcinol.

When night soil is used as fertilizer, contamination of water supplies with intestinal flukes can cause gastrointestinal infestations.

The liver fluke *(Fasciola hepatica)* is normally associated with sheep. It can also infest the human liver. Human infections with *F. hepatica* most commonly occur in Latin America and Mediterranean countries. Eggs from the fluke can be deposited in bodies of water. Snails in these water bodies can serve as intermediate hosts; only a portion of the life cycle can be accomplished in an intermediate host. *Fasciola* will pass through several life stages in snails, eventually being deposited as encysted metacercariae (a resting stage of trematodes) on vegetation. Liver flukes normally enter the body by ingestion of metacercariae on contaminated vegetation, such as watercress. The flukes then migrate from the intestine to the liver. These flukes develop within the liver, producing digestive disturbances. Fever, pain, and liver malfunction (hepatitis) result as the infestation progresses. Extensive liver damage eventually produces cirrhosis, which can be fatal. Identification of eggs in feces is important to the proper diagnosis of this disease. Several drugs, such as dehydroemetine, are used in the treatment of this disease.

Pinworms

Several roundworms (nematodes) cause human diseases (Table 23-8, p. 696). The **pinworm**, *Enterobius vermicularis,* is responsible for one of the more common worm infections of humans. The adult worms live in the large intestine, where they attach to the mucosal lining, but gravid (pregnant) female worms migrate toward the anus and emerge at night. Large numbers of worms can be seen and many eggs are extruded around the anus each night. The eggs hatch and within 6 hours the larvae are infective for the same or other individuals. Larvae may reenter the digestive tract through the anus.

Contamination of the hands with infective eggs also leads to introduction of these worms through the mouth. Hand-to-mouth contact and ingestion of contaminated food and drink are responsible for the maintenance of a high rate of infection. About one third of pinworm-infected persons suffer no symptoms. When symptoms occur, they normally include nausea, vomiting, and diarrhea. Poor sleep resulting from the irritation associated with the nocturnal migration of pregnant females often leads to nervous restlessness in children infected with this worm. Pyrantel pamoate (Antiminth) is the drug of choice to treat cases of pinworm. Good hygiene practices are also needed to prevent reinfection. Often it is necessary to treat an entire family to eliminate pinworm infestations.

Adult pinworms attach to the mucosal lining of the cecum and colon, usually producing no symptoms.

Whipworm

Whipworm is another disease caused by a nematode. In the United States this disease is prevalent in the south; it also occurs elsewhere in the world, particularly in warm climates. The causative agent, *Trichuris trichiura,* is shaped like a whip (FIG. 23-18). Adult worms live primarily within the large intestine. Hand-to-mouth contact and ingestion of food contaminated with fecal matter containing eggs is important in the transmission of this disease. Light infections often are asymptomatic. Symptoms may resemble appendicitis and result in emaciation. Bleeding may result in anemia. The broad spectrum

TABLE 23-8
Human Nematode Infections

SPECIES	TRANSMISSION	ACQUISITION	SITE OF INFESTATION
Ancylostoma duodenale (hookworm)	Person to person	Skin penetration	Small intestine
Angiostrongylus cantonensis	Animals	Ingestion of larvae in snails, crustacea	CNS (larvae)
Anisakis simplex	Animals	Ingestion of larvae in fish	Stomach, small intestine (larvae)
Ascaris lumbricoides	Person to person	Ingestion of eggs	Small Intestine
Brugia malayi	Arthropod vector	Bite of mosquito carrying infective larvae	Lymphatics (adults), blood (larvae)
Capillaria phillipinensis	Animals	Ingestion of larvae in fish	Small intestine (adults, larvae)
Enterobius vermicularis	Person to person	Ingestion of eggs	Large intestine
Loa loa	Arthropod vector	Bite of deer fly carrying infective larvae	Tissues
Necator americanus (hookworm)	Person to person	Infective larvae	Small intestine
Onchocerca volvulus	Arthropod vector	Bite of *Simulium* fly carrying infective larvae	Skin (larvae, adults), eye (larvae)
Strongyloides stercoralis	Person to person	Skin penetration by infective larvae; autoinfection	Small intestine (adults), general tissues (larvae)
Toxocara canis	Animals	Ingestion of eggs passed by dogs	Tissues, CNS (larvae)
Trichinella spiralis	Animals	Ingestion of larvae in pork, wild mammals	Small intestine (adults), muscles (larvae)
Trichuris trichiura	Person to person	Ingestion of eggs	Large intestine
Wuchereria bancrofti	Arthropod vector	Bite of mosquito carrying infective larvae	Lymphatics (adults), blood (larvae)

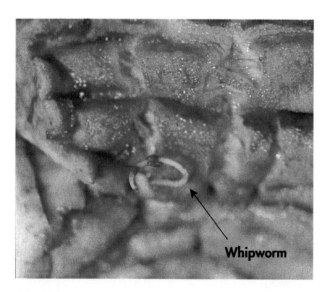

FIG. 23-18   A whipworm with its head buried in the ileal mucosa.

antihelminthic mebendazole (vermox) is considered the drug of choice for treating this disease.

Trichinosis

Trichinosis is another human disease caused by roundworms. Human infestations with the nematode *Trichinella spiralis*, the causative agent of trichinosis, generally are associated with the ingestion of undercooked pork containing larvae (FIG. 23-19). Occasionally, outbreaks of trichinosis are also associated with eating raw bear and walrus meat. Adequate cooking or freezing meat long enough to kill these larvae will prevent trichinosis. The infective larval stage of *T. spiralis* is found in an encysted form in the muscle tissues of infected pigs. When ingested, the larvae within the cyst are released and become attached to the lining of the small intestine. The larvae become adults in the intestine. The adults produce living young (larvae) that migrate into the bood-

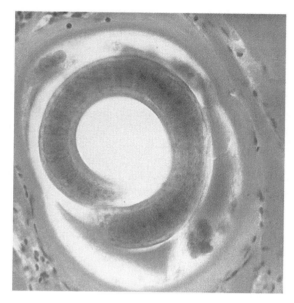

FIG. 23-19 A coiled larva of *Taenia spiralis* in fully formed cyst within striated muscle of tongue, showing hyaline capsule of the cyst.

FIG. 23-20 Large adult worms of *Ascaris lumbricoides*.

*Large masses of **Ascaris lumbricoides** can obstruct the gastrointestinal tract and must be surgically removed.*

stream and, from there, into muscles, heart, and brain. The worms reproduce within the small intestine, and larvae spread to other parts of the body.

Trichinella spiralis, the causative organism of trichinosis, is associated with undercooked pork.

Symptoms of this disease vary from one person to another. Early symptoms usually involve the digestive tract, for example, diarrhea, but as the worms are disseminated, other symptoms such as muscle pain and difficulty in breathing become prevalent. Various manifestations of trichinosis can be fatal. Cortisone is useful during the larval encystment period, and mebendazole is effective against adult worms.

Roundworm Disease

Yet another roundworm disease occurs when *Ascaris lumbricoides* infests the intestines. This infestation, which is called **roundworm disease of humans,** or ascariasis, is prevalent throughout the tropics and occurs in the southeastern United States. The eggs of this parasite can remain viable in the moist soils of these regions. Hundreds of millions of individuals are infested with this nematode throughout the world. *A. lumbricoides* is a very large nematode that can measure up to 30 centimeters (1 foot) (FIG. 23-20). Large masses of worms in the intestines can obstruct the gastrointestinal tract. Surgical removal may be necessary if the infested individual becomes too emaciated. Because it migrates, it can occlude the gallbladder. It can leave through the mouth and nose. It is known to occlude airways during surgery, resulting in death.

Transmission of this roundworm disease is often associated with fecal contamination of food and water that contains eggs. In regions where human feces (night soil) are used as fertilizer, the worms may be transmitted to crops. Ingestion of contaminated vegetables, drinking contaminated water, or oral contact with contaminated fingers are all potential routes of transmission of this parasite. Children playing or working in fields fertilized with night soil (human manure) often develop this disease. Once ingested, the worms infest the intestines. Treatment of intestinal infestations often involves use of piperazine citrate (syrup of Antepar), which paralyzes the worm. The live worm is then excreted in feces. In some cases the worm moves from the intestine to other parts of the body, for example, the lungs may also be a site of infestation. Hemorrhages occur as the worms enter the alveoli. The signs of lung infestation include dry cough, fever, and eosinophilia (increase in eosinophils, which accumulate at the site of infestation).

Hookworm

Hookworm infections are caused by the nematodes, *Necator americanus* and *Ancylostoma duodenale.* They are most common in the tropics. Transmission most frequently occurs by walking barefoot on soil contaminated with these worms. Within the body the hookworms move in a characteristic path to reach the intestines. Having burrowed through the skin, the hookworms enter the bloodstream. They move through the circulatory system to the lungs. Coughing moves the hookworms into the pharynx and swallowing brings them to the intestines. They live in the small intestine, attaching themselves firmly to the intestinal lining. Eggs are passed with the feces, and larvae develop in moist soil. The larvae sit on soil particles, surviving for up to several weeks. If they come in contact with a warmblooded animal, they penetrate the skin and are carried through the circulatory system to the lungs. They are too large to pass through the capillaries of the lungs and therefore break out of the lung. They are carried to the pharynx and are swallowed. They finally settle in the lining of the small intestine.

> **Larvae hookworms burrow through the skin, usually in the feet, and migrate through the pulmonary system and settle in the lining of the small intestine to become adults.**

Symptoms of hookworm infestations often resemble those of a duodenal ulcer. A major pathological condition of hookworm infections is anemia, caused by the worms sucking blood while attached to the intestinal mucosa. Treatment with iron will correct the anemia. Pyrantel pamoate (Antiminth) is the drug of choice for eliminating hookworm infection.

PARASITIC DISEASES OF THE CARDIOVASCULAR AND LYMPHATIC SYSTEMS

Elephantiasis

Elephantiasis is a grotesque disease caused by infestation of the lymphatic system by the nematode *Brugia malayi* or *Wuchereria bancrofti* (FIG. 23-22). The World Health Organization estimates that 250 million people have this disease. This worm is transmitted to humans by mosquito vectors and distribution is restricted to regions where there are carrier mosquitoes. It is most frequently transmitted at night when the parasite is present in peripheral blood and when mosquitoes are most active. Symptoms of this disease are associated with adult forms of the nematode. Repeated bites by infected mosquitoes result in the buildup of numerous adult worms. A hypersensitivity reaction appears to be involved in the progression of elephantiasis. Vessels break down and lymph accumulates in tissue, resulting in severe swelling. There is no suitable chemotherapy for the disease because the adult worms are resistant to drugs. The development of massive limbs and other body parts can cause severe mental problems, and psychotherapy is an important part of the supportive treatment of this disease.

> **Elephantiasis is the infestation of the lymphatic system by the nematode *Wuchereria bancrofti* that causes grotesque swelling of limbs.**

Schistosomiasis

Three species of blood flukes are responsible for infestations of the human cardiovascular system. These blood flukes, *Schistosoma mansoni*, *S. japonicum*, and *S. haematobium*, cause **schistosomiasis.** The World Health Organization reports that schistosomiasis is second only to malaria as the most prevalent disease in the tropics.

> **Blood flukes infest the cardiovascular system and cause schistosomiasis.**

The life cycle of blood flukes involves specific species of snails. This restricts the distribution of this disease (FIG. 23-21). Eggs deposited with urine or fecal matter hatch in contaminated waters. The immature larvae enter a suitable snail host where they develop to form larvae called *cercariae.* Cercariae escape from the snails and swim freely. If they contact

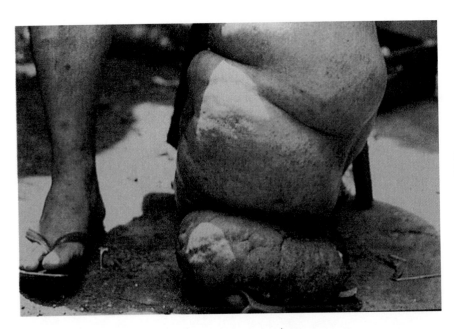

FIG. 23-21 Elephantiasis of the leg, caused by *Brugia malayi.*

humans they are able to penetrate the skin. After skin penetration, cercariae migrate through the circulatory system, moving through the heart and lungs. Cercariae settle in the hepatic portal of the liver where maturation to the adult form occurs. When they approach or reach adulthood, they migrate to the mesenteric venules of the intestines. Fertilized eggs laid in the venules move to the gastrointestinal or urinary tracts and are passed in urine or feces.

The symptoms of schistosomiasis vary with the stage of infection and the schistosome species. The penetration phase is usually asymptomatic. Most of the symptoms occur during the acute phase when the worms mature and start to lay eggs. Symptoms usually include diarrhea, fever, and discomfort. In the chronic stage there is a tissue reaction to the eggs that are deposited within the body. The eggs become walled off, forming granulomas. Cirrhosis occurs because this late stage involves the hepatic system of the liver. Diagnosis of schistosomiasis is based on the detection of eggs in the urine or feces of individuals who have been exposed to the cercariae in infested areas of the world. Various antimony-containing drugs are used for treatment of this disease.

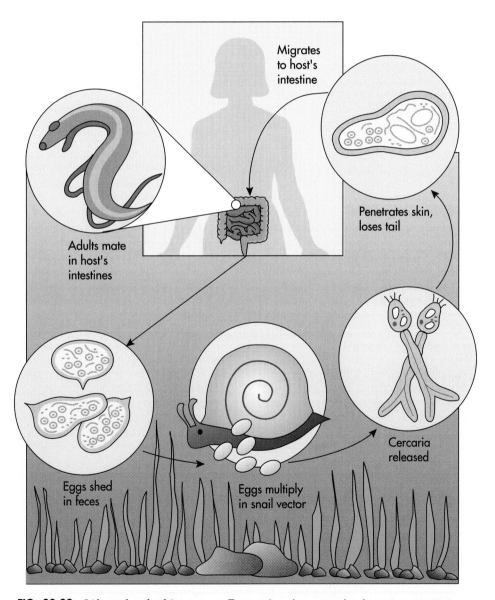

Migrates to host's intestine

Penetrates skin, loses tail

Cercaria released

Adults mate in host's intestines

Eggs shed in feces

Eggs multiply in snail vector

FIG. 23-22 Life cycle of schistosomes. Free-swimming cercariae in water penetrate unprotected skin. During penetration they lose their tails to become schistosomulae. These migrate through the blood stream via the lungs and liver to the veins of the bladder where they mature to produce eggs. The eggs then penetrate the bladder or colon, to be passed in the urine or the feces. Eggs released into fresh water are taken up by snail intermediate hosts where they mature into sporocysts. These release cercariae into the water to complete the cycle.

SUMMARY

Diseases Caused by Fungi (pp. 674-681)

Fungal Diseases of the Respiratory Tract (pp. 674-677)

- Histoplasmosis is a respiratory disease caused by the fungus *Histoplasma capsulatum*. It is transmitted by inhalation of spores that are deposited in the lungs, producing symptoms that resemble those of a mild cold. It is treated with amphotericin B.
- Coccidioidomycosis is caused by *Coccidioides immitis*, spores of which are inhaled in dust particles. Chest pain, fever, malaise, and a dry cough are symptoms of this disease. When the disease has spread systemically, it is treated with amphotericin B.
- Skin lesions are symptoms of blastomycosis that are caused by *Blastomyces dermatidis*. It is treated with amphotericin B.
- Aspergillosis is caused by infection with spores of *Aspergillus*. People with impaired immune systems, cancer, and diabetes are predisposed to this respiratory disease that is treated with amphotericin B.
- *Pneumocystis* pneumonia, a respiratory disease caused by the fungus *Pneumocystis carinii*, is associated with failures of the immune system, such as in AIDS cases. In this disease the alveoli fill with fluid, preventing gas exchange. Drugs such as pentamidine are used in treatment.

Fungal Diseases of the Gastrointestinal Tract (p. 677)

- *Aspergillus* produces aflatoxin when growing on improperly stored grains or peanuts. When ingested, aflatoxin impairs liver function and causes liver cancer.

Fungal Diseases of the Vaginal Tract (p. 677)

- Vaginitis is a disease of the vaginal tract caused by the overgrowth of *Candida*. A burning sensation and increased vaginal discharge that is thick and cheesy are symptoms of this disease. It is treated with topically applied nystatin.

Fungal Diseases of the Central Nervous System (pp. 677-679)

- Mushroom poisoning is caused by neurotoxins produced by certain poisonous mushrooms. Initial symptoms include vomiting and diarrhea; later, degenerative changes occur in liver and kidney cells and the patient becomes comatose.
- Ergotism is the result of eating grain containing ergot alkaloids produced by *Claviceps purpurea*. These toxins cause degeneration of the capillary blood vessels, producing neurological impairment due to inadequate circulation to the nervous system. Symptoms include vomiting, diarrhea, thirst, hallucinations, convulsions, and lesions.
- Cryptococcal meningitis is a disease of the central nervous system caused by the fungus *Cryptococcus neoformans*. It is spread by the inhalation of dust associated with pigeon droppings. Although it initially affects the lungs, it may spread systemically. It is treated with amphotericin B.

Fungal Diseases of the Skin (pp. 680-681)

- Dermatomycoses are infections of the skin caused by dermatophytic fungi, including *Microsporum* and *Trichophyton* species, transmission of which is enhanced by moist conditions. Tineas or ringworms are common dermatophytic diseases that are localized at particular body sites. These infections are treated with antifungal agents such as miconazole and clotrimazole.
- Sporotrichosis is a subcutaneous mycosis caused by the fungus *Sporothrix schenckii* that enters the body through an abrasion in the skin, producing lesions that may spread. When the fungus is widespread it is treated with amphotericin B.

Fungal Diseases of the Oral Cavity—Thrush (p. 681)

- Thrush is an infection of the oral cavity of newborns caused by the fungus *Candida albicans* transferred from the mother's vaginal tract during delivery. Nystatin controls this disease.

Diseases Caused by Algae—Paralytic Shellfish Poisoning (p. 682)

- Paralytic shellfish poisoning is caused by toxins produced by the dinoflagellate alga *Gonyaulax* concentrated in bivalve shellfish.

Diseases Caused by Protozoa (pp. 682-693)

Protozoan Diseases of the Gastrointestinal Tract (pp. 683-685)

- Amebiasis is caused by the contamination of water supplies by cysts of the protozoan *Entamoeba histolytica*. Cases may be asymptomatic or involve mild or severe diarrhea and abdominal pain. Antiprotozoan drugs such as metronidazole are used.
- Giardiasis is a gastrointestinal tract infection caused by the ingestion in contaminated food or water of cysts of the flagellated protozoan *Giardia lamblia*. Symptoms are diarrhea, dehydration, mucus secretion, and abdominal pain. Antiprotozoan drugs, such as metronidazole, are used.
- *Crytosporidium* is an emerging pathogen that causes waterborne infections.
- Balantidiasis is transmitted via food or water contaminated with feces containing cysts of the ciliated protozoan *Balantidium coli*. It causes symptoms of abdominal pain, nausea, vomiting, diarrhea, and loss of weight.

Protozoan Diseases of the Genital Tract (pp. 685-686)

- Vulvovaginitis is caused by excessive growth of *Trichomonas vaginalis*, producing symptoms of burning and increased vaginal discharge that is greenish and odorous. The protozoan is sexually transmitted. It is treated with metronidazole.

Protozoan Diseases of the Central Nervous System (p. 686)

- African trypanosomiasis is caused by infection with the protozoans *Trypanosoma gambiense* and *T. rhodesiense* transmitted to people by the tsetse fly. It can cause damage to heart and nerve tissues and if untreated can lead to fatal meningoencephalitis. If it is diagnosed before central nervous system involvement, it is treated with antiprotozoan agents such as suramin. Later, melarsoprol, an arsenical, is used.

Protozoan Diseases of the Cardiovascular and Lymphatic Systems (pp. 687-692)

- Malaria is caused by four species of the protozoan *Plasmodium* transmitted by the *Anopheles* mosquito. It is the most common infectious disease in the world. Early symptoms include chills, fever, headache and muscle ache. These symptoms coincide periodically with the rupture of infected erythrocytes. Malaria can be prevented by drug prophylaxis.
- Babesiois is a protozoan disease caused by the parasite *Babesia microti* that is transmitted to humans by tick vectors from rodents. It is generally a mild, self-limiting febrile disease, but people who have had their spleens removed and the elderly often experience more severe symptoms, which may include fever, sweating, joint and muscle pain, nausea, and vomiting. It is treated with chloroquine.
- Leishmaniasis is caused by infections with members of the protozoan genus *Leishmania* transmitted to humans by sand fly vectors. *L. mexicana* and *L. tropica* cause skin infections (Old World cutaneous leishmaniasis or Oriental sore). *L. donovani* multiplies throughout the mononuclear phagocyte system (kala-azar disease). These skin infections are treated with amphotericin B, and kala-azar is treated with amphotericin B or sodium stibogluconate.
- Toxoplasmosis is caused by ingestion of cysts of the protozoan *Toxoplasma gondii*. Although it is often asymptomatic, it can lead to central nervous system involvement. It is treated with antiprotozoan drugs such as aminoquinoline.
- Trypanosomiasis is caused by infections with species of the protozoan genus *Trypanosoma* transmitted to humans by cone-nosed bugs. The disease affects the mononuclear phagocyte system and later the myocardium and nervous system. It is treated with antiprotozoan drugs such as aminoquinoline.

Protozoan Diseases of the Central Nervous System (p. 693)

- *Naegleria* microencephalitis is a neurological disease caused by the protozoans *Naegleria fowleri* and *N. gruberi*. It infects the nasal mucosa and proliferates in the brain. It is almost always fatal.

Diseases Caused by Multicellular Parasites (pp. 693-699)

Parasitic Diseases of the Gastrointestinal Tract (pp. 693-698)

- Tapeworms are multicellular parasites that infect animal intestinal tracts. *Taenia saginata* is the beef tapeworm and *T. solium* is the pork tapeworm. Contaminated fecal matter or raw or inadequately cooked beef or pork containing the larval forms of these parasites transmits them to humans. Infected individuals are often asymptomatic; when there are symptoms they include loss of weight, anemia, and abdominal pain. Niclosamide is used for treatment. *T. solium* eggs hatch in the human intestine and larvae enter the

bloodsteam to localize in heart, brain, and other tissues—leading to death.

- Cysticercosis is infection throughout the body with the pork tapeworm. Larvae develop in the liver, muscles, heart, and brain. There is no treatment.
- Hydatidosis is caused by *Echinococcus granulosus* transmitted by infected dogs. The larval stages of this tapeworm form within a fluid-filled bladder called a hydatid cyst. If the cyst ruptures, the resultant hypersensitivity reaction can cause death from anaphylactic shock. Cysts can be surgically removed.
- Flukes are parasitic worms for which humans are alternate hosts. Different worms cause different diseases. Liver flukes can be accidentally ingested with contaminated vegetation and migrate from the intestine to the liver, producing digestive distress, including fever, pain, and liver malfunction. Dehydroemetine is used in treatment.
- Pinworms (*Enterobius vermicularis*) are ingested with contaminated food or drink. Hand-to-mouth contact is also common. They cause nausea, vomiting, and diarrhea, or no symptoms at all. Pyrantel pamoate (Antiminth) is the drug of choice.
- Whipworms (*Trichuris trichiura*) are nematodes. Hand-to-mouth contact and ingestion of food contaminated with fecal matter transmit this disease. The adult worms live in the large intestine. Light infestations are asymptomatic, but heavier infestations result in symptoms that resemble appendicitis and result in emaciation. The antihelminthic mebendazole (vermox) is the drug of choice.
- Trichinosis is caused by the roundworm *Trichinella spiralis* ingested in undercooked pork. Symptoms vary but usually involve the digestive tract. When the worms disseminate, other symptoms manifest, such as muscle pain and difficulty in breathing. Treatment depends on the tissues affected.
- Roundworm disease of humans is caused by infestation of the intestine with *Ascaris lumbricoides*. Large masses of worms can obstruct the gastrointestinal tract.
- Hookworms (*Necator americanus* and *Ancylostoma duodenale*) enter the body by burrowing through the skin and migrating through the pulmonary system to the intestinal tract. Symptoms resemble those of a duodenal ulcer and anemia. Iron is taken for the anemia and pyrantel pamoate (Antiminth) is taken for the infection.

Parasitic Diseases of the Cardiovascular and Lymphatic Systems (pp. 698-699)

- Elephantiasis is caused by the infestation of the lymphatic system by the nematode *Brugia malayi* or *Wuchereria bancrofti* transmitted by repeated bites of infected mosquitoes.
- Schistosomiasis is an infestation of the cardiovascular system by blood flukes. Symptoms vary with stage of infection and the specific schistosome species and usually include diarrhea, fever, and discomfort. Cirrhosis occurs in the late stages of the disease. Antimony-containing drugs are used in treatment.

CHAPTER REVIEW

1. Which diseases are caused by fungi?
2. How is histoplasmosis transmitted? Why is this disease associated with blackbirds?
3. How is coccicioidomycosis transmitted? Why is this disease called desert fever?
4. How is cryptococcal meningitis transmitted?
5. Describe ringworm, including the casuative agents and the body sites affected.

6. Which diseases are caused by protozoa?
7. How are giardiasis, amebic dysentery, and cryptosporidiosis transmittted?
8. How are malaria and leishmaniasis transmitted? How can these diseases be prevented?
9. Why is malaria so prevalent?
10. Which worms cause human infestations?
11. Which diseases are caused by *Candida?*

CRITICAL THINKING QUESTIONS

1. Why are fungal and protozoan diseases more difficult to treat than bacterial diseases? Why are antibiotics less effective against fungi and protozoa?
2. Why are protozoan infections so common in tropical countries?
3. Why is it difficult to prevent malaria? Why is there no vaccine currently available?

4. Explain the underlying causes for developing *Candida* infections. Why are newborns more likely to develop thrush than adults? Why do many women develop *Candida* vaginitis during pregnancy and when using oral contraceptives? Why do some individuals develop *Candida* infections after using broad-spectrum antibiotics?

CASE STUDY

HISTORY AND ASSESSMENT

Marla, a 53-year-old woman, came to a hospital emergency room (ER) in Philadelphia. She complained of fever, a cough of increasing severity, a sore throat, and generally feeling poorly. The main reason Marla came to the ER was because she coughed up some blood during the previous evening and in the morning.

Marla told the nurse that she had been feeling progressively worse since returning from a trip to New Mexico several weeks ago. *Why is it important for medical personnel to determine during the interview if the patient has travelled recently—and if so, where?*

Marla said that in addition to the cough, her joints were "acting up," and that she has had a fever. During the interview, Marla coughed up some sputum, noted by the nurse to contain blood. *What do you think may be the cause of blood in the sputum?* The sputum was put in a sterile container and sent to the lab for examination and culturing.

Vital signs were taken and recorded: Pulse rate 84, Respiratory rate 14, Blood pressure 122/76, and Temperature 102.4° F. *What is your assessment of these vital signs?*

PHYSICAL EXAMINATION

On EENT (eye, ear, nose, and throat) examination, the physician noted an erythematous pharynx (reddened throat) consistent with Marla's complaint of a sore throat. Auscultation (listening through a stethoscope) of

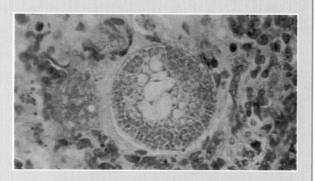

the lung fields revealed some rales and also scattered areas of dullness to percussion. *Rales are abnormal pathological lung sounds and indicate the presence of fluid in the lung airways. Dull areas to percussion suggest lung consolidation, as occurs with inflammatory exudates (pus) or other fluids or blood in the lung tissue.*

In addition, the physician also noted a somewhat decreased range of motion of Marla's right knee and left elbow joints. *From the case history and physical examination, is there any indication that Marla's elbow and knee are the sites of an active infection?*

PROCEDURES AND LABORATORY TESTS

The CBC (complete blood count) revealed a high white blood cell count (16,000 WBCs/mm3) and an increase in the relative percentage of eosinophils. *Eosinophils are a* type of white blood cell. *Under normal conditions, eosinophils make up approximately 2% of the total white blood cell count. Eosinophilia (the formation and accumulation of abnormal numbers of eosinophils in the blood) can be caused by allergies and hypersensitivity reactions, parasitic infections, immunological disorders, and malignancies.*

A chest X-ray revealed spotty granules in the lung fields. *Spotty granules in the lung fields are consistent with a nonbacterial (viral or fungal) pneumonia, as well as other noninfectious lung diseases.*

Small spherules, suggestive of a fungus infection, were seen in the stained sputum sample. Cultures would not be available for 24 hours.

DIAGNOSIS

Combining the results of the physical examination, chest X-ray, and staining, the physician determined that Marla had a pneumonia caused by a fungal infection of the lung tissue. *Some fungi are found worldwide (for example,* Candida albicans*); others are endemic to certain areas.* Histoplasma capsulatum *is most prevalent in the U.S. to the Mississippi river valley,* Coccidioides immitis *is found in the desert regions of the Americas, and* Paracoccidioides brasiliensis *is most prevalent in the area from Mexico south through Argentina.*

Since the cultures can take up to a week to grow and the physician wants to identify the organisms before then, the physician uses a special stain of hematoxylin and eosin on a sputum sample. Under the microscope, the physician identifies the characteristic spherules of *C. immitis.* (See Figure, p. 702.) *A positive histological diagnosis is as reliable as a culture for these organisms.* Sputum culture verified this finding later in the week. *A fungal infection by* C. immitis *is known as* **coccidioidomycosis.**

TREATMENT

Because of the hemoptysis (coughing up of blood), it was decided to begin treatment with amphotericin B, a very strong and potentially toxic antibiotic. *Amphotericin B can have an adverse effect on a patient's kidney tubules. The patient's creatinine level must be monitored during therapy. Creatinine is eliminated from the body by the kidneys, and serum creatinine levels are an indicator of kidney functional integrity. Hemoptysis is caused by erosion of lung tissue, internal bleeding, and coughing up blood.*

COURSE

With further drug therapy, Marla's fever and hemoptysis subsided. The patient improved dramatically and was released from the hospital after 5 days.

READINGS

Christensen CM: 1975. *Molds, Mushrooms, and Mycotoxins,* Minneapolis, University of Minnesota Press.

Covers fungi that are either toxic themselves or secrete toxic substances, the way fungi affect humans and other animals, the nature, cause, and prevention of wood decay, and the evolution of fungi.

Dubey JP and CP Beattie: 1988. *Toxoplasmosis of Animals and Man,* Boca Raton, FL; CRC Press.

Summary of current knowledge on history, structure, life cycle, transmission, control, diagnosis, and treatment of infections with Toxoplasma gondii. *Chapters on infections in livestock and humans.*

Erlandsen SL and EA Meyer (eds.): 1984. *Giardia and Giardiasis: Biology, Pathogenesis, and Epidemiology,* New York, Plenum Press.

Presents the biology of Giardia *and the correlation of structure-function relationships; the epidemiology, diagnosis, and treatment of giardiasis; and the immunological reactions that occur within the host.*

Harrison GA: 1978. *Mosquitoes, Malaria, and Man: A History of the Hostilities since 1880,* New York, Dutton.

This history of malaria and the efforts to prevent and control this disease stresses the role played by Sir Ronald Ross.

Howard DH: 1983-85. *Fungi Pathogenic for Humans and Animals,* New York, Marcel Dekker.

A three-volume set explaining the basic biology of fungi and the pathogenicity and detection of fungal diseases.

Jacobs PH and L Nall (eds.): 1990. *Antifungal Drug Therapy: A Complete Guide for the Practitioner,* New York, Marcel Dekker.

Highlights drug treatment therapies that decrease the morbidity and mortality of mycotic infections with guidelines for the diagnosis and management of superficial, systemic, and deep mycoses.

Kurstak E: 1989. *Immunology of Fungal Disease,* N Y, Marcel Dekker.

Describes the immune response to fungal infections.

Molyneux DH and RW Ashford: 1983. *The Biology of Trypanosoma and Leishmania, Parasites of Man and Domestic Animals,* New York, International Publications Service.

The modes of transmission and the methods used in the prevention and control of trypanosomiasis and leishmaniasis are discussed.

Pollock NH: 1969. *The Struggle Against Sleeping Sickness in Nyasaland and Northern Rhodesia, 1900-1922,* Athens, Ohio; Ohio University Center for International Studies.

After 20,000 deaths from sleeping sickness in Uganda between 1898 and 1906 and with the disease spreading to Nyasaland and Northern Rhodesia, British colonial administrators began to enlist their financial and scientific resources against the disease.

Ripon JW: 1988. *Medical Mycology: The Pathogenic Fungi and the Pathogenic Actinomycetes,* Philadelphia, W. B. Saunders.

A review of the mycosal infections caused by fungi and actinomycetales.

Ryley JF (ed.): 1990. *Chemotherapy of Fungal Diseases,* Berlin, Springer-Verlag.

Explains current drug therapy for the treatment of mycoses and why certain antifungal agents have therapeutic value.

Sandbach FR: 1976. The history of schistosomiasis research and policy for its control, *Medical History* 20:259-275.

Associates the history of schistosomiasis research and policy with colonialism and war as Europeans scrambled for colonies in Africa at the end of the nineteenth century.

Sarosi GA and SF Davies (eds.): 1986. *Fungal Diseases of the Lung,* Orlando, FL; Grune & Stratton.

Presents basic mycological information to enable health care workers to recognize common fungi and identify them in the clinical laboratory. Reviews specific fungal diseases and their treatment.

Williams G: 1969. *The Plague Killers,* New York, Scribner.

An account of public health campaigns against insects as carriers of malaria, hookworm, and yellow fever. Dramatically describes the work of the researchers at the Rockefeller Foundation to relieve suffering through improved public hygiene.

Wylie TD and LG Morehouse: 1977-78. *Mycotoxic Fungi, Mycotoxins, Mycotoxicoses: An Encyclopedic Handbook,* N Y, M. Dekker.

Volume 1 is "Mycotoxic Fungi and Chemistry of Mycotoxins", volume 2 is "Mycotoxicoses of Domestic and Laboratory Animals, Poultry, and Aquatic Invertebrates and Vertebrates", and volume 3 is "Mycotoxicoses of Man and Plants: Mycotoxin Control and Regulatory Practices."

UNIT SIX

Applied and

Environmental

Microbiology

CHAPTER 24

Industrial Microbiology

PREVIEW TO CHAPTER 24

In this chapter we will:
- Examine the industrial applications of microbiology.
- Learn about the various products produced by microorganisms.
- See that many foods and beverages are produced by microorganisms.
- Learn about the role of microorganisms in the production of pharmaceuticals.
- Examine the roles of microorganisms in fuel production and mineral recovery.
- Learn the following key terms and names:

alcoholic fermentation	mashing
amylases	ripening
antibiotics	scale-up
baker's yeast	screening
bioleaching	steroid hormones
biotechnology	synthetic fuels
cheeses	vaccines
distilled liquors	vitamins
fermentation	wine
fermentor	wort
industrial microbiology	yogurt
malting	

Microorganisms are used in many industrial processes to produce substances of economic value. Industrial processes exploit the enzymatic activities of microorganisms to produce substances of commercial value. Several biochemicals of commercial value are being produced by genetically engineered microorganisms, and the processes enable greater yields and greater ease of recovery and purification. Increased use of biotechnology to produce biochemicals is anticipated in the future.

The essence of an industrial process is combining the right organism, an inexpensive substrate, and the proper environment to produce high yields of a desired fermentation product. In industrial microbiology the term *fermentation* means any chemical transformation of organic compounds carried out by using microorganisms and their enzymes. In industrial fermentations, raw materials (substrates) are converted by microorganisms (specific strains or microbial enzymes) in a controlled favorable environment (created in a fermentor) to form a desired end product substance.

> **Fermentation, as used in industrial microbiology, refers to a chemical transformation of an organic compound carried out by microorganisms or their enzymes.**

Industrial microbiologists search for microorganisms whose metabolic capabilities can form substances of commercial importance. They try to find or create specific strains of microorganisms that will yield sufficient quantities of the desired product to permit commercial production on an economically favorable basis. These microbiologists also must design the optimal production process. The substrate mixture with the least expensive components that will produce the highest yield of the desired product must be defined. Fermentors must be designed and built that will optimize the environmental conditions to achieve maximal product yields (FIG. 24-1). Envi-

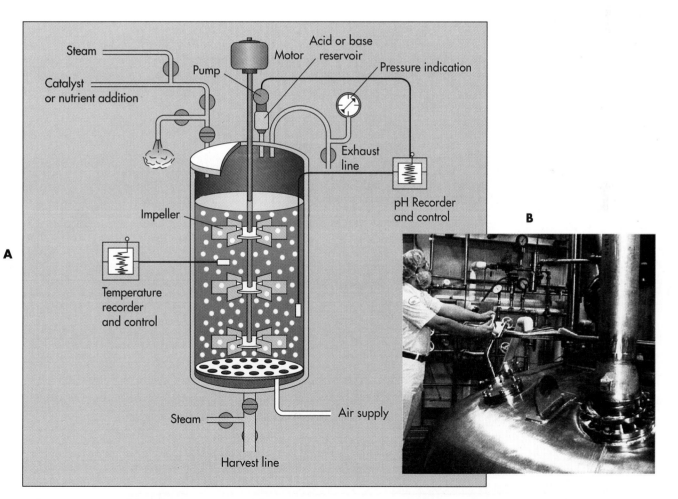

FIG. 24-1   A, A fermentor is designed to control environmental conditions so as to favor the growth of a specific microorganism and the yield of a fermentation product. The supply of oxygen and its mixing are critical in fermentor design. In a batch reactor the medium and inoculum are added, and after sufficient incubation the reaction is stopped and the fermentation product recovered. **B,** The fermentors used for commercial production of antibiotics and other fermentation products are large tanks holding hundreds or thousands of gallons of fermentation culture.

ronmental parameters must be maintained at optimal levels to maximize the yield and quality of the product. Microbial strains must retain their essential genetic and physiological features to produce high yields of compounds. Such strains are often genetically unstable. A common problem in the dairy industry is the destruction of stock cultures by lytic bacteriophage. When necessary, the strain must be replaced with a new stock culture. Recovery methods that achieve separation of the desired product from

microbial cells, residual substrate, and other metabolic products in the most economical manner must also be developed. Extraction and packaging procedures must ensure that the desired product is not contaminated.

Microorganisms with appropriate metabolic capabilities and production processes have been developed to commercially produce valuable products by fermentation.

PRODUCTION OF FOODS AND BEVERAGES

Microorganisms produce many of the foods and beverages that we consume (Table 24-1). Microbial processes in food production traditionally employ microbial enzymatic activities to transform one food into another. Many of the foods and beverages we commonly enjoy, such as wine and cheese, are the products of microbial enzymatic activity. Microbially produced food products have properties that are very different from those of the starting material. The microbial production of foods harnesses microbial biochemistry to produce desired changes in food products. The fermentative metabolism of microorganisms is most often exploited in the production of

food products. The accumulation of fermentation products, such as ethanol and lactic acid, produces characteristic flavors and other desirable properties. Trace metabolic products from specific microbial cultures give certain foods a distinctive flavor that cannot be reproduced synthetically.

FERMENTED DAIRY PRODUCTS

Several foods are made by the microbial fermentation of milk. Fermented dairy products include yogurt and cheese. The fermentation of milk is primarily carried out by lactic acid bacteria. The lactic acid fer-

TABLE 24-1	
Some Fermented Foods Produced by Microorganisms	
FOOD	**PRODUCTION MICROORGANISM**
Citron	*Saccharomyces citri medicae* and *Bacillus citri*
Cocoa	Removal of cocoa beans from pods by *Candida krusei* and lactic acid bacteria
Coffee	Removal of coffee bean pulp by pectinolytic microorganisms; lactic acid fermentation by *Leuconostoc mesenteroides, Lactobacillus plantarum, Lactobacillus brevis,* and *Streptococcus faecalis*
Green olives	*Leuconostoc mesenteroides, Lactobacillus plantarum,* and *Lactobacillus brevis*
Miso	*Aspergillus oryzae, Saccharomyces rouxii (Zygosaccharomyces rouxii), Zygosaccharomyces* spp., lactic acid bacteria, and bacilli
Pickles	Miscellaneous soil microorganisms, then *Lactobacillus plantarum, Pediococcus cerevisiae (P. acidlactici),* and *Leuconostoc mesenteroides*
Poi (Hawaiian fermented taro root)	*Pseudomonas* sp., *Lactobacillus* sp., *Streptococcus lactis,* and *Geotrichum candidum*
Sauerkraut	Mixed anaerobic population, including *Enterobacter cloacae, Leuconostoc mesenteroides,* and *Lactobacillus plantarum*
Soy sauce	*Aspergillus oryzae* and lactic acid bacteria
Tamari sauce	*Aspergillus tamarii*
Tempeh (Indonesian fermented soybean cake)	*Rhizopus* sp.
Vinegar	*Saccharomyces cerevisiae* subsp. *ellipsoideus, Acetobacter* sp., and *Gluconobacter* sp.

mentation pathway and the accumulation of lactic acid from the metabolism of the milk sugar lactose are common to the production of fermented dairy products. The accumulated lactic acid in these products acts as a natural preservative and flavoring agent. In addition, the acid lowers the pH of the milk so the protein casein coagulates with some lipids to form solid curds. The differences in the flavor and aroma of the various fermented dairy products are due to additional fermentation products that may be present in only relatively low concentrations. Some examples of fermented dairy products are listed in Table 24-2.

Lactic acid fermentation and the accumulation of lactic acid from the metabolism of lactose in milk is the basis for the production of fermented dairy products.

Yogurt

Over 550,000 pounds of yogurt are produced annually in the United States. **Yogurt** is made by fermenting milk with a mixture of *Lactobacillus bulgaricus* and *Streptococcus thermophilus* (FIG. 24-2). Yogurt fermentation is carried out at 40° C. The characteristic flavor of yogurt is due to the accumulation of lactic acid and acetaldehyde produced by *L. bulgaricus*. Because of the tart taste of acetaldehyde, most yogurt produced in the United States is flavored by adding fruit. The production of yogurt is a relatively rapid process (less than 24 hours) compared to the production of other fermented dairy products.

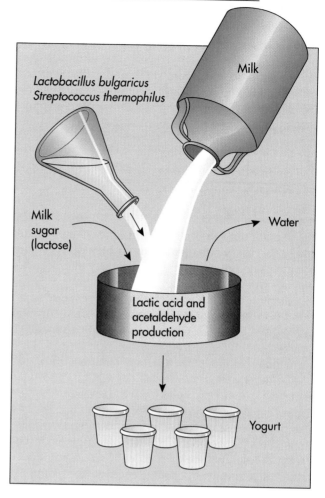

FIG. 24-2   Yogurt is produced by a lactic acid fermentation of milk by *Lactobacillus bulgaricus*.

TABLE 24-2	
Some Microbially Fermented Dairy Products	
DAIRY PRODUCT	**PRODUCTION MICROORGANISM**
Buttermilk	
Bulgarian buttermilk	*Lactobacillus delbrueckii* subsp. *bulgaricus*
Cultured buttermilk	*Streptococcus lactis* subsp. *lactis* (*Lactobacillus lactis* subsp. *lactis*) and *Leuconostoc cremoris* (*L. mesenteroides* subsp. *cremoris*)
Acidophilus milk	*Lactobacillus acidophilus*
Cheddar cheese	*Streptococcus faecium, Streptococcus thermophilus,* and *Lactococcus lactus* subsp. *lactus*
Cottage cheese	*Streptococcus lactis* subsp. *lactis* and *Leuconostoc cremoris*
Kefir (fermented milk)	*Lactobacillus brevis* and *Saccharomyces kefir*
Sour cream	*Streptococcus* (*Lactococcus*) *lactis* subsp. *diacetylactis, Streptococcus* (*Lactococcus*) *lactis* subsp. *lactis,* and *Leuconostoc cremoris*
Swiss cheese	*Propionibacterium freudenreichii* subsp. *shermanii*
Yogurt	*Streptococcus thermophilus* (*S. salivarius* subsp. *thermophilus*) and *Lactobacillus bulgaricus* (*L. delbrueckii* subsp. *bulgaricus*)

Cheese

Various cheeses are produced by microbial fermentation (FIG. 24-3). Cheeses consist of milk curds that have been separated from the liquid portion of the milk (whey). The curdling of milk is accomplished by using the enzyme rennin (casein coagulase or chymosin) and lactic acid bacterial starter cultures. Rennin is obtained from calf stomachs or by microbial production from *Mucor pussilus.*

The production of cheeses involves lactic acid fermentation. Various mixtures of *Streptococcus* and *Lactobacillus* species are used as starter cultures to initiate the fermentation. The flavors of different cheeses result from the use of different microbial starter cultures, varying incubation times and conditions, and the inclusion or omission of a secondary microorganism late in the fermentation process.

After the formation of the curds, they are separated from the whey, washed, pressed, often cooked, and salted. The cheese curd is then **ripened**—a process involving additional transformations performed by proteolytic enzymes. Different kinds of cheese are ripened at various temperatures to promote the growth of specific microorganisms. Sometimes a cheese is soaked in brine to encourage the development of selected bacterial and fungal populations during ripening. Some cheeses (for example, cream cheese) may be ripened without adding additional microorganisms. Other cheeses (for example, mozzarella) may undergo little or no ripening. Cheeses are categorized by their texture as: (1) soft cheeses—such as cottage cheese, cream cheese, and Neufchatel; (2) semisoft cheeses—such as Muenster and blue; (3) hard cheeses—such as cheddar and Swiss; or (4) very hard cheeses—such as Parmesan and Romano. The hardness of the cheese partially depends on the length of the ripening period and the amount of the water removed from the curd. Soft cheeses are ripened for 1 to 5 months, hard cheeses are ripened for 3 to 12 months, and very hard cheeses may require up to 16 months to ripen.

During ripening the curds are softened by enzymes and the cheese acquires its characteristic aroma and flavor from microbial by-products. Swiss cheese formation involves a late propionic acid fermentation, with ripening accomplished by *Propionibacterium shermanii* and *P. freudenreichii.* The propionic acid yields the characteristic aroma and flavor, and the carbon dioxide produced during this late fermentation forms the holes or eyes in the Swiss cheese.

Various fungi may also be used in the ripening of different cheeses. The unripened cheese is normally inoculated with fungal spores and incubated in a warm, moist room to promote the growth of filamentous fungi. For example, blue cheeses are produced by using *Penicillium* species. Roquefort cheese is pro-

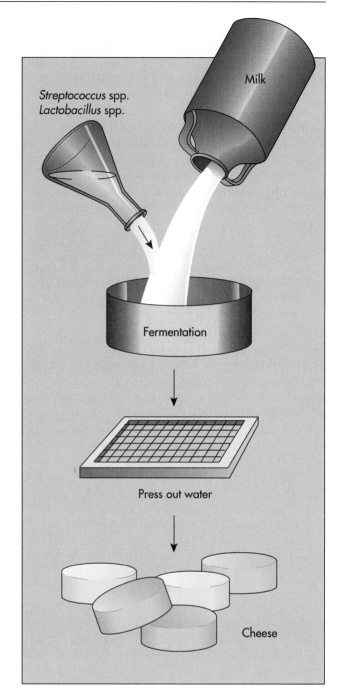

FIG. 24-3 Cheese is produced by a lactic acid fermentation using *Streptococcus* and *Lactobacillus* species. Numerous different cheeses are produced by microbial fermentation. After production of the solid cheese curd and the accumulation of fermentation products, water is removed to produce cheese.

duced by using *P. roqueforti,* and camembert and brie by using *P. camemberti* and *P. candidum.*

Streptococcus and *Lactobacillus* **species are used as starter cultures in the production of cheese.**

Propionibacterium and *Penicillium* **species are used in the ripening of some cheeses.**

FERMENTED SOYBEANS

Some Oriental foods are based on the fermentations of soybeans. **Soy,** or Japanese **shoyu,** is a brown, salty tangy sauce produced from a soybean, wheat, and wheat bran mash. Soy sauce is used as a condiment or flavoring ingredient in cooking. The starter culture for the production of soy sauce is begun by koji fermentation. This is a dry fermentation in which a mixture of soybeans and wheat is inoculated with fungal spores of *Aspergillus oryzae* (FIG. 24-4). The mixture is moistened but not immersed in liquid. The fungi grow on the surface of the soybeans and wheat and produce various enzymes, including proteases and amylases. Other microorganisms, especially lactic acid bacteria, also grow during this koji fermentation. The starter culture is then dried and extracted.

The extract is mixed with a mash containing autoclaved soybeans, crushed wheat, and steamed wheat bran. The mash and koji are incubated in flat trays at about 30° C for several days. The mixture is then soaked with concentrated brine. The resulting mixture is called *maromi.* The mash is then incubated for 10 weeks to 1 year, depending on the incubation temperature. During this incubation the enzymes of the koji are active and there is a succession of microbial populations. These include lactic acid-producing *Pediococcus soyae* and alcohol-producing *Saccharomyces (Zygosaccharomyces) rouxii, Zygosaccharomyces soyae,* and *Torulopsis* species. The lactic acid acts as a preservative and the alcohol increases the flavor of the product.

Tempeh is an Indonesian food produced from fermented soybeans. The soybeans are soaked at 25° C, dried, and inoculated with spores of various species of *Rhizopus.* The mash is incubated at 32° C for 20 hours, allowing the fungal mycelia to grow. The soybeans are then formed into a cake and salted. It is usually sliced and fried before being eaten.

BREAD

Bread is made by using yeasts that carry out an alcoholic fermentation. The substrates for this fermentation are the sugars from the carbohydrates in wheat flour that are broken down by amylase from added malt (or adding more sugar). The yeast ferments the sugar, producing carbon dioxide and ethanol. The principal yeast used in bread baking is *Saccharomyces*

A

Soybeans, wheat, wheat bran

Maromi mash — Mixing

Add inoculum — Koji fermentation

Matured mash

Pressing

Unrefined soy sauce

Pasteurizing

Refined soy sauce — Bottling

B

FIG. 24-4 **A,** Soy sauce is produced by a dry bran fermentation (koji fermentation) in which fungi grow on wheat and soybeans in a brine. **B,** Micrograph showing growth of *Aspergillus oryzae* on bran during koji fermentation.

cerevisiae, known as **baker's yeast** (FIG. 24-5). Carbon dioxide formed by baker's yeast leavens the dough and causes it to rise.

> **Yeasts that perform an alcoholic fermentation are used in making bread.**

After leavening, the bread is baked. Carbon dioxide bubbles are trapped in the dough and give rise to the honeycomb texture and increased volume of the baked bread. Although the interior of the bread does not reach 100° C, the heating is sufficient to kill the yeasts, inactivate their enzymes, expand the gas, evaporate the ethanol produced during the fermentation, and establish the structure of the bread loaf. During baking, heating the starch results in setting of the bread. Gluten, a glycoprotein in wheat that solidifies when heated, gives structural support and elasticity to dough. In the baked bread the structural support comes from the heated starch.

In addition to leavening bread, microorganisms produce the characteristic flavors of some breads. For example, the production of San Francisco sourdough bread uses a yeast *(Torulopsis holmii)* and a lactic acid bacterium *(Lactobacillus sanfrancisco)* to sour the dough and give this bread its characteristic sour flavor.

FERMENTED BEVERAGES

Microorganisms, principally yeasts in the genus *Saccharomyces,* are used to produce various types of alcoholic beverages. The production of alcoholic beverages relies on alcoholic fermentation. **Alcoholic fermentation** is the conversion of sugar to alcohol and carbon dioxide by microbial enzymes. The flavor and other characteristic differences between various types of alcoholic beverages reflect differences in the starting substrates and the production process, rather than differences in the microbial culture or the primary fermentation pathways employed in the production of alcoholic beverages.

> **The conversion of sugar to alcohol by microbial enzymes is an alcoholic fermentation that produces alcoholic beverages.**

> **The final alcoholic beverage produced is determined by the substrate and the microbial strain used.**

Beer

The worldwide production of beer is over 18 billion gallons per year. Each American consumes an average of 23.7 gallons of beer each year. The production of beer begins with the **malting** of the barley (FIG. 24-6). Malt contains a mixture of enzymes (amylases and proteinases) prepared by germinating barley grains for about a week and crushing the grains to release the plant enzymes. The malt is usually added with adjuncts in a process known as **mashing.** The

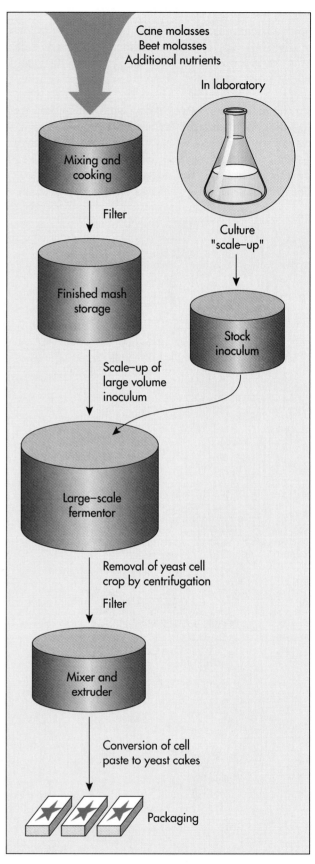

FIG. 24-5 To produce baker's yeast, cultures of *Saccharomyces* are grown on molasses. The yeasts are recovered by centrifugation and pressed into yeast cakes or freeze dried. Large quantities of yeast are grown for bread production and the brewing industry.

Water and barley

Malt

Grain

Mixing tank

Malting floor

Sugar
Hops
Malt adjucts

Brew kettle

Wort

Yeast

Fermenting vessel

Lagering tank

Canning

Atlas Lager   Atlas Lager   Atlas Lager   Atlas Lager

A

malt adjuncts—such as corn, rice, sorghum, and wheat—provide starch- and sugar-containing carbohydrate substrates for ethanol production. During the mashing process, the amylases from the barley malt hydrolyze the starches and other polysaccharides. Proteases in the malt adjunct and barley malt hydrolyze proteins. The mash is heated to temperatures of about 70° C to speed the enzymatic conversion of starch to sugars. The insoluble materials are allowed to settle from the mash. The clear liquid that is produced in this process is called **wort.**

The wort is then cooked with hops (the dried flowers of the hop plant). This cooking concentrates the mixture, inactivates the enzymes, extracts soluble flavoring compounds from the hops, and greatly reduces the number of microorganisms before the fermentation process. Additionally, compounds in the hops extract have antibacterial properties and may protect the wort from the undesirable growth of bacteria that could produce acids that would sour the beer.

The fermentation of wort to produce beer usually is carried out by the yeast *Saccharomyces carlsbergensis* or *S. cerevisiae.* The inoculation of the yeast into the cooled wort, known as **pitching,** uses a heavy inoculum, about 1 pound of yeast per barrel of beer. The wort is initially aerated to facilitate reproduction of the yeast but is then allowed to become anaerobic. This promotes the fermentative production of alcohol and carbon dioxide. At a late stage in fermentation, the yeasts aggregate and settle at the bottom. This action partially clarifies the beer. Commercially produced beer in the United States has an alcohol content of about 3.8%. In Canada it is 5%.

B

FIG. 24-6   A, Beer is produced by mixing water and barley to produce malt (active enzymes). The malt is added to grain, which produces simple fermentable sugars from the plant carbohydrates of the grain. The mash produced in this process is used as a substrate for the growth of *Saccharomyces.* The yeast produces alcohol (beer). **B,** Mash for growing yeast in beer production is typically produced in copper kettles.

HISTORICAL PERSPECTIVE

BEER IN THE UNITED STATES

Americans, while not among the nations with the greatest consumption per capita, certainly have a thirst for beer. The average person in the United States consumes 80 liters of beer each year. Since this number includes infants and toddlers who do not drink alcoholic beverages, the average adult in the United States consumes more than a six-pack of beer each week. Americans were drinking beer even before Christopher Columbus landed here. Native Americans brewed beer in earthen jugs by mixing maize (corn) and sap from the black birch tree, which served as an inoculum with naturally occurring yeasts.

The first commercial brewery was opened by the Dutch West India Company in Lower Manhattan in 1632. William Penn operated the first brewery in Pennsylvania. George Washington maintained a private brewhouse on his estate at Mount Vernon. His handwritten recipe for beer—said by his peers to be superb—is on display at the New York City Public Library. During the Revolutionary War, American soldiers received a daily quart of beer each. On one occasion, General Washington wrote to Congress complaining that the supply had run short. Both Samuel Adams and Revolutionary War General Israel Putnam were professional brewers at some point in their lives. In 1861, America's first privately endowed college for women was founded by Matthew Vassar, a brewer. America's oldest trade organization is the United States Brewers' Association, which was founded in 1862 and is still going strong.

HIGHLIGHT

PRODUCTION OF WINE

Microorganisms, principally yeasts in the genus *Saccharomyces*, are used to produce various types of alcoholic beverages. The production of alcoholic beverages relies on an alcoholic fermentation, that is, the conversion of sugar to alcohol by microbial enzymes. The flavor and other characteristic differences between various types of alcoholic beverages reflect differences in the starting substrates and the production process, rather than differences in the microbial culture or the primary fermentation pathways employed in the production of alcoholic beverages.

Initially, the grape must and yeasts are stirred to increase aeration and permit the proliferation of the yeasts. Mixing is later discontinued, creating anaerobic conditions that favor the production of alcohol. The sugar content of the grapes and the alcohol tolerance limit of the yeasts determine the final ethanol concentration. During fermentation, wine is periodically racked, that is, it is filtered through its bottom sediments and added back to the top of the fermentation vat. Carbon dioxide produced during the fermentation process forces the skins and other debris to the surface. Normally, the carbon dioxide produced during alcoholic fermentation is allowed to escape and the wine is, therefore, still. In the case of champagne and other sparkling wines, however, the carbonation is essential. In some commercially produced champagne, carbon dioxide is reinjected into the wine after fermentation. In the classic French method of champagne production, the wine is fermented in the bottle. After fermentation is complete, the bottles are inverted, and the yeast settles into the neck of the specially shaped champagne bottles. The yeasts are frozen and removed as a plug without excessive loss of carbon dioxide.

At the end of fermentation, wines typically have an alcohol content of 11% to 16% by volume. Wines are then aged to achieve their final bouquet and flavor.

Wines that are stoppered with a cork must be stored on their sides to prevent the cork from drying out; otherwise air would be permitted to enter and the alcohol would be oxidized by bacteria to form acetic acid. The spoilage, or souring, of wines through the formation of vinegar is a serious problem. In the United States, some wine bottles are sealed with plastic caps and therefore need not be stored on their sides to prevent souring of the wine.

After fermentation the beer is placed in a cool location and allowed to age. During the aging process, precipitation of proteins, yeasts, and resins occurs. This mellows the flavor of the beer. Mature beer is removed and filtered. The finished product is artificially carbonated, usually with carbon dioxide that was collected from the initial fermentation process. Most bottled or canned beers are pasteurized at 60° to 61° C or filtered to remove viable yeasts.

Light beers are low carbohydrate beers produced by using a wort prehydrolyzed with fungal glucoamylases and amylases. The prehydrolysis of carbohydrates in the wort permits the yeasts to ferment the carbohydrates completely to alcohol and carbon dioxide. This greatly reduces the concentration of residual carbohydrates in the beer.

> Beer is made from malted barley; corn, rice, or wheat provide the carbohydrate substrates for ethanol production by *Saccharomyces carlsbergensis.*

Wines

Wine is fermented from fruits, primarily grapes (FIG. 24-7). Red wines are produced by using whole red grapes. White wines are made from white grapes or from red grapes with their skins removed. The production of wine begins when the grapes are crushed to form a juice or **must.** In the classic European method of wine production, wild yeasts from the surface skins of the grapes are the only inoculum for the fermentation. In modern wine production, however, the natural microbiota associated with the grapes are inactivated by sulfur dioxide fumigation or by the addition of metabisulfite. This is done so that the wild microorganisms do not compete with the defined yeast strains used to ferment the grapes in this process. The grape must is then inoculated with a

specific strain of yeast, normally a variety of *Saccharomyces cerevisiae.* There are many strains of this yeast and each is used to make a particular type of wine. These strains of wine yeast are very different from the strains of *Saccharomyces cerevisiae* used in baking. By using specific yeast strains and controlled fermentation conditions, a product of consistent quality can be produced.

> Grape must fermented with *Saccharomyces cerevisiae* produces wine.

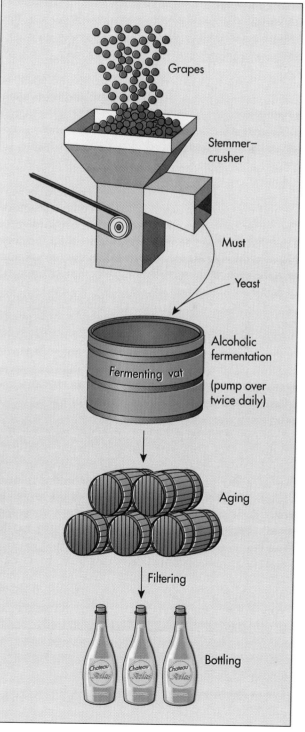

FIG. 24-7 Most wines are produced by the fermentation of grapes. Numerous types of wines are produced. Grapes are squeezed to produce grape juice called must. Yeast is added and allowed to carry out an ethanolic fermentation. The wine is then aged to remove some tart compounds, filtered to remove particulates, and bottled.

The sugar content of the grapes and the alcohol tolerance limit of the yeasts determine the final ethanol concentration. The sugar content of the grapes depends on the grape variety and its ripeness and varies from season to season. This accounts in part for the fact that some years and vineyards are better than others for the production of quality wines.

Distilled Liquors

The initial steps in the production of distilled spirits are analogous to those in beer production (FIG. 24-8). They begin with a mashing process in which the polysaccharides and proteins in a starting plant material are converted to sugars and other simple organic compounds that can be fermented by yeasts to form alcohol. These plant materials include grains such as barley, corn, wheat, or rice; fruits such as grapes, pears, apples, agave or plum; and roots such as potatoes.

After the fermentation process the alcohol is collected by distillation to produce **distilled liquors,** or spirits. The alcoholic product formed from the fermentation of wort, known as a *beer* or *wine,* is heated in a still. Ethanol has a lower boiling point than water and vaporizes while the water remains a liquid. The vapor containing the gaseous alcohol is cooled in a distillation tower so that it reforms a liquid. The liquid alcohol is collected. This permits the production of beverages with very high alcohol concentrations. In addition to alcohol, various volatile organic compounds, called fusel oils, are collected with the distillate and contribute to the characteristic flavors of the different distilled liquor products. The distilled alcohol product is normally aged in wooden casks or vats to yield a mellow-tasting alcoholic beverage. The type of wood is important because it can contribute to the flavor and smoothness of the finished product.

Various starting plant materials are used for the production of different distilled liquor products. Rum is produced by using sugar cane syrup or molasses as the initial substrate. Rye whiskey is produced from the fermentation of a rye mash, which is grain meal steeped and stirred in hot water to ferment). Bourbon or corn whiskey uses corn mash. Brandy comes from the fermentation of grapes. The yeasts used in the production of distilled liquors typically are special distiller strains of *Saccharomyces cerevisiae,* which yield relatively high concentrations of alcohol. The yeasts produced during fermentation are collected, dried, and used as animal feed.

Special distiller strains of Saccharomyces cerevisiae are used in the production of liquor.

The original plant material determines the final product in commercial alcohol production.

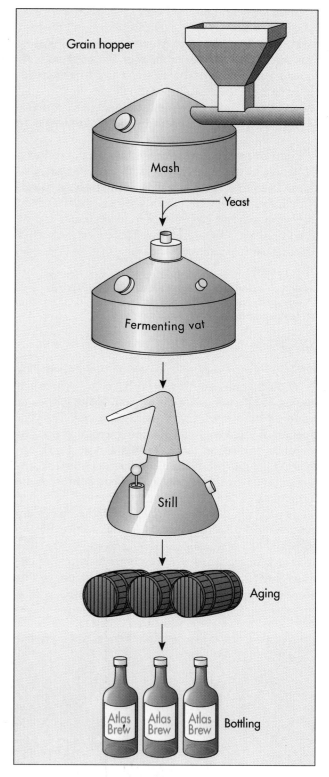

FIG. 24-8 To produce distilled spirits a beer produced by fermentation is subjected to distillation. The yeast *Saccharomyces cerevisiae* is mixed with mash formed from grain and an ethanolic fermenting occurs in a fermenting vat. The ethanol is then concentrated by distillation using a still. A concentrated alcohol is produced in this way. This removes water and produces concentrated alcohol. The alcohol can be aged to bring about additional chemical reactions. Bourbon, for example, is aged in charred oak casks. The final product (brew) is bottled.

HIGHLIGHT

SINGLE CELL PROTEIN

In addition to using microorganisms to transform substrates enzymatically into desired food products, microorganisms can be grown as a source of food. In the future one could dine on bacteria, algae, or yeasts. The food would be called single cell protein (SCP), so named because the microorganisms are single-celled organisms rich in protein. The algae *Scenedesmus* and *Spirulina* are already cultured in warm ponds and used as food sources in some nations. The production of SCP from algae is advantageous because these organisms are able to utilize solar energy, greatly reducing the amount of fuel resources required to produce SCP. Some algae currently are harvested as a source of food.

Efforts to develop large scale SCP production was begun during the 1960s when it appeared that crop production could not keep pace with the worldwide increase in human population. SCP was seen as the alternative to mass starvation. In the 1960s petroleum was inexpensive, and it was viewed as an economically attractive substrate for growing microorganisms. Massive fermentors were developed by British Petroleum and yeasts that could grow on petroleum were developed. Joint ventures were begun between oil companies and chocolate producers to make chocolate-covered microorganisms that would be palatable. However, there soon followed a dramatic increase in oil prices so that petroleum hydrocarbons could no longer be considered as the primary substrates for producing SCP; the product simply could not be economically competitive with soybean and fish meal.

Nevertheless efforts to develop SCP on other substrates have continued. Future, less expensive sources of methanol, perhaps derived from cellulose, will likely revive the prospects for large-scale production of microbial SCP. Various species of yeast, including members of the genera *Saccharomyces, Candida,* and *Torulopsis,* can be grown on waste materials, recycling these substances into useful sources of food. The growth of yeasts on waste materials serves a dual function: the removal of the unwanted substances and the production of much needed protein-rich foods. In Russia there is huge commercial production of *Candida* yeast protein from hydrolyzed peat. Approximately 1.1 million tons of yeast protein per year are being produced in a rapidly expanding Russian industry that aims to reduce Russian dependence on imported grain.

Today SCP is primarily produced as an animal feed supplement. There are problems with using SCP for direct human consumption because of the high concentrations, 6% to 11%, of nucleic acids. This high concentration of nucleic acid may result in increased serum levels of uric acid, causing kidney stone formation or gout, possible allergic reactions, and possible gastrointestinal reactions, including diarrhea and vomiting. Chickens and other animals, however, can be grown on SCP rather than on plant materials, helping to meet world food needs. Researchers are still trying to find the proper microorganism and set of production conditions to produce SCP that can be safe for human consumption.

PRODUCTION OF PHARMACEUTICALS

ANTIBIOTICS

The microbial production of antibiotics is a major industry, with annual worldwide sales of over $5 billion. **Antibiotics** are substances made in nature by various microorganisms that inhibit or kill other microorganisms. Of the thousands of different antibiotics, relatively few are produced commercially. The major antibiotics used in medicine and the microorganisms used for producing these antibiotics are shown in Table 24-3.

Antibiotics are microbially produced substances or substances synthetically derived from natural sources that inhibit or kill microorganisms.

The search for antibiotics in the pharmaceutical industry presents a good example of how screening

TABLE 24-3

Some Antibiotics Produced by Microorganisms

ANTIBIOTIC	PRODUCTION MICROORGANISM
Amphotericin B	*Streptomyces nodosus*
Bacitracin	*Bacillus licheniformis*
Chlorotetracycline	*Streptomyces aureofaciens*
Chloramphenicol	*Streptomyces venezuelae*
Erythromycin	*Streptomyces erythraeus*
Kanamycin	*Streptomyces kanamyceticus*
Neomycin	*Streptomyces fradiae*
Novobiocin	*Streptomyces niveus*
Nystatin	*Streptomyces noursei*
Penicillin	*Penicillium chrysogenum*
Polymyxin B	*Bacillus polymyxa*
Streptomycin	*Streptomyces griseus*

procedures are employed to select microorganisms for industrial applications. The discovery of new antibiotics results from laborious searches. Identification of compounds with antimicrobial activity is an essential step in the screening process. Samples from many sources, including soils from around the world, are examined as potential sources of antibiotic-producing microorganisms; countless strains of microbial isolates are tested by pharmaceutical laboratories. Of the numerous investigations, few studies yield evidence of promising new compounds of potential clinical importance.

In screening for microorganisms that possess the potential for producing industrially important substances, both naturally occurring microorganisms and genetic variants must be considered. The classic approach to find new antibiotic-producing strains has been to screen large numbers of isolates from soil samples for microorganisms that naturally produce antimicrobial substances. Additionally, mutations can be induced by exposure to radiation or mutagenic chemicals to increase genetic variability within the populations being screened. The goal is to isolate a unique microbial strain capable of producing a novel substance with the desired properties or a strain that produces large quantities of a valuable substance. Often, once a microorganism is identified as possessing the genetic information needed to produce a potentially useful substance, it is necessary to carry out successive stages of mutation before isolating a mutant strain of that organism that can be employed for commercial production (FIG. 24-9).

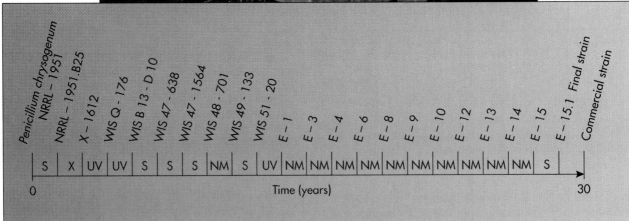

FIG. 24-9 Antibiotics inhibit the growth of microorganisms. They are produced by pharmaceutical companies and used by physicians to treat disease. **A,** Development of a commercial penicillin producing strain. **B,** Strains used for producing commercial quantities of fermentation products overexpress the formation of the products. Often, extensive genetic mutations are needed to achieve such overexpression, as in the case of penicillin production where numerous mutations, achieved spontaneously (S) and by UV light, X-radiation, and chemical mutagen (nitrogen mustard [NM]) exposure, were required to form the final strain that produces enough penicillin to permit commercial production.

METHODOLOGY

SCREENING OF ANTIBIOTIC PRODUCERS

A useful antibiotic-producing strain must produce metabolites that inhibit the growth or reproduction of pathogens. This essential property can be assayed by using test strains and examining whether the isolate being screened produces substances that inhibit the growth of these test organisms. If a suspension of the test organism is applied to the surface of an agar plate, the zone of inhibition around a colony may indicate that the organisms in that colony are producing an antibiotic. Alternatively, the crude filtrate of a broth-grown microbial culture can be added to a culture of a test organism to determine whether substances with antimicrobial activity are produced by the organism being screened.

A positive result in such a primary screening procedure in no sense ensures the discovery of an industrially useful antibiotic-producing strain. It simply identifies strains of microorganisms that have the potential for further development. Secondary screening procedures are then carried out to determine whether the organism is indeed producing a substance of industrial interest that merits further investigation and development. These procedures may include qualitative assays, aimed at identifying the nature of the substance being produced and determining whether it is a new compound not previously considered for industrial production, and quantitative assays, aimed at determining how much of the substance is being produced.

In screening for antibiotic producers, the crude filtrate from a broth culture may be separated chromatographically and the antimicrobial activities of the separated components determined. In some cases, paper chromatography is used to separate compounds for testing; in other cases, high-pressure liquid chromatography (HPLC) is employed to separate potential active compounds. The individual active components can then be isolated and used for further screening against additional test organisms to determine the microbial inhibition spectrum. This additional screening is useful in determining whether the substance has a broad or narrow range of activity and if it is particularly effective against specific pathogens. If an organism is indeed found to possess the potential for creating a useful new antibiotic, many additional tests are required to determine whether sufficient quantities of the substance can be produced to permit industrial production.

Genetic engineering provides many new possibilities for employing microorganisms to produce economically important substances. Whereas the mutation and selection approach is hit or miss, the use of recombinant DNA technology permits the purposeful manipulation of genetic information to engineer a microorganism that can produce high yields of a variety of products. Until the recent breakthroughs in the techniques of genetic engineering, a bacterium could produce only substances coded for in its bacterial genome. It is now possible to engineer bacterial strains that produce plant and animal gene products. The development of microorganisms producing high yields of new antimicrobial substances promises to revolutionize the economics of the pharmaceutical industry. The ruling of the United States Supreme Court that genetically engineered microorganisms can be patented also adds economic incentive for industrial applications of recombinant DNA technology.

Microbial strains are screened to determine those that can produce substances of industrial importance, such as antibiotics, in commercially significant quantities. Such strains may be naturally occurring or genetically engineered.

For example, the *Penicillium* species observed by Alexander Fleming to inhibit the growth of *Staphylococcus* had obvious potential for commercial development but did not produce sufficient quantities of penicillin to permit industrial production (see FIG. 24-9). Extensive screening of soil samples from around the world led to the isolation of a potentially useful strain from soil collected in Peoria, Illinois. Multiple successive mutations, though, were necessary to develop a strain of *Penicillium chrysogenum* capable of producing nearly 100 times the concentration of penicillin produced by the original strain in order to make the production of penicillin commercially feasible.

STEROIDS

The use of microorganisms to carry out biotransformations of steroids is very important in the pharmaceutical industry. **Steroid hormones** regulate various aspects of human metabolism. One such hormone, cortisone, relieves the pain associated with rheumatoid arthritis. Various cortisone derivatives are also useful in alleviating the symptoms associated with allergic and other undesired inflammatory responses of the human body. Other steroid hormones regulate human sexuality. Some regulate fertility and can be manufactured for use as oral contraceptives.

Steroid hormones regulate human metabolism.

The physiological properties of a steroid depend on the nature and the exact position of the chemical constituents on the basic steroid ring structure. The chemical synthesis of steroids is very complex because of the requirement to achieve a precise substituent location. For example, cortisone can be synthesized chemically from deoxycholic acid. This process requires 37 steps, many of which must be carried out under extreme conditions of temperature and pressure. The resulting product costs over $200 per gram. The major difficulty in the chemical synthesis of cortisone is the need to introduce an oxygen atom at the number 11 position of the steroid ring. This can be accomplished by microorganisms (FIG. 24-10).

Transformations of the steroid ring carried out by microorganisms include hydrogenations, dehydrogenations, and the removal and addition of side chains. The use of such microbial transformations in

the formation of cortisone has lowered the original cost over 400-fold, so that the price of cortisone in the United States is less than 50 cents per gram.

Microorganisms carry out biotransformations of steroids, such as hydrogenations and the addition or removal of side chains, to produce hormones.

In a typical steroid transformation process, a microorganism is grown in a fermentation tank with an appropriate growth medium and incubation conditions to achieve a high biomass. Aeration and agitation are employed to achieve rapid growth. After the growth of the microorganisms, the steroid to be transformed is added. For example, progesterone is added to a fermentor containing *Rhizopus nigricans* that has been growing for approximately 1 day, and the steroid is hydroxylated at the number 11 position to form 11-hydroxyprogesterone. The product is then recovered by extraction with a solvent, purified chromatographically, and recovered by crystallization. Numerous similar transformations are carried out to produce a variety of steroid derivatives for different medicinal uses.

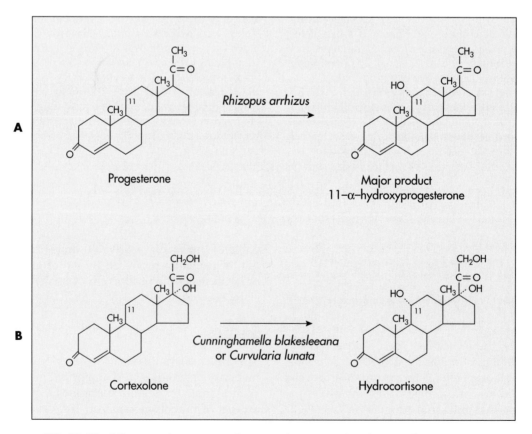

FIG. 24-10 Microorganisms are used to transform steroids. The specificity of microbial hydroxylation reaction is critical for the commercial production of various corticosteroids. **A,** Production of 11 α-hydroxyprogesterone uses *Rhizopus* to hydroxylate progesterone at the number eleven position. **B,** Production of hydrocortisone, which uses cortexolone as the substrate, depends on fungi to add a hydroxyl group at the number eleven position of the steroid ring.

HUMAN PROTEINS

Human proteins can be produced by genetically engineering bacteria and yeasts (FIG. 24-11). By using recombinant DNA technology, human DNA sequences that code for various proteins have been incorporated into the genomes of bacteria. By growing these recombinant bacteria in fermentors, human proteins are produced commercially. Human insulin, for example, is produced by a recombinant *Escherichia coli* strain. Other strains are used to produce human growth hormone, tumor necrosis factor (TNF), interferon, and interleukin-2. Human insulin is used to treat diabetics who are allergic to insulin from cattle. Human growth factor is used to treat diseases, such as dwarfism, that result from a deficiency of this hormone. Interleukin-2, interferon, and tumor necrosis factor (TNF) are important components of the natural human immune response. Their production may prove useful in treating some diseases where increased levels of these substances would be therapeutic. Interferon, for example, is important in the defense against viruses. TNF is a natural substance produced in the body in small amounts by macrophages in the blood. It appears to kill some cancer cells and infectious microorganisms without adversely affecting most normal cells. Undoubtedly, recombinant DNA technology will permit the production of other useful human proteins.

Human insulin, human growth hormone, tumor necrosis factor, interferon, and interleukin-2 are human proteins that can be produced by genetically engineered bacteria.

VACCINES

The use of **vaccines** is extremely important for preventing many serious diseases (FIG. 24-12). The development and production of these vaccines is an important function of the pharmaceutical industry. The production of vaccines involves growing microorganisms possessing the antigenic properties needed to elicit a primary immune response. Vaccines are produced by mutant strains of pathogens or by attenuating or inactivating virulent pathogens without removing the antigens necessary for eliciting the immune response.

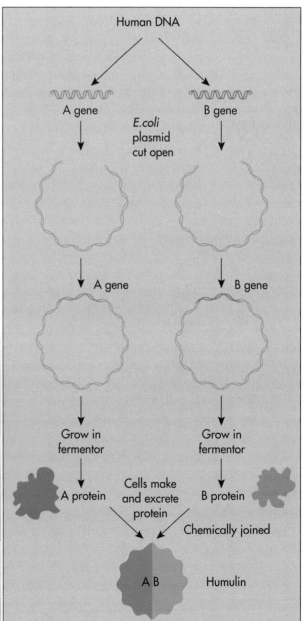

B

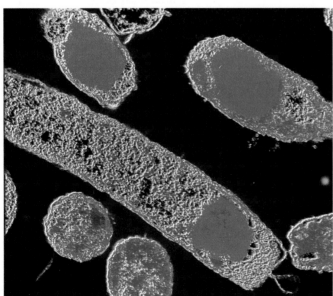

A

FIG. 24-11 **A,** This genetically engineered *Escherichia coli* is accumulating large quantities of human interleukin protein *(purple)* as shown in this colorized electron micrograph. **B,** *E. coli* strains have been genetically engineered to produce the A and B protein subunits of human insulin. The recombinant *E. coli* strains are grown in fermentors and the "human proteins" they produce are chemically combined to form humulin (human insulin).

Vaccines are produced using microorganisms with the antigenic properties to elicit a primary immune response.

Mutant, attenuated, or inactivated pathogens may be used to produce vaccines.

For the production of vaccines against viral diseases, strains of the virus often are grown by using embryonated eggs. Individuals who are allergic to eggs cannot be given such vaccine preparations. Viral vaccines may also be produced by using **tissue culture,** that is, by growing the virus in an *in vitro* culture of animal cells. For example, the older rabies vaccine, which was produced in embryonated duck eggs and had painful side effects, has been replaced with a vaccine produced in human fibroblast tissue cultures that has far fewer side effects. An entirely new approach to developing vaccines has come from recombinant DNA technology. Genes from pathogenic viruses coding for capsid proteins have been cloned and expressed in nonpathogenic bacteria or viruses.

Commercially produced vaccines must be tested and standardized before use. It is critical that the vaccine not contain active forms of a virulent pathogen, or the vaccine will transmit the disease it aims to prevent. Unfortunately, there have been several outbreaks of disease associated with improperly prepared vaccines. High standards of quality control and appropriate safety test procedures can prevent such incidents.

VITAMINS

Vitamins are essential animal nutritional factors. Some vitamins can be produced by microbial fermentation. The vitamins are produced as a result of secondary metabolism, that is, the metabolic pathways for vitamin production are not those used for the generation of cellular energy or structures. **Vitamin B$_{12}$,** for example, can be produced as a by-product of *Streptomyces* antibiotic fermentations. Vitamin B$_{12}$ can also be produced commercially by using *Propionibacterium shermanii* or *Pseudomonas denitrificans*. **Riboflavin** can also be produced as a fermentation product. It is a by-product of acetone butanol fermentation and is produced by various *Clostridium* species. Commercial production of riboflavin by fermentation uses the fungal species *Eremothecium ashbyii* or *Ashbya gossypii* with a medium containing glucose and/or corn oil.

Vitamin B$_{12}$ and riboflavin are examples of vitamins that can be produced by microbial fermentation.

FIG. 24-12 Fermentors used for production of *Haemophilus influenzae* type B (Hib) vaccine. Large fermentors are used to grow recombinant *Escherichia coli* containing the genes for human interferon and human interleukin-2 hormone for commercial production.

PRODUCTION OF ACIDS

ORGANIC ACIDS

Several organic acids can be produced by microbial fermentation. **Gluconic acid** has many commercial uses. Calcium gluconate, for example, is used as a pharmaceutical to supply calcium to the body. Ferrous gluconate is used to supply iron in the treatment of anemia. Gluconic acid in dishwasher detergents prevents spotting of glass surfaces due to the precipitation of calcium and magnesium salts. Gluconic acid is produced by various bacteria, including *Acetobacter* species, and by several fungi, including *Penicillium* and *Aspergillus* species.

Citric acid is used commercially in various ways, including as a food additive, especially in the production of soft drinks, as a metal chelating and sequestering agent, and as a plasticizer. It is produced by cultures of *Aspergillus niger*. The composition of the fermentation medium is critical for obtaining high yields of citric acid. A typical medium for the production of citric acid contains molasses, ammonium nitrate, magnesium sulfate, and potassium phosphate. It is essential to limit the growth of the fungus so that high levels of citric acid can accumulate. This can be accomplished by having a deficiency of trace metals or phosphate in the medium.

Gibberellic acid and related gibberellins are plant hormones. They enhance agricultural productivity. They are extensively used as growth-promoting substances to stimulate plant growth, flowering, and seed germination and to induce the formation of seedless fruit. Gibberellic acid is formed by the fungus *Gibberella fujikuroi* and can be produced commercially.

Lactic acid has various commercial uses. It is used in foods as a preservative, in leather production for deliming hides, and in the textile industry for fabric treatment. Various forms of lactic acid are also used for other purposes—in resins, plastics, electroplating, and in baking powder and animal feed supplements. The formation of lactic acid in making various fermented dairy products already has been discussed. *Lactobacillus delbrueckii* is widely used in the commercial production of lactic acid.

Gluconic acid, citric acid, gibberellic acid, and lactic acid are examples of organic acids that can be produced by microbial fermentation.

AMINO ACIDS

Microbial production of the amino acids lysine and glutamic acid presently accounts for over $1 billion in annual worldwide sales. L-Lysine and methionine are essential amino acids (amino acids that are not synthesized by metabolism of animal cells and therefore must be supplied) that are not present in sufficient concentrations in grains to meet animal nutritional needs. Lysine produced by microbial fermentation and methionine produced synthetically are used as animal feed supplements and as additives in cereals. L-Glutamic acid is principally made for use as monosodium glutamate (MSG), a flavor enhancer. The flavoring industry in the United States consumes more than 30,000 tons of MSG annually, some of which is imported from Japan, a major producer of amino acids by fermentation. Aspartame (L-aspartic acid-L-phenylalanine methyl ester) is used as a low-calorie sweetener.

Lysine to upgrade animal feeds is produced from carbohydrates using *Corynebacterium glutamicum*. Cane molasses is generally used as the substrate, and the pH is maintained near neutrality by adding ammonia or urea. As the sugar is metabolized, lysine accumulates in the growth medium. About 50 g/L of lysine can be produced in 2 to 3 days. L-Glutamic acid and MSG can be produced by direct fermentation, using cultures of *Corynebacterium glutamicum* and *Brevibacterium flavum* for large-scale production. The fermentation process employs a glucose-mineral salts medium with urea as a nitrogen source. The difficulty in the production of glutamic acid, as well as other amino acids, by fermentation is getting the cells to secrete sufficient quantities of the amino acid to permit commercial production.

Acids can be made by microbial fermentation using various bacterial and fungal cultures.

PRODUCTION OF ENZYMES

Enzymes have many commercial applications (FIG. 24-13). The four most extensively produced microbial enzymes are protease, glucoamylase, alpha-amylase, and glucose isomerase. Microbial production of useful industrial enzymes is advantageous because of the large number of enzymes and the virtually unlimited supply that can be produced by microorganisms.

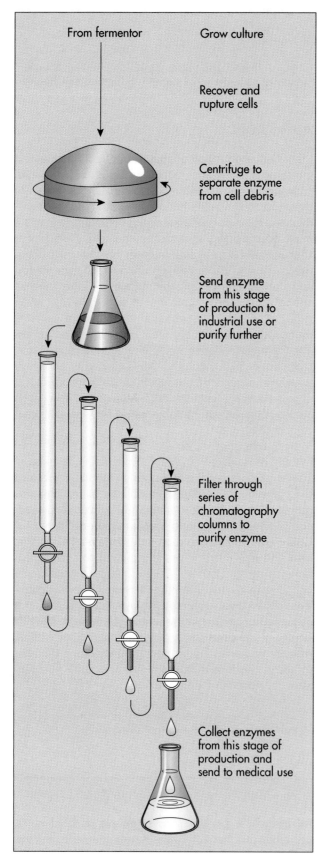

From fermentor

Grow culture

Recover and
rupture cells

Centrifuge to
separate enzyme
from cell debris

Send enzyme
from this stage
of production to
industrial use or
purify further

Filter through
series of
chromatography
columns to
purify enzyme

Collect enzymes
from this stage of
production and
send to medical use

FIG. 24-13 Enzymes are produced by recovering cells from culture and releasing the enzymes by rupturing the cells. The cell-free extract undergoes minimal purification for industrial uses. Extensive additional purification is used for enzymes used in medicine.

PROTEASES

Proteases are a class of enzymes that attack the peptide bonds of protein molecules, forming small peptides. Proteases produced by different bacterial species are used for different industrial purposes. The largest commercial application of bacterial alkaline proteases is in the laundry industry, principally in making detergent. The general trend toward the use of nonphosphate laundry detergents that function at lower wash temperatures has led to the increasing incorporation of enzymes into liquid and powdered detergents to improve their cleaning performance. In the United States, over 25% of detergents contain enzymes. Proteases are also used as spot removers in dry cleaning and as presoak treatments in laundering. The action of the enzyme degrades various proteinaceous materials such as milk and eggs, forming small peptide fragments that can be washed out readily. These protease enzymes are relatively heat stable. They remain active in warm-hot water long enough for them to degrade the proteinaceous contaminants. Currently proteases for detergents are produced largely by *Bacillus licheniformis*. The enzymes produced by these *Bacillus* strains are active against protein molecules that make up common stains such as blood and grass.

> Proteases break peptide bonds in proteins and are used as cleaning agents and in baking; bacterial proteases are produced by *Bacillus* species and fungal proteases are produced by *Aspergillus* species.

Other proteases are being developed using recombinant DNA technology. The qualities to be engineered include the ability to function over a wide pH and temperature range, to remain stable under alkaline conditions in the presence of detergent components such as bleach, and to exhibit long shelf life stability. One recombinant *Bacillus* strain secretes a protease that is highly active toward the milk protein casein. Another recombinant *Bacillus* strain produces several proteases, one of which remains active over a broad pH range, exhibits exceptional stability under highly alkaline conditions, and functions in the presence of bleach.

Recombinant DNA technology also has been employed to develop a bacterial strain that produces a specific protease, known as Kerazyme, that is used to dissolve keratin. Keratin generally resists enzymatic attack. Hair consists of the protein keratin. Kerazyme is used for opening hair-clogged drains. Kerazyme is one of the most successful products, in terms of economic profit, developed to date by genetic engineering.

Another major use of microbial proteases is in the baking industry. Proteases alter the properties of the proteins of flour. Fungal protease is added in the manufacture of most commercial bread in the United

States to reduce mixing time and improve the quality of the loaf. Either fungal or bacterial protease is used in the manufacture of crackers, biscuits, and cookies. Fungal proteases are principally obtained from *Aspergillus* species, and bacterial proteases are primarily produced using *Bacillus* species.

AMYLASES

Amylases, which are enzymes that break down starch, have many commercial applications (Table 24-4). Amylases are used for the preparation of sizing agents (agents used to fill pores in cloth and paper) and the removal of starch sizing from woven cloth. They are used in the preparation of starch-sizing pastes for use in paper coatings, the liquefaction of heavy starch pastes, in the manufacture of corn and chocolate syrups, in the production of bread, and in the removal of food spots in the dry-cleaning industry. Amylases are also sometimes used to replace or augment malt for starch hydrolysis in the brewing industry, as in the production of low-calorie beers.

α-Amylases are extracellular enzymes produced by various bacteria and fungi that cleave internal α-1,4 glycosidic bonds of starch. Some break down the starch without yielding free sugars; others form free sugars as a result of attack on the starch. *Bacillus subtilis* and *B. diastaticus* are used for the commercial production of bacterial α-amylases. Fungal production of α-amylases uses *Aspergillus* species. *A. oryzae* is used to produce amylases from wheat bran by culture on the solid bran, and *A. niger* is used to produce amylases in aerated liquid culture using a starch-mineral salts medium. *Aspergillus* and *Rhizopus* species are also used to produce glucoamylases,

which are enzymes that release free glucose units from starch.

The conversion of starch to a high-fructose corn syrup sweetener is an economically significant and relatively new industrial process that uses microbial enzymes (FIG. 24-14). It produces over 2 million tons of high-grade sweetener per year. This sweetener is produced in a three-step process using the enzymes α-amylase, glucoamylase, and glucose isomerase. Because it has a sweeter flavor, it is rapidly replacing sucrose as the primary sweetener in soft drinks. The use of mutation-screening methods combined with genetic recombination techniques has permitted the development of strains of *Bacillus subtilis* with greatly enhanced abilities to produce high yields of α-amylase. The development of such strains markedly increases the economic feasibility of producing sweeteners and similar products using microbial enzymes.

> **Amylases are used in making sizing and the removal of sizing. They are also used in the production of a high-fructose corn syrup sweetener.**
>
> **Bacterial amylase is produced by *Bacillus* species and fungal amylase is produced by *Aspergillus* species.**

OTHER ENZYMES

Various other microbial enzymes are produced for industrial applications. Rennin is used in the production of cheese. *Mucor pussilus* or *M. meihei* can be used for the commercial production of rennin for curdling milk in cheese production. Fungal pectinase enzymes are used in the clarification of fruit juices. Glucose oxidase, produced by fungi, is used for removing glu-

---

TABLE 24-4
Applications of Amylases

INDUSTRY	SOURCE *BACILLUS*	SOURCE *ASPERGILLUS*	APPLICATION
Alcohol	+	+	Liquefaction of starch before the addition of malt for saccharification
Baked goods		+	Increase in the proportion of fermentable carbohydrates
Brewing	+		Barley preparation, liquefaction of additives
Brewing		+	Improved fermentability of grains, modification of beer characteristics
Feed	+		Improvement of utilization of enzymatically treated barley in poultry and calf raising
Detergent	+		Increase in cleansing power for laundry; additive in dishwasher detergents
Paper	+		Liquefaction of starch without sugar production for sizing of paper
Textiles	+		Continuous desizing at high temperatures
Sugar	+		Improvement of filterability of cane sugar juice via breakdown of starch in juice; production of glucose, fructose and maltose

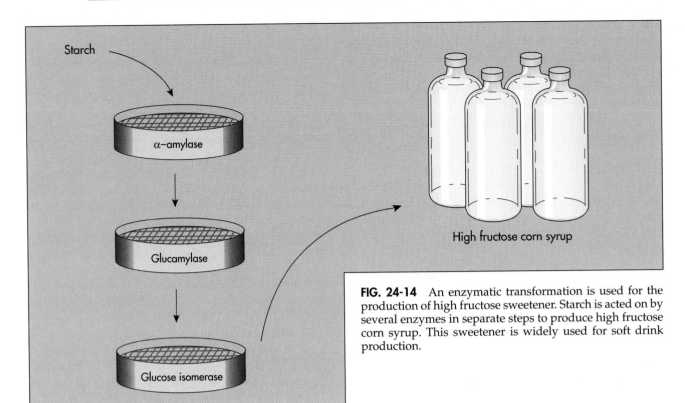

FIG. 24-14 An enzymatic transformation is used for the production of high fructose sweetener. Starch is acted on by several enzymes in separate steps to produce high fructose corn syrup. This sweetener is widely used for soft drink production.

cose from eggs prior to drying. Powdered dried eggs turn brown because of the chemical reaction of proteins with glucose. Removing the glucose stabilizes and prevents deterioration of the dried egg product. Glucose oxidase is also used to remove oxygen from various products such as soft drinks, mayonnaise, and salad dressings to prevent oxidative color and flavor changes. Furthermore, glucose oxidase is incorporated as a part of test strips used by diabetics to determine the levels of glucose in the blood or urine.

The supply of industrial enzymes produced by microorganisms is virtually unlimited.

PRODUCTION OF SOLVENTS

Several organic solvents can be produced by using microbial fermentation. Fermentation processes historically have been important in the industrial production of organic solvents. The process for producing acetone and butanol was discovered in 1915 by Chaim Weizmann (1874-1952), a Polish-born chemist then working in England. During World War I, the microbial production of acetone was very important for the production of the explosive cordite. Microbially produced butanol was converted to butadiene and used in making synthetic rubber. After the war the demand for acetone declined, but the need for *n*-butanol increased for its use in brake fluids, urea-formaldehyde resins, and the production of protective coatings such as automobile lacquers.

Microbial production of acetone and butanol uses anaerobic *Clostridium* species. The fermentation process discovered by Weizmann was based on the conversion of starch to acetone by *C. acetobutylicum*.

Other species, such as *C. saccharoacetobutylicum,* are able to convert the carbohydrates in molasses to acetone and butanol. These *Clostridium* species synthesize butyric and acetic acids, which are then converted to butanol and acetone. Yields of these neutral solvents are typically low. The production of higher concentrations is limited by the toxicity of the compounds. The solvents produced by fermentation are recovered by distillation.

> Acetone and butanol are organic solvents that are produced by microbial fermentation of *Clostridium* species.

PRODUCTION OF FUELS

Limited petroleum resources are forcing many industrialized nations to seek alternative fuel resources. Microbial production of **synthetic fuels** has the potential for helping to meet world energy demands. Useful fuels produced by microorganisms include ethanol, methane, and hydrogen. The use of microorganisms to produce commercially valuable fuels depends on finding the right strains of microorganisms that are able to produce the desired fuel efficiently. It also depends on the availability of an inexpensive supply of substrates for the fermentation process. It is obvious that the production of synthetic fuels must not consume more natural fuel resources than are produced. Microbial production of fuels can be a particularly attractive process when waste materials, such as sewage and municipal garbage, are used as the fermentation substrate.

> Microbial fermentations can be used to convert appropriate substrates, such as plant residue or waste materials, into synthetic fuels.

The microbial production of **ethanol** has become an important source of a valuable fuel, particularly in regions of the world that have abundant supplies of plant residues (FIG. 24-15). Brazil produces and uses large amounts of ethanol as an automotive fuel and plans to replace gasoline with ethanol in the next decade. At present, about 100 million gallons of ethanol per year are used as a fuel, but 12 billion gallons per year would be required to completely re-

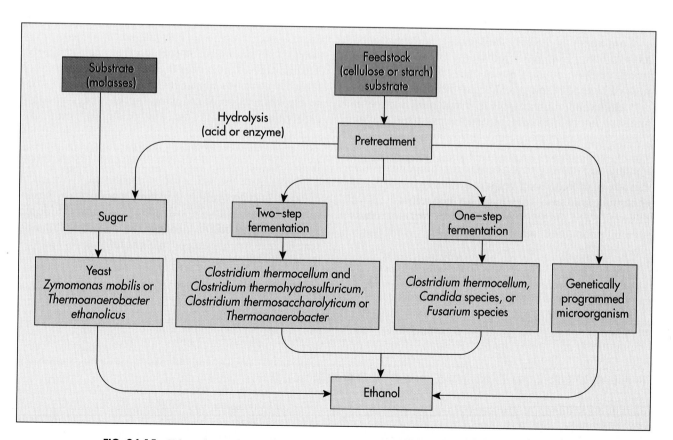

FIG. 24-15 Ethanol is commercially produced from various substrates. Yeast fermentation of molasses is commonly used for the production of alcoholic beverages and ethanol as fuel. Several bacteria, including *Clostridium* spp., and fungi can be used to produce ethanol from cellulose.

place gasoline use in the United States. Currently in the United States, ethanol is used as a fuel additive to gasoline.

There are three major limitations to the successful production of sufficient quantities of ethanol to serve as a major fuel source. Ethanol is relatively toxic to microorganisms, and therefore only limited concentrations of ethanol can accumulate in a fermentation process. Distillation to recover ethanol requires a substantial energy input. This reduces the net gain of fuel as an energy resource produced in this process. Carbohydrate substrates normally used for the production of ethanol are relatively expensive. This makes the cost of fuel produced by fermentation high. However, in tropical countries with abundant plant production the economics are more favorable for ethanol production and use as a fuel.

The bacterium *Zymomonas mobilis* ferments carbohydrates. It forms alcohol twice as rapidly as yeasts. This finding represents a significant advance in the search for a microbial strain for producing ethanol as a fuel. *Thermoanaerobacter ethanolicus*, a thermophilic bacterium, may be even more efficient than the organisms currently used for the fermentative production of ethanol. This organism could grow at a temperature above the boiling point of ethanol so that the ethanol could be recovered as a gas during the growth of the bacterium.

Corn sugar and plant starches are currently used as substrates for the production of ethanol. The prices of these substrates vary greatly and they are needed as food resources. Biomass produced by photosynthetic microorganisms is a potential source of an inexpensive substrate for ethanol production. Cellulose from wood and other plant materials is probably the most promising substrate. It is very likely that genetic engineering can create a microorganism that can efficiently convert cellulose directly to ethanol and that will also tolerate high concentrations of ethanol. Such an organism should permit the commercial production of ethanol as a fuel.

> The microbial production of ethanol uses plant residues and yeasts to ferment carbohydrates to ethanol; the bacteria *Zymomonas mobilis* and *Thermoanaerobacter ethanolicus* are potentially more efficient producers of ethanol.

NEWSBREAK

FUELS OF THE FUTURE

As the reserves of petroleum decline, scientists search for new renewable fuels. Ethanol produced by microbial fermentation already is used as a fuel supplement. In the future, gaseous fuels produced by microorganisms will likely begin to augment and then to replace gasoline as a transportation fuel. Methane (natural gas) produced by methanogenic archaebacteria and molecular hydrogen gas produced by photosynthetic algae and cyanobacteria are the most likely alternate fuels. The Japanese government is supporting a major research initiative to develop a microbial system that can produce sufficient quantities of hydrogen and an automotive engine that can efficiently use this fuel.

Methane (natural gas) produced by methanogenic bacteria is another important natural, renewable energy source. Methane can be used for the generation of mechanical, electrical, and heat energy. Large amounts of methane can be produced by anaerobic decomposition of waste materials. Many sewage treatment plants meet all or part of their own energy needs with the methane produced in their anaerobic sludge digesters. Excess methane produced in such facilities supplies power for some municipalities. Efficient generation of methane can be achieved using algal biomass (algal cells), sewage sludge, municipal refuse, plant residue, or animal waste as a substrate for the growth of methanogenic archaebacteria. Methanogens are obligate anaerobes that produce methane from the reduction of acetate and/or carbon dioxide and hydrogen. Methane production generally requires a mixed microbial community. Some bacterial populations convert the available organic carbon into low molecular weight fatty acids and hydrogen that are substrates for methanogens.

> Methane can be produced by methanogenic bacteria using algal biomass or wastes as the substrate.

RECOVERY OF MINERAL RESOURCES

Microbial mining by the process of bioleaching recovers metals from ores that are not suitable for direct smelting because of their low metal content. **Bioleaching** uses microorganisms to alter the physical or chemical properties of a metallic ore so that the metal can be extracted (FIG. 24-16). Metals can be extracted economically from low-grade sulfide or sulfide-containing ore by exploiting the metabolic activities of thiobacilli, particularly *Thiobacillus ferrooxidans*.

T. ferrooxidans is a chemolithotrophic bacterium that derives energy through the oxidation of either a reduced sulfur compound or ferrous iron. The ferric iron, in turn, chemically oxidizes the metal to be

FIG. 24-16 Large bioleaching operations are used for the recovery of copper. A solution of sulfur oxidizing bacteria *(Thiobacillus)* is sprayed over the copper-bearing rocks. The leachate is drained into a pond and the copper is recovered.

recovered to a soluble form that can be leached from the ore. The process is currently applied on a commercial scale to low-grade copper and uranium ores.

A typical low-grade copper ore contains 0.1% to 0.4% copper. In copper leaching operations, the action of *Thiobacillus* involves direct oxidation of copper sulfide (CuS) and indirect oxidation of CuS via generation of ferric ions from ferrous sulfide. This produces a solution *(leaching solution)* that contains soluble copper (1 to 3 g copper per liter of solution). Copper is recovered from the leaching solution. In the field 50% to 70% of the copper in the ore can be recovered in this process.

Uranium is required by the nuclear power generation industry. Bioleaching provides a mechanism for commercial utilization of low-grade uranium deposits and for the recovery of uranium from low-grade nuclear wastes. Recovery of uranium from radioactive wastes contributes to solving the problem of waste disposal, a major shortcoming of using nuclear power generators. Insoluble tetravalent uranium oxide (UO_2^{4+}) occurs in low-grade ores. UO_2 can be converted to the leachable hexavalent form ($UO_2SO_4^{6+}$) indirectly by *Thiobacilus ferrooxidans*. *T. ferrooxidans* oxidizes the ferrous iron in pyrite (FeS_2), which often accompanies uranium ores, to ferric iron.

The oxidized iron acts as an oxidant, converting UO_2 chemically to UO_2SO_4, which can be recovered by leaching.

In bioleaching, microorganisms are used to alter the physical or chemical properties of a metallic ore to facilitate extraction of the metal.

SUMMARY

Introduction (pp. 707-708)
- Many microorganisms have the metabolic capacity to produce substances of commercial value.
- Sufficient quantities of the desired product must be produced for the process to be economically viable.
- Inexpensive substrates must be used.
- The extraction and purification of the desired product must be efficient and cost effective.

Production of Foods and Beverages (pp. 708-717)
- Many food products are the result of microbial enzymatic activity.

Fermented Dairy Products (pp. 708-710)
- The lactic acid fermentation pathway is used in the production of fermented dairy products. Lactic acid is produced from the metabolism of lactose in milk. Differences in fermented dairy products are due to additional fermentation products.
- Yogurt is made by fermenting milk with a mixture of *Lactobacillus bulgaricus* and *Streptococcus thermophilus*.
- Cheese is made by curdling milk with the enzyme rennin and mixtures of *Streptococcus* and *Lactobacillus* species as lactic acid bacterial starter cultures; some cheeses are ripened, that is, additional microbial growth is used to add flavor and alter texture of the cheese.

Fermented Soybeans (p. 711)
- Several products, such as soy sauce, are produced by fermentation of soybeans.

Bread (pp. 711-712)
- Bread is made by yeasts, usually *Saccharomyces cerevisiae*, carrying out an alcoholic fermentation using the sugars from the carbohydrates in wheat flour as the substrate.

Fermented Beverages (pp. 712-716)
- The conversion of sugar to alcohol by microbial enzymes is an alcoholic fermentation. Differences between alcoholic beverages stem from differences in starting substrates and the production process.
- Beer is made from barley that is crushed to release its plant enzymes (malting). Malt is mixed with adjuncts (mashing). Adjuncts (corn, rice, or wheat) provide the carbohydrate substrates for ethanol production. The mash is heated. A clear liquid (wort) is produced. The wort is cooked with hops and fermented with *Saccharomyces carlsbergensis*. The beer is aged.
- Wine is fermented from grapes. Grapes are crushed to form must. *Saccharomyces cerevisiae* is inoculated into the must.
- The production of distilled liquors is similar to the production of beer. The plant substrate is mashed so that its polysaccharides and proteins are converted to sugars and simple organic compounds that can be fermented by yeasts to form alcohol. The alcohol is heated, collected, distilled, and aged.

Production of Pharmaceuticals (pp. 717-722)

Antibiotics (pp. 717-718)
- Antibiotics are substances naturally made by microorganisms that kill or inhibit other microorganisms.

- In searching for new antibiotics many microorganisms are screened to isolate a unique microbial strain capable of producing a novel metabolite with the desired properties or a strain that produces large quantities of a valuable substance.

Steroids (pp. 718-720)
- Steroids regulate human metabolism.
- The nature and position of the chemical constituents on the basic chemical determine the properties of a steroid. The precision involved in the placement of these constituents makes the chemical synthesis of steroids difficult. Microorganisms can perform these biotransformations.
- The products are recovered by extraction, purified chromatographically, and recovered by crystallization.

Human Proteins (p. 721)
- Human DNA sequences that code for the production of various human proteins have been incorporated into the genomes of bacteria using genetic engineering. These recombinant bacteria are then grown in fermentors and produce human proteins.
- Human insulin, human growth factor, interleukin-2, interferon, and tumor necrosis factor can be produced using recombinant DNA technology.

Vaccines (pp. 721-722)
- Microorganisms with the antigenic ability to elicit a primary immune response are used in the production of vaccines. Mutant, attenuated, or inactivated pathogens may be used.
- Tissue culture is the procedure used for producing viral vaccines. The virus is grown in an *in vitro* culture of animal cells.

Vitamins (p. 722)
- Vitamins are nutritional factors necessary to the health of animals. Some vitamins, such as B_{12} and riboflavin, may be produced by microbial fermentation.

Production of Acids (p. 723)

Organic Acids (p. 723)
- Gluconic acid, citric acid, gibberellic acid, and lactic acid are organic acids that can be produced by microbial fermentation.
- *Acetobacter, Penicillium, Aspergillus, Gibberella*, and *Lactobacillus* species are used in the commercial production of these acids.

Amino Acids (p. 723)
- Lysine and glutamic acid are examples of the amino acids that can be produced by microbial fermentation.
- Lysine is made by using *Corynebacterium glutamicum* with cane molasses.
- Glutamic acid is made by using *Corynebacterium glutamicum* and *Brevibacterium flavum* with a glucose-mineral salts medium.

Production of Enzymes (pp. 723-726)
- Proteases, glucoamylase, α-amylase, and glucose isomerase are commercially important industrial enzymes produced by microorganisms.

Proteases (pp. 724-725)

- Proteases attack the peptide bonds of proteins. Some proteases are used in cleaning products. Other proteases are used in baking.
- *Bacillus* species produce proteases.
- Genetic engineering is being used to create proteases that can function over a wide temperature range and a wide pH range, remain stable, and have a long shelf life.

Amylases (p. 725)

- Amylases are used in sizing agents and the removal of starch sizing.
- *Aspergillus* species and *Bacillus* species are used in the production of amylases.

Other Enzymes (pp. 725-726)

- Rennin is an enzyme used in the production of cheese. It is made using *Mucor* species.
- Fungal pectinase enzymes clarify fruit juices.
- Glucose oxidase is used in making powdered eggs, soft drinks, mayonnaise, and salad dressings. It removes oxygen and prevents oxidative color and flavor changes.

Production of Solvents (pp. 726-727)

- Organic solvents, such as acetone and butanol, can be produced by microbial fermentation using anaerobic *Clostridium* species.

Production of Fuels (pp. 727-728)

- Synthetic fuels can be produced by microorganisms. Ethanol, methane, hydrogen, and hydrocarbons can be produced.
- An inexpensive substrate, the right strain of microorganism, and an energy efficient production method must be found to make the production of synthetic fuels economically feasible.

Recovery of Mineral Resources (pp. 728-729)

- Bioleaching is microbial mining to recover metals from ore that cannot be smelted. Bioleaching alters the physical or chemical properties of the ore so that the metal can be extracted.
- *Thiobacillus ferrooxidans* is used to extract metal from low-grade sulfide or sulfide-containing ore, such as uranium and copper.

CHAPTER REVIEW

REVIEW QUESTIONS

1. What foods are produced by microbial fermentation? Describe the production of each.
2. What beverages are produced by microbial fermentation? Describe the production of each.
3. What are the differences between beer, wine, and distilled spirits?
4. What roles do microorganisms have in the production of pharmaceuticals?
5. Describe the steps involved in soy sauce production.
6. How are microorganisms used in the recovery of metals from the environment?
7. Describe the steps that are involved in the production of beer.
8. How are microorganisms used to produce enzymes? What enzymes are commercially made by microorganisms? What are their uses?
9. Which industrial processes use bacteria?
10. Which industrial processes use yeasts?
11. Why is microbial production of steroids superior to chemical synthetic production of these compounds?
12. How are microorganisms used to produce vitamins?
13. What is industrial fermentation?

CRITICAL THINKING QUESTIONS

1. How would you go about discovering a new therapeutically useful antibiotic?
2. How can microorganisms be genetically engineered to produce useful human proteins?
3. What role will microorganisms play in the future in meeting world food and fuel needs?
4. What is the difference between a commercially useful strain of microorganism and one found in nature?

READINGS

Aharonowitz Y and G Cohen: 1981. The microbiological production of pharmaceuticals, *Scientific American* 245(3):140-152.
   An excellent well-illustrated article on the use of microorganisms by the pharmaceutical industry for producing medicinals.
Ball C (ed.): 1984. *Genetics and Breeding of Industrial Microorganisms*, Boca Raton, FL; CRC Press.
   An advanced work on the genetics of organisms of biotechnological interests.

Barton JH: 1991. Patenting life, *Scientific American* 264(3):40-46.
   An article on the implications of patents for biotechnology.
Brierley CL: 1982. Microbiological mining, *Scientific American* 247(2):44-53.
   A well-written description of bioleaching to recover minerals.
Bud R: 1993. *The uses of life*, Cambridge, Cambridge University Press.
   A history of biotechnology.

Bu'Lock JD and B Kristiansen (eds.): 1987. *Basic Biotechnology*, London, Academic Press.

An introduction to biotechnology.

Bushell ME (ed.): 1984. *Modern applications of traditional biotechnologies*, Amsterdam, Elsevier.

Includes chapters on wine production, brewery fermentations, cheese and whiskey microbiology, and fermentations involved in the production of soy sauce, cocoa, coffee, and tea.

Crueger W and A Crueger: 1989. *Biotechnology: A Textbook of Industrial Microorganisms*, ed. 2, Madison, WI; Science Tech.

A comprehensive text on industrial microbiology and biotechnology.

Demain AL: 1981. Industrial microbiology, *Science* 214:987-995.

An overview of biotechnology with insight into pharmaceutical production.

Demain AL and NA Solomon: 1981. Industrial microbiology, *Scientific American* 245(3):66-76.

A well-written and well-illustrated article on the industrial uses of microorganisms.

Demain AL and NA Solomon (eds.): 1985. *Biology of Industrial Microorganisms*, Stoneham, Mass; Butterworth.

A collection of articles on industrial microorganisms.

Demain AL and NA Solomon: 1986. *Manual of Industrial Microbiology and Biotechnology*, Washington, D.C.: American Society for Microbiology.

An advanced comprehensive manual on industrial microbiology.

Ehrlick HL and CL Brierley (eds.): 1989. *Microbial Mineral Recovery*, New York, McGraw-Hill.

Complete overview of bioleaching.

Eveleigh DE: 1981. The microbiological production of industrial chemicals, *Scientific American* 245(3):154-178.

Describes the production of solvents and other industrial chemicals by microorganisms.

Franks F (ed.): 1993. *Protein biotechnology*, Totowa, NJ; Humana Press.

The biotechnological methodology in the analysis, isolation, purification, and physiology of proteins.

Frazier WC and DC Westhoff: 1988. *Food Microbiology*, ed. 4, New York, McGraw-Hill.

Classic and comprehensive text on food microbiology.

Gaman PM and KB Sherrington: 1981. *The Science of Food: An Introduction to Food Science, Nutrition, and Microbiology*, New York, Pergamon Press.

Shows how microorganisms are used to produce foods.

Hayes PR: 1985. *Food Microbiology and Hygiene*, London, Elsevier.

Discusses food handling techniques and sanitation in the food service industry.

Herrmann KM and RL Somerville (eds.): 1983. *Amino Acids: Biosynthesis and Genetic Regulation*, Reading, MA; Addison-Wesley.

Shows how microorganisms are used to produce amino acids.

Hopwood DA: 1981. The genetic programming of industrial microorganisms, *Scientific American* 245(3):91-102.

Discusses the genetic basis of biotechnology.

Hutchins S, S Davidson, J Brierly, C Brierly: 1986. Microorganisms in reclamation of metals, *Annual Review of Microbiology* 40:311-366.

Describes the physiological basis for bioleaching.

Jay JN: 1986. *Modern Food Microbiology*, ed. 4, New York, Van Nostrand.

Authoritative text on food microbiology.

Jones DT and DR Woods: 1986. Acetone-butanol fermentation revisited, *Microbiological Reviews* 50:484-524.

Describes the biochemistry and gives insight into fermentations carried out by clostridia.

Kharatyan SG: 1978. Microbes as food for humans, *Annual Review of Microbiology* 32:301-327.

Describes the production of single cell protein.

Labeda DP (ed.): 1990. *Isolation of Biotechnological Organisms from Nature*, New York, McGraw-Hill.

Methods for isolating microorganisms with industrial potential.

Microorganisms in Foods 2: 1986. Toronto, University of Toronto Press.

Discusses techniques of food microbiology and how foods themselves and the statistics they generate are analyzed.

Miller BM and W Litsky: 1976. *Industrial Microbiology*, New York, McGraw-Hill.

A classic textbook on industrial microbiology.

Montville TJ: 1987. *Food Microbiology*, Boca Raton, FL; CRC Press.

Thorough coverage of the roles of microorganisms in the food industry.

Omura S (ed.): 1992. *The Search for Bioactive Compounds from Microorganisms*, New York, Springer-Verlag.

Describes the use of microbiology and biotechnology in the discovery of biological products.

Peberdy JF (ed.): 1990. *Applied Molecular Genetics of Fungi*, Cambridge, Cambridge University Press.

Introduction to the biotechnology of fungi.

Phaff H: 1981. Industrial microorganisms, *Scientific American* 245(3):77-89.

Interesting article on the organisms used in classical industrial fermentations.

Rehm H and G Reed (eds.): 1981. *Biotechnology: A Comprehensive Treatise* (8 volumes), Deerfield Beach, FL; Verlag Chemie International.

Comprehensive coverage of biotechnology.

Rose AH: 1981. The microbiological production of food and drink, *Scientific American* 245(3):126-139.

Well-written review of microbial food and beverage production.

Rose AH (ed.): 1983. *Food Microbiology*, London, Academic Press.

Classic text on food microbiology.

Tombs MP: 1991. *Biotechnology in the Food Industry*, Philadelphia, Open University Press.

Applications of biotechnology and genetics in the food industry.

Vanek Z and Z Hostalek (eds.): 1986. *Overproduction of Microbial Metabolites: Strain Improvement and Process Control Strategies*, Stoneham, MA; Butterworth.

Discusses genetic and physiological methods for obtaining adequate microbial metabolites.

Wood BJ: 1985. *Microbiology of Fermented Foods*, New York, Elsevier Science Publishing.

Interesting coverage of uses of microorganisms to produce fermented foods.

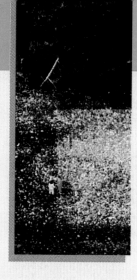

Environmental Microbiology

PREVIEW TO CHAPTER 25

In this chapter we will:
- See how microorganisms contribute to the maintenance of environmental quality.
- Review the ways in which microorganisms are used for waste disposal.
- See how microorganisms relate to drinking water quality.
- Learn how microorganisms biodegrade environmental pollutants.
- Learn the following key terms and names:

acid mine drainage	landfills
activated sludge process	nitrification
anaerobic digestors	nitrogen cycle
biodegradation	nitrogen fixation
biogeochemical cycling	oil biodegradation
biological oxygen demand	oxidation ponds
bioremediation	primary sewage treatment
carbon cycle	sanitary landfill
coliform count	secondary sewage
composting	treatment
decomposers	self-purification
detritus	septic tank
disinfection	sewage treatment
eutrophication	stabilization ponds
food web	tertiary sewage treatment
grazers	trickling filter system
indicator organisms	xenobiotics

Human activities create vast amounts of wastes and pollutants. The release of these materials into the environment sometimes causes serious health problems. It may also prevent the use of our land and water resources for some purposes. Rivers serve as habitats for fish, as a source of irrigation and drinking water, and for the disposal of sewage. The continued use of rivers for all these purposes depends on the careful management of the amounts of wastes and levels of associated pathogenic microorganisms entering the river ecosystem. Waste waters from one community flow into the potable water supply of another. There are limits to the natural capacity of an ecosystem to manage wastes and pollutants. The microorganisms in an ecosystem have a limited capacity to decompose unwanted materials without causing serious deterioration of environmental quality or an increase in the incidence of disease.

The level of wastes produced by dense human and domestic animal populations often exceeds the local ecosystem's capacity to cope with it. This can result in serious environmental pollution and epidemic outbreaks of disease. Each year more than 500 million people are afflicted with waterborne diseases, and more than 10 million of them die. Most of these disease outbreaks stem from fecal contamination of drinking water supplies. Proper treatment of wastes, employing microbial biodegradation, and disinfection of potable water supplies to kill contaminating pathogenic microorganisms can greatly improve the safety and quality of water supplies and the status of human health.

ENVIRONMENTAL ASPECTS OF MICROBIAL METABOLISM

BIOGEOCHEMICAL CYCLING

The diverse metabolic activities of microorganisms help maintain global ecological balance. Life on earth could not exist without the metabolic activities of microorganisms. Microorganisms are especially important in recycling materials. A tree that falls in the forest is attacked by microorganisms and decomposes (FIG. 25-1). The materials released from the decomposition of the tree become available to sustain the life needs of other organisms, including plants and animals. **Biogeochemical cycling** is the movement of materials via biochemical reactions through the global biosphere. The biosphere is that portion of the Earth and its atmosphere in which living organisms occur. The activities of microorganisms within the biosphere have a direct impact on the quality of human life.

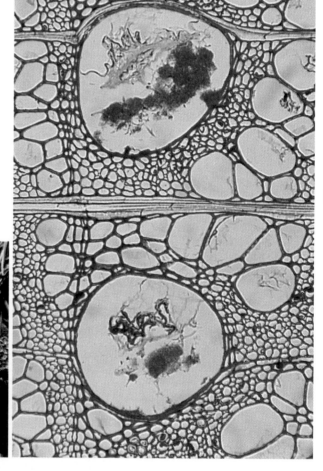

FIG. 25-1 **A,** The myxomycete *Fuligo septica* decomposing a dead tree. **B,** Micrograph showing bacteria and fungi growing within dead plant cells. These microorganisms are decomposing the dead plant matter.

Without the essential biogeochemical cycling activities of microorganisms, higher forms of life, including humans, could not exist.

The metabolism carried out by microorganisms often transfers materials from one place to another. Changes in the chemical forms of various elements can lead to the physical movement of materials, sometimes causing transfers between the atmosphere (air), hydrosphere (water), and lithosphere (land). Physical movement, for example, between soluble and insoluble states, is an important part of biogeochemical cycling.

Carbon Cycle

Carbon is actively cycled between inorganic carbon dioxide and various organic compounds that compose living organisms. The carbon cycle primarily involves the transfer of carbon dioxide and organic carbon between the atmosphere, where carbon occurs principally as inorganic CO_2, and the hydrosphere and lithosphere, which contain varying concentra-
tions of organic and inorganic carbon compounds (FIG. 25-2). The autotrophic metabolism of photosynthetic and chemolithotrophic organisms is responsible for *primary production,* the conversion of inorganic carbon dioxide to organic carbon. Once carbon is fixed (reduced) into organic compounds, it can be transferred from population to population within the biological community, supporting the growth of a wide variety of heterotrophic organisms. The respiratory and fermentative metabolism of heterotrophic organisms returns inorganic carbon dioxide to the atmosphere, completing the carbon cycle. Different microorganisms have evolved diverse metabolic capacities that utilize virtually all naturally occurring organic substances. Their biodegradative capacities have been viewed as almost limitless. Microorganisms attack a variety of organic compounds, often producing carbon dioxide and water.

Microorganisms transfer carbon between atmospheric carbon dioxide and the organic carbon compound of living organisms in soil and water.

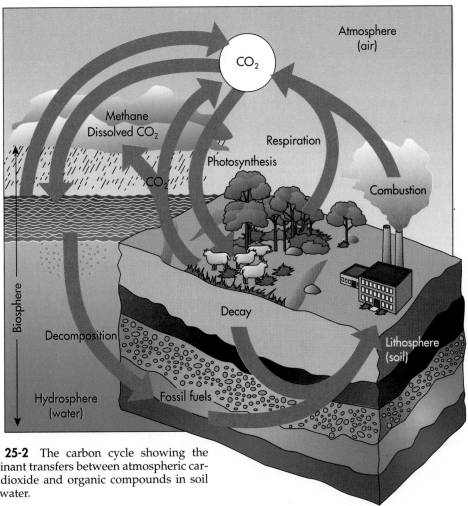

FIG. 25-2 The carbon cycle showing the dominant transfers between atmospheric carbon dioxide and organic compounds in soil and water.

The transfer of energy stored in organic compounds between the organisms in the community forms a **food web,** an integrated feeding structure (FIG. 25-3). At the base of the food web are the primary producers that form the organic matter for the system. *Grazers* are organisms that feed on primary producers. In *phytoplankton-based food webs,* algae and cyanobacteria are the primary food source for grazers. In *detrital food webs,* microbial biomass produced from growth on dead organic matter *(detritus)* serves as a primary food source for grazers. The grazers, in turn, are eaten by *predators,* which in turn may be preyed on by larger predators.

The overall feeding relationships establish a pyramid of biological populations in the food web (FIG. 25-4). Organisms that feed on others are at higher *trophic levels* (feeding levels) within the food web. The pyramid shape occurs because only a small portion of the energy stored in any trophic level is transferred to the next higher trophic level. Normally, 85% to 90% of the energy stored in the organic matter of a trophic level is consumed by respiration during transfer to the next trophic level and enters the decay portion of the food web. Consequently, the higher the trophic level, the smaller its biomass.

The transfer of matter and its stored energy occurs in food webs in which organisms at higher trophic levels feed on those at lower trophic levels.

The decay portions of food webs are dominated by microorganisms. Microbial decomposition of dead

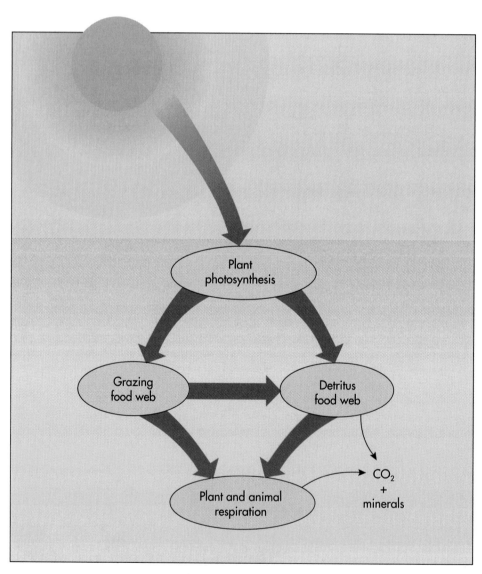

FIG. 25-3 In a food web organic compounds formed by primary producers are transferred to higher trophic (feeding) levels. Decomposers return organic biomass to carbon dioxide and minerals. In a grazing food web, primary producers feed grazers and other consumers. In a detrital food web, dead organic matter (detritus) initially feeds decomposers. The biomass of the decomposers subsequently is consumed.

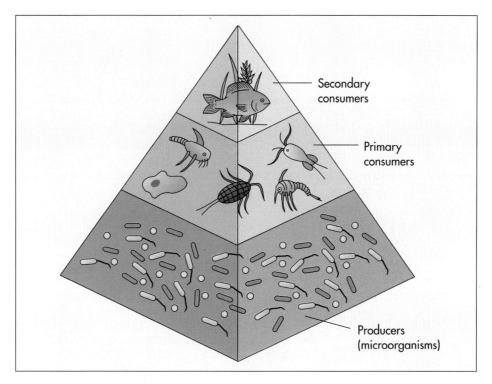

FIG. 25-4 The feeding relationships in an ecosystem form a food web pyramid. Only a portion of the stored energy is transferred, so that each successively higher trophic level has less biomass.

plants and animals and partially digested organic matter is largely responsible for the conversion of organic matter to carbon dioxide and the reinjection of inorganic CO_2 into the atmosphere. The rates of organic matter mineralization depend on various factors, including environmental conditions, such as pH, temperature, and oxygen concentration, and the chemical nature of the organic matter. Some natural organic compounds, such as lignin, cellulose, and humic acids, are relatively resistant to attack and decay only slowly. Various synthetic compounds, such as DDT, may be *recalcitrant,* that is, completely resistant to enzymatic degradation. We depend on the activities of microorganisms to decompose organic wastes, and when microbial decomposition is ineffective, organic compounds accumulate. This is evidenced by the environmental accumulation of plastic materials that are recalcitrant to microbial attack. Many modern problems relating to the accumulation of environmental pollutants reflect the inability of microorganisms to degrade rapidly enough the concentrated wastes of industrialized societies.

Microbial decomposition is essential in the recycling of carbon; some compounds are recalcitrant to microbial attack and accumulate in the environment.

The production of methane by a specialized group of methanogenic archaebacteria represents a shunt to the normal cycling of carbon because the methane that is produced cannot be used by most heterotrophic organisms and thus is lost from the biological community to the atmosphere. Normally, fossil fuels such as coal and petroleum are not actively cycled through the activities of microorganisms. Burning fossil fuels also adds CO_2 to the atmosphere, which has led to a general rise in the concentration of atmospheric CO_2 and a resultant warming of global temperatures, a phenomenon known as the *greenhouse effect.*

Nitrogen Cycle

Molecular nitrogen, the most abundant substance in the atmosphere, is not directly usable by most organisms; only a few bacteria are able to use molecular nitrogen directly. Microorganisms are able to utilize other forms of nitrogen such as NH_4^+, NO_2^- and NO_3^-, as well as organic nitrogen-containing compounds such as amino acids and proteins. The conversions of nitrogen compounds, primarily by microorganisms, changes the oxidation states of nitrogenous compounds and establishes a nitrogen cycle (FIG. 25-5).

Three processes carried out by microorganisms are critical in the nitrogen cycle: (1) nitrogen fixation, (2) nitrification, and (3) denitrification. Nitrogen fixation, which is strictly a bacterial process, converts

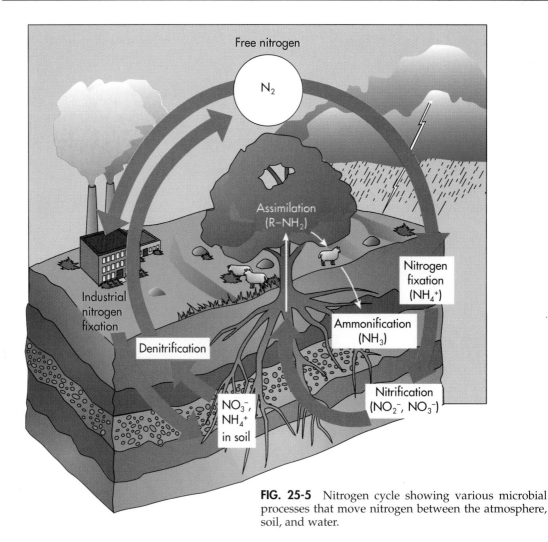

FIG. 25-5 Nitrogen cycle showing various microbial processes that move nitrogen between the atmosphere, soil, and water.

HIGHLIGHT

SOIL FERTILITY AND MANAGEMENT OF AGRICULTURAL SOILS

Microbial biogeochemical cycling activities are extremely important for the maintenance of soil fertility, that is, the ability of the soil to support plant growth. The nutrient in most limited supply normally is nitrogen, and thus the concentration of fixed forms of nitrogen in soil usually determines the potential productivity of an agricultural field. The natural availability of fixed forms of nitrogen in agricultural soils is determined by the relative balance between the rates of microbial nitrogen fixation and denitrification. Nitrogen-rich fertilizers are widely applied to soils to support increased crop yields, but proper application of nitrogen fertilizers must take into consideration the solubility and leaching characteristics of the particular chemical form of the fertilizer and the rates of microbial biogeochemical cycling activities. To avoid the losses caused by leaching and denitrification, nitrogen fertilizer is commonly applied as an ammonium salt, free ammonia, or urea. When nitrification proceeds too quickly, as

it does in some agricultural soils, wasteful losses of nitrogen fertilizer and groundwater contamination with nitrate occur. Nitrification of ammonium compounds also yields acidic products that may have to be neutralized by liming. To prevent the undesirable microbial transformation of nitrogen fertilizers, nitrification inhibitors, such as nitrapyrin, are often applied with the nitrogen fertilizer. The use of nitrification inhibitors can increase crop yields by 10% to 15% for the same amount of nitrogen fertilizer applied. In addition, by decreasing the rate of nitrification, the problem of groundwater pollution by nitrate is prevented.

Crop rotation, that is, alternating the types of crops planted in a field, is traditionally used to prevent the exhaustion of soil nitrogen and to reduce the cost of nitrogen fertilizer applications. Leguminous crops such as soybeans often are planted in rotation with other crops because of their symbiotic association with nitrogen-fixing bacteria, which reduce the soil's requirement for ex-

molecular nitrogen to ammonium ion; this is the only naturally occurring process that makes nitrogen available to living organisms. It is carried out by a few bacteria that have a nitrogenase enzyme system. It brings nitrogen from the atmosphere into the hydrosphere and lithosphere. Because plants depend on the availability of nitrogen for growth, microbial metabolism of nitrogen-containing compounds has a dramatic impact on agricultural productivity. Plants and animals rely entirely on the fixed forms of nitrogen (such as ammonium and nitrate ions) provided by bacterial nitrogen fixation. In terrestrial habitats, the microbial fixation of atmospheric nitrogen is carried out by free-living bacteria and by bacteria living in symbiotic association with plants. *Symbiotic nitrogen fixation* by *Rhizobium* or *Bradyrhizobium* is most important in agricultural fields, where these bacteria live in association with leguminous crop plants (FIG. 25-6, p. 740). *Rhizobium* and *Bradyrhizobium* species generally exhibit rates of nitrogen fixation that are two to three orders of magnitude higher than those accomplished by free-living, nitrogen-fixing soil bacteria.

Nitrification, which is a process carried out by chemolithotrophic bacteria, converts ammonium ions to nitrite and nitrate ions. Different bacteria are involved in the transformation of ammonium ions to nitrite ions and the conversion of nitrite ions to nitrate ions (Table 25-1). Because nitrite and nitrate ions are negatively charged, whereas ammonium ions are positively charged, nitrification mobilizes nitrogen (releasing the positively charged ammonium ions

from negatively charged soil particles), so that nitrogen moves from soil into groundwater (FIG. 25-7, p. 740). The leaching of nitrogen from soil reduces the fertility of the soil. The accumulation of nitrate in groundwater can be a health hazard. Denitrification, which returns molecular nitrogen to the atmosphere, is an anaerobic process in which bacteria convert nitrate ions to molecular nitrogen.

Nitrogen fixation, nitrification, and denitrification are three key steps in the biogeochemical cycling of nitrogen.

As a result of the biogeochemical cycling of nitrogen, nitrogen moves from the atmosphere through the biota, soil, and aquatic habitats.

TABLE 25-1		
Genera of Nitrifying Bacteria		
GENUS	CONVERTS	HABITAT
Nitrosomonas	Ammonia to nitrite	Soils, freshwater, marine
Nitrosospira	Ammonia to nitrite	Soils
Nitrosococcus	Ammonia to nitrite	Soils, freshwater, marine
Nitrosolobus	Ammonia to nitrite	Soils
Nitrobacter	Nitrite to nitrate	Soils, freshwater, marine
Nitrospina	Nitrite to nitrate	Marine
Nitrococcus	Nitrite to nitrate	Marine

pensive nitrogen fertilizer. Leguminous plants produce more fixed nitrogen than they require, and the excess ammonium nitrogen is released to the soil. Of more importance in terms of soil fertility is the fact that most of the combined (fixed) nitrogen is released to the soil on decomposition of the crop residues from leguminous plants that are plowed under (see Table). Soybeans and

corn are often rotated every few years in the midwestern United States because corn takes up nitrogen from the soil, substantially decreasing the concentration of soil nitrogen. During the seasons when soybeans are grown, the level of fixed nitrogen in the soil increases. In some cases, nitrogen fixation can be enhanced by inoculation of legume seeds with appropriate *Rhizobium* strains, which increases the extent of nodule formation because of the increased numbers of rhizobia that effectively initiate the infective process that leads to nodule formation. Besides increasing the extent of nodule formation, it is possible to take steps to increase the rate of nitrogen fixation within the nodules. In molybdenum-deficient soils, a dramatic improvement in the rate of nitrogen fixation can be achieved by the application of small amounts of molybdenum because this element is a constituent of the nitrogenase enzyme complex, which is required for nitrogen-fixing activities. It is important that maximal rates of nitrogen fixation be achieved by rotation with leguminous crops to successfully replenish soil nitrogen.

Nitrogen Gains in Soils in the United States Obtained by Planting Leguminous Crops	
CROP	SOIL NITROGEN INCREASE (kg NITROGEN FIXED/HECTARE/YEAR)
Alfalfa	100-280
Red clover	75-175
Pea	75-130
Soybean	60-100
Cowpea	60-120
Vetch	60-140

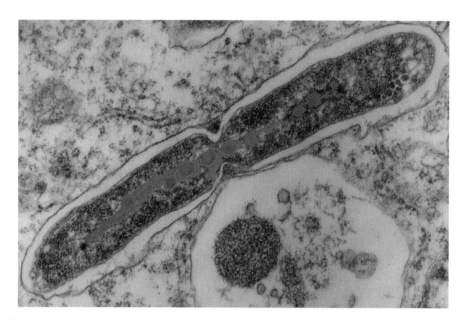

FIG. 25-6 Colorized micrograph of the nitrogen fixing bacterium *Bradyrhizobium japonicum.*

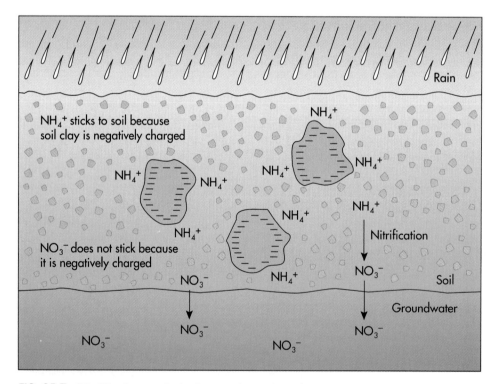

FIG. 25-7 Nitrification results in the transformation of ammonium ions (NH_4^+) to nitrate ions (NO_3^-). This process mobilizes the nitrogen in soil so that, when it rains, nitrate leaches into the groundwater.

ACID MINE DRAINAGE

Acid mine drainage is a consequence of the metabolism of sulfur and iron-oxidizing bacteria. Coal in geological deposits is often associated with pyrite (FeS_2). When coal mining activities expose pyrite ores to atmospheric oxygen, the combination of autoxidation and microbial sulfur and iron oxidation produces large amounts of sulfuric acid. When pyrites are mined as part of an ore recovery operation, oxidation may produce large amounts of acid. The acid draining from mines kills aquatic life and renders the water it contaminates unsuitable for drinking or for recreational uses. At present, approximately 10,000 miles of U.S. waterways are affected in this manner, predominantly in the states of Pennsylvania, Virginia, Ohio, Kentucky, and Indiana.

Strip mining is a particular problem of acid mine drainage because this method of coal recovery removes the overlying soil and rock, leaving a porous rubble of tailings exposed to oxygen and percolating water. Oxidation of the reduced iron and sulfur in the tailings produces acidic products, causing the pH to drop rapidly and preventing the reestablishment of vegetation and a soil cover that would seal the rubble from oxygen. The sulfuric acid produced accounts for the high acidity and the precipitated ferric hydroxide accounts for the deep brown color of the effluent. A strip-mined piece of land continues to produce acid mine drainage until most of the sulfide is oxidized and leached out; recovery of this land may take 50 to 150 years.

Sulfur Cycle

Sulfur can exist in several oxidation states within organic and inorganic compounds, and oxidation-reduction reactions—mediated by microorganisms—change the oxidation states of sulfur within various compounds, establishing the *sulfur cycle* (FIG. 25-8). Microorganisms are capable of removing sulfur from organic compounds. Under aerobic conditions, the removal of sulfur (*desulfurization*) of organic compounds results in the formation of sulfate, whereas under anaerobic conditions hydrogen sulfide is normally produced from the mineralization of organic sulfur compounds. Hydrogen sulfide may also be formed by sulfate-reducing bacteria that utilize sulfate as the terminal electron acceptor during anaerobic respiration. Hydrogen sulfide can accumulate in toxic concentrations in areas of rapid protein decomposition, is highly reactive, and is very toxic to most biological systems. It can react with metals to form insoluble metallic sulfides.

The predominant source of hydrogen sulfide in different habitats varies. In organically rich soils, most of the hydrogen sulfide is generated from the decomposition of organic sulfur-containing compounds. In anaerobic sulfate-rich marine sediments, most of it is generated from the dissimilatory reduction of sulfate by sulfate-reducing bacteria, such as members of the genus *Desulfovibrio*. Anaerobic sulfate reduction is important in corrosion processes and in the biogeochemical cycling of sulfur. Although hydrogen sulfide is toxic to many microorganisms, the photosynthetic sulfur bacteria use it during their metabolism. The anaerobic photosynthetic bacteria often occur on the surface of sediments, where there is light to support their activities and a supply of hydrogen sulfide from sulfate reduction and anaerobic degradation of organic sulfur-containing compounds.

Aerobic conversion of sulfides to sulfates is carried out by *Thiobacillus* species. Thiobacilli can be used for recovery of metals by bioleaching (see FIG. 24-16).

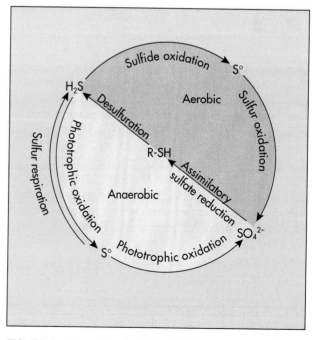

FIG. 25-8 The sulfur cycle involves conversions of sulfur in various oxidation states. Oxidation of sulfur compounds occurs in aerobic environments and reduction of sulfur compounds in anaerobic ones.

Besides their critical roles in natural biogeochemical cycling, microorganisms are essential for the destruction of wastes and pollutants. Microorganisms degrade organic wastes and pollutants, returning them to the natural biogeochemical cycling of elements. Wastes are collected and fed into systems that favor microbial degradation activities.

Urban solid waste production in the United States amounts to roughly 150 million tons per year. Much of this material is composed of glass, metal, and plastic. The rest is decomposable organic waste such as kitchen scraps, paper, and other household and industrial garbage. Sewage sludge derived from treatment of liquid wastes, animal waste from cattle feed lots, and large-scale poultry and swine farms are also major sources of solid organic waste. In traditional small family farm operations, most organic solid waste is recycled into the land as fertilizer. In highly populated urban centers and areas of large-scale agricultural production, the disposal of massive amounts of organic waste is a difficult and expensive problem.

> **Solid waste is made up of nondecomposable waste components, such as glass and metal. and decomposable organic waste, such as household and industrial garbage and sewage sludge.**

There are several options for dealing with solid waste problems. Today, many of the inert components of solid waste such as aluminum and glass are recovered and recycled. Even paper, which is relatively resistant to microbial degradation, can be recovered from solid waste. Many books and newspapers are printed on recycled paper. The remaining bulk of the solid waste may be burned in an incinerator. Incineration, however, creates air pollution problems. Alternatively, organic components can be subjected to microbial biodegradation in aquatic or terrestrial environments. In many cases, solid waste is simply dumped at sea or discarded on land. In such cases, biodegradation is allowed to occur naturally without any special treatment. Excessive dumping of organic wastes into terrestrial and marine ecosystems, however, can result in serious problems.

SANITARY LANDFILLS

The simplest and least expensive way to dispose of solid waste is to place it in **landfills** (FIG. 25-9). Solid waste is then allowed to decompose naturally. Organic and inorganic solid wastes are deposited together in low-lying land that has minimal real estate value. Exposed waste can cause aesthetic and odor problems, attract insects and rodents, and pose a fire

hazard. Each day's waste deposit, therefore, is covered over with a layer of soil. This creates a **sanitary landfill.** When the landfill is full, the site can be used for recreational purposes. It may even one day be able to provide a foundation for construction.

> **A sanitary landfill is created on a piece of low value low-lying land into which waste is dumped.**
>
> **Each waste deposit is covered with soil to minimize aesthetic, odor, rodent, and fire problems.**

For 30 to 50 years after the establishment of a landfill, the organic content of the solid waste undergoes slow, anaerobic microbial decomposition. The products of anaerobic microbial metabolism include carbon dioxide, water, methane, various low molecular weight alcohols, and acids. These products diffuse into the surrounding water and air. Extensive amounts of methane are produced during this decomposition process. The methane can be collected and sold to nearby power plants. The landfill settles slowly. Eventually, decomposition slows significantly. This signals the completion of the biodegradation of the solid waste. Settling later ceases. The land is then stabilized and suitable as a site for construction.

> **After approximately 50 years of microbial decomposition, a settled landfill may be ready for other uses.**

Although the use of sanitary landfills is simple and inexpensive, there are several problems associated with this waste disposal method. Most importantly, the number of suitable disposal sites available in urban areas is very limited. This often means long, energy-consuming hauling of the solid waste to available sites. Premature construction on a still biologically active landfill site may result in structural damage to the buildings because of movement of the land base. An explosion hazard may exist due to methane seepage into basements and other below-ground structures. Above-ground plantings may also be damaged by methane seepage.

Decomposition in a landfill is often very incomplete due to the limitation of water. As the waste material settles and compacts, it becomes more difficult for water to percolate through the landfill. Materials such as newspapers require moisture for their degradation. Other problems with landfills include the possible seepage of anaerobic decomposition products, heavy metals, and a variety of recalcitrant (nonbiodegradable) hazardous pollutants from the landfill site into underground water sources. These problems cause many municipalities to place severe restrictions on the location and operation of landfills

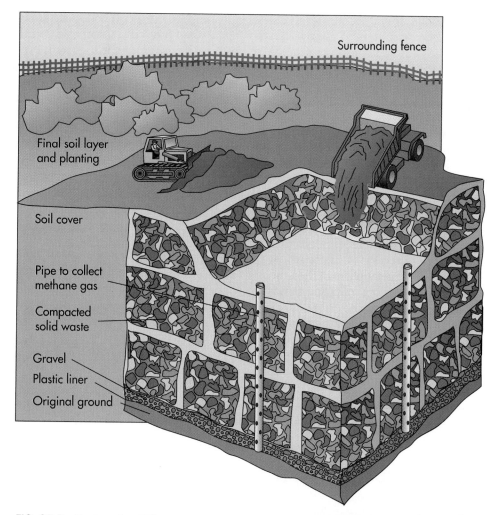

Surrounding fence

Final soil layer
and planting

Soil cover

Pipe to collect
methane gas

Compacted
solid waste

Gravel

Plastic liner

Original ground

FIG. 25-9 Sanitary landfills are an inexpensive way of decomposing solid organic wastes. Wastes are added and covered with soil. The soil is not tilled and becomes anaerobic. Decomposition by fermentation and anaerobic decomposition proceeds under these conditions. Many municipalities use this method of anaerobic decomposition for disposal of municipal waste. Methane is one of the products formed from decomposition of organic matter in landfills.

and the types of materials that can be deposited in them. Thus alternatives to the landfill technique for disposing of solid waste are being sought by many municipalities.

COMPOSTING

The organic portion of solid waste can be biodegraded by composting (FIG. 25-10). **Composting** is the process by which solid organic waste material is degraded by aerobic, mesophilic, and thermophilic microorganisms. It converts organic waste materials into a stable, sanitary, humus-like product. Reduced in bulk, it can be used for soil improvement. Composting requires that solid waste be sorted into its organic and inorganic components. Sorting can be accomplished at the source, by the separate collections of garbage (organic waste) and trash (inorganic waste).

FIG. 25-10 Compost heaps are commonly used for the aerobic decomposition of wastes. The organic matter of leaves and other dead plant matter frequently is decomposed to carbon dioxide in such compost heaps.

Solid waste can also be sorted at the receiving facility by using magnetic separators to remove ferrous metals and mechanical separators to remove glass, aluminum, and plastic materials. The remaining largely organic waste is ground up.

Aerobic, mesophilic, and thermophilic microorganisms degrade sorted organic waste material in composting.

In a compost of domestic garbage and sewage sludge, numerous microbial species that come from soil, water, and human fecal matter are present. The relatively high moisture content of the compost material favors the development of bacterial rather than fungal populations. The process is initiated by mesophilic heterotrophs. The metabolism of these organisms releases heat energy as organic compounds are decomposed, causing an increase in temperature—particularly at the interior of the pile where heat energy is trapped. As the temperature rises, mesophiles are replaced by thermophilic microorganisms.

Compost is a good soil conditioner and supplies some plant nutrients. Compost is used extensively as mulch in parks and gardens with ornamental plantings; in land reclamation, particularly after strip mining; and as part of highway beautification projects. It cannot, however, compete with synthetic fertilizers for use in agricultural production. The sale of compost can reduce the cost of the waste disposal operation but generally does not make such waste disposal operations self-supporting. When sewage sludge is used as a major component of the original compost mixture, the finished product may contain relatively high concentrations of potentially toxic heavy metals such as cadmium, chromium, and thallium. Because little is known about the behavior of these metals in agricultural soils, sewage-sludge-derived compost in agriculture is not widely used. Even though landfill operations are less expensive than composting, the long-range environmental costs in terms of groundwater contamination favor composting processes.

The compost product can be used as a soil conditioner or mulch.

TREATMENT OF LIQUID WASTES

Agricultural and industrial operations—and everyday human activities—produce **liquid wastes,** including **domestic sewage.** These liquid wastes discharge flow through natural drainage patterns or sewers, eventually entering natural bodies of water, such as groundwater, rivers, lakes, and oceans. In theory the liquid wastes disappear when they are flushed into such water bodies. The adage, "the solution to pollution is dilution" has some truth to it. Fortunately, **self-purification** is, indeed, an inherent capability of natural waters. This capacity is based on the biogeochemical cycling activities and interpopulation relationships of the indigenous microbial populations. Biogeochemical cycling is the biologically mediated transformation of elements that result in their global cycling, including transfer between the atmosphere, hydrosphere, and lithosphere. Organic nutrients in the water are metabolized and mineralized by indigenous (native) heterotrophic aquatic microorganisms. Mineralization is the breakdown of organic materials into inorganic materials brought about mainly by microorganisms. Reasonably low amounts of raw sewage can be accepted by natural waters without causing a significant decline in the level of water quality.

Biogeochemical cycling and indigenous microbial activities allow water receiving liquid wastes to be largely self-purifying.

The relative changes in some environmental parameters and populations in a river receiving sewage are illustrated in FIG. 25-11. River water receiving sewage effluents contain the filamentous aerobic bacterium *Sphaerotilus natans*, known as the sewage fungus because of its filamentous appearance. A heterogeneous microbial community also develops amid the filaments of this bacterium below a sewage outfall. *S. natans* and the associated microbial community are efficient degraders of organic matter, but they also consume oxygen at a fairly high rate. This makes it difficult for fish and other living creatures to remain in the area. Depending on the rate of sewage discharge, flow rate, river water temperature, and other environmental factors, water may reestablish an acceptable quality level at some distance downstream from the sewage outfall, typically within 24 to 60 km.

When these waters are overwhelmed by concentrated inputs of organic matter, they exhibit a high demand for oxygen. This high demand for oxygen generally indicates the presence of excessive amounts of organic carbon. The dissolved oxygen in natural waters seldom exceeds 8 mg/L because of its low solubility. It is often considerably lower because of heterotrophic microbial activity. This makes oxygen depletion a likely consequence of adding wastes with high organic content to aquatic ecosystems. Exhaustion of the dissolved oxygen content is the prin-

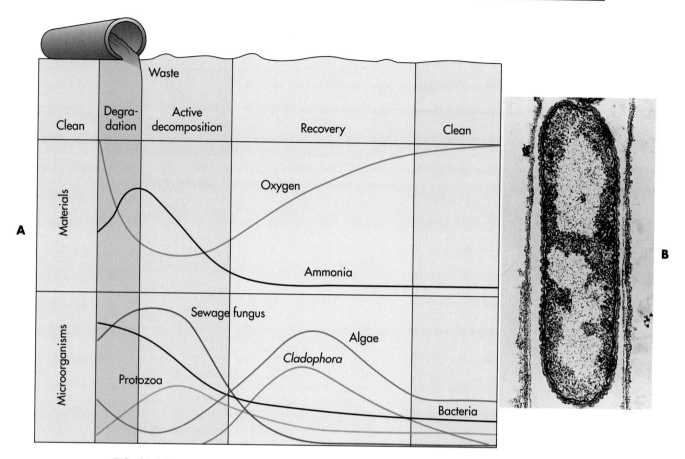

FIG. 25-11 **A,** Immediately below a sewage outfall the oxygen is depleted and the concentration of ammonia (nutrient level) is elevated. Some concentrations of microorganisms are also elevated near a sewage outfall. Further downstream, oxygen concentrations, nutrient levels, and populations of microorganisms and higher animals return to normal, indicating degradation and dilution of the sewage and the return to clean water conditions. **B,** Colorized micrograph of *Sphaerotilus natans,* a filamentous bacterium that grows in high abundance near sewage outfalls.

cipal result of sewage overload on natural waters. Oxygen deprivation kills aerobic organisms, including some microorganisms, fish, and invertebrates. The decomposition of dead organisms within the water body creates an additional oxygen demand. Fermentation products give rise to noxious odors, tastes, and colors. The water becomes putrid and septic.

High oxygen demand in water indicates the presence of excessive amounts of organic carbon, from added wastes with high organic content.

High oxygen demand depletes the oxygen supply, which kills aerobic microorganisms, fish, and invertebrates.

The maintenance of satisfactory water quality means that natural waters should not be overloaded with organic or inorganic nutrients or with toxic, noxious, or aesthetically unacceptable substances. Oxygen, temperature, salinity, turbidity, or pH levels should not be altered so significantly that these bod-

ies of water lose their ability to support fish production and recreational usage of the water body. They should not be allowed to become vehicles of disease transmission due to fecal contamination.

Modern methods of liquid waste treatment are aimed at reducing the amount of organic matter in the waste so that its oxygen demand is lessened before it is discharged into a water body. This must be done to maintain acceptable water quality. There are several different approaches to reducing the amount of organic matter, and hence the requirement for oxygen. These methods employ combinations of physical, chemical, and microbiological means. Most communities in developed countries have facilities for treating sewage. **Sewage** is the used water supply containing domestic waste, with human excrement and wash water; industrial waste, including acids, greases, oils, animal matter, and vegetable matter; and storm waters. The use of household garbage disposal units increases the organic content of domestic sewage. The treatment of liquid wastes is aimed at re-

METHODOLOGY

BIOLOGICAL OXYGEN DEMAND (BOD)

We have developed several measures of water quality that help us manage aquatic ecosystems. These measures indicate how much waste can safely be allowed to enter rivers and lakes without causing serious deterioration of water quality. One widely used measure of water quality is the biological oxygen demand (BOD) (see Figure). The BOD represents the amount of oxygen required for the microbial decomposition of the organic matter in the water. The polluting power of different sources of wastes is reflected in the BOD of the material. The BOD procedure is used extensively in monitoring water quality and biodegradation of waste materials. It is designed to determine how much oxygen is consumed by microorganisms during oxidation of the organic matter in the sample.

BOD can be easily determined in the laboratory. A water sample is incubated and the amount of oxygen consumed during a 5-day period is measured. The procedure is based on the consumption of oxygen by the microorganisms that are naturally present in the water sample. The high demand for oxygen in water that contains organic pollution is a consequence of the aerobic respiration by organisms that decompose dead organisms in the water. Aerobic respiration consumes oxygen. The oxygen remaining after 5 days of incubation can be determined chemically or, more commonly, with

the use of oxygen electrodes. The difference between the starting concentration of oxygen and the residual oxygen represents the amount of oxygen consumed by the indigenous microorganisms in degrading the organic materials in the water sample, that is, the BOD.

Incubation at 20° C for 5 days is commonly used. The test was originally developed in Great Britain, where average water temperatures are near 20° C and where it takes a maximum of 5 days for anything entering a local river to reach the ocean. Once the organic matter reaches the ocean, it is no longer considered a threat to water quality. In the United States and other large countries, it may be necessary to modify the incubation period used in the standard 5-day BOD procedure to account for the extended residence time of organic matter in the waterways receiving organic pollutants. The development of appropriate modifications to the original procedure has been slow, in part because of a lack of understanding of the original assumptions used in establishing the standard 5-day incubation procedure. Appropriate modifications of the standard BOD procedure are needed based on actual residence times in inland waterways and desirable multiple uses of water. Such modifications of the BOD test are presently being incorporated into water quality standards.

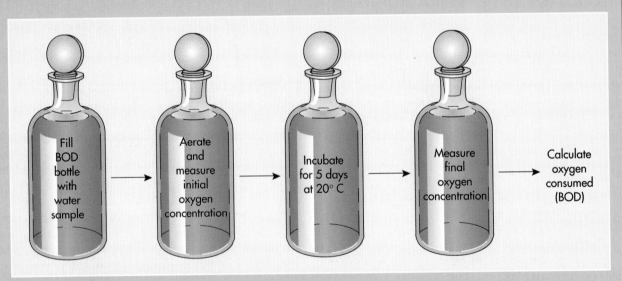

Biological oxygen demand (BOD) is determined by measuring how much oxygen is consumed by microorganisms during a 5-day incubation period. This oxygen consumption reflects the organic content of the water sample and whether the water will have a detrimental effect on animals in a receiving water. Water with too high a concentration of organic matter results in depletion of oxygen and death of the animals.

moving organic matter, human pathogens, and toxic chemicals. The treatment of sewage reduces the oxygen demand due to suspended or dissolved organics and the number of enteric pathogens so that the discharged sewage effluent will not cause unacceptable deterioration of environmental quality.

Sewage is the used water supply containing domestic, industrial, and agricultural waste.

Sewage can be subjected to different treatments (FIG. 25-12). The choice of treatment depends on the quality of the effluent deemed necessary to be achieved to permit the maintenance of acceptable water quality. **Primary treatments** rely on physical separation procedures to lower the oxygen demand. **Secondary treatments** rely on microbial biodegradation to further reduce the concentration of organic compounds in the effluent. **Tertiary treatments** use chemical methods to remove inorganic compounds. Municipal sewage treatment facilities are designed to handle organic wastes but are normally incapable of dealing with industrial wastes containing toxic chemicals, such as heavy metals. Industrial facilities frequently must operate their own treatment plants to deal with waste materials.

Primary sewage treatment lowers oxygen demand by physical means.

Secondary sewage treatments use microbial biodegradation to reduce concentrations of organic compounds.

Tertiary sewage treatments use chemical processes to remove inorganic compounds.

PRIMARY SEWAGE TREATMENT

Primary sewage treatment removes suspended solids in settling tanks or basins. Solids are drawn off from the bottom of the tank. They then may be subjected to anaerobic digestion and/or composting before final deposition in landfills or use as soil conditioner. Only a low percentage of the suspended or dissolved organic material is actually mineralized during liquid waste treatment. Most of it is removed by settling. In this method the liquid disposal problem merely becomes a solid waste problem. It is not solved. Nevertheless, this displacement is essential because of the detrimental effects of discharging effluents with high BOD into aquatic ecosystems with naturally low dissolved oxygen contents.

Suspended solids are removed by settling from liquid wastes in settling tanks or basins during primary sewage treatment.

The liquid portion of the sewage, which contains dissolved nutrients and organic matter, can be sub-

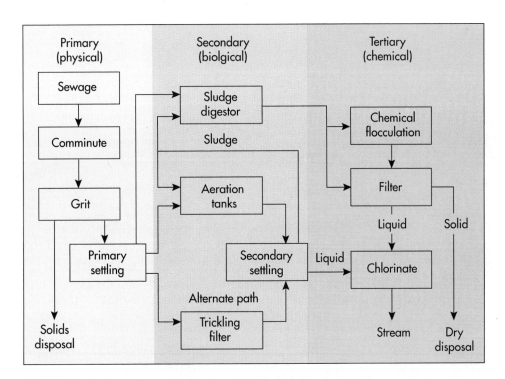

FIG. 25-12 Sewage treatment consists of a primary stage (physical treatment), secondary stage (biological treatment), and a tertiary stage (chemical treatment). Microorganisms degrade wastes in the secondary stage. Various physical, biological, and chemical treatments can be used to accomplish each respective stage.

jected to further treatment or discharged after primary treatment alone. Because liquid wastes vary in composition and may contain mainly solids and little dissolved organic matter, primary treatment may remove 70% to 80% of the BOD. It may be sufficient. For typical domestic sewage, however, primary treatment normally removes only 30% to 40% of the BOD.

SECONDARY SEWAGE TREATMENT

To achieve an acceptable reduction in BOD, secondary treatment by various means is necessary (Table 25-2). In secondary sewage treatment a small portion of the dissolved organic matter is mineralized. The larger portion is converted to removable solids. The combination of primary and secondary treatment reduces the original sewage BOD by 80% to 90%. Secondary sewage treatment relies on microbial activity. It can be aerobic or anaerobic. It can be conducted in a large variety of devices. A well-designed and efficiently operated secondary treatment unit should produce effluents with BOD and/or suspended solids of less than 20 mg/L. Because the secondary treatment of sewage is a microbial process, it is extremely sensitive to the introduction of toxic chemicals that may be contained in or accidentally added to industrial waste effluents.

> There are many secondary sewage treatment processes. All rely on microbial activity to mineralize dissolved organic matter. The remainder is converted to removable solids.

TABLE 25-2

Efficiency of Various Types of Sewage Treatment

TREATMENT	BOD (% REDUCED)	SUSPENDED SOLIDS (% REMOVED)	BACTERIA (% REDUCED)
Sedimentation	30-75	40-95	40-75
Septic tank	25-65	40-75	40-75
Trickling filter	60-90	0-80	70-85
Activated sludge	70-96	70-97	95-99

Oxidation Ponds

Oxidation ponds are also known as *stabilization ponds* or *lagoons* (FIG. 25-13). They are used for the simple secondary treatment of sewage effluents in rural communities and some industrial facilities. Heterotrophic bacteria degrade sewage organic matter within the ponds. These bacteria produce cellular material and mineral products that support the growth of algae. The algal populations produce oxygen that replenishes the oxygen depleted by the heterotrophic bacteria. This permits the continued decomposition of the organic matter. Oxygenation is usually achieved by diffusion and by the photosynthetic activity of algae. These ponds, therefore, need to be shallow with a large area able to receive the

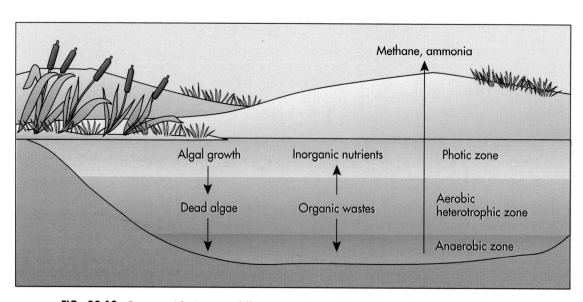

FIG. 25-13 In an oxidation pond/lagoon various microbial populations contribute to waste degradation. This is a simple system that is relatively inefficient but has low operating cost.

light necessary to support photosynthesis. Typically, oxidation ponds are less than 10 feet deep.

> **Oxidation ponds are shallow pits in which sewage effluents are subject to microbial degradation by heterotrophic bacteria. Algae grow and replenish the oxygen.**

Oxygenation is usually incomplete. This causes odor problems. The performance of oxidation ponds is strongly influenced by seasonal temperature fluctuations. Their usefulness, therefore, is largely restricted to warmer climatic regions. The bacterial and algal cells formed during the decomposition of the sewage settle to the bottom. Eventually the pond is filled. Oxidation ponds generally are low-cost operations, but they tend to be inefficient. They require large holding capacities and long retention times. The degradation of organic matter in these ponds is relatively slow. Treatment time for domestic sewage may be as long as a week. The effluents containing oxidized products are periodically removed from the ponds. The ponds are then refilled with raw sewage.

Trickling Filters

The **trickling filter system** is a simple and relatively inexpensive aerobic sewage treatment method (FIG. 25-14). The sewage is distributed by a revolving sprinkler suspended over a filter bed of porous material. The sewage slowly percolates through this porous bed. The effluent is collected at the bottom.

The porous material of the filter bed becomes coated with a dense, slimy bacterial growth. This growth is composed principally of *Zooglea ramigera* and similar slime-forming bacteria. *Zooglea ramigera* is a slime-forming Gram-negative bacterium that forms a macroscopic aggregate called *floc*. The floc supports the growth of a heterogeneous microbial community, including bacteria, fungi, protozoa, and very small animals. This microbial community absorbs and mineralizes dissolved organic nutrients in the sewage, reducing the BOD of the effluent. Aeration occurs passively as a result of the movement of air through the porous material of the bed. The sewage may be passed through two or more trickling filters or may be recirculated several times through the same filter to reduce the BOD to acceptable levels. The effluent from the trickling filters may be clarified by allowing sloughed-off biomass to settle prior to discharge.

> **Trickling filters have revolving sprinklers that dispense sewage over a porous filter bed.**
>
> **The filter bed supports a community of slime-forming bacteria that mineralizes dissolved organic nutrients in the sewage.**

A drawback of this otherwise simple and inexpensive treatment system is that a nutrient overload produces excess microbial slime. Too much slime reduces aeration and percolation rates, periodically necessitating renewal of the trickling filter bed. Also,

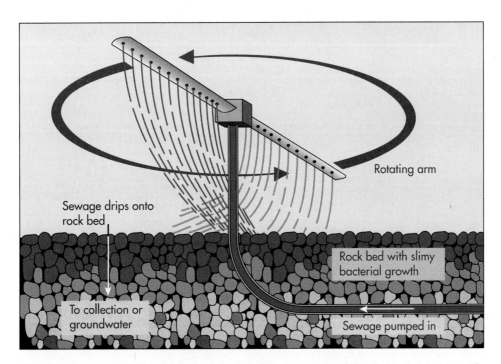

FIG. 25-14 In a trickling filter, liquid waste flows past a biofilm of microorganisms adhering to a bed of rocks. The biofilm microorganisms, many of which form slimes, aerobically degrade the organic compounds in the waste.

cold winter temperatures strongly reduce the effectiveness of such outdoor treatment facilities.

Activated Sludge

The **activated sludge process** is a very widely used aerobic suspension liquid waste treatment system (FIG. 25-15). Sewage is allowed to settle. Then it is introduced with its dissolved organic compounds into an aeration tank. Air injection and/or mechanical stirring provides aeration. Rapid development of microorganisms is also stimulated by the reintroduction of settled sludge from a previous run. The process derives its name from this inoculation with activated sludge.

Activated sludge process occurs in an aeration tank. Sewage and settled sewage sludge from a previous run are mixed in this process.

During the settling period, a large heterotrophic microbial population develops. The heterogeneous nature of the organic substrates in sewage allows the

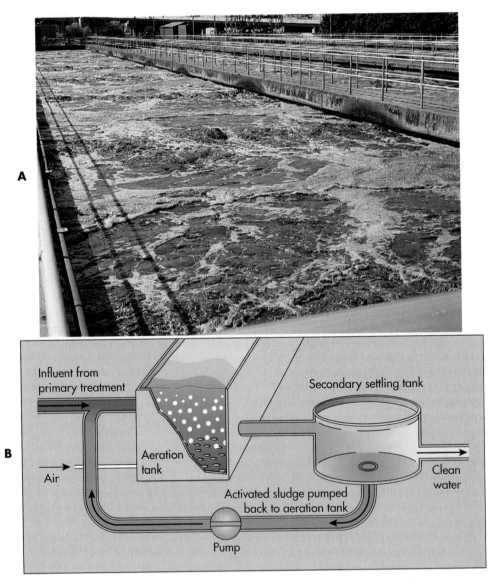

FIG. 25-15 **A,** An activated sludge treatment facility has tanks in which microorganisms degrade wastes. Extensive aeration and agitation maintain aerobic conditions that favor complete degradation of organic compounds by respiring microorganisms. **B,** An activated sludge treatment system has an aeration tank in which aerobic microorganisms actively degrade the wastes remaining after primary settling of sludge from influent sewage. This is run as a batch reaction. The biomass produced in the aeration tank settles in a secondary settling tank. A portion of this activated sludge containing massive populations of microorganisms is used as an inoculum for treatment of a new batch of sludge.

development of diverse heterotrophic bacterial populations. There are low numbers of filamentous fungi, yeasts, and protozoa, mainly ciliates. The protozoa are important predators of the bacteria. The bacteria in the activated sludge tank occur both in free suspension and as an aggregate, or floc. The flocs are composed of microbial biomass held together by bacterial slimes.

This activated sludge process tends to reduce the BOD of the effluent to 10% to 15% of that of the raw sewage. The treatment also drastically reduces the number of intestinal pathogens. This reduction is the result of the combined effects of competition, adsorption, predation, and settling.

Pathogens tend to grow poorly or not at all under the environmental conditions of an aeration tank. Nonpathogenic heterotrophs proliferate vigorously. Settling of the flocs removes additional pathogens. The number of *Salmonella, Shigella, Escherichia coli,* and enteroviruses typically is 90% to 99% lower in the effluent of the activated sludge treatment process than in the incoming raw sewage. The main removal mechanism of the enteroviruses appears to be adsorption of the virus particles onto the settling sewage sludge floc.

Septic Tank

The **septic tank** is the simplest anaerobic treatment system (FIG. 25-16). It is used extensively in rural areas that lack sewerage systems. A septic tank acts largely as a settling tank. Inside a septic tank the organic components of waste water undergo limited anaerobic degradation. The accumulated sludge is maintained under anaerobic conditions. It is degraded by anaerobic microorganisms to organic acids and hydrogen sulfide. Residual solids settle to the bottom of the septic tank. The clarified effluent is allowed to percolate into the soil. The dissolved organic compounds in the effluent undergo biodegradation in the soil. These products are distributed into the soil, along with the clarified sewage effluent. Septic tank treatment does not reliably destroy intestinal pathogens. It is important, therefore, that the soils receiving the clarified effluents not be close to drinking wells to prevent contamination of drinking water with enteric pathogens.

Septic tanks are settling tanks for sewage; anaerobic microorganisms degrade the contents to organic acids and hydrogen sulfide; the effluent seeps into the soil.

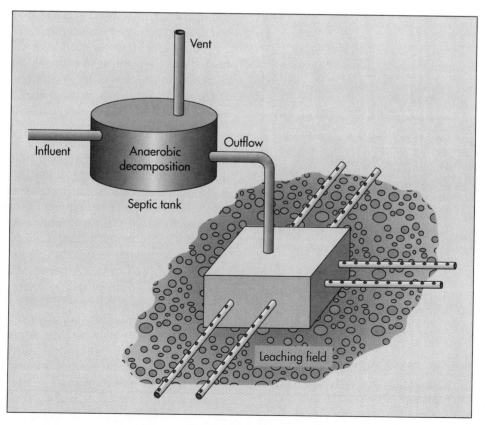

FIG. 25-16 A septic tank is a simple system that incorporates primary settling and secondary anaerobic digestion *(decomposition)* of organic compounds. The overflow liquid following anaerobic decomposition is allowed to leach into the surrounding soil.

Anaerobic Digestors

Anaerobic digestors are large fermentation tanks designed for continuous operation under anaerobic conditions (FIG. 25-17). Large-scale anaerobic digestors are used only for processing settled sewage sludge and the treatment of very high BOD industrial effluents. Anaerobic digestors contain high amounts of suspended organic matter. Much of this suspended material is bacterial biomass. Viable counts of bacteria can be as high as 10^9-10^{10} per mL. A complex bacterial community is involved in the degradation of organic matter within an anaerobic digestor.

The anaerobic digestion of wastes is a two-step process. First, complex organic materials, including microbial biomass are converted to fatty acids, CO_2 and H_2. In the next step, methane is generated. The final products obtained in an anaerobic digestor are a gas mixture, approximately 70% methane and 30% carbon dioxide, microbial biomass, and a nonbiodegradable residue. The gas can be used within the treatment plant to drive the pumps and/or to provide heat for maintaining the temperature of the digestor. If purified, the gas may be sold. Thus, in addition to its primary function in removing wastes, anaerobic digestors can produce needed fuel resources.

Large scale anaerobic digestors process settled sewage sludge and industrial effluents with very high BOD.

Complex organic materials are converted to fatty acids and methane is produced in anaerobic digestors.

A properly operating anaerobic digestor yields a greatly reduced volume of sludge compared to the starting material. This product still causes odor and water pollution problems. It must be disposed of at restricted landfill sites. Aerobic composting can be used to further consolidate the sludge. This renders it suitable for disposal in any landfill site or for use as a soil conditioner as long as the metal content is low enough to be nontoxic to plants and animals that eat those plants.

TERTIARY SEWAGE TREATMENT

Tertiary sewage treatment is defined as any practice beyond a secondary one. It is designed to remove nonbiodegradable organic pollutants and mineral nutrients, especially dissolved nitrogen and phosphorus. Secondary treatment is still required to avoid overloading this expensive treatment stage with biodegradable materials that could have been removed in more economical ways. Activated carbon filters are normally used in the removal of these materials from secondary-treated industrial effluents.

The release of sewage effluents containing phosphates and fixed forms of nitrogen can cause serious eutrophication in aquatic ecosystems (FIG. 25-18). **Eutrophication** means nutrient enrichment. Sudden nutrient enrichment by sewage discharge or agricultural runoff triggers explosive algal blooms (excessive growths of algae). Due to various causes, including mutual shading, exhaustion of micronutrients, and the presence of toxic products and/or antagonistic populations, this algal population usually dies. The subsequent decomposition of the dead algal biomass by heterotrophic microorganisms exhausts the oxygen supply in the water. This causes extensive fish kills and septic conditions. Even if the process does not proceed to this extreme, other undesirable changes due to eutrophication occur. These include the formation of algal mats, turbidity, discoloration, and shifts in the fish population from valuable species to more tolerant but less important forms.

To prevent eutrophication, phosphate is commonly removed from sewage by precipitation. This can be done as part of primary or secondary settling or in a separate facility where the precipitating agent can be recycled.

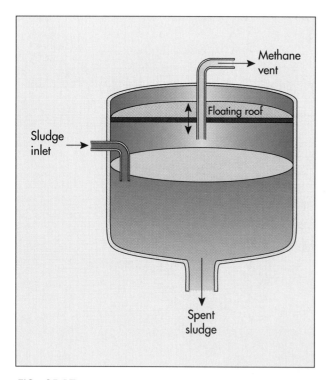

FIG. 25-17   An anaerobic sludge digestor is a tank in which anaerobic degradation processes occur, including the production of methane.

FIG. 25-18 Eutrophication of lakes occurs when high levels of inorganic nutrients, often phosphates, permit excessive growth of photosynthetic microorganisms. The growth of algae often enriches the water with organic compounds in these situations.

Eutrophication means nutrient enrichment. It is caused by sewage discharge or agricultural runoff and produces algal blooms.

The death of the algal blooms exhausts the water's oxygen supply, which causes fish kills and septic conditions.

DISINFECTION

The final step in the sewage treatment process is **disinfection.** Disinfection is designed to kill the en-teropathogenic bacteria and viruses that were not eliminated during the previous stages. It is commonly accomplished by **chlorination,** using chlorine gas (Cl_2) or hypochlorite ($Ca(ClO)_2$ or $NaClO$). Chlorine gas reacts with water to yield hypochlorous and hydrochloric acids, the actual disinfectants. Hypochlorite is a strong oxidant, which is the basis of its antibacterial action.

Chlorination is the standard disinfection process for water supplies.

TREATMENT AND SAFETY OF WATER SUPPLIES

It is important to provide safe water for drinking and other uses that is free of pathogens and toxic substances to all people. This problem is closely related to the problem of the safe disposal of liquid wastes. Fecal contamination of *potable* (clean, drinkable) water supplies by untreated or inadequately treated sewage effluents entering lakes, rivers, or groundwaters that serve as municipal water supplies creates conditions for the rapid spread of pathogens. The primary route of infection is direct ingestion of the pathogens in drinking water. Additional opportunities for infection arise when fruits, vegetables, and eating utensils are washed with contaminated water. The obvious solution for this problem is to disrupt the transmission of enteropathogenic fecal organisms to water supplies. Major sanitation efforts are required to treat and safely distribute public water supplies. Such practices have led to the virtual elimination of waterborne infections in developed countries. Such infections continue to be major causes of sickness and death in underdeveloped regions.

Water treatment facilities are designed to prevent the transmission of pathogens in drinking water.

DISINFECTION OF POTABLE WATER SUPPLIES

Most potable water supplies come from rivers, lakes, and underground wells and springs. Water from underground sources is partially purified by filtration as it passes through the soil column. This removes particulate matter and many microorganisms. This does not, however, eliminate the possibility of bacterial or viral contamination of the water supply, particularly if the source of the water is near a sewage effluent. In some rural areas, water is boiled or treated with antimicrobial chemicals to ensure its safety. In more densely populated areas, municipal water treatment facilities are designed to ensure the safety of the drinking water supply.

The principal processes of a water treatment facility are sedimentation, filtration, and disinfection. **Sedimentation** is carried out in large reservoirs, where water is held for a sufficient period of time to permit large particulate matter to settle out. Sedimentation rates can be increased by the addition of aluminum sulfate (alum). Alum forms a floc that precipitates. The floc carries microorganisms and suspended organic matter with it to the bottom of a settling basin. This flocculation treatment may be critical for the removal of microorganisms that are resistant to chlorination. In Milwaukee, during the spring of 1993, *Cryptosporidium* were found in the municipal drinking water because the flocculation procedure at the water treatment plant was inadequate to kill large numbers of this protozoan. *Cryptosporidium* presumably entered the water supply from farm run-off water that contained fecal matter from livestock. As a result, many people developed severe diarrhea.

The water then undergoes **filtration** by passage through sand filter beds. This removes up to 99% of the bacteria. The water may also be filtered through activated charcoal to remove potentially toxic organic compounds and organic compounds that give an undesirable color and/or taste to the water. The water is then subjected to disinfection to ensure that it does not contain any pathogens (FIG. 25-19). Chloramination is used for disinfecting municipal water supplies. **Chloramination** is the use of chloramines as drinking water disinfectants. This method is relatively inexpensive. The free residual chlorine in the treated water represents a built-in safety factor against pathogens surviving the treatment period and causing recontamination. The water may also be treated with fluoride to reduce dental caries. Minerals may be removed to soften the water for washing.

Water held in reservoirs undergoes sedimentation; water is filtered through sand filter beds or activated charcoal; water is then disinfected by chloramination.

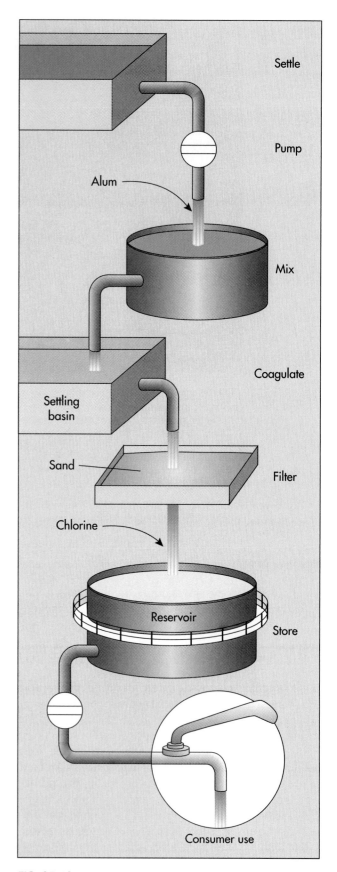

FIG. 25-19   Water purification procedure consists of coagulation, settling, and filtration to remove particulate matter and disinfection to kill microorganisms. Chlorine rapidly kills microorganisms and is widely used for the disinfection of water.

Bacterial Indicators of Water Safety

The importance of clean drinking water to public health requires objective test methods to establish high standards of water safety and to evaluate the effectiveness of treatment procedures. Routine monitoring of water to detect actual enteropathogens (pathogens of the gastrointestinal tract), such as *Salmonella* and *Shigella*, would be a difficult and uncertain undertaking. Instead, bacteriological tests of drinking water establish the degree of fecal contamination of a water sample by demonstrating the presence of **indicator organisms.** The ideal indicator organism should (1) be present whenever the pathogens concerned are present, (2) be present only when there is a real danger of pathogens being present, (3) occur in greater numbers than the pathogens to provide a safety margin, (4) survive in the environment as long as the potential pathogens, (5) be easy to detect regardless of what other organisms are present in the sample, and (6) should not multiply in water to give higher estimates of cell numbers.

> The presence of indicator organisms is used to establish the possibility of fecal contamination of a water supply.

Standards for Tolerable Levels of Fecal Contamination

The standards for tolerable limits of fecal contamination vary with the intended water use (Table 25-3). They are somewhat arbitrary and are designed with large built-in safety margins. The most stringent standards are imposed on the municipal water supplies to be used by many people. The standards for potable waters in the United States use a presence/absence test for coliforms and *E. coli*. The number of tests that are run and the number of positives that are permitted depend on the size of the water distribution system. More tests are required for large municipal water supplies. Somewhat higher coliform counts are sometimes tolerated in private wells used by only one family. Such wells would not become a source of a widespread epidemic. Mainte-

nance of a high drinking water standard does not absolutely exclude the possibility of ingesting enteropathogens with the water but helps keep this possibility to a statistically tolerable minimum. The facts that enteropathogens are very likely to be present in much lower numbers than fecal coliforms and that a few infective bacteria are unlikely to be able to overcome natural body defenses are two built-in safety factors. A minimum infectious dose of several hundred to several thousand bacteria is usually necessary for an actual infection to be established. Drinking water supplies meeting a 1/100 mL coliform standard have never been shown to be the source of a waterborne bacterial infection.

> Standards for tolerable levels of fecal contamination of water depend on the intended use of the water.

Fecal coliform counts are also used to establish the safety of water in shellfish harvesting and recreational areas. Shellfish tend to concentrate bacteria and other particles acquired through their filter-feeding activity. They are also sometimes eaten raw. They can become a source of infection by waterborne pathogens. Therefore there are relatively stringent standards for waters used for shellfishing. Clinical evidence for infection by enteropathogenic coliforms through recreational use of waters for bathing, wading, and swimming is unconvincing. In any case, as a precaution, beaches are usually closed when fecal coliform counts exceed the recreational standard of 1,000 colony forming units/100 mL. Some regional standards require that disinfected sewage discharges not exceed this limit.

Water quality standards based on fecal coliform levels do not account for the possible transmission of viruses associated with fecal matter through municipal water supplies. There is ample evidence for epidemics by enteroviruses caused by untreated drinking water in underdeveloped countries. Enteroviruses are somewhat more resistant to disinfection by chlorine or ozone than bacteria. In addition, many protozoan cysts are resistant to chloride. Occasionally, active virus particles are recovered from treated water that meets fecal coliform standards. Thus the possibility exists that water that meets accepted quality standards may still occasionally be a source of a viral infection. As many as 100 different viral types can be shed in human feces. Practical concern has been mainly with the viruses that cause infectious hepatitis, poliomyelitis, and viral gastroenteritis. Infectious hepatitis is sometimes spread by water supplies, though the more prevalent mode of infection is by the consumption of raw shellfish from fecally contaminated waters. Spread of polio infection through water supplies and/or recreational use of beaches has been suspected in many cases.

TABLE 25-3	
United States Water Standards for Coliform Contamination	
WATER USE	**MAXIMUM PERMISSIBLE COLIFORM COUNT (NUMBER/100 mL)**
Municipal drinking water	1
Waters used for shellfishing	70
Recreational waters	1,000

METHODOLOGY

COLIFORM COUNTS FOR ASSESSING WATER SAFETY

The most frequently used indicator organism is the normally nonpathogenic coliform bacterium *Escherichia coli*. (Coliform bacteria are Gram-negative rods that ferment lactose with acid and gas production; these bacteria commonly occur in the gastrointestinal tracts of warmblooded animals.) Positive tests for *E. coli* do not prove the presence of enteropathogenic organisms but do establish this possibility. *E. coli* is more numerous and easier to grow than enteropathogens. The test, therefore, has a built-in safety factor for detecting potentially dangerous fecal contamination. *E. coli* is differentiated readily from nonfecal bacteria. *E. coli*, thus, meets many of the criteria for an ideal indicator organism. There are, however, limitations to its use as an in-

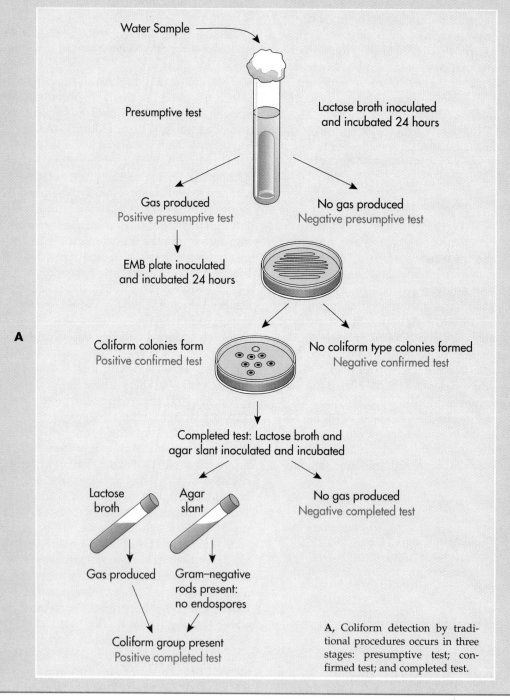

A, Coliform detection by traditional procedures occurs in three stages: presumptive test; confirmed test; and completed test.

dicator organism and various other species have been proposed as additional or replacement indicators of water safety. Water quality standards based on coliform levels do not account for possible transmission of protozoa such as *Giardia* or *Cryptosporidium* or for transmission of viruses.

The classical testing for coliform bacteria involves quantitative measures of total coliform bacteria (Gram-negative lactose fermenters that grow at 35° C) and fecal coliform bacteria (Gram-negative lactose fermenters that grow at 44.5° C). Media are employed that inhibit the growth of Gram-positive bacteria and permit the detection of lactose fermenters. A most-probable-number procedure usually is used to estimate numbers of coliform and fecal coliform bacteria. The conventional test for the detection of fecal contamination involves a three-stage test procedure (FIG. *A*). Often samples are filtered and the filters placed on a culture medium (FIG. *B*).

In a newly developed approach for coliform testing, defined substrates are used to detect total coliform bacteria and specifically the fecal indicator *E. coli*. The test detects specific enzymatic activities of coliform bacteria and *E. coli;* the products formed in these enzymatic reactions can be easily detected, greatly simplifying water quality testing (FIG. *C*).

B, Water samples (100 mL) are passed through bacteriological filters (0.2 to 0.45 μm pore size) to trap bacteria. The filters with trapped bacteria are placed on a medium containing lactose as a carbon source, an inhibitor to suppress growth of noncoliforms, and indicator substances to facilitate differentiation of coliforms. Coliform bacteria form distinct colonies on Endo medium.

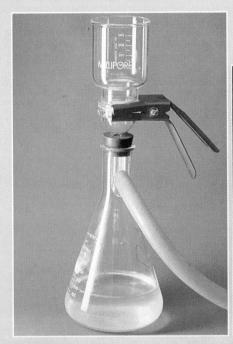
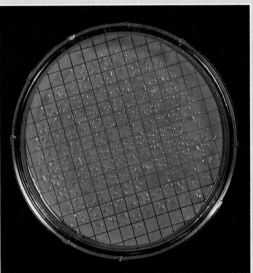

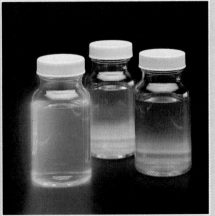

C, In the Colilert test, coliform bacteria are detected based on their production of β-galactosidase, an enzyme involved in the metabolism of lactose. A medium containing IPTG (isopropyl β-D-thiopyranogalactoside) is used to detect β-galactosidase activity of coliforms; a colored compound (*yellow liquid*) is produced. *Escherichia coli* is detected based on its specific production of β-glucuronidase. A medium containing MUG (4-methylumbelliferyl-β-D-glucuronide) is used to detect the β-glucuronidase activity of *E. coli;* a fluorescent compound (*blue liquid*) is produced. Negative test results leave a colorless fluid (*clear liquid*).

The situation with regard to viral gastroenteritis is similar. At this point, we can only say that the possibility of an occasional sporadic viral infection through drinking water adequately treated by bacteriological standards cannot be excluded. There is no hard evidence for any epidemics caused by such water.

Water quality standards do not take into account the possible presence of viruses in the water supply.

BIODEGRADATION OF ENVIRONMENTAL POLLUTANTS

Human exploitation of fossil fuel reserves and the production of many novel synthetic compounds (**xenobiotics**) in the twentieth century have introduced many new compounds into the environment. Microorganisms that normally do not encounter these compounds are not prepared to biodegrade them. Many of these compounds are toxic to living systems. Their presence in aquatic and terrestrial habitats often has serious ecological consequences, including major kills of indigenous biota. The disposal or accidental spillage of these compounds creates serious modern environmental pollution problems. Such problems can be severe when microbial biodegradation activities fail to remove these pollutants quickly enough to prevent environmental damage. Sewage treatment and water purification systems are usually incapable of removing these substances if they enter municipal water supplies, where they pose a potential human health hazard.

BIODEGRADABLE POLYMERS

In the 1970s environmentalists began to notice problems with substances that did not biodegrade. Plastic containers were accumulating along roadsides and in the bottoms of lakes. The previously reliable microorganisms were failing to attack the world's wastes. Some substances were proving to be totally resistant (recalcitrant) to microbial degradation. Some, like the pesticide DDT, were accumulating to levels that caused toxicity to birds and other animals (Table 25-4). It became clear that we would have to limit environmental contamination with such substances to avoid deterioration of environmental quality.

Some materials are recalcitrant, being totally resistant to microbial biodegradation.

Scientists recognized that substances that make their way into the environment should be biodegradable. They set out to find biodegradable substitutes for many useful products. An example of this approach is the story of alkyl benzyl sulfonates (ABS) (FIG. 25-20). ABS molecules are the major components of anionic laundry detergents. Cleaning occurs when ABS molecules form a single layer around the particles that make up the stains and dirt on clothing. An emulsion is formed that can be rinsed out of the fabric with water. The ABS molecule is a surface active molecule. It has a polar sulfonate and a nonpolar alkyl end. During laundering, ABS molecules orient their nonpolar ends toward nonpolar substances—like grease and oils—and their sulfonate ends toward the surrounding water.

The alkyl portion of the ABS molecule may be linear or branched. **Nonlinear ABS** (branched ABS) is easier to manufacture and has slightly superior detergent properties. Branched ABS has proved to be resistant to biodegradation. It causes extensive foaming of rivers receiving ABS-containing wastes. Some

TABLE 25-4	
Environmental Persistence Times of Some Pesticides	
COMMON NAME	**PERSISTENCE TIME**
Aldrin	>15 years
Chlordane	>15 years
DDT	>15 years
Heptachlor	>14 years
Lindane	15 years
Monuron	3 years
Parathion	>16 years
Picloram	>5 years
Propazine	2-3 years
Simazine	2 years
2,4,5-T	>190 days
Toxaphene	>14 years

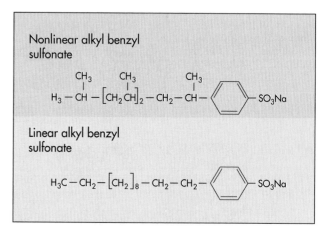

FIG. 25-20   Linear alkyl benzyl sulfates are easily biodegraded, whereas branched ABS molecules are relatively resistant to microbial attack.

communities have banned the use of these detergents because of their persistence in groundwater supplies used as sources of potable water. The methyl branching of the alkyl chain interferes with biodegradation; by changing the design of this synthetic molecule to that of **linear ABS,** the blockage can be removed.

The detergent industry has switched to linear ABS. It does not have this blockage and consequently is more easily biodegraded. The ABS story is particularly significant because it was one of the first instances in which a synthetic molecule was specifically redesigned to remove obstacles to biodegradation while preserving the useful characteristics of the compound.

> **The detergent industry changed from using branched alkyl benzyl sulfonate to linear ABS because the branched molecule cannot be biodegraded.**

In a similar manner, biodegradable plastics have been synthesized. These have been made from substances (for example, poly-β hydroxybutyric acid) that can be utilized by microorganisms. These biodegradable plastics can be safely disposed in the environment.

OIL POLLUTANTS

Over 10 million metric tons of **oil pollutants** enter the marine environment each year. This is the result of accidental spillages and the disposal of oily wastes. Periodically, pictures of dead birds floating in a sea of oil after a major oil spillage appear on the front page of the daily newspaper, evoking images of impending ecological doom. Actually, only a small proportion of all marine oil pollutants comes from major oil spills. Most oil pollution originates from minor spillages associated with routine operations.

Petroleum, with its many hydrocarbons (FIG. 25-21), has always entered the biosphere by natural seepage but at rates much slower than those involved in drilling and spillage accidents. Drilling is now estimated to produce about 2 billion metric tons of pollutant per year. The production, transportation, refining, and ultimately the disposal of

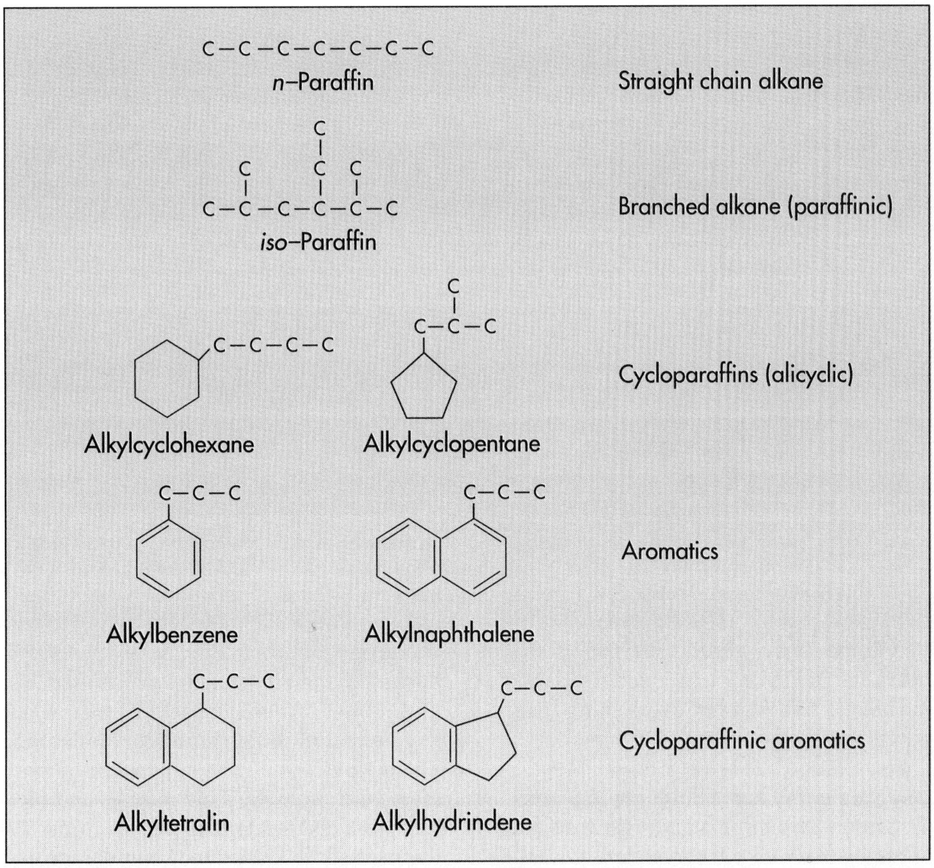

FIG. 25-21 Petroleum contains thousands of different compounds. Each class of hydrocarbon is subject to microbial degradation via different pathways. Many naturally occurring microorganisms are capable of hydrocarbon biodegradation.

used petroleum and petroleum products result in inevitable environmental pollution. Because the bulk of this load is, of course, heavily centered on offshore production sites, major shipping routes, and refineries its input frequently exceeds the self-purification capacity of the receiving waters. Petroleum pollutants in the environment are destructive to birds and marine life and, when driven ashore, cause heavy economic losses due to aesthetic damage to recreational beaches.

Petroleum is a complicated mixture of hydrocarbons (see FIG. 25-21). There are hundreds of individual compounds in every crude oil. The composition of each crude oil varies with its origin. As a result, the fate of petroleum pollutants in the environment is complex. The challenge for microorganisms to degrade all of the components of a petroleum mixture is immense. Nevertheless, microbial biodegradation of petroleum pollutants is the reason that all the oceans are not covered with oil today. In 1978, the supertanker *Amoco Cadiz* wrecked off the coast of France and spilled 223,000 tons of crude oil. As an example of the ability of microorganisms to degrade petroleum pollutants, measurements indicate that microorganisms biodegraded 10 tons of oil per day in the affected area. Microbial biodegradation represented the main process responsible for the ecological recovery of the oiled coastal region.

HIGHLIGHT

BIOREMEDIATION OF OIL SPILLS

During the 1950s and 1960s several groups of scientists examined the abilities of microorganisms to utilize petroleum hydrocarbons. Their focus was on finding microorganisms that could efficiently convert hydrocarbons into proteins. The aim was to find a new food source to feed the world's population. Petroleum was then viewed as inexpensive and in limitless supply. The threat to global existence was seen as the exponential human population increase and the inability of traditional agriculture to meet the food needs of this growing human population. Microorganisms growing on petroleum were seen as the solution. Large fermentation facilities were built for the production of microbial protein production from hydrocarbons. Then, in the mid-1960s the price of oil rose dramatically. The production of protein as a food source by growing microorganisms on petroleum was no longer economically feasible.

The global perspective during the 1960s changed from one fixated on global population explosion to one of concern with the quality of the environment. The wreck of the supertanker *Torrey Canyon* in 1967 focused that environmental concern on oil pollution of the oceans. In 1968 I began my research career in the laboratory of Richard Bartha at Rutgers University by studying the fate of petroleum in the ocean. I wanted to know how fast hydrocarbons could be degraded in the environment. I felt that if I could identify factors that limited the rates of natural hydrocarbon biodegradation that I would be able to find ways of overcoming those limitations and enhancing the microbial removal of oil pollutants.

In some early studies I examined the natural distribution of hydrocarbon-degrading microorganisms and found them to be present in all soil and water samples I examined. Knowing that nitrogen and phosphate are present in only very low concentrations and that grow-

ing microorganisms require these substances for incorporation into their cell constituents, I hypothesized that the concentrations of nitrogen and phosphorus in sea water were limiting the rates of hydrocarbon biodegradation. I set up a series of experiments to demonstrate that adding nitrogen and phosphorus could speed up the rates of microbial hydrocarbon biodegradation. When nitrogen and phosphorus were added, the microorganisms attacked the oil at increased rates. Adding cultures of microorganisms without the nitrogen and phosphorus was of little use. I then designed a fertilizer that would stick to the oil. I called this an oleophilic (oil-loving) fertilizer.

Although the U.S. Navy obtained a patent on my research findings, no use of fertilizers to enhance biodegradation rates of marine oil pollutants was made until 1989. In 1989 the *Exxon Valdez* wrecked in Prince William Sound, Alaska. The initial approach to the cleanup of the spilled oil was to wash the oiled shorelines with water under high pressure. This treatment was expensive, and cleaned shorelines became reoiled, forcing recleaning. A new method was needed.

Exxon and the U.S. Environmental Protection Agency (EPA) separately decided to consider stimulating microbial oil degradation. The use of microorganisms to degrade environmental pollutants is called bioremediation. The U.S. EPA and Exxon entered into an agreement to explore jointly the feasibility of using bioremediation. I served as a consultant on that project. The project focused on determining whether nutrient augmentation could stimulate rates of biodegradation. Three types of nutrient supplementation were considered: water soluble, slow release, and oleophilic fertilizer with chemicals that adhere to oil. Each fertilizer was tested in laboratory simulations and in field demonstration plots to show the efficacy of nutrient supplementation. Consideration was also given to po-

Because petroleum is a complex mixture of hydrocarbons, it is difficult for a microorganism to degrade all the components.

The susceptibility of petroleum hydrocarbons to biodegradation is determined by the structure and the molecular weight of the hydrocarbon molecule. The *n*-Alkanes of intermediate chain length (10 to 24 carbons) are degraded most rapidly. Short chain alkanes (less than 9 carbons) are toxic to many microorganisms, but they generally evaporate rapidly from oil slicks. As alkane chain length increases, so does resistance to biodegradation. Branching, in general,

reduces the rate of biodegradation because it interferes with the enzymes involved in hydrocarbon degradation. Aromatic compounds, especially polynuclear aromatic hydrocarbons (compounds that have multiple aromatic rings fused together), are degraded more slowly than alkanes. Alicyclic compounds (compounds that have nonaromatic rings) are frequently unable to serve as the sole carbon sources for microbial growth unless they have a sufficiently long aliphatic side chain. They can be degraded via cometabolism by two or more cooperating microbial strains with complementary metabolic capabilities.

tential adverse ecological effects, particularly eutrophication due to algal blooms and toxicity to fish and invertebrates. The application of the oleophilic fertilizer produced very dramatic results, stimulating biodegradation such that the surfaces of the oil-blackened rocks on the shoreline turned white and were essentially oil-free within 10 days after treatment (see Figure). The use of an oleophilic and slow release fertilizer was approved for

shoreline treatment and was used as a major part of the cleanup effort. A joint Exxon-U.S. EPA-State of Alaska monitoring effort followed the effectiveness of the bioremediation treatment. Bioremediation was estimated to increase the rates of biodegradation at least threefold. Due to its effectiveness, bioremediation became the major treatment method for removing oil pollutants from the impacted shorelines of Prince William Sound.

The shorelines contaminated by oil spilled by the *Exxon Valdez* were cleaned through bioremediation. Nutrients were added to stimulate the growth of indigenous oil-degrading bacteria. This test plot shows the dramatic results that demonstrated the efficacy of bioremediation.

The successful biodegradative removal of petroleum hydrocarbons from the sea depends on the enzymatic capacities of microorganisms and various abiotic factors. Microbial hydrocarbon biodegradation requires suitable growth temperatures and available supplies of fixed forms of nitrogen, phosphate, and molecular oxygen. In the oceans, temperature and nutrient concentrations often limit the rates of petroleum biodegradation. Low concentrations of nitrate and phosphate in sea water are particularly limiting to hydrocarbon biodegradation. For example, after the IXTOC I well blowout in the Gulf of Mexico, which in 1980 created the largest known oil pollution incident, little biodegradation of the oil-water emulsion (mousse) occurred in the surface waters of the Gulf because of severe nutrient limitations.

Biodegradation of petroleum depends on the enzymatic capacities of microorganisms and on temperature, nutrients, and oxygen.

Although many microorganisms can metabolize petroleum hydrocarbons, no single microorganism possesses the enzymatic capability to degrade all, or even most, of the compounds in a petroleum mixture. More rapid rates of degradation occur when there is a mixed microbial community. Apparently, the genetic information in more than one organism is required to produce the enzymes needed for extensive petroleum biodegradation. Recombinant DNA technology, however, permits the incorporation of the diverse types of genetic information extracted from several organisms into a single organism. Through genetic engineering, a "superbug" has been created that is capable of degrading many different hydrocarbon structures. It is potentially useful in oil pollution abatement programs. This hydrocarbon-degrading microorganism is the first living organism for which a patent has been granted in the United States. The ruling by the Supreme Court that genetic engineering could in essence invent microorganisms has far-reaching consequences for the future use of recombinant DNA technology in the United States to develop microorganisms of economic significance.

Microorganisms created by microbiologists have the potential to help cleanse the environment of man-made pollutants. Even with the ability to create superbugs, the usefulness of such organisms in pollution abatement depends on the compatibility with the environment. In many cases, environmental factors, rather than the genetic capability of a microorganism, limit the biodegradation of pollutants. Thus, although genetically engineered organisms are a useful addition to the arsenal of antipollution measures, there is no panacea for solving human pollution problems.

S U M M A R Y

Introduction (p. 734)
- The capacity of the microbial populations of an ecosystem to decompose waste materials is limited.
- Serious environmental pollution and epidemic outbreaks of disease may result.

Environmental Aspects of Microbial Metabolism (pp. 734-741)
- Microorganisms carry out essential biogeochemical cycling reactions.
- The carbon cycle involves conversions of inorganic carbon dioxide and various organic compounds.
- Carbon is transferred in food webs that establish a trophic structure within communities.
- Nitrogen is actively cycled by microorganisms.
- The nitrogen cycle includes nitrogen fixation, nitrification, and denitrification, which move nitrogen between the atmosphere, soil, and water.
- The sulfer cycle includes numerous sulfur compounds of differing oxidation states.

Solid Waste Disposal (pp. 742-744)
- Solid waste consists of inert materials, decomposable organic waste, and sewage sludge.
Sanitary Landfills (pp. 742-743)
- Landfills are dumpsites for solid wastes. When each day's deposit of solid waste is covered over with a layer of soil, a sanitary landfill is created. Such land can eventually be reused after the contents complete anaerobic microbial decomposition.
- Methane seepage is often a problem with landfill sites. Seepage of anaerobic decomposition products, heavy metals, and recalcitrant hazardous pollutants into underground water supplies can also occur.
Composting (pp. 743-744)
- Composting is the degradation of solid heterogeneous organic waste material by aerobic, mesophilic, and thermophilic microorganisms. It produces a stable, sanitary, humus-like product.
- Material to be composted must be sorted. Only the organic waste is ground up.
- Many microbial strains from soil, water, and fecal matter are present.

Treatment of Liquid Wastes (pp. 744-753)
- The bloom of the sewage fungus, *Sphaerotilus natans*, in a river indicates the presence of sewage in the water. *S. natans* consumes so much oxygen that other living organisms in the river cannot live.
- High oxygen demand in water indicates the presence of excessive amounts of organic carbon. The depletion of oxygen kills aerobic microorganisms, fish, and invertebrates. The decomposition of these organisms creates more oxygen demand.

- Liquid waste treatment methods reduce the oxygen demand of sewage before it is allowed into water. Sewage is the used water supply. It contains human, domestic, industrial, and agricultural waste.
- Primary treatments rely on physical separation to reduce the biological oxygen demand. Secondary treatments use microbial degradation to reduce the concentration of organic compounds. Tertiary treatments use chemical methods to remove inorganic compounds and pathogens.

Primary Sewage Treatment (pp. 747-748)

- Primary treatment removes suspended solids from sewage in settling tanks or basins. These solids are then drawn off from the bottom of the tank, creating a solid waste disposal problem.

Secondary Sewage Treatment (pp. 748-752)

- Some of the dissolved organic matter in sewage is mineralized by microbial activity in secondary treatment. The remainder of the organic matter is converted to removable solids. The oxygen demand is reduced 80% to 90%.
- Oxidation ponds (stabilization ponds or lagoons) are shallow pits where heterotrophic bacteria degrade sewage organic matter and produce minerals and materials that support the growth of algae. The algae produce oxygen, which permits continued decomposition of the organic matter.
- Trickling filters use a revolving sprinkler suspended over a filter bed of porous material through which the sewage passes several times. Slime-forming bacteria coat the filter bed and mineralize the dissolved organic nutrients in the sewage. Effluent is collected at the bottom.
- The activated sludge process uses an aeration tank in which settled sewage and sludge from a previous run are placed. Oxygen demand is reduced 85% to 90%. Numbers of intestinal pathogens are reduced.
- A septic tank is a settling tank in which the organic components of waste water undergo limited anaerobic degradation. Organic acids and hydrogen sulfide are produced. Residual solids settle on the bottom. The effluent seeps into the soil.
- Anaerobic digestors are large fermentation tanks that process settled sewage sludge and high oxygen demand effluents. Complex organic materials are first converted to fatty acids. Then methane is generated. The final products are a gas mixture, microbial biomass, and a nonbiodegradable residue.

Tertiary Sewage Treatment (pp. 752-753)

- Tertiary sewage treatment is any treatment beyond the secondary one. It removes nonbiodegradable organic pollutants and mineral salts.
- Eutrophication is nutrient enrichment. Sudden eutrophication by sewage causes algal blooms. When the algal population dies, its decomposition exhausts the oxygen supply. This leads to fish kills and septic conditions.

Disinfection (p. 753)

- Disinfection kills pathogenic bacteria and viruses not eliminated during sewage treatment. Chlorination is the standard treatment.

Treatment and Safety of Water Supplies (pp. 753-758)

- Fecal contamination of potable water permits the spread of pathogens.

Disinfection of Potable Water Supplies (p. 754)

- Municipal water treatment facilities are responsible for the safety of the drinking water supply. They use sedimentation, filtration, and disinfection.
- Large particulate matter settles out during sedimentation. Water is filtered through sand filter beds or activated charcoal. Chloramination is used for disinfection.

Bacterial Indicators of Water Safety (pp. 755-758)

- Bacteriological tests of drinking water are routinely done to determine whether there is fecal contamination of the water supply. Indicator organisms are used to determine the presence of pathogens.
- Indicator organisms must be present whenever the pathogens of concern are present, be present only when there is real danger of these pathogens being present, occur in greater numbers than the pathogens, survive as long as the pathogens, and be easy to detect.
- Standards for tolerable levels of fecal contamination vary with water intended use. Drinking water supplies meeting the 1/100 mL coliform standard are never a source of waterborne bacterial contamination.
- Because shellfish tend to concentrate bacteria and are sometimes eaten raw, fecal coliform counts are also used to establish the safety of water used for shellfishing. Beaches are closed for swimming when fecal coliform counts exceed 1000/100 mL.
- Water that meets accepted quality standards may still be a source of viral infection.

Biodegradation of Environmental Pollutants (pp. 758-762)

- Microorganisms cannot degrade all the fossil fuels and novel synthetic compounds they encounter. Such products may be toxic or cause environmental damage.

Biodegradable Polymers (pp. 758-759)

- Recalcitrant materials resist microbial degradation.
- Alkyl benzyl sulfonates are used in laundry detergents. The alkyl portion of this molecule may be linear or branched. The branched form is easier to manufacture and cleans better, but it will not biodegrade. The detergent industry switched to the linear form.

Oil Pollutants (pp. 759-762)

- Ten million metric tons of oil pollutants enter the marine environment each year. Most of the hydrocarbon components of petroleum are susceptible to biodegradation.
- Biodegradation depends on the enzymatic capabilities of the area's microbial population and such abiotic factors as temperature and availability of nitrogen, phosphorus, and oxygen.

CHAPTER REVIEW

REVIEW QUESTIONS

1. What roles do microorganisms have in the decomposition of organic compounds? How do microbial activities contribute to the maintenance of environmental quality?
2. What is the carbon cycle?
3. Describe the nitrogen cycle. How does it determine agricultural productivity?
4. What is BOD? Why is it important to reduce BOD of wastes? How does sewage treatment reduce BOD?
5. What are the different stages of sewage treatment? What is occurring in each?
6. What is an indicator organism? Why are coliform counts used to determine the safety of potable waters?
7. How are microorganisms involved in the abatement of pollution?
8. How does the chemical structure of an organic compound affect its biodegradability? How can biodegradable plastics be designed?
9. What are the differences between composting and a sanitary landfill?
10. Describe and compare the different types of sewage treatment systems.
11. How is water disinfected?
12. What procedures are used to determine water quality?

CRITICAL THINKING QUESTIONS

1. Describe how you could design a waste disposal system for a self-contained community so that water could receive wastes and still be used as a source of potable water.
2. How could microorganisms be used to reverse global warming?
3. Would it be better to replace coliform counts with direct counts of pathogens?
4. How would you replace coliform counts with direct counts of pathogens?
5. How can microorganisms be used to clean up oil spills?

READINGS

Atlas RM: 1981. Microbial degradation of petroleum hydrocarbons: An environmental perspective, *Microbiological Reviews* 45:180-209.
  Interesting review of the environmental factors that influence the fate of oil spills.
Atlas RM (ed.): 1984. *Petroleum Microbiology*, New York, McGraw-Hill.
  A collection of articles on the microbial metabolism of petroleum hydrocarbons and the fate of petroleum in the environment.
Berg G (ed.): 1978. *Indicators of Viruses in Water and Food*, Ann Arbor, MI; Ann Arbor Science.
  Explores the field of sanitary microbiology in relation to the prevention of the transmission of disease via food and water.
Bitton G and CP Gerba (eds.): 1984. *Groundwater Pollution Microbiology*, New York, Wiley.
  Authoritative articles on the microbiology of groundwater.
Brenner TE: 1969. Biodegradable detergents and water pollution, *Advances in Environmental Science and Technology* 1:147-196.
  The structures of chemicals used as detergents are considered with respect to biodegradability.
Brock T (ed.): 1986. *Thermophiles: General, Molecular, and Applied Microbiology*, New York, Wiley.
  Reviews the thermophilic bacteria in detail.
Fiksel J and VT Covello (eds.): 1986. *Biotechnology Risk Assessment: Issues and Methods for Environmental Introductions*, New York, Pergamon Press.
  Discusses techniques of health risk assessment in relation to microbial genetics and genetic engineering.
Finstein MS, FC Miller, PF Strom, ST MacGregor, KM Psarianos: 1983. Composting ecosystem management for waste treatment, *Biological Technology* 1:347-353.
  Discusses how to successfully optimize decomposition in composting

Ford TE (ed.): 1993. *Aquatic Microbiology: An Ecological Approach*, Boston, Blackwell Scientific Publications.
  Includes chapters on aquatic microbial biogeochemical cycling, microbial food webs, and the microbiology of salt marshes, wetlands, groundwater, estuaries, marine areas, and the air-water interface.
Forster CF and DAJ Wase: 1987. *Environmental Biotechnology*, New York, Halsted Press.
  Discussion of bioremediation from the point of view of the chemical engineer.
Fry JC (ed.): 1992. *Microbial Control of Pollution*, New York, Cambridge University Press.
  Collection of papers by experts presented at the Society for General Microbiology of Great Britain at its 1992 symposium.
Gaudy A and E Gaudy: 1980. *Microbiology for Environment Science Engineers*, New York, McGraw-Hill.
  An engineering perspective to environmental microbiology, including waste treatment.
Gibson DA (ed.): 1984. *Microbial Degradation of Organic Compounds*, New York, Marcel Dekker.
  Discusses the processes of microbial metabolism involved in the biodegradation of organic compounds.
Grabow WOK, M Butler, R Morris: 1988. *Health-related Water Microbiology*, Oxford, Pergamon Press.
  Discusses water microbiology, the health aspects of water pollution, viral pollution of water, and the environmental aspects of sewage.
Grainger JM and JM Lynch (eds.): 1984. *Microbiological Methods for Environmental Biotechnology*, London, Academic Press.
  Methodology for environmental microbiology and biotechnology.
Greenberg A (ed.): 1980. *Standard Methods for the Examination of Water and Wastewater*, Washington, D.C.; American Public Health Association.
  Standard methods for assessing the microbiological quality of water.

Haigler CH and PJ Weimer (eds.): 1991. *Biosynthesis and Biodegradation of Cellulose*, New York, Marcel Dekker.

   Reviews the microbial metabolism involved in the synthesis and degradation of cellulose.

Halavorson HO, D Pramer, M Rogol (eds.): 1985. *Engineered Organisms in the Environment: Scientific Issues*, Washington D.C.; American Society for Microbiology.

   Discusses the issues surrounding the deliberate environmental release of genetically engineered microorganisms.

Hollenberg CP and H Sahm (eds.): 1988. *Biosensors and Environmental Biotechnology*, Stuttgart, G. Fischer.

   Discusses the biological treatments used for the purification of sewage and air.

Horikoshi K and WD Grant (eds.): *Superbugs: Microorganisms in Extreme Environments*, New York, Springer-Verlag.

   Reviews the ecology and physiological adaptation of microorganisms to harsh living conditions.

Javor B: 1989. *Hypersaline Environments: Microbiology and Biogeochemistry*, Berlin, Springer-Verlag.

   Introduction to the microbial ecology and biogeochemistry of halophilic microorganisms.

Jenkins D and BH Olson (eds.): 1989. *Water and Wastewater Microbiology*, Oxford, Pergamon Press.

   The papers from a meeting of the International Association on Water Pollution Research and Control discuss the microbiology of water and sewage.

Kohl W (ed.): 1990. *Methanol as an Alternative Fuel Choice: An Assessment*, Washington, D. C.; Johns Hopkins University

   Proceedings of a congress sponsored by the American Petroleum Institute and the World Resources Institute on the use of methanol as a fuel.

Kulaev IS, EA Dawes, DW Tempest: 1985. *Environmental Regulation of Microbial Metabolism*, Orlando, FL; Academic Press.

   Presents the interaction between microbial ecology and metabolism.

Kundsin RB (ed.): 1988. *Architectural Design and Indoor Microbial Pollution*, New York, Oxford University Press.

   Discusses air microbiology, the health aspects of indoor air pollution, and the design and engineering of environmentally safe buildings.

Leadbetter ER and JS Poindexter (eds.): 1985. *Bacterial Activities in Perspective*, New York, Plenum Press.

   The first volume of a 10-volume set on "Bacteria in Nature." A historical treatment of the various prokaryotic trophic groups involved in ecological processes.

Leisinger T et al. (eds.): 1982. *Microbial Degradation of Xenobiotics and Recalcitrant Compounds*, New York, Academic Press.

   Series of papers discussing the biodegradation of specific types of chemical compounds.

Lidstrom ME (ed.): 1990. *Hydrocarbons and Methylotrophy*, San Diego, Academic Press.

   Discusses biodegradation by methylotrophic bacteria.

McFeters GA (ed.): 1990. *Drinking Water Microbiology: Progress and Recent Developments*, New York, Springer-Verlag.

   Introduction to microbiology of potable waters.

Metting FB (ed.): 1992. *Soil Microbial Ecology: Applications in Agricultural and Environmental Management*, New York, Marcel Dekker.

   This important, single-source reference provides detailed discussions of all the major groups of soil microorganisms, including rhizobacteria, symbiotic nitrogen-fixing bacteria, pathogenic fungi, mycorrhizal fungi, and microalgae.

Olson BH and LA Nagy: 1984. Microbiology of potable water, *Advances in Applied Microbiology* 30:73-132.

   Discusses the assessment of water quality.

Rao VC and JL Melnick: 1986. *Environmental Virology*, Washington, D.C.; American Society for Microbiology.

   Review of virology and the transmission of viral diseases.

Seppanen H and H Wihuri (eds.): 1988. *Groundwater Microbiology: Problems and Biological Treatment*, Oxford, Pergamon Press.

   Proceedings of the International Association on Water Pollution Research and Control symposium on biodegradation and groundwater pollution and purification.

Tate RL (ed.): 1986. *Microbial Autecology: A Method for Environmental Studies*, New York, Wiley.

   Monograph on microbial ecology.

Vandergrift GF, DT Reed, IR Trasker (eds.): 1992. *Environmental Remediation: Removing Organic and Metal Ion Pollutants*, Washington, D.C.; American Chemical Society.

   Discusses the use of bioremediation for the purification of groundwater and polluted soil.

Wolf RL, NR Ward, BH Olson: 1984. Inorganic chloramines as drinking water disinfectants: A review, *Journal of the American Water Works Association* 76:74-88.

   Discusses the effectiveness of chloramines for water disinfection.

Appendix

INTEGUMENTARY SYSTEM (The Skin)

STRUCTURE AND FUNCTION

The integument consists of the skin and associated structures such as nails, glands, and hair. Accessory structures are continuous with the skin and share a common origin during development of the embryo and fetus. The integument functions for protection of underlying structures from friction and microorganisms, vitamin D synthesis, body temperature regulation, immunity, sensation of various stimuli, and excretion of certain materials. Skin also acts as a permeability barrier, regulating what can and cannot diffuse through skin into the blood.

The skin is the largest organ of the human body. The average adult body is covered by about two square meters of skin. The thickness of skin varies. **Thin skin,** such as that found covering the eyelids and face, is only about 0.5 millimeters deep. **Thick skin,** such as that covering the palms and soles of the feet, can be up to 5 millimeters deep. There are two tissue layers that make up the skin—epidermis and dermis.

The **epidermis** is the superficial, avascular layer of the skin. It is composed of several stratified layers of thin cells, called *epithelial cells,* and melanocytes and Langerhans cells. The basal layer of these cells, appropriately called the *stratum basalis,* is anchored to the underlying dermis by means of a basement membrane—a glue-like substance secreted by the cells of the epidermis. Interspersed between these cells are **melanocytes**—cells that produce melanin pigment, which is responsible for suntan. The outermost layer of epidermal cells is the **stratum corneum.** This layer consists of dead cells containing large amounts of keratin protein. These cells also release lipids into the extracellular space. The stratum corneum helps waterproof the skin and serves as a physical barrier to microorganisms. Calluses and corns are areas of skin with a very thick stratus corneum. **Langerhans cells** of the epidermis are immune system cells involved in the detection of pathogens.

The **dermis** is located deep to the epidermis and is composed of large numbers of fibrocytes, protein fibers, blood vessels, lymphatic vessels, nerves, sweat and sebaceous glands, hair follicles, and arrector pili muscles. The **fibrocytes** are connective tissue cells that synthesize and secrete connective tissue proteins such as **collagen, reticulin,** and **elastin.** These proteins help give skin elasticity and strength. Blood vessels in the dermis allow the skin to function as a radiator system. Blood flow to the skin can be increased to cool the body or decreased to shunt blood to the internal body vessels to increase or maintain body core temperature. Nerves in the skin are primarily sensory nerves. They sense various stimuli—touch, temperature, vibration, pain—and relay the sensory information to the brain for processing. Some nerves to the skin stimulate small muscles in the dermis, called **arrector pili** muscles. These muscles pull on **hair follicles** located in the dermis, causing hair shafts to stand erect—producing "goose bumps." Erection of hair follicles occurs in mammals as a means of increasing hair thickness, and hence warmth, and as a behavioral signal during aggression or fear. **Sweat glands** in the dermis produce an aqueous mineral that contains water, salt, lysosome, and urea. Material from these glands is moved up ducts to the surface of the epidermis. Sweat glands function to help with evaporative cooling and excretion of small amounts of urea waste. **Sebaceous glands** are located in the dermis adjacent to hair follicles. These glands release a lipid (fatty material), called **sebum,** onto the hair shaft.

Between the dermis and underlying muscle or other tissues is often found a layer of adipose (fat) tissue called the **hypodermis.** The hypodermis is not part of the integumentary system but is relevant to the integument because it helps anchor the integument to underlying tissues. It also contains blood vessels, nerves, and lymphatic vessels that supply the integument.

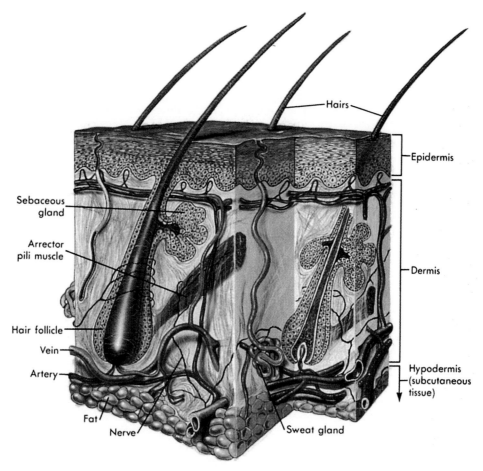

Skin and hypodermis.

Applications to Microbiology

Perspiration from sweat glands and sebum from sebaceous glands contain substances nutritive for certain microorganisms and inhibitory for other microorganisms. Ducts from these glands provide passages for microorganisms to invade the glands or deeper tissues of the skin.

The space along the hair follicles is sometimes invaded by microorganisms. **Acne** is a disease in which bacteria infect this space, feeding on the fatty acids from sebum. The inflammatory response against the bacteria leads to **folliculitis.** Most persons contain numerous harmless mites that live in their hair follicles and sebaceous gland ducts—these mites are of the genus *Demodex.*

Some bacteria secrete potent, destructive **exotoxins** that erode the integument, causing **cellulitis**—a severe, destructive, inflammation of the integument.

If not treated, cellulitis can lead to tissue death (necrosis), gangrene, and death. Strains of *Streptococcus* bacteria commonly cause cellulitis. These bacteria are known to produce collagenase—an exotoxin that destroys collagen.

Fungal infections of the skin are called **dermatomycoses.** Normal microbiota (bacteria) that inhabit the skin usually keep pathogenic bacteria and fungi from causing skin infections. Individuals who have a compromised immune system or whose skin pH is altered are susceptible to fungal infections of the skin.

Because the skin is a barrier to pathogenic microorganisms, loss of skin tissue, such as from thermal or chemical burns or trauma, can lead to invasion of deeper tissues and even the blood by microorganisms. This is serious because abscesses, deep tissue destruction, septicemia, and shock can result.

STRUCTURE AND FUNCTION

The skeletal system consists of bones and their associated connective tissues—tendons, ligaments, and cartilage. These structures collectively create a framework for the support and protection of the body. Bones provide points of attachment for skeletal muscles, thus allowing for body movement. Minerals and fat are stored in bone, and bone cavities are the site of blood cell formation. Tendons connect muscle to bone. Ligaments connect bone to bone. Cartilage provides shape for certain body parts, for example, the ear and nose, and also caps the ends of bones that articulate with each other.

Typical long bones consist of a shaft **(diaphysis),** a neck, and one or more heads **(epiphyses)** capped with cartilage, where articulation with other bones occurs. The ends of long bones in children contain regions called **epiphyseal plates,** which are crucial for bone growth. The inner core, or medulla, of bone tissue is made of **cancellous (spongy) bone.** This is the primary site of blood cell formation. Fat storage also occurs within bone marrow. The outer cortex of bone contains **dense (compact) bone** tissue and surrounds the bone marrow cavity. The space between articulat-

ing bones consists of a **joint space** that is encapsulated by connective and synovial tissue **(synovium),** creating a sterile joint cavity. Bone is surrounded by **periosteum,** a connective tissue continuous with ligaments and the joint space.

Cartilage is firm, yet softer than bone, and often caps the ends of bones to provide a suitable surface for articulation. Cartilage also provides support for structures such as the outer ear, nasal septum, and trachea, and connects the ribs to the sternum, allowing greater expansion of the rib cage during breathing.

Tendons and **ligaments** are dense, regular arrays of connective tissue containing cartilage and fibrocyte cells. Tendons connect muscle to bone. Ligaments connect bone to bone.

APPLICATIONS TO MICROBIOLOGY

Infections of skeletal system tissues are less common than infections of other bodily tissues or organs. This is a result of skeletal system tissues being located deeper (usually) than other tissue types and therefore they are less susceptible to trauma and external pathogens. Infection of ligaments and tendons is rare. Infections of bone or cartilage do occur, usually from a blood-borne microorganism, such as with septicemia or bacteremia. Deeply penetrating trauma is another cause of infection to skeletal system tissues.

Bone infection, **osteomyelitis,** typically occurs by the blood-borne spread of a microorganism to bone tissue. Blood flow through parts of a typical long bone is very sluggish, creating "pools" of blood in the bone cavities where microorganisms (usually bacteria) can get caught up in these eddies of blood and multiply locally. If this occurs, there is the risk of bone tissue destruction, pain, toxin release into the blood, fever, and dislodgement of septic emboli. Septic emboli are pieces of microbial growth (usually bacteria or fungi) that break off from a microbial growth site and travel in the bloodstream. If such emboli lodge in the lungs, local lung reflexes can lead to improper ventilation and blood oxygenation. Emboli can also travel to the heart and lodge in coronary arteries that supply heart tissue, causing heart arrythmias and death.

The joint is normally sterile but may become infected and cause **infectious arthritis.** Bacteria are usually the microorganism responsible, although viruses and fungi are also known to cause infectious arthritis. Infectious arthritis is usually caused by the spread of microorganisms through the blood. Direct contamination of the joint may also occur as a result

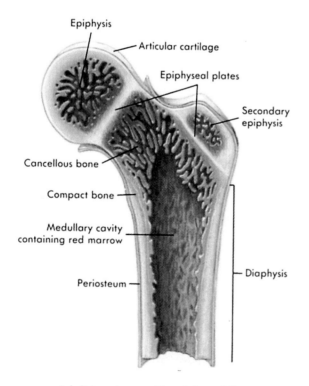

Epiphysis

— Articular cartilage

Epiphyseal plates

Secondary epiphysis

Cancellous bone

Compact bone

Medullary cavity containing red marrow

Periosteum

Diaphysis

Adult long bone with epiphyseal lines.

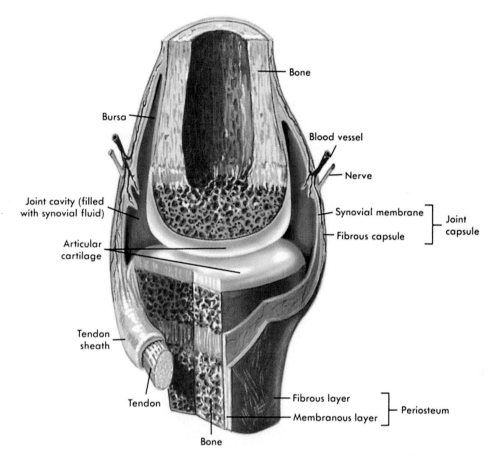

Structure of the synovial joint.

of trauma to the joint, surgical procedures, or drug inoculations to the joint (steroids, for example, to treat rheumatoid arthritis).

Cartilage has no blood supply and must receive oxygen and other nutrients by diffusion from nearby capillaries. Infection of cartilage, known as **infectious chondritis,** although not common, can be very serious. For a tissue (such as cartilage) to heal, the tissue requires blood to supply the tissue with immune system cells and regenerative cells.

MUSCULAR SYSTEM

STRUCTURE AND FUNCTION

Humans possess three classes of muscle—skeletal muscle, smooth muscle, and cardiac muscle. **Skeletal muscle,** also called **striated muscle,** is under voluntary control. As its name infers, skeletal muscle is generally attached to bone, by means of tendons. **Tendons** are actually a continuation of the connective tissue covering of a muscle with the connective tissue covering of a bone. Examples of skeletal muscles are the biceps muscle, the triceps muscle, and the diaphragm muscle (necessary for breathing). **Smooth muscle** is involuntary, which means it is under the control of the autonomic nervous system. This muscle type is found in the walls of hollow, internal or-

gans such as blood vessels, gallbladder, intestine, stomach, and lung airways. **Cardiac muscle** is located in the walls of the heart and, like smooth muscle, is under involuntary control. Cardiac muscle tissue has special cellular properties that allow very fast communication between the muscle cells.

All three classes of muscle are stimulated by **neurotransmitters**—chemicals released from nerve cell endings onto the muscle cells. Neurotransmitter molecules interact with muscle cell **receptors** at the **neuromuscular junction,** or **synapse,** eliciting a series of cellular events leading to shortening and contraction of the muscle cells. Muscle cells contract, allowing movement to occur, such as limb motion, intestinal

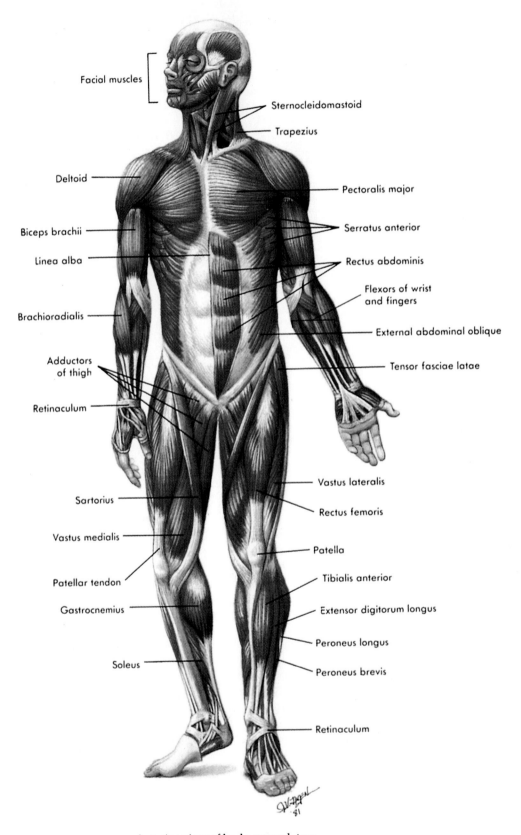

Facial muscles

Sternocleidomastoid

Trapezius

Deltoid

Pectoralis major

Serratus anterior

Biceps brachii

Linea alba

Rectus abdominis

Flexors of wrist
and fingers

Brachioradialis

External abdominal oblique

Adductors
of thigh

Tensor fasciae latae

Retinaculum

Vastus lateralis

Rectus femoris

Sartorius

Vastus medialis

Patella

Tibialis anterior

Patellar tendon

Gastrocnemius

Extensor digitorum longus

Peroneus longus

Peroneus brevis

Soleus

Retinaculum

Anterior view of body musculature.

motility, or gallbladder contraction. Shortly after a neurotransmitter performs its function, it is degraded by enzymes and thus cannot cause constant muscle contraction—a state called **tetany.** Regulation of nerve cell release of neurotransmitter and degradation of a neurotransmitter after stimulation of muscle cell receptors allows control of muscle cell contraction.

Applications to Microbiology

Toxins produced by certain bacteria can affect muscle cells. The **exotoxin** secreted by the bacterium *Clostridium botulinum* blocks the muscle cell receptor for neurotransmitter, causing a **flaccid paralysis** (a weakening or loss of muscle tone). *Clostridium tetani* produces a toxin that inhibits normal breakdown and recycling of the neurotransmitter. This toxin causes a **spastic paralysis,** characterized by involuntary contraction of one or more muscles, and is responsible for the disease commonly known as **tetanus,** or "lockjaw."

Some parasitic worm life cycle stages invade muscle tissue, causing destructive or painful muscle disease. The pork tapeworm, *Taenia solium*, lives not only in the human intestine but can spread throughout the body and invade many human organs, including muscle tissue.

Generalized muscle soreness can be a sign of virus infection. Viruses can spread by means of the blood throughout the body—a condition called **viremia.** Viremia can quickly spread virus to muscle cells, where muscle damage and soreness can result if the virus replicates inside the muscle cells. As a diagnostic aid, general muscle soreness can be of value in focusing on the cause of an infection.

NERVOUS SYSTEM

Structure and Function

The nervous system can be divided into the **central nervous system (CNS)**—consisting of the brain and spinal cord—and the **peripheral nervous system (PNS)**—consisting of the cranial nerves and the spinal nerves.

Anatomically, and somewhat functionally, the CNS and PNS are separate. The CNS consists of the brain and spinal cord, contained within the skull and vertebral column, respectively. The CNS processes incoming sensory (afferent) signals (for example, heat, touch, light) from the PNS and elicits motor (efferent) signals to muscle cells of the body. The CNS is also the site where higher neurological functions occur, such as memory, logic, and emotion. The PNS consists of nerves that communicate with the CNS but that are found, for example, in the limbs, trunk, and abdomen. Large numbers of PNS nerve cells connect with the CNS as either spinal nerves or cranial nerves.

The PNS can be further classified functionally. It can be subdivided into an **autonomic nervous system (ANS)** and a **somatic nervous system.** The ANS senses and regulates internal organ activity (for example, heart rate, bowel function, and airway diameter). The somatic nervous system is associated with the body wall and skeletal (voluntary) muscle. It senses stimuli that a person normally feels (for example, sharp pain, skin temperature, and vibration). The somatic nervous system also stimulates skeletal muscle contractions.

The CNS and PNS are enclosed by connective tissue coverings. In the CNS these coverings are given a special name—the **meninges.** The meninges occur in three layers. There is the outer **dura,** which is very tough, and it is attached to the inner layer of the skull. The middle, vascular, layer is the **arachnoid.** This layer contains the meningeal arteries. The layer attached to the brain is very thin and is called the **pia.** Between the arachnoid and pia circulates a fluid called *cerebrospinal fluid (CSF)*. The CSF is produced by cells of the **ventricles**—the hollow fluid-filled chambers deep in the brain. The CSF functions to circulate nutrients and electrolytes through brain and spinal cord tissue and to act as a shock absorber.

Blood supply to the CNS is more closely controlled than blood supply to the rest of the body. Capillaries in the CNS have tight junctions between cells, causing a decrease in permeability. In addition, CNS cells surrounding the capillaries, called astrocytes, also have tight junctions between cells. Because of this, a **blood–brain barrier (BBB)** exists in the CNS. The BBB regulates closely what can enter the brain and spinal cord tissue from the blood. Generally, molecules that are small, non-ionic, and lipid soluble cross the BBB more readily than molecules that are larger or ionic.

Applications to Microbiology

Because of the blood–brain barrier, drugs (antibiotics) used to treat diseases of the CNS should be non-ionic and lipidlike in their chemistry.

CNS infection can result from trauma or by spread through the blood or lymphatics. Head contusions, facial trauma, or vertebral injuries can result in open communication between CNS tissues and infectious microorganisms on the skin. Trauma to the nose,

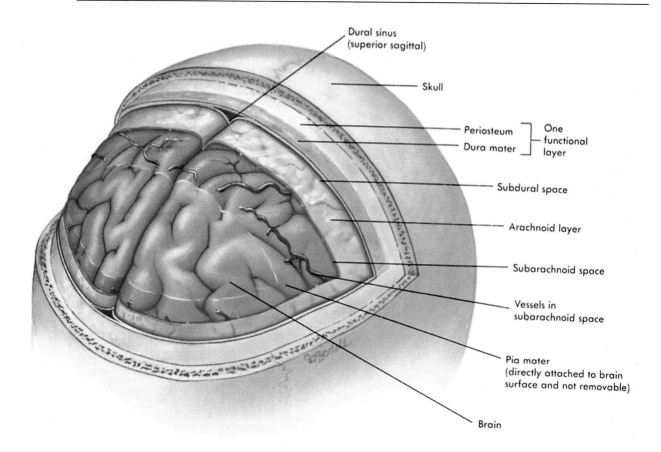

Dural sinus
(superior sagittal)

Skull

Periosteum

Dura mater

One functional layer

Subdural space

Arachnoid layer

Subarachnoid space

Vessels in subarachnoid space

Pia mater
(directly attached to brain surface and not removable)

Brain

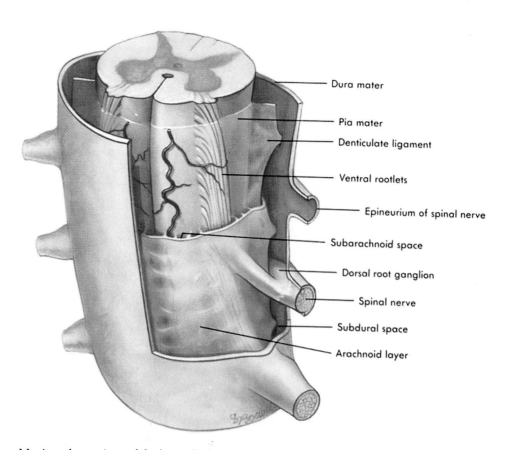

Dura mater

Pia mater

Denticulate ligament

Ventral rootlets

Epineurium of spinal nerve

Subarachnoid space

Dorsal root ganglion

Spinal nerve

Subdural space

Arachnoid layer

Meningeal coverings of the brain *(top)* and spinal cord *(bottom)*.

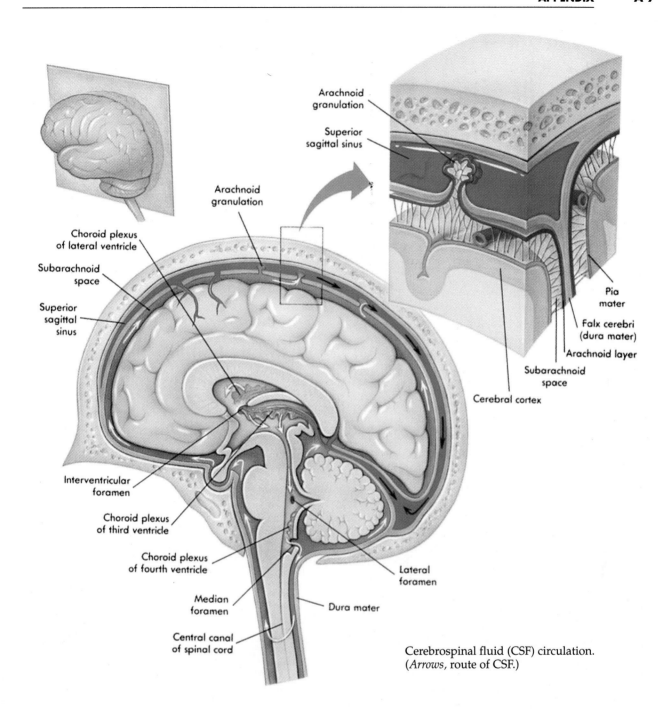

Cerebrospinal fluid (CSF) circulation.
(*Arrows*, route of CSF.)

such as from nasal fractures, can result in microorganisms gaining access to the CSF. This is usually suspected if a clear fluid (CSF) is seen dripping from the nasal cavity after trauma. Infection of the blood or lymph can lead to seeding of CNS tissue with microorganisms, with resulting infection of the meninges **(meningitis)** or the brain tissue itself **(encephalitis).** Sampling of the CSF, obtained by needle aspiration of CSF from the spinal canal through a lumbar puncture (a procedure that must be done with sterile technique, is a diagnostic method used to determine whether there is infection of the CNS. With bacterial infections of the CSF, glucose levels of the CSF will be low. With viral infections of the CSF,

glucose levels of the CSF are usually normal. Laboratory analysis of the CSF is thus important in diagnosing CNS infections.

Inflammation of the CNS, such as that resulting from microbial infections, will cause swelling of CNS tissues. This is serious because the CNS is enclosed in the nonexpandable skull and vertebral column. The skull and vertebral column leave inflamed tissue nowhere to expand. Inflamed CNS tissue will compress vital structures such as capillaries that supply neurons and ventricles that allow CSF flow and drainage. Serious neurological complications, seizures, and death can result. It is crucial to treat CNS infections aggressively.

STRUCTURE AND FUNCTION

The endocrine system consists of glands, tissues, and individual cells that secrete molecular messengers called **hormones.** Hormones travel through the body in the blood. Hormones are very potent molecules, functional at blood concentrations that are almost immeasurable. When a hormone is released into the bloodstream, it will eventually interact with a specific cell membrane or cytoplasmic **receptor** of one or more classes of **target cells.** Various cellular effects are then elicited as a result of the hormone's effect on the target cell's regulatory enzymes or even its genetic code (DNA).

Endocrine glands regulate body metabolism, reproductive function, blood electrolytes, growth, blood pressure, uterine contractions, skin pigmentation, water retention, blood nutrient levels, and many other functions. There are usually **feedback regulation** mechanisms between hormones and their target cell products and even other hormones. Some endocrine glands, such as the pituitary and adrenal glands, are closely linked to the nervous system.

The endocrine glands include the pituitary, adrenals, thyroid, parathyroids, pancreas, ovaries, testes, and thymus. The pituitary gland makes growth hormone (GH), gonadotropins (FSH and LH, which stimulate the ovary or testes), MSH (stimulates melanocyte tanning cells), prolactin (stimulates milk production by breast), TSH (stimulates thyroid gland), ACTH (stimulates adrenal glands), sommatostatin (inhibits growth), oxytocin (stimulates milk ejection from breast and uterine contractions), and antidiuretic hormone (ADH, which stimulates kidney to reabsorb water and also causes blood vessels to constrict). The adrenal glands are important for maintaining blood pressure and for stressful ("fight-or-flight") situations. Each adrenal gland is subdivided histologically into a cortex and medulla. The adrenal cortex makes steroidal hormones—glucocorticoids (involved in metabolism), mineralocorticoids (for electrolyte regulation), and small amounts of estrogens and androgens (for sexual characteristics). The adrenal medulla produces adrenaline (epinephrine)—a potent vasoconstrictor for elevating blood pressure. The thyroid gland, located in the anterior neck region, manufactures thyroid gland hormones T3, T4, and calcitonin. T3 and T4 ("thyroid hormone") increase cellular carbohydrate metabolism and basal metabolic rate (BMR). Calcitonin lowers blood calcium and increases bone density. The parathyroid glands, four in number and located on the thyroid gland, secrete parathyroid hormone (PTH, also called **parathormone**). This hormone demineralizes bone and increases blood calcium. The endocrine portion of the pancreas secretes three hormones—insulin (causes uptake of glucose into cells), glucagon (causes catabolism to increase blood glucose), and sommatostatin. The ovaries, in addition to preparing eggs for fertilization, secrete estrogen and progesterone. The male testes, in addition to making sperm cells, secrete testosterone. The thymus gland is located in the thorax and neck, superior to the heart. This gland secretes thymosin and is involved with processing of immune system cells early in body development.

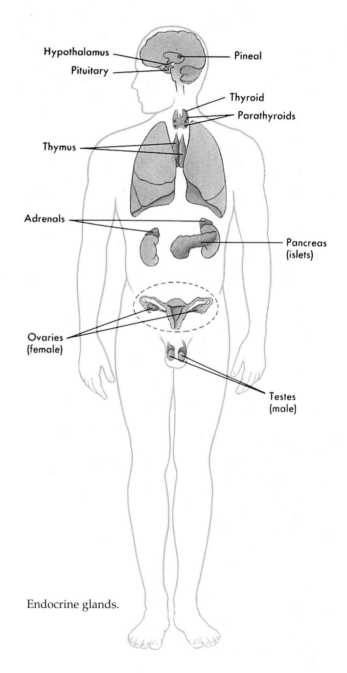

Endocrine glands.

Some hormones are secreted by individual cells instead of glands. There are endocrine system cells in the stomach and small intestine that secrete hormones that help regulate digestive function.

APPLICATIONS TO MICROBIOLOGY

Bacteremia, particularly with Gram-negative bacteria, can lead to infection and destruction of adrenal gland tissue. This can lead to severe adrenal gland insufficiency, loss of blood pressure, and electrolyte imbalance. This condition is called *Waterhouse-Friedrickson syndrome.* Hypotensive shock and cardiac arrest can occur, leading to death.

Microorganisms can affect the thyroid gland. **Thyroiditis,** an inflammation of the thyroid gland, can occur from viral infection. T3 and T4 may be elevated. The clinical history usually is that of a sudden upper respiratory infection or sore throat. A viral infection may have a possible role in a form of overactive thyroid gland disease, **hyperthyroidism.** A protein of a virus is believed to resemble the receptor on thyroid endocrine gland cells. Immune system antibody molecules formed against the viral protein may stimulate the thyroid gland, causing a form of hyperthyroidism known as *Graves disease.*

The pituitary gland, considered to be the master endocrine gland, can sometimes become inflamed, leading to **hypopituitarism.** In this condition, hormones made by the pituitary gland are not produced in sufficient amounts. The pituitary can sometimes become inflamed or infected as a result of meningitis (inflammation of the brain coverings). **Abscesses** of the pituitary (or any endocrine gland) can occur as a result of the spread of microorganisms through the blood.

CARDIOVASCULAR SYSTEM

STRUCTURE AND FUNCTION

The cardiovascular system consists of the heart and blood vessels. Vessels function to distribute the blood to the tissues and cells and then return the blood back to the heart. Arteries and arterioles are vessels that carry blood from the heart to tissues. Capillaries are vessels where diffusion of nutrients, gases, and wastes occurs between blood and cells. Venules and veins return blood to the heart.

The **heart** is a four-chambered biological pump, complete with **valves** that are designed to maintain flow in one direction. The heart is divided anatomically and functionally into a right heart and a left heart. The *right side of the heart* is designed to receive venous blood, blood that has already been distributed to tissues. The right heart pumps incoming venous blood to the lungs for oxygenation and also for expulsion of carbon dioxide from the blood. Blood re-

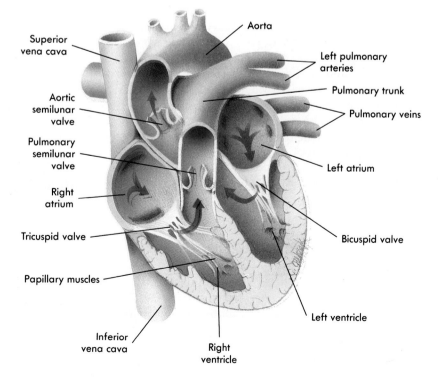

Normal blood flow. Frontal section of the heart showing the four chambers and the direction of blood flow through the heart.

turning from the lungs, rich in oxygen, flows to the *left side of the heart* to be pumped out to systemic arteries, then arterioles, and then capillaries. In this way, blood is distributed quickly and in an efficient manner to all tissues.

Arteries are larger elastic vessels with a thick wall that carry blood from the heart to organs of the body. The pulse pressure wave generated by the heart, combined with the elastic nature of large arteries, keeps blood moving through the arteries. **Arterioles,** smaller diameter vessels that branch off the arteries, are primarily responsible for alterations in blood pressure. Their diameter is controlled by nervous system signals to the smooth muscle cells encircling these vessels. Arterioles narrow to become very small vessels called **capillaries,** vessels that have a wall made of a single layer of very thin endothelial cells. Capillaries are the site where exchange of materials occurs between blood and the cells. Oxygen, sugars, amino acids, certain hormones, and other nutrients diffuse out of capillaries and into cells. Carbon dioxide, ammonia, lactic acid, and cellular products diffuse out of cells and into the blood. As capillaries enlarge, having performed their function, they become **venules,** the counterpart of arterioles. Venules unite to become larger vessels called **veins,** which ultimately bring blood back to the heart.

Blood is a mixture of noncellular elements solubilized in a liquid component of blood called **plasma** and of cellular elements. (If blood has coagulated and clotting proteins have been removed from the liquid component, the liquid component is called *serum.*) Cellular elements of blood include **leukocytes** (white blood cells), **erythrocytes** (red blood cells), and **thrombocytes** (platelets). Leukocytes, produced by bone marrow and lymphatic tissue, function in fighting off foreign invaders. **Erythrocytes** function to transport oxygen, carbon dioxide, and, to a lesser extent, hydrogen ions on a special substance called hemoglobin found in these cells. **Thrombocytes,** or platelets, are actually cell fragments of a large cell found in bone marrow called a *megakaryocyte.* When a break occurs in the wall of a blood vessel, platelets become sticky and are the first element to plug a hole in a vessel. This "buys time" for clotting proteins to form fibrous threads that mix with the platelets and erythrocytes to form a **blood clot.** The formed clot patches the break until repair cells can regenerate the damaged tissue.

Applications to Microbiology

The cardiovascular system, although designed to help maintain homeostasis in the human body by serving as a system for the distribution of nutrients and waste products, also serves as a potential distribution system for microbial agents or their toxic

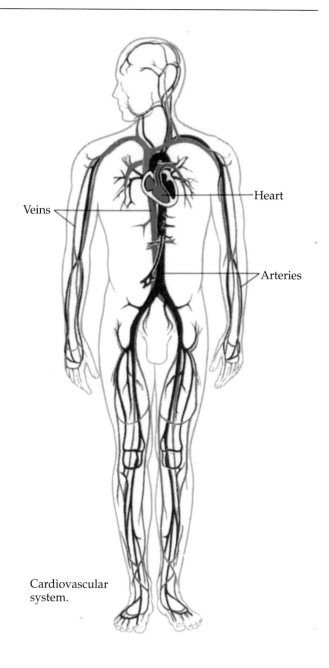

Veins

Heart

Arteries

Cardiovascular system.

products. If microorganisms contaminate the blood (a condition known as **septicemia**), the microorganisms or their products can be rapidly disseminated to tissues of the body. Widespread harm to tissues is a potential consequence of septicemia.

Sometimes microorganisms that have contaminated the blood can get caught in current eddies of blood flowing through the heart, allowing them to colonize valves in the heart, a condition known as **endocarditis.** This is a serious condition because clumps of microorganisms can dislodge from an infected valve, travel to the lungs, and lodge in lung tissue, causing pneumonia and other complications. These traveling microbial clumps are called *septic emboli.*

Toxins in certain bacteria, notably Gram-negative bacteria, can trigger activation of the clotting system of proteins in blood. This creates a life-threatening condition known as **disseminated intravascular coagulation,** or simply **DIC.** DIC is dangerous because clotting proteins are rapidly used up, leaving the body with little remaining clotting proteins for stopping microscopic bleeding that occurs in small capillaries throughout the body. Bleeding that is indicative of DIC is visible as pinpoint-size skin bleeds **(petechiae)** and later skin bruising **(ecchymoses).** Although not visible during DIC, bleeding also occurs internally, causing gastrointestinal and other systemic problems.

Because a large percent of total leukocytes in vessels are marginated (adherent to the walls of vessels), total leukocyte counts in the blood can rise significantly during times of stress or infection. The marginated leukocytes release from the vessel wall to become part of circulating blood leukocytes and part of a complete blood count (CBC). Adrenaline (epinephrine) and certain other substances will trigger the **demargination.** A sudden rise in blood leukocyte numbers during an infection is explained in part as a result of demargination, and it is a normal response of the body to foreign invaders.

LYMPHATIC SYSTEM AND IMMUNITY

STRUCTURE AND FUNCTION

The lymphatic system includes lymphatic vessels, lymph nodes, lymph nodules, tonsils, the spleen, and the thymus gland. Cells found in lymph tissue are part of the body's defense system—the immune system. Other immune system cells are found in tissues other than lymphatic tissue, such as bone marrow and blood.

The immune system cells that function to defend the body from pathogens or harmful molecules are collectively called **leukocytes.** Leukocytes defend the body by direct cellular attack or by producing cellular defense molecules—antibodies, interferon, lymphokines, histamine, and interleukins. Leukocytes are subcategorized as **granulocytes** (neutrophils, eosinophils, and basophils) and **agranulocytes** (lymphocytes and monocytes). Granulocytes are generally phagocytic and normally do not retain memory

against foreign invaders. Lymphocytic agranulocytes exist as **T lymphocytes** and **B lymphocytes.** B lymphocytes transform into **plasma cells** that produce antibodies. Lymphocytes retain memory against certain foreign invaders. Monocytes migrate from the blood to become **macrophages**—phagocytic cells that wander about in tissues.

When fluid from the blood diffuses to the tissue space before returning to the blood, some of the fluid does not return to the blood capillaries. Instead, a certain amount of this fluid enters thin-walled vessels called **lymphatic capillaries.** The lymphatic capillaries carry this fluid, called **lymph,** to larger, valve-lined vessels.

As fluid moves along the lymphatic vessels, it periodically flows through encapsulated collections of immune system white blood cells called **lymph nodes.** Lymph nodes contain macrophages, as well

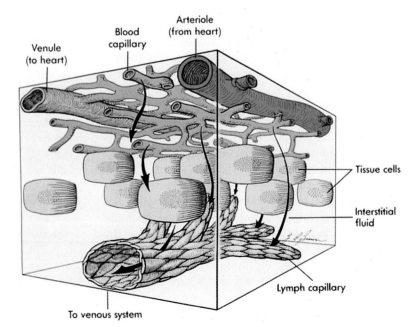

Movement of fluid from blood capillaries into tissues and from tissues into lymph capillaries.

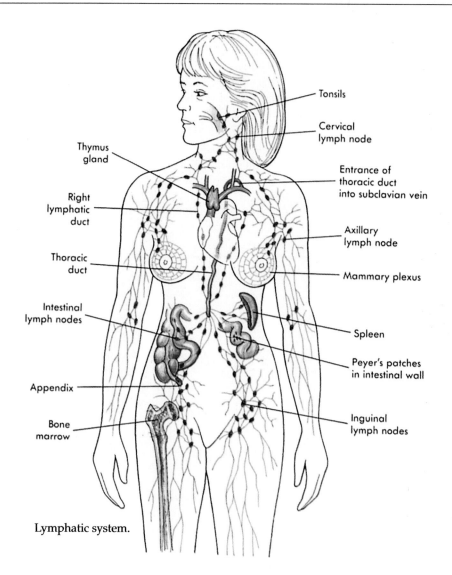

Lymphatic system.

as numerous B lymphocytes and T lymphocytes. B lymphocytes react to **antigens** (foreign cells and molecules) to produce defensive molecules of the immune system called **antibodies.** T lymphocytes react to antigens and interact with B lymphocytes to perform **cell-mediated** defensive functions against pathogens.

As lymph continues to flow through the lymphatic vessels and lymph nodes, the lymph ultimately returns to the venous system of blood vessels. In this way, fluid that exited the blood to enter the tissue space and did not return immediately to the blood eventually returns to the blood from where it originated. Along its course, this fluid can carry potential pathogens and antigens that can signal the immune system cells to react—both at the cellular and at the molecular level.

Various regions of the body contain dense, but nonencapsulated, regions of T and B lymphocytes, as well as macrophages. These regions are called **lymph nodules. Peyer's patches** in the small intestine and

lymph tissue in the spleen are examples of sites containing lymph nodules.

Tonsils are a ring of lymphoid tissue surrounding the region at the back of the throat. This tissue helps form a defense network against invading microorganisms of the mouth, throat, and nasal cavity. If the tonsils become inflamed they are sometimes removed.

The **spleen,** located on the left side of the abdominal cavity just beneath the ribs, serves several functions. It recycles red blood cells, injects red blood cells into the blood when the need arises, and responds to antigens by eliciting an immune reaction. Lymph nodules are abundant in the spleen.

The **thymus** is located in the anterior neck region, superior to the heart. It regresses after puberty so that it is almost nonexistent in adults. This gland serves the function of producing lymphocytes that migrate to other lymphatic tissues. The thymus is important for processing primitive lymphocytes to become T lymphocytes.

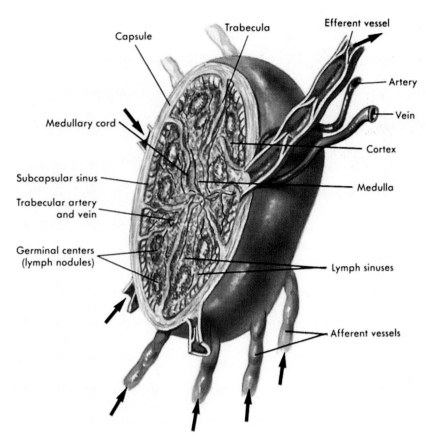

Labels on the figure:
- Capsule
- Trabecula
- Efferent vessel
- Artery
- Vein
- Medullary cord
- Cortex
- Subcapsular sinus
- Medulla
- Trabecular artery and vein
- Germinal centers (lymph nodules)
- Lymph sinuses
- Afferent vessels

Lymph node. (*Arrows,* direction of lymph flow.)

APPLICATIONS TO MICROBIOLOGY

Infection of the lymphatics provides a route for **dissemination** of a microorganism or its products through the body. The blind beginnings of the lymph capillaries are large enough so that the fluid entering the lymphatic capillaries may contain microorganisms, parts of microorganisms, or microbial products (soluble antigens; for example, toxins). As these substances circulate through the lymphatic vessels, immune system cells of the lymph nodes react to these substances and divide mitotically, increasing the size of the lymph node. Enlargement of lymph nodes, called **lymphadenopathy,** is most pronounced near the site of an infection and hence provides a clue during a physical examination as to where an individual may be infected. **Lymph node biopsies** can provide additional clues in diagnosing a pathogenic microorganism.

Certain microorganisms, such as the larval forms of some infectious parasitic worms, may replicate and grow in the lymphatics (a condition called **filariasis**), blocking lymph flow and causing tissue fluid build-up **(edema).** These parasites are called *filariae.*

Analysis of blood leukocytes, as part of complete blood count (CBC) with a "differential," can be helpful during evaluation of a patient. The term *differential* refers to a count of the relative percentage of each subtype of leukocyte in the blood. Normally, about 70% of circulating leukocytes are granulocytes and about 30% are lymphocytes. Monocytes, basophils, and eosinophils comprise very few of the circulating lymphocytes, each making up less than 5% of leukocytes. Increases in granulocytes lend evidence for a bacterial infection. Increases in eosinophilic granulocytes suggest parasitic infection or allergy. Increases in lymphocytes suggest a viral infection. Combined with a patient's presenting signs, symptoms, and clinical history, a white blood cell differential can be a valuable tool for narrowing the list of possibilities for a disease.

STRUCTURE AND FUNCTION

The respiratory system can be subdivided for convenience into upper, middle, and lower portions. The overall function of this system is to inhale, humidify, and filter air and to ventilate air to small air sacs of the lungs. During ventilation (breathing), air is moved to these air sacs for gas exchange before the exhalation of air containing waste gas (carbon dioxide [CO_2]). By regulating carbon dioxide levels in the blood, the lungs also help to regulate blood pH (acidity). Decreased lung function such as from lung disease or head trauma will result in the lowering of blood pH. Increasing lung activity can help counteract excess acid production in the body, such as occurs with exercise, fever, and shock from septicemia.

The upper portion of the respiratory system consists of the **nose** and **pharynx** (throat). The nose is separated into two nasal cavities by means of a **nasal septum.** Inside the nasal cavities are turbine- or conch-shaped bony ridges called the **turbinates** or **conchae.** These are lined by highly vascular skin tissue, helping to warm inhaled air by means of heat exchange between the blood and the air. Sinus cavities of the skull drain secretions to the nasal cavities and pharynx. Tears from the lacrimal gland of the eye also drain into the nasal cavity.

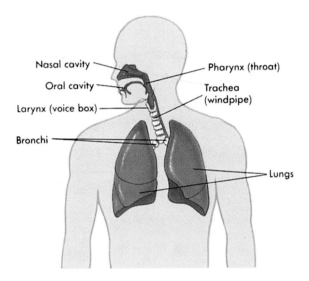

Major elements of the respiratory system.

The middle portion of the respiratory system contains the **larynx** (voice box or Adam's apple) and trachea (windpipe). The larynx contains the vocal cords. Food and beverages are prevented from entering the larynx by means of the **epiglottis.** The epiglottis is a cartilaginous flap that covers the entryway to the lar-

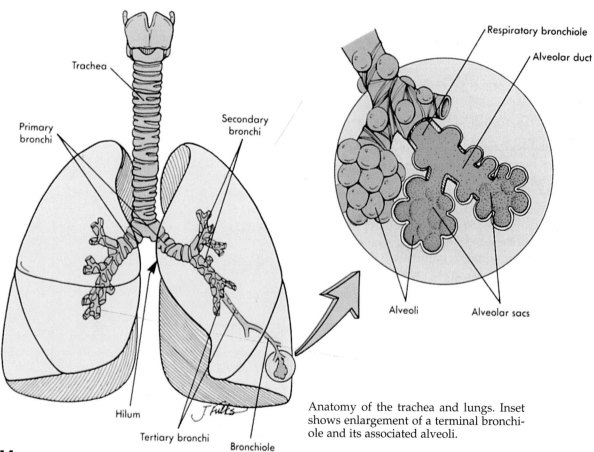

Anatomy of the trachea and lungs. Inset shows enlargement of a terminal bronchiole and its associated alveoli.

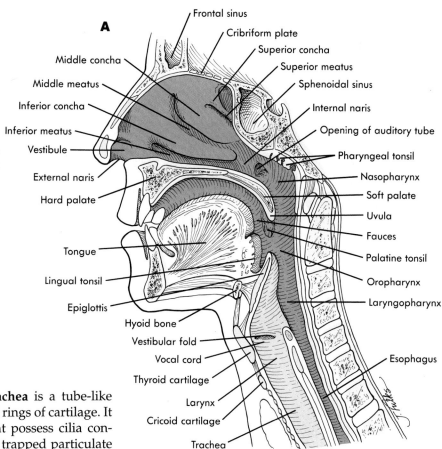

A

Frontal sinus
Cribriform plate
Superior concha
Superior meatus
Sphenoidal sinus
Internal naris
Opening of auditory tube
Pharyngeal tonsil
Nasopharynx
Soft palate
Uvula
Fauces
Palatine tonsil
Oropharynx
Laryngopharynx
Esophagus

Middle concha
Middle meatus
Inferior concha
Inferior meatus
Vestibule
External naris
Hard palate
Tongue
Lingual tonsil
Epiglottis
Hyoid bone
Vestibular fold
Vocal cord
Thyroid cartilage
Larynx
Cricoid cartilage
Trachea

Sagittal section through the nasal cavity and pharynx viewed from the medial side.

ynx during swallowing. The **trachea** is a tube-like structure, kept open by C-shaped rings of cartilage. It is lined with a layer of cells that possess cilia constantly beating upward to move trapped particulate and cellular matter out of the trachea to be swallowed. Goblet cells of the trachea produce a mucous secretion that helps trap materials, and the ciliated cells then move materials from the trachea.

The lower portion of the respiratory system consists of the lungs—composed of bronchi (large airways), bronchioles (small airways), alveoli (air sacs), and supporting connective tissue matrix. At about the level of the heart, the trachea splits into a right and left main bronchus. This branch point is called the **carina.** Each main bronchus supplies a lung. The right lung has three lobes—an upper lobe, middle lobe, and a lower lobe. The left lung has two lobes—an upper lobe and a lower lobe. The main **bronchi** branch into lobar bronchi that supply each lobe of a lung. As airways become smaller in diameter, they are called **bronchioles.** The superior region of each lung is called the **lung apex.** The apex normally receives good ventilation but receives less than optimal blood flow. The inferior region of each lung is called the **lung base.** The lung base receives good blood flow (because of gravity) but less than optimal ventilation (because of compression of alveoli from the underlying diaphragm muscle). Bronchioles eventually end as tiny air sacs—the **alveoli.** With each breath, the inhaled air is rapidly and efficiently distributed to millions of alveoli. Blood capillaries (small blood vessels) course next to each alveolar wall, ready to accept oxygen from the fresh air in the alveoli. The alveoli are lined with very thin cells, cre-

ating a thin barrier to the diffusion of gases. Oxygen diffuses from alveoli to the blood; carbon dioxide diffuses from the blood to the alveoli. Immune system phagocytic white blood cells, called **alveolar macrophages,** wander in the alveoli, cleaning up debris and attacking microorganisms. Cells lining the alveoli also serve the important function of secreting **surfactant**—a substance that prevents alveoli from collapsing and becoming nonfunctional.

Several other structures are closely associated with the respiratory system. The **diaphragm** and chest wall muscles expand the thoracic cavity, creating a negative pressure that causes the lungs to expand during inhalation. The middle ear drains secretions by way of a muscular tube, the **eustachian (auditory) tube,** to the pharynx. Rings of lymphatic tissue, the **tonsils** and **adenoids,** are found near the junction of the nose and pharynx. **Sinus cavities** (sinuses) of the skull bones and secretions of the nasolacrimal glands drain into the nasal cavity and pharynx.

Initiation, rhythm, and rate of breathing are controlled neurologically by nerve centers of the brain located in the medulla oblongata and pons—regions near the base of the brain called the brainstem. The pons and medulla give rise to phrenic nerves that innervate the diaphragm and intercostal nerves that innervate the intercostal chest wall muscles.

APPLICATIONS TO MICROBIOLOGY

Because the respiratory system is a portal of entry for microorganisms into the body, it possesses several defense mechanisms to minimize the number of microorganisms that can infect this system. The nasal cavities are lined by hairs to provide initial filtering of air. Secretions from cells of the nose, pharynx, and trachea help trap particulate matter and microorganisms. Cells of the upper and middle respiratory system have cilia to help move material out of the respiratory system. These cilia are inhibited by chemicals in cigarette smoke, making smokers more vulnerable to infections of the lower respiratory tract. Alveolar macrophages are phagocytic immune system white blood cells that wander about the lung air alveoli searching for debris or microorganisms.

Cells lining the respiratory system have different cell membrane protein **receptors** with affinities for different viruses and bacteria. This is one reason why certain microorganisms tend to attach to, and infect, certain regions of the respiratory system. Bacteria that are **encapsulated** tend to be more virulent than nonencapsulated bacteria, partly because they attach easier to human respiratory cells and partly because their capsules allow them to evade the human immune system with greater efficiency. Some pathogenic bacteria have **attachment pili** that enables them to bind to cell membrane receptors of human respiratory system cells.

Infections and inflammations of the respiratory system have specific terminology. *Pharyngitis, tracheitis, laryngitis, bronchitis,* and *pneumonitis* are terms relating to the inflammation (usually of infectious origin) of the pharynx, trachea, larynx, bronchi, and lungs, respectively. When infections involve the upper respiratory system, the patient is said to have a URI (upper respiratory infection).

"Plugged and runny noses" are a common symptom of patients with URIs. This symptom is caused by inflammatory responses in the nose to the infecting microorganisms. Because the tissue covering the turbinate bones in the nose is very vascular, vasodilation causes swelling of this tissue and a "stuffy nose." As part of the URI, cells lining the nose and pharynx secrete copious amounts of mucus. When leukocytes are added to the mucous secretion—a "runny nose" occurs, technically known as **rhinorhea.**

Pneumonias are infections of lung tissue. Whereas viruses tend to cause diffuse lung infections, bacteria tend to cause localized, lobar lung infections. The pattern of infection seen on a chest radiograph can be of diagnostic usefulness in determining the class of microorganism responsible for a patient's condition. The base of the lungs has a large number of collapsed and nonaerated alveoli. This is because of the weight of the lung pressing against the diaphragm (bedridden patients have more collapsed alveoli along the posterior lung fields). The lung base is thus susceptible to opportunistic growth by bacteria, especially anaerobes. The apices of the lungs are highly ventilated and may be susceptible to certain microorganisms that are aerobic and travel easily on air currents as a mode of transmission.

DIGESTIVE SYSTEM (The Gastrointestinal Tract)

STRUCTURE AND FUNCTION

The digestive system develops embryologically as a straight tube that enlarges, twists, and contorts at various locations. This gives rise to the definitive mouth, pharynx, esophagus, stomach, small intestine (duodenum, jejunum, and ileum), colon (large intestine), and rectum. From these essentially tubular structures develop diverticula (during embryonic and fetal development) that create the salivary glands, liver, bile ducts, gallbladder, and pancreas. Other structures that are part of the digestive tract include the teeth and the tongue.

The digestive system serves the function of taking in food, breaking it down into smaller, simpler molecules, and then absorbing useful molecules into the blood and lymph for use by cells. Unusable material that is ingested, including microorganisms, is allowed to continue along the digestive tract for elimination as feces.

The mouth, teeth, and tongue mechanically ingest and predigest food and prepare a bolus of food to be swallowed to the esophagus and stomach. Salivary digestive glands (parotid glands, sublingual glands, and submaxillary glands) produce secretions that drain into the oral cavity to assist with lubricating a food bolus and to chemically predigest food molecules (amylase enzyme in saliva begins the breakdown of starches). The esophagus is essentially a muscular tube that propels a swallowed bolus of food or liquid to the stomach by coordinated muscular contractions. The stomach performs a great deal of mechanical and acid hydrolysis digestion of food and, to a limited extent, enzymatic digestion. A special protein made by stomach cells, intrinsic factor (IF), binds to dietary vitamin B_{12} so that this molecular combination can be absorbed later by the ileum (final portion of the small intestine). Material from the stomach (chyme) is sent to the small intestine

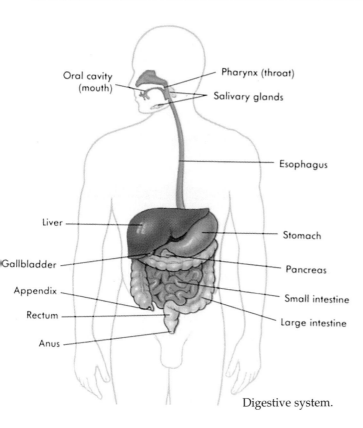

Oral cavity (mouth)
Pharynx (throat)
Salivary glands
Esophagus
Liver
Stomach
Gallbladder
Pancreas
Appendix
Small intestine
Rectum
Large intestine
Anus

Digestive system.

toxins so that blood concentrations of these substances can be closely regulated. Reabsorption of water and electrolytes (Na^+, Cl^-, K^+), absorption of bacterially synthesized vitamin K, and secretion of sodium bicarbonate occur in the colon, where formation of feces for defecation (via the rectum and anus) occurs.

APPLICATIONS TO MICROBIOLOGY

The digestive system can be affected by microorganisms at one or more locations, causing great discomfort and threatening life itself if electrolyte and nutrient losses become severe. The signs or symptoms produced by offending microorganisms can be useful in diagnosing disease. Stool consistency, stool cultures, presence of blood in the stool, and temporal sequence of events following infection are indicative of certain infections.

The oral cavity and pharynx may become colonized by fungi (*Candida albicans* [thrush]), bacteria (*Corynebacterium diphtheria,* the cause of diphtheria, or *Streptococcus,* the cause of strep throat), or other microorganisms. Mouth sores can be caused by the herpesvirus, protozoans (*Leishmania* spp.), and other microorganisms. Toxins produced by microorganisms, ingested with or without the microorganism, can irritate the stomach and intestine, causing emesis (vomiting), diarrhea, and resulting fluid and electrolyte loss and can interfere with normal digestive processes along the intestinal tract, causing malnutrition. The bile ducts drain from the liver to the small intestine and thus provide a portal of entrance to liver tissue from the intestine. Infection by liver flukes (*Schistosoma* spp. and *Taenia* spp.) occurs when fluke eggs or larval forms are ingested, mature, and then migrate up the bile ducts to invade and destroy liver tissue. An inflammation of the liver is known as *hepatitis.*

Because feces often contain a certain number of pathogens as well as normal microbiota, it is important to prevent contamination of food or drinking water with fecal material; otherwise an infectious cycle of fecal-oral spread of pathogenic microorganisms can occur.

where enzymatic digestion of nutrients occurs, as well as uptake of nutrient molecules into the blood and lymphatic vessels. Secretions from the pancreas, including many enzymes and sodium bicarbonate, drain by way of the pancreatic duct into the intestine to assist in small intestine function. Bile secretions from the liver and gallbladder drain by way of the bile duct to the small intestine and help emulsify dietary lipids for enzymatic digestion and absorption by the small intestine. Absorbed lipids enter the lymphatic vessels of the intestinal wall, then travel through the lymphatics, ultimately entering the general blood circulation. Other absorbed nutrients (and certain toxins) enter venous vessels of the intestinal wall, then travel by way of **portal venous circulation** to the liver. This transfer of material from the intestine to the portal venous blood is important because the liver can process absorbed dietary nutrients and

URINARY SYSTEM (The Urinary Tract)

STRUCTURE AND FUNCTION

The urinary system consists of two kidneys, a ureter draining each kidney, a bladder into which the ureters drain urine, and a urethra through which urine from the bladder is voided. The kidneys are located in the posterior abdominal cavity, next to the lumbar muscles. Each ureter forms a connection between a kidney and the bladder.

Each kidney serves the primary function of removing various substances from the blood. These substances include ammonia, hydrogen ions, urea, certain ingested drugs, waste organic acids, and water (needed for solubilization of these substances). The kidney also helps regulate blood electrolytes. As material from the blood moves through the kidney tubules, destined for a ureter, the kidney exchanges sodium (Na$^+$) and potassium (K$^+$) ions between blood and the fluid in the tubules. By regulating blood sodium, the kidney helps in regulating blood pressure. By regulating blood potassium, the kidney helps in regulating cellular excitability, critical for proper heart and nerve cell activity. A person can live with only one kidney. If both kidneys fail, periodic artificial dialysis is required to remove toxins from the blood.

The fluid formed by each kidney drains to the bladder by way of a narrow muscular tube called a **ureter.** The bladder is a balloon-like structure in the pelvis that collects and voids urine. The bladder is capable of significant distention so that it can hold varying volumes of urine. When stretched to a critical point, the bladder triggers neural reflexes that cause a sphincter to open between the bladder and the urethra, so that voiding of urine may occur.

The urethra is the tubular channel along which urine travels from the bladder to the external environment during urination. The female urethra is short and is located just anterior to the vagina. The male urethra is much longer, courses through the male penis, and serves a secondary function as a channel for delivery of sperm during sexual intercourse.

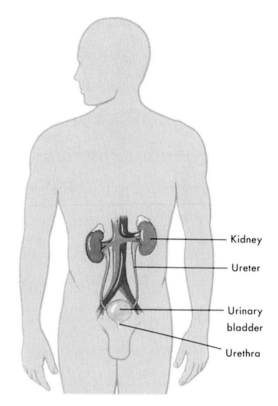

Anatomy of the urinary system.

APPLICATIONS TO MICROBIOLOGY

Kidney damage can result from several immunological mechanisms. Microbial proteins (antigens) sometimes resemble proteins of the kidney glomeruli. After an infection with some strains of *Streptococcus,* the immune system may become "confused" and attack the kidney's glomeruli, mistakenly thinking that it is attacking microorganisms. **Glomerulonephritis** and impaired kidney function results. Another type of immune-mediated glomerulonephritis occurs when microbial proteins lodge in the kidney. When soluble antigen binds to blood antibody, an **immune complex** is the result. Immune complexes can sometimes lodge in the very sinuous capillaries of the kidney glomeruli. The complexes activate serum proteins called **complement.** Activated complement attracts white blood cells. When these cells attack the immune complexes in an attempt to destroy the microbial antigen, healthy kidney tissue is also attacked and destroyed.

Urine stored in the bladder is normally devoid of microorganisms, partly because of its pH and partly because of its continual flow out of the body. How-

ever, if the pH of the urine in the bladder is altered to favor the growth of pathogens, a **urinary tract infection (UTI)** can result. Urine pH is closely linked to body metabolism and blood hydrogen ion concentration. UTIs can move retrograde, so that bladder infections move up to infect the kidneys—a condition called **pyelonephritis.**

Females are more susceptible to UTIs because of the shorter female urethra and because of the proximity of the female urethra to the microbiota of the anorectal region. *Escherichia coli* is a common microorganism present in feces; it is commonly isolated from urine cultures of females with UTI.

Because the urethra is a potential portal of entry for microorganisms into the body, indwelling urinary catheters (hollow plastic polymer tubes inserted through the urethra into the bladder) provide an open conduit for microorganisms to infect the urinary system. Catheters normally drain into a bag that collects the urine. If the bag becomes contaminated or if the catheter is contaminated during its insertion, a UTI can result.

Certain microorganisms, such as *Escherichia coli, Pseudomonas aeruginosa,* and *Serratia marcescens,* are associated with UTIs.

STRUCTURE AND FUNCTION

The female reproductive system consists of the external genitalia, clitoris, vagina, uterus, fallopian tubes, ovaries, and mammary glands. The main function of the female reproductive system in a female of reproductive age is to produce an egg each month to be fertilized by the male sperm. The fertilized egg then matures into an embryo and fetus, with ultimately the birth of an infant.

The external genitalia of the female includes the **labia majora** and **labia minora**—folds of skin that cover the entrance to the vagina and urethra. Near the point where the labia on either side of the body join, the female **clitoris** can be found. It is located just anterior and superior to the opening of the female urethra. The clitoris is erectile tissue and sensitive to tactile stimulation. The opening of the vagina is located posterior to the urethra and anterior to the rectum.

The **vagina** is a tubelike organ, approximately the length of a finger in its relaxed state. However, it is capable of significant distention longitudinally and circumferentially. This quality allows the vagina to accommodate the penis during copulation and the fetus during birth.

The **uterus** is about the size and shape of an inverted pear. The inferior part of the uterus, the **cervix,** extends down into the vagina. The main part of the uterus houses a developing embryo and fetus during pregnancy. Changes in the lining of the uterus occur each month in a female of reproductive years as a result of cyclic changes in blood levels of the hormones estrogen and progesterone.

The **fallopian tubes** (oviducts, uterine tubes) begin blindly in the lower abdominal, or pelvic, cavity, adjacent to an ovary on either side. When **ovulation** occurs (the release of an egg [ovum] from an ovary) the egg is normally carried into a fallopian tube by currents created by ciliated cells lining the tube. **Fertilization** normally occurs inside the tube as egg meets sperm to create a fertilized egg, or zygote. The **zygote** divides mitotically to produce an **embryo** that implants in the wall of the uterus.

The **ovaries** are located in the pelvic cavity. The ovaries are the female gonads and serve the principal function of maturing eggs for ovulation. Normally, one ovary ovulates each month. The ovaries manufacture and release into the bloodstream the hormones estrogen and progesterone. These are the female hormones responsible for uterus maturation,

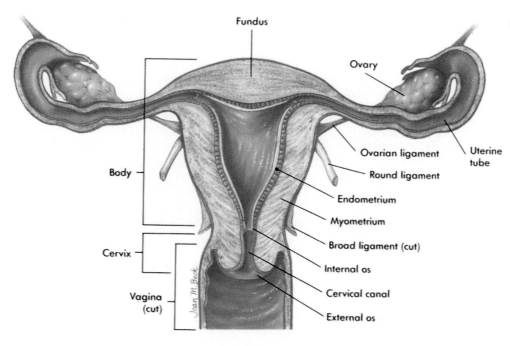

Frontal view of female reproductive organs.

menstruation, breast development, and female gender characteristics.

The **mammary glands** of the female are commonly known as breasts. Cells of the mammary glands begin to produce human milk when pregnancy nears its term. Multiple hormones, including estrogen and pituitary gland prolactin, stimulate milk production.

The male reproductive system includes the penis, testes, vas deferens, prostate gland, Cowper's glands, and seminal vesicles. The main function of the male reproductive system is to fertilize a female ovum. To accomplish this, the male reproductive system produces **sperm** cells, causes an erection of the penis for insertion into the female vagina, and moves sperm out of the penis and into the vagina.

The penis serves a dual function—the voiding of urine and coitus. The urethra passes through the penis and is surrounded in the penis by tissue composed of open spaces where blood accumulates—resulting in erection of the penis. These spaces are the **corpus spongiosum** and the **corpora cavernosa.**

Near the bladder where the urethra originates, the male urethra is surrounded by the **prostate gland.** This region of the male urethra is the **prostatic urethra.** Secretions from the prostate and two **seminal vesicle** glands are added to sperm to create semen. **Semen** is the material ejaculated from the penis following erection of the penis and orgasm.

The **testes** are the male gonads responsible for production of sperm cells. The testes are each contained within a fold of skin called the **scrotum.** The testes also contain cells that synthesize testosterone—the male hormone. Sperm cells produced by the **seminiferous tubules** of the testes exit the testes by way of the **vas deferens.** The vas is a tubular structure that carries sperm to the prostatic urethra during ejaculation.

Cowper's glands (bulbourethral glands) secrete a slightly alkaline material into the urethra. The secretions function to help neutralize any residual acidity in the urethra from urine before ejaculation. Females also have bulbourethral glands to lubricate the external genitalia.

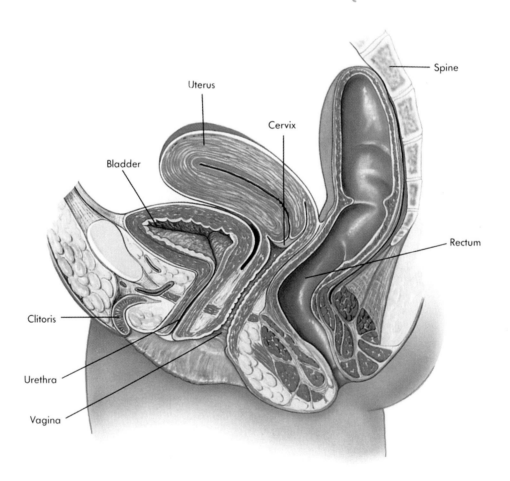

Sagittal section of the female pelvis.

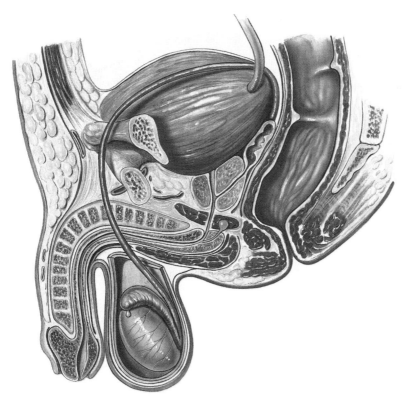

Sagittal section of the male pelvis showing male reproductive structures.

APPLICATIONS TO MICROBIOLOGY

The concept of **normal microbiota** is important in understanding infections of the male and female reproductive systems. In the male, normal microbiota of the penile urethra and the periodic voiding of urine (from a sterile bladder) help prevent infection of the urinary or reproductive tract. In the female, the vagina has a slightly acidic pH. This is partly because of normally occurring lactobacilli. This helps prevent infection of the female reproductive tract. Disruption of the normal microbiota in male or female reproductive tracts favors invasion by pathogens. Restoration of the normal microbiota inhibits pathogen growth.

Common pathogens of the female and male reproductive tract include the flagellated protozoan *Trichomonas vaginalis,* and bacteria *Gardnerella vaginalis* (*Haemophilus vaginalis*) and *Neisseria gonorrhoeae, Chlamydia,* and the yeast fungus *Candida albicans.* All of these microorganisms are classified as **sexually transmitted diseases (STDs).** STDs normally require contact between the pathogen and mucosa of the reproductive tract. **Mucosa** is a term for a non-keratinized (moist) cell lining—such as the lining of the vagina or the urethra. The shaft of the penis, although not lined with mucosa, has very thin skin and is susceptible to direct invasion by microorganisms such as the spirochete bacterium *Treponema pallidum,* the cause of syphilis. The invasive site of the spirochete is visible as a chancre (sore). The use of condoms or other physical barriers can minimize the chance of microbial infection across mucosal tissues of the genitals.

The female reproductive tract is unique compared to that of the male in that it is in open communication with the pelvic cavity. Pathogens can potentially migrate from the vagina to the uterus, to the fallopian tubes, and then to the pelvic cavity. When this happens and inflammation of the pelvis (and fallopian tubes) results secondary to the infection, the patient is said to have **pelvic inflammatory disease (PID).** *Neisseria gonorrhoeae* is commonly associated with PID. This disease must be treated aggressively.

Glossary

A

aberrations Distortions that occur in a magnifying lens

abiotic Referring to the absence of living organisms

abortive infections A viral infection in which viral replication does not occur or does not produce viral progeny that are capable of infecting other host cells

abrasion An area denuded of skin, mucous membrane, or superficial epithelium by rubbing or scraping

abscess A localized accumulation of pus

absorption The uptake, drinking in, or imbibing of a substance; the movement of substances into a cell; the transfer of substances from one medium to another, e.g., the dissolution of a gas in a liquid; the transfer of energy from electromagnetic waves to chemical bond and/or kinetic energy, e.g., the transfer of light energy to chlorophyll

accessory pigments Pigments (compounds that are colored) that harvest light energy and transfer it to the primary photosynthetic reaction centers

acellular Lacking cellular organization; not having a delimiting plasma membrane; organizational description of viruses, viroids, and prions

Acetobacter Cells ellipsoidal to rod-shaped; motile by peritrichous flagella or nonmotile; endospores not formed; young cells Gram-negative, in older cultures some strains become Gram-variable; metabolism respiratory, never fermentative; oxidize ethanol to acetic acid at neutral and acid reactions (pH 4.5); strict aerobes; optimal growth ca. 30° C; G + C 55-64 mole %

acetyl A two-carbon organic radical containing a methyl group and a carbonyl group

acetyl CoA Acetyl coenzyme A; a condensation product of coenzyme A and acetic acid; an intermediate in the transfer of 2-carbon fragments, notably in their entrance into the tricarboxylic acid cycle

acetylene reduction assay Assay for nitrogen fixation based on the conversion of acetylene to ethylene by nitrogenase, the enzyme responsible for nitrogen fixation

achievable serum levels Maximal (peak) concentrations of an antimicrobial agent that can occur in serum

achromatic lens An objective lens in which chromatic aberration has been corrected for two colors and spherical aberration for one color

acid-fast The property of some bacteria, such as mycobacteria, that retain their initial stain and do not decolorize after washing with dilute acid-alcohol

acid foods Foods with a pH value less than 4.5

acid mine drainage Consequence of the metabolism of sulfur and iron-oxidizing bacteria when coal mining exposes pyrite to atmospheric oxygen and the combination of autoxidation and microbial sulfur and iron oxidation produces large amounts of sulfuric acid, which kills aquatic life and contaminates water

acidic A compound that releases hydrogen (H^+) ions when dissolved in water; a compound that yields positive ions on dissolution; a solution with a pH value less than 7.0

acidic stains Stains with a positively charged chromophore (colored portion of the dye) that are attracted to negatively charged cells

acidophiles Microorganisms that show a preference for growth at low pH, e.g., bacteria that grow only at very low pH values, ca. 2.0

acidulant Acidic compound used as a chemical food preservative

Acinetobacter Rods, usually very short and plump, approaching coccus shape in stationary phase, predominantly in pairs and short chains; no spores formed; flagella not present; Gram-negative; oxidative metabolism; oxidase negative, catalase positive; optimal growth 30 to 32° C; G + C 40-47 mole %

acne An inflammatory disease involving the oil glands and hair follicles of the skin, occurs chiefly in adolescents and marked by papules or pustules, especially about the face

acquired immune deficiency syndrome (AIDS) An infectious disease syndrome caused by HIV retrovirus, characterized by the loss of normal immune response system functions, followed by various opportunistic infections

acquired immunity The ability of an individual to produce specific antibodies in response to antigens to which the body has been previously exposed based on the development of a memory response

acrasin Substance secreted by a slime mold that initiates aggregation to form a fruiting body (3′-5′ cyclic AMP)

Actinomyces Gram-positive, irregularly staining bacteria; non-acid-fast, non-spore-forming and nonmotile; filaments with true branching may predominate and are particularly evident in 18–48-hour microcolonies; carbohydrates are fermented with the production of acid but no gas; some species may show greening or complete lysis of rabbit red blood cells; facultative anaerobes; most are preferentially anaerobic, and one species grows well aerobically; carbon dioxide is required for maximum growth

Actinomycetales Order of bacteria characterized by the formation of branching filaments whose families are distinguished on the basis of the nature of their mycelia and spores

actinomycetes Members of an order of bacteria in which species are characterized by the formation of branching and/or true filaments

activated sludge The active microorganisms formed during the activated sludge secondary sewage treatment process that are used as an inoculum for the next batch treatment

activated sludge process An aerobic secondary sewage treatment process using sewage sludge containing active complex populations of aerobic microorganisms to break down organic matter in sewage

activation energy The energy in excess of the ground state that must be added to a molecular system to allow a chemical reaction to start

active immunity Immunity acquired as a result of the individual's own reactions to pathogenic microorganisms or their antigens; attributable to the presence of antibody or immune lymphoid cells formed in response to an antigenic stimulus

active site The site on the enzyme molecule at which the substrate binds and the catalyzed reaction actually proceeds

active transport Movement of materials across cell membranes from regions of lower to regions of higher concentration, requiring expenditure of metabolic energy

acute Referring to a disease of rapid onset, short duration, and pronounced symptoms

acute disease A disease characterized by rapid development of symptoms and signs, reaching a height of intensity, and ending fairly quickly

acute stage Stage of a disease when symptoms appear and the disease is most severe

acyclovir Antiviral agent used in the treatment of diseases caused by herpes simplex

adaptive enzymes Enzymes produced by an organism in response to the presence of a substrate or a related substance; also called *inducible enzymes*

adenine A purine base component of nucleotides, nucleosides, and nucleic acids

adenosine A mononucleoside consisting of adenine and D-ribose

adenosine diphosphate (ADP) A high-energy derivative of adenosine containing two phosphate groups, one less than ATP; formed on the hydrolysis of ATP

adenosine monophosphate (AMP) A compound composed of adenosine and one phosphate group formed on the hydrolysis of ADP

adenosine triphosphatase (ATPase) An enzyme that catalyzes the reversible hydrolysis of ATP; the membrane-bound form of this enzyme is important in catalyzing the formation of ATP from ADP and inorganic phosphate

adenosine triphosphate (ATP) A major carrier of phosphate and energy in biological systems, composed of adenosine and three phosphate groups; the free energy released from the hydrolysis of ATP is used to drive many energy-requiring reactions in biological systems

adhesins Substances involved in the attachment of microorganisms to solid surfaces; factors that increase adsorption

adhesion factors Substances involved in the attachment of microorganisms to solid surfaces; factors that increase adsorption

adhesion sites Sites of association in Gram-negative bacteria between the plasma membrane and the outer membrane; Bayer junctions

adjuncts Starchy substrates, such as corn, wheat, and rice, that provide carbohydrates for ethanol production and are added to malt during the mashing process in the production of beer

adjuvants Substances that increase the immunological response to a vaccine and, for example, can be added to vaccines to slow down absorption and increase effectiveness; substances that enhance the action of a drug or antigen

ADP Adenosine diphosphate

adrenaline Hormone secreted by the adrenal medulla in response to stress, causes a rise in blood pressure; used as a heart stimulant

adsorption A surface phenomenon involving the retention of solid, liquid, or gaseous molecules at an interface

aer- Combining form meaning air or atmosphere

aerated pile method Method of composting for the decomposition of organic waste material where the wastes are heaped in separate piles and forced aeration provides oxygen

aerial mycelia A mass of hyphae occurring above the surface of a substrate

aerobactin A hydroxamate siderophore

aerobes Microorganisms whose growth requires the presence of air or free oxygen

aerobic Having molecular oxygen present; growing in the presence of air

aerobic bacteria Bacteria requiring oxygen for growth

aerobic respiration Metabolism involving a respiration pathway in which molecular oxygen serves as the terminal electron acceptor

aerosol A fine suspension of particles or liquid droplets sprayed into the air

aflatoxin A carcinogenic poison produced by some strains of the fungus *Aspergillus flavus*

African sleeping sickness Also known as African trypanosomiasis; a protozoan disease that affects the nervous system, caused by *Trypanosoma*, a flagellate that is injected by the bite of the tsetse fly

agar A dried polysaccharide extract of red algae used as a solidifying agent in various microbiological media

agglutinating antibody Agglutinin

agglutination The visible clumping or aggregation of cells or particles due to the reaction of surface-bound antigens with homologous antibodies

agglutinin An antibody capable of causing the clumping or agglutination of bacteria or other cells

agricultural microbiology The study of the role of microorganisms in agriculture

Agrobacterium Motile Gram-negative rods; metabolism respiratory; optimal growth 25 to 30° C; G + C 59.6-62.8 mole %

AIDS Acquired immune deficiency syndrome

airborne transmission Route by which pathogens are transported to a susceptible host via the air; main route of transmission of pathogens that enter via the respiratory tract through the air

akinetes Thick-walled resting spores of cyanobacteria and green algae

Alcaligenes Cell rods, coccal rods, or cocci usually occurring singly; motile with peritrichous flagella; Gram-negative; metabolism respiratory, never fermentative; do not fix gaseous nitrogen; oxidase positive; optimal growth 20° to 37° C; G + C 57.9-70 mole %

alcoholic fermentation Conversion of sugar to alcohol by microbial enzymes; fermentation that produces alcohol (ethanol) and carbon dioxide from glucose; also known as *ethanolic fermentation*

alcohols Organic compounds characterized by one or more —OH (hydroxyl) groups

aldehydes A class of substances derived by oxidation from primary alcohols and characterized by the presence of a —CHO group

ale Alcoholic beverage produced with top-fermenting *Saccharomyces cerevisiae* and a high concentration of hops to produce a tart taste and a high alcohol concentration

algae A heterogeneous group of eukaryotic, photosynthetic, unicellular and multicellular organisms lacking true tissue differentiation

algicides Chemical agents that kill algae

alkaline A condition in which hydroxyl (OH^-) ions are in abundance; solutions with a pH greater than 7.0 are alkaline or basic

alkalophiles Bacteria that live at very high pH; bacteria that live under extremely alkaline conditions, having developed mechanisms for keeping sodium and hydroxide ions outside the cell

allele One or more alternative forms of a given gene concerned with the same trait or characteristic; one of a pair or multiple forms of a gene located at the same locus of homologous chromosomes

allelopathic substance A substance produced by one organism that adversely affects another organism

allergen An antigen that induces an allergic response, i.e., a hypersensitivity reaction

allergy An immunological hypersensitivity reaction; an antigen-antibody reaction marked by an exaggerated physiological response to a substance in sensitive individuals

allochthonous An organism or substance foreign to a given ecosystem

allochthonous population A population of allochthonous organisms

allosteric activation The acceleration of the rate of an enzymatic reaction as a result of an allosteric effector binding to an enzyme

allosteric effector Substance that can bind to the regulatory site of an allosteric enzyme, resulting in the alteration of the rate of activity of that enzyme

allosteric enzymes Enzymes with a binding and catalytic site for the substrate and a different site where a modulator (allosteric effector) acts

allosteric inhibitor An allosteric effector that results in reduced rates of activity of an allosteric enzyme

allotypes Antigenically different forms of a given type of immunoglobulin that occur in different individuals of the same species

alpha hemolysis (α hemolysis) Partial hemolysis of red blood cells as evidenced by the formation of a zone of partial clearing (greening) around certain bacterial colonies growing on blood agar

amastigotes Rounded protozoan cells lacking flagella; a form assumed by some species of Trypanosomatidae, e.g., *Plasmodium*, during a particular stage of development

amebic dysentery An inflammation of the colon caused by *Entamoeba histolytica;* also known as amebiasis

amensalism An interactive association between two populations that is detrimental to one while not adversely affecting the other

Ames test Test for the detection of chemical mutagens and potential carcinogens

amino An —NH_2 group

amino acids A class of organic compounds containing an amino (—NH_2) group and a carboxyl (—COOH) group

amino end The end of a peptide chain or protein with a free amino group, i.e., an alpha amino group not involved in forming the peptide bond

aminoacyl site Site on a ribosome where a tRNA molecule attached to a single amino acid initially binds during translation

aminoglycosides Broad-spectrum antibiotics containing an aminosugar, an amino- or guanido-inositol ring, and residues of other sugars that inhibit protein synthesis, e.g., kanamycin, neomycin, and streptomycin

ammonification The release of ammonia from nitrogenous organic matter by microbial action

ammonium ion The cation NH_4^+

Amoeba Protozoa having no distinct shape that form lobopodia and lack a skeletal structure

AMP Adenosine monophosphate

amphibolic pathway A metabolic pathway that has catabolic and anabolic functions

amphotericin B Broad-spectrum antifungal agent used for treating systemic infections

amplifying gene expression Activity by which eukaryotes can increase the amount of rRNA and thus the number of ribosomes that can be used to translate the information in a stable mRNA molecule, thus producing large amounts of the enzyme coded for by a given gene

amylases Enzymes that hydrolyze starch

anabolism Biosynthesis; the process of synthesizing cell constituents from simpler molecules, usually requiring the expenditure of energy

anaerobes Organisms that grow in the absence of air or oxygen; organisms that do not use molecular oxygen in respiration

anaerobic The absence of oxygen; able to live or grow in the absence of free oxygen

anaerobic culture chamber Enclosures designed to exclude oxygen from the atmosphere, generally by generating hydrogen, which reacts with available oxygen as a catalyst to produce water

anaerobic digester A secondary sewage treatment facility used for the degradation of sludge and solid waste

anaerobic life   Life in the absence of air

anaerobic photosynthetic bacteria   Bacteria that carry out the reactions of photosystem I

anaerobic respiration   The use of inorganic electron acceptors other than oxygen as terminal electron acceptors for energy-yielding oxidative metabolism

anaerobiosis   The state or condition characterized by the absence of air; anaerobic; lack or removal of oxygen from the atmosphere

analogously similar   Phenotypically similar

anamnestic response   A heightened immunological response in persons or animals to the second or subsequent administration of a particular antigen given some time after the initial administration; a secondary or memory immune response characterized by the rapid reappearance of antibody in the blood due to the administration of an antigen to which the subject had previously developed a primary immune response

anaphylactic hypersensitivity   An exaggerated immune response of an organism to foreign protein or other substances, involving the degranulation of mast cells and the release of histamine

anaphylactic shock   Physiological shock resulting from an anaphylactic hypersensitivity reaction, e.g., to penicillin or bee stings; in severe cases, death can result within minutes

anaplerotic sequences   Reactions in a cell that serve to replenish the supplies of key molecules

anemia   A condition characterized by having less than the normal amount of hemoglobin, reflecting a reduced number of circulating red blood cells

anergic   Unresponsive to antigens

anergic B cell clones   B cells that do not respond to antigenic stimulation

animal viruses   Viruses that multiply within animal cells

anions   Negatively charged ions

anisogametes   Gametes differing in shape, size, and/or behavior

annulus   A ring-shaped structure; a transverse groove in the cellular envelope of dinoflagellates

anode   The positive terminal of an electrolytic cell

anorexia   Absence of appetite

anoxic   Absence of oxygen; anaerobic

anoxygenic photosynthesis   Photosynthesis that takes place in the absence of oxygen and during which oxygen is not produced; photosynthesis that does not split water and evolve oxygen

anoxygenic phototrophic bacteria (anoxyphotobacteria)   Bacteria that can carry out only anoxygenic photosynthesis; a group of photosynthetic, phototrophic bacteria occurring in aquatic habitats that do not evolve oxygen

antagonism   The inhibition, injury, or killing of one species of microorganism by another; an interpopulation relationship in which one population has a deleterious (negative) effect on another

antheridium   A specialized structure where male gametes are produced; the male gametangium of oomycete fungi

anthrax   An infectious disease of animals, including humans, cattle, sheep, and pigs, caused by *Bacillus anthracis*

anti-   Combining form meaning opposing in effect or activity

antibacterial agents   Agents that kill or inhibit the growth of bacteria, e.g., antibiotics, antiseptics, and disinfectants

antibiotic resistance   The natural or acquired ability of a microorganism to overcome the inhibitory effects of an antibiotic

antibiotics   Substances of microbial origin that in very small amounts have antimicrobial activity; current usage of the term extends to synthetic and semisynthetic substances that are closely related to naturally occurring antibiotics and that have antimicrobial activity

antibodies   Glycoprotein molecules produced in the body in response to the introduction of an antigen or hapten that can specifically react with that antigen; also known as *immunoglobulins*, which are part of the serum fraction of the blood formed in response to antigenic stimulation and which react with antigens with great specificity

antibody-dependent cytotoxic hypersensitivity   Type 2 hypersensitivity; reaction in which an antigen present on the surface of a cell combines with an antibody, resulting in the death of that cell by stimulating phagocytic attack or initiating the complement pathway; examples include blood transfusions between incompatible types and Rh incompatibility

antibody-mediated immunity   Immunity produced by the activation of the B-lymphocyte population, leading to the production of several classes of immunoglobulins

anticodon   A sequence of three nucleotides in a tRNA molecule that is complementary to the codon triplet in mRNA

antifungal agents   Agents that kill or inhibit the growth and reproduction of fungi; may be fungicidal or fungistatic

antigen   Any agent that initiates antibody formation and/or induces a state of active immunological hypersensitivity and that can react with the immunoglobulins that are formed

antigen presenting cell   Cells with small polypeptide antigens attached to the MHC class 2 proteins on outer cell membrane surfaces such that the antigen is presented or shown to T_H cells

antigen-antibody complex   The molecular combination that results from the reaction between antigen and complementary antibody molecules

antigenic drift   Major antigenic changes, typically seen in influenza viruses and other pathogens, which occur because of accumulated genetic mutations and recombinations; process of antigenic change that can cause gene reassortment between an animal and a human strain; a gradual and cumulative process of antigenic changes that only become apparent with time

antigenic shift   Genetic mutation caused by the addition of new genes that produces new strains of influenza viruses

antihistamines   Compounds used for treating allergic reactions and cold symptoms that work by inactivating histamine that is released as part of the immune response

antihuman gamma globulin   Antibodies that react specifically with human antibodies

antimalarial drugs   Agents effective against the erythrocytic stage of the *plasmodium* life cycle

antimicrobial agents   Chemical or biological agents that kill or inhibit the growth of microorganisms

antimicrobial susceptibility testing   Laboratory tests designed to determine the resistance of disease-causing microorganisms to antimicrobial agents

antimicrobics   Antimicrobial agents; chemicals that inhibit or kill microorganisms and can be safely introduced into the human body; include synthetic and naturally produced antibiotics; term generally used instead of *antibiotic*

antiport   A mechanism in which different substances are actively transported across the plasma membrane simultaneously in opposite directions

antisense mRNA   An mRNA that is complementary to another mRNA, forming a double-stranded RNA that is not translated

antiseptics   Chemical agents used to treat human or animal tissues, usually skin, to kill or inactivate microorganisms capable of causing infection; not considered safe for internal consumption

antisera   Blood sera that contain antibodies

antitoxin   Antibody to a toxin capable of reacting with the poison and neutralizing the toxin

antiviral agents   Substances capable of destroying or inhibiting the reproduction of viruses

APC   Antigen presenting cell

aplanospores   Nonmotile, sexual spores

apochromatic lens   An objective microscope lens in which chromatic aberration has been corrected for three colors and spherical aberration for two colors

appendaged bacteria   Budding bacteria

appendicitis   Inflammation of the appendix, generally caused by excessive growth of indigenous microorganisms and characterized by abdominal pain, nausea, vomiting, and elevated white blood cell count

aquatic   Growing in, living in, or frequenting water; a habitat composed primarily of water

aqueous   Of, relating to, or resembling water; made from, with, or by water; solutions in which water is the solvent

arbuscules   Specialized inclusions in root cortex in the vesicular-arbuscular type of mycorrhizal association

Archaea   The kingdom of organisms that contains the archaebacteria

archaebacteria   Archaea; archaeobacteria; prokaryotes with cell walls that lack murein, having ether bonds in their membrane phospholipids; analysis of rRNA indicates that the archaebacteria represent a primary biological Kingdom related to both eubacteria and eukaryotes; considered to be a primitive group of organisms that were among the earliest living forms on Earth

armored dinoflagellates   Dinoflagellates that produce thecal plates

Arthrobacter   Cells that in complex media undergo a marked change in form during the growth cycle; older cultures (generally 2–7 days) are composed entirely or largely of coccoid cells; on transfer to fresh complex medium, growth occurs by enlargement (swelling) of the coccoid cells, followed by elongation; do not form endospores; Gram-positive, but the rods may be readily decolorized and may show only Gram-positive granules in otherwise Gram-negative cells; not acid-fast; metabolism respiratory, never fermentative; strict aerobes; optimal growth 20° to 30° C; G + C 60-72 mole %

arthropods   Animals of the invertebrate phylum Arthropoda, many of which are capable of acting as vectors of infectious diseases

arthrospores   Spores formed by the fragmentation of hyphae of certain fungi, algae, and cyanobacteria; coccoid cells present during the stationary growth phase of *Arthrobacter* species

Arthus reaction (immune-complex reaction)   Complex-mediated hypersensitivity

artifact   The appearance of something in an image or micrograph of a specimen that is due to causes within the optical system or preparation of the specimen and is not a true representation of the features of the specimen

artificially acquired active immunity   The production of antibodies and development of a memory response by the body as a result of administration of a vaccine (vaccination)

artificially acquired passive immunity   The transfer of humoral antibodies formed by one individual to a susceptible individual, accomplished by injection of antiserum

Aschelminthes   Roundworms

asci   Pleural of ascus

ascocarp   Structure on true ascomycetes that produces asci

Ascomycetes   Members of a class of fungi distinguished by the presence of an ascus, a sac-like structure containing sexually produced ascospores

ascospores   Sexual spores characteristic of ascomycetes, produced in the ascus after the union of two nuclei

ascus   The sporangium or spore case of fungi, consisting of a single terminal cell

-ase   Suffix denoting an enzyme

asepsis   State in which potentially harmful microorganisms are absent; free of pathogens

aseptic meningitis   Any infection of the meninges not caused by bacteria or fungi; a viral cause is suspected

aseptic techniques   Precautionary measures taken in microbiological work and clinical practice to prevent the contamination of cultures, sterile media, etc., and/or infection of persons, animals, or plants by extraneous microorganisms

aseptic transfer technique   The movement of microbial cultures in such a way as to minimize or eliminate exposure to the atmosphere

asexual   Lacking sex or functional sexual organs

asexual reproduction   Reproduction without union of gametes; formation of new individuals from a single individual

asexual spores   Spores produced asexually that often are involved in survival or dissemination of the microorganisms

aspirate   To draw or remove fluid by suction

assay   Analysis to determine the presence, absence, or quantity of one or more components

assembly   Stage of viral replication during which packaging of a nucleic acid genome within a protein capsid occurs

assimilation The incorporation of nutrients into the biomass of an organism

asthma Type 1 hypersensitivity reaction that primarily affects the lower respiratory tract; the condition is characterized by shortness of breath and wheezing

asymptomatic carriers Individuals infected with a pathogen who do not develop disease symptoms

ataxia The inability to coordinate muscular action

athlete's foot Disease caused by dermatophytic fungi affecting chronically wet feet with skin abrasions; tinea or ringworm of the feet

atmosphere The entire mass of air surrounding the Earth; a unit of pressure approximating 1×10^6 dynes/cm^2

atopic allergies Localized expression of type 1 hypersensitivity reaction; examples include hay fever and food allergies

ATP Adenosine triphosphate

ATPase Adenosine triphosphatase

attenuated pathogen A pathogen whose virulence has been diminished

attenuation Any procedure in which the pathogenicity of a given organism is reduced or abolished; reduction in the virulence of a pathogen; control of protein synthesis involving the translation process

attenuator site The site between the operator region and the first structural gene of the operon where transcription can be interrupted, as in the *trp* operon

attractants Chemicals that cause bacteria to move toward them; chemicals that cause phagocytic white blood cells to move toward them

atypical pneumonia Pneumonia that is not secondary to another acute infectious disease; pneumonia not treatable with penicillin; pneumonia caused by a bacterial pathogen that is not culturable on routine defined media

aureomycin 7-Chlorotetracycline; an antibiotic obtained from *Streptomyces aureofaciens*

autochthonous Microorganisms and/or substances indigenous to a given ecosystem; the true inhabitants of an ecosystem; term often used to refer to the common microbiota of the body or those species of soil microorganisms that tend to remain constant despite fluctuations in the quantity of fermentable organic matter in the soil

autochthonous populations Populations of autochthonous organisms

autoclave Apparatus in which objects or materials may be sterilized by air-free saturated steam under pressure at temperatures in excess of 100° C

autoimmune diseases Diseases that result from the failure of the immune response to recognize self-antigens so that the immune system attacks the body's own cells, resulting in the progressive degeneration of tissues

autoimmunity Immunity or hypersensitivity to some constituent in one's own body; immune reactions with self-antigens

autolysins Endogenous enzymes involved in the breakdown of certain structural components of the cell during particular phases of cellular growth and development

autolysis The breakdown of the components of a cell or tissues by endogenous enzymes, usually after the death of the cell or tissue

autospores Sexually formed, nonmotile spores resembling the parent cell morphologically

autotrophic Capable of synthesizing organic components from inorganic sources and generating ATP from light or the oxidation of inorganic compounds

autotrophic metabolism Metabolism that does not require organic carbon as a source of carbon and energy

autotrophs Organisms whose growth and reproduction are independent of external sources of organic compounds, the required cellular carbon being supplied by the reduction of CO_2 and the needed cellular energy being supplied by the conversion of light energy to ATP or the oxidation of inorganic compounds to provide the free energy for the formation of ATP

autoxidation The oxidation of a substance on exposure to air

auxotrophs Nutritional mutants that require growth factors not needed by the parental strain

avirulent Lacking virulence; a microorganism lacking the properties that normally cause disease

A$_w$ Water activity

axopodia Semipermanent pseudopodia, e.g., the pseudopodia that emanate radially from the spherical cells of heliozoans and some radiolarian species

Azobacteraceae Family of Gram-negative bacteria that exhibit pleomorphic morphology and fix molecular nitrogen

Azotobacter Large ovoid cells with marked pleomorphism; Gram-negative, with marked variability; fix atmospheric nitrogen; grow well aerobically but can also grow under reduced oxygen tension; optimal growth 20° to 30° C; G + C 63-66 mole %

B

B cell B lymphocytes

B lymphocytes A differentiated lymphocyte involved in antibody-mediated immunity; white blood cells that are able to produce specific immunoglobulins; their surfaces carry specific immunoglobulin antigen-binding receptor sites

B memory cells Specifically stimulated B lymphocytes not actively multiplying but capable of multiplication and production of plasma cells on subsequent antigenic stimuli

babesiosis Disease caused by the protozoan *Babesia microti*, which is transmitted by ticks and endemic to Nantucket Island and coastal areas of New England

Bacillaceae Family of Gram-negative, endospore-forming rod- and coccal-shaped bacteria

Bacillariophyceae A group of Chrysophycophyta algae containing the diatoms

bacilli Bacteria in the shape of cylinders

Bacillus Cells rod-shaped, straight, or nearly so; majority are motile; heat-resistant endospores formed; not more than one in a sporangial cell; Gram reaction: positive, or positive only in the early stages of growth; metabolism strictly respiratory, fermentative, or both, using various substrates; strict aerobes or facultative anaerobes; G + C 32-62 mole %

bacitracin Antibiotic that inhibits bacterial cell wall synthesis

bacteremia   Condition in which viable bacteria are present in the blood

bacteria   Members of a group of diverse and ubiquitous prokaryotic, single-celled organisms; organisms with prokaryotic cells, i.e., cells lacking a nucleus

bacterial chromosome   The single circular DNA macromolecule that contains the genetic information of bacterial cells

bacterial dysentery   Dysentery caused by *Shigella* infections; also called *shigellosis*

bacterial endotoxin   Endotoxin

bacterial meningitis   Inflammation of the meninges caused by a bacterial infection of the spinal column

bactericidal   Any physical or chemical agent able to kill some types of bacteria

bactericide   A chemical that kills microorganisms

bacteriochlorophyll   Photosynthetic pigment of green and purple anaerobic photosynthetic bacteria

bacteriocides   Chemical agents that kill bacteria

bacteriological filter   A filter with pores small enough to trap bacteria, about 0.45 μm or smaller, used to sterilize solutions by removing microorganisms during filtration

bacteriology   Science of dealing with bacteria, including their relation to medicine, industry, and agriculture

bacteriophage   A virus whose host is a bacterium; a virus that replicates within bacterial cells

bacteriostatic   An agent that inhibits the growth and reproduction of some types of bacteria but need not kill the bacteria

Bacteroidaceae   Family of Gram-negative anaerobic bacteria, many of which are important in the normal microbiota of humans

Bacteroides   Gram-negative, non-spore-forming rods; metabolize carbohydrates or peptone; fermentation products of sugar-utilizing species include combinations of succinic, lactic, acetic, formic, or propionic acids, sometimes with short-chain alcohols; obligately anaerobic; optimal growth at 37° C; G + C 40-55 mole %

bacteroids   Irregularly shaped (pleomorphic) forms that some bacteria can assume under certain conditions, e.g., *Rhizobium* in root nodules

baeocytes   Endospores of pleurocapsalean cyanobacteria

baker's yeast   *Saccharomyces cerevisiae;* yeast used in the baking industry

balantidiasis   A protozoan disease of the digestive system also known as balantidial dysentery, caused by *Balantidium coli,* typically a mild disease consisting of abdominal pain, nausea, vomiting, diarrhea, and weight loss

band cells   Stab cells

barophiles   Organisms that grow best and/or only under conditions of high pressure, e.g., in ocean depths

barotolerant   Organisms that can grow under conditions of high pressure but do not exhibit a preference for growth under such conditions

Bartonella   In stained blood films the organisms appear as rounded or ellipsoidal forms or as slender, straight, curved, or bent rods, occurring either singly or in groups within erythrocytes; within tissues they are situated within the cytoplasm of endothelial cells as isolated elements or are grouped in rounded masses; Gram-negative; not acid-fast; stain poorly or not at all with many aniline dyes but satisfactorily with Romanowsky and Giemsa stains; optimal growth at 37° C

bartonellosis   Carrión's disease; a bacterial infection of humans endemic to the Andes caused by *Bartonella baciliformis,* which attacks red blood cells

base analogs   Chemicals that structurally resemble the DNA nucleotides and therefore may substitute for them but do not function in the same manner

base substitutions   Mutations that occur when one pair of nucleotide bases in the DNA is replaced by another pair

basic stains   Dyes whose active staining parts consist of a cationic, negatively charged group that may be combined with an acid, usually inorganic, that has affinity for nucleic acids

basidia   Specialized sexual spore-producing structures found in Basidiomycetes

basidiocarps   The fruiting bodies of basidiomycetes

Basidiomycetes   A group of fungi distinguished by the formation of sexual basidiospores on a basidium

basidiomycotina   Club fungi; fungal subdivision of Amastigomycota; includes smuts, rusts, jelly fungi, shelf fungi, stinkhorns, bird's nest fungi, puffballs, and mushrooms; produce sexual basidiospores on the surfaces of basidia

basidiospores   Sexual spores formed on basidiocarps by basidiomycetes

basidium   Club-like structure of basidiomycetes on which basidiospores are borne

basophils   White blood cells containing granules (granulocytes) that readily take up basic dye

batch culture   Growth of microorganisms under conditions in which a medium in a reaction vessel is inoculated and further additions of organisms and/or growth substrates are not made

batch process   Common simple form of culture in which a fixed volume of liquid medium is inoculated and incubated for an appropriate period of time; cells grown this way are exposed to a continually changing environment; when used in industrial processes, the culture and products are harvested as a batch at appropriate times

Bauer-Kirby test   A standardized antimicrobial susceptibility procedure in which a culture is inoculated onto the surface of Mueller-Hinton agar, followed by the addition of antibiotic impregnated discs to the agar surface

Bayer junctions   Adhesion sites between the plasma membrane and outer membrane of Gram-negative bacteria

BCG vaccine   Vaccine of bacillus of Calmette-Guérin; active vaccine prepared from *Mycobacterium bovis* that is used to immunize against tuberculosis

Bdellovibrio   Cells are single, small, curved, motile rods in the parasitic state; Gram-negative; parasitic strains attach to and penetrate bacterial host cells; metabolism respiratory, not fermentative; optimal growth at 30° C; G + C 45.5-51.3 mole %

beer   Beverage produced by microbial alcoholic fermentation and brewing of cereal grains

benthos   The bottom region of aquatic habitats; collective term for the organisms living at the bottom of oceans and lakes

Bergey's Manual   Reference book describing the established status of bacterial taxonomy; describes bacterial taxa and provides keys and tables for their identification

beta-galactosidase (β-galactosidase)   An enzyme catalyzing the hydrolysis of β-linked galactose within dimers or polymers

beta-hemolysis (β-hemolysis)   Complete lysis of red blood cells, as shown by the presence of a sharply defined zone of clearing surrounding certain bacterial colonies growing on blood agar

beta-lactamase (β-lactamase)   An enzyme that attacks a β-lactam ring such as a penicillinase that attacks the lactam ring in the penicillin antimicrobials, inactivating such antibiotics

beta-oxidation (β-oxidation)   Metabolic pathway for the oxidation of fatty acids resulting in the formation of acetate and a new fatty acid that is two carbon atoms shorter than the parent fatty acid; the pathway in which fatty acids are metabolized in cells by being broken down into small 2-carbon acetyl coenzyme A units

Bifidobacterium   Rods highly variable in appearance; Gram-positive; not acid-fast; nonspore-forming; nonmotile; anaerobic; optimal growth at 36° to 38° C; G + C 57.2-64.5 mole %

bilateral parotitis   Swelling of salivary glands on both sides in a case of mumps

binary fission   A process in which two similarly sized and shaped cells are formed by the division of one cell; process by which most bacteria reproduce

binding protein transport   A mechanism for transporting substances across the Gram-negative bacterial plasma membrane that involves the cooperative activities of periplasmic-binding proteins and cytoplasmic permeases; shock-sensitive transport

binding proteins   Chemosensors in the cell envelope that bind specifically and tightly to substances in the membrane transport process, detect certain chemicals, and signal the flagella to respond

binomial nomenclature   The scientific method of naming plants, animals, and microorganisms composed of two names consisting of the species and genus

bioassay   The use of a living organism to determine the amount of a substance based on the growth or activity of the test organism under controlled conditions

biochemicals   Substances produced by and/or involved in the metabolic reactions of living organisms

biocides   Chemical agents that kill organisms, including microorganisms; can be used to sterilize materials

biodegradation   The process of chemical breakdown of a substance to smaller products caused by microorganisms or their enzymes

biodeterioration   The chemical or physical alteration of a product that decreases the usefulness of that product for its intended purpose

biodisc   Biodisc system

biodisc system   A secondary sewage treatment system employing a film of active microorganisms rotated on a disc through sewage

bioenergetics   The transfer of energy through living systems; the study of energy transformations in living systems

biogeochemical cycling   The biologically mediated transformations of elements that result in their global cycling, including transfer between the atmosphere, hydrosphere, and lithosphere

bioleaching   The use of microorganisms to transform elements so that the elements can be extracted from a material when water is filtered through it

biological control   The deliberate use of one species of organism to control or eliminate populations of other organisms; used in the control of pest populations

biological magnification   Biomagnification

biological oxygen demand (BOD)   The amount of dissolved oxygen required by aerobic and facultative microorganisms to stabilize organic matter in sewage or water; also known as *biochemical oxygen demand*

bioluminescence   The generation of light by certain microorganisms

biomagnification   An increase in the concentration of a chemical substance, such as a pesticide, as the substance is passed to higher members of a food chain

biomass   The dry weight, volume, or other quantitative estimation of organisms; the total mass of living organisms in an ecosystem

bioremediation   The use of biological agents to reclaim soils and waters polluted by substances hazardous to human health and/or the environment; it is an extension of biological treatment processes that have traditionally been used to treat wastes in which microorganisms typically are used to biodegrade environmental pollutants

biosphere   The part of the Earth in which life can exist; all living things together with their environment

biosynthesis   The production (synthesis) of chemical substances by the metabolic activities of living organisms

biotechnology   The modern use of biological systems for economic benefit

biotic   Of or relating to living organisms, caused by living things

biotype   A variant form of a given species that may be distinguished based on physiology, morphology or serology

biovar   A biotype of a given species that is differentiated based on physiology

biphasic growth curve   Growth curve reflecting diauxie, i.e., the preferential utilization of one substrate at a given rate before another substrate is metabolized at a different rate

black death   Death due to plague in which there is severe tissue necrosis so that the skin appears blackened

blastomycosis   A chronic mycosis (fungal infection) caused by *Blastomyces* in which lesions develop, e.g., in the lungs, bones, and skin

blastospores   Spores produced by budding

blight   Any plant disease or injury that results in general withering and death of the plant without rotting

blood-brain barrier   Cell membranes that control the passage of substances from the blood to the brain

blood plasma   The fluid portion of the blood minus all blood corpuscles

blood serum   The fluid expressed from clotted blood or clotted blood plasma clotted blood plasma

blood type   An immunologically distinct, genetically determined set of antigens on the surfaces of erythrocytes (red blood cells), defined as A, B, AB, and O

bloodstream   The flowing blood in a circulatory system

bloom   A visible abundance of microorganisms, generally referring to the excessive growth of algae or cyanobacteria at the surface of a body of water

BOD   Biological oxygen demand

booster vaccines   Vaccine antigens administered to elicit an anamnestic response and to maintain extended active immunity

Bordetella   Minute coccobacilli arranged singly or in pairs, more rarely in short chains; Gram-negative, bipolar; colonies on potato-glycerol-blood agar medium are smooth, convex, pearly, glistening, nearly transparent, surrounded by a zone of hemolysis without a definite periphery; metabolism respiratory, never fermentative; strict aerobes; optimal growth at 35° to 37° C

Borrelia   Cells helical, with 3–10 or more coarse, uneven, often irregular coils; Gram-negative; fermentative metabolism; strict anaerobes; optimal growth at 28° to 30° C

botulinum toxins   Neurotoxins produced by *Clostridium botulinum*

botulism   Food intoxication or poisoning that is severe and often fatal, caused by *Clostridium botulinum*

bracket fungi   Shelf fungi

bradykinin   A mediator of pain that acts by lowering the threshold for firing of nerve cells

brain abscesses   Free or encapsulated collections of pus in the brain, secondary to some other infection usually caused by pyogenic bacteria

breakpoint chlorination   Procedure for the removal and oxidation of ammonia from sewage to molecular nitrogen by the addition of hypochlorous acid

bright-field microscope   A microscope that uses visible light transmitted through a specimen to illuminate that specimen

broad-spectrum antibiotic   Antibiotics capable of inhibiting a relatively wide range of bacterial species, including Gram-negative and Gram-positive types

bronchitis   Inflammation occurring in the mucous membranes of the bronchi, often caused by *Streptococcus pneumoniae, Haemophilus influenzae,* and certain viruses

brown algae   Members of the division Phaeophycophyta; Phaeophycophyta

Brucella   Coccobacilli or short rods; mammalian parasites and pathogens; no capsules; nonmotile; do not form endospores; Gram-negative; metabolism respiratory; strict aerobes; some require 5% to 10% added CO_2 for growth, especially on initial isolation; optimal growth at 20° to 40° C; G + C 56-58 mole %

Brucellosis   A remittent febrile disease caused by infection with bacteria of the genus *Brucella*

Bruton congenital agammaglobulinemia   An immunodeficiency disease affecting males in which all immunological classes are totally or partially absent

buboes   Enlarged lymph nodes that are symptomatic of plague

bud scar   Site on a yeast cell produced by the process of fungal budding, which limits the number of progeny that can be derived from a mother cell

budding   A form of asexual reproduction in which a daughter cell develops from a small outgrowth or protrusion of the parent cell; the daughter cell is smaller than the parent cell

budding bacteria   Heterogeneous group of bacteria that form extensions or protrusions from the cell; these bacteria have reproductive or physiological functions; some reproduce by budding, others by binary fission

buffer   A solution that tends to resist the change in pH when acid or alkali is added

Burkitt lymphoma   Malignant tumorous growth on jaw or abdomen caused by the Epstein-Barr virus, usually affecting children

burst size   The average number of infectious viral units released from a single cell

butanediol fermentation pathway   Metabolic sequence during which acetoin is produced, carbon dioxide is released, and NADH is reoxidized to NAD^+; the end product is butanediol

butanol fermentation pathway   Metabolic sequence carried out by certain *Clostridium* species, with pyruvate converted to either acetone and carbon dioxide, isopropanol and carbon dioxide, butyrate, or butanol

butyric acid fermentation pathway   Butanol fermentation pathway

C

C_4 pathway   A carbon dioxide fixation pathway in heterotrophs and autotrophs that produces oxaloacetate

calcium dipicolinate   Chemical component of bacterial endospores contained within the core and involved in conferring heat resistance on endospores

Calvin cycle   The primary pathway for carbon dioxide fixation (conversion of carbon dioxide to organic matter) in photoautotrophs and chemolithotrophs

Calymmatobacterium   Pleomorphic rod-shaped, Gram-negative, facultative anaerobes; extremely fastidious; cultivation rarely successful on laboratory media

Campylobacter   Slender, Gram-negative, nonspore-forming, spirally curved rods; motile; respiratory; Methyl Red and Voges-Proskauer negative; found in the reproductive organs, intestinal tract, and oral cavity of animals and humans; G + C 30-35 mole %

cancer   A malignant, invasive cellular tumor that can spread throughout the body

candidiasis   Mild, superficial fungal infection caused by *Candida* sp., infecting the skin, nails, or mucous membranes

cankers   Plant diseases, or conditions of those diseases, that interfere with the translocation of water and minerals to the crown of the plant

canning   Method for the preservation of foodstuffs in which suitably prepared foods are placed in metal containers that are heated, exhausted, and hermetically sealed

capillary   One of a network of tiny hair-like blood vessels connecting the arteries to the veins

capnophiles Microaerophiles that grow at elevated concentrations of carbon dioxide (5% to 10% CO_2)

capsid A protein coat of a virus enclosing the naked nucleic acid

capsomere The individual protein units that form the capsid of a virus

capsule A mucoid envelope composed of polypeptides and/or carbohydrates surrounding certain microorganisms; a gelatinous or slimy layer external to the bacterial cell wall

carbohydrates A class of organic compounds consisting of many hydroxyl (-OH) group and containing a ketone or an aldehyde

carbolic acid Phenol

carbon cycle The biogeochemical cycling of carbon through oxidized and reduced forms, primarily between organic compounds and inorganic carbon dioxide

carboxyl end The terminus of a polypeptide chain with a free alpha carboxyl group not involved in forming a peptide linkage; also known as the *C terminal end*

carboxylic acid An organic chemical having a —COOH functional group

carboxysomes Inclusions within some autotrophic bacteria containing ribulose-1,5-bisphosphate carboxylase

carcinogen Cancer-causing agent

cardiovascular system The system that circulates blood throughout the body; includes the heart, arteries, veins, and capillaries

carditis Inflammation of the heart

caries Bone or tooth decay with the formation of ulceration; also known as *dental caries* or *cavities*

carotenoid pigments A class of pigments, usually yellow, orange, red, or purple, that are widely distributed among microorganisms

carpogonia The basal bodies bearing female gametes in some red algae

carpospore Red algal spore produced during fertilization

carriers Individuals who harbor pathogens but do not exhibit any signs of illness

case-control studies Studies where persons with a disease are compared with a disease-free control group for various possible risk factors

catabolic pathway A degradative metabolic pathway; a metabolic pathway in which large compounds are broken down into smaller ones

catabolism Metabolic reactions involving the enzymatic degradation of organic compounds to simpler organic or inorganic compounds with the release of free energy

catabolite repression Repression of the transcription of genes coding for certain inducible enzyme systems by glucose or other readily utilizable carbon sources

catalases Enzymes that catalyze the decomposition of hydrogen peroxide (H_2O_2) into water and oxygen and the oxidation of alcohols to aldehydes by hydrogen peroxide

catalyst Any substance that accelerates a chemical reaction but itself remains unaltered in form and amount

catalyze To subject to modification, especially an increase in the rate of chemical reaction

catheterization Insertion of a hollow tubular device (a catheter) into a cavity, duct, or vessel to permit injection or withdrawal of fluids

catheters Narrow, hollow tubes that are inserted into a body cavity to inject or to withdraw fluids

cathode The electrode at which reduction takes place in an electrolytic cell; a negatively charged electrode

cations Positively charged ions

cauldoactive bacteria Extreme thermophiles that often fail to grow at temperatures below 50° C and are able to grow at temperatures above 100° C

cDNA Complementary DNA

cell The functional and structural subunit of living organisms, separated from its surroundings by a delimiting membrane

cell envelope Structure found only in Gram-negative cell walls, extending outward from the plasma membrane to the outer membrane

cell-mediated hypersensitivity Type 4 hypersensitivity or delayed hypersensitivity; reaction involving T lymphocytes and occurring 24 to 72 hours after exposure to the antigen; contact dermatitis; poison ivy, for example

cell-mediated immune response Cell-mediated immunity

cell-mediated immunity Specific acquired immunity involving T cells, primarily responsible for resistance to infectious diseases caused by certain bacteria and viruses that reproduce within host cells

cell wall Structure outside of and protecting the cell membrane, generally containing murein in prokaryotes and composed chiefly of various other polymeric substances, e.g., cellulose or chitin, in eukaryotic microorganisms

cellular metabolism The collective process of chemical reactions that accompanies the flow of energy through a cell

cellular slime molds Members of the Acrasiales that form a sporocarp fruiting body

cellulase An extracellular enzyme that hydrolyzes cellulose

cellulose A linear polysaccharide of β-D-glucose

Centers for Disease Control Branch of the U.S. Public Health Service in Atlanta, Georgia, responsible for the generation, collection, and dissemination of epidemiological information; works to identify the causes of major public health problems; also collects and maintains samples of microbiological cultures

central metabolic pathways Metabolic sequences that have key roles in catabolism and biosynthesis

central nervous system That part of the nervous system contained within the brain and spinal cord

centrifugation Separation process in which particulate matter is sedimented from a fluid, or fluids, of different densities using a centrifuge

centrifuge An apparatus used to separate by sedimentation particulate matter suspended in a liquid by a centrifugal force

cephalosporins A heterogeneous group of natural and semisynthetic antibiotics that act against a range of Gram-positive and Gram-negative bacteria by inhibiting the formation of cross-links in peptidoglycan

cerebrospinal fluid (CSF) The fluid contained within the four ventricles of the brain, the subarachnoid space, and the central canal of the spinal cord

cestodes Tapeworms; endoparasites, adult forms of which live in the intestines of vertebrate hosts

CFU Colony forming unit

Chagas disease Also known as South American trypanosomiasis; caused by the protozoan *Trypanosoma cruzi*, which is carried by the cone-nosed bug that lives in wood or mud houses

chancre The lesion formed at the site of primary inoculation by an infecting microorganism, usually an ulcer

chancroid A lesion produced by an infection with *Haemophilus ducreyi* involving the genitalia; a sexually transmitted disease caused by *H. ducreyi*

change in free energy (ΔG) The change in the usable energy that is available for doing work

charging The attachment of an amino acid to its specific tRNA molecule

cheese Product of the microbial fermentation of milk by lactic acid bacteria

chemical mutagens Chemical substances that can modify nucleotide bases; chemicals that increase the rate of mutation

chemical oxygen demand (COD) The amount of oxygen required to oxidize completely the organic matter in a water sample

chemical preservative Chemical substances added to prevent the spoilage of a food or the biodeterioration of any substance by inhibiting microbial growth and/or activity

chemiosmosis The generation of ATP by the movement of hydrogen ions into pores in the plasma membrane that are associated with the ATPase system

chemiosmotic hypothesis The theory that the living cell establishes a proton and electrical gradient across a membrane and that, by controlled reentry of protons into the region contained by that membrane, the energy to carry out several different types of endergonic processes may be obtained, including the ability to drive the formation of ATP

chemoattractants Substances that attract motile bacteria—bacteria move through their environment toward higher concentrations of a chemoattractant

chemoautotrophic metabolism A type of metabolism in which inorganic molecules serve as electron donors and energy source and inorganic CO_2 serves as carbon source

chemoautotrophs Microorganisms that obtain energy from the oxidation of inorganic compounds and carbon from inorganic carbon dioxide; organisms that obtain energy through chemical oxidation and use inorganic compounds as electron donors; also known as *chemolithotrophs*

chemoheterotrophs Heterotrophs

chemolithotrophic metabolism Chemoautotrophic metabolism

chemolithotrophs Microorganisms that obtain energy through chemical oxidation and use inorganic compounds as electron donors and cellular carbon through the reduction of carbon dioxide; also known as chemoautotrophs

chemoorganotrophic metabolism A type of metabolism in which organic molecules serve as electron donors, energy source, and carbon source

chemoorganotrophs Organisms that obtain energy from the oxidation of organic compounds and cellular carbon from preformed organic compounds

chemorepellents Substances that repel motile bacteria—bacteria move through their environment away from higher concentrations of chemorepellents

chemostat An apparatus used for continuous-flow culture to maintain bacterial cultures in the log phase of growth, based on maintaining a continuous supply of a solution containing a nutrient in limiting quantities that controls the growth rate of the culture

chemotaxis A locomotive response in which the stimulus is a chemical concentration gradient; movement of microorganisms toward or away from a chemical stimulus

chemotherapy The use of chemical agents for the treatment of disease, including the use of antibiotics to eliminate infecting agents

chi (χ) form The joining of chromosomes at a homologous region during recombination

chickenpox Common, acute, and highly contagious infection caused by the herpesvirus varicella-zoster, occurs most frequently in children and produces a distinctive rash

childbed fever Puerperal fever

"Chinese letter" formation Snapping division

chitin A polysaccharide composed of repeating *N*-acetylglucosamine residues, abundant in arthropod exoskeletons and fungal cell walls

Chlamydia Nonmotile, spheroidal cells; obligately intracellular growth cycle; the principal developmental stages are (1) the elementary body, which is a small, electron-dense spherule containing a nucleus and numerous ribosomes surrounded by a multilaminated wall, (2) the initial body, which is a large, thin-walled, reticulated spheroid containing nuclear fibrils and ribosomal elements, and (3) an intermediate body representative of a transitional stage between the initial and elementary bodies; the elementary body is the infectious form of the organism; the initial body is the vegetative form that divides by fission intracellularly but is apparently noninfectious when separated from the host cells; Gram-negative; because these bacteria are unable to synthesize their own high-energy compounds, they have been described as "energy parasites"; optimal growth at 33° to 41° C; G + C 39-45 mole %

chlamydiae Members of the bacterial genus *Chlamydia*, all of which are obligate intracellular inhabitants

chlamydias Obligate intracellular parasites whose reproduction is characterized by a change of the small, rigid-walled, infectious form of the organism into a larger, thin-walled, noninfectious form that divides by fission

chlamydospores Thick-walled, typically spherical or ovoid resting spores produced asexually by certain types of fungi from cells of the somatic hyphae

chlor- Combining form indicating that chlorine substituted for hydrogen

chloramination The use of chloramines to disinfect water

chloramphenicol Aminoglycoside antibiotic that inhibits bacterial protein synthesis; it acts by binding to the 50S

chlorination The process of treating with chlorine, as in disinfecting drinking water or sewage

Chlorobiaceae Green sulfur bacteria; family of nonmotile, obligately phototrophic bacteria that produce green or green-brown carotenoid pigments

Chloroflexaceae Family of anaerobic phototrophic bacteria; its members have flexible walls, form filaments, and exhibit gliding motility

Chlorophycophyta Green algae; may be unicellular, colonial, or filamentous; most cells are uninucleate; some form coenocytic filaments, contain contractile vacuoles, or store starch as reserve material; their cell walls are composed of cellulose, mannans, xylans, or protein

chlorophyll The green pigment responsible for photosynthesis in plants; the primary photosynthetic pigment of algae and cyanobacteria

chloroplasts Membrane-bound organelles of photosynthetic eukaryotes where the biochemical conversion of light energy to ATP occurs; the sites of photosynthesis in eukaryotic organisms

chlorosis The yellowing of leaves and/or plant components due to bleaching of chlorophyll, often symptomatic of microbial disease

chlorosomes Vesicles that contain photosynthetic antenna pigments in some green photoautotrophic bacteria

cholera An acute infectious disease caused by *Vibrio cholerae*, characterized by severe diarrhea, delirium, stupor, and coma

choleragen Enterotoxin produced by *Vibrio cholerae* that blocks the conversion of cyclic AMP to ATP

Chromatiaceae Purple sulfur bacteria; family of phototrophic bacteria that produce carotenoid pigments, appear orange-brown, purple-red, or purple-violet, and deposit elemental sulfur

chromatic aberration An optical lens defect causing distortion of the image because light of differing wavelengths is focused at differing points instead of at a single focal point

chromatids Fibrils formed from a eukaryotic chromosome when it replicates prior to meiosis or mitosis

chromatin The deoxyribonucleic acid–protein complex that constitutes a chromosome; the readily stainable protoplasmic substance in the nuclei of cells

chromatophore Internal membranes found in some photosynthetic bacteria that contain the pigments and accessory molecules utilized in photosynthesis

chromosomes Structures that contain the nuclear DNA of a cell

chronic A disease condition in which the symptoms persist for a long time

chronic disease A persistent disease

chronometer A device for measuring time

chroococcacean cyanobacteria Subgroup of cyanobacteria, unicellular rods or cocci that reproduce by binary fission or budding; generally nonmotile

Chrysophyceae A group within the Chrysophycophyta that contains the golden algae

Chrysophycophyta Division of algae that includes the yellow-green algae, golden algae, and diatoms; all produce chrysolaminarin; most are unicellular and some are colonial

chytrids Members of the Chytridiales, which are mainly aquatic fungi that produce zoospores with a single posterior flagellum

-cide Suffix signifying a killer or destroyer, as in a chemical that kills microorganisms

CIE Countercurrent immunoelectrophoresis

cilia Threadlike appendages, having a 9 + 2 arrangement of microtubules occurring as projections from certain cells that beat rhythmically, causing locomotion or propelling fluid over surfaces

ciliated epithelial cells Cells that line the respiratory tract and act as filters by sweeping microorganisms out of the body with a wave-like motion

ciliates Members of the protozoan phylum Ciliata that use cilia for locomotion

Ciliophora Group of one subphylum of protozoa that possess simple to compound ciliary organelles in at least one stage of their life cycle; these protozoa are motile by means of cilia

circadian rhythm Daily cyclical changes that occur in an organism even when it is isolated from the natural daily fluctuations of the environment

circulatory system Vessels and organs comprising the cardiovascular and lymphatic systems of animals

cis **configuration** Genetic elements that have an effect on the same DNA molecule

cis/trans **complementation test** A test used to determine whether two mutations are in the same gene and on the same DNA molecule

cistron The functional unit of genetic inheritance; a segment of genetic nucleic acid that codes for a specific polypeptide chain; synonym for *gene*

citric acid Intermediary metabolite in the Krebs cycle; organic acid, produced by *Aspergillus niger*, used as a food additive, a metal chelating and sequestering agent, and a plasticizer

citric acid cycle Krebs cycle

Citrobacter Motile, peritrichously flagellated rods; not encapsulated; Gram-negative; can use citrate as their sole carbon source

clamp cell connections Hyphal structures in many basidiomycetes formed during cell division by dikaryotic hyphal cells, i.e., formed by hyphal cells containing two nuclei of different mating types

classical complement pathway Series of reactions initiated by the formation of a complex between an antigen and an antibody that lead to the lysis of microbial cells or the enhanced ability of phagocytic blood cells to eliminate such cells

classification The systematic arrangement of organisms in groups or categories according to established criteria

clinical immunologist A certified individual in charge of a diagnostic immunology laboratory who is responsible for the submission of diagnostic reports to physicians

clinical microbiologist A certified individual in charge of a diagnostic microbiology laboratory who is responsible for the submission of diagnostic reports to physicians

clonal deletion Elimination of clones of T cells in the thymus gland that would attack the body's own normal cells

clonal selection theory A theory that accounts for antibody formation by supposing that during fetal development a complete set of lymphocytes are developed, with each lymphocyte containing the genetic information for initiating an immune response to a single specific antigen for which it has only one type of receptor; B cells that react with self-antigens during this period are destroyed

clone A population of cells derived asexually from a single cell, often assumed to be genetically homologous; a population of genetically identical individuals

cloning vector Segment of DNA used for the replication of foreign DNA fragments

Clostridium Rods, usually motile by means of peritrichous flagella; form endospores; Gram-positive but may appear Gram-negative in the late stages of growth; fermentative; most strains are strictly anaerobic; G + C 23-43 mole %

club fungi Members of the Basidiomycotina

cluster analysis A statistical method for grouping (clustering) organisms based on their degree of similarity; method used in numerical taxonomy for defining taxonomic groups

coagglutination An enhanced agglutination reaction based on using antibody molecules whose Fc fragments are attached to cells so that a larger matrix is formed when the Fab portion reacts with other cells for which the antibody is specific

coagulase An enzyme produced by pathogenic staphylococci, causing coagulation of blood plasma

cocci Spherical or nearly spherical bacterial cells, varying in size, sometimes occurring singly, in pairs, in regular groups of 4 or more, in chains, or in irregular clusters

coccidioidomycosis A disease of humans and domestic animals caused by *Coccidioides immitis*, usually occurring via the respiratory tract

coccobacilli Bacteria that are very short oval-shaped rods

codominance Partial expression of the genetic information contained in the recessive and dominant alleles of a gene

codon A triplet of adjacent bases in a polynucleotide chain of an mRNA molecule that codes for a specific amino acid; the basic unit of the genetic code specifying an amino acid for incorporation into a polypeptide chain

coenocytic Referring to any multinucleate cell, structure, or organism formed by the division of an existing multinucleate entity or when nuclear divisions are not accompanied by the formation of dividing walls or septa; multinucleate hyphae

coenocytic hyphae Fungal filaments that lack septa and therefore do not divide into uninucleate cell-like units

coenzymes The nonprotein portions of enzymes; small, nonprotein organic chemicals that are not tightly bound to the enzymes with which they function and that act as acceptors or donors of electrons or functional groups during enzymatic reactions

cofactors Inorganic substances, such as minerals, required for enzymatic activity

cohort study The most definitive type of analytical epidemiological study based on comparisons with a defined group (cohort)

cold pack method A method of canning uncooked high-acid food by placing it in hot jars or cans and sterilizing in a bath of boiling water or steam

cold sores Also known as fever blisters; lesions that erupt on the mouth or face, caused by a recurrent infection with herpes simplex 1 virus

colicinogenic plasmids Plasmids that code for colicins

colicins Proteins produced by some bacteria that inhibit closely related bacteria

coliform count Enumeration of coliform bacteria, especially *Escherichia coli*, that is commonly used as an indicator of water quality and potential fecal contamination

coliforms Gram-negative, lactose-fermenting, enteric rods, e.g., *Escherichia coli*

colinear Two related linear information sequences arranged so that the unit may be moved from one to the other without rearrangement; RNA and DNA molecules with precisely matching base pairs

coliphage A virus that infects *Escherichia coli*

collagenase An enzyme that breaks down the proteins of collagen tissues

colonization The establishment of a site of microbial reproduction on a material, animal, or person without necessarily resulting in tissue invasion or damage

colony The macroscopically visible growth of microorganisms on a solid culture medium

colony forming units (CFUs) Number of microorganisms that can replicate to form colonies, as determined by the number of colonies that develop

colony hybridization A technique that can be used to detect the presence of a specific DNA sequence in a cell by transferring cells from a colony to a filter, lysing the cells, and identifying a target DNA sequence by hybridization with a gene probe

Colorado tick fever The only recognized tick-borne viral disease in the United States; an acute febrile disease characterized by sudden onset of fever, headache, and severe muscle pain

coma A state of unconsciousness

cometabolism The gratuitous metabolic transformation of a substance by a microorganism growing on another substrate; the cometabolized substance is not incorporated into an organism's biomass, and the organism does not derive energy from the transformation of that substance

commensalism An interactive association between two populations of different species living together in which one population benefits from the association while the other is not affected

common cold An acute self-limiting inflammation of the upper respiratory tract due to a viral infection

common source epidemic An epidemic where many individuals simultaneously acquire an infectious agent from the same source

common source outbreak Disease outbreak characterized by a sharp rise and a rapid decline in the number of cases

communicable disease A disease in which a pathogen will move with ease from one individual to the next

community Highest biological unit in an ecological hierarchy composed of interacting populations

competent　In transformation, the state of a recipient cell in which DNA can pass across its membrane, depending on environmental conditions and the cell growth phase

competent bacteria　Bacteria capable of taking up a DNA fragment from a bacterial donor cell

competition　An interactive association between two species, both of which need some limited environmental factor for growth and thus grow at suboptimal rates because they must share the growth-limiting resource

competitive exclusion　Competitive interactions tend to bring about the ecological separation of closely related populations and precludes two populations from occupying the same ecological niche

competitive inhibition　The inhibition of enzyme activity caused by the competition of an inhibitor with a substrate for the active (catalytic) site on the enzyme; impairment of the function of an enzyme due to its reaction with a substance chemically related to its normal substrate

complement　Group of proteins normally present in plasma and tissue fluids that participates in antigen-antibody reactions, allowing reactions such as cell lysis to occur

complement fixation　The binding of complement to an antigen-antibody complex so that the complement is unavailable for subsequent reactions

complement fixation test　Test that measures the degree of complement fixation for diagnostic purposes

complementary DNA (cDNA)　In cloning eukaryotic genes in bacteria, a single-stranded DNA molecule that is complementary to the complete mRNA

complementary strands　Designation used to indicate that one strand of a DNA macromolecule is the antithesis of the other strand—the juxtaposed nucleotides are complementary bases (adenine opposite thymine and guanine opposite cytosine), one strand runs from the 3′-OH hydroxyl free end to the 5′-phosphate free end and the other runs in the opposite direction

complementation　A method for determining whether mutations are in the same or different locations

completed test　In assays for assessing water safety, gas formation by subcultured colonies showing a greenish metallic sheen on eosin-methylene blue (EMB) agar grown on lactose broth incubated at 35° C; positive test for fecal coliforms

complex-mediated hypersensitivity　Type 3 hypersensitivity; reaction that occurs when excess antigens are produced during a normal inflammatory response and antibody-antigen–complement complexes are deposited in tissues

complex medium　A medium made with constituents whose compositions are not fully known and may vary

composting　Decomposition of organic matter in a heap by microorganisms; a method of solid waste disposal

computer-assisted identification　Rapid identification of microorganisms by a computer based on a large number of calculations and comparisons of data and assessment of the probability of correctly identifying a particular microorganism

concentration gradient　Condition established by the difference in concentration on opposite sides of a membrane

condenser lenses　Lenses on a microscope for focusing or directing light from the light source onto the object

conditionally lethal mutations　Mutations that cause the loss of microbial viability only under certain environmental conditions

confirmed test　In assays for assessing water safety, the formation of greenish, metallic colonies of fecal coliforms on EMB agar or brilliant green lactose–bile broth

congenital diseases　Illnesses present at birth that are the result of some *in utero* condition

conidia　Thin-walled, asexually derived spores, borne singly or in groups or clusters in specialized hyphae

conidiophores　Branches of mycelia-bearing conidia

conjugated fluorescent dyes　Fluorescent dyes coupled with antibody molecules used to tag antibodies

conjugation　The process in which genetic material is transferred from one microorganism to another, involving a physical connection or union between the two cells; a parasexual form of reproduction sometimes referred to as *mating*

conjugative plasmids　F and other plasmids that encode for self-transfer from one cell to another

conjunctivitis　Inflammation of the mucous membranes covering the eye, the conjunctiva

consensus sequence　Conserved DNA sequence that characterizes the bacterial promoter where there is a high nucleotide sequence homology among most promoters

conservation of energy　Maintenance of energy during chemical reactions: transfer of energy without destruction

conservative transposition　Excision and insertion of a transposon from one location to another without a change in copy number

consortium　An interactive association that between microorganisms generally results in combined metabolic activities

constant region　The carboxyl terminal end of an immunoglobulin molecule that has a relatively constant amino acid sequence

constipation　A condition in which the bowels are evacuated at long intervals or with difficulty; the passage of hard, dry stools

constitutive enzymes　Enzymes whose synthesis is not altered in response to changes in the environment but rather are continuously synthesized

contact dermatitis　Delayed hypersensitivity reaction resulting from exposure of the skin to chemicals; poison ivy is an example

contact diseases　Diseases caused by agents that are able to enter the subcutaneous layers of the skin through hair follicles

contagion　The process by which disease spreads from one individual to another

contagious disease　An infectious disease that is communicable to healthy, susceptible individuals by physical contact with, contact with bodily discharges from, or contact with inanimate objects contaminated by someone suffering from that disease

contamination The process of allowing the uncontrolled addition of microorganisms to an area or substance

continuous culture A method for growing microorganisms without interruption by continual addition of substrates and recovery of products

continuous feed composting process A composting process that uses a reactor to establish the environmental parameters that maximize the degradation process

continuous flow process A process for growing microorganisms without interruption by continual addition of substrates and recovery of products

continuous flow-through process Continuous flow process

continuous strand of DNA The strand of DNA that can be synthesized continuously because it runs in the appropriate direction for the continuous addition of new free nucleotide bases; also referred to as the *leading strand of DNA*

contractile vacuoles Pulsating vacuoles in certain protozoa used for the excretion of wastes and the exclusion of water for the maintenance of proper osmotic balance

contrast In microscopy, the ability to visually distinguish an object from the background

control group The reference point in a controlled experiment in which a set of conditions does not vary

controlled experiment An experiment in which results from an experimental group with variable conditions is compared with a control group with nonvariable conditions

convalescence Recovery period of a disease during which signs and symptoms disappear

copiotrophs Bacteria that grow at high nutrient concentrations, such as the nutrient concentration in most culture media

coprophagous Capable of growth on fecal matter; feeding on dung or excrement

copy DNA Complimentary DNA

cornsteep liquor The concentrated water extract byproduct resulting from the steeping of corn during the production of cornstarch; used as a medium adjunct to supply nitrogen and vitamins in industrial fermentations

cortex A layer of a bacterial endospore important in conferring heat resistance on that structure

corticosteroids Derivatives of the steroid hormones released by the adrenal glands used to treat inflammatory diseases

cortisone Hormone that relieves arthritis pain; derivatives alleviate allergy symptoms and other inflammatory responses; can be chemically synthesized

Corynebacterium Straight to slightly curved rods with irregularly stained segments and sometimes granules; frequently show club-shaped swellings; snapping division produces angular and palisade (picket fence) arrangements of cells; generally nonmotile; Gram-positive, although some species (e.g., *C. diphtheriae*) decolorize easily, especially in old cultures; not acid-fast; carbohydrate metabolism fermentative and respiratory; grow best aerobically; G + C 52-68 mole %

coryneform group Bacterial group of Gram-positive, irregularly shaped rods with a tendency to show incomplete separation following cell division and to exhibit pleomorphic morphology

coryza An inflammation of the mucous membranes of the nose, usually marked by sneezing and the discharge of watery mucus

cosmid Phage–plasmid artificial hybrids; a genetically engineered hybrid of bacteriophage lambda (λ) and plasmid genes that contains *cos* sites needed to package λ DNA into its particles

co-transporters Permeases in the plasma membrane that transport more than one type of substrate at the same time

counter immunoelectrophoresis Countercurrent immunoelectrophoresis

countercurrent immunoelectrophoresis A technique based on immunological detection of substances that relies on movement of an antibody and an antigen toward each other in an electric field, resulting in rapid formation of a detectable antigen-antibody precipitate

counterstain In microscopy, the use of a secondary stain to visualize objects differentially from those stained with a primary stain

covalent bond A strong chemical bond formed by the sharing of electrons

cowpox A mild, self-limiting disease caused by a vaccinia virus, involving the formation of vesicular lesions on the hands and arms of man and the udders of cows

Coxiella Short rods resembling organisms of the genus *Rickettsia* in their staining properties, dependence on host cells for growth, and close natural association with arthropod and vertebrate hosts; rods occur preferentially in the vacuoles of the host cell rather than in cytoplasm or nucleus, as do the species of *Rickettsia*

CPE Cytopathic effects

Creutzfeldt-Jakob disease Rapidly progressive disease that results in degeneration of the central nervous system (dementia) caused by a prion; this disease normally affects adults between ages 40 and 65

cristae Convolutions of the inner membrane that extend into the interior of the mitochondria of eukaryotic cells

critical point drying A method for removing liquids from a microbiological specimen by adjusting the temperature and pressure so that the liquid and gas phases of the liquid are in equilibrium with each other; used to minimize disruption of biological structures for viewing by scanning electron microscopy

crop rotation The alternation of the types of crops planted in a field

cross-feeding The phenomena in which the growth of an organism depends on the provision of one or more metabolic factors or nutrients by another organism growing in the vicinity; also termed syntrophism

crossing over The process in which, in effect, a break occurs in each of the two adjacent DNA strands, and the exposed 5′-P and 3′-OH ends unite with the exposed 5′-P and 3′-OH ends of the adjacent strands so that there is an exchange of homologous regions of DNA

cross-reactive antibodies Heterophile antibodies

cross walls Septa

crown gall Plant disease caused by *Agrobacterium tumefaciens*, which infects fruit trees, sugar beets, and other broadleaf plants, manifested by the formation of a tumor growth

cruciform loops   A region of DNA that forms a loop because it contains inverted repeat nucleotide sequences that form hydrogen-bonded hairpin structures

Cryptophycophyta   Group of unicellular brown algae that reproduce by longitudinal division, producing two flagella of equal length

culture   To encourage the growth of particular microorganisms under controlled conditions; the growth of particular types of microorganisms on or within a medium as a result of inoculation and incubation

culture medium   A liquid or solidified nutrient material that is suitable for the cultivation of a microorganism

curing   The loss of plasmids from a bacterial cell

curvature of field   Distortion of a microscopic field of view in which specimens in the center of the field are in clear focus, while those in the peripheral region are out of focus

cutaneous   Pertaining to the skin

cyanobacteria   Prokaryotic, photosynthetic organisms containing chlorophyll *a* that are capable of producing oxygen by splitting water; formerly known as *blue-green algae*

Cyanobacteriales   Order of Oxyphotobacteria whose primary bacterial photosynthetic pigment is chlorophyll *a*; the blue-green algae or cyanobacteria

cyclic oxidative photophosphorylation   Cyclic photophosphorylation

cyclic photophosphorylation   A metabolic pathway involved in the conversion of light energy to chemical energy, with the generation of ATP that does not produce the reduced coenzyme, NADPH

cycloserine   Antibiotic that inhibits bacterial cell wall synthesis

cyst   A dormant form assumed by some microorganisms during specific stages in their life cycles, or assumed as a response to particular environmental conditions in which the organism becomes enclosed in a thin- or thick-walled membranous structure, the function of which is protective or reproductive; a normal or pathological sac with a distinct wall containing fluid; a protozoan resting stage that has a wall

cysticercosis   Disease caused by ingestion of pork tapeworm larvae that occurs when eggs or larvae move from the infested gastrointestinal tract and develop in the liver, muscles, heart, or brain

cystites   Arthrospores of *Arthrobacter* species

cystitis   An inflammation of the urinary bladder

cytochromes   Reversible oxidation-reduction carriers in respiration

cytokines   Substances that stimulate cell growth, particularly lymphocyte proliferation

cytokinesis   The division of cytoplasm following nuclear division

cytolysis   The dissolution or disintegration of a cell

cytomegalovirus inclusion disease   Severe, often fatal disease of newborns caused by the cytomegalovirus, usually affecting salivary glands, kidneys, lungs, and liver

cytomegaly   Characteristic cellular response that results in formation of oversized cells in response to the presence of cytomegaloviruses

cytopathic effects   Generalized degenerative changes or abnormalities in the cells of a monolayer tissue culture due to infection by a virus

Cytophagales   Gliding bacteria exhibiting widely differing morphological forms and modes of metabolism that do not form fruiting bodies; however, some form filaments and others are chemolithotrophs

cytoplasm   The living substance of a cell, exclusive of the nucleus

cytoplasmic membrane   plasma membrane

cytoplasmic polyhedrosis virus   A type of virus used as a viral pesticide that develops in the cytoplasm of host midgut epithelial cells

cytoplast   The unified structure that provides the rigidity needed to hold the various structures of the eukaryotic cell in their appropriate locations

cytosine   A pyrimidine base in nucleic acids

cytosis   The movement of materials into or out of a cell, involving the engulfment and formation of a membrane-bound structure rather than passage through a membrane

cytoskeleton   Protein fibers composing the structural support framework of a eukaryotic cell

cytostomes   Mouth-like openings of some protozoa, particularly ciliates; structure found in some protozoa that acts as specialized structure for phagocytosis

cytotoxic T cells   Specialized class of T lymphocytes that are able to kill cells as part of the cell-mediated immune response

cytotoxins   Substances capable of injuring certain cells without causing cell lysis

D

D value   Decimal reduction time

dark-field microscope   A microscope in which the only light seen in the field of view is reflected from the object under examination, resulting in a light object on a dark background

deaminase   An enzyme involved in the removal of an amino group from a molecule, liberating ammonia

deamination   The removal of an amino group from a molecule, especially an amino acid

death   Permanent cessation of life functions; at the microbial level, death occurs when the microorganism loses its ability to reproduce and carry out metabolism needed to maintain cellular organization; at the human level, death is often interpreted as complete loss of brain or independent cardiovascular functions

death phase   The part of the normal growth curve that represents the inability of microorganisms to reproduce

decarboxylase   An enzyme that liberates carbon dioxide from the carboxyl group of a molecule by hydrolysis

decarboxylation   The splitting off of one or more molecules of carbon dioxide from organic acids, especially amino acids

decimal reduction time   The time required at a given temperature to heat inactivate or kill 90% of a given population of cells or spores; the time needed to reduce the number of visible microorganisms under a specified set of conditions by an order of magnitude

decimal reduction value   Decimal reduction time

decolorization Removal of a colored stain from an object

decomposers Organisms, often bacteria or fungi, in a community that convert dead organic matter into inorganic nutrients

deductive reasoning A logical process in which a conclusion drawn from a set of premises contains no more information than the premises taken collectively

defective phage A bacteriophage that carries some bacterial DNA instead of viral DNA and therefore cannot cause lysis in an infected bacterial cell

deficiencies Deletions of large numbers of base pairs that can result in the loss of genetic information for one or more complete genes

defined medium The material supporting microbial growth in which all of the constituents, including trace substances, are quantitatively known; a mixture of known composition used for culturing microorganisms

definitive hosts Organisms that harbor the adult, sexually mature form of a helminthic parasite

degenerate Describes the redundancy inherent in the genetic code that occurs because there are several codons coding for the insertion of the same amino acid into the polypeptide chain

dehydration Removal of water; drying

dehydrogenase An enzyme that catalyzes the oxidation of a substrate by removing hydrogen

delayed hypersensitivity Cell-mediated hypersensitivity

deletion mutations Mutations caused by the removal of one or more nucleotide base pairs from the DNA

Delta G (ΔG) Change in free energy

dementia Severe impairment or loss of intellectual capacity and personality integration

denaturation The alteration in the characteristics of an organic substance, especially a protein, by physical or chemical action; the loss of enzymatic activity due to modification of the tertiary protein structure

dendrograms Graphic representations of taxonomic analyses, showing the relationships between the organisms examined

dengue fever A human disease caused by togavirus and transmitted by mosquitoes, characterized by fever, rash, and severe pain in joints and muscles

denitrification The formation of gaseous nitrogen' or gaseous nitrogen oxides from nitrate or nitrite by microorganisms

dental caries Tooth decay; erosion of dental enamel and development of a cavity due to growth of lactic acid-producing bacteria

dental plaque Matrix of microbial cells and microbially produced extracellular polysaccharides that form on the tooth surface and can be removed by brushing and flossing

deoxyribonucleic acid (DNA) The carrier of genetic information; a type of nucleic acid occurring in cells, containing adenine, guanine, cytosine, and thymine, and D-2-deoxyribose linked by phosphodiester bonds

deoxyribose A 5-carbon sugar having one oxygen less than the parent sugar ribose; a component of DNA

derepress The regulation of transcription by reversibly inactivating a repressor protein

dermatitis Inflammation of the skin

dermatophytes Fungi characterized by their ability to metabolize keratin and capable of growing on the skin surface, causing disease

dermatophytic fungi Fungi characterized by their ability to metabolize keratin and capable of growing on the skin surface, causing diseases of the skin, nails, and hair called ringworm or tinea

dermis The lower layer of the skin, containing specialized nerve endings

desensitization Prevention of inflammatory allergic responses usually by a programmed course of allergen injections

desert A region of low rainfall; a dry region; a region of low biological productivity

desiccation Removal of water; drying

Desulfovibrio Curved rods; motile; Gram-negative; obtain energy by anaerobic respiration, reducing sulfates or other reducible sulfur compounds to H_2S; strict anaerobes; optimal growth at 25° to 30° C

desulfurization Removal of sulfur from organic compounds

detergent A synthetic cleaning agent containing surface-active agents that do not precipitate in hard water; a surface-active agent having a hydrophilic and a hydrophobic portion

detrital food web A food web based on the biomass of decomposers rather than on that of primary producers

detritus Waste matter and biomass produced from decompositional processes

Deuteromycetes A group of fungi with no known sexual stage; also known as *Fungi Imperfecti*

Deuteromycotina *Fungi Imperfecti*

diagnostic table A table of distinguishing features that is used as an aid in the identification of unknown organisms

diapedesis The process by which leukocytes move out of blood vessels

diarrhea A common symptom of gastrointestinal disease, characterized by increased frequency and fluid consistency of stools

diatomaceous earth A siliceous material composed largely of fossil diatoms, used in microbiological filters and industrial processes

diatoms Unicellular algae having a cell wall composed of silica, the skeleton of which persists after the death of the organism

diauxic growth Biphasic growth; growth exhibiting diauxie

diauxie The phenomenon in which, given two carbon sources, an organism preferentially metabolizes one completely before utilizing the other

dichotomous key A key for the identification of organisms, using steps with opposing choices until a final identification is achieved

dictyosomes The individual stacks of membranes in a Golgi apparatus

differential blood count Procedure for finding the ratios of various types of blood cells, used to determine the relative numbers of white blood cells as a diagnostic indication of an infectious process

differential medium Bacteriological medium on which the growth of specific types of organisms leads to readily visible changes in the appearance of the medium so that the presence of these organisms can be determined

differentially permeable membrane A membrane that selectively restricts the movement of molecules

differential stain The use of multiple staining reactions to differentiate one part of a cell from another or one cell type from another

diffraction The breaking up of a beam of light into bands of differing wavelength due to interference

diffusion The movement of molecules across a concentration gradient from the area of higher concentration to the area of lower concentration

DiGeorge syndrome Immune disorder caused by the partial or total absence of cellular immunity resulting from a deficiency of T lymphocytes because of incomplete fetal development of the thymus

digestive vacuoles Membrane-bound organelles formed when a eukaryotic cell engulfs a food source and then fuses with lysosomes, permitting digestion of the contents

dikaryotes Cells with two different nuclei resulting from the fusion of two cells

dimorphism The property of existing in two distinct structural forms, e.g., fungi that occur in filamentous and yeast-like forms under different conditions

dinitrogenase One of the two proteins that comprise nitrogenase; protein of nitrogenase that has attached iron and molybdenum or vanadium cofactor

dinitrogenase reductase One of the two proteins that comprise nitrogenase; protein of nitrogenase that has only attached iron cofactor

dinoflagellates Algae of the class Pyrrhophyta, primarily unicellular marine organisms, possessing flagella

diphtheria An acute, communicable human disease caused by strains of *Corynebacterium diphtheriae*

diphtheria toxin Cytotoxin produced by *Corynebacterium diphtheriae* that inhibits protein synthesis in mammalian cells by blocking transferase reactions during translation

diplococci Cocci occurring in pairs

diploid Having double the haploid number of chromosomes; having a duplication of genes

dipole moment The polarity resulting from the separation of electric charges; in chemical bonds resulting from an unequal distribution of electrons

direct counting procedures Methods for the enumeration of bacteria and other microorganisms that do not require the growth of cells in culture but rely on direct observation or other detection methods by which the undivided microbial cells can be counted

direct fluorescent antibody staining (FAB) Method used to detect the presence of an antigen by staining with a specific antibody linked with a fluorescent dye; the conjugated fluorescent antibody reacts directly with the antigens

disaccharides Carbohydrates formed by the condensation of two monosaccharide sugars

discharge Emission

discontinuous strand of DNA The strand of DNA that lags behind the replication of the continuous strand because DNA polymerases can add nucleotides in only one direction; therefore, synthesis of this strand can begin only after some unwinding of the double strand has occurred and takes place via the synthesis of short segments that run in the opposite direction to the overall direction of synthesis; also referred to as the *lagging strand of DNA*

disease A condition of an organ, part, structure, or system of the body in which there is incorrect functioning due to the effect of heredity, infection, diet, or environment; a physiologically impaired state of a plant or animal resulting from microbial infection, microbial products, or microbial activities; a physiological condition that occurs when microorganisms overcome host defense systems

disease syndrome Stages in the course of a disease

disinfectants Chemical agents used for disinfection

disinfection The destruction, inactivation, or removal of microorganisms likely to cause infection or produce other undesirable effects

dispersal Breaking up and spreading in various directions, e.g., the spread of microorganisms from one place to another

dissemination The scattering or dispersion of microorganisms or disease, e.g., the spread of disease associated with the dispersal of pathogens

dissociation Separation of a molecule into two or more stable fragments; a change in colony form often occurring in a new environment, associated with modified growth or virulence

distilled liquor Alcoholic beverage produced by microbial alcoholic fermentation followed by chemical distillation to achieve a high alcohol concentration

DNA Deoxyribonucleic acid

DNA double helix The two primary polynucleotide chains of DNA held together by hydrogen bonding between complementary nucleotide bases

DNA gyrase An enzyme that introduces negative supercoiling into relaxed DNA; topoisomerase II; enzyme that breaks the phosphodiester linkage of one of the strands of DNA, passing that strand through the other, which results in a localized uncoiling effect

DNA helicases Unwinding proteins that catalyze the breaking of the hydrogen bonding that hold the two strands of DNA together

DNA homology The degree of similarity of base sequences in DNA from different organisms

DNA ligase Enzyme that establishes a phosphodiester bond between the 3′-OH and 5′-P ends of chains of nucleotides—functions naturally as a repair enzyme and is used in genetic engineering to join chains of nucleotides

DNA methylases Enzymes that add methyl groups to some nucleotides of DNA after the nucleotides have been incorporated by DNA polymerases

DNA polymerases Enzymes that catalyze the phosphodiester bonds in the formation of DNA

DnaG protein RNA polymerase that makes an RNA primer of about 3-5 bases long; primase

dolipore septae The thick internal transverse openings between cell walls of basidiomycetes

domestic sewage Household liquid wastes

dominant allele The allelic form of a gene whose information is preferentially expressed

donor Any cell that contributes genetic information to another cell

donor strain Bacterial cell line that donates DNA during mating

dormant An organism or spore that exhibits minimal physical and chemical change over an extended period of time but remains alive

double bag method Disposal method for materials potentially contaminated with infectious microorganisms in which a bag containing the contaminated materials is sealed within a second bag for safety

double diffusion method Precipitin reaction technique in which an antigen and an antibody diffuse toward each other from separate wells cut into an agar gel

double helix Two primary polynucleotide strands of DNA held together by hydrogen bonds and twisted like a spiral staircase

double-stranded DNA virus Virus with a genome consisting of double-stranded DNA

double-stranded RNA virus Virus with a genome consisting of double-stranded RNA

doubling time Generation time

DPT vaccine A single vaccine used to provide active immunity against diphtheria, tetanus, and pertussis; contains diphtheria and tetanus toxoids and killed *Bordetella pertussis* cells

drugs Substances used in medicine for the treatment of disease

Durham tubes Small inverted test tubes used to detect gas production during fermentation

dust cells Macrophage cells fixed in the alveolar lining of the lungs

dwarfism Plant condition resulting from degradation or inactivation of plant growth substances by pathogens

dysentery An infectious disease marked by inflammation and ulceration of the lower part of the bowels, with diarrhea that becomes mucous and hemorrhagic; disease condition characterized by diarrhea

dysfunctional immunity An immune response that produces an undesirable physiological state, e.g., an allergic reaction, or the lack of an immune response resulting in a failure to protect the body against infectious or toxic agents

E

early proteins Proteins that are made early in viral replication

ECHO virus group Group of viruses frequently found as causative agents of gastroenteritis

eclipse period The period in the lytic reproduction cycle in which complete infective viruses are not present

ecology The study of the interrelationships between organisms and their environments

ecosystem A functional self-supporting system that includes the organisms in a natural community and their environment

ectomycorrhizae A stable, mutually beneficial (symbiotic) association between a fungus and the root of a plant where the fungal hyphae occur outside the root and between the cortical cells of the root

ectopic pregnancy Fertilization and development of an egg that occurs outside the uterus

edema The abnormal accumulation of fluid in body parts or tissues that causes swelling

effluent The liquid discharge from sewage treatment and industrial plants

Eijkman test In assays for assessing water safety, gas formation from dilutions of water samples incubated in lactose broth at 45° C demonstrates the presence of fecal coliforms

Eikenella Rods to coccobacilli; Gram-negative; acid and gas not produced from carbohydrates; facultative anaerobes; G + C 56 mole %

electromagnetic spectrum A range of energy in the form of waves of differing lengths that produces varying electric and magnetic fields as it travels through space from its source to a receiver

electron A negatively charged subatomic particle that orbits the positively charged nucleus of an atom

electron acceptors Substances that accept electrons during oxidation-reduction reactions

electron donors Substances that give up electrons during oxidation-reduction reactions

electron microscope A type of microscope with very high magnification ability that uses an electron beam; focuses by magnetic lenses instead of rays of light; the magnified image forms on a phosphorescent screen or is recorded on a photographic film

electron transport chain A series of oxidation-reduction reactions in which electrons are transported from a substrate through a series of intermediate electron carriers to a final acceptor, establishing an electrochemical gradient across a membrane that results in the formation of ATP

electrophoresis The movement of charged particles suspended in a liquid under the influence of an applied electron field

elementary body Small, rigid-walled, infectious form of chlamydias

elephantiasis Caused by the nematode *Wuchereria bancrofti*, which breeds in the tissues of the circulatory system, damaging the lymphatic vessels, causing swelling and distortion (gross enlargement) of the legs

elevated temperature Higher than normal temperature; characteristic symptom of the inflammatory response associated with the high metabolic activities of neutrophils and macrophages

ELISA Enzyme-linked immunosorbent assay

elongation factor (EF-Tu) Factor involved in placement of charged tRNA molecules into the aminoacyl site that initially forms a complex with GTP, which then binds to charged tRNA to form a ternary complex of aminoacyl-tRNA–EF-Tu–GTP

EMB agar Eosin-methylene blue agar

Embden-Meyerhof pathway A specific glycolytic pathway; a sequence of reactions in which glucose is broken down to pyruvate

Embden-Meyerhof Parnas pathway Embden-Meyerhof pathway

embryonated eggs Hen or duck eggs containing live embryos, used for culturing viruses and preparing tissue cultures

encephalitis An inflammation of the brain

end-   Combining form indicating within

end product   The chemical compound that is the final product in a particular metabolic pathway

end product inhibition   Feedback inhibition

end product repression   The process of shutting off transcription when a by-product of the metabolism coded for by the genes in that transcription region accumulates

endemic   Peculiar to a certain region, e.g., a disease that occurs regularly in an area

endemic typhus   Also known as murine, rat, or flea typhus, this milder form of epidemic typhus is caused by *Ricksettsia typhi* and is transmitted to humans by the bites of infected fleas or lice; produces fever, rash, headache, cough and muscle aches; prevention is through rat control

endergonic   A chemical reaction with a positive ΔG; a chemical reaction requiring input of free energy

endo-   Combining form indicating within

endocarditis   Infection of the endocardium or heart valves caused by bacteria or, in the cases of intravenous drug abusers, fungi

endocardium   The membrane lining the interior of the heart

endocytosis   The movement of materials into a cell by cytosis

endogenous   Produced within; due to internal causes; pertaining to the metabolism of internal reserve materials

endomycorrhizae   Mycorrhizal association in which there is fungal penetration of plant root cells

endonuclease   An enzyme that catalyzes the cleavage of DNA, normally cutting it at specific sites

endoparasitic slime molds   Plasmodiophoromycetes

endophytic   A photosynthetic organism living within another organism

endoplasmic reticulum   The extensive array of internal membranes in a eukaryotic cell involved in coordinating protein synthesis

endospore-producing bacteria   Bacteria that produce endospores; group of bacteria that includes the genera *Clostridium* and *Bacillus*; particularly important bacterial group because of the heat resistance of the endospores that are produced

endospores   Thick-walled spores formed within a parent cell; in bacteria, heat-resistant spores; spores of myxomycetes; small, coccoid reproductive cells of pleurocapsalean cyanobacteria

endosymbionts   Bacterial genera that live within the cells or tissues of other organisms without adversely affecting the other organism

endosymbiotic   A symbiotic (mutually dependent) association in which one organism penetrates and lives within the cells or tissues of another organism

endosymbiotic bacteria   Bacteria that live symbiotically within eukaryotic cells; bacteria that obligately live within protozoa

endosymbiotic evolution   Theory that bacteria living as endosymbionts within eukaryotic cells gradually evolved into organelle structures

endothelial   A single layer of thin cells lining internal body cavities; the inner layer of the seed coat of some plants

endothermic   A chemical reaction in which energy is consumed; a chemical reaction requiring an input of heat energy

endotoxins   Toxic substances found as part of some bacterial cells; the lipopolysaccharide component of the cell wall of Gram-negative bacteria

energy   The capacity to do work

energy charge   Measure of the energy status of a cell, describing its relative proportions of ATP, ADP, and AMP

energy transfer   The movement of energy from one chemical compound to another

enrichment culture   Any form of culture in a liquid medium that results in an increase in a given type of organism while minimizing the growth of any other organism present

enrichment media   Any culture medium that favors the growth of a particular microorganism

enter-   Combining form meaning the intestine

enteric   Of or pertaining to the intestines

enteric bacteria   Bacteria that live within the intestinal tract

Enterobacteriaceae   Family of Gram-negative, facultatively anaerobic rods, motile by means of peritrichous or polar flagella; divided into five tribes

enterobactin   Siderophore synthesized by enteric bacteria

enterocolitis   Infection of the lower gastrointestinal tract, lower small intestine, and colon, characterized by abdominal pain and diarrhea, often with blood in the stools

enterotoxins   Toxins specific for cells of the intestine, causing intestinal inflammation and producing the symptoms of food poisoning

enthalpy   The total heat of a system; ΔH

Entner-Doudoroff pathway   Glycolytic pathway that results in the net production of only one ATP molecule per molecule of glucose substrate metabolized

entomogenous fungi   Fungi living on insects; fungal pathogens of insects

entropy   That portion of the energy of a system that cannot be converted to work; ΔS

enumeration   Determination of the number of microorganisms

envelope   The outer covering that surrounds the capsid of some viruses

enzymatic reactions   Chemical reactions catalyzed by enzymes

enzyme kinetics   The study of the rates at which enzymatic reactions proceed

enzyme-linked immunosorbent assay (ELISA)   A technique used for detecting and quantifying specific serum antibodies based on tagging the antigen-antibody complex with a substrate that can be enzymatically converted to a readily quantifiable product by a specific enzyme

enzymes   Proteins that function as efficient biological catalysts, increasing the rate of a reaction without altering the equilibrium constant by lowering the energy of activation

eosin-methylene blue agar   A medium used for the detection of coliform bacteria; the growth of Gram-positive bacteria is inhibited on this medium, and lactose fermenters produce colonies with a green metallic sheen

eosinophil A white blood cell having an affinity for eosin or any acid stain

eosinophilia An increase above normal in the number of eosinophils in the peripheral blood

epi- Prefix meaning upon, beside, among, above, or outside

epidemic An outbreak of infectious disease among a human population in which, for a limited time, a high proportion of the population exhibits overt disease symptoms

epidemic hemorrhagic conjunctivitis An infectious disease of the eye caused by enteroviruses and characterized by subconjunctival hemorrhages

epidemic keratoconjunctivitis Acute, self-limiting adenoviral infection of the eyes characterized by redness, edema, swelling, and discomfort

epidemic typhus Also known as European, classic, or louse-borne typhus, caused by *Ricksettsia prowazekii* transmitted by the bite of infected body lice, diagnosed by the Weil-Felix reaction, and treated with antibiotics

epidemiology Study of the factors and mechanisms that govern the spread of disease within a population, including the interrelationships between a given pathogenic organism, the environment, and populations of relevant hosts

epidermis The outer layers of the skin

epifluorescence microscopy A form of microscopy employing stains that fluoresce when excited by light of a given wavelength, emitting light of a different wavelength; exciter filters are used to produce the proper excitation wavelength; barrier filters are used so that only the fluorescing specimens are visible

epigenetic Direct products derived from an organism's genome, e.g., ribosomal RNA

epilimnion The warm upper surface layer of an aquatic environment

epinephrine Adrenaline

epiphytes Organisms growing on the surface of another organism, e.g., bacteria growing on the surface of an algal cell

episomes Segments of DNA capable of existing in two alternate forms: one replicating autonomously in the cytoplasm, the other replicating as part of the bacterial chromosome

epitheca The larger of the two parts of the cell wall (frustule) of a diatom

epizootic An epidemic outbreak of infectious disease among animals other than humans

Epstein-Barr virus A member of the herpesvirus group; the causative agent of infectious mononucleosis

equilibrium A state of balance, a condition in which opposing forces equalize with one another so that no movement occurs; in a chemical reaction, the condition where forward and reverse reactions take place at equal rates so that no net change occurs; when a reaction is at equilibrium, the amounts of reactants and products remain constant

equilibrium constant The relationship among concentrations of the substances within an equilibrium system regardless of how the equilibrium condition is achieved

ergotism A condition of intoxication that results from the ingestion of grain contaminated by ergot alkaloids produced by the fungus *Claviceps purpurea*

Erwinia Cells predominantly single, straight rods; motile by peritrichous flagella; Gram-negative; fermentative; facultative anaerobes; pathogenic for plants; optimal growth at 27° to 30° C; G + C 50-58 mole %

erythema Abnormal reddening of the skin due to local congestion, symptomatic of inflammation

erythrocytes Red blood cells

erythromycin An antibiotic produced by a strain of *Streptomyces* that inhibits protein synthesis

Escherichia Straight rods; Gram-negative; motile by peritrichous flagella or nonmotile; citrate cannot be the sole carbon source; glucose and other carbohydrates are fermented, with the production of lactic, acetic, and formic acids; the formic acid is split into equal amounts of CO_2 and H_2; lactose is fermented by most strains; facultative anaerobes; indole positive; no H_2S produced from TSI agar; Methyl Red positive; Voges-Proskauer negative; optimal growth at 35° to 40° C; G + C 50-51 mole %

estuary A water passage where the ocean tide meets a river current; an arm of the sea at the lower end of a river

ethanolic fermentation A type of fermentation in which glucose is converted to ethanol and carbon dioxide

etiological agent An agent, such as a microorganism, that causes a disease

etiology The study of the causation of disease

Euascomycetidae True ascomycetes; fungi that produce asci in ascocarps that develop from dikaryotic hyphae

eubacteria Prokaryotes other than archaebacteria; prokaryotes whose plasma membranes contain phospholipids linked by ester bonds

eugenotes Theoretical primitive versions of prokaryotes

euglenoid algae Members of the Euglenophycophyta

euglenoids Members of the algal division Euglenophycophyta; unicellular organisms surrounded by a pellicle

Euglenophycophyta Division of unicellular algae that contain chlorophylls *a* and *b*, appear green, lack a cell wall, and are surrounded by a pellicle; they store paramylon as reserve material and reproduce by longitudinal division; they are widely distributed in aquatic and soil habitats

eukaryotes Cellular organisms having a membrane-bound nucleus within which the genome of the cell is stored as chromosomes composed of DNA; eukaryotic organisms include algae, fungi, protozoa, plants, and animals

eukaryotic cell A cell with a true nucleus

euphotic The top layer of water, through which sufficient light penetrates to support the growth of photosynthetic organisms

eurythermal Microorganisms that grow over a wide range of temperatures

eutrophication The enrichment of natural waters with inorganic materials, especially nitrogen and phosphorus compounds, that support the excessive growth of photosynthetic organisms

evolution The directional process of change of organisms by which descendants become distinct in form and/or function from their ancestors

evolutionary relationships The degree of relatedness of organisms based on their ancestry

excision repair A mechanism in bacteria that corrects damaged DNA by removing nucleotides and then resynthesizing the region

excystation Conversion of cysts to actively growing vegetative forms

exergonic A reaction accompanied by a liberation of free energy

exo- Prefix indicating outside, an outside layer, or out of

exocytosis Movement of materials out of a cell

exoenzymes Enzymes that occur attached to the outer surface of the cell membrane or in the periplasmic space; enzymes released into the medium surrounding a cell, including enzymes that attack extracellular polymers by sequentially removing units from one end of a polymer chain

exogenous Due to an external cause; not arising from within the organism

exon The region of a eukaryotic genome that encodes the information for protein or RNA macromolecules or regulates gene expression; a segment of eukaryotic DNA that codes for a region of RNA that is not excised during post-transcriptional processing

exonucleases Enzymes that progressively remove the terminal nucleotides of a polynucleotide chain

exothermic A chemical reaction that produces heat

exotoxins Protein toxins secreted by living microorganisms into the surrounding environment

experimental group The condition in a controlled experiment in which a factor or factors vary

exponential growth phase The period during the growth cycle of a microbial population when growth is maximal and constant and there is a logarithmic increase in population size

expression vector In gene cloning, a genetic vector that contains the desired gene and also the necessary regulatory sequences that permit control of the expression of that gene

extracellular External to the cells of an organism

extraterrestrial Originating or existing outside of the Earth or its atmosphere

extreme environments Environments characterized by extremes in growth conditions, including temperature, salinity, pH, and water availability, among others

extreme thermophiles Thermophilic archaebacteria that have optimal temperatures above 80° C

exudate Viscous fluid containing blood cells and debris that accumulates at an inflammation site

F

F pilus Attachment structure that projects from cells of certain bacteria involved in mating, found on cells that donate DNA

F plasmid Fertility plasmid coding for the donor strain that includes genes for the formation of the F pilus

F⁻ strain Bacterial strain that lacks the fertility factor; it acts as the receptor strain in conjugation

F⁺ strain Bacterial strain that has the fertility factor as a plasmid in the cytoplasm

F value The number of minutes required to heat inactivate or kill an entire population of cells or spores in an aqueous solution at 121° C

FAB (fluorescent antibody staining) Direct fluorescent antibody staining

Fab (antigen-binding fragment) Either of two identical fragments produced when an immunoglobulin is cleaved by papain; the antigen-binding portion of an antibody, including the hypervariable region

facilitated diffusion Diffusion at an enhanced rate; movement from a region of high concentration to one of low concentration that occurs more rapidly than it would on the basis of the concentration gradient

facilitator protein A plasma membrane-bound protein that carries out facilitated diffusion of substrates

facultative anaerobes Microorganisms capable of growth under aerobic or anaerobic conditions; bacteria capable of fermentative and respiratory metabolism

FAD Flavin adenine dinucleotide

FADH$_2$ Reduced flavin adenine dinucleotide

false feet Pseudopodia of some protozoa

family A taxonomic group; the principal division of an order; the classification group above a genus

fastidious An organism difficult to isolate or culture on ordinary media because of its need for special nutritional factors; an organism with stringent physiological requirements for growth and survival

fatty acids Straight chains of carbon atoms with a COOH at one end in which most of the carbons are attached to hydrogen atoms

Fc (crystallizable) fragment Remainder of the molecule when an immunoglobulin is cleaved and the Fab fragment separated; crystallizable portion of an immunoglobulin molecule containing the constant region; end of an immunoglobulin that binds with complement

feedback activation The binding of a substance to an allosteric site, thus activating the enzyme and increasing its activity

feedback inhibition A cellular control mechanism by which the end product of a series of metabolic reactions inhibits the activity of an earlier enzyme in the sequence of metabolic transformations; thus, when the end product accumulates, its further production ceases

FeMoco Cofactor for nitrogenase that contains iron and molybdenum

ferment To cause fermentation in; that which causes fermentation

fermentation A mode of energy-yielding metabolism that involves a sequence of oxidation-reduction reactions in which an organic substrate and the organic compounds derived from that substrate serve as the primary electron donor and the terminal electron acceptor, respectively; in contrast to respiration, there is no requirement for an external electron acceptor to terminate the metabolic sequence

fermentation pathways Metabolic sequences for the oxidation of organic compounds to release free energy to drive the formation of ATP in which the organic substrate acts as electron donor and a product of that substrate acts as an electron acceptor

fermented food   Food product of microbial fermentation

fermenter   An organism that carries out fermentation

fermentor   A reaction chamber in which a fermentation reaction is carried out; a reaction chamber for growing microorganisms used in industry for a batch process

fertility   Fruitfulness; the reproductive rate of a population; the ability to support life; the ability to reproduce

fertility pilus   Structure involved in cell-to-cell contact between mating bacteria; F pilus

fertility plasmid   F plasmid

fetus   Embryo after the third month of gestation in the womb

fever   The elevation of body temperature above normal

fever blisters   Cold sores

fibrin   The insoluble protein formed from fibrinogen by the proteolytic action of thrombin during normal blood clotting

fibrinogch   A protein in human plasma synthesized in the liver; it is the precursor of fibrin, which is used to increase the coagulability of blood

fibrolysin   A proteolytic enzyme capable of dissolving or preventing the formation of a fibrin clot

filament   Any elongated, threadlike bacterial cell

filamentous fungi   Fungi that develop hyphae and mycelia; also called *molds*

filterable virus   An obsolete term used to describe infectious agents that are able to pass through bacteriological filters

filtration   Separation of microorganisms from the medium in which they are suspended by passage of a fluid through a filter with pores small enough to trap the microorganisms

fire algae   Members of the Pyrrophycophyta algae

fission   A type of asexual reproduction in which a cell divides to form two or more daughter cells

fixation of carbon dioxide   The conversion of inorganic carbon dioxide to organic compounds

flagella   Flexible, relatively long appendages on cells, used for locomotion

flagella antigens   H antigens

flagellates   Organisms having flagella; one of the major divisions of protozoans, which are characterized by the presence of flagella

flagellin   Soluble, globular proteins constituting the subunits of bacterial flagella

flashlight fish   Fish of the genus *Anomalops* that have organs in which they maintain populations of luminescent bacteria as a source of light

flat-field objective   A microscope lens that provides an image in which all parts of the field are simultaneously in focus; an objective lens with minimal curvature of field

flat-sour spoilage   A type of microbially caused spoilage that occurs in canned foods in which acid but no gas is produced

flavin adenine dinucleotide   A coenzyme that is involved in transfers of electrons during oxidation-reduction reactions of the Krebs cycle and oxidative phosphorylation

floc   A mass of microorganisms caught together in a slime produced by certain bacteria, usually found in waste treatment plants

Floridean starch   Primary reserve material of Rhodophycophyta

fluid mosaic model   Currently accepted model of the structure of the plasma membrane that describes this membrane as a bilipid layer of proteins distributed in a mosaic-like pattern on the surface and in the interior of the membrane, with lateral as well as transverse movement of proteins occurring through the structure

flukes   Flatworms belonging to the class Trematoda

fluorescence   The emission of light by certain substances on absorption of an exciting radiation; the wavelength of radiation the emitted light is different from that of the excitation radiation

fluorescence microscope   A microscope in which the microorganisms are stained with a fluorescent dye and observed by illumination with short-wavelength light, e.g., ultraviolet light

fluorescent antibodies   Antibodies capable of giving off light of one color when exposed to light of another color

fluorescent antibody staining (FAB)   Direct fluorescent antibody staining

folliculitis   Bacterial infection of hair follicles that causes the formation of a pustule

fomes   Inanimate objects that can act as carriers of infectious agents

fomites   Objects and materials that have been associated with infected persons or animals and that potentially harbor pathogenic microorganisms

food additive   A substance or mixture of substances other than the basic foodstuff that is intentionally present in food as a result of an aspect of production, processing, storage, or packaging

food infection   Disease resulting from the ingestion of food or water containing viable pathogens that can establish an infectious disease, e.g., gastroenteritis from ingestion of food containing *Salmonella*

food intoxication   Disease from the ingestion of toxins produced by microorganisms that have grown in a food

food poisoning   General term for stomach or intestinal disorders due to food contaminated with certain microorganisms, their toxins, chemicals, or poisonous plant materials; disease resulting from ingestion of toxins produced by microorganisms that have grown in a food

food preservation   The prevention or delay of microbial decomposition or self-decomposition of food, and prevention of damage due to insects, animals, mechanical causes, etc.; the delay or prevention of food spoilage

food spoilage   The deterioration of a food that lessens its nutritional value or desirability, often due to the growth of microorganisms that alter the taste, smell, or appearance of the food or the safety of ingesting it

food web   An interrelationship among organisms in which energy is transferred from one organism to another; each organism consumes the preceding one and in turn is eaten by the following member in the sequence

Foraminiferans   Marine members of the protozoan class Sarcodina that form one or more chambers composed of siliceous or calcareous tests

foreign populations   Allochthonous populations

formalin   A 40% solution of formaldehyde, a pungent-smelling, colorless gas used for fixation and preservation of biological specimens and as a disinfectant

Forssman antigen   A heat-stable glycolipid; a heterophile antigen, an immunologically related antigen found in unrelated species

frameshift mutation   A type of mutation that causes a change in the three base sequences read as codons, i.e., a change in the phase of transcription arising from the addition or deletion of nucleotides in numbers other than three or multiples of three

Francisella   Very small coccoid to ellipsoidal pleomorphic rods; nonmobile; Gram-negative; strictly aerobic; optimal growth at 37° C

free energy   The energy available to do work, particularly in causing chemical reactions; ΔG

freeze etching   A technique used to examine the topography of a surface that is exposed by fracturing or cutting a deep-frozen cell, making a replica, and removing the biological material; used in transmission electron microscopy

freeze-drying   Lyophilization

freezing   Conversion of a liquid to a solid by reducing the temperature; a method used for the preservation of food by storage at −20° C, based on the fact that low temperatures restrict the rate of growth and enzymatic activities of microorganisms

freshwater habitats   Lakes, ponds, swamps, springs, streams, and rivers

fruiting body   A specialized microbial structure that bears sexually or asexually derived spores

frustules   The siliceous cell walls of a diatom

fungal gardens   Fungi grown in pure culture by insects

fungi   A group of diverse, widespread unicellular and multicellular eukaryotic organisms, lacking chlorophyll and usually bearing spores and often filaments

Fungi Imperfecti   Fungi with septate hyphae that reproduce only by means of conidia, lacking a known sexual stage; Deuteromycetes

fungicides   Agents that kill fungi

fungistasis   The active prevention or hindrance of fungal growth by a chemical or physical agent

fungus   Fungi

furuncles   Boils; painful, inflammatory sores around a central core resulting from folliculitis

G

γ-globulin   A specific class of serum proteins with antibody activity; serum fraction of blood containing antibodies; also called immunoglobulin G (IgG) or gamma-globulin

γ-hemolysis   Test result when no zone of clearing arises from the inoculation of a microorganism on blood agar; no hemolysis has occurred

galls   Abnormal plant structures formed in response to parasitic attack by certain insects or microorganisms; tumorlike growths on plants in response to an infection

gametangium   A structure that gives rise to gametes or that in its entirety functions as a gamete

gametes   Haploid reproductive cells or nuclei, the fusion of which during fertilization leads to the formation of a zygote

gamma globulin   Immunoglobulin G

gamma-hemolysis   γ-hemolysis

gamma interferon (IF-γ)   Immune interferon

gamma radiation   Short wavelength (10^{-8} to 10^{-1}) electromagnetic radiation that has high penetration power and can kill microorganisms by inducing or forming toxic free radicals

gap   Region of the double helix in which there are no complementary nucleotides opposite one of the strands

gas gangrene   A disease condition involving tissue death developing when certain species of toxin-producing bacteria grow in anaerobic wounds or necrotic tissues

gas vacuoles   Membrane-limited, gas-filled vacuoles that occur commonly in groups in the cells of several cyanobacteria and certain other bacteria

gasohol   A mixture of gasoline and ethanol used as a fuel

gastroenteritis   An inflammation of the stomach and intestine; a syndrome characterized by gastrointestinal symptoms, including nausea, vomiting, diarrhea, and abdominal discomfort

gastroenterocolitis   An inflammation of the gastrointestinal tract accompanied by the formation of pus and blood in the stools

gastrointestinal syndrome   Gastroenteritis associated with nausea, vomiting, and/or diarrhea

gastrointestinal tract   The stomach, intestines, and accessory organs

Gastromycetes   Basidiomycete group that includes the puffballs, earthstars, stinkhorns, and bird's nest fungi

gelatin   A protein that is obtained from skin, hair, bones, tendons, etc.; used in culture media for the determination of a specific proteolytic activity of microorganisms

gelatinase   A hydrolytic enzyme capable of liquefying gelatin

gene   A sequence of nucleotides that specifies a particular polypeptide chain or RNA sequence, or that regulates the expression of other genes

gene cloning   Replication of foreign DNA inserted by recombinant DNA technology

gene pool   Set of genes of an organism

gene probe   A small molecule of single-stranded RNA or DNA with a known sequence of nucleotides that is used to detect or identify a homologous complementary nucleotide sequence

generalized transduction   A form of recombination in which a phage carries bacterial DNA from a donor to a recipient cell, resulting in the exchange of any homologous genes

generation time   The time required for the cell population or biomass to double

generic drug name   Commonly used name of drug not protected by trademark registration

genetic code   Code for specific amino acids formed by three sequential nucleotides in mRNA; the 64 codons formed by sequences of three nucleotides that specify the genetic information of all organisms

genetic information of all organisms

genetic engineering   The deliberate modification of the genetic properties of an organism through the selection of desirable traits, the introduction of new information on DNA, or both; the application of recombinant DNA technology

genetic mapping   Determination of the relative positions of genes in DNA or RNA

genetics   The science dealing with inheritance

genital herpes   A sexually transmitted disease caused by a herpesvirus; an infection by herpes simplex virus marked by the eruption of groups of vesicles, often in the genital region

genital warts   Disease characterized by benign tumor development caused by human papilloma virus transmitted by sexual contact; warts most commonly develop on the moist areas of the genitalia

genitourinary tract   The combined urinary and genital systems; the combined reproductive system and urine excretion system, including the kidneys, ureters, urinary bladder, urethra, penis, prostate, testes, vagina, fallopian tubes, and uterus

genome   The complete set of genetic information contained in a haploid set of chromosomes

genomic library   A collection of clones of individual genes from a specific organism

genotype   The genetic information contained in the entire complement of alleles

genus   A taxonomic group directly above the species level, forming the principal subdivisions of the family

geometric isomers   The formation of nonequivalent structures based on the particular positions at which ligands are attached to a central atom

germ   Any microorganism, especially any of the pathogenic bacteria

germ theory of disease   Theory that infectious and contagious diseases are caused and transmitted by the activity of microorganisms

German measles   Rubella, an acute systemic infectious disease of humans caused by rubella viruses invading via the mouth or nose, characterized by a rash

germ-free animal   An animal with no normal microbiota; all of its surfaces and tissues are sterile, and it is maintained in that condition by being housed and fed in a sterile environment

germicide   A microbicidal disinfectant; a chemical that kills microorganisms

germination   A degradative process in which an activated spore becomes metabolically active, involving hydrolysis and depolymerization

giardiasis   Disease caused by the protozoan *Giardia* when it infects the human intestine

gibberellic acid   Organic acid used as a plant growth hormone, formed by the fungus *Gibberella fujikuroi* in aerated submerged culture

gingivitis   Inflammation of the gums

gliding bacteria   Bacteria that exhibit gliding motility

gliding motility   Movement that occurs when some bacteria are in contact with solid surfaces

global cycling   Biogeochemical changes in the chemical forms of various elements that lead to the physical translocations of materials, sometimes mediating transfers between the atmosphere (air), hydrosphere (water), and lithosphere (land); biogeochemical cycling that moves materials throughout global ecosystems

globular protein   The general name for a group of water-soluble proteins

glomerulonephritis   An inflammation of the filtration region of the kidneys

gluconeogenesis   The biosynthesis of glucose from noncarbohydrate substrates

gluconic acid   Organic acid produced by a submerged culture process from mycelia of *Aspergillus niger*

glucose   The monosaccharide sugar $C_6H_{12}O_6$

glutamine synthetase/glutamate synthase pathway   Pathway for the formation of L-glutamate used when ammonium concentrations are low

glycocalyx   Specialized bacterial structure with an attachment function composed of a mass of tangled fibers of polysaccharides or branching sugar molecules surrounding a cell or colony of cells

glycogen   A nonreducing polysaccharide of glucose found in many tissues and stored in the liver, where it is converted, when needed, into sugar

glycolysis   An anaerobic process of glucose dissemination by a sequence of enzyme-catalyzed reactions to a pyruvic acid

glycolytic pathways   The catabolic pathways of sugar metabolism

glycoproteins   A group of conjugated proteins that, on decomposition, yield a protein and a carbohydrate

glycosidic bonds   Bonds in disaccharides and polysaccharides formed by the elimination of water

glyoxylate cycle   A metabolic shunt within the Krebs cycle involving the intermediate glyoxylate

golden algae   Members of the Chrysophyceae

Golgi apparatus   A membranous organelle of eukaryotic organisms involved in the formation of secretory vesicles and the synthesis of complex polysaccharides

gonidia   Reproductive cells of unicellular green algae *Volvox* that lack flagella

gonococcal urethritis   An inflammation of the urethra caused by a gonococcal infection

gonorrhea   A sexually transmitted disease caused by *Neisseria gonorrhoeae*; infectious inflammation of the mucous membrane of the urethra and adjacent cavities caused by *N. gonorrhoeae*

graft-versus-host (GVH) disease   Disease that occurs when transplanted or grafted tissue contains immunocompetent cells that respond to the antigens of the recipient's tissues

Gram-negative cell wall   Bacterial cell wall composed of a thin peptidoglycan layer, lipoproteins, lipopolysaccharides, phospholipids, and proteins

Gram-positive cell wall   Bacterial cell wall composed of a relatively thick peptidoglycan layer and teichoic acids

Gram stain   Differential staining procedure in which bacteria are classified as Gram-negative or Gram-positive, depending on whether they retain or lose the primary stain when subject to treatment with a decolorizing agent; the staining procedure reflects the underlying structural differences in the cell walls of Gram-negative

grana A membranous unit formed by stacks of thylakoids

granules Small intracellular particles that usually stain selectively

granuloma An inflammatory growth composed of granulation tissue (normal and scar tissue)

granuloma inguinale The chronic destructive ulceration of external genitalia due to *Donovania granulomatis*

granulosis virus Viral pesticide that develops in the nucleus or the cytoplasm of host fat, tracheal, or epidermal cells

grazers Organisms that prey on primary producers; protozoan predators that consume bacteria indiscriminately; filter-feeding zooplanktons

green algae Members of the Chlorophycophyta algae

green beer The product of an alcoholic fermentation of grain that has not been aged or distilled

green sulfur bacteria Members of the Chlorobiaceae anoxygenic phototrophic bacteria that utilize reduced sulfur compounds as electron donors

greenhouse effect Rise in the concentrations of atmospheric CO_2 and a resulting warming of global temperatures

gross primary production Total amount of organic matter produced in an ecosystem

groundwater All subsurface water

group translocation The transfer of materials across the plasma membrane of a bacterial cell that results in chemical modification of the substance as it moves across the membrane

growth Any increase in the amount of actively metabolic protoplasm accompanied by an increase in cell number, cell size, or both

growth curve A curve obtained by plotting the increase in the size or number of microorganisms against the elapsed time

growth factors Any compound, other than the carbon and energy source, that an organism requires and cannot synthesize

growth rate Increase in the number of microorganisms per unit time

growth temperature range Established by the maximum and minimum temperatures at which a microorganism can grow

guanine A purine base that occurs naturally as a fundamental component of nucleic acids

Guillain-Barré syndrome Acute febrile polyneuritis; a diffuse neuron paresis that results from infection with a prion

Gymnomycota Slime molds; a group of protozoa that have several characteristics of fungi; their vegetative cells lack cell walls and exhibit a phagotrophic mode of nutrition

H

H⁺ hydrogen ion or proton; a hydrogen ion concentration described by the pH that is the negative logarithm of the hydrogen ion concentration

H antigen A flagellar antigen found in certain bacteria

H peplomers Hemagglutinin peplomers

habitat A location where living organisms occur

Haemophilus Coccobacillary; nonmotile; Gram-negative; strict parasites, requiring growth factors present in blood; aerobic to facultatively anaerobic; optimal growth at 37° C; G + C 38-42 mole %

hairpin loops Regions of DNA or RNA, part of which are single-stranded regions and part of which are double-stranded so that there is a loop with three-dimensional topology

halophiles Organisms requiring NaCl for growth; extreme halophiles grow in concentrated brines

haploid A single set of homologous chromosomes; having half of the normal diploid number of chromosomes

hapten A substance that elicits antibody formation only when combined with other molecules or particles but that can react with preformed antibodies

heat-labile A form that is likely to be changed or destroyed by exposure to heat

heat-resistant A form that is not likely to be changed or destroyed by exposure to heat

heat shock proteins Proteins that are synthesized at high temperature that are not otherwise expressed

heat shock response A rapid change in gene expression that occurs when there is a temperature shift to an elevated temperature

heavy chain class switching In lymphocyte maturation, the process in which the same variable region of antibodies appears in association with different heavy chain constant regions

Helicobacter Helical or curved microaerophilic Gram-negative bacteria; rapid motility; capnophilic; catalase and oxidase positive; urea hydrolyzed

Heliozoida Protozoa that produce numerous radiating axopodia, found in fresh water

helix A spiral structure

helminth A parasitic roundworm or flatworm

helper T cells T helper cells

hemagglutination The agglutination or clumping of red blood cells

hemagglutination inhibition (HI) The inhibition of hemagglutination (antibody-mediated clumping of red blood cells), usually by means of specific immunoglobulins or enzymes, used to determine whether a patient has been exposed to a specific virus

hemagglutinin peplomers (H peplomers) Peplomer spikes of an influenza virus that cause agglutination of red blood cells

hemagglutinin spikes Projections from surfaces of influenza viruses that cause agglutination of red blood cells; they increase the ability of influenza viruses to attach to human cells

heme An iron-containing porphyrin ring occurring in hemoglobin

hemocytometer A counting chamber used for estimating the number of blood cells

hemoglobin The iron-containing, oxygen-carrying molecule of red blood cells, containing four polypeptides in the heme group

hemolysin A substance that lyses erythrocytes (red blood cells)

hemolysis The lytic destruction of red blood cells and the resultant escape of hemoglobin

hemolytic disease of the newborn Disease that stems from an incompatibility of fetal (Rh-positive) and maternal (Rh-negative) blood, resulting in maternal antibody activity against fetal blood cells; also known as *erythroblastosis fetalis*

hemorrhagic Showing evidence of bleeding; the tissue becomes reddened by the accumulation of blood that has escaped from capillaries into the tissue

hepatitis Inflammation of the liver

hepatitis A Hepatitis caused by type A hepatitis virus, usually transmitted by fecal-oral route

hepatitis B Hepatitis caused by type B hepatitis virus usually transmitted via blood

hepatitis nonA–nonB Hepatitis caused by hepatitis virus lacking both A and B antigens; commonly transmitted by contaminated blood transfusions

herbicides Chemicals used to kill weeds

herd immunity The concept that an entire population is protected against a particular pathogen when 70% of the population is immune to that pathogen

heritable Any characteristic that is genetically transmissible

herpes encephalitis Form of encephalitis (inflammation of the brain) caused by herpes simplex virus; treatable with acyclovir

herpes simplex infections Localized blistery skin rash caused by herpes simplex virus, usually on the lip or the genitalia

hetero- Combining form meaning other, other than usual, different

heterocysts Cells that occur in the trichomes of some filamentous cyanobacteria that are the sites of nitrogen fixation

heteroduplex An intermediate form of DNA occurring during homologous recombination

heterogamy The conjugation of unlike gametes

heterogeneous Composed of different substances; not homologous

heterogeneous nuclear RNA (hnRNA) Heterogeneous RNA

heterogeneous RNA (hnRNA) High molecular weight RNA formed by direct transcription in eukaryotes that is then processed enzymatically to form mRNA

heterokaryotic Containing genetically different nuclei, as in some fungal hyphal cells

heterolactic fermentation Fermentation of glucose that produces lactic acid, acetic acid, and/or ethanol, and carbon dioxide, carried out by *Leuconostoc* and some *Lactobacillus* species

heterologous antigen Multivalent antigen; Forssman antigen

heterophile antibodies Antibodies that react with heterophile antigens; commonly found in sera of individuals with infectious mononucleosis

heterophile antigens Immunologically related antigens found in unrelated species; multivalent antigen; Forssman antigen

heterotrophic Incapable of utilizing carbon dioxide as sole carbon source; requiring one or more organic compounds for nutrition

heterotrophic metabolism A type of metabolism in which an organic molecule serves as carbon source; a type of metabolism in which an organic molecule serves as carbon source and energy source—often used to describe chemoheterotrophic metabolism

heterotrophs Organisms requiring organic compounds for growth and reproduction, the organic compounds serve as sources of carbon and energy

heterozygous A microorganism whose allelic forms of a gene differ

Hfr High frequency recombinant

high copy number In gene cloning, a large number of repetitive copies of a gene that are produced

high frequency recombinant A bacterial strain that exhibits a high rate of gene transfer and recombination during mating; the F plasmid is integrated into the bacterial chromosome

high temperature-short time process HTST process

histamine A physiologically active amine that has a role in the inflammatory response

histiocytes Macrophages that are located at a fixed site in a certain organ or tissue

histocompatibility antigens Genetically determined isoantigens on the lipoprotein membranes of nucleated cells of most tissues that cause an immune response when grafted onto a genetically disparate individual, and thus determine the compatibility of tissues in transplantation

histones Basic proteins rich in arginine and lysine that occur in close association with the nuclear DNA of most eukaryotic organisms

histoplasmin test Skin test designed to detect antibodies against *Histoplasma*

histoplasmosis A disease of humans and animals caused by the fungus *Histoplasma capsulatum*, characterized by fever, anemia, leukopenia, and emaciation, primarily involving the reticuloendothelial system

HIV Human immunodeficiency virus

hnRNA Heterogeneous nuclear RNA

holdfast A structure that allows certain algae and bacteria to remain attached to the substratum

holozoic nutrition Type of nutrition exhibited by some protozoa that obtain nutrients by phagocytosis of bacterial cells

homo- Combining form denoting like, common, or same

homokaryotic Containing genetically similar nuclei, as in some fungal hyphal cells

homolactic fermentation The fermentation of glucose that produces lactic acid as the sole product, carried out by certain species of *Lactobacillus*

homologous Pertaining to the structural relation between parts of different organisms due to evolutionary development of the same or a corresponding part; a substance of identical form or function

homologous recombination Recombination of regions of DNA containing alleles of the same genes

homologously dissimilar Genetically dissimilar microorganisms

homology Genetic relatedness

homozygous Microorganism whose allelic forms of a gene are identical

hookworms Roundworms that cause infestations of the small intestine

hospital Institution in which sick and injured people receive medical and surgical treatment

host A cell or organism that acts as the habitat for the growth of another organism; the cell or organism on or in which parasitic organisms live

host cell A cell within which a virus replicates

host cell range The types of cells within which replication of a virus occurs

host-range mutation Viral mutation that alters the host that the virus can infect

hot springs Thermal springs with a temperature above 98° C

HTST process High temperature-short time pasteurization process at a temperature of at least 71.5° C for at least 15 seconds; the most widely used form of commercial pasteurization

human immunodeficiency virus The retrovirus that causes AIDS

humic acids Any of the various organic acids obtained from humus, the soil matter whose origin is no longer identifiable; complex polynuclear aromatic compounds comprising the soil organic matter

humoral Referring to the body fluids

humoral immune defense system Antibody-mediated immunity

humus The organic portion of the soil remaining after microbial decomposition

hyaluronidase Enzyme that catalyzes the breakdown of hyaluronic acid

hybridization of nucleic acids Artificial construction of a double-stranded nucleic acid by complementary base pairing of two single-stranded nucleic acids

hybridomas Cells formed by fusion of lymphocytes (antibody precursors) with myeloma (tumor) cells that produce rapidly growing cells that secrete monoclonal antibodies

hydatid cysts Larval stages of the tapeworm *Echinococcus granulosus*

hydr- Combining form meaning water

hydrocarbons Compounds composed only of hydrogen and carbon that are the major compounds in petroleum

hydrogen bond A weak attraction between an atom that has a strong attraction for electrons and a hydrogen atom that is covalently bonded to another atom that attracts the electron of the hydrogen atom

hydrogen cycle The biogeochemical movement of hydrogen, usually in conjunction with carbon and oxygen; the principal transformations involve conversion between water and organic carbon compounds

hydrogenosomes Organelles containing hydrogenase in some protozoa

hydrolase Enzyme that hydrolyzes a molecule by adding water

hydrolysis The chemical process of decomposition involving the splitting of a bond and the addition of the elements of water

hydrolyze Hydrolysis

hydrophilic A substance having an affinity for water

hydrophobia Fear of water, a symptom of rabies

hydrophobic A substance lacking an affinity for water; not soluble in water

hydrosphere The aqueous envelope of the Earth, including bodies of water and aqueous vapor in the atmosphere

hydrostatic pressure Pressure exerted by the weight of a water column; it increases approximately 1 atm with every 10 m depth

3′-hydroxyl end (3′-OH) The end of a nucleic acid macromolecule (DNA or RNA) at which carbon-3 of the carbohydrate is not involved in forming the phosphate diester linkage that bonds the macromolecule; the lack of bonding to carbon-3 makes this end biochemically recognizable and confers directionality on the nucleic acid macromolecule

hydroxypropionate pathway A metabolic pathway in which two CO_2 molecules are fixed and converted into one acetyl-CoA

hyperbaric oxygen Pure oxygen under pressure

hyperchromatic shift The change in absorption of light exhibited by DNA when it is melted, forming two strands from the double helix

hyperplasia The abnormal proliferation of tissue cells, resulting in the formation of a tumor or gall

hypersensitivity An exaggerated immunological response on reexposure to a specific antigen

hypertonic A solution whose osmotic pressure is greater than that of a standard solution

hypertrophy An increase in the size of an organ, independent of natural growth, due to enlargement or multiplication of its constituent cells

hypervariable region A region of immunoglobulins that accounts for the specificity of antigen-antibody reactions; genetically specified terminal regions of the Fab fragments

hyphae Branched or unbranched filaments that constitute the vegetative form of an organism, occurring in filamentous fungi, algae, and bacteria

Hyphochytridiomycetes Class of Mastigomycota; fungi that produce uniflagellate zoospores of the tinsel type

hypolimnion The deeper, colder layer of an aquatic environment; the zone below the thermocline

hypotheca The smaller of the two parts of the cell wall of a diatom

hypothesis In the scientific method, a tentative answer to a question that has been asked

hypotonic A solution whose osmotic pressure is less than that of a standard solution

I

icosahedral virus A virus having cubical symmetry and a complex, 20-sided capsid structure

icosahedron A solid figure that is contained by 20 plane faces

identification The process of determining the closest relationship of an unknown organism to a group that has already been defined

identification key A series of questions that leads to the unambiguous identification of an organism

idiolite A secondary metabolite in the production of penicillin that is not required for the growth of the fungus

idiophase The phase of metabolism in batch culture in which secondary metabolism is dominant over primary growth-directed metabolism; the phase of antibiotic or other secondary product accumulation

idiotypes Immunoglobulin molecules with distinct variable regions determining the specificity of the antigen-antibody reaction

IF-γ Immune interferon

IgA (immunoglobulin A) An antibody that occurs primarily in mucus, semen, and secretions such as saliva, tears, and sweat; an immunoglobulin that has a major role in protecting mucous membrane surface tissues against microbial infection

IgD (immunoglobulin D) An immunoglobulin that is present on the surface of some lymphocytes, along with IgM, and appears to have a regulatory role in lymphocyte activity

IgE (immunoglobulin E) An immunoglobulin that normally is present in blood serum in very low concentrations but that becomes elevated in individuals with allergies; an immunoglobulin that attaches to mast and basophil cells and triggers an allergic response when it reacts with allergens in sensitized individuals

IgG (immunoglobulin G) The largest fraction of the body's immunoglobulins and a major antibody that circulates through the body; an immunoglobulin that has a major role in protecting the body against systemic microbial infections

IgM (immunoglobulin M) A high molecular weight immunoglobulin occurring as a pentamer that is formed prior to IgG in response to exposure to an antigen; an immunoglobulin that is important in the early response to a microbial infection

IL-2 Interleukin-2

immediate hypersensitivity Anaphylactic hypersensitivity (Type I hypersensitivity); a systemic, potentially life-threatening condition that occurs when an antigen reacts with antibody bound to mast or basophil blood cells, leading to disruption of these cells with the release of vasoactive mediators, such as histamine—it occurs shortly (5 to 30 minutes) after exposure to the antigen that triggers this response

immobilization The binding of a substance so that it is no longer reactive or able to circulate freely

immobilized enzyme An enzyme bound to a solid support

immune The condition following initial contact with a given antigen in which antibodies specific for that antigen are present in the body; the innate or acquired resistance to disease

immune adherence Opsonization

immune-complex disease Illness caused by the formation of antibodies against antigen-antibody complexes

immune defense network The integrated defense system of the body that protects against infection; the combined and interactive system that includes the B cell (antibody-mediated) and T cell (cell-mediated) immune responses

immune interferon (IF-γ) A lymphokine secreted by lymphocytes in response to a specific antigen that has antiviral activity and may kill tumor cells

immune response system The integrated mechanisms for responding to the invasion of the body by particular pathogenic microorganisms and other foreign substances; the specific immune response is characterized by specificity, memory, and the acquired ability to detect foreign substances

immunity The relative insusceptibility of a person or animal to active infection by pathogenic microorganisms or the harmful effects of certain toxins; resistance to disease by a living organism

immunization Any procedure in which an antigen is introduced into the body to produce a specific immune response

immunodeficiency The lack of an adequate immune response due to inadequate B- or T-cell recognition and/or response to foreign antigens; a lack of antibody production

immunoelectrophoresis A two-stage procedure used for the analysis of materials containing mixtures of distinguishable proteins, e.g., serum using electrophoretic separation and immunological detection

immunofluorescence Any of a variety of techniques used to detect a specific antigen or antibody by means of homologous antibodies or antigens that have been conjugated with a fluorescent dye

immunogenicity The ability of a substance to elicit an immune response

immunoglobulin A IgA

immunoglobulin D IgD

immunoglobulin E IgE

immunoglobulin G IgG

immunoglobulin M IgM

immunoglobulins (antibodies) A varied class of proteins in plasma and other body fluids, including all of the known antibodies; the antibody fraction of serum; the five classes of antibodies IgA, IgD, IgE, IgG, and IgM

immunological Referring to the immune response

immunological tolerance The unresponsiveness to self-antigens

immunology The study of immunity

immunosuppressant A drug that depresses the immune response

impetigo An acute inflammatory skin disease caused by bacteria (often *Streptococcus* and *Staphylococcus*), characterized by small blisters, weeping fluid, and crusts

IMViC test A group of tests (indole, Methyl Red, Voges-Proskauer, citrate) used in the identification of bacteria of the Enterobacteriaceae family

inclusion bodies Accumulations of reserve materials in bacteria

inclusion conjunctivitis Chlamydial infection affecting mostly newborns, causing an acute ocular inflammation, characterized by reddening of the eyelids and a thick purulent discharge; treatment is with tetracycline, erythromycin, or sulfonamide eyedrops

incompatibility group Incompatible plasmids that do not co-exist in the same cell

incompatible plasmids Pairs of plasmids that cannot be replicated with stability in the same cell

incubation The maintenance of controlled conditions to achieve the optimal growth of microorganisms

incubation period The period of time between the establishment of an infection and the onset of disease symptoms

incubators Controlled temperature chambers

indicator organism   An organism used to indicate a particular condition, commonly applied to coliform bacteria, e.g., *Escherichia coli* or *Streptococcus fecalis*, when their presence is used to indicate the degree of water pollution due to fecal contamination

indigenous   Native to a particular habitat

indigenous populations   A population that is native to a particular habitat; an autochthonous population; normal microbiota or microflora

indirect fluorescent antibody   A fluorescent antibody test to detect the presence of specific antigens associated with a microorganism, in which an immunoglobulin molecule is first reacted with the antigens to form a complex, and then a conjugated fluorescent antibody dye is added that reacts with the first unlabelled immunoglobulin that was used to form the first antigen-antibody complex; the test is indirect because the fluorescent-labelled dye actually reacts with the unlabelled immunoglobulin that was used to react with the antigens and not directly with the antigens

indirect immunofluorescence test   Test used in identifying bacteria such as *Treponema pallidum* by adding dead cells to the patient's serum and adding fluorescent anti-immunoglobulin; if the bacteria stain, the test is positive

induced mutations   Mutations that result from the exposure of the cell to exogenous DNA modifiers such as radiation or chemical substances

inducers   Substances responsible for activating certain genes, resulting in the synthesis of new proteins

inducible enzymes   Enzymes that are synthesized only in response to a particular substance in the environment

induction   An increase in the rate of synthesis of an enzyme; the turning on of enzyme synthesis in response to environmental conditions

inductive reasoning   A logical process in which a conclusion is proposed that contains more information than the observations or experience on which it is based

induration   Hardening of the skin, a positive hypersensitivity reaction

infant botulism   Caused by toxin produced by the infection of an infant's gastrointestinal tract with *Clostridium botulinum*

infantile paralysis   Poliomyelitis

infection   A condition in which pathogenic microorganisms have become established in the tissues of a host organism

infectious disease   A disease-causing agent or disease that can be transmitted from one person, animal, or plant to another

infectious dose   The number of pathogens that are needed to overwhelm host defense mechanisms and establish an infection

infectious hepatitis   Hepatitis A caused by ingestion of hepatitis A viruses in water or food

infectious mononucleosis   Glandular fever, an acute infectious disease that primarily affects the lymphoid tissues; caused by Epstein-Barr virus, which enters the body via the respiratory tract

inflammation   The reaction of tissues to injury characterized by local heat, swelling, redness, and pain

inflammatory exudate   Pussy material from blood vessels deposited in tissues or on tissue surfaces as a defensive response to injury or irritation

inflammatory response   A nonspecific immune response to injury characterized by redness, heat, swelling, and pain in the affected area; inflammation

influenza   An acute, highly communicable disease tending to occur in epidemic form; caused by influenza viruses and characterized by malaise, headache, and fever

infusoria   Archaic term for microorganisms

inhibition   Prevention of growth or multiplication of microorganisms; reduction in the rate of enzymatic activity; repression of chemical or physical activity

inhibitors   Substances that repress or stop a chemical action

initial body   Larger, thin-walled, noninfectious form of chlamydias

initiation   In protein synthesis, the stage at which the translating complex of mRNA, ribosome, and tRNA first assembles

inoculate   To deposit material, an inoculum, onto medium to initiate a culture, carried out with an aseptic technique; to introduce microorganisms into an environment that will support their growth

inoculum   The material containing viable microorganisms used to inoculate a medium

insecticides   Substances destructive to insects; chemicals used to control insect populations

insertion   A type of mutation in which a nucleotide or two or more contiguous nucleotides are added to DNA

insertion mutations   A mutation in which one or more nucleotides are inserted into a gene

insertion sequence (IS)   A transposable genetic element that can move around bacterial chromosomes, occurring at different locations on the chromosome

insertional inactivation   In gene cloning, insertion of foreign DNA at an antibiotic-resistant site, causing loss of resistance because the nucleotide sequence of the antibiotic resistance gene is disrupted

in situ   In the natural location or environment

interference microscope   A microscope that relies on destructive and/or additive interference of light waves to achieve contrast

interferons   Glycoproteins produced by animal cells that act to prevent the replication of a wide range of viruses by inducing resistance

intergenic mutations   Mutations within a single gene that affect other genes

interleukin-2 (IL-2)   A cytokine formed by T lymphocytes that stimulates the growth of T lymphocytes and cytokine production by T cells

intermediary metabolism   Intermediate steps in the cellular synthesis and breakdown of substances

intermediate hosts   Organisms that harbor the larval stage of a helminth

intermediately sensitive (I)   One of the standardized zones of inhibition used to determine the degree to which a microorganism is sensitive to a particular antibiotic

interspecies hydrogen transfer   A series of reactions that results in supplying hydrogen from complex polymers by one or more bacterial populations for the reduction of CO_2 to CH_4 by methanogenic archaebacteria

intoxication Poisoning, as by a drug, serum, alcohol, or any poison

intracellular Within a cell

intradermal Within the skin

intragenic Occurring within a gene

intragenic suppressor mutations Suppressor mutations that occur within a single gene

intramuscular Within the substance of a muscle

intravenous Within or into the vein

intron An intervening region of the DNA of eukaryotes that does not code for a known protein or a regulatory function

invasiveness The ability of a pathogen to spread through a host's tissues

inverted repeats Palindromic sequences in which the sequences of nucleotides in complementary strands of DNA are in exact opposite directions

in vitro In glass; a process or reaction carried out in a culture dish or test tube

in vivo Within the living organism

ion An atom that has lost or gained one or more orbital electrons and is thus capable of conducting electricity

ionic bond A chemical bond resulting from the transfer of electrons between metallic and nonmetallic atoms; positive and negative ions are formed and held together by electrostatic attraction

ionization The process that produces ions

ionizing radiation Radiation, such as gamma and X-radiation, that induces or forms toxic free radicals, which cause chemical reactions disruptive to the biochemical organization of microorganisms

IS Insertion sequence

iso- Combining form meaning for or from different individuals of the same species

isogamete A reproductive cell similar in form and size to the cell with which it unites; found in certain protozoas, fungi, and algae

isogamy Fertilization in which the gametes are similar in appearance and behavior

isolation Any procedure in which a given species of organism present in a sample is obtained in pure culture

isolation methods Aseptic procedures used for the establishment of pure cultures, usually involving the separation of microorganisms on a solid medium into individual cells that are then allowed to reproduce to form clones of single microorganisms

isomer One of two or more compounds having the same chemical composition but differing in the relative positions of the atoms within the molecules

isomerase Enzyme that rearranges groups within a molecule

isotope An element that has the same atomic number as another element but a different atomic weight

isotypes Antibodies differing in heavy chain constant regions associated with different classes and subclasses of immunoglobulins

itaconic acid An organic acid used as a resin in detergents; made by the transformation of citric acid by *Aspergillus terreus*

-itis Suffix denoting a disease; specifically, an inflammatory disease of a specified part

J

Jaccard coefficient A measure of similarity used in cluster analysis to show the relationship between individuals; it does not consider negative matches

jaundice Yellowness of the skin, mucous membranes, and secretions resulting from liver malfunction

juvenile dysentery Dysentery of infants, often caused by *Campylobacter fetus*

K

kala-azar disease Disease caused by the flagellate protozoan *Leishmania donovani*, generally transmitted to humans by the bite of a sandfly that has fed on an infected dog or rat; also known as visceral leishmaniasis, dumdum fever, and black fever

kappa particles Bacterial particles that occur in the cytoplasm of certain strains of *Paramecium aurelia*; such strains have a competitive advantage over other strains of *Paramecium* and are known as killer strains

karyogamy The fusion of nuclei, as of gametes in fertilization

kelp Brown algae with vegetative structures consisting of a holdfast, stem, and blade; it can form large macroscopic structures

K_{eq} Equilibrium constant

keratin A highly insoluble protein that occurs in hair, wool, horn, and skin

keratoconjunctivitis Disease caused by adenoviruses, characterized by inflammation of the eyes accompanied by redness, swelling, and discomfort

killer T cells Cytotoxic T cells

kinase Fibrinolysin

Kingdom A major taxonomic category consisting of several phyla or divisions; the primary divisions of living organisms

Kirby-Bauer test Bauer-Kirby test

Klebsiella Nonmotile, Gram-negative, encapsulated rods; can use citrate and glucose as their sole carbon sources; glucose is fermented, with the production of acid and gas; Voges-Proskauer positive; Methyl Red negative; H_2S not produced from TSI agar; catalase positive, oxidase negative; optimal growth at 35° to 37° C; G + C 52-56 mole %

K_m The Michaelis constant; describes the affinity of an enzyme for a substrate; the substrate concentration at half of the maximal velocity of an enzyme

Koch's postulates A process for elucidating the etiological (causative) agent of an infectious disease

Koji fermentation Dry fermentation in which a soybean/wheat mixture is inoculated with spores of *Aspergillus oryzae* and moistened, not submerged in liquid, so that fungi grow on the surface; used in soy sauce production

Koplik's spots Small red spots surrounded by white areas occurring on the mucous membranes of the mouth during the early stages of measles

Krebs cycle The tricarboxylic acid cycle; the citric acid cycle; the metabolic pathway in which acetate derived from pyruvic acid is converted to carbon dioxide and reduced coenzymes are produced

Kupffer cells   Macrophages lining the sinusoids of the liver

kuru   Disease caused by a prion affecting the central nervous system; observed among cannibals in New Guinea

L

labile   Unstable, readily changed by physical, chemical, or biological processes

lac **operon**   Inducible enzyme system of *Escherichia coli* for the utilization of lactose

lactam   An organic compound containing an -NH-CO- group in ring form

lactamase   An enzyme that breaks a lactam ring

lactic acid   Organic acid with antimicrobial activity produced by lactic acid bacteria (*Lactobacillus*, *Streptococcus*, *Leuconostoc*) involved in antagonistic relationships among microorganisms and used as a preservative

lactic acid fermentation   Fermentation that produces lactic acid as the primary product

Lactobacillaceae   Family of Gram-positive, asporogenous, regularly shaped rods that produce lactic acid as a major fermentation product

Lactobacillus   Rods; chain formation common; do not produce spores; Gram-positive but become Gram-negative with increasing age; metabolism fermentative even though growth generally occurs in the presence of air; some are strict anaerobes; lactic acid is the major end product of fermentation; optimal growth at 30° to 40° C; G + C 33.3-53.9 mole %

lactoferrin   An iron-containing compound that binds the iron necessary for microbial growth, resulting in a slight antimicrobial action

lactose   A disaccharide in milk; when hydrolyzed, it yields glucose and galactose

lag phase   A period following inoculation of a medium during which the number of microorganisms does not increase

lagging strand of DNA   Discontinuous strand of DNA

lagoons   Ponds used for the secondary treatment of sewage and industrial effluents

laminar flow   The flow of air currents in which streams do not intermingle; the air moves along parallel flow lines; used in a laminar flow hood to provide air free of microorganisms over a work area

landfill   A site where solid waste is dumped and allowed to decompose; a process in which solid waste containing organic and inorganic material is added to soil and allowed to decompose

late proteins   Proteins coded for late in the developmental sequence of a virus

late syphilis   Tertiary phase of syphilis that can damage any body organ, occurring several years after the initial infection

latent   Potential; not manifest; present but not visible or active

latent period   The period of time following infection of a cell by a virus before new viruses are assembled

late-onset hypogammaglobulinemia   Immunodeficiency disorder characterized by a shortage of circulating B cells and/or B cells with IgG surface receptors

leach   To wash or extract soluble constituents from insoluble materials

leader sequence   The beginning sequence of nucleotides in an mRNA molecule involved in the initiation of protein synthesis at the ribosomes

leading strand of DNA   Continuous strand of DNA

leaf spots   Plant diseases in which infection of the foliage interferes with photosynthesis

leavening   Substance used to produce fermentation in dough or liquid; the production of CO_2 that results in the rising of dough

lecithinases   Extracellular phospholipid-splitting enzymes

Legionella   Cells rod-shaped; Gram-negative; weakly oxidase positive, catalase positive; nonmotile; fastidious, with narrow optimal temperature and pH ranges; do not utilize carbohydrates; urea not utilized; aerobic; G + C 39 mole %

Legionellaceae   Family of Gram-negative, fermentative, rod-shaped bacteria that require iron and cysteine as growth factors

Legionnaire's disease   A form of pneumonia caused by *Legionella pneumophila*

leishmaniasis   Disease caused by protozoa of the genus *Leishmania*

leprosy   A chronic contagious disease affecting humans and armadillos, caused by *Mycobacterium leprae*; also known as Hansen disease

leptospirosis   Disease of humans or animals caused by *Leptospira*

lesion   A region of tissue mechanically damaged or altered by any pathological process

lethal dose   The amount of a toxin that results in the death of an organism

lethal mutations   Mutations that result in the death of a microorganism or its inability to reproduce

leukemia   Type of cancer characterized by the malignant proliferation of abnormally high numbers of leukocytes; treatment is with chemotherapy; in some cases, treatment is with bone marrow transplant

leukocidin   An extracellular bacterial product that can kill leukocytes

leukocyte   A type of white blood cell characterized by a beaded, elongated nucleus

leukocytosis   An increase above the normal upper limits of the leukocyte count

leukopenia   A decrease below the normal lower limit of the leukocyte count

lichens   A large group of composite organisms, each consisting of a fungus in symbiotic association with an alga or cyanobacterium

life   A state that characterizes living systems, encompassing the complex series of physicochemical processes essential for maintaining the organization of the system and the ability to reproduce that organization

ligases   Enzymes that catalyze reactions in which a bond is formed between 2 substrate molecules using energy obtained from cleavage of a pyrophosphate bond; enzyme that joins 2 molecules using energy from ATP

light beer   Beer with a low calorie content produced with fungal enzymes to ensure that simple substrates are available for alcoholic fermentation

light microscope A microscope in which visible light is used to illuminate the specimen; often referred to as a bright-field microscope

light scattering Dispersion of light when it strikes particles; used in some instruments for estimating quantities of suspended particles, including microorganisms

lignins A class of complex polymers in the woody material of higher plants

Limulus amoebocyte assay Assay that uses aqueous extracts from the blood cells (amoebocytes) of the horseshoe crab (*Limulus*) to detect endotoxin

linear alkyl benzyl sulfonate (ABS) Synthetic molecule with a straight hydrocarbon chain, benzene ring, and sulfate group designed as a component of anionic laundry detergent that is easily biodegraded

lipases Fat-splitting enzymes; enzymes that break down lipids

lipids Fats or fat-like substances that are insoluble in water and soluble in nonpolar solvents

lipophilic Preferentially soluble in lipids or nonpolar solvents

lipopolysaccharide toxin (LPS toxin) Endotoxin

lipopolysaccharides Molecules consisting of covalently linked lipids and polysaccharides; a component of Gram-negative bacterial cell walls

liquid diffusion method A method for detecting a substance based on the diffusion of that substance into a medium to achieve the appropriate concentration for a reaction to occur; a method used in serology for detection of antigens and antibodies based on its ability to achieve a zone of equivalence in which antigen-antibody reactions can occur

liquid wastes Waste material in liquid form, the result of agricultural, industrial, and all other human activities

Listeria Small, coccoid, Gram-positive rods; do not produce spores or capsules; not acid-fast; motile by peritrichous flagella; acid but no gas produced from glucose and several other carbohydrates; esculin is hydrolyzed; aerobic to microaerophilic; optimal growth at 20° to 30° C; G + C 38 mole %, except one species with G + C 56 mole %

liter A metric unit of volume equal to 1,000 milliliters; approximately equal in volume to a quart

lithosphere The solid part of the Earth

lithotrophs Microorganisms that live in and obtain energy from the oxidation of inorganic matter; autotrophs

litmus Plant extract dye used as an indicator of pH and of oxidation or reduction

living system A system separated from its surroundings by a semipermeable barrier; composed of macromolecules, including proteins and nucleic acids, having lower entropy than its surroundings and thus requiring inputs of energy to maintain its high degree of organization, capable of self-replication and normally based on cells as the primary functional and structural unit

lobopodia False feet that are extensions of ectoplasm, which includes the flow of endoplasm

lockjaw Tetanus

locus The point on a chromosome occupied by a gene

logarithmic growth phase Exponential phase

long-term immunity Acquired immunity that establishes a bank of memory cells that persist within the body and permit recognition of specific antigens to which the body has been previously exposed, thereby ensuring long-term protection (immunity) against disease

low-acid food Food with a pH above 4.5

low-nutrient bacteria Oligotrophic bacteria; bacteria that grow at low nutrient concentrations

low temperature-long-time pasteurization process (LTH) Pasteurization process at 63° C for 30 minutes

LPS Lipopolysaccharides

LTH process Low temperature-long-time pasteurization process

luciferase Enzyme that catalyses light-producing reaction in bioluminescent bacteria

lumbar puncture The removal of cerebrospinal fluid from the vertebral canal

luminescence The emission of light without production of heat sufficient to cause incandescence, produced by physiological processes or by friction, chemical, or electrical action

luminescent bacteria Bacteria that carry out light-producing metabolism

ly-, lys-, lyt- Combining forms meaning to loosen or dissolve

lyase Enzyme that removes groups from a molecule to form double bonds or adds groups to double bonds

Lyme disease An infectious disease that produces arthritis caused by a spirochete and transmitted by a tick

lymph A plasma filtrate that circulates through the body

lymph nodes An aggregation of lymphoid tissues surrounded by a fibrous capsule found along the course of the lymphatic system

lymphatic system The widely spread system of capillaries, nodes, and ducts that collects, filters, and returns tissue fluid, including protein molecules, to the blood

lymphocytes Lymph cells

lymphogranuloma venereum A sexually transmitted disease caused by a *Chlamydia*, characterized by an initial lesion, usually on the genitalia, followed by regional lymph node enlargement and systemic involvement

lymphokines A varied group of biologically active extracellular proteins formed by activated T lymphocytes involved in cell-mediated immunity

lymphomas Cancers of the lymph glands and other lymphoid tissues

lyophilization The process of rapidly freezing a substance at low temperature and then dehydrating the frozen mass in a high vacuum; a process in which water is removed by sublimation, moving from the solid to the gaseous phase

lysins Antibodies or other entities that under appropriate conditions are capable of causing the lysis of cells

lysis The rupture of cells

lysogenic conversion The process in which the genes of temperate bacterial viruses (viruses capable of lysogeny) can be expressed by the bacterial host, with the bacterium producing proteins that are coded for by the viral genes

lysogeny The nondisruptive infection of a bacterium by a bacteriophage

lysosomes Organelles containing hydrolytic enzymes involved in autolytic and digestive processes

lysozymes Enzymes that hydrolyze peptidoglycan; they act as bactericidal agents when they degrade the bacterial cell walls

lytic Of or relating to lysis or a lysin; viruses that cause lysis of cells within which they reproduce

lytic phage Phage that kill host bacterial cells when they are released; bacteriophage the replication of which causes lysis of the host bacterial cell

M

MacConkey agar A solid medium used for the growth of enteric bacteria

macro- Combining form meaning long or large

macromolecules Very large organic molecules having polymeric chain structures, as in proteins, polysaccharides, and other natural and synthetic polymers

macronucleus Larger nucleus in multinucleate protozoa

macroorganisms Organisms that are large enough to be visible to the naked eye

macrophages Mononuclear phagocytes; large, actively phagocytic cells in spleen, liver, lymph nodes, and blood; important factors in nonspecific immunity

macroscopic Of a size visible to the naked eye

macular rash Small red dots on the skin

"magic bullets" Term used to describe early synthetic drug compounds, particularly those from Paul Ehrlich's laboratory, that were portrayed as being able to seek out and destroy disease-causing pathogens

magnetosomes Dense inclusion bodies within bacterial cells that contain iron granules and act as magnetic compasses, permitting bacteria to move in response to the Earth's magnetic field

magnetotaxis Motility that is directed by a geomagnetic field

magnification The extent to which the image of an object is larger than the object itself

major histocompatibility complex (MHC) Proteins found on almost all cells in the body that are responsible for showing processed foreign protein antigens to T_H cells or cytotoxic T cells—they were first identified as the main determinants of tissue or graft rejection when tissue from one individual is transplanted to a second individual

malaise A general feeling of illness, accompanied by restlessness and discomfort

maltase An enzyme that converts maltose to glucose

malting Enzymatic conversion of barley by plant amylases and proteases that is used to prepare grain for microbial alcoholic fermentation

maltose A disaccharide formed on hydrolysis of starch or glycogen and metabolized by a wide range of fungi and bacteria

manganese nodules Nodules (round, irregular mineral masses) produced by microbial oxidation of manganese oxides

marine Of or relating to the oceans

maromi A mixture of a starter culture and a mash consisting of autoclaved soybeans, autoclaved crushed wheat, and steamed wheat bran after incubation and soaking in concentrated brine

mash The crushed malt or grain meal steeped and stirred in hot water with amylases to produce wort as a substrate for microorganisms

mashing Process in the production of beer in which adjuncts are added to malt

mast cells Cells that contain granules of histamine, serotonin, and heparin, especially in connective tissues involved in hypersensitivity reactions

Mastigomycota True fungi; some are unicellular, whereas others form extensive filamentous, coenocytic mycelia to produce motile cells with flagella; asexual reproduction involves zoospores, nutrition provided by the absorption of nutrients

Mastigophora A subclass of protozoans characterized by the presence of flagella

mastitis Inflammation of the breast

mating The meeting of individuals for sexual reproduction

maturation The process in the replication of some viruses in which an envelope is added

MBC Minimal bactericidal concentration

MCPs Methyl-accepting chemotaxis proteins

measles An acute, contagious systemic human disease caused by a paramyxovirus that enters via the oral and nasal routes, characterized by the presence of Koplik spots

media Plural form of medium

medical microbiology The study of medical science as it relates to microorganisms

medical technologist An allied health professional trained and certified to perform tests used in the diagnosis of disease

medium The material that supports the growth/reproduction of microorganisms

meiosis Cell division that results in a reduction of the state of ploidy, normally from diploid to haploid during the formation of the germ cells

melting temperature of DNA The midpoint temperature of a denaturation curve used in the analysis of DNA composition in which DNA is heated and the double-stranded helix is converted to single-stranded DNA

membrane filter A cellulose-ester membrane used for microbiological filtrations

memory cells Clones of lymphocytes with receptors of high affinity for a particular antigenic molecule

memory response Anamnestic response

meninges The membranes covering the brain and spinal cord

meningitis Inflammation of the membranes of the brain or spinal cord

merozoites Progeny cells of a protozoan formed from a sporozoite by schizogony

mesophiles Organisms whose optimum growth is in the temperature range of 20° to 45° C

mesosomes Intracellular membranous structures observed as infoldings of bacterial cell membranes in electron microscopy; their function is unknown, and in fact they now appear to be artifacts of specimen preparation

messenger RNA (mRNA) The RNA that specifies the amino acid sequence for a particular polypeptide chain

metabolic pathway   A sequence of biochemical reactions that transforms a substrate into a useful product for carbon assimilation or energy transfer

metabolism   The total of all chemical reactions by which energy is provided for the vital processes and new cell substances are assimilated

metabolites   Chemicals participating in metabolism; nutrients

metabolize   To transform by means of metabolism

metachromatic granules   Cytoplasmic granules of polyphosphate occurring in the cells of certain bacteria that stain intensively with basic dyes but appear a different color

metastasis   The spread of cancer from a primary tumor to other parts of the body

methane monooxygenase   Enzyme that catalyzes the initial step in the utilization of methane, namely its oxidation by reaction with O_2

methanogenesis   A type of anaerobic metabolism that results in methane production

methanogenic bacteria   Methanogens

methanogens   Methane-producing prokaryotes; a group of archaebacteria capable of reducing carbon dioxide or low molecular weight fatty acids to produce methane

methanotrophs   Bacteria that have the ability to use methane as their sole carbon source

Methyl Red test (MR)   A diagnostic test used to detect significant acid production by bacterial metabolism, particularly by mixed-acid fermentations

methyl-accepting chemotaxis proteins (MCPs)   Plasma membrane-bound proteins in bacteria that are involved in transmitting signals to the flagellum that dictate its direction of rotation and therefore the movement of the cell—these proteins are alternately methylated and demethylated by specific enzymes

methylation   The process of substituting a methyl group for a hydrogen atom

Methylonionadaceae   A family of Gram-negative bacteria that can utilize carbon monoxide, methane, or methanol as the sole source of carbon; they also utilize respiratory metabolism

methylotrophs   Bacteria that can utilize organic C-1 compounds other than methane as their sole source of carbon

MHC   Major histocompatibility complex

MIC   Minimum inhibitory concentration

Michaelis-Menten equation   Mathematical description of the relationship between the rate of an enzymatic reaction and the substrate concentration

micro-   Combining form meaning small

microaerophiles   Aerobic organisms that grow best in an environment with less than atmospheric oxygen levels; oxygen-requiring microorganisms that grow only at reduced oxygen concentrations

microbes   Microscopic organisms; microorganisms

microbial ecology   The field of study that examines the interactions of microorganisms with their biotic and abiotic surroundings

microbial mining   A mineral recovery method that uses bioleaching to recover metals from ores not suitable for direct smelting

microbial pesticides   Preparations of populations of pathogenic or predatory microorganisms that are antagonistic toward a particular pest population

microbicidal   Any agent capable of destroying, killing, or inactivating microorganisms so that they cannot replicate

microbiology   The study of microorganisms and their interactions with other organisms and the environment

microbiostatic   Chemical agents that inhibit the growth of microorganisms but do not kill them; when the agent is removed, growth is resumed

microbiota   The microorganisms normally associated with a given environment; the microorganisms associated with a particular tissue

microbodies   Organelles within a cell containing specialized enzymes whose functions involve hydrogen peroxide

microcidal   Microbicidal

Micrococcaceae   Family of Gram-positive cocci whose cells occur singly or as irregular clusters

Micrococcus   Cells spherical, occurring singly or in pairs and characteristically dividing in more than one plane to form a regular cluster, tetrads, or cubical packets; no resting stages known; Gram-positive; metabolism strictly respiratory; aerobes; optimal growth at 25° to 30° C; G + C 66-75 mole %

microcysts   Refractile, encapsulated myxospores

microfibrils   Threadlike structures in the cell walls of filamentous fungi, consisting of chitin

microfilament   An elongated structure composed of protein subunits

microflora   Microbiota

microglia   Macrophages of the central nervous system

microhabitat   The location where microorganisms live defined on a small scale

micro-ID system   A miniaturized commercial identification system

micrometer   One millionth (10^{-6}) of a meter; one thousandth (10^{-3}) of a millimeter

micronuclei   Smaller nuclei in multinucleate protozoa

microorganisms   Microscopic organisms, including algae, bacteria, fungi, protozoa, and viruses

microscope   An optical or electronic instrument for viewing objects too small to be visible to the naked eye

microtome   An instrument used for cutting thin sheets or sections of tissues or individual cells for examination by light or electron microscopy

microtubules   Cylindrical protein tubes that occur within all eukaryotic organisms; they aid in maintaining cell shape, comprise the structure of organelles of cilia and flagella, and serve as spindle fibers in mitosis

microwave radiation   Long wavelength radiation having poor penetrating power that apparently is unable to kill microorganisms directly

mildew   Any of various of plant diseases in which the mycelium of the parasitic fungus is visible on the affected plant; biodeterioration of a fabric due to fungal growth

millipore filter   A specific commercial brand of membrane filters

mineralization   The microbial breakdown of organic materials into inorganic materials brought about mainly by microorganisms

miniaturized commercial identification systems   Small devices containing multicompartmentalized chambers that each perform separate biochemical tests, used for the identification of bacterial species

minimal bactericidal concentration (MBC)   The lowest concentration of an antibiotic that will kill a defined proportion of viable organisms in a bacterial suspension during a specified exposure period

minimum inhibitory concentration (MIC)   The concentration of an antimicrobial drug necessary to inhibit the growth of a particular strain of microorganism

Minitek system   Miniaturized commercial identification system

mismatch correction enzyme   The gene products of *mut*H, *mut*L, *mut*S and *mut*U that form an enzyme that recognizes and excises improperly inserted nucleotides in a DNA double helix

mismatch repair   A mechanism in bacterial cells that corrects incorrectly matched base pairs in the DNA

misó   Product of Koji fermentation of rice with *Aspergillus oryzae*, it is ground into a paste and combined with other foods to be eaten

missense mutations   Type of base substitution that results in the change in the amino acid inserted into the polypeptide chain specified by the gene in which the mutation occurs

mitochondrion   A semiautonomous organelle in eukaryotic cells, the site of respiration and other cellular processes; consists of an outer membrane and an inner one that is convoluted

mitosis   The sequence of events resulting in the division of the nucleus into two genetically identical cells during asexual cell division; each of the daughter nuclei has the same number of chromosomes as the parent cell

mixed acid fermentation   A type of fermentation carried out by members of the Enterobacteriaceae that converts glucose to acetic, lactic, succinic, and formic acids

mixed amino acid fermentation pathway   Metabolism of amino acids resulting in their deamination and decarboxylation

mixotrophic   Capable of utilizing autotrophic and heterotrophic metabolic processes, e.g., the concomitant use of organic compounds as sources of carbon and light as a source of energy

MMR vaccine   A single vaccine designed to provide immunity against measles, mumps, and rubella

modification   The methylation of nucleotide residues in DNA; modification of newly synthesized DNA by specific enzymes in a manner characteristic of the particular bacterial strain

moiety   A part of a molecule having a characteristic chemical property

mold   A type of fungus having a filamentous structure

mole % G + C   The proportion of guanine and cytosine in a DNA macromolecule

Mollicutes   A class of prokaryotic organisms that do not form cell walls, e.g., *Mycoplasma*

Monera   Prokaryotic protists with a unicellular, simple colonial organization; bacteria

mono-   Combining form meaning single, one, or alone

monocistronic   mRNA that contains the information for only one polypeptide sequence

monoclonal antibody   An antibody produced from a clone of cells making only that antibody

monocytes   Ameboid, agranular, phagocytic leukocytes derived from the bone marrow

monokaryotic   Containing one nucleus per cell in fungal septate hyphae

mononuclear   Having only one nucleus

mononuclear phagocyte system   The macrophage system of the body, including all phagocytic white blood cells except granular white blood cells; the reticuloendothelial system

monosaccharide   Any carbohydrate whose molecule cannot be split into simpler carbohydrates; a simple sugar

Montoux test   Test for tuberculosis in which an appropriate dilution of purified protein derivative is injected intradermally into the superficial layers of the skin of the forearm

Moraxella   Rods, usually very short and plump (coccobacilli); Gram-negative; oxidative metabolism; a limited number of organic acids, alcohols, and amino acids serve as carbon and energy sources; carbohydrates not utilized; oxidase positive, catalase usually positive; H_2S not produced; strict aerobes; optimal growth at 32° to 35° C; G + C 40-46 mole %

morbidity   The state of being diseased; the ratio of the number of sick individuals to the total population of the community; the conditions inducing disease

morbidity rate   Number of diseased individuals per unit population per unit time

mordant   A substance that fixes the dyes used in staining tissues or bacteria; a substance that increases the affinity of a stain for a biological specimen

morphogenesis   Morphological changes, including growth and differentiation of cells and tissues during development; the transformations involved in the growth and differentiation of cells and tissues

morphology   The study of the shape and structure of microorganisms

morphovar   A biotype of a given species that is differentiated based on morphology

mortality   Death; the proportion of deaths within a population

mortality rate   Death rate; number of deaths per unit population per unit time

mosaics   A plant disease in which a patchy pattern of symptoms develop

most probable number (MPN)   The statistical estimate of a bacterial population through the use of dilution and multiple tube inoculations

motility   The capacity for independent locomotion

MPN   Most probable number

MR test   Methyl Red test

mRNA   Messenger RNA

mucociliary escalator system   Defense system that lines the upper respiratory tract and protects it against pathogens; the system consists of mucous membranes and cilia; mucous secretions trap microorganisms and cilia beat with an upward wave-like motion to expel microorganisms from the respiratory tract

mucopeptide   Peptidoglycan component of bacterial cell walls

mucosa   A mucous membrane, the lining of body cavities that communicate with the exterior

mucous membrane   The type of membrane lining body cavities and canals that have communication with air

mucus   A viscid fluid secreted by mucous glands consisting of mucin, water, inorganic salts, epithelial cells, and leukocytes

multicellular   Composed of or containing more than one cell

multilateral budding   In fungi, budding that occurs all around the mother cell

multiple antibiotic resistance   The ability to resist the effects of two or more unrelated antibiotics by bacterial strains generally containing R plasmids

mumps   An acute infectious disease caused by a virus, characterized by swelling of the salivary glands

murein   Peptidoglycan; the repeating polysaccharide unit comprising the backbone of the cell walls of eubacteria

mushroom poisoning   Food poisoning caused by the ingestion of toxin-producing fungi in which the toxin accumulates in the fruiting body (mushroom) that is eaten; often mushroom poisoning affects the central nervous system and in many cases is fatal; *Amanita* mushrooms are often called death-angel mushrooms because of their beauty and production of deadly toxins

mushrooms   Fungi that are members of the Agaricales; the basidiocarps of basidiomycetes

must   The fluid extracted from crushed grapes; the ingredients, e.g., fruit pulp or juice, used as substrate for fermentation in wine making

mutagen   Any chemical or physical agent that promotes the occurrence of mutation; a substance that increases the rate of mutation above the spontaneous rate

mutant   Any organism that differs from the naturally occurring type because its base DNA has been modified, resulting in an altered protein that gives the cell properties different from those of its parent

mutation   A stable, heritable change in the nucleotide sequence of the genetic nucleic acid, resulting in an alteration in the products coded for by the gene

mutation rate   The average number of mutations per cell generation

mutualism   A stable condition in which two organisms of different species live in close physical association, each organism deriving some benefit from the association; symbiosis

myc-   Combining form meaning fungus

mycelium (plural, **mycelia**)   The interwoven mass of discrete fungal hyphae

mycetoma   A chronic infection usually involving the foot, characterized by the presence of pussy nodules and caused by a wide variety of fungi or bacteria; also known as madura foot

Mycobacterium   Slightly curved or straight rods; filamentous or mycelium-like growth may occur; acid-alcohol fast at some stage of growth; Gram-positive, but not readily stained by Gram's method; nonmotile; no endospores, conidia, or capsules; growth slow to very slow; optimal growth at ca. 40° C; G + C 62-70 mole %

mycobiont   The fungal partner in a lichen

mycolic acids   Fatty acids found in the cell walls of *Mycobacterium* and several other bacteria related to the actinomycetes

mycology   The study of fungi

mycoplasmas   Members of the group of bacteria composed of cells lacking cell walls, bounded by a single triple-layered membrane, exhibiting various shapes; the smallest organisms capable of self-reproduction

mycorrhizae   A stable, symbiotic association between a fungus and the root of a plant; the term also refers to the root–fungus structure itself

mycosis   Any disease in which the causal agent is a fungus

mycotoxins   Toxic substances produced by fungi, including aflatoxin, amatoxin, and ergot alkaloids

mycovirus   A virus that infects fungi

myocarditis   Infection of the myocardium; can result from viral, bacterial, helminthic, or parasitic infections, hypersensitivity immune reactions, radiation therapy, or chemical poisoning

myocardium   The muscular tissue of the heart wall

myx-   Combining form meaning mucus

myxamoebae   Nonflagellated ameboid cells that occur in the life cycle of the slime molds and members of the Plasmodiophorales

Myxobacterales   Fruiting myxobacteria; gliding, small, rod-shaped bacteria normally embedded in a slime layer; under appropriate conditions they aggregate to form fruiting bodies

myxobacteria   Myxobacterales

Myxomycetes   True slime molds, class of Plasmodiogymnomycotina; some form myxamoebae, others swarm cells; their classification is based on the structure of the fruiting body

myxospores   Resting cells in the fruiting bodies of members of the Myxobacteriales

N

N peplomers   Neuraminidase peplomers

NAD⁺   Oxidized nicotinamide adenine dinucleotide

NADH   Reduced nicotinamide adenine dinucleotide

NADP⁺   Oxidized nicotinamide adenine dinucleotide phosphate

NADPH   Reduced nicotinamide adenine dinucleotide phosphate

narrow-spectrum antibiotics   Antibiotics that are relatively selective and are usually targeted at a particular pathogen

nasopharyngeal swabs   Culture taken from the nasopharynx by means of a polyester attached to and wrapped around a thin stick

nasopharynx   The upper part of the pharynx continuous with the nasal passages

natto   Food product from the Orient; the fermentation product of boiled soybeans and *Bacillus subtilis*

NADH   Reduced nicotinamide adenine dinucleotide

NADP⁺   Oxidized nicotinamide adenine dinucleotide phosphate

NADPH   Reduced nicotinamide adenine dinucleotide phosphate

narrow-spectrum antibiotics   Antibiotics that are relatively selective and are usually targeted at a particular pathogen

nasopharyngeal swabs   Culture taken from the nasopharynx by means of a polyester attached to and wrapped around a thin stick

nasopharynx   The upper part of the pharynx continuous with the nasal passages

natto   Food product from the Orient; the fermentation product of boiled soybeans and *Bacillus subtilis*

natural killer cells   A special subset of lymphocytes that are neither T nor B cells; natural killer (NK) cells that are responsible for lysis of tumor cells

necrosis   The pathological death of a cell or group of cells in contact with living cells

negative interactions   Interactions between populations that act as feedback mechanisms and limit population densities

negative stain   The treatment of cells with dye so that the background, rather than the cell itself, is made opaque; used to demonstrate bacterial capsules or the presence of parasitic cysts in fecal samples; a stain with a positively charged chromophore

negative staining   The treatment of cells with dye so that the background, rather than the cell itself, is made opaque; used to demonstrate bacterial capsules or the presence of parasitic cysts in fecal samples

negatively supercoiled   Underwound DNA that is twisted in the opposite direction than the helix turns

Negri bodies   Acidophilic, intracytoplasmic inclusion bodies that develop in cells of the central nervous system in cases of rabies

Neisseria   Cocci, occurring singly but often in pairs, with adjacent sides flattened; endospores not produced; nonmotile; capsules may be present; Gram-negative; complex growth requirements; few carbohydrates utilized; aerobic or facultatively anaerobic; catalase positive; oxidase positive; optimal growth at ca. 37° C; G + C 47-52 mole %

Neisseriaceae   Family of Gram-negative cocci and coccobacilli, including the genera *Neisseria*, *Branhamella*, *Moraxella*, and *Acinetobacter*

nematodes   Worms of the class Nematoda

neoplasm   The result of the abnormal and excessive proliferation of the cells of a tissue; if the progeny cells remain localized, the resulting mass is called a *tumor*

nephrons   The microscopic functional units of the kidneys that control the concentration and volume of blood by removing and adding selected amounts of water and solutes and excreting wastes

net primary production   Amount of organic carbon in the form of biomass and soluble metabolites available for heterotrophic consumers in terrestrial and aquatic habitats

neuraminidase peplomers (N peplomers)   Projections from surfaces of influenza viruses that split neuraminic acid from polysaccharides and are involved in the release of viruses from infected cells following viral replication

neurotoxin   A toxin capable of destroying nerve tissue or interfering with neural transmission

neutralism   The relationship between two different microbial populations characterized by the lack of a recognizable interaction

neutralization   Reaction of antitoxin with toxin that renders the toxin harmless

neutralization of toxic materials   Conversion of toxic materials to nontoxic forms

neutralophiles   Bacteria that tend to thrive under neutral pH conditions

neutropenia   A decrease below the normal standard in the number of neutrophils in the peripheral blood

neutrophilia (neutrophilic leukocytosis)   An increase above the normal standard in the number of neutrophils in the peripheral blood

neutrophils   Large granular leukocytes with highly variable nuclei consisting of three to five lobes and cytoplasmic granules that stain with neutral dyes and eosin

NGU   Nongonococcal urethritis

niche   The functional role of an organism within an ecosystem; the combined description of the physical habitat, functional role, and interactions of the microorganisms occurring at a given location

nicotinamide adenine dinucleotide (NAD⁺)   A coenzyme used as an electron acceptor in oxidation-reduction reactions

nicotinamide adenine dinucleotide phosphate (NADP⁺)   The phosphorylated form of NAD⁺ formed when NADPH serves as an electron donor in oxidation-reduction reactions

nif **genes**   Genes that code for nitrogenase; genes that code for nitrogen fixation

nine + two (9 + 2) system   The arrangement of microtubules in eukaryotic flagella and cilia, consisting of nine peripheral pairs of microtubules surrounding two single central microtubules

nitrate   A salt of nitric acid, NO_3^-

nitrate reduction   The reduction of nitrate to reduced forms; for example, under anaerobic and microaerophilic conditions, bacteria use nitrate as a terminal electron acceptor for respiratory metabolism

nitrification   The process in which ammonia is oxidized to nitrite and nitrite to nitrate; a process primarily carried out by the strictly aerobic, chemolithotrophic bacteria of the family Nitrobacteraceae

nitrifying bacteria   Nitrobacteraceae; Gram-negative, obligately aerobic, chemolithotrophic bacteria occurring in fresh and marine waters and in soil that oxidize ammonia to nitrite or nitrite to nitrate

nitrite   A salt of nitrous acid; NO_2^-; nitrites of sodium and potassium are used as food additives and preservatives

nitrite ammonification   Reduction of nitrite to ammonium ions by bacteria; does not remove nitrogen from the soil

Nitrobacter   Cells or short rods; Gram-negative; chemolithotrophs that oxidize nitrite to nitrate and fix CO_2; strictly aerobic; temperature range for growth 5° to 40° C; G + C 60.7-61.7 mole %

nitrogenase The enzyme that catalyzes biological nitrogen fixation

nitrogenous Containing nitrogen

nitrogen-rich fertilizers Products containing fixed forms of nitrogen that can serve as plant nutrients when applied to crop fields to support increased production

Nitrosomonas Cells ellipsoidal or short rods; motile or nonmotile; occur singly, in pairs, or in short chains; Gram-negative; chemolithotrophic; oxidize ammonia to nitrite and fix CO_2; strictly aerobic; optimal growth at 5° to 30°C; G + C 47.4-51 mole %

NK cells Natural killer cells

Nocardia Produce true mycelia, but mycelia production may be rudimentary; Gram-positive; some species acid-fast to partially acid-fast; obligate aerobes; nonmotile; pigments are produced by several species; G + C 60-72 mole %

nodules Tumor-like growths formed by plants in response to infections with specific bacteria within which the infecting bacteria fix atmospheric nitrogen; a rounded, irregularly shaped mineral mass

nodulin genes The genes involved in root nodule formation

Nomarski differential interference microscope A specialized type of interference microscope that produces high-contrast images of unstained specimens with a three-dimensional appearance; its special features are a polarizing filter, an interference contrast condenser, and a prism analyzer plate

nomenclature The naming of organisms, a function of taxonomy governed by codes, rules, and priorities laid down by committees

nonA–nonB hepatitis Hepatitis nonA–nonB

noncompetitive inhibition Inhibition of enzyme activity by a substance that does not compete with the normal substrate for the active site and thus cannot be reduced by increasing the substrates concentration

noncyclic photophosphorylation A metabolic pathway involved in the conversion of light energy for the generation of ATP in which an electron is transferred from an electron donor, normally water, by a series of electron carriers, with the eventual formation of a reduced coenzyme, normally $NADPH_2$

nongonococcal urethritis Any inflammation of the urethra not caused by *Neisseria gonorrhoeae*

nonhomologous recombination Recombination involving little or no homology between the donor DNA and the region of the DNA in the recipient where insertion occurs

nonlinear alkyl benzyl sulfonate (ABS) A component of anionic laundry detergent that contains a braided alkane chain, is resistant to biodegradation, and causes foaming of receiving waters; banned because of its persistence in groundwater

nonperishable foods Food products that are not subject to spoilage by microorganisms under normal storage conditions and consequently have an extended shelf life if those conditions are maintained

nonreciprocal recombination Nonhomologous recombination

nonself-antigens Foreign antigens; antigens not found as part of the normal cells and tissues of the body

nonsense codon A codon that does not specify an amino acid but acts as a punctuator of mRNA

nonsense mutation A mutation in which a codon specifying an amino acid is altered to a nonsense codon

nonspecific defense system Host resistance that tends to afford protection against various pathogens; host resistance that is innate

nonspecific urethritis (NSU) A sexually transmitted disease that results in inflammation of the urethra caused by bacteria other than *Neisseria gonorrhoeae*

normal growth curve Characteristic growth curve exhibited by bacteria when inoculated on fresh media, obtained by plotting increases in the numbers of microorganisms against elapsed time; consisting of the lag, log, stationary, and death phases

normal microbiota Microbial populations most frequently found in association with particular tissues that typically do not cause disease; also known as indigenous microbial populations

normal microflora Normal microbiota

Northern blotting A technique that permits the separation and identification of specific RNA sequences

Norwalk agent Small DNA virus responsible for an outbreak of winter vomiting disease in Norwalk, Ohio, in 1968

nosocomial infection An infection acquired while in the hospital

NSU Nonspecific urethritis

nuclear membrane A double layer with a distinct space between the two membranes surrounding the genomes of eukaryotic cells

nuclear polyhedrosis virus Viral pesticide that develops in host-cell nuclei

nuclease An enzyme capable of splitting nucleic acids to nucleotides, nucleosides, or their components

nucleic acid A large, acidic, chain-like macromolecule containing phosphoric acid, sugar, and purine and pyrimidine bases; the nucleotide polymers RNA and DNA

nucleocapsid The combined viral genome and capsid

nucleoid region The region of a prokaryotic cell in which the genome occurs

nucleolus An RNA-rich intranuclear body not bounded by a limiting membrane that is the site of rRNA synthesis in eukaryotes

nucleoprotein A conjugated protein closely associated with nucleic acid

nucleoside A class of compound in which a purine or pyrimidine base is linked to a pentose sugar

nucleosome The fundamental structural unit of DNA in eukaryotes, having approximately 190 base pairs folded and held together by histones

nucleotide The combination of a purine or pyrimidine base with a sugar and phosphoric acid; the basic structural unit of nucleic acid

nucleus An organelle of eukaryotes in which the cell's genome occurs; the differentiated protoplasm of a cell that is rich in nucleic acids and is surrounded by a membrane

numerical aperture (NA) The property of a lens that describes the amount of light that can enter it

numerical profile   Used in commercial systems for the identification of clinical isolates; calculates and compares the test pattern of an unknown with that of a defined group to determine the probability that the test results could represent a member of that taxon

numerical taxonomy   A system that uses overall degrees of similarity and large numbers of characteristics to determine the taxonomic position of an organism; allows organisms of unknown affiliation to be identified as members of established taxa

nurse   Person trained to take care of the ill and infirm

nutrient   A growth-supporting substance

nutrition   Requirements of living organisms for growth and sustenance

nutritional mutations   Mutations that alter the nutritional requirements of the progeny of a microorganism

nutritional requirements   The essential growth substances needed for metabolism and reproduction

nystatin   Polyene antibiotic used in the treatment of topical *Candida* infections

O

O antigens   Lipopolysaccharide–protein antigens occurring in the cell walls of Gram-negative bacteria

objective lens   The microscope lens closest to the object

obligate aerobes   Organisms that grow only under aerobic conditions, i.e., in the presence of air or oxygen

obligate anaerobes   Organisms that cannot use molecular oxygen; organisms that grow only under anaerobic conditions, i.e., in the absence of air or oxygen; organisms that cannot carry out respiratory metabolism

obligate intracellular parasites   Organisms that can live and reproduce only within the cells of other organisms, such as viruses, all of which must find suitable host cells for their replication

obligate thermophiles   Organisms restricted to growth at high temperatures

occluded   Closed or shut up

oceans   The body of salt water that covers nearly three fourths of the Earth's surface

ocular lens   The eyepiece of a microscope; the lens closest to the eye

3′-OH free end   Unattached hydroxyl group at the 3-carbon position at one end of a nucleic acid molecule

-oid   Combining form meaning resembling

oil immersion lens   A high-power objective lens of a microscope designed to work with the space between the objective and the specimen, filled with oil to enhance resolution

oil pollutants   Petroleum hydrocarbons that contaminate the environment

Okazaki fragments   The short segments of newly synthesized DNA along the trailing or discontinuous strand that are linked by a ligase to form the completed DNA

oligodynamic action   The ability of a small amount of a heavy metal compound to act as an antimicrobial agent

oligotrophic   Pertaining to lakes and other bodies of water that are poor in those nutrients that support the growth of aerobic, photosynthetic organisms; microorganisms that grow at very low nutrient concentrations

oligotrophic bacteria   Bacteria that possess physiological properties that permit them to grow at low nutrient concentrations

oncogenes   Genes that can lead to malignant transformations of animal cells

oncogenic viruses   Viruses capable of inducing tumor formation, i.e., animal cell transformations

one gene–one enzyme hypothesis   An hypothesis developed in the 1940s, which states that one gene codes for a specific protein

one-step growth curve   Curve that describes the lytic reproduction cycle that releases a large number of phage simultaneously

oogamy   A form of fertilization that involves a motile male gamete and a relatively large, nonmotile female gamete or gametangial contact in which the gametangia are morphologically different

oogonium   A specialized structure where female gametes are produced; the female gametangium of oomycete water molds

oomycetes   Water molds, class of Mastigomycota; fungi that reproduce using flagellated zoospores

oospores   Thick-walled, resting spores of fungi

operator region   A section of an operon involved in the control of the synthesis of the gene products encoded within that region of DNA; a regulatory gene that binds with a regulatory protein to turn on and off transcription of a specified region of DNA

operon   A group or cluster of structural genes whose coordinated expression is controlled by a regulator gene

operon model   A model that explains the control of the expression of structural genes, such as for lactose metabolism, by the regulation of the transcription of the mRNA directing the synthesis of the products of those structural genes

opportunistic pathogens   Organisms that exist as part of the normal body microbiota but that may become pathogenic under certain conditions, e.g., when the normal antimicrobial body defense mechanisms have been impaired; organisms that are not normally considered pathogens but that cause disease under some conditions

opsonization   The process by which a cell becomes more susceptible to phagocytosis and lytic digestion when a surface antigen combines with an antibody and/or other serum component

optical isomers   Compounds having the same number and kind of atoms and grouping of atoms but differing in their configurations or arrangements in space; specifically their structures are not superimposable

optimal growth temperature   The temperature at which microorganisms exhibit the maximal growth rate

optimal oxygen concentration   The oxygen concentration at which microorganisms exhibit the maximal growth rate with maximal product yield

oral cavity   Mouth

orally   Ingestion into the gastrointestinal tracts

orchitis   Inflammation of the testes

organelle   A membrane-bound structure that forms part of a microorganism and that performs a specialized function

Orleans process   Method for the production of vinegar in which raw vinegar from a previous run provides the active inoculum; classic slow process for producing vinegar that relies on a microbial surface film

oscillatorian cyanobacteria   Subgroup of cyanobacteria that form filamentous structures composed of straight or helical vegetative cells

-ose   Combining form denoting a sugar

-osis   Combining form meaning disease of

osmophiles   Organisms that grow best or only in or on media of relatively high osmotic pressure

osmosis   The passage of a solvent through a membrane from a dilute solution into a more concentrated one

osmotic pressure   The force resulting from differences in solute concentrations on opposite sides of a semipermeable membrane

osmotic shock   Any disturbance or disruption in a cell or subcellular organelle that occurs when it is transferred to a significantly hypertonic or hypotonic medium, with lysis of cells resulting from osmotic pressure

osmotolerant   Capable of withstanding high osmotic pressures and growth in solutions of high solute concentrations

otitis externa   Also known as swimmer's ear, inflammation of the skin of the external ear canal and auricle

otitis media   Inflammation of the inner ear

outer membrane   A structure found in Gram-negative cell walls that acts as a coarse molecular sieve and allows the diffusion of hydrophilic and hydrophobic molecules

overgrowths   A plant disease condition characterized by excessive growth

oxidase   An enzyme (oxidoreductase) that catalyzes a reaction in which electrons removed from a substrate are donated directly to molecular oxygen

oxidation   An increase in the positive valence or a decrease in the negative valence of an element resulting from the loss of electrons that are taken on by some other element

oxidation pond   A method of aerobic waste disposal employing biodegradation by aerobic and facultative microorganisms growing in a standing water body

oxidation-reduction potential   A measure of the tendency of a given oxidation-reduction system to donate electrons (i.e., to behave as a reducing agent) or to accept electrons (i.e., to act as an oxidizing agent); determined by measuring the electrical potential difference between the given system and a standard system

oxidation-reduction reactions   Coupled reactions in which one substrate loses an electron (oxidation) and a second substrate gains that electron (reduction)

oxidative phosphorylation   A metabolic sequence of reactions occurring within a membrane in which an electron is transferred from a reduced coenzyme by a series of electron carriers, establishing an electrochemical gradient across the membrane that drives the formation of ATP from ADP and inorganic phosphate by chemiosmosis

oxidative photophosphorylation   A metabolic sequence of reactions occurring within a membrane in which light initiates the transfer of an electron by a series of electron carriers, establishing an electrochemical gradient across the membrane that drives the formation of ATP from ADP and inorganic phosphate by chemiosmosis

oxidize   To produce an increase in the positive valence through the loss of electrons

oxidoreductase   Enzyme that carries out oxidation-reduction reactions

oxygen cycle   The biogeochemical cycle by which oxygen is exchanged and distributed throughout the biosphere

oxygenic photosynthesis   A type of photosynthesis carried out by plants, algae, and cyanobacteria in which oxygen is produced from water

oxygenic phototrophic bacteria (oxyphotobacteria)   Subclass of Photobacteria; bacteria capable of splitting water to form oxygen as part of photosynthetic metabolism; bacteria capable of producing oxygen during photosynthesis

oxyphotobacteria   Bacteria capable of evolving oxygen during photosynthesis

ozonation   The killing of microorganisms by exposure to ozone

P

5′-P free end   Unattached phosphate ester group at the 5-carbon position at one end of a nucleic acid molecule

5′-phosphate end (5′-P)   The end of a nucleic acid macromolecule (DNA or RNA) at which carbon-5 of the carbohydrate is not involved in forming the phosphate diester linkage that bonds the macromolecule; the lack of bonding to carbon-5 makes this end biochemically recognizable and confers directionality on the nucleic acid macromolecule

pain   Characteristic of the inflammatory response, an unpleasant sensation due to lysis of blood cells, triggering the production of bradykinin and prostaglandins that alter the threshold and intensity of the nervous system's response to pain

palindrome (palindromic sequence)   A word reading the same backward and forward; a base sequence the complement of which has the same sequence; a nucleotide sequence that is the same when read in the antiparallel direction

pandemic   An outbreak of disease that affects large numbers of people in a major geographical region or that has reached epidemic proportions simultaneously in different parts of the world

papilloma viruses   Small, icosahedral DNA viruses that cause warts

papular rash   A skin rash characterized by raised spots

paralytic shellfish poisoning   Caused by toxins produced by the dinoflagellates *Gonyaulax*, which concentrate in shellfish such as oysters and clams

parasites   Organisms that live on or in the tissues of another living organism, the host, from which they derive nutrients

parasitism   An interactive relationship between two organisms or populations in which one is harmed and the other benefits; generally, the population that benefits (the parasite) is smaller than the population that is harmed

parenteral route   Route of infection when microorganisms are deposited directly into tissues beneath the skin and mucous membranes

parfocal   Pertaining to microscopic oculars and objectives that are so constructed or so mounted that in changing from one to another the image remains in focus

parotitis   Inflammation of the parotid gland, as in mumps

passive agglutination   A procedure in which the combination of an antibody with a soluble antigen is made readily detectable by the prior adsorption of the antigen to erythrocytes or to minute particles of organic or inorganic materials

passive diffusion   Unassisted movement of molecules from areas of high concentration to areas of low concentration

passive immunity   Short-term immunity brought about by the transfer of preformed antibody from an immune subject to a nonimmune subject

Pasteur effect   The slower rate of glucose utilization by a microorganism growing aerobically by respiratory metabolism than by the same organism growing anaerobically, reflecting feedback inhibition; in organisms capable of both fermentative and respiratory metabolism, the inhibition of glucose utilization in anaerobically grown cells on exposure to oxygen

Pasteurella   Cells ovoid or rod-shaped; nonmotile; do not produce endospores; Gram-negative, but bipolar staining is common; metabolism fermentative; Methyl Red negative; Voges-Proskauer negative; aerobic to facultatively anaerobic; optimal growth at 37° C; G + C 36.5-43.0 mole %

pasteurization   Reduction in the number of microorganisms by exposure to elevated temperatures but not necessarily the killing of all microorganisms in a sample; a form of heat treatment that is lethal for the causal agents of a number of milk-transferable diseases, as well as for a proportion of normal milk microbiota, which also inactivates certain bacterial enzymes that may cause deterioration in milk

pathogenicity   The ability of an organism to cause disease in the host it infects

pathogens   Organisms capable of causing disease in animals, plants, or microorganisms

pathology   The study of the nature of disease through the study of its causes, processes, and effects, along with the associated alterations of structure and function

PBP   Penicillin-binding protein

pellicle   A thin protective membrane occurring around some protozoa, also known as a *periplast*; a continuous or fragmentary film that sometimes forms at the surface of a liquid culture; it consists entirely of cells or may be largely extracellular products of the cultured organisms

pelvic inflammatory disease (PID)   Inflammation that results from a generalized bacterial infection of the uterus, pelvic organs, uterine tubes, and ovaries

penetration   Entry of the phage genome into the host cell

penicillinase   A β-lactamase that hydrolyzes the beta-lactam linkage of many penicillins, rendering it ineffective as an antibiotic

penicillin-binding proteins   Bacterial plasma membrane-bound proteins that are involved in some of the reactions in peptidoglycan biosynthesis—these proteins irreversibly bind penicillins

penicillins   A group of natural and semisynthetic antibiotics with a β-lactam ring that are active against Gram-positive bacteria inhibiting the formation of cross-links in the peptidoglycan of growing bacteria

pentose   A class of carbohydrates containing five carbon atoms

pentose phosphate pathway   A metabolic pathway that involves the oxidative decarboxylation of glucose 6-phosphate to ribulose 5-phosphate, followed by a series of reversible, nonoxidative sugar interconversions

peplomers   Protruding peptide spikes of a virus that affect the pathogenicity and antigenicity of the particular viral strain

PEP:PTS   Phosphoenolpyruvate:phosphotransferase system

pepsin   A proteolytic enzyme

peptidase   An enzyme that splits peptides to form amino acids

peptide bond   A bond in which the carboxyl group of one amino acid is condensed with the amino group of another amino acid

peptides   Compounds of two or more amino acids containing one or more peptide bonds

peptidoglycan   The rigid component of the cell wall in most bacteria, consisting of a glycan (sugar) backbone of repetitively alternating *N*-acetylglucosamine and *N*-acetylmuramic acid with short, attached, cross-linked peptide chains containing unusual amino acids; also called murein

peptidyl site   The site on the ribosome where the growing peptide chain is moved during protein synthesis

Peptococcaceae   Family of Gram-positive cocci with complex nutritional requirements whose cells occur singly or in pairs, or in regular or irregular masses; they are obligately anaerobic, producing low molecular weight fatty acids, carbon dioxide, hydrogen, and ammonia

peptones   A water-soluble mixture of proteases and amino acids produced by the hydrolysis of natural proteins by an enzyme or by an acid

perfringens food poisoning   Food poisoning by ingestion of *Clostridium perfringens* type A, a self-limiting condition characterized by abdominal pain and diarrhea

peripheral nervous system   That part of the nervous system outside the brain and spinal cord; the ganglia and the cranial and spinal nerves

period of decline   Stage of disease after the period of illness during which the signs and symptoms of the disease disappear

period of illness   Acute phase of disease during which the patient experiences characteristic symptoms

period of infection   Time period during which viable disease-causing microorganisms are present in the body

periodontal disease   Disease of the tissues surrounding the teeth

periodontal pockets   Holes in the gums deepened by periodontal disease

periodontitis   Inflammation of the periodontium, the tissues surrounding a tooth

periodontosis   Juvenile periodontitis, noninflammatory degeneration of the periodontium leading to bone regression

periplasm   The region between the plasma membrane and the outer cell wall membrane of Gram-negative bacteria

periplasmic space   In Gram-negative bacterial cells, the area between outer cell wall membrane and the plasma membrane

periplast Pellicle of cryptomonad algae

perishable foods Food products that are readily subject to spoilage by microorganisms and consequently have a short shelf life

peritonitis Inflammation of the peritoneum

peritrichous flagella Referring to the arrangement of a cell's flagella in a more or less uniform distribution over the surface of the cell

permeability The property of cell membranes that permits transport of molecules and ions in solution across the membrane

permease An enzyme that increases the rate of transport of a substance across a membrane

peroxidase An oxidoreductase that catalyzes a reaction in which electrons removed from a substrate are donated to hydrogen peroxide

peroxide The anion O_2^- or HO_2^-, or a compound containing one of these anions

peroxisomes Microbodies that contain D-amino acid oxidase, α-hydroxy acid oxidase, catalase, and other enzymes, found in yeasts and certain protozoa

person-to-person epidemic Epidemiological disease pattern characterized by a relatively slow, prolonged rise and decline in the number of cases; an epidemic where there is a chain of transmission from one infected individual to an uninfected individual

person-to-person transmission Spread of infectious disease from one person to another

pest A population that is an annoyance for economic, health, or aesthetic reasons

pesticides Substances destructive to pests, especially insects

Petri plate A round, shallow, flat-bottomed dish with a vertical edge and a similar, slightly larger structure that forms a loosely fitting lid, made of glass or plastic, widely used as receptacles for various types of solid media

pH An expression of the hydrogen ion (H^+) concentration; the logarithm to the base 10 of the reciprocal of the hydrogen ion concentration

Phaeophycophyta Brown algae that produce xanthophylls; algae where the primary reserve materials are laminarin and mannitol and the cell wall is two-layered cell and composed of alginic acid

phage Bacteriophage

phagocytes Any of a variety of cells that ingest and break down certain categories of particulate matter

phagocytosis The process in which particulate matter is ingested by a cell, involving the engulfment of that matter by the cell's membrane

phagosomes Membrane-bound vesicles in phagocytes formed by the invagination of the cell membrane and the phagocytized material

phagotrophic Referring to the ingestion of nutrients in particulate form by phagocytosis

pharmaceutical A drug used in the treatment of disease

pharyngitis Inflammation of the pharynx

phase contrast microscope A microscope that achieves enhanced contrast of the specimen by altering the phase of light that passes through the specimen relative to the phase of light that passes through the background, eliminating the need for staining to view microorganisms and making the viewing of live specimens possible

PHB Poly-β hydroxybutyric acid

phenetic Pertaining to the physical characteristics of an individual without consideration of its genetic makeup; in taxonomy, a classification system that does not take evolutionary relationships into consideration; a classification system that assesses similarity based on appearance

phenol coefficient A number that expresses the antibacterial power of a substance relative to that of the disinfectant phenol

phenolics Class of antiseptics and disinfectants derived from carbolic acid

phenotype The totality of observable structural and functional characteristics of an individual organism, determined jointly by its genotype and the environment

phenotypic characteristics The observable qualities of an organism

-phile Combining form meaning similar to or having an affinity for

philopodia False feet that are filamentous projections composed entirely of ectoplasm

-phobic Combining form meaning having an aversion for or lacking affinity for

phosphatases Enzymes that hydrolyze esters of phosphoric acid

phosphate diester linkage Bond that links nucleotides of DNA and RNA, consisting of a phosphate group bonded via ester linkages between carbon-3 of one carbohydrate and carbon-5 of the next carbohydrate portion of the macromolecule; the bonding of two moieties through a phosphate group in which each moiety is held to the phosphate by an ester linkage

phosphatidylcholine phosphohydrolase The alpha toxin of *Clostridium perfringens* that is a lecithinase

phosphodiester bond The bonding of two moieties by a phosphate group; each moiety is held to the phosphate by an ester linkage

phosphoenolpyruvate:phosphotransferase system (PEP: PTS) A type of group translocation in which a phosphate group is added to a sugar as it is transported through the plasma membrane of some bacteria

phosphofructokinase An enzyme that mediates the addition of a phosphate group to glucose 6-phosphate, with the formation of glucose-1,6-bisphosphate, a key step during glycolysis

phospholipase An enzyme that catalyzes the hydrolysis of a phospholipid

phospholipase C The alpha toxin of *Clostridium perfringens* that is a lecithinase

phospholipid A lipid compound that is an ester of phosphoric acid and also contains one or two molecules of fatty acid, an alcohol, and sometimes a nitrogenous base

phosphorylation The esterification of compounds with phosphoric acid; the conversion of an organic compound into an organic phosphate

photo- Combining form meaning light

photoautotrophic metabolism A type of metabolism in which inorganic molecules serve as electron donors and carbon source, and light serves as energy source

photoautotrophs Organisms whose source of energy is light and whose source of carbon is carbon dioxide; characteristic of algae and some prokaryotes

photoheterotrophs Organisms that obtain energy from light but require exogenous organic compounds for growth

photolithotrophic metabolism Photoautotrophic metabolism

photolithotrophs Photoautotrophs

photolysis Liberation of oxygen by splitting of water during photosynthesis

photoorganotrophic metabolism A type of metabolism in which organic molecules serve as electron donors and carbon source, and light serves as energy source

photophosphorylation A metabolic sequence by which light energy is trapped and converted to chemical energy, with the formation of ATP

photoreactivation A mechanism whereby the effects of ultraviolet radiation on DNA may be reversed by exposure to radiation of wavelengths in the range 320–500 nm; an enzymatic repair mechanism of DNA present in many microorganisms

photosynthesis The process in which radiant (light) energy is absorbed by specialized pigments of a cell and is subsequently converted to chemical energy; the ATP formed in the light reactions is used to drive the fixation of CO_2, with the production of organic matter

photosynthetic membranes Specialized membranes in photosynthetic bacteria that are the anatomical sites where light energy is converted to chemical energy in the form of ATP during photosynthesis

photosynthetic metabolism Photosynthesis

photosystem I Cyclic photophosphorylation

photosystem II Noncyclic photophosphorylation

photosystems Pathways of electron transfer initiated by light energy; pathways of ATP synthesis in photosynthetic bacteria used to convert light energy to chemical energy

phototaxis The ability of bacteria to detect and respond to differences in light intensity, moving toward or away from light

phototrophic bacteria Phototrophs

phototrophs Organisms whose sole or principal primary source of energy is light; organisms capable of photophosphorylation

phycobilisomes Granules found in cyanobacteria and some algae on the surface of their thylakoids

phycobiont The algal partner of a lichen

phycocyanin Type of pigment in cyanobacteria and some algae that confers blue color

phycoerythrin Type of pigment in cyanobacteria and red algae that confers red color

phycology The study of algae

phycomycete A group of true fungi that lack regularly spaced septae in the actively growing portions of the fungus and produce sporangiospores by cleavage as the primary method of asexual reproduction

phycovirus Any virus whose host cell within which it replicates is a cyanobacterium or alga

phylogenetic Referring to the evolution of a species from simpler forms; in taxonomy, a classification based on evolutionary relationships

phylum A taxonomic group composed of groups of related classes

physician Doctor of medicine

physiology Study of functions of living organisms and their physicochemical parts and metabolic reactions

phytoalexin Polyaromatic antimicrobial substances produced by higher plants in response to a microbial infection

phytoplankton Passively floating or weakly motile photosynthetic aquatic organisms, primarily cyanobacteria and algae

phytoplankton food web A food web in aquatic habitats based on the grazing of primary producers

PID Pelvic inflammatory disease

pili Filamentous appendages that project from the cell surface of certain Gram-negative bacteria apparently involved in adsorption phenomena; filamentous appendages involved in bacterial mating

pilin A chain of proteins, the subunits of pili

pilot plant Facility with small-intermediate size fermentors that is used during scale up to determine appropriate fermentation conditions for full-scale commercial production

pinkeye Infection of the eye caused by *Haemophilus aegyptius*, characterized by swelling of the eyelids, discharge from the eye, and bleeding within the conjunctiva, as well as redness and itching characteristic of many eye inflammations

pinworm Disease caused by the nematode *Enterobius vermicularis,* which end their life cycles in human hosts, particularly the large intestine

pinworms *Enterobius vermicularis* nematodes

pitching The inoculation of yeast into cooled wort or grape must during the production of beer or wine

plague A contagious disease often occurring as an epidemic; an acute infectious disease of humans and other animals, especially rodents, caused by *Yersinia pestis* that is transmitted by fleas

planapochromatic lens A flat-field apochromatic objective microscope lens

plankton Collectively, all microorganisms that passively drift in the pelagic zone of lakes and other bodies of water, chiefly microalgae and protozoans

plant pathogens Microorganisms that cause plant diseases

plant pathology The study of the diseases of plants

plant pests Plant pathogenic bacteria

plant viruses Viruses that replicate within plant cells

plaques Clearings in areas of bacterial growth due to lysis by phage; also, the accumulation of bacterial cells within a polysaccharide matrix on the surfaces of teeth; also known as *dental plaque*

plasma cells Cells that are able to synthesize a specific antibody and secondary B cells

plasma membrane The selectively permeable membrane that forms the outer limit of the protoplast, bordered externally by the cell wall in most bacteria

plasmids Extrachromosomal genetic structures that can replicate independently within a bacterial cell

Plasmodiogymnomycotina Subdivision of Gymnomycota; includes two classes, Protostetliomycetes and Myxomycetes

plant viruses   Viruses that replicate within plant cells

plaques   Clearings in areas of bacterial growth due to lysis by phage; also, the accumulation of bacterial cells within a polysaccharide matrix on the surfaces of teeth; also known as *dental plaque*

plasma cells   Cells that are able to synthesize a specific antibody and secondary B cells

plasma membrane   The selectively permeable membrane that forms the outer limit of the protoplast, bordered externally by the cell wall in most bacteria

plasmids   Extrachromosomal genetic structures that can replicate independently within a bacterial cell

Plasmodiogymnomycotina   Subdivision of Gymnomycota; includes two classes, Protostetliomycetes and Myxomycetes

Plasmodiophoromycetes   Endoparasitic slime molds, class of Mastigomycota; protozoa that are obligate parasites of plants, algae, and fungi, forming a plasmodium within host cells

Plasmodium   Genus of malaria-causing protozoa; the life stage of acellular slime molds, characterized by a motile, multinucleate body

plasmogamy   Fusion of cells without nuclear fusion to form a multinucleate mass

plastids   A class of membrane-bound organelles found within cells of higher plants and algae, containing pigments and/or certain products of the cell, e.g., chloroplasts

plate counting   Method of estimating numbers of microorganisms by diluting samples, culturing on solid media, and counting the colonies that develop to estimate the number of viable microorganisms in the sample

Platyhelminthes   Flatworms

pleomorphic   Exhibiting pleomorphism

pleomorphism   The variation in size and form among cells in a clone or a pure culture

pleurocapsalean cyanobacteria   Unicellular subgroup of cyanobacteria, exhibiting multiple fission to produce coccoid reproductive cells that fail to separate completely following binary fission, forming multicellular aggregates

ploidy   In a eukaryotic nucleus or cell, the number of complete sets of chromosomes

plus (+) strand viruses   Viruses whose RNA genomes can serve as mRNAs

PMNs   Neutrophils

pneumonia   Inflammation of the lungs

poi   Hawaiian fermented food product made from the root of the taro plant

polar   Located at an end

polar budding   In fungi, budding that occurs at only one end of the mother cell

polar flagella   Flagella emanating from one or both polar ends of a cell

polar mutations   Mutations that prevent the translation of subsequent polypeptides coded for in the same mRNA molecule

polarized light   Light vibrating in a defined pattern

poliomyelitis   Inflammation of the gray matter of the spinal cord, caused by a picornavirus

pollutant   A material that contaminates air, soil, or water; substances—often harmful—that foul water or soil, reducing their purity and usefulness

poly-β-hydroxybutyric acid   A polymeric storage product formed by some bacteria

polycistronic   Coding for multiple cistrons; mRNA molecules that code for the synthesis of several proteins, often the proteins are functionally related and under the control of a specific operon

polyene antibiotics   A group of antibiotics used to treat fungal diseases; they act by altering the permeability properties of plasma membranes

polyenes   Antimicrobial agents capable of altering sterols in eukaryotic plasma membranes

polymerase   An enzyme that catalyzes the formation of a polymer

polymers   The products of the combination of two or more molecules of the same substance

polymorph   A leukocyte with granules in the cytoplasm; also known as a polymorphonuclear leukocyte (PMN)

polymorphonuclear   Having a nucleus that resembles lobes connected by thin strands of nuclear substance

polymorphonuclear leukocytes   Polymorphs

polymorphonuclear neutrophils (PMNs)   Neutrophils

polymorphs   Leukocytes having granules in the cytoplasm, also known as polymorphonuclear leukocytes, or PMNs

polymyxins   Antibiotics whose effectiveness is based on their ability to cause the disintegration of phospholipids

polypeptide   A chain of amino acids linked by peptide bonds, but of lower molecular weight than a protein

polyphosphate   Reserves of organic phosphate that can be used in the synthesis of ATP

polysaccharides   Carbohydrates formed by the condensation of monosaccharides, e.g., starch and cellulose, that have multiple monosaccharide subunits

polysomes   Complexes of ribosomes bound together by a single mRNA molecule; also known as polyribosomes

population   An assemblage of organisms of the same type living at the same location; a clone of organisms

porins   Proteins found in the outer membranes of Gram-negative cells in groups of three, they form cross-membrane channels through which small molecules can diffuse

portals of entry   The sites through which pathogens can gain access and entry to the body

positional isomers   Molecules with identical molecular formulas but differing in the locations of substituents; isomers with functional groups located at different positions

positive interactions   Between biological populations, interactions that enhance the ability of the interacting populations to survive within the community or a particular habitat

positive stain   A stain with a positively charged chromophore; a stain that is attracted to negatively charged cells

positive staining procedures   Use of a basic, positively charged chromophore to stain a negatively charged structure

positively supercoiled   Overwound DNA that is twisted in the same direction that the helix turns

postpartum sepsis   Puerperal fever

predation A mode of life in which food is primarily obtained by killing and consuming animals; an interaction between organisms in which one benefits and one is harmed, based on the ingestion of the smaller organism (the prey) by the larger organism (the predator)

predators Organisms that practice predation

preemptive colonization Alteration of environmental conditions by pioneer organisms in a way that discourages further succession

preservation The use of physical and/or chemical means to kill or retard the growth of microorganisms that cause spoilage

preservatives Chemicals used for preservation

presumptive diagnosis Preliminary diagnosis of a disease based on signs and symptoms reported to the physician

presumptive test Test that points toward a probable diagnosis but does not definitely identify the etiology of a disease—generally used to aid in the preliminary diagnosis so that treatment can be initiated while additional tests are run to define the best and most specific treatment; in assays for assessing water safety, gas formation in Durham tubes containing lactose broth and water samples if positive evidence of fecal contamination exists

prey An animal taken by a predator for food

Pribnow sequence A sequence within nucleotide bases in DNA that determines the site of transcription initiation

primary atypical pneumonia Pneumonia caused by *Mycoplasma pneumoniae*

primary immune response The first immune response to a particular antigen that has a characteristically long lag period and a relatively low titer of antibody production

primary producers Organisms capable of converting carbon dioxide to organic carbon, including photoautotrophs and chemoautotrophs

primary production The autotrophic conversion of inorganic carbon dioxide to organic carbon

primary sewage treatment The removal of suspended solids from sewage by physical settling in tanks or basins

primary staining Use of the first or primary stain in a differential staining procedure

principle of energy conservation Principle stating that energy involved in chemical reactions is never created or destroyed

prions Infectious protein substances that reproduce within living systems—they appear to be proteinaceous, based on degradation by proteases, and to lack nucleic acids, based on resistance to digestion by nucleases

probabilistic identification matrices Combinations of characteristics of organisms used to characterize large numbers of strains of a taxonomic group to establish the variability of a particular feature within a group; data matrices used to allow organisms of unknown affiliation to be identified as members of established taxa

processed cheeses Cheese products to which water has been added, thereby diluting their nutritional value; blends of various cheeses

processivity A mechanism in which an enzyme or enzyme complex that copies a long message maintains an uninterrupted contact with the template until the copying is terminated

Prochlorales Order of Oxyphotobacteria; the primary photosynthetic pigments are chlorophyll *a* and *b*, only members of the genus *Prochloron* occur as green, single-celled, extracellular symbionts of marine invertebrates

prodigiosin A red pigment produced by some *Serratia* species

prodromal stage Time period in the infectious process following incubation when the symptoms of the illness begin to appear

productive infection An infection that results in viral replication, with the production of viruses that can infect other compatible host cells

progenotes Theoretical primitive, self-replicating, protein-containing, cell-like structures

progeny Offspring

projector lens The lens of an electron microscope that focuses the beam on the film or viewing screen

prokaryotes Cells whose genomes are not contained within a nucleus; the bacteria and archaebacteria

prokaryotic cells Eubacterial and archaebacterial cells

promastigote An elongated, flagellated form assumed by many species of the Trypanosomatidae during a particular stage of development

promoter region Specific initiation site of DNA where the RNA polymerase enzyme binds for transcription on the DNA

proofreading The $3' \rightarrow 5'$ exonuclease activity of DNA polymerases; excision of improperly inserted nucleotides by DNA polymerases during DNA replication

propagated transmission Person-to-person transmission

propagules The reproductive units of microorganisms

prophage The integrated phage genome formed when the genome becomes integrated with the host's chromosome and is replicated as part of the bacterial chromosome during subsequent cell division

prophylaxis The measures taken to prevent the occurrence of disease

Propionibacteriaceae Family of Gram-positive rods that produce propionic acid, acetic acid, or mixtures of organic acids by fermentation; consists of the genera *Propionibacterium* and *Eubacterium*

Propionibacterium Gram-positive, non-spore-forming, nonmotile rods. Usually pleomorphic, diphtheroid, or club-shaped; metabolize carbohydrates, peptone, pyruvate, or lactate; fermentation products include combinations of propionic and acetic acids; anaerobic to aerotolerant; optimal growth at 30° to 37° C; G + C 59-66 mole %

propionic acid fermentation pathway Metabolic sequence carried out by the propionic bacteria that produces propionic acid

prostaglandins Naturally occurring fatty acids that circulate in the blood, stimulate the contraction of smooth muscles, and have the ability to lower blood pressure and affect the action of certain hormones

prostatitis Inflammation of the prostate gland, usually caused by a Gram-negative bacterial infection

prosthecae A cell wall–limited appendage forming a narrow extension of a prokaryotic cell

proteases Exoenzymes that break down proteins into their component amino acids

protein A class of high molecular weight polymers composed of amino acids joined by peptide linkages

protein toxins Exotoxins of bacteria; proteins secreted by bacteria that act as poisons

proteinase One of the subgroups of proteases or proteolytic enzymes that act directly on native proteins in the first step of their conversion to simpler substances

proteolytic enzymes Enzymes that break down proteins

Proteus Straight rods; motile by peritrichous flagella; nonpigmented; acid produced from glucose; Methyl Red test usually positive; Gram-negative; optimal growth at 37° C; G + C 38-42 mole %

Protista In one proposed classification system, a Kingdom of organisms lacking true tissue differentiation, i.e., the microorganisms; in another classification system, a Kingdom that includes many of the algae and protozoa

protobionts Progenotes

proto-cooperation Synergism; a nonobligatory relationship between two microbial populations in which both populations benefit

protonmotive force Potential chemical energy in a gradient of hydrogen ions and electrical energy across the bacterial plasma membrane

protoplasm The viscid material constituting the essential substance of living cells on which all the vital functions of nutrition, secretion, growth, reproduction, irritability, and locomotion depend

protoplasts Spherical, osmotically sensitive structures formed when cells are suspended in an isotonic medium and their cell walls are completely removed; a bacterial protoplast consists of an intact cell membrane and the cytoplasm it contains

prototrophs Parental strains of microorganisms that give rise to nutritional mutants known as auxotrophs

protozoa A group of diverse eukaryotic, typically unicellular, nonphotosynthetic microorganisms generally lacking a rigid cell wall

protozoology The study of protozoa

provirus A viral genome that becomes integrated with the host genome

Pseudomonadaceae Family of Gram-negative, straight or curved rods that are motile by means of polar flagella; most strains carry out obligately aerobic respiration, unable to fix atmospheric nitrogen; nutritionally versatile; some produce characteristic fluorescent pigments; widely distributed in soil and water

Pseudomonas Cells single, straight, or curved rods; motile by polar flagella; monotrichous or multitrichous; no resting stages known; Gram-negative; metabolism respiratory, never fermentative; some are facultative chemolithotrophs; G + C 58-70 mole %

pseudopeptidoglycan Component of archaebacterial cell walls

pseudoplasmodium Structure formed by swarming together or aggregation of myxamoebae; undergoes a developmental sequence to form a sporocarp

pseudopodia False feet formed by protoplasmic streaming in protozoa; used for locomotion and the capture of food

psittacosis An infectious disease of parrots, other birds, and humans, caused by *Chlamydia psittaci*; also called parrot fever

psychro- Combining form meaning cold

psychroduric Capable of surviving but not of growing at low temperatures

psychrophile An organism that has an optimum growth temperature below 20° C

psychrotroph A mesophile that can grow at low temperatures

puerperal fever An acute febrile condition following childbirth caused by infection of the uterus and/or adjacent regions by streptococci

puntae Holes in the silica walls of diatoms that allow exchange of nutrients and metabolic wastes between the cell and its surroundings

pure culture A culture that contains cells of one kind; the progeny of a single cell

purine $C_5H_4N_4$, a cyclic nitrogenous compound, the parent of several nucleic acid bases

purple membrane The portion of the plasma membrane that contains bacteriorhodopsin, found in *Halobacterium*

purple nonsulfur bacteria Members of the Chromatiaceae anoxygenic phototrophic bacteria that do not oxidize sulfur to sulfate

purulent Full of pus; containing or discharging pus

pus A semifluid, creamy yellow or greenish-yellow product of inflammation composed mainly of leukocytes and serum

pustule A small elevation of the skin containing pus

putrefaction The microbial breakdown of protein under anaerobic conditions

pyelonephritis Inflammation of the kidneys

pyknosis A condition in which the nucleus is contracted

pyoderma A pus-producing skin lesion

pyogenic Pus producing

pyorrhea Periodontitis

pyrimidine A six-membered cyclic compound containing four carbon and two nitrogen atoms in a ring; the parent compound of several nucleotide bases

pyrite A common mineral containing iron disulfite

pyrogenic Fever producing

pyrogens Fever-producing substances

Pyrrophycophyta Fire algae; generally brown or red because of xanthophyll pigments; unicellular and biflagellate; store starch or oils as the reserve material; the cell walls contain cellulose

Q

Q_{10} describes the actual change in the rate at which a reaction proceeds when the temperature is increased by 10° C; for enzymatic reactions the Q_{10} usually is about 2

Q fever An acute disease in humans characterized by sudden onset of headache, malaise, fever, and muscular pain, caused by *Coxiella burnetii*; the reservoirs of infection are cattle, sheep, and ticks

quality control A system for verifying and maintaining a desired level of quality in a product or process by careful planning, use of proper equipment, continued inspection, and corrective action when required; in fermentation processes, quality is determined by the yield and purity of the product

quarantine The isolation of persons or animals suffering from an infectious disease to prevent transmission of the disease to others

quaternary ammonium compounds Group of cationic detergents that disrupt bacterial cell membranes used as antiseptics and disinfectants

quats Quaternary ammonium compounds

quick freezing Subjecting cooked or uncooked foods to rapid refrigeration, permitting them to be stored almost indefinitely at freezing temperatures

quinolones Antimicrobial agents that act by blocking normal DNA replication by interfering with DNA gyrase

R

R plasmid A plasmid encoding for antibiotic resistance

rabies An acute and usually fatal disease of humans, dogs, cats, bats, and other animals, caused by the rabies virus and commonly transmitted in saliva by the bite of a rabid animal

racking A step in the fermentation of wine in which the wine is filtered through the bottom sediments and added back to the top of the fermentation vat

radappertization Reduction in the number of microorganisms by exposure to ionizing radiation

radiation The process in which energy is emitted in particles or waves

radioimmunoassay (RIA) A highly sensitive serological technique used to assay specific antibodies or antigens, employing a radioactive label to tag the reaction

radioisotopes Radioactive isotopes; isotopes emitting radioactivity

radiolaria Free-living protozoa occurring almost exclusively in marine habitats; they contain axopodia, with a skeleton of silicon or strontium sulfate

radurization Sterilization by exposure to ionizing radiation

rancid Having the characteristic odor of decomposing fat, chiefly due to the liberation of butyric and other volatile fatty acids

raphe A slit or pore in the cell wall of a diatom

rDNA Recombinant DNA

reading frame Groups of three nucleotide sequences

reagins A group of antibodies in serum that react with the allergens responsible for the specific manifestations of human hypersensitivity; a heterophile antibody formed during syphilis infections

reaneal To reestablish double-stranded DNA

rearrangement of genes Change in the relative positions of genes within the chromosome, thus altering the expression of the information contained in those genes

rec genes Recombination genes

recalcitrant A chemical that is totally resistant to microbial attack

receptor The binding constituent on a surface

recessive allele The allelic form of a gene whose information is not expressed

recipient strain Any strain that receives genetic information from another strain

reciprocal recombination Occurs as a result of crossing-over in which a symmetrical exchange of genetic material takes place, i.e., the genes lost by one chromosome are gained by the other, and vice versa

recombinant Any organism whose genotype has arisen as a result of recombination; also, any nucleic acid that has arisen as a result of recombination

recombinant DNA technology Genetic engineering

recombination The exchange and incorporation of genetic information into a single genome, resulting in the formation of new combinations of alleles

recombination genes Genes that code for heteroduplex formation during homologous recombination

recombination repair A mechanism in bacteria that is used to repair damaged DNA that involves cutting and splicing a piece of template DNA from a complementary strand

recovery The end of a disease syndrome

red algae Members of the Rhodophycophyta algae

red eyespot The stigma or pigmented region of many unicellular green algae

red tides Aquatic phenomenon caused by toxic blooms of *Gonyaulax* and other dinoflagellates that color the water and kill invertebrate organisms; the toxins concentrate in the tissues of filter-feeding molluscs, causing food poisoning

redness Characteristic of the inflammatory response resulting from capillary dilation

reduced flavin adenine dinucleotide The reduced form of the coenzyme flavin adenine dinucleotide; $FADH_2$

reducing power The capacity to bring about reduction

reduction An increase in the negative valence or a decrease in the positive valence of an element resulting from the gain of electrons

reduction potential The relative susceptibility of a substrate to oxidation or reduction

reductive amination The reaction of an α-carboxylic acid with ammonia to produce an amino acid

reductive tricarboxylic acid cycle A metabolic pathway in some photoautotrophs for the fixation of carbon dioxide in which oxaloacetate is reduced to malate, converted to fumarate, and then reduced again to succinate—the succinate is converted to α-ketoglutarate, and a second molecule of CO_2 is reductively added to the α-ketoglutarate to form isocitric acid and then citric acid, which is split into oxaloacetate and acetyl-CoA

refraction The deviation of a ray of light from a straight line in passing obliquely from one transparent medium to another of different density

refractive index An index of the change in velocity of light when it passes through a substance causing a deviation in the path of the light

refrigeration Method used for the preservation of food by storage at 5° C, based on the fact that low temperatures restrict the rates of growth and enzymatic activities of microorganisms

regulatory genes Genes that serve a regulatory function; genes that do not code for specific peptides but instead regulate the expression of structural genes

relapsing fever A human disease characterized by recurrent fever, caused by a *Borrelia* sp. and transmitted by ticks and lice

relative humidity (RH) The availability of water in the atmosphere

relaxed DNA A circular double-helix of DNA that does not have additional supercoiling and can lie flat on a planar surface without being contorted

relaxed strains Bacterial strains that have mutations in the *rel*A gene and, therefore, do not exhibit a stringent response under conditions of amino acid starvation

release factors Bacterial proteins, RF1 and RF2, that help catalyze the termination of peptide bond formation and end translation

renin Enzyme obtained from a calf's stomach that can hydrolyze proteins

Rep protein An unwinding protein in bacteria that catalyzes breaking the hydrogen bonding that holds the two strands of DNA together

repellents Chemicals that push substances away from them; chemicals that cause microorganisms to move away from them

replica plating A technique by which various types of mutants can be isolated from a population of bacteria grown under nonselective conditions, based on plating cells from each colony onto multiple plates and noting the positions of inoculation

replication Multiplication of a microorganism; duplication of a nucleic acid from a template; the formation of a replica mold for viewing by electron microscopy

replication fork The Y-shaped region of a chromosome that is the growing point during replication of DNA

replicative form DNA (RF DNA) Doubled-stranded DNA that is formed during the replication of a single-stranded DNA virus and that serves as a template for the formation of new viral genomes

replicative RNA strands Templates for the synthesis of new viral genomes produced by RNA polymerase

replicative transposition Insertion of a transposon from one location to another by copying the original sequence and inserting the copied sequence at another site—the source transposon is retained and does not move from its site

replicon Segments of a DNA macromolecule having their own origin and termini; a nucleic acid molecule that possesses an origin and is therefore capable of initiating its own replication

replisome The complex of DNA polymerase and accessory proteins that are involved in DNA replication

reporter genes Genes that code for an easily detectable trait in the cell in which they are placed and can be used to identify recombinant DNA

repressible A characteristic of enzymes that allows them to be made unless stopped by the presence of a specific repression substance

repressible operons Regulatory genes that can be shut off under specific conditions

repression The blockage of gene expression

repressor protein A protein that binds to the operator and inhibits the transcription of structural genes

reproduction A fundamental property of living systems by which organisms give rise to other organisms of the same kind

reservoirs The constant sources of infectious agents found in nature

resistance plasmids R plasmid

resistant (R) In antimicrobial sensitivity testing, the standardized zone of inhibition shows that the antibiotic disc has little or no effect on the microorganism

resistant crop varieties Species of agricultural plants that are not susceptible to particular plant pathogens

resolution The fineness of detail observable in the image of a specimen

resolving power A quantitative measure of the closest distance between two points that can still be seen as distinct points when viewed in a microscope field; depends largely on the characteristics of the microscope's objective lens and the optimal illumination of the specimen

respiration A mode of energy-yielding metabolism that requires a terminal electron acceptor for substrate oxidation; oxygen is frequently used as the terminal electron acceptor

respiration pathways Metabolic sequences for the oxidation of organic compounds to release free energy to drive the formation of ATP that require an external electron acceptor

respiratory therapist An allied health professional who aids patients with respiratory problems and in alleviating or preventing pneumonias

respiratory tract In animals, the structures and passages involved in the intake of oxygen and the expulsion of carbon dioxide

restriction endonuclease A bacterial enzyme that cuts double-stranded DNA at specific locations

restriction enzymes Restriction endonucleases; enzymes capable of cutting DNA macromolecules

restriction map A map of a genome indicating sites where specific restriction endonucleases will cut

restrictive infections Viral infections that occur when the host cell is transiently permissive, so that infective viral progeny are sometimes produced and at other times the virus persists in the infective cell without the production of infective viral progeny

reticuloendothelial system Mononuclear phagocyte system

Retroviruses Family of enveloped RNA animal viruses that use reverse transcriptase to form a DNA macromolecule needed for their replication

reverse transcriptase An enzyme that synthesizes a complementary DNA from an RNA template

reverse transcription Mechanism for RNA synthesis in which the RNA viruses use their RNA genome as a template for an RNA-directed DNA polymerase; RNA-directed synthesis of DNA that is the reversal of normal informational flow within a cell

reverse tricarboxylic acid cycle Reductive tricarboxylic acid cycle

reversion mutations Genotypically double mutants that appear phenotypically like wild type cells because the second mutation cancels out the first

Reye syndrome A neurological disease that sometimes occurs after a viral infection

RF DNA Replicative form DNA

RH Relative humidity

Rh antigen (Rhesus antigen) An antigen that occurs on the surfaces of some red blood cells; individuals with the antigen have Rh-positive blood and those lacking this antigen have Rh-negative blood

Rh incompatibility Type 2 hypersensitivity reaction that occurs when a mother is Rh negative and the father and fetus are Rh positive; the Rh-negative mother develops Rh antibodies during the birth of an Rh-positive infant that may cross the placenta and cause anemia in her next Rh-positive fetus

rheumatic fever A febrile disease, characterized by painful migratory arthritis and a predilection to heart damage, leading to chronic valvular disease, that results from the systemic spread of *Streptococcus pyogenes* toxins

rhinoviruses Causal agents of 25% of all common colds in adults

Rhizobiaceae Gram-negative family of rod-shaped bacteria capable of fixing atmospheric nitrogen

Rhizobium Rods, commonly pleomorphic; Gram-negative; metabolism respiratory; characteristically able to invade root hairs of leguminous plants and initiate production of root nodules; within nodules bacteria are pleomorphic (bacteroids); nodule bacteroids characteristically involved in fixing molecular nitrogen; optimal growth at 25° to 30° C; G + C 59.1-65.5 mole %

rhizopodia Rootlike pseudopodia of some protozoa

rhizosphere An ecological niche that comprises the surfaces of plant roots and the region of the surrounding soil in which the microbial populations are affected by the presence of the roots

rhizosphere effect Evidence of the direct influence of plant roots on bacteria, demonstrated by the fact that microbial populations usually are higher within the rhizosphere (the region directly influenced by plant roots) than in root-free soil

rho (ρ) protein Protein required to interrupt transcription; a factor that is involved in some types of termination (ρ-dependent) of transcription in bacteria

Rhodophycophyta Red algae that occur in marine habitats and contain phycocyanin, phycoerythrin, and chlorophyll pigments; the primary reserve material is Floridean starch; exhibit a specialized type of oogamous sexual reproduction; some produce tetraspores and have a bilayered cell wall

Rhodospirillaceae Purple, nonsulfur bacteria; family of phototrophic bacteria that produce red-purple carotenoid pigments; consist of the genera *Rhodospirillum*, *Rhodopseudomonas*, and *Rhodomicrobium*; carry out photoheterotrophic metabolism, converting carbon dioxide to organic matter by the Calvin cycle

RIA Radioimmunoassay

ribonucleic acid (RNA) A linear polymer of ribonucleotides in which the ribose residues are linked by 3'–5'-phosphodiester bridges; the nitrogenous bases attached to each ribose residue may be adenine, guanine, uracil, or cytosine

ribosomal RNA (rRNA) RNA of various sizes that makes up part of the ribosomes, constituting up to 90% of the total RNA of a cell; single-stranded RNA, but with helical regions formed by base pairing between complementary regions within the strand

ribosomes Cellular structures of rRNA and protein; the sites where protein synthesis occurs within cells

70S ribosomes Sites of protein synthesis in bacterial cells, mitochondria, and chloroplasts

80S ribosomes Sites of protein synthesis in the cytoplasm of eukaryotic cells

ribozyme RNA molecule that is capable of catalyzing a reaction

ribulose 1,5-bisphosphate carboxylase (RuBisCo) Enzyme that determines the rates of the Calvin cycle

ribulose monophosphate cycle A metabolic pathway in which formaldehyde initially reacts with ribulose-5-phosphate to form hexulose-6-phosphate, and the hexulose-6-phosphate is then split to form glyceraldehyde-3-phosphate

Rickettsia Parasitic bacteria occurring intracellularly or intimately in association with tissue cells other than erythrocytes or with certain organs in arthropods; growth generally occurs in the cytoplasm of host cells; transmitted by arthropod vectors; short rods; no flagella; Gram-negative; G + C 30-32.5 mole %

rickettsialpox A disease caused by *Rickettsia akari*, characterized by enlargement of the lymph nodes, fever, chills, headache, secondary rash, and leukopenia following an initial papule at the locale of the bite of a mite

rickettsias Members of the family Rickettsiaceae; Gram-negative bacterial parasites or pathogens of vertebrates and arthropods that reproduce within host cells by binary fission

ringspots Symptom of viral plant disease characterized by the appearance of chlorotic or necrotic rings on the leaves

ringworm Any mycosis of the skin, hair, or nails in humans or other animals in which the causal agent is a dermatophyte; also called *tinea*

ripen To bring to completeness or perfection; to age or cure, as in cheese; to develop a characteristic flavor, odor, texture, and color

ripened cheeses Cheeses that have undergone additional microbial growth beyond a single fermentation step to achieve a characteristic taste, texture, or aroma

RNA Ribonucleic acid

RNA polymerase An enzyme that catalyzes the formation of RNA macromolecules

RNA replicase RNA-dependent RNA polymerase used in replication of some RNA viruses

Rocky Mountain spotted fever A tick-borne human disease caused by *Rickettsia rickettsii* that occurs in parts of North America

rods Bacteria in the shape of cylinders; bacilli

roll tube method Technique used to create anaerobic conditions in which a prereduced, sterilized medium is rolled during cooling so that it covers the inside of the test tube and inoculation is accomplished under a stream of carbon dioxide or nitrogen

rolling circle model Replication pattern of viral DNA in which a circular DNA model is used to spin off unidirectionally a linear DNA molecule

rot Any of various unrelated plant diseases characterized by primary decay and disintegration of host tissue

rotating biological contactor Biodisc system

rotavirus A large DNA virus, the common etiological agent for diarrhea in infants

rough endoplasmic reticulum (ER) A network of interconnected closed internal membrane vesicles in eukaryotic cells where ribosomes synthesize certain membrane and secreted proteins

roundworm disease Any infestation of the body by roundworms

roundworms Members of the phylum Aschelminthes, commonly called helminths; parasitic animals that spend part or all of their life cycle in human hosts

rRNA Ribosomal ribonucleic acid

rubella German measles

RuBiCo Ribulose-1,5-bisphosphate carboxylase

rumen One of the four compartments that form the stomach of a ruminant animal where anaerobic microbial degradation of plant residues occurs, producing nutrients that can be metabolized by the animal

runs Straight-line movements by motile bacteria

rusts Plant diseases caused by fungi of the order Uredianales, so-called because of the rust-colored spores formed by many of the causal agents on the surfaces of the infected plants

S

S layer In some bacteria, a crystalline protein layer surrounding the cell

Sabin vaccine Attenuated live viral antigenic preparation administered for the prevention of polio

sac fungi Members of the Ascomycotina

sacculus The cross-linked peptidoglycan molecule that forms a little sac around the eubacterial cell

saki Yellow rice beer made in Japan

salinity The concentration of salts dissolved in a solution

Salk vaccine Inactivated viral antigenic preparation administered for the prevention of polio

Salmonella Rods, usually motile by peritrichous flagella; can utilize citrate as their carbon source; Gram-negative; indole negative; H_2S produced on TSI agar; Methyl Red positive; Voges-Proskauer negative; catalase positive; oxidase negative; nitrate reduced; optimal growth at 37° C; G + C 50-53 mole %

salmonellosis Any disease of humans or animals in which the causal agent is a species of *Salmonella*, including typhoid and paratyphoid fever, but most frequently referring to a gastroenteritis

salpingitis Inflammation of the Fallopian tubes

salt lake An inland water body with a high salt concentration normally approaching saturation

salt tolerant Organisms that can grow at elevated salt concentrations (greater than 5%) but that do not require added sodium chloride for growth

salt-tolerant bacteria Bacteria that can grow at concentrations of NaCl of 3% to 15%, which most bacteria cannot tolerate

sanitary engineering The science dealing with the removal of waste materials

sanitary landfill A method for disposal of solid wastes in low-lying areas, with wastes covered with a layer of soil each day

sanitary methods Techniques that prevent contamination of food or objects with pathogenic and spoilage organisms, including washing, sanitizing, heating, and packaging

sanitary practices Any practice that produces sanitary conditions, such as by cleaning and/or sterilizing, or removes microorganisms and/or the substances that support microbial growth

sanitize To make sanitary, as by cleaning or sterilizing

sanitizers Substances capable of sanitizing

sanitizing agents Compounds that reduce the number of microorganisms without necessarily killing them or inhibiting their growth

saprophytes Organisms (e.g., bacteria and fungi) whose nutrients are obtained from dead and decaying plant or animal matter in the form of organic compounds in solution

saprozoic nutrition Type of nutrition exhibited by some protozoa that obtain nutrients by diffusion, active transport, or pinocytosis from nonliving sources

Sarcodina A major taxonomic group of protozoa characterized by the formation of pseudopodia

sarcomastigophora Phylum of protozoans, that includes the Sarcodina and the Mastigophora; members are motile by means of flagella, pseudopodia, or both; reproduction is by syngamy

saturation Phenomenon in enzyme kinetics in which raising the concentration of a substrate does not continue to increase the rate of the reaction; the maximal concentration of a substance that will dissolve in a given solvent

scale-up A stepwise process of going from small laboratory flasks to large production fermentors

scanning electron microscope (SEM) An electron microscope in which a beam of electrons systematically sweeps over the specimen, and the intensity of secondary electrons generated at the specimen's surface where the beam's impact is measured and the resulting signal is used to determine the intensity of a signal viewed on a cathode ray tube that is scanned in synchrony with the scanning of the specimen

scanning electron microscopy A form of electron microscopy in which the image is formed by a beam of electrons that has been reflected from the surface of a specimen

scar tissue Tissue that forms after a wound, composed of normal functional and supportive tissues

scarlet fever Infection caused by *Streptococcus pyogenes* transmitted by inhalation and direct contact, most common in children 2–10 years of age, characterized by sore throat, nausea, vomiting, fever, rash, and strawberry tongue; treated with penicillin or erythromycin

schistosomiasis Slowly progressive disease caused by trematodes that infect the urinary or intestinal tracts

schizogamy A form of asexual reproduction characteristic of certain groups of sporozoan protozoa; coincident with cell growth; nuclear division occurs several or numerous times, producing a schiziont that then further segments into other cells

schizontocidal action Effect of antimalarial drugs that rapidly interrupts schizogony within red blood cells

Schlicter test A direct method for determining the antibacterial activity of serum of patients receiving antimicrobial drugs, that is, the dilution of a patient's serum that kills the infecting organism is determined, also known as the serum bactericidal test

SCID Severe combined immunodeficiency

scientific method A method of research in which a problem is identified, relevant data are gathered, a hypothesis is formulated from these data, and the hypothesis is empirically tested

sclerotia Hard resting bodies that are resistant to unfavorable conditions and may remain dormant for prolonged periods

scolex The head of a tapeworm, containing suckers and sometimes hooks

SCP Single cell protein

screening A discovery process of searching for microorganisms with desired metabolic capabilities that can be used in a commercial process

screening methods Diagnostic tests used to determine the likelihood that an individual has an infectious disease; initial tests used to direct the course of further diagnosis; tests used to identify a microorganism with a desired metabolic feature

scrub typhus Acute infection caused by *Rickettsia tsutsugamushi* and transmitted to humans by mite larvae; also known as tsutsugamushi fever

sebum The secretion of the sebaceous gland containing unsaturated free fatty acids that act as antimicrobics

secondary B lymphocytes Memory B-lymphocyte cells that are capable of initiating the antibody-mediated immune response for which they are genetically programmed

secondary immune response Anamnestic response; the response of an individual to the second or subsequent contact with a specific antigen, characterized by a short lag period and the production of a high antibody titer

secondary productivity The heterotrophic recapture of dilute nutrients; formation of bacterial biomass from utilization of nutrients at low concentrations

secondary sewage treatment The treatment of the liquid portion of sewage containing dissolved organic matter, using microorganisms to degrade the organic matter that is mineralized or converted to removable solids

secretory Pertaining to the act of exporting a fluid from a cell or organism

secretory vesicles Vesicles containing proteins destined for secretion that bud off of the Golgi apparatus in eukaryotic cells

sedimentation The process of settling, commonly of solid particles from a liquid

segmented genome A viral genome composed of several separate RNA molecules

selective culture medium An inhibitory medium or one designed to encourage the growth of certain types of microorganisms in preference to any others that may be present

selective toxicity The toxic effect of some antimicrobial agents on some microorganisms but not on others

self-antigens Antigens associated with the normal cells and tissues of the body; one's own antigens, in contrast to nonself, or foreign, antigens

self-limiting A disease that normally does not result in mortality even without medical intervention; an infection that is eliminated by natural host immune defenses prior to mortality and without the need for antimicrobics to curtail progression of the infection

self-purification Inherent capability of natural waters to cleanse themselves of pollutants based on biogeochemical cycling activities and interpopulation relationships of indigenous microbial populations

SEM Scanning electron microscope

semiconservative replication The production of double-stranded DNA containing one new strand and one parental strand

semidiscontinuous replication A term used to describe the mechanism of DNA replication because the leading strand is replicated continuously and the lagging strand is replicated discontinuously

semiperishable foods Food products that are not readily subject to spoilage by microorganisms and consequently have a long shelf life unless improperly handled

sense strand The strand of DNA that codes for the synthesis of RNA

sensitive (S) In antimicrobial sensitivity testing, the standardized zone of inhibition shows that the antibiotic disc is effective on the microorganism

sensitization A process in which specific IgE antibodies are synthesized in response to an allergen, move through the bloodstream to mast cells in connective tissue, and become firmly fixed to receptors so that the next time the individual is exposed to the same allergen, that allergen can react directly with the IgE fixed to the mast cells

septate Separated by septa, or cross walls

septic tank A simple anaerobic treatment system for waste water where residual solids settle to the bottom of the tank and the clarified effluent is distributed over a leaching field

septicemia A condition in which an infectious agent is distributed throughout the body via the bloodstream; blood poisoning, the condition attended by severe symptoms in which the blood contains large numbers of bacteria

septum In bacteria, the partition or cross wall formed during cell division that divides the parent cell into two daughter cells; in filamentous organisms (e.g., fungi) one of a number of internal transverse cross walls that occur at intervals within each hypha

septum formation In binary fission, the inward movement of the plasma membrane and cell wall that establishes the separation of the two complete bacterial chromosomes

serine pathway A metabolic pathway in type II methanotrophs, the first step of which is the reaction of formaldehyde with glycine to form serine—the serine is then deaminated to form pyruvate, which is subsequently reduced to form glycerate

serology The *in vitro* study of antigens and antibodies and their interactions; immunological (antigen-antibody) reactions carried out *in vitro*

serotypes The antigenically distinguishable members of a single species; serovar

serotyping Tests to identify microorganisms based on serological procedures that detect the presence of specific characteristic antigens

serovar A biotype of a given species that is differentiated based on serology (antigenic characteristics)

Serratia Motile, peritrichously flagellated rods; many strains produce pink, red, or magenta pigments; glucose is fermented, with or without the production of a small volume of gas; Methyl Red negative; Voges-Proskauer positive; Gram-negative; facultatively anaerobic; catalase positive; oxidase negative; G + C 53-59 mole %

serum The fluid fraction of coagulated blood

serum hepatitis A form of viral hepatitis transmitted by the parenteral injection of human blood or blood products contaminated by hepatitis viruses; hepatitis B and hepatitis nonA–nonB

serum killing power The antimicrobial activity of the serum of a patient receiving antibiotics; an *in vivo* measure of antibiotic activity

serum sickness A hypersensitivity reaction that occurs 8 to 12 days after exposure to a foreign antigen; symptoms caused by the formation of immune complexes include a rash, joint pain, and fever

severe combined immunodeficiency (SCID) A genetically determined type of immunodeficiency caused by the failure of stem cells to differentiate properly; victims are incapable of any immunological response

sewage The refuse liquids or waste matter carried by sewers

sewage fungus The bacterium *Sphaerotilus natans*, which grows beneath sewage outfalls and forms filaments—giving it a fungal-like appearance

sewage treatment The treatment of sewage to reduce its biological oxygen demand and to inactivate the pathogenic microorganisms present

sex pilus F pilus

sexual reproduction Reproduction involving the union of gametes from two individuals

sexual spore A spore resulting from the conjugation of gametes or nuclei from individuals of different mating type or sex

sexually transmitted diseases (STDs) Diseases whose transmission occurs primarily or exclusively by direct contact during sexual intercourse

sheath A tubular structure formed around a filament or bundle of filaments, occurring in some bacteria

sheathed bacteria Bacteria whose cells occur within a filamentous sheath that permits attachment to solid surfaces and affords protection

shelf fungi Members of the order Aphyllophorales; fungi that grow on trees with tough leathery fruiting bodies

shelf life The period of time during which a stored product remains effective, useful, or suitable for consumption

shift to the left An increase in stab cells, indicative of neutrophilia, that refers to a blood cell classification system in which immature blood cells are positioned on the left side of a standard reference chart and mature blood cells are placed on the right

Shigella Nonmotile rods; not encapsulated; cannot use citrate or malonate as their sole carbon source; H$_2$S is not produced; glucose and other carbohydrates are fermented, with the production of acid but not gas; Gram-negative; generally catalase positive; oxidase negative

shigellosis Bacillary dysentery caused by bacteria of the genus *Shigella*

Shine-Dalgarno sequence A polypurine consensus sequence on bacterial mRNA that helps position the mRNA on the 30S ribosomal subunit by forming a base-paired region to a complementary sequence on the 16S rRNA

shingles An acute inflammation of the peripheral nerves caused by reactivation of an infection with the herpes varicella virus that remained latent since causing chickenpox in that individual, characterized by painful, small, red nodular skin lesions; also known as herpes zoster

shipyard eye Epidemic keratoconjunctivitis

shock-sensitive transport Binding-protein transport

short-term immunity Immunity that only lasts a relatively short time because it does not result in the formation of long-lasting B and T memory cells; a type of immunity that can be conferred by passive immunization and the transfer of immunoglobulins from mother to fetus

shotgun cloning A technique used to randomly clone DNA when the sequence is unknown by breaking an entire genome into fragments that are individually cloned

shoyu Soy sauce

shunt A diversion from the normal path as an alternative pathway in metabolism

shuttle vectors Cloning vectors that permit the transfer of recombinant DNA from one cell type to another

Siderocapsaceae Unicellular family of chemolithotrophic bacteria that oxidize iron or manganese, depositing iron and/or manganese oxides in capsules or extracellular material

siderophores Iron chelators that solubilize ferric hydroxide, making soluble iron available

sigma unit A subunit of RNA polymerase that helps to recognize the promoter site

signal sequence A region of nucleotides at the beginning of an mRNA molecule and the corresponding sequence of amino acids in the synthesized protein that indicates that the protein is an exoprotein and is responsible for initiating the export of that protein across the plasma membrane

signs Observable and measurable changes in a patient caused by a disease

silent mutations Mutations that do not alter the phenotype of an organism and therefore go undetected

simple matching coefficient A similarity measure used in taxonomic analysis that includes both negative and positive matches in its calculation

simple stain Procedure in which bacteria are stained with a single basic dye before viewing under a bright-field microscope

simple staining procedure A method using a single stain that does not differentiate parts of a cell or different types of cells

single cell protein (SCP) Protein produced by microorganisms and primarily composed of microbial cells; sources of this protein include bacteria, fungi, and algae

single diffusion method Precipitin reaction technique in which an antigen is allowed to diffuse unidirectionally into a tube containing a uniform concentration of soluble antibody so that the antigen establishes a concentration gradient through the tube

simple stain   Procedure in which bacteria are stained with a single basic dye before viewing under a bright-field microscope

simple staining procedure   A method using a single stain that does not differentiate parts of a cell or different types of cells

single cell protein (SCP)   Protein produced by microorganisms and primarily composed of microbial cells; sources of this protein include bacteria, fungi, and algae

single diffusion method   Precipitin reaction technique in which an antigen is allowed to diffuse unidirectionally into a tube containing a uniform concentration of soluble antibody so that the antigen establishes a concentration gradient through the tube

single-stranded binding proteins   Bacterial proteins that preferentially bind to regions of single-stranded DNA and prevent association of the strand to other strands

single-stranded DNA viruses   Viruses with genomes consisting of single-stranded DNA

single-stranded RNA viruses   Viruses with genomes consisting of single-stranded RNA

singlet oxygen   Form of oxygen in which two of the electrons in the valence shell have antiparallel spins chemically reactive with and lethal to microorganisms

site-directed mutagenesis   A technique in which a single and specific base is altered in a gene sequence, producing a mutation at a desired site

site-specific recombination   Nonhomologous recombination

SJ   The Jaccard coefficient

skin   The external covering of the body, consisting of the dermis and epidermis

skin rash   Cutaneous eruption; sign of a disease condition

skin surfaces   An environment characterized by lack of available water, high salt concentrations, low water activity, and the presence of antimicrobial agents; generally an unfavorable habitat for microbial growth

skin testing   Testing procedure based on delayed hypersensitivity reactions useful in the presumptive diagnosis of some diseases

SLE   Systemic lupus erythematosus

slime layer   An external polysaccharide layer surrounding microbial cells composed of diffuse secretions that adhere loosely to the cell surface

slime molds   Members of the gymnomycota fungi

slow-reacting substance of anaphylaxis (SRS-A)   A mixture of leukotrienes that acts as a potent bronchial constrictor

sludge   The solid portion of sewage

smallpox   An extinct disease caused by the variola virus that in humans caused an acute, highly communicable disease characterized by cutaneous lesions on the face and limbs

smooth endoplasmic reticulum (ER)   A network of interconnected closed internal membrane vesicles in eukaryotic cells where fatty acids and phospholipids are synthesized and metabolized

smuts   Plant diseases caused by fungi of the order Ustilaginales; typically involve the formation of masses of dark-colored teliospores on or within the tissues of the host plant

snapping division   After binary fission, cells do not completely separate; they appear to form groups resembling Chinese ideographs

sneeze   A sudden, noisy, spasmodic expiration through the nose, caused by the irritation of nasal nerves

sodium   A metal of the alkali group (Na)

sodium-potassium pump   A mechanism in eukaryotic cells by which Na^+ is pumped out of the cell and K^+ is pumped into the cell by the enzyme Na^+/K^+-ATPase

soft spots   Evidence of microbial spoilage of fruits and vegetables resulting from the action of microbially produced pectinesterases and polygalacturanases

sofu   Chinese word for tofu; tofu

soil fungistasis   The inhibition of fungi by soil, which is believed by some to be due to microbial activities

solenoid   A structure in eukaryotic cells in which chromatin is wound into a secondary helix with about six nucleosomes per turn

solid waste refuse   Waste material composed largely of inert materials—glass, plastic, and metal—and some decomposable organic wastes, including paper and kitchen scraps

somatic antigens   Anitgens that form part of the main body of a cell, usually at the cell surface; distinguishable from antigens that occur on the flagella or capsule

somatic cell gene therapy   Introduction of genes into somatic cells of an individual through genetic engineering that overcomes a genetic defect

somatic cells   Any cell of the body of an organism except the specialized reproductive germ cell

sonti   Indian rice beer made with *Rhizopus sonti*

sore throat   Inflammation of the mucous membrane lining the throat, producing pain and redness; pharyngitis

SOS system   Radical, complex multifunctional system for repairing DNA damage

Southern blotting   A technique that permits the separation and identification of specific DNA sequences

soy sauce   Brown, salty, tangy sauce made in Japan from soybeans, wheat, and wheat bran fermented with *Aspergillus oryzae*

sp.   Species singular

specialized transduction   Form of gene transfer and recombination accomplished by the transmission of bacterial DNA from a donor to a recipient cell by a temperate phage in which only a small amount of genetic information is transferred; the transferred genes occur at specific locations

species   A taxonomic category ranking just below a genus; includes individuals that display a high degree of mutual similarity and that actually or potentially inbreed

specific immune response   Defense system of the body characterized by a high degree of specificity to different antigens, the ability to distinguish self from nonself and the development of memory to recognize and react with foreign substances

specificity   The restrictiveness of interaction; of an antibody, refers to the range of antigens with which an antibody may combine; of an enzyme, refers to the substrate that is acted on by that enzyme; of a pathogen or parasite, refers to the range of hosts

spectrophotometer   An instrument that measures the transmission of light as a function of wavelength, allow-

spherical aberration A form of distortion of a microscope lens based on the differential refraction of light passing through the thick central portion of a convex-convex lens and the light passing through the thin peripheral regions of the lens

spheroplasts Spherical structures formed from bacteria, yeasts, and other cells by weakening or partially removing the rigid component of the cell wall

spiral bacteria Bacterial group characterized by the presence of helically curved rods, motile by means of polar flagella

spirilli Bacteria in the shape of spirals

Spirillum Rigid, helical cells; motile by means of polar, multitrichous flagella; generally exhibit bipolar flagellation; Gram-negative; strictly respiratory metabolism; aerobic-microaerophilic; optimal growth at 30° C; G + C 38-65 mole %

spirochetes Bacterial group characterized by the presence of helically coiled rods wound around one or more central axial filaments; mobile by a flexing motion of the cell.

split genes Genes coded for by noncontiguous segments of the DNA so that the mRNA and the DNA for the protein product of that gene are not colinear; genes with intervening nucleotide sequences not involved in coding for the gene product

spontaneous generation Formation of living organisms from nonliving entities by natural processes, a now proven impossibility

spontaneous mutations Naturally occurring changes in the DNA sequence in cells

sporadic Occurrences of cases of a disease in areas geographically remote from each other or temporally separated, implying that the occurrences are not related

sporangiospores Asexual fungal spores formed within a sporangium

sporangium A sac-like structure within which numbers of motile or nonmotile asexually derived spores are formed

spore An asexual reproductive or resting body that is resistant to unfavorable environmental conditions, capable of generating viable vegetative cells when conditions are favorable; resistant and/or disseminative forms produced asexually by certain types of bacteria by a process that involves differentiation of vegetative cells or structures; characteristically formed in response to adverse environmental conditions

sporicidal Capable of killing bacterial spores; any agent capable of killing bacterial spores

sporocarp Special type of fruiting body that bears a mucoid droplet at the tip of each branch containing spores with cell walls

sporotrichosis Chronic disease caused by the fungus *Sporotrix schenckii* that usually enters the body through breaks in the skin, producing characteristic skin lesions on the fingers and hands; treated with applications of potassium iodide solution

Sporozoa A subphylum of parasitic protozoa in which mature organisms lack cilia and flagella, characterized by the formation of spores

sporozoite A motile infective stage of a protozoan; the cells produced by the division of the zygote of a sporozoan

sporulation The process of spore formation

spp. Species plural

spreading factor Hyaluronidase, an enzyme that allows pathogens to spread through the body

spread plate technique A method of microbial inoculation whereby a small volume of liquid inoculum is dispersed with a glass spreader over the entire surface of an agar plate

sputum The material discharged from the surface of the air passages, throat, or mouth, consisting of saliva, mucus, pus, microorganisms, fibrin, and/or blood

SRS-A Slow reactive substance of anaphylaxis

SSM The simple matching coefficient

stab cells Immature leukocytes

stabilization ponds Ponds used for the secondary treatment of sewage and industrial effluents

stain A substance used to treat cells or tissues to enhance contrast so that specimens and their details may be detected by microscopy

staining Coloring cells or tissues with dyes

stalked bacteria Bacteria with appendages (stalks) that are relatively wide; the stalks can attach to substrate or to other cells, or may serve to increase the efficiency of nutrient acquisition

stalks Relatively wide bacterial appendages that can attach to a substrate or to other cells; may serve to increase the efficiency of nutrient acquisition

staphylococcal food poisoning An acute, nonfebrile condition caused by the enterotoxins of certain strains of *Staphylococcus*

staphylococcal scalded skin syndrome (SSSS) Severe skin disorder most often in infants, caused by *Staphylococcus aureus* marked by erythema, peeling, and necrosis that give the skin a scalded appearance

Staphylococcus Cells spherical, occurring singly, in pairs, and characteristically dividing in more than one plane to form irregular clusters; nonmotile; no resting stages known; Gram-positive; metabolism respiratory and fermentative; catalase positive; a wide range of carbohydrates may be utilized, with the production of acid; facultative anaerobes; optimal growth at 35° to 40° C; G + C 30-40 mole %

staphylokinase A fibrinolytic enzyme that catalyzes the lysis of fibrin clots produced by *Staphylococcus* sp.

starter cultures A pure or mixed culture of microorganisms used to initiate a desired fermentation process; used in the microbial production of food

stationary growth phase A growth phase during which the death rate equals the rate of reproduction, resulting in a zero growth rate in batch cultures

statospore A resting spore of some algae, consisting of two pieces

STDs Sexually transmitted diseases

stem cell A formative cell; a blood cell capable of giving rise to various differentiated types of blood cells

stenothermophiles Microorganisms that grow only at temperatures near their optimal growth temperature

sterilization Process that results in a condition totally free of microorganisms and all other living forms

sterilize To render incapable of reproducing or free from microorganisms

sterilizing agents Substances capable of yielding a sterile condition

steroids Lipid molecules with four fused carbon rings that include cholesterol and hormones

sterol A polycyclic alcohol, e.g. cholesterol or ergosterol

Stickland reaction The coupling of oxidation-reduction reactions between pairs of amino acids

stigma Red eyespot, a pigmented region in the chloroplasts of many unicellular green algae

stock culture A culture that is maintained as a source of authentic subcultures; a culture whose purity is ensured and from which working cultures are derived

storage vacuoles Membrane-bound organelles involved in maintaining accumulated reserve materials segregated from the cytoplasm within eukaryotic cells

strain A population of cells derived by asexual reproduction from a single parental cell; a cell or population of cells that has the general characteristics of a given type of organism (e.g., a bacterium or fungus) or of a particular genus, species, and serotype

streak plate technique A method of microbial inoculation whereby a loopful of culture is scratched across the surface of a solid culture medium so that single cells are deposited at a given location

strep throat Disease caused by *Streptococcus pyogenes* transmitted by direct contact and inhalation, characterized by sore throat with pain and difficulty swallowing; treated with penicillin or erythromycin, rest, and isolation; also known as streptococcal pharyngitis

Streptococcaceae Family of Gram-positive cocci whose cells occur as pairs or chains, exhibiting facultative, anaerobic, fermentative metabolism

Streptococcus Cells spherical, occurring in pairs or chains; Gram-positive; metabolism fermentative; predominant end product of glucose fermentation is lactic acid; catalase negative; facultative anaerobes; minimal nutritional requirements generally complex; optimal growth at ca. 37° C; G + C 33-42 mole %

streptokinase A fibrinolytic enzyme that catalyzes the lysis of fibrin clots produced by *Streptococcus* species

Streptomyces Produces true mycelia; slender hyphae; the aerial mycelium at maturity forms chains of three to many spores; reproduction by germination of the aerial spores, sometimes by growth of fragments of the vegetative mycelium; Gram-positive; produce a wide variety of pigments; highly oxidative heterotrophs; aerobes; optimal growth at 25° to 35° C; G + C 69-73 moles %

streptomycin An aminoglycoside antibiotic produced by *Streptomyces griseus*, affecting protein synthesis by inhibiting polypeptide chain initiation

strict anaerobes Microorganisms that cannot tolerate molecular oxygen and are inhibited or killed in its presence; microorganisms that cannot use oxygen or survive in its presence

stringent response A response in bacteria that enables them to shut down a number of energy-draining processes (RNA synthesis and protein synthesis) during poor growth conditions (amino acid starvation)

stroma The interior compartment of the chloroplast where carbon dioxide fixation occurs during photosynthesis

structural gene A gene whose product is an enzyme, structural protein, tRNA, or rRNA, as opposed to a regulator gene whose product regulates the transcription of structural genes; genes that code for polypeptides

structural RNA Ribosomal RNA

subcutaneous Beneath the skin

submerged culture reactors Used for the commercial production of vinegar, using forced aeration to maximize the rate of acetic acid production, with bacteria growing in a fine suspension created by the air bubbles and the fermenting liquid

subspecies Division of species that describes a specific clone of cells

substrate A substance on which an enzyme acts

substrate-level phosphorylation Reaction in which ATP is formed from ADP by the direct transfer of a high-energy phosphate group from an intermediate substrate in a metabolic pathway, as opposed to chemiosmotic generation of ATP

substrate specificity A characteristic of enzymes reflecting the fact that the enzyme and substrate must fit together in a specific way for the enzyme to lower the activation energy

succession The replacement of populations by other populations that are better adapted to fill the ecological niche

sulfate-reducing bacteria Bacteria that can utilize sulfate as a final electron acceptor, thereby converting it to sulfide; they are important in the cycling of sulfur compounds in soil, sediment, and water

sulfide A compound of sulfur with an element or basic radical

sulfide stinker A type of microbially caused spoilage that occurs in canned foods, producing the noxious odor of hydrogen sulfide from putrefying proteins

sulfur cycle Biogeochemical cycle mediated by microorganisms that changes the oxidation state of sulfur within various compounds

sulfur granules Internal or external deposits of elemental sulfur formed by some photosynthetic bacteria

superoxide dismutase An enzyme that catalyzes the reaction between superoxide anions and protons, the products being hydrogen peroxide and oxygen

superoxide radical A toxic free radical of oxygen (OH_2^-)

suppressor mutation A mutation that alleviates the effects of an earlier mutation at a different locus

suppressor T cells T cells usually with CD8 that tend to suppress the immune response

surface antigens Antigens associated with cell surfaces

surfactant A surface-active agent

susceptibility The likelihood that an individual will acquire a disease if exposed to the causative agent

Svedberg unit (S) The unit in which the sedimentation coefficient of a particle is commonly expressed; when values are given in seconds, the basic unit is 10 to 13 seconds

swan-necked flasks Flasks whose necks were curved by Pasteur for use in his experiments disproving the theory of spontaneous generation

swarm cells Flagellated cells of Myxomycetes that fuse to form a true plasmodium

swelling Characteristic of the inflammatory response associated with the accumulation of fluids in the bases surrounding tissue cells

swimmer's ear Otitis externa

symbiosis   An obligatory interactive association between members of two populations, producing a stable condition in which the two organisms live together in close physical proximity, to their mutual advantage

symbiotic nitrogen fixation   Fixation of atmospheric nitrogen by bacteria living in mutually dependent associations with plants

symport   A mechanism in which different substances are actively transported across the plasma membrane simultaneously in the same direction

symptom   A physiological disorder that results in a detectable deviation from the normal healthy state and is usually indicated by complaints from a patient

symptomatology   All symptoms of a disease

synchronous growth   Growth that occurs when all cells divide at the same time

synchrony   A state or condition of a culture in which all cells are dividing at the same time

synergism   In antibiotic action, when two or more antibiotics are acting together, the production of inhibitory effects on a given organism that are greater than the additive effects of those antibiotics acting independently; an interactive but nonobligatory association between two populations in which each population benefits

synergy   An additive interaction between two drugs; in antibiotic action, when two or more antibiotics are acting together, the production of inhibitory effects on a given organism that are greater than the additive effects of the antibiotics acting independently

syngamy   The union of gametes to form a zygote

synthetic fuels   Fuels, such as ethanol, methane, hydrogen, and hydrocarbons, produced by microorganisms

syntrophism   A phenomenon in which the extent of growth of an organism depends on the provision of one or more metabolic factors or nutrients by another organism growing in the vicinity

syphilis   A sexually transmitted disease of humans caused by *Treponema pallidum* that is characterized by a variety of lesions and stages

systematics   A system of taxonomy; the range of theoretical and practical studies involved in the classification of organisms

systemic infections   Infections that are disseminated throughout the body via the circulatory system

systemic lupus erythematosus (SLE)   Autoimmune disease resulting from the failure of the immune response to recognize self-antigens; results in kidney failure

T

T$_C$ cells   Cytotoxic T cells

T cell receptor (TCR)   A receptor on the surface of a T cell that in association with CD4 or CD8 is responsible for MHC-restricted antigen recognition—the TCR on most cells is a heterodimer of two polypeptide chains that are anchored to the T cell membrane and contain immunoglobulin-like constant domains and amino-terminal variable domains

T cells   T lymphocytes; lymphocyte cells that are differentiated in the thymus and are important in cell-mediated immunity, as well as in the modulation of antibody-mediated immunity

T helper cells   (T$_H$)   A class of T cells with CD4 markers that enhance the activities of B cells in antibody-mediated immunity

T killer cells   Cytotoxic T cells

T lymphocytes   T cells

T suppressor cells   (T$_s$)   A class of T cells that depress the activities of B cells in antibody-mediated immunity

TATA box   A conserved consensus sequence in eukaryotic cells that is recognized by TFIID and assists RNA polymerase II in initiating transcription

taxis   A directional locomotive response to a given stimulus exhibited by certain motile organisms or cells

taxon   A taxonomic group, e.g., genus, family, or order

taxonomic hierarchy   Organizational levels used to group living things; the levels are Kingdom, phylum, class, order, family, genus, and species

taxonomy   The science of biological classification; the grouping of organisms according to their mutual affinities or similarities

TCA cycle   Krebs cycle

TCR   T cell receptor

teichoic acids   Polymers of ribitol or glycerol phosphate in the cell walls of Gram-positive bacteria

teichuronic acids   Polymers containing uronic acids and *N*-acetylglucosamine in the cell walls of Gram-positive bacteria that are growing at limiting phosphate concentrations

teliospores   Thick-walled, binucleate resting spores of rusts and smuts

TEM   Transmission electron microscope

tempeh   Food from Indonesia made from soybeans fermented with spores of *Rhizopus*

temperate phage   Bacteriophage with the ability to form a stable, nondisruptive relationship within a bacterium; a prophage in which the phage DNA is incorporated into the bacterial chromosome

temperature   Degree of heat or coldness of a body or substance, as measured by a thermometer or other graduated scale; environmental parameter that influences the rates of chemical reactions and the three-dimensional configuration of proteins

temperature growth range   The range between the maximum and minimum temperatures at which a microorganism can grow

temperature sensitive mutations   Mutations that alter the range of temperatures over which a microorganism may grow, using specific substrates

template   A pattern that acts as a guide for directing the synthesis of new macromolecules

termination   Cessation of strand elongation as in DNA replication, RNA transcription, or protein synthesis

termination codons   Three codons, UGA, UAG, and UAA, that do not code for a particular aminoacyl–tRNA anticodon and bring about the release of a nascent polypeptide from the ribosome (termination of protein synthesis)

termination sites   Sequences of nucleotides in the DNA that act as signals to stop transcription

terrestrial   Relating to or consisting of land, as distinct from water or air

tertiary recovery of petroleum   The use of biological and chemical means to enhance oil recovery

tertiary sewage treatment A sewage treatment process that follows a secondary process, aimed at removing nonbiodegradable organic pollutants and mineral nutrients

test Algal cell wall structure containing calcium or silicon; the outer protective covering or shell formed by some protozoa

tetanospasmin Neurotoxin produced by *Clostridium tetani* that interferes with the ability of peripheral nerves of the spinal column to properly transmit signals to the muscle cells

tetanus A disease of humans and other animals in which the symptoms are due to a powerful neurotoxin formed by the causal agent, *Clostridium tetani*, present in an anaerobic wound or other lesion, characterized by sustained involuntary contraction of the muscles of the jaw and neck; also known as lockjaw

tetracyclines A group of natural and semisynthetic antibiotics that have in common a modified naphthalene ring; bacteriostatic, with a broad spectrum of activity

TFs Transcription factors

thallus The vegetative body of a fungus or alga

theca A layer of flattened, membranous vesicles beneath the external membrane of a dinoflagellate; an open or perforated shell-like structure that houses part or all of a cell

theory of spontaneous generation Nonscientific theory that held that living organisms could arise without external cause from nonliving matter

therapeutic value A measure of the usefulness of an antimicrobial agent for use in treating disease based on the degree of effectiveness and selectivity against an infecting pathogen compared to the toxicity toward the host organism

thermal death time The time required at a given temperature for the thermal inactivation or killing of a specified number of microorganisms

thermal vents Hot areas located at depths 800 to 1000 m on the sea floor, where spreading allows sea water to percolate deeply into the crust and react with hot core materials; life around the vents is supported energetically by chemoautotrophic oxidation of reduced sulfur

thermoacidophiles Microorganisms in the Kingdom Archaea that grow optimally at low pH and high temperatures

thermocline Zone of water characterized by a rapid decrease in temperature, with little mixing of water across it

thermoduric Microorganisms capable of surviving but not growing at high temperatures

thermodynamics The basic relationships between properties of matter, especially those affected by changes in temperature, and a description of the conversion of energy from one form to another

thermophiles Microorganisms with optimal growth temperatures above 45° C

theta (θ) structure A structure formed during replication of circular DNA molecules that appears like the Greek letter, θ

thin sectioning Preparation of specimens for viewing in a transmission electron microscope by cutting them into thin slices with a microtome

Thiobacillus Small, rod-shaped cells; motile by means of a single polar flagellum; no resting stages known; Gram-negative; energy derived from the oxidation of one or more reduced or partially reduced sulfur compounds; final oxidation product is sulfate; obligate aerobes; optimal growth at 28° to 30° C; G + C 50-68 mole %

thrush Candidiasis of the mucous membranes of the mouth of infants, characterized by the formation of whitish spots

thylakoids Flattened, membranous vesicles that occur in the photosynthetic apparatus of cyanobacteria and algae; the thylakoid membrane contains chlorophylls, accessory pigments, and electron carriers and is the site of light reaction in photosynthesis

thymine A pyrimidine component of DNA

thymine dimers Structures formed by base substitutions creating covalent linkages between pyrimidine bases on the same strand of the DNA, caused by exposure to ultraviolet light—they cannot act as templates for DNA polymerase and so prevent the proper functioning of polymerases

thymocytes T cells

Ti plasmid Tumor-inducing plasmid of *Agrobacterium tumefaciens*

Tine test Test for tuberculosis in which a mechanical device makes multiple punctures in the skin to expose the individual to the antigen

tinea The lesions of dermatophytosis; also called ringworm

tinsel flagellum Flagellum of eukaryotic organisms that bear fine, filamentous appendages along their lengths

tissue culture The maintenance or culture of isolated tissues and of plant or animal cell lines *in vitro*

tissue toxicity test A test in which germicides are examined for their ability to kill bacteria and their toxicity to chick heart tissue cells

tissues In plants and animals, a group of similar cells performing the same function

titer The concentration in a solution of a dissolved substance or particulate substance

tofu Japanese cheese-like food product made by fermenting soybeans with *Mucor* species

tonsillitis Inflammation of the tonsils, commonly caused by *Streptococcus pyogenes*

topoisomerase I An enzyme that converts negatively supercoiled DNA into relaxed DNA by uncoiling the helix

topoisomerase II An enzyme that introduces negative supercoiling into relaxed DNA

toxic enzymes Enzymes produced by microorganisms that adversely affect human cells or tissues; the production of toxic enzymes contributes to the virulence of some pathogens

toxicity The quality of being toxic; the kind and quantity of a poison produced by a microorganism or possessed by a nonbiological chemical

toxicity index A relative measure of the ability of a chemical to kill microorganisms and its toxicity to mammalian cells that is useful for determining the suitability of antiseptics for use on human tissues

toxic shock syndrome A disease caused by the release of toxins from *Staphylococcus*, resulting in a physiological state of shock—major outbreaks of the disease have been associated with the use of tampons during menstruation

toxigenicity The ability to produce toxins

toxin Any organic microbial product or substance that is harmful or lethal to cells, tissue cultures, or organisms; a poison

toxoid A modified protein exotoxin that has lost its toxicity but has retained its specific antigenicity

toxoplasmosis An acute or chronic disease of humans and other animals, caused by the intracellular pathogen *Toxoplasma gondii*; transmission is by ingestion of insufficiently cooked meats containing tissue cysts

trachoma A communicable disease of the eye, caused by *Chlamydia trachomatis*

trans **configuration** Genetic elements that have an effect on different DNA molecules

transamination The transfer of one or more amino groups from one compound to another; the formation of a new amino acid by the transfer of the amino group from another amino acid

transcription The synthesis of mRNA, rRNA, and tRNA from a DNA template

transcription factors Eukaryotic proteins that bind to DNA and are responsible for binding the correct RNA polymerases to their correct promoters

transduction The transfer of bacterial genes from one bacterium to another by bacteriophage; transfer of DNA from a donor to a recipient cell by a viral carrier

transfer RNA (tRNA) A type of RNA involved in carrying amino acids to the ribosomes during translation; for each amino acid there are one or more corresponding tRNAs that can bind it specifically

transferase Enzyme that transfers a part of one molecule to another molecule

transferrin Serum beta-globulin that binds and transports iron

transformation A mode of genetic transfer in which a naked DNA fragment derived from one microbial cell (typically bacterial) is taken up by another and subsequently undergoes recombination with the recipient's chromosome; transfer of a free DNA molecule from a donor to a recipient bacterium; in tissue culture, the conversion of normal cells to cells that exhibit some or all of the properties typical of tumor cells; morphological and other changes that occur in B and T lymphocytes on exposure to antigens to which they are specifically reactive

transformed cells Cells produced *in vitro* that have altered surface properties and continue to grow even when they contact a neighboring cell; microbial cells that have undergone transformation; cancer cells; malignant cells; bacterial cells that have incorporated DNA by transformation

transfusion incompatibility Because blood contains red cells with surface antigens and antibodies to the antigens that are not present, transfused blood must be compatible between donor and recipient to prevent deleterious antigen-antibody reactions; the compatibility of blood types is primarily determined by the presence or absence of A and B antigens and anti-A and anti-B antibodies

transgenic plants Plants that have gained new genetic information from foreign sources

transition A point mutation in which one purine or one pyrimidine is replaced by another

translation The assembly of polypeptide chains with mRNA serving as the template, a process that occurs at the ribosomes

translocation Nonhomologous recombination

transmission electron microscope (TEM) An electron microscope in which the specimen transmits an electron beam focused on it; image contrasts are formed by the scattering of electrons out of the beam, and various magnetic lenses perform functions analogous to those of ordinary lenses in a light microscope

transplant To graft tissues from the same body or another, or move organs from one body to another

transposable genetic elements Specific segments of DNA that can undergo nonreciprocal recombination and thus move from one location to another

transposons Translocatable genetic elements; genetic elements that move from one locus to another by nonhomologous recombination, allowing them to move around a genome

transtracheal aspiration Procedure used to obtain samples from lower respiratory tract free of upper respiratory tract microorganisms by collecting fluid via a tube passed through the trachea

transversion A point mutation in which a purine is replaced by a pyrimidine or a pyrimidine is replaced by a purine

Treponema Unicellular, helical rods with tight regular or irregular spirals; cells have one or more axial fibrils inserted at each end of the protoplasmic cylinder; motile; Gram-negative; metabolism fermentative, using amino acids and/or carbohydrates; strict anaerobes; G + C 32-50 mole %

treponemes Small spirochetes, such as *Treponema pallidum*, the etiologic agent of syphilis

tricarboxylic acid cycle Krebs cycle

trichinosis *Trichinella spiralis* infection acquired by eating encysted larvae in pork or other meat

trichome A chain or filament of cells that may or may not include one or more resting spores

trickling filter system A simple, film-flow aerobic sewage treatment system; the sewage is distributed over a porous bed coated with bacterial growth that mineralizes the dissolved organic nutrients

triplet code Describes the genetic code because three sequential nucleotides in mRNA are needed to code for a specific amino acid

trismus Tetanus; the name given to tetanus indicating that the jaw and neck contract convulsively and that the mouth remains locked closed, making swallowing difficult

tRNA Transfer RNA

-troph Combining form indicating a relation to nutrition or nourishment

trophic level Steps in the transfer of energy stored in organic compounds from one organism to another

trophic structure Steps in the transfer of energy stored in organic compounds from one organism to another

trophophase The growth phase during a fermentation process when biomass forms, but during which a desired secondary metabolite is not yet accumulating; during batch culture of a fungus, that phase in which growth-directed metabolism is dominant over secondary metabolism

trophozoite   A vegetative or feeding stage in the life cycle of certain protozoa

trp **operon**   Contains the structural genes that code for the enzymes required for the biosynthesis of the amino acid tryptophan and the regulatory genes that control the expression of these structural genes

trypanosomiasis   Any of a number of human and animal diseases in which the causal organism is a member of the genus *Trypanosoma*

tuberculin reaction   Classic skin test for detecting probable cases of tuberculosis in which a purified protein derivative of *Mycobacterium tuberculosis* is injected subcutaneously and the area near the injection site is observed for evidence of a delayed hypersensitivity reaction

tuberculosis   An infectious disease of humans and other animals, caused in humans by *Mycobacterium tuberculosis* and *M. bovis* and may affect any organ or tissue of the body, usually the lungs

tularemia   An acute or chronic systemic disease characterized by malaise, fever, and an ulcerative granuloma at the site of infection, caused by *Francisella tularensis*

tumbles   Turning movements that occur when bacteria stop traveling in a straight line

tumor-inducing (Ti) plasmid   Plasmid in *Agrobacterium tumefaciens* that codes for tumorous plant growths (galls) when this bacterium infects plants

turbidity   Cloudiness or opacity of a solution

turbidostat   A system in which an optical sensing device measures the turbidity of the culture in a growth vessel and generates an electrical signal that regulates the flow of fresh medium into the vessel and the release of spent medium and cells

twiddles   Tumbles

tyndallization   A sterilization process designed to eliminate endospore formers in which the material is heated to 80° to 100° C for several minutes on each of 3 successive days and incubated at 37° C during the intervening periods

type   A subdivision of a species; subspecies

type culture   Collections centralized storage depositories for the preservation of all microbial species

type 1 hypersensitivity   Anaphylactic hypersensitivity

type 2 hypersensitivity   Antibody-dependent hypersensitivity

type 3 hypersensitivity   Complex-mediated hypersensitivity; Arthus hypersensitivity; serum sickness

type 4 hypersensitivity   Cell-mediated hypersensitivity; delayed type hypersensitivity

type strain   Specific microbial strain deposited in a culture collection

typhoid fever   An acute, infectious disease of humans caused by *Salmonella typhi* that invades via the oral route; the symptoms include fever, and skin, intestinal, and lymphoid lesions

typhus fever   An acute infectious disease of humans characterized by a rash, high fever, and marked nervous symptoms, caused by rickettsia and transmitted by vectors

U

UHT sterilization   Ultra high-temperature sterilization

ultra high-temperature sterilization   Sterilization using very high temperatures and short exposure times, such as 141° C for 2 seconds

ultracentrifuge   A high-speed centrifuge that produces centrifugal fields up to several hundred thousand times the force of gravity; used for the study of proteins and viruses, for the sedimentation of macromolecules, and for the determination of molecular weights

ultraviolet light   Short wavelength electromagnetic radiation in the range 100-400 nm

uncoating   Stage in viral replication in which the nucleic acid is released from the capsid; the removal of a viral capsid

unicellular   Having the form and characteristics of a single cell

uniporters   Permeases that transport only one kind of molecule

universal donors   Persons with type O blood

universal recipients   Individuals with type AB blood

unripened cheeses   Cheeses produced by a single-step fermentation

uracil   A pyrimidine base, a component of nucleic acids

urea $CO(NH_2)_2$ A product of protein degradation

Ureaplasma   Cells coccoid; lack a true cell wall; cells bounded by a triple-layered membrane; colonies exhibit a cauliflower head or fried egg appearance; urea hydrolyzed, with simultaneous production of carbon dioxide and ammonia; catalase negative; glucose fermented aerobically and anaerobically; optimal growth at 35° to 37° C; G + C 26.9-29.8 mole %

ureases   Enzymes that split urea into carbon dioxide and ammonia

urethra   The canal through which urine is discharged

urethritis   Inflammation of the urethra

urinary tract   The system that functions in the elaboration and excretion of urine

urkaryote   The proposed progenitor of prokaryotic and eukaryotic cells; the primordial living cell

use-dilution method   Method of the Association of Official Analytical Chemists (AOAC) for evaluating the effectiveness of disinfectants that establishes appropriate dilutions of a germicide for actual conditions—this method gauges the effects of disinfectants by comparing them to each other, not to phenol, and it tests nonphenol-like disinfectants

UV   Ultraviolet light

V

V_{max} The maximal velocity of an enzymatic reaction occurring when the enzyme is saturated with substrate

VA mycorrhizae   Vesicular-arbuscular mycorrhizae

vaccination   The administration of a vaccine to stimulate the immune response to protect an individual from a pathogen or toxin

vaccine   A preparation of antigens that is used for vaccination

vacuole   A membrane-bound cavity within a cell that may function in digestion, secretion, storage, or excretion

vaginal tract A region of the female genital tract, the canal that leads from the uterus to the external orifice of the genital canal

vaginitis Inflammation of the vagina, marked by pain and a purulent discharge

valves Frustules of diatoms

vancomycin Antibiotic that inhibits bacterial cell wall synthesis

variable region The amino terminal end of an immunoglobulin molecule that is characterized by a high degree of variability

variant A strain that differs in some way from a particular organism

variolation A historically old procedure used by some cultures to protect individuals from smallpox; inoculation of an individual with smallpox virus

vasodilators Substances that cause the dilation or enlargement of blood vessels

vector vaccines Vaccines that act as carriers for antigens associated with pathogens other than the one from which the vaccine was derived; created through recombinant DNA technology

vectors Organisms that act as carriers of pathogens and are involved in the spread of disease from one individual to another

vegetative cells Cells that are engaged in nutrition and growth; they do not act as specialized reproductive or dormant forms

vegetative growth Production of a new organism from a portion of an existing organism exclusive of sexual reproduction

vegetative structures Structures involved in nutrition and growth that are not specialized reproductive or dormant forms

Veillonella Small cocci; nonmotile; Gram-negative; carbohydrates and alcohols not fermented; produce acetate, propionate, CO_2, and H_2 from lactate; oxidase negative; complex nutritional requirements; carbon dioxide required; anaerobic; optimal growth at 30° to 37° C; G + C 40-44 mole %

venereal disease Sexually transmitted disease

venous puncture Medical procedure in which a hypodermic needle is passed through the skin and into a vein; used for intravenous injection of drugs and for obtaining blood samples

vesicle A membrane-bounded sphere; specialized inclusion in root cortex in the vesicular-arbuscular type of mycorrhizal association

vesicular arbuscular mycorrhizae A common type of mycorrhizae characterized by the formation of vesicles and arbuscules

viability The ability to grow and reproduce

viable plate count method Procedure for the enumeration of bacteria whereby serial dilutions of a suspension of bacteria are plated onto a suitable solid growth medium, the plates are incubated, and the number of colony-forming units is counted

Vibrio Short rods, axis curved or straight; motile by a single polar flagellum; Gram-negative; not acid-fast; endospores not produced; nonencapsulated; metabolism respiratory and fermentative; fermentation of carbohydrates produces mixed products but no CO_2 or H_2; oxidase positive; facultative anaerobes; G + C 40-50 mole %

Vibrionaceae Family of Gram-negative, facultatively anaerobic rods consisting of the genera *Vibrio, Aeromonas, Plesiomonas*, and *Photobacterium*

vinegar A condiment prepared by the microbial oxidation of ethanol to acetic acid

vir **genes** Genes that code for proteins required for the transfer of T-DNA (transforming DNA) of *Agrobacterium tumefaciens*

viral Of or pertaining to a virus

viral gastroenteritis Gastroenteritis caused by adeno-, echo-, or coxsackieviruses

viremia Viral infection of the bloodstream

viricides Chemicals capable of inactivating viruses so that they lose their ability to replicate

virion A single, structurally complete, mature virus

viroids The causal agents of certain diseases, resembling viruses in many ways but differing in their apparent lack of a virus-like structural organization and their resistance to a wide variety of treatments to which viruses are sensitive; naked infective RNA

virology The study of viruses and viral diseases

virucides Chemicals capable of killing viruses

virulence The capacity of a pathogen to cause disease, broadly defined in terms of the severity of the disease in the host

virulence factors Special inherent properties of disease-causing microorganisms that enhance their pathogenicity, allowing them to invade human tissue and disrupt normal body functions

virulent pathogen An organism with specialized properties that enhance its ability to cause disease

virus A noncellular entity that consists minimally of protein and nucleic acid and that can replicate only after entry into specific types of living cells; it has no intrinsic metabolism, and its replication depends on the direction of cellular metabolism by the viral genome; within the host cell, viral components are synthesized separately and are assembled intracellularly to form mature, infectious viruses

visible light Radiation in the wavelength range of 400–800 nm that is required for photosynthesis but can be lethal to nonphotosynthetic microorganisms

vital force The force that animates and perpetuates living organisms

vitamins A group of unrelated organic compounds, some or all of which are necessary in small quantities for the normal metabolism and growth of microorganisms

Voges-Proskauer test (VP) A diagnostic test used to detect acetoin production by bacterial butanediol fermentation

volutin Metachromatic granules

volva A cup-shaped remnant of the universal veil that surrounds the base of the stalk in mature fruiting bodies of certain fungi

vomiting The forcible ejection of the contents of the stomach through the mouth

VP test Voges- Proskauer test

vulvovaginitis Inflammation of the vulva and vagina, usually caused by *Candida albicans*, herpes viruses, *Trichomonas vaginalis*, or *Neisseria gonorrhoeae*

W

wandering cells Cells capable of ameboid movement, including free macrophages, lymphocytes, mast cells, and plasma cells

wandering macrophage Wandering cells

warts Small, benign tumors of the skin, caused in humans by the human papilloma virus

Wasserman test A classic complement fixation test for the diagnosis of syphilis that has been replaced by more accurate methods

water activity (A$_w$) A measure of the amount of reactive water available, equivalent to the relative humidity; the percentage of water saturation of the atmosphere

water molds Members of the oomycetes fungi

Weil-Felix test Serological test for the diagnosis of some diseases caused by *Rickettsia* species, especially typhus fever, using heterophile antibodies

whey The fluid portion of milk that separates from the curd

whiplash flagella Smooth flagella of algae and fungi

whipworms Nematode *Trichuris truchiura*, which infest the first part of the human large intestine

whooping cough An acute, respiratory tract disease mainly in children, caused by *Bordetella pertussis*, characterized by paroxysms of coughing that usually end in loud whooping inspirations; also known as pertussis

Widal test Agglutination test for the diagnosis of typhoid fever, using antigens from *Salmonella* species

wild type Cells that contain the most common form of DNA sequences

wilts Plant diseases characterized by a reduction in host tissue turgidity, commonly affecting the vascular system; common causal agents are species of the fungi *Fusarium* and *Verticillium* and the bacteria *Erwinia* and *Pseudomonas*

windrow method A slow composting process that requires turning and covering with soil or compost

wine An alcoholic beverage produced by microbial fermentation of grapes and other fruit

wobble hypothesis Proposed by Frances Crick, this hypothesis accounts for the observed pattern of degeneracy in the third base of a codon and says that this base can undergo unusual base pairing with the corresponding first base in the anticodon

wort In brewing, the liquor that results from the mixture of mash and water held at 40° to 65° C for 1 to 2 hours, during which the starch is broken down by amylases to glucose, maltose, and dextrins, and proteins are degraded to amino acids and polypeptides

wound botulism Infection of a wound with *Clostridium botulinum*, diffusions of toxins from the site of infection can affect the nervous system

X

X-rays Type of ionizing radiation with wavelengths from 10^{-3} to 10^2 nm that may be used for sterilization

Xanthophyceae Members of the chrysophycophyta algae; yellow-green algae

xanthophyll A pigment containing oxygen and derived from carotenes; a yellow photosynthetic accessory pigment in some algae

xenobiotic A synthetic product that is not formed by natural biosynthetic processes; a foreign substance or poison

xerophiles Organisms that grow under conditions of low water activity

xerotolerant Able to withstand dryness; an organism capable of growth at low water activity

Y

yeasts A category of fungi defined in terms of morphological and physiological criteria; typically, unicellular, saprophytic organisms that characteristically ferment a range of carbohydrates and in which asexual reproduction occurs by budding

yellow fever An acute, systemic, infectious disease that affects humans and other primates, caused by a togavirus, transmitted to humans by mosquitoes; in its severe form it is marked by fever, jaundice, hemorrhage, and renal damage

yellow-green algae Members of the Xanthophyceae

Yersinia Gram-negative, ovoid cells or rods; nonmotile or motile with peritrichous flagella; nonencapsulated; various carbohydrates fermented, without production of gas; Methyl Red positive; Voges-Proskauer negative; optimal growth at 30° to 37° C; G + C 45.8-46.8 mole %

yersiniosis Infection caused by *Yersinia enterolytica* in which symptoms resemble those of appendicitis

yogurt The product of the fermentation of milk by *Lactobacillus bulgaricus* and *Streptococcus thermophilus*

Z

Z pathway of oxidative photophosphorylation The combination of the cyclic and noncyclic photophosphorylation pathways in oxygenic photosynthetic organisms describing the metabolic reactions accounting for the trapping of light energy, and the generation of ATP, oxygen, and NADPH during photosynthesis

z value The number of degrees Fahrenheit required to reduce the thermal death time tenfold

zone of greening Area of green discoloration with partial clearing around the colony resulting from alpha hemolysis

zone of inhibition The area of no bacterial growth around an antimicrobial agent in an agar diffusion test for antimicrobial sensitivity

zoology The study of animal life, including its origin, development, structure, function, and classification

zoonoses Diseases of lower animals

zoospores Motile, flagellated spores

zooxanthellae Symbiotic relationships between dinoflagellates and other marine invertebrates

Zygomycotina Fungal subdivision of Amastigomycota; its members have coenocytic mycelia and form zygospores, exhibit sexual reproduction, or produce asexual sporangiospores

zygospore Thick-walled resting spores formed after gametangial fusion by members of the zygomycetes

zygote A single diploid cell formed from two haploid parental cells during fertilization; diploid reproductive form produced by union of haploid gametes

zymogenous Term used to describe soil microorganisms that grow rapidly on exogenous substrates

Illustration Credits

Chapter 1

Chapter opener From Terry J. Beveridge, University of Guelph/Biological Photo Service.

Box figures, p. 6, Biological Photo Service and The Bettmann Archive

Figure 1-4, **A,** John J. Cardamone, Jr., University of Pittsburgh/Biological Photo Service; **C,** William L. Dentler, University of Kansas/Biological Photo Service

Figure 1-5, **B,** Terry J. Beveridge, University of Guelph/Biological Photo Service

Box figure, p. 14, **B,** Courtesy of Anheuser-Busch Companies, Inc.

Box figures, p. 15, **C,** John NA Lott, McMaster University/Biological Photo Service; **D,** Science Vu-SIM/Visuals Unlimited.

Figure 1-6, **A,** Courtesy the Library of Congress; **B,** From Thompson JM, McFarland GK, Hirsch JE, Tucker SM: *Mosby's Clinical Nursing,* ed 3, St. Louis, Mosby, 1993; **C,** Yoav Levy/Phototake.

Figure 1-8, **B,** UPI/Bettmann

Figure 1-9 Ken Edward/Science Source/Photo Researchers

Figure 1-10 The Bettmann Archive

Figure 1-11, **B,** Science VU-Genetech/Visuals Unlimited; **C,** Hank Morgan/Photo Researchers.

Chapter 2

Chapter opener From Baron EJ, Peterson LR, Finegold SM: Bailey & Scott's Diagnostic Microbiology, ed 9, St. Louis, Mosby, 1994.

Figure 2-5, **A,** Courtesy Bion Enterprises Ltd., Park Ridge, IL; **B,** Habif TP: *Clinical Dermatology: Color Guide to Diagnosis and Therapy,* St. Louis, Mosby, 1985.

Figure 2-6, **B** and **D,** GW Willis, Ochsner Medical Institution/Biological Photo Service; C, Paul W. Johnson/Biological Photo Service

Box figures, p. 40, (top) Paul W. Johnson/Biological Photo Service; (bottom) GW Willis, Ochsner Medical Institution/Biological Photo Service

Box figures, p. 41, (top) from Baron EJ, Peterson LR, Finegold SM: *Bailey & Scott's Diagnostic Microbiology,* ed 9, St. Louis, Mosby 1994; (bottom) GW Willis, Ochsner Medical Institution/Biological Photo Service

Box figures, p. 42, (top) **A,** John J. Cardamone, Jr., University of Pittsburgh and BK Pugashetti, University of Pittsburgh/Biological Photo Service; **B,** Visuals Unlimited/Christine Case; (bottom) EH Newcomb and SR Tandon, University of Wisconsin/Biological Photo Service

Box figures, p. 43, (top) Courtesy William Merrill, Pennsylvania State University; (bottom) from Baron EJ, Peterson LR, Finegold SM: *Bailey & Scott's Diagnostic Microbiology,* ed 9, St. Louis, Mosby, 1994.

Box figures, p. 44, (top) Linda S. Thomashow, Washington State University/Biological Photo Service; (bottom) from Baron EJ, Peterson LR, Finegold SM: *Bailey & Scott's Diagnostic Microbiology,* ed 9, St. Louis, Mosby, 1994.

Box figures, p. 45, (top) Richard L. Moore/Biological Photo Service; (bottom) Judith FM Hoeniger, University of Toronto/Biological Photo Service

Box figures, p. 46, (top) Karen Stephens, University of Washington/Biological Photo Service; (bottom) **A,** Visual Unlimited/Science Vu; **B,** Arthur M. Siegelman/Visuals Unlimited

Box figures, p. 47, (top and bottom) GW Willis, Ochsner Medical Institution/Biological Photo Service

Box figures, p. 48, (top) From Baron EJ, Peterson LR, Finegold SM: *Bailey & Scott's Diagnostic Microbiology,* ed 9, St. Louis, Mosby, 1993. Courtesy Ellena Patterson, University of California, Irvine; (bottom) From Baron EJ, Peterson LR, Finegold SM: *Bailey & Scott's Diagnostic Microbiology,* ed 9, St. Louis, Mosby, 1994.

Box figures, p. 49, (top) Stanley C. Holt, University of Texas Health Center, San Antonio/Biological Photo Service; (bottom) Paul W. Johnson/Biological Photo Services

Box figures, p. 50, (top) Recolorized from an image Science Vu-SW Watson/Visual Unlimited; (bottom) Terry J. Beveridge, University of Guelph and GD Sprott, National Research Council of Canada/Biological Service

Figure 2-8 Courtesy CP Kurtzman/US Department of Agriculture

Figure 2-9 From Baron EJ, Peterson LR, Finegold SM: *Bailey & Scott's Diagnostic Microbiology,* ed 9, St. Louis, Mosby, 1994.

Figure 2-10 John NA Lott, McMaster University/Biological Photo Service

Figure 2-12 Courtesy Orson K. Miller, Jr., Virginia Polytechnic Institute

Figure 2-13 J. Robert Waaland, University of Washington/Biological Photo Service

Figure 2-14, **A,** Alfred Oxczarzak/Biological Photo Service; **B,** Paul W. Johnson/Biological Photo Service

Figure 2-15 Robert Brons/Biological Photo Service

Figure 2-16 Biophoto Associations/SS/Photo Researchers

Figure 2-17 JJ Pauline, University of Georgia/Biological Photo Service

Chapter 3

Chapter opener Courtesy Esther Angert, Indiana University

Figure 3-4 The Bettmann Archive

Figure 3-6, **A,** John J. Cardamone Jr., University of Pittsburgh/Biological Photo Service; **B,** Courtesy William Ghiorse, Cornell University

Box figure, p. 65, Courtesy Esther R. Angert, Indiana University

Figure 3-11 GW Willis, Ochsner Medical Institution/Biological Photo Service

Figure 3-13, **B** and **C,** GW Willis, Ochsner Medical Institution/Biological Photo Service

Figure 3-14 From Farrar WE, Woods MJ, Innes JA: *Infectious Diseases: Text and Color Atlas,* ed 2, London, Mosby Europe, 1992. Courtesy I Farrell.

Figure 3-15, **B,** GW Willis, Ochsner Medical Institution/Biological Photo Service

Figure 3-17 From Baron EJ, Peterson LR, Finegold SM: *Bailey & Scott's Diagnostic Microbiology,* ed 9, St. Louis, Mosby, 1994.

Figure 3-18 From Baron EJ, Peterson LR, Finegold SM: *Bailey & Scott's Diagnostic Microbiology,* ed 9, St. Louis, Mosby, 1994.

Figure 3-20 From Immunology slide set, UpJohn, Kalamazoo, MI

Figure 3-21 Robert Brons/Biological Photo Service

Figure 3-22 J. Robert Waaland, University of Washington/Biological Photo Service

Figure 3-23 Kennedy/Biological Photo Service

Figure 3-24, **B**, Garry T. Cole, University of Texas at Austin/ Biological Photo Service; C, Lara Hartley/Terraphotographics/ Biological Photo Service

Figure 3-25 Stanley C. Holt, University of Texas Health Center, San Antonio/ Biological Photo Service

Box figure, p. 77, Terry J. Beveridge, University of Guelph/ Biological Photo Service

Figure 3-27 Cathy M. Pringle/Biological Photo Service

Figure 3-28 Photo courtesy of Anheuser-Busch Companies, Inc.

Figure 3-29 Terry J. Beveridge, University of Guelph/Biological Photo Service

Figure 3-30, **A**, Stewart Halperin; **B**, From Baron EJ, Peterson LR, Finegold SM: *Bailey & Scott's Diagnostic Microbiology*, ed 9, St. Louis, Mosby, 1994.

Figure 3-33 From Baron EJ, Peterson LR, Finegold SM: *Bailey & Scott's Diagnostic Microbiology*, ed 9, St. Louis, Mosby, 1994.

Figure 3-34, **B**, From Baron EJ, Peterson LR, Finegold SM: *Bailey & Scott's Diagnostic Microbiology*, ed 9, St. Louis, Mosby, 1994.

Box figure, p. 85, From Baron EJ, Peterson LR, Finegold SM: *Bailey & Scott's Diagnostic Microbiology*, ed 9, St. Louis, Mosby 1994.

Box figures, p. 86, (top) Christine Case/Visuals Unlimited; (bottom) from Olds JK: *Color Atlas of Microbiology*, London, Mosby, 1975

Box figures, p. 87, (top and bottom) RM Atlas, University of Louisville

Chapter 4

Figure 4-11 D. Tronrud, University of Oregon/Biological Photo Service

Figure 4-17 Stewart Halperin

Figure 4-19 Stewart Halperin

Chapter 5

Figure 5-1, **B**, Terry J. Beveridge, University of Guelph/Biological Photo Service

Figure 5-7 H. Stuart Pankratz, Michigan State University/ Biological Photo Service

Figure 5-8 Terry J. Beveridge, University of Guelph/Biological Photo Service

Figure 5-13 Terry J. Beveridge, University of Guelph/Biological Photo Service

Figure 5-14 Terry J. Beveridge, University of Guelph/Biological Photo Service

Figure 5-15 From Baron EJ, Peterson LR, Finegold SM: *Bailey & Scott's Diagnostic Microbiology*, ed 9, St. Louis, Mosby, 1994.

Figures 5-16 Terry J. Beveridge, University of Guelph/Biological Photo Service

Figure 5-17 Max Listergarten, University of Pennsylvania/ Biological Photo Service

Figure 5-18 S. Abraham and EH Beachey, VA Medical Center, Memphis, TN/Biological Photo Service

Figure 5-19 From Mims CA, Playfair JHL, Roitt IM et al: *Medical Microbiology*, St. Louis, Mosby, 1993. A, Courtesy ET Nelson: **B**, Courtesy T Yamamoto.

Figure 5-20, **A**, Paul W. Johnson/Biological Photo Service; **B**, William L. Dentler, University of Kansas/Biological Photo Services

Figure 5-23 From Baron EJ, Peterson LR, Finegold SM: *Bailey & Scott's Diagnostic Microbiology*, ed 9, St. Louis, Mosby, 1994.

Figure 5-24 William L. Dentler, University of Kansas/Biological Photo Service

Box figure, p. 138, Terry J. Beveridge, University of Guelph and Y. Gorby and D. Blakemore, University of New Hampshire/ Biological Photo Service

Figure 5-25, **B**, Paul W. Johnson/Biological Photo Service

Figure 5-26 Courtesy R. Dourmashkin, New York University

Figure 5-27 Terry J. Beveridge, University of Guelph/Biological Photo Service

Figure 5-28, **B**, Richard Rodewald, University of Virginia/ Biological Photo Service

Figure 5-29, **B**, Ada L. Olins, University of Tennessee/Biological Photo Service

Figure 5-32 Recolorized from an image by Science Vu-RGE Murray/Visuals Unlimited

Figure 5-33 Eldon H. Newcomb, University of Wisconsin/ Biological Photo Service

Figure 5-34, **B**, Barry F King, University of California School of Medicine/Biological Photo Service

Figure 5-35, **B**, Paul W. Johnson/Biological Photo Service

Figure 5-36, **A**, Terry J. Beveridge, University of Guelph and J. Ingram McGill University/Biological Photo Service; **B**, Paul W. Johnson and John Sierburth, University of Rhode Island/ Biological Photo Service

Figure 5-38, **B**, Paul W. Johnson/Biological Photo Service

Figure 5-39 Charles L. Sanders/Biological Photo Service

Figure 5-41 Robert Brons/Biological Photo Service

Figure 5-42, **B**, Terry J. Beveridge, University of Guelph/ Biological Photo Service

Chapter 6

Chapter opener From Courtesy Holger Jannasch, Woodshole Oceanographic Institute

Figure 6-8 From Baron EJ, Peterson LR, Finegold SM: *Bailey & Scott's Diagnostic Microbiology*, ed 9, St. Louis, Mosby, 1994.

Figure 6-17 Stewart Halperin

Figure 6-19 Stewart Halperin

Figure 6-20 Stewart Halperin

Figure 6-21 Stewart Halperin

Figure 6-24 J. Robert Waaland, University of Washington/ Biological Photo Service

Figure 6-28 Courtesy Holger Jannasch, Woodshole Oceanographic Institute

Figure 6-30 Paul W. Johnson/Biological Photo Service

Figure 6-31, **B**, Phil Gates, University of Durham/Biological Photo Service

Figure 6-32 Terry J. Beveridge, University of Guelph and GD Sprott, National Research Council of Canada/Biological Photo Service

Chapter 7

Figure 7-6, Courtesy Richard J. Feldmann, National Institutes of Health

Box figures, p. 197, From Cold Springs Laboratory Archives

Figure 7-11, **B**, Barbara J. Miller/Biological Photo Service

Box figure, p. 204, Jack Bostrack/Visuals Unlimited

Chapter 8

Figure 8-3, **B**, Courtesy Huntington Potter and David Dressler Harvard Medical School

Figure 8-8 R. Welch, University of Wisconsin Medical School/ Biological Photo Service

Figure 8-14 David P. Allison, Oak Ridge National Laboratory/ Biological Photo Service

Figure 8-19 From Gottfried SS: *Human Biology*, St. Louis, Mosby, 1994; Photos from CA Hasenkamp, University of Toronto/ Biological Photo Service; Illustrations by Barbara Cousins.

Box figures, pp. 246-247, Courtesy Steve Lindow, University of California, Berkeley

Chapter 9

Chapter opener From Mims CA, Playfair JHL, Roitt IM et al: *Medical Microbiology*, St. Louis, Mosby, 1993. Courtesy D. Hockley.

Figure 9-3 Lee Simon/Photo Researchers
Box figure, p. 259, Visual Unlimited/Cabisco
Figure 9-6 Lee Simon/Photo Researchers
Figure 9-10, **B,** Leon J. Le Beau/Biological Photo Service
Box figures, p. 273, From Baron EJ, Peterson LR, Finegold SM:
Bailey & Scott's Diagnostic Microbiology, ed 9, St. Louis, Mosby,
1994.
Figure 9-16, C. Garon and J. Rose, National Institute of
Health/Biological Photo Service
Figure 9-19, **A,** John J. Cardamone, Jr, University of
Pittsburgh/Biological Photo Service
Figure 9-20, **A,** John J. Cardamone, Jr, University of
Pittsburgh/Biological Photo Service
Figure 9-21 Centers for Disease Control/Biological Photo Service
Figure 9-22 Petit Format Institut Pasteur/Charles Dauget/Photo
Researchers
Figure 9-23 From Mims CA, Playfair JHL, Roitt IM et al: *Medical
Microbiology,* St. Louis, Mosby, 1993. Courtesy D. Hockley.

Chapter 10

Chapter opener Courtesy Holger Jannasch, Woodshole
Oceanographic Institute.

Figure 10-1, **A,** Stanley C Holt, University of Texas Health Center,
San Antonio/Biological Photo Service
Figure 10-2 Terry J. Beveridge, University of Guelph/Biological
Photo Service
Figure 10-6, **B,** From Baron EJ, Peterson LR, Finegold SM: *Bailey
& Scott's Diagnostic Microbiology,* ed 9, St. Louis, Mosby, 1994.
Figure 10-7 Jack M. Bostrack/Visuals Unlimited
Box figure, p. 297, Courtesy Exxon Research and Engineering
Company
Box figures, pp. 299-300, Courtesy Holger Jannasch, Woodshole
Oceanographic Institute
Figure 10-16, Paul A. Zahl, National Geographic Society In
National Geographic, August 1967, pp. 258-259.

Chapter 11

Box figure, p. 321, The Bettmann Archive
Figure 11-4 Courtesy American Can Company, Barrington, IL
Figure 11-10 Leon J. Le Beau/Biological Photo Service
Figure 11-12 Rich Humbert/Biological Photo Service

Chapter 12

Chapter opener From Farrar WE, Woods MJ, Innes JA: *Infectious
Diseases: Text Color Atlas,* ed 2, London, Mosby-Europe, 1992,
Courtesy J. Newman.

Figure 12-1 The Bettmann Archive
Figure 12-3 From Farrar WE, Woods MJ, Innes JA: I*nfectious
Diseases: Text Color Atlas,* ed 2, London, Mosby-Europe, 1992,
Courtesy J. Newman.
Figure 12-5 Leon J. Le Beau/Biological Photo Service
Figure 12-9 From Baron EJ, Peterson LR, Finegold SM: *Bailey &
Scott's Diagnostic Microbiology,* ed 9, St. Louis, Mosby, 1994.
Figure 12-13 (background) from Whaley DL: *Whaley and
Wong's Essentials of Pediatric Nursing,* ed 4, St. Louis, Mosby,
1993.
Figure 12-14, A, From Thompson JM, McFarland GK, Hirsch JE,
Tucker SM: *Mosby's Clinical Nursing,* ed 3, St. Louis, Mosby,
1993; B, Leon J. Le Beau/Biological Photo Service
Figure 12-16 RM Atlas, University of Louisville
Figure 12-19 NIH/Science Source/Photo Researchers

Chapter 13

Chapter opener From Roitt IM, Brostoff J, Male DK: *Immunology,*
ed 3, St. Louis, Mosby, 1993. Courtesy D. McLaren.

Box figure, p. 391, The Bettmann Archive

Figure 13-15 From Roitt IM, Brostoff J, Male DK: *Immunology,* ed
3, St. Louis, Mosby, 1993. Courtesy D. McLaren.
Figure 13-16 From Olds JK: *Color Atlas of Microbiology,* London,
Mosby, 1975.

Chapter 15

Chapter opener From Cerio R and Jackson WF: *A Colour Atlas of
Allergic Skin Disorders,* London, Mosby-Europe, 1992.

Figure 15-1 The Bettmann Archive
Box figures, p.439, The Bettmann Archive
Box figure, p.442, Courtesy World Health Organization
Figure 15-7 UPI/The Bettmann Archive
Figure 15-9 James Snyder, University of Louisville Hospital
Box figures, p. 456, From Cerio R and Jackson WF: *A Colour Atlas
of Allergic Skin Disorders,* London, Mosby-Europe, 1992.

Chapter 16

Figure 16-1 From Farrar WE, Woods MJ, Innes JA: *Infectious
Diseases: Text and Color Atlas,* ed 2, London, Mosby-Europe, 1992.
Figure 16-3 From Farrar WE, Woods MJ, Innes JA: *Infectious
Diseases: Text and Color Atlas,* ed 2, London, Mosby-Europe, 1992.
Figure 16-4 From Olds JK: *Color Atlas of Microbiology,* London,
Mosby, 1975.
Figure 16-7 From Baron EJ, Peterson LR, Finegold SM: *Bailey &
Scott's Diagnostic Microbiology,* ed 9, St. Louis, Mosby, 1994.
Figure 16-8 From Olds JK: *Color Atlas of Microbiology,* London,
Mosby, 1975. Courtesy MH Gleeson-White.
Figure 16-9 From Baron EJ, Peterson LR, Finegold SM: *Bailey &
Scott's Diagnostic Microbiology,* ed 9, St. Louis, Mosby, 1994.
Figure 16-10, **B,** From Baron EJ, Peterson LR, Finegold SM: *Bailey
& Scott's Diagnostic Microbiology,* ed 9, St. Louis, Mosby, 1994.
Figure 16-11 From Olds JK: *Color Atlas of Microbiology,* London,
Mosby, 1975.
Figure 16-12 From Baron EJ, Peterson LR, Finegold SM: *Bailey &
Scott's Diagnostic Microbiology,* ed 9, St. Louis, Mosby, 1994.
Figure 16-13, **A,** From Baron EJ, Peterson LR, Finegold SM: *Bailey
& Scott's Diagnostic Microbiology,* ed 9, St. Louis, Mosby, 1994; **B,**
From Olds JK: *Color Atlas of Microbiology,* London, Mosby, 1975.
Courtesy GRE Naylor.
Figure 16-14 From Olds JK: *Color Atlas of Microbiology,* London,
Mosby, 1975.
Figure 16-15 From Olds JK: *Color Atlas of Microbiology,* London,
Mosby, 1975.
Figure 16-16, **B,** From Baron EJ, Peterson LR, Finegold SM: *Bailey
& Scott's Diagnostic Microbiology,* ed 9, St. Louis, Mosby, 1994.
Box figures, p. 489, From Baron EJ, Peterson LR, Finegold SM:
Bailey & Scott's Diagnostic Microbiology, ed 9, St. Louis, Mosby,
1994.
Figure 16-18 From Mims CA, Playfair JHL, Roitt IM et al: *Medical
Microbiology,* St. Louis, Mosby, 1993. Courtesy DK Banerjee.
Figure 16-20, B, From Olds JK: *Color Atlas of Microbiology,* London,
Mosby, 1975. Courtesy P. Walker.

Chapter 17

Chapter opener Courtesy RM Atlas, University of Louisville

Figure 17-1 Yoav Levy/Phototake
Box figures, pp. 504-505, The Bettmann Archive
Figure 17-5 RM Atlas, University of Louisville
Figure 17-6, **B,** Leon J. Le Beau/Biological Photo Service

Chapter 18

Figure 18-1 Courtesy Robert Dourmashkin
Figure 18-3 RM Atlas, University of Louisville

Figure 18-4 John J. Cardamone, Jr., University of Pittsburgh/ Biological Photo Service

Figure 18-10 From Emond RT and Rowland HAK: *A Colour Atlas of Infectious Diseases,* ed 2, Mosby-Wolfe, London, 1987.

Figure 18-12 Centers for Disease Control/Biological Photo Service

Figure 18-13 Centers for Disease Control, Atlanta/Biological Photo Service

Figure 18-14 Bernard Roizman, University of Chicago/Biological Photo Service

Figure 18-15, **A** and **B,** From Mims CA, Playfair JHL, Roitt IM et al: *Medical Microbiology,* St. Louis, Mosby, 1993. Courtesy JS Bingham.

Figure 18-16, **A** and **B,** From Mims CA, Playfair JHL, Roitt IM et al: *Medical Microbiology,* St. Louis, Mosby, 1993. Courtesy JS Bingham.

Case study figure, From Farrar WE, Woods MJ, Innes JA: *Infectious Diseases: Text Color Atlas,* ed 2, London, Mosby-Europe, 1992. Courtesy Regional Virus Laboratory, East Birmingham Hospital.

Chapter 19

Figure 19-3 Mims CA, Playfair JHL, Roitt IM et al: *Medical Microbiology,* St. Louis, Mosby, 1993. Courtesy P. Garen.

Figure 19-5 The Bettmann Archive

Figure 19-10 From Farrar WE, Woods MJ, Innes JA: *Infectious Diseases: Text and Color Atlas,* ed 2, London, Mosby-Europe, 1992. Courtesy E. Sahn.

Figure 19-15, **A** to **C,** From Mims CA, Playfair JHL, Roitt IM et al: *Medical Microbiology,* St. Louis, Mosby, 1993. Courtesy World Health Organization.

Figure 19-17, **A** and **B,** From Emond RT and Rowland HAK: *A Colour Atlas of Infectious Diseases,* ed 2, Mosby-Wolfe, London, 1987.

Figure 19-18 From Farrar WE, Woods MJ, Innes JA: *Infectious Diseases: Text and Color Atlas,* ed 2, London, Mosby-Europe, 1992.

Figure 19-19 The Bettmann Archive

Figure 19-20 From Farrar WE, Woods MJ, Innes JA: *Infectious Diseases: Text and Color Atlas,* ed 2, London, Mosby-Europe, 1992.

Figure 19-21 From Farrar WE, Woods MJ, Innes JA: *Infectious Diseases: Text and Color Atlas,* ed 2, London, Mosby-Europe, 1992.

Figure 19-23 From Farrar WE, Woods MJ, Innes JA: *Infectious Diseases: Text and Color Atlas,* ed 2, London, Mosby-Europe, 1992.

Figure 19-24 From Farrar WE, Woods MJ, Innes JA: *Infectious Diseases: Text and Color Atlas,* ed 2, London, Mosby-Europe, 1992.

Case study figure, From Farrar WE, Woods MJ, Innes JA: *Infectious Diseases: Text Color Atlas,* ed 2, London, Mosby-Europe, 1992. Courtesy Regional Virus Laboratory, East Birmingham Hospital.

Chapter 20

Chapter opener From Mims CA, Playfair JHL, Roitt IM et al: *Medical Microbiology,* St. Louis, Mosby, 1993. Courtesy I Farrell.

Figure 20-2 Leodocia M. Pope, University of Texas/Biological Photo Service

Figure 20-3 From Mims CA, Playfair JHL, Roitt IM et al: *Medical Microbiology,* St. Louis, Mosby, 1993.

Figure 20-4 From Olds JK: *Color Atlas of Microbiology,* London, Mosby, 1975.

Figure 20-5 GW Willis, Ochsner Medical Institution/Biological Photo Service

Figure 20-6 From Farrar WE, Woods MJ, Innes JA: *Infectious Diseases: Text and Color Atlas,* ed 2, London, Mosby-Europe, 1992.

Figure 20-9 James Snyder, University of Louisville Hospital

Figure 20-10 From Baron EJ, Peterson LR, Finegold SM: *Bailey & Scott's Diagnostic Microbiology,* ed 9, St. Louis, Mosby, 1994.

Figure 20-11 Courtesy James Snyder, University of Louisville Hospital

Figure 20-12 From Olds JK: *Color Atlas of Microbiology,* London, Mosby, 1975.

Figure 20-13 Mims CA, Playfair JHL, Roitt IM et al: *Medical Microbiology,* St. Louis, Mosby, 1993. Courtesy K. Nye.

Figure 20-15 From Baron EJ, Peterson LR, Finegold SM: *Bailey & Scott's Diagnostic Microbiology,* ed 9, St. Louis, Mosby, 1994.

Figure 20-17 RB Morrison, MD, Austin, TX/Biological Photo Service

Figure 20-18 From Mims CA, Playfair JHL, Roitt IM et al: Medical Microbiology, St. Louis, Mosby, 1993. Courtesy I. Farrell.

Figure 20-19 From Mims CA, Playfair JHL, Roitt IM et al: *Medical Microbiology,* St. Louis, Mosby, 1993. Courtesy J Newman.

Figure 20-20 The Bettmann Archive

Figure 20-22 From Mims CA, Playfair JHL, Roitt IM et al: *Medical Microbiology,* St. Louis, Mosby, 1993. Courtesy AM Geddes.

Case study figure, From Mims CA, Playfair JHL, Roitt IM et al: *Medical Microbiology,* St. Louis, Mosby, 1993. Courtesy ID Starke and ME Hodson.

Chapter 21

Chapter opener AP/Wide World Photos

Figure 21-1 From Olds JK: *Color Atlas of Microbiology,* London, Mosby, 1975. Courtesy MH Gleeson-White.

Figure 21-2 From Emond RT and Rowland HAK: *A Colour Atlas of Infectious Diseases,* ed 2, Mosby Wolfe, London, 1987. **A,** courtesy GDW McKendrick; **B,** courtesy JA Forbes.

Figure 21-6 Courtesy Theodore Khaukin, Interfon Science, New Brunswick

Figure 21-7 From Olds JK: *Color Atlas of Microbiology,* London, Mosby, 1975. Courtesy HM Gilles.

Figure 21-8 AP/Wide World Photos

Figure 21-9 Farrar WE, Woods MJ, Innes JA: *Infectious Diseases: Text and Color Atlas,* ed 2, London, Mosby-Europe, 1992. Courtesy JR Cantey.

Figure 21-10 Farrar WE, Woods MJ, Innes JA: *Infectious Diseases: Text and Color Atlas,* ed 2, London, Mosby-Europe, 1992. Courtesy E. Taylor.

Figure 21-11 From Olds JK: *Color Atlas of Microbiology,* London, Mosby, 1975. Courtesy GM Gilles.

Figure 21-12 Mims CA, Playfair JHL, Roitt IM et al: *Medical Microbiology,* St. Louis, Mosby, 1993. Courtesy TF Sellers, Jr.

Figure 21-16 Farrar WE, Woods MJ, Innes JA: *Infectious Diseases: Text and Color Atlas,* ed 2, London, Mosby-Europe, 1992. Courtesy E. Sahn.

Case Study figure from Murray PR, Kobayashi GS, Pfaller MA, Rosenthal KS: *Medical Microbiology,* ed 2, Mosby, 1994.

Chapter 22

Chapter opener From Mims CA, Playfair JHL, Roitt IM et al: *Medical Microbiology,* St. Louis, Mosby, 1993. Courtesy MJ Wood.

Figure 22-2 From Olds JK: *Color Atlas of Microbiology,* London, Mosby, 1975. Courtesy TJ Alexander.

Figure 22-3 From Baron EJ, Peterson LR, Finegold SM: *Bailey & Scott's Diagnostic Microbiology,* ed 9, St. Louis, Mosby, 1994.

Figure 22-5 From Mims CA, Playfair JHL, Roitt IM et al: *Medical Microbiology,* St. Louis, Mosby, 1993. Courtesy J. Clay.

Figure 22-8 From Farrar WE, Woods MJ, Innes JA: *Infectious Diseases: Text and Color Atlas,* ed 2, London, Mosby-Europe, 1992.

Figure 22-9 From Farrar WE, Woods MJ, Innes JA: *Infectious Diseases: Text and Color Atlas,* ed 2, London, Mosby-Europe, 1992. Courtesy RD Catterall.

Figure 22-10 Farrar WE, Woods MJ, Innes JA: *Infectious Diseases: Text and Color Atlas*, ed 2, London, Mosby-Europe, 1992. Courtesy GL Ridgway.

Figure 22-11 From Farrar WE, Woods MJ, Innes JA: *Infectious Diseases: Text and Color Atlas*, ed 2, London, Mosby-Europe, 1992. Courtesy L. Brown.

Figure 22-12 From Mims CA, Playfair JHL, Roitt IM et al: *Medical Microbiology*, St. Louis, Mosby, 1993. Courtesy MJ Wood.

Figure 22-13 Camera MD Studios

Figure 22-14 From Mims CA, Playfair JHL, Roitt IM et al: *Medical Microbiology*, St. Louis, Mosby, 1993. Courtesy DA Lewis.

Figure 22-15 From Mims CA, Playfair JHL, Roitt IM et al: *Medical Microbiology*, St. Louis, Mosby, 1993. Courtesy WE Farrar.

Figure 22-16 From Mims CA, Playfair JHL, Roitt IM et al: *Medical Microbiology*, St. Louis, Mosby, 1993. Courtesy A. du Vivier.

Figure 22-17 From Farrar WE, Woods MJ, Innes JA: *Infectious Diseases: Text and Color Atlas*, ed 2, London, Mosby-Europe, 1992. Courtesy S. Harding.

Figure 22-18 Leon J. Le Beau/Biological Photo Service

Figure 22-19 From Mims CA, Playfair JHL, Roitt IM et al: *Medical Microbiology*, St. Louis, Mosby, 1993. Courtesy M. Chaput de Saitonge.

Figure 22-20 Max Listergarten, University of Pennsylvania/Biological Photo Service

Case Study figure, From Baron EJ, Peterson LR, Finegold SM: *Bailey & Scott's Diagnostic Microbiology*, ed 9, St. Louis, Mosby, 1994.

Chapter 23

Chapter opener From Farrar WE, Woods MJ, Innes JA: *Infectious Diseases: Text and Color Atlas*, ed 2, London, Mosby-Europe, 1992. Courtesy SE Thompson.

Figure 23-1 From Mims CA, Playfair JHL, Roitt IM et al: *Medical Microbiology*, St. Louis, Mosby, 1993. Courtesy TF Sellers.

Figure 23-3 Farrar WE, Woods MJ, Innes JA: *Infectious Diseases: Text and Color Atlas*, ed 2, London, Mosby-Europe, 1992. Courtesy K. Nye.

Figure 23-4 R. Calentine/Visuals Unlimited

Figure 23-5, **A** and **B,** Farrar WE, Woods MJ, Innes JA: *Infectious Diseases: Text and Color Atlas*, ed 2, London, Mosby-Europe, 1992. Courtesy AE Prevost.

Figure 23-6 Mims CA, Playfair JHL, Roitt IM et al: *Medical Microbiology*, St. Louis, Mosby, 1993. Courtesy JA Innes.

Figure 23-7 Gordon Leedale/Photo Researchers

Figure 23-8 Biological Photo Service

Figure 23-9 From Mims CA, Playfair JHL, Roitt IM et al: *Medical Microbiology*, St. Louis, Mosby, 1993.

Figure 23-10 From Mims CA, Playfair JHL, Roitt IM et al: *Medical Microbiology*, St. Louis, Mosby, 1993. Courtesy DK Banerjee.

Figure 23-11 Farrar WE, Woods MJ, Innes JA: *Infectious Diseases: Text and Color Atlas*, ed 2, London, Mosby-Europe, 1992. Courtesy SE Thompson.

Figure 23-12 Farrar WE, Woods MJ, Innes JA: *Infectious Diseases: Text and Color Atlas*, ed 2, London, Mosby-Europe, 1992.

Figure 23-15 Farrar WE, Woods MJ, Innes JA: *Infectious Diseases: Text and Color Atlas*, ed 2, London, Mosby-Europe, 1992.

Figure 23-16 Farrar WE, Woods MJ, Innes JA: *Infectious Diseases: Text and Color Atlas*, ed 2, London, Mosby-Europe, 1992.

Figure 23-18 Farrar WE, Woods MJ, Innes JA: *Infectious Diseases: Text and Color Atlas*, ed 2, London, Mosby-Europe, 1992.

Figure 23-19 Farrar WE, Woods MJ, Innes JA: *Infectious Diseases: Text and Color Atlas*, ed 2, London, Mosby-Europe, 1992.

Figure 23-20 Farrar WE, Woods MJ, Innes JA: *Infectious Diseases: Text and Color Atlas*, ed 2, London, Mosby-Europe, 1992.

Figure 23-21 From Mims CA, Playfair JHL, Roitt IM et al: *Medical Microbiology*, St. Louis, Mosby, 1993. Courtesy AE Bianco.

Case study figure from Farrar WE, Woods MJ, Innes JA: *Infectious Diseases: Text Color Atlas*, ed 2, London, Mosby-Europe, 1992. Courtesy AE Provost.

Chapter 24

Chapter opener Courtesy Dhigeomi Ushijima, Kikkoman Corporation, Noda City, Chiba Prefecture, Japan.

Figure 24-1, **B,** Visuals Unlimited/Science Vu-URSCIM

Figure 24-3 Stewart Halperin

Figure 24-4, **B,** Courtesy Dhigeomi Ushijima, Kikkoman Corporation, Noda City, Chiba Prefecture, Japan.

Figure 24-6, **B,** Photo courtesy of Anheuser-Busch Companies, Inc.

Figure 24-9 Richard Norwitz/Phototake

Figure 24-11 Secchi-Lecaque-Roussel-UCLAF/CNRI/Science Photo Library/Photo Researchers

Figure 24-26 Hank Morgan/Photo Researchers

Chapter 25

Chapter opener Exxon Research and Engineering Company

Chapter 25-1, **A,** William E Schadel, Small World Enterprise/Biological Photo Service

Figure 25-1, **B,** Orson K Miller, Jr., Virginia Polytechnic Institute

Figure 25-6, **B,** L. Evans Roth, University of Tennessee/Biological Photo Service

Figure 25-10 Stewart Halperin

Figure 25-11, **B,** Judith FM Hoeniger, University of Toronto/Biological Photo Service

Figure 25-15 John NA Lott, McMaster University/Biological Photo Service

Figure 25-18 John NA Lott, McMaster University/Biological Photo Service

Box figures, p. 757, **B** (right), Visuals Unlimited/Christine L. Case; C, Courtesy Ideex Laboratories, Westbrook, ME, Food and Environmental Division

Box figure, p. 761, Exxon Research and Engineering Company

Appendix

Figure, p. A-1, Ronald J. Ervin

Figure, p. A-2, John V. Hagen

Figure, p. A-3, David J. Mascaro and Associates

Figure, p. A-4, John V. Hagen

Figures, p. A-6, Scott Bodell

Figure, p. A-7, Barbara Cousins

Figure, p. A-8, Joan Beck

Figure, p. A-9, Christine Oleksyk

Figure, p. A-10, Joan Beck

Figure, p. A-11, G. David Brown

Figure, p. A-12, Barbara Cousins

Figure, p. A-13, From Seeley RR, Stephens TD, Tate P: *Anatomy and Physiology*, ed 2, St. Louis, Mosby, 1992.

Figures, p. A-14, (top) Joan Beck; (bottom) Jody L. Fulks

Figure, p. A-15, Jody L. Fulks

Figure, p. A-17, Joan Beck

Figure, p. A-18, Joan Beck

Figure, p. A-19, Joan Beck

Figure, p. A-20, Ronald J. Ervin

Figure, p. A-21 Ronald J. Ervin

Index

Italic letters refer to *tables* and *figures*.

Emerging Infectious Diseases

Disease	Deaths
Pneumonia	4,300,000
Diarrheal diseases	3,200,000
Tuberculosis	3,000,000
Hepatitis B	1–2,000,000
Malaria	1,000,000
Measles	880,000
Neonatal tetanus	600,000
AIDS	550,000

AIDS cases

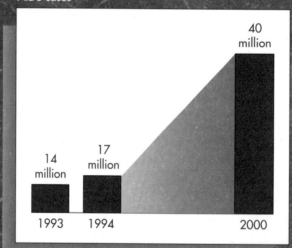

HIV infections

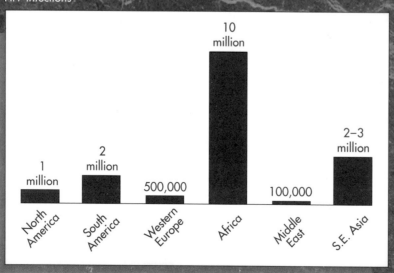